Understanding and fostering soil carbon sequestration

It is widely recognised that agriculture is a significant contributor to global warming and climate change. Agriculture needs to reduce its environmental impact and adapt to current climate change whilst still feeding a growing population, i.e. become more 'climate-smart'. Burleigh Dodds Science Publishing is playing its part in achieving this by bringing together key research on making the production of the world's most important crops and livestock products more sustainable.

Based on extensive research, our publications specifically target the challenge of climate-smart agriculture. In this way we are using 'smart publishing' to help achieve climate-smart agriculture.

Burleigh Dodds Science Publishing is an independent and innovative publisher delivering high quality customer-focused agricultural science content in both print and online formats for the academic and research communities. Our aim is to build a foundation of knowledge on which researchers can build to meet the challenge of climate-smart agriculture.

For more information about Burleigh Dodds Science Publishing simply call us on +44 (0) 1223 839365, email info@bdspublishing.com or alternatively please visit our website at www.bdspublishing.com.

Related titles:

Managing soil health for sustainable agriculture Volume 1: Fundamentals
Print (ISBN 978-1-78676-188-0) Online (ISBN 978-1-78676-191-0; 978-1-78676-190-3)

Managing soil health for sustainable agriculture Volume 2: Monitoring and management
Print (ISBN 978-1-78676-192-7) Online (ISBN 978-1-78676-195-8; 978-1-78676-194-1)

Advances in Conservation Agriculture Volume 2: Practice and Benefits
Print (ISBN 978-1-78676-268-9) Online (ISBN 978-1-78676-270-2; 978-1-78676-271-9)

Climate change and agriculture
Print (ISBN 978-1-78676-320-4) Online (ISBN 978-1-78676-323-5; 978-1-78676-322-8)

Improving soil health
Print (ISBN 978-1-78676-670-0) Online (ISBN 978-1-78676-673-1; 978-1-78676-672-4)

Chapters are available individually from our online bookshop:
https://shop.bdspublishing.com

climate•SMART•publishing
IN AGRICULTURAL SCIENCE

BURLEIGH DODDS SERIES IN AGRICULTURAL SCIENCE
NUMBER 121

Understanding and fostering soil carbon sequestration

Edited by Dr Cornelia Rumpel, CNRS, Sorbonne University, Institute of Ecology and Environmental Sciences Paris, France

burleigh dodds
SCIENCE PUBLISHING

Published by Burleigh Dodds Science Publishing Limited
82 High Street, Sawston, Cambridge CB22 3HJ, UK
www.bdspublishing.com

Burleigh Dodds Science Publishing, 1518 Walnut Street, Suite 900, Philadelphia, PA 19102-3406, USA

First published 2023 by Burleigh Dodds Science Publishing Limited

Library of Congress Control Number: 2022934581

British Library Cataloguing in Publication Data
A catalogue record for this book is available from the British Library

ISBN 978-1-78676-969-5 (print)
ISBN 978-1-78676-972-5 (PDF)
ISBN 978-1-78676-971-8 (ePub)
ISSN 2059-6936 (print)
ISSN 2059-6944 (online)

DOI 10.19103/AS.2022.0106

Typeset by Deanta Global Publishing Services, Dublin, Ireland

Contents

Series list

Title	Series number
Achieving sustainable cultivation of maize - Vol 1 From improved varieties to local applications Edited by: Dr Dave Watson, CGIAR Maize Research Program Manager, CIMMYT, Mexico	001
Achieving sustainable cultivation of maize - Vol 2 Cultivation techniques, pest and disease control Edited by: Dr Dave Watson, CGIAR Maize Research Program Manager, CIMMYT, Mexico	002
Achieving sustainable cultivation of rice - Vol 1 Breeding for higher yield and quality Edited by: Prof. Takuji Sasaki, Tokyo University of Agriculture, Japan	003
Achieving sustainable cultivation of rice - Vol 2 Cultivation, pest and disease management Edited by: Prof. Takuji Sasaki, Tokyo University of Agriculture, Japan	004
Achieving sustainable cultivation of wheat - Vol 1 Breeding, quality traits, pests and diseases Edited by: Prof. Peter Langridge, The University of Adelaide, Australia	005
Achieving sustainable cultivation of wheat - Vol 2 Cultivation techniques Edited by: Prof. Peter Langridge, The University of Adelaide, Australia	006
Achieving sustainable cultivation of tomatoes Edited by: Dr Autar Mattoo, USDA-ARS, USA and Prof. Avtar Handa, Purdue University, USA	007
Achieving sustainable production of milk - Vol 1 Milk composition, genetics and breeding Edited by: Dr Nico van Belzen, International Dairy Federation (IDF), Belgium	008
Achieving sustainable production of milk - Vol 2 Safety, quality and sustainability Edited by: Dr Nico van Belzen, International Dairy Federation (IDF), Belgium	009
Achieving sustainable production of milk - Vol 3 Dairy herd management and welfare Edited by: Prof. John Webster, University of Bristol, UK	010
Ensuring safety and quality in the production of beef - Vol 1 Safety Edited by: Prof. Gary Acuff, Texas A&M University, USA and Prof. James Dickson, Iowa State University, USA	011
Ensuring safety and quality in the production of beef - Vol 2 Quality Edited by: Prof. Michael Dikeman, Kansas State University, USA	012
Achieving sustainable production of poultry meat - Vol 1 Safety, quality and sustainability Edited by: Prof. Steven C. Ricke, University of Arkansas, USA	013
Achieving sustainable production of poultry meat - Vol 2 Breeding and nutrition Edited by: Prof. Todd Applegate, University of Georgia, USA	014

Achieving sustainable cultivation of sorghum - Vol 1 031
Genetics, breeding and production techniques
Edited by: Prof. William Rooney, Texas A&M University, USA

Achieving sustainable cultivation of sorghum - Vol 2 032
Sorghum utilization around the world
Edited by: Prof. William Rooney, Texas A&M University, USA

Achieving sustainable cultivation of potatoes - Vol 2 033
Production, storage and crop protection
Edited by: Dr Stuart Wale, Potato Dynamics Ltd, UK

Achieving sustainable cultivation of mangoes 034
Edited by: Prof. Víctor Galán Saúco, Instituto Canario de Investigaciones Agrarias (ICIA), Spain and Dr Ping Lu, Charles Darwin University, Australia

Achieving sustainable cultivation of grain legumes - Vol 1 035
Advances in breeding and cultivation techniques
Edited by: Dr Shoba Sivasankar et al., formerly International Crops Research Institute for the Semi-Arid Tropics (ICRISAT), India

Achieving sustainable cultivation of grain legumes - Vol 2 036
Improving cultivation of particular grain legumes
Edited by: Dr Shoba Sivasankar et al., formerly International Crops Research Institute for the Semi-Arid Tropics (ICRISAT), India

Achieving sustainable cultivation of sugarcane - Vol 1 037
Cultivation techniques, quality and sustainability
Edited by: Prof. Philippe Rott, University of Florida, USA

Achieving sustainable cultivation of sugarcane - Vol 2 038
Breeding, pests and diseases
Edited by: Prof. Philippe Rott, University of Florida, USA

Achieving sustainable cultivation of coffee 039
Edited by: Dr Philippe Lashermes, Institut de Recherche pour le Développement (IRD), France

Achieving sustainable cultivation of bananas - Vol 1 040
Cultivation techniques
Edited by: Prof. Gert H. J. Kema, Wageningen University and Research, The Netherlands and Prof. André Drenth, University of Queensland, Australia

Global Tea Science 041
Current status and future needs
Edited by: Dr V. S. Sharma, formerly UPASI Tea Research Institute, India and Dr M. T. Kumudini Gunasekare, Coordinating Secretariat for Science Technology and Innovation (COSTI), Sri Lanka

Integrated weed management 042
Edited by: Emeritus Prof. Rob Zimdahl, Colorado State University, USA

Achieving sustainable cultivation of cocoa 043
Edited by: Prof. Pathmanathan Umaharan, Cocoa Research Centre – The University of the West Indies, Trinidad and Tobago

Robotics and automation for improving agriculture 044
Edited by: Prof. John Billingsley, University of Southern Queensland, Australia

Water management for sustainable agriculture 045
Edited by: Prof. Theib Oweis, ICARDA, Jordan

Improving organic animal farming 046
Edited by: Dr Mette Vaarst, Aarhus University, Denmark and Dr Stephen Roderick, Duchy College, UK

Acknowledgements

We wish to acknowledge the following for their help in reviewing particular chapters:

- Chapter 6: Dr David Whitehead, Manaaki Whenua - Landcare Research, New Zealand
- Chapter 18: Dr Gilles Lemaire, formerly INRAE-Lusignan, France
- Chapter 24: Professor John M. Galbraith, Virginia Polytechnic Institute and State University, USA and Dr Klaus Lorenz, Ohio State University, USA

Chapter 1

Introduction: soil carbon sequestration – a process linking soils to humanity

C. Rumpel, CNRS, Sorbonne University, Institute of Ecology and Environmental Sciences Paris, France

1 Introduction

Human survival is threatened by perturbations of the earth system, which are affecting nine key processes, defined as planetary boundaries (Steffen et al., 2015). Most threatening are the perturbations affecting global biogeochemical cycling and, in particular, the carbon cycle. Indeed, fossil fuel combustion and land use change have led to increasing atmospheric CO_2 concentrations with severe consequences for the Earth's climate (IPCC, 2022). Climate change is expected to affect production on agricultural soils, which, however, needs to be maintained or even increased in view of a growing global population, because soils deliver as much as 98.8% of our food calories along with multiple other ecosystem services. However, intensification of soil use through modern agriculture is also an important contributor to the perturbations affecting the planetary boundaries (Kopittke et al., 2021).

It has recently been pointed out that, in view of the challenges faced by human societies in particular, soil organic carbon (SOC) plays a central role as a crucial indicator of soil functioning, determining many of the soil's complex and interconnected processes at several scales (Lal et al., 2021; Kopittke et al., 2022). In meeting the objective of providing solutions to global challenges, SOC sequestration has received particular attention. This attention is related to the fact that soils can store huge amounts of carbon and, as part of the global carbon cycle, can influence atmospheric CO_2 concentrations responsible for climate change (Fuss et al., 2018). As a consequence, carbon sequestration in soils has been proposed as one of the nature-based solutions to remove CO_2 from the atmosphere (Smith, 2016; IPCC, 2022). Indeed, only small increases

http://dx.doi.org/10.19103/AS.2022.0106.01

in total global SOC stocks may be necessary to mitigate fossil fuel emissions (Balesdent and Arrouays, 1999). Although SOC sequestration may be needed to combat climate change, it clearly should not be used as an excuse to not reduce emissions because it cannot be the only climate solution (Chabbi et al., 2017; Anderson et al., 2019). SOC sequestration may be a win-win strategy (Bossio et al., 2020) because increasing SOC is also required to improve all other soil functions with beneficial implications for human societies (Chabbi et al., 2017).

Indeed, SOC sequestration can be seen as underpinning several UN Sustainable Development Goals (SDGs) (Lal et al., 2021; Rumpel et al., 2022) through the provision of food and fiber production, purification of drinking water, biodiversity and climate regulation (FAO et al., 2020; Kopittke et al., 2022). The provision of these ecosystem services relies on the good functioning of soil, in particular its microbial processes responsible for biogeochemical cycling (Lemanceau et al., 2015). In this context, SOC sequestration is an important process, which captures CO_2 via photosynthesis and transfers organic materials via plant activity belowground where they can be used by the soil (micro-) organisms. Microbial processes involve the decomposition, transformation and turnover of plant-derived organic materials, providing nutrients, carbon and energy for microbial activity and growth and more generally SOC accumulation. This microbial functioning leads to nutrient provision to plants and aggregate formation, a structure-giving process allowing for water infiltration and storage thereby reducing soil loss through erosion. Moreover, microbial carbon use also leads to microbial necromass formation, a process largely responsible for SOC stabilization (Cotrufo et al., 2013).

2 Key issues in understanding soil carbon sequestration

Despite the vital importance of biogeochemical cycling, understanding the processes occurring within the soil matrix is still incomplete. One of the reasons is that soil organic matter is one of nature's most versatile materials comprising a continuum of microbial and plant compounds in various stages of decomposition, which need to be characterized by a combination of methods (Kögel-Knabner, 2002, 2017). In recent years, it has become apparent that a complete understanding of microbial utilization of soil organic compounds is only possible by taking into account the physical organization of the soil matrix determining microbial access (Vogel et al., 2022). The understanding of soil functioning and the processes responsible for ecosystem service provision thus requires the combination of concepts derived from biological, chemical and physical sciences. Information needs to be collected at several scales, from the microscale, where the processes are occurring through to the profile and landscape scales, where soil management takes place (O'Rourke et al., 2015).

Integration and upscaling through different modeling approaches is also needed to understand the importance and consequences of specific processes for biogeochemical cycling at the global scale (Jungkunst et al., 2022). However, global carbon modeling is characterized by a huge uncertainty, making climate predictions at the global scale variable (Jones and Frielingstein, 2020). This is related to the poor understanding of land-based processes to which soil and in particular SOC cycling makes a major contribution. Biogeochemical and biophysical understanding of soil biogeochemical cycling may help to reduce the uncertainty in modeling the global carbon cycle (Bradford et al., 2016).

However, SOC is not the only carbon type in soil. Especially in arid regions and soils with limestone materials, inorganic carbon contributes greatly to soil carbon. It has been estimated that inorganic carbon represents about one-third of the total amount of carbon stored in the soil, amounting to between 700 and 940 GT (Monger et al., 2015). Particularly in regions with poor vegetation growth, where weathering processes are dominant, mineral weathering may be one process which could lead to inorganic carbon sequestration through chemical weathering of Ca-silicate and carbonate rocks (Liu et al., 2011). The cycling of organic and inorganic carbon is affected by fundamentally contrasting processes. While organic carbon is mainly cycled by biological processes relatively rapidly, chemical processes control the slow cycling of inorganic carbon. Interactions between the two carbon pools may occur through biologically enhanced weathering (Song et al., 2018) and pedogenic carbonate formation (Zamanian et al., 2016), two processes, which have poorly been described up to now. Management of soils for organic carbon sequestration may lead to loss of soil inorganic carbon, which therefore needs to be taken into account (Raza et al., 2021).

In assessing global SOC stocks, it must be considered that the up to 3000 GT of carbon stored in global soils (Scharlemann et al., 2014) are not homogenously distributed. More than one-fifth of the world's SOC is stored in organic soils such as peatlands (Leifeld and Menichetti, 2018). These organic soil types are characterized by the occurrence of completely different processes compared to those occurring in mineral soils (Rumpel, 2019). On the other hand, organic soils occupy only about 3% of the total land surface (Yu et al., 2010). Most agricultural production occurs on mineral soil types, which are generally impoverished in SOC and therefore present a huge potential for additional SOC sequestration, especially in deep soil horizons which have been poorly studied (Rumpel and Kögel-Knabner, 2011) but present possibilities of increasing soil carbon with appropriate management interventions (Button et al., 2022).

Understanding the processes affecting the quantity, nature and turnover of organic matter in soils, but also of inorganic carbon and biogeochemical cycling more generally, is thus of crucial importance to find solutions to reducing carbon loss and increasing its sequestration. Specific management practices are needed to influence the different processes affecting contrasting carbon forms in the various soil types (Kögel-Knabner and Amelung, 2021). Moreover, the scope of research on soil carbon has evolved in recent years because soil carbon sequestration relates to land management practices and thus to socioeconomic and political factors. Soil carbon sequestration concerns management decisions affecting biogeochemical processes and is intimately linked to human societies. Its understanding and implementation requires a transdisciplinary approach bringing together different scientific disciplines and various stakeholders (Chabbi et al., 2021; Rumpel et al., 2022). Knowledge gaps are manifold. They range from factors controlling biophysical processes in different pedoclimatic environments to socioeconomic factors influencing the behavior of land managers and political decision-makers at different levels.

To develop adequate policies to enhance good soil management, leading to maintenance and accumulation of soil carbon, the effect of management practices on (bio-)geochemical cycling needs to be taken into account. In view of the numerous soil types and pedoclimatic conditions in the Earth system, the practices to be employed need to be region specific, taking into account not only biophysical constraints and trade-offs but also the socioeconomic, legal and cultural environment (Amelung et al., 2020).

3 This book

This book is a first attempt to gather the knowledge of the global (soil) science community working on the different aspects of soil carbon sequestration. It is divided into four parts.

The first part describes the biological, physical and chemical processes determining soil carbon sequestration and their implications for soil management. The second part is devoted to measuring, reporting and verification (MRV) of carbon sequestration in soils. These aspects are crucial for developing management strategies with positive effects on soil carbon. Only if something is measurable can it be influenced by human activity and taken into account by policymakers. However, due to its heterogeneous nature, distribution and residence time, MRV and modeling of soil carbon are quite complex.

In the third part of the book, management practices influencing the different carbon types present in various ecosystems are presented and discussed, while the fourth part of the book is devoted to socioeconomic, legal and policy aspects of soil carbon sequestration. The latter presents global (1) and national policy frameworks for carbon credits related to soil management, (2) the approaches needed to take into account farmers' practices and (3) legal requirements for implementing SOC sequestration.

I would like to thank the authors of the 29 chapters of the book for their valuable contributions. Special acknowledgments go to the reviewers, David Whitehead, Gilles Lemaire, John M. Galbraith and Klaus Lorenz. I hope that this book will be of use to students, scientists and practitioners and will lead to better management of global (soil) carbon resources.

4 References

Amelung, W., Bossio, D., de Vries, W., Kögel-Knabner, I., Lehmann, J., Amundson, R., Bol, R., Collins, C., Lal, R., Leifeld, J., Minasny, B., Pan, G., Paustian, K., Rumpel, C., Sanderman, J., van Groenigen, J. W., Mooney, S., van Wesemael, B., Wander, M. and Chabbi, A. 2020. Towards a global-scale soil climate mitigation strategy. *Nature Communications* 11(1), 5427.

Anderson, C. M., Defries, R. S., Utterman, R., Matson, P. A., Nepstad, D. C., Pacala, S., Schlesinger, W. H., Shaw, M. R., Smith, P., Weber, C. and Field, C. B. 2019. Natural climate solutions are not enough. *Science* 363(6430), 933–934.

Balesdent, J. and Arrouays, D. 1999. Usage des terres et stockage de carbone dans les sols du territoire français. Une estimation des flux nets annuels pour la période 1900–1999. C. R. Acad. *Agriculturists* 85, 265–277.

Bossio, D. A., Cook-Patton, S. C., Ellis, P. W., Fargione, J., Sanderman, J., Smith, P., Wood, S., Zomer, R. J., von Unger, M., Emmer, I. M. and Griscom, B. W. 2020. The role of soil carbon in natural climate solutions. *Nature Sustainability* 3(5), 391–398.

Bradford, M., Wieder, W., Bonan, G., Fierer, N., Raymond, P. A. and Crowther, T. W. 2016. Managing uncertainty in soil carbon feedbacks to climate change. *Nature Climate Change* 6, 751–758.

Button, E. S., Pett-Ridge, J., Murphy, D. V., Kuzyakov, Y., Chadwick, D. R. and Jones, D. L. 2022. Deep-C storage: Biological, chemical and physical strategies to enhance carbon stocks in agricultural subsoils. *Soil Biology and Biochemistry* 170, 108697.

Chabbi, A., Kögel-Knabner, I. and Rumpel, C. 2021. Soil Science in transition – Re-defining its role under the 4p1000 initiative. *Geoderma* 385, 114891.

Chabbi, A., Lehmann, J., Ciais, P., Loescher, H. W., Cotrufo, M. F., Don, A., SanClements, M., Schipper, L., Six, J., Smith, P. and Rumpel, C. 2017. Aligning agriculture and climate policy. *Nature Climate Change* 7(5), 307–309.

Cotrufo, M. F., Wallenstein, M. D., Boot, C. M., Denef, K. and Paul, E. 2013. The microbial efficiency-matrix stabilization (MEMS) framework integrates plant litter decomposition with soil organic matter stabilization: Do labile plant inputs form stable soil organic matter? *Global Change Biology* 19(4), 988–995.

FAO, ITPS, GSBI, SCBD, EC. 2020. *State of Knowledge of Soil Biodiversity - Status, Challenges and Potentialities*. Rome: Food and Agriculture Organization.

Fuss, S., Lamb, W. F., Callaghan, M. W., Hilaire, J., Creutzig, F., Amann, T., Beringer, T., de Oliveira Garcia, W., Hartmann, J., Khanna, T., Luderer, G., Nemet, G. F., Rogelj, J., Smith, P., Vicente, J. L. V., Wilcox, J., del Mar Zamora Dominguez, M. and Minx, J. C. 2018. Negative emissions–Part 2: Costs, potentials and side effects. *Environmental Research Letters* 13(6), 063002.

IPCC. 2022. Mitigation of climate change. Contribution of Working Group III to the sixth assessment report of the intergovernmental panel on climate change [P.R. Shukla, J.

Skea, R. Slade, A. Al Khourdajie, R. van Diemen, D. McCollum, M. Pathak, S. Some, P. Vyas, R. Fradera, M. Belkacemi, A. Hasija, G. Lisboa, S. Luz, J. Malley, (eds.)]. Summary for policymakers. In: *Climate Change*. Cambridge, UK and New York, NY: Cambridge University Press.

Jones, C. D. and Friedlingstein, P. 2020. Quantifying process-level uncertainty contributions to TCRE and carbon budgets for meeting Paris Agreement climate targets. *Environmental Research Letters* 15(7), 074019.

Jungkunst, H. F., Goepel, J., Horvath, T., Ott, S. and Brunn, M. 2022. Global soil organic carbon-climate interactions: Why scales matter. *WIREs Climate Change* 13(4). DOI: 10.1002/wcc.780.

Kögel-Knabner, I. 2002. The macromolecuar organic composition of plant and microbial residues as inputs to soil organic matter. *Soil Biology and Biochemistry* 34(2), 139–162.

Kögel-Knabner, I. 2017. The macromolecular organic composition of plant and microbial residues as inputs to soil organic matter: Fourteen years on. *Soil Biology and Biochemistry* 105, A3–A8.

Kögel-Knabner, I. and Amelung, W. 2021. Soil organic matter in major pedogenic soil groups. *Geoderma* 384, 114785.

Kopittke, P. M., Berhe, A. A., Carrillo, Y., Cavagnaro, T. R., Chen, D., Chen, Q., Román Dobarco, M., Dijkstra, F. A., Field, D. J., Grundy, M. J., He, J., Hoyle, F. C., Kögel-Knabner, I., Lam, S. K., Marschner, P., Martinez, C., McBratney, A. B., McDonald-Madden, E., Menzies, N. W., Mosley, L. M., Mueller, C. W., Murphy, D. V., Nielsen, U. N., O'Donnell, A. G., Pendall, E., Pett-Ridge, J., Rumpel, C., Young, I. M. and Minasny, B. 2022. Ensuring planetary survival: The centrality of organic carbon in balancing the multifunctional nature of soils. *Critical Reviews in Environmental Science and Technology*, 1–17. DOI: 10.1080/10643389.2021.2024484.

Kopittke, P. M., Menzies, N. W., Dalal, R. C., McKenna, B. A., Husted, S., Wang, P. and Lombi, E. 2021. The role of soil in defining planetary boundaries and the safe operating space for humanity. *Environment International* 146, 106245.

Lal, R., Bouma, J., Brevik, E., Dawson, L., Field, D. J., Glaser, B., Hatano, R., Hartemink, A. E., Kosaki, T., Lascelles, B., Monger, C., Muggler, C., Ndzana, G. M., Norra, S., Pan, X., Paradelo, R., Reyes-Sánchez, L. B., Sandén, T., Singh, B. R., Spiegel, H., Yanai, J. and Zhang, J. 2021. Soils and sustainable development goals of the United Nations (New York, USA): An IUSS perspective. *Geoderma Regional* 25, e00398.

Leifeld, J. and Menichetti, L. 2018. The underappreciated potential of peatlands in global climate change mitigation strategies. *Nature Communications* 9(1), 1071.

Lemanceau, P., Maron, P.-M., Mazurier, S., Mougel, C., Pivato, B., Plassart, P., Ranjard, L., Revellin, C., Tardy, V. and Wipf, D. 2015. Understanding and managing soil biodiversity: A major challenge in agroecology. *Agronomy for Sustainable Development* 35(1), 67–81.

Liu, Z., Dreybrodt, W. and Liu, H. 2011. Atmospheric CO2 sink: Silicate weathering or carbonate weathering? *Applied Geochemistry* 26 (Suppl.), S292–S294.

Monger, H. C., Kraimer, R. A., Khresat, S., Cole, D. R., Wang, X. and Wang, J. 2015. Sequestration of inorganic carbon in soil and groundwater. *Geology* 43(5), 375–378.

O'Rourke, S. M., Angers, D. A., Holden, N. M. and McBratney, A. B. 2015. Soil organic carbon across scales. *Global Change Biology* 21(10), 3561–3574.

Raza, S., Zamanian, K., Ullah, S., Kuzyakov, Y., Virto, I. and Zhou, J. 2021. Inorganic carbon losses by soil acidification jeopardize global efforts on carbon sequestration and climate change mitigation. *Journal of Cleaner Production* 315, 128036.

Rumpel, C. 2019. Soils linked to climate change. *Nature* 572(7770), 442–443.

Rumpel, C. and Kögel-Knabner, I. 2011. Deep soil organic matter - a key but poorly understood component of terrestrial C cycle. *Plant and Soil*, 338, 143-158.

Rumpel, C., Amiraslai, F., Bossio, D., Chenu, C., Henry, B., Fuentes Espinoza, A., Koutika, L.-K., Ladha, J., Madari, B., Minasni, B., Olaleye, A. O., Shirato, Y., Sall, N. S., Soussana, J.-F. and Varela-Ortega, C. 2022. The role of soil carbon sequestration in enhancing human resilience in tackling global crises including pandemics. *Soil Security*, 8, 100069.

Scharlemann, J. P. W., Tanner, E. V. J., Hiederer, R. and Kapos, V. 2014. Global soil carbon: Understanding and managing the largest terrestrial carbon pool. *Carbon Management* 5(1), 81-91.

Smith, P. 2016. Soil carbon sequestration and biochar as negative emission technologies. *Global Change Biology* 22(3), 1315–1324.

Song, Z., Liu, H., Strömberg, C. A. E., Wang, H., Strang, P. J., Yang, X. and Wu, Y. 2018. Contribution of forests to the carbon sink via biologically-mediated silicate weathering: A case study of China. *Science of the Total Environment* 615, 1–8.

Steffen, W., Richardson, K., Rockström, J., Cornell, S. E., Fetzer, I., Bennett, E. M., Biggs, R., Carpenter, S. R., de Vries, W., de Wit, C. A., Folke, C., Gerten, D., Heinke, J., Mace, G. M., Persson, L. M., Ramanathan, V., Reyers, B. and Sörlin, S. 2015. Sustainability. Planetary boundaries: Guiding human development on a changing planet. *Science* 347(6223), 1259855.

Vogel, H.-J., Balseiro-Romero, M., Kravchenko, A., Otten, W., Pot, V., Schlüter, S., Weller, U. and Baveye, P. C. 2022. A holistic perspective on soil architecture is needed as a key to soil functions. *European Journal of Soil Science* 73(1), e13152.

Yu, Z. C., Loisel, J., Brosseau, D. P., Beilman, D. W. and Hunt, S. J. 2010. Global peatland dynamics since the last glacial maximum. *Geophysical Research Letters* 37(13), L13402.

Zamanian, K., Pustovoytov, K. and Kuzyakov, Y. 2016. Pedogenic carbonates: Forms and formation processes. *Earth-Science Reviews* 157, 1–17.

Part 1

Understanding carbon sequestration in soils

Chapter 2

Mechanisms of soil organic carbon sequestration and implications for management

Ingrid Kögel-Knabner, Chair of Soil Science, TUM School of Life Sciences and Institute for Advanced Study, Technical University of Munich, Germany; Martin Wiesmeier, TUM School of Life Sciences, Technical University of Munich and Bavarian State Research Center for Agriculture, Institute for Organic Farming, Soil and Resource Management, Germany; and Stefanie Mayer, Chair of Soil Science, TUM School of Life Sciences, Germany

1 Introduction

1.1 Introduction and definitions

A major potential for increasing carbon sequestration in mineral soils is in agricultural systems under cropland use (Amelung et al., 2020). Understanding organic carbon (OC) sequestration in (mineral) soils requires considering the pathways and the associated different types of organic matter (OM) input. As pointed out in Box 1, OC sequestration refers to 'the process of transferring CO_2 from the atmosphere into the soil of a land unit, through plants, plant residues and other organic solids which are stored or retained in the unit as part of the soil organic matter (humus)' (Chenu et al., 2019; Olson et al., 2014). If we accept this definition, all processes are relevant that lead to a storage or retention of OC in soils. A number of mechanisms have been described that lead to the retention of OC in soils. As the OC that enters the soil is in dynamic equilibrium, all the different OM pools that are retained in a soil must be considered. SOC sequestration implies raising soil organic carbon (SOC) levels, where they are currently undersaturated, and to maintain maximal OC levels in well-managed soil systems (Lehmann et al., 2020).

http://dx.doi.org/10.19103/AS.2022.0106.02

Olson et al. (2014) pointed out that it is essential to strictly differentiate between the application of any of OM to soils from sources external or outside a land unit (e.g. amendments like manure, compost, biochar) and OC sequestration sensu strictu. Sequestration of OC in soils as defined here (Box 1) requires that atmospheric CO_2 is fixed through photosynthesis and stored in the soil. No atmospheric CO_2 is converted and stored as a result of amendment transfer and it does not add to reducing atmospheric CO_2 levels. Therefore, we will not consider the application of organic amendments in this chapter. As organic amendments may in specific cases influence SOC sequestration through their impact on plant growth and soil microbial functioning, their management is discussed in Chapter 9 of this book.

Box 1: Definitions

'We generally use the comprehensive term stabilization of OM for processes or mechanisms that lead to prolonged turnover times in soil. Stabilization is defined as protection of OM from mineralization. Stability is the integrated effect of recalcitrance, organo-mineral interactions and accessibility. Generally, stabilized OM is older than unstabilized OM, a fact that is demonstrated in longer turnover and mean residence times. Turnover times of stabilized OM cannot be given as absolute values because they are dependent on environmental conditions and soil properties (e.g. climate, soil mineralogy)' (von Lützow et al., 2006).

OC sequestration is defined as 'the process of transferring CO_2 from the atmosphere into the soil of a land unit, through plants, plant residues and other organic solids which are stored or retained in the unit as part of the soil organic matter (humus)'. Retention time of sequestered OC in the soil (terrestrial pool) can range from short-term (not immediately released back to atmosphere) to long-term (millennia) storage. Sequestration can therefore be quantified for a given duration (Chenu et al., 2019; Olson et al., 2014).

'OC storage is broader as it is defined as the increase in SOC stocks over time in the soils of a given land unit, not necessarily associated with a net removal of CO_2 from the atmosphere' (Chenu et al., 2019).

'The OC storage potential of a given soil unit may be defined as the maximum gain in SOC stock attainable under a given climate and a given timeline (e.g. time required to attain a new equilibrium or a given time period, such as 20 years for IPCC)' (Chenu et al., 2019).

'Similarly, the OC sequestration potential of a given soil would be the maximum gain in SOC allowing a net removal of CO_2 from the atmosphere under a given climate and for a specified timeline' (Chenu et al., 2019).

1.2 Organic matter input to soils

Organic C enters the soil mainly as:

- aboveground litter or crop residues,
- belowground litter or crop residues, and
- rhizodeposition.

Both above and belowground litter or crop residues are mainly composed of OC bound in large polymers (celluloses, hemicelluloses, lignin, cutin, suberin) in leaves, stems, twigs and other woody debris, or roots, with only a small contribution of low-molecular weight organic components (Kögel-Knabner, 2017). They are either deposited on the mineral soil surface, or in different soil depths as root litter. Incorporation of OC from aboveground litter occurs via bioturbation or leaching of soluble components. In contrast, rhizodeposition consists mainly of low molecular weight compounds released from roots into the surrounding soil at various depths (Fig. 1).

Long-term OC storage in soils occurs primarily when OC derived from plant biomass is stabilized in soils as soil OM. Plant biomass makes up the majority of OC input also to agricultural soils. But we have to take into

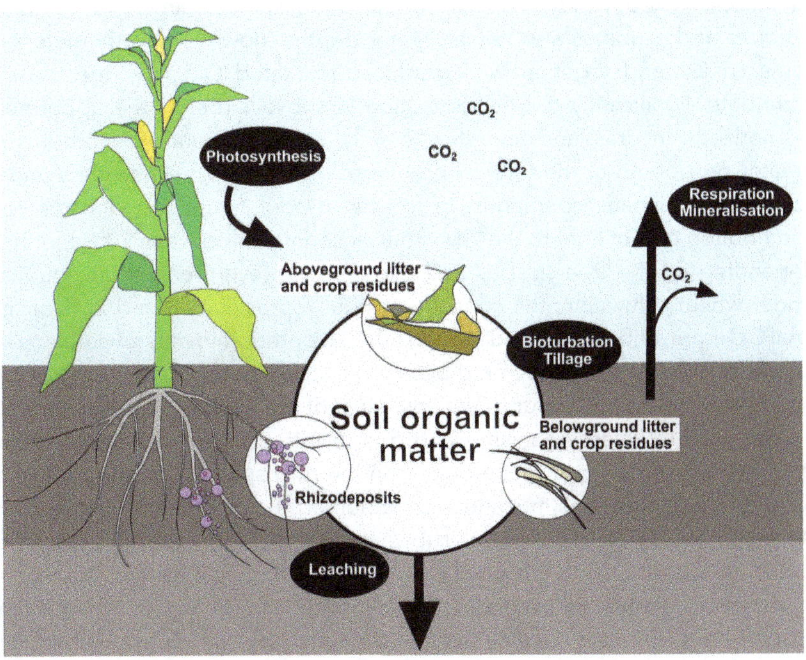

Figure 1 Aboveground and belowground OM input to soils.

Published by Burleigh Dodds Science Publishing Limited, 2023.

account that OM is also added to cultivated soils through fertilization and waste disposal (e.g. liquid manure, compost, sludge, animal excreta, biochar, biogas digestate), which contribute significant amounts (Jacobs et al., 2020). Soils are often also contaminated with organic constituents from the petroleum and coal chemistry/industry, as well as from coal combustion, e.g. tar oil, coal dusts, black carbon, specifically in industrial-urban areas (Kiem and Kögel-Knabner, 2003; Schmidt and Noack, 2000), as well as plastics (Rillig et al., 2021). Geogenic C such as kerogen or black shale can also be inherited from the parent material (Fox et al., 2020 and references therein). This short listing demonstrates the large diversity of OM input to soils. Lehmann et al. (2020) suggest that the molecular diversity of the organic compounds rather than the material properties of individual compounds controls decomposition in soils. As pointed out above, the amendments help to increase the OC content and stocks of a soil, but may not help sequestering OC in soils. At the same time, it is important to return organic residue materials to soils, rather than burning them or using them otherwise, e.g. for energy production or production of chemicals.

A considerable portion of biomass OM reaches the soil as subsurface input, i.e. as root litter and rhizodeposition. About 30%, 50%, and 75% of the root biomass are found in the topmost 10 cm, 20 cm and 40 cm of the soil, respectively; however, the maximum rooting depth, and thus also the range influenced by root residues and rhizodeposition, is much larger. Root debris/residues and exudates also reach greater depths (down to several meters in some cases), and together with displaced/relocated OM from the topsoil, contribute significantly to SOC formation in subsoils (below the uppermost (A) horizon). In grassland and steppe soils, root-C generally contributes in greater proportion to the total carbon input than in forest ecosystems under comparable climatic conditions. In forests of cool temperate climates, the contribution of root litter to the OM input, accounts for between 20% and 50% depending on the tree species and the life form (evergreen or deciduous), Under wheat cultivation, the subsurface input represents around 25% of the total OC input, under grassland around 40%. The predominantly low-molecular, usually N-rich OM that is secreted by plant roots (rhizodeposition) contributes significantly to the OC input to the soil. Most of the root exudates are rapidly metabolized by microorganisms, and due to this constant OC source, there is a high (microbial) population density in the rhizosphere. The proportion of belowground input flux (much less documented than inputs from leaves and stems) is considered to contribute substantially to soil OM through dead roots and rhizodeposition, but the exact proportions and quantities remain a matter of debate. Quantitative estimates are complicated by the fact that a major proportion of the root exudates is transferred to the soil via symbiotic fungi (mycorrhiza) (Dijkstra et al., 2021; Zhou et al., 2020).

1.3 Resynthesis of organic compounds in soil: Formation of microbial necromass and exudates

When the OM is deposited on or in the soil and is broken down by the soil decomposer community, it is rapidly mineralized with the liberation of CO_2, while the majority of the mineralized N is incorporated into the microbial biomass.

As described in Fig. 2, a portion of the plant OC is used by the soil (micro-) organisms for build-up of their biomass (anabolism). Microbial biomass is the mass of living microorganisms in soil. He et al. (2020) estimated the global stocks of living microbial biomass C to be 12.6 (6.6–16.4) Pg C for fungal biomass and 4.3 (0.5–10.3) Pg C for bacterial biomass in topsoil (0–30 cm), but most of the microbes are dormant or barely active. Microbial products are generally composed of the same type of (macro)molecules as plants, except for lignin, which is exclusively found in plant tissue. Microorganisms also introduce new monomers that build up the cell-wall biopolymers composed

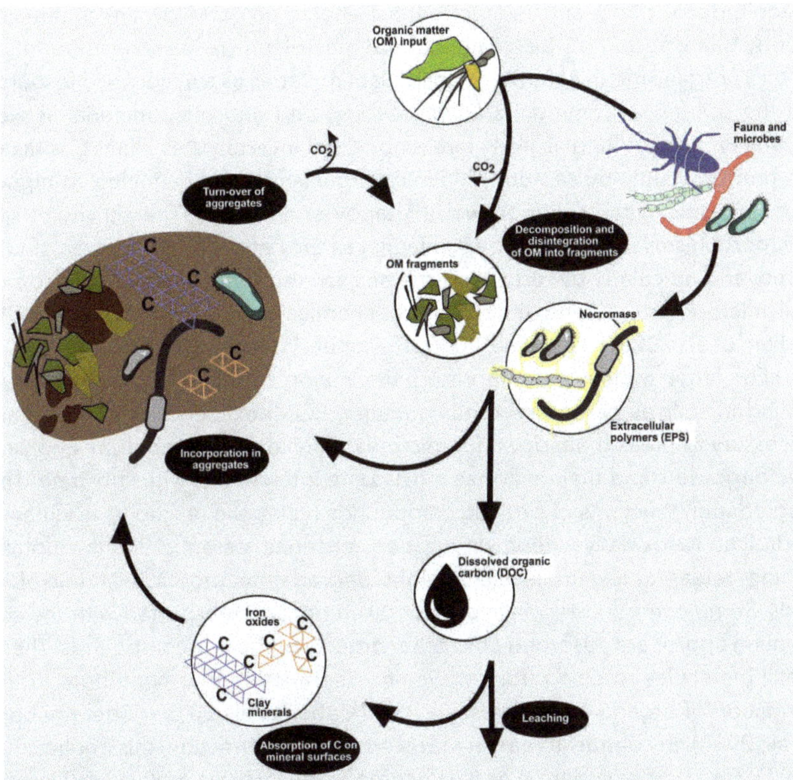

Figure 2 Flow of OC through different binding forms in soils.

of amino sugars (Kögel-Knabner, 2002). This means that the C/N ratio of decomposing plant residues decreases during the course of decomposition. The microbial necromass, which is further decomposed in soils has a narrow C:N ratio, ranging often between 4 and 8. The OC assimilated by microbes and forming their biomass is thus recycled several times through the microbial biomass (Gleixner et al., 2002; Buckeridge et al., 2020). This fact is obvious in the isotopic enrichment of SOM with ^{13}C as the microbial fractionation preferentially transfers the lighter isotope (^{12}C) into the product (CO_2) and leaves the substrate (decomposing plant tissues or individual organic compounds) enriched in ^{13}C. The proportion of OC that is used for building the microbial biomass and its exudates (so-called extracellular polymeric substances, EPS) is the C use efficiency (CUE) of the soil microbial biomass. It is often considered to be between 0.4 and 0.5, but varies in a large range (Manzoni et al., 2018; Qiao et al., 2019).

The decomposition and release of organic residues introduced to soils is to a large extend through the activity of microorganisms. The direct contribution of soil fauna to the release of CO_2 from soils is comparably small, but the magnitude of direct and indirect animal effects on SOM dynamics remains poorly quantified. The ^{13}C signature of soil macrofauna reveals the importance of microorganisms, in contrast to dead plant material, as a major food resource for the soil faunal community. Direct feeding on living plant material as well as on mycorrhizal fungi is likely rare among soil invertebrates. Plant C is taken up predominantly by saprotrophic microorganisms and channeled to higher trophic levels of the soil food web (Potapov et al., 2019). The activity of soil microorganisms is regulated by soil fauna, as they affect many aspects of OM turnover, particularly by regulating the activity and functional composition of soil microorganisms and their physical-chemical connectivity with soil OM (Erktan et al., 2020; Filser et al., 2016; Thakur and Geisen, 2019).

For large molecules comprising the major part of the plant residues (found in soils as particulate organic matter, POM) extracellular enzymes are necessary to break them down for microbial uptake and metabolism, implying that degraders and their enzymes must come into contact with substrate. The degradation reactions of organic compounds (catabolic reactions) are mainly hydrolytic or oxidative depolymerization reactions, essential for the uptake of the released low molecular weight degradation products in microbial cells. Some compounds, which are not taken up, are transported with the soil solution and/or are adsorbed onto other organic or mineral compounds. These small molecules resulting from enzymatic degradation thus contribute to the formation of organo-mineral associations (Kleber et al., 2015; Kögel-Knabner et al., 2008). In contrast to earlier views on soil OM formation, this implies that soil OM is also derived from easily decomposable organic molecules (Cotrufo et al., 2013).

Published by Burleigh Dodds Science Publishing Limited, 2023.

The mineralization of SOC is thus a two-stage process: first, non-bioavailable forms are converted to bioavailable forms, which, only then, undergo the second process, i.e. mineralization after uptake in the cell. It is also considered that other unknown nonenzymatic processes contribute to the first step (Brookes et al., 2017). Such depolymerization reactions are not necessary for the exudates released from plant roots in the rhizosphere, which are low molecular weight compounds.

The capacity of a soil to effectively retain OC is described by the CUE, i.e. the ratio of OC remaining in a system to OC entering that system. Microbial CUE is the fraction of OC taken up allocated to biosynthesis, forming new biomass and biological products, including exudates (Manzoni et al., 2018). The efficiency of long-term OC storage in soils depends on how much OC enters protected compartments or is removed from the system by abiotic transport in vertical and/or horizontal direction (e.g. leaching or erosion).

1.4 Protective mechanisms leading to storage of organic carbon in soils

Plant and microbial residues and their transformation products can be protected from further microbial degradation through various mechanisms. These stabilization processes have an effect on the mixture of fresh and transformed plant residues resulting from decomposition processes, as well as on newly formed microbial residues. They lead to a significant deceleration of the initial rapid mineralization rate. As synthesized recently by Angst et al. (2021), OM stabilized in aggregates and mineral-associated OM is with about equal contributions of plant and microbial origin.

The action of decomposers on their organic substrates takes place in soil pores. As most bacteria are in the size range between 0.5 and 2 μm (Liang et al., 2019), the soil structure and its heterogeneity controls accessibility and biodegradation at the scale of the microbial habitat (Juarez et al., 2013). Large tracts of the soil-pore network are practically devoid of decomposer organisms, and the distribution and forms of OM are equally patchy. Physical separation between decomposer and substrate is therefore an important stabilization mechanism (Lehmann et al., 2020). The structural arrangement of the soil particles creates microsites with limited O_2 diffusion, enzymatic and microbial access (Chevallier et al., 2010; Keiluweit et al., 2017; Mbé et al., 2021; Zimmerman et al., 2004). Different microdomains were identified within an aggregate associated with either occluded POM or dissolved organic matter (DOM), and different spatial and functional properties (Steffens et al., 2017). Domains with POM are densely packed, having low porosity, and are considered to store and protect OC, whereas other domains are porous and consist of mineral grains covered with thin OM films, serving as sites of microbial growth and OM turnover (Steffens et al., 2017). Organic compounds can be stabilized and protected from microbial degradation through interactions with minerals. This mainly involves interactions with particles in the fine fraction and leads to the formation of 'organo-mineral associations, the OM being present in the form of mineral-associated OM. If isolated from soil they are often small organo-mineral microaggregates, rather than primary particles with sorbed OM' (Chenu and Plante, 2006; Totsche et al., 2018), the differentiation between microaggregates and mineral-associated OM being impossible at the smallest scales. Depending on the pedogenetic environment controlled mainly by parent material and climate, different pedogenic minerals form (Kögel-Knabner and Amelung, 2021), which are highly reactive and provide a large specific surface area for interaction with different organic compounds. In addition, dissolution of metals, and the synthesis of organo-metallic phases and their association with clays may also add to protection of OM from decomposition (Basile-Doelsch et al., 2015; Wagai et al., 2020). The contribution of these different processes in combination with protection mechanisms, such as the associations with mineral surfaces, and the physical isolation of decomposers and substrates, strongly depend on soil properties and site conditions, leading to a wide range of OC turnover rates in different soil types.

1.4.1 Particulate organic matter

Plant residues in or transported into the mineral soil are an easily available carbon and nutrient source and are thus immediately attacked and degraded by soil fauna and microorganisms. They thereby act as nuclei for the formation of soil aggregates (Bucka et al., 2019; Witzgall et al., 2021) and at the same time are stabilized against degradation. One way of stabilizing otherwise readily decomposable substrates in the soil can result from the spatial separation of the substrate from the decomposers, which thus prevents degradation (see above). They can be isolated from the mineral soil as POM, existing in different stages of decomposition. Although all organic compounds of plant or animal origin entering the soil are prone to decomposition, the more recalcitrant compounds are relatively enriched in older/stabilized POM fractions while the polysaccharides are decomposed (Baldock et al., 1992; Leifeld and Kögel-Knabner, 2005; Mueller and Koegel-Knabner, 2009). Bacteria and fungi attack the plant particles already comminuted by soil fauna and remain attached to their surfaces, together with fine mineral particles (Chenu and Cosentino, 2011; Tisdall and Oades, 1982). The proliferating microbes excrete EPS during growth and metabolism (Costa et al., 2018; Flemming and Wingender, 2010). Those patches of microbes and their EPS are sticky and adhere as gluing agents to surrounding minerals leading to soil structure formation by cross-linking of particles (Chenu and Jaunet, 1992; Gaillard et al., 1999; Rabbi et al.,

2020; Young and Crawford, 2004). This leads to the formation of aggregated organo-mineral compounds inside the aggregates and as the pores are filled with growing organisms and their exudates, degradation ceases, and the partly degraded POM particles are inaccessible to other microorganisms (McCarthy et al., 2008). This is supported by results from (Steffens et al., 2017). Different microdomains were identified within an aggregate associated with either occluded POM or DOM, and different spatial and functional properties. Domains with POM are densely packed, having low porosity, and are considered to store and protect OC, whereas other domains are porous and consist of mineral grains covered with thin OM films, serving as sites of microbial growth and OM turnover. Smaller POM particles have a larger specific surface area compared to larger ones, which provides a higher potential contact area with mineral particles and colonizing microbes. With increasing degree of decomposition, the POM isolated has a more narrow C:N ratio, due to the accumulation of N-rich microbes and their necromass and exudates, acting as glue between the surface of the POM particles and fine-grained soil minerals. This induces particle assembly around the POM particles and the formation of aggregates (Witzgall et al., 2021). Within time frame of decades and turnover of the aggregates, the particulate plant residues become available again for microbial decay and thus get more and more degraded. The amount of OC stored in a soil as POM is the net result of stabilizing and destabilizing processes. Release of POM from physical occlusion can occur through processes such as tillage, bioturbation, or freeze-thaw and wetting-drying cycles (Bailey et al., 2019).

1.4.2 Interactions of organic matter with mineral surfaces

The predominating minerals forming associations with OM in soil environments include metal oxides, hydroxides, and oxyhydroxides, phyllosilicates, and short-range ordered aluminosilicates. A detailed assessment of the mineral surfaces found in soils and the mechanisms of interaction is given by Kleber et al. (2015, 2021).

Phyllosilicates provide permanently charged surfaces in external and interlayer siloxane surfaces with significant isomorphic substitution of lattice cations, resulting in a pH-independent and permanent net negative charge of the siloxane surfaces. Pedogenic Fe and Al can be present in monomeric form by chelating with organic ligands or in polymeric form as polynuclear complexes and as secondary minerals. The latter includes Fe and Al oxides, hydroxides, and oxyhydroxides as well as short-range-order aluminosilicates (allophane, imogolite, and proto-imogolite) that have high sorptive capacity for OM due to their small size (down to several nanometers) and high surface reactivity via surface hydroxyl groups. Since phyllosilicates possess variable-charge surfaces at crystallite edges, which behave similarly to the ones found on metal oxides, inner- and outersphere adsorption as well as H-bonding of organic ligands may also contribute to OM adsorption to phyllosilicates (Kleber et al., 2015). The extent of permanent negative charge of siloxane surfaces varies with the amount of isomorphic substitution, phyllosilicates thus differ considerably in their ability to bind OM. Protective effects of phyllosilicate clays and pedogenic metal phases on OM are likely to occur concurrently, and possibly synergistically (Wagai et al., 2020).

The organic compounds associated with minerals range from low-molecular weight compounds to large polymeric molecules or cell wall residues. They originate from both plant and microbial origin. Data compiled by Angst et al. (2021) show that plant-derived compounds account for ~50% of the SOM in aggregates and mineral-associated OM in cropland soils, and even more in forest soils. They found that soil types less conducive to the proliferation of microbes, such as Podzols or Alisols, may contain relatively more plant compounds than soil types with favorable conditions for microbial growth, such as Chernozems and Luvisols. Grasslands, in particular, with a continuous input of plant material, low C:N ratios and ratios of lignin to N in plants, and a higher soil pH, seem to favor the accumulation of microbial residues in stabilized SOM as compared to forest or agricultural soils. Thus, if the clay fraction is isolated from the soil and its composition is investigated, mainly alkyl compounds and polysaccharides are found, however, only low concentrations of aromatic compounds, such as strongly transformed lignin fragments. The composition of mineral-associated material may vary not only in different soil types but also with soil depth (Rumpel et al., 2015). This is most probably associated with the microbial processing and exchange of DOM with increasing depth (Roth et al., 2019). Bound OM is also rich in carboxyl and N-containing groups, consistent with the low C:N ratio and a microbial origin of its components.

Low-molecular weight compounds include alcohols, sugars, amino acids, and amines, as well as simple aromatic compounds such as benzene, phenols, etc. that are bound via sorption processes. EPS are released by microbes and strongly associate with mineral surfaces. They are highly variable, hydrated polymers composed of polysaccharides, proteins, nucleic acids, and lipids (Costa et al., 2018). EPS are responsible for the cohesion of microorganisms and adhesion of (groups of) microorganisms to surfaces. It is important to understand that many microorganisms can form supracellular structures comprised of surface-associated microbial cells embedded in hydrated EPS that facilitates adhesion and survival (Cai et al., 2019; Flemming and Wuertz, 2019). Foster (1988) already described small groups of cells, sometimes coated with clay particles, and embedded in partially degraded OM or surrounded by EPS. A planktonic lifestyle (i.e. free swimming) is not possible in the unsaturated zone and most bacterial cells are restricted in their motility and found as small colonies attached to surfaces, protecting themselves from changing hydration conditions using a matrix of EPS (Kleber et al., 2015). This implies that microbial residues are often already attached to mineral surfaces in soils when building necromass. Buckeridge et al. (2020) describe that necromass added to soil strongly binds to mineral surfaces. According to the estimates by Liang et al. (2019) up to around half of the total SOC is microbial necromass. How much of this is associated with mineral surfaces is presently not known.

The binding between minerals and organic compounds can take place through various mechanisms and depends on the one hand on the type of mineral and its surface charge, and on the other, on the type and charge of the functional groups in the OM. Therefore, the pH value and the base saturation are of great importance for the development of organo-mineral compounds (Rasmussen et al., 2018). The binding modes that contribute to the stabilization of organic ligands at mineral surfaces include specific and nonspecific adsorption, H-bond formation, van der Waals interactions, and hydrophobic interactions (see Kleber et al. (2015) for a detailed description). Due to the different mineral composition of soils and the associated large variability of properties several binding mechanisms operate simultaneously. The decomposability of OM can also be strongly reduced by complex formation with metal cations. In this case, the bonding of metal cations (Ca, Al, Fe, heavy metals) onto the OM inhibits attack by enzymes. Soluble complexes of Fe and Al with organic ligands can be precipitated especially in acidic, OM-rich environments such as volcanic and podzolic soils (Wagai et al., 2020). These secondary mineral phases such as allophanes, Fe and Al oxyhydroxides form in the presence of OM via coprecipitation. The size of these mineral phases ranges from one to a few dozen nanometers (Eusterhues et al., 2008; Kleber et al., 2015; Levard et al., 2012; Tamrat et al., 2019; Torn et al., 1997), as the recrystallization ('aging') of such pedogenic Al and Fe oxides can be delayed or prevented if adsorbed or coprecipitated organic compounds block the growth positions (crystallization nuclei) of the crystals or form very stable complexes with the Al and Fe ions.

Evidence of the high stability of organo-mineral compounds is reflected by the fact that the temperature of thermal decomposition of OM increases after bonding onto the mineral phase, and at the same time, the decomposition rates of these bound organic substances are clearly lower compared to free OM. The radiocarbon data compiled by Kleber et al. (2015) show that turnover times of OC stored in association with minerals are on average four times longer than those of OC in free or occluded OM, and range from decades to millennia. Mineral-associated OM is in a state of dynamic equilibrium with DOM, resulting in a constant exchange between the soil solution and mineral surfaces. Thus, a significant fraction of the mineral-associated OM is released into the soil solution and becomes available for microbial uptake and mineralization. This process has been suggested as regulatory gate for microbial activity in soils (Kemmitt et al., 2008).

The formation of organo-mineral associations depends on the presence, but also the accessibility of mineral surfaces for interaction with OM. Interaction of mineral surfaces with DOM is constrained by soil texture and the associated soil-pore system, as preferential flow paths restrict the direct interaction of DOM with mineral surfaces. Changes to the soil environment that affect the solubility of minerals or surface charges of minerals including increased pH or in the reductive dissolution of Fe-bearing minerals can lead to a release of sorbed OM (Kleber et al., 2015). Turbation processes and specifically bioturbation may enhance the exposure of mineral surfaces for interaction with OM. Earthworms induce the formation of organo-mineral associations and at the same time create closer association of partly degraded OM and iron oxides (Barthod et al., 2020; Vidal et al., 2019).

1.4.3 Isolation of particulate organic matter and organo-mineral associations

Physical fractionation, based on a separation resulting from differences in density or particle size, is used to separate plant residues from organo-mineral compounds (Poeplau et al., 2018). It is generally performed following pretreatment consisting of agitation or ultrasound to destroy aggregates. The obtained fractions are then separated using sieves, sedimentation, and/or density fractionation (Christensen, 2001; Oades, 1988). During density fractionation, the sample is suspended in a liquid with a density between 1.6 and 2.4 g cm^{-3} (organic solvent mixtures or inorganic salts), so that the light, floating organic material can be separated. For particle size fractionation, the material in the >63 μm fraction (sand fraction) is separated, which can be further fractionated by sieving. The organo-mineral compounds <63 μm are separated using sedimentation in water, and the <2 μm fraction can be further separated via centrifugation. Fractionation according to the particle size and density are often combined. Such combined particle size and density fractionation approaches are used to isolate free, and occluded POM as well as mineral-associated OM. The POM is found in the light or coarse fraction, occluded POM after dispersion of aggregates. If the aggregation is selectively destroyed, the occluded POM, mainly plant residues at an advanced state of decomposition, can be isolated. It is most often not possible to isolate the 'pure' POM particles, as some mineral material is strongly attached to the free, but especially to the occluded POM fraction. This leads to the fact that the OC concentration of occluded POM ranges between 200 mg g^{-1} and 300 mg g^{-1} fraction. The heavy and fine fractions mainly contain mineral-associated OM. Recently, simple fractionation protocols for agricultural topsoils have been proposed and tested for routine analysis of large sample sets to obtain indicators for soil management and OC input strategies (Just et al., 2021; van Wesemael et al., 2019).

1.5 Organic carbon sequestration in soil compartments with high and specific input: detritusphere and rhizosphere

The detritusphere and the rhizosphere are soil microenvironments characterized by a high input of OM. The detritusphere is defined as a zone of recognizable plant and animal detritus undergoing decay and the soil in their vicinity. It represents a small volume of the soil, as it is limited to the soil adjacent to the plant residues, where most of the microbial activity and biogeochemical processes take place. The detritusphere constitutes an area a few millimeters thick around decomposing residues, and most of the OC transferred from the residues to the soil via diffusive or advective transport is found in this zone (Gaillard et al., 1999; Marschner and Rengel, 2012; Védère et al., 2020). In the first stages of decomposition, most of the residue OC mineralization takes place on the residues themselves (Védère et al., 2020). The incorporation of plant residues into soil leads to the sequestration of OM and the formation of mineral-associated OM in aggregates within weeks of decomposition (Bucka et al., 2019). Detritus can be located on top of the soil or within the mineral soil. These fresh litter surfaces serve as hotspots of microbial activity driving the formation of organo-mineral associations together with comprising a nucleus for aggregate formation. Both processes induced by plant detritus, i.e. the occlusion of POM in soil aggregates and the association of OM with mineral surfaces, strongly rely on the spatial proximity of particulate litter and its surfaces, microbial residues, and fine-sized mineral particles (Witzgall et al., 2021). Detritus is distributed in the mineral soil via bioturbation, peloturbation in Vertisols or cryoturbation in permafrost soils. In cropland soils, the incorporation of detritus is supported by ploughing operations. Most often this is limited to the topsoil (0–30 cm), but deep ploughing may introduce topsoil and detritus to larger depths (Alcántara et al., 2017, 2016). Ploughing converts subsoil material with low OC contents into topsoil with higher OC concentrations and renders a larger soil volume and its associated mineral surfaces available for storing OM (Wiesmeier et al., 2014b). Thus, extending the detritusphere into larger soil depths may enhance OC sequestration. This is in contrast to the concept of no or reduced tillage which relies on bioturbation for incorporation of plant residues into the mineral soil. Here the OC increase in the uppermost topsoil is most often counterbalanced by a release of stored OC in deeper soil layers, leading to a low to nonsignificant effect on SOC stocks over the soil profile (Chenu et al., 2019). However, it is important to note that ploughing also leads to tillage erosion (Chapter 5 of this book), the disruption of aggregates and release of CO_2 to the atmosphere, until a new vegetation is established. These processes must be carefully balanced for a given soil and agricultural system to enhance SOC sequestration or preventing SOC loss.

Plant roots affect OC dynamics in soils by providing OC primarily in the forms of root litter and rhizodeposition (Fig. 1), thus promoting OC sequestration (Dijkstra et al., 2021; Rasse et al., 2005). Root depth distribution is thus an important plant trait that influences OC sequestration, specifically in the subsoil (Jobbagy and Jackson, 2000; Poirier et al., 2018; Thorup-Kristensen et al., 2020). However, roots can also promote SOC mineralization, exposing previously protected OC to microbial decomposition and leading to a loss of SOC. The role of plant roots contributing to SOC stabilization has been often studied in isolation from their role in contributing to destabilization. As pointed out by Dijkstra et al. (2021), there are results from studies on SOC stabilization by plant roots and their rhizodeposition suggesting that roots stabilize or increase SOC, and reduce SOC loss, whereas, other studies show that plant roots or their exudates accelerate SOC mineralization (Keiluweit et al., 2015; Li et al., 2021), often referred to as rhizosphere priming. The contribution of roots to OC sequestration has to be differentiated for root litter and rhizodeposition (Kögel-Knabner, 2000, 2017). The OC inputs from these sources may be retained preferentially in different SOC pools and by different mechanisms (Villarino et al., 2021). Root necromass in soils is a major OC input to the soil (Rasse et al., 2005; Sokol and Bradford, 2019), creating new detritusphere regions with all implications for OC sequestration resulting from the formation and storage of POM year after year. Rhizodeposition is easily taken up by microorganisms and its OC remaining in the soil ends up as microbial necromass and microbial EPS, finally contributing to the mineral-associated OM (Villarino et al., 2021), after internal microbial recycling. Increasing OC allocation to roots and rhizodeposition may thus be an important tool for increasing SOC storage.

2 Potential of different soils for carbon sequestration

The efficiency of OC sequestration in soils is largely affected by site conditions. The build-up and decomposition of SOM occur as nonlinear processes and the turnover of SOM depends on the amount of SOM (Chenu et al., 2019; Poeplau et al., 2011). This relationship is site- and soil-specific: SOM turnover is generally higher in coarse-textured soils and under warmer climates than in fine-textured soils and at lower temperatures. Therefore, the SOC storage potential clearly differs among different soils and depends on several site characteristics such as parent material, soil type, texture, mineralogy, and climate.

At the same time, it is relevant to identify soils with high SOC stocks and to maintain or even improve their SOC sequestering capacity through increasing OC inputs and reducing OC loss from these soils. Similarly, a high

OC sequestration potential is identified for soils that are depleted in OM, due to use as grazing land or cropland (Sanderman et al., 2017b).

2.1 Identifying soils with high organic carbon sequestration potential

In general, SOC accumulation depends on pedogenic soil differentiation in terms of the formation of reactive mineral surfaces, soil aggregation, and translocation processes (Kögel-Knabner and Amelung, 2021). Soil texture plays a decisive role as fine-textured soils with high silt and clay contents can have considerably higher SOC stocks than coarse-textured soils (Arrouays et al., 2006; Hassink, 1997; Zinn et al., 2007). Moreover, baseline SOC stocks are important, as the efficiency of OC sequestration is highest in soils with a relatively low SOC level and a corresponding high SOC storage potential (Chenu et al., 2019).

Soil type/soil group is important for understanding and quantifying the sequestration of OM (Keyvanshokouhi et al., 2019; Mayer et al., 2019; Soucémarianadin et al., 2018; Wiesmeier et al., 2015). The distribution, content, and stocks of OM differ between major soil groups as differentiated according to the World Reference Base for Soil Resources (WRB) (IUSS Working Group WRB, 2015). As already described by Batjes (1996) large proportions of SOC are stored below 1 m, which requires consideration in relation to pedogenic OM formation and translocation mechanisms. Yost and Hartemink (2020) report that many investigations do not report soil depth and do not consider the subsoil. As estimations of the areas covered by the Reference Soil Groups are still vague, our understanding on the OC state of soil groups and their changes has to rely on very rough estimates of soil changes (FAO, 2015).

Whereas some mechanisms of OM accumulation occur in many soils, some are specific for a soil group (Kögel-Knabner and Amelung, 2021). While sorptive preservation mechanisms occur more or less in all soil orders, though largely in those with larger portions of reactive clay minerals and oxides, physical protection mechanisms dominate in the steppe soils (Chernozems, Phaeozems, Kastanozems) and are relevant also in Nitisols, Ferralsols, Cambisols, and Luvisols. Some mechanisms are related to specific pedogenic processes of OC accumulation and translocation. This refers to Cryosols, Vertisols, Luvisols, Podzols, but also some other soil groups have specific translocation mechanisms into the subsoil (Kögel-Knabner and Amelung, 2021). These pedogenic OM formation mechanisms and their effects with respect to climate change are only considered for some soil groups (e.g. for Cryosols) in the current discussions on the role of soils in the global C cycle. In addition, we are lacking information on the concentration and stocks of OM in many soils, and specifically for depths below 30 cm. There is increasing evidence that OC storage in subsoils is controlled by soil group or associated parent material, while land use type and soil management often only affects topsoil OC stocks (Chen et al., 2019; Mayer et al., 2019; Poeplau et al., 2020). Moreover, most studies on OM stabilisation mechanisms have been carried out in temperate regions with acid soils, while tropical and arid region with alkaline soils received less attention. The specific processes occurring in these soils and in particular the interaction between organic and inorganic carbon compounds are less known (Chapters 7 and 23 of this book).

Building and maintaining high SOM levels in soils has to rely on high or increasing plant inputs (Amelung et al., 2020). High plant inputs foster microbial activity, and in turn accumulation of microbially derived OC through the accumulation of microbial-derived compounds, including necromass. This process complex implies that SOC sequestration can be improved via increasing productivity in soils (Sanderman et al., 2017a), which is in turn specifically promising in soils with presently low yields. It is therefore essential to provide soil information containing localized information on soil group, degradation status, crop yield gap, and the associated OC sequestration potentials (Amelung et al., 2020). As summarized by Kögel-Knabner and Amelung (2021) only for about 52% of the total area, soil group-specific OC stock data estimates are available. From these considerations, it becomes evident that there may also be a large global potential to sequester OC in soil groups that cover large areas, even if their capacity to store OC is limited and they are only partially suitable for plant production. Arenosols and Leptosols cover 8.7% and 11.1% of the total soil area, but their SOC stocks are not known (Kögel-Knabner and Amelung, 2021) (see, e.g. review by Yost and Hartemink (2020) for sandy soils). It also becomes obvious that soil-specific land-use (e.g. cropland versus afforestation) and management practices for OC sequestration need to be evaluated and adjusted. Degraded soils with low initial SOC stocks could be consequently targeted to foster C sequestration (Lal, 2018). Positive feedbacks of measures to enhance SOC on soil fertility and thus yields and OC inputs in degraded soils could in turn contribute to SOC accrual. Moreover, fine-textured subsoils with a high protective capacity and relatively low SOC contents could efficiently store additional SOC (Balesdent et al., 2018; Chenu et al., 2019).

2.2 Organic carbon saturation concept

In the course of global efforts to foster C sequestration in soils, information on the potential of soils to sequester additional C is required with regard to a prioritization of land management practices to increase SOC stocks (Minasny et al., 2017). There is growing evidence that soils have a finite capacity to store C as their potential to stabilize SOM against microbial mineralization may be limited by the amount of fine mineral particles (Baldock and Skjemstad, 2000; Chung et al., 2008; Six et al., 2002; Stewart et al., 2007). The OC saturation concept is based on the assumption that there is an upper limit of stable OC storage via interaction of SOM with mineral surfaces as quantitatively most important stabilization mechanism in most soils (Arrouays et al., 2006; Oades, 1988; Sollins et al., 1996; von Lützow et al., 2006). The capacity of soils to preserve SOC may be determined by the amount of the fine mineral fraction (<20/<53 μm, medium/fine silt and clay) as proxy for mineral surfaces. In his pioneer work, Hassink (1997) proposed an equation to estimate the maximum OC storage capacity related to the fine fraction <20 μm based on a range of uncultivated and grassland soils in temperate and tropical regions. According to this equation, the OC sequestration potential (corresponding to the OC saturation deficit) can be calculated as the difference between the maximum OC storage capacity and the actual amount of OC in the fine fraction, which can be determined by fractionation. The OC saturation concept was further expanded to other land uses and climatic regions and specific equations were proposed for cropland, grassland, and forest soils, different clay types (1:1 vs. 2:1 clays) and size definitions of the fine fraction (<20 μm vs. <53 μm) (Beare et al., 2014; Feller and Beare, 1997; Feng et al., 2013; Six et al., 2002; Wiesmeier et al., 2015; Zhao et al., 2006). Further refinements of the OC saturation approach were proposed with regard to the incorporation of additional parameters such as specific surface area, extractable Al and pH as well as boundary line or quantile regression analysis of the data (Beare et al., 2014; Feng et al., 2013; McNally et al., 2017).

The OC saturation concept was frequently used to estimate the OC sequestration potential of agricultural soils (Angers et al., 2011; Chan, 2001; Chen et al., 2019; Six et al., 2002; Sparrow et al., 2006; Wiesmeier et al., 2014a; Zhao et al., 2006). However, the suitability of this concept was discussed controversially. A critical aspect of the OC saturation concept is the fact that it is not suitable to estimate the total OC sequestration potential of soils as it only refers to mineral-protected OC and does not include POM, which may potentially accumulate indefinitely (Cotrufo et al., 2019) and OC in the sand fraction (Barré et al., 2017; Chen et al., 2019). Thus, a considerable proportion of total SOC may be not taken into account, although most agriculturally used soils contain relatively large proportions (often >80% of total SOC) of OC in the fine fraction (Beare et al., 2014; Chen et al., 2019; Wiesmeier et al., 2014a). Moreover, the OC saturation concept refers to the biophysically determined SOC storage capacity independent from management aspects and economic or policy constraints and may thus not be suitable to estimate the 'effective' OC sequestration potential of soils that can actually be achieved by means of improved management (Castellano et al., 2015). Therefore, alternative 'data- or model-driven' approaches were proposed that provide references for maximum SOC storage derived from highest SOC stocks that can be achieved in specific regions/soils under given land use and management practices (Barré et al., 2017; Chenu et al., 2019). In order to assess the applicability of the OC saturation approach for an estimation of the SOC sequestration potential, field studies on managed soils with varying degrees of OC saturation are needed.

3 Soil management options for organic carbon sequestration

3.1 *Increasing organic carbon input vs. decreasing organic carbon output*

The storage of SOC depends on a variety of pedogenic, biological, topographic, and climatic properties depending on the spatial scale considered (Wiesmeier et al., 2019; Chapter 5 of this book). Under long-term constant environmental conditions, SOC stocks approach a dynamic equilibrium of OC input (e.g. harvest residues, organic fertilizers, roots, rhizodeposition) and OC output (by mineralization or erosion). In order to foster OC sequestration in soils, the soil management can be changed in a way to either increase OC inputs to the soil by improved land use/management practices or to decrease OC outputs by increasing the retention time of OM through addition of stable OM (e.g. biochar) or increasing the interaction with soil minerals. What is crucial in this regard is to avoid or at least reduce trade-offs with other soil functions/ ecosystem services, e.g. reduced biomass production/food security or emission of greenhouse gases such as N_2O (Guenet et al., 2021) that potentially offsets SOC gains (Chapter 6 of this book).

3.1.1 Management practices to increase organic carbon input

As the type of land use/management largely controls OC input in soils, land use changes or improved management of agricultural soils proofed to be successful strategies to increase OC inputs and SOC sequestration in mineral soils.

As there are substantial differences of SOC stocks among main land use types, a land use change from intensively managed land use systems such as croplands to more extensively managed land uses such as grasslands or forests normally results in a long-term increase of SOC stocks (Guo and Gifford, 2002; Poeplau et al., 2011; Post and Kwon, 2000; Schulp and Veldkamp, 2008). The conversion of cropland to permanent grassland is an efficient way to sequester C in the long-term (Freibauer et al., 2004; Post and Kwon, 2000; Smith, 2004; Soussana et al., 2004). Converting cropland to forest normally results in lower OC sequestration than its conversion to grassland, particularly in temperate environments, and may lead to a relative shift from stable to labile SOC, particularly when coniferous species are used (Paul et al., 2002; Poeplau and Don, 2013; Post and Kwon, 2000; Smith, 2004). For afforestation of grassland, no globally consistent effect on SOC stocks was observed, while in temperate environments SOC tended to decrease (Guo and Gifford, 2002; Paul et al.,

Published by Burleigh Dodds Science Publishing Limited, 2023.

2002; Poeplau et al., 2011; Post and Kwon, 2000). While OC sequestration by land use extensification was mainly attributed to increased OC inputs, other factors that increase OC retention time such as reduced tillage/soil disturbance probably also play a role. In fact, evidence was found that total OC inputs do not differ between cropland and grassland soils in temperate regions and that OM quality (more root-derived OC inputs in grassland soils) rather than quantity drives differences in SOC stocks between land use types (Jacobs et al., 2020).

For cropland soils, there are different management practices that increase OC inputs and proofed to sequester OC in the long-term (Chapter 16 of this book). In this regard, the design of crop rotations is decisive to maximise OC inputs. Generally, increasing the diversity of cropping systems has been shown to be beneficial for the build-up of SOC (Jarecki and Lal, 2003; McDaniel et al., 2014; Tiemann and Grandy, 2015). In particular, the integration of legumes, ley grassland, cover crops, undersown crops, deep-rooting crops, and perennial crops into cropping systems leads to sequestration of C, mainly by increased belowground OC inputs (Bolinder et al., 2012; Ledo et al., 2020; Martin et al., 2020; Paustian et al., 2019; Poeplau and Don, 2015; West and Post, 2002). An increased retention of crop residues is another strategy to enhance the OC input into agricultural soils and their SOC stocks. In particular, the retention of aboveground harvest residues such as stubble, beet leaves, and straw or their return via organic fertilizers (slurry, manure, or digestate) is an efficient way to increase OC inputs and thus SOC stocks (Lal, 2005; Liu et al., 2014; Mann et al., 2002; Powlson et al., 2008). In this regard, it should be mentioned that the application of external organic fertilizers does not result in OC sequestration if they return to land anyway, as in this case, OC sources are only relocated and locally concentrated. Organic fertilizers would only directly contribute to soil carbon sequestration, if the alternative were burning as in the case of straw in many region of the world. They would indirectly contribute to OC sequestration, when crop yields and thus OC inputs increased because of improved chemical and physical soil properties (Paustian et al., 2019; Sykes et al., 2019).

Agroforestry systems that integrate woody species in agricultural land provide diverse ecosystem services and increase SOC stocks mainly due to increased above and belowground OC inputs (Chapter 20 of this book). Several global meta-analyses provided evidence for OC sequestration by agroforestry systems, particularly in tropical and subtropical regions, but also in temperate environments (Cardinael et al., 2018; Chatterjee et al., 2018; De Stefano and Jacobson, 2018; Feliciano et al., 2018; Hübner et al., 2021; Kim et al., 2016; Mayer et al., 2022; Shi et al., 2018). The introduction of trees in agricultural systems increases the overall OC input by pruning residues, litterfall, root turnover, and rhizodeposition within the sphere of influence of trees (Cardinael et al., 2018; De Stefano and Jacobson, 2018). Indications were found that root-derived OC inputs are deposited even in deep subsoils as tree roots may grow deeper in agroforestry system compared to forest ecosystems (Cardinael et al., 2015; Germon et al., 2016). Moreover, SOC increases can be attributed to OC inputs of understorey vegetation within tree rows of agroforestry systems as well as increased yields on adjacent agricultural land due to improved microclimate conditions (Cardinael et al., 2018; Lorenz and Lal, 2014). Although increased OC inputs may be the most important factor for OC sequestration in agroforestry systems, other processes leading to better microbial functioning due to enhanced soil quality (Guillot et al., 2021), and/or an increased retention of OM through reduced soil disturbance and erosion and reduced decomposition of recalcitrant tree litter may be relevant.

For grassland ecosystems, several improved grassland management practices are discussed in order to increase SOC stocks mainly via increased root-derived OC inputs (Chapter 18 of this book). Most importantly, improved grazing management such as optimized grazing intensity or holistic grazing systems may increase SOC stocks (Conant et al., 2017), although effects of grazing on SOC are highly context-specific and grazing management has to be locally adapted in order to contribute to C sequestration (McSherry and Ritchie, 2013). Particularly in semiarid regions, reduced grazing intensity or spatial/temporal grazing exclusion leads to increased above and belowground OC inputs and increased SOC stocks (Hobley et al., 2018; Wang et al., 2018). In temperate regions, optimized grassland management intensity in terms of the frequency of cutting and fertilization may affect SOC storage, but evidence for that is scarce (Soussana et al., 2004; Ward et al., 2016). Moreover, plant diversity in temperate grasslands seems to be positively correlated with SOC storage, mainly due to increased rhizosphere OC input (Lange et al., 2015; Yang et al., 2019). Potential positive effects of management practices such as sward renewal and integration of deep rooting and/or leguminous species on belowground OC inputs and SOC storage have to be further studied (Whitehead, 2020; Whitehead et al., 2018). Moreover, these positive effects must be balanced with trade-offs in form of N_2O and/or CH_4 emissions from lifestock to evaluate the contribution SOC sequestration in grazing systems can make to GHG emission mitigation (Chapters 6 and 18 of this book).

3.1.2 Approaches to decrease organic carbon loss

Besides strategies to increase OC inputs into agricultural soils, there are several management practices that can contribute to OC sequestration by reducing the mineralization of OM and increasing its retention time in mineral soils. This can generally be achieved by increasing the spatial extend of the detritusphere-mineral soil interface through incorporation of OM deeper in the mineral soil by inversion tillage, or by increasing the available mineral surface for interaction with OM by amending the soil with fine-sized minerals. Another approach is the addition of stable OM, such as biochar.

In general, a deepening of cropland topsoils by tillage may increase OC inputs in deeper parts of the mineral soil and promote the formation of organo-mineral associations, as it was indicated by relatively high proportions of mineral-associated OM in cropland topsoils compared to other land uses (Stemmer et al., 1999; Wiesmeier et al., 2014a). As a result, SOC stocks may be increased in the long term, as it was observed in temperate cropland soils (Don et al., 2009; Nieder and Richter, 1986; Wiesmeier et al., 2012). Moreover, deep inversion tillage in soils with compacted subsoil layers (e.g. hard or plough pan) or for pasture renewal may be an effective measure to foster C sequestration (Alcántara et al., 2016, 2017; Lawrence-Smith et al., 2021; Schiedung et al., 2019). The burial of OC-rich topsoils (usually below 60 cm depth) results in reduced decomposition of OM, while fresh OC inputs in exposed subsoils depleted in C may be effectively stabilized. Reduced tillage or no-till systems are practiced around the world to control erosion and improve soil structure, water holding capacity, and soil macrofauna by increased soil cover and decreased soil disturbance. Moreover, it is commonly assumed that under reduced tillage/no-till systems soil aggregation and thus the physical protection of OM is improved, while conventional tillage leads to a disruption of (macro) aggregates and increased mineralization of OM (Balesdent et al., 2000; Beare et al., 1994; Six et al., 1998). Although some studies reported increased SOC stocks in topsoils as a result of reduced tillage/no till (West and Post, 2002), there is growing evidence from several studies and meta-analyses that when subsoil layers are considered, reduced tillage/no-till systems do not necessarily contribute to a net OC sequestration but rather induce a vertical redistribution of C in the soil profile (Haddaway et al., 2017; Luo et al., 2010; Meurer et al., 2018; Powlson et al., 2014). In reduced tillage/no-till systems, SOC stocks are often higher in the upmost topsoil (usually in 0–10 cm) compared to conventional tillage systems, as input of OC is concentrated in this depth. However, in the lower topsoil SOC stocks are lower as the incorporation of OM in this depth is reduced (except for root-derived OC inputs).

Amending soils with minerals to increase their protective capacity is a promising strategy to enhance the retention of OC inputs, particularly in sandy soils with a naturally low SOC storage potential. In Australia, sandy topsoils are often amended with clay to improve their chemical and physical properties, reduce soil erosion and nutrient leaching, and increase SOC stocks (Cann, 2000; Churchman et al., 2014; Churchman et al., 2020; Hall et al., 2010; Schapel et al., 2018). Besides the incorporation of external clay-rich material into agricultural topsoils, mixing subsoil clay into sandy topsoils ('delving') is a common practice (Betti et al., 2015; Churchman et al., 2020). The amendment of soils with crushed silicate rocks such as basalt is discussed as option to foster the sequestration of inorganic C, as enhanced weathering of silicate rocks in soils liberates Ca and/or Mg that delivers stable carbonates that either remain in the soil or are leached to the oceans (Beerling et al., 2020, 2018). Besides silicate rocks, also silicate by-products from industrial processes such as steel slag can be used (Reddy et al., 2019). Despite potential risks with regard to the release of metals and persistent organic compounds, further benefits in terms of crop production can be expected due to soil pH increases and nutrient additions that in turn may increase organic OC sequestration (Lehmann and Possinger, 2020). However, evidence for efficient organic and/or inorganic C sequestration by mineral amendments is still poor and additional GHG emissions for mining/excavation, grinding, transport, and application of minerals have to be taken into account (Beerling et al., 2020; Chenu et al., 2019).

The application of thermally transformed OM (biochar and hydrochar) to agricultural soils is a strategy to contribute to OC sequestration by adding OM with a high stability against microbial degradation and thus low turnover (Lehmann et al., 2006). Thermal conversion of organic material at high temperatures may occur at high temperatures (>450°C) through pyrolysis or though heating at lower temperatures (around 200°C) in the presence of water. The latter procedure transforms substrates with high residual water into lignite-like products by hydrothermal carbonization (HTC) under high pressure. The stability of thermally transformed products and thus their potential contribution to SOC sequestration largely depends on biomass parent material and production conditions, particularly the pyrolysis temperature (Spokas, 2010). Generally, biochars produced by pyrolysis at temperatures >450°C are characterized by an overall high stability (Crombie and Mašek, 2015), whereas materials produced at lower temperatures such as HTC coals have a much lower stability against decomposition (Naisse et al., 2015).

3.2 Organic carbon sequestration in soils due to historic management techniques

On the other hand, specific soil management techniques have been developed already in historic times that produce significant OC accumulation, while at the same time strongly modifying the soils, such as in the Terra Preta, paddy soils, or plaggen soils (Kögel-Knabner and Amelung, 2021). The Terra Preta or Amazonian Dark Earths are anthropic soils formed on land surfaces enriched by inputs resulting from pre-Columbian settlers. These inputs resulted in soils characterized by a dark thick A horizon, enriched in OM, and elevated contents of plant-available nutrients (P and Ca), a less acidic pH value and a higher cation exchange capacity than the adjacent tropical soils. Terra Preta is one of the rare but prominent examples, how the addition and alteration of OM to (oxidic) soils may change their properties so dramatically that they turn into fertile soils for centuries. Similarly, the plaggen management established in medieval times in NW Europe resulted in the improvement of sandy soils with respect to OC

Published by Burleigh Dodds Science Publishing Limited, 2023.

contents and stocks. The addition of plaggen material to the topsoils induced a volume increase as well as higher OC concentrations in the Ap horizons, resulting in significantly higher SOC stocks and fertility in the Plaggic Anthrosols compared to surrounding soils. The increase in yields, in turn resulted in higher in-situ OM input, most probably also contributing to high SOC stocks (Urbanski et al., 2022). Paddy soils occurring widely in Asia are also highly modified by specific paddy management operations, especially artificial submergence and drainage, ploughing and puddling (= ploughing and levelling the surface layer of a submerged soil), organic manuring (animal manure, rice straw, and other crop residues, often fermented with sediments taken from the river or channel), liming, and fertilization. Paddy soil management has a clear effect on the accumulation of OM and the OM storage in paddy soils exceeded OM storage in corresponding non-paddy soils (Kalbitz et al., 2013; Wissing et al., 2011). The large accumulation of OM observed in paddy soils is considered to be due to high input of plant residues coupled to alternating redox conditions (Chen et al., 2021; Kögel-Knabner et al., 2010; Kölbl et al., 2014).

These are examples for historic soil management that resulted in higher OC contents and stocks in combination with higher fertility, increased yields and thus in turn a further increase in OC sequestration. The amount of OM input materials available is a major constraint for increasing OC sequestration in soils, specifically when taking into account the higher amounts of OM input necessary to comply for the increased turnover of OM due to increasing temperatures (Bruni et al., 2021; Riggers et al., 2021; Wiesmeier et al., 2016). If OM input materials are available, it seems worthwhile to consider such 'extreme' management options to increase OC sequestration in soils. This calls for judicious management of organic waste materials, which should be considered as raw materials for transformation into soil amendment instead of allowing their destruction and rapid release of CO_2 to the atmosphere.

4 Acknowledgements

The authors acknowledge funding by the German Federal Ministry of Education and Research (BMBF) in the framework of the funding measure 'Soil as a Sustainable Resource for the Bioeconomy–BonaRes', project 'BonaRes (Module B): BonaRes Centre for Soil Research (grant 031B0511C)'.

5 References

Alcántara, V., Don, A., Vesterdal, L., Well, R. and Nieder, R. 2017. Stability of buried carbon in deep-ploughed forest and cropland soils - Implications for carbon stocks. *Scientific Reports* 7(1), 5511.

Alcántara, V., Don, A., Well, R. and Nieder, R. 2016. Deep ploughing increases agricultural soil organic matter stocks. *Global Change Biology* 22(8), 2939-2956.

Amelung, W., Bossio, D., de Vries, W., Kogel-Knabner, I., Lehmann, J., Amundson, R., Bol, R., Collins, C., Lal, R., Leifeld, J., Minasny, B., Pan, G., Paustian, K., Rumpel, C., Sanderman, J., van Groenigen, J. W., Mooney, S., van Wesemael, B., Wander, M. and Chabbi, A. 2020. Towards a global-scale soil climate mitigation strategy. *Nature Communications* 11(1), 5427.

Angers, D. A., Arrouays, D., Saby, N. P. A. and Walter, C. 2011. Estimating and mapping the carbon saturation deficit of French agricultural topsoils. *Soil Use and Management* 27(4), 448–452.

Angst, G., Mueller, K. E., Nierop, K. G. J. and Simpson, M. J. 2021. Plant- or microbial-derived? A review on the molecular composition of stabilized soil organic matter. *Soil Biology and Biochemistry* 156, 108189.

Arrouays, D., Saby, N., Walter, C., Lemercier, B. and Schvartz, C. 2006. Relationships between particle-size distribution and organic carbon in French arable topsoils. *Soil Use and Management* 22(1), 48–51.

Bailey, V. L., Pries, C. H. and Lajtha, K. 2019. What do we know about soil carbon destabilization? *Environmental Research Letters* 14(8), 083004.

Baldock, J. A., Oades, J. M., Waters, A. G., Peng, X., Vassallo, A. M. and Wilson, M. A. 1992. Aspects of the chemical structure of soil organic materials as revealed by solid-state13C NMR spectroscopy. *Biogeochemistry* 16(1), 1–42.

Baldock, J. A. and Skjemstad, J. O. 2000. Role of the soil matrix and minerals in protecting natural organic materials against biological attack. *Organic Geochemistry* 31(7–8), 697–710.

Balesdent, J., Basile-Doelsch, I., Chadoeuf, J., Cornu, S., Derrien, D., Fekiacova, Z. and Hatté, C. 2018. Atmosphere–soil carbon transfer as a function of soil depth. *Nature* 559(7715), 599–602.

Balesdent, J., Chenu, C. and Balabane, M. 2000. Relationship of soil organic matter dynamics to physical protection and tillage. *Soil and Tillage Research* 53(3–4), 215–230.

Barré, P., Angers, D. A., Basile-Doelsch, I., Bispo, A., Cécillon, L., Chenu, C., Chevallier, T., Derrien, D., Eglin, T. K. and Pellerin, S. 2017. Ideas and perspectives: Can we use the soil carbon saturation deficit to quantitatively assess the soil carbon storage potential, or should we explore other strategies? *Biogeosciences* Discuss. [preprint], https://doi.org/10.5194/bg-2017-395, 2017.

Barthod, J., Dignac, M.-F., Le Mer, G., Bottinelli, N., Watteau, F., Kögel-Knabner, I. and Rumpel, C. 2020. How do earthworms affect organic matter decomposition in the presence of clay-sized minerals? *Soil Biology and Biochemistry* 143, 107730.

Basile-Doelsch, I., Balesdent, J. and Rose, J. 2015. Are interactions between organic compounds and nanoscale weathering minerals the key drivers of carbon storage in soils? *Environmental Science and Technology* 49(7), 3997–3998.

Batjes, N. H. 1996. Total carbon and nitrogen in the soils of the world. *European Journal of Soil Science* 47(2), 151–163.

Beare, M. H., Hendrix, P. F., Cabrera, M. L. and Coleman, D. C. 1994. Aggregate-protected and unprotected organic matter pools in conventional- and no-tillage soils. *Soil Science Society of America Journal* 58(3), 787–795.

Beare, M. H., McNeill, S. J., Curtin, D., Parfitt, R. L., Jones, H. S., Dodd, M. B. and Sharp, J. 2014. Estimating the organic carbon stabilisation capacity and saturation deficit of soils: A New Zealand case study. *Biogeochemistry* 120(1–3), 71–87.

Beerling, D. J., Kantzas, E. P., Lomas, M. R., Wade, P., Eufrasio, R. M., Renforth, P., Sarkar, B., Andrews, M. G., James, R. H., Pearce, C. R., Mercure, J. F., Pollitt, H., Holden, P. B.,

Edwards, N. R., Khanna, M., Koh, L., Quegan, S., Pidgeon, N. F., Janssens, I. A., Hansen, J. and Banwart, S. A. 2020. Potential for large-scale CO_2 removal via enhanced rock weathering with croplands. *Nature* 583(7815), 242-248.

Beerling, D. J., Leake, J. R., Long, S. P., Scholes, J. D., Ton, J., Nelson, P. N., Bird, M., Kantzas, E., Taylor, L. L., Sarkar, B., Kelland, M., DeLucia, E., Kantola, I., Müller, C., Rau, G. and Hansen, J. 2018. Farming with crops and rocks to address global climate, food and soil security. *Nature Plants* 4(3), 138-147.

Betti, G., Grant, C., Churchman, G. and Murray, R. 2015. Increased profile wettability in texture-contrast soils from clay delving: Case studies in South Australia. *Soil Research* 53(2), 125-136.

Bolinder, M. A., Kätterer, T., Andrén, O. and Parent, L. E. 2012. Estimating carbon inputs to soil in forage-based crop rotations and modeling the effects on soil carbon dynamics in a Swedish long-term field experiment. *Canadian Journal of Soil Science* 92(6), 821-833.

Brookes, P. C., Chen, Y., Chen, L., Qiu, G., Luo, Y. and Xu, J. 2017. Is the rate of mineralization of soil organic carbon under microbiological control? *Soil Biology and Biochemistry* 112, 127-139.

Bruni, E., Guenet, B., Huang, Y., Clivot, H., Virto, I., Farina, R., Kätterer, T., Ciais, P., Martin, M. and Chenu, C. 2021. Additional carbon inputs to reach a 4 per 1000 objective in Europe: Feasibility and projected impacts of climate change based on century simulations of long-term arable experiments. *Biogeosciences* 18(13), 3981-4004.

Bucka, F. B., Kölbl, A., Uteau, D., Peth, S. and Kögel-Knabner, I. 2019. Organic matter input determines structure development and aggregate formation in artificial soils. *Geoderma* 354, 113881.

Buckeridge, K. M., Mason, K. E., McNamara, N. P., Ostle, N., Puissant, J., Goodall, T., Griffiths, R. I., Stott, A. W. and Whitaker, J. 2020. Environmental and microbial controls on microbial necromass recycling, an important precursor for soil carbon stabilization. *Communications Earth & Environment* 1(1), 36.

Cai, P., Sun, X., Wu, Y., Gao, C., Mortimer, M., Holden, P. A., Redmile-Gordon, M. and Huang, Q. 2019. Soil biofilms: Microbial interactions, challenges, and advanced techniques for ex-situ characterization. *Soil Ecology Letters* 1(3-4), 85-93.

Cann, M. A. 2000. Clay spreading on water repellent sands in the south east of South Australia–Promoting sustainable agriculture. *Journal of Hydrology* 231-232, 333-341.

Cardinael, R., Chevallier, T., Barthès, B. G., Saby, N. P. A., Parent, T., Dupraz, C., Bernoux, M. and Chenu, C. 2015. Impact of alley cropping agroforestry on stocks, forms and spatial distribution of soil organic carbon–A case study in a Mediterranean context. *Geoderma* 259-260, 288-299.

Cardinael, R., Guenet, B., Chevallier, T., Dupraz, C., Cozzi, T. and Chenu, C. 2018. High organic inputs explain shallow and deep SOC storage in a long-term agroforestry system - Combining experimental and modeling approaches. *Biogeosciences* 15(1), 297-317.

Castellano, M. J., Mueller, K. E., Olk, D. C., Sawyer, J. E. and Six, J. 2015. Integrating plant litter quality, soil organic matter stabilization, and the carbon saturation concept. *Global Change Biology* 21(9), 3200-3209.

Chan, K. Y. 2001. Soil particulate organic carbon under different land use and management. *Soil Use and Management* 17(4), 217-221.

Chatterjee, N., Nair, P. K. R., Chakraborty, S. and Nair, V. D. 2018. Changes in soil carbon stocks across the forest-agroforest-agriculture/pasture continuum in various agroecological regions: A meta-analysis. *Agriculture, Ecosystems and Environment* 266, 55–67.

Chen, S., Arrouays, D., Angers, D. A., Martin, M. P. and Walter, C. 2019. Soil carbon stocks under different land uses and the applicability of the soil carbon saturation concept. *Soil and Tillage Research* 188, 53–58.

Chen, X., Hu, Y., Xia, Y., Zheng, S., Ma, C., Rui, Y., He, H., Huang, D., Zhang, Z., Ge, T., Wu, J., Guggenberger, G., Kuzyakov, Y. and Su, Y. 2021. Contrasting pathways of carbon sequestration in paddy and upland soils. *Global Change Biology* 27(11), 2478–2490.

Chenu, C., Angers, D. A., Barré, P., Derrien, D., Arrouays, D. and Balesdent, J. 2019. Increasing organic stocks in agricultural soils: Knowledge gaps and potential innovations. *Soil and Tillage Research* 188, 41–52.

Chenu, C. and Cosentino, D. 2011. Microbial regulation of soil structural dynamics. In: *The Architecture and Biology of Soils: Life in Inner Space*, Edited by K. Ritz and I. Young. Wallingford, UK: CABI (2011), pp. 416.

Chenu, C. and Jaunet, A. M. 1992. Cryoscanning electron microscopy of microbial extracellular polysaccharides and their association with minerals. *Scanning* 14(6), 360–364.

Chenu, C. and Plante, A. F. 2006. Clay-sized organo-mineral complexes in a cultivation chronosequence: Revisiting the concept of the 'primary organo-mineral complex'. *European Journal of Soil Science* 57(4), 596–607.

Chevallier, T., Woignier, T., Toucet, J. and Blanchart, E. 2010. Organic carbon stabilization in the fractal pore structure of andosols. *Geoderma* 159(1–2), 182–188.

Christensen, B. T. 2001. Physical fractionation of soil and structural and functional complexity in organic matter turnover. *European Journal of Soil Science* 52(3), 345–353.

Chung, H., Grove, J. H. and Six, J. 2008. Indications for soil carbon saturation in a temperate agroecosystem. *Soil Science Society of America Journal* 72(4), 1132–1139.

Churchman, G. J., Noble, A., Bailey, G., Chittleborough, D. and Harper, R. 2014. Clay addition and redistribution to enhance carbon sequestration in soils. In: Hartemink, A. E. and McSweeney, K. (Eds.), *Soil Carbon*. Cham: Springer International Publishing, pp. 327–335.

Churchman, G. J., Singh, M., Schapel, A., Sarkar, B. and Bolan, N. 2020. Clay minerals as the key to the sequestration of carbon in soils. *Clays and Clay Minerals* 68(2), 135–143.

Conant, R. T., Cerri, C. E. P., Osborne, B. B. and Paustian, K. 2017. Grassland management impacts on soil carbon stocks: A new synthesis. *Ecological Applications* 27(2), 662–668.

Costa, O. Y. A., Raaijmakers, J. M. and Kuramae, E. E. 2018. Microbial extracellular polymeric substances: Ecological function and impact on soil aggregation. *Frontiers in Microbiology* 9(1636), 1636.

Cotrufo, M. F., Ranalli, M. G., Haddix, M. L., Six, J. and Lugato, E. 2019. Soil carbon storage informed by particulate and mineral-associated organic matter. *Nature Geoscience* 12(12), 989–994.

Cotrufo, M. F., Wallenstein, M. D., Boot, C. M., Denef, K. and Paul, E. 2013. The Microbial Efficiency-Matrix Stabilization (MEMS) framework integrates plant litter decomposition with soil organic matter stabilization: Do labile plant inputs form stable soil organic matter? *Global Change Biology* 19(4), 988–995.

Crombie, K. and Mašek, O. 2015. Pyrolysis biochar systems, balance between bioenergy and carbon sequestration. *GCB Bioenergy* 7(2), 349-361.

De Stefano, A. and Jacobson, M. G. 2018. Soil carbon sequestration in agroforestry systems: A meta-analysis. *Agroforestry Systems* 92(2), 285-299.

Dijkstra, F. A., Zhu, B. and Cheng, W. 2021. Root effects on soil organic carbon: A double-edged sword. *New Phytologist* 230(1), 60-65.

Don, A., Scholten, T. and Schulze, E.-D. 2009. Conversion of cropland into grassland: Implications for soil organic-carbon stocks in two soils with different texture. *Journal of Plant Nutrition and Soil Science* 172(1), 53-62.

Erktan, A., Or, D. and Scheu, S. 2020. The physical structure of soil: Determinant and consequence of trophic interactions. *Soil Biology and Biochemistry* 148, 107876.

Eusterhues, K., Wagner, F. E., Häusler, W., Hanzlik, M., Knicker, H., Totsche, K. U., Kögel-Knabner, I. and Schwertmann, U. 2008. Characterization of ferrihydrite-soil organic matter coprecipitates by X-ray diffraction and Mössbauer spectroscopy. *Environmental Science and Technology* 42(21), 7891-7897.

FAO. 2015. *Status of the World's Soil Resources (SWSR) - Main Report*. Rome, Italy: Food and Agriculture Organization of the United Nations and Intergovernmental Technical Panel on Soils.

Feliciano, D., Ledo, A., Hillier, J. and Nayak, D. R. 2018. Which agroforestry options give the greatest soil and above ground carbon benefits in different world regions? *Agriculture, Ecosystems and Environment* 254, 117-129.

Feller, C. and Beare, M. H. 1997. Physical control of soil organic matter dynamics in the tropics. *Geoderma* 79(1-4), 69-116.

Feng, W., Plante, A. F. and Six, J. 2013. Improving estimates of maximal organic carbon stabilization by fine soil particles. *Biogeochemistry* 112(1-3), 81-93.

Filser, J., Faber, J. H., Tiunov, A. V., Brussaard, L., Frouz, J., De Deyn, G., Uvarov, A. V., Berg, M. P., Lavelle, P., Loreau, M., Wall, D. H., Querner, P., Eijsackers, H. and Jiménez, J. J. 2016. Soil fauna: Key to new carbon models. *Soil* 2(4), 565-582.

Flemming, H. C. and Wingender, J. 2010. The biofilm matrix. *Nature Reviews. Microbiology* 8(9), 623-633.

Flemming, H. C. and Wuertz, S. 2019. Bacteria and archaea on Earth and their abundance in biofilms. *Nature Reviews Microbiology* 17(4), 247-260.

Foster, R. C. 1988. Microenvironments of soil microorganisms. *Biology and Fertility of Soils* 6(3), 189-203.

Fox, J. F., Campbell, J. E. and Acton, P. M. 2020. Carbon sequestration by reforesting legacy grasslands on coal mining sites. *Energies* 13(23).

Freibauer, A., Rounsevell, M. D. A., Smith, P. and Verhagen, J. 2004. Carbon sequestration in the agricultural soils of Europe. *Geoderma* 122(1), 1-23.

Gaillard, V., Chenu, C., Recous, S. and Richard, G. 1999. Carbon, nitrogen and microbial gradients induced by plant residues decomposing in soil. *European Journal of Soil Science* 50(4), 567-578.

Germon, A., Cardinael, R., Prieto, I., Mao, Z., Kim, J., Stokes, A., Dupraz, C., Laclau, J.-P. and Jourdan, C. 2016. Unexpected phenology and lifespan of shallow and deep fine roots of walnut trees grown in a silvoarable Mediterranean agroforestry system. *Plant and Soil* 401(1-2), 409-426.

Gleixner, G., Poirier, N., Bol, R. and Balesdent, J. 2002. Molecular dynamics of organic matter in a cultivated soil. *Organic Geochemistry* 33(3), 357-366.

Guenet, B., Gabrielle, B., Chenu, C., Arrouays, D., Balesdent, J., Bernoux, M., Bruni, E., Caliman, J. P., Cardinael, R., Chen, S., Ciais, P., Desbois, D., Fouche, J., Frank, S., Henault, C., Lugato, E., Naipal, V., Nesme, T., Obersteiner, M., Pellerin, S., Powlson, D. S., Rasse, D. P., Rees, F., Soussana, J. F., Su, Y., Tian, H., Valin, H. and Zhou, F. 2021. Can N_2O emissions offset the benefits from soil organic carbon storage? *Global Change Biology* 27(2), 237–256.

Guillot, E., Bertrand, I., Rumpel, C., Gomez, C., Arnal, D., Abadie, J. and Hinsinger, P. 2021. Spatial heterogeneity of soil quality within a Mediterranean alley cropping agroforestry system: Comparison with a monocropping system. *European Journal of Soil Biology* 105, 103330.

Guo, L. B. and Gifford, R. M. 2002. Soil carbon stocks and land use change: A meta analysis. *Global Change Biology* 8(4), 345-360.

Haddaway, N. R., Hedlund, K., Jackson, L. E., Kätterer, T., Lugato, E., Thomsen, I. K., Jørgensen, H. B. and Isberg, P.-E. 2017. How does tillage intensity affect soil organic carbon? A systematic review. *Environmental Evidence* 6(1), 30.

Hall, D. J. M., Jones, H. R., Crabtree, W. L. and Daniels, T. L. 2010. Claying and deep ripping can increase crop yields and profits on water repellent sands with marginal fertility in southern Western Australia. *Soil Research* 48(2), 178–187.

Hassink, J. 1997. The capacity of soils to preserve organic C and N by their association with clay and silt particles. *Plant and Soil* 191(1), 77–87.

He, X., Hou, E., Veen, G. F., Ellwood, M. D. F., Dijkstra, P., Sui, X., Zhang, S., Wen, D. and Chu, C. 2020. Soil microbial biomass increases along elevational gradients in the tropics and subtropics but not elsewhere. *Global Ecology and Biogeography* 29(2), 345–354.

Hobley, E., Garcia-Franco, N., Hübner, R. and Wiesmeier, M. 2018. Reviewing our options: Managing water-limited soils for conservation and restoration. *Land Degradation and Development* 29(4), 1041–1053.

Hübner, R., Kühnel, A., Lu, J., Dettmann, H., Wang, W. and Wiesmeier, M. 2021. Soil carbon sequestration by agroforestry systems in China: A meta-analysis. *Agriculture, Ecosystems and Environment* 315, 107437.

IUSS Working Group. 2015. *International Soil Classification System for Naming Soils and Creating Legends for Soil Maps. World Soil Resources Reports No. 106. WRB, 2015. World Reference Base for Soil Resources 2014. Update*. Rome: Food and Agriculture Organization.

Jacobs, A., Poeplau, C., Weiser, C., Fahrion-Nitschke, A. and Don, A. 2020. Exports and inputs of organic carbon on agricultural soils in Germany. *Nutrient Cycling in Agroecosystems* 118(3), 249-271.

Jarecki, M. K. and Lal, R. 2003. Crop management for soil carbon sequestration. *Critical Reviews in Plant Sciences* 22(6), 471-502.

Jobbagy, E. G. and Jackson, R. B. 2000. The vertical distribution of soil organic carbon and its relation to climate and vegetation. *Ecological Applications* 10(2), 423–436.

Juarez, S., Nunan, N., Duday, A.-C., Pouteau, V. and Chenu, C. 2013. Soil carbon mineralisation responses to alterations of microbial diversity and soil structure. *Biology and Fertility of Soils* 49(7), 939–948.

Just, C., Poeplau, C., Don, A., van Wesemael, B., Kögel-Knabner, I. and Wiesmeier, M. 2021. A simple approach to isolate slow and fast cycling organic carbon fractions in central European soils – Importance of dispersion methods. *Frontiers in Soil Science* 1, 692583.

Kalbitz, K., Kaiser, K., Fiedler, S., Kolbl, A., Amelung, W., Brauer, T., Cao, Z., Don, A., Grootes, P., Jahn, R., Schwark, L., Vogelsang, V., Wissing, L. and Kogel-Knabner, I. 2013. The carbon count of 2000 years of rice cultivation. *Global Change Biology* 19(4), 1107-1113.

Keiluweit, M., Bougoure, J. J., Nico, P. S., Pett-Ridge, J., Weber, P. K. and Kleber, M. 2015. Mineral protection of soil carbon counteracted by root exudates. *Nature Climate Change* 5(6), 588-595.

Keiluweit, M., Wanzek, T., Kleber, M., Nico, P. and Fendorf, S. 2017. Anaerobic microsites have an unaccounted role in soil carbon stabilization. *Nature Communications* 8(1), 1771.

Kemmitt, S. J., Lanyon, C. V., Waite, I. S., Wen, Q., Addiscott, T. M., Bird, N. R. A., O'Donnell, A. G. and Brookes, P. C. 2008. Mineralization of native soil organic matter is not regulated by the size, activity or composition of the soil microbial biomass - A new perspective. *Soil Biology and Biochemistry* 40(1), 61-73.

Keyvanshokouhi, S., Cornu, S., Lafolie, F., Balesdent, J., Guenet, B., Moitrier, N., Moitrier, N., Nougier, C. and Finke, P. 2019. Effects of soil process formalisms and forcing factors on simulated organic carbon depth-distributions in soils. *Science of the Total Environment* 652, 523-537.

Kiem, R. and Kögel-Knabner, I. 2003. Contribution of lignin and polysaccharides to the refractory carbon pool in C-depleted arable soils. *Soil Biology and Biochemistry* 35(1), 101-118.

Kim, D.-G., Kirschbaum, M. U. F. and Beedy, T. L. 2016. Carbon sequestration and net emissions of CH_4 and N_2O under agroforestry: Synthesizing available data and suggestions for future studies. *Agriculture, Ecosystems and Environment* 226, 65-78.

Kleber, M., Bourg, I. C., Coward, E. K., Hansel, C. M., Myneni, S. C. B. and Nunan, N. 2021. Dynamic interactions at the mineral-organic matter interface. *Nature Reviews Earth and Environment* 2(6), 402-421.

Kleber, M., Eusterhues, K., Keiluweit, M., Mikutta, C., Mikutta, R. and Nico, P. S. 2015. Mineral-organic associations: Formation, properties, and relevance in soil. *Environments* 130, 1-140.

Kögel-Knabner, I. 2000. Analytical approaches for characterizing soil organic matter. *Organic Geochemistry* 31(7-8), 609-625.

Kögel-Knabner, I. 2002. The macromolecular organic composition of plant and microbial residues as inputs to soil organic matter. *Soil Biology and Biochemistry* 34(2), 139-162.

Kögel-Knabner, I. 2017. The macromolecular organic composition of plant and microbial residues as inputs to soil organic matter: Fourteen years on. *Soil Biology and Biochemistry* 105, A3-A8.

Kögel-Knabner, I. and Amelung, W. 2021. Soil organic matter in major pedogenic soil groups. *Geoderma* 384, 114785.

Kögel-Knabner, I., Amelung, W., Cao, Z., Fiedler, S., Frenzel, P., Jahn, R., Kalbitz, K., Kölbl, A. and Schloter, M. 2010. Biogeochemistry of paddy soils. *Geoderma* 157(1-2), 1-14.

Kögel-Knabner, I., Guggenberger, G., Kleber, M., Kandeler, E., Kalbitz, K., Scheu, S., Eusterhues, K. and Leinweber, P. 2008. Organo-mineral associations in temperate soils: Integrating biology, mineralogy, and organic matter chemistry. *Journal of Plant Nutrition and Soil Science* 171(1), 61-82.

Kölbl, A., Schad, P., Jahn, R., Amelung, W., Bannert, A., Cao, Z. H., Fiedler, S., Kalbitz, K., Lehndorff, E., Müller-Niggemann, C., Schloter, M., Schwark, L., Vogelsang, V.,

Wissing, L. and Kögel-Knabner, I. 2014. Accelerated soil formation due to paddy management on marshlands (Zhejiang Province, China). *Geoderma* 228-229, 67-89.

Lal, R. 2005. World crop residues production and implications of its use as a biofuel. *Environment International* 31(4), 575-584.

Lal, R. 2018. Digging deeper: A holistic perspective of factors affecting soil organic carbon sequestration in agroecosystems. *Global Change Biology* 24(8), 3285-3301.

Lange, M., Eisenhauer, N., Sierra, C. A., Bessler, H., Engels, C., Griffiths, R. I., Mellado-Vázquez, P. G., Malik, A. A., Roy, J., Scheu, S., Steinbeiss, S., Thomson, B. C., Trumbore, S. E. and Gleixner, G. 2015. Plant diversity increases soil microbial activity and soil carbon storage. *Nature Communications* 6(1), 6707.

Lawrence-Smith, E. J., Curtin, D., Beare, M. H., McNally, S. R., Kelliher, F. M., Calvelo Pereira, R. and Hedley, M. J. 2021. Full inversion tillage during pasture renewal to increase soil carbon storage: New Zealand as a case study. *Global Change Biology* 27(10), 1998-2010.

Ledo, A., Smith, P., Zerihun, A., Whitaker, J., Vicente-Vicente, J. L., Qin, Z., McNamara, N. P., Zinn, Y. L., Llorente, M., Liebig, M., Kuhnert, M., Dondini, M., Don, A., Diaz-Pines, E., Datta, A., Bakka, H., Aguilera, E. and Hillier, J. 2020. Changes in soil organic carbon under perennial crops. *Global Change Biology* 26(7), 4158-4168.

Lehmann, J., Gaunt, J. and Rondon, M. 2006. Bio-char sequestration in terrestrial ecosystems - A review. *Mitigation and Adaptation Strategies for Global Change* 11(2), 403-427.

Lehmann, J., Hansel, C. M., Kaiser, C., Kleber, M., Maher, K., Manzoni, S., Nunan, N., Reichstein, M., Schimel, J. P., Torn, M. S., Wieder, W. R. and Kögel-Knabner, I. 2020. Persistence of soil organic carbon caused by functional complexity. *Nature Geoscience* 13(8), 529-534.

Lehmann, J. and Possinger, A. 2020. *Removal of Atmospheric CO_2 by Rock Weathering Holds Promise for Mitigating Climate Change*. Nature Publishing Group.

Leifeld, J. and Kögel-Knabner, I. 2005. Soil organic matter fractions as early indicators for carbon stock changes under different land-use? *Geoderma* 124(1-2), 143-155.

Levard, C., Doelsch, E., Basile-Doelsch, I., Abidin, Z., Miche, H., Masion, A., Rose, J., Borschneck, D. and Bottero, J. -Y. 2012. Structure and distribution of allophanes, imogolite and proto-imogolite in volcanic soils. *Geoderma* 183-184, 100-108.

Li, H., Bölscher, T., Winnick, M., Tfaily, M. M., Cardon, Z. G. and Keiluweit, M. 2021. Simple plant and microbial exudates destabilize mineral-associated organic matter via multiple pathways. *Environmental Science and Technology* 55(5), 3389-3398.

Liang, C., Amelung, W., Lehmann, J. and Kästner, M. 2019. Quantitative assessment of microbial necromass contribution to soil organic matter. *Global Change Biology* 25(11), 3578-3590.

Liu, C., Lu, M., Cui, J., Li, B. and Fang, C. 2014. Effects of straw carbon input on carbon dynamics in agricultural soils: A meta-analysis. *Global Change Biology* 20(5), 1366-1381.

Lorenz, K. and Lal, R. 2014. Soil organic carbon sequestration in agroforestry systems. A review. *Agronomy for Sustainable Development* 34(2), 443-454.

Luo, Z., Wang, E. and Sun, O. J. 2010. Can no-tillage stimulate carbon sequestration in agricultural soils? A meta-analysis of paired experiments. *Agriculture, Ecosystems and Environment* 139(1-2), 224-231.

Mann, L., Tolbert, V. and Cushman, J. 2002. Potential environmental effects of corn (*Zea mays* L.) stover removal with emphasis on soil organic matter and erosion. *Agriculture, Ecosystems and Environment* 89(3), 149-166.

Manzoni, S., Čapek, P., Porada, P., Thurner, M., Winterdahl, M., Beer, C., Brüchert, V., Frouz, J., Herrmann, A. M., Lindahl, B. D., Lyon, S. W., Šantrůčková, H., Vico, G. and Way, D. 2018. Reviews and syntheses: Carbon use efficiency from organisms to ecosystems - Definitions, theories, and empirical evidence. *Biogeosciences* 15(19), 5929-5949.

Marschner, P. and Rengel, Z. 2012. Chapter 12: Nutrient availability in soils. In: Marschner, P. (Ed.), *Marschner's Mineral Nutrition of Higher Plants* (3rd edn.). San Diego: Academic Press, pp. 315-330.

Martin, G., Durand, J.-L., Duru, M., Gastal, F., Julier, B., Litrico, I., Louarn, G., Médiène, S., Moreau, D., Valentin-Morison, M., Novak, S., Parnaudeau, V., Paschalidou, F., Vertès, F., Voisin, A.-S., Cellier, P. and Jeuffroy, M.-H. 2020. Role of ley pastures in tomorrow's cropping systems: A review. *Agronomy for Sustainable Development* 40(3), 17.

Mayer, S., Kühnel, A., Burmeister, J., Kögel-Knabner, I. and Wiesmeier, M. 2019. Controlling factors of organic carbon stocks in agricultural topsoils and subsoils of Bavaria. *Soil and Tillage Research* 192, 22-32.

Mayer, S., Wiesmeier, M., Sakamoto, E., Hübner, R., Cardinael, R., Kühnel, A. and Kögel-Knabner, I. 2022. Soil organic carbon sequestration in temperate agroforestry systems - A meta-analysis. *Agriculture, Ecosystems and Environment* 323, 107689.

Mbé, B., Monga, O., Pot, V., Otten, W., Hecht, F., Raynaud, X., Nunan, N., Chenu, C., Baveye, P. C. and Garnier, P. 2021. Scenario modelling of carbon mineralization in 3D soil architecture at the microscale: Toward an accessibility coefficient of organic matter for bacteria. *European Journal of Soil Science* 73(1), e13144.

McCarthy, J. F., Ilavsky, J., Jastrow, J. D., Mayer, L. M., Perfect, E. and Zhuang, J. 2008. Protection of organic carbon in soil microaggregates via restructuring of aggregate porosity and filling of pores with accumulating organic matter. *Geochimica et Cosmochimica Acta* 72(19), 4725-4744.

McDaniel, M. D., Grandy, A. S., Tiemann, L. K. and Weintraub, M. N. 2014. Crop rotation complexity regulates the decomposition of high and low quality residues. *Soil Biology and Biochemistry* 78, 243-254.

McNally, S. R., Beare, M. H., Curtin, D., Meenken, E. D., Kelliher, F. M., Calvelo Pereira, R., Shen, Q. and Baldock, J. 2017. Soil carbon sequestration potential of permanent pasture and continuous cropping soils in New Zealand. *Global Change Biology* 23(11), 4544-4555.

McSherry, M. E. and Ritchie, M. E. 2013. Effects of grazing on grassland soil carbon: A global review. *Global Change Biology* 19(5), 1347-1357.

Meurer, K. H. E., Haddaway, N. R., Bolinder, M. A. and Kätterer, T. 2018. Tillage intensity affects total SOC stocks in boreo-temperate regions only in the topsoil–A systematic review using an ESM approach. *Earth-Science Reviews* 177, 613-622.

Minasny, B., Malone, B. P., McBratney, A. B., Angers, D. A., Arrouays, D., Chambers, A., Chaplot, V., Chen, Z.-S., Cheng, K., Das, B. S., Field, D. J., Gimona, A., Hedley, C. B., Hong, S. Y., Mandal, B., Marchant, B. P., Martin, M., McConkey, B. G., Mulder, V. L., O'Rourke, S., Richer-de-Forges, A. C., Odeh, I., Padarian, J., Paustian, K., Pan, G., Poggio, L., Savin, I., Stolbovoy, V., Stockmann, U., Sulaeman, Y., Tsui, C.-C., Vågen, T.-G., van Wesemael, B. and Winowiecki, L. 2017. Soil carbon 4 per mille. *Geoderma* 292, 59-86.

Mueller, C. W. and Koegel-Knabner, I. 2009. Soil organic carbon stocks, distribution, and composition affected by historic land use changes on adjacent sites. *Biology and Fertility of Soils* 45(4), 347-359.

Naisse, C., Girardin, C., Lefèvre, R., Pozzi, A., Maas, R., Stark, A. and Rumpel, C. 2015. Effect of physical weathering on the carbon sequestration potential of biochars and hydrochars in soil. *GCB Bioenergy* 7(3), 488–496.

Nieder, R. and Richter, J. 1986. C- und N-Festlegung in Böden Südostniedersachsens nach Krumenvertiefung. *Zeitschrift für Pflanzenernährung und Bodenkunde* 149(2), 189-201.

Oades, J. M. 1988. The retention of organic-matter in soils. *Biogeochemistry* 5(1), 35–70.

Olson, K. R., Al-Kaisi, M. M., Lal, R. and Lowery, B. 2014. Experimental consideration, treatments, and methods in determining soil organic carbon sequestration rates. *Soil Science Society of America Journal* 78(2), 348–360.

Paul, K. I., Polglase, P. J., Nyakuengama, J. G. and Khanna, P. K. 2002. Change in soil carbon following afforestation. *Forest Ecology and Management* 168(1-3), 241–257.

Paustian, K., Larson, E., Kent, J., Marx, E. and Swan, A. 2019. Soil C sequestration as a biological negative emission strategy. *Frontiers in Climate* 1(8).

Poeplau, C. and Don, A. 2013. Sensitivity of soil organic carbon stocks and fractions to different land-use changes across Europe. *Geoderma* 192, 189-201.

Poeplau, C. and Don, A. 2015. Carbon sequestration in agricultural soils via cultivation of cover crops – A meta-analysis. *Agriculture, Ecosystems and Environment* 200, 33–41.

Poeplau, C., Don, A., Six, J., Kaiser, M., Benbi, D., Chenu, C., Cotrufo, M. F., Derrien, D., Gioacchini, P., Grand, S., Gregorich, E., Griepentrog, M., Gunina, A., Haddix, M., Kuzyakov, Y., Kühnel, A., Macdonald, L. M., Soong, J., Trigalet, S., Vermeire, M.-L., Rovira, P., van Wesemael, B., Wiesmeier, M., Yeasmin, S., Yevdokimov, I. and Nieder, R. 2018. Isolating organic carbon fractions with varying turnover rates in temperate agricultural soils – A comprehensive method comparison. *Soil Biology and Biochemistry* 125, 10–26.

Poeplau, C., Don, A., Vesterdal, L., Leifeld, J., Van Wesemael, B. A. S., Schumacher, J. and Gensior, A. 2011. Temporal dynamics of soil organic carbon after land-use change in the temperate zone – Carbon response functions as a model approach. *Global Change Biology* 17(7), 2415–2427.

Poeplau, C., Jacobs, A., Don, A., Vos, C., Schneider, F., Wittnebel, M., Tiemeyer, B., Heidkamp, A., Prietz, R. and Flessa, H. 2020. Stocks of organic carbon in German agricultural soils—Key results of the first comprehensive inventory. *Journal of Plant Nutrition and Soil Science* 183(6), 665–681.

Poirier, V., Roumet, C. and Munson, A. D. 2018. The root of the matter: Linking root traits and soil organic matter stabilization processes. *Soil Biology and Biochemistry* 120, 246-259.

Post, W. M. and Kwon, K. C. 2000. Soil carbon sequestration and land-use change: Processes and potential. *Global Change Biology* 6(3), 317-327.

Potapov, A. M., Tiunov, A. V. and Scheu, S. 2019. Uncovering trophic positions and food resources of soil animals using bulk natural stable isotope composition. *Biological Reviews of the Cambridge Philosophical Society* 94(1), 37–59.

Powlson, D. S., Riche, A. B., Coleman, K., Glendining, M. J. and Whitmore, A. P. 2008. Carbon sequestration in European soils through straw incorporation: Limitations and alternatives. *Waste Management* 28(4), 741–746.

Powlson, D. S., Stirling, C. M., Jat, M. L., Gerard, B. G., Palm, C. A., Sanchez, P. A. and Cassman, K. G. 2014. Limited potential of no-till agriculture for climate change mitigation. *Nature Climate Change* 4(8), 678–683.

Published by Burleigh Dodds Science Publishing Limited, 2023.

Qiao, Y. J., Liu, Q., Gao, Z. Q., Ge, J. Y., Feng, X. M., Xue, J. F. and Sun, M. 2019. Impacts of long-term fertilization on bacterial community structure, soil microbial biomass, and grain yields in red paddy soil in China. *Applied Ecology and Environmental Research* 17(2), 3353-3370.

Rabbi, S. M. F., Minasny, B., McBratney, A. B. and Young, I. M. 2020. Microbial processing of organic matter drives stability and pore geometry of soil aggregates. *Geoderma* 360, 114033.

Rasmussen, C., Heckman, K., Wieder, W. R., Keiluweit, M., Lawrence, C. R., Berhe, A. A., Blankinship, J. C., Crow, S. E., Druhan, J. L., Hicks Pries, C. E., Marin-Spiotta, E., Plante, A. F., Schädel, C., Schimel, J. P., Sierra, C. A., Thompson, A. and Wagai, R. 2018. Beyond clay: Towards an improved set of variables for predicting soil organic matter content. *Biogeochemistry* 137(3), 297-306.

Rasse, D. P., Rumpel, C. and Dignac, M.-F. 2005. Is soil carbon mostly root carbon? Mechanisms for a specific stabilisation. *Plant and Soil* 269(1-2), 341-356.

Reddy, K. R., Gopakumar, A. and Chetri, J. K. 2019. Critical review of applications of iron and steel slags for carbon sequestration and environmental remediation. *Reviews in Environmental Science and Bio/Technology* 18(1), 127-152.

Riggers, C., Poeplau, C., Don, A., Frühauf, C. and Dechow, R. 2021. How much carbon input is required to preserve or increase projected soil organic carbon stocks in German croplands under climate change? *Plant and Soil* 460(1-2), 417-433.

Rillig, M. C., Leifheit, E. and Lehmann, J. 2021. Microplastic effects on carbon cycling processes in soils. *PLoS Biology* 19(3), e3001130.

Roth, V.-N., Lange, M., Simon, C., Hertkorn, N., Bucher, S., Goodall, T., Griffiths, R. I., Mellado-Vázquez, P. G., Mommer, L., Oram, N. J., Weigelt, A., Dittmar, T. and Gleixner, G. 2019. Persistence of dissolved organic matter explained by molecular changes during its passage through soil. *Nature Geoscience* 12(9), 755-761.

Rumpel, C., Baumann, K., Remusat, L., Dignac, M.-F., Barré, P., Deldicque, D., Glasser, G., Lieberwirth, I. and Chabbi, A. 2015. Nanoscale evidence of contrasted processes for root-derived organic matter stabilization by mineral interactions depending on soil depth. *Soil Biology and Biochemistry* 85, 82-88.

Sanderman, J., Creamer, C., Baisden, W. T., Farrell, M. and Fallon, S. 2017a. Greater soil carbon stocks and faster turnover rates with increasing agricultural productivity. *Soil* 3(1), 1-16.

Sanderman, J., Hengl, T. and Fiske, G. J. 2017b. Soil carbon debt of 12,000 years of human land use. *Proceedings of the National Academy of Sciences of the United States of America* 114(36), 9575-9580.

Schapel, A., Marschner, P. and Churchman, J. 2018. Clay amount and distribution influence organic carbon content in sand with subsoil clay addition. *Soil and Tillage Research* 184, 253-260.

Schiedung, M., Tregurtha, C. S., Beare, M. H., Thomas, S. M. and Don, A. 2019. Deep soil flipping increases carbon stocks of New Zealand grasslands. *Global Change Biology* 25(7), 2296-2309.

Schmidt, M. W. I. and Noack, A. G. 2000. Black carbon in soils and sediments: Analysis, distribution, implications, and current challenges. *Global Biogeochemical Cycles* 14(3), 777-793.

Schulp, C. J. E. and Veldkamp, A. 2008. Long-term landscape - Land use interactions as explaining factor for soil organic matter variability in Dutch agricultural landscapes. *Geoderma* 146(3-4), 457-465.

Shi, L., Feng, W., Xu, J. and Kuzyakov, Y. 2018. Agroforestry systems: Meta-analysis of soil carbon stocks, sequestration processes, and future potentials. *Land Degradation and Development* 29(11), 3886-3897.

Six, J., Conant, R. T., Paul, E. A. and Paustian, K. 2002. Stabilization mechanisms of soil organic matter: Implications for C-saturation of soils. *Plant and Soil* 241(2), 155-176.

Six, J., Elliott, E. T., Paustian, K. and Doran, J. W. 1998. Aggregation and soil organic matter accumulation in cultivated and native grassland soils. *Soil Science Society of America Journal* 62(5), 1367-1377.

Smith, P. 2004. Carbon sequestration in croplands: The potential in Europe and the global context. *European Journal of Agronomy* 20(3), 229-236.

Sokol, N. W. and Bradford, M. A. 2019. Microbial formation of stable soil carbon is more efficient from belowground than aboveground input. *Nature Geoscience* 12(1), 46-53.

Sollins, P., Homann, P. and Caldwell, B. A. 1996. Stabilization and destabilization of soil organic matter: Mechanisms and controls. *Geoderma* 74(1-2), 65-105.

Soucémarianadin, L. N., Cécillon, L., Guenet, B., Chenu, C., Baudin, F., Nicolas, M., Girardin, C. and Barré, P. 2018. Environmental factors controlling soil organic carbon stability in French forest soils. *Plant and Soil* 426(1-2), 267-286.

Soussana, J.-F., Loiseau, P., Vuichard, N., Ceschia, E., Balesdent, J., Chevallier, T. and Arrouays, D. 2004. Carbon cycling and sequestration opportunities in temperate grasslands. *Soil Use and Management* 20(2), 219-230.

Sparrow, L. A., Belbin, K. C. and Doyle, R. B. 2006. Organic carbon in the silt+clay fraction of Tasmanian soils. *Soil Use and Management* 22(2), 219-220.

Spokas, K. A. 2010. Review of the stability of biochar in soils: Predictability of O:C molar ratios. *Carbon Management* 1(2), 289-303.

Steffens, M., Rogge, D. M., Mueller, C. W., Höschen, C., Lugmeier, J., Kölbl, A. and Kögel-Knabner, I. 2017. Identification of distinct functional microstructural domains controlling C Storage in soil. *Environmental Science and Technology* 51(21), 12182-12189.

Stemmer, M., Von Lützow, M., Kandeler, E., Pichlmayer, F. and Gerzabek, M. H. 1999. The effect of maize straw placement on mineralization of C and N in soil particle size fractions. *European Journal of Soil Science* 50(1), 73-85.

Stewart, C. E., Paustian, K., Conant, R. T., Plante, A. F. and Six, J. 2007. Soil carbon saturation: Concept, evidence and evaluation. *Biogeochemistry* 86(1), 19-31.

Sykes, A. J., Macleod, M., Eory, V., Rees, R. M., Payen, F., Myrgiotis, V., Williams, M., Sohi, S., Hillier, J., Moran, D., Manning, D. A. C., Goglio, P., Seghetta, M., Williams, A., Harris, J., Dondini, M., Walton, J., House, J. and Smith, P. 2019. Characterising the biophysical, economic and social impacts of soil carbon sequestration as a greenhouse gas removal technology. *Global Change Biology* 26(3), 1085-1108.

Tamrat, W. Z., Rose, J., Grauby, O., Doelsch, E., Levard, C., Chaurand, P. and Basile-Doelsch, I. 2019. Soil organo-mineral associations formed by co-precipitation of Fe, Si and Al in presence of organic ligands. *Geochimica et Cosmochimica Acta* 260, 15-28.

Thakur, M. P. and Geisen, S. 2019. Trophic regulations of the soil microbiome. *Trends in Microbiology* 27(9), 771-780.

Thorup-Kristensen, K., Halberg, N., Nicolaisen, M., Olesen, J. E., Crews, T. E., Hinsinger, P., Kirkegaard, J., Pierret, A. and Dresbøll, D. B. 2020. Digging deeper for agricultural resources, the value of deep rooting. *Trends in Plant Science* 25(4), 406-417.

Tiemann, L. K. and Grandy, A. S. 2015. Mechanisms of soil carbon accrual and storage in bioenergy cropping systems. *GCB Bioenergy* 7(2), 161-174.

Tisdall, J. M. and Oades, J. M. 1982. Organic-matter and water-stable aggregates in soils. *Journal of Soil Science* 33(2), 141-163.

Torn, M. S., Trumbore, S. E., Chadwick, O. A., Vitousek, P. M. and Hendricks, D. M. 1997. Mineral control of soil organic carbon storage and turnover. *Nature* 389(6647), 170-173.

Totsche, K. U., Amelung, W., Gerzabek, M. H., Guggenberger, G., Klumpp, E., Knief, C., Lehndorff, E., Mikutta, R., Peth, S., Prechtel, A., Ray, N. and Kögel-Knabner, I. 2018. Microaggregates in soils. *Journal of Plant Nutrition and Soil Science* 181(1), 104-136.

Urbanski, L., Schad, P., Kalbitz, K., Van Mourik, J., Gehrt, E. and Kögel-Knabner, I. 2022. Legacy of Plaggen agriculture: High soil organic carbon stocks as result from high carbon input and volume increase. *Geoderma* 406, 115513.

van Wesemael, B., Chartin, C., Wiesmeier, M., von Lützow, M., Hobley, E., Carnol, M., Krüger, I., Campion, M., Roisin, C., Hennart, S. and Kögel-Knabner, I. 2019. An indicator for organic matter dynamics in temperate agricultural soils. *Agriculture, Ecosystems and Environment* 274, 62-75.

Védère, C., Vieublé Gonod, L., Pouteau, V., Girardin, C. and Chenu, C. 2020. Spatial and temporal evolution of detritusphere hotspots at different soil moistures. *Soil Biology and Biochemistry* 150, 107975.

Vidal, A., Watteau, F., Remusat, L., Mueller, C. W., Nguyen Tu, T.-T., Buegger, F., Derenne, S. and Quenea, K. 2019. Earthworm cast formation and development: A shift From plant litter to mineral associated organic matter. *Frontiers in Environmental Science* 7, 55.

Villarino, S. H., Pinto, P., Jackson, R. B. and Piñeiro, G. 2021. Plant rhizodeposition: A key factor for soil organic matter formation in stable fractions. *Science Advances* 7(16), eabd3176.

von Lützow, Mv, Kogel-Knabner, I., Ekschmitt, K., Matzner, E., Guggenberger, G., Marschner, B. and Flessa, H. 2006. Stabilization of organic matter in temperate soils: Mechanisms and their relevance under different soil conditions - A review. *European Journal of Soil Science* 57(4), 426-445.

Wagai, R., Kajiura, M. and Asano, M. 2020. Iron and aluminum association with microbially processed organic matter via meso-density aggregate formation across soils: Organo-metallic glue hypothesis. *Soil* 6(2), 597-627.

Wang, L., Gan, Y., Wiesmeier, M., Zhao, G., Zhang, R., Han, G., Siddique, K. H. M. and Hou, F. 2018. Grazing exclusion–An effective approach for naturally restoring degraded grasslands in Northern China. *Land Degradation and Development* 29(12), 4439-4456.

Ward, S. E., Smart, S. M., Quirk, H., Tallowin, J. R., Mortimer, S. R., Shiel, R. S., Wilby, A. and Bardgett, R. D. 2016. Legacy effects of grassland management on soil carbon to depth. *Global Change Biology* 22(8), 2929-2938.

West, T. O. and Post, W. M. 2002. Soil organic carbon sequestration rates by tillage and crop rotation. *Soil Science Society of America Journal* 66(6), 1930-1946.

Whitehead, D. 2020. Management of grazed landscapes to increase soil carbon stocks in temperate, dryland grasslands. *Frontiers in Sustainable Food Systems* 4, 197.

Whitehead, D., Schipper, L. A., Pronger, J., Moinet, G. Y. K., Mudge, P. L., Calvelo Pereira, R., Kirschbaum, M. U. F., McNally, S. R., Beare, M. H. and Camps-Arbestain, M. 2018. Management practices to reduce losses or increase soil carbon stocks in temperate grazed grasslands: New Zealand as a case study. *Agriculture, Ecosystems and Environment* 265, 432-443.

Wiesmeier, M., Hubner, R., Sporlein, P., Geuss, U., Hangen, E., Reischl, A., Schilling, B., von Lutzow, M. and Kogel-Knabner, I. 2014a. Carbon sequestration potential of soils in southeast Germany derived from stable soil organic carbon saturation. *Global Change Biology* 20(2), 653–665.

Wiesmeier, M., Schad, P., von Lützow, M., Poeplau, C., Spörlein, P., Geuß, U., Hangen, E., Reischl, A., Schilling, B. and Kögel-Knabner, I. 2014b. Quantification of functional soil organic carbon pools for major soil units and land uses in southeast Germany (Bavaria). *Agriculture, Ecosystems and Environment* 185, 208–220.

Wiesmeier, M., Lützow, M. v., Spörlein, P., Geuß, U., Hangen, E., Reischl, A., Schilling, B. and Kögel-Knabner, I. 2015. Land use effects on organic carbon storage in soils of Bavaria: The importance of soil types. *Soil and Tillage Research* 146, 296–302.

Wiesmeier, M., Poeplau, C., Sierra, C. A., Maier, H., Frühauf, C., Hübner, R., Kühnel, A., Spörlein, P., Geuß, U., Hangen, E., Schilling, B., von Lützow, M. and Kögel-Knabner, I. 2016. Projected loss of soil organic carbon in temperate agricultural soils in the 21st century: Effects of climate change and carbon input trends. *Scientific Reports* 6(1), 32525.

Wiesmeier, M., Spörlein, P., Geuß, U., Hangen, E., Haug, S., Reischl, A., Schilling, B., Lützow, M. and Kögel-Knabner, I. 2012. Soil organic carbon stocks in southeast Germany (Bavaria) as affected by land use, soil type and sampling depth. *Global Change Biology* 18(7), 2233–2245.

Wiesmeier, M., Urbanski, L., Hobley, E., Lang, B., von Lützow, M., Marin-Spiotta, E., van Wesemael, B., Rabot, E., Ließ, M., Garcia-Franco, N., Wollschläger, U., Vogel, H.-J. and Kögel-Knabner, I. 2019. Soil organic carbon storage as a key function of soils - A review of drivers and indicators at various scales. *Geoderma* 333, 149–162.

Wissing, L., Kölbl, A., Vogelsang, V., Fu, J.-R., Cao, Z.-H. and Kögel-Knabner, I. 2011. Organic carbon accumulation in a 2000-year chronosequence of paddy soil evolution. *CATENA* 87(3), 376–385.

Witzgall, K., Vidal, A., Schubert, D. I., Höschen, C., Schweizer, S. A., Buegger, F., Pouteau, V., Chenu, C. and Mueller, C. W. 2021. Particulate organic matter as a functional soil component for persistent soil organic carbon. *Nature Communications* 12(1), 4115.

Yang, Y., Tilman, D., Furey, G. and Lehman, C. 2019. Soil carbon sequestration accelerated by restoration of grassland biodiversity. *Nature Communications* 10(1), 718.

Yost, J. L. and Hartemink, A. E. 2020. How deep is the soil studied - An analysis of four soil science journals. *Plant and Soil* 452(1–2), 5–18.

Young, I. M. and Crawford, J. W. 2004. Interactions and self-organization in the soil-microbe complex. *Science* 304(5677), 1634–1637.

Zhao, L., Sun, Y., Zhang, X., Yang, X. and Drury, C. F. 2006. Soil organic carbon in clay and silt sized particles in Chinese mollisols: Relationship to the predicted capacity. *Geoderma* 132(3–4), 315–323.

Zhou, Y., Coventry, D. R., Gupta, V. V. S. R., Fuentes, D., Merchant, A., Kaiser, B. N., Li, J., Wei, Y., Liu, H., Wang, Y., Gan, S. and Denton, M. D. 2020. The preceding root system drives the composition and function of the rhizosphere microbiome. *Genome Biology* 21(1), 89.

Zimmerman, A. R., Chorover, J., Goyne, K. W. and Brantley, S. L. 2004. Protection of mesopore-adsorbed organic matter from enzymatic degradation. *Environmental Science and Technology* 38(17), 4542–4548.

Zinn, Y. L., Lal, R., Bigham, J. M. and Resck, D. V. S. 2007. Edaphic controls on soil organic carbon retention in the Brazilian cerrado: Texture and mineralogy. *Soil Science Society of America Journal* 71(4), 1204–1214.

Published by Burleigh Dodds Science Publishing Limited, 2023.

Chapter 3

Plant influences on soil organic carbon dynamics

Xiaojuan Feng, Institute of Botany, Chinese Academy of Sciences and College of Resources and Environment, University of the Chinese Academy of Sciences, China

1 Introduction

As the primary producers in terrestrial ecosystems, plants transform carbon dioxide (CO_2) into organic molecules via photosynthesis. Plant-synthesized organic matter is subsequently transferred into soils as litter or rhizodeposition, partly decomposed or utilized by soil biota, with the remnants contributing to soil organic matter (Kögel-Knabner and Amelung, 2014). Globally, about 60 Pg of carbon (or half of the plant assimilates) is estimated to enter soils annually (Lal, 2008). Plant inputs are hence crucial drivers for soil organic carbon accumulation, especially during the initial stages of soil formation (Khedim et al., 2021). Early studies put much emphasis on plant-derived moieties that survive decomposition processes owing to either chemical recalcitrance (such as lignin) or physiochemical protection as the primary contributor to soil organic carbon (Berg, 2000; Schmidt et al., 2011). Recent studies, however, find that heterotrophic microbes feeding on plant-derived or soil-derived organic matter also contribute significantly to soil organic carbon by converting labile substrates into relatively stable microbial residues and byproducts (Cotrufo

http://dx.doi.org/10.19103/AS.2022.0106.03

et al., 2013; Liang et al., 2017). Regardless of the pathways, it is doubtless that plants are the ultimate primary source of soil organic carbon, either directly preserved as plant molecules or converted into microbial products. The composition and delivery pathways of plant-derived organic matter into soils (such as rhizodeposition vs. aboveground litter) directly affect its transformation or preservation in the soil and therefore contribution to soil organic carbon. Furthermore, plants regulate soil carbon dynamics by influencing microbial community structure and activity via organic substrate supply and/or interaction with root-associated microbes (such as mycorrhizae). Plant roots, exudates and the associated mycorrhizae also play a vital role in soil aggregation (see also Chapter 2 of this book) and organo-mineral interactions, which further affect soil organic carbon preservation and stabilization (Poirier et al., 2018). These processes differ substantially among ecosystems (grassland vs. forest) and between soil horizons, some of which remain poorly resolved or a topic of debate. This chapter aims to introduce plant influences on soil organic carbon dynamics by summarizing existing knowledge and recent advances pertaining to the above aspects and highlighting potential areas for future research.

2 Pathways and composition of plant inputs into soils

Plant-derived organic matter enters the soil from both above and below ground in the form of litter or rhizodeposition (Fig. 1). Aboveground inputs mainly consist of fallen litter such as leaves, twigs, stems and fruits, but also include dissolved organic matter (DOM) originating from plant surfaces (such as forest canopy), decomposing litter and partly decomposed materials in the forest floor (Kaiser and Kalbitz, 2012). In forests overlain by organic layers, the downward transfer of DOM derived from organic horizons is a significant source of soil organic matter in the mineral horizons (Fröberg et al., 2007; Kalbitz et al., 2007). By comparison, belowground inputs include root litter, root-associated fungal (mycorrhizal) turnover and rhizodeposition (e.g. sloughed root cells, passive exudation and active secretion; Pausch et al., 2012). While an accurate estimate of plant belowground carbon inputs is challenging and highly variable depending on vegetation type, climate and edaphic properties, on average nearly 50% of primary production is considered to enter the soil via roots (Freschet et al., 2013; Gherardi and Sala, 2020). Belowground carbon allocation may be higher in boreal forests (~65% of gross primary productivity) with a low nitrogen (N) availability than in tropical forests (~30%; Gill and Finzi, 2016) and can be even higher in grasslands with a high root-to-shoot ratio (71% of net primary productivity; Martinez et al., 2016). Hence, belowground inputs are considered to be a major source of soil organic carbon because they may be preferentially affected by several stabilization processes than aboveground

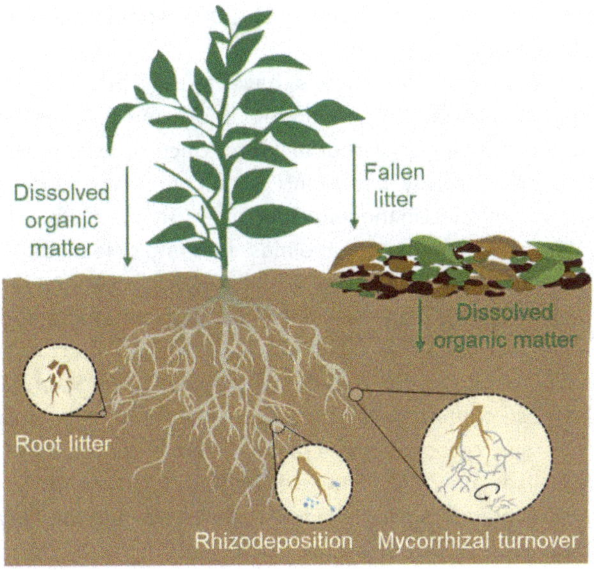

Figure 1 Pathways of plant carbon inputs into soils.

plant material (Rasse et al., 2005; Keller et al., 2021). In some climate regions, root-derived and root-associated (microbial) carbon is estimated to contribute up to 50–70% of total soil organic carbon stocks, e.g. boreal forests (Clemmensen et al., 2013).

The composition of plant-derived organic matter varies considerably depending on the pathway and form of inputs into soils. Plant tissues or litters are composed of a wide array of biomolecules including cellulose, hemicellulose, lignin, polyphenols, lipids, proteins, pigments and storage compounds such as non-structural carbohydrates (Kögel-Knabner, 2002). Among these, cellulose and hemicellulose are the main structural components of plant cell walls and are made up of polysaccharides (Kögel-Knabner, 2002). They account for 40–60% and 20–40% of dried plant biomass, respectively, with variations among species and plant tissues (McKendry, 2002; Yang et al., 2006). Lignin, the second most abundant biopolymer in vascular plants, is composed of phenylpropane units and accounts for ~15–30% of dried plant biomass (Kirk and Farrell, 1987; Kögel-Knabner, 2002; Kögel-Knabner and Rumpel, 2018). It provides rigidity as well as protection against pathogens to plants (Thevenot et al., 2010; Santhanam et al., 2012). Other plant biomolecules occur in different abundances in various plant tissues and function as cell wall structure units, storage compounds, pigments, regulators, nutrient transporters and enzymes (Kögel-Knabner, 2002). These plant molecules vary in degradability and environmental behavior (such as affinity to soil minerals and propensity to associate with soil aggregates; Lützow et al., 2006; Kögel-Knabner et al., 2008; Schmidt et al., 2011).

Numerous field litterbag studies have found that soluble substrates (including organic acids, carbohydrates and amino acids) are rapidly exhausted during the early phase of leaf litter decomposition while cellulose and hemicellulose are slowly lost during the second stage of decomposition, leaving lignin concentrated in the remaining litter (Berg and Staaf, 1980; Melillo et al., 1989; Berg, 2000). Moreover, the mass loss rate of leaf litter is negatively correlated with the ratio of lignin to N in the starting material (Melillo et al., 1982; Zhang et al., 2008). These results contributed to a long-standing, former paradigm that lignin, as an aromatic structural macromolecule in plant cell walls, is a recalcitrant component of plant litter (Berg and Staaf, 1980; Melillo et al., 1989), whose content may play a decisive role on litter decomposition rate as well as soil carbon sequestration (Schimel et al., 1994; Thevenot et al., 2010; Hall et al., 2020). However, emerging evidence suggests that biochemical recalcitrance is probably most relevant during the initial stages of (litter) decomposition (Marschner et al., 2008), while lignin or other chemically recalcitrant compounds may not be as stable in the long term (Kögel-Knabner et al., 2008; Preston et al., 2009; Klotzbücher et al., 2016). Hence, the paradigm is now shifting, de-emphasizing the predominant role of chemical recalcitrance or lignin as control of soil organic matter persistence especially in mineral soils (Schmidt et al., 2011; Lehmann and Kleber, 2015) (see Section 3), although some studies indicate that stabilized carbon may contain plant litter compounds along with microbial material (Barré et al., 2018). Nevertheless, it is doubtless that the chemical composition of plant-derived organic matter is greatly modified before incorporation into soils and plays a role in the fate of plant carbon, at least during the early phase of decomposition.

Compared with leaf litter rich in non-structural carbohydrates and cellulose, plant roots (especially for woody species) have higher contents of structural biomolecules and condensed tannins that are relatively difficult to decompose (Cusack et al., 2009; Xia et al., 2015). Hence, root litter decomposes more slowly than leaf litter across a range of terrestrial ecosystems (Gholz et al., 2000; Abiven et al., 2005; Vivanco and Austin, 2006; Freschet et al., 2013). The biochemical traits that control the decomposition rate of root litter may differ from that of leaf litter. For instance, in a decomposition study of first-order roots and leaf litter from 35 woody plant species in temperate forests, Sun et al. (2018) found that non-lignin (rather than lignin) root chemistry (including non-structural carbohydrates and polyphenols) best explains the large inter-species variation of root decomposition.

Root exudates are another important form of plant inputs into soil, representing around 5–21% of photosynthetically fixed carbon that is transferred into the rhizosphere (Lynch and Whipps, 1990; Nguyen, 2003; Derrien et al.,

2004). They are often divided into two categories of compounds: (1) low-molecular weight compounds such as amino acids, organic acids, sugars and other secondary metabolites and (2) high-molecular weight exudates, such as mucilage (polysaccharides) and proteins (Badri and Vivanco, 2009; Haichar et al., 2014). Compared to root litter, root exudates (especially the low-molecular weight compounds) are easily taken up by soil microbes when not protected via mineral associations (although the direct stabilization and microbial uptake of root exudates are rarely studied simultaneously; Sokol et al., 2019a), and hence may not directly contribute to soil organic matter (Hütsch et al., 2002; de Graaff et al., 2014; Haichar et al., 2014). However, root exudates that fuel microbial turnover may promote the production of microbial necromass (dead cells) and extracellular compounds, which in turn enhances soil organic carbon accrual relative to inputs of chemically recalcitrant litter (Cotrufo et al., 2015). Some recent studies showed that microbial formation of soil carbon is more efficient using root deposits than root litter or aboveground input (Sokol and Bradford, 2018; Sokol et al., 2019a). Some root exudates may also function as an inhibitor of microbial activity (e.g. phenolic compounds; Murray et al., 1996; Baetz and Martinoia, 2014) or as organic chelators that may potentially disrupt organo-mineral interactions (e.g. oxalic acid; Keiluweit et al., 2015) (see Section 4). Hence, root exudates play an important role in regulating soil organic carbon dynamics as both sources and mediators of soil carbon processes (Rasse et al., 2005; Jones et al., 2009).

3 Fate of plant carbon in the soil

Once released into the soil environment, plant-derived organic matter is subject to biotic and abiotic decomposition as well as preservation via organo-mineral interactions and soil aggregation (Fig. 2). These processes may occur simultaneously or independently, resulting in different fates of plant carbon including complete mineralization (to CO_2 or methane in anoxic conditions), transformation (to decomposition or microbial products) and preservation (in mineral-associated organic matter, organic complexes or aggregates). The environmental fate also varies for different forms of plant organic matter, depending on its morphology, chemical property, point of entry and physical environment into the soil (i.e. accessibility to microbes) (Sokol and Bradford, 2018; Sokol et al., 2019a).

Low-molecular weight compounds such as simple sugars and amino acids from root exudates or soil DOM may be directly taken up by soil microbes when physically accessible in the soil, mineralized to CO_2 or partially converted into microbial biomass and byproducts (Cotrufo et al., 2013; Sokol et al., 2019a). The majority of plant-derived organic matter is, however, deposited in the form of biopolymers. The decomposition of plant biopolymers usually initiates

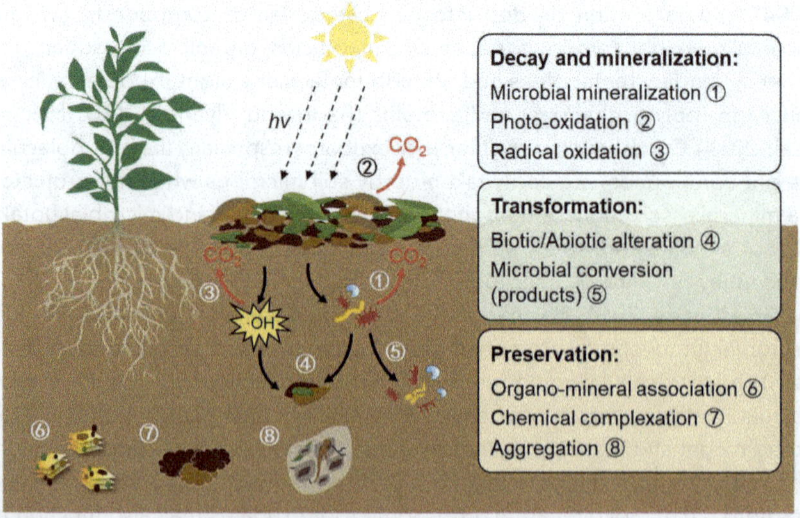

Figure 2 Fate of plant-derived organic matter in the soil. Numbers in the figure refer to different transformation pathways listed in the right boxes.

with hydrolysis or oxidation by extracellular enzymes and/or abiotic processes to smaller units that can be transported into microbial cells for metabolism (Ratledge, 1994). For instance, cellulose and hemicellulose, the most abundant constituents of plant litter, are depolymerized by hydrolytic enzymes before their constituting units, monomers and dimers can be utilized by microbes. Hydrolytic enzymes also liberate much of the organic N and phosphorus (P), which is ultimately assimilated by plants and microbes. While hydrolytic enzymes operate on a lock-and-key mechanism, oxidative enzymes (such as peroxidases) are not substrate-specific and may attack a variety of chemical bonds including aromatics (Allison, 2006). They are therefore known to exert a strong control on the decomposition of lignin and polyphenols that are also abundant in terrestrial plants (Burns et al., 2013). White-rot fungi (predominantly *Basidiomycota*) are considered to be the main microorganisms that are able to mineralize lignin using an array of oxidative enzymes while other wood-degrading fungi such as brown rot (*Ascomycota* and *Basidiomycota*) and soft rot (*Ascomycota*) may also produce some oxidative enzymes and catalyze partial oxidation of lignin (Burns et al., 2013; Ayuso-Fernández et al., 2018).

Abiotic processes such as ultraviolet radiation may also induce the oxidation of organic matter, which proceeds in chain reactions involving reactive oxygen species (free radicals; Hofrichter, 2002). In semiarid systems, degradation associated with photo-oxidation may contribute to up to 60% of litter mass loss (Austin and Vivanco, 2006). Light-absorbing compounds such as lignin and tannin are preferentially oxidized by radiation (Austin and Ballare,

2010). In mineral soils, redox-active metals such as iron (Fe) and manganese (Mn) or redox-active minerals such as hematite, goethite and ferrihydrite may induce Fenton or Fenton-like reactions in the presence of hydrogen peroxide, which also produce radicals to oxidize organic matter (Lin and Gurol, 1998; Garrido-Ramirez et al., 2010). Radical attack, mediated by either oxidizing enzymes or abiotic processes, is considered to be particularly important for the decomposition of chemically recalcitrant compounds including lignin (Waggoner et al., 2017; Shi et al., 2020). However, compared with microbial and enzymatic processes, the prevalence and magnitude of abiotic decomposition of plant-derived organic matter remain relatively poorly constrained, especially in mineral soils.

If not decomposed, plant-derived organic matter may be in total or partially preserved in the soil under environmental conditions that constrain microbial decomposition activities through low temperatures (permafrost soils), lack of oxygen leading to anaerobic conditions (wetlands) or lack of moisture (drylands) (Fenner and Freeman, 2011; Schmidt et al., 2011). It has been shown that permafrost soils contain high abundance of labile (plant-derived) substrates despite storage over thousands of years, mainly due to temperature-constrained microbial activity (Sistla et al., 2013). These substrates are rapidly utilized upon permafrost thawing (Schuur et al., 2009; Hicks Pries et al., 2015). Similarly, anaerobic environments such as wetlands greatly inhibit the activity of microbes and oxidative enzymes that are essential for the decomposition of chemically recalcitrant compounds and hence promote the preservation of lignin and other plant-derived phenolics (Freeman et al., 2001; Fenner and Freeman, 2011). The accumulation of phenolics constrains the activity of hydrolases, further enhancing organic matter accumulation in wetlands (Freeman et al., 2001). The above chain reactions, known as 'enzyme latch' on organic matter decomposition in anaerobic environments (Freeman et al., 2001, 2004), contribute to the persistence of tremendous carbon stocks in wetlands, which mainly take the form of plant-derived molecules (Chen et al., 2021). Upon wetland drainage or drought, oxygen exposure may greatly alleviate the 'enzyme latch' and stimulate decomposition (Fenner and Freeman, 2011).

Plant-derived organic matter may also be protected against decomposition through interactions with soil minerals (Torn et al., 1997; Hemingway et al., 2019) or organic matter and encapsulation within aggregates (Sanaullah et al., 2011; Cotrufo et al., 2015; Angst et al., 2021). Mineral association via sorption, complexation and co-precipitation is considered to be an effective pathway for preserving otherwise degradable organic matter against microbial decomposition (Schmidt et al., 2011; Lehmann and Kleber, 2015; Hemingway et al., 2019). Clays (in particular, smectites) and reactive metal (including Fe and aluminum, Al) oxides with large specific surface areas exhibit high affinity for organic matter and have been shown to strongly govern soil organic carbon contents and turnover (Torn et al., 1997; Doetterl et al., 2015; Hemingway et al., 2019). Similarly, organic matter may also help to preserve organic molecules through forming chemical bonds and complexation (Tombacz et al., 2004; Feng et al., 2005), but this mechanism is much less studied by comparison.

Minerals show selectivity towards organic molecules during interaction. For instance, N-rich, microbial-derived biomolecules like amino sugars and amino acids tend to bind strongly to mineral surfaces due to their amphiphilicity (Sollins et al., 2006; Kopittke et al., 2018), leading to a relative enrichment of microbial-derived organic matter in the mineral-associated fraction of soils (indicated by low C/N ratios, enriched ^{13}C and high abundance of microbial residues; Kopittke et al., 2018; Sokol et al., 2019a). However, plant-derived aromatic and phenolic moieties such as lignin also appear to be selectively retained on clay surfaces during mineral sorption of DOM (Kaiser et al., 2004; Feng et al., 2005) and exhibit preferential association with reactive Fe and Al oxides in soils and aquatic systems (Riedel et al., 2012; Wang et al., 2017). In addition, plant-derived hydrophobic lipids such as cutin, suberin and waxes are known to bind strongly to mineral surfaces and relatively concentrate in mineral-associated organic matter (Xu et al., 2009). Hence, both plant- and microbial-derived organic matter may form organo-mineral complexes, but the occurrence and abundance of plant-derived molecules in mineral-associated organic matter are less well known relative to microbial moieties (Angst et al., 2021). The likelihood for a plant molecule to be preserved through mineral association may depend on its 'point of entry' into the soil, in addition to its chemical properties (Sokol and Bradford, 2018; Sokol et al., 2019a). In the microbe- and organics-depleted bulk soil, for instance, plant carbon has a higher chance to directly interact with minerals upon entering the soil, while in the rhizosphere with high abundance of microorganisms, plant carbon (especially that with a small molecular weight) may be used by decomposers before it interacts with soil minerals (Sokol et al., 2019a).

While organo-mineral association takes place at submicron scales, soil aggregation may protect organic matter at micron to millimeter scales by constraining the diffusion of substrate, enzymes and oxygen as well as the microbial accessibility within aggregates (Christensen, 2001; Totsche et al., 2018). There is a wide array of organic carbon functionalities associated with aggregates so that aggregation seems to be less selective towards organic molecules compared with organo-mineral associations (Lehmann et al., 2008; Solomon et al., 2012). Fine roots, plant rhizodeposition, mycorrhizal fungi, bacteria as well as cementing agents such as clay, carbonate and metal oxides are all known to promote aggregate formation (Totsche et al., 2018). However, it is commonly reported that aggregates (in particular, macroaggregates; >250 μm) contain more plant detritus relative to mineral-associated fractions in grassland soils (Amelung et al., 1999), suggesting that aggregation is an important way of preserving plant-derived carbon. It is also noteworthy that soil faunal activity (e.g. earthworms) is an important driver of plant-derived organic matter incorporation into aggregates (see Chapter 4 of this book).

4 Rhizosphere processes

The rhizosphere is defined as the volume of soil around living roots (Hiltner, 1904; Hartmann et al., 2008), which may range from a few millimeters of the root surface to the entire rooting zone of the soil system, depending on root density (Dijkstra et al., 2021). Root litter and rhizodeposition, including sloughed root cells, passive exudation and active secretion (Pausch and Kuzyakov, 2012), lead to a proliferation of soil microbes in the rhizosphere containing up to 10^6–10^9 bacteria, 10^4 protozoa, 10–10^2 nematodes and 10^5–10^6 fungi per gram of rhizosphere soil (Watt et al., 2006; Hinsinger et al., 2009; Mendes et al., 2013). The interplays of plant roots-soil-microbes, i.e. rhizosphere processes, underpin essential functions of terrestrial ecosystems, including nutrient cycling and carbon sequestration (Finzi et al., 2015). It has been shown that root inputs into the rhizosphere are a more important contributor to soil organic matter than aboveground plant inputs (Rasse et al., 2005; Jackson et al., 2017; Sokol et al., 2019b), whereas rhizosphere processes may release 30–60% of global primary production on land as CO_2 (Friedlingstein et al., 2019). Hence, the rhizosphere is a hotspot for both soil carbon accumulation and decomposition (Hartley et al., 2012; Cheng et al., 2014; Dijkstra et al., 2021).

Rhizosphere processes may enhance soil carbon accumulation and stabilization in several ways (Poirier et al., 2018; Adamczyk et al., 2019; Dijkstra et al., 2021). First, inputs of root litter containing high amounts of chemically recalcitrant structural components such as suberin and lignin (Cusack et al., 2009; Xia et al., 2015) may promote the buildup of particulate organic matter (Angst et al., 2016) and enhance short-term soil carbon stabilization by slowing decomposition (Fulton-Smith and Cotrufo, 2019). Alternatively, inputs of labile substrates including low-molecular weight root exudates are beneficial for microbial growth, which fuels the 'microbial carbon pump' (Liang et al., 2017) driving the accumulation of microbial necromass and extracellular compounds preferentially preserved in the mineral-associated organic matter (Cotrufo et al., 2019). Recent studies have suggested that microbial formation of mineral-associated soil carbon is more efficient in the rhizosphere than bulk soil (Sokol and Bradford, 2018; Sokol et al., 2019b), highlighting the pivotal role of rhizodeposition in soil organic carbon accumulation. Second, plant roots, rhizodeposition and the associated microbes (in particular, mycorrhizae) play a vital role in soil aggregation and structure development (Wei et al., 2019). For instance, root-derived extracellular polymeric substances as well as fungal hyphae associated with roots may enhance soil aggregation (Sher et al., 2020). More microaggregates are therefore found in the rhizosphere than bulk soil of a beech forest system (Angst et al., 2018; De Feudis et al., 2019).

In addition to the prominent role of rhizosphere in soil carbon accumulation, it may also be a place of enhanced degradation of native SOC. Rhizodeposition may stimulate or suppress the decomposition of native soil organic carbon. This phenomenon, known as rhizosphere priming effect (RPE; Kuzyakov, 2002), is mainly attributed to the input of plant-derived labile substrates, which changes microbial activity and overall decomposition (Dalenberg and Jager, 1989; Kuzyakov et al., 2000). The magnitude of RPE varies widely from 380% enhancement to 50% reduction relative to decomposition rates in the non-rhizosphere (Zhu and Cheng, 2011; Cheng et al., 2014) and is considered to play a pivotal role in regulating soil carbon turnover in general (Finzi et al., 2015; Street et al., 2020).

Both positive RPE (i.e. enhanced decomposition) and negative RPE (i.e. suppressed decomposition) can be induced by root effects on microbial processes (Fig. 3). For instance, labile substrates (e.g. from rhizodeposition) may increase microbial biomass and its activity by inducing increased microbial utilization of soil organic carbon (microbial activation hypothesis; Drake et al., 2012). Alternatively, root uptake of mineral N often significantly reduces mineral N in the rhizosphere, inducing microbial N limitation and in turn promoting microbial decomposition of soil organic matter to acquire N (microbial N-mining hypothesis; Dijkstra et al., 2013; Chen et al., 2014). However, microbial growth and decomposition may also be inhibited due to strong competition

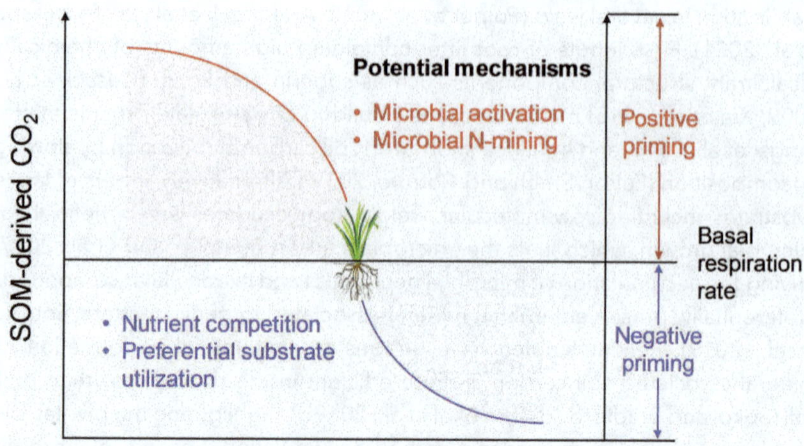

Figure 3 Concept of priming effect and potential mechanisms. Positive and negative priming refer to increased and decreased mineralization of native soil organic matter (SOM) relative to basal respiration rate following labile substrate input (e.g. from rhizodeposition), respectively.

for mineral nutrients with plant roots in the rhizosphere (nutrient competition hypothesis; Cheng and Bledsoe, 2005; van der Heijden et al., 2008; Kuzyakov and Xu, 2013). Moreover, soil microbes may prefer labile rhizodeposition to native soil organic matter when mineral nutrients are abundant, leading to negative RPE (preferential substrate utilization hypothesis; Cheng, 1999). Hence, the direction and magnitude of RPE depend on the composition and dose of rhizodeposition as well as microbial status in the soil.

The priming effect is considered to occur shortly after substrate input, lasting over a period from hours to months, as most studies on priming effect rely on short-term experiments (see Blagodatskaya and Kuzyakov, 2008). However, as rhizodeposition occurs continuously or frequently in the field, RPE is thought to play a vital role in the long-term stability and sequestration of soil organic carbon. For instance, Keuper et al. (2020) combined depth-resolved datasets of plant and permafrost properties with empirical relationships of RPE and estimated that RPE amplified soil respiration in permafrost-affected ecosystems by ~12% overall. By incorporating priming effect into the global land surface model, ORCHIDEE, Guenet et al. (2018) demonstrated that priming decreased soil carbon sequestration by 51% (12 ± 3 Pg C) during the period 1901–2010.

Other than directly mediating microbial responses, plant roots may also promote soil organic matter decomposition through destructing soil aggregates and/or organo-mineral associations. For instance, by taking up soil water, roots intensify drying-rewetting cycles, destructing existing macroaggregates, particularly in heavy-textured soils (He et al., 2020). Aggregate disruption may expose protected organic matter to microbial decomposition, resulting in a positive RPE (Lu et al., 2019; Wang et al., 2020). Furthermore, root exudates contain organic acids and ligands (e.g. oxalate acid, citrate) that may disrupt chemical associations between organic matter and soil minerals, especially those including polyvalent metals such as Fe, Al and Mn, thereby destabilizing and releasing mineral-associated organic matter for microbial decomposition (Clarholm et al., 2015; Keiluweit et al., 2015).

Hence, rhizosphere processes may have contrasting effects on soil organic carbon dynamics, inducing both stabilization and destabilization of soil carbon (Adamczyk et al., 2019; Dijkstra et al., 2021). The trade-off between the two effects may depend on many factors including the composition and dose of rhizodeposition (Phillips et al., 2011; Yin et al., 2014; Meier et al., 2017), soil physiochemical properties affecting aggregate distribution and organo-mineral interactions (Huo et al., 2017; Jilling et al., 2018; He et al., 2020) and symbiotic relationships between plants and microbes (e.g. mycorrhizal types; Craig et al., 2018). The outcome of the trade-off may vary on different timescales, as aggregate and organo-mineral disruptions may be short-lasting, followed by new formation of soil aggregates and mineral-associated organic matter in the longer term (Jia et al., 2019). The root effect on soil carbon processes may also differ in the top- versus subsoils, given the divergent root distribution, microbial activity and organic matter composition at different soil depths (Fontaine et al., 2007; Rumpel et al., 2012; Ma et al., 2019). These aspects of rhizosphere processes continue to be a hot topic of research, which have important impacts on soil carbon dynamics in general.

5 Mycorrhizae

Mycorrhiza refers to a mutualistic relationship between a plant and a fungus, where the fungus colonizes host roots, either intra- or extracellularly (Frey, 2019). Approximately 94% of terrestrial vascular plants (including grasses and trees) are associated with mycorrhiza (Brundrett, 2009), where plant trades carbon for nutrients (especially N and P) acquired through filamentous fungal hyphae that extend outward from mycorrhizal roots into the soil (known as extraradical hyphae or extrametrical mycelium; Frey, 2019). Two most commonly known mycorrhizae are the arbuscular mycorrhizal fungi (AMF) (Hooper et al., 2012) and ectomycorrhizal fungi (EMF). AMF is an ancient and widespread mycorrhizal symbiosis where a monophyletic clade of fungi (*Glomeromycota*) obligately associates with plants in temperate grasslands and tropical forests and is also common in temperate forests (Frey, 2019; Steidinger et al., 2019). EMF is formed by a diverse group of fungi in symbiosis with primarily woody plants and is the dominant mycorrhiza in boreal and temperate forests and is also common in some tropical forests (Frey, 2019; Steidinger et al., 2019). Up to 35% of net primary productivity has been estimated to allocate to mycorrhizae in temperate forests (Ouimette et al., 2019), although the number may vary greatly across climates, ecosystem types (Gill and Finzi, 2016), stand age (Wallander et al., 2010) and nutrient availability (Vicca et al., 2012; Brzostek et al., 2015). Nonetheless, it is widely accepted that mycorrhizae constitute a considerable fraction of plant photosynthates and belowground carbon inputs into the soil (Frey, 2019).

As an important conduit of rhizodeposition inhabiting the interface between plant roots and soil, mycorrhizae affect soil organic carbon dynamics in three ways. First, mycorrhizae promote soil carbon stabilization by enhancing soil aggregation via the production of extracellular compounds (such as glomalin; Rillig, 2004; Wei et al., 2019) and extraradical hyphae that entangle and enmesh soil particles (Lehmann et al., 2017; Sher et al., 2020). Mycorrhizal hyphae can represent up to half of the standing mycelial biomass (Bååth et al., 2004), thereby making mycorrhizae effective promotors for soil aggregate formation.

Second, mycorrhiza is an important player in the soil 'microbial carbon pump' (Liang et al., 2017) not only as a direct contributor of mycorrhizal necromass and extracellular compounds which may stabilize on mineral surfaces (Kleber et al.,

2015; Jilling et al., 2018) but also as an important conduit delivering labile plant photosynthates to soil microbes in the rhizosphere (Kaiser et al., 2015). Since (1) mycorrhiza is directly produced from plant carbon, (2) has a high proximity to soil mineral surfaces and (3) contains melanin, a complex nonhydrolyzable structure that may benefit the preservation of fungal necromass (Clemmensen et al., 2013), mycorrhiza potentially serves as an efficient pathway converting photosynthates to relatively stable soil organic carbon as mineral-associated mycorrhizal necromass (Frey, 2019). For instance, over half of the newly formed soil organic matter was attributed to extraradical hyphae in EMF-dominated systems (Godbold et al., 2006; Clemmensen et al., 2013). Yet, the efficiency of mycorrhizal necromass accumulation relative to other microbial groups has not been examined.

Third, although mycorrhizae do not act as saprotrophs to directly decompose organic matter for carbon acquisition (Tisserant et al., 2013; Martin et al., 2016), they may mediate soil organic matter decay through their nutrient mining activities and interactions with saprotrophs (Read, 1991; Frey, 2019). For instance, some widespread and abundant EMF genera (e.g. *Cortinarius, Russula, Lactarius*) can produce hydrolytic and oxidative enzymes to access and mobilize organic N (Talbot et al., 2015; Nicolas et al., 2019). Some EMF taxon (*P. involutus*) may also employ radicals produced via Fenton reactions to degrade organic matter-protein complexes for N acquisition (Op De Beeck et al., 2018), releasing carbon as a byproduct in the process (Talbot et al., 2008). In contrast, the AMF genera have largely lost the genetic machinery to enzymatically break down organic macromolecules (Tisserant et al., 2013) and hence rely heavily on saprotrophs for nutrient acquisition by providing labile carbon (as plant photosynthates, exudates and/or mycorrhizal turnover products) and thereby stimulating their decomposition activity (i.e. rhizosphere priming; Dijkstra et al., 2013; Bukovska et al., 2018). Alternatively, mycorrhizae may suppress organic matter decomposition through antagonistic interactions with free-living saprotrophs for limited nutrients and/or water resources (Averill et al., 2014; Leifheit et al., 2015). This mechanism, known as the Gadgil effect (Gadgil and Gadgil, 1971, 1975), has been observed in soil trenching or tree girdling experiments, where declines in living mycorrhizal hyphae were accompanied by increased litter decay (Gadgil and Gadgil, 1975; Sterkenburg et al., 2018) and enhanced C and N mineralization (Moore et al., 2015; Averill and Hawkes, 2016). However, the prevalence and magnitude of the Gadgil effect remain uncertain, particularly in the presence of confounding factors such as moisture alterations after roots removal (Frey, 2019).

Given the different nutrient acquisition strategies and distribution of AMF versus EMF, nutrient cycling and soil organic carbon dynamics differ significantly between EMF- and AMF-dominated soils. Compared with AMF-associated trees, the EMF-associated trees typically have higher ratios of C:N in the leaf litter (Read, 1991; Lin et al., 2017; Zhu et al., 2018) and slower decomposition rates for both leaf and root litter (Cornelissen et al., 2001; Midgley et al., 2015; Jacobs et al., 2018; Keller and Phillips, 2019), resulting in slow cycling of carbon and nutrients and a greater proportion of nutrients in organic forms (Read and Perez-Moreno, 2003). Following the soil 'microbial carbon pump' process (Liang et al., 2017), it can be deduced that AMF, associated with fast carbon and nutrient cycling, promotes the formation of microbial residues potentially stabilized as mineral-associated organic matter. This hypothesis is consistent with the positive correlations between AMF tree dominance and microbial residue concentrations observed in temperate forests and grasslands soils (Craig et al., 2018; Cotrufo et al., 2019), whereas high concentrations of particulate organic carbon (>53 μm), containing high contributions of plant detritus, are found in EMF forest soils (Cotrufo et al., 2019). However, how AMF and EMF affect the overall storage and persistence of soil organic carbon remains unclear. Conflicting results have been reported with higher soil organic carbon concentrations or stocks in EMF-dominated (Averill et al., 2014; Taylor et al., 2016; Lin et al., 2017) or AMF-dominated (Craig et al., 2018; Jo et al., 2019) sites, probably related to the soil conditions and plant-mycorrhiza-saprotroph interactions within the specific ecosystem. Hence, it is crucial to elucidate mechanisms regulating the mycorrhizal effect on soil carbon dynamics and its controls.

6 Plant influences on soil organic carbon dynamics in different environments

Plants' impact on soil organic carbon dynamics may differ significantly among different ecosystems, given the large variability of plant carbon allocation, litter composition, mycorrhizal association, climate, land-use and edaphic conditions. The difference is well manifested in the comparison between grassland and forest soil organic carbon composition and dynamics. Compared with forests that are dominated by woody plants, grasslands have a higher ratio of belowground to aboveground carbon allocation in plant biomass (Jackson et al., 1996) and less litter with lower C/N ratios (Rowe et al., 2006; Cotrufo et al., 2019). In addition, grasslands are mostly associated with AMF in contrast to forests with prominent EMF association (Terrer et al., 2021). These characteristics of grasslands largely promote microbial utilization of plant photosynthates (as rhizodeposition and high-quality litter), resulting in a higher proportion of microbial residues (indicated by amino sugars; Ma et al., 2018) and mineral-associated organic matter dominated by microbial inputs (Cotrufo et al., 2019) in total soil organic carbon. By contrast, particulate organic matter dominated by plant detritus seems to comprise a larger fraction of soil organic carbon in forests, especially in those dominated by species with chemically recalcitrant litter (such as pine; Cotrufo et al., 2019). Hence, it appears that microbial-mediated stabilization plays an important role in soil organic carbon accumulation in grasslands (Ma et al., 2018) while direct preservation of plant carbon is more crucial for forest soil carbon increases (Cotrufo et al., 2019).

The grassland-forest contrast is further reflected in the divergent relationships between plant and soil carbon responses to elevated CO_2 (Ainsworth and Long, 2005). Elevated CO_2 tends to increase plant biomass and litter production in grasslands and, to a larger extent, also in forests (Bazzaz, 1990; Ainsworth and Long, 2005). However, a recent meta-analysis showed that the increased plant-derived organic matter inputs under elevated CO_2 lead to increased soil organic carbon (especially mineral-associated organic carbon) stocks in grasslands while they decreased in forests (Terrer et al., 2021). These contrasting responses may be explained by increased rhizodeposition and microbial necromass production in grassland soils and priming effects through microbial mining of nutrients locked in soil organic matter to meet the demands of growing biomass in forest soils (Terrer et al., 2016, 2021). Therefore, there seems to be a trade-off between plant and soil organic carbon pools under elevated CO_2, largely related to the contrasting responses of different ecosystems. It is conceivable that plants' impact on soil organic carbon dynamics under other environmental change effects (e.g. elevated temperature and/or extreme events) are also dependent on plant-microbe interactions regulated by soil nutrient and moisture availability, which warrant further investigations.

The top-subsoil contrast presents another conspicuous demonstration of different plant influences on soil organic carbon dynamics in contrasting soil environments. Subsoil (often referred to as the soil residing >20 cm below ground) contains more than half of global soil organic carbon stocks (Harrison et al., 2011; Rumpel et al., 2012) and differs substantially from topsoil in organic matter composition, microbial communities, organo-mineral associations, substrate and oxygen availability (Rumpel et al., 2012). It has been widely shown that plant-derived compounds dominate soil organic matter in the topsoil characterized by higher abundance of plant roots and aboveground litter inputs than the subsoil (Ma et al., 2019). By comparison, subsoil contains relatively more microbial-derived organic matter indicated by low C/N ratios, enriched ^{13}C contents and microbial biomarkers (Rumpel et al., 2012; Ma et al., 2019), likely owing to microbial conversion of DOM percolating through the soil (Kammer et al., 2012) and/or root carbon input, despite lower microbial biomass and activity in the carbon- and energy-constrained subsoil relative to topsoil (Fontaine et al., 2007; Kramer et al., 2013; Lee et al., 2013). Similarly, particulate organic matter is relatively more abundant in the topsoil whereas mineral-associated form represents a higher fraction of soil organic carbon in the organics-lean subsoil (Li et al., 2013). Moreover, plant material may show different tendencies to be stabilized directly by mineral interactions or via transformation into microbial compounds in the topsoil versus subsoil (Rumpel et al., 2015).

When plant inputs increase (e.g. due to warming or N deposition), topsoil and subsoil organic carbon may show divergent changes depending on microbial responses. Given the energy and carbon constraints of subsoil microbes (Fontaine et al., 2007), enhanced decomposition (i.e. positive priming) is often observed following increased plant inputs of labile carbon into the subsoil (Wang et al., 2014; Jia et al., 2017). For instance, warming-induced root growth was found to enhance the decomposition of plant-derived compounds (such as lignin) in alpine grassland (Jia et al., 2019), while such effects were absent in the topsoil. On the other hand, newly synthesized plant carbon may also accumulate at higher rates in the subsoil, where carbon-unsaturated mineral surfaces may be relatively prevailing (Kaiser and Guggenberger, 2003; Poirier et al., 2014) and soil microbes are less active in the short term relative to the topsoil (Zhu et al., 2021). Indeed, an enhanced accumulation of new carbon was observed in the mineral-associated fraction of subsoil but not topsoil of an alpine grassland witnessing increased root growth under warming (Jia et al., 2019).

7 Implications for soil management

Plant photosynthates are the primary source of soil organic carbon. While soil organic carbon stocks typically increase with plant productivity during the early stage of soil development (Guelland et al., 2012; Khedim et al., 2021), increasing plant input does not always enhance soil carbon accumulation in developed soils (Fontaine et al., 2004) owing to the complex interactions of plants, microbes and soil environments (e.g. priming effect). At the global scale, the most productive ecosystems (such as tropical forests) do not have the highest soil organic carbon stocks (Davidson and Janssens, 2006) due to elevated ecosystem carbon turnover (Carvalhais et al., 2014). At the site scale, plant biomass increases induced by elevated CO_2 or N deposition are sometimes accompanied by soil organic carbon decreases following enhanced microbial decomposition of organic matter to meet up nutrient demands (van Groenigen et al., 2014; Volk et al., 2018). Therefore, attempts to enhance soil carbon sequestration via plant inputs must consider and assess the nutrient demands invoked by plant growth to minimize potential microbial priming effect associated with nutrient mining activities (Chen et al., 2014) or to avoid increased soil carbon turnover. This consideration may be particularly important for agricultural soils when soil nutrients are partially removed during harvests and in nutrient-limited ecosystems. In this regard, cultivation of N_2-fixing leguminous plants has been shown to enhance soil carbon sequestration via increasing soil N content in symbiosis with the soil bacteria rhizobia and promoting plant productivity, particularly in N-limited systems (Christopher and Lal, 2007; Jensen et al., 2012). The potential use of leguminous plants to replace inorganic N fertilizers in agro-ecosystems may also reduce the emission of nitrous oxide, which is often stimulated by inorganic N applications (Jensen et al., 2012).

Plant-induced changes to soil structures (such as porosity and aggregation; Angers and Caron, 1998) or physiochemical processes (such as exudate-assisted desorption and soil erosion; Keiluweit et al., 2015) may also affect soil carbon turnover and storage. Plant cover, roots and the litter layer protect surface soils from erosion by promoting soil particle trapping (Gyssels et al., 2005). Alternatively, root growth and turnover can increase soil porosity and affect the diffusion of oxygen, substrates and extracellular enzymes (Angers and Caron, 1998; Dijkstra et al., 2021), potentially increasing soil carbon turnover in compacted soils (Adamczyk et al., 2019). Root exudates (e.g. organic acids) may also lower soil pH in the rhizosphere and lead to inorganic carbon releases (see Chapter 22 of this book). Such abiotic processes should also be considered for soil carbon management.

The form of plant inputs is another important factor to consider for promoting plant-mediated soil carbon sequestration. Historically, soil management practices focused on aboveground litter manipulations to enhance soil carbon storage (Lajtha et al., 2014). However, in recent years, management of rhizodeposition is regarded as a more effective way for soil carbon formation and accumulation compared to aboveground litter (Villarino et al., 2021). Increasing belowground inputs or rhizodeposition by growing plants with large root masses (especially fine roots with high labile carbon inputs) may be particularly beneficial for the soil microbial carbon pump and the accumulation of mineral-associated organic matter (Liang et al., 2017; Jia et al., 2019; Villarino et al., 2021). This organic carbon pool is vital for soil carbon sequestration on decadal to centennial timescales, given its relatively longer turnover time compared with particulate organic matter (Kögel-Knabner et al., 2008; Lavallee et al., 2020) and the location of roots in deep soil, where soil organic matter turnover may be much slower than in topsoil. Hence, although the accumulation of mineral-associated organic carbon may be counteracted by enhanced decomposition of non-protected carbon pools following increased root inputs initially (Adamczyk et al., 2019; Jia et al., 2019), the long-term effect of rhizodeposition-induced soil carbon sequestration deserves assessment. Similarly, manipulating mycorrhizal associations with plant roots in managed ecosystems may not only mediate soil nutrient cycling but also adjust carbon allocation and turnover in different soil carbon pools (Averill et al., 2014; Dignac et al., 2017). As such, commercial AMF inoculum products have been used to increase crop yield or nutrition (Berruti et al., 2015; Elliott et al., 2020), but their effect on soil carbon dynamics is less well known and the quantitative importance of mycorrhiza-derived soil carbon deserves attention.

The above considerations should not be limited to surface soils only but also include subsoils with unsaturated mineral surfaces, contrasting microbial communities and edaphic properties (Rumpel et al., 2012). Divergent responses of soil carbon storage, composition and turnover have been observed in the surface versus subsoils following altered plant growth or community composition under global changes or land-use alterations (Jobbagy and Jackson, 2000; Jia et al., 2019; Ofiti et al., 2021). For instance, elevated root growth under warming resulted in an accumulation of newly synthesized carbon in the mineral-associated fraction of subsoil (30–40 cm) but not topsoil (0–10 cm) in an alpine grassland (Jia et al., 2019). Keidel et al. (2018) observed a depth-dependent response of macroaggregation to elevated CO_2 in a temperate grassland ecosystem where macroaggregation as well as the organic carbon content of large macroaggregates increased in the subsoil (15–45 cm) but not in the topsoil (0–15 cm). Also, 20 years of straw addition to a farmland with fertilizers mainly changed soil organic carbon content in the deep (0.6–3 m) rather than surface soil (Li et al., 2013). By contrast, for the use of cover crops during 19 years, soil carbon stock decreases were observed in the subsoil whereas those of topsoil increased (Tautges et al., 2019). Given the tremendous carbon storage in subsoils (Jobbagy and Jackson, 2000) and shifting plant root distribution under global and land-use changes (Liu et al., 2018a), subsoils must be included in assessing and planning soil management practices for carbon sequestration.

8 Conclusion and future trends in research

In summary, plants play a pivotal role in governing soil organic carbon dynamics as (1) a direct contributor of soil organic matter, (2) a mediator and fuel for microbial decomposition and stabilization of organic carbon and (3) a regulator of soil physiochemical processes. Many of the above processes may occur independently or interactively, sometimes inducing opposite effects on soil carbon turnover or storage depending on ecosystem and soil conditions. In the pledge of enhancing soil carbon sequestration for climate mitigations, elucidating plant influences on soil carbon storage and stability underpins an accurate assessment of soil carbon sink potentials and successful efforts of soil carbon sequestration. However, mechanisms regulating plant-microbe-soil interactions and the associated effects on soil carbon dynamics are not fully understood yet. Here I summarize, although not exhaustively, several key aspects that should be prioritized for future research to improve our understanding and utilization of the plant effects on soil carbon storage.

8.1 Transformation and persistence of plant molecules in the soil

Although recent studies have demonstrated and emphasized microbial conversion and stabilization of plant-derived carbon in the form of microbial residues in the soil (Miltner et al., 2011; Cotrufo et al., 2013; Kallenbach et al., 2016; Liang et al., 2017; Ma et al., 2018), the fate, persistence and storage of soil carbon in the form of plant-derived molecules remain unclear. For instance, the disappearance of lignin in mineral soils is quite puzzling, given its chemical recalcitrance and high abundance in decaying litter. Current analytical methods (such as copper oxide-assisted oxidation of lignin) cannot fully recover and quantify the lignin macromolecule in complex environmental matrices (Hernes et al., 2013; Wang et al., 2017), let alone its degradation products. Similar analytical constraints apply to other transformed, complexed or physiochemically protected plant molecules, making them difficult (if not impossible) to trace in the soil (Kögel-Knabner, 2000; Lehmann and Kleber, 2015). Leveraging recent analytical advances, such as high-resolution mass spectrometry (Zhang et al., 2021) and isotope tracing (Huang et al., 2019), to identify transformed or 'hidden' plant molecules in the soil will help decipher the fate and persistence of plant-derived organic matter in the soil. This effort will improve our understanding of soil carbon stabilization mechanisms and soil management strategies.

8.2 Plant and soil carbon responses to global changes and soil management

Manipulating vegetation or plant litter for soil carbon sequestration is an appealing strategy. However, soil carbon does not always synchronize with plant responses to environmental changes, making it difficult to predict and manage ecosystem carbon stocks. Comparing plant and soil carbon responses to environmental changes at different temporal-spatial scales will help disentangle the underlying connections and improve soil carbon modeling. Recent advances in vegetation phenology and carbon flux monitoring may aid in this aspect (Xin et al., 2020). In addition, exploring connections between plant functional traits (especially belowground traits) and soil organic carbon dynamics may help identify species that are most efficient in promoting soil carbon accumulations. Such efforts should include subsoils, which sometimes show contrasting responses from the surface soil.

8.3 Effects of plant diversity on soil carbon dynamics

Plant diversity has been shown to exert a positive effect on soil organic carbon storage in both grasslands (Flombaum and Sala, 2008; Steinbeiss et al., 2008; Lange et al., 2015) and forests (Zhang et al., 2011; Liu et al., 2018b) owing to diversity-enhanced plant productivity and resource use efficiency (Hooper et al., 2005, 2012; Huang et al., 2018). However, the prevalence, persistence and effectiveness of the relationship between plant diversity and soil organic carbon accumulation are not clear. Nor are the mechanisms affecting the effectiveness of this relationship. Protecting and promoting plant diversity is important for multiple ecosystem functions and services other than carbon sequestration (Loreau et al., 2001; Isbell et al., 2011; Hooper et al., 2012; Pennekamp et al., 2018). Revealing mechanisms and factors regulating soil organic carbon responses to plant diversity changes will help identify 'hot spots' for promoting plant diversity for soil carbon sequestration.

8.4 Mycorrhizal influences on rhizosphere processes

The past decades have witnessed many advances in mycorrhizal research that greatly improve our understanding of plant-mycorrhiza-soil interactions. However, how AMF and EMF affect the overall storage and persistence of soil organic carbon remains unclear, especially in different types of soil. As mycorrhizal association plays a key role in regulating plant growth, nutrient cycling as well as carbon dynamics, investigating mechanisms regulating mycorrhizal effect on soil carbon dynamics and the operating conditions are much needed to illuminate the utility of mycorrhizal associations for sustainability and climate-smart agriculture.

9 Where to look for further information

The International '4 per 1000' Initiative, launched on 1 December 2015 at COP 21, aims to show that agriculture and forestry, and in particular agricultural and forestry soils, can play a crucial role in food security and climate change. The initiative consists of federating all voluntary public and private stakeholders within the framework of the Lima-Paris Action Plan (LPAP) to increase soil carbon content of the top 30–40 cm by 0.4% per year to reduce the annual increase of CO_2 in the atmosphere. The initiative recommends or entails many practices involving vegetation management to enhance soil carbon sequestration. More information about '4 per 1000' initiative can be found at: https://4p1000.org/.

Individual organizations or groups also focus their research on plant effects on soil carbon storage. For instance, the international Detrital Input and Removal Treatment (DIRT) network was established to assess the long-term impacts of altered plant litter inputs on SOC dynamics in forest ecosystems (https://dirtnet.wordpress.com/). Bangor Wetlands Group led by Prof. Chris Freeman, a group based in the Wolfson Carbon Capture Laboratories at Bangor University in North Wales, UK, mainly focuses on increasing carbon storage in peatland (https://wetlands.bangor.ac.uk/index.php.en).

10 References

Abiven, S., Recous, S., Reyes, V. and Oliver, R. (2005). Mineralisation of C and N from root, stem and leaf residues in soil and role of their biochemical quality, *Biol. Fertil. Soils* 42(2): 119-28.

Adamczyk, B., Sietio, O. M., Strakova, P., Prommer, J., Wild, B., Hagner, M., Pihlatie, M., Fritze, H., Richter, A. and Heinonsalo, J. (2019). Plant roots increase both decomposition and stable organic matter formation in boreal forest soil, *Nat. Commun.* 10(1): 3982.

Ainsworth, E. A. and Long, S. P. (2005). What have we learned from 15 years of free-air CO_2 enrichment (FACE)? A meta-analytic review of the responses of photosynthesis, canopy properties and plant production to rising CO_2, *New Phytol.* 165(2): 351-71.

Allison, S. D. (2006). Brown ground: A soil carbon analogue for the green world hypothesis?, *Am. Nat.* 167(5): 619-27.

Amelung, W., Flach, K. -W. and Zech, W. (1999). Lignin in particle-size fractions of native grassland soils as influenced by climate, *Soil Sci. Soc. Am. J.* 63(5): 1222-8.

Angers, D. A. and Caron, J. (1998). Plant-induced changes in soil structure: Processes and feedbacks, *Biogeochemistry* 42(1-2): 55-72.

Angst, G., Kögel-Knabner, I., Kirfel, K., Hertel, D. and Mueller, C. W. (2016). Spatial distribution and chemical composition of soil organic matter fractions in rhizosphere and non-rhizosphere soil under European beech (*Fagus sylvatica* L.), *Geoderma* 264: 179-87.

Angst, G., Messinger, J., Greiner, M., Häusler, W., Hertel, D., Kirfel, K., Kögel-Knabner, I., Leuschner, C., Rethemeyer, J. and Mueller, C. W. (2018). Soil organic carbon stocks in topsoil and subsoil controlled by parent material, carbon input in the rhizosphere, and microbial-derived compounds, *Soil Biol. Biochem.* 122: 19-30.

Angst, G., Mueller, K. E., Nierop, K. G. J. and Simpson, M. J. (2021). Plant- or microbial-derived? A review on the molecular composition of stabilized soil organic matter, *Soil Biol. Biochem.* 156: 108189.

Austin, A. T. and Ballare, C. L. (2010). Dual role of lignin in plant litter decomposition in terrestrial ecosystems, *Proc. Natl. Acad. Sci. U. S. A.* 107(10): 4618-22.

Austin, A. T. and Vivanco, L. (2006). Plant litter decomposition in a semi-arid ecosystem controlled by photodegradation, *Nature* 442(7102): 555-8.

Averill, C. and Hawkes, C. V. (2016). Ectomycorrhizal fungi slow soil carbon cycling, *Ecol. Lett.* 19(8): 937-47.

Averill, C., Turner, B. L. and Finzi, A. C. (2014). Mycorrhiza-mediated competition between plants and decomposers drives soil carbon storage, *Nature* 505(7484): 543-5.

Ayuso-Fernández, I., Ruiz-Dueñas, F. J. and Martínez, A. T. (2018). Evolutionary convergence in lignin-degrading enzymes, *Proc. Natl. Acad. Sci. U. S. A.* 115(25): 6428-33.

Bååth, E., Nilsson, L. O., Göransson, H. and Wallander, H. (2004). Can the extent of degradation of soil fungal mycelium during soil incubation be used to estimate ectomycorrhizal biomass in soil?, *Soil Biol. Biochem.* 36(12): 2105-9.

Badri, D. V. and Vivanco, J. M. (2009). Regulation and function of root exudates, *Plant Cell Environ.* 32(6): 666-81.

Baetz, U. and Martinoia, E. (2014). Root exudates: The hidden part of plant defense, *Trends Plant Sci.* 19(2): 90-8.

Barré, P., Quénéa, K., Vidal, A., Cécillon, L., Christensen, B. T., Kätterer, T., Macdonald, A., Petit, L., Plante, A. F., van Oort, F. and Chenu, C. (2018). Microbial and plant-derived compounds both contribute to persistent soil organic carbon in temperate soils, *Biogeochemistry* 140(1): 81-92.

Bazzaz, F. A. (1990). The response of natural ecosystems to the rising global CO_2 levels, *Annu. Rev. Ecol. Syst.* 21(1): 167-96.

Berg, B. (2000). Litter decomposition and organic matter turnover in northern forest soils, *Forest Ecol. Manag.* 133(1-2): 13-22.

Berg, B. and Staaf, H. (1980). Decomposition rate and chemical changes of Scots pine needle litter. II. Influence of chemical composition, *Ecol. Bull.* 32: 373-90.

Berruti, A., Lumini, E., Balestrini, R. and Bianciotto, V. (2015). Arbuscular mycorrhizal fungi as natural biofertilizers: Let's benefit from past successes, *Front. Microbiol.* 6: 1559.

Blagodatskaya, E and Kuzyakov, Y. (2008). Mechanisms of real and apparent priming effects and their dependence on soil microbial biomass and community structure: Critical review, *Biol. Fertil. Soils* 45(2): 115-31.

Brundrett, M. C. (2009). Mycorrhizal associations and other means of nutrition of vascular plants: Understanding the global diversity of host plants by resolving conflicting information and developing reliable means of diagnosis, *Plant Soil* 320(1-2): 37-77.

Brzostek, E. R., Dragoni, D., Brown, Z. A. and Phillips, R. P. (2015). Mycorrhizal type determines the magnitude and direction of root-induced changes in decomposition in a temperate forest, *New Phytol.* 206(4): 1274-82.

Bukovska, P., Bonkowski, M., Konvalinkova, T., Beskid, O., Hujslová, M., Püschel, D., Řezáčová, V., Gutiérrez-Núñez, M. S., Gryndler, M. and Jansa, J. (2018). Utilization of organic nitrogen by arbuscular mycorrhizal fungi-is there a specific role for protists and ammonia oxidizers?, *Mycorrhiza* 28(3): 269-83.

Burns, R. G., DeForest, J. L., Marxsen, J., Sinsabaugh, R. L., Stromberger, M. E., Wallenstein, M. D., Weintraub, M. N. and Zoppini, A. (2013). Soil enzymes in a changing environment: Current knowledge and future directions, *Soil Biol. Biochem.* 58: 216-34.

Carvalhais, N., Forkel, M., Khomik, M., Bellarby, J., Jung, M., Migliavacca, M., Mu, M., Saatchi, S., Santoro, M., Thurner, M., Weber, U., Ahrens, B., Beer, C., Cescatti, A., Randerson, J. T. and Reichstein, M. (2014). Global covariation of carbon turnover times with climate in terrestrial ecosystems, *Nature* 514(7521): 213-7.

Chen, R., Senbayram, M., Blagodatsky, S., Myachina, O., Dittert, K., Lin, X., Blagodatskaya, E. and Kuzyakov, Y. (2014). Soil C and N availability determine the priming effect: Microbial N mining and stoichiometric decomposition theories, *Glob. Change Biol.* 20(7): 2356-67.

Chen, X., Hu, Y., Xia, Y., Zheng, S., Ma, C., Rui, Y., He, H., Huang, D., Zhang, Z., Ge, T., Wu, J., Guggenberger, G., Kuzyakov, Y. and Su, Y. (2021). Contrasting pathways of carbon sequestration in paddy and upland soils, *Glob. Change Biol.* 27(11): 2478-90. DOI: 10.1111/gcb.15595.

Cheng, W., Parton, W. J., Gonzalez-Meler, M. A., Phillips, R., Asao, S., McNickle, G. G., Brzostek, E. and Jastrow, J. D. (2014). Synthesis and modeling perspectives of rhizosphere priming, *New Phytol.* 201(1): 31-44.

Cheng, W. X. (1999). Rhizosphere feedbacks in elevated CO_2, *Tree Physiol.* 19(4_5): 313-20.

Cheng, X. and Bledsoe, C. S. (2005). Effects of annual grass senescence on $^{15}NH_4^+$ and ^{15}N-glycine uptake by blue oak (*Quercus douglasii*) seedlings and soil microorganisms in California oak woodland, *Soil Biol. Biochem.* 37(3): 551-9.

Christensen, B. T. (2001). Physical fractionation of soil and structural and functional complexity in organic matter turnover, *Eur. J. Soil Sci.* 52(3): 345–53.

Christopher, S. F. and Lal, R. (2007). Nitrogen management affects carbon sequestration in North American cropland soils, *Crit. Rev. Plant Sci.* 26(1): 45–64.

Clarholm, M., Skyllberg, U. and Rosling, A. (2015). Organic acid induced release of nutrients from metal-stabilized soil organic matter - The unbutton model, *Soil Biol. Biochem.* 84: 168–76.

Clemmensen, K. E., Bahr, A., Ovaskainen, O., Dahlberg, A., Ekblad, A., Wallander, H., Stenlid, J., Finlay, R. D., Wardle, D. A. and Lindahl, B. D. (2013). Roots and associated fungi drive long-term carbon sequestration in boreal forest, *Science* 339(6127): 1615–8.

Cornelissen, J., Aerts, R., Cerabolini, B., Werger, M. and van der Heijden, M. (2001). Carbon cycling traits of plant species are linked with mycorrhizal strategy, *Oecologia* 129(4): 611–9.

Cotrufo, M. F., Ranalli, M. G., Haddix, M. L., Six, J. and Lugato, E. (2019). Soil carbon storage informed by particulate and mineral-associated organic matter, *Nat. Geosci.* 12(12): 989–94.

Cotrufo, M. F., Soong, J. L., Horton, A. J., Campbell, E. E., Haddix, M., Wall, D. H. and Parton, W. J. (2015). Formation of soil organic matter via biochemical and physical pathways of litter mass loss, *Nat. Geosci.* 8(10): 776–9.

Cotrufo, M. F., Wallenstein, M. D., Boot, C. M., Denef, K. and Paul, E. (2013). The microbial efficiency-matrix stabilization (MEMS) framework integrates plant litter decomposition with soil organic matter stabilization: Do labile plant inputs form stable soil organic matter?, *Glob. Change Biol.* 19(4): 988–95.

Craig, M. E., Turner, B. L., Liang, C., Clay, K., Johnson, D. J. and Phillips, R. P. (2018). Tree mycorrhizal type predicts within-site variability in the storage and distribution of soil organic matter, *Glob. Change Biol.* 24(8): 3317–30.

Cusack, D. F., Chou, W. W., Yang, W. H., Harmon, M. E. and Silver, W. L. (2009). Controls on long-term root and leaf litter decomposition in Neotropical forests, *Glob. Change Biol.* 15(5): 1339–55.

Dalenberg, J. W. and Jager, G. (1989). Priming effect of some organic additions to ^{14}C-labelled soil, *Soil Biol. Biochem.* 21(3): 443–8.

Davidson, E. A. and Janssens, I. A. (2006). Temperature sensitivity of soil carbon decomposition and feedbacks to climate change, *Nature* 440(7081): 165–73.

De Feudis, M., Cardelli, V., Massaccesi, L., Trumbore, S. E., Vittori Antisari, L., Cocco, S., Corti, G. and Agnelli, A. (2019). Small altitudinal change and rhizosphere affect the SOM light fractions but not the heavy fraction in European beech forest soil, *CATENA* 181: 104091.

de Graaff, M.-A., Jastrow, J. D., Gillette, S., Johns, A. and Wullschleger, S. D. (2014). Differential priming of soil carbon driven by soil depth and root impacts on carbon availability, *Soil Biol. Biochem.* 69: 147–56.

Derrien, D., Marol, C. and Balesdent, J. (2004). The dynamics of neutral sugars in the rhizosphere of wheat. An approach by ^{13}C pulse-labelling and GC/C/IRMS, *Plant Soil* 267(1-2): 243–53.

Dignac, M. F., Derrien, D., Barre, P., Barot, S., Cécillon, L., Chenu, C., Chevallier, T., Freschet, G. T., Garnier, P., Guenet, B., Hedde, M., Klumpp, K., Lashermes, G., Maron, P., Nunan, N., Roumet, C. and Basile-Doelsch, I. (2017). Increasing soil carbon storage:

Mechanisms, effects of agricultural practices and proxies. A review, *Agron. Sustain. Dev.* 37(2): 14.

Dijkstra, F. A., Carrillo, Y., Pendall, E. and Morgan, J. A. (2013). Rhizosphere priming: A nutrient perspective, *Front. Microbiol.* 4: 216.

Dijkstra, F. A., Zhu, B. and Cheng, W. (2021). Root effects on soil organic carbon: A double-edged sword, *New Phytol.* 230(1): 60–5.

Doetterl, S., Stevens, A., Six, J., Merckx, R., Van Oost, K., Casanova Pinto, M., Casanova-Katny, A., Muñoz, C., Boudin, M., Zagal Venegas, E. and Boeckx, P. (2015). Soil carbon storage controlled by interactions between geochemistry and climate, *Nat. Geosci.* 8(10): 780–3.

Drake, J. E., Oishi, A. C., Giasson, M.-A., Oren, R., Johnsen, K. H. and Finzi, A. C. (2012). Trenching reduces soil heterotrophic activity in a loblolly pine (*Pinus taeda*) forest exposed to elevated atmospheric [CO_2] and N fertilization, *Agr. Forest Meteorol.* 165: 43–52.

Elliott, A. J., Daniell, T. J., Cameron, D. D. and Field, K. J. (2020). A commercial arbuscular mycorrhizal inoculum increases root colonization across wheat cultivars but does not increase assimilation of mycorrhiza-acquired nutrients, *Plants People Planet* 3: 588–99.

Feng, X., Simpson, A. J. and Simpson, M. J. (2005). Chemical and mineralogical controls on humic acid sorption to clay mineral surfaces, *Org. Geochem.* 36(11): 1553–66.

Fenner, N. and Freeman, C. (2011). Drought-induced carbon loss in peatlands, *Nat. Geosci.* 4(12): 895–900.

Finzi, A. C., Abramoff, R. Z., Spiller, K. S., Brzostek, E. R., Darby, B. A., Kramer, M. A. and Phillips, R. P. (2015). Rhizosphere processes are quantitatively important components of terrestrial carbon and nutrient cycles, *Glob. Change Biol.* 21(5): 2082–94.

Flombaum, P. and Sala, O. E. (2008). Higher effect of plant species diversity on productivity in natural than artificial ecosystems, *Proc. Natl. Acad. Sci. U. S. A.* 105(16): 6087–90.

Fontaine, S., Bardoux, G., Abbadie, L. and Mariotti, A. (2004). Carbon input to soil may decrease soil carbon content, *Ecol. Lett.* 7(4): 314–20.

Fontaine, S., Barot, S., Barre, P., Bdioui, N., Mary, B. and Rumpel, C. (2007). Stability of organic carbon in deep soil layers controlled by fresh carbon supply, *Nature* 450(7167): 277–80.

Freeman, C., Ostle, N. and Kang, H. (2001). An enzymic 'latch' on a global carbon store, *Nature* 409(6817): 149.

Freeman, C., Ostle, N. J., Fenner, N. and Kang, H. (2004). A regulatory role for phenol oxidase during decomposition in peatlands, *Soil Biol. Biochem.* 36(10): 1663–7.

Freschet, G. T., Cornwell, W. K., Wardle, D. A., Elumeeva, T. G., Liu, W., Jackson, B. G., Onipchenko, V. G., Soudzilovskaia, N. A., Tao, J. and Cornelissen, J. H. C. (2013). Linking litter decomposition of above- and below-ground organs to plant-soil feedbacks worldwide, *J. Ecol.* 101(4): 943–52.

Frey, S. D. (2019). Mycorrhizal fungi as mediators of soil organic matter dynamics, *Annu. Rev. Ecol. Evol. Syst.* 50(1): 237–59.

Friedlingstein, P., Jones, M. W., O'Sullivan, M., Andrew, R. M., Hauck, J., Peters, G. P., Peters, W., Pongratz, J., Sitch, S., Le Quéré, C., Bakker, D. C. E., Canadell, J. G., Ciais, P., Jackson, R. B., Anthoni, P., Barbero, L., Bastos, A., Bastrikov, V., Becker, M., Bopp, L., Buitenhuis, E., Chandra, N., Chevallier, F., Chini, L. P., Currie, K. I., Feely, R. A., Gehlen,

M., Gilfillan, D., Gkritzalis, T., Goll, D. S., Gruber, N., Gutekunst, S., Harris, I., Haverd, V., Houghton, R. A., Hurtt, G., Ilyina, T., Jain, A. K., Joetzjer, E., Kaplan, J. O., Kato, E., Klein Goldewijk, K., Korsbakken, J. I., Landschützer, P., Lauvset, S. K., Lefèvre, N., Lenton, A., Lienert, S., Lombardozzi, D., Marland, G., McGuire, P. C., Melton, J. R., Metzl, N., Munro, D. R., Nabel, J. E. M. S., Nakaoka, S., Neill, C., Omar, A. M., Ono, T., Peregon, A., Pierrot, D., Poulter, B., Rehder, G., Resplandy, L., Robertson, E., Rödenbeck, C., Séférian, R., Schwinger, J., Smith, N., Tans, P. P., Tian, H., Tilbrook, B., Tubiello, F. N., van der Werf, G. R., Wiltshire, A. J. and Zaehle, S. (2019). Global carbon budget 2019, *Earth Syst. Sci. Data* 11(4): 1783–838.

Fröberg, M., Jardine, P. M., Hanson, P. J., Swanston, C. W., Todd, D. E., Tarver, J. R. and Garten, C. T. (2007). Low dissolved organic carbon input from fresh litter to deep mineral soils, *Soil Sci. Soc. Am. J.* 71(2): 347–54.

Fulton-Smith, S. and Cotrufo, M. F. (2019). Pathways of soil organic matter formation from above and belowground inputs in a *Sorghum bicolor* bioenergy crop, *GCB Bioenergy* 11(8): 971–87.

Gadgil, R. and Gadgil, P. (1975). Suppression of litter decomposition by mycorrhizal roots of *Pinus radiata*, *N. Z. J. For. Sci.* 5: 35–41.

Gadgil, R. L. and Gadgil, P. D. (1971). Mycorrhiza and litter decomposition, *Nature* 233(5315): 133.

Garrido-Ramirez, E. G., Theng, B. K. G. and Mora, M. L. (2010). Clays and oxide minerals as catalysts and nanocatalysts in Fenton-like reactions - A review, *Appl. Clay Sci.* 47(3-4): 182–92.

Gherardi, L. A. and Sala, O. E. (2020). Global patterns and climatic controls of belowground net carbon fixation, *Proc. Natl. Acad. Sci. U. S. A.* 117(33): 20038–43.

Gholz, H. L., Wedin, D. A., Smitherman, S. M., Harmon, M. E. and Parton, W. J. (2000). Long-term dynamics of pine and hardwood litter in contrasting environments: Toward a global model of decomposition, *Glob. Change Biol.* 6(7): 751–65.

Gill, A. L. and Finzi, A. C. (2016). Belowground carbon flux links biogeochemical cycles and resource-use efficiency at the global scale, *Ecol. Lett.* 19(12): 1419–28.

Godbold, D. L., Hoosbeek, M. R., Lukac, M., Cotrufo, M. F., Janssens, I. A., Ceulemans, R., Polle, A., Velthorst, E. J., Scarascia-Mugnozza, G., De Angelis, P., Miglietta, F. and Peressotti, A. (2006). Mycorrhizal hyphal turnover as a dominant process for carbon input into soil organic matter, *Plant Soil* 281(1–2): 15–24.

Guelland, K., Hagedorn, F., Smittenberg, R. H., Göransson, H., Bernasconi, S. M., Hajdas, I. and Kretzschmar, R. (2012). Evolution of carbon fluxes during initial soil formation along the forefield of Damma glacier, Switzerland, *Biogeochemistry* 113(1–3): 545–61.

Guenet, B., Camino-Serrano, M., Ciais, P., Tifafi, M., Maignan, F., Soong, J. L. and Janssens, I. A. (2018). Impact of priming on global soil carbon stocks, *Glob. Change Biol.* 24(5): 1873–83.

Gyssels, G., Poesen, J., Bochet, E. and Li, Y. (2005). Impact of plant roots on the resistance of soils to erosion by water: A review, *Prog. Phys. Geog.* 29(2): 189-217.

Haichar, F. el. Z., Santaella, C., Heulin, T. and Achouak, W. (2014). Root exudates mediated interactions belowground, *Soil Biol. Biochem.* 77: 69-80.

Hall, S. J., Huang, W., Timokhin, V. I. and Hammel, K. E. (2020). Lignin lags, leads, or limits the decomposition of litter and soil organic carbon, *Ecology* 101(9): e03113.

Harrison, R. B., Footen, P. W. and Strahm, B. D. (2011). Deep soil horizons: Contribution and importance to soil carbon pools and in assessing whole-ecosystem response to management and global change, *For. Sci.* 57(1): 67-76.

Hartley, I. P., Garnett, M. H., Sommerkorn, M., Hopkins, D. W., Fletcher, B. J., Sloan, V. L., Phoenix, G. K. and Wookey, P. A. (2012). A potential loss of carbon associated with greater plant growth in the European Arctic, *Nat. Clim. Change* 2(12): 875-9.

Hartmann, A., Rothballer, M. and Schmid, M. (2008). Lorenz Hiltner, a pioneer in rhizosphere microbial ecology and soil bacteriology research, *Plant Soil* 312(1-2): 7-14.

He, Y., Cheng, W., Zhou, L., Shao, J., Liu, H., Zhou, H., Zhu, K. and Zhou, X. (2020). Soil DOC release and aggregate disruption mediate rhizosphere priming effect on soil C decomposition, *Soil Biol. Biochem.* 144: 107787.

Hemingway, J. D., Rothman, D. H., Grant, K. E., Rosengard, S. Z., Eglinton, T. I., Derry, L. A. and Galy, V. V. (2019). Mineral protection regulates long-term global preservation of natural organic carbon, *Nature* 570(7760): 228-31.

Hernes, P. J., Kaiser, K., Dyda, R. Y. and Cerli, C. (2013). Molecular trickery in soil organic matter: Hidden lignin, *Environ. Sci. Technol.* 47(16): 9077-85.

Hicks Pries, C. E., Schuur, E. A. G., Natali, S. M. and Crummer, K. G. (2015). Old soil carbon losses increase with ecosystem respiration in experimentally thawed tundra, *Nat. Clim. Change* 6(2): 214-8.

Hiltner, L. (1904). Über neuere Erfahrungen und Probleme auf dem Gebiete der Bodenbakteriologie unter besonderer Berücksichtigung der Gründüngung und Brache, *Arb. D. L. G.* 98: 59-78.

Hinsinger, P., Bengough, A. G., Vetterlein, D. and Young, I. M. (2009). Rhizosphere: Biophysics, biogeochemistry and ecological relevance, *Plant Soil* 321(1-2): 117-52.

Hofrichter, M. (2002). Review: Lignin conversion by manganese peroxidase (MnP), *Enzyme Microb. Technol.* 30(4): 454-66.

Hooper, D. U., Adair, E. C., Cardinale, B. J., Byrnes, J. E., Hungate, B. A., Matulich, K. L., Gonzalez, A., Duffy, J. E., Gamfeldt, L. and O'Connor, M. I. (2012). A global synthesis reveals biodiversity loss as a major driver of ecosystem change, *Nature* 486(7401): 105-8.

Hooper, D. U., Chapin, F. S., Ewel, J. J., Hector, A., Inchausti, P., Lavorel, S., Lawton, J. H., Lodge, D. M., Loreau, M., Naeem, S., Schmid, B., Setälä, H., Symstad, A. J., Vandermeer, J. and Wardle, D. A. (2005). Effects of biodiversity on ecosystem functioning: A consensus of current knowledge, *Ecol. Monogr.* 75(1): 3-35.

Huang, W., Hammel, K. E., Hao, J., Thompson, A., Timokhin, V. I. and Hall, S. J. (2019). Enrichment of lignin-derived carbon in mineral-associated soil organic matter, *Environ. Sci. Technol.* 53(13): 7522-31.

Huang, Y., Chen, Y., Castro-Izaguirre, N., Baruffol, M., Brezzi, M., Lang, A., Li, Y., Härdtle, W., von Oheimb, G., Yang, X., Liu, X., Pei, K., Both, S., Yang, B., Eichenberg, D., Assmann, T., Bauhus, J., Behrens, T., Buscot, F., Chen, X. Y., Chesters, D., Ding, B. Y., Durka, W., Erfmeier, A., Fang, J., Fischer, M., Guo, L. D., Guo, D., Gutknecht, J. L. M., He, J. S., He, C. L., Hector, A., Hönig, L., Hu, R. Y., Klein, A. M., Kühn, P., Liang, Y., Li, S., Michalski, S., Scherer-Lorenzen, M., Schmidt, K., Scholten, T., Schuldt, A., Shi, X., Tan, M. Z., Tang, Z., Trogisch, S., Wang, Z., Welk, E., Wirth, C., Wubet, T., Xiang, W., Yu, M., Yu, X. D., Zhang, J., Zhang, S., Zhang, N., Zhou, H. Z., Zhu, C. D., Zhu, L., Bruelheide, H., Ma, K., Niklaus, P. A. and Schmid, B. (2018). Impacts of species richness on productivity in a large-scale subtropical forest experiment, *Science* 362(6410): 80-3.

Huo, C., Luo, Y. and Cheng, W. (2017). Rhizosphere priming effect: A meta-analysis, *Soil Biol. Biochem.* 111: 78-84.

Hütsch, B. W., Augustin, J. and Merbach, W. (2002). Plant rhizodeposition–An important source for carbon turnover in soils, *J. Plant Nutr. Soil Sci.* 165(4): 397-407.

Isbell, F., Calcagno, V., Hector, A., Connolly, J., Harpole, W. S., Reich, P. B., Scherer-Lorenzen, M., Schmid, B., Tilman, D., van Ruijven, J., Weigelt, A., Wilsey, B. J., Zavaleta, E. S. and Loreau, M. (2011). High plant diversity is needed to maintain ecosystem services, *Nature* 477(7363): 199-202.

Jackson, R. B., Canadell, J., Ehleringer, J. R., Mooney, H. A., Sala, O. E. and Schulze, E. D. (1996). A global analysis of root distributions for terrestrial biomes, *Oecologia* 108(3): 389-411.

Jackson, R. B., Lajtha, K., Crow, S. E., Hugelius, G., Kramer, M. G. and Piñeiro, G. (2017). The ecology of soil carbon: Pools, vulnerabilities, and biotic and abiotic controls, *Annu. Rev. Ecol. Evol. Syst.* 48(1): 419-45.

Jacobs, L. M., Sulman, B. N., Brzostek, E. R., Feighery, J. J. and Phillips, R. P. (2018). Interactions among decaying leaf litter, root litter and soil organic matter vary with mycorrhizal type, *J. Ecol.* 106(2): 502-13.

Jensen, E. S., Peoples, M. B., Boddey, R. M., Gresshoff, P. M., Hauggaard-Nielsen, H., J.R. Alves, B. and Morrison, M. J. (2012). Legumes for mitigation of climate change and the provision of feedstock for biofuels and biorefineries: A review, *Agron. Sustain. Dev.* 32(2): 329-64.

Jia, J., Cao, Z., Liu, C., Zhang, Z., Lin, L., Wang, Y., Haghipour, N., Wacker, L., Bao, H., Dittmar, T., Simpson, M. J., Yang, H., Crowther, T. W., Eglinton, T. I., He, J. and Feng, X. (2019). Climate warming alters subsoil but not topsoil carbon dynamics in alpine grassland, *Glob. Change Biol.* 25(12): 4383-93.

Jia, J., Feng, X. J., He, J. S., He, H., Lin, L. and Liu, Z. (2017). Comparing microbial carbon sequestration and priming in the subsoil versus topsoil of a Qinghai-Tibetan alpine grassland, *Soil Biol. Biochem.* 104: 141-51.

Jilling, A., Keiluweit, M., Contosta, A. R., Frey, S., Schimel, J., Schnecker, J., Smith, R. G., Tiemann, L. and Grandy, A. S. (2018). Minerals in the rhizosphere: Overlooked mediators of soil nitrogen availability to plants and microbes, *Biogeochemistry* 139(2): 103-22.

Jo, I., Fei, S., Oswalt, C. M., Domke, G. M. and Phillips, R. P. (2019). Shifts in dominant tree mycorrhizal associations in response to anthropogenic impacts, *Sci. Adv.* 5(4): eaav6358.

Jobbagy, E. G. and Jackson, R. B. (2000). The vertical distribution of soil organic carbon and its relation to climate and vegetation, *Ecol. Appl.* 10(2): 423-36.

Jones, D. L., Nguyen, C. and Finlay, R. D. (2009). Carbon flow in the rhizosphere: Carbon trading at the soil-root interface, *Plant Soil* 321(1-2): 5-33.

Kaiser, C., Kilburn, M. R., Clode, P. L., Fuchslueger, L., Koranda, M., Cliff, J. B., Solaiman, Z. M. and Murphy, D. V. (2015). Exploring the transfer of recent plant photosynthates to soil microbes: Mycorrhizal pathway vs direct root exudation, *New Phytol.* 205(4): 1537-51.

Kaiser, K. and Guggenberger, G. (2003). Mineral surfaces and soil organic matter, *Eur. J. Soil Sci.* 54(2): 219-36.

Kaiser, K., Guggenberger, G. and Haumaier, L. (2004). Changes in dissolved lignin-derived phenols, neutral sugars, uronic acids, and amino sugars with depth in forested Haplic Arenosols and Rendzic Leptosols, *Biogeochemistry* 70(1): 135-51.

Kaiser, K. and Kalbitz, K. (2012). Cycling downwards - Dissolved organic matter in soils, *Soil Biol. Biochem.* 52: 29-32.

Kalbitz, K., Meyer, A., Yang, R. and Gerstberger, P. (2007). Response of dissolved organic matter in the forest floor to long-term manipulation of litter and throughfall inputs, *Biogeochemistry* 86(3): 301–18.

Kallenbach, C. M., Frey, S. D. and Grandy, A. S. (2016). Direct evidence for microbial-derived soil organic matter formation and its ecophysiological controls, *Nat. Commun.* 7: 13630.

Kammer, A., Schmidt, M. W. I. and Hagedorn, F. (2012). Decomposition pathways of C-13-depleted leaf litter in forest soils of the Swiss Jura, *Biogeochemistry* 108(1-3): 395–411.

Keidel, L., Lenhart, K., Moser, G. and Müller, C. (2018). Depth-dependent response of soil aggregates and soil organic carbon content to long-term elevated CO_2 in a temperate grassland soil, *Soil Biol. Biochem.* 123: 145-54.

Keiluweit, M., Bougoure, J. J., Nico, P. S., Pett-Ridge, J., Weber, P. K. and Kleber, M. (2015). Mineral protection of soil carbon counteracted by root exudates, *Nat. Clim. Change* 5(6): 588–95.

Keller, A. B., Brzostek, E. R., Craig, M. E., Fisher, J. B. and Phillips, R. P. (2021). Root-derived inputs are major contributors to soil carbon in temperate forests, but vary by mycorrhizal type, *Ecol. Lett.* 24(4): 626–35.

Keller, A. B. and Phillips, R. P. (2019). Leaf litter decay rates differ between mycorrhizal groups in temperate, but not tropical, forests, *New Phytol.* 222(1): 556–64.

Keuper, F., Wild, B., Kummu, M., Beer, C., Blume-Werry, G., Fontaine, S., Gavazov, K., Gentsch, N., Guggenberger, G., Hugelius, G., Jalava, M., Koven, C., Krab, E. J., Kuhry, P., Monteux, S., Richter, A., Shahzad, T., Weedon, J. T. and Dorrepaal, E. (2020). Carbon loss from northern circumpolar permafrost soils amplified by rhizosphere priming, *Nat. Geosci.* 13(8): 560–5.

Khedim, N., Cecillon, L., Poulenard, J., Barré, P., Baudin, F., Marta, S., Rabatel, A., Dentant, C., Cauvy-Fraunié, S., Anthelme, F., Gielly, L., Ambrosini, R., Franzetti, A., Azzoni, R. S., Caccianiga, M. S., Compostella, C., Clague, J., Tielidze, L., Messager, E., Choler, P. and Ficetola, G. F. (2021). Topsoil organic matter build-up in glacier forelands around the world, *Glob. Change Biol.* 27(8): 1662–77.

Kirk, T. K. and Farrell, R. L. (1987). Enzymatic "combustion": The microbial degradation of lignin, *Annu. Rev. Microbiol.* 41(1): 465–505.

Kleber, M., Eusterhues, K., Keiluweit, M., Mikutta, C., Mikutta, R. and Nico, P. S. (2015). Mineral-organic associations: Formation, properties, and relevance in soil environments. In: Donald, L. S. (Ed.), *Advances in Agronomy*. San Diego, CA: Academic Press.

Klotzbücher, T., Kalbitz, K., Cerli, C., Hernes, P. J. and Kaiser, K. (2016). Gone or just out of sight? The apparent disappearance of aromatic litter components in soils, *Soil* 2(3): 325–35.

Kögel-Knabner, I. (2000). Analytical approaches for characterizing soil organic matter, *Org. Geochem.* 31(7–8): 609–25.

Kögel-Knabner, I. (2002). The macromolecular organic composition of plant and microbial residues as inputs to soil organic matter, *Soil Biol. Biochem.* 34(2): 139–62.

Kögel-Knabner, I. and Amelung, W. (2014). Dynamics, chemistry, and preservation of organic matter in soils. In: Holland, H. D. and Turekian, K. K. (Eds.), *Treatise on Geochemistry* (2nd edn.). Oxford: Elsevier.

Kögel-Knabner, I., Guggenberger, G., Kleber, M., Kandeler, E., Kalbitz, K., Scheu, S., Eusterhues, K. and Leinweber, P. (2008). Organo-mineral associations in temperate soils: Integrating biology, mineralogy, and organic matter chemistry, *J. Plant Nutr. Soil Sci.* 171(1): 61-82.

Kögel-Knabner, I. and Rumpel, C. (2018). Advances in molecular approaches for understanding soil organic matter composition, origin, and turnover: A historical overview. In: Sparks, D. L. (Ed.), *Advances in Agronomy*. San Diego, CA: Academic Press.

Kopittke, P. M., Hernandez-Soriano, M. C., Dalal, R. C., Finn, D., Menzies, N. W., Hoeschen, C. and Mueller, C. W. (2018). Nitrogen-rich microbial products provide new organo-mineral associations for the stabilization of soil organic matter, *Glob. Change Biol.* 24(4): 1762-70.

Kramer, S., Marhan, S., Haslwimmer, H., Ruess, L. and Kandeler, E. (2013). Temporal variation in surface and subsoil abundance and function of the soil microbial community in an arable soil, *Soil Biol. Biochem.* 61: 76-85.

Kuzyakov, Y. (2002). Review: Factors affecting rhizosphere priming effects, *J. Plant Nutr. Soil Sci.* 165(4): 382-96.

Kuzyakov, Y., Friedel, J. K. and Stahr, K. (2000). Review of mechanisms and quantification of priming effects, *Soil Biol. Biochem.* 32(11-12): 1485-98.

Kuzyakov, Y. and Xu, X. (2013). Competition between roots and microorganisms for nitrogen: Mechanisms and ecological relevance, *New Phytol.* 198(3): 656-69.

Lajtha, K., Bowden, R. D. and Nadelhoffer, K. (2014). Litter and root manipulations provide insights into soil organic matter dynamics and stability, *Soil Sci. Soc. Am. J.* 78(S1): S261-SS9.

Lal, R. (2008). Sequestration of atmospheric CO_2 in global carbon pools, *Energy Environ. Sci.* 1(1): 86-100.

Lange, M., Eisenhauer, N., Sierra, C. A., Bessler, H., Engels, C., Griffiths, R. I., Mellado-Vázquez, P. G., Malik, A. A., Roy, J., Scheu, S., Steinbeiss, S., Thomson, B. C., Trumbore, S. E. and Gleixner, G. (2015). Plant diversity increases soil microbial activity and soil carbon storage, *Nat. Commun.* 6: 6707.

Lavallee, J. M., Soong, J. L. and Cotrufo, M. F. (2020). Conceptualizing soil organic matter into particulate and mineral-associated forms to address global change in the 21st century, *Glob. Change Biol.* 26(1): 261-73.

Lee, S. H., Jang, I., Chae, N., Choi, T. and Kang, H. (2013). Organic layer serves as a hotspot of microbial activity and abundance in arctic tundra soils, *Microb. Ecol.* 65(2): 405-14.

Lehmann, A., Leifheit, E. F. and Rillig, M. C. (2017). Mycorrhizas and soil aggregation. In: Johnson, N. C., Gehring, C. and Jansa, J. (Eds.), *Mycorrhizal Mediation of Soil*. Amsterdam: Elsevier.

Lehmann, J. and Kleber, M. (2015). The contentious nature of soil organic matter, *Nature* 528(7580): 60-8.

Lehmann, J., Solomon, D., Kinyangi, J., Dathe, L., Wirick, S. and Jacobsen, C. (2008). Spatial complexity of soil organic matter forms at nanometre scales, *Nat. Geosci.* 1(4): 238-42.

Leifheit, E. F., Verbruggen, E. and Rillig, M. C. (2015). Arbuscular mycorrhizal fungi reduce decomposition of woody plant litter while increasing soil aggregation, *Soil Biol. Biochem.* 81: 323-8.

Li, C. H., Li, Y. and Tang, L. S. (2013). The effects of long-term fertilization on the accumulation of organic carbon in the deep soil profile of an oasis farmland, *Plant Soil* 369(1-2): 645-56.

Liang, C., Schimel, J. P. and Jastrow, J. D. (2017). The importance of anabolism in microbial control over soil carbon storage, *Nat. Microbiol.* 2: 17105.

Lin, G., McCormack, M. L., Ma, C. and Guo, D. (2017). Similar below-ground carbon cycling dynamics but contrasting modes of nitrogen cycling between arbuscular mycorrhizal and ectomycorrhizal forests, *New Phytol.* 213(3): 1440-51.

Lin, S. S. and Gurol, M. D. (1998). Catalytic decomposition of hydrogen peroxide on iron oxide: Kinetics, mechanism, and implications, *Environ. Sci. Technol.* 32(10): 1417-23.

Liu, H., Mi, Z., Lin, L., Wang, Y., Zhang, Z., Zhang, F., Wang, H., Liu, L., Zhu, B., Cao, G., Zhao, X., Sanders, N. J., Classen, A. T., Reich, P. B. and He, J. S. (2018a). Shifting plant species composition in response to climate change stabilizes grassland primary production, *Proc. Natl. Acad. Sci. U. S. A.* 115(16): 4051-6.

Liu, X., Trogisch, S., He, J. S., Niklaus, P. A., Bruelheide, H., Tang, Z., Erfmeier, A., Scherer-Lorenzen, M., Pietsch, K. A., Yang, B., Kühn, P., Scholten, T., Huang, Y., Wang, C., Staab, M., Leppert, K. N., Wirth, C., Schmid, B. and Ma, K. (2018b). Tree species richness increases ecosystem carbon storage in subtropical forests, *Proc. Biol. Sci.* 285(1885): 20181240.

Loreau, M., Naeem, S., Inchausti, P., Bengtsson, J., Grime, J. P., Hector, A., Hooper, D. U., Huston, M. A., Raffaelli, D., Schmid, B., Tilman, D. and Wardle, D. A. (2001). Biodiversity and ecosystem functioning: Current knowledge and future challenges, *Science* 294(5543): 804-8.

Lu, J., Dijkstra, F. A., Wang, P. and Cheng, W. (2019). Roots of non-woody perennials accelerated long-term soil organic matter decomposition through biological and physical mechanisms, *Soil Biol. Biochem.* 134: 42-53.

Lützow, M. v., Kögel-Knabner, I., Ekschmitt, K., Matzner, E., Guggenberger, G., Marschner, B. and Flessa, H. (2006). Stabilization of organic matter in temperate soils: Mechanisms and their relevance under different soil conditions - A review, *Eur. J. Soil Sci.* 57(4): 426-45.

Lynch, J. M. and Whipps, J. M. (1990). Substrate flow in the rhizosphere, *Plant Soil* 129(1): 1-10.

Ma, T., Dai, G., Zhu, S., Chen, D., Chen, L., Lü, X., Wang, X., Zhu, J., Zhang, Y., He, J., Bai, Y., Han, X. and Feng, X. (2019). Vertical variations in plant- and microbial-derived carbon components in grassland soils, *Plant Soil* 446(1-2): 441-55.

Ma, T., Zhu, S., Wang, Z., Chen, D., Dai, G., Feng, B., Su, X., Hu, H., Li, K., Han, W., Liang, C., Bai, Y. and Feng, X. (2018). Divergent accumulation of microbial necromass and plant lignin components in grassland soils, *Nat. Commun.* 9(1): 3480.

Marschner, B., Brodowski, S., Dreves, A., Gleixner, G., Gude, A., Grootes, P. M., Hamer, U., Heim, A., Jandl, G., Ji, R., Kaiser, K., Kalbitz, K., Kramer, C., Leinweber, P., Rethemeyer, J., Schäffer, A., Schmidt, M. W. I., Schwark, L. and Wiesenberg, G. L. B. (2008). How relevant is recalcitrance for the stabilization of organic matter in soils?, *J. Plant Nutr. Soil Sci.* 171(1): 91-110.

Martin, F., Kohler, A., Murat, C., Veneault-Fourrey, C. and Hibbett, D. S. (2016). Unearthing the roots of ectomycorrhizal symbioses, *Nat. Rev. Microbiol.* 14(12): 760-73.

Martinez, C., Alberti, G., Cotrufo, M. F., Magnani, F., Zanotelli, D., Camin, F., Gianelle, D., Cescatti, A. and Rodeghiero, M. (2016). Belowground carbon allocation patterns as determined by the in-growth soil core ^{13}C technique across different ecosystem types, *Geoderma* 263: 140-50.

McKendry, P. (2002). Energy production from biomass (part 1): Overview of biomass, *Bioresour. Technol.* 83(1): 37-46.

Meier, I. C., Finzi, A. C. and Phillips, R. P. (2017). Root exudates increase N availability by stimulating microbial turnover of fast-cycling N pools, *Soil Biol. Biochem.* 106: 119-28.

Melillo, J. M., Aber, J. D., Linkins, A. E., Ricca, A., Fry, B. and Nadelhoffer, K. J. (1989). Carbon and nitrogen dynamics along the decay continuum - Plant litter to soil organic-matter, *Plant Soil* 115(2): 189-98.

Melillo, J. M., Aber, J. D. and Muratore, J. F. (1982). Nitrogen and lignin control of hardwood leaf litter decomposition dynamics, *Ecology* 63(3): 621-6.

Mendes, R., Garbeva, P. and Raaijmakers, J. M. (2013). The rhizosphere microbiome: Significance of plant beneficial, plant pathogenic, and human pathogenic microorganisms, *FEMS Microbiol. Rev.* 37(5): 634-63.

Midgley, M. G., Brzostek, E. and Phillips, R. P. (2015). Decay rates of leaf litters from arbuscular mycorrhizal trees are more sensitive to soil effects than litters from ectomycorrhizal trees, *J. Ecol.* 103(6): 1454-63.

Miltner, A., Bombach, P., Schmidt-Brücken, B. and Kästner, M. (2011). SOM genesis: Microbial biomass as a significant source, *Biogeochemistry* 111(1-3): 41-55.

Moore, J. A. M., Jiang, J., Patterson, C. M., Mayes, M. A., Wang, G. and Classen, A. T. (2015). Interactions among roots, mycorrhizas and free-living microbial communities differentially impact soil carbon processes, *J. Ecol.* 103(6): 1442-53.

Murray, A. H., Iason, G. R. and Stewart, C. (1996). Effect of simple phenolic compounds of heather (*Calluna vulgaris*) on rumen microbial activity in vitro, *J. Chem. Ecol.* 22(8): 1493-504.

Nguyen, C. (2003). Rhizodeposition of organic C by plants: Mechanisms and controls, *Agronomie* 23(5-6): 375-96.

Nicolas, C., Martin-Bertelsen, T., Floudas, D., Bentzer, J., Smits, M., Johansson, T., Troein, C., Persson, P. and Tunlid, A. (2019). The soil organic matter decomposition mechanisms in ectomycorrhizal fungi are tuned for liberating soil organic nitrogen, *ISME J.* 13(4): 977-88.

Ofiti, N. O. E., Zosso, C. U., Soong, J. L., Solly, E. F., Torn, M. S., Wiesenberg, G. L. B. and Schmidt, M. W. I. (2021). Warming promotes loss of subsoil carbon through accelerated degradation of plant-derived organic matter, *Soil Biol. Biochem.* 156: 108185.

Op De Beeck, M., Troein, C., Peterson, C., Persson, P. and Tunlid, A. (2018). Fenton reaction facilitates organic nitrogen acquisition by an ectomycorrhizal fungus, *New Phytol.* 218(1): 335-43.

Ouimette, A. P., Ollinger, S. V., Lepine, L. C., Stephens, R. B., Rowe, R. J., Vadeboncoeur, M. A., Tumber-Davila, S. J. and Hobbie, E. A. (2019). Accounting for carbon flux to mycorrhizal fungi may resolve discrepancies in forest carbon budgets, *Ecosystems* 23(4): 715-29.

Pausch, J. and Kuzyakov, Y. (2012). Soil organic carbon decomposition from recently added and older sources estimated by $\delta^{13}C$ values of CO_2 and organic matter, *Soil Biol. Biochem.* 55: 40-7.

Pausch, J., Tian, J., Riederer, M. and Kuzyakov, Y. (2012). Estimation of rhizodeposition at field scale: Upscaling of a ^{14}C labeling study, *Plant Soil* 364(1-2): 273-85.

Pennekamp, F., Pontarp, M., Tabi, A., Altermatt, F., Alther, R., Choffat, Y., Fronhofer, E. A., Ganesanandamoorthy, P., Garnier, A., Griffiths, J. I., Greene, S., Horgan, K., Massie, T. M., Mächler, E., Palamara, G. M., Seymour, M. and Petchey, O. L. (2018). Biodiversity increases and decreases ecosystem stability, *Nature* 563(7729): 109-12.

Phillips, R. P., Finzi, A. C. and Bernhardt, E. S. (2011). Enhanced root exudation induces microbial feedbacks to N cycling in a pine forest under long-term CO_2 fumigation, *Ecol. Lett.* 14(2): 187–94.

Poirier, V., Angers, D. A. and Whalen, J. K. (2014). Formation of millimetric-scale aggregates and associated retention of ^{13}C-^{15}N-labelled residues are greater in subsoil than topsoil, *Soil Biol. Biochem.* 75: 45-53.

Poirier, V., Roumet, C. and Munson, A. D. (2018). The root of the matter: Linking root traits and soil organic matter stabilization processes, *Soil Biol. Biochem.* 120: 246-59.

Preston, C. M., Nault, J. R. and Trofymow, J. A. (2009). Chemical changes during 6 years of decomposition of 11 litters in some Canadian forest sites. Part 2. ^{13}C abundance, solid-state ^{13}C NMR Spectroscopy and the meaning of "lignin", *Ecosystems* 12(7): 1078-102.

Rasse, D. P., Rumpel, C. and Dignac, M.-F. (2005). Is soil carbon mostly root carbon? Mechanisms for a specific stabilisation, *Plant Soil* 269(1-2): 341-56.

Ratledge, C. (1994). *Biochemistry of Microbial Degradation*. Netherland: Springer.

Read, D. J. (1991). Mycorrhizas in ecosystems, *Experientia* 47(4): 376-91.

Read, D. J. and Perez-Moreno, J. (2003). Mycorrhizas and nutrient cycling in ecosystems - A journey towards relevance?, *New Phytol.* 157(3): 475-92.

Riedel, T., Biester, H. and Dittmar, T. (2012). Molecular fractionation of dissolved organic matter with metal salts, *Environ. Sci. Technol.* 46(8): 4419-26.

Rillig, M. C. (2004). Arbuscular mycorrhizae, glomalin, and soil aggregation, *Can. J. Soil Sci.* 84(4): 355–63.

Rowe, E. C., Evans, C. D., Emmett, B. A., Reynolds, B., Helliwell, R. C., Coull, M. C. and Curtis, C. J. (2006). Vegetation type affects the relationship between soil carbon to nitrogen ratio and nitrogen leaching, *Water Air Soil Pollut.* 177(1-4): 335–47.

Rumpel, C., Baumann, K., Remusat, L., Dignac, M., Barré, P., Deldicque, D., Glasser, G., Lieberwirth, I. and Chabbi, A. (2015). Nanoscale evidence of contrasted processes for root-derived organic matter stabilization by mineral interactions depending on soil depth, *Soil Biol. Biochem.* 85: 82-8.

Rumpel, C., Chabbi, A. and Marschner, B. (2012). Carbon storage and sequestration in subsoil horizons: Knowledge, gaps and potentials. In: *Recarbonization of the Biosphere*. New York: Springer.

Sanaullah, M., Chabbi, A., Leifeld, J., Bardoux, G., Billou, D. and Rumpel, C. (2011). Decomposition and stabilization of root litter in top-and subsoil horizons: What is the difference?, *Plant Soil* 338(1-2): 127–41.

Santhanam, N., Badri, D. V., Decker, S. R., Manter, D. K., Reardon, K. F. and Vivanco, J. M. (2012). Lignocellulose decomposition by microbial secretions. In: Vivanco, J. M. and Baluška, F. (Eds.), *Secretions and Exudates in Biological Systems*. Berlin, Heidelberg: Springer.

Schimel, D. S., Braswell, B. H., Holland, E. A., McKeown, R., Ojima, D. S., Painter, T. H., Parton, W. J. and Townsend, A. R. (1994). Climatic, edaphic, and biotic controls over storage and turnover of carbon in soils, *Global Biogeochem. Cycles* 8(3): 279-93.

Schmidt, M. W., Torn, M. S., Abiven, S., Dittmar, T., Guggenberger, G., Janssens, I. A., Kleber, M., Kögel-Knabner, I., Lehmann, J., Manning, D. A., Nannipieri, P., Rasse, D. P., Weiner, S. and Trumbore, S. E. (2011). Persistence of soil organic matter as an ecosystem property, *Nature* 478(7367): 49-56.

Schuur, E. A., Vogel, J. G., Crummer, K. G., Lee, H., Sickman, J. O. and Osterkamp, T. E. (2009). The effect of permafrost thaw on old carbon release and net carbon exchange from tundra, *Nature* 459(7246): 556–9.

Sher, Y., Baker, N. R., Herman, D., Fossum, C., Hale, L., Zhang, X., Nuccio, E., Saha, M., Zhou, J., Pett-Ridge, J. and Firestone, M. (2020). Microbial extracellular polysaccharide production and aggregate stability controlled by switchgrass (*Panicum virgatum*) root biomass and soil water potential, *Soil Biol. Biochem.* 143: 107742.

Shi, Y., Zhu, K., Dai, Y., Zhang, C. and Jia, H. (2020). Evolution and stabilization of environmental persistent free radicals during the decomposition of lignin by laccase, *Chemosphere* 248: 125931.

Sistla, S. A., Moore, J. C., Simpson, R. T., Gough, L., Shaver, G. R. and Schimel, J. P. (2013). Long-term warming restructures Arctic tundra without changing net soil carbon storage, *Nature* 497(7451): 615–8.

Sokol, N. W. and Bradford, M. A. (2018). Microbial formation of stable soil carbon is more efficient from belowground than aboveground input, *Nat. Geosci.* 12(1): 46–53.

Sokol, N. W., Sanderman, J. and Bradford, M. A. (2019a). Pathways of mineral-associated soil organic matter formation: Integrating the role of plant carbon source, chemistry, and point of entry, *Glob. Change Biol.* 25(1): 12–24.

Sokol, N. W., Kuebbing, S. E., Karlsen-Ayala, E. and Bradford, M. A. (2019b). Evidence for the primacy of living root inputs, not root or shoot litter, in forming soil organic carbon, *New Phytol.* 221(1): 233–46.

Sollins, P., Swanston, C., Kleber, M., Filley, T., Kramer, M., Crow, S., Caldwell, B. A., Lajtha, K. and Bowden, R. (2006). Organic C and N stabilization in a forest soil: Evidence from sequential density fractionation, *Soil Biol. Biochem.* 38(11): 3313–24.

Solomon, D., Lehmann, J., Harden, J., Wang, J., Kinyangi, J., Heymann, K., Karunakaran, C., Lu, Y., Wirick, S. and Jacobsen, C. (2012). Micro- and nano-environments of carbon sequestration: Multi-element STXM-NEXAFS spectromicroscopy assessment of microbial carbon and mineral associations, *Chem. Geol.* 329: 53–73.

Steidinger, B. S., Crowther, T. W., Liang, J., Van Nuland, M. E., Werner, G. D. A., Reich, P. B., Nabuurs, G. J., de-Miguel, S., Zhou, M., Picard, N., Herault, B., Zhao, X., Zhang, C., Routh, D., Peay, K. G. and GFBI Consortium. (2019). Climatic controls of decomposition drive the global biogeography of forest-tree symbioses, *Nature* 569(7756): 404–8.

Steinbeiss, S., Bessler, H., Engels, C., Temperton, V. M., Buchmann, N., Roscher, C., Kreutziger, Y., Baade, J., Habekost, M. and Gleixner, G. (2008). Plant diversity positively affects short-term soil carbon storage in experimental grasslands, *Glob. Change Biol.* 14(12): 2937–49.

Sterkenburg, E., Clemmensen, K. E., Ekblad, A., Finlay, R. D. and Lindahl, B. D. (2018). Contrasting effects of ectomycorrhizal fungi on early and late stage decomposition in a boreal forest, *ISME J.* 12(9): 2187–97.

Street, L. E., Garnett, M. H., Subke, J. A., Baxter, R., Dean, J. F. and Wookey, P. A. (2020). Plant carbon allocation drives turnover of old soil organic matter in permafrost tundra soils, *Glob. Change Biol.* 26(8): 4559–71.

Sun, T., Hobbie, S. E., Berg, B., Zhang, H., Wang, Q., Wang, Z. and Hättenschwiler, S. (2018). Contrasting dynamics and trait controls in first-order root compared with leaf litter decomposition, *Proc. Natl. Acad. Sci. U. S. A.* 115(41): 10392–7.

Talbot, J. M., Allison, S. D. and Treseder, K. K. (2008). Decomposers in disguise: Mycorrhizal fungi as regulators of soil C dynamics in ecosystems under global change, *Funct. Ecol.* 22(6): 955–63.

Talbot, J. M., Martin, F., Kohler, A., Henrissat, B. and Peay, K. G. (2015). Functional guild classification predicts the enzymatic role of fungi in litter and soil biogeochemistry, *Soil Biol. Biochem.* 88: 441–56.

Tautges, N. E., Chiartas, J. L., Gaudin, A. C. M., O'Geen, A. T., Herrera, I. and Scow, K. M. (2019). Deep soil inventories reveal that impacts of cover crops and compost on soil carbon sequestration differ in surface and subsurface soils, *Glob. Change Biol.* 25(11): 3753–66.

Taylor, M. K., Lankau, R. A. and Wurzburger, N. (2016). Mycorrhizal associations of trees have different indirect effects on organic matter decomposition, *J. Ecol.* 104(6): 1576–84.

Terrer, C., Phillips, R. P., Hungate, B. A., Rosende, J., Pett-Ridge, J., Craig, M. E., van Groenigen, K. J., Keenan, T. F., Sulman, B. N., Stocker, B. D., Reich, P. B., Pellegrini, A. F. A., Pendall, E., Zhang, H., Evans, R. D., Carrillo, Y., Fisher, J. B., Van Sundert, K., Vicca, S. and Jackson, R. B. (2021). A trade-off between plant and soil carbon storage under elevated CO_2, *Nature* 591(7851): 599–603.

Terrer, C., Vicca, S., Hungate, B. A., Phillips, R. P. and Prentice, I. C. (2016). Mycorrhizal association as a primary control of the CO_2 fertilization effect, *Science* 353(6294): 72–4.

Thevenot, M., Dignac, M. F. and Rumpel, C. (2010). Fate of lignins in soils: A review, *Soil Biol. Biochem.* 42(8): 1200–11.

Tisserant, E., Malbreil, M., Kuo, A., Kohler, A., Symeonidi, A., Balestrini, R., Charron, P., Duensing, N., Frei dit Frey, N., Gianinazzi-Pearson, V., Gilbert, L. B., Handa, Y., Herr, J. R., Hijri, M., Koul, R., Kawaguchi, M., Krajinski, F., Lammers, P. J., Masclaux, F. G., Murat, C., Morin, E., Ndikumana, S., Pagni, M., Petitpierre, D., Requena, N., Rosikiewicz, P., Riley, R., Saito, K., San Clemente, H., Shapiro, H., van Tuinen, D., Bécard, G., Bonfante, P., Paszkowski, U., Shachar-Hill, Y. Y., Tuskan, G. A., Young, J. P., Sanders, I. R., Henrissat, B., Rensing, S. A., Grigoriev, I. V., Corradi, N., Roux, C. and Martin, F. (2013). Genome of an arbuscular mycorrhizal fungus provides insight into the oldest plant symbiosis, *Proc. Natl. Acad. Sci. U. S. A.* 110(50): 20117–22.

Tombacz, E., Libor, Z., Illes, E., Majzik, A. and Klumpp, E. (2004). The role of reactive surface sites and complexation by humic acids in the interaction of clay mineral and iron oxide particles, *Org. Geochem.* 35(3): 257–67.

Torn, M. S., Trumbore, S. E., Chadwick, O. A., Vitousek, P. M. and Hendricks, D. M. (1997). Mineral control of soil organic carbon storage and turnover, *Nature* 389(6647): 170–3.

Totsche, K. U., Amelung, W., Gerzabek, M. H., Guggenberger, G., Klumpp, E., Knief, C., Lehndorff, E., Mikutta, R., Peth, S., Prechtel, A., Ray, N. and Kögel-Knabner, I. (2018). Microaggregates in soils, *J. Plant Nutr. Soil Sci.* 181(1): 104–36.

van der Heijden, M. G., Bardgett, R. D. and van Straalen, N. M. (2008). The unseen majority: Soil microbes as drivers of plant diversity and productivity in terrestrial ecosystems, *Ecol. Lett.* 11(3): 296–310.

van Groenigen, K. J., Qi, X., Osenberg, C. W., Luo, Y. and Hungate, B. A. (2014). Faster decomposition under increased atmospheric CO_2 limits soil carbon storage, *Science* 344(6183): 508–9.

Vicca, S., Luyssaert, S., Penuelas, J., Campioli, M., Chapin, F. S., Ciais, P., Heinemeyer, A., Högberg, P., Kutsch, W. L., Law, B. E., Malhi, Y., Papale, D., Piao, S. L., Reichstein,

M., Schulze, E. D. and Janssens, I. A. (2012). Fertile forests produce biomass more efficiently, *Ecol. Lett.* 15(6): 520–6.

Villarino, S. H., Pinto, P., Jackson, R. B. and Piñeiro, G. (2021). Plant rhizodeposition: A key factor for soil organic matter formation in stable fractions, *Sci. Adv.* 7(16): eabd3176.

Vivanco, L. and Austin, A. T. (2006). Intrinsic effects of species on leaf litter and root decomposition: A comparison of temperate grasses from North and South America, *Oecologia* 150(1): 97–107.

Volk, M., Bassin, S., Lehmann, M. F., Johnson, M. G. and Andersen, C. P. (2018). ^{13}C isotopic signature and C concentration of soil density fractions illustrate reduced C allocation to subalpine grassland soil under high atmospheric N deposition, *Soil Biol. Biochem.* 125: 178–84.

Waggoner, D. C., Wozniak, A. S., Cory, R. M. and Hatcher, P. G. (2017). The role of reactive oxygen species in the degradation of lignin derived dissolved organic matter, *Geochim. Cosmochim. Acta* 208: 171–84.

Wallander, H., Johansson, U., Sterkenburg, E., Brandström Durling, M. and Lindahl, B. D. (2010). Production of ectomycorrhizal mycelium peaks during canopy closure in Norway spruce forests, *New Phytol.* 187(4): 1124–34.

Wang, Q., Wang, Y., Wang, S., He, T. and Liu, L. (2014). Fresh carbon and nitrogen inputs alter organic carbon mineralization and microbial community in forest deep soil layers, *Soil Biol. Biochem.* 72: 145–51.

Wang, X., Yin, L., Dijkstra, F. A., Lu, J., Wang, P. and Cheng, W. (2020). Rhizosphere priming is tightly associated with root-driven aggregate turnover, *Soil Biol. Biochem.* 149: 107964.

Wang, Y., Wang, H., He, J. S. and Feng, X. (2017). Iron-mediated soil carbon response to water-table decline in an alpine wetland, *Nat. Commun.* 8: 15972.

Watt, M., Silk, W. K. and Passioura, J. B. (2006). Rates of root and organism growth, soil conditions, and temporal and spatial development of the rhizosphere, *Ann. Bot.* 97(5): 839–55.

Wei, L., Vosátka, M., Cai, B., Ding, J., Lu, C., Xu, J., Yan, W., Li, Y. and Liu, C. (2019). The role of arbuscular mycorrhiza fungi in the decomposition of fresh residue and soil organic carbon: A mini-review, *Soil Sci. Soc. Am. J.* 83(3): 511–7.

Xia, M., Talhelm, A. F. and Pregitzer, K. S. (2015). Fine roots are the dominant source of recalcitrant plant litter in sugar maple-dominated northern hardwood forests, *New Phytol.* 208(3): 715–26.

Xin, F. F., Xiao, X. M., Cabral, O. M. R., White, P. M., Guo, H., Ma, J., Li, B. and Zhao, B. (2020). Understanding the land surface phenology and gross primary production of sugarcane plantations by eddy flux measurements, MODIS images, and data-driven models, *Remote Sens.* 12(14): 2186.

Xu, C. H., Guo, L. D., Ping, C. L. and White, D. M. (2009). Chemical and isotopic characterization of size-fractionated organic matter from cryoturbated tundra soils, northern Alaska, *J. Geophys. Res.* 114(G3): G03002.

Yang, H., Yan, R., Chen, H., Zheng, C., Lee, D. H. and Liang, D. T. (2006). In-depth investigation of biomass pyrolysis based on three major components: Hemicellulose, cellulose and lignin, *Energy Fuels* 20(1): 388–93.

Yin, H., Wheeler, E. and Phillips, R. P. (2014). Root-induced changes in nutrient cycling in forests depend on exudation rates, *Soil Biol. Biochem.* 78: 213–21.

Zhang, D., Hui, D., Luo, Y. and Zhou, G. (2008). Rates of litter decomposition in terrestrial ecosystems: Global patterns and controlling factors, *J. Plant Ecol.* 1(2): 85–93.

Zhang, X., Chen, Z., Huo, X., Kang, J., Zhao, S., Peng, Y., Deng, F., Shen, J. and Chu, W. (2021). Application of Fourier transform ion cyclotron resonance mass spectrometry in deciphering molecular composition of soil organic matter: A review, *Sci. Total Environ.* 756: 144140.

Zhang, Y., Duan, B., Xian, J., Korpelainen, H. and Li, C. (2011). Links between plant diversity, carbon stocks and environmental factors along a successional gradient in a subalpine coniferous forest in Southwest China, *Forest Ecol. Manag.* 262(3): 361–9.

Zhu, B. and Cheng, W. (2011). Rhizosphere priming effect increases the temperature sensitivity of soil organic matter decomposition, *Glob. Change Biol.* 17(6): 2172–83.

Zhu, E., Cao, Z., Jia, J., Liu, C., Zhang, Z., Wang, H., Dai, G., He, J. S. and Feng, X. (2021). Inactive and inefficient: Warming and drought effect on microbial carbon processing in alpine grassland at depth, *Glob. Change Biol.* 27(10): 2241–53.

Zhu, K., McCormack, M. L., Lankau, R. A., Egan, J. F. and Wurzburger, N. (2018). Association of ectomycorrhizal trees with high carbon-to-nitrogen ratio soils across temperate forests is driven by smaller nitrogen not larger carbon stocks, *J. Ecol.* 106(2): 524–35.

Chapter 4

Biological basis of soil organic carbon sequestration: a complex set of interactive processes

Patrick Lavelle, Institute of Ecology and Environmental Sciences Paris, Sorbonne University, France

1 Introduction

Soil management is increasingly seen as a way to combat global warming (FAO, 2019). Soils contain twice as much carbon (C) as the atmosphere. Poor management has caused them to lose between 115 Gt and 154 Gt of C, contributing between 1/4 and 1/3 of overall atmospheric C (FAO, 2019). Sustainable management methods could reconstitute soil organic carbon (SOC) stocks with considerable co-benefits in the form of increased ecosystem services (FAO, 2019). Sixteen of the 24 ecosystem services listed by the Millennium Ecosystem Assessment are directly linked to the soil (MEA, 2005).

The general management options that allow this recovery in soil carbon have been identified: avoid tillage, maintain soils with permanent plant cover, add organic fertilizers and create more diverse landscapes. But SOC recovery is often slow and unpredictable, and there is still a significant lack of understanding of the details of the processes involved and how to optimize

http://dx.doi.org/10.19103/AS.2022.0106.04

them. Progress in understanding would certainly be a significant step forward toward the implementation of sustainable management practices.

The accumulation of carbon in soils is a predominantly biological process, mediated by a number of biotic and abiotic conditions (Tisdall and Oades, 1982). SOC sequestration is an emergent property of the decomposition process. The increase or loss and stabilization of organic matter (OM) stocks in the soil are the difference between (Rees et al., 2005):

- the amount of OM added to the soil system by plants and secondary production and/or organic amendments, and
- the amount lost through decomposition as dissolved OM, mineral nutrients and/or gases (CO_2, CH_4, NH_4 and N_2O).

The residence time of decomposing residues depends on the speed of the carbon flow from the initial state of the residues to their complete mineralization. The mechanisms influencing SOC stabilization (see Chapter 2 of this book) will thus determine the amount of accumulated SOC (Schmidt et al., 2011; Lehmann and Kleber, 2015). Differences in SOC accumulation rates and persistence time are interpreted as 'a property of the interactions between organic matter and the surrounding soil ecosystem' (Schmidt et al., 2011).

In this chapter, I propose a conceptual framework to guide understanding of the complex nature of the underlying biological processes involved in SOC dynamics. I analyze, step by step, the conditions that promote an accumulation of OM in the soil and the maintenance of large SOC stocks. The hierarchical model proposed by Lavelle et al., (1993) identifies three nested drivers:

- climate as an overarching control,
- general soil conditions (pH, redox status, nutrient contents and stoichiometry, types and concentrations of clay-sized minerals), and
- organisms.

Organisms, in turn, develop strong interactions with the environment and among themselves. They create self-organized structures, pores and aggregates. These become hot spots of microbial activity and the ultimate microsites where C is conserved or actively mineralized (Lavelle et al., 2016, 2020).

I will focus on the process of soil aggregation, which allows the local creation of conditions unfavorable to microbial activity or disconnection among decomposers and their organic resources, thus contributing to the temporary storage of SOC at different levels of decomposition and different depths in soils. Taking the earthworm drilosphere as an example, I consider the biotic and abiotic factors in their genesis, their diversity and conditions for the stabilization or disaggregation of these structures.

I finally discuss the theoretical and technical bottlenecks that prevent research to provide a clearer understanding needed to design better management practices that would enhance carbon sequestration in natural and managed ecosystems.

2 Decomposition: a set of complex processes that drives soil organic carbon dynamics

Decomposition may be defined as the digestion by soils of all dead organic elements that are deposited at their surface in the litter layer or in the soil itself as dead roots and microbial and invertebrate dead biomass. Decomposition is a progressive transformation of dead OM toward mineral, solid or gaseous, end products (Swift et al., 1979). It is driven by general rules of ecosystem complexity where hierarchies, self-organization and panarchies regulate processes at different scales (Allen and Starr, 1982; Perry, 1995; Gunderson and Holling, 2002). Decomposition patterns and rates are, ultimately, emergent properties of a self-organization process driven by a hierarchy of determining factors operating at several nested scales (Lavelle et al., 1993, 2016).

2.1 Chemical aspects

Decomposition has been traditionally presented as a combination of two processes:

- mineralization, and
- humification.

Mineralization is clearly a catalytic process in which organic molecules progressively lose C as CO_2 or CH_4 and are finally transformed into purely mineral ions such as NH_{4+}, NO_3– or PO_4–. These can be directly taken up by plants and microbes to produce new biomass.

Humification is described as a partly anabolic process in which high-molecular-weight molecules are accumulated and even synthesized as a result of microbial digestion. The humification index, a ratio of non-humified compounds to the sum of fulvic and humic acids, is used to measure this process (Roletto et al., 1985; Sequi et al., 1986). Recently, authors have argued that humic acids are not really an obstacle to decomposition or accumulated end products. This is because priming effects on microorganisms occurring in invertebrate guts and other hot spots of microbial activity allow the mineralization of such recalcitrant organic compounds (Martin et al., 1992; Fontaine et al., 2007;

Schmidt et al., 2011; Lehmann and Kleber, 2015). In the absence of chemical restrictions on decomposition, physical protection is now considered to be the main process that slows down the decomposition rate (Oades, 1984; Feller and Beare, 1997; Six et al., 2000). Unfavorable stoichiometric conditions can also slow down the mineralization process when nutrient contents are unbalanced (Sterner and Elser, 2002; Marichal et al., 2011; Hill et al., 2018; Garcia-Palacios *et* al., 2016).

2.2 Physical transformation and transfer of organic residues to protected sites

Physical protection can occur in three different ways:

1. the inclusion of organic debris in compact structures which do not offer good aeration and/or moisture conditions to microorganisms and/or favor physical disconnection between OM and enzymes,
2. the association of organic molecules and/or microbial cells with clay by sorption mechanisms in the clay-humic complexes of soil microaggregates, and
3. the localization of SOC at great depths in the soil where the physical conditions and/or the dilution of the OM protect it from microbial activities.

Decomposition is accompanied by physical transformations – fragmentation and solubilization – and transfers. Anecic organisms like some earthworms and termites, which inhabit the soil but come to the surface to feed on litter, are mainly responsible for the transfer of OM from leaf litter to the mineral soil layers (Lavelle and Spain, 2001). In their absence, organic layers (humus) of various morphologies and thicknesses form on the surface of the soil (Zanella et al., 2018). In such humus types, fine roots and their associated mycorrhizal networks are often seen to develop at the base of this edifice, where the well-advanced decomposition process releases nutrients. The association of roots with mycorrhizae, which directly extract nutrients from organic residues, is then a predominant feature. In such systems, the accumulation of OM, kept at sometimes rather early stages of decomposition, depends on climate and soil conditions. Low temperatures at high latitudes or acid pH and low nutrient concentrations in tropical areas generate this type of accumulation processes (see Chapter 5 of this book).

2.3 Biological processes

Decomposition is a digestion process dominated by microbial enzymatic activities. While over 90% of the overall soil respiration activity is microbial, invertebrates and roots also have an essential role in creating suitable conditions for microbial activities (Lavelle and Spain, 2001). As ecosystem engineers, invertebrates and roots fragment the litter, disperse microbial propagules and directly stimulate microbial activity in their guts and rhizospheres. Mutualistic interactions among microorganisms and the larger ecosystem engineers provide a common response to two major constraints facing life in soils (Lavelle and Spain, 2001):

- Difficulties associated with moving and living in a compact environment, an obvious limitation to microbial activity.
- Using organic resources that are either abundant but of low quality (leaf litter and soil OM) or highly energetic but with a sporadic and unpredictable spatial distribution. The latter include such materials as root exudates, earthworm intestinal and cutaneous mucus, termite saliva and others.

Macroinvertebrates and roots create suitable conditions for microorganisms, benefiting from the digestive capacities that only microorganisms possess (Barois and Lavelle, 1986; Lattaud et al., 1998).

Invertebrates are able to carry out the physical transformation of OM: its fragmentation and burial in the soil. Responsible for creating a high proportion of the porosity in the surface horizons of the soil, they create habitats and favorable conditions for aeration, infiltration and water retention for microorganisms and the smaller fauna that cannot move or have a physical impact on the soil. On the other hand, these ecosystem engineers have limited digestive capacities. Invertebrates do not have the enzymes that would allow efficient digestion of complex compounds such as phenol protein complexes, which represent more than 85% of the litter nitrogen and are precursor molecules of very simple mineral or organic forms required for root uptake. In this context, the mutualist digestion systems of invertebrates, the microbial loops in the rhizosphere or the networks of mycorrhizal filaments connected to the roots facilitate the release of plant nutrients. These processes allow the transport, the activation and the selection of microbial decomposer populations in exchange for digestion with benefits for both partners.

The networks of micropredators, e.g. protists, nematodes and acari, which live in the habitats created by the engineering organisms operate as an additional control of microbial populations and the transformation of OM and nutrients contained in the biomass of their prey.

3 A general conceptual framework for understanding soil biological activity

While the hierarchical organization of general drivers of decomposition sets the general patterns of the dynamics of OM, self-organization processes optimize biological activities, while positive feedback processes make soils truly living entities.

3.1 Hierarchy of drivers of the decomposition process

Decomposition is driven by a suite of hierarchically organized determinants that influence the chemical, physical and biological processes involved at different scales of time and space. In hierarchical systems, upper levels have properties, called emergent properties, which lower levels do not have. On the other end, the explanation of mechanisms that generate these properties is to be found at the lower level (Allen and Starr, 1982).

Climate is by far the dominant driver of decomposition at the upper level of the hierarchical model, Fig. 1). The effect of temperature and moisture on microbial activity and the resulting decomposition rates have been largely demonstrated and documented (Heal et al., 1981; Wall et al., 2008 but see Faber et al., 2018).

At the next scale down (measured temporally in hundreds of years and spatially in hectares), soil parameters greatly influence decomposition,

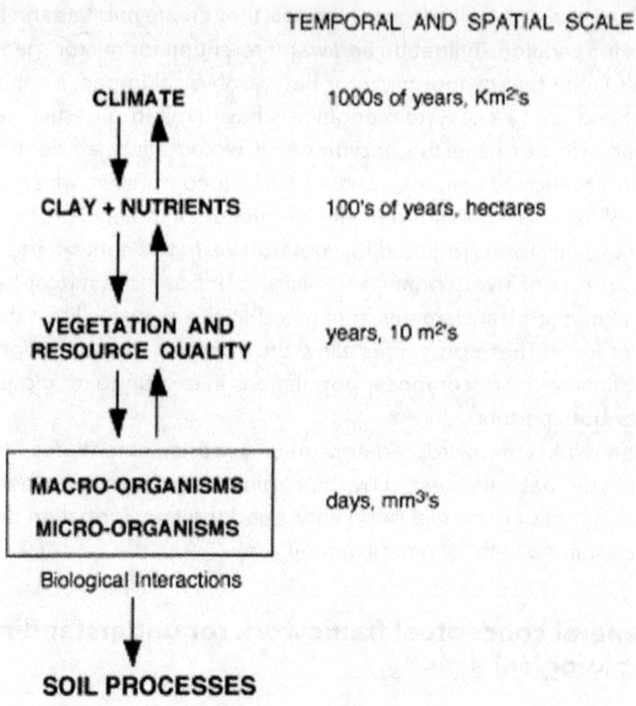

Figure 1 A hierarchical model for drivers of decomposition and other processes in soil (Lavelle et al., 1993).

principally via the nutrient concentrations and stoichiometric balances of the original bedrock and the presence and nature of clay minerals involved in sorption–desorption processes. While nutrient concentrations and their relative proportions may limit the growth of microorganisms, clay minerals can protect organic molecules from decomposition by:

- forming organomineral complexes that are difficult to digest, and
- coating organic debris, thereby restricting the access of microorganisms to these resources.

In soils where neither climatic nor edaphic conditions are an obstacle to the activity of decomposing microorganisms, the quality of the OM supplied to the system can generate important differences in the rates of decomposition (Swift et al., 1979). The stoichiometry of the resources (in particular, their C:N and C:P ratios), and their physical structure, may represent important chemical and physical limitations to microbial activity (Sterner and Elser, 2002).

Finally, at the lowest level of this hierarchy (at scales measured temporally in hours to days and spatially in millimeters to centimeters), microbial activity is dependent on the effects of plant root activity and macroinvertebrates as ecosystem engineers that create hot spots of intense microbial activity. These hot spots may include the rhizospheric environment, the invertebrate gut and the structures created by such invertebrates as earthworms, termites and others, where physical and trophic conditions for microbial activity are optimal.

3.2 Decomposition and self-organization of soils

While communities of ecosystem engineers are largely shaped by climate and soil conditions, self-organization processes allow them to influence soil and microclimate conditions with positive feedbacks on their activity. Self-organization is a general property of natural systems whereby interactions, at discrete scales, among biotic and abiotic elements, result in the formation of structures that have positive feedback on the whole system (Perry, 1995). The system is then in a metastable state, far from equilibrium, 'on the edge of chaos' and susceptible to sudden changes and threshold effects. The plant rhizosphere is a well-known example of a self-organized system in soil (Fig. 2).

The release of large amounts of root exudates triggers massive activation and selection of soil microbial communities that were mostly inactive before (Mueller et al., 2019). Bacteria activated in the root zone digest root litter and soil OM. They are then ingested by a community of micropredators that release part of the nutrients of the microbial biomass for the benefit of plants and for themselves (Clarholm, 1985). Mycorrhizae associated with roots also enormously extend the volume of soil explored by the root system and

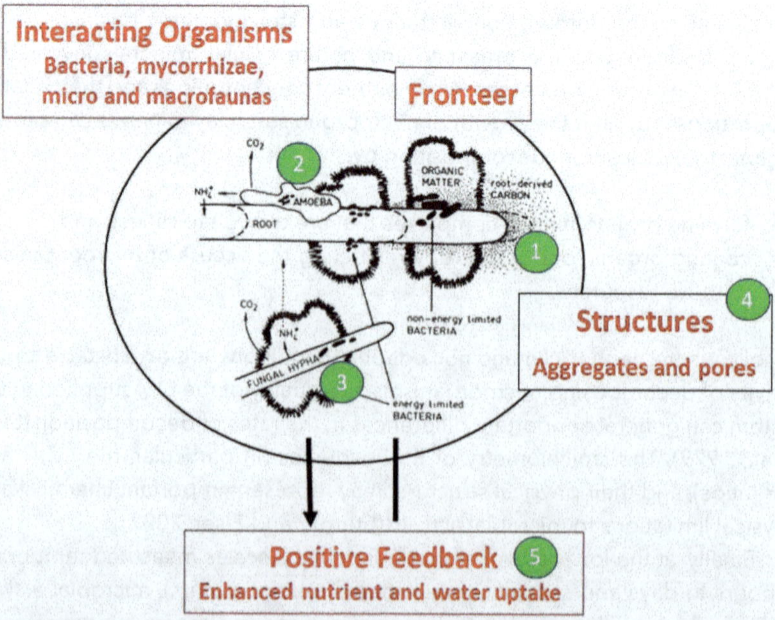

Figure 2 Plant rhizosphere. 1. Roots produce highly assimilable and energy-rich exudates that create a hot spot of bacterial activity around the growing root tip. 2. Bacteria that have multiplied, extracting nutrients from the root neighborhood, are preyed upon by protists and nematodes. These micropredators excrete mineral N as NH_4+ that is taken up by the plant. 3. This microbial loop (Coleman et al., 1983; Clarholm, 1985) is completed by an association with mycorrhizae that receive C directly from the root and explore a vast volume of soil where they uptake water and nutrients that they provide to the plant. 4. In addition, mycorrhizae form networks of hyphae that enmesh and glue soil particles forming macroaggregates (Rillig, 2004; Wilson et al., 2009).

significantly increase the nutrient and water supply of plants. At the same time, soil compression by the growing root and microbial activities form stable macroaggregates inside which highly stable microaggregates are created. This system, in a metastable state of equilibrium, is active as long as the root tip is present at the microsite. Once the root tip has been displaced by plant growth, microbial activity decreases and the aggregates formed enter a maturation stage and a further disaggregation process while the root-derived OM within the aggregates is decomposed.

Self-organization involves interactions among all soil organisms and the creation of structures that improve the conditions of their habitats. Soil as a self-organized system has a hierarchical structure with a suite of discrete scales (Fig. 3):

1 microaggregates colonized by microorganisms,

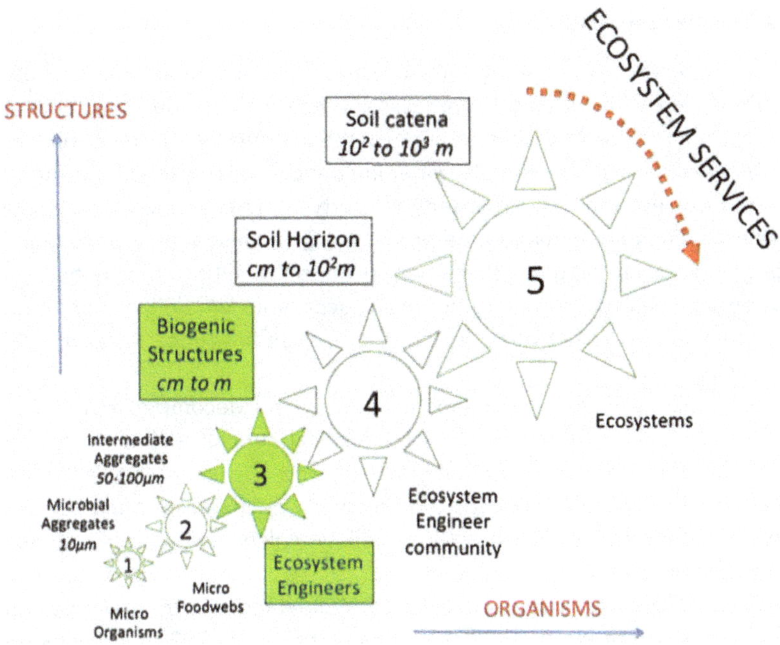

Figure 3 Soil as a self-organized system of biological entities. Organisms of increasing sizes, from microorganisms to microinvertebrates (microfoodwebs), ecosystem engineers and ecosystems (horizontal axis), create structures of different sizes (vertical axis) and organize soil at different nested scales from 1 (microbial aggregates inhabited by microbial communities) to 5 (landscape viewed as a mosaic of different ecosystems). Most soil-based ecosystem services are perceived and delivered at the landscape (5) scale with a general positive feedback effect on plant nutrient and water uptakes (Lavelle et al., 2006).

2 intermediate aggregates and pores - the habitat of micropredators that control microbial communities,
3 millimetric to centimetric biogenic structures - pores and aggregates produced by soil ecosystem engineers (roots, earthworms, ants, termites and others),
4 mosaics of functional domains of a community of engineers that correspond to a soil horizon in a patch of an ecosystem, and
5 landscapes viewed as a mosaic of different ecosystem types.

Most ecosystem services, including carbon storage in ecosystems, are provided at the landscape level as a result of a combination of processes operating at the mentioned suite of discrete scales (Fig. 3).

3.3 The soil panarchy

If soil is considered a self-organized system, the decomposition process is itself coupled with other soil processes. These include the construction of a physical structure favorable to life and the production of other soil-based ecosystem services (for example, water infiltration and storage and protection of soil from erosion), an emergent property of soils at the largest scales. Understanding and managing the process of C sequestration in soils requires taking into consideration all these patterns and the full complexity of the soil environment without forgetting the social, economic and political dimensions that guide human activities which themselves influence self-organizing processes of soils.

In parallel with the hierarchies that determine decomposition and self-organized systems, there are other hierarchical structures that interact with them: the social and economic context of land use and/or its institutional structures. This complex conceptual framework, defined as a panarchy in the sense of Gunderson and Holling (2001), allows linking soil science to practice and public policies.

Such holistic systems are difficult to understand, and modeling them is a real challenge (Allen et al., 2014; Garmestani et al., 2020). However, only modeling makes it possible to reach the objective of increasing the stock of C in the soils and thus reducing the load of greenhouse gases in the atmosphere.

4 From theory to practice: how could we manage soil self-organized engineering to enhance carbon sequestration?

At the crossroads of physical and biological processes and sensitive to land use and soil management options, manipulation of the aggregation process is probably the most practical way to change the conditions of carbon storage in soil.

4.1 Aggregation and carbon conservation

The relationship between the stabilization of soil aggregates and carbon, and the impact of aggregation on carbon sequestration, has now been widely documented (Table 1). Various papers describe the diversity of physical and biogenic aggregates present in soils as well as their origin and evolution in time (Lavelle et al., 2020).

From 3336 papers selected from the ISI Web of Science database with soil and aggregation used as key words, 28.1% had the word stabilization in the title and 18.9% the word carbon. In contrast, biological processes and the effect

Table 1 Number of papers available on ISI Web (July 2021) on soil aggregation associated with different key words

Soil + aggregation + word	N counts	% counts
Stabilization	939	28.1
Carbon	630	18.9
Biological processes	224	6.7
Bacteria	76	2.3
Mycorrhizae	70	2.1
Earthworm	48	1.4
Sequestration	46	1.4
Root	44	1.3
Ant	14	0.4
Termite	2	0.1

of ecosystem engineers on soil aggregation are still poorly studied, although the number of related publications is increasing steadily.

In a meta-analysis of the effects of 279 different species of soil microorganisms and invertebrates of 11 different phyla, Lehmann et al. (2017) showed a significantly higher aggregation of soils in the presence of biota. The effect, however, although significant globally with a 24% ± 6% more aggregation, was highly variable among and within the different groups. Macroaggregation was, on average, more enhanced than microaggregation, especially by fungi. This study confirms the major role of microorganisms, especially bacteria, in the stabilization of macroaggregates, as suggested by many different authors (Tisdall and Oades, 1982; Chenu, 1993).

However, it is important not to generalize due to context dependency of the soil aggregation process and idiosyncratic effects within different species. For example, earthworms, which generally produce enormous amounts of casts, may have rather different effects on global soil aggregation. Morphological differences in the anatomy and physiology of the last few intestinal segments result in the production of two different classes of casts, compact and globulous, or loose and granulous. These may result in contrasting 'compacting' and 'decompacting' effects on soil (Rossi, 2003; Blanchart et al., 1997).

Aggregates may be classified according to their formation process. A first group is formed as a consequence of secretory processes (exobiopolymers) produced by roots, soil bacteria, ectomycorrhizal hyphae, Arbuscular Mycorrhizal fungal hyphae (including the secretion of glomalin-associated protein) and/or soil saprophytic fungi. Other aggregates are formed by ingestion/egestion processes of macroinvertebrates (earthworms, termites and Coleoptera larvae) or microarthropods.

Manipulation of soil by social insects, dung beetles, Cicada and others is also a source of aggregates. A final category comprises smaller structures resulting from the breaking down of macroaggregates. Based on these observations, I propose to classify soil macroaggregates into seven main categories that significantly differ in their morphologies, chemical and biochemical compositions and spectral NIR signatures (Table 2) (Plante and Mc Gill, 2002., Velasquez et al., 2007; Decaëns et al., 1999; Hedde et al., 2005; Dominguez-Haydar, 2018).

Microaggregates form a very diverse component, and six main processes have been identified that contribute to their formation and stabilization (Tötsche et al., 2018). A minimal classification into three groups has been proposed by Blanchart et al. (unp. data) (Fig. 4):

- organomineral aggregates that associate clay particles with amorphous OM,
- organomineral aggregates with visible plant debris and fungal hyphae, and
- bacterial microaggregates.

While the first group probably corresponds to purely physical and chemical processes, the second one might be formed during the maturation stage of macroaggregates formed in the root zone, while the last group is macroaggregates formed by earthworms and other macroinvertebrates. Land use may influence the contribution of the three aggregate types (Fig. 4).

The simplest type of organomineral microaggregates may be associated with microorganisms, rootlets or hyphae that stick particles to their bodies with gluing agents to build soil blocks at their own size scales, i.e. at a scale of <53 µm (Chenu, 1993) down to nanometer scale (Rumpel et al., 2015).

Chemical processes may also be involved in the process of microaggregate stabilization, although there are no reports of the formation of specific microaggregates through chemical processes alone. Formation of secondary microaggregates within macroaggregates may occur during maturation. This is particularly the case of earthworm cast formation with several successive phases. First, during the gut transit, microaggregates that may have occurred in the structure of the ingested soil are completely dispersed (Barois et al., 1993). Occluded particulate OM released in the gut is partly decomposed, favoring its association with clay minerals in newly formed microaggregates. While freshly egested, casts are highly dispersible. A first drying cycle allows the structure to consolidate. Microorganisms that are still active for some time after the egestion of the cast may build microaggregates inside the aging cast (Shipitalo and Protz, 1988; Six et al., 2000; Bossuyt et al., 2005).

4.2 Carbon dynamics in the earthworm drilosphere

A practical way to enhance SOC sequestration could be the activation of mechanisms leading to self-organization associated with ecosystem engineer

Table 2 A general classification and characterization of macroaggregates based on their origin (after Totsche et al., 2018; Lavelle et al., 2021)

Macroaggregate type	Main forming process	Macroaggregate binding agent	Microaggregate type	Global effect on soil bulk density
Physical	Soil drying and rewetting	Sorption–desorption	Organomineral	+
	Soil freeze and thaw	Sorption–desorption	Organomineral	+
Fungal	Entangling of soil particles by hyphae glomalin		Enmeshed	+
Root	Enmeshment of particles in fine roots	Root exudates Bacterial polysaccharides	Enmeshed	+
	Compression of soil by growing roots	Physical processes Root exudates Bacterial polysaccharides	Bacterial	+
Earthworms	Casting + drying	Physicochemical	Bacterial	+
		Mucus residues	Bacterial	–
Ants	Rubble		Organomineral	–
	Nests		Organomineral	+/–
Termites	Sheathings		All types	–
	Nests		Bacterial	+
Mixed structures	Roots in invertebrate structures			+/–

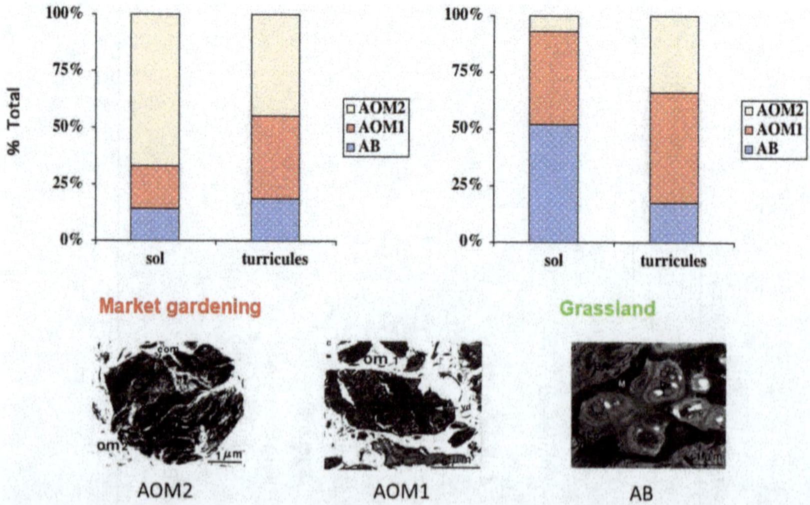

Figure 4 Distribution of microaggregates among three different dominant types in vertisols subjected to different management options: market gardening (1.5% C, 32% stable macroaggregates, macroaggregate stability 23 µm mean weight diameter, 1 g m^{-2} earthworm biomass) and 10-year pasture (3.5% C, 78% stable macroaggregates, macroaggregate stability 180 µm mean weight diameter, 90 g m^{-2} earthworm biomass). AOM1: organomineral; AOM2: organomineral with root debris; AB: bacterial microaggregates (Blanchart et al., 2000). Distribution was calculated as the percentage of the number of a given type of microaggregates on a definite surface of ultra-thin sections of 2–20 µm soil fractions observed with an electronic microscope (×3400 magnification) (Blanchart et al., 2000, unpublished data).

activities. While rhizosphere processes have been fairly well described (Fig. 2, Chapter 3 of this book), I explore here processes that control C abundance in the earthworm drilosphere.

Earthworm drilosphere effects operate at four different scales (Martin, 1991, Lavelle and Spain, 2001). Gut contents are hot spots of microbial activity where the ingested soil undergoes digestion. Gut transit may last from 20 min to several hours, depending on feeding habits and temperature. Fresh casts are microsites with largely available nutrient contents and unstable physical structure. In a short time period (ranging from hours to days), their structure consolidates and microbial activity decreases. In the next phase of aging, cast structure is highly stable and microbial activity is very low. At the scale of the drilosphere functional domain, carbon dynamics is the result of these three different phases occurring in casts deposited by the whole earthworm community.

4.2.1 Soil ingestion

The amount of soil ingested by an individual of a given species is a function of the individual weight of the earthworm, soil moisture, soil nutritional quality linked to the depth at which the individual feeds and temperature (Lavelle, 1978). In temperate conditions, endogeic earthworms may ingest daily up to between 0.5 and 3.0 times their own weight of soil and OM (Boström, 1988; Curry et al., 1995; Zhang et al., 2009). Recently born tropical endogeic earthworms may ingest up to 20–30 times their own weight daily when the soil has low SOC content and temperature and moisture conditions are optimal (Lavelle, 1978). Earthworms mostly ingest small soil particles and aggregates that have higher SOC contents than bulk soil. Total ingested soil in Mg ha^{-1} $year^{-1}$ is obtained by multiplying these figures by the average biomass of earthworms and the number of days of activity. An absolute maximum value of 1250 Mg ha^{-1} $year^{-1}$ has been recorded in a humid African savanna of Côte d'Ivoire (Lavelle, 1978). This is equivalent to a 10-cm thick soil layer passed each year through the earthworm guts. In temperate grasslands, values of several hundreds Mg have been recorded (Curry et al., 1995).

Mechanical activity required by ingestion has an elevated energy cost evaluated at 95% of the assimilated OM (86% of the ingested OM in the African endogeic *Millsonia anomala*) or 1.2 Mg ha^{-1} OM mineralized (Lavelle, 1978).

4.2.2 Gut transit

An average of 10% of the ingested soil C is lost to respiration and, to a much lower extent, to earthworm biomass production and reproduction. During gut transit, soil structure is dispersed by water and mucus addition and intense mixing in the gizzard (Shipitalo and Protz, 1988; Marinissen and Dexter, 1990; Barois et al., 1993; Botinelli et al., 2020). With the addition of water (up to 100%) and intestinal mucus (5–40% of the dry weight of the ingested soil), the gut content becomes a hot spot of high microbial activity where dormant soil bacterial populations are selectively activated (Barois and Lavelle, 1986; Trigo and Lavelle, 1993, 1994; Brown et al., 2000). Intestinal mucus is no longer found in the second half of the gut. Activated microbes that have recuperated their enzymatic activities (Lattaud et al., 1998) digest a fraction evaluated at 3-19% of the soil OM in endogeic species and 8-70% of leaf litter in litter-feeding epígeics and anecics (Lavelle and Spain, 2001).

4.2.3 Fresh casts

These are characterized by high respiration rates at the moment of their deposition. In time periods of hours to a few days, their nutrient and microbial biomass contents drop. They are deposited at the soil surface or inside the mineral soil, depending on the species and seasons although, in most cases, surface deposits do not exceed 10-20% of total cast production. They are highly dispersible structures that require a drying cycle to harden and become stable macroaggregates (Shipitalo and Protz, 1988; Blanchart et al., 1993).

These casts may have a wide diversity of structures, ranging from fragile granular casts that can be easily dispersed by water to highly compact globular casts, with bulk densities of up to 1.8 g cm^{-3} or more. Fresh casts generally have high contents of water and mineral nutrients, N, P and other cations and can be colonized by small roots. A meta-analysis of 405 case studies showed a positive difference of 40–48% between the SOC content and various nutrient elements of the castings of species of the different ecological categories and the surrounding soil (Van Groenigen et al., 2019).

This average value measured on casts with roughly identified ages (> or <7 days) probably underestimates nutrient contents in freshly deposited casts that steadily decrease as a function of the age of the casts. In freshly deposited casts of *Pontoscolex corethrurus*, mineral NH_4-N contents of 114.7 μg g^{-1}, equivalent to 18.5 times the control soil value, rapidly decreased during the maturation process down to less than half this value (47.3) 24 h later and 21.1 μg g^{-1} after 10 days while microbial biomass followed a similar trend (Lavelle et al., 1992). Observations of Decaëns et al. (1999) on the large tropical anecic earthworm *Martiodrilus carimaguensis* show similar figures and dynamics and suggest that a great part of the losses in mineral N and microbial biomass may be due to a migration of these elements with water leaching out of the drying cast. Elevated concentrations of mineral elements and increased root production below these casts support this hypothesis. Fresh casts also contain higher amounts of SOC than the surrounding soil, which may be used for microbial activity inside the casts after their deposition. A recent study showed that fresh casts are enriched in unprotected as well as protected SOC pools and that all of these pools decrease during the aging process (Bottinelli et al., 2021).

4.2.4 Aging casts

These are structures that have experienced a drying cycle and thus acquire physical stability. They will persist until physical and/or biological disaggregation processes disperse them into non-macroaggregated soil or transform them into other structures (Blanchart et al., 1997; Jouquet et al., 2008). This may be the case when decompacting invertebrates (Oligochaeta, ants or termites) break them down. Globular earthworm casts are highly resistant structures that can persist for several years in the soil if no specific process of mechanical dispersion occurs. They are the main components of the biogenic macroaggregate fraction that often represents 50% or more of the volume of the upper layers of soils that have active earthworm communities (Blanchart et al., 1997; Velasquez and Lavelle, 2019).

Microbial activity may be significantly reduced in their compact structure. In aging casts of *Millsonia anomala*, microbial respiration almost stopped (Martin, 1991) to the point that 100 days after deposition, they have the same SOC content as non-ingested soil submitted to the same conditions of laboratory incubation. Decaëns et al. (1999) even report a 100% increase in SOC contents of casts of the tropical anecic species *M. carimaguensis*. They explain this spectacular and unexpected result by an enrichment of the cast structure in root products and fine root biomass. This effect of aging casts on carbon conservation, although confirmed in other studies, is rather site specific and depends on the initial soil texture and management (Zhang and Schräder, 1993; Jouquet et al., 2007; Van Groenigen et al., 2019; Angst et al., 2017).

The maturation phase is also a period when microaggregates form inside macroaggregates, thus increasing SOC stability (Six et al., 2000; Bossuyt et al., 2004; Pulleman et al., 2005). This formation is enhanced by the effect of earthworms on clay OM interactions (Le Mer et al., 2020; Barthod et al., 2020). In addition, co-constructions of mixed structures with roots seem to occur frequently, although experiments suggest that not all plants associate with all earthworms in this process (Decaëns et al., 1999; Zangerlé et al., 2011; Botinelli et al., 2021). The overall effect of soil aggregation by earthworms will ultimately depend on the time of persistence of mature casts in the soil (Lavelle et al., 2020). The age of casts can be determined by their near-infrared spectral signature that changes as the chemical and physical properties of casts are modified (Zangerlé et al., 2014; Dominguez-Haydar et al., 2020; Botinelli et al., 2020, 2021) (Fig. 5).

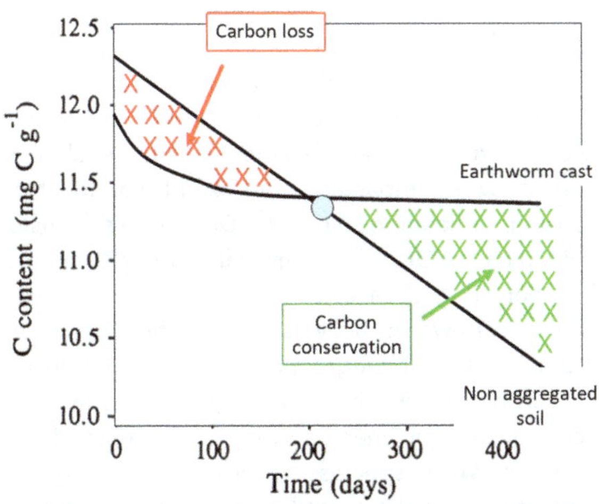

Figure 5 Variation of C contents in aging casts of *Millsonia anomala* and a disaggregated control soil (after Martin, 1991).

4.2.5 Drilosphere-aggregatusphere

The community of earthworms and other ecosystem engineers produce aggregates of diverse structures and resistances in their respective drilospheres (the part of the soil controlled by earthworms) and, more generally, in the aggregatusphere, defined as the sum of aggregates created by soil ecosystem engineers (Beare et al., 1995). The pattern of the aggregatusphere created and its ability to store SOC in stable forms will obviously depend on the combination of climate, soil and plant resource quality that shapes the community of ecosystem engineers. This includes roots and their associated mycorrhizae and other fungal and bacterial components and macroinvertebrates that can either compact the soil by creating structures with low porosity, or have a decompacting effect in breaking down compact structures into smaller ones.

At that stage, roots may have colonized fresh casts in search of assimilable nutrients and added their effects to the aggregation process, as revealed by their near infrared spectrometric pattern (Zangerlé et al., 2011). When compact casts (that better protect OM from microbial attacks than porous macrostructures) accumulate, the soil gets increasingly more compact (Blanchart et al., 1997; Alegre et al., 1996). In steady-state equilibrium, compacting and decompacting species combine their activities. In the Lamto African savanna, there is evidence that decompacting earthworm species may break down the compact aggregates, probably enriching this soil with root exudates, a mixture of soluble OM compounds released in the rooting zone (Rossi, 2003). Conversely, compacting species that never reingest their own casts seem to feed on small casts of the decompacting species that have a rather high SOC content. In other ecosystems, Enchytraeidae, ants or termites can have a similar effect (Didden, 1990; Dawod and Fitzpatrick, 1993).

Unbalancing compacting to decompacting functions may induce severe soil compaction. This is for example the case when the compacting earthworm *P. corethrurus* invades soils submitted to severe disturbances where the decompacting community has been depleted (Chauvel et al., 1999). This study showed that the soil was compacted to an extent that anaerobic conditions developed, causing vegetation dieback. During this transformation, the equivalent of 18 Mg ha^{-1} of OM was lost from the upper 20 cm of soil, probably in the form of methane (Fig. 6; Lavelle et al., 2001).

At this scale, earthworm activities also significantly enhance plant production, especially the aboveground part of it. This increased fixation of atmospheric carbon in plant biomass may be important and offsets the C respiration cost of building macroaggregates. In a set of 16 experiments conducted on the effects of 13 species of tropical earthworms on the production of 14 different plant species, the overall average increase in biomass of shoot and grain due to earthworms was 56.9%, 39.3% and 35.88%, respectively.

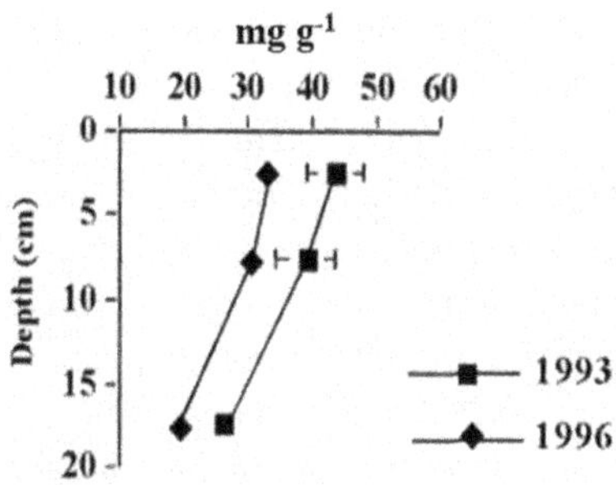

Figure 6 Changes in carbon contents at different depths of a pasture soil before (1993) and after (1996) invasion of the endogeic earthworm *P. corethrurus* (Glossoscolecidae) (Manaus, Brazil) (Barros, 1999).

Highest values were obtained for soils with a sandy texture, low OM contents and moderately acid pH. Legumes responded much less than grasses (Brown et al., 1999). Similar experiments in temperate conditions show similar trends (Scheu, 2003; Van Groenigen et al., 2014).

4.2.6 Landscape

The overall effect of a landscape on soil aggregation, in a given set of climate, soil and plant resource quality conditions and biodiversity, will depend on the spatial organization and allocation to land use of the different ecosystem patches that comprise it. Soil macro invertebrate biodiversity is clearly linked to plant cover and the history of land use (Lavelle et al., 2022). In Amazonia, for example, deforestation and the progressive elimination of the remaining tree cover have a general depleting effect on all kinds of biodiversity and on the different elements of soil quality (Grimaldi et al., 2014).

4.3 Manipulating earthworm and plant communities to enhance carbon storage in biomass and soil: three experiments and an accident

4.3.1 The earthworm carbon trap

A large number of short- and medium-term (up to 2-3 years) laboratory and field experiments have been conducted with controlled numbers of earthworms and plant cover (Villenave et al., 1999; Coq et al., 2007; Zhang et al., 2013; Lubbers et al., 2017). The aim was to estimate the potential of earthworm management as a means to enhance plant production and SOC storage in soil. Results were highly variable, given the diversity of conditions at all four stages of the hierarchical suite of determinants. There was a general increase in CO_2 and sometimes NO_x emissions resulting from earthworm activities. This phase corresponds with the digestion and incorporation to the soil of the OM displayed at the soil surface or to the digestion of OM in the mineral soil.

At the same time, soil aggregation increased and microaggregates formed within the macroaggregates, gradually increasing the protection of the OM they contain. The extrapolation to 30 years using the CENTURY model, of the trend outlined in an experiment carried out in the Peruvian Amazon (Pashanasi et al., 1996), showed two phases with contrasting earthworm effects (Villenave et al., 1999). Immediately after the cultivation of the recently deforested soil, inoculation of the worms led to accelerated mineralization of OM in the forest soil. This loss of C was compensated by an intense aggregation of the soil, which protected the OM.

It would take 10 years under the conditions of this experiment for the soil carbon content to reach that of the soil without earthworms. At the end of the 30-year simulated period, the soil inoculated with earthworms had 28% more carbon and had higher amounts of SOC associated with the fraction of the 'slow' SOC pool protected in the macroaggregates (+70%). When the soil was not initially macroaggregated, as may be the case for heavy tillage, on a sandy textured soil, the protective effect can be immediate. The long-term effect described in this study has been confirmed by Zhang et al. (2013) and is known as the earthworm carbon trap.

Although verified in a number of cases, conservation of SOC in macroaggregates is highly variable and depends on the properties of the macroaggregates, their porosity and nature of OM occluded, whether in particulate or organomineral forms. The type of clay minerals present in soils is also important. In soils with dominant smectitic clay minerals, roots are the major agents of C sequestration in macroaggregates, while earthworms play a dominant role in kaolinitic soils (Blanchart et al., 2004).

Finally, there are reasons to believe that this carbon trap is not an exclusive feature of earthworm structures. All soil-based ecosystem engineers that produce compact macroaggregates, starting with roots, probably participate in this process (Jouquet et al., 2011; Angers and Caron, 1998).

4.3.2 Biodiversity accidents

The cohabitation of invertebrates with a compacting or decompacting effect, and their possible cooperation with roots in creating mixed macroaggregated structures, generates spatial heterogeneity in soils. Indeed, the patchy distribution of earthworm populations observed in many sites (Rossi et al., 1997; Decaëns and Rossi, 2001; Decaëns et al., 2009; Jimenez et al., 2006, 2014) leads them to form separate areas where soil physical conditions are different (Rossi, 2003). In all cases, only a small proportion of the earthworm distribution pattern was explained by soil texture or chemical properties, supporting the hypothesis of predominantly biotic controls and self-organization processes. The results of a modeling exercise suggest that the spots dominated by compacting or decompacting species succeed each other over time, with each of the two groups feeding on the castings of the other group (Barot et al., 2007). However, the field data collected at time scales of less than 2 or 3 years, in other situations (Jimenez et al., 2006), did not confirm this hypothesis, suggesting that if they are occurring, they operate at larger time scales.

The cohabitation of functionally different groups in a spatial mosaic stable at time scales of at least 2 years allows the regulation of the physical properties of the soil, mainly water infiltration and bulk density.

The disturbance of this spatial arrangement can seriously affect the physical quality of the soil. As an example, the introduction of the invasive earthworm species *P. corethrurus* most often leads to soil compaction due to its exceptional bioturbation activity combined with the weakening of the activity of natural decompactants: termites, ants or other species of earthworms (Chauvel et al., 1999). Remarkably, however, this situation can be reverted by soil macroinvertebrate communities in 1 year, as was shown for a block of this compacted soil, which was inserted into the nearby natural soil without the invasive species (Barros et al., 2001).

5 Discussion

After describing the extreme complexity of processes involved at all scales, from the micrometer to the landscape, and the interactions with other biophysical, social, economic and institutional elements of the panarchy constituting (agro) ecosystems, one can legitimately ask if we could do more than follow general recommendations.

My analysis has identified soil aggregation involving plants, invertebrates and microbial communities as a natural player, as a major process in carbon sequestration. Loss of SOC from managed soils is generally associated with degraded physical structures. Their reconstruction by ecosystem engineers and protection with permanent soil cover seems to be a way to enhance carbon sequestration. Biodiversity at all levels, from the micrometer to the landscape scale, is needed to foster SOC sequestration. Our analysis has identified two major knowledge gaps, which need to be addressed to allow better control of the processes involved. One is the cooperation among the different natural actors and, in particular, ecosystem engineers that would allow the production of more stable macroaggregates. The other one is a quantitative knowledge of the amounts of macroaggregates produced and their permanence time in soil.

5.1 Who made this aggregate? A general diagnosis of macroaggregation status

Our analysis has highlighted two important facts:

- the need for a minimum level of functional diversity in the engineer community to build equilibrated macroaggregated structures (aggregatuspheres), and
- cooperation among engineers and microorganisms in the construction of macroaggregates.

Biodiversity is necessary to promote stable macroaggregation. While compacting species are necessary to create macroaggregated structures, the regulation of this process by decompacting species prevents any excess in the accumulation of compact structures that would give the soil excessive bulk density.

A problem here is that analytical methods used to assess macroaggregation only focus on their resistance to physical stresses, not on their origin and persistence. The origin of macroaggregates can be defined using their near or mid-infrared spectroscopic signal. Spectral signatures provided by these methods are extremely sensitive to the nature and distribution of organic and inorganic aggregate components. They are able to separate macroaggregates according to the organism that created them, their large taxonomic group and even the specific earthworm species involved and indicate if the structure was produced by a single actor, a root or an earthworm, or by a combination of at least two of them. A careful manual separation allows extracting all macroaggregated structures that will be individually submitted to infrared spectroscopy (Hedde et al., 2005; Zangerlé et al., 2016). A cluster analysis of the spectral data allows the grouping of macroaggregates produced by the same ecosystem.

Laboratory conditions show significant differences among spectral signals of earthworm casts and root macroaggregates. Intermediate signals were recorded for mixed macroaggregates that had been formed by the growing of plant rootlets into a fresh earthworm cast (Zangerlé et al., 2011).

The spectral signal of fresh macroaggregates changes with time, as N and/or C losses occur in dissolved form or as NO_x, CO_2 and CH_4 emissions, or microbial biomass. This change allows evaluating the time elapsed since the deposition of a given earthworm cast in laboratory experiments (Zangerlé et al., 2014; Dominguez-Haydar et al., 2020). The infrared signals could also be used to differentiate different cast aging stages under field conditions (Botinelli et al., 2021).

These analyses tend to show that after some time, all macroaggregate structures evolve toward a 'bulk soil' signature showing no further signs of earthworm activity. Although these results are very promising, much research is still needed to develop and implement a standard methodology to describe the macroaggregation status of a given soil.

5.2 Macroaggregate production rates and persistence: key elements of the carbon balance of the aggregation process

The determination of aggregate production by specific ecosystem engineers under field conditions is extremely time consuming and tedious. Methodological developments based on infrared spectroscopy might be a way to overcome these limitations. In an initial diagnosis phase, studies should provide a database of identified spectral signals. Rates of macroaggregate production by the various ecosystem engineers could be derived from spectral analyses of the origin of individual macroaggregates extracted from a standard soil sample. Monitoring the production along a seasonal cycle of fresh structures of all types would provide production rates differentiated by different ecosystem engineers.

In a steady-state situation, the amount of macroaggregates formed will be equal to the amount destroyed, thus providing a global average value of the persistence in time of these structures. Finally, measurement of the respirometric activity of each class of macroaggregates will make it possible to establish a general model of C flux through the aggregated and non-aggregated parts of the soil (Fig. 7).

6 Conclusion

The amount of carbon stored in soils is the result of complex interactions among different sets of self-organized systems operating at nested scales in time and space. To foster organic carbon accumulation in soil, a rigorous scientific analysis is necessary to understand the underlying mechanisms, identify the processes that can be managed, and the ones that cannot, and refine management options to accelerate the process.

In the meantime, simple implementation of sustainable management options proposed by the FAO and other institutions should be applied without delay. Based on a considerable body of evidence, they all favor the development of diverse self-organized systems that allow soils to evolve toward a state of metastable equilibrium with improved characteristics. Improving and accelerating these dynamics require a precise monitoring of a few key biological parameters, soil macrofauna communities, root systems and a precise diagnosis of the macroaggregate compartments and of their dynamics. The creation of databases gathering such datasets will allow for a better definition of the conditions for faster and greater C storage in soils. They will also serve to

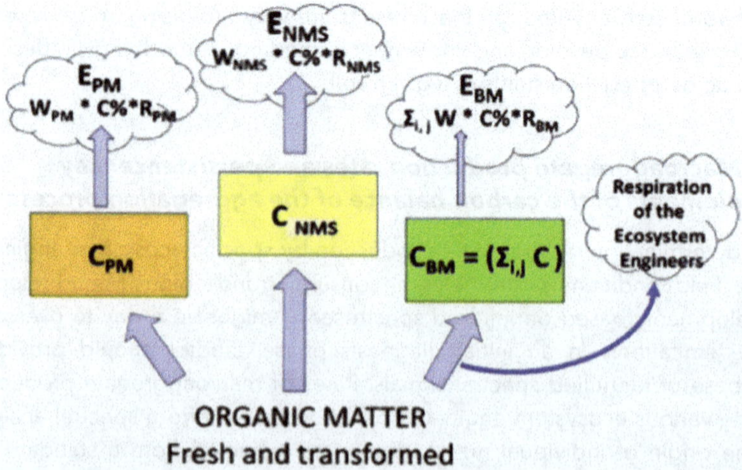

Figure 7 Flow of organic matter and gas emissions through the macroaggregated and non-macroaggregated soil compartments (Lavelle et al., 2020). Carbon in physical macroaggregates (C_{PM}), non-macroaggregated soil (C_{NMS}) and biogenic macroaggregates (C_{BM}), E_{PM}, E_{NMS} and E_{BM} are gases released by respiratory activities from the C_{PM}, C_{NSM} and C_{BM} pools, respectively. W: respective weights of the C_{PM}, C_{NMS} and C_{BM} pools. C%: carbon contained in the respective pools; R_{PM}, R_{NMS} and R_{BM} emission rates in % in the different pools. Note that respiration of soil ecosystem engineers when they produce biogenic aggregates is an additional source of gas emissions that can be compensated by increased plant production.

create simulation models to propose and test novel management options and/or to improve existing empirical systems.

7 Where to look for further information

General methodology to develop research and implementation is proposed in Lavelle et al. (2020). The FAO GSBI (Global Biodiversity Initiative) and FAO Global Soil Partnership programs might provide information on appropriate institutional support.

8 References

Alegre, J. C., Pashanasi, B. and Lavelle, P. 1996. Dynamics of soil physical properties in Amazonian agroecosystems inoculated with earthworms. *Soil Science Society of America Journal* 60(5): 1522–1529.

Allen, C. R., Angeler, D. G., Garmestani, A. S., Gunderson, L. H. and Holling, C. S. 2014. Panarchy: theory and application. *Ecosystems* 17(4): 578–589.

Allen, T. F. H. and Starr, T. B. 1982. *Hierarchy. Perspectives for Ecological Complexity*. Chicago: The University of Chicago Press.

Angers, D. A. and Caron, J. 1998. Plant-induced changes in soil structure: processes and feedbacks. *Biogeochemistry* 42(1–2): 55–72.

Angst, Š, Mueller, C. W., Cajthaml, T., Angst, G., Lhotakova, Z., Bartuska, M., Spaldonova, A. and Frouz, J. 2017. Stabilization of soil organic matter by earthworms is connected with physical protection rather than with chemical changes of organic matter. *Geoderma* 289: 29–35.

Barois, I. and Lavelle, P. 1986. Changes in respiration rate and some physicochemical properties of a tropical soil during transit through *Pontoscolex corethrurus* (Glossoscolecidae, Logichatea). *Soil Biology and Biochemistry* 18(5): 539–541.

Barois, I., Villemin, G., Lavelle, P. and Toutain, F. 1993. Transformation of the soil structure through *Pontoscolex corethrurus* (Oligochaeta) intestinal tract. *Geoderma* 56(1–4): 57–66.

Barot, S., Rossi, J.-P. and Lavelle, P. 2007. Self-organization in a simple consumer-resource system, the example of earthworms. *Soil Biology and Biochemistry* 39(9): 2230–2240.

Barros, E., Curmi, P., Hallaire, V., Chauvel, A. and Lavelle, P. 2001. The role of macrofauna in the transformation and reversibility of soil structure of an oxisol in the process of forest to pasture conversion. *Geoderma* 100(1–2): 193–213.

Barthod, J., Dignac, M. -F., Le Mer, G., Bottinelli, N., Watteau, F., Koegel-Knabner, I. and Rumpel, C. 2020. How do earthworms affect organic matter decomposition in the presence of clay-sized minerals? *Soil Biology and Biochemistry* 143: 107730.

Beare, M. H., Coleman, D. C., Crossley, D. A., Hendrix, P. F. and Odum, E. P. 1995. Preface. *Plant and Soil* 170(1): 1–2.

Blanchart, E., Bruand, A. and Lavelle, P. 1993. The physical structure of casts of Millsonia-Anomala (Oligochaeta, Megascolecidae) in shrub savanna soils (cote-Divoire). *Geoderma* 56(1–4): 119–132.

Blanchart, E., Lavelle, P., Braudeau, E., Le Bissonnais, Y. and Valentin, C. 1997. Regulation of soil structure by geophagous earthworm activities in humid savannas of Cote d'Ivoire. *Soil Biology and Biochemistry* 29(3–4): 431–439.

Blanchart, E., Albrecht, A., Brown, G., Decaëns, T., Duboisset, A., Lavelle, P., Mariani, L. and Roose, E. 2004. Effects of tropical endogeic earthworms on soil erosion. *Agriculture, Ecosystems and Environment* 104(2): 303–315.

Bossuyt, H., Six, J. and Hendrix, P. F. 2005. Protection of soil carbon by microaggregates within earthworm casts. *Soil Biology and Biochemistry* 37(2): 251–258.

Boström, U. 1988. Ecology of earthworms in arable land. Population dynamics and activity in four cropping systems, Phd thesis. Uppsala.

Bottinelli, N., Kaupenjohann, M., Maerten, M., Jouquet, P., Soucemarianadin, L., Baudin, F., Tran, T. M. and Rumpel, C. 2020. Age matters: fate of soil organic matter during ageing of earthworm casts produced by the anecic earthworm *Amynthas khami*. *Soil Biology and Biochemistry* 148: 107906.

Bottinelli, N., Jouquet, P., Tran, T. M., Aroui Boukbida, H. and Rumpel, C. 2021. Mid-infrared spectroscopy to trace biogeochemical changes of earthworm casts during ageing under field conditions. *Geoderma* 383: 114811.

Brown, G. G., Pashanasi, B., Villenave, C., Patron, J. C., Senapati, B. K., Giri, S., Barois, I., Lavelle, P., Blanchart, E., Blakemore, R. J., Spain, A. V. and Boyer, J. 1999. Effects of earthworms on plant production in the tropics. In: *Earthworm Management in Tropical Agroecosystems*. Lavelle, P., Brussaard, L. and Hendrix, P. (Eds). Wallingford, UK: CAB International, 87–147.

Brown, G. G., Barois, I. and Lavelle, P. 2000. Regulation of soil organic matter dynamics and microbial activity in the drilosphere and the role of interactions with other edaphic functional domains. *European Journal of Soil Biology* 36(3–4): 177–198.

Chauvel, A., Grimaldi, M., Barros, E., Blanchart, E., Desjardins, T., Sarrazin, M. and Lavelle, P. 1999. Pasture damage by an Amazonian earthworm. *Nature* 398(6722): 32–33.

Chenu, C. 1993. Clay-or sand-polysaccharide association as models for the interface between microorganisms and soil water related properties and microstructure. *Geoderma* 56: 142–156

Clarholm, M. 1985. Interactions of bacteria, protozoa and plant leading to mineralization of soil nitrogen. *Soil Biology and Biochemistry* 17(2): 181–187.

Coq, S., Barthes, B. G., Oliver, R., Rabary, B. and Blanchart, E. 2007. Earthworm activity affects soil aggregation and organic matter dynamics according to the quality and localization of crop residues - an experimental study (Madagascar). *Soil Biology and Biochemistry* 39(8): 2119-2128.

Curry, J. P., Byrne, D. and Boyle, K. E. 1995. The earthworm population of a winter cereal field and its effects on soil and nitrogen turnover. *Biology and Fertility of Soils* 19(2-3): 166-172.

Dawod, V. and Fitzpatrick, E. A. 1993. Some population sizes and effects of the Enchytraeidae (Oligochaeta) on soil structure in a selection of Scottish soils. *Geoderma* 56(1-4): 173-178.

Decaëns, T., Rangel, A. F., Asakawa, N. and Thomas, R. J. 1999. Carbon and nitrogen dynamics in ageing earthworm casts in grasslands of the eastern plains of Colombia. *Biology and Fertility of Soils* 30(1-2): 20-28.

Decaëns, T. and Rossi, J. P. 2001. Spatio-temporal structure of earthworm community and soil heterogeneity in a tropical pasture. *Ecography* 24(6): 671–682.

Decaëns, T., Jimenez, J. J. and Rossi, J.-P. 2009. A null-model analysis of the spatio-temporal distribution of earthworm species assemblages in Colombian grasslands. *Journal of Tropical Ecology* 25(4): 415–427.

Didden, W. A. M. 1990. Involvement of Enchytraeidae (Oligoquaeta) in soil structure evolution in agricultural fields. *Biology and Fertility of Soils* 9(2): 152–158.

Dominguez-Haydar, Y., Velasquez, E., Zangerle, A., Lavelle, P., Gutierrez-Eisman, S. and Jimenez, J. J. 2020. Unveiling the age and origin of biogenic aggregates produced by earthworm species with their NIRS fingerprint in a subalpine meadow of Central Pyrenees. *PLoS ONE* 15(8): e0237115.

Faber, J., Quadros, A. F. and Zimmer, M. 2018. A Space-For-Time approach to study the effects of increasing temperature on leaf litter decomposition under natural conditions. *Soil Biology and Biochemistry* 123: 250–256.

FAO 2019. Recarbonization of global soils. A tool to support the implementation of the Koronivia joint work on agriculture. Available at: http://www.fao.org/3/ca6522en/CA6522EN.pdf.

Feller, C. and Beare, M. H. 1997. Physical control of soil organic matter dynamics in the tropics. *Geoderma* 79(1-4): 69–116.

Fontaine, S., Barot, S., Barré, P., Bdioui, N., Mary, B. and Rumpel, C. 2007. Stability of organic carbon in deep soil layers controlled by fresh carbon supply. *Nature* 450(7167): 277–280.

Garcia-Palacios, P., McKie, B. G., Handa, I. T., Frainer, A. and Hattenschwiler, S. 2016. The importance of litter traits and decomposers for litter decomposition: a comparison of aquatic and terrestrial ecosystems within and across biomes. *Functional Ecology* 30(5): 819–829.

Garmestani, A., Twidwell, D., Angeler, D. G., Sundstrom, S., Barichievy, C., Chaffin, B. C., Eason, T., Graham, N., Granholm, D., Gunderson, L., Knutson, M., Nash, K. L., Nelson,

R. J., Nystrom, M., Spanbauer, T. L., Stow, C. A. and Allen, C. R. 2020. Panarchy: opportunities and challenges for ecosystem management. *Frontiers in Ecology and the Environment* 18(10): 576–583.

Grimaldi, M., Oszwald, J., Dolédec, S., Hurtado, MdP., de Souza Miranda, I., Arnauld de Sartre, X., Assis, W. Sd, Castañeda, E., Desjardins, T., Dubs, F., Guevara, E., Gond, V., Lima, T. T. S., Marichal, R., Michelotti, F., Mitja, D., Noronha, N. C., Delgado Oliveira, M. N., Ramirez, B., Rodriguez, G., Sarrazin, M., Silva, M. Ld, Costa, L. G. S., Souza, S. Ld, Veiga, I., Velasquez, E. and Lavelle, P. 2014. Ecosystem services of regulation and support in Amazonian pioneer fronts: searching for landscape drivers. *Landscape Ecology* 29(2): 311-328.

Gunderson, L. H. and Holling, C. S. 2002. *Panarchy. Understanding Transformations in Human and Natural Systems*. Washington, D.C.: Island Press.

Heal, O. W., Flannagan, P. W., French, D. D. and MacLean, S. F. 1981. Decomposition and accumulation of organic matter in tundra. Tundra ecosystems: a comparative analysis. Bliss, L C., Heal, O. W. and Moore, J. J. (Eds.). Cambridge: Cambridge University Press.

Hedde, M., Lavelle, P., Joffre, R., Jimenez, J. J. and Decaëns, T. 2005. Specific functional signature in soil macro-invertebrate biostructures. *Functional Ecology* 19(5): 785–793.

Hill, B. H., Elonen, C. M., Herlihy, A. T., Jicha, T. M. and Serenbetz, G. 2018. Microbial ecoenzyme stoichiometry, nutrient limitation, and organic matter decomposition in wetlands of the conterminous United States. *Wetlands Ecology and Management* 26(3): 425–439.

Jimenez, J. J., Decaëns, T. and Rossi, J. P. 2006. Stability of the spatio-temporal distribution and niche overlap in Neotropical earthworm assemblages. *Acta Oecologica* 30(3): 299–311.

Jimenez, J. J., Decaëns, T., Lavelle, P. and Rossi, J. P. 2014. Dissecting the multi-scale spatial relationship of earthworm assemblages with soil environmental variability. *BMC Ecology* 14: 26.

Jouquet, P., Bernard-Reversat, F., Bottinelli, N., Orange, D., Rouland-Lefevre, C., Duc, T. T. and Podwojewski, P. 2007. Influence of changes in land use and earthworm activities on carbon and nitrogen dynamics in a steepland ecosystem in Northern Vietnam. *Biology and Fertility of Soils* 44(1): 69–77.

Jouquet, P., Bottinelli, N., Podwojewski, P., Hallaire, V. and Duc, T. T. 2008. Chemical and physical properties of earthworm casts as compared to bulk soil under a range of different land-use systems in Vietnam. *Geoderma* 146(1–2): 231–238.

Jouquet, P., Traoré, S., Choosai, C., Hartmann, C., Bignell, D. 2011. Influence of termites on ecosystem functioning. Ecosystem services provided by termites. *European Journal of Soil Biology* 47: 215–222.

Lattaud, C., Locati, S., Mora, P., Rouland, C., Lavelle, P. 1998. The diversity of digestive systems in tropical geophagous earthworms. *Applied Soil Ecology* 9(1998): 189–195.

Lavelle, P. 1978. Les vers de terre de la savane de Lamto (Côte d' Ivoire). Peuplements, populations et fonctions dans l´ecosysteme. PhD thesis. Université P. et M. Curie. Publications du Laboratoire de Zoologie de l´ENS 12. 301p.

Lavelle, P., Barros, E., Blanchart, E., Brown, G., Desjardins, Th., Mariani, L. and Rossi, J. P. 2001. SOM management in the tropics: why feeding the soil macrofauna? *Nutrient Cycling in Agroecosystems* 61: 53–61.

Lavelle, P., Blanchart, E., Martin, A., Martin, S. and Spain, A. 1993. A hierarchical model for decomposition in terrestrial ecosystems: application to soils of the humid tropics. *Biotropica* 25: 130–150.

Lavelle, P. and Spain, A. V. 2001. *Soil Ecology*. Amsterdam: Kluwer Publishers Scientific Publications.

Lavelle, P., Spain, A., Blouin, M., Brown, G., Decaëns, T., Grimaldi, M., Jimenez, J. J., McKey, D., Mathieu, J., Velasquez, E. and Zangerle, A. 2016. Ecosystem engineers in a self-organized soil: a review of concepts and future research questions. *Soil Science* 181(3/4): 91–109.

Lavelle, P., Spain, A., Fonte, S., Camilo Bedano, J. C., Blanchart, E., Galindo, V., Grimaldi, M., Jimenez, J. J., Velasquez, E. and Zangerle, A. 2020. Soil aggregation, ecosystem engineers and the C cycle. *Acta Oecologica* 105: 103561.

Lavelle, P., Mathieu, J., Spain, A., Brown, G., Fragoso, C., Lapied, E., De Aquino, A., Barois, I., Barrios, E., Barros, M. E., Bedano, J. C., Blanchart, E., Caulfield, M., Chagueza, Y., Dai, J., Decaëns, T., Dominguez, A., Dominguez, Y., Feijoo, A., Folgarait, P., Fonte, S. J., Gorosito, N., Huerta, E., Jimenez, J. J., Kelly, C., Loranger, G., Marchão, R., Marichal, R., Praxedes, C., Rodriguez, L., Rousseau, G., Rousseau, L., Ruiz, N., Sanabria, C., Suarez, J. C., Tondoh, J. E., De Valença, A., Vanek, S. J., Vasquez, J., Velasquez, E., Webster, E. and Zhang, C. 2022. Soilmacroinvertebrate communities: a worldwide assessment. *Global Ecology and Biogeography* 31(7): 31.1261–1276.

Lehmann, J. and Kleber, M. 2015. The contentious nature of soil organic matter. *Nature* 528(7580): 60–68.

Lehmann, A., Zheng, W. and Rillig, M. 2017. Soil biota contributions to soil aggregation. *Nature Ecology and Evolution* 1: 1828–1835.

Le Mer, G., Barthod, J., Dignac, M.-F., Barre, P., Baudin, F. and Rumpel, C. 2020. Inferring the impact of earthworms on the stability of organo-mineral associations, by Rock-Eval thermal analysis and 13 C NMR spectroscopy. *Organic Geochemistry* 144: 104016.

Lubbers, I. M., Pulleman, M. M. and Van Groenigen, J. W. 2017. Can earthworms simultaneously enhance decomposition and stabilization of plant residue carbon? *Soil Biology and Biochemistry* 105: 12–24.

Marichal, R., Mathieu, J., Couteaux, M. -M., Mora, P., Roy, J. and Lavelle, P. 2011. Earthworm and microbe response to litter and soils of tropical forest plantations with contrasting C:N:P stoichiometric ratios. *Soil Biology and Biochemistry* 43(7): 1528–1535.

Marinissen, J. C. Y. and Dexter, A. R. 1990. Mechanisms of stabilization of earthworm casts and artificial casts. *Biology and Fertility of Soils* 9(2): 163–167.

Martin, A. 1991. Short- and long-term effects of the endogeic earthworm *Millsonia anomala* (Omodeo) (Megascolecidae, Oligochaeta) of tropical savannas, on soil organic matter. *Biology and Fertility of Soils* 11(3): 234–238.

Martin, A., Mariotti, A., Balesdent, J. and Lavelle, P. 1992. Soil organic matter assimilation by a geophagous tropical earthworm based on ð^{13}C measurements. *Ecology* 73(1): 118–128.

MEA 2005. Millennium Ecosystem Assessment. Available at: www.millenniumassessment.org.

Mueller, C. W., Carminati, A., Kaiser, C., Subke, J.-A. and Gutjahr, C. 2019. Editorial: rhizosphere functioning and structural development as complex interplay Between plants, microorganisms and soil minerals. *Frontiers in Environmental Science* 7: 130. doi: 10.3389/fenvs.2019.00130.

Oades, J. M. 1984. Soil organic matter and structural stability: mechanisms and implication for management. *Plant and Soil* 76(1-3): 319-337.

Pashanasi, B., Lavelle, P., Alegre, J. and Charpentier, F. 1996. Effect of the endogenic earthworm Pontoscolex corethrurus on soil chemical characteristics and plant growth in a low-input tropical agroecosystem. *Soil Biology and Biochemistry* 28(6): 801-810.

Perry, D. A. 1995. Self-organizing systems across scales. *Trends in Ecology and Evolution* 10(6): 241-244.

Plante, A. F. and McGill, W. B. 2002. Soil aggregate dynamics and the retention of organic matter in laboratory-incubated soil with differing simulated tillage frequencies. *Soil and Tillage Research* 66(1): 79-92.

Pulleman, M. M., Six, J., Uyl, A., Marinissen, J. C. Y. and Jongmans, A. G. 2005. Earthworms and management affect organic matter incorporation and microaggregate formation in agricultural soils. *Applied Soil Ecology* 29(1): 1-15.

Rees, R. M., Bingham, I. J., Baddeley, J. A. and Watson, C. A. 2005. The role of plants and land management in sequestering soil carbon in temperate arable and grassland ecosystems. *Geoderma* 128(1-2): 130-154.

Rillig, M. C. 2004. Arbuscular mycorrhizae, glomalin, and soil aggregation. *Canadian Journal of Soil Science* 84(4): 355-363.

Roletto, E., Barberis, R., Consiglio, M. and Jodice, R. 1985. Chemical parameters for evaluating compost maturity. *BioCycle* 26(2): 46-48.

Rossi, J. P. 2003. The spatiotemporal pattern of a tropical earthworm species assemblage and its relationship with soil structure. *Pedobiologia* 47(5-6): 497-503.

Rossi, J. P., Lavelle, P. and Albrecht, A. 1997. Relationships between spatial pattern of the endogeic earthworm *Polypheretima elongata* and soil heterogeneity. *Soil Biology and Biochemistry* 29(3-4): 485-488.

Rumpel, C., Baumann, K., Remusat, L., Dignac, M.-F., Barré, P., Deldicque, D., Glasser, G., Lieberwirth, I. and Chabbi, A. 2015. Nanoscale evidence of contrasted processes for root-derived organic matter stabilization by mineral interactions depending on soil depth. *Soil Biology and Biochemistry* 85: 82-88.

Scheu, S. 2003. Effects of earthworms on plant growth: patterns and perspectives. *Pedobiologia* 47(5-6): 846-856.

Schmidt, M. W. I., Torn, M. S., Abiven, S., Dittmar, T., Guggenberger, G., Janssens, I. A., Kleber, M., Koegel-Knabner, I., Lehmann, J., Manning, D. A. C., Nannipieri, P., Rasse, D. P., Weiner, S. and Trumbore, S. E. 2011. Persistence of soil organic matter as an ecosystem property. *Nature* 478(7367): 49-56.

Sequi, P., De Nobile, M., Leita, G. y Cercignani, G. 1986. A new index of humification. *Agrochimica* 30: 1-2.

Shipitalo, M. J. and Protz, R. 1988. Factors influencing the dispersibility of clay in worm casts. *Soil Science Society of America Journal* 52(3): 764-769.

Six, J., Elliott, E. T. and Paustian, K. 2000. Soil macroaggregate turnover and microaggregate formation: a mechanism for C sequestration under no-tillage agriculture. *Soil Biology and Biochemistry* 32(14): 2099-2103.

Sterner, R. W. and Elser, J. J. 2002. *Ecological Stoichiometry. The Biology of Elements from Molecules to the Biosphere*. Princeton, NJ: Princeton University Press.

Swift, M. J., Heal, O. W. and Anderson, G. M. 1979. *Decomposition in Terrestrial Ecosystems*. Oxford: Blackwell Publishing, Scientific Publications.

Tisdall, J. M. and Oades, J. M. 1982. Organic matter and water-stable aggregates in soils. *Journal of Soil Science* 33(2): 141–163.

Tötsche, K. U., Amelung, W., Gerzabek, M. H., Guggenberger, G., Klumpp, E., Knief, C., Lehndorff, E., Mikutta, R., Peth, S., Prechtel, A., Ray, N. and Kögel-Knabner, I. 2018. Micro-aggregates in soils. *Journal of Plant Nutrition and Soil Science* 181(1): 104–136.

Trigo, D. and Lavelle, P. 1993. Changes in respiration rate and some physicochemical properties of soil during gut transit through Allolobophora-Molleri (Lumbricidae, Oligochaeta). *Biology and Fertility of Soils* 15(3): 185–188.

Trigo, D. and Lavelle, P. 1994. Soil changes during gut transit through *Octolasion lacteum* Oerley (Lumbricidae, Oligochaeta). *Acta Zoologica Fennica* 194: 1–3.

Van Groenigen, J. W., Lubbers, I. M., Vos, H. M. J., Brown, G. G., De Deyn, G. B. and van Groenigen, K. J. 2014. Earthworms increase plant production: a meta-analysis. *Scientific Reports* 4: 6365.

Van Groenigen, J. W., Van Groenigen, K. J., Koopmans, G. F., Stokkermans, L., Vos, H. M. J. and Lubbers, I. M. 2019. How fertile are earthworm casts? A meta-analysis. *Geoderma* 338: 525–535.

Velasquez, E., Pelosi, C., Brunet, D., Grimaldi, M., Martins, M., Rendeiro, A., Barrios, E., Lavelle, P. 2007. This ped is my ped: visual separation and near infrared spectra allow determination of the origins of soil macro-aggregates. *Pedobiologia* 51: 75–87.

Velasquez, E. and Lavelle, P. 2019. Soil macrofauna as an indicator for evaluating soil-based ecosystem services in agricultural landscapes. *Acta Oecologica* 100: 1034446.

Villenave, C., Charpentier, F., Lavelle, P., Feller, C., Brussaard, L., Pashanasi, B., Barois, I., Albrecht, A. and Patron, J. C. 1999. Effects of earthworms on soil organic matter and nutrient dynamics following earthworm inoculation in field experimental situations. In: *Earthworm Management in Tropical Agroecosystems*. Lavelle, P., Brussaard, L. and Hendrix, P. (Eds). Wallingford: CAB International, 173–197.

Wall, D. H., Bradford, M. A., John, M. G. St., Trofymow, J. A., Behan-Pelletier, V., Bignell, D. E., Dangerfield, J. M., Parton, W. J., Rusek, J., Voigt, W., Wolters, V., Gardel, H. Z., Ayuke, F. O., Bashford, R., Beljakova, O. I., Bohlen, P. J., Brauman, A., Flemming, S., Henschel, J. R., Johnson, D. L., Jones, T. H., Kovarova, M., Kranabetter, J. M., Kutny, L., Lin, K. C., Maryati, M., Masse, D., Pokarzhevskii, A., Rahman, H., Sabara, M. G., Salamon, J. A., Swift, M. J., Varela, A., Vasconcelos, H. L., White, D. and Zou, X. M. 2008. Global decomposition experiment shows soil animal impacts on decomposition are climate-dependent. *Global Change Biology* 14(11): 2661–2677.

Wilson, G. W. T., Rice, C. W., Rillig, M. C., Springer, A. and Harnet, D. C. 2009. Soil aggregation and carbon sequestration are tightly correlated with the abundance of arbuscular mycorrhizal fungi: results from long-term field experiments. *Ecology Letters* 12(5): 452–461.

Zanella, A., Ponge, J. F. and Briones, M. J. I. 2018. Humusica 1, article 8: terrestrial humus systems and forms – biological activity and soil aggregates, space-time dynamics. *Applied Soil Ecology* 122: 103–137.

Zangerlé, A., Pando, A. and Lavelle, P. 2011. Do earthworms and roots cooperate to build soil macro-aggregates? A microcosm experiment. *Geoderma* 167: 303–309.

Zangerlé, A., Hissler, C., Blouin, M. and Lavelle, P. 2014. Near infrared spectroscopy (NIRS) to estimate earthworm cast age. *Soil Biology and Biochemistry* 70: 47–53.

Zangerlé, A., Hissler, C., McKey, D. and Lavelle, P. 2016. Using near infrared reflectance spectroscopy (NIRS) to identify the contribution of earthworms to soil macro-aggregation in field conditions. *Applied Soil Ecology* 104: 138–147.

Zhang, C., Langlest, R., Velasquez, E., Pando, A., Brunet, D., Dai, J. and Lavelle, P. 2009. Cast production and NIR spectral signatures of *Aporrectodea caliginosa* fed soil with different amounts of half-decomposed *Populus nigra* litter. *Biology and Fertility of Soils* 45(8): 839–844.

Zhang, H. and Schräder, S. 1993. Earthworm effects on selected physical and chemical properties of soil aggregates. *Biology and Fertility of Soils* 15(3): 229-234.

Zhang, W. X., Hendrix, P. F., Dame, L. E., Burke, R. A., Wu, J. P., Neher, D. A., Li, J. X., Shao, Y. H. and Fu, S. L. 2013. Earthworms facilitate carbon sequestration through unequal amplification of carbon stabilization compared with mineralization. *Nature Communications* 4: 2576.

Chapter 5

Understanding soil organic carbon dynamics at larger scales

Sebastian Doetterl, ETH Zurich, Switzerland; Rose Abramoff, Oak Ridge National Laboratory, USA; Jean-Thomas Cornelis, University of British Columbia, Canada; Aline Frossard, Swiss Federal Institute for Forest, Snow and Landscape Research WSL, Switzerland; Peter Fiener, Institute of Geography, Augsburg University, Germany; Gina Garland, ETH Zurich and Soil Quality and Use Group, Agroscope, Switzerland; Michael Kaiser, University of Nebraska-Lincoln, USA; Moritz Laub, ETH Zurich, Switzerland; Sophie Opfergelt, Earth and Life Institute, UCLouvain, Belgium; Marijn Van de Broek and Sarah van den Broek, ETH Zurich, Switzerland; and Sophie F. von Fromm, ETH Zurich, Switzerland and Max Planck Institute for Biogeochemistry, Germany

1 Introduction

In this chapter, we focus on the effects of biotic and abiotic factors controlling soil organic carbon dynamics at continental to global scales. On the side of natural effects, we highlight processes that can control carbon inputs, turnover, and stabilization in soils. On the side of anthropogenic effects, we focus on the role of climate change as well as historic and modern land conversion. We hereby divide anthropogenic effects into direct and indirect disturbances done by humans. Both overarching sections close with a short synthesis.

2 Natural processes affecting soil carbon dynamics[1]

The cycling of carbon (C) between plants, soil and the atmosphere is undoubtedly one of the most important fields of soil research in the twenty-first

[1] This section was written by Sebastian Doetterl.

http://dx.doi.org/10.19103/AS.2022.0106.05

century. At global scales, the amount of C involved in this cycle is primarily a function of organic matter entering soil through plant growth, its transformation within and incorporation into soil, and the specific nature of interactions with the soil (mineral) matrix under varying environmental conditions. Given the importance of healthy soils for supporting diverse ecosystems and the delivery of a multitude of ecosystem services including global climate regulation, it is still surprising to see that, despite thousands of studies on soil properties, plant-soil interactions and soil biogeochemistry, only relatively few soil types (namely those in wealthy countries in temperate climate zones) are well studied (see Kögel-Knabner and Amelung, 2021). While these studies have greatly improved our understanding of fundamental physicochemical soil organic carbon (SOC) stabilization and destabilization processes in soils, the understanding of biogeochemical cycles in many other soil systems cannot be simply derived from findings of specific, well-studied soil types. Too diverse are the controls of processes in soil types of contrasting pedoclimatic environments that ultimately control (SO)C sequestration. Consequently, biogeochemical models developed for specific soil types are not necessarily representative for other soil systems and may only be applied with great uncertainties and potentially result in false estimates of responses of soil C cycling to global change. Lack of understanding of spatial patterns of and links between biogeochemical cycles is thus one of the largest sources of uncertainty for predicting terrestrial C cycling in global models overall (Friedlingstein et al., 2020). With examples from different climate zones, we highlight in the following chapter some key processes and gaps in our understanding of C cycling at larger scales and how closing those gaps may help to better predict SOC sequestration of soils with global significance (Fig. 1).

2.1 Controls on carbon input, turnover and sequestration[2]

2.1.1 Global carbon reservoirs

Carbon is cycling within and across different earth spheres, being stored, transformed, and moved at different rates through sedimentary rocks, rivers, lakes, oceans, the soil system (mineral, organic, and permafrost soil), terrestrial vegetation, and the atmosphere. Besides the total concentration, C stored in different earth systems can be characterized according to the mean residence time (i.e. time needed to exchange each C atom of the considered system or subsystem at least once). Based on the C mean residence times, different systems can be subdivided into active and inactive drivers of C dynamics. Under undisturbed conditions, almost the entire C stored in sediments or sedimentary

[2] This section was written by Gina Garland and Sebastian Doetterl.

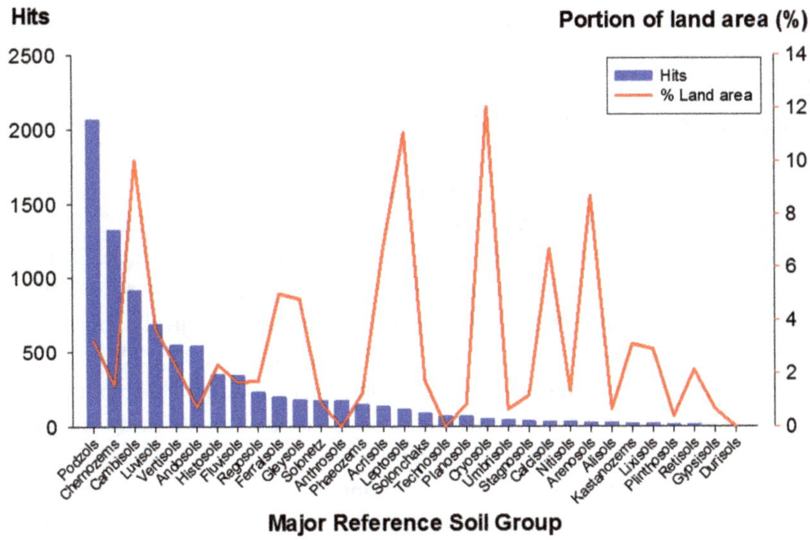

Figure 1 Database hits for joint search of the key words 'soil organic matter and Podzols' or with any other major Reference Soil Group according to IUSS Working Group WRB (2015), using the literature search program SciFinder Scholar (www.scifinder.cas.org; March 14, 2020; blue bars). The red lines represent the contribution of each soil group to total land area, as given in IUSS Working Group WRB (2015; Annex 1 - Description, distribution, use and management of Reference Soil Groups). The figure illustrates the frequency of organic matter research for a given soil group. It is not designed to give a full honest overview, because the search did neither include the key words soil organic carbon, nor any soil group as translated to other classification systems. Reprint of Kögel-Knabner and Amelung (2021).

rocks, for example, is considered to be inactive with mean residence times of up to millions of years (Mackenzie et al., 2004). The C stored in the deep ocean waters and sediments is also considered to be inactive, with mean residence times of more than 1000 years. In contrast, C stored in the atmosphere, in the surface oceans, in living vegetation, and in soil organic matter is considered to be relatively active, with mean residence times ranging from seconds to centuries (e.g. Killough and Emanuel, 1981; Trumbore, 2000). However, fossil fuel C from sedimentary deposits is an example of how an inactive C reservoir can be rapidly transformed into a highly active driver of global C dynamics due its anthropogenic uses.

Besides geologic systems (upper mantle, continental crust, oceanic crust), sediments at the surface of the earth containing calcite and dolomite deposits and rocks as well as organic matter are by far the largest C reservoir (7.78×10^7 Gt) with fossil fuel-C accounting for less than 0.5% of the total 5000 Gt sedimentary organic C (Mackenzie et al., 2004). The world oceans store about 38 000 Gt of C whereby the surface ocean waters (first 100 m) that are directly interacting with the atmosphere contain about 700 Gt dissolved C and 3 Gt of C stored in the oceanic biota (e.g. Houghton, 2007; Mackenzie, et al., 2004). The living vegetation of terrestrial ecosystems stores 450–650 Gt organic C whereas soils contain 1500–3000 Gt organic C (first 2 m) (Sanderman et al., 2017; Friedlingstein et al., 2020).

Current global carbon models estimate that soils store approximately 81% of terrestrial C stocks amounting to approximately 1500 Gt C, which is more than the atmosphere (800 Gt C) and terrestrial vegetation (500 Gt C) combined (Erb et al., 2018; Friedlingstein et al., 2020; FAO and ITPS, 2015; IPCC, 2013). However, this number does not include soil in the permafrost region. Data from Tarnocai et al. (2009) and Hugelius et al. (2013, 2014) suggest that up to 1700 Gt of organic C can be found in shallow and deep permafrost soils. As such, the importance of soils as sources and sinks for atmospheric CO_2 is increasingly recognized at both the scientific and policy-making levels. Thus, on-the-ground measures to protect and increase this important C stock are currently being explored at a vigorous pace. Strong expectations are placed on soils as a critical element for combatting the current global climate crisis (Richter and Yaalon, 2012). This has in turn spurned the continuous improvement of Earth system models to determine accurate estimates of SOC stocks, the main drivers of the dynamic processes contributing to these stocks, and how they may change over time under varying climatic conditions and human activities (Chapin et al., 2009; Todd-Brown et al., 2014; Kawamiya et al., 2020).

2.1.2 Plant-soil relationships as a key component to understand soil organic carbon dynamics

One important factor to consider as a regulator for SOC stocks is vegetation (Chapter 3 of this book). Although this pool accounts for only 19% of terrestrial C stocks, it is one of the key drivers of C input into soil systems and thus the overall significance of plants in maintaining SOC stocks must not be overlooked. Indeed, it is this plant-soil feedback (Laliberte et al., 2013; Pugnaire et al., 2019), or the process by which plants modify the biotic and abiotic soil properties through feedback mechanisms, which drive not only the input of C into the system, but are also one of the key components of pedogenesis in general, along with parent material, climate, topography, and time (Dokuchaev, 1886, 1899; Jenny, 1941). In fact, through time, soil-plant interactions evolve from initial soil formation stages, where the composition and size of the soil organic matter stock is primarily determined by plant-derived inputs (Khedim et al., 2021) to later stages, where soil properties such as mineralogy and texture play an increasingly dominant role in stabilizing these different organic matter pools accumulating in the soil (Chorover et al., 2004; Mikutta et al., 2006, 2009;

Kleber et al., 2021). Ultimately, soil properties determine the capacity of soils to accumulate C (Paul, 2016). However, even at these later soil developmental stages, vegetation still plays an important role for soil carbon sequestration (Chapters 2 and 3 of this book), as the main transporters of atmospheric C into the soil system.

The net amount of SOC sequestered in a given system depends not only on the initial C introduced via plants but also on a variety of biotic and abiotic processes including rates of decomposition and the capacity of soils to adsorb organic matter onto soil minerals (Kleber et al., 2021). However, in general terms, the higher the net primary productivity (NPP), or the total vegetative biomass produced within a given ecosystem or land use type, the higher carbon inputs to soils (Smith, 2008) and subsequently the higher potential SOC sequestration. As such, the specific type of vegetation in a given area plays an important role in overall SOC dynamics, including distribution of SOC pools at different soil depths (Jobbagy and Jackson, 2000). For example, current estimates of carbon stored in different vegetation types show that forests contain the largest proportion of terrestrial C pools, with 359 Gt C stored in the vegetation and 787 Gt C in forest soils (IPCC, 2022). This is in large part because growth often occurs year round, and also because the forest litter C sources tend to be more recalcitrant than other vegetation types (Smith, 2008). Though less than forests, tropical savannas (66 Gt C in the vegetation and 264 Gt C in the soil) and natural grasslands (9 Gt C in the vegetation and 295 Gt C in the soil) are also known to sequester large quantities of C, much of which is through decomposition from their proportionally large root systems (IPCC, 2022; Soussana et al., 2006; Lorenz and Lal, 2018; Ma et al., 2021). Croplands, on the other hand, tend to provide relatively few C input (3 Gt C in the vegetation and 128 Gt C in the soil) due to the removal of biomass C during crop harvest (IPCC, 2022). However, the specific amount for a given area depends greatly on the management practices employed (see Sections 2.2–2.4 and Chapters 15–24 of this book).

2.1.3 Impact of climate on plant species distribution and soil organic carbon cycling

While the integral role of plant growth in driving both pedogenesis and SOC dynamics is unquestioned, various abiotic factors influencing these biological communities at the global scale must be considered in addition. It is well known, for example, that climatic patterns can drive and shape soil formation and SOC accrual through both physical and chemical weathering of parent material (Ehlers and Goss, 2016; Torn et al., 1997). This impacts SOC dynamics not only directly, but also indirectly through effects on geomorphology, vegetation patterns, and soil (geo)chemistry. Indeed, both historic and current temperature and precipitation patterns, coupled with parent material and topography, have a direct influence on global aboveground and belowground biodiversity patterns, including plant species distribution and productivity (Adler et al., 2013; Zuckerberg et al., 2020). As access to global databases and advanced modeling and experimental approaches expands, it is increasingly clear that such climatically driven changes in the composition and functioning of biological communities have a significant impact on the global carbon cycle (Peng et al., 1998; Gerten et al., 2008; Crowther et al., 2019; Lu et al., 2019; Eglinton et al., 2021).

For example, in colder high latitude and high altitude climate regions, low temperatures can limit both plant growth and soil microbial activity, which in turn reduces decomposition of organic material and therefore the turnover and availability of key plant nutrients such as nitrogen and phosphorus (Hobbie, 1992; Paul, 2016). In natural ecosystems, such strong controls on nutrient availability in turn significantly influence the capacity of different plant species to grow (Hobbie, 1992) and thus significantly shape global plant distribution patterns (Aerts and Chapin, 1999). Similarly, it has often been noted that plant diversity tends to decrease from the equatorial region toward the poles due to the interactive effect of temperature and precipitation patterns (Hillebrand, 2004; Laliberte et al., 2020). Yet even within a given ecosystem type, regional climatic differences can impact SOC dynamics as well. For example, global temperature patterns in forests are known to directly influence the allocation of new carbon supplies to different tissue types. For example, trees in colder climates allocate more new C to roots in comparison to shoots (Reich et al., 2014; Ma et al., 2021). This difference in C allocation not only influences C uptake through photosynthesis, but also decomposition rates and nutrient availability, which in turn impacts C input into the soil, its turnover and potential sequestration as SOC (Kirschbaum, 1995, 2000; Davidson and Janssens, 2006). Furthermore, as vegetation patterns are strongly influenced by anthropogenic factors (i.e. management practices), the relative impact of climate on SOC sequestration must be assessed considering these other important drivers (Smith, 2008; Morris et al., 2020).

In addition to changes in temperature, moisture availability as driven by precipitation patterns are also known to greatly impact plant species distribution and biological activity (Schaffer and Rodriguez-Iturbe, 2018), and thus both directly and indirectly SOC dynamics. For example, in arid and semiarid regions, rates of microbial decomposition decrease drastically during the dry season, thus reducing nutrient turnover and availability to plants and ultimately reducing NPP (Manzoni et al., 2012a). As new soil C inputs are primarily plant derived, this impacts not only the total SOC content, but also the transport of more mobile SOC compounds (i.e. dissolved organic compounds) to greater soil depths and thus the soils' capacity to accumulate SOC (Kaiser and Kalbitz, 2012). At the same time, under these dry and nutrient poor conditions, microbial C decomposition is reduced as well, leading to sometimes contrasting feedbacks between low C input but high SOC accrual. Conclusively, these processes highlight the interacting and complex abiotic controls on SOC storage under different climates, changing in importance as pedogenesis progresses (Rumpel et al., 2002; Quideau et al., 2001).

2.2 Soil microorganisms as agents of carbon turnover at the global scale[3]

2.2.1 The effect of soil microbiome composition on carbon dynamics at global scales

Belowground ecosystems are hosting very diverse microbial communities, with the number of species present in a gram of soil typically exceeding a thousand (Gans et al., 2005; Torsvik et al., 1990), and the number of individuals ranging from millions to billions (Raynaud and Nunan, 2014; Torsvik et al., 2002). Despite their microscopic size, soil microbes are important drivers of terrestrial biogeochemical cycles as they are responsible for the formation and turnover of soil organic matter (Chapter 2 of this book), and through SOC mineralization also control the amount of nutrients available to plants.

Globally, the amount of microbial biomass carbon has been estimated at 16.7 Gt C in the top 0.3 m of soils (ca. 2.3 % of SOC) and 23.2 Gt C in the top meter (ca. 1.5 % of SOC) (Xu et al., 2013). Soil fungi constitute the largest biomass (ca. 12 Pg C), followed by bacteria (ca. 7 Gt) and Archaea (ca. 0.5 Gt) (Bar-On et al., 2018). The largest stock of soil microbial biomass carbon is located in tundra, while desert ecosystems contain the lowest amount of microbes (Xu et al., 2013). In general, managed soils (i.e. grassland and agriculture) contain less microbial biomass carbon compared to soil under natural vegetation (Xu et al., 2013). At the global scale, the mass of soil microbes thus follows the opposite trend of aboveground biomass, with the greatest abundance in Arctic and sub-Arctic ecosystems (Crowther et al., 2019). This seems counterintuitive, as larger microbial abundances lead to faster rates of soil organic matter processing and thus lower SOC stocks. However, cold temperatures at high latitudes reduce the rate of soil organic matter mineralization, leading to large SOC stocks, which support large microbial abundances.

Not only the abundance of soil microbes affects soil biogeochemical cycles, also the microbial community composition is important. Bacteria are the phylogenetically most diverse group on earth (Hug et al., 2016) with the numbers of species in the millions. Similarly while the number of described fungal species is ca. 100 000 (Blackwell, 2011), their total number is estimated to

[3] This section was written by Marijn Van de Broek and Aline Frossard.

be as high as 5.1 million species (O'Brien et al., 2005). The immense diversity of belowground microorganisms has been shown to have a significant impact on biogeochemical cycles in terrestrial ecosystems (Bardgett and van der Putten, 2014). Microbial diversity is driven by spatial heterogeneity (Bardgett and van der Putten, 2014), which varies from the micrometer to millimeter scale in soils (Vos et al., 2013) to the continental scale (Fierer et al., 2009). While in the past the great local diversity in soil microorganisms has obscured spatial patterns in their community composition, recent meta-analyses of high-throughput sequencing data have shown that at the global scale, microbial community composition is related to soil pH (Lauber et al., 2009; Tedersoo et al., 2014), aridity (Maestre et al., 2015), plant diversity (Prober et al., 2015), and plant productivity (Delgado-Baquerizo et al., 2018). In addition, there appears to be no clear relationship between latitude and belowground species richness as is the case for aboveground species (Bardgett and van der Putten, 2014). The ratio of plant-to-fungi richness, for example, decreases exponentially from the equator toward the poles (Tedersoo et al., 2014). The effect of the composition of the soil microbiome on soil organic matter transformation processes is more important for specific processes which can only be performed by specialized microbial groups, such as N transformation processes and lignin decomposition, compared to more general processes, such as mineralization of labile plant litter compounds such as sugars, which can be performed by a broad range of soil microorganisms (Crowther et al., 2019; Baumann et al., 2013). The latter, referred to as functional redundancy, seems to be a common characteristic of many microbial systems (Louca et al., 2018), although reduced microbial diversity was shown to affect specific decomposition functions (Baumann et al., 2013). However, a common paradigm is that not changes in microbial species diversity level, but rather community composition, are more important for soil organic matter cycling (Bardgett and van der Putten, 2014).

The two microbial functional groups whose effect on soil organic matter cycling has been studied most intensively at the global scale are arbuscular mycorrhizal fungi (AMF) and ectomycorrhizal mycorrhizal fungi (EMF). While both form symbioses with plant roots to facilitate plant nutrient uptake, at least some EMF have the capacity to produce oxidative and hydrolytic extracellular enzymes to mineralize organic matter and obtain nutrients (Chalot and Brun, 1998; Courty et al., 2010). In contrast, AMF lack this capacity and rely on saprotrophic microbes for organic matter mineralization (Smith and Smith, 2011). While ca. 2% of plant species associate with EMF, EMF-associated trees represent ca. 60% of all trees on earth (Steidinger et al., 2019). There are two main mechanisms by which the type of mycorrhizal symbiont associated to trees has an impact on SOC dynamics. The first one is related to the ability of EMF to produce extracellular enzymes, thereby affecting nutrient availability for microorganisms and plants. This may lead to decreasing microbial activity and soil organic matter mineralization, while on the other hand facilitating plant NPP, and thus organic carbon inputs to the soil (Orwin et al., 2011). The second mechanism is related to the quality of plant leaf litter produced by AMF and EMF-associated trees. It has been shown for mature forests in the United States that AMF-associated trees produce litter of high quality, which decomposes faster compared to litter of lower quality produced by EMF-associated trees (Phillips et al., 2013). This has a substantial effect on SOC stocks, as showed that soils in ecosystems dominated by AMF-associated trees store 70% less SOC per unit nitrogen compared to soils in ecosystems dominated by EMF dominated trees (Averill et al., 2014). In addition, Averill et al. (2014) showed that mycorrhizal association with trees has a larger effect on SOC stocks compared to climatic or edaphic factors. This knowledge has been used to successfully improve the representation of the soil microbial communities in SOC models, both at the local (Brzostek et al., 2014; Fatichi et al., 2019; Fisher et al., 2010) and global scale (Sulman et al., 2019).

2.2.2 The role of soil microorganisms in the formation and loss of soil organic carbon

While historically input quantity and quality of plant material was considered as the main driver of soil C storage and stabilization, recent paradigms have recognized soil microbial metabolism, through assimilation, growth, and necromass production, as a key control for both soil C losses and stabilization in soils (e.g. Cotrufo et al., 2013, Chapter 2 of this book). In fact, a recent review concluded that SOC stabilized through mineral interactions consists of some plant but mostly microbial-derived compounds (Angst et al., 2022). Many soil microbes utilize this soil organic matter as their main source of energy by assimilating organic molecules, after which a portion of carbon is lost in the form of CO_2 produced during microbial maintenance, growth, and enzyme production with the rest being incorporated into microbial biomass or released into mineral soil again. Since most biomolecules are polymers and too large for direct microbial uptake, many soil microbes have the ability to produce extracellular enzymes which mediate the depolymerization of large organic molecules. This microbial extracellular enzyme activity (EEA) is then directly linked to resource availability (e.g. Burns et al., 2013). The induction of enzyme synthesis and secretion by microbial cells is hereby mainly governed by local substrate concentration. Low availability of carbon or nutrient compounds will trigger the synthesis of the corresponding enzyme to facilitate nutrient release into the soil solution for uptake by plants and microorganisms. However, enzyme production is costly for cells and EEA is usually reflecting cellular resource and energy economics by maximizing the benefit while minimizing the costs. When cells endure pressure to lower the energetic demands of extracellular enzyme production, a sufficient concentration of the reaction products need to be maintained to support cellular functions and maintain viability (Burns et al., 2013). Modeling results indicate that a minimum of 2% of microbially assimilated C must be allocated to enzyme production to sustain microbial biomass of decomposer communities at constant levels (Schimel and Weintraub, 2003). However, the percentage of C allocated to enzyme production sufficient to maintain cell viability is likely to vary according to soil environmental factors (pH and temperature) as well as quality and substrate availability, diffusion conditions, and extracellular enzyme survival and sorption to soil minerals (Burns et al., 2013). Globally, extracellular enzyme stoichiometry is strongly constrained, with an averaged C:N:P ratio of 1:1:1, based on log-transformed potential activities of C-, N-, and P-acquiring enzymes. Although mechanisms are currently mostly unknown, data gathered from many different soils confirmed that the regulation of extracellular enzymes contribute to match the microbial stoichiometric demand. As an example, an increased N availability through the excretion of N acquiring enzymes (e.g. aminopeptidases) in soil was shown to stimulate C mobilizing enzymes (Allison and Vitousek, 2005) and to enhance the activity of the phosphatase (Mooshammer et al., 2012). In this way, the investment of microorganisms to acquire N also includes mining for P to maintain an elemental stoichiometric ratio. Moreover, once excreted from cells, several extracellular enzymes can be stabilized by binding to clay minerals and organic matter compounds to retain their activity, though often reduced in comparison to free enzymes, for prolonged periods of time (Burns et al., 2013). It has been suggested that stabilized enzymes associated with organic and inorganic colloids could contribute through extracellular metabolism substantially to soil respiration (Kandeler, 1990; Maire et al., 2013) but quantitative assessments of this contribution are still lacking.

In addition to being the main agents of SOC mineralization, microorganisms play a key role in soil organic matter formation and ultimately stabilization (Chapter 2 of this book), depending on their location in the soil matrix. In the rhizosphere surrounding living roots, microbes receive carbon either from rhizodeposition or from decaying roots (Chapter 3 of this book). In mineral soil, however, microbes utilize SOC that is located in patches on soil minerals and which is often not directly plant derived but already processed through one or several microbial life cycles before being stabilized on mineral surfaces (see also Section 1.3 for more detail). While the composition of the soil microbiome will have a substantial effect on the rate of carbon mineralization in the rhizosphere, spatial (in)accessibility of SOC to microbes can be become a substantial additional limitation for carbon mineralization in mineral soil (Schimel and Schaeffer, 2012). Ultimately, the fraction of carbon utilized by soil microorganisms will depend on their carbon use efficiency (CUE). The microbial CUE refers to the balance between SOC retained in microbial biomass versus the amount of C respired and released to the atmosphere (Sinsabaugh et al., 2016, 2013; Spohn et al., 2016). CUE can also be represented as a function of stoichiometric environmental resource supply in relation to biomass composition (Sinsabaugh et al., 2016). Microbial CUE can be higher when SOC availability is limiting and microbial catabolic and anabolic metabolism is coupled. In contrast, nutrient limitation can reduce CUE by uncoupling catabolism and anabolism which increases the cost for extracellular enzyme production (Blagodatskaya and Kuzyakov, 2008). Thus, CUE is expected to vary greatly along gradients of resource availability (Sinsabaugh et al., 2017). Moreover, environmental variation in soil temperature and moisture alter the microbial metabolism thereby affecting CUE by shifting the balance between microbial growth and respiration (Allison et al., 2010; Manzoni et al., 2012b).

Thus, microbial mediation of soil organic matter formation can occur via two major mechanisms: (1) the *ex vivo* one by which extracellular enzymes attack and transform plant residue, resulting in plant-derived SOC and (2) the *in vivo* turnover of organic substrates via cell uptake, biosynthesis, growth, and death resulting in microbial-derived SOC (Liang et al., 2017). The necromass left by dead microorganisms is then either recycled as a new substrate for microbial growth or adsorbed to mineral surfaces and stabilized in the soil (Miltner et al., 2012). The balance between microbial catabolic activity, releasing CO_2 to the atmosphere and anabolic activity, which contributes to necromass formation, controls largely the magnitude of soil C sequestration (e.g. Liang et al., 2017). Necromass mostly includes organic matter from fragmented cell membranes and former intracellular compounds such as enzymes, ribosomes, and small biopolymers not recycled by the active microbial pool. In a comprehensive review, it was reported that about 30% of stable SOC is derived from microbial necromass in temperate forests, while microbial necromass contributes to more than half of the total SOC in grassland and agricultural soils (Liang et al., 2019). Moreover, the authors of that study observed that the proportion of the fungal necromass (more than 70% of total necromass) is much higher than the bacterial one (less than 30% of total necromass), independently of land use type.

In conclusion, the heterogeneous nature of soil and its chemically diverse compounds produce large variations of elemental stoichiometry, which microorganisms have to manage. Moreover, the C:N:P ratio of soil microbial biomass was observed to be significantly smaller and less variable than that of their substrates in the surrounding soil. This implies a certain degree of homeostasis (i.e. stoichiometric invariance) of microbial cells (Cleveland and Liptzin, 2007), which can constrain microbial activities. The resulting stoichiometric imbalance between microbial cells and their resource (i.e. the surrounding soil) is very important for carbon and nutrient cycling and microbial growth and functioning (Mooshammer et al., 2012, 2014). However, it is not clear yet how soil microbes regulate C:N:P homeostasis (i.e. stoichiometric imbalance toward the surrounding soil). Four complementary mechanisms allowing microbial communities to adapt to environmental constraints have been proposed (Mooshammer et al., 2014): (1) microorganisms reduce their stoichiometric resource imbalance by adjusting C:N:P ratios internally (i.e. cellular stoichiometry), (2) microorganisms adjust the production of extracellular enzymes in order to maximize the mobilization of substrates, which are rich in limiting nutrients, (3) microorganisms ingest any available elements in their surrounding and excrete nutrient in excess to their demand, while keeping the limited ones in their cells, and (4) specialization of microorganisms, such as nitrogen-fixing prokaryotes or saprophytic fungi, which increase the availability of limiting nutrients (N and P), which are then available to other microorganisms. However, the relative importance of these mechanisms to overcome stoichiometric resource imbalances and how these imbalances affect the metabolization of organic matter and SOC sequestration via the composition and functioning of microbial communities is poorly understood.

2.3 Geochemical perspective on the properties of soils to store carbon[4]

2.3.1 Soil-forming processes at larger scales

Climate has important direct and indirect consequences on soil C cycling and ultimately sequestration. Globally, SOC stocks are generally positively correlated with moisture and negatively correlated with temperature (Post et al., 1982), whereas the longest SOC turnover rates are usually found in cold or dry regions where biological activity is constrained (Carvalhais et al., 2014; Shi et al., 2020). However, large variability in SOC stocks and turnover rates still exist within one climate zone, due to variation in soil internal properties that evolve in response to the soil-forming factors, namely soil physicochemical properties (Torn et al., 1997; Rasmussen et al., 2018; von Fromm et al., 2021).

At global scales, geology, climate, vegetation, and their interactions drive soil biogeochemical cycles and soil development (Fig. 2). Direct and indirect effects of human-related activities on soils have been recognized for long as a sixth soil-forming factor with local to global impacts on soil functions and development (Amundson and Jenny, 1997). Among the number of ecosystem services provided by soils, the biogeochemical cycling of key nutrients and water is one of the most important functions for regulating environmental systems. For example, water input and percolation rates in

[4] This section was written by Jean-Thomas Cornelis, Sophie Opfergelt and Sebastian Doetterl.

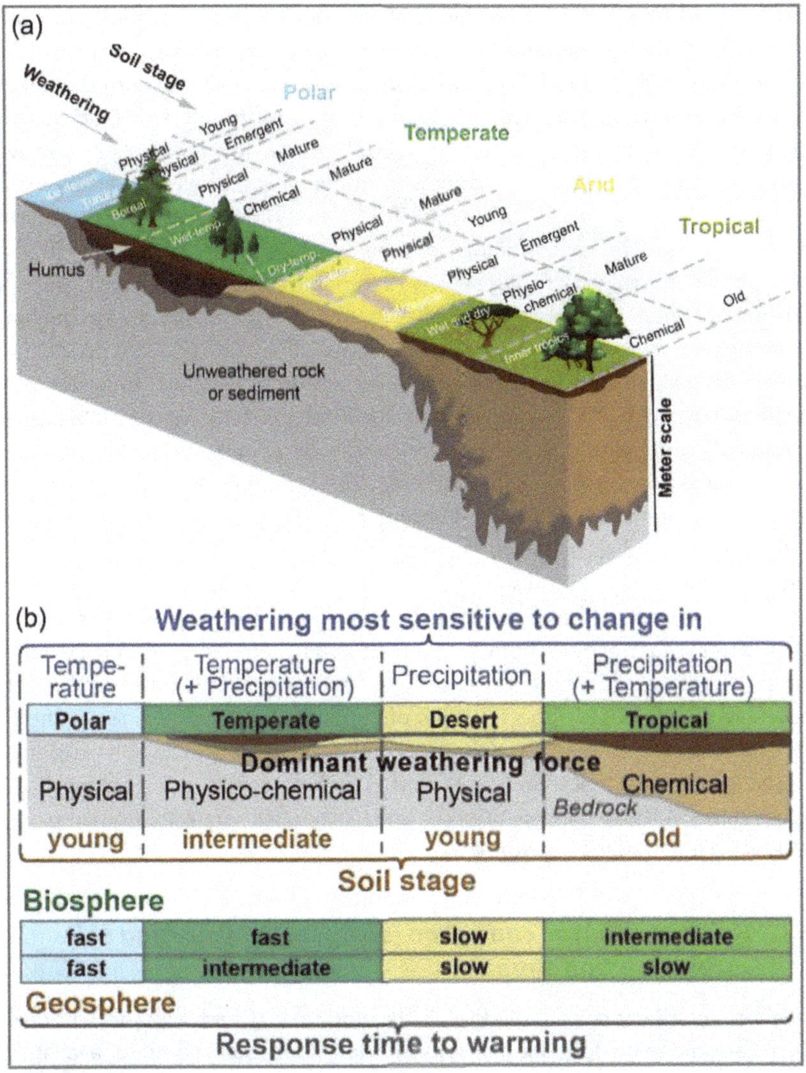

Figure 2 (a) Distribution of weathering forces and soil development stages at the global scale. (b) Conceptual design of sensitivity of bio- and geosphere to climate change in relation to weathering and rate of expected change (modified after Doetterl et al., 2018).

soils are key drivers of thermodynamics and kinetics of soil biogeochemical reactions (Chadwick and Chorover, 2001), which in turn govern the ability of soils to retain nutrients and to stabilize organic matter against decomposition (Witzgall et al., 2021). With the rising human influence on soils, we entered an era in which anthropogenic soil disturbance needs to be considered as the main driver of soil development in many ecosystems. Thus, there is an urgent need in revisiting our understanding of the main controls on the upper and lower boundaries of soils and the development direction of key soil properties (Cline, 1961; Richter and Billings, 2015). For example, soil moisture is spatially variable across climate zones as a function of atmospheric and land surface conditions (Lawrence and Hornberger, 2007). From a geochemical perspective, the transition from alkaline to acidic soil domains is globally driven by water balances (Slessarev et al., 2016). Shifts in hydrology and soil saturation/aeration regulate redox conditions potentially altering carbon storage in soils through variations in microbial activity (Lehmann et al., 2020) and/or electron transfer between minerals and organic matter (Kleber et al., 2021). Similarly, hydrology controls carbon inputs through its effect on biomass production and microbial decomposer communities controlling SOC turnover (see Sections 1.1 and 1.2). However, human disturbances of the water cycle are not adequately represented at scales at which soils differ in their development and geochemical characteristics (Vörösmarty and Sahagian, 2000). Furthermore, most of our mechanistic understanding of soil processes and warming is derived from studies in temperate zones (Haaf et al., 2021; Kögel-Knabner and Amelung, 2021). A holistic, global assessment on factors controlling soil respiration except for basic variables that integrate various processes at once (clay content) has not been done yet (Davidson and Janssens, 2006). In addition, current terrestrial biosphere models (TBM) use highly simplified approaches to represent carbon (C), nitrogen (N), and phosphorus (P) dynamics in soils and do not consider hydrology or geochemistry-related differences in soil properties and soil formation that affect C, N, and P cycling. Soil is not mechanistically represented in these models but is treated as a black box with a given mostly budgetary function for key nutrients.

2.3.2 Soil formation as control on (bio)geochemistry and soil organic carbon storage

Environmental factors controlling soil development (Chadwick and Chorover, 2001) can lead to the formation of soils with very contrasting contents, availability and speciation of major (Ca, Mg, K, P, Si) and trace elements (Fe, Mn, Mo, Zn) (Jobbagy and Jackson, 2001). Changing soil-forming conditions can affect the geochemistry of such elements and thereby the capacity of soils to store organic carbon. The chemistry of these elements is also sensitive to changes in water-related soil-forming conditions (change in drainage and redox conditions). At global scale, SOC retention in terrestrial ecosystems is forced by climate-driven thresholds, which govern prevailing mineral-related carbon stabilization mechanisms in which a range of elements play a key role (Kramer and Chadwick, 2018). For example, in humid regions, it is well-established that Al, Fe, and Mn geochemistry is central with regard to the ability of soils to store SOC (Kleber et al., 2021; Allard et al., 2017; Johnson et al., 2015; Stuckey et al., 2018). In the warming Arctic, permafrost thaw is not only associated with direct losses of labile organic matter but also with changing conditions for Fe-bearing mineral dissolution (Monhonval et al., 2021; Opfergelt, 2020; Patzner et al., 2020), which are intimately linked with redox-sensitive elements and a lowering of the potential for soils to store SOC through organo-mineral associations. Similarly, decreasing soil drainage in warming high-latitude peatlands is considered to decrease Fe export from soils as Fe-organic complexes, which in turn has potential implications for the part of organic matter stored in soils as chelated with Fe ions (Opfergelt et al., 2017). In dry regions with alkaline and less acidic soils conditions, other elements such as calcium may take over the role of the main binding partner for organic matter (Rowley et al., 2018). Thus, shifts in hydrology and soil saturation/aeration shape the biogeochemical cycling of key elements (Si, Al, and Fe) involved in secondary mineral formation (clay minerals and Fe-, Al-oxyhydroxides) or redox-sensitive elements (Fe, Mn, or Mo) with key ecological functions (Herndon et al., 2020). The distribution and mobility of both cations and anions in soils shows stark contrasts between tropical (Fritsch et al., 2009) and temperate regions (Fekiacova et al., 2017; Otero et al., 2009) due to the varying development and weathering stages of soils, with often sharply contrasting pH and redox conditions. In tropical soils, hematite (Fe_2O_3) is the main soil constituent involved in organic matter stabilization while soil-forming processes in temperate zones lead to the formation of goethite (crystalline FeOOH) and/or ferrihydrite (amorphous $Fe_2O_3.0.5H_2O$), which are key constituents to SOC stabilization (Kleber et al., 2015). Beyond soil redox controls, it is worth pointing out that soil aeration can also have indirect effects by moderating the bioavailability of trace elements (Fe, Mo, Mn, Zn) that are key in biomass production and in microbial enzymatic ability to process soil organic matter.

In conclusion, these examples remind us that the control of environmental conditions on soil development and soil properties will be highly influential on biogeochemical cycles and thus SOC sequestration through direct (adsorption of organic molecules) or indirect (interaction with trace elements acting as enzymatic cofactors) effects on organic matter storage between contrasting climate regions. Furthermore, the relative importance of various environmental controls (Fig. 3) on the carbon cycle changes over time as soils develop. Thus, soil processes as controls on geochemistry and the intertwined interactions with trace elements should certainly deserve further investigations with regard to SOC stabilization at a global scale. The role of trace elements such as Mo certainly should receive more attention. This is exemplified by the role of Mo in biomass production as a cofactor in the N-fixing enzyme nitrogenase that is associated with roots (Barron et al., 2009; Bellenger et al., 2008). Another

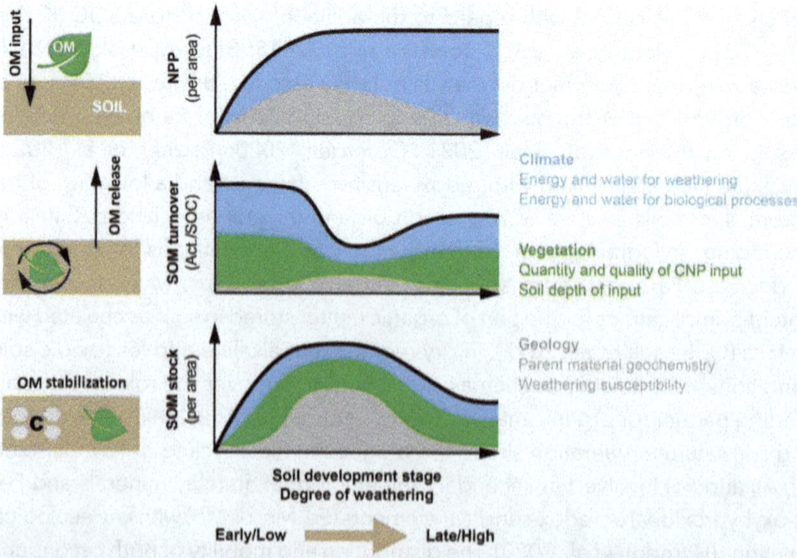

Figure 3 Representation of biological (vegetation and microbial) processes in focus and their hypothesized trajectories along timelines of soil development. Upper panels: NPP and OM input. Middle panels: OM matter release and microbial OM cycling in soils. Lower panels: OM stabilization with minerals and SOM stock. Y-axis represents size of stock or flux. X-axis represents the degree of soil development. Colored area in plots represent the effect size of key environmental controls (climate, vegetation, and geology) on the respective process in focus at various soil development stages.

element is Mn, a cofactor of Mn peroxidase, strongly controlling the later stage of soil organic matter decomposition (Berg et al., 2007).

3 Carbon cycling in the Anthropocene: direct and indirect human disturbance of soil and its impact on soil organic carbon

The global influence of human activity on soils and the biogeochemical cycles therein is an important factor determining our influence on the environment as a species with reach to almost all terrestrial ecosystems. For thousands of years, humans have altered soils directly through changing land use and the deliberate modification of soil properties to satisfy their needs. With climate change, humans now also alter soils indirectly and in regions where humans are not present in large numbers or where no direct interest in the use of soils exists. In the light of the rapidly increasing global impact of human activity on soils and ecosystems, the following section aims at illustrating the main processes and drivers for historic and future patterns of SOC gains and losses through examples of anthropogenic activity, which have altered soils at large scales.

3.1 The role of human-induced climate change[5]

3.1.1 Temperature effects on soil organic carbon cycling

Temperature describes the amount of thermal energy present in a system, with higher temperatures typically leading to increased reaction rates of processes such as microbial decomposition of soil organic matter and greenhouse gas (GHG) fluxes to the atmosphere (Carey et al., 2016). The relationship between reaction rates and temperature is generally positive and nonlinear (e.g. exponentially increasing). However, not all warming results in increased C loss from soils. For example, warming temperatures during the winter in temperate ecosystems can melt the insulating snow layer and paradoxically result in soil freezing (Reinmann and Templer, 2016). Winter soil freezing can reduce SOC loss due to decreases in microbial biomass and exoenzyme activity (Sorensen et al., 2016). At high temperature extremes decomposition may also be slowed, either due to the denaturing of enzymes involved in decomposition (Schipper et al., 2014), microbial acclimation (Bradford et al., 2010; Bárcenas-Moreno et al., 2009), community shifts (DeAngelis et al., 2015), or interactions between temperature and soil moisture. At higher temperatures, surfaces tend to increase their evaporation, resulting in potential water stress, which can inhibit soil microbial activity (e.g. warming-induced drying decreases biological activity) (Hicks Pries et al., 2017). Contrastingly, in cold regions, warming can lead to higher water availability (e.g. water is less frozen throughout the year) and stimulate plant growth and microbial activity alike. This increased activity could potentially lead to higher C input to soils (see the following section for details).

Most of our hypotheses for how soils will respond to warming temperatures come from field warming experiments. These experiments typically measure total SOC stock over a period of 1–25 years, though most data from warming experiments are the result of manipulations of <5 years (Crowther et al., 2016; van Gestel et al., 2018). Crowther et al. (2016) found that the majority of soil warming experiments measured a loss in SOC stocks. However, a later study by van Gestel et al. (2018) found that when including more data points than in the original study, the SOC stock warming response was not significantly different from zero. Indeed, soil responses in field warming experiments are very heterogeneous. This may be partially accounted for by the difficulty of

[5] This section was written by Rose Abramoff and Sophie von Fromm.

measuring a relatively small change to a large pool, and by changes to plant inputs induced by warming. Field-scale warming experiments consistently report increased rates of C lost as CO_2 from the soil surface (Carey et al., 2016; Haaf et al., 2021), SOC losses from subsoil (Soong et al., 2021) but also increased plant inputs, which may offset to some extent these losses.

Soil warming may not affect all SOC pools equally. Historically and for the purpose of modeling, soils have been divided into conceptual pools based on their turnover time such as *Active*, *Slow*, and *Passive* (Parton et al., 1987). More recent conceptual and simulation models consider SOC pools that are measurable by fractionation or extraction (Abramoff et al., 2018; Robertson et al., 2019). Two common SOC pools measured by density fractionation are particulate organic matter (POM) that represents incompletely decomposed plant material, and mineral-associated organic matter (MAOM) that consists of organic matter that is complexed in some way with mineral surfaces (Kleber et al., 2021). Warming temperatures typically cause loss of total SOC in the absence of changes to plant inputs, yet the amount of C in some functional fractions of soil such as POM and/or MAOM may increase (Rocci et al., 2021). Recent studies have highlighted the differences in the vulnerability of POM and MAOM to warming (Rocci et al., 2021; Lugato et al., 2021; Soong et al., 2021), with POM generally found to be more vulnerable to warming than MAOM. For example, Lugato et al. (2021) used a space-for-time approach to infer the vulnerability of POM and MAOM to global warming, showing that estimates of POM and MAOM were lower in warmer than in colder locations. This inferred sensitivity was greater for POM than it was for MAOM, although there were differences across land use types (e.g. more absolute POM loss in forests, more absolute MAOM loss in croplands). Similarly, Soong et al. (2021) noted that there was a significant decrease in POM concentrations at 1 m depth after 3.5 years of soil warming in a conifer forest, while MAOM concentrations increased in the subsoil. The increase in MAOM concentrations may have been due to an increase in microbial activity and dissolved organic C (DOC) concentrations, which were also observed at this depth, and the more rapid and selective decomposition of the easily available POM relative to the total SOC stock.

While there is still little evidence available on the effects of temperature on POM and/or MAOM fractions at the field level, there are theoretical (Conant et al., 2011) and modeling (Sulman et al., 2018; Abramoff et al., 2019) studies that make different assumptions about the temperature sensitivity of MAOM (from negative to neutral to positive) depending on the bond type and whether the majority of bonds involved in the organo-mineral associations are expected to be endo- or exothermic. These studies find large potential variations in the modeled SOC stock as a result of these assumptions.

The large variation in SOC temperature response observed in field experiments and for different soil functional fractions underlines that temperature is not the only driver for the response of soils to warming, and it may not even be the main driver. Rather, soil responses to warming interact with the local vegetation, climate, geochemistry, and land use. Haaf et al. (2021) found that climate was not the dominant control over the temperature sensitivity of soil respiration, especially in temperate ecosystems. Rather, a wide variety of environmental characteristics explained the temperature sensitivity, such as soil solution chemistry, plant growth conditions, texture, clay, and weathering. In particular, the temperature sensitivity of organic compound decomposition depends on soil to resist microbial decomposition is depending on the mechanisms that control their stabilization with minerals, rather than their chemical composition (von Lützow and Kögel-Knabner, 2009; Moinet et al., 2018, 2020). However, climate also has a strong influence on the temperature sensitivity of decomposition in extreme environments, such as cold and dry regions. Thus, using large datasets to assess SOC turnover should always be combined with field and laboratory experiments aiming to disentangle relevant stabilization mechanisms. In doing so we will be able to better understand the major factors controlling temperature sensitivity of SOC turnover in different regions of the world, and their interaction with moisture availability as main controls of SOC sequestration and loss.

3.1.2 Moisture effects on soil organic carbon cycling

Water is the main driver of weathering, mobilization of nutrients, leaching, microbial activity, and redox potentials in soils. The effects of changes in soil water levels can depend on soil texture, porosity, bulk density, and drainage, and often effects are nonlinear and strongly interact with temperature (Xu et al., 2004). Many model representations of soil moisture and SOC loss assume that soil-water relations follow an optimum curve, with specific soil moisture values for which SOC loss is maximized (Moyano et al., 2013). At other moisture conditions, SOC loss from heterotrophic respiration may be lower, either due to anaerobic conditions at high moisture levels (Skopp et al., 1990), or a decrease in effective substrate availability due to low diffusion rates and pore connectivity under dry conditions (Kleber et al., 2021; Peyton Smith et al., 2017).

Experiments that manipulate soil water are often difficult to compare due to the numerous types of manipulations that are conducted (e.g. amount and frequency of watering). A recent review of more than 500 drought studies found that only a third of the studies explicitly reported how 'drought' was defined, and there were many different types of definitions, ranging from reduced rainfall, low water depth, plant stress, soil moisture, natural occurrence of the dry season, or a difference from ambient conditions (Slette et al., 2019). Nevertheless, field and laboratory experiments that manipulate soil moisture have provided many insights into the biological and chemical effects of soil water.

Drought and rewetting experiments are often observed to affect SOC cycling indirectly via changes to plant growth and subsequent C allocation belowground (Bredemeier et al., 1998; Ingrisch et al., 2020). However, direct effects of soil moisture conditions on SOC cycling have also been observed. For, example, dry conditions were shown to affect soil enzyme activity (Sanaullah et al., 2011) and directly suppress CO_2 emission from soils in seasonally dry climates such as the Mediterranean forests (Asensio et al., 2007; Hicks Pries et al., 2017) or well-drained temperate soils (Borken et al., 2006). However, in moist soils, drying and draining may increase the aerobic pore space in the soil, which may both increase CO_2 emissions and decrease CH_4 emissions, respectively increasing CH_4 uptake. This is commonly observed in both cool moist soils such as peatlands (Qiu et al., 2021; Huang et al., 2021) and in warm moist soils such as tropical soils (Wood and Silver, 2012; Cleveland et al., 2010). Therefore, the overall effect of the moisture regime on SOC sequestration depends on the balance of CO_2, CH_4, and N_2O emissions, SOC accumulation via plant inputs and microbial transformation, and DOC leaching.

3.1.3 Integrating biosphere responses to human-induced climate change

Anthropogenic GHG emissions are causing rapid increases in soil temperature and changes to precipitation regimes (IPCC, 2021). Under climate change, the effects of temperature and moisture on SOC cycling are becoming increasingly important, but the magnitude of their effects and internal feedbacks remain uncertain across different regions and scales (Wiesmeier et al., 2019). While temperatures are expected to warm globally, predicted changes in precipitation are much more variable. Both, the frequency and intensity of heavy precipitation and droughts have increased since the 1950s and are predicted to continue to increase in the future (IPCC, 2021). These indirect disturbances of ecosystems may interact with direct disturbances to soil such as those related to land cover change, which can alter local and regional temperature and moisture regimes.

While the effects of typical climatic trends on vegetation patterns and carbon cycling have been well-described, it is important to note that our rapidly changing climate has introduced unprecedented and unpredictable climatic events, of which the complex environmental consequences are just beginning to be examined. Intense shifts in precipitation have resulted in simultaneous flooding and droughts, with drastically fluctuating temperatures further exacerbating these changing conditions (Seneviratne et al., 2021). For example, although wildfires are considered typical environmental disturbances that have been recorded both historically and contemporarily, both the frequency, intensity, and even regions of fire occurrence have changed dramatically over the past decades (Westerling et al., 2006; Canadell et al., 2021). Such drastic changes are estimated to affect not only vegetation dynamics, but the global carbon cycle as well. For example, it was shown that the global average of soil C and N contents from fire-affected ecosystems decreased by 12.1% and 10.4%, respectively, with savanna grasslands and broadleaf forests showing the highest decreases (Pellegrini et al., 2018). Such significant changes are hypothesized to limit future plant growth and thus further exacerbate the carbon losses predicted from such events (Pellegrini et al., 2018). Furthermore, changes in fire regime have also been shown to change the actual structure and composition of vegetation species over time, with more herbaceous and grass species becoming dominant in fire-prone areas compared to woody species (Reich et al., 2001) and appearing in ecosystems where they were formerly less frequent or even unknown (Hu et al., 2015; Bowman et al., 2020). As forests have been shown to sequester more C than any other ecosystem type (IPCC, 2022), it is likely that reducing global forest cover in such a way will significantly reduce the capacity for terrestrial C storage as we know it.

In addition, legacy SOC stored in organic soils under boreal soil is likely to be affected by increasing fire return intervals thereby turning boreal forests from a small sink into a C source (Walker et al., 2020). One changing climatic factor undoubtedly contributing to these more frequent fire events is the increasingly widespread occurrence of prolonged droughts brought about by increasing temperatures and outcompeting evapotranspiration rates relative to water recharge (Westerling et al., 2006; IPCC, 2013). Not only are such extreme droughts becoming more frequent in the past decades, they are also more widely spread than previously recorded, a trend which is predicted to increase even further in the coming years (IPCC, 2013). Such changes in temperature and water availability have been shown to significantly shape plant species distribution at both local and regional scales through mechanisms such as niche differentiation (Engelbrecht et al., 2007) as well as through changes in the spatial distribution of soil nutrient availability (John et al., 2007). Together, these direct changes in both plant species composition as well as soil nutrient composition impact rates of organic matter decomposition, which in turn influence the amount of nutrients available for plant uptake, SOC sequestration onto soil minerals, and release into the atmosphere through microbial respiration (Davidson and Janssens, 2006).

However, one key factor which must not be overlooked is the strong change in precipitation patterns that have occurred in the past decades. For example, in addition to droughts, flooding has also become more frequent in recent years (Doocy et al., 2013). Such events can lead to soil erosion and landslides, which have a clear and direct impact on vegetation patterns and thus subsequently SOC dynamics. Yet even mild flooding can increase the soil water content to the extent that oxygen diffusion into the soil system is drastically reduced, inducing changes in the soil redox potential with effects on electron transfer and organo-mineral complexes (Keiluweit et al., 2017) and thus decreasing potential

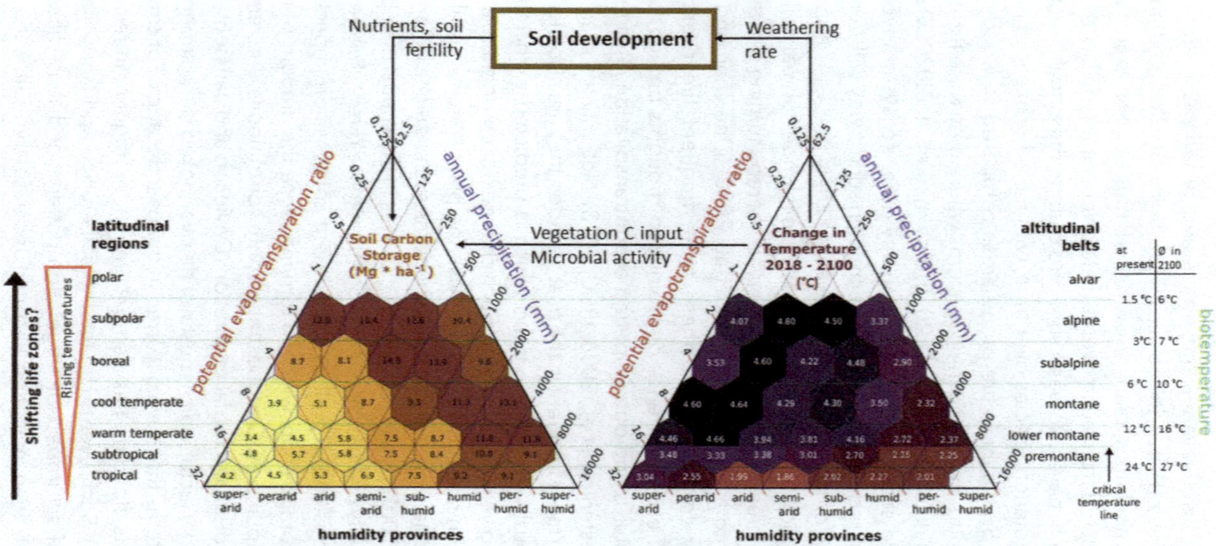

Figure 4 Holdridge life zones (HLZ) represented in hexagons for vegetated climatic regions of the world. Eight humidity provinces based on the mean annual precipitation are distinguished and plotted against annual mean precipitation and potential evapotranspiration leading to distinct bioclimatic (life) zones. Left panel: current SOC stocks following Jungkunst et al. (2021). Right panel: potential temperature increase in each HLZ according to the CMIP6 SSP3 RCP7.0 scenario following IPCC (2014). As a consequence of global warming, short- and long-term responses of biological processes (vegetation and soil microbial activity) as well as of soil weathering rates are to be expected with unknown consequences for the global SOC budget.

metabolic processes leading to a slowdown of decomposition rates (Davidson and Janssens, 2006). While it is estimated that such changes in climate will decrease land carbon uptake and SOC sequestration in future years, the exact magnitude of such changes is difficult to accurately predict due to the complex interconnected factors influencing these biogeochemical processes (Canadell et al., 2021). A way to illustrate the potential effects of changing climates on global SOC stocks and potential feedback to soil development among Holdridge life zones (Holdridge, 1947) is given below, modified following Jungkunst et al. (2021) (Fig. 4). In this study, Holdridge life zones system as a global bioclimatic scheme for the classification of land areas are used to map distinct soil and (climax) vegetation patterns as important factors for assessing SOC stocks.

3.2 The historic and modern role of global land use change on biogeochemical cycling and soil organic carbon sequestration[6]

3.2.1 Human impact on soil organic carbon dynamics through land use change and intensive agriculture

Among the active C reservoirs, especially soils and the living above- and belowground biomass show distinct regional variations in the amounts of C stored, with both components being highly vulnerable to human-induced disturbances. For example, boreal forests currently store about 559 Gt of C globally, of which 84% is found in soils, whereas tropical forests store about 428 Gt of C with 50% in soils (Janzen, 2004). On the lower end of the spectrum are cropland agro-ecosystems that contain about 131 Gt of C with more than 90% of the C found in soil (Janzen, 2004). Both soil and biomass store C mainly as organic C that can be lost directly to the atmosphere in the form of CO_2 or CH_4 or to aquatic ecosystems in various organic or inorganic forms. Such losses are accelerated and magnified by anthropogenically induced disturbances of intact terrestrial ecosystems by, for example, destroying, removing, or degrading the above ground vegetation and using the soil for agricultural production. The intensity and rate of such disturbances, expressed as area disturbed per unit time is thereby directly related to the increasing demand of a growing global population for food and other commodities. Over the last centuries, the human population expanded globally from about 253 million in 500 AD to about 7.3 billion in 2015 (Klein Goldewijk et al., 2017). This growth resulted in an intensified and increased rate of land use to provide food, feed, energy, fiber, construction materials, consumer products, space for infrastructure and urbanization, as well as resources and raw materials for industrial production.

[6] This section was written by Michael Kaiser, Gina Garland and Sarah van den Broek.

From 500 to 2015, the areas globally used as cropland and grassland (pasture and rangeland), for example, increased from 133 to 1591 million hectares and from 249 to 3241 million hectare (Klein Goldewijk et al., 2017; see also Fig. 5), respectively. This increase came mainly at the cost of primary or intact terrestrial ecosystems (Hurtt et al., 2020).

Studies dealing with temporal aspects of global land use change patterns consider the majority of terrestrial ecosystems to prevail in their primary undisturbed conditions until 1800–1850 (e.g. Sanderman et al., 2017; Friedlingstein et al., 2020) suggesting that the majority of SOC lost due to system transformations appeared in the last 150–200 years. Based on Hurtt et al. (2020), global trends in the distribution of major land use classes between 850 and 2015 can be summarized as a decrease in the areas covered by primary non-forest and primary forest and an increase in the areas covered by secondary forest as well as managed pasture and rangeland, crop land (annual

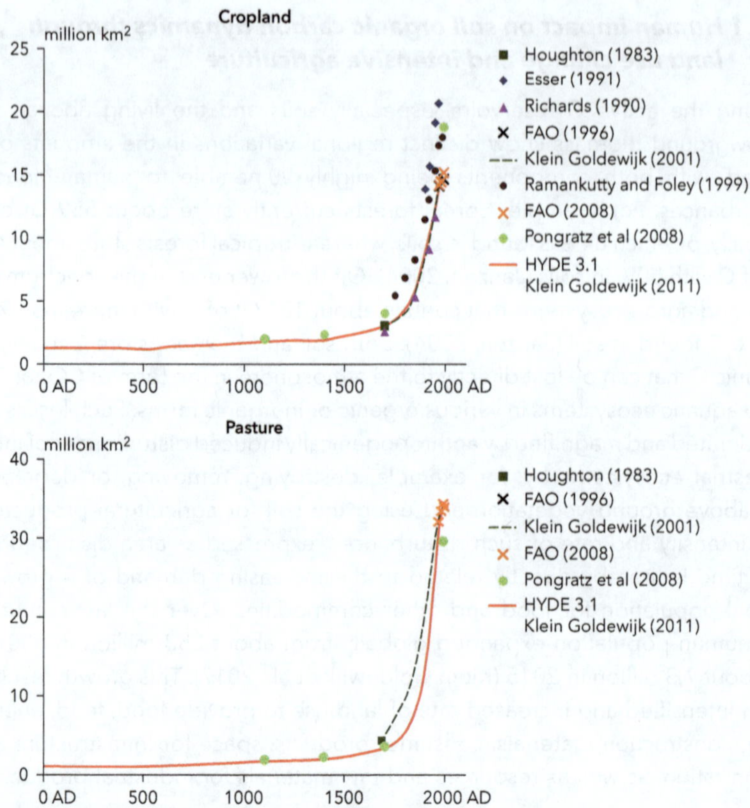

Figure 5 Overview of historical changes in global agricultural land area, divided in cropland and pastures. Published in Klein Goldewijk et al. (2011).

C3 and C4, perennial C3 and C4, C3 legumes), and urban land. The percentage of agricultural land (crop land and grazing-pasture/rangeland) increased from 2.9% of the total land area in 500 to 37.1% in 2015 (Klein Goldewijk et al., 2017). For estimating historical and actual losses of C from terrestrial ecosystems to the atmosphere due to land use change, both compartments biomass-C (living and dead) and SOC need to be considered.

As summarized by Friedlingstein et al. (2020; see also Fig. 6), land use change-induced CO_2 emissions and uptake fluxes are calculated based on

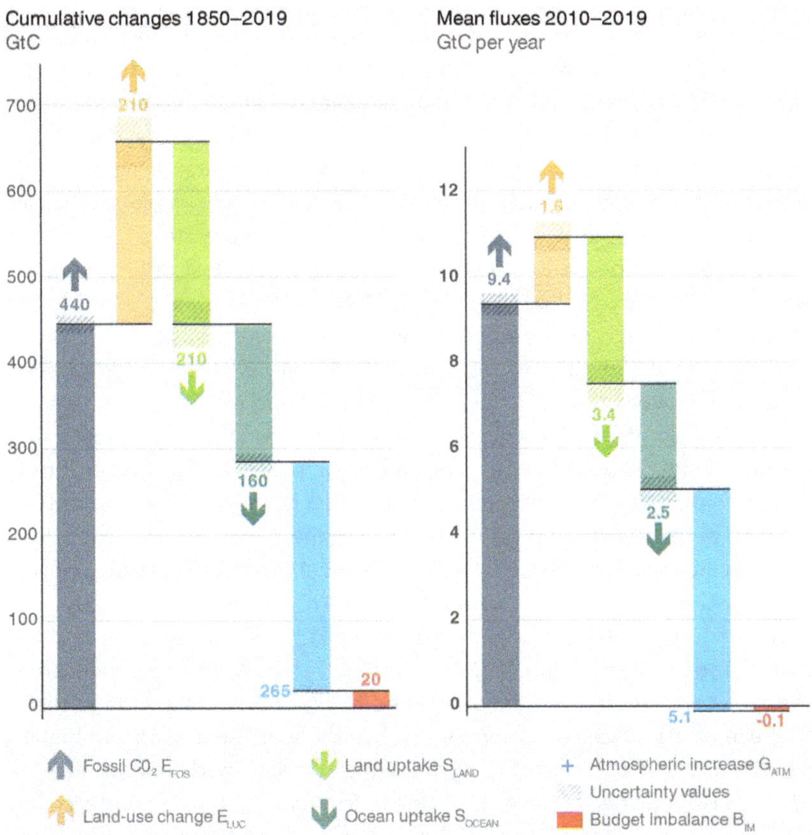

Figure 6 Cumulative changes during 1850-2019 and mean fluxes during 2010-2019 for the atmospheric perturbation as defined in the legend. The components of the CO_2 budget that are reported annually include the following separate estimates for the CO_2 emissions: (1) fossil fuel combustion and oxidation from all energy and industrial processes, also including cement production and carbonation (*E*FOS); (2) the emissions resulting from deliberate human activities on land, including those leading to land-use change (*E*LUC); (3) their partitioning among the growth rate of atmospheric CO_2 concentration (*G*ATM; Gt C year[1]); (4) the sink of CO_2 in the ocean (*S*OCEAN); and (5) the sink of CO_2 on land (*S*LAND). Published in Friedlingstein et al. (2020).

bookkeeping models, which go back to the original bookkeeping approach of Houghton (2003) that includes C stored in vegetation and soils before and after land use change expressed as the transitions between various natural vegetation types to cropland and pasture. Land use change related CO_2 emissions have also been estimated using dynamic global vegetation models (DGVMs) that account for deforestation and regrowth but do not represent all processes resulting directly from human activities on land such as peat fires and fires as a management tool, N fertilization, or irrigation (Friedlingstein et al., 2020). Based on calculated yearly emissions, Gasser et al. (2020) estimated a cumulative total loss of 206 ± 57 Pg C due to land cover change for the period from 1750 to 2018. A comparison to the cumulative losses of 235 ± 75 Pg C for the period of 1750–2018 and of 205 ± 60 Pg C for the period of 1850–2018 as calculated by Friedlingstein et al. (2020) highlights that the majority of C losses resulting from land use change took place over the last 150 years. The land use change over the last 250 years therefore transformed vegetation types that can provide a large carbon sink per area unit (i.e. forests, native grasslands) into less efficient land use types such as cropland or degraded grassland. Gasser et al. (2020) accounted for a land use change-induced global 'loss of additional sink capacity' of 0.68 ± 0.57 Pg C per year (2009–2018) and 32 ± 23 Pg C in total (1750–2018), which is excluded from the calculated emissions.

Gasser et al. (2020) provided a breakdown into different model based C pools, which revealed mean annual C fluxes from and to the atmosphere over the last decade of −3.66 ± 0.96 Pg C year^{1} for vegetation C due to biomass regrowth, 2.84 ± 0.85 Pg C year^{1} for SOC mainly derived from heterotrophic respiration of SOC and dead biomass, and 2.18 ± 0.65 Pg C year^{1} emitted from oxidation of harvest wood products (HWP). For the period of 1750–2018, these pool-specific annual fluxes summed to −443 ± 155 Pg C for vegetation, 373 ± 137 Pg C for soil, and 276 ± 84 Pg C for HWP. Using a different approach, Sanderman et al. (2017) modeled the loss of SOC due to land use and land cover change during the last 12 000 years (Fig. 7) based on a machine learning model that was fitted using a global compilation of SOC data and the History Database of the Global Environment (HYDE) for land use data in combination with climatic, landform, and lithology covariates. In their study, Sanderman et al. (2017, 2018) reported a lower C emission from soil due to land use change than previously assumed. The model results indicated a cumulative loss of SOC with about 116 Pg C from the top 2 m due to agricultural activities over the last 12 000 years with a strongly increased loss rate over the last 200 years. The modeled data thereby compared well with global datasets from paired sites where each site had one plot with undisturbed native vegetation and one adjacent plot under agricultural land use but with similar soil conditions for both plots per specific site (Sanderman et al., 2017). The authors showed that regions with the highest losses of SOC are often associated with major

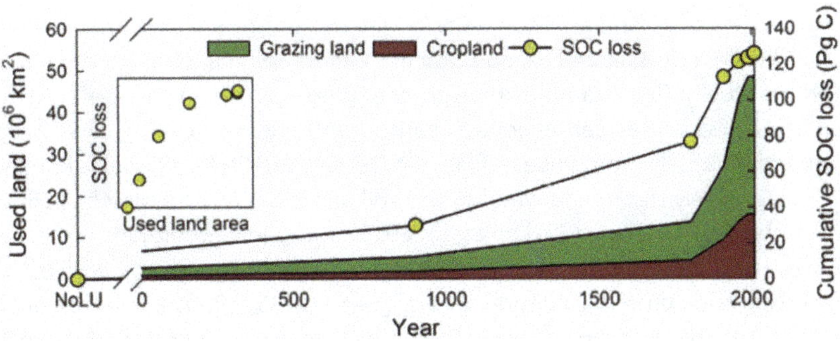

Figure 7 Historic reconstruction of loss in soil organic C (SOC) relative to 10 000 BC (assumed no land use [NoLU]). Temporal evolution of cropland and grazing land is given in stacked area plots. (Inset) Biplot of SOC loss (Pg C) v. total used land area (106 km^2) for each predicted time interval. Published in Sanderman et al. (2017).

cropping regions of intensified agriculture and degraded grazing lands. These regions are also the most likely ones to respond positively with increased SOC sequestration when managed differently (Paustian et al., 2019), which should in turn be targeted first for restoration efforts rebuilding SOC stocks as a means to develop strategies for climate change mitigation by using well-known sustainable and emerging technologies and practices.

3.2.2 Impacts of agricultural management practices on soil microbial communities and plant productivity

Soil fauna and microbial communities transform nutrients (Van Groenigen et al., 2015), provide plant nutrition (Bossolani et al., 2020) and maintain soil fertility by contribute to soil structure (Lehmann and Kleber, 2015) and cycling carbon and nitrogen (Cotrufo et al., 2013; Schmidt et al., 2019). Beneficial bacteria in these communities can inhibit plant pathogens (De Tender et al., 2019; Lupatini et al., 2017), promote plant growth (Aasfar et al., 2021), protect against plant parasitic weeds (Masteling et al., 2019), and help to suppress disease in soils (Jayaraman et al., 2021). Changes in microbial activity affect plant nutrient uptake (Leite et al., 2017) but also the interactions between soil microbes that determine microbial community functioning (Cordero and Datta, 2016), affecting nutrient turnover and plant productivity (Andreote and Pereira e Silva, 2017).

However, the increasing demand for food security due to a rising world population leads to an intensification of agriculture to improve plant productivity. Farmers often reach to inorganic fertilization because it is readily available and easy to apply. Although the number of organic farms is on the rise, the majority of agriculture is conventional with continuous cropping systems (Willer et al., 2021). Continous cropping increases pest prevalence

(Thomine et al., 2022) and often relies on the use of pesticides (Cernava et al., 2019). While inorganic fertilization practices affect soil microbial communities more indirectly (for example through changing soil pH) (Pan et al., 2014), the use of pesticides can affect soil carbon and nitrogen cycling directly by altering microbial communities (Ullah and Dijkstra, 2019). Additionally, farming practices largely determine whether the soil and plant microbiome benefit or harm plant productivity (Bakker et al., 2012). Inorganic fertilization practices hinder the plants' ability to recruit beneficial microorganisms (Reid et al., 2021) which increase plant productivity (Jones et al., 2019; Schulz-Bohm et al., 2018). Contrarily, using organic amendments increases soil microbial diversity (Hartmann et al., 2015) and can increase the abundance of plant growth promoting bacteria (de Carvalho et al., 2021).

While intensification of agricultural systems has undoubtedly contributed to the vast increases in worldwide crop production over the past decades, it has unfortunately also furthered numerous negative environmental impacts such as GHG emissions, eutrophication of waterways, and reduction of biodiversity (Tilman et al., 2011). For example, the use of crop protection products (i.e. fungicides, herbicides, and insecticides) has increased exponentially in the past decades and it is currently estimated that four million tons of pesticides are applied to agricultural systems globally (FAO Stat). As only a fraction of these chemicals actually reach the target plants, this practice is known to cause significant direct and indirect pollution of air, water, and soils (Sharma et al., 2019). As soils are known to harbor some of the most diverse communities of microorganisms on earth (Bardgett and van der Putten, 2014), such pollution has potentially severe consequences for soil biodiversity. For example, application of pesticides has been shown to reduce the diversity and activity of certain microbial populations (Bünemann et al., 2006; Al-Ani et al., 2019), which in turn impacts soil nutrient and carbon cycling (Pelosi et al., 2014), and even can inhibit nitrogen-fixing and phosphorus-solubilizing bacteria (Hussain et al., 2009). They may thus affect soil health and ultimately also SOC sequestration. Even more alarming, pesticide residues have been found in soils of organic agricultural systems (Riedo et al., 2021), showing the strong persistence of such chemicals in the environment even after decades of discontinued use.

Similarly, fertilization of agricultural systems has also increased greatly in the past decades, from as estimated 11.3 Tg N year^{-1} in 1961 to 107.6 Tg N year^{-1} in 2013 (Lu and Tian, 2017). Such inputs have also been shown to impact soil microbial communities, although the general trends are difficult to predict as it seems to depend on the fertilizer type (i.e. organic or mineral), specific amount applied, duration of application, soil type, climate, and cropping system (Geisseler and Scow, 2014). For example, while some studies have shown that increasing N fertilization decreases soil microbial diversity and biomass (Lu et al., 2011), others have shown that the application of mineral fertilizers actually increases soil microbial biomass and SOC content (Geisseler and Scow, 2014; Geisseler et al., 2017). Similarly, while it has been shown that organic fertilization can increase SOC stocks, it may also stimulate soil microbial communities responsible for N_2O emissions (Lazcano et al., 2021), and thus site-specific best management practices must be determined prior to fertilization regime in order to balance tradeoffs between increased crop production and C storage with other negative environmental impacts.

Soil emissions strongly depend on the type of farming practices, because some practices increase these emissions (Lourenço et al., 2021), while others mitigate them (Shu et al., 2021). The soil microbial community structure can help predict these GHG emissions (Morales et al., 2010). Thus, increased understanding of the soil microbial communities helps to lay hold of how changes in these communities affect SOC sequestration and release (de Bruijn et al., 2015; Fierer, 2017). It is essential to increase our research efforts to unravel how novel fertilization practices affect soil emissions and plant productivity to increase agricultural sustainability while increasing or maintaining the SOC pool.

3.3 The consequences and dimension of lateral soil (and soil organic carbon) movement through human activity[7]

3.3.1 Human-induced redistribution of soil within landscapes

What is often ignored in assessments of land cover and land use change-induced alterations of the C cycle and SOC stocks (Fig. 7; Section 2.2) is the fact that considerable amounts of soil and SOC are redistributed laterally across landscape units. The lateral movement of soil driven by water and wind erosion is a natural process that can drastically alter biogeochemical cycling in the plant-soil-atmosphere continuum. Soil redistribution is in part responsible for the small-scale variability of soil properties, often following typical patterns along slope profiles (Fig. 8). Naturally occurring erosion is most pronounced in environments with minimal soil cover, e.g. arid to semiarid areas, higher altitudes or very steep slopes limiting plant growth, and those exposed to high wind speeds or heavy rainfall events in combination with steep slopes. However, human activities substantially accelerated erosion processes and hence lateral soil movement. By far the most important change in land use affecting soil erosion is a shift toward arable use, as soil tillage introduces periods of more or less bare soils. Since the beginning of soil cultivation during the Neolithic Revolution (Weisdorf, 2005), not only water and wind erosion dramatically

[7] This section was written by Peter Fiener.

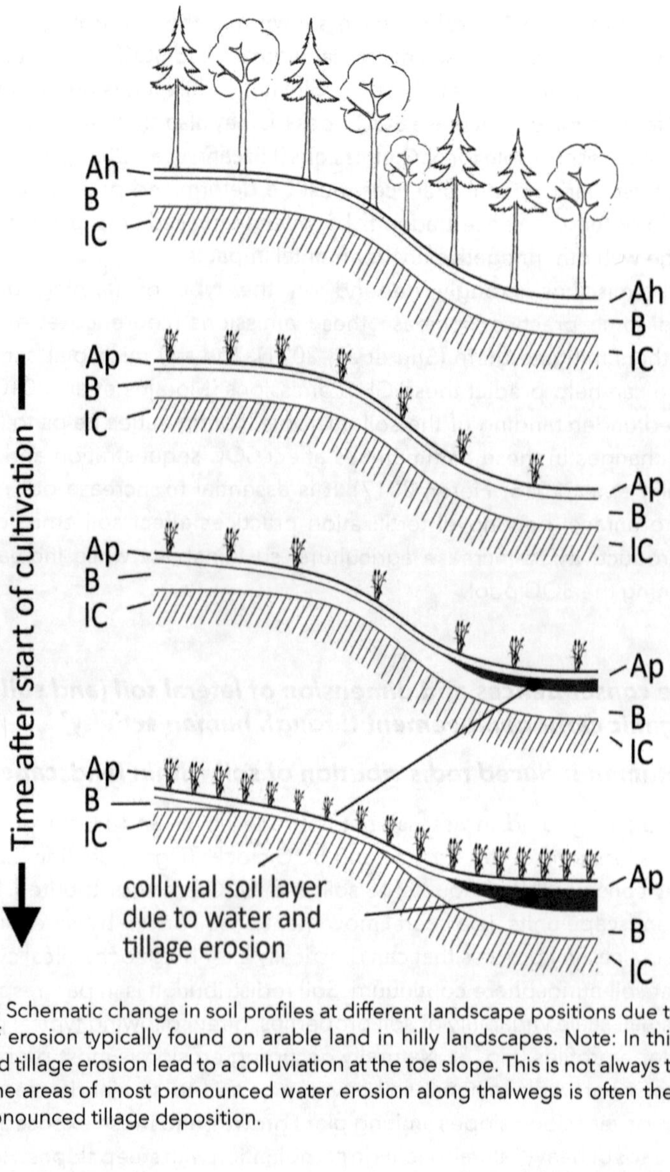

Figure 8 Schematic change in soil profiles at different landscape positions due to a long phase of erosion typically found on arable land in hilly landscapes. Note: In this figure, water and tillage erosion lead to a colluviation at the toe slope. This is not always the case, as e.g. the areas of most pronounced water erosion along thalwegs is often the area of most pronounced tillage deposition.

increased, but also other erosion processes came into play. Namely, tillage erosion, moving soils from slope convexities to concavities (Govers et al., 1994), and harvest erosion, removing soil attached to root crops (Ruysschaert et al., 2007). Furthermore, anthropogenic modifications of landscapes are often associated with forest burning and/or clearing for different purposes (hunting, cultivation, etc.), and declines in soil cover for increasingly large areas (Kaplan et al., 2009). In consequence, today large areas are globally affected by erosion processes, which under natural conditions would hardly be affected at all. Therefore, in order to prevent organic matter loss and enhance SOC sequestration, one of the principal measures is to reduce soil loss through erosion by establishing permanent soil cover (see, for example, Chenu et al., 2019).

Recent global estimates regarding the importance of different erosion types suggest that water erosion is the dominant process, while tillage erosion moves roughly a fifth of the amount of mass compared to water erosion, but over twice the amount moved by wind erosion (Quinton et al., 2010). Whether water and wind erosion increased or decreased from historic to modern agriculture is open to debate. Estimates of global soil erosion rates under recent land use are challenging and associated with large uncertainties (Table 1), and our knowledge regarding the long-term dimension of global lateral soil redistribution and loss and its effects on soils and their SOC sequestration potential is very limited, and mostly based on small to regional scale case studies. Especially if soil cultivation takes place for centuries or even millennia, detailed, spatially distributed data regarding land use and management over such long periods of time are hardly available.

Table 1 Estimates of global soil erosion

Global soil erosion (Pg year^{1})	Method	Processes	Source
74	Extrapolated for plot measurements	Interill and rill erosion (no deposition)	Pimentel et al. (1995)
120			Hooke (2000)
24-65	Estimated from sediment fluxes in large rivers and storage in lakes and reservoirs	All erosion and deposition processes are estimated from sediment delivery	Stallard (1998)
75			Lal (2003)
60			Wilkinson and McElroy (2007)
75			Berhe et al. (2007)
81	Modeling with universal soil loss equation-based model types (only Van Oost et al. (2007) include tillage erosion modeling)	Interill and rill erosion (no deposition)	Yang et al. (2003)
140			Ito (2007)
20.5 ± 10.3			Doetterl et al. (2012)
47.6 ± 10			Naipal et al. (2018)
34.9			Guerra et al. (2020)
43 ± 9/7			Borrelli et al. (2020)
30-40		Interill and rill erosion (no deposition) and tillage erosion	Van Oost et al. (2007)

Depending on literature information, ranges or variability of means (standard deviation) are given. Note: all plot and modeling approaches focus on human-induced erosion (mostly on arable land), while the estimates from sediment fluxes in rivers and lakes and reservoirs also include natural erosion.

Clearly, tillage erosion increased since cultivation started as rates of soil movement from hand tools and plows pulled by animals are much lower than those associated with mechanized agriculture and these rates have especially accelerated in recent decades as agriculture has intensified and machinery has increased in size and power (Schjønning et al., 2015). In consequence, the following evaluation of human-induced lateral soil movement is mostly dedicated to water and tillage erosion. In this context, it is important to note that tillage erosion solely redistributes soils within arable land, while in the case of water erosion, soil (and associated C) is at least partly moved into other ecosystems, mostly inland waters.

3.3.2 Effects of lateral soil redistribution on soil organic carbon cycling

In (agricultural) landscapes with severe erosion and soil degradation, soil profiles exhibit a large variability in soil properties (Fig. 9) resulting from the removal of topsoil at erosional sites and burial of these soils at depositional sites. With the loss and gain of soil at different landscape positions, large amounts of SOC are also redistributed (Van Oost et al., 2007; Doetterl et al., 2016). However, the effects of soil redistribution on soil C cycling are complex, as redistribution may also affect SOC sequestration and its mineralization through their impact on soil physical properties and microbial communities. Effects can change with (1) severity of erosion, (2) the time since cultivation began, or (3) the amount of soil that is deposited and buried in colluvial and alluvial valleys and floodplains (Wang et al., 2017; Schrumpf et al., 2021).

Briefly, soil redistribution with regard to SOC can be structured into several compartments. First, at the beginning of erosional processes (see 1A and 3A in Fig. 9), the most important process affecting landscape scale C balance is the transport of SOC-rich topsoil from slopes to valleys and the partial SOC loss to adjacent ecosystems, mostly inland waters. In this context, it is important to note that SOC is often preferentially transported and hence lost to other ecosystems since it is tied to smaller, easily transportable particles (clay sized) or transported as particulate or dissolved organic matter of low density. This enrichment of SOC in general and in particular its labile and low density compounds in sediments was shown numerous times in many experimental studies (Jacinthe et al., 2004; Rumpel et al., 2006; Schiettecatte et al., 2008b; Juarez et al., 2011), However, it is difficult to generalize for larger areas (e.g. catchments), where soil redistribution takes place through a wide mix of selective and nonselective erosion and deposition processes (Kuhn et al., 2012; Schiettecatte et al., 2008a). Second, during erosional transport and shortly after deposition of soil, mineralization of SOC might occur following aggregate breakdown (Six et al., 2000a, 2001) (see 2A and 2B in Fig. 9). The breakdown of aggregates and the

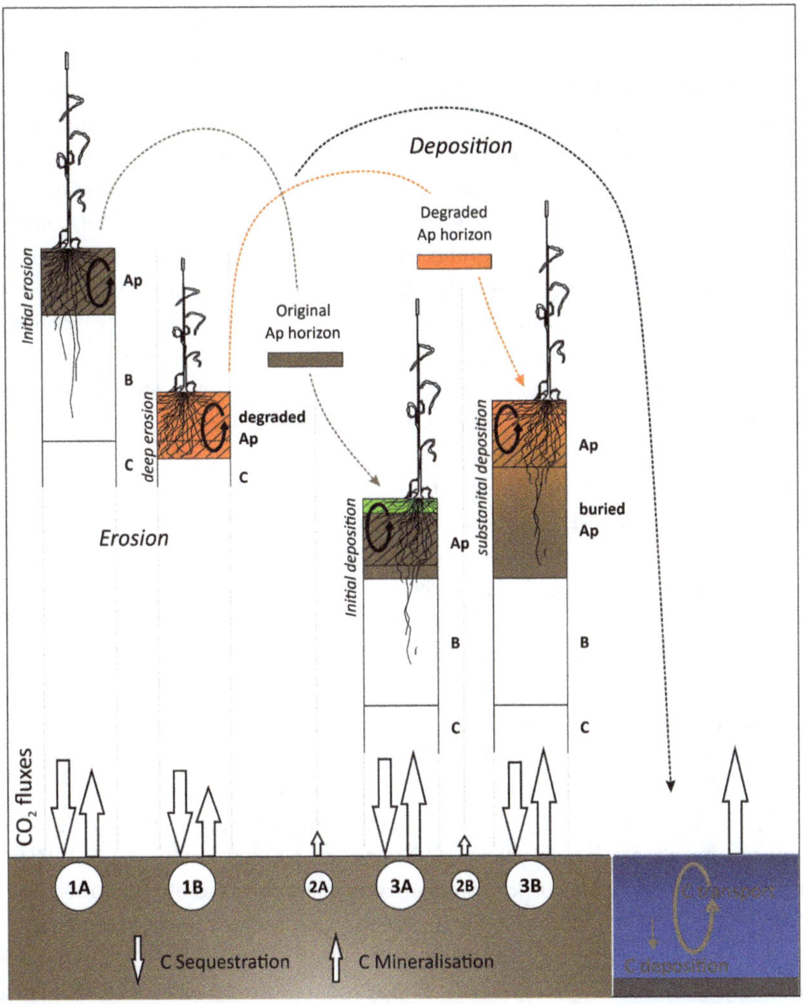

Figure 9 Schematic overview of soil and C redistribution and its effect on C sequestration and mineralization at different landscape positions. 1A-B and 3A-B refer to fluxes in case of different erosion and deposition intensity. 2A-B refers to C mineralization during transport.

associated CO_2 effluxes from mobilized SOC hereby depends predominantly on the kinetic energy of raindrop impact. Several studies have demonstrated this relationship and quantified an increase in SOC mineralization under laboratory (Jacinthe et al., 2004; Polyakov and Lal, 2008) or field conditions (Van Hemelryck et al., 2011). However, the increase in CO_2 fluxes is short lived and only recognizable for several days after an erosion event (Bremenfeld et al., 2013; Jacinthe et al., 2002; Van Hemelryck et al., 2010). Third, with increasing changes in soil horizonation other processes than the above are becoming increasingly important for erosion-induced changes of SOC cycling (see 1B and 3B in Fig. 9). At eroding landscape positions, two contrasting processes need to be considered: (1) plant productivity and hence C input tends to decrease if soils are getting thinner due to prolonged erosion (Öttl et al., 2021), because degraded soils often show a reduced water storage capacity and lower nutrient availability. Soil surface removal experiments show that the effect of soil loss upon biomass production is significantly more pronounced in case of zero or low fertilizer inputs (Bakker et al., 2004). (2) The eroded SOC is replaced by new C inputs via new photosynthates, which was first termed 'dynamic replacement' by Harden et al. (1999). This replacement is driven by a number of processes: surface soil at eroded sites is mixed with subsurface soil via tillage. As long as C inputs from agricultural crops are not substantially reduced, SOC can be replaced to reach the previous equilibrium between C inputs and outputs. Additionally, chemical weathering of former subsoils may lead to the formation of new organo-mineral complexes when minerals stabilize newly sequestered SOC at reactive surfaces that are not yet saturated with sorbed SOC (Harden et al., 1999; Van Oost et al., 2007; Remus et al., 2018;). These processes are very site specific, as they depend strongly on the biogeochemical properties of mineral soil and the environmental conditions under which weathering occurs.

At depositional landscape positions (Fig. 9, see 3B), soil-eroded upslope accumulates leading to the burial of transported and in-situ sequestered SOC. At the same time, deposition of fertile, former topsoil can increase plant productivity at depositional landscape positions (Bouchoms et al., 2018; Öttl et al., 2021). Overall our knowledge regarding the scale of SOC burial in depositional sites is still very limited (Rumpel and Kögel-Knabner, 2011; Wang et al., 2017) as well as our knowledge regarding the stability of buried SOC (Zhang et al., 2021) and its stability under changing climatic and environmental conditions (Soong et al., 2021; Zosso et al., 2021). On average, it is estimated that only between 10% and 30% of topsoil material redistributed due to agricultural (water) erosion is subsequently transported into lakes and oceans via major rivers (de Vente et al., 2007; Walling, 1996). The rest is deposited in different soil environments, ranging from footslope and colluvial deposits to alluvial flood plains. In conclusion, the effects of soil redistribution can lead to both SOC losses and SOC gains depending on the overall ability of soils affected by erosion to sustain plant growth, and the amount of SOC buried over time in depositional sites. The efficiency at which SOC can be conserved at depositional sites largely depends on the rate of burial, the time since burial, the quality of the buried SOC and very importantly the environmental conditions, where SOC is buried (Doetterl et al., 2016). In well-aerated colluvial soils, often strong depth gradients in SOC concentrations can be found with lower SOC concentrations in the deepest and therefore oldest layers, whereas in alluvial soils under oxygen limited conditions soils can be stored for centuries without substantial changes (Hoffmann et al., 2009, 2013a,b; Steger et al., 2019). River regulation and the associated drainage of alluvial plains can pose therefore a serious risk of increased mineralization of buried SOC when soils become increasingly aerated (e.g. for the areas, exposing alluvial soil SOC to better aerated conditions).

3.3.3 Global consequences and dimensions of lateral soil redistribution on soil organic carbon cycling

The approaches used to estimate the global effects of erosion on SOC cycling can be roughly subdivided in two subgroups: (1) soil-centered approaches upscaling results from plot to small catchment scale while using parsimonious modeling approaches (Naipal et al., 2018; Van Oost et al., 2007; Wang et al., 2017) and (2) sediment-centered approaches estimating the importance of soil redistribution on SOC cycling from sediment in globally important river systems (Smith et al., 2001; Lal, 2001). While the first appropriately address the catchment internal erosion and deposition and its effect on SOC fluxes, the latter integrate more processes such as erosion, deposition, transport, and mineralization on land as well as mineralization and burial in inland waters, but lack detail in catchment internal dynamics. Thus, at the local to regional scale, the SOC sink or source effect of this redistribution is well determined. However, we still lack data, methods, and process understanding to appropriately analyze the global C cycle effect of SOC, which is moving through landscapes into oceans. Global estimates of the dimensions of lateral soil redistribution on SOC cycling are challenging and associated with large uncertainty, as briefly illustrated while describing the different processes involved. A recent study in Australia (Chappell et al., 2016) indicates that SOC loss from arable land via erosion is often incorrectly attributed to a loss via soil respiration. In the case of the Australian National Greenhouse Gas inventory, this incorrect attribution leads to an overestimation of the net C flux from cropland by up to 40% and an overestimation of the potential (100 year) C sink (if erosion is reduced) by up to 17% (Chappell et al., 2016). Overall, common approaches used to estimate the global effects of soil erosion on SOC cycling result in contradicting estimates, which were widely discussed within the last two decades (for reviews, see Doetterl et al., 2016; Kirkels et al., 2014). These global estimates range roughly from a net source of 1 Pg SOC year^{-1} to a net sink of 1 Pg SOC year^{-1}, whereas later studies indicate more moderate effects, mostly assuming a slight global C sink introduced via erosion processes (Van Oost et al., 2007; Quinton et al., 2010) (Fig. 10).

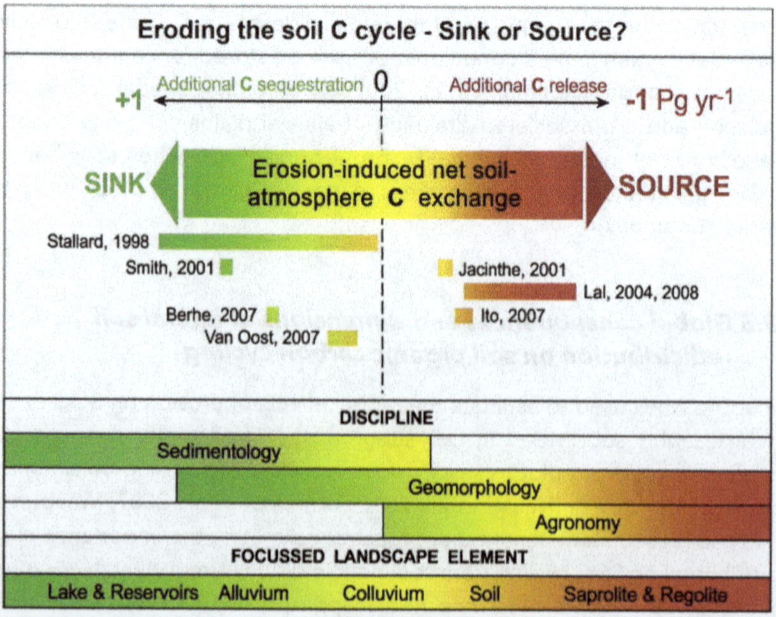

Figure 10 Net effect of soil redistribution on SOC cycling as determined by current global-scale studies. Colors and position along the X-axis show the relationship between the disciplines and landscape elements focused on in these studies. Republished from Doetterl et al. (2016).

3.4 The soil organic carbon sequestration (and loss) potential of human-modified landscapes and land use systems[8]

3.4.1 Recovery of soil organic carbon stocks at global scales

The literature on the connection between human management and SOC provides global estimates on the average efficiency of sequestering SOC by different management, mostly by meta-analyses. Present global SOC stock estimates help to put sequestration into perspective. The amount of SOC stored until 2 m soil depth was estimated as 2400 Pg C (Batjes, 1996; Minasny et al., 2017). With a global land surface area of 149 million km^2, an average of about 100 t SOC ha^{-1} can be found in the first meter of soil. Tang et al. (2019) reported in their meta-analysis that the global average loss of SOC stocks to 60 cm depth after conversion of grassland to arable land was about 18% of initial stocks. Wei et al. (2014) reported globally differentiated SOC losses following conversion of forest to agricultural land: about 50% for temperate forest, 40% for tropical and 30% for boreal forests. Historical losses in SOC have been highest, where land was under cultivation for many centuries. Global estimates state that in

[8] This section was written by Moritz Laub and Sebastian Doetterl.

cultivated soils overall, between 40% and 60% of initial SOC stocks have been lost due to agriculture (Lal, 2004; Wei et al., 2014). These historic losses are the theoretical maximum potential for SOC sequestration when soils are put out of management to approach their natural state. In practice, land available to be given up for this purpose is strongly constrained by the human need for food and biomass. The fact that gains in SOC occur much slower than losses (Attard et al., 2016) also limits the sequestration potential in the near future, indicating that maintenance of intact ecosystems should be a priority. However, despite many uncertainties and limitations (e.g. Poulton et al., 2018), improved management of soils by, for example, reducing SOC losses through measures limiting erosion, reducing tilling depths and frequencies, the use of deep rooting or perennial crops as well as the active addition of stable organic C compounds to soils (biochar) are discussed as possible ways for soils to remain under agricultural land use while continuously sequestering SOC (Minasny et al., 2017; Paustian et al., 2019; Chapters 16–24 of this book).

Despite the SOC sequestration potential of (agricultural) soils, rebuilding historic stocks is tedious and the effects of improved management practices on the reversal some of these losses at decadal time scales are uncertain. For example, soil tillage disrupts soil aggregates and speeds up soil C turnover (Six et al., 2000a,b). The difference between plowing and a no-till management has been estimated to be about 0–0.2 t C ha^1 year^1 under humid climate conditions (Six et al., 2004; Virto et al., 2012), and very little potential for additional SOC sequestration in dry conditions. Exception may be highly weathered soils in tropical regions (Chapter 22 of this book). In addition, it was noted that due to an increase in N_2O emissions in the first 10 years, a net benefit for the reduction of GHG-related global warming potential would only be possible if no-till is maintained for at least two decades. Similarly, mineral N application on previously unfertilized agricultural land also shows potential to increase SOC stocks between 5% and 10%, mostly by increasing plant biomass inputs into the soil (Han et al., 2016; Yue et al., 2016). Yet, to fully assess the influence of such measures on global warming, tradeoffs such as increased N_2O emissions must be accounted for (Chapter 6 of this book), at best normalized per unit of crop yield (Clark and Tilman, 2017). Others have provided estimates of global average SOC sequestration for a number of management practices (Freibauer et al., 2004; Smith, 2004) such as no-tillage, 0.4 t C ha^1 year^1; application of animal manure, 0.4 t C ha^1 year^1; planting of permanent or deep rooting crops, 0.6 t C ha^1 year^1; extensification or organic farming, 0.5 t C ha^1 year^1, and conversion of cropland to woodland, 0.6 t C ha^1 year^1; or to grassland, 1.5 t C ha^1 year^1. As part of the 4 per 1000 initiative, Minasny et al. (2017) estimated that a targeted annual sequestration of 0.4% of current stocks would be possible for a few decades on agricultural soils, offsetting 20–35% of global GHG emissions. They assumed the largest potential for SOC accumulation for afforestation

(+0.6 t C ha^1 year^1), pasture implementation (+0.5 t C ha^1 year^1), residue incorporation (+0.35 t C ha^1 year^1), and reduced tillage (+0.3 t C ha^1 year^1) but acknowledge that initial SOC stocks determine the locally achievable potential. Also in the context of 4 per 1000, Poulton et al. (2018) compared several long-term experiments in United Kingdom and concluded that by adding about 3 t C ha^1 year^1 of manure, it is theoretically possible to sequester between 0.1 and 0.4 t C ha^1 year^1, with a decrease in sequestration rates to zero after around 150 years. At the same time, several practical limitations such as sourcing such high amounts of manure, economic and food security tradeoffs arise that need to be taken into consideration when optimizing land use for SOC sequestration.

Similar to these examples from temperate regions, small effects of higher C inputs to soil to increase SOC stocks can also be confirmed in tropical environments, where turnover of soil C is high due to higher temperatures and more heavily weathered soils with limited capacity to stabilize C by association with minerals (Doetterl et al., 2021, Chapter 22 of this book). In recent studies, the strongest influences of improved management on SOC stocks were related to the amount of C inputs (0.45 ± 0.14 t C ha^1 year^1), while type of management practices (0.32 ± 0.06 t C ha^1 year^1), such as reduced tillage and residue retention, was less important (Fujisaki et al., 2018). Fujisaki et al. (2018) also reported that on average, only 8% of carbon inputs into the soil were translated into SOC. Another meta-analysis in the sub-Saharan Africa context looking at relative SOC sequestration rates, found that the implementation of agroforestry or conservation agriculture applying a combination of reduced tillage, mulching and crop rotation/intercropping could help to achieve the goal of the 4 per 1000 initiative of an annual rate of 0.4% C increase of SOC stocks (Corbeels et al., 2019).

It is also important to note that the land use history of a site determines its future potential to sequester SOC by improved management. For example, if cultivation of a site has started relatively recently, with the status of soils still being similar to their natural equivalent, the introduction of an improved management scheme may slow SOC losses, but not sequester SOC. In contrast, if a site has been under long-term cultivation and is close to reach the new steady state for low SOC input, improved management shows potential to sequester SOC (Fig. 11). This concept also shows why heavily eroded soils may sequester SOC (Wang et al., 2017), as SOC stocks of the new topsoil are below the level of steady state. In the long run, regardless of the current state of soil, a specific balance between C inputs and outputs to soil after agricultural land use has stopped should, in theory, lead to the same levels of SOC before conversion, if soil fertility (and the resulting plant productivity) remains similar. Yet, the timescale to reach this is expected to be in the order of centuries to millennia and is thus several magnitudes slower than the loss of SOC. Furthermore, many soils under agricultural land use are formed from parent

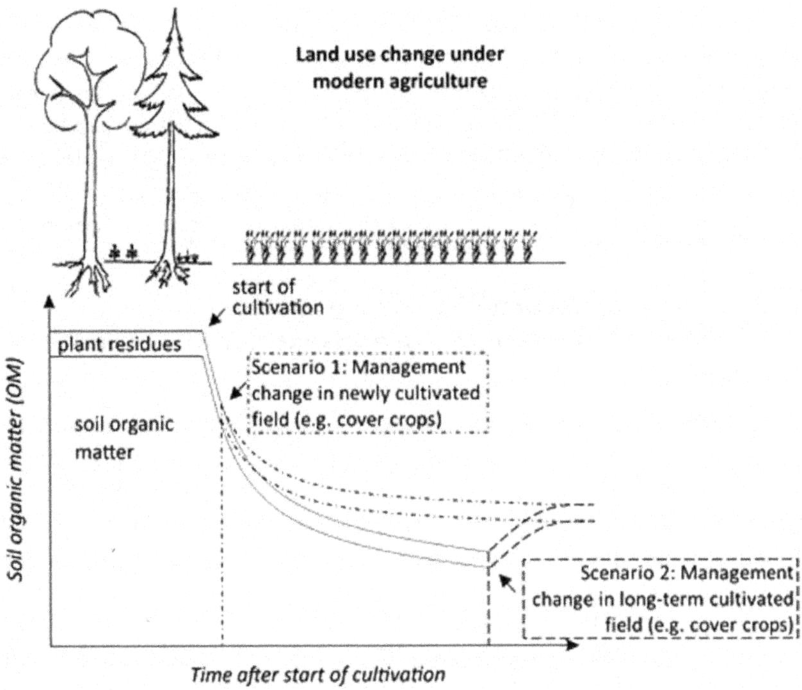

Figure 11 Schematic figure of potential changes in different C pools following the change to improved management. The time passed since the transformation of the natural to an agricultural ecosystem determines how close the system is to a new low input steady state. This determines whether improved management can only reduce soil C losses (Scenario 1) or lead to soil C sequestration (Scenario 2).

material with limited availability and no active mechanism for replacement. For example, some of the most fertile soils across the northern hemisphere are formed from aeolian late periglacial deposits (loess) with often only a few cm to several meters thickness (IUSS Working Group WRB, 2015; Kögel-Knabner and Amelung, 2014; Rejman et al., 2014). Once this resource has been eroded, underlying material (often bedrock or tertiary, deeply weathered sediments) does often have less suitable properties to form productive and fertile soils. With the recovery of loess resources being tied to stadial/interstadial cycles, the recovery of loess-derived soils is therefore practically not possible even at millennial timescales (Roberts et al., 2013).

An exception to the slow recovery of SOC stocks with immediate effects of increasing SOC sequestration is the management of peatland and wetland soils by ending and reversing draining measures (Chapter 21 of this book). Being comparatively small in their spatial extent and rich in SOC, drained peatlands experience yearly CO_2 losses of a factor of 20 and more compared to mineral soils. For example, a significant correlation between depth of the water table and yearly peat SOC losses was found in meta-analysis for Southeast Asia (Carlson et al., 2015), estimating yearly losses of 20 t SOC ha^{1} year^{1} at a water table depth of 70 cm, losses being reduced by 0.2 t SOC ha^{1} year^{1} with each cm of increased water table. Equally important is therefore the conservation of remaining peatlands. Peatland soils, while making up less than 3% of global soil covered land (Leifeld and Menichetti, 2018), can annually lose as much SOC, as several optimistic scenarios promise to sequester in all mineral soils combined.

3.4.2 Synthesis of the potential soil organic carbon sequestration potential on a human dominated planet Earth

Due to the heterogeneity of global soils and a lack of high-resolution soil data in many areas, estimates of global SOC losses and sequestration potential are yet rather rough and uncertainties are high. However, as more than 40% of the earth surface is covered by croplands and pastures combined (Foley et al., 2005), the potential for SOC sequestration by improved management, while being slow and in parts hard to implement, is still substantial. For example, Doetterl et al. (2012) and Sanderman et al. (2017) compared natural sites to agricultural sites globally and created maps of estimated SOC losses due to agricultural activity. They estimate that as a result of agriculture, ~116–150 Pg SOC have been lost in the last 12 000 years, globally. Recovering and rebuilding only a fraction of this C mass through SOC sequestration can buy important time for transforming the world's leading industrial nations into carbon neutral economies (Rogelj et al., 2018).

Studies on the global potential of SOC sequestration are still scarce and based on simplified global assumptions, such as that all soils could additionally sequester a certain percentage of their current SOC stocks. Yet, region and land use-specific estimates provide an initial basis of where to focus. Due to the land-use history effect, it will be important to target the most promising sites for SOC sequestration - sites with depleted SOC and poor management at the moment (Rumpel et al., 2020). Spatially diversified strategies that acknowledge difference in storage potentials and availability of C inputs for the soil will therefore be needed in the future (Amelung et al., 2020). These strategies should ideally be based on measured soil properties, as recent evidence suggests the human influence on SOC could be overestimated by global models, finding a mismatch between global soil carbon model outputs and observed global soil profiles data (Georgiou et al., 2021). For example, the work by Chen et al. (2018) provided spatially explicit estimates of SOC sequestration potential across France, based on the difference between measured SOC and SOC at an assumed, texture-related saturation level. While they estimated total sequestration potential to be 2.4 Pg SOC for top- and subsoils overall, huge variations in saturation deficits existed for all land types, highlighting the importance for spatially resolved maps (Chapter 12 of this book). Such spatially explicit approaches can also recognize the potential of pedogenetic drivers of SOC storage potential, which can in some cases explain more variation in SOC than the land use (von Fromm et al., 2021). Therefore, a better global coverage of soil and land use data will be highly relevant to better estimate local SOC sequestration potential and thereby improve the global estimates.

4 Future trends in research

4.1 Climate change and a biogeochemical perspective on soil properties to store carbon

Part of this chapter illustrates that the effects of climate change on soil C cycling and SOC sequestration are not yet fully understood. They will vary across regions and scales and are dependent on other factors like soil mineralogy and plant community dynamics. Among the soil properties controlling carbon storage, future research from a geochemical perspective should better constrain the influence of the ongoing climate change on soil water dynamics and their controls on organo-mineral interactions. To move this field forward, analysis of large data sets should always be combined with (long-term) field and laboratory experiments to understand the mechanisms behind observed patterns and improve our understanding of the main factors controlling the temperature sensitivity and the effect of moisture and drought on SOC sequestration in different regions of the world. In particular, it is important to consider regional differences rather than trying to find universal, global-scale relationships. Being able to quantitatively assess the local impact of this shift on carbon storage would be a major step forward to improve predictions of the capacity of soil to store carbon by Earth System Models. As climate change accelerates, it will also become important to make newly collected measurements and existing data available under FAIR (Findable, Accessible, Interoperable, Reusable) principles so that researchers, stakeholders, and policymakers have access to the information needed for climate mitigation and adaptation.

4.2 Soil microbial community responses and functions

The link between soil microbial diversity and functions (i.e. who is doing what?) has been a central theme of investigation for a long time, yet much remains unclear. However, recently developed molecular techniques, such as metagenomics and metatranscriptomics, can help to shed light on this important question. With the introduction of next generation sequencing, there has been an increasing number of studies characterizing the soil microbial communities in various settings using meta-omics methods. Combined with well-designed experiments, these techniques might also help to understand the impact of the environmental factors on the microbial carbon use efficiency and the functioning of microbial communities at local to global scales. Furthermore, these studies help to disentangle these complex soil microbial communities and their response to climate change as well as changes in agricultural management and its consequences for soil carbon dynamics. To be able to better include soil microbial community responses in climate models, it is also essential to fill spatial gaps by measuring soil microbial communities and plant productivity in previously uncharacterized soils that are representative for more remote areas. Closing these gaps and integrating the knowledge gained from existing studies has the potential to enable us to determine drivers of changes to carbon dynamics and investigate how to utilize agricultural management to create climate-smart soil microbial communities (for instance to stimulate soil carbon sequestration).

4.3 Future soil carbon-sequestration potential in anthropogenically modified landscapes

As described in this chapter, the interactions between soil organic carbon mineralization and sequestration and soil redistribution driven by human activities that modify agricultural and natural landscapes for millennia through a wide range of processes are not yet fully understood. Gaps in knowledge range from detailed process understanding regarding erosion-affected carbon fluxes, e.g. changes in crop yields and crop root-to-shoot ratios following severe erosion, to landscape-scale effects on erosion-affected C balance taking several erosion processes into account. As soil and soil organic carbon redistribution is mostly associated with short extreme events occurring in local scales, upscaling to regional or even global assessments is still challenging. Furthermore, for targeting modern farming techniques and management practices to achieve additional C sequestration in soil, further research on the site-specific limitations and potentially achievable C increases are needed. A second concern is what rates are realistic under real farming conditions to reach C increase in soils in a specific timeframe. This information then will help to assess the best techniques and practices to achieve fast gains in soil C while not jeopardizing other important soil functions (soil health, biodiversity, crop yields, water storage, etc.).

Finally, with new data collected and studies being published everyday at an increasing rate, it will be of growing importance that the knowledge gained from these efforts is shared and analyzed in a more synergistic way than currently done. One way to achieve this is to motivate researchers (and create the infrastructure and conditions for them) to store data in open-access databases, promote open science, and in general decrease the barriers for knowledge transfer between fields to stimulate interdisciplinary and transdisciplinary systems research. One way to deal with the sometimes contrasting results that new studies deliver on 'hot' subjects in relation to soil C research or studies from remote regions is to increase our efforts to conduct meta-analyses. Lastly, it is important to remember that global model predictions on C release and sequestration potential fundamentally rely on good data. This involves not only current-day observations at fine resolution at a global scale but also a solid understanding of past processes and changes and the fact that soils change and develop at different time scales than most of the changes to the C cycle that we would like to gain knowledge on. Estimates of C emissions from land use change can be compromised by poor quality of historical land-cover and land-use change maps, the rudimentary representation of management processes in most models, and the confusion in methodologies and boundary conditions used across methods. For example, the very coarse land-use categories used in the different upscaling approaches do not represent specific soil management strategies that might reduce or even reverse the emission of C from agricultural soils. Consequently, improved model parametrization is recognized as a major factor to lower uncertainties in estimates of land use change derived C emissions. Improving future land use scenarios and related C dynamics also desperately need better links to socioeconomic models that include the effects of global trade, consumer behavior, and population growth on agricultural production and the expansion of agricultural land into natural landscapes (especially in the Tropics). Together with climatic change, land degradation, and the overuse of soils that typically lowers the capacity of soils for cropping, economic and societally driven short-term and globally variable developments are ultimately the factors that will define future land-use patterns and the exchange and storage of C in soils.

5 References

Aasfar, A., Bargaz, A., Yaakoubi, K., Hilali, A., Bennis, I., Zeroual, Y. and Meftah Kadmiri, I. 2021. Nitrogen fixing Azotobacter species as potential soil biological enhancers for crop nutrition and yield stability. *Frontiers in Microbiology* 12, 628379. doi: 10.3389/fmicb.2021.628379.

Abramoff, R., Xu, X., Hartman, M., O'Brien, S., Feng, W., Davidson, E., Finzi, A., Moorhead, D., Schimel, J., Torn, M. and Mayes, M. A. 2018. The Millennial model: In search of measurable pools and transformations for modeling soil carbon in the new century. *Biogeochemistry* 137(1-2), 51-71. doi: 10.1007/s10533-017-0409-7.

Abramoff, R. Z., Torn, M. S., Georgiou, K., Tang, J. and Riley, W. J. 2019. Soil organic matter temperature sensitivity cannot be directly inferred from spatial gradients. *Global Biogeochemical Cycles* 33(6), 761-776. doi: 10.1029/2018GB006001.

Adler, P. B., Byrne, K. M. and Leiker, J. 2013. Can the past predict the future? Experimental tests of historically based population models. *Global Change Biology* 19(6), 1793-1803. doi: 10.1111/gcb.12168.

Aerts, R. and Chapin, F. S. 1999. The mineral nutrition of wild plants revisited: A re-evaluation of processes and patterns. *Advances in Ecological Research* 30, 1-67. Available at: https://www.sciencedirect.com/science/article/pii/S0065250408600161.

Al-Ani, M., Hmoshi, R., Kanaan, I. and Thanoon, A. 2019. Effect of pesticides on soil microorganisms. *Journal of Physics: Conference Series* 1294(7), 072007. doi: 10.1088/1742-6596/1294/7/072007.

Allard, S., Gutierrez, L., Fontaine, C., Croué, J. P. and Gallard, H. 2017. Organic matter interactions with natural manganese oxide and synthetic birnessite. *Science of the Total Environment* 583, 487-495. doi: 10.1016/j.scitotenv.2017.01.120.

Allison S. D., Vitousek P. M. 2005. Responses of extracellular enzymes to simple and complex nutrient inputs. *Soil Biology and Biochemistry* 37, 937-944. doi: 10.1016/j.soilbio.2004.09.014.

Allison, S. D., Wallenstein, M. D. and Bradford, M. A. 2010. Soil-carbon response to warming dependent on microbial physiology. *Nature Geoscience* 3(5), 336-340. doi: 10.1038/ngeo846.

Amelung, W., Bossio, D., de Vries, W., Kögel-Knabner, I., Lehmann, J., Amundson, R., Bol, R., Collins, C., Lal, R., Leifeld, J., Minasny, B., Pan, G., Paustian, K., Rumpel, C., Sanderman, J., van Groenigen, J. W., Mooney, S., can Wessemael, B., Wander, M. and Chabbi, A. 2020. Towards a global-scale soil climate mitigation strategy. *Nature Communications* 11, 5427.

Amundson, R. and Jenny, H. 1997. On a state factor model of ecosystems. *BioScience* 47(8), 536-543. doi: 10.2307/1313122.

Andreote, F. D. and Pereira e Silva, M. de C. 2017. Microbial communities associated with plants: Learning from nature to apply it in agriculture. *Current Opinion in Microbiology* 37, 29-34. doi: 10.1016/j.mib.2017.03.011.

Angst, G., Frouz, J., van Groenigen, J. W., Scheu, S., Kögel-Knabner, I. and Eisenhauer, N. 2022. Earthworms as catalysts in the formation and stabilization of soil microbial necromass. *Global Change Biology* 28(16), 4775-4782. doi: 10.1111/gcb.16208.

Asensio, D., Peñuelas, J., Llusià, J., Ogaya, R. and Filella, I. 2007. Interannual and interseasonal soil CO_2 efflux and VOC exchange rates in a Mediterranean holm oak forest in response to experimental drought. *Soil Biology and Biochemistry* 39(10), 2471-2484. doi: 10.1016/j.soilbio.2007.04.019.

Attard, E., Le Roux, X., Charrier, X., Delfosse, O., Guillaumaud, N., Lemaire, G. and Recous, S. 2016. Delayed and asymmetric responses of soil C pools and N fluxes to grassland/cropland conversions. *Soil Biology and Biochemistry* 97, 31-39. doi: 10.1016/j.soilbio.2016.02.016.

Averill, C., Turner, B. L. and Finzi, A. C. 2014. Mycorrhiza-mediated competition between plants and decomposers drives soil carbon storage. *Nature* 505(7484), 543-545. doi: 10.1038/nature12901.

Bakker, M. G., Manter, D. K., Sheflin, A. M., Weir, T. L. and Vivanco, J. M. 2012. Harnessing the rhizosphere microbiome through plant breeding and agricultural management. *Plant and Soil* 360(1-2), 1-13. doi: 10.1007/s11104-012-1361-x.

Bakker, M. M., Govers, G. and Rounsevell, M. D. A. 2004. The crop productivity-erosion relationship: An analysis based on experimental work. *CATENA* 57(1), 55-76. doi: 10.1016/j.catena.2003.07.002.

Bárcenas-Moreno, G., Gómez-Brandón, M., Rousk, J. and Bááth, E. 2009. Adaptation of soil microbial communities to temperature: Comparison of fungi and bacteria in a laboratory experiment. *Global Change Biology* 15(12), 2950–2957. doi: 10.1111/j.1365-2486.2009.01882.x.

Bardgett, R. D. and van der Putten, W. H. 2014. Belowground biodiversity and ecosystem functioning. *Nature* 515(7528), 505–511. doi: 10.1038/nature13855.

Bar-On, Y. M., Phillips, R. and Milo, R. 2018. The biomass distribution on Earth. *Proceedings of the National Academy of Sciences of the United States of America* 115(25), 6506–6511. doi: 10.1073/pnas.1711842115.

Barron, A. R., Wurzburger, N., Bellenger, J. P., Wright, S. J., Kraepiel, A. M. L. and Hedin, L. O. 2009. Molybdenum limitation of asymbiotic nitrogen fixation in tropical forest soils. *Nature Geoscience* 2(1), 42–45. doi: 10.1038/ngeo366.

Batjes, N. H. 1996. Total carbon and nitrogen in the soils of the world. *European Journal of Soil Science* 47(2), 151–163. doi: 10.1111/j.1365-2389.1996.tb01386.x.

Baumann, K., Dignac, M.-F., Rumpel, C., Bardoux, G., Sarr, A., Steffens, M. and Maron, P.-A. 2013. Soil microbial diversity affects soil organic matter decomposition in a silty grassland soil. *Biogeochemistry* 114(1–3), 201–212. doi: 10.1007/s10533-012-9800-6.

Bellenger, J. P., Wichard, T., Kustka, A. B. and Kraepiel, A. M. L. 2008. Uptake of molybdenum and vanadium by a nitrogen-fixing soil bacterium using siderophores. *Nature Geoscience* 1(4), 243–246. doi: 10.1038/ngeo161.

Berg, B., Steffen, K. T. and McClaugherty, C. 2007. Litter decomposition rate is dependent on litter Mn concentrations. *Biogeochemistry* 82(1), 29–39. doi: 10.1007/s10533-006-9050-6.

Berhe, A. A., Harte, J., Harden, J. W. and Torn, M. S. 2007. The significance of the erosion-induced terrestrial carbon sink. *BioScience* 57(4), 337–346. doi: 10.1641/B570408.

Blackwell, M. 2011. The fungi: 1, 2, 3 ... 5.1 million species? *American Journal of Botany* 98(3), 426–438. doi: 10.3732/ajb.1000298.

Blagodatskaya, E. and Kuzyakov, Y. 2008. Mechanisms of real and apparent priming effects and their dependence on soil microbial biomass and community structure: Critical review. *Biology and Fertility of Soils* 45(2), 115–131. doi: 10.1007/s00374-008-0334-y.

Borken, W., Savage, K., Davidson, E. A. and Trumbore, S. E. 2006. Effects of experimental drought on soil respiration and radiocarbon efflux from a temperate forest soil. *Global Change Biology* 12(2), 177–193. doi: 10.1111/j.1365-2486.2005.001058.x.

Borrelli, P., Robinson, D. A., Panagos, P., Lugato, E., Yang, J. E., Alewell, C., Wuepper, D., Montanarella, L. and Ballabio, C. 2020. Land use and climate change impacts on global soil erosion by water (2015-2070). *Proceedings of the National Academy of Sciences of the United States of America* 117(36), 21994–22001. doi: 10.1073/pnas.2001403117.

Bossolani, J. W., Crusciol, C. A. C., Merloti, L. F., Moretti, L. G., Costa, N. R., Tsai, S. M. and Kuramae, E. E. 2020. Long-term lime and gypsum amendment increase nitrogen fixation and decrease nitrification and denitrification gene abundances in the rhizosphere and soil in a tropical no-till intercropping system. *Geoderma* 375, 114476. doi: 10.1016/j.geoderma.2020.114476.

Bouchoms, S., Wang, Z., Vanacker, V. and Van Oost, K. 2018. Evaluating the interaction between sediment fluxes, carbon dynamics and biomass production using an integrated model. *Soil Discuss.* 15, 4459–4480. doi: 10.5194/bg-15-4459-2018.

Bowman, D. M. J. S., Kolden, C. A., Abatzoglou, J. T., Johnston, F. H., van der Werf, G. R. and Flannigan, M. 2020. Vegetation fires in the Anthropocene. *Nature Reviews Earth and Environment* 1(10), 500–515. doi: 10.1038/s43017-020-0085-3.

Bradford, M. A., Watss, B. W. and Davies, C. A. 2010. Thermal adaptation of heterotrophic soil respiration in laboratory microcosms. *Global Change Biology* 16(5), 1576–1588. doi: 10.1111/j.1365-2486.2009.02040.x.

Bredemeier, M., Blanck, K., Dohrenbusch, A., Lamersdorf, N., Meyer, A. C., Murach, D., Parth, A. and Xu, Y.-J. 1998. The Solling roof project - Site characteristics experiments and results. *Forest Ecology and Management* 101(1-3), 281-293. doi: 10.1016/S0378-1127(97)00143-6.

Bremenfeld, S., Fiener, P. and Govers, G. 2013. Effects of interrill erosion, soil crusting and soil aggregate breakdown on in situ CO_2 effluxes. *CATENA* 104, 14-20. doi: 10.1016/j.catena.2012.12.011.

Brzostek, E. R., Fisher, J. B. and Phillips, R. P. 2014. Modeling the carbon cost of plant nitrogen acquisition: Mycorrhizal trade-offs and multipath resistance uptake improve predictions of retranslocation. *Journal of Geophysical Research: Biogeosciences* 119(8), 1684-1697. doi: 10.1002/2014JG002660.

Bünemann, E., Schwenke, G. and Van Zwieten, L. 2006. Impact of agricultural inputs on soil organisms - A review. *Soil Research* 44(4), 379-406. doi: 10.1071/SR05125.

Burns, R. G., DeForest, J. L., Marxsen, J., Sinsabaugh, R. L., Stromberger, M. E., Wallenstein, M. D., Weintraub, M. N. and Zoppini, A. 2013. Soil enzymes in a changing environment: Current knowledge and future directions. *Soil Biology and Biochemistry* 58, 216-234. doi: 10.1016/j.soilbio.2012.11.009.

Canadell, J. G., Monteiro, P. M. S., Costa, M. H., Cotrim da Cunha, L., Cox, P. M., Eliseev, A. V., Henson, S., Ishii, M., Jaccard, S., Koven, C., Lohila, A., Patra, P. K., Piao, S., Rogelj, J., Syampungani, S., Zaehle, S. and Zickfeld, K. 2021. Global carbon and other biogeochemical cycles and feedbacks. In: Masson-Delmotte, V., Zhai, P., Pirani, A., Connors, S. L., Péan, C., Berger, S., Caud, N., Chen, Y., Goldfarb, L., Gomis, M. I., Huang, M., Leitzell, K., Lonnoy, E., Matthews, J. B. R., Maycock, T. K., Waterfield, T., Yelekçi, O., Yu, R. and Zhou, B. (Eds.), Climate Change 2021: The Physical Science Basis. Contribution of Working Group I to the Sixth Assessment Report of the Intergovernmental Panel on Climate Change. Cambridge University Press.

Carey, J. C., Tang, J., Templer, P. H., Kroeger, K. D., Crowther, T. W., Burton, A. J., Dukes, J. S., Emmett, B., Frey, S. D., Heskel, M. A., Jiang, L., Machmuller, M. B., Mohan, J., Panetta, A. M., Reich, P. B., Reinsch, S., Wang, X., Allison, S. D., Bamminger, C., Bridgham, S., Collins, S. L., de Dato, G., Eddy, W. C., Enquist, B. J., Estiarte, M., Harte, J., Henderson, A., Johnson, B. R., Larsen, K. S., Luo, Y., Marhan, S., Melillo, J. M., Peñuelas, J., Pfeifer-Meister, L., Poll, C., Rastetter, E., Reinmann, A. B., Reynolds, L. L., Schmidt, I. K., Shaver, G. R., Strong, A. L., Suseela, V. and Tietma, A. 2016. Limited acclimation of soil respiration to warming. *PNAS* 113(48), 13797–13802. doi: 10.1073/pnas.1605365113.

Carlson, K. M., Goodman, L. K. and May-Tobin, C. C. 2015. Modeling relationships between water table depth and peat soil carbon loss in Southeast Asian plantations. *Environmental Research Letters* 10(7), 074006. doi: 10.1088/1748-9326/10/7/074006.

Carvalhais, N., Forkel, M., Khomik, M., Bellarby, J., Jung, M., Migliavacca, M., Mu, M., Saatchi, S., Santoro, M., Thurner, M., Weber, U., Ahrens, B., Beer, C., Cescatti, A., Randerson, J. T. and Reichstein, M. 2014. Global covariation of carbon turnover times with climate in terrestrial ecosystems. *Nature* 514(7521), 213–217. doi: 10.1038/nature13731.

Cernava, T., Chen, X., Krug, L., Li, H., Yang, M. and Berg, G. 2019. The tea leaf microbiome shows specific responses to chemical pesticides and biocontrol applications. *Science of the Total Environment* 667, 33–40. doi: 10.1016/j.scitotenv.2019.02.319.

Chadwick, O. A. and Chorover, J. 2001. The chemistry of pedogenic thresholds. *Geoderma, Developments and Trends in Soil Science* 100(3–4), 321–353. doi: 10.1016/S0016-7061(01)00027-1.

Chalot, M. and Brun, A. 1998. Physiology of organic nitrogen acquisition by ectomycorrhizal fungi and ectomycorrhizas. *FEMS Microbiology Reviews* 22(1), 21–44. doi: 10.1111/j.1574-6976.1998.tb00359.x.

Chapin, F., McFarland, J., McGuire, A. D., Euskirchen, E., Ruess, R. and Kielland, K. 2009. The changing global carbon cycle: Linking plant-soil carbon dynamics to global consequences. *Journal of Ecology*, 97(5), 840--850. doi: doi10.1111/j.1365-2745.2009.01529.x: Available at: https://besjournals.onlinelibrary.wiley.com/doi/10.1111/j.1365-2745.2009.01529.x.

Chappell, A., Baldock, J. and Sanderman, J. 2016. The global significance of omitting soil erosion from soil organic carbon cycling schemes. *Nature Climate Change* 6(2), 187–191. doi: 10.1038/nclimate2829.

Chen, S., Martin, M. P., Saby, N. P. A., Walter, C., Angers, D. A. and Arrouays, D. 2018. Fine resolution map of top- and subsoil carbon sequestration potential in France. *Science of the Total Environment* 630, 389–400. doi: 10.1016/j.scitotenv.2018.02.209.

Chenu, C., Angers, D. A., Barré, P., Derrien, D., Arrouayss, D. and Balesdent, J. 2019. Increasing organic stocks in agricultural soils: Knowledge gaps and potential innovations. *Soil and Tillage Research* 188, 41–52. doi: 10.1016/j.still.2018.04.011.

Chorover, J., Amistadi, M. K. and Chadwick, O. A. 2004. Surface charge evolution of mineral-organic complexes during pedogenesis in Hawaiian basalt. *Geochimica et Cosmochimica Acta* 68(23), 4859–4876. doi: 10.1016/j.gca.2004.06.005.

Clark, M. and Tilman, D. 2017. Comparative analysis of environmental impacts of agricultural production systems, agricultural input efficiency, and food choice. *Environmental Research Letters* 12(6), 064016. doi: 10.1088/1748-9326/aa6cd5.

Cleveland, C. C. and Liptzin, D. 2007. C:N:P stoichiometry in soil: Is there a "Redfield ratio" for the microbial biomass? *Biogeochemistry* 85(3), 235–252. doi: 10.1007/s10533-007-9132-0.

Cleveland, C. C., Wieder, W. R., Reed, S. C. and Townsend, A. R. 2010. Experimental drought in a tropical rain forest increases soil carbon dioxide losses to the atmosphere. *Ecology* 91(8), 2313–2323. doi: 10.1890/09-1582.1.

Cline, M. G. 1961. The changing model of soil. *Soil Science Society of America Journal* 25(6), 442–446.

Conant, R. T., Ryan, M. G., Ågren, G. I., Birge, H. E., Davidson, E. A., Eliasson, P. E., Evans, S. E., Frey, S. D., Giardina, C. P., Hopkins, F. M., Hyvönen, R., Kirschbaum, M. U. F., Lavallee, J. M., Leifeld, J., Parton, W. J., Megan Steinweg, J., Wallenstein, M. D., Martin Wetterstedt, J. Å and Bradford, M. A. 2011. Temperature and soil organic matter decomposition rates – Synthesis of current knowledge and a way forward. *Global Change Biology* 17(11), 3392–3404. doi: 10.1111/j.1365-2486.2011.02496.x.

Corbeels, M., Cardinael, R., Naudin, K., Guibert, H. and Torquebiau, E. 2019. The 4 per 1000 goal and soil carbon storage under agroforestry and conservation agriculture systems in sub-Saharan Africa. *Soil and Tillage Research* 188, 16–26. doi: 10.1016/j.still.2018.02.015.

Cordero, O. X. and Datta, M. S. 2016. Microbial interactions and community assembly at microscales. *Current Opinion in Microbiology* 31, 227–234. doi: 10.1016/j.mib.2016.03.015.

Cotrufo, M. F., Wallenstein, M. D., Boot, C. M., Denef, K. and Paul, E. 2013. The Microbial Efficiency-Matrix Stabilization (MEMS) framework integrates plant litter decomposition with soil organic matter stabilization: Do labile plant inputs form stable soil organic matter? *Global Change Biology* 19(4), 988–995. doi: 10.1111/gcb.12113.

Courty, P.-E., Buée, M., Diedhiou, A. G., Frey-Klett, P., Le Tacon, F., Rineau, F., Turpault, M.-P., Uroz, S. and Garbaye, J. 2010 The role of ectomycorrhizal communities in forest ecosystem processes: New perspectives and emerging concepts. *Soil Biology and Biochemistry* 42(5), 679–698. doi: 10.1016/j.soilbio.2009.12.006.

Crowther, T. W., Todd-Brown, K. E., Rowe, C. W., Wieder, W. R., Carey, J. C., Machmuller, M. B., Snoek, B. L., Fang, S., Zhou, G., Allison, S. D., Blair, J. M., Bridgham, S. D., Burton, A. J., Carrillo, Y., Reich, P. B., Clark, J. S., Classen, A. T., Dijkstra, F. A., Elberling, B., Emmett, B. A., Estiarte, M., Frey, S. D., Guo, J., Harte, J., Jiang, L., Johnson, B. R., Kröel-Dulay, G., Larsen, K. S., Laudon, H., Lavallee, J. M., Luo, Y., Lupascu, M., Ma, L. N., Marhan, S., Michelsen, A., Mohan, J., Niu, S., Pendall, E., Peñuelas, J., Pfeifer-Meister, L., Poll, C., Reinsch, S., Reynolds, L. L., Schmidt, I. K., Sistla, S., Sokol, N. W., Templer, P. H., Treseder, K. K., Welker, J. M. and Bradford, M. A. 2016. Quantifying global soil carbon losses in response to warming. *Nature* 540(7631), 104–108. doi: 10.1038/nature20150.

Crowther, T. W., van den Hoogen, J., Wan, J., Mayes, M. A., Keiser, A. D., Mo, L., Averill, C. and Maynard, D. S. 2019. The global soil community and its influence on biogeochemistry. *Science* 365(6455), eaav0550. doi: 10.1126/science.aav0550.

Davidson, E. A. and Janssens, I. A. 2006. Temperature sensitivity of soil carbon decomposition and feedbacks to climate change. *Nature* 440(7081), 165–173. doi: 10.1038/nature04514. Available at: https://www.nature.com/articles/nature04514.

DeAngelis, K. M., Pold, G., Topçuoğlu, B. D., van Diepen, L. T. A., Varney, R. M., Blanchard, J. L., Melillo, J. and Frey, S. D. 2015. Long-term forest soil warming alters microbial communities in temperate forest soils. *Frontiers in Microbiology* 6, 104. doi: 10.3389/fmicb.2015.00104.

de Bruijn, I., Cheng, X., de Jager, V., Expósito, R. G., Watrous, J., Patel, N., Postma, J., Dorrestein, P. C., Kobayashi, D. and Raaijmakers, J. M. 2015. Comparative genomics and metabolic profiling of the genus Lysobacter. *BMC Genomics* 16(1), 991. doi: 10.1186/s12864-015-2191-z.

de Carvalho, L. A. L., Teheran-Sierra, L. G., Funnicelli, M. I. G., da Silva, R. C., Campanari, M. F. Z., de Souza, R. S. C., Arruda, P., Soares, M. A. and Pinheiro, D. G. 2021. Farming systems influence the compositional, structural, and functional characteristics of the sugarcane-associated microbiome. *Microbiological Research* 252, 126866. doi: 10.1016/j.micres.2021.126866.

De Tender, C., Mesuere, B., Van der Jeugt, F., Haegeman, A., Ruttink, T., Vandecasteele, B., Dawyndt, P., Debode, J. and Kuramae, E. E. 2019. Peat substrate amended with chitin modulates the N-cycle, siderophore and chitinase responses in the lettuce rhizobiome. *Scientific Reports* 9(1), 9890. doi: 10.1038/s41598-019-46106-x.

de Vente, J., Poesen, J., Arabkhedri, M. and Verstraeten, G. 2007. The sediment delivery problem revisited. *Progress in Physical Geography: Earth and Environment* 31(2), 155–178. doi: 10.1177/0309133307076485.

Delgado-Baquerizo, M., Oliverio, A. M., Brewer, T. E., Benavent-González, A., Eldridge, D. J., Bardgett, R. D., Maestre, F. T., Singh, B. K. and Fierer, N. 2018. A global atlas of the dominant bacteria found in soil. *Science* 359(6373), 320–325. doi: 10.1126/science.aap9516.

Doetterl, S., Asifiwe, R. K., Baert, G., Bamba, F., Bauters, M., Boeckx, P., Bukombe, B., Cadisch, G., Cooper, M., Cizungu, L. N., Hoyt, A., Kabaseke, C., Kalbitz, K., Kidinda, L., Maier, A., Mainka, M., Mayrock, J., Muhindo, D., Mujinya, B. B., Mukotanyi, S. M., Nabahungu, L., Reichenbach, M., Rewald, R., Six, J., Stegmann, A., Summerauer, L., Unseld, R., Vanlauwe, B., Van Oost, K., Verheyen, K., Vogel, C., Wilken, F. and Fiener, P. 2021. Organic matter cycling along geochemical, geomorphic, and disturbance gradients in forest and cropland of the African Tropics – Project TropSOC database version 1.0. *Scientific Data* 13, 4133–4153. doi: 10.5194/essd-13-4133-2021.

Doetterl, S., Berhe, A. A., Arnold, C., Bodé, S., Fiener, P., Finke, P., Fuchslueger, L., Griepentrog, M., Harden, J. W., Nadeu, E., Schnecker, J., Six, J., Trumbore, S., Van Oost, K., Vogel, C. and Boeckx, P. 2018. Links among warming, carbon and microbial dynamics mediated by soil mineral weathering. *Nature Geoscience* 11(8), 589–593. doi: 10.1038/s41561-018-0168-7.

Doetterl, S., Berhe, A. A., Nadeu, E., Wang, Z. G., Sommer, M. and Fiener, P. 2016. Erosion, deposition and soil carbon: A review of process-level controls, experimental tools and models to address C cycling in dynamic landscapes. *Earth-Science Reviews* 154, 102–122. doi: 10.1016/j.earscirev.2015.12.005.

Doetterl, S., Van Oost, K. and Six, J. 2012. Towards constraining the magnitude of global agricultural sediment and soil organic carbon fluxes. *Earth Surface Processes and Landforms* 37(6), 642–655. doi: 10.1002/esp.3198.

Dokuchaev, V. V. 1886. Materials on land evaluation of the Nizhni Novgorod governorate. Natural and historical part: Report ot the Nizhni Novgorod governorate zemstvo. In: *Key Points in the History of Land Evaluation in the European Russia, with Classification of Russian Soils* 1. Evdokimova, St. Petersburg: Tipogr [in Russian].

Dokuchaev, V. V. 1899. *Report to the Transcaucasian Statistical Committee on Land Evaluation in General and Especially for the Transcaucasia. Horizontal and Vertical Soil Zones*. Tipogr. Kantselyarii Glavnonachal'stvuyushchego Grazhdanskoi chast'yu Na Kavkaze, Tiflis [in Russian].

Doocy, S., Daniels, A., Murray, S. and Kirsch, T. D. 2013. The human impact of floods: A historical review of events 1980–2009 and systematic literature review. *PLoS Currents* 5, 5. doi: 10.1371/currents.dis.f4deb457904936b07c09daa98ee8171a.

Eglinton, T., Galy, V., Hemingway, J., Feng, X., Bao, J., Blattmann, T., Dickens, A. F., Gies, H., Giosan, L., Haghipour, N., Hou, P., Lupker, M., McIntyre, C. P., Montluçon, D. B., Peucker-Ehrenbrink, B., Ponton, C., Schefuß, E., Schwab, M. S., Voss, B. M., Wacker, L., Wu, Y. and Zhao, M. 2021. Climate control on terrestrial biospheric carbon turnover. *PNAS* 118(8), e2011585118. Available at: https://www.pnas.org/content/pnas/118/8/e2011585118.full.pdf.

Ehlers, W. and Goss, M. 2016. *Water Dynamics in Plant Production* (2nd edn.). Wallingford, UK: CABI Publishing, 344 p.

Engelbrecht, B. M., Comita, L. S., Condit, R., Kursar, T. A., Tyree, M. T., Turner, B. L. and Hubbell, S. P. 2007. Drought sensitivity shapes species distribution patterns in tropical forests. *Nature* 447(7140), 80–82. doi: 10.1038/nature05747.

Erb, K. H., Kastner, T., Plutzar, C., Bais, A. L. S., Carvalhais, N., Fetzel, T., Gingrich, S., Haberl, H., Lauk, C., Niedertscheider, M., Pongratz, J., Thurner, M. and Luyssaert, S. 2018. Unexpectedly large impact of forest management and grazing on global vegetation biomass. *Nature* 553(7686), 73–76. doi: 10.1038/nature25138.

FAO. 1996. *FAOSTAT. PC Software*. Rome, Italy: Food and Agriculture Organization of the United Nations (FAO).

FAO. 2008. *FAOSTAT*. Rome, Italy: Food and Agriculture Organization of the United Nations (FAO).

FAO. 2017. Soil organic carbon: The hidden potential. Rome, Italy: Food and Agriculture Organization of the United Nations (FAO). Available at: http://www.fao.org/3/i6937e/i6937e.pdf.

FAO and ITPS 2015. *Status of the World's Soil Resources (SWSR)–Main Report*. Food and Agriculture Organization of the United Nations and Intergovernmental Technical Panel on Soils, Rome, Italy. Available at: http://www.fao.org/3/a-i5199e.pdf.

Fatichi, S., Manzoni, S., Or, D. and Paschalis, A. 2019. A mechanistic model of microbially mediated soil biogeochemical processes: A reality check. *Global Biogeochemical Cycles* 33(6), 620–648. doi: 10.1029/2018GB006077.

Fekiacova, Z., Vermeire, M. L., Bechon, L., Cornelis, J. T. and Cornu, S. 2017. Can Fe isotope fractionations trace the pedogenetic mechanisms involved in podzolization? *Geoderma* 296, 38–46. doi: 10.1016/j.geoderma.2017.02.020.

Fierer, N. 2017. Embracing the unknown: Disentangling the complexities of the soil microbiome. *Nature Reviews. Microbiology* 15(10), 579–590. doi: 10.1038/nrmicro.2017.87.

Fierer, N., Strickland, M. S., Liptzin, D., Bradford, M. A. and Cleveland, C. C. 2009. Global patterns in belowground communities. *Ecology Letters* 12(11), 1238–1249. doi: 10.1111/j.1461-0248.2009.01360.x.

Fisher, J. B., Sitch, S., Malhi, Y., Fisher, R. A., Huntingford, C. and Tan, S.-Y. 2010. Carbon cost of plant nitrogen acquisition: A mechanistic, globally applicable model of plant nitrogen uptake, retranslocation, and fixation. *Global Biogeochemical Cycles* 24(1). doi: 10.1029/2009GB003621.

Foley, J. A., Defries, R., Asner, G. P., Barford, C., Bonan, G., Carpenter, S. R., Chapin, F. S., Coe, M. T., Daily, G. C., Gibbs, H. K., Helkowski, J. H., Holloway, T., Howard, E. A., Kucharik, C. J., Monfreda, C., Patz, J. A., Prentice, I. C., Ramankutty, N. and Snyder, P. K. 2005. Global consequences of land use. *Science* 309(5734), 570–574. doi: 10.1126/science.1111772.

Freibauer, A., Rounsevell, M. D. A., Smith, P. and Verhagen, J. 2004. Carbon sequestration in the agricultural soils of Europe. *Geoderma* 122(1), 1–23. doi: 10.1016/j.geoderma.2004.01.021.

Friedlingstein, P., O'Sullivan, M., Jones, M. W., Andrew, R. M., Hauck, J., Olsen, A., Peters, G. P., Peters, W., Pongratz, J., Sitch, S., Le Quéré, C., Canadell, J. G., Ciais, P., Jackson, R. B., Alin, S., Aragão, L. E. O. C., Arneth, A., Arora, V., Bates, N. R., Becker, M., Benoit-Cattin, A., Bittig, H. C., Bopp, L., Bultan, S., Chandra, N., Chevallier, F., Chini, L. P., Evans, W., Florentie, L., Forster, P. M., Gasser, T., Gehlen, M., Gilfillan, D., Gkritzalis, T., Gregor, L., Gruber, N., Harris, I., Hartung, K., Haverd, V., Houghton, R. A., Ilyina, T., Jain, A. K., Joetzjer, E., Kadono, K., Kato, E., Kitidis, V., Korsbakken, J. I., Landschützer, P., Lefèvre, N., Lenton, A., Lienert, S., Liu, Z., Lombardozzi, D., Marland, G., Metzl, N., Munro, D. R., Nabel, J. E. M. S., Nakaoka, S., Niwa, Y., O'Brien, K., Ono, T., Palmer, P. I., Pierrot, D., Poulter, B., Resplandy, L., Robertson, E., Rödenbeck, C., Schwinger, J., Séférian, R., Skjelvan, I., Smith, A. J. P., Sutton, A. J., Tanhua, T., Tans, P. P., Tian, H., Tilbrook, B., van der Werf, G., Vuichard, N., Walker, A. P., Wanninkhof, R., Watson, A. J., Willis, D., Wiltshire, A. J., Yuan, W., Yue, X. and Zaehle, S. 2020. Global carbon budget 2020. *Earth System Science Data* 12(4), 3269–3340. doi: 10.5194/essd-12-3269-2020.

Fritsch, E., Allard, Th., Benedetti, M. F., Bardy, M., do Nascimento, N. R., Li, Y. and Calas, G. 2009. Organic complexation and translocation of ferric iron in podzols of the Negro River watershed. Separation of secondary Fe species from Al species. *Geochimica et Cosmochimica Acta* 73(7), 1813–1825. doi: 10.1016/j.gca.2009.01.008.

Fujisaki, K., Chevallier, T., Chapuis-Lardy, L., Albrecht, A., Razafimbelo, T., Masse, D., Ndour, Y. B. and Chotte, J.-L. 2018. Soil carbon stock changes in tropical croplands are mainly driven by carbon inputs: A synthesis. *Agriculture, Ecosystems and Environment* 259, 147–158. doi: 10.1016/j.agee.2017.12.008.

Gans, J., Wolinsky, M. and Dunbar, J. 2005. Computational improvements reveal great bacterial diversity and high metal toxicity in soil. *Science* 309(5739), 1387–1390. doi: 10.1126/science.1112665.

Gasser, T., Crepin, L., Quilcaille, Y., Houghton, R. A., Ciais, P. and Obersteiner, M. 2020. Historical CO_2 emissions from land use and land cover change and their uncertainty. *Biogeosciences* 17(15), 4075–4101. doi: 10.5194/bg-17-4075-2020.

Geisseler, D., Linquist, B. A. and Lazicki, P. A. 2017 Effect of fertilization on soil microorganisms in paddy rice systems – A meta-analysis. *Soil Biology and Biochemistry* 115, 452–460. doi: 10.1016/j.soilbio.2017.09.018.

Geisseler, D. and Scow, K. M. 2014. Long-term effects of mineral fertilizers on soil microorganisms – A review. *Soil Biology and Biochemistry* 75, 54–63. doi: 10.1016/j.soilbio.2014.03.023.

Georgiou, K., Malhotra, A., Wieder, W. R., Ennis, J. H., Hartman, M. D., Sulman, B. N., Berhe, A. A., Grandy, A. S., Kyker-Snowman, E., Lajtha, K., Moore, J. A. M., Pierson, D. and Jackson, R. B. 2021. Divergent controls of soil organic carbon between observations and process-based models. *Biogeochemistry* 156(1), 5–17. doi: 10.1007/s10533-021-00819-2.

Gerten, D., Luo, Y., LeMaire, G., Parton, W. J., Keough, C., Weng, E., Beier, C., Ciais, P., Cramer, W., Dukes, J. S., Hanson, P. J., Knapp, A. A. K., Linder, S., Nepstad, D., Rustad, L. and Sowerby, A. 2008. Modelled effects of precipitation on ecosystem carbon and water dynamics in different climatic zones. *Global Change Biology* 14(10), 2365–2379. doi: 10.1111/j.1365-2486.2008.01651.x. Available at: https://onlinelibrary.wiley.com/doi/full/10.1111/j.1365-2486.2008.01651.x.

Goldberg, S., Forster, H. S. and Godfrey, C. L. 1996. Molybdenum adsorption on oxides, clay minerals, and soils. *Soil Science Society of America Journal* 60(2), 425–432. doi: 10.2136/sssaj1996.03615995006000020013x.

Govers, G., Vandaele, K., Desmet, P., Poesen, J. and Bunte, K. 1994. The role of tillage in soil redistribution on hillslopes. *European Journal of Soil Science* 45(4), 469–478. doi: 10.1111/j.1365-2389.1994.tb00532.x.

Guerra, C. A., Rosa, I. M. D., Valentini, E., Wolf, F., Filipponi, F., Karger, D. N., Xuan, A. N., Mathieu, J., Lavelle, P. and Eisenhauer, N. 2020. Global vulnerability of soil ecosystems to erosion. *Landscape Ecology* 35, 823–842. doi: 10.1007/s10980-020-00984-z.

Haaf, D., Six, J. and Doetterl, S. 2021. Global patterns of geo-ecological controls on the response of soil respiration to warming. *Nature Climate Change* 11(7), 623–627. doi: 10.1038/s41558-021-01068-9.

Han, P., Zhang, W., Wang, G., Sun, W. and Huang, Y. 2016. Changes in soil organic carbon in croplands subjected to fertilizer management: A global meta-analysis. *Scientific Reports* 6, 27199. doi: 10.1038/srep27199.

Harden, J. W., Sharpe, J. M., Parton, W. J., Ojima, D. S., Fries, T. L., Huntington, T. G. and Dabney, S. M. 1999. Dynamic replacement and loss of soil carbon by eroding cropland. *Global Biogeochemical Cycles* 13(4), 885–901. doi: 10.1029/1999GB900061.

Hartmann, M., Frey, B., Mayer, J., Mäder, P. and Widmer, F. 2015. Distinct soil microbial diversity under long-term organic and conventional farming. *The ISME Journal* 9(5), 1177–1194. doi: 10.1038/ismej.2014.210.

Herndon, E., Kinsman-Costello, L. and Godsey, S. 2020. Biogeochemical cycling of redox-sensitive elements in permafrost-affected ecosystems. In:. *Geophysical Monograph Series*. American Geophysical Union (AGU), pp. 245–265. doi: 10.1002/9781119413332.ch12.

Hicks Pries, C. E., Castanha, C., Porras, R. C. and Torn, M. S. 2017. The whole-soil carbon flux in response to warming. *Science* 355(6332), 1420–1423. doi: 10.1126/science.aal1319.

Hillebrand, H. 2004. On the generality of the latitudinal diversity gradient. *The American Naturalist* 163(2), 192–211. doi: 10.1086/381004.

Hobbie, S. E. 1992. Effects of plant species on nutrient cycling. *Trends in Ecology and Evolution* 7(10), 336–339. doi: 10.1016/0169-5347(92)90126-V.

Hoffmann, T., Glatzel, S. and Dikau, R. 2009. A carbon storage perspective on alluvial sediment storage in the Rhine catchment. *Geomorphology* 108(1–2), 127–137. doi: 10.1016/j.geomorph.2007.11.015.

Hoffmann, T., Mudd, S. M., van Oost, K., Verstraeten, G., Erkens, G., Lang, A., Middelkoop, H., Boyle, J., Kaplan, J. O., Willenbring, J. and Aalto, R. 2013a. Short Communication: Humans and the missing C-sink: Erosion and burial of soil carbon through time. *Earth Surface Dynamics* 1(1), 45–52.

Hoffmann, T., Schlummer, M., Notebaert, B., Verstraeten, G. and Korup, O. 2013b. Carbon burial in soil sediments from Holocene agricultural erosion, Central Europe. *Global Biogeochemical Cycles* 27(3), 828–835. doi: 10.1002/gbc.20071.

Holdridge, L. R. 1947. Determination of world plant formations from simple climatic data. *Science* 105(2727), 367–368. doi: 10.1126/science.105.2727.367.

Hooke, R. L. 2000. On the history of human as geomorphic agents. *Geology* 28(9), 843–846. doi: 10.1130/0091-7613(2000)28<843:OTHOHA>2.0.CO;2.

Houghton, R. A. 2003. Revised estimates of the annual net flux of carbon to the atmosphere from changes in land use and land management 1850–2000. *Tellus B* 55(2), 378–390. doi: 10.1034/j.1600-0889.2003.01450.x.

Houghton, R. A. 2007. Balancing the global carbon budget. *Annual Review of Earth and Planetary Sciences* 35(1), 313–347. doi: 10.1146/annurev.earth.35.031306.140057.

Hu, F. S., Higuera, P. E., Duffy, P., Chipman, M. L., Rocha, A. V., Young, A. M., Kelly, R. and Dietze, M. C. 2015. Arctic tundra fires: Natural variability and responses to climate change. *Frontiers in Ecology and the Environment* 13(7), 369–377. doi: 10.1890/150063.

Huang, Y., Ciais, P., Luo, Y., Zhu, D., Wang, Y., Qiu, C., Goll, D. S., Guenet, B., Makowski, D., De Graaf, I., Leifeld, J., Kwon, M. J., Hu, J. and Qu, L. 2021. Tradeoff of CO_2 and CH_4 emissions from global peatlands under water-table drawdown. *Nature Climate Change* 11(7), 618–622. doi: 10.1038/s41558-021-01059-w.

Hug, L. A., Baker, B. J., Anantharaman, K., Brown, C. T., Probst, A. J., Castelle, C. J., Butterfield, C. N., Hernsdorf, A. W., Amano, Y., Ise, K., Suzuki, Y., Dudek, N., Relman, D. A., Finstad, K. M., Amundson, R., Thomas, B. C. and Banfield, J. F. 2016. A new view of the tree of life. *Nature Microbiology* 1, 16048. doi: 10.1038/nmicrobiol.2016.48.

Hugelius, G., Bockheim, J. G., Camill, P., Elberling, B., Grosse, G., Harden, J. W., Johnson, K., Jorgenson, T., Koven, C. D., Kuhry, P., Michaelson, G., Mishra, U., Palmtag, J., Ping, C.-L., O'Donnell, J., Schirrmeister, L., Schuur, E. A. G., Sheng, Y., Smith, L. C., Strauss, J. and Yu, Z. 2013. A new data set for estimating organic carbon storage to 3 m depth in soils of the northern circumpolar permafrost region. *Earth System Science Data* 23, 393–402. doi: 10.5194/essd-5-393-2013.

Hugelius, G., Strauss, J., Zubrzycki, S., Harden, J. W., Schuur, E. A. G., Ping, C. -L., Schirrmeister, L., Grosse, G., Michaelson, G. J., Koven, C. D., O'Donnell, J. A., Elberling, B., Mishra, U., Camill, P., Yu, Z., Palmtag, J. and Kuhry, P. 2014. Estimated stocks of circumpolar permafrost carbon with quantified uncertainty ranges and identified data gaps. *Biogeosciences* 11(23), 6573–6593. doi: 10.5194/bg-11-6573-2014.

Hurtt, G. C., Chini, L., Sahajpal, R., Frolking, S., Bodirsky, B. L., Calvin, K., Doelman, J. C., Fisk, J., Fujimori, S., Klein Goldewijk, K., Hasegawa, T., Havlik, P., Heinimann, A., Humpenöder, F., Jungclaus, J., Kaplan, J. O., Kennedy, J., Krisztin, T., Lawrence, D., Lawrence, P., Ma, L., Mertz, O., Pongratz, J., Popp, A., Poulter, B., Riahi, K., Shevliakova, E., Stehfest, E., Thornton, P., Tubiello, F. N., van Vuuren, D. P. and Zhang, X. 2020. Harmonization of global land use change and management for the period 850–2100 (LUH2) for CMIP6. *Geoscientific Model Development* 13(11), 5425–5464. doi: 10.5194/gmd-13-5425-2020.

Hussain, S., Siddique, T., Saleem, M., Arshad, M. and Khalid, A. 2009. Chapter 5: Impact of pesticides on soil microbial diversity, enzymes, and biochemical reactions. *Advances in Agronomy* 102, 159–200. doi: 10.1016/S0065-2113(09)01005-0.

Ingrisch, J., Karlowsky, S., Hasibeder, R., Gleixner, G. and Bahn, M. 2020. Drought and recovery effects on belowground respiration dynamics and the partitioning of recent carbon in managed and abandoned grassland. *Global Change Biology* 26(8), 4366–4378. doi: 10.1111/gcb.15131.

IPCC. 2013. The physical science basis. Summary for policymakers. Contribution of Working Group I to the 5th assessment report of the intergovernmental panel on climate change. In: *Climate Change*. Cambridge and New York, NY: Cambridge University Press, pp. 1–30.

IPCC. 2014. Climate Change 2014: Synthesis Report. Contribution of Working Groups I, II and III to the Fifth Assessment Report of the Intergovernmental Panel on Climate Change [Core Writing Team, Pachauri, R. K. and Meyer, L. A. (Eds)]. IPCC, Geneva, Switzerland, 151 pp.

IPCC. 2021. 2021: The physical science basis. Contribution of working Group I to the sixth assessment report of the Intergovernmental Panel on Climate Change. In: Masson-Delmotte, V., Zhai, P., Pirani, A., Connors, S. L., Péan, C., Berger, S., Caud, N., Chen, Y., Goldfarb, L., Gomis, M. I. and Huang, M. (Eds.), *Climate Change*. Cambridge: Cambridge University Press.

IPCC. 2022. Climate Change 2022: Impacts, adaptation, and vulnerability. Contribution of Working Group II to the Sixth Assessment Report of the Intergovernmental Panel on Climate Change Pörtner, H.-O., Roberts, D. C., Tignor, M., Poloczanska, E. S., Mintenbeck, K., Alegría, A., Craig, M., Langsdorf, S., Löschke, S., Möller, V., et al. (Eds). Cambridge University Press, Cambridge, UK.

Ito, A. 2007. Simulated impacts of climate and land-cover change on soil erosion and implication for the carbon cycle, 1901 to 2100. *Geophysical Research Letters* 34(9), L09403. doi: 10.1029/2007GL029342.

IUSS Working Group WRB. 2015. World reference base for soil resources 2014. Update 2015. World Soil Resources Reports No. 106. Food and Agriculture Organization, Rome.

Jacinthe, P.-A., Lal, R. and Kimble, J. M. 2002. Carbon dioxide evolution in runoff from simulated rainfall on long-term no-till and plowed soils in southwestern Ohio. *Soil and Tillage Research* 66(1), 23–33.

Jacinthe, P.-A., Lal, R., Owens, L. B. and Hothem, D. L. 2004. Transport of labile carbon in runoff as affected by land use and rainfall characteristics. *Soil and Tillage Research* 77(2), 111–123.

Janzen, H. H. 2004. Carbon cycling in earth systems—A soil science perspective. *Agriculture, Ecosystems and Environment* 104(3), 399–417. doi: 10.1016/j.agee.2004.01.040.

Jayaraman, S., Naorem, A. K., Lal, R., Dalal, R. C., Sinha, N. K., Patra, A. K. and Chaudhari, S. K. 2021. Disease-suppressive soils - Beyond food production: A critical review. *Journal of Soil Science and Plant Nutrition* 21(2), 1437-1465. doi: 10.1007/s42729-021-00451-x.

Jenny, H. 1941. *Factors of Soil Formation: a System of Quantitative Pedology*. Dover, New York, republished in 1994.

Jobbagy, E. G. and Jackson, R. B. 2000. The vertical distribution of soil organic carbon and its relation to climate and vegetation. *Ecological Applications* 10(2), 423–436. doi: 10.1890/1051-0761(2000)010[0423:TVDOSO]2.0.CO;2.

Jobbágy, E. G. and Jackson, R. B. 2001. The distribution of soil nutrients with depth: global patterns and the imprint of plants. *Biogeochemistry* 53(1), 51-77. doi: 10.1023/A:1010760720215.

John, R., Dalling, J. W., Harms, K. E., Yavitt, J. B., Stallard, R. F., Mirabello, M., Hubbell, S. P., Valencia, R., Navarrete, H., Vallejo, M. and Foster, R. B. 2007. Soil nutrients influence spatial distributions of tropical tree species. *Proceedings of the National Academy of Sciences of the United States of America* 104(3), 864-869. doi: 10.1073/pnas.0604666104.

Johnson, K., Purvis, G., Lopez-Capel, E., Peacock, C., Gray, N., Wagner, T., März, C., Bowen, L., Ojeda, J., Finlay, N., Robertson, S., Worrall, F. and Greenwell, C. 2015. Towards a mechanistic understanding of carbon stabilization in manganese oxides. *Nature Communications* 6, 7628. doi: 10.1038/ncomms8628.

Jones, P., Garcia, B. J., Furches, A., Tuskan, G. A. and Jacobson, D. 2019. Plant host-associated mechanisms for microbial selection. *Frontiers in Plant Science* 10, 862. doi: 10.3389/fpls.2019.00862.

Juarez, S., Rumpel, C., Mchunu, C. and Chaplot, V. 2011. Carbon mineralization and lignin content of erodedd sediments from a grazed watershed of South-Africa. *Geoderma* 167-168, 247–253. doi: 10.1016/j.geoderma.2011.09.007.

Jungkunst, H. F., Goepel, J., Horvath, T., Ott, S. and Brunn, M. 2021. New uses for old tools: Reviving Holdridge Life Zones in soil carbon persistence research. *Journal of Plant Nutrition and Soil Science* 184(1), 5-11. doi: 10.1002/jpln.202100008.

Kaiser, K. and Kalbitz, K. 2012. Cycling downwards - Dissolved organic matter in soils. *Soil Biology and Biochemistry* 52, 29-32. doi: 10.1016/j.soilbio.2012.04.002.

Kandeler, E. 1990. Characterization of free and adsorbed phosphatases in soils. *Biology and Fertility of Soils* 9(2), 199–202. doi: 10.1007/BF00335808.

Kaplan, J. O., Krumhardt, K. M. and Zimmermann, N. 2009. The prehistoric and preindustrial deforestation of Europe. *Quaternary Science Reviews* 28(27–28), 3016-3034. doi: 10.1016/j.quascirev.2009.09.028.

Kawamiya, M., Hajima, T., Tachiiri, K., Watanabe, S. and Yokohata, T. 2020. Two decades of Earth system modeling with an emphasis on Model for Interdisciplinary Research on Climate (MIROC). *Progress in Earth and Planetary Science* 7(1), 64. doi: 10.1186/s40645-020-00369-5.

Keiluweit, M., Wanzek, T., Kleber, M., Nico, P. and Fendorf, S. 2017. Anaerobic microsites have an unaccounted role in soil carbon stabilization. *Nature Communications* 8(1), 1771. doi: 10.1038/s41467-017-01406-6.

Khedim, N., Cécillon, L., Poulenard, J., Barré, P., Baudin, F., Marta, S., Rabatel, A., Dentant, C., Cauvy-Fraunié, S., Anthelme, F., Gielly, L., Ambrosini, R., Franzetti, A., Azzoni, R. S., Cacciangia, M. S., Compostella, C., Clague, J., Tielidze, L., Messenger, E., Choler, P. and Ficetola, G. F. 2021. Topsoil organic matter build-up in glacier forelands around the world. *Global Change Biology* 27(8), 1662–1677. doi: 10.1111/gcb.15496.

Killough, G. G. and Emanuel, W. R. 1981. A comparison of several models of carbon turnover in the ocean with respect to their distributions of transit time and age, and responses to atmospheric CO_2 and ^{14}C. *Tellus* 33(3), 274–290. doi: 10.3402/tellusa.v33i3.10715.

Kirkels, F. M. S. A., Cammeraat, L. H. and Kuhn, N. J. 2014. The fate of soil organic carbon upon erosion, transport and deposition in agricultural landscapes – A review of different concepts. *Geomorphology* 226, 94–105. doi: 10.1016/j.geomorph.2014.07.023.

Kirschbaum, M. U. F. 1995. The temperature dependence of soil organic matter decomposition, and the effect of global warming on soil organic C storage. *Soil Biology and Biochemistry* 27(6), 753–760. doi: 10.1016/0038-0717(94)00242-S.

Kirschbaum, M. U. F. 2000. Will changes in soil organic carbon act as a positive or negative feedback on global warming? *Biogeochemistry* 48(1), 21–51. doi: 10.1023/A:1006238902976.

Kleber, M., Bourg, I. C., Coward, E. K., Hansel, C. M., Myneni, S. C. B. and Nunan, N. 2021. Dynamic interactions at the mineral–organic matter interface. *Nature Reviews Earth and Environment* 2(6), 402–421. doi: 10.1038/s43017-021-00162-y.

Kleber, M., Eusterhues, K., Keiluweit, M., Mikutta, C., Mikutta, R. and Nico, P. S. 2015. Mineral-organic associations: formation, properties, and relevance in soil environments. *Advances in Agronomy* 130, 1–140. doi: 10.1016/bs.agron.2014.10.005.

Klein Goldewijk, K., Beusen, A., Doelman, J. and Stehfest, E. 2017. Anthropogenic land use estimates for the Holocene-HYDE 3.2. *Earth System Science Data* 9(2), 927–953. doi: 10.5194/essd-9-927-2017.

Klein Goldewijk, K., Beusen, A., Van Drecht, G. and De Vos, M. 2011. The HYDE 3.1 spatially explicit database of human-induced global land-use change over the past 12,000 years. *Global Ecology and Biogeography* 20(1), 73–86. doi: 10.1111/j.1466-8238.2010.00587.x.

Kögel-Knabner, I. and Amelung, W. 2014. Dynamics, chemistry, and preservation of organic matter in soils. In: Holland, H. D. and Turekian, K. K. (Eds.), *Treatise on Geochemistry* (2nd edn.). Amsterdam, Netherlands, Oxford, UK, Waltham: Elsevier Publishers, p. 9144.

Kögel-Knabner, I. and Amelung, W. 2021. Soil organic matter in major pedogenic soil groups. *Geoderma* 384, 114785, doi: 10.1016/j.geoderma.2020.114785.

Kramer, M. G. and Chadwick, O. A. 2018. Climate-driven thresholds in reactive mineral retention of soil carbon at the global scale. *Nature Climate Change* 8(12), 1104–1108. doi: 10.1038/s41558-018-0341-4.

Kuhn, N. J., Armstrong, E. K., Ling, A. C., Connolly, K. L. and Heckrath, G. 2012. Interrill erosion of carbon and phosphorus from conventionally and organically farmed Devon silt soils. *CATENA* 91, 94–103. doi: 10.1016/j.catena.2010.10.002.

Lal, R. (Ed.) 2001. *Soil Carbon Sequestration and the Greenhouse Effect*. SSSA, Madison, pp. 173-182.

Lal, R. 2003. Soil erosion and the global carbon budget. *Environment International* 29(4), 437–450. doi: 10.1016/S0160-4120(02)00192-7.

Lal, R. 2004. Soil carbon sequestration to mitigate climate change. *Geoderma* 123(1-2), 1-22. doi: 10.1016/j.geoderma.2004.01.032.

Laliberte, E., Grace, J. B., Huston, M. A., Lambers, H., Teste, F. P., Turner, B. L. and Wardle, D. A. 2013. How does pedogenesis drive plant diversity? *Trends in Ecology and Evolution* 28(6), 331-340. doi: 10.1016/j.tree.2013.02.008.

Laliberté, E., Schweiger, A. K. and Legendre, P. 2020. Partitioning plant spectral diversity into alpha and beta components. *Ecology Letters* 23(2), 370-380. doi: 10.1111/ele.13429.

Lauber, C. L., Hamady, M., Knight, R. and Fierer, N. 2009. Pyrosequencing-based assessment of soil pH as a predictor of soil bacterial community structure at the continental scale. *Applied and Environmental Microbiology* 75(15), 5111-5120. doi: 10.1128/AEM.00335-09.

Lawrence, J. E. and Hornberger, G. M. 2007. Soil moisture variability across climate zones. *Geophysical Research Letters* 34(20). doi: 10.1029/2007GL031382.

Lazcano, C., Zhu-Barker, X. and Decock, C. 2021. Effects of organic fertilizers on the soil microorganisms responsible for N_2O emissions: A review. *Microorganisms* 9(5), 983. doi: 10.3390/microorganisms9050983.

Lehmann, J. and Kleber, M. 2015. The contentious nature of soil organic matter. *Nature* 528(7580), 60–68. doi: 10.1038/nature16069.

Lehmann, J., Hansel, C.M., Kaiser, C., Kleber, M., Maher, K., Manzoni, S., Nunan, N., Reichstein, M., Schimel, J.P., Torn, M.S., Wieder, W.R., Kögel-Knabner, I., 2020. Persistence of soil organic carbon caused by functional complexity. *Nat. Geosci.* 1-6. https://doi.org/10.1038/s41561-020-0612-3

Leifeld, J. and Menichetti, L. 2018. The underappreciated potential of peatlands in global climate change mitigation strategies. *Nature Communications* 9(1), 1071. doi: 10.1038/s41467-018-03406-6.

Leite, M. F. A., Pan, Y., Bloem, J., ten Berge, H. T. and Kuramae, E. E. 2017. Organic nitrogen rearranges both structure and activity of the soil-borne microbial seedbank. *Scientific Reports* 7, 42634. doi: 10.1038/srep42634.

Liang, C., Amelung, W., Lehmann, J. and Kästner, M. 2019. Quantitative assessment of microbial necromass contribution to soil organic matter. *Global Change Biology* 25(11), 3578–3590. doi: 10.1111/gcb.14781.

Liang, C., Schimel, J. P. and Jastrow, J. D. 2017. The importance of anabolism in microbial control over soil carbon storage. *Nature Microbiology* 2, 17105. doi: 10.1038/nmicrobiol.2017.105.

Lorenz, K. and Lal, R. 2018. Carbon sequestration in grassland soils. In: *Carbon Sequestration in Agricultural Ecosystems*. Cham: Springer. doi: 10.1007/978-3-319-92318-5_4.

Louca, S., Polz, M. F., Mazel, F., Albright, M. B. N., Huber, J. A., O'Connor, M. I., Ackermann, M., Hahn, A. S., Srivastava, D. S., Crowe, S. A., Doebeli, M. and Parfrey, L. W. 2018.

Function and functional redundancy in microbial systems. *Nature Ecology and Evolution* 2(6), 936–943. doi: 10.1038/s41559-018-0519-1.

Lourenço, K. S., Cantarella, H., Soares, J. R., Gonzaga, L. C. and Menegale, P. LdC. 2021. DMPP mitigates N_2O emissions from nitrogen fertilizer applied with concentrated and standard vinasse. *Geoderma* 404, 115258. doi: 10.1016/j.geoderma.2021.115258.

Lu, C. and Tian, H. 2017. Global nitrogen and phosphorus fertilizer use for agricultural production in the past half century: Shifted hot spots and nutrient imbalance. *Earth System Science Data* 9, 181–192. doi: 10.1594/PANGAEA.863323.

Lu, M., Yang, Y., Luo, Y., Fang, C., Zhou, X., Chen, J., Yang, X. and Li, B. 2011. Responses of ecosystem nitrogen cycle to nitrogen addition: A meta-analysis. *New Phytologist* 189(4), 1040–1050. doi: 10.1111/j.1469-8137.2010.03563.x.

Lu, Z., Miller, P. A., Zhang, Q., Warlind, D., Nieradzik, L., Sjolte, J., Li, Q. and Smith, B. 2019. Vegetation patterns and terrestrial carbon variation in past warm and cold climates. *Geophysical Research Letters* 46(14), 8133–8143. doi: 10.1029/2019GL083729.

Lugato, E., Lavallee, J. M., Haddix, M. L., Panagos, P. and Cotrufo, M. F. 2021. Different climate sensitivity of particulate and mineral-associated soil organic matter. *Nature Geoscience* 14(5), 295–300. doi: 10.1038/s41561-021-00744-x.

Lupatini, M., Korthals, G. W., de Hollander, M., Janssens, T. K. S. and Kuramae, E. E. 2017. Soil microbiome is more heterogeneous in organic than in conventional farming system. *Frontiers in Microbiology* 7, 2064. doi: 10.3389/fmicb.2016.02064.

Ma, H., Mo, L., Crowther, T. W., Maynard, D. S., van den Hoogen, J., Stocker, B. D., Terrer, C. and Zohner, C. M. 2021. The global distribution and environmental drivers of aboveground versus belowground plant biomass. *Nature Ecology and Evolution* 5(8), 1110–1122. doi: 10.1038/s41559-021-01485-1.

Mackenzie, F. T., Lerman, A. and Andersson, A. J. 2004. Past and present of sediment and carbon biogeochemical cycling models. *Biogeosciences* 1(1), 11–32. doi: 10.5194/bg-1-11-2004.

Maestre, F. T., Delgado-Baquerizo, M., Jeffries, T. C., Eldridge, D. J., Ochoa, V., Gozalo, B., Quero, J. L., García-Gómez, M., Gallardo, A., Ulrich, W., Bowker, M. A., Arredondo, T., Barraza-Zepeda, C., Bran, D., Florentino, A., Gaitán, J., Gutiérrez, J. R., Huber-Sannwald, E., Jankju, M., Mau, R. L., Miriti, M., Naseri, K., Ospina, A., Stavi, I., Wang, D., Woods, N. N., Yuan, X., Zaady, E. and Singh, B. K. 2015. Increasing aridity reduces soil microbial diversity and abundance in global drylands. *Proceedings of the National Academy of Sciences of the United States of America* 112(51), 15684–15689. doi: 10.1073/pnas.1516684112.

Maire, V., Alvarez, G., Colombet, J., Comby, A., Despinasse, R., Dubreucq, E., Joly, M., Lehours, A.-C., Perrier, V., Shazad, T. and Fontaine, S. 2013. An unknown oxidative metabolism substantially contributes to soil CO_2 emissions. *Biogeosciences* 10(2), 1155-1167. doi: 10.5194/bg-10-1155-2013.

Manzoni, S., Schimel, J. P. and Porporato, A. 2012a. Responses of soil microbial communities to water stress: Results from a meta-analysis. *Ecology* 93(4), 930–938. doi: 10.1890/11-0026.1.

Manzoni, S., Taylor, P., Richter, A., Porporato, A. and Ågren, G. I. 2012b. Environmental and stoichiometric controls on microbial carbon-use efficiency in soils. *New Phytologist* 196(1), 79-91. doi: 10.1111/j.1469-8137.2012.04225.x.

Masteling, R., Lombard, L., de Boer, W., Raaijmakers, J. M. and Dini-Andreote, F. 2019. Harnessing the microbiome to control plant parasitic weeds. *Current Opinion in Microbiology* 49, 26-33. doi: 10.1016/j.mib.2019.09.006.

Mikutta, R., Kleber, M., Torn, M. S. and Jahn, R. 2006. Stabilization of soil organic matter: Association with minerals or chemical recalcitrance? *Biogeochemistry* 77(1), 25–56. doi: 10.1007/s10533-005-0712-6.

Mikutta, R., Schaumann, G. E., Gildemeister, D., Bonneville, S., Kramer, M. G., Chorover, J., Chadwick, O. A. and Guggenberger, G. 2009 Biogeochemistry of mineral-organic associations across a long-term mineralogical soil gradient (0.3–4100kyr), Hawaiian Islands. *Geochimica et Cosmochimica Acta* 73(7), 2034–2060. doi: 10.1016/j.gca.2008.12.028.

Miltner, A., Bombach, P., Schmidt-Brücken, B. and Kästner, M. 2012. SOM genesis: Microbial biomass as a significant source. *Biogeochemistry* 111(1-3), 41-55. doi: 10.1007/s10533-011-9658-z.

Minasny, B., Malone, B. P., McBratney, A. B., Angers, D. A., Arrouays, D., Chambers, A., Chaplot, V., Chen, Z.-S., Cheng, K., Das, B. S., Field, D. J., Gimona, A., Hedley, C. B., Hong, S. Y., Mandal, B., Marchant, B. P., Martin, M., McConkey, B. G., Mulder, V. L., O'Rourke, S., Richer-de-Forges, A. C., Odeh, I., Padarian, J., Paustian, K., Pan, G., Poggio, L., Savin, I., Stolbovoy, V., Stockmann, U., Sulaeman, Y., Tsui, C.-C., Vågen, T.-G., van Wesemael, B. and Winowiecki, L. 2017. Soil carbon 4 per mille. *Geoderma* 292, 59–86. doi: 10.1016/j.geoderma.2017.01.002.

Moinet, G. Y. K., Hunt, J. E., Kirschbaum, M. U. F., Morcom, C. P., Midwood, A. J. and Millard, P. 2018. The temperature sensitivity of soil organic matter decomposition is constrained by microbial access to substrates. *Soil Biology and Biogeochemistry* 116, 333–339. doi: 10.1016/j.soilbio.2017.10.031.

Moinet, G. Y. K., Moinet, M., Hunt, J. E. Rumpel, C., Chabbi, A. and Millard, P. 2020. Temperature sensitivity of decomposition decreases with increasing soil organic matter stability. *Science of the Total Environment* 704(135460), Elsevier.

Monhonval, A., Strauss, J., Mauclet, E., Hirst, C., Bemelmans, N., Grosse, G., Schirrmeister, L., Fuchs, M. and Opfergelt, S. 2021. Iron redistribution upon thermokarst processes in the Yedoma domain. *Frontiers in Earth Science* 9. doi: 10.3389/feart.2021.703339.

Mooshammer, M., Wanek, W., Schnecker, J., Wild, B., Leitner, S., Hofhansl, F., Blöchl, A., Hämmerle, I., Frank, A. H., Fuchslueger, L., Keiblinger, K. M., Zechmeister-Boltenstern, S. and Richter, A. 2012. Stoichiometric controls of nitrogen and phosphorus cycling in decomposing beech leaf litter. *Ecology* 93(4), 770–782.

Mooshammer, M., Wanek, W., Zechmeister-Boltenstern, S. and Richter, A. A. 2014. Stoichiometric imbalances between terrestrial decomposer communities and their resources: mechanisms and implications of microbial adaptations to their resources. *Frontiers in Microbiology* 5, 22, doi: 10.3389/fmicb.2014.00022.

Morales, S. E., Cosart, T. and Holben, W. E. 2010. Bacterial gene abundances as indicators of greenhouse gas emission in soils. *ISME Journal* 4(6), 799–808. doi: 10.1038/ismej.2010.8.

Morris, W. F., Ehrlen, J., Dahlgren, J. P., Loomis, A. K. and Louthan, A. M. 2020. Biotic and anthropogenic forces rival climatic/abiotic factors in determining global plant population growth and fitness. *Proceedings of the National Academy of Sciences of the United States of America* 117(2), 1107–1112. doi: 10.1073/pnas.1918363117.

Moyano, F. E., Manzoni, S. and Chenu, C. 2013. Responses of soil heterotrophic respiration to moisture availability: An exploration of processes and models. *Soil Biology and Biochemistry* 59, 72-85. doi: 10.1016/j.soilbio.2013.01.002.

Naipal, V., Ciais, P., Wang, Y. L., Lauerwald, R., Guenet, B. and Van Oost, K. 2018. Global soil organic carbon removal by water erosion under climate change and land use change during AD 1850-2005. *Biogeosciences* 15(14), 4459-4480. doi: 10.5194/bg-15-4459-2018.

O'Brien, H. E., Parrent, J. L., Jackson, J. A., Moncalvo, J. M. and Vilgalys, R. 2005. Fungal community analysis by large-scale sequencing of environmental samples. *Applied and Environmental Microbiology* 71(9), 5544-5550. doi: 10.1128/AEM.71.9.5544-5550.2005.

Opfergelt, S. 2020. The next generation of climate model should account for the evolution of mineral-organic interactions with permafrost thaw. *Environmental Research Letters* 15(9), 091003. doi: 10.1088/1748-9326/ab9a6d.

Opfergelt, S., Williams, H. M., Cornelis, J. T., Guicharnaud, R. A., Georg, R. B., Siebert, C., Gislason, S. R., Halliday, A. N. and Burton, K. W. 2017. Iron and silicon isotope behaviour accompanying weathering in Icelandic soils, and the implications for iron export from peatlands. *Geochimica et Cosmochimica Acta* 217, 273-291. doi: 10.1016/j.gca.2017.08.033.

Orwin, K. H., Kirschbaum, M. U. F., John, M. G. S. and Dickie, I. A. 2011. Organic nutrient uptake by mycorrhizal fungi enhances ecosystem carbon storage: A model-based assessment. *Ecology Letters* 14(5), 493-502. doi: 10.1111/j.1461-0248.2011.01611.x.

Otero, X. L., Ferreira, T. O., Huerta-Díaz, M. A., Partiti, C. S. M., Souza, V., Vidal-Torrado, P. and Macías, F. 2009. Geochemistry of iron and manganese in soils and sediments of a mangrove system, Island of Pai Matos (Cananeia - SP, Brazil). *Geoderma* 148(3-4), 318-335. doi: 10.1016/j.geoderma.2008.10.016.

Öttl, L. K., Wilken, F., Auerswald, K., Sommer, M., Wehrhan, M. and Fiener, P. 2021. Tillage erosion as an important driver of in-field biomass patterns in an intensively used hummocky landscape. *Land Degradation and Development* 32(10), 3077-3091. doi: 10.1002/ldr.3968.

Pan, Y., Cassman, N., de Hollander, M., Mendes, L. W., Korevaar, H., Geerts, R. H. E. M., van Veen, J. A. and Kuramae, E. E. 2014. Impact of long-term N, P, K, and NPK fertilization on the composition and potential functions of the bacterial community in grassland soil. *FEMS Microbiology Ecology* 90(1), 195-205. doi: 10.1111/1574-6941.12384.

Parton, W. J., Schimel, D. S., Cole, C. V. and Ojima, D. S. 1987. Analysis of factors controlling soil organic matter levels in Great Plains grasslands. *Soil Science Society of America Journal* 51(5), 1173-1179. doi: 10.2136/sssaj1987.03615995005100050015x.

Patzner, M. S., Mueller, C. W., Malusova, M., Baur, M., Nikeleit, V., Scholten, T., Hoeschen, C., Byrne, J. M., Borch, T., Kappler, A. and Bryce, C. 2020. Iron mineral dissolution releases iron and associated organic carbon during permafrost thaw. *Nature Communications* 11(1), 6329. doi: 10.1038/s41467-020-20102-6.

Paul, E. A. 2016. The nature and dynamics of soil organic matter: Plant inputs, microbial transformations, and organic matter stabilization. *Soil Biology and Biochemistry* 98, 109-126. doi: 10.1016/j.soilbio.2016.04.001.

Paustian, K., Larson, E., Kent, J., Marx, E. and Swan, A. 2019. Soil C sequestration as a biological negative emission strategy. *Frontiers in Climate* 1(8), doi: 10.3389/fclim.2019.00008.

Pellegrini, A. F. A., Ahlstrom, A., Hobbie, S. E., Reich, P. B., Nieradzik, L. P., Staver, A. C., Scharenbroch, B. C., Jumpponen, A., Anderegg, W. R. L., Randerson, J. T. and Jackson, R. B. 2018. Fire frequency drives decadal changes in soil carbon and nitrogen and ecosystem productivity. *Nature* 553(7687), 194–198. doi: 10.1038/nature24668.

Pelosi, C., Barot, S., Capowiez, Y., Hedde, M. and Vandenbulcke, F. 2014. Pesticides and earthworms: A review. *Agronomy for Sustainable Development* 34(1), 199–228. doi: 10.1007/s13593-013-0151-z.

Peng, C. H., Guiot, J. and Van Campo, E. 1998. Estimating changes in terrestrial vegetation and carbon storage. *Quaternary Science Reviews* 17(8), 719–735. doi: 10.1016/S0277-3791(97)00045-0.

Peyton Smith, A. P., Bond-Lamberty, B., Benscoter, B. W., Tfaily, M. M., Hinkle, C. R., Liu, C. and Bailey, V. L. 2017. Shifts in pore connectivity from precipitation versus groundwater rewetting increases soil carbon loss after drought. *Nature Communications* 8(1), 1335. doi: 10.1038/s41467-017-01320-x.

Phillips, R. P., Brzostek, E. and Midgley, M. G. 2013. The mycorrhizal-associated nutrient economy: A new framework for predicting carbon-nutrient couplings in temperate forests. *New Phytologist* 199(1), 41–51. doi: 10.1111/nph.12221.

Pimentel, D., Harvey, C., Resosudarmo, P., Sinclair, K., Kurz, D., McNair, M., Crist, S., Shpritz, L., Fitton, L., Saffouri, R. and Blair, R. 1995. Environmental and economic costs of soil erosion and conservation benefits. *Science* 267(5201), 1117–1123. doi: 10.1126/science.267.5201.1117.

Polyakov, V. O. and Lal, R. 2008. Soil organic matter and CO_2 emission as effected by water erosion on field runoff plots. *Geoderma* 143(1-2), 216–222. doi: 10.1016/j.geoderma.2007.11.005.

Post, W. M., Emanuel, W. R., Zinke, P. J. and Stangenberger, A. G. 1982. Soil carbon pools and world life zones. *Nature* 298(5870), 156–159. doi: 10.1038/298156a0.

Poulton, P., Johnston, J., Macdonald, A., White, R. and Powlson, D. 2018. Major limitations to achieving "4 per 1000" increases in soil organic carbon stock in temperate regions: Evidence from long-term experiments at Rothamsted Research, United Kingdom. *Global Change Biology* 24(6), 2563–2584. doi: 10.1111/gcb.14066.

Prober, S. M., Leff, J. W., Bates, S. T., Borer, E. T., Firn, J., Harpole, W. S., Lind, E. M., Seabloom, E. W., Adler, P. B., Bakker, J. D., Cleland, E. E., DeCrappeo, N. M., DeLorenze, E., Hagenah, N., Hautier, Y., Hofmockel, K. S., Kirkman, K. P., Knops, J. M. H., La Pierre, K. J., MacDougall, A. S., McCulley, R. L., Mitchell, C. E., Risch, A. C., Schuetz, M., Stevens, C. J., Williams, R. J. and Fierer, N. 2015. Plant diversity predicts beta but not alpha diversity of soil microbes across grasslands worldwide. *Ecology Letters* 18(1), 85–95. doi: 10.1111/ele.12381.

Pugnaire, F. I., Morillo, J. A., Penuelas, J., Reich, P. B., Bardgett, R. D., Gaxiola, A., Wardle, D. A. and van der Putten, W. H. 2019. Climate change effects on plant-soil feedbacks and consequences for biodiversity and functioning of terrestrial ecosystems. *Science Advances* 5(11), eaaz1834. doi: 10.1126/sciadv.aaz1834.

Qiu, C., Ciais, P., Zhu, D., Guenet, B., Peng, S., Petrescu, A. M. R., Lauerwald, R., Makowski, D., Gallego-Sala, A. V., Charman, D. J. and Brewer, S. C. 2021. Large historical carbon emissions from cultivated northern peatlands. *Science Advances* 7(23), eabf1332. doi: 10.1126/sciadv.abf1332: Available at: https://doi.org./10.1126/sciadv.abf1332.

Quideau, S. A., Chadwick, O. A., Benesi, A., Graham, R. C. and Anderson, M. A. 2001. A direct link between forest vegetation type and dsoil organic matter composition. *Geoderma* 104(1-2), 41–60. doi: 10.1016/S0016-7061(01)00055-6.

Quinton, J. N., Govers, G., Van Oost, K. and Bardgett, R. D. 2010. The impact of agricultural soil erosion on biogeochemical cycling. *Nature Geoscience* 3(5), 311–314. doi: 10.1038/ngeo838.

Rasmussen, C., Heckman, K., Wieder, W. R., Keiluweit, M., Lawrence, C. R., Berhe, A. A., Blankinship, J. C., Crow, S. E., Druhan, J. L., Pries, C. E. H., Marin-Spiotta, E., Plante, A. F., Schädel, C., Schimel, J. P., Sierra, C. A., Thompson, A. and Wagai, R. 2018. Beyond clay: Towards an improved set of variables for predicting soil organic matter content. *Biogeochemistry* 137(3), 297–306. doi: 10.1007/s10533-018-0424-3.

Raynaud, X. and Nunan, N. 2014. Spatial ecology of bacteria at the microscale in soil. *PLoS ONE* 9(1), e87217. doi: 10.1371/journal.pone.0087217.

Reich, P. B., Luo, Y., Bradford, J. B., Poorter, H., Perry, C. H. and Oleksyn, J. 2014. Temperature drives global patterns in forest biomass distribution in leaves, stems, and roots. *Proceedings of the National Academy of Sciences of the United States of America* 111(38), 13721–13726. doi: 10.1073/pnas.1216053111.

Reich, P. B., Peterson, D. W., Wedin, D. A. and Wrage, K. 2001. Fire and vegetation effects on productivity and nitrogen cycling across a forest-grassland continuum. *Ecology* 82(6), 1703–1719. doi: 10.1890/0012-9658(2001)082[1703:FAVEOP]2.0.CO;2.

Reid, T. E., Kavamura, V. N., Abadie, M., Torres-Ballesteros, A., Pawlett, M., Clark, I. M., Harris, J. and Mauchline, T. H. 2021. Inorganic chemical fertilizer application to wheat reduces the abundance of putative plant growth-promoting rhizobacteria. *Frontiers in Microbiology* 12, 642587. doi: 10.3389/fmicb.2021.642587.

Reinmann, A. B. and Templer, P. H. 2016. Reduced winter snowpack and greater soil frost reduce live root biomass and stimulate radial growth and stem respiration of red maple (Acer rubrum) trees in a mixed-hardwood forest. *Ecosystems* 19(1), 129–141. doi: 10.1007/s10021-015-9923-4.

Rejman, J., Iglik, I., Paluszek, J. and Rodzik, J. 2014. Soil redistribution and crop productivity in loess areas (Lublin, Upland, Poland). *Soil and Tillage Research* 143, 77–84. doi: 10.1016/j.still.2014.05.011.

Remus, R., Kaiser, M., Kleber, M., Augustin, J. and Sommer, M. 2018. Demonstration of the rapid incorporation of carbon into protective, mineral-associated organic carbon fractions in an eroded soil from the CarboZALF experimental site. *Plant and Soil* 430(1–2), 329–348. doi: 10.1007/s11104-018-3724-4.

Richter, D. D. and Billings, S. A. 2015. "One physical system": Tansley's ecosystem as Earth's critical zone. *New Phytologist* 206(3), 900–912. doi: 10.1111/nph.13338.

Richter, D. D. and Yaalon, D. H. 2012. "The changing model of soil" revisited. *Soil Science Society of America Journal* 76(3), 766–778. doi: 10.2136/sssaj2011.0407.

Riedo, J., Wettstein, F. E., Rösch, A., Herzog, C., Sanerjee, S., Büchi, L., Charles, R., Wächter, D., Martin-Laurent, F., Bucheli, T. D., Walder, F. and van der Heijden, M. G. A. 2021 Widespread occurrence of pesticides in organically managed agricultural soils - The ghost of a conventional agricultural past? *Environmental Science and Technology* 55(5), 2919–2928. doi: 10.1021/acs.est.0c06405.

Roberts, H. M., Muhs, D. R. and Bettis III, E. A. 2013. Loess records|North America. In: Elias, S. A. and Mock, C. J. (Eds.), *Encyclopedia of Quarternary Science* (2nd edn.). Amsterdam, Netherlands: Elsevier Publishers, pp. 620–628. doi: 10.1016/B978-0-444-53643-3.00159-X.

Robertson, A. D., Paustian, K., Ogle, S., Wallenstein, M. D., Lugato, E. and Cotrufo, M. F. 2019. Unifying soil organic matter formation and persistence frameworks: The MEMS model. *Biogeosciences* 16(6), 1225–1248. doi: 10.5194/bg-16-1225-2019.

Rocci, K. S., Lavalle, J. M., Stewart, C. E. and Cotrufo, M. F. 2021. Soil organic carbon response to global environmental change depends on its distribution between mineral-associated and particulate organic matter: A meta-analysis. *Science of the Total Environment* 793, 148569. doi: 10.1016/j.scitotenv.2021.148569.

Rogelj, J., Shindell, D., Jiang, K., Fifita, S., Forster, P., Ginzburg, V., Handa, C., Kheshgi, H., Kobayashi, S., Kriegler, E., Mundaca, L., Séférian, R. and Vilariño, M. V. 2018. Mitigation pathways compatible with 1.5°C in the context of sustainable development. In: *Global Warming of 1(5)°C*. An IPCC Special Report on the impacts of global warming of 1.5°C above pre-industrial levels and related global greenhouse gas emission pathways, in the context of strengthening the global response to the threat of climate change, sustainable development, and efforts to eradicate poverty [Masson-Delmotte, V., Zhai, P., Pörtner, H.-O., Roberts, D., Skea, J., Shukla, P. R., Pirani, A., Moufouma-Okia, W., Péan, C., Pidcock, R., Connors, S., Matthews, J. B. R., Chen, Y., Zhou, X., Gomis, M. I., Lonnoy, E., Maycock, T., Tignor, M. and Waterfield, T. (Eds.)].

Rowley, M. C., Grand, S. and Verracchia, É. P. 2018. Calcium-mediated stabilisation of soil organic carbon. *Biogeochemistry* 137(1–2), 27–49. doi: 10.1007/s10533-017-0410-1.

Rumpel, C., Amiraslani, F., Chenu, C., Garcia Cardenas, M., Kaonga, M., Koutika, L. S., Ladha, J., Madari, B., Shirato, Y., Smith, P., Soudi, B., Soussana, J. F., Whitehead, D. and Wollenberg, E. 2020. The 4p1000 initiative: Opportunities, limitations and challenges for implementing soil organic carbon sequestration as a sustainable development strategy. *Ambio* 49(1), 350–360. doi: 10.1007/s13280-019-01165-2.

Rumpel, C., Chaplot, V., Planchon, O., Bernadou, J., Valentin, C. and Matiottit, A. 2006. Preferential erosion of black carbon on steep slopes with slash and burn agriculture. *CATENA* 65(1), 30–40. doi: 10.1016/j.catena.2005.09.005.

Rumpel, C. and Kögel-Knabner, I. 2011. Deep soil organic matter—A key but poorly understood component of terrestrial C cycle. *Plant and Soil* 338(1–2), 143–158. doi: 10.1007/s11104-010-0391-5.

Rumpel, C., Kögel-Knabner, I. and Bruhn, F. 2002. Vertical distribution, age and chemical composition of organic carbon in two forest soils of different pedogenesis. *Organic Geochemistry* 33(10), 1131–1142. doi: 10.1016/S0146-6380(02)00088-8.

Ruysschaert, G., Poesen, J., Verstraeten, G. and Govers, G. 2007. Soil loss due to harvesting of various crop types in contrasting agro-ecological environments. *Agriculture, Ecosystems and Environment* 120(2–4), 153–165. doi: 10.1016/j.agee.2006.08.012.

Sanaullah, M., Blagodatskaya, E., Chabbi, A., Rumpel, C. and Kuzyakov, Y. 2011. Drought effects on microbial biomass and enzyme activities in the rhizosphere of grasses depend on plant community composition. *Applied Soil Ecology* 48(1), 38–44. doi: 10.1016/j.apsoil.2011.02.004.

Sanderman, J., Hengl, T. and Fiske, G. J. 2017. Soil carbon debt of 12,000 years of human land use. *Proceedings of the National Academy of Sciences of the United States of America* 114(36), 9575–9580. doi: 10.1073/pnas.1706103114.

Sanderman, J., Hengl, T. and Fiske, G. J. 2018. Correction for Soil carbon debt of 12,000 years of human land use. *PNAS* 115(7), E1700. doi: 10.1073/pnas.1800925115.

Schaffer, B. E. and Rodriguez-Iturbe, I. 2018. Water-limited vegetated ecosystems driven by stochastic rainfall: Feedbacks and bimodality. *Proceedings: Mathematical, Physical, and Engineering Sciences* 474(2214), 20170649, doi: 10.1098/rspa.2017.0649.

Schiettecatte, W., Gabriels, D., Cornelis, W. M. and Hofman, G. 2008a. Impact of deposition on the enrichment of organic carbon in eroded sediment. *CATENA* 72(3), 340–347. doi: 10.1016/j.catena.2007.07.001.

Schiettecatte, W., Gabriels, D., Cornelis, W. M. and Hofman, G. 2008b. Enrichment of organic carbon in sediment transport by interrill and rill erosion processes. *Soil Science Society of America Journal* 72(1), 50–55. doi: 10.2136/sssaj2007.0201.

Schimel, J. P. and Schaeffer, S. M. 2012. Microbial control over carbon cycling in soil. *Frontiers in Microbiology* 3, 348. doi: 10.3389/fmicb.2012.00348.

Schimel, J. P. and Weintraub, M. N. 2003. The implications of exoenzyme activity on microbial carbon and nitrogen limitation in soil: A theoretical model. *Soil Biology and Biochemistry* 35(4), 549–563. doi: 10.1016/S0038-0717(03)00015-4.

Schipper, L. A., Hobbs, J. K., Rutledge, S. and Arcus, V. L. 2014. Thermodynamic theory explains the temperature optima of soil microbial processes and high Q_{10} values at low temperatures. *Global Change Biology* 20(11), 3578–3586. doi: 10.1111/gcb.12596.

Schjønning, P., Akker, J. J. H., Keller, T., Greve, M. H., Lamandé, M., Simojoki, A., Stettler, M., Arvidsson, J. and Breuning-Madsen, H. 2015. Driver-Pressure-State-Impact-Response (DPSIR) analysis and risk assessment for soil compaction – A European perspective. *Advances in Agronomy* 133, 183–237. doi: 10.1016/bs.agron.2015.06.001.

Schmidt, J. E., Kent, A. D., Brisson, V. L. and Gaudin, A. C. M. 2019. Agricultural management and plant selection interactively affect rhizosphere microbial community structure and nitrogen cycling. *Microbiome* 7(1), 146. doi: 10.1186/s40168-019-0756-9.

Schrumpf, M., Kaiser, K., Mayer, A., Hempel, G. and Trumbore, S. 2021. Age distribution, extractability, and stability of mineral-bound organic carbon in central European soils. *Biogeosciences* 18(3), 1241–1257. doi: 10.5194/bg-18-1241-2021.

Schulz-Bohm, K., Gerards, S., Hundscheid, M., Melenhorst, J., De Boer, W. and Garbeva, P. 2018. Calling from distance: Attraction of soil bacteria by plant root volatiles. *ISME Journal* 12(5), 1252–1262. doi: 10.1038/s41396-017-0035-3.

Seneviratne, S. I., Zhang, X., Adnan, M., Badi, W., Dereczynski, C., Di Luca, A., Ghosh, S., Iskandar, I., Kossin, J., Lewis, S., Otto, F., Pinto, I., Satoh, M., Vicente-Serrano, S. M., Wehner, M. and Zhou, B. 2021. The Physical Science Basis. Contribution of Working Group I to the Sixth Assessment Report of the Intergovernmental Panel on Climate Change [Masson-Delmotte, V., P. Zhai, A. Pirani, S. L. Connors, C. Péan, S. Berger, N. Caud, Y. Chen, L. Goldfarb, M. I. Gomis, M. Huang, K. Leitzell, E. Lonnoy, J. B. R. Matthews, T. K. Maycock, T. Waterfield, O. Yelekçi, R. Yu and B. Zhou (Eds.)]. Weather and climate extreme events in a changing climate. In: *Climate Change*. Cambridge: Cambridge University Press.

Sharma, A., Kumar, V., Shahzad, B., Tanveer, M., Sidhu, G. P. S., Handa, N., Kohli, S. K., Yadav, P., Bali, A. S., Parihar, R. D., Dar, O. I., Singh, K., Jasrotia, S., Bakshi, P., Ramakrishnan, M., Kumar, S., Bhardwaj, R. and Thukral, A. K. 2019. Worldwide pesticide usage and its impacts on ecosystem. *SN Applied Sciences* 1(11), 1446. doi: 10.1007/s42452-019-1485-1.

Shi, Z., Allison, S. D., He, Y., Levine, P. A., Hoyt, A. M., Beem-Miller, J., Zhu, Q., Wieder, W. R., Trumbore, S. and Randerson, J. T. 2020. The age distribution of global soil carbon inferred from radiocarbon measurements. *Nature Geoscience* 13(8), 555–559. doi: 10.1038/s41561-020-0596-z.

Shu, X., Wang, Y., Wang, Y., Ma, Y., Men, M., Zheng, Y., Xue, C., Peng, Z. and Noulas, C. 2021. Response of soil N_2O emission and nitrogen utilization to organic matter in the wheat and maize rotation system. *Scientific Reports* 11(1), 4396. doi: 10.1038/s41598-021-83832-7.

Siebert, C., Pett-Ridge, J. C., Opfergelt, S., Guicharnaud, R. A., Halliday, A. N. and Burton, K. W. 2015. Molybdenum isotope fractionation in soils: influence of redox conditions, organic matter, and atmospheric inputs. *Geochimica et Cosmochimica Acta* 162, 1-24. doi: 10.1016/j.gca.2015.04.007.

Sinsabaugh, R. L., Manzoni, S., Moorhead, D. L. and Richter, A. 2013. Carbon use efficiency of microbial communities: Stoichiometry, methodology and modelling. *Ecology Letters* 16(7), 930-939. doi: 10.1111/ele.12113.

Sinsabaugh, R. L., Moorhead, D. L., Xu, X. and Litvak, M. E. 2017. Plant, microbial and ecosystem carbon use efficiencies interact to stabilize microbial growth as a fraction of gross primary production. *New Phytologist* 214(4), 1518-1526. doi: 10.1111/nph.14485.

Sinsabaugh, R. L., Turner, B. L., Talbot, J. M., Waring, B. G., Powers, J. S., Kuske, C. R., Moorhead, D. L. and Shah, J. J. F. 2016. Stoichiometry of microbial carbon use efficiency in soils. *Ecological Monographs* 86(2), 172-189. doi: 10.1890/15-2110.1.

Six, J., Elliott, E. T. and Paustian, K. 2000a. Soil macroaggregate turnover and microaggregate formation: A mechanism for C sequestration under no-tillage agriculture. *Soil Biology and Biochemistry* 32(14), 2099-2103. doi: 10.1016/S0038-0717(00)00179-6.

Six, J., Paustian, K., Elliot, E. T. and Combrink, C. 2000b. Soil structure and organic matter: I. Distribution of aggregate-size classes and aggregate-associated carbon. *Soil Science Society of America Journal* 64(2), 681-689. doi: 10.2136/sssaj2000.642681x.

Six, J., Guggenberger, G., Paustian, K., Haumaier, L., Elliot, E. T. and Zech, W. 2001. Sources and composition of soil organic matter fractions between and within soil aggregates. *European Journal of Soil Science* 52(4), 607-618. doi: 10.1046/j.1365-2389.2001.00406.x.

Six, J., Ogle, S. M., Breidt, F. J., Conant, R. T., Mosier, A. R. and Paustian, K. 2004. The potential to mitigate global warming with no-tillage management is only realized when practised in the long term. *Global Change Biology* 10(2), 155-160. doi: 10.1111/j.1529-8817.2003.00730.x.

Skopp, J., Jawson, M. D. and Doran, J. W. 1990. Steady-state aerobic microbial activity as a function of soil water content. *Soil Science Society of America Journal* 54(6), 1619-1625. doi: 10.2136/sssaj1990.03615995005400060018x.

Slessarev, E. W., Lin, Y., Bingham, N. L., Johnson, J. E., Dai, Y., Schimel, J. P. and Chadwick, O. A. 2016. Water balance creates a threshold in soil pH at the global scale. *Nature* 540(7634), 567-569. doi: 10.1038/nature20139.

Slette, I. J., Post, A. K., Awad, M., Even, T., Punzalan, A., Williams, S., Smith, M. D. and Knapp, A. K. 2019. How ecologists define drought, and why we should do better. *Global Change Biology* 25(10), 3193-3200. doi: 10.1111/gcb.14747.

Smith, P. 2004. Carbon sequestration in croplands: The potential in Europe and the global context. *European Journal of Agronomy* 20(3), 229-236. doi: 10.1016/j.eja.2003.08.002.

Smith, P. 2008. Land use change and soil organic carbon dynamics. *Nutrient Cycling in Agroecosystems* 81(2), 169-178. doi: 10.1007/s10705-007-9138-y.

Smith, S. E. and Smith, F. A. 2011. Roles of arbuscular mycorrhizas in plant nutrition and growth: New paradigms from cellular to ecosystem scales. *Annual Review of Plant Biology* 62, 227-250. doi: 10.1146/annurev-arplant-042110-103846.

Smith, S. V., Renwick, W. H., Buddemeier, R. W. and Crossland, C. J. 2001. Budgets of soil erosion and deposition for sediments and sedimentary organic carbon across the conterminous United States. *Global Biogeochemical Cycles* 15(3), 697-707. doi: 10.1029/2000GB001341.

Soong, J. L., Castanha, C., Hicks Pries, C. E., Ofiti, N., Porras, R. C., Riley, W. J., Schmidt, M. W. I. and Torn, M. S. 2021. Five years of whole-soil warming led to loss of subsoil carbon stocks and increased CO_2 efflux. *Science Advances* 7(21), eabd1343. doi: 10.1126/sciadv.abd1343.

Sorensen, P. O., Templer, P. H. and Finzi, A. C. 2016. Contrasting effects of winter snowpack and soil frost on growing season microbial biomass and enzyme activity in two mixed-hardwood forests. *Biogeochemistry* 128(1-2), 141-154. doi: 10.1007/s10533-016-0199-3.

Soussana, J.-F., Loiseau, P., Vuichard, N., Ceschia, E., Balesdent, J., Chevallier, T. and Arrouays, D. 2006. Carbon cycling and sequestration opportunities in temperate grasslands. *Soil Use and Management* 20(2), 219-230. doi: 10.1111/j.1475-2743.2004.tb00362.x.

Spohn, M., Klaus, K., Wanek, W. and Richter, A. 2016. Microbial carbon use efficiency and biomass turnover times depending on soil depth - Implications for carbon cycling. *Soil Biology and Biochemistry* 96, 74-81. doi: 10.1016/j.soilbio.2016.01.016.

Stallard, R. F. 1998. Terrestrial sedimentation and the carbon cycle: Coupling weathering and erosion to carbon burial. *Global Biogeochemical Cycles* 12(2), 231-257.

Steger, K., Fiener, P., Marvin-DiPasquale, M., Viers, J. H. and Smart, D. R. 2019. Human-induced and natural carbon storage in floodplains of the Central Valley of California. *Science of the Total Environment* 651(1), 851-858. doi: 10.1016/j.scitotenv.2018.09.205.

Steidinger, B. S., Crowther, T. W., Liang, J., Nuland, M. E. V., Werner, G. D. A., Reich, P. B., Nabuurs, G. J., De-Miguel, S., Zhou, M., Picard, N., Herault, B., Zhao, X., Zhang, C., Routh, D., Peay, K. G. and GFBI consortium 2019. Climatic controls of decomposition drive the global biogeography of forest-tree symbioses. *Nature* 569(7756), 404-408. doi: 10.1038/s41586-019-1128-0.

Stuckey, J. W., Goodwin, C., Wang, J., Kaplan, L. A., Vidal-Esquivel, P., Beebe, T. P. and Sparks, D. L. 2018. Impacts of hydrous manganese oxide on the retention and lability of dissolved organic matter. *Geochemical Transactions* 19(1), 6. doi: 10.1186/s12932-018-0051-x.

Sulman, B. N., Moore, J. A. M., Abramoff, R., Averill, C., Kivlin, S., Gorgeou, K., Sridhar, B., Hartman, M. D., Wang, G., Wieder, W. R., Bradford, M. A., Luo, Y., Mayes, M. A., Morrison, E., Riley, W. J., Salazar, A., Schimel, J. P., Tang, J. and Classen, A. T. 2018. Multiple models and experiments underscore large uncertainty in soil carbon dynamics. *Biogeochemistry* 141(2), 109-123. doi: 10.1007/s10533-018-0509-z.

Sulman, B. N., Shevliakova, E., Brzostek, E. R., Kivlin, S. N., Malyshev, S., Menge, D. N. L. and Zhang, X. 2019. Diverse mycorrhizal associations enhance terrestrial C Storage in a global model. *Global Biogeochemical Cycles* 33(4), 501-523. doi: 10.1029/2018GB005973.

Tang, S., Guo, J., Li, S., Li, J., Xie, S., Zhai, X., Wang, C., Zhang, Y. and Wang, K. 2019. Synthesis of soil carbon losses in response to conversion of grassland to agriculture land. *Soil and Tillage Research* 185, 29-35. doi: 10.1016/j.still.2018.08.011.

Tarnocai, C., Canadell, J. G., Schuur, E. A. G., Kuhry, P., Mazhitova, G. and Zimov, S. 2009. Soil organic carbon pools in the northern circumpolar permafrost region. *Global Biogeochemical Cycles* 23(2), GB2023. doi: 10.1029/2008GB003327.

Tedersoo, L., Bahram, M., Põlme, S., Kõljalg, U., Yorou, N. S., Wijesundera, R., Ruiz, L. V., Vasco-Palacios, A. M., Thu, P. Q., Suija, A., Smith, M. E., Sharp, C., Saluveer, E., Saitta, A., Rosas, M., Riit, T., Ratkowsky, D., Pritsch, K., Põldmaa, K., Piepenbring, M., Phosri, C., Peterson, M., Parts, K., Pärtel, K., Otsing, E., Nouhra, E., Njouonkou, A. L., Nilsson, R. H., Morgado, L. N., Mayor, J., May, T. W., Majuakim, L., Lodge, D. J., Lee, S. S., Larsson, K. H., Kohout, P., Hosaka, K., Hiiesalu, I., Henkel, T. W., Harend, H., Guo, L. D., Greslebin, A., Grelet, G., Geml, J., Gates, G., Dunstan, W., Dunk, C., Drenkhan, R., Dearnaley, J., De Kesel, A., Dang, T., Chen, X., Buegger, F., Brearley, F. Q., Bonito, G., Anslan, S., Abell, S. and Abarenkov, K. 2014. Fungal biogeography. Global diversity and geography of soil fungi. *Science* 346(6213), 1256688. doi: 10.1126/science.1256688.

Thomine, E., Mumford, J., Rusch, A. and Desneux, N. 2022. Using crop diversity to lower pesticide use: Socio-ecological approaches. *Science of the Total Environment* 804, 150156. doi: 10.1016/j.scitotenv.2021.150156.

Tilman, D., Balzer, C., Hill, J. and Befort, B. L. 2011. Global food demand and the sustainable intensification of agriculture. *Proceedings of the National Academy of Sciences of the United States of America* 108(50), 20260-20264. doi: 10.1073/pnas.1116437108.

Todd-Brown, K. E. O., Randerson, J. T., Hopkins, F., Arora, V., Hajima, T., Jones, C., Shevliakova, E., Tjiputra, J., Volodin, E., Wu, T., Zhang, Q. and Allison, S. D. 2014. Changes in soil organic carbon storage predicted by Earth system models during the 21st century. *Biogeosciences* 11(8), 2341-2356. doi: 10.5194/bg-11-2341-2014.

Torn, M. S., Trumbore, S. E., Chadwick, O. A., Vitousek, P. M. and Hendricks, D. M. 1997. Mineral control of soil organic carbon storage and turnover. *Nature* 389(6647), 170-173. doi: 10.1038/38260.

Torsvik, V., Goksøyr, J. and Daae, F. L. 1990. High diversity in DNA of soil bacteria. *Applied and Environmental Microbiology* 56(3), 782-787. doi: 10.1128/aem.56.3.782-787.1990.

Torsvik, V., Øvreås, L. and Thingstad, T. F. 2002. Prokaryotic diversity–Magnitude, dynamics, and controlling factors. *Science* 296(5570), 1064-1066. doi: 10.1126/science.1071698.

Trumbore, S. 2000. Age of soil organic matter and soil respiration: Radiocarbon constraints on belowground C dynamics. *Ecological Applications* 10(2), 399-411. doi: 10.1890/1051-0761(2000)010[0399:AOSOMA]2.0.CO;2.

Ullah, M. and Dijkstra, F. 2019. Fungicide and bactericide effects on carbon and nitrogen cycling in soils: A meta-analysis. *Soil Systems* 3(2), 23. doi: 10.3390/soilsystems3020023.

van Gestel, N., Shi, Z., van Groenigen, K. J., Osenberg, C. W., Andresen, L. C., Dukes, J. S., Hovenden, M. J., Luo, Y., Michelsen, A., Pendall, E., Reich, P. B., Schuur, E. A. G. and Hungate, B. A. 2018. Predicting soil carbon loss with warming. *Nature* 554(7693), E4-E5. doi: 10.1038/nature25745.

Van Groenigen, J. W., Huygens, D., Boeckx, P., Kuyper, T. W., Lubbers, I. M., Rütting, T. and Groffman, P. M. 2015. The soil N cycle: New insights and key challenges. *SOIL* 1(1), 235-256. doi: 10.5194/soil-1-235-2015.

Van Hemelryck, H., Fiener, P., Van Oost, K., Govers, G. and Merckx, R. 2010. The effect of soil redistribution on soil organic carbon: An experimental study. *Biogeosciences* 7(12), 3971–3986. doi: 10.5194/bg-7-3971-2010.

Van Hemelryck, H., Govers, G., Van Oost, K. and Merckx, R. 2011. Evaluating the impact of soil redistribution on the in situ mineralization of soil organic carbon. *Earth Surface Processes and Landforms* 36(4), 427–438. doi: 10.1002/esp.2055.

Van Oost, K., Quine, T. A., Govers, G., De Gryze, S., Six, J., Harden, J. W., Ritchie, J. C., McCarty, G. W., Heckrath, G., Kosmas, C., Giraldez, J. V., Marques da Silva, J. R. and Merckx, R. 2007. The impact of agricultural soil erosion on the global carbon cycle. *Science* 318(5850), 626–629. doi: 10.1126/science.1145724.

Virto, I., Barré, P., Burlot, A. and Chenu, C. 2012. Carbon input differennces as the main factor explaining the variability in soil organic C storage in no-tilled compared to inversion tilled agrosystems. *Biogeochemistry* 108(1-3), 17–26. doi: 10.1007/s10533-011-9600-4.

von Fromm, S. F., Hoyt, A. M., Lange, M., Acquah, G. E., Aynekulu, E., Berhe, A. A., Haefele, S. M., McGrath, S. P., Shepherd, K. D., Sila, A. M., Six, J., Towett, E. K., Trumbore, S. E., Vågen, T.-G., Weullow, E., Winowiecki, L. A. and Doetterl, S. 2021. Continental-scale controls on soil organic carbon across sub-Saharan Africa. *Soil* 7(1), 305–332. doi: 10.5194/soil-7-305-2021.

von Lützow, M. and Kögel-Knabner, I. 2009. Temperature sensitivity of soil organic matter decomposition – What do we know? *Biology and Fertility of Soils* 46(1), 1–15. doi: 10.1007/s00374-009-0413-8.

Vörösmarty, C. J. and Sahagian, D. 2000. Anthropogenic disturbance of the terrestrial water cycle. *BioScience* 50(9), 753–765.

Vos, M., Wolf, A. B., Jennings, S. J. and Kowalchuk, G. A. 2013. Micro-scale determinants of bacterial diversity in soil. *FEMS Microbiology Reviews* 37(6), 936–954. doi: 10.1111/1574-6976.12023.

Walker, X. J., Baltzer, J. L., Cumming, S. G. Day, N. J., Ebert, C., Goetz, S., Johnstone, J. F., Potter, S., Rogers, B. M., Schuur, E. AG., and et. al. 2019. Increasing wildfires threaten historic carbon sink of boreal forest soils. *Nature,* Nature Publishing Group, 572(7770), 520–523.

Walker, T. W. N., Janssens, I. A., Weedon, J. T., Sigurdsson, B. D., Richter, A., Peñuelas, J., Leblans, N. I. W., Bahn, M., Bartrons, M., De Jonge, C., Fuchslueger, L., Gargallo-Garriga, A., Gunnarsdóttir, G. E., Marañón-Jiménez, S., Oddsdóttir, E. S., Ostonen, I., Poeplau, C., Prommer, J., Radujković, D., Sardans, J., Sigurðsson, P., Soong, J. L., Vicca, S., Wallander, H., Ilieva-Makulec, K. and Verbruggen, E. 2020. A systemic overreaction to years versus decades of warming in a subarctic grassland ecosystem. *Nature Ecology and Evolution* 4(1), 101–108. doi: 10.1038/s41559-019-1055-3.

Walling, D. E. 1996. Suspended sediment transport by rivers: A geomorphological and hydrological perspective. *Arch. Hydrobiol. Spec. Issues Advanc. Limnol.* 47, 1–27.

Wang, Z., Hoffmann, T., Six, J., Kaplan, J. O., Govers, G., Doetterl, S. and Van Oost, K. 2017. Human-induced erosion has offset one-third of carbon emissions from land cover change. *Nature Climate Change* 7(5), 345–349. doi: 10.1038/nclimate3263.

Wei, X., Shao, M., Gale, W. and Li, L. 2014. Global pattern of soil carbon losses due to the conversion of forests to agricultural land. *Scientific Reports* 4, 4062. doi: 10.1038/srep04062.

Weisdorf, J. L. 2005. From foraging to farming: Explaining the neolithic revolution. *Journal of Economic Surveys* 19(4), 561–586. doi: 10.1111/j.0950-0804.2005.00259.x.

Westerling, A. L., Hidalgo, H. G., Cayan, D. R. and Swetnam, T. W. 2006. Warming and earlier spring increase Western U.S. forest wildfire activity. *Science* 313(5789), 940–943. doi: 10.1126/science.1128834.

Wiesmeier, M., Urbanski, L., Hobley, E., Lang, B., von Lützow, M., Marin-Spiotta, E., van Wesemael, B., Rabot, E., Ließ, M., Garcia-Franco, N., Wollschläger, U., Vogel, H.-J. and Kögel-Knabner, I. 2019. Soil organic carbon storage as a key function of soils – A review of drivers and indicators at various scales. *Geoderma* 333, 149–162. doi: 10.1016/j.geoderma.2018.07.026.

Wilkinson, B. H. and McElroy, B. J. 2007. The impact of humans on continental erosion and sedimentation. *Geological Society of America Bulletin* 119(1-2), 140–156. doi: 10.1130/B25899.1.

Willer, H., Trávníček, J., Meier, C. and Schlatter, B. 2021. *The World of Organic Agriculture: Statistics and Emerging Trends 2021*. Research Institute of Organic Agriculture FiBL. Frick and IFOAM. 2009. Organics international, Bonn (v20210301). *International Journal of Sustainability in Higher Education* 10(1). doi: 10.1108/ijshe.2009.24910aae.004.

Witzgall, K., Vidal, A., Schubert, D. I., Höschen, C., Schweizer, S. A., Buegger, F., Pouteau, V., Chenu, C. and Mueller, C. W. 2021. Particulate organic matter as a functional soil component for persistent soil organic carbon. *Nature Communications* 12(1), 4115. doi: 10.1038/s41467-021-24192-8.

Wood, T. E. and Silver, W. L. 2012. Strong spatial variability in trace gas dynamics following experimental drought in a humid tropical forest. *Global Biogeochemical Cycles* 26(3), GB3005. doi: 10.1029/2010GB004014.

Xu, L., Baldocchi, D. D. and Tang, J. 2004. How soil moisture, rain pulses, and growth alter the response of ecosystem respiration to temperature. *Global Biogeochemical Cycles* 18(4), GB4002. doi: 10.1029/2004GB002281.

Xu, X., Thornton, P. E. and Post, W. M. 2013. A global analysis of soil microbial biomass carbon, nitrogen and phosphorus in terrestrial ecosystems. *Global Ecology and Biogeography* 22(6), 737–749. doi: 10.1111/geb.12029.

Yang, D., Kanae, S., Oki, T., Koike, T. and Musiake, K. 2003. Global potential soil erosion with reference to land use and climate changes. *Hydrological Processes* 17(14), 2913–2928. doi: 10.1002/hyp.1441.

Yue, K., Peng, Y., Peng, C., Yang, W., Peng, X. and Wu, F. 2016. Stimulation of terrestrial ecosystem carbon storage by nitrogen addition: A meta-analysis. *Scientific Reports* 6, 19895. doi: 10.1038/srep19895.

Zhang, W. C., Gregory, A. S., Whalley, W. R., Ren, T. S. and Gao, W. D. 2021. Characteristics of soil organic matter within an erosional landscape under agriculture in Northeast China: stock, source, and thermal stability. *Soil and Tillage Research* 209, 104927. doi: 10.1016/j.still.2020.104927.

Zosso, C. U., Ofiti, N. O. E., Soong, J. L., Solly, E. F., Torn, M. S., Huguet, A., Wiesenberg, G. L. B. and Schmidt, M. W. I. 2021. Whole-soil warming decreases abundance and modifies the community structure of microorganisms in the subsoil but not in surface soil. *Soil* 7(2), 477–494. doi: 10.5194/soil-7-477-2021.

Zuckerberg, B., Strong, C., LaMontagne, J. M., St. George, S., Betancourt, J. L. and Koenig, W. D. 2020. Climate dipoles as continental drivers of plant and animal populations. *Trends in Ecology and Evolution* 35(5), 440–453. doi: 10.1016/j.tree.2020.01.010.

Chapter 6

Benefits and trade-offs of soil organic carbon sequestration

C. Rumpel, CNRS, Sorbonne University, Institute of Ecology and Environmental Sciences Paris, France; B. Henry, Queensland University of Technology, Australia; C. Chenu, AgroParisTech, UMR Ecosys INRA, AgroParisTech, Université Paris-Saclay, France; and F. Amiraslani, Ulster University, UK

1 Introduction

Soil organic carbon (SOC) sequestration has received increasing attention due to its numerous co-benefits, particularly its crucial role in enhancing important soil functions and ecosystem services (Lal, 2004). SOC is stored in the soil as organic compounds in the form of soil organic matter (SOM), which influences plant growth and soil (micro-)biological activity in numerous ways related to its positive impacts on biological, chemical and physical soil functions. In this context, SOC concentrations rather than stock is a critical soil quality variable positively correlated with most soil functions sustaining supporting, regulating and provisioning ecosystem services (Rumpel and Chabbi, 2021). As a result, SOC sequestration, which mostly implies increases in its concentration, has been indicated as important for soil fertility and agricultural yields (Lal, 2016). Following

http://dx.doi.org/10.19103/AS.2022.0106.06

the identification of its potential role in climate change mitigation (Balesdent and Arrouays, 1999), SOC sequestration received policy attention. Indeed, removal of atmospheric CO_2 through carbon storage in soils is attractive relative to other negative emissions technologies because the technology is available, readily accessible, and because the concomitant SOC concentration increases are often associated with co-benefits for enhanced food security (Paustian et al., 2016; Frank et al., 2017) and ecosystem services (Bossio et al., 2020). Recent estimates showed that SOC could make up 9%, 72% and 47% of the mitigation potential of forest, wetland, agriculture and grassland biomes, equivalent to protection or sequestration of 1.2 Gt CO_2e year^{-1}, 2.0 Gt CO_2e year^{-1} and 2.3 Gt CO_2e year^{-1}, respectively, for these biome soils (Bossio et al., 2020).

Due to potential co-benefits, it has been proposed that SOC sequestration may be a win-win-win strategy (Lal, 2008) to be implemented with 'no regrets' (Bossio et al., 2020). However, in order to fulfill these expectations, trade-offs in terms of other greenhouse gas (GHG) emissions, water and nutrient requirements and other potential drawbacks of this nature-based solution (Fig. 1) must be taken into account and addressed by applying sustainable management practices that also restore and protect ecosystems. These trade-offs can be either a consequence of the increased SOM content or consequences of the practices implemented to increase SOC. The former include potential increased nitrogen (N) leaching from SOM-rich soils. The latter are related to environmental and economic resources needed to sequester SOC and also to changes in the energy balance of the system. Without rigorous quantification of these trade-offs, the levels of climate change mitigation that

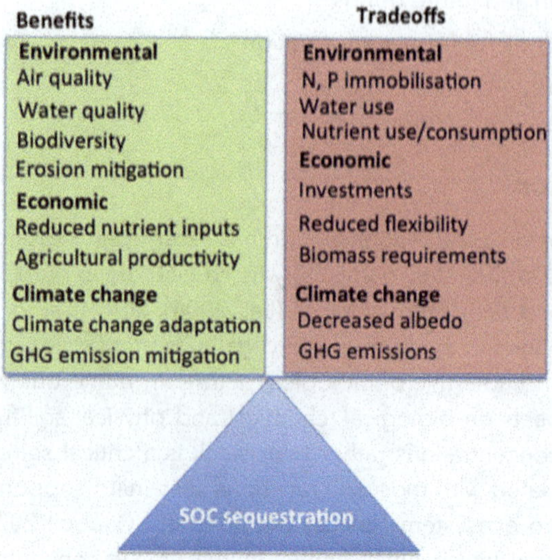

Figure 1 Potential benefits and trade-offs of SOC sequestering practices.

Published by Burleigh Dodds Science Publishing Limited, 2023.

can be achieved from the implementation of SOC sequestering practices are uncertain. Accounting for changes in carbon dioxide (CO_2), nitrous oxide (N_2O) and methane (CH_4) emissions is essential to estimate net abatement due to SOC sequestration at field or project scale. Moreover, this estimation is necessary to ensure consistency with IPCC guidelines (IPCC, 2019) and reporting obligations under the UNFCCC (see Chapter 10 of this book). For example, a modeling study of soil and livestock emissions and removals for cropping and grazing management options in Australia showed that N_2O emissions have the potential to partially or fully offset SOC sequestration benefits in cropping and permanent pasture systems (Meier et al., 2020). The study also indicated significant variation in outcomes depending on site and management factors because both influence the impact of the practices on the soil-plant system.

While it is important to consider the potential trade-offs of SOC sequestration, it is also necessary to carefully evaluate and quantify the benefits. The general paradigm is that increasing SOC has positive effects on soil water (quality and quantity), soil physical properties, i.e. bulk density and aggregation, nutrient storage, biological activity and biodiversity. Together these benefits enhance soil functioning and the ecosystem services derived from the soil. While SOC sequestration trade-offs other than GHG emissions received little attention, its benefits are widely accepted. This is probably mainly related to the fact that soils rich in organic matter are highly fertile. However, quantitative assessments of the relationship between SOC increase and its benefits for physical, chemical and biological functions are scarce. The reason may be that benefits and trade-offs are both variable and difficult to quantify. The effects depend on location factors, such as soil and climate characteristics and the form of sequestered SOC, and they also depend on whether SOC is reported as the carbon (C) concentration of soil (expressed as percentage C in dry matter) or as the stock (mass) of C in a given volume of soil.

In this chapter, we will identify and discuss the benefits and trade-offs of sequestering SOC related to soil processes and ecosystem functions and the factors influencing them. We present literature evidence for SOC sequestration benefits and trade-offs and discuss the role of specific carbon forms. Moreover, we collated published data on changes in CO_2, N_2O and CH_4 emissions (or removals) as the main trade-offs for several land management options that are widely seen as prospective for increasing C storage in soils.

2 Should soil organic carbon concentrations or stocks be considered to evaluate the benefits and trade-offs of soil organic carbon sequestration?

SOC sequestration, i.e. the transfer of atmospheric CO_2 into terrestrial reservoirs, implies that this process needs to consider changes in the amount of stored SOC, i.e. its stock. Assessing SOC stock changes is thus particularly important for evaluating the contribution SOC sequestration can make to climate change mitigation. SOC stocks are measured by taking into account SOC concentration, soil depth and bulk density (see Chapter 11 of this book). Thus, a higher SOC concentration may or may not be associated with an increase in SOC stocks, and, therefore, with CO_2 transfer from atmosphere to the soil. A specific example is the implementation of the no-till practice, which commonly leads to increases in SOC concentrations in the topsoil. However, because of associated bulk density changes, this increase does not necessarily translate into corresponding increases in SOC stocks, but may rather reflect re-distribution of SOC between different layers of the soil profile (Virto et al., 2012; Angers and Eriksen-Hamel, 2008). As a result, SOC sequestration due to introducing no-till may be less important for climate change mitigation than previously thought. However, these processes may be soil type and climate dependent (see Chapters 16 and 22 of this book). In addition, increasing SOC content in the upper soil horizon through no-till practices may have benefits resulting from reduced erosion risk (Skaalsveen et al., 2019) and more favorable biogeochemical cycling compared to regular tillage (Mbuthia et al., 2015).

When evaluating SOC benefits regarding soil physical, chemical and biological properties, such as water holding capacity, plant nutrient availability, erosion control and biodiversity, but also mitigation of GHG emissions through decreased albedo, SOC concentration is frequently a more appropriate variable than SOC stocks. This can be illustrated using the example of deep soil horizons, which may contain high carbon stocks at low concentrations because of their high bulk density and depth (Rumpel et al., 2002). Because of the dilution of SOM in subsoil horizons, the biological activity of most subsoil horizons is low and restricted to the root zone, whereas SOC-rich topsoils with low bulk density and lower SOC stocks may be more biologically active and productive because of more favorable physical conditions that are prone to support plant growth and microbial functioning. SOC concentration may thus be a useful variable to evaluate benefits and trade-offs in ecosystem services related to soil fertility and biological activity, whereas SOC stock changes need to be assessed to quantify SOC sequestration. However, N stocks calculated based on SOC stocks may be useful to predict N availability for plant growth and to manage N fertilization (e.g. Thorup-Kristensen and Nielsen, 1998; Chalhoub et al., 2013).

3 Quantitative evidence of benefits related to soil carbon sequestration

In this section, we review quantitative evidence for the effects of increasing SOC concentrations and stocks on soil functionality and ecosystem services other than GHG emission mitigation (Table 1). It is generally accepted that

SOC increases improve most soil functions and the soil properties determining them, and here we discuss three examples of related ecosystem service benefits: water infiltration and retention, plant growth and yields (related to food provision), and biodiversity.

Correlations between SOC content and water infiltration rate as well as aggregate mean diameter have been reported (Singh Brar et al., 2015). SOC correlates positively with aggregate stability (Blanco-Canqui and Lal, 2007; Chaney and Swift, 1984; Chenu et al., 2000; Pikul et al., 2009), and the presence of SOM was found to reduce the maximum compactability of soils (e.g. Diaz-Zorita and Grosso, 2000). Regarding the role of soils in infiltrating and retaining water, bulk density is generally negatively correlated with SOC, which indicates its important contribution to porosity (e.g. Rawls, 1983). Although specific studies commonly indicate strong relationships between SOC content and physical, chemical and biological variables, these relationships were not confirmed in other (general) studies using larger datasets. For example, in a global meta-analysis, Minasny and McBratney (2018) showed that increasing

Table 1 Studies providing quantitative evidence for benefits between SOC and soil properties that are relevant for ecosystem services provision

Ecosystem service	Effect of SOC	Reference
Nutrients and metals retention	Increased cation exchange capacity	*van Erp et al. (2001); Curtin and Rostad (1997); Bigorre et al. (2000); Krogh et al. (2000)*
Infiltrating and retaining water	Decreased bulk density (increased soil porosity)	*Rawls (1983); Lettens et al. (2005)*
	Increased infiltration rate	*Singh Brar et al. (2015)*
	Very small increase in soil water content	*Minasny and McBratney (2018)*
Supporting traffic	Reduces soil compactability	*Diaz-Zorita and Grosso (2000); Quiroga et al. (1999)*
Food provision	Higher yields	Pan et al. (2009); *Oldfield et al. (2019)*
	Lower yield variability	Pan et al. (2009)
	Lower amounts of N fertilizer needed for similar yields	*Oldfield at al. (2019)*
Climate change resilience	Drought tolerance	Iizumi and Wagai (2019)
Protection against erosion	Mean weight diameter	*Singh Brar et al. (2015)*
	Aggregate stability	*Blanco-Canqui and Lal (2007); Chaney and Swift (1984); Chenu et al. (2000); Pikul et al. (2009)*
Biodiversity	Higher insect diversity	Flores-Rios et al. (2020)

Studies reporting concentrations are printed in *italic*, those reporting stocks in normal fonts.

SOC content had only a small effect on soil water content and available water holding capacity, depending on soil texture. As a result, they concluded that SOC changes following the adoption of agricultural practices may have negligible impacts on soil hydrological properties. In contrast, a meta-analysis investigating the effect of organic amendments on soil properties using long-term experiments concluded that there is a positive effect of SOC on water retention, but that the main effect is on water content at saturation (Eden et al., 2017). It therefore seems that the impact of SOC on soil (hydrological) properties cannot be generalized and that the controlling factors leading to a quantifiable effect of SOC increases are not completely understood. The effect of SOC on soil properties is largely controlled by soil texture, and increases as the clay content of soils decreases (e.g. Curtin and Rostad, 1997 for CEC; Kay et al., 1997 for water retention).

Many studies have investigated the beneficial effect of SOC for plant growth and yields as often sustainable soil management options that enable an increase in SOC are associated with increased yields (e.g. Han et al., 2018; Yeboah et al., 2016; Olson et al., 2010). For example, Pan et al. (2009) and Zhang et al. (2016) showed that higher SOC stocks in Chinese croplands led to higher yields and greater yield stability. In contrast, Oelofse et al. (2015) and Hijbeek et al. (2016) observed non-significant effects of SOC contents on yields in long-term European experiments. They explained their results by the sufficient nutrient levels supplied by mineral fertilizers in these trials, indicating that the effects of SOC on crop yields may be influenced by other confounding factors. However, a meta-analyses of a global dataset indicated that, regardless of initial SOC stocks, increasing the SOC concentration up to a critical level of 2% (20 mg g^{-1} soil) could enhance agricultural yields (Oldfield et al., 2019). The authors noted that two-thirds of global croplands are below this level. Moreover less N fertilizer may be required to achieve similar yields in soils with higher SOC concentrations (Oldfield et al., 2019). In terms of climate change impact mitigation (e.g. improved resilience to extreme events), increasing the SOC stocks in the topsoil seems to decrease the drought-related yield gap, in particular in arid regions, as shown by Iizumi and Wagai (2019) based on a global data set.

While SOC is often higher in systems with high biodiversity (e.g. Lange et al., 2015), the effect of SOC on biodiversity has rarely been quantified. Flores-Rios et al. (2020) observed that insect diversity was positively correlated with SOC stocks. This may be a benefit in terms of system stability, which, in turn, could prevent pathogenic insect invasions. However, to the best of our knowledge, no studies have analyzed the effects of SOC on biodiversity at other trophic levels (soil microorganisms, plant communities, fauna) with a quantitative approach.

4 Quantitative evidence for trade-offs related to soil carbon sequestration

Fostering SOC storage requires carbon input, usually in the form of plant-derived organic matter. Due to stoichiometric constraints because of the low C:N:P ratios of SOM, nutrient inputs are also necessary. If nutrient sources are not provided, organic matter addition to soils can lead to SOC decrease due to nutrient mining (Fontaine et al., 2014). It was suggested that to meet stoichiometric requirements, some 73 kg ha^{-1}, 17 kg ha^{-1} and 11 kg ha^{-1} of nitrogen, phosphorus and sulfur are necessary to store one additional t of SOC per ha (Richardson et al., 2014). This nutrient cost of SOC storage may constitute a significant trade-off and must be considered in large-scale SOC sequestration programs (van Groeningen et al., 2017).

Moreover, water requirements for enhancing biomass inputs via plant growth could constitute a significant trade-off, especially under climate change. For example, the high SOC stocks in paddy rice and organic soils rely on waterlogged conditions (see Chapters 17 and 21 of this book). In addition, with the frequency of drought projected to increase in the future, more irrigation will likely be required to maintain agricultural growth and hence organic matter input in (rainfed) agricultural systems. The effect of this practice on SOC sequestration has been found to be positive (Kelliher et al., 2015), negative (Condron et al., 2014), or neutral (Hunt et al., 2016). The mechanisms leading to these effects are poorly understood. They may be related to changes in photosynthate allocation (Carmona et al., 2020) and/or changes in soil texture due to clay translocation, depending on the type of irrigation water (Warrington et al., 2007). A global meta-analysis indicated that irrigated agriculture could increase SOC stocks in the soil profile by on average 5.9%, depending on climate, soil type and irrigation method (Emde et al., 2021). The highest increases of 11–35% were noted for semi-arid sites (Trost et al., 2013). In contrast, the amount of water needed to store 1 kg of SOC in different pedoclimatic conditions has not yet been determined.

Moreover, it has been suggested that the beneficial effects of SOC sequestration on reducing atmospheric CO_2 concentrations may be offset by increased emissions of other GHGs, i.e. N_2O and CH_4. The global warming potential of these emissions is 273 and 27.9 times higher than that of CO_2 over 100 years, respectively (IPCC, 2021). Although the processes leading to N_2O emissions from soil are not entirely understood (Lugato et al., 2018; Pärn et al., 2018; Tian et al., 2020), it seems generally accepted that increased soil N supply, decreased soil pH, high C availability, high water content and soil compaction lead to increased emissions (Skinner et al., 2014; Smith, 2016). Moreover, the input of labile plant litter and/or organic amendments can lead to loss of existing SOC and enhanced CO_2 emissions via priming effects (Guenet et al., 2018) and, if reactive N is present, also to N_2O emissions (Guenet et al., 2020). However, the drivers of N_2O emissions are insufficiently understood to trigger appropriate soil management to reduce this loss. In terms of CH_4 emissions, rice paddies and other waterlogged anaerobic systems are net emitters of large amounts of this GHG (see Chapters 17 and 21 of this book), produced by methanogenic *Archaea*, during the decomposition of root exudates or dead roots from recently assimilated CO_2 (Conrad, 2009). The controls of these emissions, i.e. the microbial methanogenesis pathway, need to be better understood to predict net abatement from SOC sequestration practices and to inform the development of mitigation strategies (Conrad, 2020). The impact of soil CH_4 emissions in many systems will be less predictable than that of N_2O. A New Zealand study found that methane uptake ranged from 10 to 11 kg CH_4 ha^{-1} $year^{-1}$ in beech forest soils down to <1 kg CH_4 ha^{-1} $year^{-1}$ in most pasture soils (Saggar et al., 2008), while other studies indicated net soil CH_4 emissions from pasture sites (Allen et al., 2009). In grazing systems, CH_4 emissions from ruminants greatly exceed those from soil (see Chapter 18 of this book). Fluxes of GHG other than CO^2 can thus be important under certain conditions and should be included in estimates of the net GHG mitigation benefits of SOC stock changes.

Sustainable practices adopted to increase SOC in agricultural land may also alter surface albedo thereby influencing the energy balance of the system. This may be a benefit or a trade-off. Higher albedo results in more reflection of solar energy back to space, thereby reducing climate warming. In the case of cover crops, often used to enhance SOC, surface albedo and, therefore, reflection of solar energy may be increased because vegetation has a higher albedo than bare soil (Carrer et al., 2018). In Europe, the beneficial albedo effect of cover crops may enhance their climate change mitigation potential in addition to their positive effect on SOC sequestration (Carrer et al., 2018). In contrast, soil application of biochar may result in a darker surface, less effective at reflecting solar radiation to space, therefore leading to increased surface temperature and evaporation (Genesio et al., 2012) when vegetation cover is low. As higher SOC concentrations generally darken the soils' color, a change in albedo is a trade-off when the soil is bare. For example, an assessment of the albedo impact of biochar systems indicated that the overall GHG emission mitigation effect may be reduced by 13-22% due to the albedo effect (Meyer et al., 2012). Despite their potential importance, these albedo effects are rarely considered when assessing the climate change mitigation potential of agricultural practices, because the focus is generally on the GHG balance.

Published by Burleigh Dodds Science Publishing Limited, 2023.

5 Different carbon forms may have contrasting benefits and trade-offs

The SOM is a continuum ranging from fresh plant material consisting of intact plant litter compounds, wholly or partly degraded material, and microbial-processed carbon, which may be contributing to stabilized SOC compounds (Kogel-Knabner, 2002; Fig. 2). In particular for modeling purposes, SOM has been partitioned into three pools with contrasting turnover time and thus stability (Parton et al., 1987). Labile OM consisting of fresh OM inputs is distinguished from an intermediate pool comprising SOC stabilized within soil aggregates and a stable pool composed of mineral-associated SOC and combusted and/or pyrolyzed SOC compounds, such as black carbon and/or biochar (Krull et al., 2003). The labile SOC pool may be degraded rapidly and/or incorporated into the mineral soil, thereby fueling microbial activity necessary for promoting soil aggregation (e.g. Cosentino et al., 2006; Abiven et al., 2009). Labile SOC compounds may be incorporated into aggregates, and stabilized until the disintegration of such structures lead to microbial access and their mineralization. After degradation, microbial necromass and/or partly degraded plant material may be stabilized by mineral interactions (Barre et al., 2018; Angst et al., 2021, Chapter 2 of this book). The earlier conceptualization of soil particulate organic

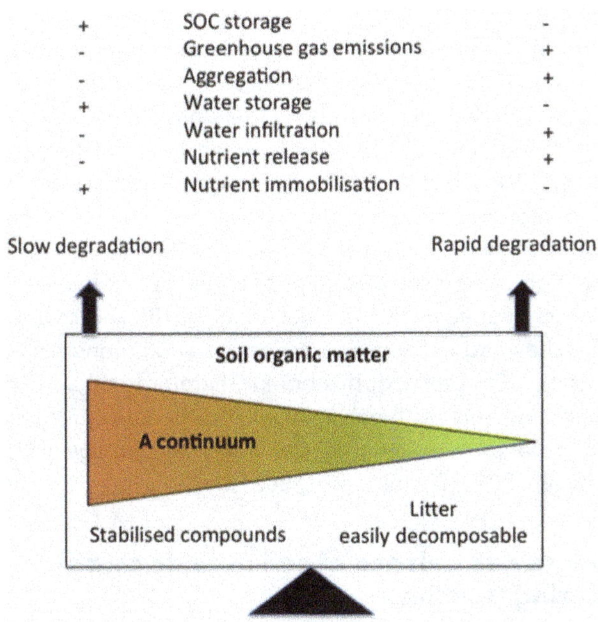

Figure 2 Effects of specific SOM compounds on benefits and trade-off.

matter (POM) and mineral-associated organic matter (MAOM) fractions into SOC kinetic pools (Balesdent, 1996) and into SOC functional pools (Gregorich et al., 1995, 2006; Feller et al., 2001) has been recently promoted as a useful framework to assess SOC dynamics (Lavallee et al., 2020) and therefore SOC sequestration.

While the major benefits derived from labile SOC may be related to its degradability, which results in readily available energy and nutrient release for soil microorganisms and plants, the same properties can also result in significant trade-offs as their rapid degradation may lead to losses of nitrate, and release of CO_2 and other GHGs. The availability of nutrients and energy for microbial activity decreases with increasing SOC stability, while its benefits for other soil functions, such as carbon and water storage may increase. The benefits of very stable SOM compounds may be mostly related to their SOC storage function, although, they may also improve physical soil functions and microbial habitat (Janzen, 2006, Fig. 2).

It has been pointed out that labile and stable SOC and, in particular OM decay processes, are needed for good soil functioning and to secure the multiple roles of SOM for plant growth and the provision of ecosystem services derived from soil (Janzen, 2006). It seems that the balance between both SOC types is essential. However, at the moment, the most beneficial ratio between labile and stable SOM compounds to optimize ecosystem services from the soil is not known.

The ratio of SOC forms may be managed to some extent through addition of organic amendments. Indeed, the chemical composition of OM amendments was found to control their fate in soil and therefore their contribution to SOC sequestration (Peltre et al., 2012) through enhancing SOC storage in organo-mineral fractions (Paetsch et al., 2018). While high microbial use efficiency may be required to stabilize plant-derived compounds, aromatic carbon types such as those produced by pyrolysis, i.e. biochar, may show high stability after soil addition. The high carbon sequestration potential of biochar is relatively well documented (Schmidt et al., 2019; Woolf et al., 2010), and its addition to soil has also been shown to enhance soil properties and agronomic performance in some situations (Chapter 9 of this book). Therefore, soil addition of stable carbon compounds, such as biochar to soil, has been suggested in addition to SOC sequestration as a negative emission technology with numerous co-benefits (Smith, 2016; Schmidt et al., 2021).

6 Greenhouse gas balance of soil organic carbon sequestering practices

To assess the SOC sequestration benefits in climate change mitigation, its effects on all GHG emissions need to be quantified. We therefore collated

Published by Burleigh Dodds Science Publishing Limited, 2023.

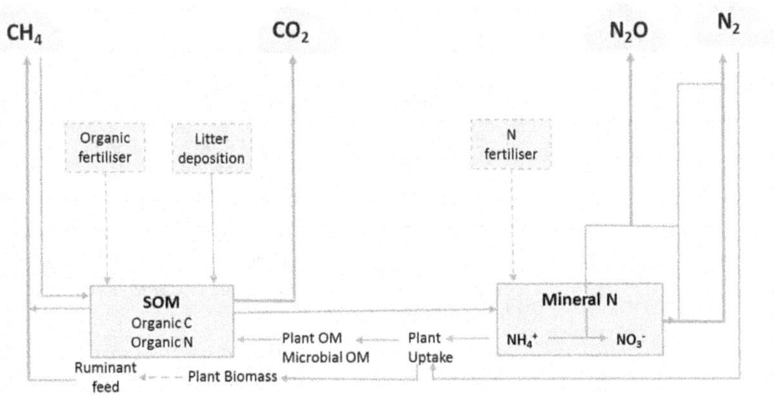

Figure 3 Simplified representation of key linkages between C and N cycling in soils. Sources and sinks of GHG emissions affected by practices that influence SOC stocks are shown.

published data for changes in CO_2, N_2O and CH_4 emissions (or removals) for several land management practice options that are widely promoted for increasing SOC sequestration. These data are helpful for understanding the likely GHG emission benefits and trade-offs from implementing practices that affect C and N cycling in agricultural lands (Fig. 3).

Recent research in response to growing global interest in the potential for SOC sequestration as a climate change mitigation strategy has provided some understanding of the balance between increasing C storage in soil and changes in other GHG net emissions resulting from selected management practices (Table 2). However, C and N cycle interactions are highly dynamic and generalizations of changes to net emissions are problematic. The relative effects tend to be dynamic and site and management specific and we discuss the processes responsible for GHG emissions other than CO_2, generated mainly by agricultural activities.

6.1 Nitrous oxide (N_2O)

The close linkages between C and N cycling in soils mean that changes in SOC turnover affect N transformations and consequently N_2O emissions (Soussana et al., 2019). The agricultural sector is a major contributor to these emissions, but due to their substantial spatial and temporal variability, robust scaling of estimates of N_2O emissions are scarce.

The complex and variable nature of the links between the C and N cycles and changing relationships over time make it difficult to generalize on the extent of any trade-off or synergy between N_2O emissions and C sequestration actions (Xu et al., 2021). However, the risk of significant offsets of the contribution of

Table 2 Potential impacts on GHG emissions of practices commonly undertaken to increase carbon stocks in agricultural and forestry soils

Activity to increase SOC stocks	GHG flux changed	SOC impact	Possible GHG effects relative to baseline
Reduced- or no-till	CO_2 N_2O	Reduced loss of SOC	Reduced fuel use; higher denitrification
Synthetic fertilizer	CO_2 N_2O CH_4	Higher C inputs (higher yields)	Fossil fuel use for application; higher N availability; higher N_2O loss; urea can also increase CO_2 emissions
Liming	CO_2	Higher C inputs (neutralize acid soils)	Carbonates react to release CO_2
Irrigation	CO_2 N_2O CH_4	Higher C inputs (higher yields)	Fossil fuel use (diesel); higher soil moisture can increase N_2O and CH_4
Organic amendments	CO_2 N_2O CH_4	Higher C inputs (higher yields)	Higher N availability; higher denitrification; if organic amendment not from a waste stream; for biochar-enhanced adsorption of N may lower nitrification
Cover crops	CO_2 N_2O	Higher C inputs + lower C loss	Lower denitrification due to higher plant uptake (non-legume); higher N availability (legume) may increase loss
Agroforestry	CO_2 N_2O CH_4	Higher C inputs; lower C loss; increased stability (aggregates and increased depth)	Lower denitrification (lower bulk density; lower soil moisture); lower denitrification due to higher plant uptake; higher CH_4 sink (lower bulk density)
Grazing management	CO_2 N_2O CH_4	Higher C inputs; added N through dung, urine	Changed SOM dynamics through trampling; dung and urine C, N inputs and N_2O, CH_4 emissions; enteric CH_4 emissions

Sources: Baldock and Burgess (2017); Guenet et al. (2020); Meier et al. (2020). Effects on each GHG need to be quantified to calculate the net climate change mitigation due to a practice change.

N_2O to climate change mitigation benefits is expected to increase as the rate of sequestration slows toward a new equilibrium SOC concentration with the adoption of new management practices (Liu and Greaver, 2009; Lugato et al., 2018). This will be exacerbated if nutrients become limiting and fertilizer (synthetic or organic) containing N is applied to achieve a constant increase in SOC stocks over a long time (Kirkby et al., 2014). In these conditions, achieving reduced net GHG emissions becomes more difficult (van Groenigen et al., 2017), unless N_2O emissions can be reduced. Applying organic materials to soil (Lehmann et al., 2006) or adopting cover crop or agroforestry practices (Poeplau and Don, 2015; Feliciano et al., 2018) that increase C inputs and/or reduce C losses tend to enhance the GHG mitigation benefit of SOC sequestration. These materials and practices have been shown to lower N_2O emissions. This is likely a result of decreased denitrification in soil with lower bulk density and lower soil moisture and, in the case of biochar, due to adsorption of mineral N (Borchard et al., 2019; Guenet et al., 2020; see Table 2). Higher plant productivity in agroforestry systems with the use of cover crops will also reduce N_2O emissions due to increased N uptake. The exception is where N-fixing species are used, in which case soil N availability and therefore available substrate increases (Kim et al., 2016; Lugato et al., 2018). Table 3 provides indicative directions of these emissions from recent literature, but many are derived from short-term studies.

Accounting for GHG emissions from grasslands with livestock grazing may need to include additional N_2O sources. Practices that increase SOC sequestration in grasslands include management options that reduce periods with low or no ground cover and increase pasture organic matter inputs to soil (Chapter 18 of this book). In grazing systems, the reduction of grazing animals will have a major impact on reducing N_2O and CH_4 emissions. The intensity and timing of livestock grazing affect N_2O emissions directly through dung and urine deposition, offsetting SOC sequestration or potentially enhancing it through overcoming nutrient limitations for pasture growth. In ruminant production systems, improving pasture quality and feeding supplements can increase dietary N and reduce N losses from manure and also decrease ruminants' CH_4 emissions (Chapter 18 of this book). However, the animal feed and soil interactions are complex and the capacity to estimate by measurement or modeling the impact of management change in grazing systems on the GHG and SOC sequestration balance is currently limited.

6.2 Methane (CH_4)

Changes in soil management to increase SOC sequestration can also affect ecosystem methane emissions and removals. While wetland soils and paddy rice soils are CH_4 sources (Chapters 17 and 21 of this book), well-drained and aerated agricultural soils are generally thought to be a small net sink for CH_4 (Conrad, 2009), mediated by the action of methanotrophic bacteria. There are gaps in knowledge of how the biogeochemical methane cycle is regulated in both terrestrial aerated land and anaerobic systems. Studies show that a range of trade-offs or co-benefits are possible due to changes in CH_4 pathways with practice change (Table 3). A comparison of agroforestry and adjacent agricultural lands revealed only minor differences in net CH_4 and N_2O emissions, with no clear overall change direction (Kim et al., 2016). Improved fallow systems showed a CH_4 uptake of -0.109 ± 0.034 t CO_2-e ha^{-1} y^{-1}, but agroforestry practices increased the CH_4 sink by only 0.003 ± t CO_2-e ha^{-1} y^{-1}

Table 3 Estimates of impact on nitrous oxide and methane emissions of practices recognized as likely to increase SOC storage in agricultural lands, expressed as the change relative to the 'without practice' scenario

Activity to increase SOC stocks	Change in emissions (t CO_2e ha^{-1} year^{-1})[a]	Reference	Indicative balance with SOC
N_2O			
Conservation tillage	0.167	Mei et al. (2018)	Little or no offset of SOC increase
Conservation tillage	0.7 ± 0.13	Van Kessel et al. (2013)[b]	N_2O approx. unchanged or slightly decreased
Cover crops	0.08[c]	Abdalla et al. (2019)	On av. N_2O emissions partly but do not fully offset the SOC climate change benefits
Organic fertilizer (relative to synthetic)	−0.492 ± 0.16	Skinner et al. (2014)	Enhancement to SOC climate change benefit
Agroforestry	2.3	Kim et al. (2016)[b]	Agroforestry increases SOC storage; only minor change to N_2O (higher if N-fixing trees)
CH_4			
Conservation tillage	0	Abdalla et al. (2013)	No effect to minor enhancement to SOC climate change benefit
Cover crops	0	Abdalla et al. (2019)	Negligible effect on SOC climate change benefit
Organic fertilizer (relative to synthetic)	−0.003 ± 0.003	Skinner et al. (2014)	Enhancement to SOC climate change benefit (arable soils)
Organic fertilizer (relative to synthetic)	0.95 ± 0.415	Skinner et al. (2014)	Enhancement to SOC climate change benefit (rice paddies)
Agroforestry	−0.054 ± 0.034	Kim et al. (2016)	Minor enhancement to SOC climate change benefit

Effects on emissions of these GHGs are highly variable and depend on site-specific factors. The indicative trade-off or enhancement outcome is an average from published reviews or meta-analyses representing a range of timescales.

[a] Positive values indicate net emissions to the atmosphere; negative values are net removal from the atmosphere.

[b] Change (± SE, if reported) in N_2O emissions relative to without practice scenario, and relationship with SOC climate change benefit from meta-analyses (Guenet et al., 2020, and Supporting Information).

[c] Net of direct (−0.08 t CO_2e ha^{-1} year^{-1}) and indirect (0.16 t CO_2e ha^{-1} year^{-1}) emissions.

compared to adjacent conventional agricultural land (Kim et al., 2016). The increase in the CH_4 sink under agroforestry, which is very small relative to other GHG fluxes in these systems, may be due to larger soil pore spaces and lower bulk density which facilitates the rate of oxidation of CH_4 (Christiansen and Gundersen 2011).

In livestock production systems, CH_4 emissions from urine patches and from enteric fermentation in ruminants can be high, while in aerated cropping soils, CH_4 fluxes are generally minor. Therefore, in upland grazing systems, in particular, stocking rates should be monitored to maximize SOC sequestration benefits and avoid offsets due to livestock emissions (Chapter 18 of this book).

7 Socioeconomic benefits and trade-offs of soil organic carbon sequestration

While we target 2030 sustainable development goals (SDGs), we have to consider that the world population will surpass 8.5 billion (UN, 2017). This population growth will require increased food production and exerts enormous pressure on existing soil and water resources. SOC sequestration, especially with co-benefits in yield enhancement and stability and climate change adaptation, may be beneficial at the farm scale. The biophysical benefits and trade-offs of SOC sequestration may affect farmers' economic situation, especially if SOC sequestration enhances soil fertility and/or if carbon markets are operating. In such cases, direct economic benefits may include higher income and indirect benefits may be related to reduced risks of crop failure (Chapters 26 and 27 of this book). These socioeconomic benefits of SOC sequestration may contribute to agricultural development and poverty reduction particularly in low-income countries (Graff-Zivin and Lipper, 2008), as in these countries most of the complex social and economic challenges occur at the farm scale. Most of these farms are small and insufficiently fertile to provide income for the stakeholders. The situation has worsened amid the COVID-19 pandemic, so there is a global call to improve the structural settings of rural areas (Amiraslani, 2021). Consequently, implementation at scale is necessary, not only to take advantage of the GHG emission mitigation benefits of SOC sequestration, but also because of its potential broader impacts on the economic situation and well-being of societies (Chabbi et al., 2017). Social benefits of SOC sequestration may include the prevention of social unrest and mass migration. When SOC sequestration leads to biodiversity enhancement and healthier food production due to reduction of agrochemical input, and also to increasing recreation value of landscapes, health benefits can be expected (Rumpel et al., 2022).

Adoption of SOC sequestering practices requires in most cases changing farming practices, which is a potential risk for the farmers. Decisions by farmers to adopt (new) sustainable SOC sequestering practices depend on many factors, related to the farmers personal situation (e.g. age, income, training), and to characteristics of their biophysical, political and institutional environment (Piñeiro et al., 2020). In many cases, capacity building and investments are required, potentially constituting an economic trade-off.

In addition, to have a global impact on climate change mitigation, changes need to be implemented by millions of farmers globally in very different pedoclimatic, socioeconomic and cultural contexts. For instance, dryland soils contain the largest pool of inorganic C (Plaza-Bonilla et al., 2015), and loss of inorganic C must be considered as well as the risk of salinity development (Chapters 7 and 23 of this book). Therefore, it is important to employ region-specific strategies to overcome biophysical, economic, social, legal and political barriers (Bossio et al., 2020; Amelung et al., 2020; Demenois et al., 2020).

Trade-offs concerning the enhancement of uptake of sustainable practices may include funding to pay for (1) changes in agricultural practices, (2) outreach and teaching of benefits of improved practices for all stakeholders in order to convince policymakers to remove legislative barriers and (3) investment in SOC sequestration strategies. Such funding will be important where the benefits of SOC sequestration require up-front costs and/or are not immediately profitable. Moreover, long-term solutions must be found in order to maintain sustainable management. To acknowledge the value of SOC sequestration in a global context, we suggest that socioeconomic benefits and trade-offs should be included in monitoring, reporting and verification (MRV) schemes.

8 Conclusion and future trends

SOC sequestration is a potential sustainable development strategy, which results in benefits and trade-offs for mitigating GHG emissions and climate change impacts. These include positive effects on soil properties and negative effects in form of GHG emissions, water requirements and/or costs of the practices employed to enhance soil carbon. Positive effects on soil properties are mostly related to the enhancement of SOC concentrations, while mitigation of GHG emissions requires SOC sequestration, measured as increase in SOC stocks. Both are linked but not synergistic in all cases.

In terms of GHG emission mitigation, trade-offs need to be quantified accurately. More efforts are necessary to consider the spatial and temporal variability of soil CH_4 and N_2O emissions. Other trade-offs such as water requirements and/or albedo changes have rarely been considered in the past, but they should be taken into account when assessing SOC sequestering practices. Factors controlling GHG emissions other than CO_2 should further be elucidated. This must include GHG emissions from grazing animals in addition to soil processes.

In terms of co-benefits, relationships between SOC and positive impacts on ecosystem services have rarely been quantified. In particular, the relationship between yield increases and SOC stocks is poorly understood and should be elucidated in various pedoclimatic environments. Moreover, quantitative relationships between SOC concentrations and soil properties need to be established and their confounding factors elucidated.

(Socio)economic benefits and trade-offs are related to the farmers' income and may be directly related to agricultural yields and indirectly to stability of the system and/or investments needed to change agricultural practices. Development of policy frameworks to encourage the implementation of strategies need to be regionspecific. These policies need to overcome barriers due to trade-offs, including (socio)economic ones. The latter should be included in monitoring, reporting and verification programs.

9 Where to look for further information

Benefits and trade-offs of specific agricultural practices are further detailed in the following two articles:

- Sykes, A. J., Macleod, M., Eory, V. Rees, R. M., Payen, F., Myrgiotis, V., Williams, M., Sohi, S., Hillier, J., Moran, D., Manning, D. A. C., Goglio, P., Seghetta, M., Williams, A., Harris, J., Dondini, M., Walton, J., House, J. and Smith, P. 2020. Characterising the biophysical, economic and social impacts of soil carbon sequestration as a greenhouse gas removal technology. *Global Change Biology*, 26:1085–1108.
- Tiefenbacher, A., Sandén, T., Haslmayr, H.-P., Miloczki, J., Wenzel, W., and Spiegel, H. 2021: Optimizing Carbon Sequestration in Croplands: A Synthesis. *Agronomy*, 11, 882.

10 References

Abdalla, M., Hastings, A., Cheng, K., Yue, Q., Chadwick, D., Espenberg, M., Truu, J., Rees, R. M. and Smith, P. 2019. A critical review of the impacts of cover crops on nitrogen leaching, net greenhouse gas balance and crop productivity. *Global Change Biology* 25(8), 2530–2543.

Abdalla, M., Osborne, B., Lanigan, G., Forristal, D., Williams, M., Smith, P. and Jones, M. B. 2013. Conservation tillage systems: a review of its consequences for greenhouse gas emissions. *Soil Use and Management* 29(2), 199–209.

Abiven, S., Menasseri, S. and Chenu, C. 2009. The effects of organic inputs over time on soil aggregate stability - a literature analysis. *Soil Biology and Biochemistry* 41(1), 1–12.

Allen, D. E., Mendham, D. S., Bhupinderpal, S., Cowie, A., Wang, W., Dalal, R. C. and Raison, R. J. 2009. Nitrous oxide and methane emissions from soil are reduced following afforestation of pasture lands in three contrasting climatic zones. *Soil Research* 47(5), 443–458.

Amelung, W., Bossio, D., de Vries, W., Kögel-Knabner, I., Lehmann, J., Amundson, R., Bol, R., Collins, C., Lal, R., Leifeld, J., Minasny, B., Pan, G., Paustian, K., Rumpel, C., Sanderman, J., van Groenigen, J. W., Mooney, S., van Wesemael, B., Wander, M.

and Chabbi, A. 2020. Towards a global-scale soil climate mitigation strategy. *Nature Communications* 11(1), 5427.

Amiraslani, F. 2021. Tackling rural health, energy, … and technological issues all at once: a call for a global interdisciplinary platform for strengthening rural setting Amid the COVID-19 pandemic. *Challenges* 12(2). DOI: 10.3390/challe12020016.

Angers, D. A. and Eriksen-Hamel, N. S. 2008. Full-inversion tillage and organic carbon distribution in soil profiles: a meta-analysis. *Soil Science Society of America Journal* 72(5), 1370–1374.

Angst, G., Mueller, K. E., Nierop, K. G. J. and Simpson, M. J. 2021. Plant- or microbial-derived? A review on the molecular composition of stabilized soil organic matter. *Soil Biology and Biochemistry* 156, 108189.

Baldock, J. and Burgess, R. (2017). Soil Carbon Accounting-the Australian example. In: Workshop on "Soil Organic Carbon Measurement for Management and Management for Measurement". Available at: https://jahnresearchgroup.cals.wisc.edu/wp-content/uploads/sites/200/2017/08/3.-Soil-carbon-accounting-in-Australia-1.pdf [Accessed July 2021].

Balesdent, J. 1996. The significance of organic separates to carbon dynamics and its modelling in some cultivated soils. *European Journal of Soil Science* 47(4), 485–493.

Balesdent, J. and Arrouays, D. 1999. Usage des terres et stockage de carbone dans les sols du territoire français. Une estimation des flux nets annuels pour la période 1900–1999. *C. R. Acad. Agriculturists* 85, 265–277.

Barre, P., Quenea, K., Vidal, A., Cecillon, L., Christensen, B. T., Katterer, T., Macdonald, A. J., Petit, L., Plante, A. F., Van Oort, F. and Chenu, C. 2018. Microbial and plant-derived compounds both contribute to persistent soil organic carbon in temperate soils. *Biogeochemistry* 140(1), 81–92.

Bigorre, F., Tessier, D. and Pédro, G. 2000. Contribution des argiles et des matières organiques à la rétention de l'eau dans les sols. Signification et rôle fondamental de la capacité d'échange en cations. Comptes Rendus de l'Académie des Sciences-Series IIA-Earth and Planetary Sciences.

Blanco-Canqui, H. and Lal, R. 2007. Soil structure and organic carbon relationships following 10 years of wheat straw management in no-till. *Soil and Tillage Research* 95(1–2), 240–254.

Borchard, N., Schirrmann, M., Cayuela, M. L., Kammann, C., Wrage-Mönnig, N., Estavillo, J. M., Fuertes-Mendizábal, T., Sigua, G., Spokas, K., Ippolito, J. A. and Novak, J. 2019. Biochar, soil and land-use interactions that reduce nitrate leaching and N2O emissions: a meta-analysis. *Science of the Total Environment* 651(2), 2354–2364.

Bossio, D. A., Cook-Patton, S. C., Ellis, P. W., Fargione, J., Sanderman, J., Smith, P., Wood, S., Zomer, R. J., von Unger, M., Emmer, I. M. and Griscom, B. W. 2020. The role of soil carbon in natural climate solutions. *Nature Sustainability* 3(5), 391–398.

Carmona, C. R., Clough, T. J., McNally, S. R., Beare, M. H., Tregurtha, C. S. and Hunt, J. E. 2020. Seasonal irrigation affects the partitioning of new photosynthate carbon in soil. *Soil Biology and Biochemistry* 143, 107751.

Carrer, D., Pique, G., Fericoq, M., Ceamanos, X. and Ceschia, E. 2018. What is the potential of cropland albedo management in the fight against global warming? A case study based on the use of cover crops. *Environmental Research Letters* 13(4), 044030.

Chabbi, A., Lehmann, J., Ciais, P., Loescher, H. W., Cotrufo, M. F., Don, A., SanClements, M., Schipper, L., Six, J., Smith, P. and Rumpel, C. 2017. Aligning agriculture and climate policy. *Nature Climate Change* 7(5), 307–309.

Chalhoub, M., Garnier, P., Coquet, Y., Mary, B., Lafolie, F. and Houot, S. 2013. Increased nitrogen availability in soil after repeated compost applications: use of the PASTIS model to separate short and long-term effects. *Soil Biology and Biochemistry* 65, 144–157.

Chaney, K. and Swift, R. S. 1984. The influence of organic matter on aggregate stability in some British soils. *Journal of Soil Science* 35(2), 223–230.

Chenu, C., Le Bissonnais, Y. and Arrouays, D. 2000. Organic matter influence on clay wettability and soil aggregate. *Soil Science Society of America Journal* 64(4), 1479–1486.

Christiansen, J. R. and Gundersen, P. 2011. Stand age and tree species affect N2O and CH4 exchange from afforested soils. *Biogeosciences* 8(9), 2535–2546.

Condron, L. M., Hopkins, D. W., Gregorich, E. G., Black, A. and Wakelin, S. A. 2014. Long-term irrigation effects on soil organic matter under temperate grazed pasture. *European Journall of Soil Sciencee* 65, 741–750.

Conrad, R. 2009. The global methane cycle: recent advances in understanding the microbial processes involved. *Environmental Microbiology Reports* 1(5), 285–292.

Conrad, R. 2020. Importance of hydrogenotrophic, aceticlastic and methylotrophic methanogenesis for methane production in terrestrial, aquatic and other anoxic environments: a mini review. *Pedosphere* 30(1), 25–39.

Cosentino, D., Chenu, C. and Le Bissonnais, Y. 2006. Aggregate stability and microbial community dynamics under drying-wetting cycles in a silt loam soil. *Soil Biology and Biochemistry* 38(8), 2053–2062.

Curtin, D. and Rostad, H. P. W. 1997. Cation exchange and buffer potential of Saskatchewan soils estimated from texture, organic matter and pH. *Canadian Journal of Soil Science* 77(4), 621–626. 10.4141/S97-015.

Demenois, J., Torquebiau, E., Arnoult, M. H., Eglin, T., Masse, D., Assouma, M. H., Blanfort, V., Chenu, C., Chapuis-Lardy, L., Medoc, J.-M. and Sall, S. N. 2020. Barriers and strategies to boost soil carbon sequestration in agriculture. *Frontiers in Sustainable Food Systems* 4. 10.3389/fsufs.2020.00037.

Díaz-Zorita, M. and Grosso, G. A. 2000. Effect of soil texture, organic carbon and water retention on the compactability of soils from the Argentinean pampas. *Soil and Tillage Research* 54(1–2), 121–126.

Eden, M., Gerke, H. H. and Houot, S. 2017. Organic waste recycling in agriculture and related effects on soil water retention and plant available water: a review. *Agronomy for Sustainable Development* 37(2), 11.

Emde, D., Hannam, K. D., Most, I., Nelson, L. M. and Jones, M. D. 2021. Soil organic carbon in irrigated agricultural systems: a meta-analysis. *Global Change Biology* 27(16), 3898–3910.

Feliciano, D., Ledo, A., Hillier, J. and Nayak, D. R. 2018. Which agroforestry options give the greatest soil and above ground carbon benefits in different world regions? *Agriculture, Ecosystems and Environment* 254, 117–129.

Feller, C., Balesdent, J., Nicolardot, B. and Cerri, C. 2001. Approaching "functional" soil organic matter pools through particle-size fractionation: examples for tropical soils. In: Lal, R., Kimble, J. M. and Stewart, B. A. (Eds), *Assessment Methods for Soil Carbon*. CRC Press, Boca Raton, pp. 53–67.

Flores-Rios, A., Thomas, E., Peri, P. P., Amelung, W., Duarte-Guardia, S., Borchard, N., Lizárraga-Travaglini, A., Vélez-Azañero, A., Sheil, D., Tscharntke, T., Steffan-Dewenter,

I. and Ladd, B. 2020. Co-benefits of soil carbon protection for invertebrate conservation. *Biological Conservation* 252, 108859.

Fontaine, S., Bardoux, G., Abbadie, L. and Mariotti, A. 2014. Carbon input to soil may decrease soil carbon content. *Ecology Letters* 7(4), 314–320.

Frank, S., Havlík, P., Soussana, J. F., Levesque, A., Valin, H., Wollenberg, E., Kleinwechter, U., Fricko, O., Gusti, M., Herrero, M. and Smith, P. (2017). Reducing greenhouse gas emissions in agriculture without compromising food security? *Environmental Research Letters* 12(10), 105004.

Genesio, L., Miglietta, F., Lugato, E., Baronti, S., Pieri, M. and Vaccari, F. P. 2012. Surface albedo following biochar application in durum wheat Surface albedo following biochar application in durum wheat. *Environmental Research Letters* 1623(7), 149–139.

Graff-Zivin, J. and Lipper, L. 2008. Poverty, risk, and the supply of soil carbon sequestration. *Environment and Development Economics* 13(3), 353–373.

Gregorich, E. G., Beare, M. H., McKim, U. F. and Skjemstad, J. O. 2006. Chemical and biological characteristics of physically uncomplexed organic matter. *Soil Science Society of America Journal* 70(3), 975–985.

Gregorich, E. G., Carter, M. R., Angers, D. A., Monreal, C. M. and Ellert, B. H. 1995. Towards a minimum data set to assess soil organic matter quality in agricultural soils. *Canadian Journal of Soil Science* 74(4), 367–385.

Guenet, B., Camino-Serrano, M., Ciais, P., Tifafi, M., Maignan, F., Soong, J. L. and Janssens, I. A. 2018. Impact of priming on global soil carbon stocks. *Global Change Biology*, 24(5), 1873–1883.

Guenet, B., Gabrielle, B., Chenu, C., Arrouays, D., Balesdent, J., Bernoux, M., Bruni, E., Caliman, J. P., Cardinael, R., Chen, S., Ciais, P., Desbois, D., Fouche, J., Frank, S., Henault, C., Lugato, E., Naipal, V., Nesme, T., Obersteiner, M., Pellerin, S., Powlson, D. S., Rasse, D. P., Rees, F., Soussana, J. F., Su, Y., Tian, H., Valin, H. and Zhou, F. 2020. Can N_2O emissions offset the benefits from soil organic carbon storage? *Global Change Biology* 27(2), 237–256.

Han, X., Xu, C., Dungait, J. A. J., Bol, R., Wang, X. J., Wu, W. L. and Meng, F. Q. 2018. Straw incorporation increases crop yield and soil organic carbon sequestration but varies under different natural conditions and farming practices in China: a system analysis. *Biogeosciences* 15(7), 1933–1946.

Hijbeek, R., van Ittersum, M. K., Ten Berge, H. F. M., Gort, G., Spiegel, H. and Whitmore, A. P. 2016. Do organic inputs matter? A meta-analysis of additional yield effects for arable crops in Europe. *Plant and Soil* 411(1–2), 293–303.

Hunt, J. E., Laubach, J., Bartel, M., Fraser, A. and Phillips, R. L. 2016. Carbon budgets for an irrigated intensively grazed dairy pasture and an unirrigated winter-grazed pasture. *Biogeosciences* 13(10), 2927–2944.

Iizumi, T. and Wagai, R. 2019. Leveraging drought risk reduction for sustainable food, soil and climate via soil organic carbon sequestration. *Scientific Reports* 9(1), 19744.

IPCC. 2019. *Climate Change 2021: The Physical Science Basis. Contribution of Working Group I to the Sixth Assessment Report of the Intergovernmental Panel on Climate Change*. Shukla, P. R., Skea, J., Calvo Buendia, E., Masson-Delmotte, V., Pörtner, H.-O., Roberts, D. C., Zhai, P., Slade, R., Connors, S., van Diemen, R., Ferrat, M., Haughey, E., Luz, S., Neogi, S., Pathak, M., Petzold, J., Portugal Pereira, J., Vyas, P., Huntley,

E., Kissick, K., Belkacemi, M. and Malley, J. (Eds.). Cambridge University Press, Cambridge.

IPCC. 2021. Climate change 2021: the physical science basis. Contribution of working group I to the sixth assessment report of the Intergovernmental Panel on Climate Change Masson-Delmotte, V., Zhai, P., Pirani, A., Connors, S. L., Péan, C., Berger, S., Caud, N., Chen, Y., Goldfarb, L., Gomis, M. I., et al. (Eds.). Cambridge University Press, Cambridge.

Janzen, H. 2006. The soil carbon Idilemma: Schall we hoard it or use It. *Soil Biology and Biochemistry* 28, 419–424.

Kay, B. D., da Silva, A. Pd and Baldock, J. A. 1997. Sensitivity of soil structure to changes in organic carbon content: predictions using pedotransfer functions. *Canadian Journal of Soil Science* 77(4), 655–667.

Kelliher, F. M., West, P. S. J. and Moir, J. L. 2015. Soil carbon stocks beneath an established irrigated pasture grazed by dairy cattle. *New Zealand Journal of Agricultural Research* 58(1), 78–83.

Kim, D. G., Kirschbaum, M. U. F. and Beedy, T. L. 2016. Carbon sequestration and net emissions of CH4 and N2O under agroforestry: synthesizing available data and suggestions for future studies. *Agriculture, Ecosystems and Environment* 226, 65–78.

Kirkby, C. A., Richardson, A. E., Wade, L. J., Passioura, J. B., Batten, G. D., Blanchard, C. and Kirkegaard, J. A. 2014. Nutrient availability limits carbon sequestration in arable soils. *Soil Biology and Biochemistry* 68, 402–409.

Koegel-Knabner, I. 2002. The macromolecular organic composition of plant and microbial residues as inputs to soil organic matter. *Soil Biology and Biochemistry* 34(2), 139–162.

Krogh, L., Breuning-Madsen, H. and Greve, M. H. 2000. Cation-exchange capacity pedotransfer functions for Danish soils. *Acta Agriculturae Scandinavica, Section B – Soil and Plant Science* 50(1), 1–12.

Krull, E. S., Baldock, J. A. and Skjemstad, J. O. 2003. Importance of mechanisms and processes of the stabilisation of soil organic matter for modelling carbon turnover. *Functional Plant Biology : FPB* 30(2), 207–222.

Lal, R. 2004. Soil carbon sequestration impacts on global climate change and food security. *Science* 304(5677), 1623–1627.

Lal, R. 2008. Carbon sequestration. *Philosophical Transactions of the Royal Society of London. Series B, Biological Sciences* 363(1492), 815–830.

Lal, R. 2016. Soil health and carbon management. *Food and Energy Security* 5(4), 212–222.

Lavallee, J. M., Soong, J. L. and Cotrufo, M. F. 2020. Conceptualizing soil organic matter into particulate and mineral-associated forms to address global change in the 21st century. *Global Change Biology* 26(1), 261–273.

Lange, M., Eisenhauer, N., Sierra, C. A., Bessler, H., Engels, C., Griffiths, R. I., Mellado-Vázquez, P. G., Malik, A. A., Roy, J., Scheu, S., Steinbeiss, S., Thomson, B. C., Trumbore, S. E. and Gleixner, G. 2015. Plant diversity increases soil microbial activity and soil carbon storage. *Nature Communications* 6, 6707.

Lehmann, J., Gaunt, J. and Rondon, M. 2006. Bio-char sequestration in terrestrial ecosystems–a review. *Mitigation and Adaptation Strategies for Global Change* 11(2), 403–427.

Lettens, S., Van Orshoven, J., van Wesemael, B., De Vos, B. and Muys, B. 2005. Stocks and fluxes of soil organic carbon for landscape units in Belgium derived from heterogeneous data sets for 1990 and 2000. *Geoderma* 127(1–2), 11–23.

Liu, L. and Greaver, T. L. 2009. A review of nitrogen enrichment effects on three biogenic GHGs: the CO2 sink may be largely offset by stimulated N2O and CH4 emission. *Ecology Letters* 12(10), 1103-1117.

Lugato, E., Leip, A. and Jones, A. 2018. Mitigation potential of soil carbon management overestimated by neglecting N2O emissions. *Nature Climate Change* 8(3), 219–223.

Mbuthia, L. W., Acosta-Martinez, V., DeBryun, J., Schaeffer, S., Tyler, D., Odoi, E., Mpheshea, M., Walker, F. and Eash, N. 2015. Long term tillage, cover crop, and fertilization effects on microbial community structure, activity: implications for soil quality. *Soil Biology and Biochemistry* 89, 24–34.

Mei, K., Wang, Z., Huang, H., Zhang, C., Shang, X., Dahlgren, R. A., Zhang, M. and Xia, F. 2018. Stimulation of N2O emission by conservation tillage management in agricultural lands: a meta-analysis. *Soil and Tillage Research* 182, 86-93.

Meier, E. A., Thorburn, P. J., Bell, L. W., Harrison, M. T. and Biggs, J. S. 2020. Greenhouse gas emissions from cropping and grazed pastures are similar: a simulation analysis in Australia. *Frontiers in Sustainable Food Systems* 3, 121.

Meyer, S., Bright, R. M., Fischer, D., Schulz, H. and Glaser, B. 2012. Albedo impacts on the suitability of biochar systems to mitigate global climate warming. *Environmental Science and Technology* 46(22), 12726-12734.

Minasny, B. and McBratney, A. B. 2018. Limited effect of organic matter on soil available water capacity. *European Journal of Soil Science* 69(1), 39-47.

Oelofse, M., Markussen, B., Knudsen, L., Schelde, K., Olesen, J. E., Stoumann, L. S., Bruun, S., Jensen, L. S. and Bruun, S. 2015. Do soil organic carbon levels affect potential yields and nitrogen use efficiency? An analysis of winter wheat and spring barley field trials. *European Journal of Agronomy* 66, 62–73.

Oldfield, E. E., Bradford, M. A. and Wood, S. A. 2019. Global meta-analysis of the relationship between soil organic matter and crop yields. *SOIL* 5(1), 15–32.

Olson, K. R., Ebelhar, S. A. and Lang, J. M. 2010. Cover crop effects on crop yields and soil organic carbon content. *Soil Science* 175(2), 89-98.

Paetsch, L., Mueller, C. W., Kögel-Knabner, I., von Lützow, M., Girardin, C. and Rumpel, C. 2018. Effect of *in-situ* aged and fresh biochar on soil water holding capacity and microbial C use under drought conditions. *Scientific Reports* 8(1), 6852.

Pan, G., Smith, P. and Pan, W. 2009. The role of soil organic matter in maintaining the productivity and yield stability of cereals in China. *Agriculture, Ecosystems and Environment* 129(1-3), 344-348.

Pärn, J., Verhoeven, J. T. A., Butterbach-Bahl, K., Dise, N. B., Ullah, S., Aasa, A., Egorov, S., Espenberg, M., Järveoja, J., Jauhiainen, J., Kasak, K., Klemedtsson, L., Kull, A., Laggoun-Défarge, F., Lapshina, E. D., Lohila, A., Lõhmus, K., Maddison, M., Mitsch, W. J., Müller, C., Niinemets, Ü., Osborne, B., Pae, T., Salm, J. O., Sgouridis, F., Sohar, K., Soosaar, K., Storey, K., Teemusk, A., Tenywa, M. M., Tournebize, J., Truu, J., Veber, G., Villa, J. A., Zaw, S. S. and Mander, Ü. 2018. Nitrogen-rich organic soils under warm well-drained conditions are global nitrous oxide emission hotspots. *Nature Communications* 9(1), 1135.

Parton, W. J., Schimel, D. S., Cole, C. V. and Ojima, D. S. 1987. Analysis of factors controlling soil organic levels of grasslands in the Great Plains. *Soil Science Society of America Journal* 51:1173-1179.

Paustian, K., Lehmann, J., Ogle, S., Reay, D., Robertson, G. P. and Smith, P. 2016. Climate-smart soils. *Nature* 532(7597), 49-57.

Peltre, C., Christensen, B. T., Dragon, S., Icard, C., Kätterer, T. and Houot, S. 2012. RothC simulation of carbon accumulation in soil after repeated application of widely different organic amendments. *Soil Biology and Biochemistry* 52, 49-60.

Pikul, J. L., Chilom, G., Rice, J., Eynard, A., Schumacher, T. E., Nichols, K., Johnson, J. M. F., Wright, S., Caesar, T. and Ellsbury, M. 2009. Organic matter and water stability of field aggregates affected by tillage in South Dakota. *Soil Science Society of America Journal* 73(1), 197-206.

Piñeiro, V., Arias, J., Dürr, J., Elverdin, P., Ibáñez, A. M., Kinengyere, A., Opazo, C. M., Owoo, N., Page, J. R., Prager, S. D. and Torero, M. 2020. A scoping review on incentives for adoption of sustainable agricultural practices and their outcomes. *Nature Sustainability* 3(10), 809-820.

Plaza-Bonilla, D., Arrúe, J. L., Cantero-Martínez, C., Fanlo, R., Iglesias, A. and Álvaro-Fuentes, J. 2015. Carbon management in dryland agricultural systems: a review. *Agronomy for Sustainable Development* 35(4), 1319-1334, DOI: 10.1007/s13593-015-0326-x.

Poeplau, C. and Don, A. 2015. Carbon sequestration in agricultural soils via cultivation of cover crops-a meta-analysis. *Agriculture, Ecosystems and Environment* 200, 33-41.

Quiroga, A. R., Buschiazzo, D. E. and Peinemann, N. 1999. Soil compaction is related to management practices in the semi-arid Argentine pampas. *Soil and Tillage Research* 52(1-2), 21-28.

Rawls, W. J. 1983. Estimating soil bulk density from particle size analysis and organic matter content. *Soil Science* 135, 124-124.

Richardson, A. E., Kirkby, C. A., Bnerjee, S. and Kirkgaard, J. A. 2014. The inorganic nutrient cost of building soil carbon. *Carbon Management* 5(3), 265-268.

Rumpel, C., Amiraslai, F., Bossio, D., Chenu, C., Henry, B., Fuentes Espinoza, A. F., Koutika, L.-K., Ladha, J., Madari, B., Minasni, B., Olaleye, A. O., Shirato, Y., Sall, S. N., Soussana, J.-F. and Varela-Ortega, C. 2022. The role of soil carbon sequestration in enhancing human resilience in tackling global crises including pandemics. *Soil Security* 8, 100069.

Rumpel, C. and Chabbi, A. 2021. Managing soil organic carbon for mitigating climate change and increasing food security. *Agronomy* 11(8), 1553.

Rumpel, C., Kögel-Knabner, I. and Bruhn, F. 2002. Vertical distribution, age, and chemical composition of organic carbon in two forest soils of different pedogenesis. *Organic Geochemistry* 33(10), 1131-1142.

Saggar, S., Tate, K. R., Giltrap, D. L. and Singh, J. 2008. Soil-atmosphere exchange of nitrous oxide and methane in New Zealand terrestrial ecosystems and their mitigation options: a review. *Plant and Soil* 309(1-2), 25-42.

Schmidt, H. P., Anca-Couze, A., Hagemann, N., Werner, C., Gerten, D., Lucht, W. and Kammann, C. 2019. Pyrogenic carbon capture and storage. *GCB Bioenergy* 11(4), 573-591.

Schmidt, H. P., Kamman, C., Hagemann, N., Leifeld, J., Buecheli, T., Sanchez Mandeno, M. A. and Cayuela, M. A. 2021. Biochar in agriculture. A systematic reivew of 26 metaanalyses. *Global Change Biology* 13, 1708-1730.

Singh Brar, B., Sing, J., Singh, G. and Kaur, G. 2015. Effects of long-term application of inorganic and organic fertilizers on soil organic carbon and physical properties in maize-wheat rotations. *Agronomy* 5(2), 220-238.

Skaalsveen, K., Ingram, J., Julie, L. E. and Clarke, L. E. 2019. The effect of no-till farming on the soil functions of water purification and retention in northwestern Europe: a literature review. *Soil and Tillage Research* 189, 98-109.

Skinner, C., Gattinger, A., Muller, A., Mäder, P., Flieβbach, A., Stolze, M., Ruser, R. and Niggli, U. 2014. Greenhouse gas fluxes from agricultural soils under organic and non-organic management–A global meta-analysis. *Science of the Total Environment* 468–469, 553–563.

Smith, P. 2016. Soil carbon sequestration and biochar as negative emission technologies. *Global Change Biology* 22(3), 1315–1324.

Soussana, J., Lutfalla, S., Ehrhardt, F., Rosenstock, T., Lamanna, C., Havlík, P., Richards, M., Wollenberg, E., Chotte, J., Torquebiau, E., Ciais, P., Smith, P. and Lal, R. 2019. Matching policy and science: rationale for the '4 per 1000-soils for food security and climate' initiative. *Soil and Tillage Research* 188, 3–15.

Thorup-Kristensen, K. and Nielsen, N. E. 1998. Modelling and measuring the effect of nitrogen catch crops on the nitrogen supply for succeeding crops. *Plant and Soil* 203(1), 79–89.

Tian, H., Xu, R., Canadell, J. G., Thompson, R. L., Winiwarter, W., Suntharalingam, P., Davidson, E. A., Ciais, P., Jackson, R. B., Janssens-Maenhout, G., Prather, M. J., Regnier, P., Pan, N., Pan, S., Peters, G. P., Shi, H., Tubiello, F. N., Zaehle, S., Zhou, F., Arneth, A., Battaglia, G., Berthet, S., Bopp, L., Bouwman, A. F., Buitenhuis, E. T., Chang, J., Chipperfield, M. P., Dangal, S. R. S., Dlugokencky, E., Elkins, J. W., Eyre, B. D., Fu, B., Hall, B., Ito, A., Joos, F., Krummel, P. B., Landolfi, A., Laruelle, G. G., Lauerwald, R., Li, W., Lienert, S., Maavara, T., MacLeod, M., Millet, D. B., Olin, S., Patra, P. K., Prinn, R. G., Raymond, P. A., Ruiz, D. J., van der Werf, G. R., Vuichard, N., Wang, J., Weiss, R. F., Wells, K. C., Wilson, C., Yang, J. and Yao, Y. 2020. A comprehensive quantification of global nitrous oxide sources and sinks. *Nature* 586(7828), 248–256.

Trost, B., Prochnow, A., Drastig, K., Meyer-Aurich, A., Ellmer, F. and Baumecker, M. 2013. Irrigation, soil organic carbon and N_2O emissions: a review. *Agronomy for Sustainable Development* 33(4), 733–749.

United Nations 2017. World population prospects: the 2017 revision, key findings and advance tables. Working Paper No. ESA/P/WP/248, Department of Economic and Social Affairs, Population Division (2017).

van Erp, P. J., Houba, V. J. G. and van Beusichem, M. L. 2001. Actual cation exchange capacity of agricultural soils and its relationship with pH and content of organic carbon and clay. *Communications in Soil Science and Plant Analysis* 32(1-2), 19-31.

van Groenigen, J. W., Van Kessel, C., Hungate, B. A., Oenema, O., Powlson, D. S. and Van Groenigen, K. J. 2017. Sequestering soil organic carbon: a nitrogen dilemma. *Environmental Science and Technology* 51(9), 4738–4739.

Van Kessel, C., Venterea, R., Six, J., Adviento-Borbe, M. A., Linquist, B. and van Groenigen, K. J. 2013 Climate, duration, and N placement determine N2O emissions in reduced tillage systems: a meta-analysis. *Global Change Biology* 19(1), 33–44.

Virto, I., Barré, P., Burlot, A. and Chenu, C. 2012. Carbon input differences as the main factor explaining the variability in soil organic C storage in no-tilled compared to inversion tilled agrosystems. *Biogeochemistry* 108(1-3), 17–26.

Warrington, D. N., Goldstein, D. and Levy, G. J. 2007. Clay translocation Within the soil profile as affected by intensive irrigation with treated wastewater. *Soil Science* 172(9), 692–700.

Woolf, D., Amonette, J. E., Street-Perrott, F. A., Lehmann, J. and Joseph, S. 2010. Sustainable biochar to mitigate global climate change. *Nature Communications* 1, 56.

Xu, C., Han, X., Zhuge, Y., Xiao, G., Ni, B., Xu, X. and Meng, F. 2021. Crop straw incorporation alleviates overall fertilizer-N losses and mitigates N2O emissions per unit applied N from intensively farmed soils: an in situ 15N tracing study. *Science of the Total Environment* 764, 142884.

Yeboah, S., Zhang, R., Cai, L., Li, L., Xie, J., Luo, Z., Liu, J. and Wu, J. 2016. Tillage effect on soil organic carbon, microbial biomass carbon and crop yield in spring wheat-field pea rotation. *Plant, Soil and Environment* 62(6), 279–285.

Zhang, X., Sun, N., Wu, L., Xu, M., Bingham, I. J. and Li, Z. 2016. Effects of enhancing soil organic carbon sequestration in the topsoil by fertilization on crop productivity and stability: evidence from long-term experiments with wheat-maize cropping systems in China. *Science of the Total Environment* 562, 247–259.

Chapter 7

Soil inorganic carbon: stocks, functions, losses and their consequences

Kazem Zamanian, University of Hannover, Germany and Nanjing University of Information Science and Technology (NUIST), China; and Yakov Kuzyakov, University of Göttingen, Germany and RUDN University, Russia

1 Introduction

Soil inorganic carbon (SIC) comprises:

1. all carbonate-containing minerals, most commonly in the form of calcite ($CaCO_3$) but also $MgCO_3$, $FeCO_3$ etc. in the solid phase (Chesworth et al., 2008; Zamanian et al., 2016);
2. dissolved CO_2 and ions $H_2CO_3^{\circ}$, HCO_3^- and CO_3^{2-} in the liquid phase - so-called dissolved inorganic carbon (DIC) (Li et al., 2015; Liu et al., 2010).

Solid SIC can be further divided into three groups (Fig. 1) based on their origin, morphology and formation mechanisms (Zamanian et al., 2016):

1. **Geogenic carbonates** are inherited from soil parent materials such as limestone blocks, particles and crystals, as in loess deposits.

http://dx.doi.org/10.19103/AS.2022.0106.07

Figure 1 Examples and forms of geogenic, biogenic and pedogenic carbonates most common in soils. See more details and examples in Zamanian et al. (2016).

2 **Biogenic carbonates** are formed via activities of organisms such as earthworm biospheroliths and/or from parts of an organism's skeleton, for example, land-snail shells and eggshells.

3 **Pedogenic carbonates** are *authigenic* carbonates formed in the soil matrix via precipitation of carbonate ions with cations (mainly Ca^{2+}) from soil solution when the solubility constant of carbonates is achieved. The source of carbonate ions in pedogenic carbonates is from soil respiration (Rovira and Vallejo, 2008; Zamanian et al., 2016), but cation sources can include (i) mineral weathering including dissolution of any type of carbonate minerals (Beerling et al., 2020; Zamanian et al., 2016), (ii) groundwater (Guo et al., 2016), (iii) dust (Gile, 1999), (iv) organic matter decomposition (Mikhailova et al., 2009), (v) rain water (Goddard et al., 2007; Mikhailova et al., 2013) and (vi) irrigation water (de Soto et al., 2017).

Under natural conditions there is an equilibrium between the solid phase (hereafter SIC for simplicity) and the liquid phase of SIC, i.e. DIC (Eq. 1). Dissolution (when Eq. 1 goes to the right) or precipitation (when Eq. 1 goes to the left) of calcite (a_{CaCO3}) controls the activity of calcium ions in soil solution (m_{Ca2+}), the partial pressure of CO_2 in soil gas (pCO_2) (Fig. 2) and temperature-sensitive

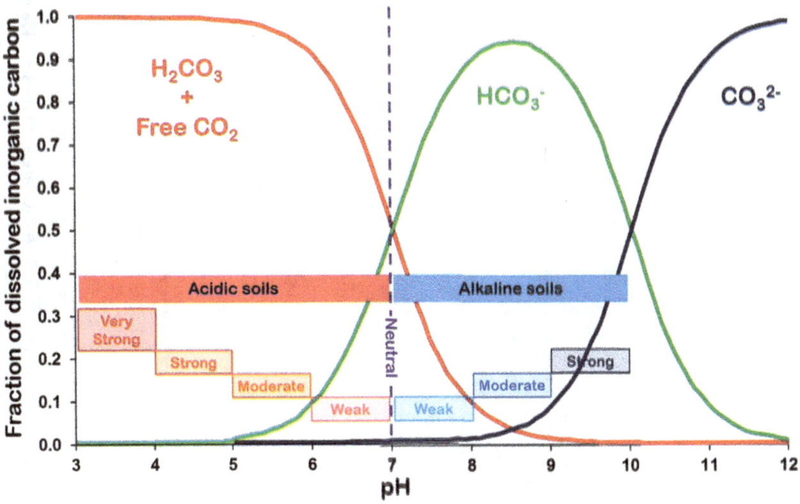

Figure 2 Speciation of dissolved CO_2 in aqueous solutions as a function of acidity and the range of soil pH values from very strong acidic (i.e. acid sulfate soils) to strong alkaline soils (i.e. Solonetz).

equilibrium constants for the dissociation of carbonic acid (K_1), bicarbonate (K_2) and calcite (K_{cal}), and for the hydration of CO_2 (K_{CO2}) (Eq. 2) (Drever, 1982).

$$CaCO_{3(s)} + H_2O_{(l)} + CO_{2(aq)} \leftrightarrow Ca^{2+}_{(aq)} + 2HCO^-_{3(aq)} \tag{1}$$

$$a_{CaCO_3} = \frac{4m^3_{Ca^{2+}}}{pCO_2}\left[\frac{K_2}{K_1 K_{cal} K_{CO_2}}\right] \tag{2}$$

The presence of SIC and its dissolution/re-precipitation processes have various ecosystem functions, both directly via particle cementation and providing Ca^{2+} ions and indirectly by controlling soil pH.

2 Importance of soil inorganic carbon in global carbon cycle and ecosystem functions

2.1 Soil inorganic carbon stocks and global carbon cycle

The presence of SIC is highly correlated with climatic conditions, and SIC content decreases with increasing mean annual precipitation (MAP) (Wang et al., 2010; Wu et al., 2009; Yang et al., 2018). Hence, the largest SIC stocks are present in arid to sub-humid regions (Fig. 3). Globally, about 54% of land surface area (Zamanian et al., 2018) stores ca. 960 (Batjes, 1996) to 1783 (Eswaran et al., 1995) Pg (10^{15} g) C as SIC in the top 100 cm. It is estimated that another 1000 Pg SIC is stored between 100 cm and 200 cm (Zamanian

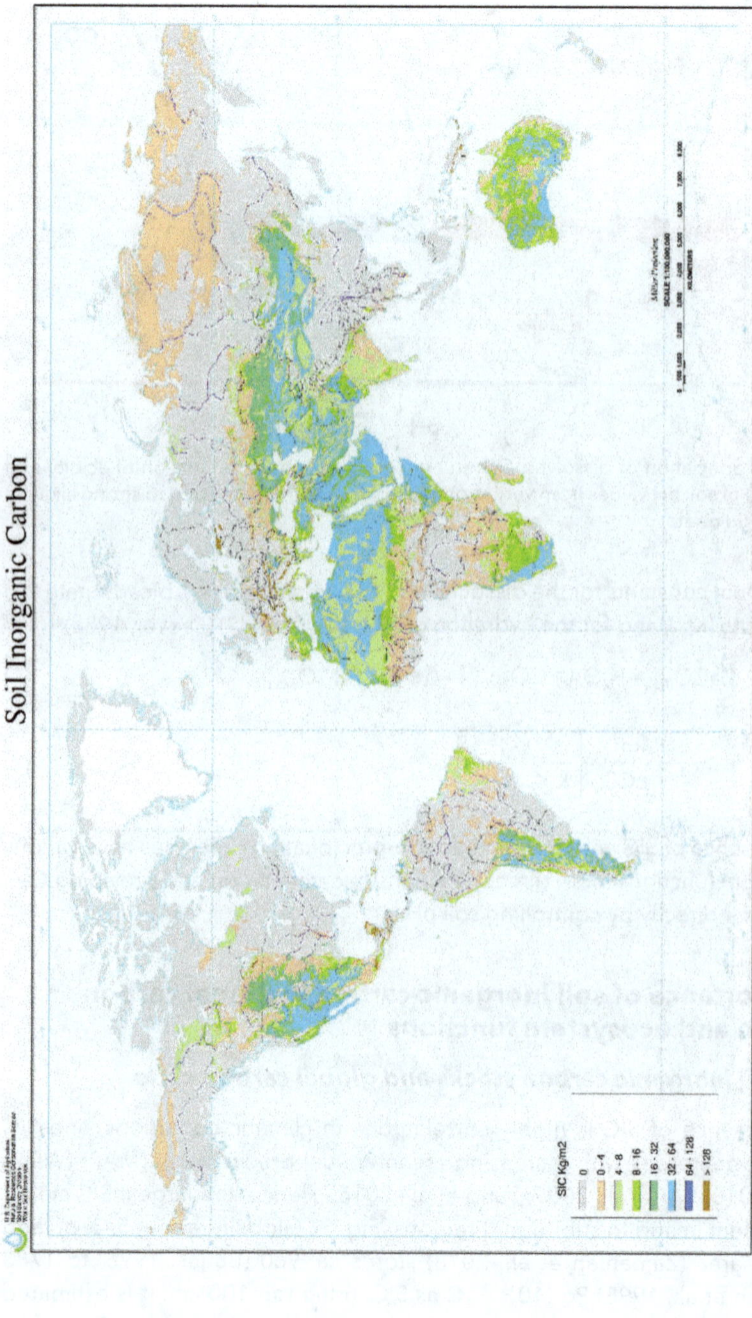

Figure 3 Global distribution of soil inorganic carbon (SIC) in the topsoil (USDA-NRCS, https://www.nrcs.usda.gov/). The SIC area covers ~54% of the land surface (Zamanian et al., 2018).

et al., 2021). Neglecting deeper layers may lead to underestimating the soil C stock by as much as ~50% (Díaz-Hernández, 2010; Li et al., 2007; Wang et al., 2010; You et al., 2020; Zhang et al., 2015a). Global SIC stocks in the top 200 cm of soil are comparable to soil organic carbon (SOC) stocks (ca. 2400 Pg C) (Batjes, 1996). The addition of around 1404 Pg C as DIC stocks in soil and groundwater (Monger et al., 2015) makes the total SIC stocks even greater. It is estimated that 0.71 Pg C can be stored globally per year in soil water and in aquatic bodies as DIC (Li et al., 2015; Liu et al., 2010). Such huge SIC stocks point not only to the importance of SIC as part of the global C cycle (Yang et al., 2010, 2012; Zamanian et al., 2016) but also as an effective regulator of global warming over both long-term (Lerman et al., 2007) and short-term timescales (Yang et al., 2010; Zamanian et al., 2021; Zamanian and Kuzyakov, 2019). Carbon sequestration as SIC is of particular importance in dry regions, where SIC is the dominant form of soil C (Raheb et al., 2016, 2017) and where SOC accumulation, even when using irrigation, is slow (Lal and Kimble, 2000).

2.2 Soil buffering capacity

Soil acidification via natural processes such as organic acid exudation and H^+ release in exchange of cationic nutrients uptake by roots or (micro)organisms respiration, and through anthropogenic activities such as burning fossil fuels and fertilization, alters the biogeochemistry of ecosystems (Luo et al., 2015). While natural processes may take place over millennia, anthropogenic acidification affects ecosystems within a few years to a few decades (Fig. 4). Acidification results in the gradual elimination of acid-sensitive species including some bacterial groups, plants and animals, which ultimately changes the structure and functions of terrestrial ecosystems ((Rengel, 2003) and references therein). Furthermore, acidity is partially transported from terrestrial ecosystems to water bodies and endangers aquatic ecosystems.

Soils are, to a varying degree, resistant to pH changes. Various buffering systems are active within specific pH ranges and prevent pH decrease to certain thresholds (Fig. 4) (Raza et al., 2020). Among them, SIC is the most effective, reactive and largest buffer system (Berner and Berner, 2012; Van Breemen et al., 1983) that can neutralize large quantities of acid (20 000 mmol+ $kg_{calcite}^{-1}$) (Bloom et al., 2005). As long as free SIC is present, the importance of other buffering systems is negligible (Fig. 5) and the pH value remains above 6.5. The pH value drops only after SIC has been completely consumed.

SIC dissolution takes place when acidity is added to the soil, for example, via carbonic acid and when Eq. 1 goes to the right. SIC dissolution is a reversible process if:

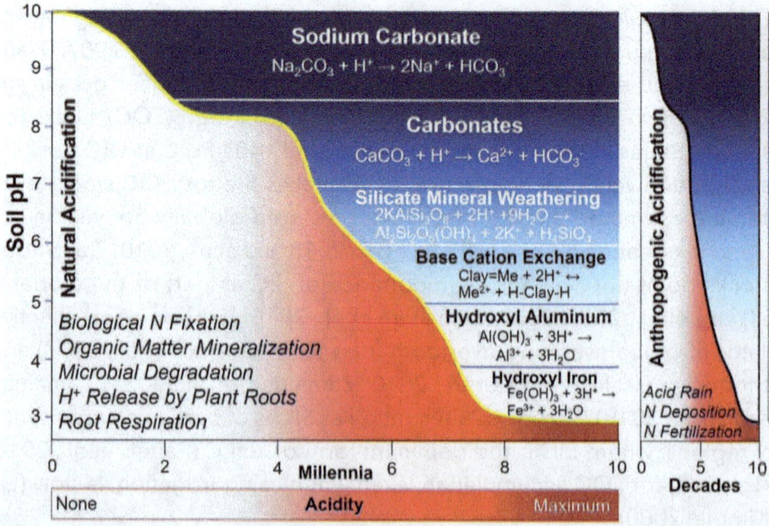

Figure 4 Various acid buffering mechanisms in soils, which are active within specific pH ranges. While soil's acid neutralizing capacity decreases over millennia under natural conditions, anthropogenic activities may acidify soils over a few decades (modified from Raza et al., 2020).

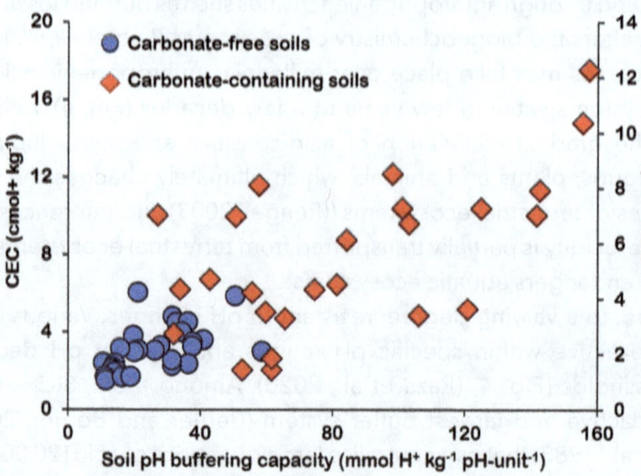

Figure 5 The relation of pH buffering capacity either to cation exchange capacity (CEC) in carbonate-free soils (blue circles) or to SIC content in carbonate-containing soils (red diamonds) along a 3600 km long transect across northern China (Luo et al., 2015). While CEC is the most determinant control in carbonate-free soils, pH buffering capacity is generally controlled by SIC content in carbonate-containing soils.

1 carbonic acid or organic acids are the sources of acidity, because CO_2 degassing (Eq. 1) and microbial decomposition of organic anions (Eq. 3), respectively, will lead to re-precipitation of SIC;
2 water is removed by transpiration or evaporation;
3 the temperature increases, consequently decreasing CO_2 solubility and degassing.

If the source of acidity is mineral acids, such as sulfuric or nitric acids (e.g. mining lands, acid rain), then the reaction (Eq. 4) is irreversible as long as the inorganic anions remain in the solution. Either way, the neutralization of acidity via SIC dissolution leads to a release of Ca^{2+} (and/or other basic) ions into the solution that may be transported through the soil profile or ultimately completely washed out of the soil (Yin et al., 2016). Such Ca^{2+} removal and consequently the depletion of SIC stocks increase the vulnerability of soil to further acidification (Shaofei and Wang, 2018):

$$CaCO_{3(s)} + R-COOH_{(aq)} \leftrightarrow Ca^{2+}_{(aq)} + HCO^{-}_{3(aq)} + R-COO^{-}_{(aq)} \tag{3}$$

$$CaCO_{3(s)} + 2H^{+}_{(aq)} + 2NO^{-}_{3(aq)} \rightarrow Ca^{2+}_{(aq)} + 2NO^{-}_{3(aq)} + H_2CO_{3(aq)} \tag{4}$$

2.3 Soil organic carbon stability

Soil organic carbon content is the main characteristic controlling the state of many other biological, chemical and physical properties (Schmidt et al., 2011). Despite inconsistent reports due to differences in analyzed soil types (e.g. calcite content), climatic conditions and land management (Zhang et al., 2020), SOC content can be directly and indirectly controlled by SIC (Clough and Skjemstad, 2000; Guo et al., 2016; Zhang et al., 2020). SIC precipitation around fresh organic matter directly removes SOC from microbial access (Virto et al., 2018). SIC precipitation results in soil aggregation (see Section 2.5), which also protects SOC via entrapment of organic compounds within aggregates (Clough and Skjemstad, 2000; Garcia-Franco et al., 2020).

SIC controls plant growth and microbial activities through buffering soil pH and providing nutrients (Lal and Kimble, 2000). However, SIC content may also have negative consequences for plant growth, due to deficiencies in phosphorus, iron and various micronutrients, as well as physiological deficiency of potassium and magnesium. Nonetheless, lower SIC content can also increase plant resistance against pathogens or drought, which will increase SOC inputs, for example, via more leaf and root litter and exudates (Cater and Ogrinc, 2011; Paradelo et al., 2015). On the other hand, microbial activity increases in neutral to slightly alkaline soils (compared to acidic soils), which favors organic matter mineralization and could decrease SOC content (Chan and Heenan, 1999;

Whittinghill and Hobbie, 2012). However, SOC stored in microbial biomass and necromass (i.e. dead microbial biomass) is more stable in comparison to the C in plant residues. Therefore, increased SOC input may sometimes (over) compensate for SOC loss via microbial mineralization (Paradelo et al., 2015).

SIC dissolution controls the concentration of Ca^{2+} ions in the soil solution (Eq. 1). The presence of Ca^{2+} ions increases the adsorption of organic compounds on mineral surfaces via formation of SOC–Ca-mineral complexes (Mikutta et al., 2007; Paradelo et al., 2015; Rowley et al., 2018). Ca^{2+} ions may participate in the formation of multilayer organo-mineral complexes of mineral-Ca-SOC-Ca-SOC (Gao et al., 2020; Kleber et al., 2015). Therefore, increasing the Ca^{2+} concentration increases the amount of adsorbed organic compounds (Barreto et al., 2021; Huang et al., 2019). Ca^{2+} ions also increase the thermal stability of organic compounds by increasing the bonding strength with mineral surfaces (Barreto et al., 2021). Furthermore, Ca^{2+} ions increase preferential adsorption of low molecular weight organic compounds with a high density of carboxylic groups on mineral surfaces (Kang and Xing, 2008; Sowers et al., 2018), which changes the physico-chemical properties of the sorbed organic compounds (Barreto et al., 2021). Furthermore, Ca^{2+} ions are effective aggregating agents (see Section 2.5), which improve soil structure (Huang et al., 2019; Virto et al., 2018). An improved structure not only increases SOM stability via occlusion but also provides a better medium for water storage and plant growth, which in turn increases SOC input to the soil (Caires et al., 2015; Rowley et al., 2020; Virto et al., 2011).

2.4 Activity of soil organisms: microorganisms and animals

Microorganisms are essential for the sustainable functioning of ecosystems. Soil microorganisms are involved in the decomposition of organic compounds, cycling of elements, nutrient availability and improving resilience of ecosystems to natural and anthropogenic changes (Dai et al., 2021; Girvan et al., 2005; Kao and Ko, 1986a). SIC content not only controls the abundance and composition of microorganisms (Shen et al., 2013) but also affects the abundance and diversity of meso- and macroorganisms (Pabian and Brittingham, 2012) as well as higher plants (Bigelow and Canham, 2002). The directions and rates of biochemical and chemical processes, as well as SOC stability (see Section 2.3), therefore depend on SIC content (Bakina et al., 2014).

SIC controls pH in the neutral and alkaline range (Kim et al., 2016; Shen et al., 2013; Vestergard et al., 2018) and, consequently, has direct and indirect effects on ecosystem food systems (Vestergard et al., 2018). Increasing microbial activity, for example, increases the availability of mineral nitrogen (N), which benefits plant growth. The nitrification rate strongly increases at pH values above six (Fig. 6), where SIC is one of the main regulators of microbial

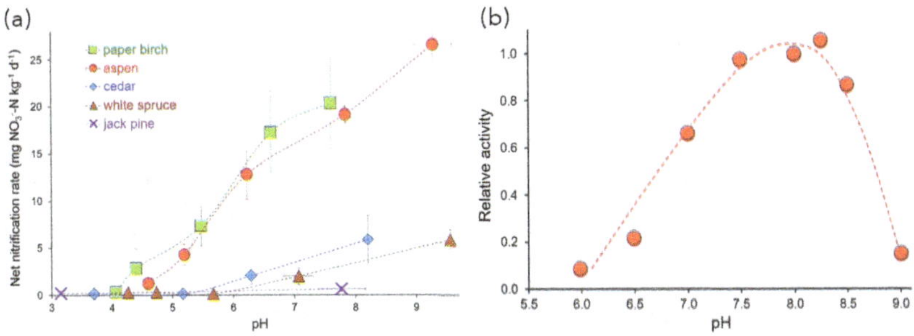

Figure 6 Net nitrification rate in soils under various tree species (a) and the relative activity of Nitrobacter (b) as a function of soil pH value. Re-drawn, respectively, from Ste-Marie and Paré (1999) and Blackburne et al. (2007).

activities (Ste-Marie and Paré, 1999). Denitrification processes, in particular the reduction of N_2O to N_2, is also highly pH dependent (Cuhel and Simek, 2011). Efficient reduction of N_2O to N_2 at pH values above 6.4 - common in SIC-containing soils - prevents N_2O emission (Goulding, 2016; Hénault et al., 2019).

Furthermore, Ca is an essential element for plant growth (Marschner, 2011; McLaughlin and Wimmer, 1999; White and Broadley, 2003) as well as for elongation of fungal hyphae, for example, *Histoplasma capsulatum* (Batanghari and Goldman, 1997). Ca availability increases N content in plant tissues and crop yield (Kao and Ko, 1986b; Tiritan et al., 2016) and increases plant resistance to diseases (e.g. cucumber seedlings by *Pythium splendens*) (Kao and Ko, 1986a) and environmental extremes such as drought or cold (Hosseini et al., 2019; Song et al., 2008; Yuan et al., 2018). SIC *per se* is also the source of Ca necessary for breeding egg-laying animals and snails (Graveland and van der Wal, 1996). Low Ca availability results in thin eggshells or even the absence of eggshells altogether (Graveland and van der Wal, 1996).

Despite all these positive effects, excessive amounts of Ca may limit plant growth (Li et al., 2018a) due to hardening of cell walls, disturbance of plant energy metabolism and antagonistic effects for magnesium (White and Broadley, 2003) and potassium. Nevertheless, 'Calcium depletion [has] the potential to disrupt entire ecosystems' (Pabian and Brittingham, 2012).

2.5 Soil aggregation and aggregate stability

SIC availability increases aggregation and aggregate stability (Wuddivira and Camps-Roach, 2007) via various mechanisms. SIC precipitation, i.e. formation of pedogenic carbonates, can directly cement soil particles (Vargas et al., 2019). Accumulation of micro-crystals of pedogenic carbonate thus decreases micro- and macro-porosity and changes the geometry of soil pores (Virto et al., 2018). Higher concentrations of Ca^{2+} ions (as well as other divalent cations such

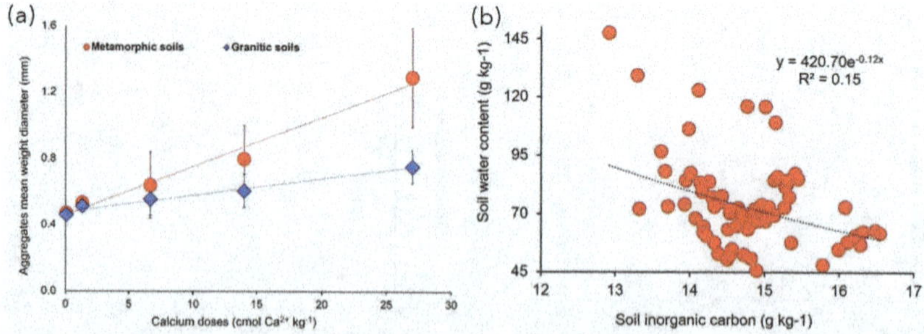

Figure 7 (a) Effect of Ca^{2+} ions added as $Ca(OH)_2$ on aggregates' mean weight diameter in soils developed on metamorphic (circles) and granitic (diamonds) parent materials (Vargas et al., 2019). (b) Soil water content as a function of SIC content (Zhao et al., 2016).

as Mg^{2+} ions) in carbonate-containing soils also enable aggregation (Fig. 7a). Ca^{2+} ions may act as cationic bridges among clay minerals bonding mineral and/or organic particles together (Rowley et al., 2018; Wuddivira and Camps-Roach, 2007).

The effects of SIC as well as Ca^{2+} ions on aggregation, however, are dependent on the amount of SOC (Boix-Fayos et al., 2001; Bronick and Lal, 2005) and clay mineralogy (Wuddivira and Camps-Roach, 2007). For example, SIC is the main aggregating agent when SOC content is below 5% in Mediterranean soils (Boix-Fayos et al., 2001). Furthermore, when more than about 15% (w_{CaCO3}:w_{soil}) of pedogenic carbonates accumulate in soil, porosity decreases considerably, which subsequently has negative effects on water storage as well as root prolongation (Fig. 7b). Nevertheless, the improved structure from a smaller content of SIC gives soils greater water storage (Fernández-Ugalde et al., 2014) and better infiltration rates (Groshans et al., 2018; Virto et al., 2018). Thus, carbonate-containing soils generally become resistant to various types of physico-chemical degradation (Fernández-Ugalde et al., 2014) and have lower erodability (Ostovari et al., 2016).

3 Soil inorganic carbon response to land management and climate change

3.1 Land use and land management

Land management practices such as irrigation, tillage and fertilization (de Soto et al., 2017; Dong et al., 2017; Eshel et al., 2007; Lal, 2009), as well as grazing (Reeder et al., 2004) and plantation (Gao et al., 2017; Plaza-Bonilla et al., 2015), can significantly change SIC levels over short periods of time (de Soto et al., 2019; Kim et al., 2020; Liu et al., 2014; Raza et al., 2020; Tao et al., 2022; Wang et al., 2016). For example, irrigation with Ca-containing water, whether with surface water (Bughio et al., 2016; de Soto et al., 2019; Zhang et al., 2015b) or groundwater (Schlesinger, 2000), increases SIC content via the formation of pedogenic carbonates (Wang et al., 2016). Irrigation may also lead to SIC dissolution and loss, or relocation in the soil profile, i.e. dissolution at the topsoil and re-precipitation at deeper depths (An et al., 2019; Kim et al., 2020; Magaritz and Amiel, 1981). Thus, the quality and quantity of irrigation water determine if SIC precipitates or dissolves (Artiola and Walworth, 2009).

Plowing deep soils formed on calcareous parent materials brings SIC particles to the surface (de Soto et al., 2017), common in Calcaric Leptosols (Terra fusca, Terra rossa). The addition of straw on the surface may lead to the formation of pedogenic carbonates as Ca^{2+} ions may be released following the decomposition of organic compounds (Lu et al., 2020; Wang et al., 2015; Zhao et al., 2018). On the other hand, straw decomposition, especially organic compounds with a relatively higher amount of N (Poss and Saragoni, 1992), may decrease soil pH (Rahman et al., 2008), which increases SIC dissolution at the surface. Fertilizers (and in particular N fertilizers) are the main sources of acidity in croplands (Caires et al., 2015; Goulding, 2016; You et al., 2020), which directly dissolve SIC (Poss and Saragoni, 1992) and provide a significant source of atmospheric CO_2 (Cater and Ogrinc, 2011; Zamanian et al., 2018; Zamanian and Kuzyakov, 2019).

Inconsistent responses of SIC content to agricultural practices, such as accumulation (Bughio et al., 2016; de Soto et al., 2019), losses (Raza et al., 2020), relocation (You et al., 2020) or no effect (Dong et al., 2017), have been addressed in the literature, but long-term investigations (over decades) might be necessary to understand the ultimate effects of cropping on SIC dynamics. Nevertheless, SIC content and distribution can be used as an indicator of agrogenic soil development (Kuznetsova et al., 2010; Kuzyakov and Zamanian, 2019). For details regarding the effects of agricultural practices on SIC stock, readers are referred to Chapter 22 of this volume.

Liming (Bramble et al., 2020a,b; Haynes and Naidu, 1998), or the addition of Ca-containing minerals such as Ca-silicates (Beerling et al., 2020; Haque et al., 2020), as well as building materials, steel slag (Renforth et al., 2009), wood ash (Zhao et al., 2018) and biochar (Ghorbani et al., 2021) are other sources of SIC accumulation in agricultural soils. Ca-containing minerals (mostly lime) are added to soil to increase pH and nutrient availability for plant growth (Haque et al., 2020), which may subsequently result in C sequestration as inorganic (from lime) and organic (from litter and roots) carbon (Haque et al., 2020; Zhao et al., 2018). The released Ca^{2+} ions via mineral weathering may bind to soil/atmospheric CO_2 and precipitate as pedogenic carbonates leading to C sequestration (Beerling et al., 2020). However, this depends very much on the acidity source. If acidity originates from inorganic acids, i.e. acids stronger than carbonic acid, then C sequestration decreases compared to mineral weathering by carbonic acid (Beaulieu et al., 2011; Lerman et al., 2007). In the case of

SIC as a liming material, liming is a net source of atmospheric CO_2 (Ali and Atekwana, 2011; Beaulieu et al., 2011; Zamanian et al., 2018, 2021). Thus, to understand the effects of liming and the addition of Ca-containing minerals on SIC dynamics, weathering should be taken into account (Beaulieu et al., 2011). Furthermore, the relation between SIC and SOC should be known (Bramble et al., 2019).

Other land management practices, such as grazing, also affect SIC content and distribution in soil via changes in dominant vegetation community and water balance. For example, domination of a C_4 grass, blue grama (*Bouteloua gracilis*), by heavy grazing in the steppes of northeastern Colorado increased SIC content in the subsoil compared to non-grazed soil covered mostly by C_3 grasses, particularly prickly pear cactus (*Opuntia polycantha*) (Reeder et al., 2004). This has been related to the better water use efficiency of blue grama, leading to higher water availability in soil and enhanced weathering by secreting more organic acids via blue grama roots, as well as Ca^{2+} migration from topsoil to the subsoil. The released Ca was then transported to deeper horizons and partly precipitated as pedogenic SIC (Reeder et al., 2004).

3.2 Climate change

Climate change modifies the concentration of greenhouse gases in the atmosphere, especially CO_2, acidifies surface and subsurface waters, and alters precipitation amounts and frequencies. Climate change therefore affects soil water content, vegetation community and net primary production, which are all determinant factors of SIC solubility. Alterations in vegetation community due to climate change may increase SIC dissolution due to:

1. introduction of plant species with a different rooting system than native species including changes in rooting depth, N demand and belowground productivity (Raich and Tufekciogul, 2000), root exudates and organic acid inputs, H^+ releasing potential;
2. changes in belowground respiration due to alterations in aboveground biomass, quality and quantity of organic residues (Song et al., 2016) and canopy photosynthesis (Bremer et al., 1998) as well as changes to rhizosphere microbiome resulting in residue decomposition;
3. modification of chemical and physical soil properties such as solutes concentrations, bulk density (Song et al., 2016; Yang and Yang, 2020), moisture content and moisture distribution ((Wilsey et al., 2020) and references therein).

Invasive species (Jobbágy and Jackson, 2000), for example, have a shallower rooting depth (Wilsey and Polley, 2006) than perennial grasses. Therefore, higher rhizosphere respiration at the topsoil due to invasive species, as well as larger amounts of organic acids, increases SIC dissolution, which increases CO_2 emissions and reduces SIC content (Wilsey et al., 2020). The presence of larger amounts of finer roots with lower residence time in the topsoil under oak (*Quercus liaotungensis*) forest compared to soils under black locust (*Robinia pseudoacacia*) resulted in higher root and microbial respiration and consequently increased SIC dissolution (Song et al., 2016). Furthermore, finer roots and higher root exudation improve soil structure and decrease bulk density, which facilitate water percolation and subsequent leaching of dissolved ions (Song et al., 2016).

Shrub communities in temperate regions have deeper rooting depth and density compared with grasslands and forests (Jackson et al., 1996). This results in a lower water content throughout the soil profile under shrubs, which decreases SIC dissolution. Decomposition of shrub residues releases more Ca^{2+} ions into the solution, which results in more SIC deposition on the surface (Zhao et al., 2016). Invasion of *Spartina alterniflora* (Loisel) also causes SIC relocation in coastal ecosystems as a result of changes in soil water content, salinity and pH as well as SOC content (Yang and Yang, 2020). The shallow rooting system of *S. alterniflora* changes the SIC profile (i.e. depletion in the topsoil and re-precipitation in the subsoil) due to increased respiration, loosened soil structure and stimulated leaching of dissolved Ca^{2+} and bicarbonate ions in the top 30 cm of soil (Yang and Yang, 2020). Afforestation at the Loess Plateau has also changed SIC distribution depth following enhanced biological activity as well as increased leaching from the topsoil and decreased water content in the subsoil (Chang et al., 2012).

In general, SIC can be an active C pool (Wu et al., 2009; Yang et al., 2012) that substantially differs between biomes (Rasmussen, 2006; Yang et al., 2010). Changing the soil water balance, either directly by climate change, such as by increased precipitation (Schindlbacher et al., 2019), or indirectly by altering the vegetation community and the corresponding evapotranspiration, significantly affects SIC dissolution and DIC concentrations leading to depletion or redistribution of SIC stocks at depth. Thus a 'substantial stock of soil carbon may be cycling unknowingly' (Wilsey et al., 2020), especially if only the topsoil is considered. Understanding SIC cycling further points to the fact that protecting native vegetation communities may prevent C loss via SIC dissolution (Wilsey et al., 2020). For further details regarding agricultural and anthropogenic effects on SIC stock, see Chapter 22 in this volume.

4 Consequences of soil inorganic carbon loss

SIC losses due to land-use change, particularly intensive agricultural management and N fertilization, have significant consequences regarding

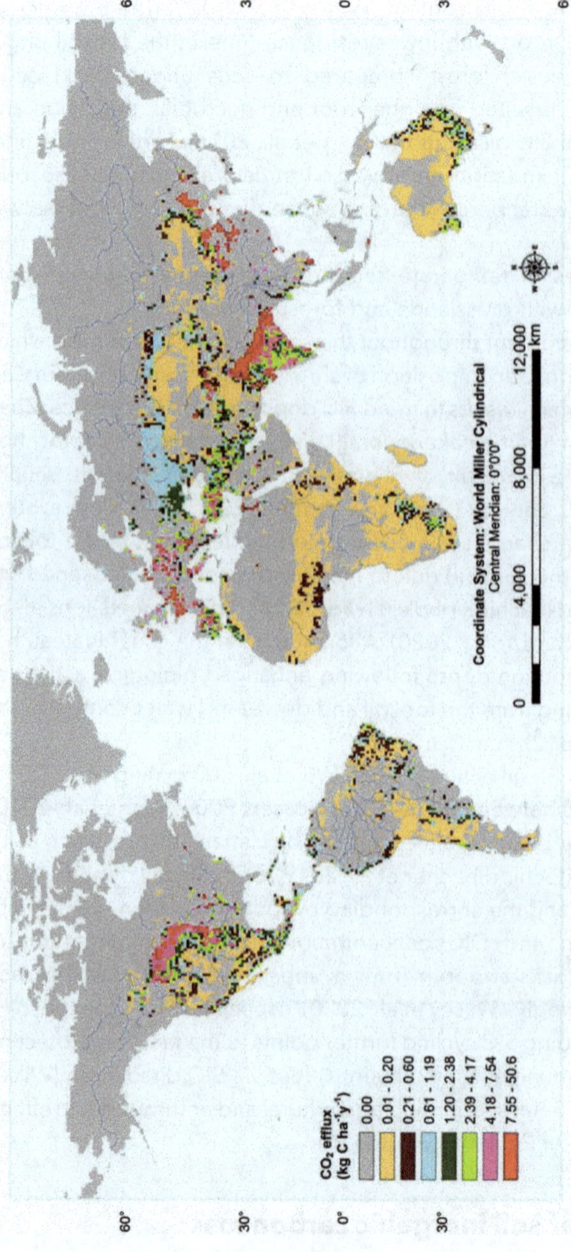

Figure 8 Global map of annual CO_2 emissions from SIC due to neutralization of acidity induced by nitrogen fertilization (Zamanian et al., 2018).

land sustainability and ecosystem functions, as well as global warming. All the positive functions of SIC discussed earlier will be lost with soil acidification (Pabian and Brittingham, 2012; Rowley et al., 2020; Zamanian et al., 2021). SIC depletion makes soils vulnerable to acidification and aluminum toxicity (Kuzyakov et al., 2021; Raza et al., 2020; Tao et al., 2022), which results in decreasing root growth and crop production (Azura et al., 2011). Furthermore, SIC losses due to N fertilization-induced acidification provide an unexplored net source of atmospheric CO_2 as much as 25 10^{12} g C y^{-1} (Fig. 8) (Zamanian et al., 2018). This source of atmospheric CO_2 increases with SIC content (Fig. 9) and can reach 90% of soil-emitted CO_2 in dry regions and as high as 50% in humid regions (Ramnarine et al., 2012; Tamir et al., 2011).

In order to increase profitability, cultivation of acid soils may need liming, i.e. (Ca, Mg) CO_3, which provides a significant source of atmospheric CO_2 with an estimated rate of 270 10^{12} g C y^{-1} (Zamanian et al., 2018, 2021). This estimated value for liming contributing to atmospheric CO_2 is conservative (Zamanian and Kuzyakov, 2019) as it has been assumed that:

1 about 40% of the dissolved lime (West and McBride, 2005) is re-precipitated in soil or water bodies;
2 other acidity sources are not considered;
3 N fertilization is solely in the form of urea.

Furthermore, the CO_2 emitted by burning fossil fuels to provide energy for lime extraction, grinding, transportation and implication is not considered.

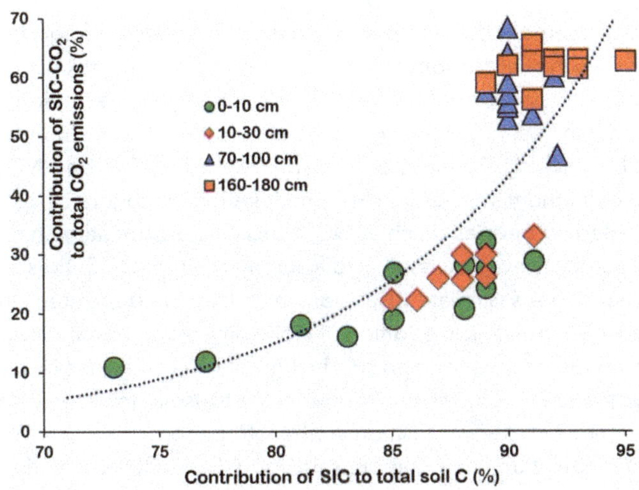

Figure 9 Contribution of SIC to CO_2 efflux from soil depending on the share of SIC as total soil C (Cardinael et al., 2020). Soil samples from four depths were incubated to determine the contribution of SIC in the emitted CO_2.

The significance of carbonates as a source of atmospheric CO_2 is not limited to soils and the solid phase of SIC. Acidification of surface and groundwater also releases CO_2 to the atmosphere from dissolved HCO_3^- and CO_3^{2-} ions (Perrin et al., 2008). Therefore, reducing soil CO_2 emissions and subsequently controlling climate change requires a profound understanding of C sources and their relative magnitudes as well as the driving mechanisms and factors (e.g. fertilization amount, crop species, SOM content and quality, soil water content and redox condition) that control their dynamics (Dong et al., 2014; Li et al., 2018b; Pan et al., 2011; Rey, 2015). Nonetheless, the significance of carbonates and abiotic CO_2 exchange with the atmosphere remains an underexplored aspect of the C cycle (Serrano-Ortiz et al., 2010).

5 Best management practices to protect soil inorganic carbon and to avoid acidification

Feeding a large, continuously increasing population without fertilization and in particular N fertilizers is unrealistic (Vlek, 1990) The focus should therefore be on sustainable land management such as managing fertilization strategies, employing agroforestry systems and avoiding crop production on marginal lands. Proper fertilization management will avoid soil acidification and subsequent loss of SIC and basic cations and will prevent CO_2 emissions from SIC. Precision agriculture in this regard is one of the most promising approaches to avoid over-fertilization by providing plant-available nutrients at the right time (Conyers et al., 1996; Poss and Saragoni, 1992).

Agroforestry appears to be another promising approach. Although no studies have been conducted to solely detect the effects of agroforestry on SIC dissolution and re-precipitation, it can be predicted that taking water from deeper depths by deep-rooting trees in agroforestry will result in partial soil desiccation (Tian et al., 2016), which decreases nutrient loss including Ca^{2+} and NO_3^- ions by leaching. Furthermore, deeper roots take Ca^{2+} and NO_3^- ions at depth and uplift them, resulting in SIC precipitation in the topsoil, as Ca^{2+} ions are not needed by trees as much as NO_3^- ions. SIC precipitation in the topsoil will subsequently prevent subsoil acidification and improve SOM stability. On the other hand, the assimilated Ca^{2+} ions may later be deposited on the soil surface via tree litter fall. The addition of labile organic compounds via litter fall triggers microbial respiration and production of organic acids (Adeleke et al., 2017), which results in SIC destabilization in the topsoil. However, the released Ca^{2+} ions following decomposition and bioturbation due to an increasing earthworm population may compensate for SIC dissolution at the surface, especially if leaching is limited.

Cropping should be avoided on marginal lands, especially in semiarid and arid regions where water is scarce. Keeping natural vegetation as well as planting deep-rooted, indigenous species that are adapted to such harsh conditions is a practical method for long-term belowground sequestration of organic C (Plaza-Bonilla et al., 2015; Zhang et al., 2015a) and avoiding inorganic C losses. Plantations in semiarid regions can globally sequester 5–20% of current anthropogenic CO_2 emissions in the form of pedogenic carbonates (Gao et al., 2017; Plaza-Bonilla et al., 2015). This, however, needs proper financial support by policymakers as well as incentives to farmers to prevent land-use change on marginal soils (Plaza-Bonilla et al., 2015).

6 Conclusion

In this overview we have shown the importance of SIC stocks (960–1780 Pg C to 100 cm, and ~2500 Pg C to 200 cm) and dissolved inorganic carbon (1400 Pg C) for the global C cycle and functioning of the ecosystem. The assumed long-term stability of SIC has been challenged by the full range of global change processes (elevated CO_2, distribution of invasive plants and global warming), but mainly by soil acidification induced by continuously increasing N fertilization of crops. SIC losses and Ca^{2+} leaching out of the soil can have dramatic consequences for soil properties and processes, including reduced plant growth due to damage to aggregate structure, decreased water permeability, decreased nutrient availability, a decrease in microbial activities for nutrient mineralization, reduced organic matter stabilization and an increasing susceptibility to pathogens, drought, cold or other environmental stresses. Consequently, SIC is crucial not only as a buffering system for pH regulation but also contributes many other soil functions and health. The losses of SIC are irrecoverable, because Ca^{2+} leaching out of soil cannot be fulfilled by weathering or dust input over the human lifetime. Therefore, to protect soil cover and functionality, it is crucial to maintain SIC stocks to resist degradation factors such as erosion, compaction, acidification and nutrient depletion. In conclusion, SIC is an important soil constituent responsible for a broad range of soil properties and processes as well as for ecosystem services such as cycles of C, N and other elements.

7 Acknowledgements

The authors would like to thank the German Research Foundation (DFG; grant number: ZA 1068/4-1), the 'RUDN University program 5-100', Russian Science Foundation (project 19-77-30012), the 'CarboRus' project and the NSFC (grant number: 42050410320) for their support.

8 References

Adeleke, R., Nwangburuka, C. and Oboirien, B. 2017. Origins, roles and fate of organic acids in soils: A review. *South African Journal of Botany* 108, 393–406. https://doi.org/10.1016/j.sajb.2016.09.002.

Ali, H. N. and Atekwana, E. A. 2011. The effect of sulfuric acid neutralization on carbonate and stable carbon isotope evolution of shallow groundwater. *Chemical Geology* 284(3–4), 217–228. https://doi.org/10.1016/j.chemgeo.2011.02.023.

An, H., Wu, X., Zhang, Y. and Tang, Z. 2019. Effects of land-use change on soil inorganic carbon: A meta-analysis. *Geoderma* 353, 273–282. https://doi.org/10.1016/j.geoderma.2019.07.008.

Artiola, J. F. and Walworth, J. L. 2009. Irrigation water quality effects on soil carbon fractionation and organic carbon dissolution and leaching in a semiarid calcareous soil. *Soil Science* 174(7), 356–371. https://doi.org/10.1097/SS.0b013e3181aea7b4.

Azura, A. E., Shamshuddin, J. and Fauziah, C. I. 2011. Root elongation, root surface area and organic acid by rice seedling under Al^{3+} and/or H^+ stress. *American Journal of Agricultural and Biological Sciences* 6, 324–331. https://doi.org/10.3844/ajabssp.2011.324.331.

Bakina, L. G., Chugunova, M. V., Zaitseva, T. B. and Nebol'sina, Z. P. 2014. The effect of liming on the complex of soil microorganisms and the humus status of a soddy-podzolic soil in a long-term experiment. *Eurasian Soil Science* 47(2), 110–118. https://doi.org/10.1134/S1064229314020021.

Barreto, M. S. C., Elzinga, E. J., Ramlogan, M., Rouff, A. A. and Alleoni, L. R. F. 2021. Calcium enhances adsorption and thermal stability of organic compounds on soil minerals. *Chemical Geology* 559, 119804. https://doi.org/10.1016/j.chemgeo.2020.119804.

Batanghari, J. W. and Goldman, W. E. 1997. Calcium dependence and binding in cultures of Histoplasma capsulatum. *Infection and Immunity* 65(12), 5257-5261. https://doi.org/10.1128/IAI.65.12.5257-5261.1997.

Batjes, N. H. 1996. Total carbon and nitrogen in the soils of the world. *European Journal of Soil Science* 47(2), 151–163. https://doi.org/10.1111/j.1365-2389.1996.tb01386.x.

Beaulieu, E., Goddéris, Y., Labat, D., Roelandt, C., Calmels, D. and Gaillardet, J. 2011. Modeling of water-rock interaction in the Mackenzie Basin: Competition between sulfuric and carbonic acids. *Chemical Geology* 289(1–2), 114–123. https://doi.org/10.1016/j.chemgeo.2011.07.020.

Beerling, D. J., Kantzas, E. P., Lomas, M. R., Wade, P., Eufrasio, R. M., Renforth, P., Sarkar, B., Andrews, M. G., James, R. H., Pearce, C. R., Mercure, J. F., Pollitt, H., Holden, P. B., Edwards, N. R., Khanna, M., Koh, L., Quegan, S., Pidgeon, N. F., Janssens, I. A., Hansen, J. and Banwart, S. A. 2020. Potential for large-scale CO_2 removal via enhanced rock weathering with croplands. *Nature* 583(7815), 242–248. https://doi.org/10.1038/s41586-020-2448-9.

Berner, E. K. and Berner, R. A. 2012. *Global Environment: Water, Air, and Geochemical Cycles*. Princeton, NJ: Princeton University Press.

Bigelow, S. W. and Canham, C. D. 2002. Community organization of tree species along soil gradients in a north-eastern USA forest. *Journal of Ecology* 90(1), 188–200. https://doi.org/10.1046/j.0022-0477.2001.00655.x.

Blackburne, R., Vadivelu, V. M., Yuan, Z. G., Keller, J. 2007. Determination of growth rate and yield of nitrifying bacteria by measuring carbon dioxide uptake rate. *Water Environment Research* 79(12), 2437–2445. doi: 10.2175/106143007X212139.

Bloom, P. R., Skyllberg, U. L. and Sumner, M. E. 2005. Soil acidity: Chemical processes in soils. In: Tabatabai, M. A. and Sparks, D. L. (Eds.), *Soil Science of America Book Series* (vol. 8). Madison, WI: Madison Book Company, pp. 411–460.

Boix-Fayos, C., Calvo-Cases, A., Imeson, A. C. and Soriano-Soto, M. D. 2001. Influence of soil properties on the aggregation of some Mediterranean soils and the use of aggregate size and stability as land degradation indicators. *CATENA* 44(1), 47–67. https://doi.org/10.1016/S0341-8162(00)00176-4.

Bramble, D. S. E., Gouveia, G. A. and Ramnarine, R. 2019. Organic residues and ammonium effects on CO_2 emissions and soil quality indicators in limed acid tropical soils. *Soil Systems* 3(1), 16. https://doi.org/10.3390/soilsystems3010016.

Bramble, D. S. E., Gouveia, G. A., Ramnarine, R. and Farrell, R. E. 2020a. Short-term effects of aglime on inorganic- and organic-derived CO_2 emissions from two acid soils amended with an ammonium-based fertiliser. *Journal of Soils and Sediments* 20(1), 52–65. https://doi.org/10.1007/s11368-019-02407-2.

Bramble, D. S. E., Gouveia, G. A., Ramnarine, R. and Farrell, R. E. 2020b. Organic residue and agricultural lime interactions on CO_2 emissions from two contrasting soils: Implications for carbon management in acid soils. *Journal of Soils and Sediments* 21(1), 172–188. https://doi.org/10.1007/s11368-020-02736-7.

Bremer, D. J., Ham, J. M., Owensby, C. E. and Knapp, A. K. 1998. Responses of soil respiration to clipping and grazing in a Tallgrass prairie. *Journal of Environmental Quality* 27(6), 1539–1548. https://doi.org/10.2134/jeq1998.00472425002700060034x.

Bronick, C. J. and Lal, R. 2005. Soil structure and management: A review. *Geoderma* 124(1-2), 3-22. https://doi.org/10.1016/j.geoderma.2004.03.005.

Bughio, M. A., Wang, P., Meng, F., Qing, C., Kuzyakov, Y., Wang, X. and Junejo, S. A. 2016. Neoformation of pedogenic carbonates by irrigation and fertilization and their contribution to carbon sequestration in soil. *Geoderma* 262, 12–19. https://doi.org/10.1016/j.geoderma.2015.08.003.

Caires, E. F., Haliski, A., Bini, A. R. and Scharr, D. A. 2015. Surface liming and nitrogen fertilization for crop grain production under no-till management in Brazil. *European Journal of Agronomy* 66, 41–53. https://doi.org/10.1016/j.eja.2015.02.008.

Cardinael, R., Chevallier, T., Guenet, B., Girardin, C., Cozzi, T., Pouteau, V. and Chenu, C. 2020. Organic carbon decomposition rates with depth and contribution of inorganic carbon to CO_2 emissions under a Mediterranean agroforestry system. *European Journal of Soil Science* 71, 909–923. https://doi.org/10.1111/ejss.12908.

Cater, M. and Ogrinc, N. 2011. Soil respiration rates and [image omitted] in natural beech forest (Fagus sylvatica L.) in relation to stand structure. *Isotopes in Environmental and Health Studies* 47(2), 221–237. https://doi.org/10.1080/10256016.2011.578214.

Chan, K. Y. and Heenan, D. P. 1999. Lime-induced loss of soil organic carbon and effect on aggregate stability. *Soil Science Society of America Journal* 63(6), 1841–1844. https://doi.org/10.2136/sssaj1999.6361841x.

Chang, R., Fu, B., Liu, G., Wang, S. and Yao, X. 2012. The effects of afforestation on soil organic and inorganic carbon: A case study of the Loess Plateau of China. *CATENA* 95, 145–152. https://doi.org/10.1016/j.catena.2012.02.012.

Chesworth, W., Camps Arbestain, M., Macías, F., Spaargaren, O., Spaargaren, O., Mualem, Y., Morel-Seytoux, H. J., Horwath, W. R., Almendros, G. and Chesworth, W. 2008. Carbonates. In: Chesworth, W. (Ed.), *Encyclopedia of Soil Science*. Dordrecht: Springer Netherlands, pp. 99–101. https://doi.org/10.1007/978-1-4020-3995-9_90.

Clough, A. and Skjemstad, J. O. 2000. Physical and chemical protection of soil organic carbon in three agricultural soils with different contents of calcium carbonate. *Soil Research* 38(5), 1005–1016. https://doi.org/10.1071/SR99102.

Conyers, M. K., Heenan, D. P., Poile, G. J., Cullis, B. R. and Helyar, K. R. 1996. Influence of dryland agricultural management practices on the acidification of a soil

profile. *Soil and Tillage Research* 37(2–3), 127–141. https://doi.org/10.1016/0167-1987(95)01005-X.
Cuhel, J. and Simek, M. 2011. Effect of pH on the denitrifying enzyme activity in pasture soils in relation to the intrinsic differences in denitrifier communities. *Folia Microbiologica (Praha)* 56(3), 230–235. https://doi.org/10.1007/s12223-011-0045-x.
Dai, H., Zhang, H., Li, Z., Liu, K. and Zamanian, K. 2021. Tillage practice impacts on the carbon sequestration potential of topsoil microbial communities in an agricultural field. *Agronomy* 11(1), 60. https://doi.org/10.3390/agronomy11010060.
de Soto, I. S., Virto, I., Barré, P. and Enrique, A. 2019. Effect of irrigation on carbonate dynamics in a calcareous soil using isotopic determinations. *Spanish Journal of Soil Science* 9, 9. https://doi.org/10.3232/SJSS.2019.V9.N2.06.
de Soto, I. S., Virto, I., Barré, P., Fernández-Ugalde, O., Antón, R., Martínez, I., Chaduteau, C., Enrique, A. and Bescansa, P. 2017. A model for field-based evidences of the impact of irrigation on carbonates in the tilled layer of semi-arid Mediterranean soils. *Geoderma* 297, 48–60. https://doi.org/10.1016/j.geoderma.2017.03.005.
Díaz-Hernández, J. L. 2010. Is soil carbon storage underestimated? *Chemosphere* 80(3), 346–349. https://doi.org/10.1016/j.chemosphere.2010.04.038.
Dong, X., Hao, Q., Li, G., Lin, Q. and Zhao, X. 2017. Contrast effect of long-term fertilization on SOC and SIC stocks and distribution in different soil particle-size fractions. *Journal of Soils and Sediments* 17(4), 1054-1063. https://doi.org/10.1007/s11368-016-1615-y.
Dong, Y., Cai, M. and Zhou, J. 2014. Effects of moisture and carbonate additions on CO_2 emission from calcareous soil during closed-jar incubation. *Journal of Arid Land* 6(1), 37–43. https://doi.org/10.1007/s40333-013-0195-6.
Drever, J. I. 1982. *The Geochemistry of Natural Waters*. Canada: Pearson Education.
Eshel, G., Fine, P. and Singer, M. J. 2007. Total soil carbon and water quality: An implication for carbon sequestration. *Soil Science Society of America Journal* 71(2), 397–405. https://doi.org/10.2136/sssaj2006.0061.
Eswaran, H., Van den Berg, E., Reich, P. and Kimble, J. 1995. Global soil carbon resources. In: Kimble, J. M., Levine, E. R. and Stewart, B. A. (Eds.), *Soils and Global Change*. Boca Raton, FL: CRC Press, pp. 27–44.
Fernández-Ugalde, O., Virto, I., Barré, P., Apesteguía, M., Enrique, A., Imaz, M. J. and Bescansa, P. 2014. Mechanisms of macroaggregate stabilisation by carbonates: Implications for organic matter protection in semi-arid calcareous soils. *Soil Research* 52(2), 180-192. https://doi.org/10.1071/SR13234.
Gao, J., Mikutta, R., Jansen, B., Guggenberger, G., Vogel, C. and Kalbitz, K. 2020. The multilayer model of soil mineral–organic interfaces–A review. *Journal of Plant Nutrition and Soil Science* 183(1), 27–41. https://doi.org/10.1002/jpln.201900530.
Gao, Y., Tian, J., Pang, Y. and Liu, J. 2017. Soil inorganic carbon sequestration following afforestation is probably induced by pedogenic carbonate formation in Northwest China. *Frontiers in Plant Science* 8, 1282. https://doi.org/10.3389/fpls.2017.01282.
Garcia-Franco, N., Walter, R., Wiesmeier, M., Hurtarte, L. C. C., Berauer, B. J., Buness, V., Zistl-Schlingmann, M., Kiese, R., Dannenmann, M. and Kögel-Knabner, I. 2020. Biotic and abiotic controls on carbon storage in aggregates in calcareous alpine and prealpine grassland soils. *Biology and Fertility of Soils* 57(2), 203–218. https://doi.org/10.1007/s00374-020-01518-0.
Ghorbani, M., Amirahmadi, E. and Zamanian, K. 2021. In-situ biochar production in paddies: Direct involvement of farmers in greenhouse gases reduction policies

besides increasing nutrients availability and rice production. *Land Degradation and Development*. https://doi.org/10.1002/ldr.4006.

Gile, L. H. 1999. Eolian and associated pedogenic features of the Jornada Basin floor, Southern New Mexico. *Soil Science Society of America Journal* 63(1), 151-163. https://doi.org/10.2136/sssaj1999.03615995006300010022x.

Girvan, M. S., Campbell, C. D., Killham, K., Prosser, J. I. and Glover, L. A. 2005. Bacterial diversity promotes community stability and functional resilience after perturbation. *Environmental Microbiology* 7(3), 301-313. https://doi.org/10.1111/j.1462-2920.2005.00695.x.

Goddard, M. A., Mikhailova, E. A., Christopher, J. P. and Schlautman, M. A. 2007. Atmospheric Mg2+ wet deposition within the continental United States and implications for soil inorganic carbon sequestration. *Tellus B Chem Phys Meteorol* 59(1), 50-56. https://doi.org/10.1111/j.1600-0889.2006.00228.x.

Goulding, K. W. T. 2016. Soil acidification and the importance of liming agricultural soils with particular reference to the United Kingdom. *Soil Use and Management* 32(3), 390-399. https://doi.org/10.1111/sum.12270.

Graveland, J. and van der Wal, R. 1996. Decline in snail abundance due to soil acidification causes eggshell defects in forest passerines. *Oecologia* 105(3), 351-360. https://doi.org/10.1007/BF00328738.

Groshans, G. R., Mikhailova, E. A., Post, C. J. and Schlautman, M. A. 2018. Accounting for soil inorganic carbon in the ecosystem services framework for United Nations sustainable development goals. *Geoderma* 324, 37-46. https://doi.org/10.1016/j.geoderma.2018.02.009.

Guo, Y., Wang, X., Li, X., Wang, J., Xu, M. and Li, D. 2016. Dynamics of soil organic and inorganic carbon in the cropland of upper Yellow River Delta, China. *Scientific Reports* 6, 36105. https://doi.org/10.1038/srep36105.

Haque, F., Santos, R. M. and Chiang, Y. W. 2020. Optimizing inorganic carbon sequestration and crop yield With wollastonite soil amendment in a microplot study. *Frontiers in Plant Science* 11, 1012. https://doi.org/10.3389/fpls.2020.01012.

Haynes, R. J. and Naidu, R. 1998. Influence of lime, fertilizer and manure applications on soil organic matter content and soil physical conditions: A review. *Nutrient Cycling in Agroecosystems* 51(2), 123-137. https://doi.org/10.1023/A:1009738307837.

Hénault, C., Bourennane, H., Ayzac, A., Ratié, C., Saby, N. P. A., Cohan, J. P., Eglin, T. and Gall, C. L. 2019. Management of soil pH promotes nitrous oxide reduction and thus mitigates soil emissions of this greenhouse gas. *Scientific Reports* 9(1), 20182. https://doi.org/10.1038/s41598-019-56694-3.

Hosseini, S. A., Réthoré, E., Pluchon, S., Ali, N., Billiot, B. and Yvin, J. C. 2019. Calcium application enhances drought stress tolerance in sugar beet and promotes plant biomass and beetroot sucrose concentration. *International Journal of Molecular Sciences* 20(15), 3777. https://doi.org/10.3390/ijms20153777.

Huang, X., Jia, Z., Guo, J., Li, T., Sun, D., Meng, H., Yu, G., He, X., Ran, W., Zhang, S., Hong, J. and Shen, Q. 2019. Ten-year long-term organic fertilization enhances carbon sequestration and calcium-mediated stabilization of aggregate-associated organic carbon in a reclaimed cambisol. *Geoderma* 355, 113880. https://doi.org/10.1016/j.geoderma.2019.113880.

Jackson, R. B., Canadell, J., Ehleringer, J. R., Mooney, H. A., Sala, O. E., Schulze, E. D. 1996. A global analysis of root distributions for terrestrial biomes. *Oecologia* 108(3), 389-411. doi: 10.1007/BF00333714.

Jobbágy, E. G. and Jackson, R. B. 2000. The vertical distribution of soil organic carbon and its relation to climate and vegetation. *Ecological Applications* 10(2), 423–436. https://doi.org/10.1890/1051-0761(2000)010[0423:TVDOSO]2.0.CO;2.

Kang, S. and Xing, B. 2008. Humic acid fractionation upon sequential adsorption onto goethite. *Langmuir* 24(6), 2525–2531. https://doi.org/10.1021/la702914q.

Kao, C. W. and Ko, W. 1986a. Suppression of pythium-splendens in a Hawaiian soil by calcium and microorganisms. *Phytopathology* 76(2), 215–220. https://doi.org/10.1094/Phyto-76-215.

Kao, C. W. and Ko, W. 1986b. The role of calcium and microorganisms in suppression of cucumber damping-off caused by pythium-splendens in a Hawaiian soil. *Phytopathology* 76(2), 221–225. https://doi.org/10.1094/Phyto-76-221.

Kim, J. H., Jobbágy, E. G., Richter, D. D., Trumbore, S. E. and Jackson, R. B. 2020. Agricultural acceleration of soil carbonate weathering. *Global Change Biology* 26(10), 5988–6002. https://doi.org/10.1111/gcb.15207.

Kim, J. M., Roh, A. S., Choi, S. C., Kim, E. J., Choi, M. T., Ahn, B. K., Kim, S. K., Lee, Y. H., Joa, J. H., Kang, S. S., Lee, S. A., Ahn, J. H., Song, J. and Weon, H. Y. 2016. Soil pH and electrical conductivity are key edaphic factors shaping bacterial communities of greenhouse soils in Korea. *Journal of Microbiology* 54(12), 838–845. https://doi.org/10.1007/s12275-016-6526-5.

Kleber, M., Eusterhues, K., Keiluweit, M., Mikutta, C., Mikutta, R. and Nico, P. S. 2015. Chapter one: Mineral-organic associations: Formation, properties, and relevance in soil environments. In: Sparks, D. L. (Ed.), *Advances in Agronomy*. Cambridge: Academic Press, pp. 1–140. https://doi.org/10.1016/bs.agron.2014.10.005.

Kuznetsova, A. M., Khokhlova, O. S., Chendev, Y. G. and Aleksandrovskii, A. L. 2010. Evolution of the carbonate state of Agrogenically transformed dark gray forest soils in the central Forest-Steppe. *Eurasian Soil Science* 43(13), 1527–1534. https://doi.org/10.1134/S1064229310130119.

Kuzyakov, Y., Kuzyakova, I., Raza, S., Zhou, J. and Zamanian, K. 2021. Letter-to-the-editor: Does acidification really increase soil carbon in croplands? How statistical analyses of large datasets might mislead the conclusions. *Geoderma* 384, 114806. https://doi.org/10.1016/j.geoderma.2020.114806.

Kuzyakov, Y. and Zamanian, K. 2019. Reviews and syntheses: Agropedogenesis - Humankind as the sixth soil-forming factor and attractors of agricultural soil degradation. *Biogeosciences* 16(24), 4783–4803. https://doi.org/10.5194/bg-16-4783-2019.

Lal, R. 2009. Myths and facts about soils and the greenhouse effect. In: Lal, R. and Follett, R. F. (Eds.), *Soil Carbon Sequestration and the Greenhouse Effect*. ASA-CSSA-SSSA.

Lal, R. and Kimble, J. M. 2000. Pedogenic carbonates and the global carbon cycle. In: Lal, R., Kimble, J. M., Eswaran, H. and Stewart, B. A. (Eds.), *Global Climate Change and Pedogenic Carbonates*. Boca Raton, FL: CRC Press, pp. 1–14.

Lerman, A., Wu, L. and Mackenzie, F. T. 2007. CO_2 and H_2SO_4 consumption in weathering and material transport to the ocean, and their role in the global carbon balance. *Marine Chemistry* 106(1–2), 326–350. https://doi.org/10.1016/j.marchem.2006.04.004.

Li, F., Zhang, X., Gong, J., Liu, L. and Yi, Y. 2018a. Specialized core bacteria associate with plants adapted to adverse environment with high calcium contents. *PLoS ONE* 13(3), e0194080. https://doi.org/10.1371/journal.pone.0194080.

Li, H., Wang, S., Bai, X., Luo, W., Tang, H., Cao, Y., Wu, L., Chen, F., Li, Q., Zeng, C. and Wang, M. 2018b. Spatiotemporal distribution and national measurement of the global carbonate carbon sink. *Science of the Total Environment* 643, 157–170. https://doi.org/10.1016/j.scitotenv.2018.06.196.

Li, Y., Wang, Y.-G., Houghton, R. A. and Tang, L.-S. 2015. Hidden carbon sink beneath desert. *Geophysical Research Letters* 42(14), 5880–5887. https://doi.org/10.1002/2015GL064222.

Li, Z. P., Han, F. X., Su, Y., Zhang, T. L., Sun, B., Monts, D. L. and Plodinec, M. J. 2007. Assessment of soil organic and carbonate carbon storage in China. *Geoderma* 138(1-2), 119–126. https://doi.org/10.1016/j.geoderma.2006.11.007.

Liu, W., Wei, J., Cheng, J. and Li, W. 2014. Profile distribution of soil inorganic carbon along a chronosequence of grassland restoration on a 22-year scale in the Chinese Loess Plateau. *CATENA* 121, 321-329. https://doi.org/10.1016/j.catena.2014.05.019.

Liu, Z., Dreybrodt, W. and Wang, H. 2010. A new direction in effective accounting for the atmospheric CO_2 budget: Considering the combined action of carbonate dissolution, the global water cycle and photosynthetic uptake of DIC by aquatic organisms. *Earth-Science Reviews* 99(3–4), 162–172. https://doi.org/10.1016/j.earscirev.2010.03.001.

Lu, T., Wang, X., Xu, M., Yu, Z., Luo, Y. and Smith, P. 2020. Dynamics of pedogenic carbonate in the cropland of the North China plain: Influences of intensive cropping and salinization. *Agriculture, Ecosystems and Environment* 292, 106820. https://doi.org/10.1016/j.agee.2020.106820.

Luo, W. T., Nelson, P. N., Li, M. -H., Cai, J. P., Zhang, Y. Y., Zhang, Y. G., Yang, S., Wang, R. Z., Wang, Z. W., Wu, Y. N., Han, X. G. and Jiang, Y. 2015. Contrasting pH buffering patterns in neutral-alkaline soils along a 3600 km transect in northern China. *Biogeosciences* 12(23), 7047–7056. https://doi.org/10.5194/bg-12-7047-2015.

Magaritz, M. and Amiel, A. J. 1981. Influence of intensive cultivation and irrigation on soil properties in the Jordan Valley, Israel: Recrystallization of carbonate minerals. *Soil Science Society of America Journal* 45(6), 1201–1205. https://doi.org/10.2136/sssaj1981.03615995004500060038x.

Marschner, P. 2011. *Marschner's Mineral Nutrition of Higher Plants* (3rd edn.). Oxford, London; Waltham, MA: Elsevier Ltd.

McLaughlin, S. B. and Wimmer, R. 1999. Tansley review no. 104. *New Phytologist* 142(3), 373–417. https://doi.org/10.1046/j.1469-8137.1999.00420.x.

Mikhailova, E., Post, C., Cihacek, L. and Ulmer, M. 2009. Soil inorganic carbon sequestration as a result of cultivation in the mollisols. In: *Geophysical Monograph Series*. American Geophysical Union (American Geophysical Union), Washington, DC, pp. 129-133. https://doi.org/10.1029/2005GM000313.

Mikhailova, E. A., Goddard, M. A., Post, C. J., Schlautman, M. A. and Galbraith, J. M. 2013. Potential contribution of combined atmospheric Ca^{2+} and Mg^{2+} wet deposition within the continental U.S. to soil inorganic carbon sequestration. *Pedosphere* 23(6), 808–814. https://doi.org/10.1016/S1002-0160(13)60072-7.

Mikutta, R., Mikutta, C., Kalbitz, K., Scheel, T., Kaiser, K. and Jahn, R. 2007. Biodegradation of forest floor organic matter bound to minerals via different binding mechanisms. *Geochimica et Cosmochimica Acta* 71(10), 2569–2590. https://doi.org/10.1016/j.gca.2007.03.002.

Monger, H. C., Kraimer, R. A., Khresat, S., Cole, D. R., Wang, X. and Wang, J. 2015. Sequestration of inorganic carbon in soil and groundwater. *Geology* 43(5), 375–378. https://doi.org/10.1130/G36449.1.

Ostovari, Y., Ghorbani-Dashtaki, S., Bahrami, H.-A., Naderi, M., Dematte, J. A. M. and Kerry, R. 2016. Modification of the USLE K factor for soil erodibility assessment on calcareous soils in Iran. *Geomorphology* 273, 385–395. https://doi.org/10.1016/j.geomorph.2016.08.003.

Pabian, S. E. and Brittingham, M. C. 2012. Soil calcium and forest birds: Indirect links Between nutrient availability and community composition. *Ecosystems* 15(5), 748–760. https://doi.org/10.1007/s10021-012-9543-1.

Pan, Y., Birdsey, R. A., Fang, J., Houghton, R., Kauppi, P. E., Kurz, W. A., Phillips, O. L., Shvidenko, A., Lewis, S. L., Canadell, J. G., Ciais, P., Jackson, R. B., Pacala, S. W., McGuire, A. D., Piao, S., Rautiainen, A., Sitch, S. and Hayes, D. 2011. A large and persistent carbon sink in the World's forests. *Science* 333(6045), 988–993. https://doi.org/10.1126/science.1201609.

Paradelo, R., Virto, I. and Chenu, C. 2015. Net effect of liming on soil organic carbon stocks: A review. *Agriculture, Ecosystems and Environment* 202, 98–107. https://doi.org/10.1016/j.agee.2015.01.005.

Perrin, A.-S., Probst, A. and Probst, J.-L. 2008. Impact of nitrogenous fertilizers on carbonate dissolution in small agricultural catchments: Implications for weathering CO_2 uptake at regional and global scales. *Geochimica et Cosmochimica Acta* 72(13), 3105–3123. https://doi.org/10.1016/j.gca.2008.04.011.

Plaza-Bonilla, D., Arrúe, J. L., Cantero-Martínez, C., Fanlo, R., Iglesias, A. and Álvaro-Fuentes, J. 2015. Carbon management in dryland agricultural systems. A review. *Agronomy for Sustainable Development* 35(4), 1319–1334. https://doi.org/10.1007/s13593-015-0326-x.

Poss, R. and Saragoni, H. 1992. Leaching of nitrate, calcium and magnesium under maize cultivation on an oxisol in Togo. *Fertilizer Research* 33(2), 123–133. https://doi.org/10.1007/BF01051167.

Raheb, A., Heidari, A. and Mahmoodi, S. 2016. Storage of organic and inorganic carbon in arid-semihumid soils: A case study of the rangelands of Northwestern Iran. *Soil Science* 181(11/12), 473–486.

Raheb, A., Heidari, A. and Mahmoodi, S. 2017. Organic and inorganic carbon storage in soils along an arid to dry sub-humid climosequence in northwest of Iran. *CATENA* 153, 66–74. https://doi.org/10.1016/j.catena.2017.01.035.

Rahman, M. H., Okubo, A., Sugiyama, S. and Mayland, H. F. 2008. Physical, chemical and microbiological properties of an andisol as related to land use and tillage practice. *Soil and Tillage Research* 101(1–2), 10–19. https://doi.org/10.1016/j.still.2008.05.006.

Raich, J. W. and Tufekciogul, A. 2000. Vegetation and soil respiration: Correlations and controls. *Biogeochemistry* 48(1), 71–90. https://doi.org/10.1023/A:1006112000616.

Ramnarine, R., Wagner-Riddle, C., Dunfield, K. E. and Voroney, R. P. 2012. Contributions of carbonates to soil CO_2 emissions. *Canadian Journal of Soil Science* 92(4), 599–607. https://doi.org/10.4141/cjss2011-025.

Rasmussen, C. 2006. Distribution of soil organic and inorganic carbon pools by biome and soil taxa in Arizona. *Soil Science Society of America Journal* 70(1), 256–265. https://doi.org/10.2136/sssaj2005.0118.

Raza, S., Miao, N., Wang, P., Ju, X., Chen, Z., Zhou, J. and Kuzyakov, Y. 2020. Dramatic loss of inorganic carbon by nitrogen-induced soil acidification in Chinese croplands. *Global Change Biology* 26(6), 3738–3751. https://doi.org/10.1111/gcb.15101.

Reeder, J. D., Schuman, G. E., Morgan, J. A. and Lecain, D. R. 2004. Response of organic and inorganic carbon and nitrogen to long-term grazing of the shortgrass steppe. *Environmental Management* 33(4), 485–495. https://doi.org/10.1007/s00267-003-9106-5.

Renforth, P., Manning, D. A. C. and Lopez-Capel, E. 2009. Carbonate precipitation in artificial soils as a sink for atmospheric carbon dioxide. *Applied Geochemistry* 24(9), 1757–1764. https://doi.org/10.1016/j.apgeochem.2009.05.005.

Rengel, Z. 2003. *Handbook of Soil Acidity*. Boca Raton, FL: CRC Press.

Rey, A. 2015. Mind the gap: Non-biological processes contributing to soil CO_2 efflux. *Global Change Biology* 21(5), 1752–1761. https://doi.org/10.1111/gcb.12821.

Rovira, P. and Vallejo, V. R. 2008. Changes in δ13C composition of soil carbonates driven by organic matter decomposition in a Mediterranean climate: A field incubation experiment. *Geoderma* 144(3-4), 517–534. https://doi.org/10.1016/j.geoderma.2008.01.006.

Rowley, M. C., Grand, S., Adatte, T. and Verrecchia, E. P. 2020. A cascading influence of calcium carbonate on the biogeochemistry and pedogenic trajectories of subalpine soils, Switzerland. *Geoderma* 361, 114065. https://doi.org/10.1016/j.geoderma.2019.114065.

Rowley, M. C., Grand, S. and Verrecchia, É. P. 2018. Calcium-mediated stabilisation of soil organic carbon. *Biogeochemistry* 137(1-2), 27–49. https://doi.org/10.1007/s10533-017-0410-1.

Schindlbacher, A., Beck, K., Holzheu, S. and Borken, W. 2019. Inorganic carbon leaching From a warmed and irrigated carbonate forest soil. *Frontiers in Forests and Global Change* 2. https://doi.org/10.3389/ffgc.2019.00040.

Schlesinger, W. H. 2000. Carbon sequestration in soils: Some cautions amidst optimism. *Agriculture, Ecosystems and Environment* 82(1-3), 121–127. https://doi.org/10.1016/S0167-8809(00)00221-8.

Schmidt, M. W. I., Torn, M. S., Abiven, S., Dittmar, T., Guggenberger, G., Janssens, I. A., Kleber, M., Kögel-Knabner, I., Lehmann, J., Manning, D. A. C., Nannipieri, P., Rasse, D. P., Weiner, S. and Trumbore, S. E. 2011. Persistence of soil organic matter as an ecosystem property. *Nature* 478(7367), 49–56. https://doi.org/10.1038/nature10386.

Serrano-Ortiz, P., Roland, M., Sanchez-Moral, S., Janssens, I. A., Domingo, F., Goddéris, Y. and Kowalski, A. S. 2010. Hidden, abiotic CO_2 flows and gaseous reservoirs in the terrestrial carbon cycle: Review and perspectives. *Agricultural and Forest Meteorology* 150(3), 321–329. https://doi.org/10.1016/j.agrformet.2010.01.002.

Shaofei, J. and Wang, H. 2018. Relationships between soil pH and soil carbon in China's carbonate soils. *Fresenius Environmental Bulletin* 27, 605–611.

Shen, C., Xiong, J., Zhang, H., Feng, Y., Lin, X., Li, X., Liang, W. and Chu, H. 2013. Soil pH drives the spatial distribution of bacterial communities along elevation on Changbai Mountain. *Soil Biology and Biochemistry* 57, 204–211. https://doi.org/10.1016/j.soilbio.2012.07.013.

Song, B.-L., Yan, M.-J., Hou, H., Guan, J.-H., Shi, W.-Y., Li, G.-Q. and Du, S. 2016. Distribution of soil carbon and nitrogen in two typical forests in the semiarid region of the Loess Plateau, China. *CATENA* 143, 159–166. https://doi.org/10.1016/j.catena.2016.04.004.

Song, W. Y., Zhang, Z. B., Shao, H. B., Guo, X. L., Cao, H. X., Zhao, H. B., Fu, Z. Y. and Hu, X. J. 2008. Relationship between calcium decoding elements and plant abiotic-stress resistance. *International Journal of Biological Sciences* 4(2), 116–125. https://doi.org/10.7150/ijbs.4.116.

Sowers, T. D., Stuckey, J. W. and Sparks, D. L. 2018. The synergistic effect of calcium on organic carbon sequestration to ferrihydrite. *Geochemical Transactions* 19(1), 4. https://doi.org/10.1186/s12932-018-0049-4.

Ste-Marie, C. and Paré, D. 1999. Soil, pH and N availability effects on net nitrification in the forest floors of a range of boreal forest stands. *Soil Biology and Biochemistry* 31(11), 1579–1589. https://doi.org/10.1016/S0038-0717(99)00086-3.

Tamir, G., Shenker, M., Heller, H., Bloom, P. R., Fine, P. and Bar-Tal, A. 2011. Can soil carbonate dissolution lead to overestimation of soil respiration? *Soil Science Society of America Journal* 75(4), 1414–1422. https://doi.org/10.2136/sssaj2010.0396.

Tao, J., Raza, S., Zhao, M., Cui, J., Wang, P., Sui, Y., Zamanian, K., Kuzyakov, Y., Xu, M., Chen, Z., Zhou, J. 2022. Vulnerability and driving factors of soil inorganic carbon stocks in Chinese croplands. *Science of The Total Environment* 825, 154087. https://doi.org/10.1016/j.scitotenv.2022.154087.

Tian, F., Feng, X., Zhang, L., Fu, B., Wang, S., Lv, Y. and Wang, P. 2016. Effects of revegetation on soil moisture under different precipitation gradients in the Loess Plateau, China. *Hydrology Research* 48(5), 1378–1390. https://doi.org/10.2166/nh.2016.022.

Tiritan, C. S., Büll, L. T., Crusciol, C. A. C., Carmeis Filho, A. C. A., Fernandes, D. M. and Nascente, A. S. 2016. Tillage system and lime application in a tropical region: Soil chemical fertility and corn yield in succession to degraded pastures. *Soil and Tillage Research* 155, 437–447. https://doi.org/10.1016/j.still.2015.06.012.

Van Breemen, N. van, Mulder, J. and Driscoll, C. T. 1983. Acidification and alkalinization of soils. *Plant and Soil* 75(3), 283–308. https://doi.org/10.1007/BF02369968.

Vargas, G., Verdejo, J., Rivera, A., Suarez, D., Youlton, C., Celis-Diez, J. L., Le Bissonnais, Y., Dovletyarova, E. A. and Neaman, A. 2019. The effect of four calcium-based amendments on soil aggregate stability of two sandy topsoils. *Journal of Plant Nutrition and Soil Science* 182(2), 159–166. https://doi.org/10.1002/jpln.201700562.

Vestergard, M., Bang-Andreasen, T., Buss, S. M., Cruz-Paredes, C., Bentzon-Tilia, S., Ekelund, F., Kjoller, R., Mortensen, L. H. and Ronn, R. 2018. The relative importance of the bacterial pathway and soil inorganic nitrogen increase across an extreme wood-ash application gradient. *GCB Bioenergy* 10(5), 320–334. https://doi.org/10.1111/gcbb.12494.

Virto, I., Antón, R., Apesteguía, M. and Plante, A. 2018. Chapter 9: Role of carbonates in the physical stabilization of soil organic matter in agricultural Mediterranean soils. In: Muñoz, M. Á. and Zornoza, R. (Eds.), *Soil Management and Climate Change*. Academic Press, pp. 121-136. https://doi.org/10.1016/B978-0-12-812128-3.00009-4.

Virto, I., Gartzia-Bengoetxea, N. and Fernández-Ugalde, O. 2011. Role of organic matter and carbonates in soil aggregation estimated using laser diffractometry. *Pedosphere* 21(5), 566-572. https://doi.org/10.1016/S1002-0160(11)60158-6.

Vlek, P. L. G. 1990. The role of fertilizers in sustaining agriculture in sub-Saharan Africa. *Fertilizer Research* 26(1–3), 327–339. https://doi.org/10.1007/BF01048771.

Wang, J., Monger, C., Wang, X., Serena, M. and Leinauer, B. 2016. Carbon sequestration in response to grassland–shrubland–Turfgrass conversions and a test for carbonate biomineralization in desert soils, New Mexico, USA. *Soil Science Society of America Journal* 80(6), 1591–1603. https://doi.org/10.2136/sssaj2016.03.0061.

Wang, X., Wang, J., Xu, M., Zhang, W., Fan, T. and Zhang, J. 2015. Carbon accumulation in arid croplands of northwest China: Pedogenic carbonate exceeding organic carbon. *Scientific Reports* 5, 11439. https://doi.org/10.1038/srep11439.

Wang, Y., Li, Y., Ye, X., Chu, Y. and Wang, X. 2010. Profile storage of organic/inorganic carbon in soil: From forest to desert. *Science of the Total Environment* 408(8), 1925-1931. https://doi.org/10.1016/j.scitotenv.2010.01.015.

West, T. O. and McBride, A. C. 2005. The contribution of agricultural lime to carbon dioxide emissions in the United States: Dissolution, transport, and net emissions. *Agriculture, Ecosystems and Environment* 108(2), 145-154. https://doi.org/10.1016/j.agee.2005.01.002.

White, P. J. and Broadley, M. R. 2003. Broadly, M.R. *Annals of Botany* 92(4), 487-511. https://doi.org/10.1093/aob/mcg164.

Whittinghill, K. A. and Hobbie, S. E. 2012. Effects of pH and calcium on soil organic matter dynamics in Alaskan tundra. *Biogeochemistry* 111(1-3), 569-581. https://doi.org/10.1007/s10533-011-9688-6.

Wilsey, B., Xu, X., Polley, H. W., Hofmockel, K. and Hall, S. J. 2020. Lower soil carbon stocks in exotic vs. native grasslands are driven by carbonate losses. *Ecology* 101(7), e03039. https://doi.org/10.1002/ecy.3039.

Wilsey, B. J. and Polley, H. W. 2006. Aboveground productivity and root-shoot allocation differ between native and introduced grass species. *Oecologia* 150(2), 300-309. https://doi.org/10.1007/s00442-006-0515-z.

Wu, H., Guo, Z., Gao, Q. and Peng, C. 2009. Distribution of soil inorganic carbon storage and its changes due to agricultural land use activity in China. *Agriculture, Ecosystems and Environment* 129(4), 413-421. https://doi.org/10.1016/j.agee.2008.10.020.

Wuddivira, M. N. and Camps-Roach, G. 2007. Effects of organic matter and calcium on soil structural stability. *European Journal of Soil Science* 58(3), 722-727.

Yang, F., Huang, L., Yang, R., Yang, F., Li, D., Zhao, Y., Yang, J., Liu, F. and Zhang, G. 2018. Vertical distribution and storage of soil organic and inorganic carbon in a typical inland river basin, Northwest China. *Journal of Arid Land* 10(2), 183-201. https://doi.org/10.1007/s40333-018-0051-9.

Yang, R.-M. and Yang, F. 2020. Impacts of Spartina alterniflora invasion on soil inorganic carbon in coastal wetlands in China. *Soil Science Society of America Journal* 84(3), 844-855. https://doi.org/10.1002/saj2.20073.

Yang, Y., Fang, J., Ji, C., Ma, W., Mohammat, A., Wang, S., Wang, S., Datta, A., Robinson, D. and Smith, P. 2012. Widespread decreases in topsoil inorganic carbon stocks across China's grasslands during 1980s-2000s. *Global Change Biology* 18(12), 3672-3680. https://doi.org/10.1111/gcb.12025.

Yang, Y., Fang, J., Ji, C., Ma, W., Su, S. and Tang, Z. 2010. Soil inorganic carbon stock in the Tibetan alpine grasslands. *Global Biogeochemical Cycles* 24(4), n/a-n/a. https://doi.org/10.1029/2010GB003804.

Yin, J., Wang, R., Liu, H., Feng, X., Xu, Z. and Jiang, Y. 2016. Nitrogen addition alters elemental stoichiometry within soil aggregates in a temperate steppe. *Solid Earth* 7(6), 1565-1575. https://doi.org/10.5194/se-7-1565-2016.

You, M., Han, X., Hu, N., Du, S., Doane, T. A. and Li, L.-J. 2020. Profile storage and vertical distribution (0-150 cm) of soil inorganic carbon in croplands in northeast China. *CATENA* 185, 104302. https://doi.org/10.1016/j.catena.2019.104302.

Yuan, P., Yang, T. and Poovaiah, B. W. 2018. Calcium signaling-mediated plant response to cold stress. *International Journal of Molecular Sciences* 19(12). https://doi.org/10.3390/ijms19123896.

Zamanian, K. and Kuzyakov, Y. 2019. Contribution of soil inorganic carbon to atmospheric CO_2: More important than previously thought. *Global Change Biology* 25(1), e1–e3. https://doi.org/10.1111/gcb.14463.

Zamanian, K., Pustovoytov, K. and Kuzyakov, Y. 2016. Pedogenic carbonates: Forms and formation processes. *Earth-Science Reviews* 157, 1–17. https://doi.org/10.1016/j.earscirev.2016.03.003.

Zamanian, K., Zarebanadkouki, M. and Kuzyakov, Y. 2018. Nitrogen fertilization raises CO_2 efflux from inorganic carbon: A global assessment. *Global Change Biology* 24(7), 2810–2817. https://doi.org/10.1111/gcb.14148.

Zamanian, K., Zhou, J. and Kuzyakov, Y. 2021. Soil carbonates: The unaccounted, irrecoverable carbon source. *Geoderma* 384, 114817. https://doi.org/10.1016/j.geoderma.2020.114817.

Zhang, F., Wang, X., Guo, T., Zhang, P. and Wang, J. 2015a. Soil organic and inorganic carbon in the loess profiles of Lanzhou area: Implications of deep soils. *CATENA* 126, 68–74. https://doi.org/10.1016/j.catena.2014.10.031.

Zhang, Y., Zhou, A., Zhou, J., Liu, C., Cai, H., Liu, Y. and Xu, W. 2015b. Evaluating the sources and fate of nitrate in the alluvial aquifers in the Shijiazhuang rural and suburban area, China: Hydrochemical and multi-isotopic approaches. *Water* 7(12), 1515–1537. https://doi.org/10.3390/w7041515.

Zhang, W., Wang, X., Lu, T., Shi, H. and Zhao, Y. 2020. Influences of soil properties and hydrological processes on soil carbon dynamics in the cropland of North China Plain. *Agriculture, Ecosystems and Environment* 295, 106886. https://doi.org/10.1016/j.agee.2020.106886.

Zhao, H., Zhang, H., Shar, A. G., Liu, J., Chen, Y., Chu, S. and Tian, X. 2018. Enhancing organic and inorganic carbon sequestration in calcareous soil by the combination of wheat straw and wood ash and/or lime. *PLoS ONE* 13(10), e0205361. https://doi.org/10.1371/journal.pone.0205361.

Zhao, W., Zhang, R., Huang, C., Wang, B., Cao, H., Koopal, L. K. and Tan, W. 2016. Effect of different vegetation cover on the vertical distribution of soil organic and inorganic carbon in the Zhifanggou watershed on the loess plateau. *CATENA* 139, 191–198. https://doi.org/10.1016/j.catena.2016.01.003.

Chapter 8

Soil organic carbon sequestration and climate change

M. Sanaullah and T. Afzal, University of Agriculture Faisalabad, Pakistan; T. Shahzad, Government College University Faisalabad, Pakistan; and A. Wakeel, University of Agriculture Faisalabad, Pakistan

1 Introduction

Climate change is one of the most threatening global issues, putting agricultural production and food security at risk, which are already under pressure due to population growth and poverty. Adapting to the changing climate is the only way to ensure sustainable food production. However, the general public, including farmers, has little access to scientific knowledge on the impacts of climate change on food systems and technical solutions for adaptation, although they face the consequences of climate change and its implications in their daily lives (Brulle, 2018).

It is generally accepted that increasing greenhouse gas (GHG) emissions are the result of the widespread use of fossil fuels and land-use changes, particularly the massive conversion of perennial vegetation systems into arable cropping systems. These lead to higher CO_2 concentrations in the atmosphere and global warming (IPCC, 2020). Although essential to life on earth, the concentration of atmospheric CO_2 has increased from pre-industrial levels of 280 ppm to over 400 ppm (IPCC, 2020). As a result, average surface temperature records are now higher when compared with the last five centuries. An average global increase in temperature of 0.74°C has been recorded over the last hundred years. If sufficient actions are not undertaken to mitigate GHGs emissions and/or drawdown GHGs from the atmosphere, the global average temperature will

http://dx.doi.org/10.19103/AS.2022.0106.08

increase by 2.8°C compared to pre-industrial levels by the end of this century (IPCC, 2021).

Climate change directly affects food security and agriculture through changes in precipitation patterns, elevated temperature, and occurrence of extreme events (Flannigan et al., 2009; Pautasso et al., 2012). Higher temperatures and extreme events may affect soil biogeochemical processes, inducing changes in soil function and soil organic carbon (SOC) sequestration. As this process leads to the removal of atmospheric CO_2 and its long-term storage in the soil through plant residues and other organic solids retained as soil organic matter (SOM) (Olson, 2013), changing climatic conditions may have positive and negative consequences for atmospheric CO_2 concentrations. Indeed, climate change may affect *soil carbon input* through its effect on biomass production and also *soil carbon output* by its effect on soil water status and microbial functioning (Chapter 5 of this book).

Increasing CO_2 concentrations, climate change, and SOC sequestration interact in multiple ways. Soil respiration is strongly related to soil temperature and moisture availability and even a slight change in soil temperature and moisture can alter this process (Liu et al., 2016). Drought and heat waves have an immediate and direct effect on the moisture and temperature of the soil. Heat waves can have contradictory effects on soil respiration, including (1) accelerated soil respiration due to elevated temperatures, resulting in accelerated enhanced microbial activities, and (2) negative effects on soil microbes by reducing the soil moisture through enhanced evapotranspiration, leading to a decline in soil respiration (Hoover et al., 2016). Elevated global temperatures affect soil microbial activity and are hypothesized to lead to SOC loss. Moreover, extreme events such as drought, flooding, hail and/or frost may lead to crop failure and therefore reduce biomass input, thereby preventing SOC accumulation. High rainfall intensity may lead to water erosion and soil loss, removing SOC along with fertile topsoil (Chapter 5 of this book). On the other hand, increased SOC may reduce the yield gap related to such extreme events (Iizumi and Wagai, 2019). Globally, there is growing interest in increasing the SOC levels of terrestrial ecosystems to counteract global GHG emissions and adapt to climate change. Consequently, the approach has been integrated into the IPCC's climate change mitigation actions. Recently, the French '4 per mille' initiative was introduced as an international multistakeholder platform aimed at increasing SOC sequestration globally with the aspirational goal of 0.4% (4 per mille) per year as a nature-based solution to mitigate climate change and to adapt to its consequences (Rumpel et al., 2020). This initiative is closely linked with agricultural soils that have low SOC levels as a result of persistent cultivation and regularly unsustainable crop production practices following intensification (Pingali, 2012). These soils are especially vulnerable to extreme events and have high GHG emissions due to the high input of agrochemicals.

In this chapter, we will address the causes of climate change and its effects on soil processes affecting SOC storage. The impacts of climate change variables such as elevated CO_2, elevated temperature, drought, floods and other extreme events on C storage in the terrestrial ecosystem will be the main focus of this chapter.

2 Causes of climate change

Over the past few decades, increased amounts of GHGs in the atmosphere have caused changes in the earth's climate, leading to a rise in global temperatures (IPCC, 2021). An increase in global temperatures changes precipitation regimes with effects on the intensity, frequency, and geographical distribution of extreme events (Houghton et al., 2001). Soils play a vital role in global biogeochemical cycles. Three important GHGs (CO_2, N_2O, and CH_4) are generated from the soil and contribute to global warming. Before detailing the impacts of climate change on SOC sequestration, we will present the main causes of climate change. A large number of studies using statistical modeling and methods have confirmed that anthropogenic activities have contributed substantially to cause climatic changes (Stern and Kaufmann, 2014). While anthropogenic GHG emissions related to fossil fuel use are leading to increased atmospheric CO_2 concentrations, agricultural activities are also among the major causes responsible for climate change.

2.1 Deforestation

Deforestation and conversion of forest land to agriculture gained strength in the last part of the twentieth century and noticeable global deforestation was seen between 1990 and 2010, particularly in tropical regions (Mather, 2004). It is a global concern, even if tropical forest area decreases in predominantly developing countries (Barraclough and Ghimire, 2000). Deforestation affects the climate both locally and globally, which leads to increased CO_2 emissions as forests are replaced by marginal land and crops (Bala et al., 2007). Forests contain over three-quarters of C stored by terrestrial flora and 40% of the soil C (Gorte and Sheikh, 2010). Larcher (2003) reported that, from a climate perspective, the increase in atmospheric CO_2 is one of the most important consequences of deforestation and land use change. Assessments indicate that deforestation has resulted in a release of around 5.9 peta grams (Pg) of CO_2 into the atmosphere, equivalent to about 17% of all annual anthropogenic GHG emissions during the 1990s (IPCC, 2007). Put another way, global deforestation has released CO_2 equivalent to approximately 25% of emissions originating from the combustion of fossil fuels. Moreover, biophysical impacts of deforestation on the climate system include an increase in the albedo of the land surface, which affects the radiation budget of the area (Bounoua et al., 2002; Gupta et al., 2005). Deforestation has an impact on wind flow, water vapor flow, and solar energy absorption, affecting the local and global climate (Buys, 2007). Kindermann et al. (2008) reported that a 10% decrease in deforestation between 2005 and 2030 could lower emissions by 0.3 GT to 0.6 GT of CO_2 per year. Forests are the primary terrestrial C sinks and deforestation enhances global warming by increasing concentrations of GHGs (Houghton, 2005).

2.2 Agricultural practices

Intensive agricultural practices significantly contribute to GHG emissions through two processes: (1) ongoing cultivation of land during farm operations, and (2) manufacturing of farm inputs such as fertilizers and pesticides (Desjardins and Riznek, 2000).

Regional climates have changed through land-use changes, further affecting the water balance and surface energy. Additionally, anthropogenic perturbations of the hydrological cycle through fresh water use in industry, for irrigation and domestic consumption, add to these changes (Vörösmarty and Sahagian, 2000; Kalnay and Cai, 2003).

Furthermore, intensive agriculture leads to biodiversity loss induced by pesticide use, soil degradation, water pollution, loss of habitat, and overexploitation of native species. An increased global population has put significant pressure on the biosphere. Land-use practices also affect the air by changing atmospheric conditions as the tropospheric ozone has a significant impact on the biogenic emission and vegetative cover (Kalnay and Cai, 2003).

2.3 Edaphic factors

Soil organic matter contains a mixture of substances, primarily plant and animal residues, microbial biomass and various substances released by microbes into the soil (Miltner et al., 2012). It plays an important role in the global carbon cycle because soils act as a potential source and/or sink for C. Edaphic and climatic factors such as temperature, precipitation, pH, and soil content of C, N, and P along with soil texture significantly affect the C and N content of soils and their exchange rates with the atmosphere (Hosseini Bai et al., 2012). The individual organism responds to local climate change instead of the regional and global climate. Climate stress leads to a potential risk of extinction for small populations dependent on edaphic factors. Their alteration may increase the climatic stress for local species thereby favoring the invasion of alien species (Cartwright, 2019).

2.4 Biological factors

Soil (micro) biota typically play a pivotal role in the stabilization of SOM (Wolters, 2000). Soil faunal activities through ingestion of plant litter can favor carbon mineralization but they also promote the mixing of OM with the mineral soil (Frouz et al., 2020). Microbial decomposition of plant-derived C regulates its partitioning into SOM through the addition of microbial necromass and partly decomposed OM compounds and losses of CO_2 from soil (Paterson et al., 2008).

Climate change can strongly affect these processes as it may influence (1) biodiversity mainly by altering phenological traits, species distribution, (2) ecosystem functioning, and (3) the prevalence of invasive species (Hopkins et al., 2007). For example, earthworms tend to be sensitive to harsh environmental conditions. During periods of high temperature and drought conditions, they usually aestivate themselves by migrating from upper horizons to deeper ones where they become dormant (Edwards and Bohlen, 1996). Earthworm densities have also been observed to be in decline (Eggleton et al., 2009). Climate change can cause notable variations in soil properties, particularly soil temperature and moisture, which are known to pose severe effects on earthworm densities (Perreault and Whalen, 2006; Wever et al., 2001) since they are characterized by limited dispersal ability and intimate connection with the soil's ambient environment. For instance, Eggleton et al. (2009) reckoned that earthworm populations normally decline in summer due to hot and dry weather while they likely surge in the winter season and wet conditions.

Similarly, microbial activity may be affected by various environmental factors, including temperature, CO_2 effect, and precipitation (Metcalfe et al., 2011; Paterson et al., 2008). Among these ecological factors, soil warming has been observed to cause variable effects on SOM mineralization (Conant et al., 2011). Application of chemical kinetic theory predicts that soil warming will cause an increase in SOM mineralization, which would be of a higher magnitude at low temperatures with wide effects on the turnover of relatively recalcitrant SOC pools (Conant et al., 2011). However, the associated effects of soil warming on the biological drivers of mineralization, as well as the complex nature of the soil matrix, confound the application of this basic theory to SOM mineralization. Changes in community structure (microbial acclimatization) can favor microorganisms adapted to warm conditions that can cause amplification of the sensitivity of SOM mineralization to warm temperatures. Indeed, soil warming leads to a rapid depletion of labile C-pools and thus can slow down mineralization rates over time (Kirschbaum, 2004), but this might be different with fresh plant inputs as they cannot be replenished as fast. Thus, the sensitivity of priming effects to future warming may be linked with shifts in microbial community structure and altered plant inputs to soil (Balser and Wixon, 2009;

Ostle et al., 2009). However, basal soil respiration, contrary to the priming effect, is observed to have increased significantly in soils amended with water only under increasing temperatures (Ghee et al., 2013).

2.5 Temperature sensitivity of soil organic matter decomposition

Variations in temperature sensitivity of SOC decomposition can cause uncertainties in forecasting climate-C cycle response. Uncertainties in predicting temperature sensitivity of SOC decay partly depend on how different models parameterize the Q_{10} factor, which is, in short, the temperature dependence of SOC decay, and describes a relative increase in SOC decomposition for each 10°C rise in temperature (Zhou et al., 2018). Additionally, earth system models (ESMs) assume that all SOC fractions respond identically to climate warming, excluding the SOC stability, the tenacity of which is due to its recalcitrant molecular structure, physiochemical, and biological protectivity, etc. (Friedlingstein et al., 2006; Schmidt et al., 2011). Previous studies report highly inconsistent results, showing either more or less sensitivity of old SOC to temperature than young components (Arora et al., 2013; Friedlingstein et al., 2006). Zhou et al. (2018) examined the apparent sensitivity of SOC decay to temperature in a long-term field incubation using deep soil collars to exclude root C input under warming and ambient conditions and predicted temperature sensitivity and C residence times of three SOC fractions (active, slow, and passive) using data assimilation techniques. They found ancient SOC to be more sensitive to temperature than young SOC content, which may make old SOC highly vulnerable to future warmer climatic events. Since the ancient SOC represents a considerable part of the total amount of SOC (50–90%) with long stay periods spanning several decades (Schmidt et al., 2011), its vulnerability to temperature will largely ascertain the global trend and magnitude of the future soil C stock in the long-term (Zhou et al., 2018). Enhanced oxidation of SOC compounds under warming can increase SOC decomposition and respiration (Rustad et al., 2001). However, Zhou et al. (2018) reported a reduction in SOC decomposition and SOC content in deep soil, which might be linked with the depletion of an active C pool with short residence at first and relative deposition of ancient SOC fractions (slow and passive C pools) with long stay periods thereafter. Additionally, in the deep soil collars, Cheng et al. (2017) found that microbial communities in the warmed soil showed an abundance of key functional genes involved in decomposing ancient SOC relative to control and estimated an increase of about 73 g C m^{-2} $year^{-1}$ in annual CO_2 emissions from the deep collars during the study period (2000 to 2010). They indicated microbial respiration as the dominant cause of C losses from the old SOM pools in the subsoil during a decade of experimental warming.

The kinetic theory envisages that ancient SOM, if more chemically recalcitrant, tends to be highly sensitive to temperature variability compared to new and young SOM pools (Knorr et al., 2005; Davidson and Janssens, 2006). Soil C losses under warming can be ascribed to changes in biotic (e.g. microbial respiration) and abiotic (e.g. freeze-thaw and leaching) factors (Davidson and Janssens, 2006). However, these theories need to be revisited with regards to the changing paradigm that SOC persistence is controlled by microbial accessibility rather than chemical recalcitrance (Cotrufo et al., 2013). Indeed, a recent study showed that physicochemical protection of SOM constrains the temperature sensitivity of SOM decomposition under field conditions (Moinet et al., 2020). It may be referred to as apparent temperature sensitivity of SOM decomposition, which relates to the environmental protection of SOC pools by means of physical, chemical, and biological factors (also known as environmental constraints), such as lack of enzymatic contact, controlled access to soil microbes, and exposure to external temperature (Xiong et al., 2014; Pang et al., 2015). The intrinsic temperature sensitivity of SOC decomposition depends on the molecular structure of SOC, its activation energy, and ambient temperature, which can be observed under laboratory conditions with unlimited substrate availability (Manzoni et al., 2012).

The addition of fresh plant inputs with higher net primary productivity (NPP) can stimulate the microbial degradation of SOC and thus constrain SOC accrual (Fontaine et al., 2004; Hungate et al., 2013; Kuzyakov et al., 2000; Phillips et al., 2011; Wieder et al., 2014) via priming of SOC (Neill and Gignoux, 2006; Carney et al., 2007; Cheng et al., 2014). Todd-Brown et al. (2014) estimated a rise in heterotrophic respiration from 7 Pg C year^{-1} to 43 Pg C year^{-1} by studying outcomes from historical and the most extreme IPCC RCP 8.5 (radiative forcing scenario) between 1850 and 2100 using ESMs, which were primarily due to increases in soil inputs and shortening of SOC turnover times, leading to an unchanged net SOC balance.

3 Carbon sequestration and climate change in terrestrial ecosystems

Climate change affects SOC buildup by its impact on biotic and abiotic processes related to above-ground and below-ground productivity and decomposition processes (Chapter 5 of this book). Terrestrial ecosystems, including forests, grasslands, croplands, and drylands, are affected by a variety of climate change drivers, such as CO_2 fertilization, elevated temperature, altered precipitation, drought, heat waves, and biotic stresses. Since these ecosystems have the potential to accumulate a massive amount of SOC, they have received considerable research attention globally. Soils are the largest terrestrial sink for C on the planet (Sharma et al., 2012). Estimates of these reserves range from 1100 Pg C to

1700 Pg C, which is twice as much as vegetation or atmosphere (Scharlemann et al., 2014). The impacts of climate change drivers on C sequestration in four types of terrestrial ecosystems are discussed in (Fig. 1) in detail.

3.1 Forests

Forest ecosystems have been identified as the largest land C sink and account for more than half of C stored in the terrestrial ecosystem (Hui et al., 2017). The total amount of C sequestered in forest vegetation is approximately 359 Pg (Allen et al., 2010). Forests contain an important amount of SOC (about 0.9 Pg C $year^{-1}$) (Yude et al., 2011). A major part of the land-based climate mitigation measures envisioned in the 2015 Paris Agreement is the global enrichment of forest biomass C stores via the preservation of existing mature forests and the establishment of new ones (Jiang et al., 2020). However, forest ecosystems can be heavily affected by climate change due to their longevity and limited ability to adapt to altered temperature and precipitation regimes (Lindner et al., 2010). Globally, especially in boreal regions, forests are heavily affected by warming, alterations in hydrological cycling, and their biodiversity (Bonan, 2008). They are also vulnerable to extreme climate events, which in turn could weaken their SOC sink potential or even change them into a net C source in some areas (Kurz et al., 2008). The impacts of climate change variables on SOC stocks in forest ecosystems are given as follows.

3.1.1 Elevated carbon dioxide

Forests can capture substantial amounts of CO_2 from the air due to the abundance of vegetation. Higher atmospheric CO_2 concentration and atmospheric nutrient addition can have a fertilization effect on increasing plant primary productivity (Wenzel et al., 2016; Campbell et al., 2017; Peters et al., 2013). However, this additional uptake may not accumulate in the soil to enhance the SOC stock at the ecosystem level (Fig. 1). Utilizing ESMs, Todd-Brown et al. (2014) showed that cumulative global NPP is likely to increase from 11% to 59%, but with a concurrent decline in SOC turnover times (15–28%) over the twenty-first century. This illustrates soil microbial response to increasing temperatures, with microbial activity being the most important pathway to reducing the SOC sequestration potential of soils (Todd-Brown et al., 2014; Naipal et al., 2018). This calls into question that elevated CO_2 in the atmosphere can increase the forests' SOC sink potential. However, the obscured constraints on NPP can limit the conversion of plant biomass into SOC (Todd-Brown et al., 2014). Talhelm et al. (2014) reported that elevated CO_2 promoted an overall 39% increase in NPP in three young northern temperate forest communities in the central north US in an Aspen FACE experiment, leading to an 11% rise in ecosystem

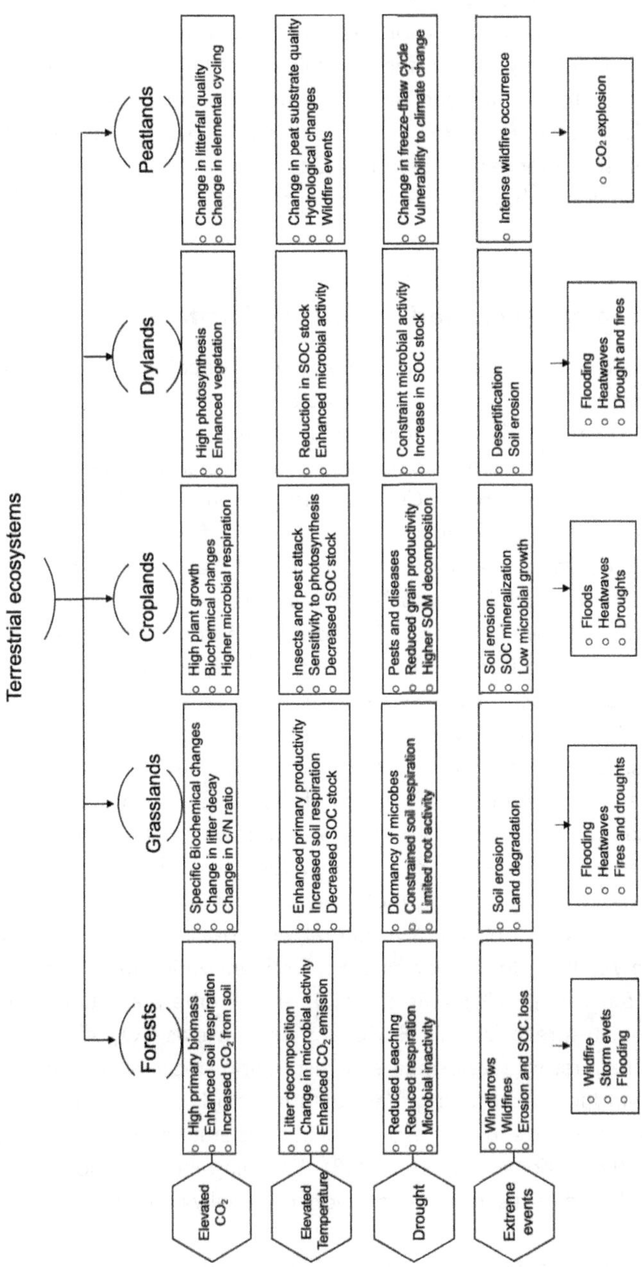

Figure 1 The impact of climate change drivers on the carbon storage potential of terrestrial ecosystems.

C content; but this increase tended to conflict with the effects of enhanced CO_2 on SOC stock. In temperate forests, high C allocation to root exudates under elevated CO_2 stimulates microbial activity, leading to an acceleration in soil organic matter decomposition and the release of N bound to the SOM and its uptake by plants, which can result in long-term enhancement of a C sink in woody biomass but not in the mineral soil (Drake et al., 2011). Thus, the implication of the subterranean exchange of C for N is a sustained sink for C in forest biomass, but it may have only a minor effect on SOC (Bader et al., 2013; Talhelm et al., 2009).

3.1.2 Elevated temperature and altered precipitation

Due to global rises in temperature, an increase in litter decomposition in forest ecosystems is likely to be observed due to increased microbiological activity. Indeed, according to a study conducted in Canadian forests, an increase of 4-7% in litter degradation is predicted, due to an increase in temperature and precipitation under the double CO_2 scenario (Moore et al., 1999). Prietzel and Christophel (2014) determined the SOC stock from 150 forest soil profiles covering the entire German Alps. They found that at sites with low air temperature and high precipitation, SOC stocks were particularly large, as in most German forests. There is a risk of SOC depletion from these forests due to future temperature rises. Beniston et al. (1997) and Beniston (2005) reported that a temperature increase of 2 Kelvin (K) might cause a reduction in SOC stocks by 1.5 kg SOC m^{-2}, corresponding to 14% of the total forest SOC stock of the German Alps. Forest floor SOC stocks are found to be negatively correlated with air temperature (Prietzel and Christophel, 2014). Elevated atmospheric temperatures may be reflected by soil temperature, which can considerably affect soil biochemical processes (Kirschbaum, 2004). Xu et al. (2010) assessed the effects of artificial warming using the open top chamber (OTC) method in two different types of forests (planted vs natural) in the eastern Tibetan Plateau of China and showed that, due to warming, SOC mineralization rates increased and led to a 10.6% and 15.4% rise in soil CO_2 efflux in the plantation and in the natural forest sites, respectively. However, this C loss as CO_2 efflux is not associated with respiratory rates since soil respiration declined, presumably due to warming-induced soil drying. Various other studies also reported no increase or even a decrease in soil respiration in temperate and boreal forests under warming (Lellei-Kovács et al., 2008). Warming-induced soil drying could explain such decreases in soil respiration, especially in artificial warming experiments.

There is a strong link between soil temperature and moisture, and alterations in either of these variables can cause changes in SOC due to their significant effect on microbial activities (Chapter 5 of this book). Borken et al. (1999) examined the effects of rainfall and its distribution on CO_2 emissions and the DOC pool in temperate forest soil. They did not notice any effect on DOC leaching under drought below the rhizosphere on an annual basis. Drought tended to reduce soil respiration but not to a significant degree. Rewetting enhanced CO_2 emissions by 48% and 144% in consecutive years (1993 and 1994), and emissions from the soil were more pronounced at higher temperatures upon rewatering, leading to enhanced soil respiration. Rewetting following drought enhanced annual CO_2 emissions from 2981 kg C ha^{-1} yr^{-1} in 1993 to 4813 kg C ha^{-1} yr^{-1} in 1994, a 51% increase (Borken et al., 1999). Campo and Merino (2016) reported changes in SOC stock due to precipitation and drought in 12 mature seasonally dry tropical forests. The reduction in rainfall from 1240 mm yr^{-1} to 642 mm yr^{-1} only slightly reduced the litterfall by up to 10% while decomposition was greatly reduced by up to 56%. It was concluded that low precipitation in these forests can increase SOC stocks in mineral and organic layers, due to low decomposition rates and the recalcitrant nature of organic matter. These results indicate that elevated temperatures and drought can increase SOC in dry tropical forests, assuming primary production is constant.

3.1.3 Other extreme events

The current process of climate change is leading to an increase in the occurrence of extreme events, which are expected to occur at even higher frequencies in the future. Extreme events include windstorms, heat waves, and wildfires, which can particularly affect forest ecosystems. These climatic extremes can indirectly influence SOC storage by influencing the above-ground biomass and thereby organic matter input, which may affect the underground SOC equilibrium. Storms are a major disturbing agent in forests. Wind damage, or windthrow, occurs when the force applied by the wind overcomes the anchorage of a tree or the strength of its stem, resulting in uprooting or stem breakage (Mitchell, 2012). Saad et al. (2017) assessed the effect of climate change on windthrow risks in forests in Eastern Canada using a modeling approach and predicted an increased risk of windthrow (3-30%) due to unfrozen soil conditions. However, windthrow does not really have the potential to affect forest SOC as forest soils tend to respond slowly to changes in above-ground biomass C. For example, it was shown that a frequency increase of 30% only slightly affected the national SOC budget of Swiss forests (Thürig et al., 2005).

Changes to the climate can lead to more frequent occurrences of heat waves and severe drought events (Anderegg et al., 2015). In Europe, severe drought periods in 2005 and 2010, and the 2003 heat wave, reversed four years of continental C uptake by terrestrial ecosystems (Marinov et al., 2010). The drought also caused three to four years of reduced growth. These legacy effects were more prominent in dry ecosystems, among *Pinaceae* and other species with limited hydraulic activity (Anderegg et al., 2015), which might have an indirect effect on SOC balance due to reduced C inputs into soil. Fast turnover of leaves and roots following drought events can also induce a reduction in ecosystem-level C storage (Somero, 1995), with negative indirect effects on SOC sequestration.

Climate variables, particularly temperature and soil moisture, are intertwined with climate change, and thus play a role in wildfire risk. Climate warming promotes the drying of organic matter in forests (the fuel that burns and causes wildfires), which has doubled the incidence of severe forest fires in the western US between 1984 and 2015. Disturbance from fire is a significant factor in the SOC dynamics of forest ecosystems. In a boreal forest of the Greater Higgnan Mountains region, northeastern China, Han et al. (2021) explored how SOC changed following a wildfire. Their study showed that the SOC storage of boreal forest soil was reduced due to wildfire disturbances, which could not recover to the levels of the control site at the decadal scale. Wildfire-induced reduction in SOC levels was most apparent in surface soil horizons. The observed SOC differences and dynamics were caused by changes in physical and biochemical properties of organic inputs, re-vegetation, and canopy after forest fires. Wildfire turned the SOC into a small amount of black C resistant to any further disturbance and resulted in a large CO_2 release (Chen and Shrestha, 2012), with ultimate declines in the overall SOC concentration in the soil. Furthermore, the forest fire destroyed a part of the aggregate-associated SOC, which did not recover to its initial values during 25 years of plant succession. Magnani et al. (2007) found that although wildfire events first change the forest to a net C source, they may later be transformed into a significant SOC sink through increased primary production, especially in mature forests. Organic matter transformation to black carbon under wildfire events is a crucial aspect to be considered, as it confers resistance to further breakdown. However, only a small percentage of ecosystem carbon is returned to the soil as black carbon, while the major part of the biomass is lost as CO_2 (Alexis et al., 2010).

In recent studies, it has been shown that smoldering buried biomass causes additional losses and destabilization of SOC (Rein and Huang, 2021). Smoldering is the most persistent type of burning process under low temperature and without a flame. These fires spread slowly and transversely via the soil's organic layers, gradually burning everything in their way. In boreal forests, growing on organic soils, this type of fire may especially lead to SOC release. Despite heavy rain, weather changes, and firefighting efforts, smoldering fires are very hard to control once they are ignited. They turn organic soil first into char and then into ash until only a thin coating of ash and very small char fragments remain, leaving a void inside. In boreal forests, this may lead to the loss of legacy carbon. With increased frequency, these fires may turn these ecosystems from small carbon sinks into carbon sources (Walker et al., 2019).

Under a changing climate, extreme climate events such as flooding and wind storms can cause soil erosion, which ultimately affects SOC stocks (Ma et al., 2016). Forest soils covered with vegetation are less prone to soil erosion because debris lying on the surface of the forest reduces the impact of raindrops and wind velocity, and at the same time, covers the soil, protecting its surface against erosion (Pimentel and Burgess, 2013). The rate of erosion in mature forest ecosystems is limited to just 0.004 tons ha^{-1} yr^{-1} to 0.05 tons ha^{-1} yr^{-1} (Heywood and Watson, 1995). However, this situation can change when forests are disturbed by windthrow or fire, removing the tree cover (Daily, 2013).

3.2 Grasslands

Grassland ecosystems encompass above 29% of ice-free land and accordingly are principal components of global C stocks below-ground and SOC sequestration (Hovenden et al., 2019; Sharma et al., 2012). Several studies have shown that grassland ecosystems are vulnerable to climate change, but the causes and implications of their susceptibility are debatable, particularly for different management practices. The terrestrial ecosystem's C, water, and energy fluxes are strongly dependent on regional climate and soil moisture feedbacks (Gentine et al., 2019), determining the grasslands' functioning and SOC sequestration potential.

Grasslands can significantly increase SOC reserves and their sequestering ability has been recognized for a long time (Chapter 18 of this book). They contain up to 331 Mg ha^{-1} SOM (Schlesinger, 1977), more SOC than other terrestrial ecosystems. Grasslands, along with croplands, are deemed to have high potential for increasing SOC sequestration, especially in areas where soils have been historically degraded and SOC pools are depleted. However, these enhanced SOC stocks can only be achieved if proper management practices are adopted (Chapter 18 of this book). In line with the uniform adoption of management practices, croplands and grasslands could store up to 0.32–0.86 Pg SOC yr^{-1} and 0.60–1.01 Pg SOC yr^{-1}, respectively. According to recent estimates, grasslands absorbed 10% of annual CO_2 emissions released from fossil fuels and industry between 2005 and 2015 (Batjes, 2019). However, this may be affected by climate change.

3.2.1 Elevated carbon dioxide

Elevated atmospheric CO_2 may impact C sequestration in grassland soils by changing their biochemical composition, including the C:N ratio of grassland species that, in return, affect the litter decomposition in soil (Franck et al., 1997). This change in the C:N ratio, due to elevated $CO_{2,}$ varies with plant species. Franck et al. (1997) demonstrated in a laboratory microcosm that litter quality and its subsequent decomposition under elevated CO_2 concentration is species-specific. They reported that *V. microstachys* was the only species for which increased CO_2 growth steadily reduced the rate of decomposition and associated N mineralization, despite the fact that elevated CO_2 did decrease the litter C:N ratio in this species. Furthermore, elevated CO_2 can also affect the physical and biotic aspects of the ecosystem, which in turn modify the allocation patterns of CO_2 inside the plants and affect litter decomposition. Seibert et al. (2021) examined the long-term *in situ* effects of elevated CO_2 on different forage qualities of grassland biomass in a managed temperate grassland. They reported that 18 years of increased CO_2 enrichment resulted in lower forage quality, showing decreased crude protein and fat content and an increase in crude fiber content. For most forages, the forbs showed a significant increase in CO_2-induced effect, whereas the grasses demonstrated a significant adverse effect of elevated CO_2, mainly on ash content.

Along with the direct effects of elevated CO_2 on litter quality, there are also indirect effects of elevated CO_2, including changes in species composition that may be equal to or even greater than the direct effects. In this way, changes in species dominance induced by higher CO_2 concentration may have the potential to affect decomposition rates and SOC sequestration (Dukes and Hungate, 2002).

Elevated CO_2 can stimulate grassland ecosystem productivity due to its fertilizing effect on plant species, either through enhancing photosynthesis (Reich et al., 2014) or controlling stomatal conductance (Ainsworth and Rogers, 2007) and reduced respiration (Haworth et al., 2015). Yuan et al. (2018) studied the combined effects of extreme events (e.g. heat waves, extreme dry events, and hard frost events) alongside elevated CO_2 by utilizing the long-range data from a C3 grassland FACE trial in Germany. They characterized the CO_2 fertilization effect as enhanced photosynthesis and water-use efficiency, which tended to be fairly different between plots treated with CO_2 enrichment and ambient CO_2. Their results implied that elevated CO_2 can enhance the adaptability of grassland ecosystems to extreme climatic events, presumably by expanding the growing period. Likewise, elevated CO_2 can enhance productivity by diminishing stomatal aperture, enhancing plant water-use efficiency, and retaining higher soil moisture in drier ecosystems.

3.2.2 Elevated temperature and altered precipitation

High temperatures and precipitation together are important factors and can have a significant impact on SOC reserves in grassland ecosystems. Chen et al. (2020) investigated the response of SOM to future climate change scenarios in an alpine grassland and found that increased precipitation enhanced the SOC stock due to the positive effects on vegetation. Conversely, warming-induced drought stress, resulting in low biomass production and low SOC stock. Jackson et al. (2002) evaluated the change in vegetation on SOC stock at six grasslands and woody sites in south-western US along a precipitation gradient and ascertained the relationship between biomass C and SOC levels under the invasion of grassland by woody plants. They found the SOC buildup under woody plants invasion in the grassland ecosystem to be negatively correlated with precipitation, showing drier areas gaining SOC and wetter ones losing SOC. Tian et al. (2010) suggested that with an increase of 2°C in temperature, higher microbial activities will lead to losses of 10% of SOC stocks in Qinghai-Tibetan grasslands. However, a meta-analysis to analyze the ecosystem-level responses to warming, precipitation, and their combination reported that experimental warming and high precipitation tend to promote plant productivity while ecosystem C fluxes and plant productivity were more sensitive to reduced precipitation. They suggested that warming may cause an increase in above-ground and below-ground productivity by 19% and 52%, respectively (Wu et al., 2011). Although experimental warming and precipitation can increase net plant productivity, positive effects on SOC balance might be offset by the stimulation of microbial activities.

Drought may have an impact on carbon allocation from shoot to root and can affect microbial activities and SOC stock in grassland soils (Sanaullah et al., 2012). Tietjen et al. (2017) used the ecohydrological simulation model SOILWAT to separate the direct and indirect effects of climate change on soil moisture conditions. They found that vegetation change under drought conditions will likely worsen the soil water availability. Craine et al. (2012) studied the seasonal impacts of climate change on SOC sequestration by considering grass productivity under temperate humid conditions for 27 years and found that drought and high-intensity precipitation reduced grass productivity in 110 days, while high temperatures reduced productivity in just 25 days. Warming not only increased the primary productivity of the alpine meadow but also increased the SOM decomposition and thus enhanced soil respiration, which led to a significant decline in SOC stocks in the alpine grassland ecosystem.

A high level of biodiversity may confer stability to grassland ecosystems against climatic changes. Liu et al. (2018) explored the effects of climate change on plant community structure and ecosystem functioning of grasslands. They showed that 32 years of warming caused an increased abundance of herbal grass species and decreased the richness of sedge species, without any effect on above-ground NPP. Changes in the composition of plants due to climate change could sustain primary production and change the proportion of above-ground to below-ground productivity. Sanaullah et al. (2014) demonstrated that extreme climatic events such as high temperatures and drought may alter the stoichiometry of grassland species, i.e. *D. glomerata* and *F. arundinancea,* which differentially affected SOC concentrations after their addition to soil. They reported that the plant biomass of both grassland species developed under drought conditions mineralized readily, presumably due to higher N content and lower C/N and lignin/N ratios of moisture-stressed plants.

3.2.3 Other extreme events

Extreme events due to climate change (such as heat waves) can severely affect C sequestration in grasslands. Yuan et al. (2018) documented a down-regulation in above-ground biomass due to extreme climatic events, i.e. intense and extended heat waves, in both summer and spring periods under elevated CO_2 levels. They found that higher temperatures and drier conditions can negate the CO_2-fertilization effect on growth, presumably by restricting plant growth through reducing enzyme activity, enhancing susceptibility to pathogens and herbivores, and via intensifying respiratory losses. Fires are part of the natural cycle in many biomes and are notably dominant in grassland ecosystems. In grassland ecosystems, fires appear to constrain C sequestration, due to returning C fixed by vegetation directly to the atmosphere and thereby curbing its storage in soil (Farage et al., 2003). However, stable carbon may be returned to the soil through the production of black carbon resulting from fire. Moreover, the SOC loss can be subsequently replaced by enhanced photosynthesis and vegetative growth (Svejcar and Browning, 1988).

3.3 Croplands

Almost 34% of the global land area is linked to food production, and over the past few centuries, there has been a historical loss of SOC reserves in agroecosystems due to intensified agricultural production following the Green revolution (Godfray et al., 2011). Indeed, agroecosystems' soils have lost around 25–75% of their SOC pool (Lal, 1999) due to intensive agricultural practices such as extensive tillage, mono-cropping, use of synthetic fertilizers, and other agro-chemicals. This significant depletion in SOC stocks implies that these systems have the potential to re-build their SOC if properly managed (Chapter 16 of this book). The global capacity of SOC sequestration in agroecosystems' soils is anticipated to be roughly 2.1 Pg C yr^{-1} (Lal, 2010). If the global SOC pool can be enriched by 10% (+250 Pg) throughout the twenty-first century, it can cause a diminution of around 110 ppm of atmospheric CO_2 (Lal, 2010). The SOC sequestering potential of agricultural soils is about 25% to 75% less than its undisturbed counterparts and natural ecosystems (Lal, 2004). This not only reduces the amount of CO_2 in the atmosphere but also enhances soil fertility by improving SOC concentrations in addition to stocks (Rumpel and Chabbi, 2021). While it is crucial to improve SOC stocks, ongoing climate change is further affecting these reserves in agroecosystems. Climate and agriculture are inextricably linked, and climate has a profound effect on crop development and productivity. The impacts of climatic variables on SOC stocks in croplands are discussed as follows.

3.3.1 Elevated carbon dioxide

Like in other ecosystems, elevated atmospheric CO_2 is the most important driver of climate change, which can affect crop growth, biomass production, and consequently SOC stocks. Muluneh (2020) predicted a 7% increase in maize yield and crop growth due to the CO_2 fertilization effect, which is likely due to the higher assimilation of C under elevated CO_2. Ren et al. (2012) reported that CO_2 fertilization improved the NPP by 18% and SOC by 22.6%. Free Air CO_2 Enrichment (FACE) studies anticipated around a 14% increase in average productivity of C3 crops such as wheat, rice, and soybean with the CO_2 enrichment from 367 ppm to 583 ppm, showing that the 47 ppm rise in CO_2 since 1980 has raised yield by almost 3% (Ainsworth et al., 2008). Models relating the yield of the four largest commodity crops to climatic trends suggest a decline of about 3.8% and 5.5% in the global productivity of maize and rice, respectively (Lobell et al., 2011). However, the effect on SOC reserves is not yet clear since the processes taking place in the soil are highly complex. Van Groenigen et al. (2014) reported that this CO_2 enrichment tends to stimulate soil microbial respiration and in turn greater decomposition of plant detritus and thus cannot increase SOC stocks as hoped. Indeed, elevated CO_2 was found to maintain the SOC pool while process rates could be increased, thereby accelerating soil biogeochemical cycling and organic matter turnover (Kuzyakov et al., 2019).

The gains in NPP by elevated CO_2 can be constrained by nutrient deficiencies, as indicated by the theory of progressive nutrient limitation, which limits plant growth (Nowak et al., 2004; Norby and Zak, 2011; Norby et al., 2010; Todd-Brown et al., 2014). However, increased nutrient mineralization due to soil warming and higher nutrient inputs from atmospheric deposition could compensate for these limitations (Bai et al., 2013). Hungate et al. (2009) reported a sizeable increase in SOC accumulation due to enhanced CO_2 with N fertilization after analyzing four meta-analyses. However, the SOC accrual was either poorly associated with elevated CO_2 levels or even missing in unmanaged ecosystems lacking N (van Groenigen et al., 2006). The effectiveness of elevated CO_2 as a natural fertilizer to enhance the SOC storage capacity is reinforced by N inputs, such as N deposition and mineral fertilizer (Reich et al., 2006). Nitrogen addition positively affects the SOC buildup indirectly by stimulating plant growth and thus biomass inputs to soil (Alvarez and Alvarez, 2005). Depending on the soil's characteristics, P fertilization may also be required.

In cropland ecosystems, paddy lands very often provide environmental constraints to SOC decomposition relative to aerobic soil, and thus somehow limit the impact of climate change (Dash et al., 2019). For instance, in the Philippines, submergence imparted protection to SOC against decomposition and increased SOC content, regardless of fertilizer and crop residue addition in a long-term (15-year) fertilizer experiment. However, sequential wet and dry cycles due to climatic perturbations have been shown to cause significant loss of SOC from soil (Shrestha et al., 2002).

3.3.2 Elevated temperature and altered precipitation

Elevated temperatures due to global warming can change cropping seasons and their life cycles. Rising temperatures can greatly reduce agricultural production by increasing harmful pests and diseases (Wienhold et al., 2018). Similarly, warm temperatures can decrease grain productivity by affecting pollination and photosynthesis since each crop requires a certain temperature and even the slightest deviation can affect plant growth. Lobell et al. (2011) demonstrated different responses by crop types to warming. Whereas a 1°C increase in temperature may lead to reduced yields by up to 10% in high latitude regions, rice yields in low latitude regions may increase after warming. Impact on crop growth may have a significant effect on SOC stocks since it is related to the addition of plant residues to the soil.

It has been suggested that the combination of temperature, precipitation, evaporation, and soil properties causes changes in SOC, and these factors and their combination vary from region to region (Wan et al., 2011). Overall trends show that a changing climate will have a negative effect on SOC stocks in cropland soils unless proper measures are adopted to increase OM, such as crop straw return, manure application, and introducing conservative tillage methods.

3.4 Drylands

In terrestrial ecosystems, drylands cover over 47% of the earth's surface (Koutroulis, 2019). The global potential rate of soil C sequestration in drylands is estimated to range from 0.5 Pg C year^{-1} to 1.4 Pg C year^{-1} (Lal, 2019) and their SOC stock is estimated to total 241 Pg (Eswaran et al., 2000), which is about 27% of the global SOC stock (Millennium Ecosystem Assessment, 2005). Drylands are globally vulnerable and at risk of degradation due to various environmental conditions, which will likely be compounded by the consequences of global climate change. At the same time, a high amount of atmospheric CO_2 might decrease climate change impacts by boosting plant WUE and productivity (Tietjen et al., 2010). However, this increased production may not show any significant increase in the SOC stock since it accelerates soil respiration by stimulating microbial activities. Dryland ecosystems are characterized by low vegetation cover and low SOC density. Regional and global C cycling may be significantly impacted by dryland ecosystem response to climate change due to their huge size and sensitivity to climate change (Schimel, 2010).

Elevated CO_2 is an important driver that can affect all other current climate change factors. Rising CO_2 levels can affect the SOC stock in drylands by changing other climatic factors, which include temperature, precipitation, and drought events. In drylands, however, water supply is, specifically, the main limitation that alters productivity potential (Lal, 2019), with implications of less C fixation by plant communities in these ecosystems exerted by water scarcity (Farage et al., 2003). Song et al. (2018) estimated that vegetation biomass may increase under enhanced CO_2 levels, but this increase could be affected by other climatic factors such as elevated temperatures and uncertain precipitation. In addition, these climate changes can affect the availability of soil moisture and cause land desertification (Costa and Soares, 2012). Burrell et al. (2020) estimate that 6% of the world's drylands became deserts between 1982 and 2015. It is estimated that 11.2% of the total land area has turned into drylands due to changes in precipitation patterns and evapotranspiration since the Industrial Revolution (Koutroulis, 2019). The expansion of drylands could reduce SOC stocks in the future and contribute to global warming by converting stored SOC into CO_2 in the atmosphere.

It is generally accepted that rainfall increases the assimilation of C, which has a positive effect on plant growth (Berger et al., 2019). However, it can also increase CO_2 emissions by stimulating microbial activities and thus soil respiration. By contrast, less precipitation can reduce emissions of CO_2 by inducing drought, which can increase SOC stock. Munjonji et al. (2020) examined the effects of 5 years of severe drought on SOC stock, CO_2 emissions and soil characteristics in drylands. Severe drought caused a significant change of up to 32% in CO_2 emissions, from 43.6 tons CO_2 ha^{-1} $year^{-1}$ under normal conditions to 30 tons CO_2 ha^{-1} $year^{-1}$ during drought. Consequently, SOC stocks were also higher under drought, ranging from 1.118 kg C m^{-2} to 0.999 kg C m^{-2} in normal conditions. In addition, low rainfall and high temperatures can cause wildfires in drylands (Middleton and Sternberg, 2013), which ultimately reduces the SOC stock over the short term. Effects can be different in the long run, such as organic compounds, which resist degradation, also being formed. The low rainfall under current climate change conditions is also causing desertification of drylands, which can induce dramatic fluctuations in SOC pools (Sharma et al., 2012). For instance, soil C density declined during the decade-long drought of 1998-2008, inducing loss of SOC stock by 0.46 Pg from 1979 to 2011. This loss was more pronounced in Kazakhstan, which faced a decline in rainfall of about 90 mm/decade (Li et al., 2015). There is also evidence that drylands are less likely to lose sequestered C than other terrestrial ecosystems (Glenn et al., 1993) since the moisture limitation restricts mineralization in the soil and thereby the flux of C to the atmosphere, implying that the residence times of C in these ecosystems is much higher even than in forests (Gifford et al., 1992).

Drylands are at high risk of soil erosion, which may be due to the lack of vegetative cover (Lal, 2019). Flooding could make it worse. Future projections for climate change show that the rate of erosion could increase from 25% to 55% in the current century (Delgado-Baquerizo et al., 2013). Lal et al. (2007) evaluated that soil erosion in drylands could induce emissions of 0.23–0.29 Pg C yr^{-1}. Soil desertification due to drought and heat wave events could further aggravate SOC losses by soil erosion. The loss of SOC is intensified due to erosion since it specifically displaces the finer soil fractions and SOC is largely bound to these (Quiroga et al., 1996).

The effect of erosion on SOC loss from surface soil is obvious as it has high concentrations of SOC. However, soil erosion can expose carbonates and/or other minerals in the soil to climatic elements, which may cause their weathering and subsequent CO_2 release (Yang et al., 2012) and/or sequestration through the process of silicate weathering (Chapter 5 of this book). Maintaining vegetative cover either via planting or adding crop residues can play an important role in reducing existing soil erosion in drylands (Delgado-Baquerizo et al., 2013) and may also enhance weathering processes.

Climate change therefore has the potential to have a severe impact on dryland ecosystems via a variety of climate drivers, affecting the NPP and SOC storage capacity of these ecosystems. Additionally, dryland ecosystems could also be turned into deserts if climate change advances. This process should be counteracted by good management practices and human intervention. Management of extreme events (drought, heat waves, and wildfires) is an important loophole to preserve the SOC stock of drylands (Lal, 2019). SOC storage in drylands may be an important negative feedback to the global C cycle since they faced a historic loss of SOC, and therefore have a great sink capacity (Sharma et al., 2012). If eroded and degraded drylands are rehabilitated, they could potentially sequester up to 0.4–0.6 Gt C yr^{-1} (Lal, 2001).

3.5 Peatlands

Although peatlands cover less than 3% of the terrestrial surface land globally (Bridgham et al., 2000), they store a tremendous amount of C. One estimate puts this at ~455 Pg of C to 644 Pg of C (Dargie et al., 2017), which is approximately one-third (21%) of the total C stock of other terrestrial ecosystems (~3000 Pg across 0–3 m) (Scharlemann et al., 2014). Organic soils in peatlands are quite distinct from mineral soils, based on their high C and N densities and characterized by organic matter content greater than 90%, several meters deep (Frolking et al., 2001; Tarnocai et al., 2009). Although the direct effects of climate change on peatlands are obvious, short-term climatic perturbations could also cause changes in peat substrate quality (Chapin et al., 1995; Moore and Dalva, 1997), either by altering the plant community composition and quality of litterfall, or by changing the decomposition process through changes in soil quality. Climate change may cause a change to C (CO_2 and CH_4 emissions) and N mineralization dynamics in peatlands. The CO_2 production from different peatlands appears to have converged with time under different climate conditions, which supports the notion that a relatively small labile C pool consumes rapidly.

Global temperature rises are expected to cause changes in hydrology and spread wildfire events in peatlands as well as escalate the degradation of organic material (Hengeveld, 2000). Tarnocai (2006) ascertained the sensitivity of Canadian peatlands to climate change and predicted subsequent changes in their distribution and SOC dynamics. The peatland sensitivity model suggests that climate change will affect the greatest portion (about 85%) of the peatland area, particularly the Subarctic wetland region, which could trigger the release of approximately 78% (33.96 Gt) of stored organic C mass in conjunction with the thawing of perennially frozen peatlands. Water-saturated conditions will develop due to the thawing of ice-rich peat and underlying mineral soil, which will probably result in anaerobic decomposition, triggering CH_4 emission, which is 21 times more potent a GHG compared to CO_2 (Houghton and Meira, 1996). Under unsaturated conditions, peatlands can emit large quantities of CO_2 due to aerobic decomposition (Christensen, 1991). Drier conditions could aggravate the occurrence and intensity of wildfires on peatlands, which could lead to high CO_2 release with severe consequences for the earth's environment.

4 Conclusion

Climate change is one of the most serious problems facing the world today and may have consequences for SOC sequestration. In most global ecosystems, rising levels of CO_2 in the atmosphere may lead to increased biomass production and consequently carbon input into soil. The resulting increasing temperatures may stimulate microbial activities and SOC mineralization in organic soil layers, while the temperature sensitivity of SOM decomposition in mineral soils may be constrained by substrate accessibility. In addition, climate change is also increasing the occurrence of extreme events such as drought, heat waves, high rainfall and flooding, which may affect SOC storage under contrasting land-use systems. Wildfires following heat waves, and soil erosion as a result of flooding events, have gained much attention in terms of changes in SOC pools in several terrestrial ecosystems. Overall, current climate change effects are having a significant impact on SOC reserves of terrestrial ecosystems under different land uses and the associated processes need to be taken into account by ESMs.

5 References

Ainsworth, E. A., Leakey, A. D. B., Ort, D. R. and Long, S. P. (2008). FACE-ing the facts: Inconsistencies and interdependence among field, chamber and modeling studies of elevated CO_2 impacts on crop yield and food supply. *New Phytologist* 179(1): 5-9.

Ainsworth, E. A. and Rogers, A. (2007). The response of photosynthesis and stomatal conductance to rising CO_2: Mechanisms and environmental interactions. *Plant, Cell and Environment* 30(3): 258-270.

Alexis, M. A., Rumpel, C., Knicker, H., Leifeld, J., Rasse, D. P., Péchot, N., Bardoux, G. and Mariotti, A. (2010). Thermal alteration of organic matter during a shrubland fire: A field study. *Organic Geochemistry* 41(7): 690-697.

Allen, C. D., Macalady, A. K., Chenchouni, H., Bachelet, D., McDowell, N., Vennetier, M., Kitzberger, T., Rigling, A., Breshears, D. D., Hogg, E. H., Gonzalez, P., Fensham, R., Zhang, Z., Castro, J., Demidova, N., Lim, J., Allard, G., Running, S. W., Semerci, A. and Cobb, N. (2010). A global overview of drought and heat-induced tree mortality reveals emerging climate change risks for forests. *Forest Ecology and Management* 259(4): 660-684.

Alvarez, R. and Alvarez, R. (2005). A review of nitrogen fertilizer and conservation tillage effects on soil organic carbon storage. *Soil Use and Management* 21(1): 38-52.

Anderegg, W. R. L., Schwalm, C., Biondi, F., Camarero, J. J., Koch, G., Litvak, M., Ogle, K., Shaw, J. D., Shevliakova, E., Williams, A. P., Wolf, A., Ziaco, E. and Pacala, S. (2015). Forest ecology. Pervasive drought legacies in forest ecosystems and their implications for carbon cycle models. *Science* 349(6247): 528-532.

Arora, V. K., Boer, G. J., Friedlingstein, P., Eby, M., Jones, C. D., Christian, J. R., Bonan, G., Bopp, L., Brovkin, V., Cadule, P., Hajima, T., Ilyina, T., Lindsay, K., Tjiputra, J. F. and Wu, T. (2013). Carbon-concentration and carbon-climate feedbacks in CMIP5 earth system models. *Journal of Climate* 26(15): 5289-5314.

Bader, M. K.-F., Leuzinger, S., Keel, S. G., Siegwolf, R. T. W., Hagedorn, F., Schleppi, P. and Körner, C. (2013). Central European hardwood trees in a high-CO_2 future: Synthesis of an 8-year forest canopy CO_2 enrichment project. *Journal of Ecology* 101(6): 1509-1519.

Bai, E., Li, S., Xu, W., Li, W., Dai, W. and Jiang, P. (2013). A meta-analysis of experimental warming effects on terrestrial nitrogen pools and dynamics. *New Phytologist* 199(2): 441-451.

Bala, G., Caldeira, K., Wickett, M., Phillips, T. J., Lobell, D. B., Delire, C. and Mirin, A. (2007). Combined climate and carbon-cycle effects of large-scale deforestation. *Proceedings of the National Academy of Sciences of the United States of America* 104(16): 6550-6555.

Balser, T. C. and Wixon, D. L. (2009). Investigating biological control over soil carbon temperature sensitivity. *Global Change Biology* 15(12): 2935-2949.

Barraclough, S. L. and Ghimire, K. B. (2000). *Agricultural Expansion and Tropical Deforestation: Poverty, International Trade and Land Use*. London and Sterling, VA: Earthscan Publications.

Batjes, N. H. (2019). Technologically achievable soil organic carbon sequestration in world croplands and grasslands. *Land Degradation and Development* 30(1): 25-32.

Beniston, M. (2005). Mountain climates and climatic change: An overview of processes focusing on the European Alps. *Pure and Applied Geophysics* 162(8-9): 1587-1606.

Beniston, M., Diaz, H. and Bradley, R. (1997). Climatic change at high elevation sites: An overview. *Climatic Change* 36(3): 233–251.

Berger, A. W., Valenca, R., Miao, Y., Ravi, S., Mahendra, S. and Mohanty, S. K. (2019). Biochar increases nitrate removal capacity of woodchip biofilters during high-intensity rainfall. *Water Research* 165: 115008.

Bonan, G. B. (2008). Forests and climate change: Forcings, feedbacks, and the climate benefits of forests. *Science* 320(5882): 1444–1449.

Borken, W., Xu, Y.-J., Brumme, R. and Lamersdorf, N. (1999). A climate change scenario for carbon dioxide and dissolved organic carbon fluxes from a temperate forest soil drought and rewetting effects. *Soil Science Society of America Journal* 63(6): 1848–1855.

Bounoua, L., DeFries, R., Collatz, G. J., Sellers, P. and Khan, H. (2002). Effects of land cover conversion on surface climate. *Climatic Change* 52(1): 29–64.

Bridgham, S. D., Ping, C.-L., Richardson, J. and Updegraff, K. (2000). Soils of northern peatlands: Histosols and gelisols. In: *Wetland Soils: Genesis, Hydrology, Landscapes and Classification*. Boca Raton, FL: CRC Press, pp. 343–370.

Brulle, R. J. (2018). The climate lobby: A sectoral analysis of lobbying spending on climate change in the USA, 2000 to 2016. *Climatic Change* 149(3–4): 289–303.

Burrell, A. L., Evans, J. P. and De Kauwe, M. G. (2020). Anthropogenic climate change has driven over 5 million km^2 of drylands towards desertification. *Nature Communications* 11(1): 3853.

Buys, P. (2007). At loggerheads?: Agricultural expansion, poverty reduction, and environment in the tropical forests.

Campbell, J. E., Berry, J. A., Seibt, U., Smith, S. J., Montzka, S. A., Launois, T., Belviso, S., Bopp, L. and Laine, M. (2017). Large historical growth in global terrestrial gross primary production. *Nature* 544(7648): 84–87.

Campo, J. and Merino, A. (2016). Variations in soil carbon sequestration and their determinants along a precipitation gradient in seasonally dry tropical forest ecosystems. *Global Change Biology* 22(5): 1942–1956.

Carney, K. M., Hungate, B. A., Drake, B. G. and Megonigal, J. P. (2007). Altered soil microbial community at elevated CO_2 leads to loss of soil carbon. *Proceedings of the National Academy of Sciences* 104(12): 4990–4995.

Cartwright, J. (2019). Ecological islands: Conserving biodiversity hotspots in a changing climate. *Frontiers in Ecology and the Environment* 17(6): 331–340.

Chapin, F. S., Shaver, G. R., Giblin, A. E., Nadelhoffer, K. J. and Laundre, J. A. (1995). Responses of arctic tundra to experimental and observed changes in climate. *Ecology* 76(3): 694–711.

Chen, H. Y. H. and Shrestha, B. M. (2012). Stand age, fire and clearcutting affect soil organic carbon and aggregation of mineral soils in boreal forests. *Soil Biology and Biochemistry* 50: 149–157.

Chen, Q., Niu, B., Hu, Y., Luo, T. and Zhang, G. (2020). Warming and increased precipitation indirectly affect the composition and turnover of labile-fraction soil organic matter by directly affecting vegetation and microorganisms. *Science of the Total Environment* 714: 136787.

Cheng, L., Zhang, N., Yuan, M., Xiao, J., Qin, Y., Deng, Y., Tu, Q., Xue, K., Van Nostrand, J. D., Wu, L., He, Z., Zhou, X., Leigh, M. B., Konstantinidis, K. T., Schuur, E. A., Luo, Y., Tiedje, J. M. and Zhou, J. (2017). Warming enhances old organic carbon decomposition through altering functional microbial communities. *The ISME Journal* 11(8): 1825–1835.

Cheng, W., Parton, W. J., Gonzalez-Meler, M. A., Phillips, R., Asao, S., McNickle, G. G., Brzostek, E. and Jastrow, J. D. (2014). Synthesis and modeling perspectives of rhizosphere priming. *New Phytologist* 201(1): 31–44.

Christensen, T. (1991). Arctic and sub-arctic soil emissions: Possible implications for global climate change. *Polar Record* 27(162): 205–210.

Conant, R. T., Ryan, M. G., Ågren, G. I., Birge, H. E., Davidson, E. A., Eliasson, P. E., Evans, S. E., Frey, S. D., Giardina, C. P., Hopkins, F. M., Hyvönen, R., Kirschbaum, M. U. F., Lavallee, J. M., Leifeld, J., Parton, W. J., Megan Steinweg, J., Wallenstein, M. D., Martin Wetterstedt, J. Å. and Bradford, M. A. (2011). Temperature and soil organic matter decomposition rates-synthesis of current knowledge and a way forward. *Global Change Biology* 17(11): 3392–3404.

Costa, A. C. and Soares, A. (2012). Local spatiotemporal dynamics of a simple aridity index in a region susceptible to desertification. *Journal of Arid Environments* 87: 8–18.

Cotrufo, M. F., Wallenstein, M. D., Boot, C. M., Denef, K. and Paul, E. (2013). The Microbial Efficiency-Matrix Stabilization (MEMS) framework integrates plant litter decomposition with soil organic matter stabilization: Do labile plant inputs form stable soil organic matter? *Global Change Biology* 19(4): 988–995.

Craine, J. M., Nippert, J. B., Elmore, A. J., Skibbe, A. M., Hutchinson, S. L. and Brunsell, N. A. (2012). Timing of climate variability and grassland productivity. *Proceedings of the National Academy of Sciences of the United States of America* 109(9): 3401–3405.

Daily, G. C. (2013). *Nature's Services: Societal Dependence on Natural Ecosystems (1997)*. New Haven, CT: Yale University Press.

Dargie, G. C., Lewis, S. L., Lawson, I. T., Mitchard, E. T., Page, S. E., Bocko, Y. E. and Ifo, S. A. (2017). Age, extent and carbon storage of the central Congo Basin peatland complex. *Nature* 542(7639): 86–90.

Dash, P. K., Bhattacharyya, P., Roy, K. S., Neogi, S. and Nayak, A. K. (2019). Environmental constraints' sensitivity of soil organic carbon decomposition to temperature, management practices and climate change. *Ecological Indicators* 107: 105644.

Davidson, E. A. and Janssens, I. A. (2006). Temperature sensitivity of soil carbon decomposition and feedbacks to climate change. *Nature* 440(7081): 165–173.

Delgado-Baquerizo, M., Maestre, F. T., Gallardo, A., Bowker, M. A., Wallenstein, M. D., Quero, J. L., Ochoa, V., Gozalo, B., García-Gómez, M., Soliveres, S., García-Palacios, P., Berdugo, M., Valencia, E., Escolar, C., Arredondo, T., Barraza-Zepeda, C., Bran, D., Carreira, J. A., Chaieb, M., Conceição, A. A., Derak, M., Eldridge, D. J., Escudero, A., Espinosa, C. I., Gaitán, J., Gatica, M. G., Gómez-González, S., Guzman, E., Gutiérrez, J. R., Florentino, A., Hepper, E., Hernández, R. M., Huber-Sannwald, E., Jankju, M., Liu, J., Mau, R. L., Miriti, M., Monerris, J., Naseri, K., Noumi, Z., Polo, V., Prina, A., Pucheta, E., Ramírez, E., Ramírez-Collantes, D. A., Romão, R., Tighe, M., Torres, D., Torres-Díaz, C., Ungar, E. D., Val, J., Wamiti, W., Wang, D. and Zaady, E. (2013). Decoupling of soil nutrient cycles as a function of aridity in global drylands. *Nature* 502(7473): 672–676.

Desjardins, R. L. and Riznek, R. (2000). Agricultural greenhouse gas budget. Report.

Drake, J. E., Gallet-Budynek, A., Hofmockel, K. S., Bernhardt, E. S., Billings, S. A., Jackson, R. B., Johnsen, K. S., Lichter, J., McCarthy, H. R., McCormack, M. L., Moore, D. J., Oren, R., Palmroth, S., Phillips, R. P., Pippen, J. S., Pritchard, S. G., Treseder, K. K., Schlesinger, W. H., Delucia, E. H. and Finzi, A. C. (2011). Increases in the flux of carbon below-ground stimulate nitrogen uptake and sustain the long-term enhancement of forest productivity under elevated CO_2. *Ecology Letters* 14(4): 349-357.

Dukes, J. S. and Hungate, B. A. (2002). Elevated carbon dioxide and litter decomposition in California annual grasslands: Which mechanisms matter? *Ecosystems* 5(2): 171-183.

Edwards, C. and Bohlen, P. (1996). *Biology and Ecology of Earthworms*. London: Chapman & Hall, p. 426.

Eggleton, P., Inward, K., Smith, J., Jones, D. T. and Sherlock, E. (2009). A six year study of earthworm (Lumbricidae) populations in pasture woodland in southern England shows their responses to soil temperature and soil moisture. *Soil Biology and Biochemistry* 41(9): 1857-1865.

Eswaran, H., Reich, P. and Kimble, J. (2000). Global carbon stocks. In: Lal, R. (Ed.), *Global Climate Change and Pedogenic Carbonates*. Boca Raton, FL: Lewis Publishers, pp. 15-26.

Farage, P., Pretty, J. and Ball, A. (2003). *Biophysical Aspects of Carbon Sequestration in Drylands*. University of Essex, UK.

Flannigan, M., Stocks, B., Turetsky, M. and Wotton, M. (2009). Impacts of climate change on fire activity and fire management in the circumboreal forest. *Global Change Biology* 15(3): 549-560.

Fontaine, S., Bardoux, G., Abbadie, L. and Mariotti, A. (2004). Carbon input to soil may decrease soil carbon content. *Ecology Letters* 7(4): 314-320.

Franck, V. M., Hungate, B. A., Chapin, F. S. and Field, C. B. (1997). Decomposition of litter produced under elevated CO_2: Dependence on plant species and nutrient supply. *Biogeochemistry* 36(3): 223-237.

Friedlingstein, P., Cox, P., Betts, R., Bopp, L., von Bloh, W., Brovkin, V., Cadule, P., Doney, S., Eby, M., Fung, I., Bala, G., John, J., Jones, C., Joos, F., Kato, T., Kawamiya, M., Knorr, W., Lindsay, K., Matthews, H. D., Raddatz, T., Rayner, P., Reick, C., Roeckner, E., Schnitzler, K. -G., Schnur, R., Strassmann, K., Weaver, A. J., Yoshikawa, C. and Zeng, N. (2006). Climate-carbon cycle feedback analysis: Results from the C4MIP model intercomparison. *Journal of Climate* 19(14): 3337-3353.

Frolking, S., Roulet, N. T., Moore, T. R., Richard, P. J. H., Lavoie, M. and Muller, S. D. (2001). Modeling northern peatland decomposition and peat accumulation. *Ecosystems* 4(5): 479-498.

Frouz, J., Novotná, K., Čermáková, L. and Pivokonský, M. (2020). Soil fauna reduce soil respiration by supporting N leaching from litter. *Applied Soil Ecology* 153: 103585.

Gentine, P., Green, J. K., Guérin, M., Humphrey, V., Seneviratne, S. I., Zhang, Y. and Zhou, S. (2019). Coupling between the terrestrial carbon and water cycles–A review. *Environmental Research Letters* 14(8): 083003.

Ghee, C., Neilson, R., Hallett, P. D., Robinson, D. and Paterson, E. (2013). Priming of soil organic matter mineralisation is intrinsically insensitive to temperature. *Soil Biology and Biochemistry* 66: 20-28.

Gifford, R. M., Cheney, N. P., Noble, J. C., Russell, J. S., Wellington, A. B. and Zammit, C. (1992). *Australian Land Use, Primary Production of Vegetation and Carbon Pools in Relation to Atmospheric Carbon Dioxide Levels* (vol. 14). Bureau of Rural Resources Proceeding, 14: 151-187.

Glenn, E., Squires, V., Olsen, M. and Frye, R. (1993). Potential for carbon sequestration in the drylands. In: *Terrestrial Biospheric Carbon Fluxes Quantification of Sinks and Sources of Co_2* 70(1-4): 341-355. Dordrecht: Springer.

Godfray, H. C., Pretty, J., Thomas, S. M., Warham, E. J. and Beddington, J. R. (2011). Linking policyon climate and food. *Science* 331(6020): 1013–1014.

Gorte, R. W. and Sheikh, P. A. (2010). *Deforestation and Climate Change*. Washington, DC: Congressional Research Service.

Gupta, A., Thapliyal, P., Pal, P. and Joshi, P. C. (2005). Impact of deforestation on Indian monsoon - A GCM sensitivity study. *J. Indian Geophys. Union* 9(2): 97–104.

Han, C.-L., Sun, Z.-X., Shao, S., Wang, Q., Libohova, Z. and Owens, P. R. (2021). Changes of soil organic carbon after wildfire in a boreal forest, northeast China. *Agronomy* 11(10): 1925.

Haworth, M., Moser, G., Raschi, A., Kammann, C., Grünhage, L. and Müller, C. (2015). Carbon dioxide fertilisation and supressed respiration induce enhanced spring biomass production in a mixed species temperate meadow exposed to moderate carbon dioxide enrichment. *Functional Plant Biology* 43(1): 26–39.

Hengeveld, H. (2000). *Projections for Canada's Climate Future - A Discussion of Recent Simulations With the Canadian Climate Model*. Meteorological Service of Canada, Minister of Public Works and Government Services Canada.

Heywood, V. H. and Watson, R. T. (1995). *Global Biodiversity Assessment*. Cambridge: Cambridge University Press.

Hoover, D. L., Knapp, A. K. and Smith, M. D. (2016). The immediate and prolonged effects of climate extremes on soil respiration in a mesic grassland. *Journal of Geophysical Research: Biogeosciences* 121(4): 1034–1044.

Hopkins, J., Allison, H., Walmsley, C., Gaywood, M. and Thurgate, G. (2007). *Conserving Biodiversity in a Changing Climate: Guidance on Building Capacity to Adapt*. London: Defra, p. 26.

Hosseini Bai, S., Blumfield, T. J., Xu, Z., Chen, C. and Wild, C. (2012). Soil organic matter dynamics and nitrogen availability in response to site preparation and management during revegetation in tropical Central Queensland, Australia. *Journal of Soils and Sediments* 12(3): 386–395.

Houghton, J. T. and Meira, L. G. (1996). 1995: The science of climate change: Contribution of working group I to the second assessment report of the Intergovernmental Panel on Climate Change. In: *Climate Change*. Cambridge: Cambridge University Press.

Houghton, R. A. (2005). Above-ground forest biomass and the global carbon balance. *Global Change Biology* 11(6): 945–958.

Houghton, R. A., Ding, Y. and Griggs, D. J. (2001). The scientific basis. In: *Climatic Change* 48(4): 525–534. Cambridge, UK: The Press Syndicate of the University of Cambridge.

Hovenden, M. J., Leuzinger, S., Newton, P. C. D., Fletcher, A., Fatichi, S., Lüscher, A., Reich, P. B., Andresen, L. C., Beier, C., Blumenthal, D. M., Chiariello, N. R., Dukes, J. S., Kellner, J., Hofmockel, K., Niklaus, P. A., Song, J., Wan, S., Classen, A. T. and Langley, J. A. (2019). Globally consistent influences of seasonal precipitation limit grassland biomass response to elevated CO_2. *Nature Plants* 5(2): 167–173.

Hui, D., Deng, Q., Tian, H. and Luo, Y. (2017). Climate change and carbon sequestration in forest ecosystems. In: Chan, W. C. (Ed.), *Handbook of Climate Change Mitigation and Adaptation*, pp. 1–40. Springer, Cham. https://doi.org/10.1007/978-3-319-14409-2_13.

Hungate, B. A., Dijkstra, P., Wu, Z., Duval, B. D., Day, F. P., Johnson, D. W., Megonigal, J. P., Brown, A. L. P. and Garland, J. L. (2013). Cumulative response of ecosystem carbon and nitrogen stocks to chronic CO_2 exposure in a subtropical oak woodland. *New Phytologist* 200(3): 753–766.

Hungate, B. A., Van Groenigen, K. J., Six, J., Jastrow, J. D., Luo, Y., de Graaff, M., van Kessel, C. and Osenberg, C. W. (2009). Assessing the effect of elevated carbon dioxide on soil carbon: A comparison of four meta-analyses. *Global Change Biology* 15(8): 2020-2034.

Iizumi, T. and Wagai, R. (2019). Leveraging drought risk reduction for sustainable food, soil and climate via soil organic carbon sequestration. *Scientific Reports* 9(1): 19744.

IPCC. (2007). The physical science basis. Contribution of Working Group I to the Fourth Assessment Report of the Intergovernmental Panel on Climate Change. Summary for policymakers. In: Solomon, S., Qin, D., Manning, M., Chen, Z., Marquis, M., Averyt, K. B., Tignor, M. and Miller, H. L. (Eds.), *Climate Change*. Cambridge, UK and New York, NY: Cambridge University Press.

IPCC. (2020). *Climate Change and Land: An IPCC Special Report on Climate Change, Desertification, Land Degradation, Sustainable Land Management, Food Security, and Greenhouse Gas Fluxes in Terrestrial Ecosystems*.

IPCC. (2021). The physical science basis. Contribution of working group I to the sixth assessment report of the Intergovernmental Panel on Climate Change. In: Masson-Delmotte, V., Zhai, P., Pirani, A., Connors, S. L., Péan, C., Berger, S., Caud, N., Chen, Y., Goldfarb, L. and Gomis, M. I. (Eds.), *Climate Change*. Cambridge: Cambridge University Press.

Jackson, R. B., Banner, J. L., Jobbágy, E. G., Pockman, W. T. and Wall, D. H. (2002). Ecosystem carbon loss with woody plant invasion of grasslands. *Nature* 418(6898): 623-626.

Jiang, M., Medlyn, B. E., Drake, J. E., Duursma, R. A., Anderson, I. C., Barton, C. V. M., Boer, M. M., Carrillo, Y., Castañeda-Gómez, L., Collins, L., Crous, K. Y., De Kauwe, M. G., Dos Santos, B. M., Emmerson, K. M., Facey, S. L., Gherlenda, A. N., Gimeno, T. E., Hasegawa, S., Johnson, S. N., Kännaste, A., Macdonald, C. A., Mahmud, K., Moore, B. D., Nazaries, L., Neilson, E. H. J., Nielsen, U. N., Niinemets, Ü, Noh, N. J., Ochoa-Hueso, R., Pathare, V. S., Pendall, E., Pihlblad, J., Piñeiro, J., Powell, J. R., Power, S. A., Reich, P. B., Renchon, A. A., Riegler, M., Rinnan, R., Rymer, P. D., Salomón, R. L., Singh, B. K., Smith, B., Tjoelker, M. G., Walker, J. K. M., Wujeska-Klause, A., Yang, J., Zaehle, S. and Ellsworth, D. S. (2020). The fate of carbon in a mature forest under carbon dioxide enrichment. *Nature* 580(7802): 227-231.

Kalnay, E. and Cai, M. (2003). Impact of urbanization and land-use change on climate. *Nature* 423(6939): 528-531.

Kindermann, G., Obersteiner, M., Sohngen, B., Sathaye, J., Andrasko, K., Rametsteiner, E., Schlamadinger, B., Wunder, S. and Beach, R. (2008). Global cost estimates of reducing carbon emissions through avoided deforestation. *Proceedings of the National Academy of Sciences of the United States of America* 105(30): 10302-10307.

Kirschbaum, M. U. F. (2004). Soil respiration under prolonged soil warming: Are rate reductions caused by acclimation or substrate loss? *Global Change Biology* 10(11): 1870-1877.

Knorr, W., Prentice, I. C., House, J. I. and Holland, E. A. (2005). Long-term sensitivity of soil carbon turnover to warming. *Nature* 433(7023): 298-301.

Koutroulis, A. G. (2019). Dryland changes under different levels of global warming. *Science of the Total Environment* 655: 482-511.

Kurz, W. A., Dymond, C. C., Stinson, G., Rampley, G. J., Neilson, E. T., Carroll, A. L., Ebata, T. and Safranyik, L. (2008). Mountain Pine beetle and forest carbon feedback to climate change. *Nature* 452(7190): 987-990.

Kuzyakov, Y., Friedel, J. K. and Stahr, K. (2000). Review of mechanisms and quantification of priming effects. *Soil Biology and Biochemistry* 32(11-12): 1485-1498.

Kuzyakov, Y., Horwath, W. R., Dorodnikov, M. and Blagodatskaya, E. (2019). Review and synthesis of the effects of elevated atmospheric CO_2 on soil processes: No changes in pools, but increased fluxes and accelerated cycles. *Soil Biology and Biochemistry* 128: 66-78.

Lal, R. (1999). Soil management and restoration for C sequestration to mitigate the accelerated greenhouse effect. *Progress in Environmental Science* 1(4): 307-326.

Lal, R. (2001). Potential of desertification control to sequester carbon and mitigate the greenhouse effect. *Climatic Change* 51(1): 35-72.

Lal, R. (2004). Soil carbon sequestration impacts on global climate change and food security. *Science* 304(5677): 1623-1627.

Lal, R. (2010). Managing soils and ecosystems for mitigating anthropogenic carbon emissions and advancing global food security. *BioScience*. *BioScience* 60(9): 708-721.

Lal, R. (2019). Carbon cycling in global drylands. *Current Climate Change Reports* 5(3): 221-232.

Lal, R., Follett, R. F., Stewart, B. A. and Kimble, J. M. (2007). Soil carbon sequestration to mitigate climate change and advance food security. *Soil Science* 172(12): 943-956.

Larcher, W. (2003). *Physiological Plant Ecology: Ecophysiology and Stress Physiology of Functional Groups*. Heidelberg: Springer-Verlag.

Lellei-Kovács, E., Kovács-Láng, E., Kalapos, T., Botta-Dukát, Z., Barabás, S. and Beier, C. (2008). Experimental warming does not enhance soil respiration in a semiarid temperate forest-steppe ecosystem. *Community Ecology* 9(1): 29-37.

Li, C., Zhang, C., Luo, G., Chen, X., Maisupova, B., Madaminov, A. A., Han, Q. and Djenbaev, B. M. (2015). Carbon stock and its responses to climate change in Central Asia. *Global Change Biology* 21(5): 1951-1967.

Lindner, M., Maroschek, M., Netherer, S., Kremer, A., Barbati, A., Garcia-Gonzalo, J., Seidl, R., Delzon, S., Corona, P., Kolström, M., Lexer, M. J. and Marchetti, M. (2010). Climate change impacts, adaptive capacity, and vulnerability of European forest ecosystems. *Forest Ecology and Management* 259(4): 698-709.

Liu, H., Mi, Z., Lin, L., Wang, Y., Zhang, Z., Zhang, F., Wang, H., Liu, L., Zhu, B., Cao, G., Zhao, X., Sanders, N. J., Classen, A. T., Reich, P. B. and He, J. S. (2018). Shifting plant species composition in response to climate change stabilizes grassland primary production. *Proceedings of the National Academy of Sciences of the United States of America* 115(16): 4051-4056.

Liu, L., Wang, X., Lajeunesse, M. J., Miao, G., Piao, S., Wan, S., Wu, Y., Wang, Z., Yang, S., Li, P. and Deng, M. (2016). A cross-biome synthesis of soil respiration and its determinants under simulated precipitation changes. *Global Change Biology* 22(4): 1394-1405.

Lobell, D. B., Schlenker, W. and Costa-Roberts, J. (2011). Climate trends and global crop production since 1980. *Science* 333(6042): 616-620.

Ma, W., Li, Z., Ding, K., Huang, B., Nie, X., Lu, Y. and Xiao, H. (2016). Soil erosion, organic carbon and nitrogen dynamics in planted forests: A case study in a hilly catchment of Hunan Province, China. *Soil and Tillage Research* 155: 69-77.

Magnani, F., Mencuccini, M., Borghetti, M., Berbigier, P., Berninger, F., Delzon, S., Grelle, A., Hari, P., Jarvis, P. G., Kolari, P., Kowalski, A. S., Lankreijer, H., Law, B. E., Lindroth, A., Loustau, D., Manca, G., Moncrieff, J. B., Rayment, M., Tedeschi, V., Valentini, R. and

Grace, J. (2007). The human footprint in the carbon cycle of temperate and boreal forests. *Nature* 447(7146): 848–850.

Manzoni, S., Taylor, P., Richter, A., Porporato, A. and Ågren, G. I. (2012). Environmental and stoichiometric controls on microbial carbon-use efficiency in soils. *New Phytologist* 196(1): 79–91.

Marinov, I., Doney, S. C. and Lima, I. D. (2010). Response of ocean phytoplankton community structure to climate change over the 21st century: Partitioning the effects of nutrients, temperature and light. *Biogeosciences* 7(12): 3941–3959.

Mather, A. S. (2004). Forest transition theory and the reforesting of Scotland. *Scottish Geographical Journal* 120(1-2): 83-98.

Metcalfe, D. B., Fisher, R. A. and Wardle, D. A. (2011). Plant communities as drivers of soil respiration: Pathways, mechanisms, and significance for global change. *Biogeosciences* 8(8): 2047-2061.

Middleton, N. J. and Sternberg, T. (2013). Climate hazards in drylands: A review. *Earth-Science Reviews* 126: 48-57.

Millennium Ecosystem Assessment. (2005). *Millennium Ecosystem Assessment. Ecosystems and Human Well-Being: Desertification Synthesis*. Report.

Miltner, A., Bombach, P., Schmidt-Brücken, B. and Kästner, M. (2012). SOM genesis: Microbial biomass as a significant source. *Biogeochemistry* 111(1-3): 41-55.

Mitchell, S. J. (2012). Wind as a natural disturbance agent in forests: A synthesis. *International Journal of Forestry Research* 86(2): 147-157.

Moinet, G. Y. K., Moinet, M., Hunt, J. E., Rumpel, C., Chabbi, A. and Millard, P. (2020). Temperature sensitivity of decomposition decreases with increasing soil organic matter stability. *Science of the Total Environment* 704: 135460.

Moore, T. R. and Dalva, M. (1997). Methane and carbon dioxide exchange potentials of peat soils in aerobic and anaerobic laboratory incubations. *Soil Biology and Biochemistry* 29(8): 1157-1164.

Moore, T. R., Trofymow, J. A., Taylor, B., Prescott, C., Camiré, C., Duschene, L., Fyles, J., Kozak, L., Kranabetter, M., Morrison, I., Siltanen, M., Smith, S., Titus, B., Visser, S., Wein, R. and Zoltai, S. (1999). Litter decomposition rates in Canadian forests. *Global Change Biology* 5(1): 75-82.

Muluneh, A. (2020). Impact of climate change on soil water balance, maize production, and potential adaptation measures in the Rift Valley drylands of Ethiopia. *Journal of Arid Environments* 179: 104195.

Munjonji, L., Ayisi, K. K., Mudongo, E. I., Mafeo, T. P., Behn, K., Mokoka, M. V. and Linstädter, A. (2020). Disentangling drought and grazing effects on soil carbon stocks and CO_2 fluxes in a semi-arid African savanna. *Frontiers in Environmental Science* 8: 207.

Naipal, V., Ciais, P., Wang, Y., Lauerwald, R., Guenet, B. and Van Oost, K. (2018). Global soil organic carbon removal by water erosion under climate change and land use change during AD 1850-2005. *Biogeosciences* 15(14): 4459-4480.

Neill, C. and Gignoux, J. (2006). Soil organic matter decomposition driven by microbial growth: A simple model for a complex network of interactions. *Soil Biology and Biochemistry* 38(4): 803-811.

Norby, R. J., Warren, J. M., Iversen, C. M., Medlyn, B. E. and McMurtrie, R. E. (2010). CO_2 enhancement of forest productivity constrained by limited nitrogen availability. *Proceedings of the National Academy of Sciences of the United States of America* 107(45): 19368-19373.

Norby, R. J. and Zak, D. R. (2011). Ecological lessons from free-air CO_2 enrichment (FACE) experiments. *Annual Review of Ecology, Evolution, and Systematics* 42(1): 181-203.
Nowak, R. S., Ellsworth, D. S. and Smith, S. D. (2004). Functional responses of plants to elevated atmospheric CO_2 - Do photosynthetic and productivity data from FACE experiments support early predictions? *New Phytologist* 162(2): 253-280.
Olson, K. R. (2013). Soil organic carbon sequestration, storage, retention and loss in U.S. croplands: Issues paper for protocol development. *Geoderma* 195-196: 201-206.
Ostle, N. J., Smith, P., Fisher, R., Ian Woodward, F., Fisher, J. B., Smith, J. U., Galbraith, D., Levy, P., Meir, P., McNamara, N. P. and Bardgett, R. D. (2009). Integrating plant-soil interactions into global carbon cycle models. *Journal of Ecology* 97(5): 851-863.
Pang, X., Zhu, B., Lü, X. and Cheng, W. (2015). Labile substrate availability controls temperature sensitivity of organic carbon decomposition at different soil depths. *Biogeochemistry* 126(1-2): 85-98.
Paterson, E., Thornton, B., Midwood, A. J., Osborne, S. M., Sim, A. and Millard, P. (2008). Atmospheric CO_2 enrichment and nutrient additions to planted soil increase mineralisation of soil organic matter, but do not alter microbial utilisation of plant- and soil C-sources. *Soil Biology and Biochemistry* 40(9): 2434-2440.
Pautasso, M., Döring, T. F., Garbelotto, M., Pellis, L. and Jeger, M. J. (2012). Impacts of climate change on plant diseases–Opinions and trends. *European Journal of Plant Pathology* 133(1): 295-313.
Perreault, J. M. and Whalen, J. K. (2006). Earthworm burrowing in laboratory microcosms as influenced by soil temperature and moisture. *Pedobiologia* 50(5): 397-403.
Peters, E. B., Wythers, K. R., Zhang, S., Bradford, J. B. and Reich, P. B. (2013). Potential climate change impacts on temperate forest ecosystem processes. *Canadian Journal of Forest Research* 43(10): 939-950.
Phillips, R. P., Finzi, A. C. and Bernhardt, E. S. (2011). Enhanced root exudation induces microbial feedbacks to N cycling in a pine forest under long-term CO_2 fumigation. *Ecology Letters* 14(2): 187-194.
Pimentel, D. and Burgess, M. (2013). Soil erosion threatens food production. *Agriculture* 3(3): 443-463.
Pingali, P. L. (2012). Green revolution: Impacts, limits, and the path ahead. *Proceedings of the National Academy of Sciences of the United States of America* 109(31): 12302-12308.
Prietzel, J. and Christophel, D. (2014). Organic carbon stocks in forest soils of the German Alps. *Geoderma* 221-222: 28-39.
Quiroga, A. R., Buschiazzo, D. E. and Peinemann, N. (1996). Soil organic matter particle size fractions in soils of the semiarid Argentinian Pampas. *Soil Science* 161(2): 104-108.
Reich, P. B., Hobbie, S. E., Ellsworth, D. S., West, J. B., Tilman, D., Knops, J. M. H., Naeem, S. and Trost, J. (2006). Nitrogen limitation constrains sustainability of ecosystem response to CO_2. *Nature Letters* 440(7086): 922-925. doi: 10.1038/nature004486.
Reich, P. B., Hobbie, S. E. and Lee, T. D. (2014). Plant growth enhancement by elevated CO_2 eliminated by joint water and nitrogen limitation. *Nature Geoscience* 7(12): 920-924.
Rein, G. and Huang, X. (2021). Smouldering wildfires in peatlands, forests and the arctic: Challenges and perspectives. *Current Opinion in Environmental Science and Health* 24: None.
Ren, W., Tian, H., Tao, B., Huang, Y. and Pan, S. (2012). China's crop productivity and soil carbon storage as influenced by multifactor global change. *Global Change Biology* 18(9): 2945-2957.

Ren, W., Banger, K., Tao, B., Yang, J., Huang, Y. and Tian, H. (2020) Global pattern and change of cropland soil organic carbon during 1901-2010: Roles of climate, atmospheric chemistry, land use and management. *Geography and Sustainability* 1(1): 59-69.

Rumpel, C., Amiraslani, F., Chenu, C., Garcia Cardenas, M., Kaonga, M., Koutika, L. S., Ladha, J., Madari, B., Shirato, Y., Smith, P., Soudi, B., Soussana, J. F., Whitehead, D. and Wollenberg, E. (2020). The 4p1000 initiative: Opportunities, limitations and challenges for implementing soil organic carbon sequestration as a sustainable development strategy. *Ambio* 49(1): 350-360.

Rumpel, C. and Chabbi, A. (2021). Managing soil organic carbon for mitigating climate change and increasing food security. *Agronomy* 11(8): 1553.

Rustad, L., Campbell, J., Marion, G., Norby, R., Mitchell, M., Hartley, A., Cornelissen, J., Gurevitch, J. and GCTE-NEWS (2001). A meta-analysis of the response of soil respiration, net nitrogen mineralization, and above-ground plant growth to experimental ecosystem warming. *Oecologia* 126(4): 543-562.

Saad, C., Boulanger, Y., Beaudet, M., Gachon, P., Ruel, J. and Gauthier, S. (2017). Potential impact of climate change on the risk of windthrow in eastern Canada's forests. *Climatic Change* 143(3-4): 487-501.

Sanaullah, M., Chabbi, A., Girardin, C., Durand, J., Poirier, M. and Rumpel, C. (2014). Effects of drought and elevated temperature on biochemical composition of forage plants and their impact on carbon storage in grassland soil. *Plant and Soil* 374(1-2): 767-778.

Sanaullah, M., Chabbi, A., Rumpel, C. and Kuzyakov, Y. (2012). Carbon allocation in grassland communities under drought stress followed by ^{14}C pulse labeling. *Soil Biology and Biochemistry* 55: 132-139.

Scharlemann, J. P., Tanner, E. V., Hiederer, R. and Kapos, V. (2014). Global soil carbon: Understanding and managing the largest terrestrial carbon pool. *Carbon Management* 5(1): 81-91.

Schimel, D. S. (2010). Climate: Drylands in the earth system. *Science* 327(5964): 418-419.

Schlesinger, W. H. (1977). Carbon balance in terrestrial detritus. *Annual Review of Ecology and Systematics* 8(1): 51-81.

Schmidt, M. W., Torn, M. S., Abiven, S., Dittmar, T., Guggenberger, G., Janssens, I. A., Kleber, M., Kögel-Knabner, I., Lehmann, J., Manning, D. A., Nannipieri, P., Rasse, D. P., Weiner, S. and Trumbore, S. E. (2011). Persistence of soil organic matter as an ecosystem property. *Nature* 478(7367): 49-56.

Seibert, R., Donath, T. W., Moser, G., Laser, H., Grünhage, L., Schmid, T. and Müller, C. (2021). Effects of long-term CO_2 enrichment on forage quality of extensively managed temperate grassland. *Agriculture, Ecosystems and Environment* 312: 107347.

Sharma, P., Abrol, V., Abrol, S. and Kumar, R. (2012). Climate change and carbon sequestration in dryland soils. In: *Resource Management for Sustainable Agriculture*, pp. 139-164. IntechOpen, London.

Shrestha, R. K., Ladha, J. K. and Lefroy, R. D. (2002). Carbon management for sustainability of an intensively managed rice-based cropping system. *Biology and Fertility of Soils* 36(3): 215-223.

Somero, G. N. (1995). Proteins and temperature. *Annual Review of Physiology* 57(1): 43-68.

Song, X. P., Hansen, M. C., Stehman, S. V., Potapov, P. V., Tyukavina, A., Vermote, E. F. and Townshend, J. R. (2018). Global land change from 1982 to 2016. *Nature* 560(7720): 639-643.

Stern, D. I. and Kaufmann, R. K. (2014). Anthropogenic and natural causes of climate change. *Climatic Change* 122(1-2): 257-269.

Svejcar, T. J. and Browning, J. A. (1988). Growth and gas exchange of Andropogon gerardii as influenced by burning. *Journal of Range Management/Journal of Range Management Archives* 41(3): 239-244.

Talhelm, A. F., Pregitzer, K. S., Kubiske, M. E., Zak, D. R., Campany, C. E., Burton, A. J., Dickson, R. E., Hendrey, G. R., Isebrands, J. G., Lewin, K. F., Nagy, J. and Karnosky, D. F. (2014). Elevated carbon dioxide and ozone alter productivity and ecosystem carbon content in northern temperate forests. *Global Change Biology* 20(8): 2492-2504.

Talhelm, A. F., Pregitzer, K. S. and Zak, D. R. (2009). Species-specific responses to atmospheric carbon dioxide and tropospheric ozone mediate changes in soil carbon. *Ecology Letters* 12(11): 1219-1228.

Tarnocai, C. (2006). The effect of climate change on carbon in Canadian peatlands. *Global and Planetary Change* 53(4): 222-232.

Tarnocai, C., Canadell, J. G., Schuur, E. A. G., Kuhry, P., Mazhitova, G. and Zimov, S. (2009). Soil organic carbon pools in the northern circumpolar permafrost region. *Global Biogeochemical Cycles* 23(2): 1-11.

Thürig, E., Palosuo, T., Bucher, J. and Kaufmann, E. (2005). The impact of windthrow on carbon sequestration in Switzerland: A model-based assessment. *Forest Ecology and Management* 210(1-3): 337-350.

Tian, Y., Liu, J., Zhang, X. and Gao, L. (2010). Effects of summer catch crop, residue management, soil temperature and water on the succeeding cucumber rhizosphere nitrogen mineralization in intensive production systems. *Nutrient Cycling in Agroecosystems* 88(3): 429-446.

Tietjen, B., Jeltsch, F., Zehe, E., Classen, N., Groengroeft, A., Schiffers, K. and Oldeland, J. (2010). Effects of climate change on the coupled dynamics of water and vegetation in drylands. *Ecohydrology and Hydrobiology* 3(2): 226-237.

Tietjen, B., Schlaepfer, D. R., Bradford, J. B., Lauenroth, W. K., Hall, S. A., Duniway, M. C., Hochstrasser, T., Jia, G., Munson, S. M., Pyke, D. A. and Wilson, S. D. (2017). Climate change-induced vegetation shifts lead to more ecological droughts despite projected rainfall increases in many global temperate drylands. *Global Change Biology* 23(7): 2743-2754.

Todd-Brown, K. E. O., Randerson, J. T., Hopkins, F., Arora, V., Hajima, T., Jones, C., Shevliakova, E., Tjiputra, J., Volodin, E., Wu, T., Zhang, Q. and Allison, S. D. (2014). Changes in soil organic carbon storage predicted by Earth system models during the 21st century. *Biogeosciences* 11(8): 2341-2356.

Van Groenigen, K. J., Qi, X., Osenberg, C. W., Luo, Y. and Hungate, B. A. (2014). Faster decomposition under increased atmospheric CO_2 limits soil carbon storage. *Science* 344(6183): 508-509.

van Groenigen, K. J., Six, J., Hungate, B. A., de Graaff, M. A., van Breemen, N. and van Kessel, C. (2006). Element interactions limit soil carbon storage. *Proceedings of the National Academy of Sciences of the United States of America* 103(17): 6571-6574.

Vörösmarty, C. J. and Sahagian, D. (2000). Anthropogenic disturbance of the terrestrial water cycle. *BioScience* 50(9): 753-765.

Walker, X. J., Baltzer, J. L., Cumming, S. G., Day, N. J., Ebert, C., Goetz, S., Johnstone, J. F., Potter, S., Rogers, B. M., Schuur, E. A. G., Turetsky, M. R. and Mack, M. C. (2019). Increasing wildfires threaten historic carbon sink of boreal forest soils. *Nature* 572(7770): 520-523.

Wan, Y., Lin, E., Xiong, W., Li, Y. and Guo, L. (2011). Modeling the impact of climate change on soil organic carbon stock in upland soils in the 21st century in China. *Agriculture, Ecosystems and Environment* 141(1-2): 23-31.

Wenzel, S., Cox, P. M., Eyring, V. and Friedlingstein, P. (2016). Projected land photosynthesis constrained by changes in the seasonal cycle of atmospheric CO_2. *Nature* 538(7626): 499-501.

Wever, L. A., Lysyk, T. J. and Clapperton, M. J. (2001). The influence of soil moisture and temperature on the survival, aestivation, growth and development of juvenile Aporrectodea tuberculata (Eisen)(Lumbricidae). *Pedobiologia* 45(2): 121-133.

Wieder, W. R., Grandy, A. S., Kallenbach, C. M. and Bonan, G. B. (2014). Integrating microbial physiology and physio-chemical principles in soils with the MIcrobial-MIneral Carbon Stabilization (MIMICS) model. *Biogeosciences* 11(14): 3899-3917.

Wienhold, B. J., Vigil, M. F., Hendrickson, J. R. and Derner, J. D. (2018). Vulnerability of crops and croplands in the US Northern Plains to predicted climate change. *Climatic Change* 146(1-2): 219-230.

Wolters, V. (2000). Invertebrate control of soil organic matter stability. *Biology and Fertility of Soils* 31(1): 1-19.

Wu, Z., Dijkstra, P., Koch, G. W., Peñuelas, J. and Hungate, B. A. (2011). Responses of terrestrial ecosystems to temperature and precipitation change: A meta-analysis of experimental manipulation. *Global Change Biology* 17(2): 927-942.

Xiong, X., Grunwald, S., Myers, D. B., Ross, C. W., Harris, W. G. and Comerford, N. B. (2014). Interaction effects of climate and land use/land cover change on soil organic carbon sequestration. *Science of the Total Environment* 493: 974-982.

Xu, Z., Wan, C., Xiong, P., Tang, Z., Hu, R., Cao, G. and Liu, Q. (2010). Initial responses of soil CO_2 efflux and C, N pools to experimental warming in two contrasting forest ecosystems, Eastern Tibetan Plateau, China. *Plant and Soil* 336(1-2): 183-195.

Yang, Y., Fang, J., Ji, C., Ma, W., Mohammat, A., Wang, S., Wang, S., Datta, A., Robinson, D. and Smith, P. (2012). Widespread decreases in topsoil inorganic carbon stocks across China's grasslands during 1980s-2000s. *Global Change Biology* 18(12): 3672-3680.

Yuan, N., Moser, G., Mueller, C., Obermeier, W. A., Bendix, J. and Luterbacher, J. (2018). Extreme climatic events down-regulate the grassland biomass response to elevated carbon dioxide. *Scientific Reports* 8(1): 17758.

Yude, P., Richard, A.B., Jingyun, F., Pan, Y., Birdsey, R. A., Fang, J., Houghton, R., Kauppi, P. E., Kurz, W. A., Phillips, O. L., Shvidenko, A., Lewis, S. L., Canadell, J. G., Ciais, P., Jackson, R. B., Pacala, S. W., McGuire, A. D., Piao, S., Rautiainen, A., Sitch, S. and Hayes, D. (2011). A large and persistent carbon sink in the world's forests. *Science* 333(6045): 988--993.

Zhou, X., Xu, X., Zhou, G. and Luo, Y. (2018). Temperature sensitivity of soil organic carbon decomposition increased with mean carbon residence time: Field incubation and data assimilation. *Global Change Biology* 24(2): 810-822.

Chapter 9

Innovative agriculture management to foster soil organic carbon sequestration

María de la Luz Mora, Jorge Medina, Patricia Poblete-Grant, Rolando Demanet, Paola Durán, Patricio Barra, Cecilia Paredes and Marcela Calabi-Floody, Universidad de La Frontera, Chile*

1 Introduction

The largest carbon (C) pool in terrestrial ecosystems is in soils (Batjes, 1996) which contain approximately 75% of the total terrestrial C pool, three times the amount of atmospheric CO_2. Soils are dynamic reservoirs of organic and inorganic C, they can act as a source or a sink for the atmospheric CO_2 (Lal, 2010), and play a key role in the global C cycle (Schlesinger, 1986). To adequately manage this reservoir, we need to understand the key knowledge gaps in understanding soil C turnover processes and management options affecting these processes (Paustian et al., 2016). Here lies the importance of visualizing the soil systems as a 'Natural soil bioreactor', taking into account different factors that could affect soil C turnover (Chapters 2 and 5 of this book), such as:

* Marcela Calabi-Floody is the corresponding author

http://dx.doi.org/10.19103/AS.2022.0106.30

- its allocation to specific fractions (particle size, density or aggregates), and organo-mineral associations (SOM-Al/Fe; SOM-Ca complexes);
- microbial processes, particularly the MCP (MCP; Liang et al., 2017);
- the combined effects of these two factors.

These factors may be altered by soil management and, in particular, fertilizer input which may affect the accessibility of nutrients (especially N) and soil reactions. If they are provided as organic amendments, fertilizers may also directly affect soil organic matter (SOM) content, texture, and porosity, among other physicochemical parameters, thereby influencing the structure and functionality of the microbiome. External inputs may offer an opportunity to influence the soil–plant microbiome to enhance SOC sequestration (Paustian et al., 2016; Lal, 2018; Bai et al., 2019; Chowdhury et al., 2020; Liang and Zhu, 2021).

However, since the green revolution, intensive agricultural management using chemical fertilizers, herbicides, tillage, and other energy-based inputs have been used to achieve high yields but with a long-term reduction in the quality of soils. Soil degradation has a detrimental impact on the C, N, and phosphorus (P) cycles, with for example losses of 1.16 tons of SOC, 52.3 kg of N, and 0.28 kg of P per hectare in arable land in the UK (Muhammed et al., 2018). According to Kopittke et al. (2019), cropping for several years reduces SOC stocks by 30% to 60%, while N, P, and sulfur stocks decrease by 42%, 27%, and 33%, respectively. SOC losses will inevitably reduce soil fertility, microbial biomass, soil aggregate formation, and stability, resulting in decreased crop production, and therefore decreased soil C sink capacity (Obalum et al., 2017), and increased greenhouse gas (GHG) emissions (Lal, 2015). Nutrient exportation through plant harvesting without residue incorporation is a significant issue that contributes to chemical soil degradation and the risk of soil nutrient depletion (Bashagaluke et al., 2018).

Continuous applications of inorganic fertilizers are now required to maintain a high crop yield. Different types of fertilizers may have contrasting effects on SOC. Li et al. (2017) demonstrated that inorganic fertilizer application, although promoting C inputs from crops, led to SOC stock decline after 12 years whereas, in soils with manure inputs, SOC levels were maintained. Depending on soil type (see Chapter 2 of this book), restoring organic matter to account for harvesting losses may be necessary to maintain SOC stocks in agricultural soils. While organic amendments may be used for this purpose, they may not lead to additional SOC sequestration as, in most cases, only translocation of C input into soils occurs. However, SOC sequestration can be achieved indirectly by enhancement of soil productivity and consequently higher (plant-derived) organic matter inputs. Soil type strongly influences the effect of organic matter additions through specific soil reactions, physicochemical, and biological properties (Poblete-Grant et al., 2019). These should be considered before the incorporation of any kind of amendment.

In this chapter we present the novel concept of soil as a 'natural bioreactor' and the factors influencing its functionality (Fig. 1) by using Andisols as a model (for assessing other soil groups). We also discuss how organic fertilizers may be used to manage the natural soil bioreactor to enhance its SOC sequestration potential. We also discuss innovations in organic amendment production to increase their stability and nutritional quality.

We suggest that sustainable management of organic amendments can promote enhanced SOC sequestration and improve nutrient use efficiency via the soil microbial biomass. We also argue that the transformation of organic residues into more stable amendments, and their utilization as smart fertilizers to maximize nutrient uptake and SOC sequestration, is key to meeting the increasing demand for food worldwide while mitigating GHG emissions and optimizing agroecosystem function.

2 Microbial activity and biogeochemical cycling in soils: the concept of the 'natural soil bioreactor'

The concept of soil as a 'natural bioreactor' refers to the interaction of organic matter (OM)-mineral soil complexes and microbiome associated in a soil-plant-microorganism continuum (Fig. 1). These factors interact to enhance SOC sequestration via the MCP (Liang et al., 2017).

The dynamics of SOC are mediated by microorganisms which may affect both SOC stabilization and destabilization (Liang et al., 2017, 2020). As biogeochemical cycles are ultimately a consequence of bacterial growth and activity, the mechanistic basis for SOM decomposition, transformation, and stabilization in soils depends on a detailed understanding of the bacterial microbiome composition and its functionality (Liang and Balser, 2012). To this end, the recently proposed concept of the MCP emphasizes the importance of the soil microbiome metabolism in the production, stabilization and sequestration of SOC (Liang and Balser, 2012; Liang and Zhu, 2021). The MCP states that bacteria can play a dominant role in 'pumping' bio-available C into a pool of persistent organic molecules (Liang et al., 2017). The MCP transports microbially produced chemicals into soil, where they can be maintained by physicochemical interactions with soil minerals or aggregates, known as the 'entombing effect'. Liang (2020) suggests that MCP is modulated by external factors such as edaphic, climate, faunal, and land use changes. Microbial physiology, chemical recalcitrance and physical protection are also internal factors regulating MCP

We can understand the factors affecting MCP by viewing soil as a natural bioreactor. This can be illustrated using Andisols as an example. These are SOC-rich soil types derived from volcanic ash (Parfitt et al., 1983; Wada, 1989; Calabi-Floody et al., 2011). Andisols occur along the Circum-Pacific belt (also

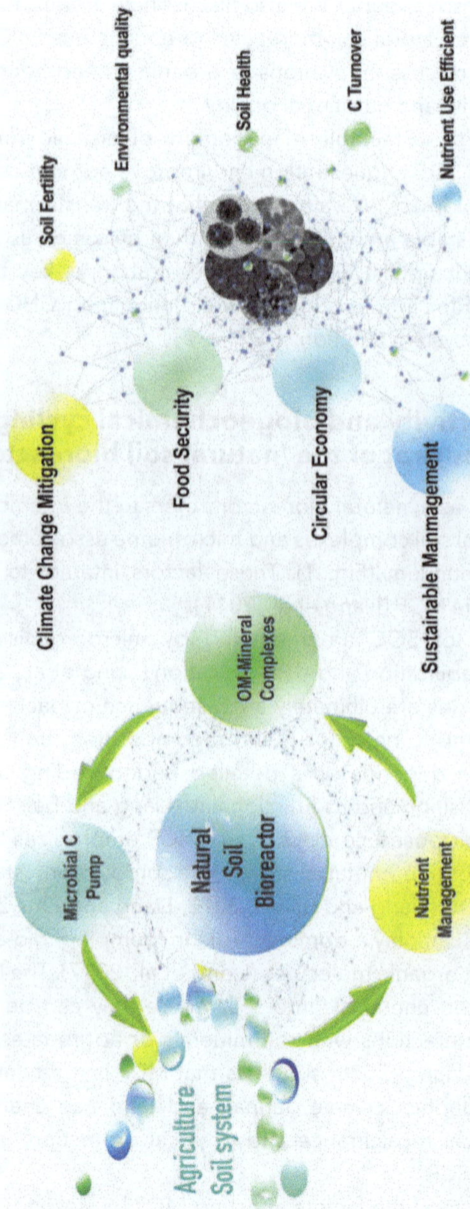

Figure 1 Visualization of the 'natural soil bioreactor'.

called 'Ring of Fire') in countries including Chile, New Zealand, and Japan among others (Brady and Weil, 2002). Because of the inherently favorable conditions of Andisols for SOC storage, we suggest that appropriate agronomic management of these soils could improve SOC sequestration.

From the point of view of SOC stocks, Andisols are major global C sinks. They represent only 0.8% of the global land area but store 5% of the total SOC in the terrestrial biosphere (Eswaran et al., 1993; FAO, 2001). These soils are naturally acidic (pH <5.5), with high exchangeable aluminum (Al) (and free Al; Al^{3+}), high levels of significantly modified and stabilized SOM and phosphorus (P) retention capacity (Matus et al., 2014; Redel et al., 2016; Borie et al., 2019). Several studies have shown that the main traits of these soils are a consequence of the presence of allophane and imogolite together with Al- and iron (Fe)-organic complexes (humus complexes or OM-Al/Fe soil complexes), which have many reactive surfaces (Calabi-Floody et al., 2011; Panichini et al., 2012; Merino et al., 2015; Velásquez et al., 2016; Borie et al., 2019).

Allophanic clays are present in Andisols as natural nanoclay materials belonging to the sesquioxide group with unique physicochemical characteristics. They have a very high specific surface area (700–1500 m^2g^{-1}) and able to retain much higher amounts of SOM than layered silicate clay minerals (Calabi-Floody et al., 2011, 2015; Alekseeva, 2011; Saidy et al., 2013), and can be considered as natural gels (Woignier et al., 2007). These minerals are also able to take part in specific chemical interactions (e.g. ligand exchange) with SOM and prevent microbial mineralization (Kleber et al., 2015).

Bacterial activities are driven primarily by readily available or labile C provided by root turnover and exudation (Dijkstra et al., 2013). Competition between microorganisms and soil reactive particles for nutrients and dissolved organic C in the rhizosphere (Chapter 3 of this book) is often neglected as a factor in understanding soil processes. It has been assumed that microorganisms can easily access and consume labile C released into soil although, in fact, a number of biotic and abiotic processes regulate C concentration and availability in the soil solution (Merino et al., 2015).

In addition to microbial uptake, available C can decrease in the rhizosphere due to its adsorption onto soil particles with the formation of soil (micro-) aggregates (Albalasmeh and Ghezzehei, 2014). Allophane minerals have a large affinity for organic substances, particularly in the presence of cations. Organic matter (OM) adsorption influences water-holding capacity, soil density (both particle and bulk densities), and aggregate stability (Borie et al., 2019; Vistoso et al., 2009, 2012). Their capacity to retain OM is reflected by a high mean residence time of SOM stabilized in Andisols, which is generally much longer than those from other soil types, ranging between 1000 years and 5000 years (Torn et al., 1997; Borie et al., 2019). Andisols can therefore be expected to have a higher potential for SOM sequestration than non-Andisols (Matus et al., 2014; Calabi-Floody et al., 2015).

3 Soil processes and fertilization

Although it is evident that Andisols have several favorable physicochemical properties, their allophanic nature leading to P fixation and Al toxicity also negatively affects SOC sequestration by reducing biomass production of croplands and pastures (Mora et al., 2002, 2006; Vistoso et al., 2012, 2021; Demanet et al., 2015) and therefore C input into agricultural soils. As an example, to cope with low P availability in Andisols, the application of P fertilizers exceeding crop requirements has led to the accumulation of P in the soil in various inorganic (Pi) and organic (Po) forms (Seguel et al., 2017; Nash et al., 2019). This has resulted in the subsequent accumulation of a significant soil 'P bank', commonly referred as 'Legacy P' (Condron et al., 2013; Zhu et al., 2018). Much of this P may have accumulated as Po forms differing in availability (Velásquez et al., 2016; Borie et al., 2019). Velásquez et al. (2016) recently showed that residual P in Andisols is 45–63% of the total P, with 53–77% accumulated as Pi, while Po forms represented 21–42% of total P, most of which occurred as myo- (16%) and scyllo-inositol hexakisphosphate (10%). As the accumulation of legacy P occurs not only in Andisols but also affects P availability in many agricultural soils worldwide, the development of technologies that allow mobilizing at least a portion of this legacy P is key to achieving sustainable agricultural production, thereby reducing the requirement of additional P fertilization.

It seems evident that the dynamics of P release from mineral soil and its subsequent availability for plant nutrition could be closely related to the availability of essential nutrients for bacterial metabolism, whose availability is, in turn, closely related to the soils' inherent physicochemical properties. Although knowledge about P dynamics, and the role of organic amendments such as manures, in the cycling of nutrients has increased in recent decades, the physicochemical and biological mechanisms that control the processes of release-retention of nutrients and their effect on SOC storage are far from being fully understood. The important role microorganisms play in P release from recalcitrant P forms may be dependent on the availability of other elements such as C and N. This is suggested by Fuentes et al. (2009), reporting a significant increase in the level of water-soluble P (from 570 mg kg^{-1} to 1890 mg kg^{-1}) and biomass P (from 390 mg kg^{-1} to 870 mg kg^{-1}) during the initial 40 days of cattle manure incubation in a laboratory-scale reactor. After this period, water-soluble P remained constant (around 1500 mg kg^{-1}) and biomass P decreased (around 220 mg kg^{-1}) probably due to the exhaustion of easily available C.

At the global scale, N is an essential element for plant nutrition and is typically the most growth-limiting factor (Mora et al., 2007). In the soil-plant-microorganism continuum, N availability mainly depends on the microbial-mediated mineralization of organic-N forms (such as crop residues, SOM, manures or sludge amendments and urea-based fertilizers) in the inorganic form of ammonium (NH_4^+) and its subsequent nitrification to nitrate (NO_3^-) (Cartes et al., 2009; Ollivier et al., 2011). A considerable portion of the N input comes from biological nitrogen fixation, i.e. from the action of the nitrogenase enzyme from N-fixing bacteria.

Although most biological nitrogen fixation is carried out by symbiotic bacteria in root nodules of legumes, N-fixation by free-living diazotrophic bacteria is a natural source for biological N inputs in non-leguminous crops (Ouyang et al., 2018). This can reach 40–100 kg N $ha^{-1}y^{-1}$, or 40–60% of the crops' requirement in wheat-sorghum-oat rotation systems and sugarcane, respectively (Gupta et al., 2006; Yeoh et al., 2016). However, this biological N input is generally not considered by farmers, who often apply N fertilizer in excess (Gupta et al., 2014; Jorquera et al., 2014).

Although the application of high rates of N increases crop yields, this practice is unsustainable considering that crop N use efficiency (NUE) in modern agro-ecosystems is notoriously low. More than 50% of N fertilizer applied is lost to the environment through ammonia volatilization, nitrate leaching, and emissions of nitrous oxide (denitrification), the third most important GHG (Núñez et al., 2010; Cartes et al., 2009; Hu and He, 2018, Chapter 6 of this book). When urea is added to the soil, it is first hydrolyzed to NH_4^+-N by urease enzymes before being oxidized to NO_3^--N during the nitrification process. The pool of net mineralized N thus comprises the sum of NH_4^+-N and NO_3^--N in the soil (Cartes et al., 2009).

Soil processes driving N turnover (nitrification, denitrification, N-fixation and mineralization) are highly influenced by (Ollivier et al., 2011; Hu and He, 2018):

- abiotic factors (oxygen availability, water temperature, acidity and alkalinity, soil texture, and climatic condition);
- biotic factors (vegetation type, plant diversity, grazing animals, macro-decomposers and root exudation, C/N ratio, and SOM content);
- anthropogenic factors (agricultural management, N deposition, climate change, grazing, human disturbance, land-use change).

In addition to environmental factors and soil properties, N availability also affects soil physicochemical and biological processes, which, in turn, influence other biogeochemical cycles and plant performance (Hu and He, 2018; Wei et al., 2018; Liang and Zhu, 2021).

However, the role N plays in SOC sequestration and SOM mineralization, especially via the priming effect, continues to be a controversial topic since the net N effects on SOC pools can either be neutral, negative, or positive depending on the amount C input and output controlled by N in different environments (Jung et al., 2011a; Liang and Zhu, 2021). On the one hand, N fertilization generally increases biomass production, with positive effects on SOC pools by microbial transformation of plant-derived C, including both root exudates and decomposing plant biomass (Yan et al., 2018; Liang and Zhu, 2021). On average between 11% and 17% of net C fixed by plants is released to rhizosphere by root exudates and rhizodeposits, providing labile C easily consumed by rhizobacteria (Richardson et al., 2009a; Dijkstra et al., 2013). The addition of N can also retard the decomposition of recalcitrant SOM by inducing a switch of N sources for bacteria from recalcitrant SOM to readily available N.

In contrast, according to the N mining theory, low N availability will lead to the decomposition of recalcitrant SOM to acquire N (Jung et al., 2011a; Chen et al., 2014). However, exogenous N application in N-limited agroecosystems can alleviate microbial N-limitation and increase soil microbial biomass respiration (Wei et al., 2018), increasing N fertilizer application rates can cause increased total GHG emissions and a higher carbon footprint for several crops (Hamer et al., 2009; Jiang et al., 2019).

Similar responses have been described for P turnover in response to N fertilization. The increase of microbial biomass under exogenous N addition may exacerbate P immobilization by microorganisms, resulting in the reduction of soil available P (He et al., 2016a). Meanwhile, other studies have found that exogenous N addition can stimulate bacteria to enhance soil phosphatase activity and accelerate the soil organic P mineralization process, leading to inorganic P release (Marklein and Houlton, 2012; Wei et al., 2018). While N and P availability in soils are crucial to ensure food security, both nutrients also interact with SOC. Recently, Vermeiren et al. (2021) indicated that SOM molecules can effectively prevent P diffusion into soil reactive mineral micropores by spreading out across the sorption surface, thereby reducing P accumulation in the long term.

4 Using organic amendments to increase soil organic carbon sequestration in soils: the role of manure

It has been suggested that long-term application of organic amendments increases SOC concentration and P bioavailability in Andisols as compared to unamended controls (Poblete-Grant et al., 2020). The increased effect has been attributed to:

- the direct supply of P and SOC from the amendment;

- the mobilization of P from the accumulated P source in soil;
- the decrease of soil adsorption sites.

These findings highlight the potential role that organic amendments can play in SOC sequestration. As an example, Calabi-Floody et al. (2019) have explored the use of wheat straw as a material in co-composting. This can also be seen in the fact that between 2000 and 2021 there was an 11.8-fold increase in the number of published articles related to SOC sequestration and organic residues according to the SCOPUS database (Fig. 2).

This chapter reviews the literature discussing whether soil as a 'natural bioreactor' can be managed using organic amendment additions to foster SOC sequestration and nutrient release. Sustainable soil management strategies in agricultural systems will be critical in addressing the triple challenges of food security, environmental degradation, and global warming (Xia et al., 2017). We suggest that harvested residues and animal waste need to be appropriately transformed into organic amendments. These amendments could increase SOC sequestration through their impact on different elements of the plant-soil system. This section discusses manure and manure-based amendments.

As a consequence of the growing global population and a switch to meat-based diets with growing levels of affluence, animal production has increased substantially with high corresponding amounts of animal waste. According to OECD/FAO (2018), beef production in developing countries is projected to be 21% higher by 2027 while sheep meat consumption worldwide has also increased substantially. The global number of sheep and goats increased by 60% from 1961 to 2016, while swine increased by 140% and poultry production showed a remarkable fivefold increase.

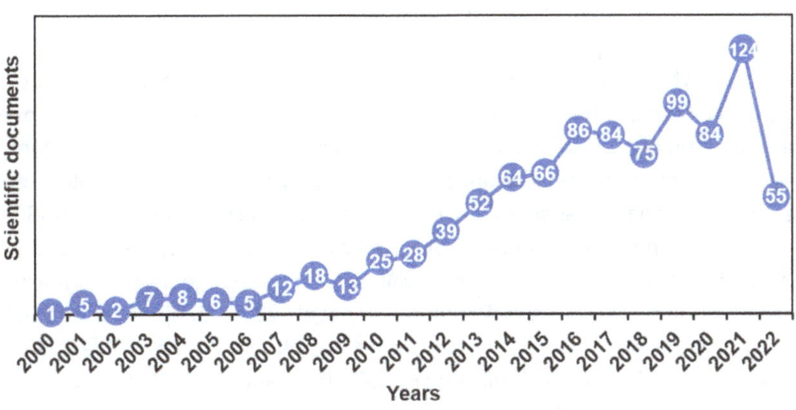

Figure 2 Results from searches of the SCOPUS database of scientific documents with the title-abstract-keywords 'Organic amendments AND soil organic carbon sequestration' (https://www.scopus.com, accessed 12.12.21).

The livestock industry produces 9×10^9 Mg of manure annually with a high nutrient content (He et al., 2016b) which can potentially be reused as a soil amendment and source of macro and micronutrients (Fuentes et al., 2006; He et al., 2016b; Poblete-Grant et al., 2019). Several studies have demonstrated the potential of biotechnological tools to increase nutrient availability from organic molecules contained in manure to enhance its beneficial effects on plant growth and maintain agricultural production (Calabi-Floody et al., 2012; Fuentes et al., 2009; Menezes-Blackburn et al., 2014, 2013).

Manure composition varies widely depending on manure type, animal physiology, species and age, composition of diets, and moisture content (Fuentes et al., 2006). For example, the Pi pool of manures has been reported to vary between 35% and 60%, being quite soluble and readily available to plants (Bolan et al., 2010; He and Honeycutt, 2001; Pagliari and Laboski, 2012; Sato et al., 2005). Manure can potentially increase plant-available P pools in soil (Richardson et al., 2009b). However, when used without pre-treatment, manure application can lead to soil acidification, nutrient loss, and GHG emissions. To avoid these negative effects, composting can transform organic (animal) waste into more stable soil organic amendments to increase soil fertility, biomass production, and consequently SOM content (Ngo et al., 2011, 2014; Chabbi et al., 2017).

Several studies have shown the impact of composted manures on SOC sequestration through (Table 1): Al/Fe soil complex formation (Poblete-Grant et al., 2020; Yin et al., 2016); microbial biomass (Poblete-Grant et al., 2019); N and P availability (Poblete-Grant et al., 2020, 2019).

In addition, OM addition by compost application was found to influence plant-derived carbon flow toward protected fractions (Vidal et al., 2020; Poblete-Grant et al., 2022). This can help maintain long-term SOC stocks more efficiently than inorganic fertilizers (Li et al., 2017).

However, OM application also leads to tradeoffs in form of CO_2 and N_2O emissions, which must be accounted for when evaluating their net effect on SOC sequestration (Barthod et al., 2016) (Chapter 6 of this book). Moreover, the evaluation of SOC sequestration potential must consider both the potential of amendments to increase plant productivity and therefore CO_2 fixation, and the soils' potential for successfully incorporating it in stabilized pools of SOM e.g. organo-mineral associations (Wiesmeier et al., 2019). The nature of these interactions may also control nutrient availability in soils subjected to contrasting types of organic amendments (Poblete-Grant et al., 2020; Guajardo et al., 2020).

Manure-based organic amendment application may affect SOC sequestration indirectly via altering microbial activity, and N and P biogeochemical cycles. Several studies have highlighted that fresh and composted manure significantly increased microorganism abundance, depending on soil types and their physicochemical characteristics (Barra et al., 2018; Bol et al., 2003; Li

Table 1 Organic amendments effect on factors that modulate 'Natural Soil Bioreactor'

Country	Soil type	Manure	Type	Composition			Condition	Natural soil bioreactor factors				Reference
				Total N (g kg^{-1})	Total P (g kg^{-1})	OC (g kg^{-1})		Al/Fe complexes	Microbial C pump	N/P available for plants	SOC storage	
China	Anthrosol	Horse	Unprocessed	-	-	-	Field	↑Fepy	-	-	-	Yin et al. (2016)
China	Cambisol/ Luvisol	Swine	Unprocessed	-	-	-	Field	-	-	↑N	-	Duan et al. (2011)
China	Anthrosol	Poultry	Unprocessed	-	-	-	Field	-	↑MBC	-	-	Atere et al. (2020)
China	Ultisol	Swine	Fresh	33.1	23.7	376	Field	-	-	-	↑SOC	Zhang et al. (2015a)
China	Vertisol	Swine	Composted	-	-	-	Field	-	-	-	↑SOC	Guo et al. (2018)
China	Mollisol	Cattle/ Swine	Fresh	-	-	-	Field	-	-	-	↑SOC	Li et al. (2017)
China	-	Poultry	Unprocessed	19.9	-	233	Field	-	↑MBC	-	↑SOC	Li et al. (2018)
Japan	Andisol	Cattle/ Horse	Composted	-	-	-	Field	-	-	-	↑SOC	Kitamura et al. (2021)
Japan	Andisol	Poultry/ Cattle	Composted	27.0/46.2	31.3/10.2	193/329	Greenhouse	-	↑MBP	↑P	-	Wickramatilake et al. (2010)
Pakistan	Inceptisol	Poultry	Air dried	21.3	9.80	321	Field	-	-	↑N and P	↑	Abbasi et al. (2013)
Pakistan	Aridisol/ Entisol	Poultry/ Cattle	Unprocessed	8.70/8.60	5.10/4.00	143/112	Incubation	-	↑MBC/ MBN/ MBP	↑P	-	Malik et al. (2013)

(Continued)

Table 1 *(Continued)*

				Composition				Natural soil bioreactor factors				
Country	Soil type	Manure	Type	Total N (g kg^{-1})	Total P (g kg^{-1})	OC (g kg^{-1})	Condition	Al/Fe complexes	Microbial C pump	N/P available for plants	SOC storage	Reference
India	Inceptisol	Poultry/ Cattle	Unprocessed	25.0/11.6	17.9/5.80	225/385	Field	-	-	↑N and P	↑SOC	Antil and Singh (2007)
Canada	Regosol	Poultry	Composted	73.0	32.0	520	Field	-	-	↑N and P	-	Warman and Cooper (2000)
Chile	Andisol	Cattle	Fresh	28.6	-	-	Incubation	-	-	-	-	Aguilera et al. (2010)
Chile	Andisol	Poultry	Composted	37.0	25.0	268	Field sampling	↑Al/Fepy ↓Al/Feox	-	↑N/P	↑SOC	Poblete-Grant et al. (2020)
United States	-	Poultry	Unprocessed	-	-	-	Field	-	-	↑N/P	↑SOC	Hoover et al. (2019)
Chile	Andisol	Poultry	Oven dried	33.6	9.50	335	Incubation	-	-	↑N/P	-	Hirzel et al. (2010)
France	Cambisol/ Luvisol	Poultry	Composted	-	13.2	348	Greenhouse	-	↑MBP	↑N/P	↑ SOC	Poblete-Grant et al. (2019)
Chile	Andisol	Cattle	Air dried	-	-	-	Greenhouse	-	↑MBC/ MBP/ MBN	↑ P	-	Paredes et al. (2022)
Chile	Andisol	Cattle	Air dried	21.4	-	410	Incubation/ Greenhouse	-	-	↑ P	-	Paredes et al. (2021)
Chile	Andisol	Cattle	Air dried	12.0-14.0	23.4-44.0	150-227	Incubation/ Greenhouse	-	-	↑ P	-	Menezes-Blackburn et al. (2014)

Chile	Andisol	Cattle	Air dried	-	7.60	-	Incubation	-	-	↑ P	-	Calabi-Floody et al. (2012)
Chile	Andisol	Cattle	Air dried	-	7.60	-	Incubation	-	-	↑ P	-	Fuentes et al. (2009)
Brazil	Latossol	Swine	Fresh	2.63	3.70	-	Field	-	-	↑ N	-	Costa et al. (2014)
France	Luvisol, Neoluvisol	Poultry	Composted	45.0	13.2	288	Greenhouse	-	-	↑ N	↑SOC	Poblete-Grant et al. (2022)
United States	Spodosol	Poultry	Fresh	40.3	19.5	335	Greenhouse	-	-	↑ P	-	Waldrip et al. (2011)
Chile	Andisol	Poultry	Fresh	28.5	9.73	337	Field	-	-	↑ P ↑ N	-	Hirzel and Walter (2008)

MBC: microbial biomass carbon; MBP: microbial biomass phosphorus; MBN: microbial biomass nitrogen; SOC: soil organic carbon; Al/Fe py: aluminum or iron pyrophosphate; Al/Fe ox: aluminum or iron oxalate.

et al., 2018; Poblete-Grant et al., 2019-2022; Rahman et al., 2017) The effect of different types of organic amendments on processes occurring in the 'natural soil bioreactor' must be considered (Table 1) and their effects on biogeochemical functioning of contrasting soil types need to be quantified.

Organic amendments usually contain high amounts of C, ranging from 25 g kg^{-1} to 460 g kg^{-1} (Table 1), which can be potentially incorporated and stabilized in soil (Poblete-Grant et al., 2022). Hoover et al. (2019) found an improvement of 6.1–6.7 g kg $soil^{-1}$ of the occluded OM fraction after 20 years of poultry manure application as compared to 4.6 g kg $soil^{-1}$ of conventional management (N urea fertilizer). In soils receiving organic amendments (straw and manure) over nine years, Li et al. (2018) found that SOC was increased by 143% as compared to inorganic fertilizers, which did not affect SOC. In addition, Guajardo et al. (2020) showed that up to six years of cattle slurry application of 15 m^3 ha^{-1} increased SOM from 11% to 16%. OC from plant residues contributes to SOC sequestration where SOC retention in the soil can be twice as high as observed for manure-derived C (Zhang et al., 2015a).

Further research (Poblete-Grant et al., 2022) on the impact of poultry manure compost on C allocation in a Cambisol under ryegrass growth showed that SOC increased and that plant-derived C was associated with the most active SOM fraction (free and occluded particulate OM). However, the same study also showed that organic amendments were able to stimulate plant-carbon transfer into the most-stable SOM fractions indicating its potential to enhance SOC sequestration in Cambisol in the short term (Poblete-Grant et al., 2022). The long-term effect of organic amendment addition on C allocation to specific soil fractions in contrasting soils still needs to be investigated, considering the high capacity of some soil types to sequester C in this context (Calabi-Floody et al., 2015).

Soil type is an important consideration when developing guidelines on organic amendment input (Shepherd and Newell-Price, 2016). Animal species with contrasting feed requirements (Barthram et al., 2005) and type of digestion (ruminant or non-ruminant) (Jha et al., 2019), and transformation (composting, thermal treatment and digestion), also influence differences in nutrient concentration of organic amendments.

However, to prevent tradeoffs in the form of GHG emissions after organic amendment application, biotechnological processes transforming organic amendments to environmentally friendly fertilizers must be implemented. These are discussed in the next section.

5 Stabilizing treatments for organic amendments

Agricultural residues, such as harvesting residues and manure, constitute an important labile source of carbon and nutrients, which can positively influence biological, chemical, and physical properties when incorporated into soils (Zhang et al., 2015b). However, they are also easily decomposed and lost from the system (Plaza and Senesi, 2009). Instead of applying agricultural residues without modification, they can also be transformed into more stable organic amendments using several processes.

5.1 Composting

Aerobic composting is one of the most frequently used technologies for organic waste transformation into organic amendments and an important route to C stabilization through partial degradation and chemical alterations of OM (Scotti et al., 2015). During the process of composting, successive activities of diverse microorganisms can produce more stable and complex OM under aerobic conditions (de Bertoldi et al., 1983; Medina et al., 2015). The resulting humified OM is a potentially suitable amendment for agricultural (Castaldi et al., 2005) and degraded soils (Park et al., 2011).

The addition of compost to soil accelerates the growth of plants, thereby increasing CO_2 uptake and C storage within plant-soil systems (Paustian et al., 2016). Composts can improve tillage and workability of soil via changes in physical, chemical, and biological properties, thereby contributing to reducing GHG emissions (Barthod et al., 2018b). Aerobic composting produces a stabilized material that can valorize residues, recycle crop nutrients and reduce reliance on synthetic fertilizers (Cogger, 2005; Favoino and Hogg, 2008). Long-term observations show that frequent compost application increases soil C content, although a large part of the applied OM may be subject to quick decomposition (Paetsch et al., 2016).

However, as noted, even stabilized composts may show a high rate of decomposition once incorporated into soils and may produce GHG emissions during their production and after being incorporated into soils (Barthod et al., 2020, 2021). Gao et al. (2019) described different levels of stability and chemical transformation in mixtures composted under the same conditions, given the diverse microbes and raw material composition involved. This has focused interest in practices for reducing GHG emissions during the composting process and increasing the mean residence time of compost once applied to the soil.

5.2 Composting with additives

The addition of organic and inorganic materials and inoculants during organic waste composting is used to improve physical, chemical, and biological conditions during the composting processes (Dias et al., 2010). The primary function of these additives is providing structural support and increasing air voids to allow aeration of the pile (Haug, 1993). In other cases, the additives may act as a primary C and energy source for microorganisms (Adhikari et al., 2009). The use of additives in composting of organic residues can also facilitate the process and improve the quality of end products (Fig. 3).

As an example, metallic oxides and clays, which are the main constituents of the inorganic soil matrix and well known to have an important role in SOM stabilization (Mikutta and Kaiser, 2011), are being used as additives during composting processes to reduce the emission of GHGs (Ren et al., 2019). Clays or nanoclays are more often used to reduce GHG emissions and N losses during composting and to enhance C stabilization by promoting organo-mineral associations (i.e. physical protection) (Barthod et al., 2018a; Chowdhury et al., 2016; Mao et al., 2019; Medina et al., 2020). Bolan et al. (2012) reported that oxides and clay minerals such as goethite and allophane can enhance C stabilization and end-product quality in composting mixtures, increasing the half-life ($t_{1/2}$) of the amendments in soils. The incorporation of additives such as zeolite or vermiculite and other materials can also promote the retention of nutrients and reduce ammonia volatilization and nitrogen loss which is a common problem in composting mixtures from animal manures (Pisa et al., 2020; Wang et al., 2018). More importantly in terms of SOC sequestration is the positive effect of co-composts produced with clay minerals on plant biomass and OC flow from plants to soil as compared to regular compost (Vidal et al., 2020; Medina et al., 2021).

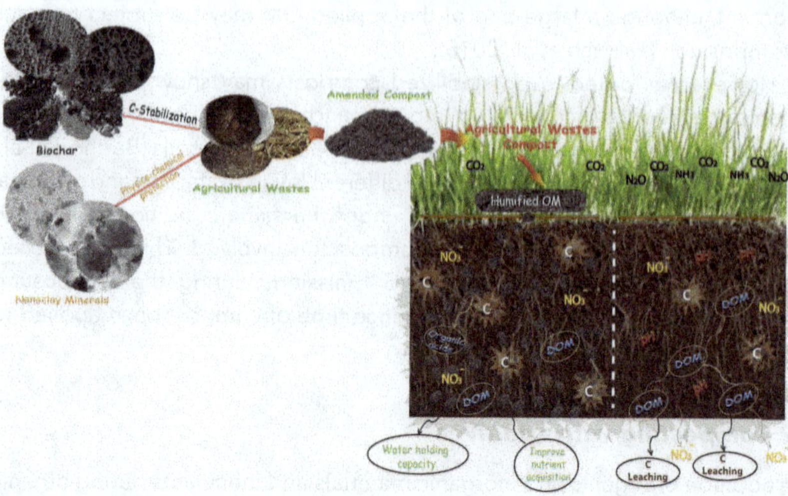

Figure 3 Utilization of organic and inorganic additives in the stabilization of agricultural residues to produce innovative amendments with potential C sequestration.

5.3 Thermal treatment: biochar

Thermal treatment of organic waste through pyrolysis to obtain biochar, an aromatic C-rich product which is highly stable against microbial decomposition, has been suggested as a possible contribution to carbon sequestration in the biosphere (Smith, 2016). Biochar has attracted attention for its benefits as a soil amendment (Godlewska et al., 2017), because it enhances soil properties (water and nutrient retention, health, fertility), and its SOC sequestration capacity (Jeffery et al., 2011; Lehmann and Joseph, 2015; Shin et al., 2019).

The combination of biochar with other materials can have beneficial effects on soil properties. Cardelli et al. (2018) reported that the addition of pruning residues, biochar, and digestates (obtained from anaerobic digestion of animal wastewater and other agro-industrial materials) appeared to enrich the soil C sink and reduce CO_2 emissions, potentially promoting SOC sequestration. The authors reported that the addition of biochar and digestate reduced soluble organic C, limiting microbial activity and reducing mineralization of organic materials (Cardelli et al., 2018).

Recent studies have suggested that biochar could be also useful as composting additive, for example during composting of poultry manure, where it has been found to improve the stability and quality of mature composts (Wang et al., 2014; Curaqueo et al., 2016; Meier et al., 2015). Biochar addition to compost can reduce N and C losses (Awasthi et al., 2020). Studies conducted by Prost et al. (2013) and Vandecasteele et al. (2016) showed that biochar reduced and adsorbed leachates produced during conventional composting processes. In general, the combination of biochar and compost has shown synergistic effects, improving SOC sequestration and some soil properties (Ngo et al., 2016).

However, biochar's use as soil amendment involves tradeoffs. It can increase soil dust emissions of particles <10 μm (PM 10) during soil application, or after it has been incorporated as the result of natural wind-driven erosion or through mechanical tillage events (Gelardi et al., 2019), with a potential negative impact on human health. Conventional biochar application can result in losses of around 25%. Recent studies have demonstrated that pelletized biochar could reduce these problems on site (Shin et al., 2019, 2021). Pyrolyzed material might also be a way to produce slow-release fertilizers (see below). The combination of technologies and additives (e.g. Fig. 3) for the treatment of organic waste provides an effective way to stabilize organic amendments and to promote SOC sequestration, both by improving the waste treatment process and the properties of the resulting organic amendments.

6 Smart fertilizers to optimize nutrient uptake and soil organic carbon sequestration

A combination of biotechnology and nanotechnology has the potential to contribute to the development of a new generation of fertilizer which improves plant nutrient use efficiency (Usman et al., 2020; Mora et al., 2021). This can be achieved through controlled nutrient release (Calabi-Floody et al., 2018; Nguyen, 2021) together with bioformulations based on bacteria or enzymes (Trenkel, 1997; Manjunatha et al., 2016; Naz and Sulaiman, 2016; DeRosa et al., 2010).

Controlled-release fertilizers are those containing a nutrient in a form which delays its availability for plant uptake and use after application. These fertilizers can also be in a form which is available to plants significantly longer than a reference 'rapidly available nutrient fertilizer' such as ammonium or urea (AAPFCO, 1995; Trenkel, 1997).

These fertilizers may be designed with materials such as nanomaterials (Calabi-Floody et al., 2018; Usman et al., 2020). Hydroxyapatite (HA, $(Ca_{10}(PO_4)_6(OH)_2)$ nanoparticles have been widely studied for biomedical applications, there has been less research on nano-fertilizer applications, in spite of its high P content (Kottegoda et al., 2017; Marchiol et al., 2019). Research on HA nanoparticles has focused on developing slow-release urea-based N fertilizer, partly to reduce N losses through leaching (Kottegoda et al., 2011, 2017).

The re-utilization of C-rich agricultural wastes as materials to develop smart fertilizers to foster SOC sequestration promotes the objectives of a sustainable, circular bioeconomy, and improved food security (Calabi-Floody et al., 2019; Nguyen, 2021; Shin et al., 2021). It combines improved soil fertility, SOC sequestration, and reduced environmental impacts from agriculture. A good example of these synergies is biochar pellets blended with pig manure compost and fertilizer which induced higher NUE compared to a traditional inorganic fertilizer (Shin et al., 2019). Biochar from crop residues can also provide a favorable microbial habitat for promoting microbial C-use efficiency and thus SOC stabilization (Majumder et al., 2019). Shin et al. (2021) reported that activated biochar pelletized fertilizers reduced the need for inorganic N fertilizer in rice production by 60% and sequestered around 40% more C, achieving 1.23 ton ha^{-1} compared to 0.89 ton ha^{-1} in the control.

Waste treatment, composting and co-composting strategies open up huge possibilities to use C-rich agricultural waste for smart fertilizer development to foster SOC sequestration. As an example, Calabi-Floody et al. (2019) reported the option of producing C-rich carrier material from wheat straw. This combines high C content (>40%), an aromatic polycondensated structure and high water-holding capacity (>91%) with the potential to increase surface reactivity and sorption properties, suitable for bioformulation to design smart fertilizers.

The development of smart fertilizers must take place within a broader context of coupling biogeochemical cycling of carbon and nutrients to provide environmental benefits at the same time as improving nutrient use efficiency and crop yields. This requires smart soil management that increases SOM and optimizes biogeochemical cycles both to reduce adverse environmental impacts while improving the quality of agricultural soils (Paustian et al., 2016; Lal, 2018; Bai et al., 2019; Chowdhury et al., 2020; Liang and Zhu, 2021).

7 Conclusion

Soils largely sustain global food production which provides more than 95% of the food calories needed by the human population. However, agricultural soils are often characterized by degradation of their physicochemical and biological properties in systems based on intensive mineral fertilizer use. Sustainable management of agricultural systems should include the use of organic amendments and a new generation of fertilizers to establish closed biogeochemical cycles and to foster SOC sequestration and to re-build soil's organic carbon stocks. Soil should be considered as a natural bioreactor. This requires understanding the factors driving changes in C, N, and P cycles and the need to develop innovative and eco-friendly strategies to meet plant requirements adapted to a specific pedoclimatic context.

New fertilization strategies should be based on the transformation of organic waste into organic amendments using composting and/or thermal treatments. Co-composting processes with organic, mineral, and/or biological additives should be developed to increase fertilizer value and environmental benefits of the end products. Biochar application to soil may help foster SOC sequestration. However, tradeoffs such as emission of greenhouse gases and nutrients (e.g. P leaching to water bodies) should be considered when using organic amendments. Innovative alternative nutrient sources may include smart fertilizers based on associations of carrier materials with slow-release nutrient sources. Their development is critical to increasing productivity in degraded agricultural soils and reducing the environmental impact of conventional fertilization strategies.

8 Where to look for further information

A recent overview is provided by: Mora, M. L., Calabi-Floody, M., Rumpel, C., 2021. Smart Fertilizers and Innovative Organic Amendments for Sustainable Agricultural Systems. MDPI Agronomy Basel, 173, pp. and Calabi-Floody, M., Medina, J., Rumpel, C., Condron, L. M., Hernandez, M., Dumont, M., Mora, M. L.

2018. Chapter Three - Smart fertilizers as a strategy for sustainable agriculture. In: Donald LS, editor. Advances in Agronomy. Volume 147: Academic Press. p. 119-157.

To complement the readers knowledge about the impact of the climate change on food security please visit Climate Change 2022: Impacts, Adaptation and Vulnerability specifically read Chapter 5 Food, fibre, and other ecosystem products (https://www.ipcc.ch/report/ar6/wg2/).

For more information on Chilean volcanic soil please read Borie, F., Aguilera, P., Castillo, C. et al. (2019), Revisiting the Nature of Phosphorus Pools in Chilean Volcanic Soils as a Basis for Arbuscular Mycorrhizal Management in Plant P Acquisition, J. Soil Sci. Plant Nutr., 19, 390–401.

An interesting way to understand the inherent capacity of soils to store carbon, and its potential for sequestering carbon inputs from the atmosphere, it is the RECOIL program:

(https://www.fao.org/documents/card/en/c/ca6522en).

To improve the reader knowledge about animal manure procedures, storage, and characterization the following references could be revised in: Waldrip, H., Pagliari, P., He, Z., 2020. Animal Manure: Production, Characteristics, Environmental Concerns, and Management. ASA Special Publication 67, 448, pp.

In addition, please visit the following links to find further information related to Chilean organic farmers experiences using manure as sustainable management:

- The experience of Tierra Viva AG, Chile's first Organic Family Farmer's Organization (https://www.organicwithoutboundaries.bio/2019/12/11/organic-walnuts-and-sheep-cheese-a-model-of-regenerative-agriculture-in-chile/).
- An article investigating the indigenous smallholder's agricultural production in southern Chile performed by Mariola, M. 2018. Limited Fertility, Limited Land: Barriers to Sustainability in a Chilean Agrarian Community (https://www.intechopen.com/chapters/60267).

Countries such as Chile also provide their guidance related to composting operations. In this sense, information related to waste management and design and development of composting facilities and processes can be found in some manuals based on Scientific information collected for the Environmental Minister.

- Programa Reciclo Orgánicos. (2021). Manual de compostaje: Una herramienta para combatir el cambio climático. https://reciclorganicos.com/wpcontent/uploads/2021/03/Manual-de-compostaje.pdf.

- Fuente: ImplementaSur. (2019). Asesoría sobre el manejo de residuos orgánicos generados a nivel municipal en Chile. https://rechile.mma.gob.cl/wpcontent/uploads/2020/07/Informe-3-Sistemas-de-manejo-e-instrumentosaplicables-en-Chile.pdf.

9 Acknowledgements

We gratefully acknowledge ANID (National Research and Development Agency) for the financial support under FONDECYT projects Regular N° 1201375, Regular N° 1181050, Iniciación N° 11201107, and Postdoctoral N° 3210228.

10 References

AAPFCO (1995). Official Publication No. 48. Association of American Plant Food Control Officials. Association of American Plant Food Control Officials (AAPFCO). Inc., West Lafayette, IN.

Abbasi, M. K., Mansha, S., Rahim, N. and Ali, A. (2013). Agronomic effectiveness and phosphorus utilization efficiency of rock phosphate applied to winter wheat, *Agron. J.* 105(6), 1606-1612.

Adhikari, B. K., Barrington, S., Martinez, J. and King, S. (2009). Effectiveness of three bulking agents for wood waste composting, *Waste Manag.* 29(1), 197-203.

Aguilera, P., Briceño, G., Mora, M. L., Demanet, R. and Palma, G. (2010). Effect of liquid cow manure on chemical and biological properties in an andisol, *J. Soil Sci. Plant Nutr.* 10(2), 158 -169.

Albalasmeh, A. A. and Ghezzehei, T. A. (2014). Interplay between soil drying and root exudation in rhizosheath development, *Plant Soil* 374(1-2), 739-751.

Alekseeva, T. V. (2011). Clay minerals and organo-mineral associates, in Gliński, J., Horabik, J. and Lipiec, J. (Eds), *Encyclopedia of Agrophysics*, pp. 117-122. Dordrecht, The Netherlands: Springer Netherlands.

Antil, R. S. and Singh, M. (2007). Effects of organic manures and fertilizers on organic matter and nutrients status of the soil, *Arch. Agron. Soil Sci.* 53(5), 519-528.

Atere, C. T., Gunina, A., Zhu, Z., Xiao, M., Liu, S., Kuzyakov, Y., Chen, L., Deng, Y., Wu, J. and Ge, T. (2020). Organic matter stabilization in aggregates and density fractions in paddy soil depending on long-term fertilization: tracing of pathways by 13C natural abundance, *Soil Biol. Biochem.* 149, 107931.

Awasthi, M. K., Duan, Y., Awasthi, S. K., Liu, T. and Zhang, Z. (2020). Effect of biochar and bacterial inoculum additions on cow dung composting, *Bioresour. Technol.* 297, 122407.

Bai, X., Huang, Y., Ren, W., Coyne, M., Jacinthe, P. A., Tao, B., Hui, D., Yang, J. and Matocha, C. (2019). Responses of soil carbon sequestration to climate smart agriculture practices: a meta-analysis, *Glob. Change Biol.* 25(8), 2591-2606.

Barra, P. J., Viscardi, S., Jorquera, M. A., Duran, P. A., Valentine, A. J. and de la Luz Mora, M. (2018). Understanding the strategies to overcome phosphorus-deficiency and aluminum-toxicity by ryegrass endophytic and rhizosphere phosphobacteria, *Front. Microbiol.* 9, 1155.

Barthod, J., Dignac, M.-F., Le Mer, G., Bottinelli, N., Watteau, F., Kögel-Knabner, I. and Rumpel, C. (2020). How do earthworms affect organic matter decomposition in the presence of clay-sized minerals?, *Soil Biol. Biochem.* 143, 107730.

Barthod, J., Dignac, M. F. and Rumpel, C. (2021). Effect of microbial decomposition products produced in the presence or absence of epigeic earthworms and minerals on soil carbon stabilization, *Soil Biol. Biochem.* 160, 108308.

Barthod, J., Rumpel, C., Calabi-Floody, M., Mora, M. L., Bolan, N. S. and Dignac, M. F. (2018a). Adding worms during composting of organic waste with red mud and fly ash reduces CO2 emissions and increases plant available nutrient contents, *J. Environ. Manage.* 222, 207-215.

Barthod, J., Rumpel, C. and Dignac, M. F. (2018b). Composting with additives to improve organic amendments: a review, *Agron. Sustain. Dev.* 38(2), 17.

Barthod, J., Rumpel, C., Paradelo, R. and Dignac, M. F. (2016) The effects of worms, clay and biochar on CO_2 emissions during production and soil application of co-composts. *Soil* 2, 673-683.

Barthram, G. T., Duff, E. I., Elston, D. A., Griffiths, J. H., Common, T. G. and Marriott, C. A. (2005). Frequency distributions of sward height under sheep grazing, *Grass Forage Sci.* 60(1), 4-16.

Bashagaluke, J. B., Logah, V., Opoku, A., Sarkodie-Addo, J. and Quansah, C. (2018). Soil nutrient loss through erosion: impact of different cropping systems and soil amendments in Ghana, *PLoS ONE* 13(12), e0208250.

Batjes, N. H. (1996). Total carbon and nitrogen in the soils of the world, *Eur. J. Soil Sci.* 47(2), 151-163.

Bol, R., Kandeler, E., Amelung, W., Glaser, B., Marx, M. C., Preedy, N. and Lorenz, K. (2003). Short-term effects of dairy slurry amendment on carbon sequestration and enzyme activities in a temperate grassland, *Soil Biol. Biochem.* 35(11), 1411-1421.

Bolan, N. S., Kunhikrishnan, A., Choppala, G. K., Thangarajan, R. and Chung, J. W. (2012). Stabilization of carbon in composts and biochars in relation to carbon sequestration and soil fertility, *Sci. Total Environ.* 424, 264-270.

Bolan, N. S., Szogi, A. A., Chuasavathi, T., Seshadri, B., Rothrock, M. J. and Panneerselvam, P. (2010). Uses and management of poultry litter, *Worlds Poult. Sci. J.* 66(4), 673-698.

Borie, F., Aguilera, P., Castillo, C., Valentine, A., Seguel, A., Barea, J. M. and Cornejo, P. (2019). Revisiting the nature of phosphorus pools in Chilean volcanic soils as a basis for arbuscular mycorrhizal management in plant P acquisition, *J. Soil Sci. Plant Nutr.* 19(2), 390-401.

Brady, N. C. and Weil, R. R. (2002). *The Nature and Properties of Soils* (13th edn.), New Jersey: Prentice Hall.

Calabi-Floody, M., Bendall, J. S., Jara, A. A., Welland, M. E., Theng, B. K. G., Rumpel, C. and Mora, MdlL. (2011). Nanoclays from an andisol: extraction, properties and carbon stabilization, *Geoderma* 161(3-4), 159-167.

Calabi-Floody, M., Medina, J., Rumpel, C., Condron, L. M., Hernandez, M., Dumont, M. and Mora, M. L. (2018). Chapter 3: Smart fertilizers as a strategy for sustainable agriculture, in Donald, L. S. (Ed) *Advances in Agronomy*, (vol. 147), pp. 119-157. Academic Press.

Calabi-Floody, M., Medina, J., Suazo, J., Ordiqueo, M., Aponte, H., Mora, M. L. and Rumpel, C. (2019). Optimization of wheat straw co-composting for carrier material development, *Waste Manag.* 98, 37-49.

Calabi-Floody, M., Rumpel, C., Velásquez, G., Violante, A., Bol, R., Condron, L. M. and Mora, M. L. (2015). Role of nanoclays in carbon stabilization in andisols and cambisols, *J. Soil Sci. Plant Nutr.* 15, 587-604.

Calabi-Floody, M., Velásquez, G., Gianfreda, L., Saggar, S., Bolan, N., Rumpel, C. and Mora, M. L. (2012). Improving bioavailability of phosphorous from cattle dung by using phosphatase immobilized on natural clay and nanoclay, *Chemosphere* 89(6), 648-655.

Cardelli, R., Giusani, G., Marchini, F. and Saviozzi, A. (2018). Short-term effects of biogas digestate, biochar and their combination, *Soil Res.* 56(6), 623-631.

Cartes, P., Jara, A. A., Demanet, R. and Mora, MdlL. (2009). Urease activity and nitrogen mineralization kinetics as affected by temperature and urea input rate in southern Chilean andisols, *RC Suelo Nutr. Veg.* 9(1), 69-82.

Castaldi, P., Santona, L. and Melis, P. (2005). Heavy metal immobilization by chemical amendments in a polluted soil and influence on white lupin growth, *Chemosphere* 60(3), 365-371.

Chabbi, A., Lehmann, J., Ciais, P., Loescher, H. W., Cotrufo, M. F., Don, A., SanClements, M., Schipper, L., Six, J., Smith, P. and Rumpel, C. (2017). Aligning agriculture and climate policy, *Nat. Clim. Change* 7(5), 307-309.

Chen, R., Senbayram, M., Blagodatsky, S., Myachina, O., Dittert, K., Lin, X., Blagodatskaya, E. and Kuzyakov, Y. (2014). Soil C and N availability determine the priming effect: microbial N mining and stoichiometric decomposition theories, *Glob. Chang. Biol.* 20(7), 2356-2367.

Chowdhury, S., Bolan, N., Farrell, M., Sarkar, B., Sarker, J. R., Kirkham, M. B., Hossain, M. Z. and Kim, G. (2020). Role of cultural and nutrient management practices in carbon sequestration in agricultural soil, *Adv. Agron.* 166, 131-196.

Chowdhury, S., Bolan, N. S., Seshadri, B., Kunhikrishnan, A., Wijesekara, H., Xu, Y., Yang, J., Kim, G. H., Sparks, D. and Rumpel, C. (2016). Co-composting solid biowastes with alkaline materials to enhance carbon stabilization and revegetation potential, *Environ. Sci. Pollut. Res. Int.* 23(8), 7099-7110.

Condron, L. M., Spears, B. M., Haygarth, P. M., Turner, B. L. and Richardson, A. E. (2013). Role of legacy phosphorus in improving global phosphorus-use efficiency, *Environ. Dev.* 8, 147-148.

Cogger, C. G. (2005). Potential compost benefits for restoration of soils disturbed By urban development, *Compost. Sci. Util.* 13(4), 243-251.

Costa, M., Shigaki, F., Alves, B., Kleinman, P. and Pereira, M. (2014). Swine manure application methods effects on ammonia volatilization, forage quality, and yield in the Pre-Amazon Region of Brazil, *Chil. J. Agric. Res.* 74(3), 311-318.

Curaqueo, G., Sánchez-Monedero, M., Meier, S., Medina, J., Panichini, M., Borie, Fernando and Navia, R. (2016). Characterization of a soil amendment derived from co-composting of agricultural wastes and biochar, EGU General Assembly 2016. Austria, Vol. 18. 18184.

de Bertoldi, M., Vallini, G. and Pera, A. (1983). The biology of composting: a review, *Waste Manag. Res.* 1, 157-176.

Demanet, R., Mora, M. L., Herrera, M. Á., Miranda, H. and Barea, J. M. (2015). Seasonal variation of the productivity and quality of permanent pastures in Adisols of temperate regions, *J. Soil Sci. Plant Nutr.* 15, 111-128.

DeRosa, M. C., Monreal, C., Schnitzer, M., Walsh, R. and Sultan, Y. (2010). Nanotechnology in fertilizers, Nat, *Nanotechnol. J.* 5, 91.

Dias, B. O., Silva, C. A., Higashikawa, F. S., Roig, A. and Sánchez-Monedero, M. A. (2010). Use of biochar as bulking agent for the composting of poultry manure: effect on organic matter degradation and humification, *Bioresour. Technol.* 101(4), 1239–1246.

Dijkstra, F. A., Carrillo, Y., Pendall, E. and Morgan, J. A. (2013). Rhizosphere priming: A nutrient perspective, *Front. Microbiol.* 4, 216.

Duan, Y., Xu, M., He, X., Li, S. and Sun, X. (2011). Long-term pig manure application reduces the requirement of chemical phosphorus and potassium in two rice-wheat sites in subtropical China, *Soil Use Manag.* 27(4), 427–436.

Eswaran, H., Van Den Berg, E. and Reich, P. (1993). Organic carbon in soils of the world, *Soil Sci. Soc. Am. J.* 57(1), 192–194. doi:10.2136/sssaj1993.03615995005700010034x.

FAO (2001). Driessen, P. M., Deckers, J. and Spaargaren, O. Lecture notes on the major soils of the world. (World Soil Resources Reports : FAO; Vol. 94). Food and Agriculture Organization of the United Nations (FAO). https://www.fao.org/3/y1899e/y1899e.pdf.

Favoino, E. and Hogg, D. (2008). The potential role of compost in reducing greenhouse gasses, *Waste Manag. Res.* 26(1), 61–69.

Fuentes, B., Bolan, N., Naidu, R. and Mora, MdlL. (2006). Phosphorus in organic waste-soil systems, *RC Suelo Nutr. Veg.* 6(2), 64–83.

Fuentes, B., Jorquera, M. and Mora, Mde L. (2009). Dynamics of phosphorus and phytate-utilizing bacteria during aerobic degradation of dairy cattle dung, *Chemosphere* 74(2), 325–331.

Gao, X., Tan, W., Zhao, Y., Wu, J., Sun, Q., Qi, H., Xie, X. and Wei, Z. (2019). Diversity in the mechanisms of humin formation during composting with different materials, *Environ. Sci. Technol.* 53(7), 3653–3662.

Gelardi, D. L., Li, C. and Parikh, S. J. (2019). An emerging environmental concern: biochar-induced dust emissions and their potentially properties, *Sci. Total Environ.* 678, 813–820.

Godlewska, P., Schmidt, H. P., Ok, Y. S. and Oleszczuk, P. (2017). Biochar for composting improvement and contaminants reduction. A review, *Bioresour. Technol.* 246, 193–202.

Guajardo, C., Recio-Espejo, J. M., Sandoval, M. A., Olmo, F. Dd, Bustamante, M. and García-Ferrer, A. (2020). Anthropogenic alteration of available, amorphous, and total iron in an andisol from dairy slurry applications over a 12-year period, *Chil. J. Agric. Res.* 80(1), 108–117.

Guo, Z., Zhang, Z., Zhou, H., Wang, D. and Peng, X. (2018). The effect of 34-year continuous fertilization on the SOC physical fractions and its chemical composition in a vertisol. *Sci. Rep.* 9(1), 2505.

Gupta, V. V. S. R., Kroker, S. J., Hicks, M., Davoren, C. W., Descheemaeker, K. and Llewellyn, R. (2014). Nitrogen cycling in summer active perennial grass systems in South Australia: non-symbiotic nitrogen fixation, *Crop Pasture Sci.* 65(10), 1044–1056.

Gupta, V. V. S. R., Roper, M. M. and Roget, D. K. (2006). Potential for non-symbiotic N_2-fixation in different agroecological zones of southern Australia, *Soil Res.* 44(4), 343–354.

Hamer, U., Potthast, K. and Makeschin, F. (2009). Urea fertilisation affected soil organic matter dynamics and microbial community structure in pasture soils of southern Ecuador, *Appl. Soil Ecol.* 43(2–3), 226–233.

Haug, R. T. (1993). *The Practical Handbook of Compost Engineering*. Boca Raton: Lewis Publishers.

He, Z. and Honeycutt, C. W. (2001). Enzymatic characterization of organic phosphorus in animal manure, *J. Environ. Qual.* 30(5), 1685-1692.

He, Z., Pagliari, P. H. and Waldrip, H. M. (2016a). Applied and Environmental Chemistry of Animal Manure: a review, *Pedosphere* 26(6), 779-816.

He, P., Wan, S. Z., Fang, X. M., Wang, F. and Chen, F. (2016b). Exogenous nutrients and carbon resource change the responses of soil organic matter decomposition and nitrogen immobilization to nitrogen deposition, *Sci. Rep.* 6(1), 23717.

Hirzel, J., Undurraga, P. and Walter, I. (2010). Nitrogen mineralization and released nutrients in a volcanic soil amended with poultry litter, *Chil. J. Agric. Res.* 70(1), 113-121.

Hirzel, J. and Walter, I. (2008). Availability of nitrogen, phosphorus and potassium from poultry litter and conventional fertilizers in a volcanic soil cultivated with silage corn, *Chil. J. Agric. Res.* 68(3), 264-273.

Hoover, N. L., Law, J. Y., Long, L. A. M., Kanwar, R. S. and Soupir, M. L. (2019). Long-term impact of poultry manure on crop yield, soil and water quality, and crop revenue, *J. Environ. Manage.* 252, 109582.

Hu, H. W. and He, J. Z. (2018). Manipulating the soil microbiome for improved nitrogen management, *Microbiol. Aust.* 39(1), 2-5.

Jeffery, S., Verheijen, F. G. A., VanderVelde, M. and Bastos, A. C. (2011). A quantitative review of the effect of biochar application to soils on crop productivity using meta-analysis, *Agric. Ecosyst. Environ.* 144(1), 175-187.

Jha, R., Fouhse, J. M., Tiwari, U. P., Li, L. and Willing, B. P. (2019). Dietary fiber and intestinal health of monogastric animals, *Front. Vet. Sci.* 6, 48.

Jiang, Z., Zhong, Y., Yang, J., Wu, Y., Li, H. and Zheng, L. (2019). Effect of nitrogen fertilizer rates on carbon footprint and ecosystem service of carbon sequestration in rice production, *Sci. Total Environ.* 670, 210-217.

Jorquera, M. A., Martínez, O. A., Marileo, L. G., Acuña, J. J., Saggar, S. and Mora, M. L. (2014). Effect of nitrogen and phosphorus fertilization on the composition of rhizobacterial communities of two Chilean andisol pastures, *World J. Microbiol. Biotechnol.* 30(1), 99-107.

Jung, J. Y., Lal, R. and Ussiri, D. A. N. (2011). Changes in CO_2, 13C abundance, inorganic nitrogen, β-glucosidase, and oxidative enzyme activities of soil during the decomposition of switchgrass root carbon as affected by inorganic nitrogen additions, *Biol. Fertil. Soils* 47(7), 801-813.

Kitamura, R., Sugiyama, C., Yasuda, K., Nagatake, A., Yuan, Y., Du, J., Yamaki, N., Taira, K., Kawai, M. and Hatano, R. (2021). Effects of three types of organic fertilizers on greenhouse gas emissions in a grassland on andosol in Southern Hokkaido, Japan, *Front. Sustain. Food Syst.* 5, 649613.

Kleber, M., Eusterhues, K., Keiluweit, M., Mikutta, C., Mikutta, R. and Nico, P. S. (2015). Mineral-organic associations: formation, properties, and relevance in soil environments, *Adv. Agron.* 130, 1-140.

Kopittke, P. M., Menzies, N. W., Wang, P., McKenna, B. A. and Lombi, E. (2019). Soil and the intensification of agriculture for global food security, *Environ. Int.* 132, 105078.

Kottegoda, N., Munaweera, I., Adasooriya, N. M. and Karunaratne, V. (2011). A green slow-release fertilizer composition based on urea-modified hydroxyapatite nanoparticles encapsulated wood, *Curr. Sci.* 101(1), 73-78.

Kottegoda, N., Sandaruwan, C., Priyadarshana, G., Siriwardhana, A., Rathnayake, U. A., Berugoda Arachchige, D. M., Kumarasinghe, A. R., Dahanayake, D., Karunaratne, V. and Amaratunga, G. A. (2017). Urea-hydroxyapatite nanohybrids for slow release of nitrogen, *ACS Nano* 11(2), 1214–1221.

Lal, R. (2010). Terrestrial sequestration of carbon dioxide (CO2). In: *Developments and Innovation in Carbon Dioxide (CO2) Capture and Storage Technology: Carbon Dioxide (CO2) Storage and Utilization* (vol. 2). Maroto-Valer, M. (Ed.). Woodhead Publishing, pp. 217-298.

Lal, R. (2015). Restoring soil quality to mitigate soil degradation, *Sustainability* 7(5), 5875–5895.

Lal, R. (2018). Digging deeper: A holistic perspective of factors affecting soil organic carbon sequestration in agroecosystems, *Glob. Change Biol.* 24(8), 3285-3301.

Lehmann, J. and Joseph, S. (2015). *Biochar for Environmental Management: Science, Technology and Implementation*. New York: Routledge.

Li, H., Feng, W. T., He, X. H., Zhu, P., Gao, H., Sun, N. and Xu, M. (2017). Chemical fertilizers could be completely replaced by manure to maintain high maize yield and soil organic carbon (SOC) when SOC reaches a threshold in the Northeast China Plain, *J. Integr. Agric.* 16(4), 937-946.

Li, J., Wu, X., Gebremikael, M. T., Wu, H., Cai, D., Wang, B., Li, B., Zhang, J., Li, Y. and Xi, J. (2018). Response of soil organic carbon fractions, microbial community composition and carbon mineralization to high-input fertilizer practices under an intensive agricultural system, *PLoS ONE* 13(4), e0195144.

Liang, C. (2020). Soil microbial carbon pump: mechanism and appraisal, *Soil Ecol. Lett.* 2(4), 241–254.

Liang, C. and Balser, T. C. (2012). Warming and nitrogen deposition lessen microbial residue contribution to soil carbon pool, *Nat. Commun.* 3, 1222.

Liang, C., Schimel, J. P. and Jastrow, J. D. (2017). The importance of anabolism in microbial control over soil carbon storage, *Nat. Microbiol.* 2, 17105.

Liang, C. and Zhu, X. (2021). The soil Microbial Carbon Pump as a new concept for terrestrial carbon sequestration, *Sci. China Earth Sci.* 64(4), 545–558.

Majumder, S., Neogi, S., Dutta, T., Powel, M. A. and Banik, P. (2019). The impact of biochar on soil carbon sequestration: meta-analytical approach to evaluating environmental and economic advantages, *J. Environ. Manage.* 250, 109466.

Malik, M. A., Khan, K. S., Marschner, P. and Ali, S. (2013). Organic amendments differ in their effect on microbial biomass and activity and on P pools in alkaline soils, *Biol. Fertil. Soils* 49(4), 415 -425.

Manjunatha, S. B., Biradar, D. P. and Aladakatti, Y. R. (2016). Nanotechnology and its applications in agriculture: a review, *J. Farm Sci.* 29(1), 1-13.

Mao, H., Zhang, H., Fu, Q., Zhong, M., Li, R., Zhai, B., Wang, Z. and Zhou, L. (2019). Effects of four additives in pig manure composting on greenhouse gas emission reduction and bacterial community change, *Bioresour. Technol.* 292, 121896.

Marchiol, L., Filippi, A., Adamiano, A., Degli Esposti, L., Iafisco, M., Mattiello, A., Petrussa, E. and Braidot, E. (2019). Influence of hydroxyapatite nanoparticles on germination and plant metabolism of tomato (*solanum Lycopersicum L.*): preliminary evidence, *Agronomy* 9(4), 161.

Marklein, A. R. and Houlton, B. Z. (2012). Nitrogen inputs accelerate phosphorus cycling rates across a wide variety of terrestrial ecosystems, *New Phytol.* 193(3), 696–704.

Matus, F., Rumpel, C., Neculman, R., Panichini, M. and Mora, M. L. (2014). Soil carbon storage and stabilisation in andic soils: a review, *CATENA* 120, 102–110.

Medina, J., Calabi-Floody, M., Aponte, H., Santander, C., Paneque, M., Meier, S., Panettieri, M., Cornejo, P., Borie, F. and Knicker, H. (2021). Utilization of inorganic nanoparticles and biochar as additives of agricultural waste composting: effects of end-products on plant growth, C and nutrient stock in soils from a Mediterranean region, *Agronomy* 11(4), 767.

Medina, J., Monreal, C. M., Antilén, M., Calabi-Floody, M., Velasco-Molina, M., Meier, S., Borie, F., Cornejo, P. and Knicker, H. (2020). Influence of inorganic additives on wheat straw composting: characterization and structural composition of organic matter derived from the process, *J. Environ. Manage.* 260, 110137.

Medina, J., Monreal, C., Barea, J. M., Arriagada, C., Borie, F. and Cornejo, P. (2015). Crop residue stabilization and application to agricultural and degraded soils: a review, *Waste Manag.* 42, 41–54.

Meier, S., Curaqueo, G., Khan, N., Bolan, N., Rilling, J., Vidal, C., Fernández, N., Acuña, J., González, M., Cornejo, P. and Borie, F. (2015). Effects of biochar on copper immobilization and soil microbial communities in a metal-contaminated soil, *J. Soils Sediments* 17(5), 1237–1250.

Menezes-Blackburn, D., Jorquera, M. A., Gianfreda, L., Greiner, R. and de la Luz Mora, M. (2014). A novel phosphorus biofertilization strategy using cattle manure treated with phytase - nanoclay complexes, *Biol. Fertil. Soils* 50, 583–592.

Menezes-Blackburn, D., Jorquera, M. A., Greiner, R., Gianfreda, L. and de la Luz Mora, M. (2013). Phytases and phytase-labile organic phosphorus in manures and soils, *Crit. Rev. Environ. Sci. Technol.* 43(9), 916–954.

Merino, C., Nannipieri, P. and Matus, F. (2015). Soil carbon controlled by plant, microorganism and mineralogy interactions, *J. Soil Sci. Plant Nutr.* 15(2), 321–332.

Mikutta, R. and Kaiser, K. (2011). Organic matter bound to mineral surfaces: resistance to chemical and biological oxidation, *Soil Biol. Biochem.* 43(8), 1738–1741.

Mora, M. L., Alfaro, M. A., Jarvis, S. C., Demanet, R. and Cartes, P. (2006). Soil aluminium availability in andisols of southern Chile and its effect on forage production and animal metabolism, *Soil Use Manag.* 22(1), 95–101.

Mora, M. L., Calabi-Floody, M. and Rumpel, C. (2021). Closing biogeochemical cycles and meeting plant requirements by smart fertilizers and innovative organic amendments, *Agronomy* 11(6), 1158.

Mora, M. L., Cartes, P., Demanet, P and Cornforth, I. S. (2002). Effects of lime and gypsum on pasture growth and composition on an acid andisol in Chile, South America, *Commun. Soil Sci. Plant Anal.* 33(13–14), 2069–2081.

Mora, M. L., Cartes, P., Núñez, P., Salazar, M and Demanet, R. (2007). Movement of NO_3^-N and NH_4^+-N in an andisol and its influence on ryegrass production in a short term study, *J. Soil Sci. Plant Nutr.* 7, 46–64.

Muhammed, S. E., Coleman, K., Wu, L., Bell, V. A., Davies, J. A. C., Quinton, J. N., Carnell, E. J., Tomlinson, S. J., Dore, A. J., Dragosits, U., Naden, P. S., Glendining, M. J., Tipping, E. and Whitmore, A. P. (2018). Impact of two centuries of intensive agriculture on soil carbon, nitrogen and phosphorus cycling in the UK, *Sci. Total Environ.* 634, 1486–1504.

Nash, D. M., McDowell, R. W., Condron, L. M. and McLaughlin, M. J. (2019). Direct exports of phosphorus from fertilizers applied to grazed pastures, *J. Environ. Qual.* 48(5), 1380–1396.

Naz, M. Y. and Sulaiman, S. A. (2016). Slow release coating remedy for nitrogen loss from conventional urea: a review, *J. Control. Release* 225(10), 109-120.

Ngo, P. T., Rumpel, C., Dignac, M.-F., Billou, D., Duc, T. T. and Jouquet, P. (2011). Transformation of buffalo manure by composting or vermicomposting to rehabilitate degraded tropical soils, *Ecol. Eng.* 37(2), 269-276.

Ngo, P. T., Rumpel, C., Doan, T. T., Thierry H.-D.-T., Ce D.-K. and Jouquet, P. (2014). Use of organic substrates for increasing soil organic matter quality and carbon sequestration of tropical degraded soil (a 3 years mesocosms experiment), *Carbon Manag.* 5, 155-168.

Ngo, P. T., Rumpel, C., Janeau, J. L., Dang, D., Doan, T. T. and Jouquet, P. (2016). Mixing of biochar with organic amendments reduces carbon removal after field exposure under tropical conditions, *Ecol. Eng.* 91, 378-380.

Nguyen, M. N. (2021). Potential use of silica-rich biochar for the formulation of adaptively controlled release fertilizers: a mini review, *J. Clean. Prod.* 307, 127188.

Núñez, P. A., Demanet, R., Misselbrook, T. H. et al. and Jouquet, P. (2010). Nitrogen losses under different cattle grazing frequencies and intensities in a volcanic soil of Southern Chile, *Chil. J. Agric. Res.* 70, 237-250.

Obalum, S. E., Chibuike, G. U., Peth, S. and Ouyang, Y. (2017). Soil organic matter as sole indicator of soil degradation, *Environ. Monit. Assess.* 189(4), 176.

OECD/FAO (2018). OECD-FAO Agricultural Outlook 2018-2027, OECD Publishing, Paris/ Food and Agriculture Organization of the United Nations, Rome, Italy.

Ollivier, J., Töwe, S., Bannert, A., Hai, B., Kastl, E. M., Meyer, A., Su, M. X., Kleineidam, K. and Schloter, M. (2011). Nitrogen turnover in soil and global change, *FEMS Microbiol. Ecol.* 78(1), 3-16.

Ouyang, Y., Evans, S. E., Friesen, M. L. and Tiemann, L. K. (2018). Effect of nitrogen fertilization on the abundance of nitrogen cycling genes in agricultural soils: a meta-analysis of field studies, *Soil Biol. Biochem.* 127, 71-78.

Paetsch, L., Mueller, C. W., Rumpel, C., Houot, S. and Kögel-Knabner, I. (2016). Urban waste composts enhance OC and N stocks after long-term amendment but do not alter organic matter composition, *Agric. Ecosyst. Environ.* 223, 211-222.

Pagliari, P. H. and Laboski, C. A. M. (2012). Investigation of the inorganic and organic phosphorus forms in animal manure, *J. Environ. Qual.* 41(3), 901-910.

Panichini, M., Matus, F., Mora, M. L., Godoy, R., Bolan, N. S., Rumpel, C. and Borie, F. (2012). Carbon distribution in top- and subsoil horizons of two contrasting andisols under pasture or forest, *Eur. J. Soil Sci.* 63(5), 616-624.

Paredes, C., Staunton, S., Durán, P., Rodríguez, R. and de la Luz Mora, M. (2022). Assessment of the combined effects of beef cattle manure and lemon peel waste on soil-plant biochemical properties and phosphorus uptake by ryegrass, *Appl. Soil Ecol.* 169, 104207.

Paredes, C., Staunton, S., Simeone, G. D. R., Rao, M. A., Violante, A. and de la Luz Mora, M. (2021). Citrus residue enhances the effectiveness of beef cattle manure improving the phosphorus availability in acidic andisol, *Commun. Soil Sci. Plant Anal.* 52(21), 2767-2781.

Parfitt, R. L., Russell, M. and Orbell, G. E. (1983). Weathering sequence of soils from volcanic ash involving allophane and halloysite, New Zealand, *Geoderma* 29(1), 41-57.

Park, J. H., Lamb, D., Paneerselvam, P., Choppala, G., Bolan, N. and Chung, J. W. (2011). Role of organic amendments on enhanced bioremediation of heavy metal(loid) contaminated soils, *J. Hazard. Mater.* 185(2-3), 549-574.

Paustian, K., Lehmann, J., Ogle, S., Reay, D., Robertson, G. P. and Smith, P. (2016). Climate-smart soils, *Nature* 532(7597), 49-57.

Pisa, C., Wuta, M. and Muchaonyerwa, P. (2020). Effects of incorporation of vermiculite on carbon and nitrogen retention and concentration of other nutrients during composting of cattle manure, *Bioresour. Technol. Rep.* 9, 100383.

Plaza, C. and Senesi, N. (2009). Chapter 4: The effect of organic matter amendment on native soil humic substances, in Senesi, P.M. and Huang, X. B. (Eds), *Biophysico-Chemical Processes Involving Natural Nonliving Organic Matter in Environmental System. IUPAC Series on Biophysico-Chemical Processes in Environmental Systems (N)* pp. 147-181, (vol. 2). Hoboken, NJ: J. Wiley & Sons.

Poblete-Grant, P., Biron, P., Bariac, T., Cartes, P., Mora, MdL. L. and Rumpel, C. (2019). Synergistic and antagonistic effects of poultry manure and phosphate rock on soil P availability, ryegrass production, and P uptake, *Agronomy* 9(4), 191.

Poblete-Grant, P., Cartes, P., Pontigo, S., Biron, P., de La Luz Mora, M. and Rumpel, C. (2022). Phosphorus fertiliser source determines the allocation of root-derived organic carbon to soil organic matter fractions, *Soil Biol. Biochem.* 167, 108614.

Poblete-Grant, P., Suazo-Hernández, J., Condron, L., Rumpel, C., Demanet, R., Malone, S. L. and Mora, MdL. L. (2020). Soil available P, soil organic carbon and aggregation as affected by long-term poultry manure application to andisols under pastures in Southern Chile, *Geoderma Reg.* 21, e00271.

Prost, K., Borchard, N., Siemens, J., Kautz, T., Séquaris, J. M., Möller, A. and Amelung, W. (2013). Biochar affected by composting with farmyard manure, *J. Environ. Qual.* 42(1), 164-172.

Rahman, M. T., Zhu, Q. H., Zhang, Z. B., Zhou, H. and Peng, X. (2017). The roles of organic amendments and microbial community in the improvement of soil structure of a vertisol, *Appl. Soil Ecol.* 111, 84-93.

Redel, Y., Cartes, P., Velásquez, G., Poblete-Grant, P., Bol, R. and Mora, M. L. (2016). Assessment of phosphorus status influenced by Al and Fe compounds in volcanic grassland soils, *J. Soil Sci. Plant Nutr.* 16, 490-506.

Ren, X., Wuang, Q., Awasthi, M. K., Zhao, J., Wang, J., Liu, T., Li, R. and Zhang, Z. (2019). Improvement of cleaner composting production by adding diatomite: from the nitrogen conservation and greenhouse gas emission, *Bioresour. Technol.* 286, 121377.

Richardson, A. E., Barea, J. M., McNeill, A. M. and Prigent-Combaret, C. (2009a). Acquisition of phosphorus and nitrogen in the rhizosphere and plant growth promotion by microorganisms, *Plant Soil* 321(1-2), 305-339.

Richardson, A. E., Hocking, P. J., Simpson, R. J. and George, T. S. (2009b). Plant mechanisms to optimise access to soil phosphorus, *Crop Pasture Sci.* 60(2), 124-143.

Saidy, A. R., Smernik, R. J., Baldock, J. A., Kaiser, K. and Sanderman, J. (2013). The sorption of organic carbon onto differing clay minerals in the presence and absence of hydrous iron oxide, *Geoderma* 209-210, 15-21.

Sato, S., Solomon, D., Hyland, C., Ketterings, Q. M. and Lehmann, J. (2005). Phosphorus speciation in manure and manure-amended soils using XANES spectroscopy, *Environ. Sci. Technol.* 39(19), 7485-7491.

Schlesinger, W. H. (1986). Changes in soil carbon storage and associated properties with disturbance and recovery, in *The Changing Carbon Cycle: A Global Analysis*, pp. 194-220, Trabalka, J. R. and Reichle, D. E. (Eds). New York: Springer-Verlag.

Scotti, R., Bonanomi, G., Scelza, R., Zoina, A. and Rao, M. A. (2015). Organic amendments as sustainable tool to recovery fertility in intensive agricultural systems, *J. Soil Sci. Plant Nutr.* 15(2), 333-352.

Seguel, A., Cornejo, P., Ramos, A., Von Baer, E., Cumming, J. and Borie, F. (2017). Phosphorus acquisition by three wheat cultivars contrasting in aluminium tolerance growing in an aluminium-rich volcanic soil, *Crop Pasture Sci.* 68(4), 305-316.

Shepherd, M. A. and Newell-Price, J. P. (2016). The effects of pig manure type and application timing and frequency on nitrate leaching from a seven-course arable rotation on a retentive alluvial soil, *Soil Use Manag.* 32, 117-126.

Shin, J., Jang, E., Park, S., Ravindran, B. and Chang, S. W. (2019). Agro-environmental impacts, carbon sequestration, and profit analysis of blended biochar pellet application in the paddy soil-water system, *J. Environ. Manage.* 244, 92-98.

Shin, J., Park, D., Hong, S., Jeong, C., Kim, H. and Chung, W. (2021). Influence of activated biochar pellet fertilizer application on greenhouse gas emissions and carbon sequestration in rice (*Oryza sativa L.*) production, *Environ. Pollut.* 285, 117457.

Smith, P. (2016). Soil carbon sequestration and biochar as negative emission technologies, *Glob. Change Biol.* 22(3), 1315-1324.

Torn, M. S., Trumbore, S. E., Chadwick, O. A., Vitousek, P. M. and Hendricks, D. M. (1997). Mineral control of soil organic carbon storage and turnover, *Nature* 389(6647), 170-173.

Trenkel, M. E. (1997). *Improving Fertilizer Use Efficiency: Controlled-Release and Stabilized Fertilizers in Agriculture*. The International Fertilizer Industry Association, Paris. International Fertilizer Industry Association (IFA), Paris, France. pp. 151.

Usman, M., Farooq, M., Wakeel, A., Nawaz, A., Cheema, S. A., Rehman, H. U., Ashraf, I. and Sanaullah, M. (2020). Nanotechnology in agriculture: current status, challenges and future opportunities, *Sci. Total Environ.* 721, 137778.

Vandecasteele, B., Sinicco, T., D'Hose, T., Vanden Nest, T. and Mondini, C. (2016). Biochar amendment before or after composting affects compost quality and N losses, but not plant uptake, *J. Environ. Manage.* 168, 200-209.

Velásquez, G., Ngo, P. T., Rumpel, C., Calabi-Floody, M., Redel, Y., Turner, B. L., Condron, L. M. and Mora, MdlL. (2016). Chemical nature of residual phosphorus in andisols, *Geoderma* 271, 27-31.

Vermeiren, C., Kerckhof, P., Reheul, D. and Smolders, E. (2021). Increasing soil organic carbon content can enhance the long-term availability of phosphorus in agricultural soils, *Eur. J. Soil Sci.* 73(1), 1-18.

Vidal, A., Lenhart, T., Dignac, M. F., Biron, P., Höschen, C., Barthod, J., Vedere, C., Vaury, V., Bariac, T. and Rumpel, C. (2020). Promoting plant growth and carbon transfer to soil with organic amendments produced with mineral additives, *Geoderma* 174, 114454.

Vistoso, E., Bolan, N. S., Theng, B. K. and Mora, M. L. (2009). Kinetics of molybdate and phosphate sorption by some Chilean andisols, *J. Soil Sci. Plant Plant Nutr.* 9, 55-68.

Vistoso, E., Iraira, S. and Sandaña, P. (2021). Phosphorus use efficiency in permanent pastures in andisols, *J. Soil Sci. Plant Nutr.* 21(4), 2587-2599.

Vistoso, E., Theng, B. K. G., Bolan, N. S., Parfitt, R. L. and Mora, M. L. (2012). Competitive sorption of molybdate and phosphate in andisols, *J. Soil Sci. Plant Nutr.* 12(1), 59-72.

Wada, K. (1989). Allophane and imogolite. In: *Minerals in Soil Environments* (2nd edn.) Dixon, J. B. and Weed, S. B. (Eds). Soil Science Society of America, Madison, WI, pp. 1051-1087.

Waldrip, H. M., He, Z. and Erich, M. S. (2011). Effects of poultry manure amendment on phosphorus uptake by ryegrass, soil phosphorus fractions and phosphatase activity, *Biol. Fertil. Soils* 47(4), 407–418.

Wang, Q., Awasthi, M. K., Ren, X., Zhao, J., Li, R., Wang, Z., Wang, M., Chen, H. and Zhang, Z. (2018). Combining biochar, zeolite and wood vinegar for composting of pig manure: the effect on greenhouse gas emission and nitrogen conservation, *Waste Manag.* 74, 221–230.

Wang, C., Tu, Q., Dong, D., Strong, P. J., Wang, H., Sun, B. and Wu, W. (2014). Spectroscopic evidence for biochar amendment promoting humic acid synthesis and intensifying humification during composting, *J. Hazard. Mater.* 280, 409–416.

Warman, P. R. and Cooper, J. M. (2000). Fertilization of a mixed forage crop with fresh and composted chicken manure and NPK fertilizer: effects on dry matter yield and soil and tissue N, P and K, *Can. J. Soil Sci.* 80(2), 337–344.

Wei, K., Sun, T., Tian, J., Chen, Z. and Chen, L. (2018). Soil microbial biomass, phosphatase and their relationships with phosphorus turnover under mixed inorganic and organic nitrogen addition in a *Larix gmelinii* plantation, *For. Ecol. Manag.* 422, 313–322.

Wickramatilake, A. R. P., Kouno, K. and Nagaoka, T. (2010). Compost amendment enhances the biological properties of andosols and improves phosphorus utilization from added rock phosphate, *Soil Sci. Plant Nutr.* 56(4), 607–616.

Wiesmeier, M., Urbanski, L., Hobley, E., Lang, B., von Lützow, M., Marin-Spiotta, E., van Wesemael, B., Rabot, E., Ließ, M., Garcia-Franco, N., Wollschläger, U., Vogel, H. and Kögel-Knabner, I. (2019). Soil organic carbon storage as a key function of soils - a review of drivers and indicators at various scales, *Geoderma* 333, 149–162.

Woignier, T., Pochet, G., Doumenc, H., Dieudonné, P. and Duffours, L. (2007). Allophane: a natural gel in volcanic soils with interesting environmental properties, *J. Sol Gel Sci. Technol.* 41(1), 25–30.

Xia, L., Lam, S. K., Yan, X. and Chen, D. (2017). How does recycling of livestock manure in agroecosystems affect crop productivity, reactive nitrogen losses, and soil carbon balance?, *Environ. Sci. Technol.* 51(13), 7450–7457.

Yan, J., Wang, L., Hu, Y., Tsang, Y. F., Zhang, Y., Wu, J., Fu, X. and Sun, Y. (2018). Plant litter composition selects different soil microbial structures and in turn drives different litter decomposition pattern and soil carbon sequestration capability, *Geoderma* 319, 194–203.

Yeoh, Y. K., Paungfoo-Lonhienne, C., Dennis, P. G., Robinson, N., Ragan, M. A., Schmidt, S. and Hugenholtz, P. (2016). The core root microbiome of sugarcanes cultivated under varying nitrogen fertilizer application, *Environ. Microbiol.* 18(5), 1338–1351.

Yin, Y., Wang, L., Liang, C., Xi, F., Pei, Z. and Du, L. (2016). Soil aggregate stability and iron and aluminium oxide contents under different fertiliser treatments in a long-term solar greenhouse experiment, *Pedosphere* 26(5), 760–767.

Zhang, W., Liu, K., Wang, J., Shao, X., Xu, M., Li, J., Wang, X. and Murphy, D. V. (2015a). Relative contribution of maize and external manure amendment to soil carbon sequestration in a long-term intensive maize cropping system, *Sci. Rep.* 5, 10791.

Zhang, L., Zheng, J., Chen, L., Shen, M., Zhang, X., Zhang, M., Bian, X., Zhang, J. and Zhang, W. (2015b). Integrative effects of soil tillage and straw management on crop yield sand greenhouse gas emissions in a rice-wheat cropping system, *Eur. J. Agron.* 63, 47–54.

Zhu, J., Li, M. and Whelan, M. (2018). Phosphorus activators contribute to legacy phosphorus availability in agricultural soils: a review, *Sci. Total Environ.* 612, 522–537.

Part 2

Measuring carbon sequestration in soils

Chapter 10

Measuring and monitoring soil carbon sequestration

Matthias Kuhnert, Sylvia H. Vetter and Pete Smith, Institute of Biological & Environmental Sciences, University of Aberdeen, UK

1 Introduction

Soils have recently received attention in the context of climate change, because they are the largest terrestrial carbon pool (Batjes, 1996, Lal, 2004), and small changes in this large reservoir may affect atmospheric CO_2 concentrations. To preserve current atmospheric CO_2 levels and to limit climate change impacts, it is thus important to protect the carbon stored in soils and prevent its release into the atmosphere. Carbon may be stored in soils at long timescales, and soil carbon sequestration may therefore have a large potential as negative emission technology (Paustian et al., 2016, Minasny et al., 2017, Paustian et al., 2019). Such technologies need to be employed to meet the climate targets of the Paris Agreement or the different national net zero targets (Climate Ambition Alliance: Net Zero 2050). Increasing soil organic carbon (SOC) sequestration will remove atmospheric carbon and store it in the soil. This process can have a range of positive side effects (Smith et al., 2021). Rumpel et al. (2018) describe eight steps to make the soil more resilient and more productive and improve the storage of carbon. One central part of awarding land managers for improving soils and for the application of SOC storage as a large-scale climate mitigation strategy is the availability of a system for measuring, reporting and verification (MRV) of the effect of land management practice. This includes the quantification of SOC changes over time. However, SOC sequestration is a complex and slow process affected by a wide range of factors (see Chapters 2–9 of this book). This makes the

http://dx.doi.org/10.19103/AS.2022.0106.09

measurement and monitoring of SOC challenging. While point measurements are associated with errors and uncertainties, large-scale quantification presents even more challenges, as it requires either large amounts of samples, which are costly and labour intensive, or upscaling methods based on assumptions.

Soils are used for different purposes and are often managed by a variety of users with their own interests. Activities to increase SOC are not necessarily the main interest of the users and/or landowners. Farmers, for example, need to rely on constant harvest to provide food and make a living. Incentives or legal obligations are required to introduce changes to maintain and/or increase SOC (see Chapters 26–29 of this book), and their implementation depends on a functional monitoring system. This system will be based on the available and future development of tools and approaches to quantify SOC changes (measurements, modelling, etc.). Available tools show a wide range of complexity and accuracy, with a general trend towards the application of simpler options with easy, cheap and rapid methods.

To account for varying complexity of methodologies, the Intergovernmental Panel on Climate Change (IPCC) introduced a three-step tier system, with Tier 1 indicating a basic method with an equation and default emission factors, Tier 2 using the same equation but country-/region-specific emission factors and Tier 3 any more complex method, ranging from alternative equations to process-based models. The equation and default factor for Tier 1 describes a linear relation between an activity (e.g. fertilizer application on the field) and the related estimated greenhouse gas (GHG) emissions. Scientific data build the basis for this relation, which is a simplified but effective and standardised approach to estimating GHG emissions. While the used values in Tier 1 are more generic (based on global data), the Tier 2 approach is similar to Tier 1 but uses country-specific values that provide a more accurate estimate for the target country. Available emission factors are summarised in the emission factor database (https://www.ipcc-nggip.iges.or.jp/EFDB/main.php).

There is an increasing interest in the economics of carbon sequestration. While there is already an interest in an investment in more sustainable companies (Kareiva et al., 2015), carbon accounting and trading of carbon units generated by SOC sequestration has started already. However, all of these different interests rely on a functional and applicable MRV system. Recent research focused on MRV applications and has developed suitable frameworks (Paustian et al., 2019, Smith et al., 2020, FAO, 2020). In this chapter, we will present available systems and discuss their advantages and limitations.

2 Measurement/monitoring, reporting and verification

In this chapter, we distinguish MRV frameworks from MRV applications. While a framework provides a more theoretical description of an optimum MRV system, the application describes an applied MRV system, using protocols that include concrete definitions of used models, measurement approaches and other required details (e.g. responsibilities for the different actions). Smith et al. (2020) outlined a generic concept for an MRV framework as a combination of different approaches to quantify SOC change over time. Generally, management changes are applied on a field or farm level. The farm level is the scale for the beneficiaries of subsidies or carbon trading. Therefore, field and farm scales are most relevant to an MRV scheme. A central part of quantifying SOC changes on this scale is field measurements, but these are costly and labour intensive. Additionally, MRV protocols often lack clear measurement standards (Bispo et al., 2017). While some aspects are clarified (depth of the top 30 cm, 1 m if possible; FAO, 2020), other specifications are missing (number of samples, date of sampling relative to the management practices, spatial distribution of sampling, etc.). Therefore, alternative approaches to measurements need to be considered. Modelling is a very attractive alternative, as all problems and limitations of the measurements are resolved by using a model. But the quality of the simulation result needs to be considered, especially in comparison to measurements. Models include errors and uncertainty based on the assumptions used and the underlying concepts, and they require data for calibration and validation, in addition to those needed for running the models. MRV frameworks, as outlined by the FAO (2020) and Smith et al. (2020), specify that only calibrated models can contribute to SOC quantification, but further specification of the models is not provided. A combination of both (measurement and modelling) will compensate for the disadvantages of each other and improve the result (Smith et al., 2020). Overall, a combination of different approaches secures the optimum quantification of SOC changes over time. Smith et al. (2020) list seven components of an MRV framework: long-term experimental sites, field experiments (short term), field-specific modelling, spatial data analysis combined with modelling, collection and aggregation of activity data (e.g. conventional and intervention management), remote sensing and spatial re-sampling. The different components complement each other to allow an optimum framework for measuring and verifying SOC changes over time. There are advantages and disadvantages of all different components, and all show some limitations. Only a combination of all, or at least several of these methods, will provide good MRV outcomes.

2.1 Measurements

Direct measurements of SOC content involve quantifying the fine earth and coarse earth fraction, the organic carbon concentration and soil bulk density or fine earth mass (FAO, 2019). Estimating the rock content of sample soils can be a challenge but will significantly affect soil bulk density (Poeplau et al.,

2017, Throop et al., 2012). Another challenge is that a change in management (whole practice as well as depth at which that practice is applied) will not only impact the bulk density of the soil but also the amount of soil in a soil sample at a certain depth (Haynes and Naidu, 1998). Therefore, corrections and use of the equivalent mass approach may be necessary (Chapter 11 of this book). As soils are characterised by high spatial variability, direct measurements rely on appropriate study designs and sampling protocols (Minasny et al., 2017, Chapter 12 of this book). At the field scale, a large number of soil samples are usually required to give reliable SOC stock estimates with an acceptable error margin (Garten and Wullschleger, 1999, Vanguelova et al., 2016).

The IPCC recommends a sample depth of 30 cm, but several methods for increasing SOC content require deeper sampling to confirm the expected effect (Smith et al., 2020). For example, the effect of a no-tillage practice on the SOC content may be overestimated if the measuring depth is insufficient (Angers and Eriksen-Hamel, 2008, Blanco-Canqui and Lal, 2008).

A change in SOC stocks can also be estimated through indirect measurements and by presenting the full carbon budget. This approach uses the net balance of carbon fluxes measured through chamber measurements or the eddy covariance (EC) method (Baldocchi, 2003). From the carbon fluxes, the initial uptake of carbon through photosynthesis and its subsequent partial loss through respiration (from soil, plant and litter) are estimated to give net ecosystem exchange or net ecosystem production and further C inputs (organic fertilization) and outputs (harvest) to and from the system (Smith et al., 2010, Soussana et al., 2010). Through this complied carbon budget, a change in SOC can be estimated. This approach indirectly measures the change in SOC for larger landscapes but can only be used under horizontal homogeneity of the footprint area and under sufficient air turbulences (Aubinet et al., 1999). The maintenance of most measurement systems is costly and time-consuming. The post-processing of the measured data also needs time and expert knowledge about flux corrections for density and gap filling (Falge et al., 2001, Reichstein et al., 2005). In an MRV application, EC provides landscape-specific data, which can be used as baseline data or for model optimisation purposes (calibration and validation).

Long-term study sites are crucial for the implementation of the MRV framework (Smith et al., 2020). Study sites for different management combinations allow a long-term observation and quantification of all relevant parameters and variables that affect SOC sequestration. 'Long-term' is relative and not defined as a fixed duration. The IPCC suggests 20 years as the default period to observe SOC changes because SOC sequestration rates are fast at the beginning but slow down over time until they approach zero (Sommer and Bossio, 2014). Measurements are impractical for a generic implementation of an MRV process (too costly and labour intensive), and other solutions that replace or at least reduce the sampling intensity in the field are required. Besides modelling and remote sensing, long-term study sites in combination with short-term field experiments can complement field measurements. These study sites provide data for re-assessment of potential impacts, reference for expected changes or baseline for a particular management practice. Additionally, these data will be the basis for the development, calibration and validation of models and remote sensing approaches. Ideally, land cover, soil, climate, management and environmental conditions are represented by available study sites or at least a reasonable number of combinations (good representation of all climate zones, soil types, crop species, etc.). A standard protocol for the acquisition of these data would be beneficial, because differences in the set-up of the measurement approaches could introduce uncertainty. Organizing and providing these data on accessible platforms is the best way for an open and transparent handling of the data. Two platforms for long-term experimental sites were initiated by the SOMNET (Smith et al., 2002) and the EuroSOMNET (https://www.ufz.de/somnet/, Franko et al., 2002) platforms. The SOMNET platform evolved later into an online, real-time inventory project, including a webpage with Long-Term Soil-Ecosystems Experiments. The database contains meta-data of more than 200 long-term experiments and is hosted by the International Soil Carbon Network (http://iscn.fluxdata.org/network/partner-networks/ltse/). More than 80% of the long-term experimental sites concern agricultural systems (Smith et al., 2012). However, the majority of the sites are in the temperate climate zone with a focus on Europe and North America, under-representing tropical and sub-tropical regions and the Southern hemisphere (Smith et al., 2012). For good coverage of the variability of global agricultural systems, more long-term sites in other parts of the world need to be established. The better the representation of different management options, soil and climate zones by experimental sites, the better the data basis for MRV application. This requires immediate action, as study sites that are established today, will be able to be used to assess long-term effects on SOC in 20 years (Smith et al., 2012). Special funding is required to initiate long-term monitoring sites, as project funding for 3–5 years duration is insufficient.

2.2 Remote sensing

Besides *in-situ* measurements, remote sensing can support the monitoring of SOC changes and/or provide data for the verification of measured SOC changes. This technology allows non-invasive measurements, including at a large scale. Remote sensing can be applied in the lab or in the field by handheld or transportable systems or by airborne or satellite device devices (Chabrillat et al., 2019). As part of an MRV application, the latter two options are more useful, as these systems allow a wider coverage and delivery of large-scale data globally. Considering the wide application and availability of the data, this would reduce costs for monitoring SOC changes (Nocita et al., 2015) once the approach is established. There are different approaches used for SOC estimation, and two of them are highlighted below.

One established remote sensing approach is reflectance spectroscopy. It uses characteristic spectra that are reflected from the soil surface for quantitative and quantitative analysis of soil properties. The recommended wavelength range for these measurements is the visible near infrared–shortwave infrared (700–2500 nm), as it shows a good signal-to-noise ratio and is a cost- and time-effective option for spectroscopy (Mohamed et al., 2018). The characteristic spectra are reflected by the bonds in the SOC molecules (O-H, N-H, C-H), which allow a qualitative and quantitative analysis of SOC. This method provides soil-type-specific quantitative SOC estimates (Grinand et al., 2012). To secure a wider application without site specific measurements, spectral libraries are required that contain several thousand soil types with varying soil properties as a reference. This is a cost- and time-effective alternative to other traditional measurement options in the laboratory, such as wet digestion or dry combustion (Nayak et al., 2019).

The introduced high spectroscopy measures for fixed wavelength using multispectral sensors, which is associated with some limitations, especially for quantitative measurements on SOC (Ben-Dor et al., 2018). Therefore, recent developments in hyperspectral sensors show an improved approach with higher capability for quantitative data over large areas. Hyperspectral remote sensing (also called image spectroscopy) provides a continuous spectrum for each pixel, using 100 or more contiguous spectral bands. However, Ben-Dor et al. (2018) also list a wide range of drawbacks with the signal-to-noise ratio as a major problem (caused, e.g. by the non-transparent atmosphere, problems with sensor calibration) and more problems with changing conditions (e.g. changes in soil particle size). Further developments are required to improve the approach. For large-scale applications, there is again a demand for developing new libraries for the new approach.

The advantage of large-scale remote sensing using airborne devices and/or satellites is that it can provide additional information, e.g. land-use change (Winkler et al., 2021), primary production (Zhao et al., 2005) or different soil properties (Viscarra Rossel et al., 2006). Nevertheless, remote sensing has limitations. The availability of images is affected by cloud cover, measurements are affected by plant cover on the ground and only the top centimetre can be measured (Smith et al., 2020). Despite good results in different studies and the availability of spectral libraries, the measurement is still uncertain, which renders remote sensing as the sole MRV method unsuitable. In contrast, it is an excellent additional approach to complement other methods and should therefore be used only in combination.

The latest developments in multispectral systems to quantify SOC have shown great progress (Aldana-Jague et al., 2016). These kinds of measurements have the potential to reduce uncertainty (Chabrillat et al., 2019), new libraries have to be built for the new approach. Chabrillat et al. (2019) refer also to studies using hyperspectral systems (Gomez et al., 2008, Lu et al., 2013) but rates the performance as moderate. Similar to the other approaches, remote sensing shows good potential to complement measurements and reduced costs but is not able to replace field measurements completely.

2.3 Modelling

As there are limitations to field measurements and remote sensing, modelling becomes the most prominent supplement to provide data for MRV applications. Models can contribute in different ways to MRV: (1) provide baseline information, (2) interpolate measurements (temporally and spatially), (3) extrapolate measurements for projections or for an ex-ante assessment, (4) estimate SOC changes, and (5) provide information for an optimised measurement plan. Different models with different complexity and accuracy can be used in MRV applications. However, there are no standards defining the quality of a model used in an MRV application. Choosing the right model depends on the objective, data availability and modelling skills of the user (Table 1), as different models vary in their characteristics, complexity and accuracy (Table 1).

Biogeochemical models seem to be most suitable for MRV approaches, as they are able to simulate SOC with the highest accuracy and provide additional information about impacts on yield and GHG emissions (Camino-Serrano et al., 2018, Campbell and Paustian, 2015). However, these models are sometimes impractical, as they require a large amount of data and expert knowledge to use them. Emission factors and simple equations are often used by carbon trading platforms, but these models are developed for large-scale (country scale) applications and might show large errors on the field scale. Most suitable seems to be decision support tools for carbon accounting, like the Cool Farm Tool (CFT, https://coolfarmtool.org/, Hillier et al., 2011) or COMET-Farm (http://comet-farm.com/, Paustian et al., 2017), as they address or show the potential to address the aforementioned problems of the emission factors and process-based models (Whittaker et al., 2013). These tools use different routines of different complexities (Tier 1 to Tier 3, depending on the tool). Unfortunately, the most popular options also show limitations, as the SOC component of the CFT uses the Tier 1 approach of the IPCC 2006 guidelines (although this is under review at the moment) and the COMET-Farm Tool uses a Tier 3 approach in combination with a database that only covers the United States. Further developments of both tools are ongoing, and in the future, these may be reasonable options. Both tools are developed for stakeholders with

Table 1 Specifications for different model categories

	Emission factors	Decision support tools	SOC models	Biogeochemical models (Tier 3)
Data requirement	Low	High (farm-specific data)	High (environmental data)	High (environmental data)
Calibration requirement	Low	Low	High	High
Required expertise	Low	Medium	High	High
Management options	Medium (categories)	Medium-high	No-high	High
Targeted scale	Country and larger	Field-farm	Point/site	Point/site
Uncertainty/expected error for field scale	High	Medium-high	Low	Low
Example models	UNFCCC models, Tier 1, Tier 2	Cool Farm Tool, Comet Farm Tool	RothC	EPIC, DAYCENT, DNDC

The category emission factors also include simple empirical equations (Tier 2 approaches). The categories SOC and biogeochemical models (Tier 3 approaches) are separated to indicate if a model only includes SOC dynamics or also addresses processes (e.g. N cycle, plant growth). The category decision support includes tools that are designed to provide information on GHG emissions, SOC changes or both (these tools use mainly Tier 1 and Tier 2 approaches but can also include Tier 3 routines).

an easy-to-use interface. The CFT was developed with an interface usable by farmers and the input information they have at hand. The CFT calculates the GHG emissions on a farm level for a specific site and specific management. The methods used within CFT range from emission factors to a model approach considering region-specific parameters and farm-level data (e.g. management, soil, climate) on an annual basis. Therefore, the CFT can be seen as a Tier 2 or simple Tier 3 model. COMET-Farm calculates the carbon footprint of a farm. The tool provides the opportunity to test different management interventions and explore their mitigation potential, i.e. the potential reduction of GHG emissions. GHG estimates for crops are calculated using the DayCent dynamic model (Del Grosso et al., 2010, Parton et al., 1998) - a process-based model - and follow the official USDA GHG inventory guidelines for entity-scale reporting (Eve et al., 2014). Both tools consider soil carbon sequestration and calculate the SOC change for a land use change or change in soil management.

3 Existing monitoring, reporting and verification protocols

The FAO (2020) published a protocol that provides concrete guidelines about the structure and steps to apply an MRV application (Fig. 1). The FAO protocol differentiates reporting into four different categories: (1) pre-implementation report, (2) initial report, (3) biannual report and (4) final report. These reports describe the different stages of MRV frameworks as outlined in Fig. 1. The protocol also provides suggestions and guidelines for the responsibilities of MRV framework. It is suggested that the reporting can be organised by the farmer but needs to be done in consultation with a relevant expert. An independent person or entity must verify the reports. The monitoring and verification require expert knowledge, which can be secured by the accreditation of independent experts specialised in these activities and by external expert reviews, respectively. The accreditation and certification can be organised by governmental institutions (e.g. for subsidies) and other large organisations (e.g. the FAO or entities). Certain specifications are not included (e.g. measurement approaches, suitable models) as the protocol is a blueprint for global use and might require local adaptation.

In addition to the FAO's standardised framework, two other examples of MRV protocols in Alberta, Canada and Australia are already in place. The Government of Alberta published an MRV protocol to quantify the impacts of tillage management on GHG emissions and SOC. It differs from the FAO protocol by specifying the target area and the management to be applied, which allows some aspects to be considered in more detail. For example, some reversal events are allowed for natural farms. Conventional tillage is allowed in less than 10% of the farm area for weed control. Another protocol has been published by

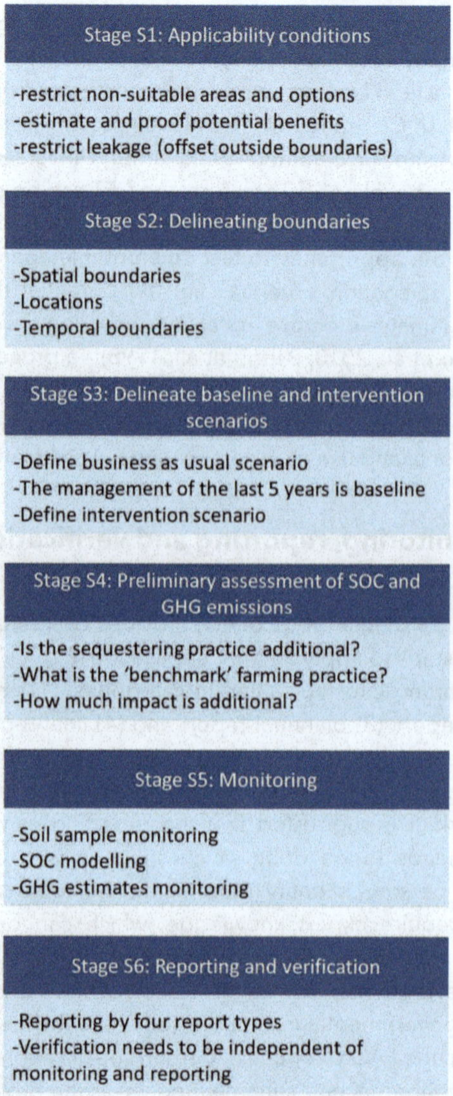

Figure 1 The MRV framework of the FAO (2020) follows a six-stage approach to set up an MRV protocol.

the Australian Government (Smith et al., 2020), which includes bare land and pastures alongside croplands for baseline conditions. This protocol has a similar structure and content as the FAO framework, but it is more specific in defining some details in the management options that are allowed but differs in some other details (e.g. review every 5 years).

Carbon accounting platforms have also started to trade carbon based on SOC gains. The good standards of the protocols are undermined by their implementation. One example of the actual protocols is the Verified Carbon Standard (Shoch et al., 2020). Measurements are very limited (one measurement suggested), SOC changes are quantified by Tier 1 models and the project time is restricted to a short period (e.g. 10 years). The implementation of a simple MRV application by businesses is economically motivated. Even though there is a demand for cost reduction, the methods applied need to be improved to provide an adequate data for carbon trading. Nevertheless, improvements in this sector would provide a business solution that will improve mitigation actions once a functional system is established.

4 Outlook of the use of monitoring, reporting and verification applications

MRV applications are essential to the implementation of strategies to mitigate climate change (Smith et al., 2020). It is also a requirement for subsidies from governmental institutions, carbon trading or the initiatives of companies with net zero targets (FAO, 2020, Kareiva et al., 2015, Paustian et al., 2019). In contrast to MRV applications for other processes or variables, monitoring of SOC stock changes has additional challenges: (1) the slow rate of change in soil carbon against the large background stock, (2) the heterogeneous distribution in space and depth and (3) the complexity of measurements and the reversibility of the gains, which make the requirements of MRV complex. MRV protocols overcome these problems by applying a combination of different methods, to compensate for the limitations of individual quantification methods. The implementation of MRV applications requires an integrated approach but barriers exist. The combined approach can be costly, labour intensive and/or requires a wider skill set (or even expert knowledge). In summary, current MRV methods are often impractical for stakeholders.

In the near future, the challenge for science is to reduce complexity and to remove these barriers in order to provide practical solutions. One relatively easy target could be the development or improvement of models that are easily applicable by stakeholders but that provide robust results at the field and farm scale. More challenging and more time intensive will be the further development of remote sensing approaches. Remote sensing will never be able to replace field measurements, but it will improve the quality of the measurements and might allow for a reduction of the number of samples to save labour time and lower costs.

Other approaches like digital mapping will also contribute to an improved understanding and quantification of SOC changes. Such developments have the potential to improve the measurements in MRV applications.

The following chapters will further detail some aspects of MRV approaches. Chapter 11 of this book will give an overview of methods for quantifying SOC stocks and characterising its turnover times at the profile scale. Chapter 12 of this book will introduce digital soil mapping as an additional option to quantify SOC on a farm level (De Gruijter et al., 2015). The chapter will indicate the advantages and limitations of this approach, including the measurement demand and the associated uncertainty. Chapter 13 of this book will give a detailed overview of SOC modelling approaches with a special focus on its permanency. Finally, Chapter 14 of this book will outline digital stock taking, with a focus on the field scale. This will include an analysis of knowledge gaps in field-specific digital stock taking and new approaches, such as the application of smartphones to quantify SOC stocks. These methods will be discussed in the context of an application in MRV applications, which would bring down measurement costs and potentially improve accuracy.

5 Conclusion

Sequestering atmospheric carbon through increased SOC stocks requires a functional MRV application to monitor the impacts of management practices on the soil. A single quantification approach is not sufficient; instead a combination of different methods is necessary to monitor SOC changes over time and to provide appropriate verification methods. For simplified MRV applications, there is a risk of errors and uncertainty. Developments of the available tools do not meet all the demands for an MRV application applied for different purposes. There is an imbalance between complexity and accuracy (for modelling) as well as in costs and accuracy (measurements). The chapters in this section describe currently available approaches and future developments that might provide effective solutions to be applied in MRV applications. The approaches presented do not target the implementation in MRV applications, but they are measurement tools, which can be used in MRV applications.

6 Where to look for further information

The carbon trading market for soil organic carbon developed very recently. Therefore, there are no agreed standards for MRV systems. The number of trading platforms is increasing, and the protocols of the existing platforms are diverse and still changing. These platforms usually provide the MRV protocols as reports on their webpages, which provide detailed information about the accounting process. However, as these are businesses, no platform should be highlighted here. In a very recent publication, different accounting protocols were compared by Oldfield et al. (2022).

As mentioned, there are no agreed standards and policymakers acknowledged this gap. Therefore, there are a couple of calls for upcoming projects (starting 2023) that will tackle this problem (e.g. projects funded by the HORIZON program of the European Union). These projects as well as some ongoing projects (e.g. ClieNFarms (European), RETINA (UK)) will provide information on this topic in future. The results of former projects and existing governmental initiatives on this topic are listed in the list of references and introduced in the text.

Beside the key publication by Smith et al. (2020), the Food and Agricultural Organization of the United Nations provides a global standard for the application of an MRV system (https://www.fao.org/documents/card/en/c/cb0509en/).

The European Soil Data Centre published in 2015 a sampling protocol to verify changes of the organic carbon stock in the soil for a region in Italy (https://esdac.jrc.ec.europa.eu/content/validation-eu-soil-sampling-protocol-verify-changes-organic-carbon-stock-mineral-soil). Additionally, the European Commission had recently published a working document on Sustainable Carbon Cycles - Carbon Farming (https://ec.europa.eu/clima/system/files/2021-12/swd_2021_450_en_0.pdf).

7 References

Aldana-Jague, E., Heckrath, G., Macdonald, A., van Wesemael, B. and Van Oost, K. 2016. UAS-based soil carbon mapping using VIS-NIR (480–1000 nm) multi-spectral imaging: potential and limitations, *Geoderma* 275, 55–66.

Angers, D. A. and Eriksen-Hamel, N. S. 2008. Full-inversion tillage and organic carbon distribution in soil profiles: a meta-analysis, *Soil Science Society of America Journal* 72(5), 1370–1374.

Aubinet, M., Grelle, A., Ibrom, A., Rannik, Ü., Moncrieff, J., Foken, T., Kowalski, A. S., Martin, P. H., Berbigier, P., Bernhofer, C., Clement, R., Elbers, J., Granier, A., Grünwald, T., Morgenstern, K., Pilegaard, K., Rebmann, C., Snijders, W., Valentini, R. and Vesala, T. 1999. Estimates of the annual net carbon and water exchange of forests: the EUROFLUX methodology, *Advances in Ecological Research* 30, 113–175.

Baldocchi, D. D. 2003. Assessing the eddy covariance technique for evaluating carbon dioxide exchange rates of ecosystems: past, present and future, *Global Change Biology* 9(4), 479–492.

Batjes, N. H. 1996. Total carbon and nitrogen in the soils of the world, *European Journal of Soil Science* 47(2), 151–163.

Ben-Dor, E., Chabrillat, S. and Dematte, J. A. M. 2018. Characterization of soil properties using reflectance spectroscopy. In: *Hyperspectral Remote Sensing of Vegetation* (2nd edn.), Thenkabail, P., Lyon, J. and Huete, A. (Eds), CRC Press, Boca Raton, pp. 187–161.

Bispo, A., Andersen, L., Angers, D. A., Bernoux, M., Brossard, M., Cecillon, L., Comans, R. N. J., Harmsen, J., Jonassen, K., Lame, F., Lhuillery, C., Maly, S., Martin, E., Mcelnea,

A. E., Sakai, H., Watabe, Y. and Eglin, T. K. 2017. Accounting for carbon stocks in soils and measuring GHGs emission fluxes from soils: do we have the necessary standards?, *Frontiers in Environmental Science* 5, 41.

Blanco-Canqui, H. and Lal, R. 2008. No-tillage and soil-profile carbon sequestration: an on-farm assessment, *Soil Science Society of America Journal* 72(3), 693-701.

Camino-Serrano, M., Guenet, B., Luyssaert, S., Ciais, P., Bastrikov, V., De Vos, B., Gielen, B., Gleixner, G., Jornet-Puig, A., Kaiser, K., Kothawala, D., Lauerwald, R., Penuelas, J., Schrumpf, M., Vicca, S., Vuichard, N., Walmsley, D. and Janssens, I. A. 2018. ORCHIDEE-SOM: modeling soil organic carbon (SOC) and dissolved organic carbon (DOC) dynamics along vertical soil profiles in Europe, *Geoscientific Model Development* 11(3), 937-957.

Campbell, E. E. and Paustian, K. 2015. Current developments in soil organic matter modeling and the expansion of model applications: a review, *Environmental Research Letters* 10(12), 123004.

Chabrillat, S., Ben-Dor, E., Cierniewski, J., Gomez, C., Schmid, T. and van Wesemael, B. 2019. Imaging spectroscopy for soil mapping and monitoring, *Surveys in Geophysics* 40(3), 361-399.

Climate Ambition Alliance: Net Zero 2050 2021. Available at: https://climateaction.unfccc.int/views/cooperative-initiative-details.html?id=94.

De Gruijter, J. J., Minasny, B. and McBratney, A. B. 2015. Optimizing stratification and allocation for design-based estimation of spatial means using predictions with error, *Journal of Survey Statistics and Methodology* 3(1, March), 19-42.

Del Grosso, S. J., Ogle, S. M., Parton, W. J. and Breidt, F. J. 2010. Estimating uncertainty in N2O emissions from US cropland soils, *Global Biogeochemical Cycles* 24(1), GB1009.

Eve, M., Flugge, M. and Pape, D. 2014. Chapter 2: considerations when estimating agriculture and forestry GHG emissions and removals. In: *Quantifying Greenhouse Gas Fluxes in Agriculture and Forestry: Methods for Entity-Scale Inventory*, Eve, M., Pape, D., Flugge, M., et al. (Eds), Office of the Chief Economist, U.S. Department of Agriculture, Washington, DC, pp. 2-1-2-25.

Falge, E., Baldocchi, D., Olson, R., Anthoni, P., Aubinet, M., Bernhofer, C., Burba, G., Ceulemans, R., Clement, R., Dolman, H., Granier, A., Gross, P., Grunwald, T., Hollinger, D., Jensen, N. O., Katul, G., Keronen, P., Kowalski, A., Lai, C. T., Law, B. E., Meyers, T., Moncrieff, J., Moors, E., Munger, J. W., Pilegaard, K., Rannik, Ü., Rebmann, C., Suyker, A., Tenhunen, J., Tu, K., Verma, S., Vesala, T., Wilson, K. and Wofsy, S. 2001. Gap filling strategies for defensible annual sums of net ecosystem exchange, *Agricultural and Forest Meteorology* 107(1), 43-69.

FAO. 2020. *A Protocol for Measurement, Monitoring, Reporting and Verification of Soil Organic Carbon in Agricultural Landscapes-GSOC MRV Protocol*, Food and Agriculture Organization of the United Nations, Rome.

FAO. 2019. *Measuring and Modelling Soil Carbon Stocks and Stock Changes in Livestock Production Systems: Guidelines for Assessment (1st version). Livestock Environmental Assessment and Performance (LEAP) Partnership*, FAO, Rome, Italy.

Franko, U., Schramm, G., Rodionova, V., Korschens, M., Smith, P., Coleman, K., Romanenkov, V. and Shevtsova, L. 2002. EuroSOMNET - a database for long-term experiments on soil organic matter in Europe, *Computers and Electronics in Agriculture* 33(3), 233-239.

Garten, C. T. and Wullschleger, S. D. 1999. Soil carbon inventories under a bioenergy crop (switchgrass): measurement limitations, *Journal of Environmental Quality* 28(4), 1359-1365.

Gomez, C., Rossel, R. A. V. and McBratney, A. B. 2008. Soil organic carbon prediction by hyperspectral remote sensing and field vis-NIR spectroscopy: an Australian case study, *Geoderma* 146(3-4), 403-411.

Grinand, C., Barthes, B. G., Brunet, D., Kouakoua, E., Arrouays, D., Jolivet, C., Caria, G. and Bernoux, M. 2012. Prediction of soil organic and inorganic carbon contents at a national scale (France) using mid-infrared reflectance spectroscopy (MIRS), *European Journal of Soil Science* 63(2), 141-151.

Haynes, R. J. and Naidu, R. 1998. Influence of lime, fertilizer and manure applications on soil organic matter content and soil physical conditions: a review, *Nutrient Cycling in Agroecosystems* 51(2), 123-137.

Hillier, J., Walter, C., Malin, D., Garcia-Suarez, T., Mila-i-Canals, L. and Smith, P. 2011. A farm-focused calculator for emissions from crop and livestock production, *Environmental Modelling and Software* 26(9), 1070-1078.

Kareiva, P. M., McNally, B. W., McCormick, S., Miller, T. and Ruckelshaus, M. 2015. Improving global environmental management with standard corporate reporting, *Proceedings of the National Academy of Sciences of the United States of America* 112(24), 7375-7382.

Lal, R. 2004. Soil carbon sequestration impacts on global climate change and food security, *Science* 304(5677), 1623-1627.

Lu, P., Wang, L., Niu, Z., Li, L. and Zhang, W. 2013. Prediction of soil properties using laboratory VIS-NIR spectroscopy and Hyperion imagery, *Journal of Geochemical Exploration* 132, 26-33.

Minasny, B., Malone, B. P., McBratney, A. B., Angers, D. A., Arrouays, D., Chambers, A., Chaplot, V., Chen, Z., Cheng, K., Das, B. S., Field, D. J., Gimona, A., Hedley, C. B., Hong, S. Y., Mandal, B., Marchant, B. P., Martin, M., McConkey, B. G., Mulder, V. L., O'Rourke, S., Richer-de-Forges, A. C., Odeh, I., Padarian, J., Paustian, K., Pan, G., Poggio, L., Savin, I., Stolbovoy, V., Stockmann, U., Sulaeman, Y., Tsui, C., Vågen, T., van Wesemael, B. and Winowiecki, L. 2017. Soil carbon 4 per mille, *Geoderma* 292, 59-86.

Mohamed, E. S., Saleh, A. M., Belal, A. B. and Gad, A. 2018. Application of near-infrared reflectance for quantitative assessment of soil properties, *Egyptian Journal of Remote Sensing and Space Science* 21(1), 1-14.

Nayak, A. K., Rahman, M. M., Naidu, R., Dhal, B., Swain, C. K., Nayak, A. D., Tripathi, R., Shahid, M., Islam, M. R. and Pathak, H. 2019. Current and emerging methodologies for estimating carbon sequestration in agricultural soils: a review, *Science of the Total Environment* 665, 890-912.

Nocita, M., Stevens, A., van Wesemael, B., Aitkenhead, M., Bachmann, M., Barthès, B., Ben Dor, E., Brown, D. J., Clairotte, M., Csorba, A., Dardenne, P., Demattê, J. A. M., Genot, V., Guerrero, C., Knadel, M., Montanarella, L., Noon, C., Ramirez-Lopez, L., Robertson, J., Sakai, H., Soriano-Disla, J. M., Shepherd, K. D., Stenberg, B., Towett, E. K., Vargas, R. and Wetterlind, J. 2015. Soil spectroscopy: an alternative to wet chemistry for soil monitoring, *Advances in Agronomy* 132, 139-159.

Oldfield, E. E., Eagle, A. J., Rubin, R. L., Rudek, J., Sanderman, J., Gordon, D. R. 2022. Crediting agricultural soil carbon sequestration, *Science* 375(6586), 1222-1225. doi: 10.1126/science.abl7991. Epub 2022 Mar 17. PMID: 35298251.

Parton, W. J., Hartman, M., Ojima, D. and Schimel, D. 1998. DAYCENT and its land surface submodel: description and testing, *Global and Planetary Change* 19(1-4), 35-48.

Paustian, K., Collier, S., Baldock, J., Burgess, R., Creque, J., DeLonge, M., Dungait, J., Ellert, B., Frank, S., Goddard, T., Govaerts, B., Grundy, M., Henning, M., Izaurralde, R.

C., Madaras, M., McConkey, B., Porzig, E., Rice, C., Searle, R., Seavy, N., Skalsky, R., Mulhern, W. and Jahn, M. 2019. Quantifying carbon for agricultural soil management: from the current status toward a global soil information system, *Carbon Management* 10(6), 567–587.

Paustian, K., Easter, M., Brown, K., Chambers, A., Eve, M., Huber, A., Marx, E., Layer, M., Stermer, M., Sutton, B., Swan, A., Toureene, C., Verlayudhan, S. and Williams, S. 2017. Field- and farm-scale assessment of soil greenhouse gas mitigation using COMET-Farm, *Agronomy Monographs*, 341–359.

Paustian, K., Lehmann, J., Ogle, S., Reay, D., Robertson, G. P. and Smith, P. 2016. Climate-smart soils, *Nature* 532(7597), 49–57.

Poeplau, C., Vos, C. and Don, A. 2017. Soil organic carbon stocks are systematically overestimated by misuse of the parameters bulk density and rock fragment content, *SOIL* 3(1), 61-66.

Reichstein, M., Falge, E., Baldocchi, D., Papale, D., Aubinet, M., Berbigier, P., Bernhofer, C., Buchmann, N., Gilmanov, T., Granier, A., Grunwald, T., Havrankova, K., Ilvesniemi, H., Janous, D., Knohl, A., Laurila, T., Lohila, A., Loustau, D., Matteucci, G., Meyers, T., Miglietta, F., Ourcival, J. M., Pumpanen, J., Rambal, S., Rotenberg, E., Sanz, M., Tenhunen, J., Seufert, G., Vaccari, F., Vesala, T., Yakir, D. and Valentini, R. 2005. On the separation of net ecosystem exchange into assimilation and ecosystem respiration: review and improved algorithm, *Global Change Biology* 11(9), 1424–1439.

Rumpel, C., Amiraslani, F., Koutika, L. S., Smith, P., Whitehead, D. and Wollenberg, E. 2018. Put more carbon in soils to meet Paris climate pledges, *Nature* 564(7734), 32–34.

Shoch, D., Swails, E. and indigo 2020. *Methodology for Improved Agricultural Land Management*. Verified, Carbon Standard. VCS Methodology, version 1. Report VW0042.

Smith, P., Davies, C. A., Ogle, S., Zanchi, G., Bellarby, J., Bird, N., Boddey, R. M., McNamara, N. P., Powlson, D., Cowie, A., Noordwijk, M., Davis, S. C., Richter, D. D. B., Kryzanowski, L., Wijk, M. T., Stuart, J., Kirton, A., Eggar, D., Newton-Cross, G., Adhya, T. K. and Braimoh, A. K. 2012. Towards an integrated global framework to assess the impacts of land use and management change on soil carbon: current capability and future vision, *Global Change Biology* 18(7), 2089-2101.

Smith, P., Falloon, P., Korschens, M., Shevtsova, L., Franko, U., Romanenkov, V., Coleman, K., Rodionova, V., Smith, J. and Schramm, G. 2002. EuroSOMNET - a European database of long-term experiments on soil organic matter: the WWW metadatabase, *Journal of Agricultural Science* 138(2), 123–134.

Smith, P., Keesstra, S. D., Silver, W. L., Adhya, T. K., De Deyn, G. B., Carvalheiro, L. G., Giltrap, D. L., Renforth, P., Cheng, K., Sarkar, B., Saco, P. M., Scow, K., Smith, J. U., Morel, J. C., Thiele-Bruhn, S., Lal, R. and McElwee, P. 2021. Soil-derived nature's contributions to people and their contribution to the UN sustainable development goals, *Philosophical Transactions of the Royal Society of London. Series B, Biological Sciences* 376(1834), 20200185.

Smith, P., Lanigan, G., Kutsch, W. L., Buchmann, N., Eugster, W., Aubinet, M., Ceschia, E., Beziat, P., Yeluripati, J. B., Osborne, B., Moors, E. J., Brut, A., Wattenbach, M., Saunders, M. and Jones, M. 2010. Measurements necessary for assessing the net ecosystem carbon budget of croplands, *Agriculture, Ecosystems and Environment* 139(3), 302–315.

Smith, P., Soussana, J. F., Angers, D., Schipper, L., Chenu, C., Rasse, D. P., Batjes, N. H., van Egmond, F., McNeill, S., Kuhnert, M., Arias-Navarro, C., Olesen, J. E., Chirinda,

N., Fornara, D., Wollenberg, E., Álvaro-Fuentes, J., Sanz-Cobena, A. and Klumpp, K. 2020. How to measure, report and verify soil carbon change to realize the potential of soil carbon sequestration for atmospheric greenhouse gas removal, *Global Change Biology* 26(1), 219-241.

Sommer, R. and Bossio, D. 2014. Dynamics and climate change mitigation potential of soil organic carbon sequestration, *Journal of Environmental Management* 144, 83-87.

Soussana, J. F., Tallec, T. and Blanfort, V. 2010. Mitigating the greenhouse gas balance of ruminant production systems through carbon sequestration in grasslands, *Animal: An International Journal of Animal Bioscience* 4(3), 334-350.

Throop, H. L., Archer, S. R., Monger, H. C. and Waltman, S. 2012. When bulk density methods matter: implications for estimating soil organic carbon pools in rocky soils, *Journal of Arid Environments* 77, 66-71.

Vanguelova, E. I., Bonifacio, E., De Vos, B., Hoosbeek, M. R., Berger, T. W., Vesterdal, L., Armolaitis, K., Celi, L., Dinca, L., Kjonaas, O. J., Pavlenda, P., Pumpanen, J., Puttsepp, Ü., Reidy, B., Simoncic, P., Tobin, B. and Zhiyanski, M. 2016. Sources of errors and uncertainties in the assessment of forest soil carbon stocks at different scales-review and recommendations, *Environmental Monitoring and Assessment* 188(11), 630.

Viscarra Rossel, R. A., Walvoort, D. J. J., McBratney, A. B., Janik, L. J. and Skjemstad, J. O. 2006. Visible, near infrared, mid infrared or combined diffuse reflectance spectroscopy for simultaneous assessment of various soil properties, *Geoderma* 131(1-2), 59-75.

Whittaker, C., McManus, M. C. and Smith, P. 2013. A comparison of carbon accounting tools for arable crops in the United Kingdom, *Environmental Modelling and Software* 46, 228-239.

Winkler, K., Fuchs, R., Rounsevell, M. and Herold, M. 2021. Global land use changes are four times greater than previously estimated, *Nature Communications* 12(1), 2501.

Zhao, M., Heinsch, F. A., Nemani, R. R. and Running, S. W. 2005. Improvements of the MODIS terrestrial gross and net primary production global data set, *Remote Sensing of Environment* 95(2), 164-176.

Chapter 11

Advances in measuring soil organic carbon stocks and dynamics at the profile scale

Christopher Poeplau, Thünen Institute of Climate-Smart Agriculture, Germany; and Edward Gregorich, Agriculture and Agri-Food Canada, Canada

1 Introduction

Quantification of soil organic carbon (SOC) stocks and dynamics is required for and applied to different frameworks, situations and scales, ranging from an experimental plot (for evaluating treatment effects) to whole fields (e.g. in SOC-accounting or crediting systems) and inventories of whole countries or continents (Orgiazzi et al., 2018, Saby et al., 2008, Bolinder et al., 2010, Viscarra Rossel and Brus, 2018). Accurate estimates of belowground stocks of carbon are necessary for understanding ecosystem responses to changes in land use, management, climate and other environmental stresses. Monitoring of SOC stock responses to management has gained interest in recent years due to the call to raise ambitions for countries to include SOC storage in climate change negotiations (Wiese et al., 2019). At the same time, the opposite is also the case: understanding the impact of these drivers and underlying processes is key for designing appropriate frameworks for monitoring, reporting and verification (MRV) of SOC stock changes (Smith et al., 2020).

The estimation of SOC stocks is challenging at any scale because it requires measurement of several key parameters including SOC content, bulk density and rock fragments, some of which are highly variable in space and time, and are prone to sampling biases and other pitfalls. It is therefore even more

http://dx.doi.org/10.19103/AS.2022.0106.10

challenging to measure and verify changes in SOC stock, which are usually small compared to the total SOC stock and its spatial heterogeneity (Smith, 2004). Of all the parameters mentioned, the one with the largest spatial heterogeneity (both vertically and horizontally) is usually SOC content (Goidts et al., 2009). An appropriate sampling design, which accounts for the scale-dependent *in situ* variability, is thus crucial for representative sampling and quantification of SOC changes (VandenBygaart and Angers, 2006, VandenBygaart, 2006). Detecting change in SOC stock usually requires a large number of samples to account for spatial variability, that are collected at sufficiently long intervals to allow for changes to be detectable (Saby et al., 2008, Schrumpf et al., 2011, Heikkinen et al., 2020, de Gruijter et al., 2018, Schöning et al., 2006, Viscarra Rossel et al., 2016). Our focus in this chapter will be to describe the methods, techniques and approaches used to measure SOC stocks and dynamics in the soil profile. The spatial variability of SOC stocks and their associated parameters and the sampling designs to best tackle this issue will not be discussed in detail. Instead, our objective here is to provide a general introduction to traditional, state-of-the-art and novel techniques and frameworks used to quantify SOC stocks and their dynamics.

Characterizing SOC dynamics refers not only to quantitative changes in bulk SOC over time, but also to the partitioning of SOC in fractions with distinct turnover kinetics, their residence times and stabilization pathways. The concept of fractionating SOC to derive these functional pools is introduced, as well as the use of C isotopes to accurately trace inputs in the soil profile. We also discuss techniques used to characterize and predict changes in SOC stocks for evaluating environmental or management changes.

2 Accurate estimation of parameters needed to calculate soil organic carbon stocks

2.1 Sampling and sensing methods

Three parameters are required to estimate SOC stocks at a given soil depth: (1) SOC content, which is usually measured in the fine soil fraction < 2 mm; (2) bulk density (mass of soil per volume); and (3) the fraction of rock fragments > 2 mm, which are usually considered to be free of SOC. An appropriate sampling design for the determination of these parameters for a particular spatial unit depends primarily on the size and properties of that unit (Goidts et al., 2009). Collecting a sample that contains a representative amount of SOC for the spatial unit of interest is not only the major goal, but also a major challenge of soil sampling. Here, we focus on the smallest spatial unit of field sampling – the soil profile – to introduce some basic techniques and pitfalls for sampling soils for subsequent measurements of the required parameters.

Published by Burleigh Dodds Science Publishing Limited, 2023.

The nature of the 'soil profile' and adequate sampling depth are important considerations when measuring and evaluating SOC stocks. Globally, the thickness of the layer that is usually defined as soil, that is, loose material above the bedrock that underwent (initial) pedogenesis and is biologically active, mostly exceeds 1 m (https://www.isric.org/explore/wise-databases). Only in regions characterized by strong erosion processes, such a mountain areas or deserts, as well as regions of initial soil development, can the soil thickness be much lower. Accordingly, the upper 30 cm of soil, which is the standard sampling depth in many experiments and inventories, stores only half of the total SOC stock on a global scale (Batjes, 2014), which implies that SOC storage and storage potential are strongly underestimated when only topsoils are considered. Furthermore, neglecting SOC in deeper layers may also lead to false conclusions about SOC dynamics (Luo et al., 2010, Hicks Pries et al., 2017).

The ability to detect either temporal or spatial changes in SOC stocks depends on the depth of sampling because deeper sampling usually increases the variability of the measurement (Ellert et al., 2008). However, increasing the depth of sampling increases the cost and effort to do so. Thus, a primary consideration with regard to the appropriate sampling depth relates to the soil management or process under investigation. Factors such as land use, cropping system, tillage management, rooting depth, landscape position, drainage class and other soil-forming factors affect the quantity and distribution of SOC with depth, and will therefore influence decisions on appropriate sampling depth. For example, in studies comparing contrasting tillage systems, the depth of sampling required to determine difference in SOC stocks may be just below the depth of plowing (e.g. 30-40 cm). In contrast, an assessment of SOC in systems with deep-rooted crops or those that receive additions of large amounts of organic amendments may require sampling below this depth (Tautges et al., 2019).

The sampling techniques and methods used to determine the SOC content can be used for any other variable to be measured in 'disturbed' soil samples (as opposed to intact soil cores). However, consideration of the depth gradients within the soil profile is of particular importance for elements (e.g., C, N and P) contained in soil organic matter. In the vast majority of soils, the distribution of SOC along the soil profile shows a clear depth dependency, with decreasing SOC at increasing soil depth. This is primarily related to the OC inputs to the soil, which are highest close to the surface and exponentially decline with depth (Jackson et al., 1996). In agricultural soils, particularly in tilled arable soils, the depth distribution of SOC differs from that of soils under natural vegetation, due to the homogenizing effect of tillage. The depth gradient within the plow layer can thus be expected to be relatively small, which, however, causes a sharp decline of SOC at the border between plow layer and the underlying subsoil (Fig. 1a). In any case, the depth gradient of

Figure 1 Example profile pictures of (a) a plowed arable soil with a sharp decline in soil organic carbon (SOC) below the plow layer; (b) a gleysol with a more complex depth distribution of SOC; and (c) a Stagnosol with hydromorphic properties. Photos belong to the Thünen Institute of Climate-Smart Agriculture, Germany.

SOC needs to be considered during soil sampling to obtain reliable estimates of SOC content in a given volume of soil. When sampling a particular vertical distribution, it is necessary that the whole depth of the increment is covered in approximately equal shares. This might be the most basic, but is also the most important rule of soil sampling; disregarding this rule can cause large biases in estimates of SOC. There are exceptions to the typical sharp gradients in SOC with depth (e.g. soils with large amounts of allochthonous SOC, such as those observed in fluvisols or those characterized by pedogenesis that involves SOC redistribution with depth, such as podsols (Schneider et al., 2020)). For these soils, it is even more important to ensure that each layer or soil horizon is well represented in the sample used to estimate SOC stocks (Fig. 1b).

The simplest and cheapest technique to sample a soil is to use a spade to extract a slice of topsoil material from above ground (Fernández-Ugalde et al., 2020b). However, this method can be particularly prone to a sampling bias, since the spade does not ensure collection of a homogeneously distributed sample for a given depth increment. Moreover, the sampling depth is not easily controllable and reproducible with a spade, since sampling usually occurs in a V-shape (http://kvkcachar.nic.in/soil.html). Finally, sampling multiple depth increments becomes almost impossible, unless a full soil pit is excavated first. The latter will be discussed as a separate technique in the following section. Spade sampling has the advantage of being cheap, quick and applicable to a wide range of situations. Therefore it has also been used in continental scale soil inventories such as the European LUCAS Soil Survey (Orgiazzi et al., 2018). However, it is not a volumetric sampling method, which precludes deriving the

Published by Burleigh Dodds Science Publishing Limited, 2023.

other two parameters (i.e., bulk density and rock fragment fraction) needed for SOC stock estimation from the same sample. Therefore, this technique is used mainly when only the SOC content of a single layer of topsoil is of interest, not the SOC stock of a whole profile.

A hand-operated gouge auger or a core sampler of any kind is often used for soil sampling because it allows for controlled depth sampling at multiple depths, and to obtain a representative sample of each depth increment. Depending on the diameter of the core, an undisturbed core of soil can be obtained with relatively low effort. For these reasons, a gouge auger or a core sampler are the most common tools used for soil sampling. A machine-driven hydraulic system is usually preferred if the diameter of the core exceeds 4–5 cm and the desired sampling depth exceeds ~30 cm. The disadvantages of the auger or core sampler include deformations of the core (compression and stretching) as well as a partial mixing of topsoil and subsoil (Walter et al., 2016). Wet, coarse-textured soil is specifically prone to compaction. Also, a high rock fragment fraction, a dense network of coarse roots in forest topsoils, frozen or extremely dry soil can aggravate or impede auger or core sampling. In specific situations, e.g. in permafrost soils, a drilling auger is indispensable to extract cores of sufficient depth and quality. The fact that the extracted cores derived from auger sampling or soil coring resemble relatively undisturbed soil samples, allows for simultaneous determination of physical (e.g. bulk density) and chemical soil properties in the same sample, which helps to reduce the uncertainty of SOC stock estimation. Also, the fraction of rock fragments, at least those not exceeding the diameter of the auger/core sampler, can be estimated. Thus, if the soil core has a sufficient diameter (e.g. 4–10 cm) for volumetric and relatively undisturbed soil sampling, all parameters for estimating SOC stocks can be derived from one sample.

Another commonly used method to sample and describe soils is the soil pit method. This is usually the most labor-intensive method, but it is also a low-tech method which enables the most unbiased description of the soil profile and thus provides information related to pedogenesis (Vadeboncoeur et al., 2012). For example, hydromorphic properties, such as concretions of pedogenic oxides, are an important indicator for the presence of partial oxygen limitation which is directly coupled to SOC cycling (Poeplau et al., 2020). These properties are often only visible at the profile scale (see Fig. 1c). In addition, the abundance of macropores, the proportion of larger rock fragments, root distribution, as well as exact borders of pedogenic horizons can be best determined by a direct view of the entire soil profile. The length and depth of a soil pit can range from a few decimeters to several meters and are adjusted to the purpose of sampling and the site properties. Representative soil sampling of a profile pit for SOC content determination can best be achieved by (i) taking a thin slice of the whole depth increment or horizon, or (ii) using small metal cylinders to extract a known volume of soil. The latter has the advantage that bulk density can also be measured in the same sample. Due to the strong negative relationship between bulk density and SOC content (Kaur et al., 2002), the sampling distance for both parameters should generally be minimized.

The preference of sampling by horizon or depth increment is context-specific. Usually, when SOC stocks are targeted, depth-specific values are of interest. However, the profile pit sampling provides the opportunity of a mixed approach, that is, a general sampling by fixed depth increments, with additional increments if horizon borders vary strongly from depth increment borders (Jacobs et al., 2018). Finally, the profile pit method is also of advantage in more stony soils, or soils with a high share of larger stones that will not be included in auger sampling or soil coring. The volumetric proportion of the rock fragment fraction can be visually estimated on the profile wall. If many spatial replicates are required, the profile pit sampling is less suitable than taking soil cores. It should be noted that two different sampling techniques will not yield identical results. In the German Agricultural Soil Inventory, profile pit sampling has been accompanied by eight core samples around the pit. Figure 2 shows that depth-specific sampling biases were detected for bulk density, SOC content and stocks. In the topsoil (0–30 cm), all parameters tended to be higher in the soil pit, while in the subsoils (50–100 cm), the opposite was true. A change in sampling method between sampling dates should be avoided.

Representative sampling of larger spatial units, for example, the field or landscape scale, requires greater sampling effort and ideally some knowledge about the spatial distribution of SOC (Viscarra Rossel and Brus, 2018, Heikkinen et al., 2020, VandenBygaart, 2006). In recent years, cost-effective indirect measures such as proximal sensing or even mobile phone-based image analysis tools have been developed for non-destructive 'sampling' of soils to estimate SOC and bulk density (Aitkenhead et al., 2016, Viscarra Rossel et al., 2016, Wills et al., 2007). Mobile phone applications,

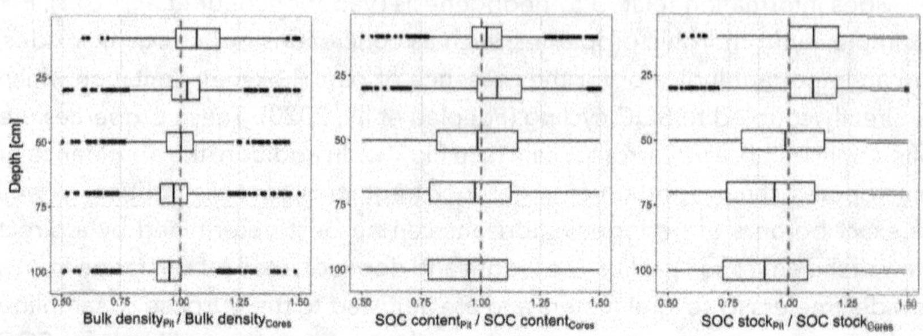

Figure 2 Ratios of bulk density, SOC content and stock, determined via profile pit (n = 1 per site) or core sampler (n = 8 per site), for a total of 1236 sampling sites of the German Agricultural Soil Inventory. Dashed red line indicates no change.

Published by Burleigh Dodds Science Publishing Limited, 2023.

which are usually based on soil color sensing and a statistical model in the background (machine learning), might have the advantage that citizen science projects can easily be designed and big datasets can be gathered. However, among currently available proximal sensing techniques, visible near-infrared spectroscopy for SOC content estimation and active gamma-ray attenuation for bulk density have been found to be the most accurate measures (England and Viscarra Rossel, 2018). Claims have been made that these indirect methods might be as accurate, or even more accurate than composite sampling (Viscarra Rossel and Brus, 2018), and could be used for SOC stock estimation and even SOC stock change detection. For this reason, mobile proximal-sensing techniques are already used in official carbon accounting schemes such as the Emissions Reduction Fund of the Australian Government (https://www.industry.gov.au/regulations-and-standards/methods-for-the-emissions-reduction-fund/measurement-of-soil-carbon-sequestration-in-agricultural-systems-method). The accuracy of aerial means of proximal sensing might be higher than simple random composite sampling for determining aerial means of SOC in certain situations, since the spatial coverage is higher and more information about SOC heterogeneity can be obtained. Therefore, it can be a way forward for soil mapping, deriving field-landscape-scale baseline SOC stocks and the development of sampling designs (see Chapter 11 of this book). However, it is doubtful that proximal sensing, or even remote sensing, will be accurate enough to verify changes in SOC stocks. Even with a large spectral library, Jaconi et al. (2017) used stationary (not mobile) near-infrared spectroscopy to determine SOC contents in dried and homogenized soil, and the uncertainty was 3.2 g C kg^{-1}, which exceeds the expected changes - as for example, induced by agricultural management - in reasonable timescales. The analytical error of, for example, elemental analyses, is one order-of-magnitude lower (Saby et al., 2008). Mobile field sensors are even less accurate (England and Viscarra Rossel, 2018). The cost efficiency of sensor-based estimation of soil properties is often highlighted (Viscarra Rossel and Brus, 2018). However, this applies only to mobile sensors that are used in the field (with high uncertainty), while for stationary laboratory spectrometers, the costs of sampling and sample preparation for spectroscopy are equal to those of actual measurements. Furthermore, sensors for determining the rock fragment fraction are in an early stage of development (Jiang et al., 2021). Therefore, estimates of SOC contents or stocks should not rely on proximal sensing or image analysis methods alone. However, the major advantage and the great potential of these techniques are that the easily and rapidly obtained spectra contain information on a wide range of chemical, physical and even biological soil properties (Cécillon et al., 2009, Jaconi et al., 2019, Vohland et al., 2014, Hermansen et al., 2017). The development of global comprehensive spectral libraries is a pre-condition for the routine use of infrared spectroscopy, for example, in soil monitoring (Cécillon et al., 2009).

As indicated above, sampling soil cores is more rapid and less labor-intensive than excavation of soil pits. Research involving spectroscopic analysis of soil cores has shown that this approach allows for rapid and data-rich investigation of the vertical distribution of soil properties, including SOC, at the scale of centimeters (Doetterl et al., 2013). Research using hyperspectral imaging of soil cores in the laboratory illustrates that even more detailed information can be obtained at even finer microscales (Steffens and Buddenbaum, 2013). The very fine resolution of this type of imaging can provide data about the hot- and cold-spots of SOC vertically through the profile (Hobley et al., 2018), potentially allowing for detailed assessments of subsoil changes in SOC.

In summary, all of these sampling techniques have their strengths and weaknesses and should be selected depending on the scientific question (including, for example, the desired accuracy and spatial extent of the area of interest), labor and resources available for sampling, expected soil properties, sampling depth and of course, cost-benefit considerations. However, especially in repeated samplings, there should not be a change in sampling methodology, since this can severely bias all three parameters of SOC stock estimation (Walter et al., 2016) and thus impede the interpretation of temporal trends or at least add significant noise to the data.

2.2 Calculation of soil organic carbon stocks and the equivalent-soil-mass concept

Once the required parameters are obtained, the calculation of SOC stocks based on the amount of fine soil present and its SOC content should be straightforward. However, careful consideration must be given to all of the parameters.

2.2.1 Case study 1

It has been shown recently that up to 87 out of 100 randomly picked studies overestimated SOC stocks due to misuse of the parameters 'bulk density' and 'rock fragment fraction' in various ways (Poeplau et al., 2017). In brief, the following three biases (M1–M3) were found (summarized in Fig. 3):

M1: Bulk density of the whole soil (including stones with a bulk density of ~2.6 g cm^{-3}) was determined, and the volume occupied by fine soil was not corrected for the volume of stones. This method, which basically neglects the presence of stones, is correct when the soil has no rock fragments at all (e.g., eolian deposits). In all other cases, such as those with

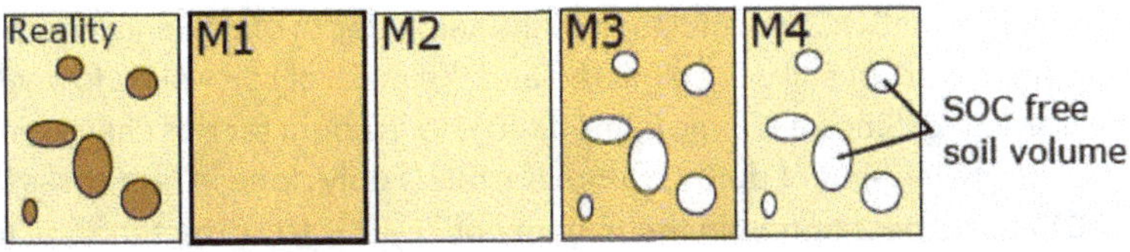

Figure 3 Schematic figure of four methods to calculate SOC stocks (Poeplau et al., 2017) in samples with rock fragments. The different shades of brown indicate the density of the material. The highest bulk density is found in stones (oval) and the lowest in fine soil.

missing data on the rock fragments fraction, the method will overestimate the SOC stock (i.e. fine soil bulk density is overestimated and stones are considered to be filled with fine soil). Sampling with a spade, a narrow auger or by using proximal sensing all lead to this type of error, since no reliable data on the rock fragment fraction can be obtained and it is not possible to estimate this parameter using other soil properties (e.g. by using pedotransfer functions).

M2: The rock fragment fraction was estimated and used to calculate the bulk density of the fine soil, but was not used to correct the total volume of soil. Therefore, the amount of fine soil was overestimated, as was the SOC stock.

M3: The opposite happened in this method: the bulk density of the whole soil was used (which is an overestimation) while the volume of the soil was corrected for the rock fragment fraction.

Only when the rock fragment fraction was used to determine the correct fine soil mass and the correct soil volume (M4), unbiased SOC stock estimates were obtained. It is particularly relevant to apply M4 when investigating stony soils. More details and equations for all methods, as well as a simplified version of M4 can be found in (Poeplau et al., 2017). This method was applied to the dataset of the German agricultural soil inventory, which contained relatively few soil profiles with a high rock fragment fraction; the average bias caused by methods M1-M3 ranged from 2.1% to 10.1%, depending on depth increment and method.

For unbiased comparison of SOC stocks in time series or the valuation of treatment effects, the SOC stock estimates must be calculated for equal masses of soil (von Haden et al., 2020). The equivalent-soil-mass approach has been proposed under various names for a long time (Ellert and Bettany, 1995, Sposito et al., 1976), but is often overlooked or solved in different ways. Due to the fact that SOC content and bulk density are negatively correlated, the effect can be quite severe and usually leads to underestimation of SOC stock differences when fixed depth increments are compared. When a soil accumulates SOC in a specific depth increment (e.g. 0–30 cm), its bulk density decreases. In the case of resampling after SOC accumulation to the same depth of the initial sampling, several centimeters of soil will be missed and the mass of soil will be lower than at initial sampling. Since it is almost impossible to account for this change in soil mass per depth increment during sampling, this is only done in rare cases (Don et al., 2019). The best option in the majority of cases is to correct the soil mass arithmetically. In the above-mentioned example with a single-layer sampling, no data is available for the depth increment below 30 cm. Therefore, the only possible correction is to 'cut' the heavier soil to a depth that equals the mass of the lighter soil in 0–30 cm. In single-layer sampling, this is straightforward, especially when this layer does not cross a pedogenic horizon border in which there is a strong shift in SOC content and bulk density. However, when multiple layers are sampled, the problem of equivalent soil mass to a given depth becomes more complex and can be solved (approximately) using post hoc non-linear models which are fitted to cumulative SOC stocks, to derive specific SOC stocks for any desired soil mass (Wendt and Hauser, 2013, Rovira et al., 2015). As discussed by Rovira et al. (2015), such a correction has limitations in stony soils, in which differences in fine soil mass can be driven by random differences in the rock fragment fraction.

For comparison of cumulative SOC stocks, such approximations are a reduction of the fixed depth bias, and in most cases, reasonably accurate. However, a serious problem related to the equivalent soil mass remains unsolved: if a depth increment includes the border between two pedogenic horizons (A and B horizons), a change in bulk density within one horizon (e.g. the Ap layer in tilled arable soils), will lead to a shift in the proportions of A and B horizon material, characterized by distinct SOC contents (Fig. 1a). Such 'dilution effects' can mask actual treatment effects and increase the noise in SOC stock data. This is another argument for exceeding the typical sampling depth of 30 cm and estimating the SOC stock in deeper layers.

In summary, even if all parameters for SOC stock quantification are estimated with high accuracy, major biases can still be introduced when calculating SOC stocks.

3 Understanding soil organic carbon dynamics

3.1 Long-term experiments, repeated inventories and space-for-time substitution

Changes in SOC that occur in response to alterations in land use, management or environmental factors are slow (often on the timescale of decades) and occur at relatively low rates compared to the total SOC stock and its variability (Smith, 2004). Therefore, detecting significant changes in SOC can take many years, depending on the effects of a given treatment and the background variability in SOC. The effects of agricultural management, such as tillage, fertilization, residue management or crop rotation on the SOC stock are relatively small, usually ranging from <0.1–0.5 Mg C ha^{-1} yr^{-1} (Freibauer et al., 2004, Bolinder et al., 2020). The management effects are therefore mostly studied in long-term field experiments (Debreczeni and Körschens, 2003), in which treatment effects are compared to a reference or control treatment and spatial variability is ideally tackled by random, block-wise replication. Having a control treatment is crucial to distinguish between treatment effects and changes in storage that represent the absolute transfer of C between soil and atmosphere (Fig. 4). In this case, the average initial SOC stocks in the plots of treatment A and B were equal. However, in reality, this is rarely the case because many of the existing older experiments were not designed for rigorous statistical analysis. For example, many old, long-term experiments have only two replicates per treatment. When plots are laid out in a field having a gradient in SOC, it is likely that two treatments may differ in initial SOC stocks, which would be a major pitfall in quantifying treatment effects on SOC dynamics. For unbiased quantification of treatment effects, this difference should be accounted for if there is information on the initial SOC contents available. In ten Swedish long-term experiments on mineral fertilization with a total of 1116 treatment pairs, we found that initial absolute SOC content differences between plots were on average 1.3 g C kg^{-1}, while the final differences after about 50 years were 1.5 g C kg^{-1}, with 47% of the

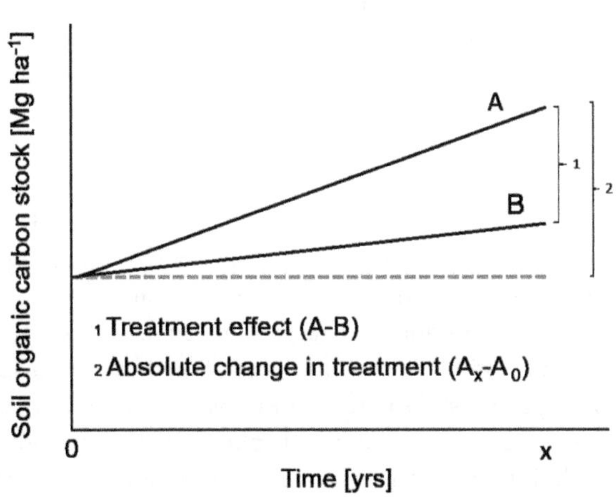

Figure 4 An example of SOC changes between two treatments, A and B. For treatment A, the absolute change is the difference in SOC at time x, compared to that at time 0. The net change is the difference between SOC in treatment A and that in treatment B, at time x, assuming that SOC was the same in both treatments at time 0. The latter approach is often used to measure the effect on SOC of a proposed treatment (e.g. no-till) compared to a standard 'control' (e.g. conventional tillage). Figure content adapted from (Ellert et al., 2001).

treatments having larger initial differences than final differences (Poeplau et al., 2016). We also found that the initial and final differences were significantly correlated, indicating a long-lasting legacy effect of different starting conditions across plots. Ideally, treatment effects at time x should thus be calculated as $(A_x - A_0) - (B_x - B_0)$, rather than $A_x - B_x$.

On larger spatial scales, repeated soil inventories are useful to evaluate SOC stock changes over large areas and long periods of time (Bellamy et al., 2005). It has been suggested that a period of ten years is necessary for detecting changes in average SOC stocks on a national scale (Schrumpf et al., 2011), while the continental-scale soil monitoring network, LUCAS Soil, was able to detect significant SOC losses (decline in SOC contents) in European croplands between 2009 and 2015 (Fernández-Ugalde et al., 2020a). For a correct interpretation of the regional trend in SOC stocks, careful records of land use and management practices are important. For example, Bellamy et al. (2005) suggested that SOC losses in Great Britain were mainly climate change-driven, while Smith et al. (2007) later showed that the reported losses were most likely overestimated and mainly attributable to changes in agricultural management. Other difficulties arise if the types of soil are not fully characterized across the sampling area or if the sampling is not conducted uniformly in space and time. If SOC inventories are carried out over long periods of time, factors that significantly affect the soil profile depth (e.g., erosion and deposition) or bulk density (e.g., compaction) should also be quantified repeatedly. A closer look at those inventories that have repeatedly sampled soils also reveals that even if the exact same sampling plots are resampled, the reproducibility of site-specific SOC stocks can be limited (Heikkinen et al., 2013). Figure 5 depicts the distribution of observed SOC changes in European topsoils (0–20 cm) between 2009 and 2015. Only 25% of the 16145 resampled soils had a relative deviation in SOC content between +/-10%, a range that could be related to changes caused by management or even climate change within six years. The other 75% of soils were out of this range, indicating that resampling caused a strong random error, masking potential management effects. It is not known what causes such large deviations in repeated sampling, but the reasons could be manifold, including slight shifts in sampling position, sampling depth, sample preparation or the above-mentioned problem on changes in bulk density and a related 'dilution' of depth increments. Although this effect is not only observed in the LUCAS dataset (Heikkinen et al., 2013), the relatively large deviations in this case could be related to the low reproducibility of the spade sampling method, as described above. Soil monitoring networks are currently set up in many parts of the world. The issue of the random resampling error should be further evaluated.

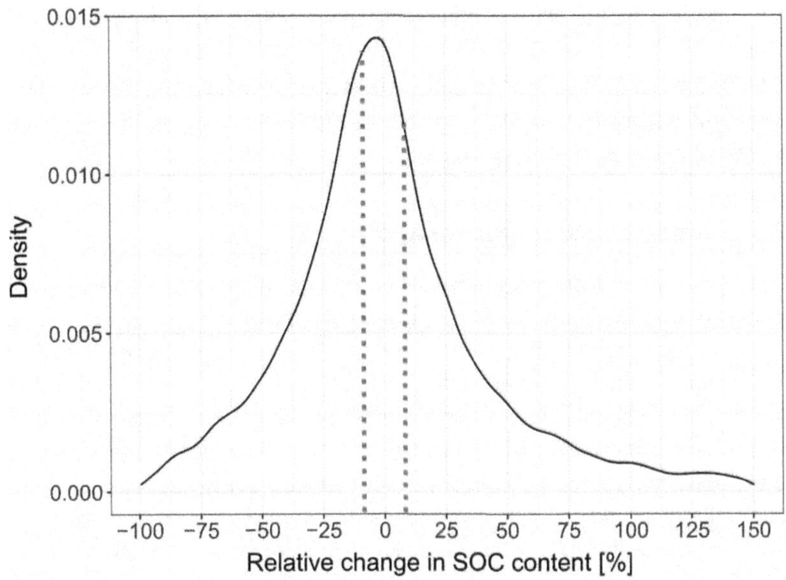

Figure 5 Distribution of relative deviations in SOC content between the LUCAS Soil sampling campaigns in 2009 and 2015. A total of 16 145 sampling locations could be compared, while only 15 406 are displayed here, due to even more extreme deviations of 739 sampling locations. The area between the red dotted lines indicates the range of deviations that could potentially be attributed to actual changes in SOC over time (+/-10%).

The space-for-time substitution approach is a common and powerful way to overcome the problem that SOC dynamics occur relatively slowly, while research questions often need to be answered in the short term (Pickett, 1989). This approach is based on the extrapolation of a temporal trend from a series of differently aged sites, also known as a chronosequence. For example, forest sites of different ages (i.e. time since afforestation) are compared to each other and to a long-term cropland to estimate the temporal dynamic of SOC since afforestation (Vesterdal et al., 2002, Quezada et al., 2019). Of course, the major assumption is that all sites had the same initial SOC stock at the time of conversion. This may rarely occur (see above), but site selection should account for this by minimizing pedological or topographical differences as well as different management across sites. Complementary to chronosequences, other naturally occurring gradients can be useful to study SOC dynamics. One example is thermosequences – a series of plots in close proximity to a geothermal source that are used to study soil-warming effects on SOC dynamics (O'Gorman et al., 2014, Sigurdsson et al., 2016). In addition, translocation experiments, along altitudinal gradients for example, can be used to mimic climate change effects on SOC dynamics (Luan et al., 2014).

In summary, SOC dynamics are assessed in a wide range of scenarios, frameworks and experimental approaches, all of which require a high degree of standardization and reproducibility.

3.2 Soil organic carbon fractionation

SOC is the main constituent element of soil organic matter, a highly complex material consisting of countless compounds of various chemical properties and biogeochemical stability. For a better understanding, predicting and modeling of its dynamics and turnover in the soil, SOC is often conceptualized into different functional pools with distinct turnover kinetics (von Lützow et al., 2007). SOC models distinguish roughly between 2 and 7 partly interrelated pools, and experimental attempts have been made to quantify these pools by various fractionation schemes. Those approaches developed over time, following the prevailing paradigms of SOC stabilization in soils or vice versa (Dynarski et al., 2020). Starting from pure and harsh chemical extractions (Waksman, 1925), SOC fractionation has become more related to physical separation methods, such as size or density fractionation (Carter et al., 2003). A problem with SOC fractionation is that a myriad of different fractionation methods and method variations have been used, which usually preclude cross-study comparisons. A total of 20 different fractionation methods were evaluated by Poeplau et al. (2018) to investigate their efficiency to isolate pools with different turnover times. It turned out that methods separating clay or clay and silt-associated SOC from coarse and free particulate SOC were among the most effective. These methods were only outperformed by methods that additionally included a step to obtain an oxidation-resistant SOC fraction. How meaningful such chemical fractionation methods are with respect to *in situ* biogeochemical stability (Poeplau et al., 2019, Lutfalla et al., 2014) or regarding their representativeness of model pools has often been debated (Elliott et al., 1996, Luo et al., 2014). It has recently been proposed that the mere separation of particulate organic matter (POM) and mineral-associated organic matter (MAOM), two pools that have been shown to be substantially different in their mean residence time and response to changes, should be a possible way forward toward an improved understanding and prediction of SOC dynamics (Lavallee et al., 2020). This suggestion will be examined by many researchers around the globe, since this efficient simplification strongly reduces the workload compared to other, more complex methods and increases the comparability across studies at the same time. However, the choice of a method should always suit the purpose, and a reduction to the mentioned two fractions might not always be the optimal solution.

3.3 Use of carbon isotopes to quantify soil organic carbon turnover

3.3.1 ^{13}C natural abundance (discrimination with age)

Well-drained soils under stable vegetation (i.e. forest or grassland) are characterized by regular inputs of litter and root material which are gradually decomposed and mixed as they move down through the soil profile. Thus, a gradient is established in these profiles as the undecomposed organic material at the surface is transformed by decomposition into smaller particles/compounds that become more mobile and are transported deeper in the profile. The soil profile under forest or grassland contains an undisturbed, long-term record of decomposition and mixing processes useful for evaluating the rate of SOC turnover and determining the factors that regulate it. This evaluation can be conducted by coupling measurements of the stable C isotope composition ($\delta^{13}C$) and SOC at different depths through the soil profile.

The observation that the natural abundance of $\delta^{13}C$ increases as SOC concentration decreases from the surface litter to deep layers in forest and grassland soils has been widely reported in different ecosystems and climates (Powers and Schlesinger, 2002, Wang et al., 2018, Garten Jr. et al., 2000). These observations show that $\delta^{13}C$ increases vertically in the profile, until it reaches a maximum steady-state value (Acton et al., 2013, Kohl et al., 2015). This increase in $\delta^{13}C$ with depth is attributed to several factors, including the mixing of two isotopically different sources of C (Garten et al., 2000). The change in $\delta^{13}C$ values with depth may reflect the depletion of atmospheric $\delta^{13}C$-CO_2 due to combustion of $\delta^{13}C$-depleted fossil fuels (a decrease of about 2‰ since the pre-industrial era) (Keeling et al., 2017). The change in isotopic composition with depth may also be attributed to the mixing of above- and below-ground vegetative sources of C because of the isotopic differences in, for example, root inputs compared to leaf litter inputs. Soil C may also be enriched with ^{13}C due to isotopic discrimination against the heavier isotope in biological processes.

Soils formed under similar conditions exhibit similar trends in $\delta^{13}C$ with depth, and the generally observed opposing trends of $\delta^{13}C$ increases and SOC concentration decreases with depth result in a negative linear relationship between $\log_{10}$-transformed SOC and $\delta^{13}C$. The slope of the gradient, β, has been proposed as a proxy for SOC turnover (Garten Jr. et al., 2000). More recently, litter quality and mean annual temperature (MAT) were observed to be related to β (Campbell et al., 2009), as are soil texture and drainage (Powers and Schlesinger, 2002). It should be noted that these observations hold true only for well-drained forest and grassland soils, because it has been observed that the gradient in $\delta^{13}C$ with depth was reversed in soils with very high clay content

(Krull and Skjemstad, 2003) and in water-logged soils with poor drainage due to the production of ^{13}C-depleted methane (Krull and Retallack, 2000).

Recently a meta-analysis was conducted with data from studies that measured the vertical gradients of stable C isotopes and SOC in 176 forest and grassland soil profiles to evaluate the controls of climate, soil properties and nutrients on SOC turnover in different biomes worldwide (Wang et al., 2018). A significant relationship between β and modeled SOC turnover rates was observed. Also, relationships observed between β and MAT, MAP, soil N concentration and clay content suggest that these data could be used at a regional level to estimate SOC turnover. This would help to clarify the role of SOC in the climate system and could reduce modeling uncertainty.

3.3.2 C_3-C_4 vegetation change

The natural ^{13}C abundance of SOC is also useful as a quantitative tracer in systems when two plant species sources differ in their isotopic composition. For example, introduction of C4 plants to soil previously developed under C3 vegetation (or vice versa) results in the SOC containing two isotopically different sources of C. The $\delta^{13}C$ signature of the soil organic matter integrates the relative contribution of the fresh C4 plant-derived carbon and the C3-derived SOC. In this situation, a two-end-member mixing model (Chalk et al., 2020) is used to partition the SOC to origin, thereby allowing for an *in situ* estimate of SOC turnover:

$$\delta M = f_A \delta_A + f_B \delta_B$$

where f_A and f_B are the fractions or proportions of A (e.g. C4-C) and B (e.g. C3-C) in M, and δ_M, δ_A and δ_B represent the mean ^{13}C signatures for the mixture M and sources A and B, respectively. The estimate of the proportion of C4-C derived from this model best characterizes turnover for the period of time following the vegetation change. The turnover rate of bulk soil and physical fractions of soil can also be determined using this mixing model (Derrien and Amelung, 2011).

The natural $\delta^{13}C$ abundance method has been widely used to estimate SOC turnover in soils and soil fractions under different climates and types of vegetation (Balesdent, 1987, Gregorich et al., 1995). The strength of the method is evident in its use as a soil profile indicator of both known and presumed changes in vegetation and/or climate, in timescales ranging from years to centuries (depending on the approach). In a comprehensive review, Chalk et al. (2020) summarized studies using $\delta^{13}C$ measurements at different depths/horizons in the soil profile to indicate shifts in ecotone boundaries and changes in vegetation (e.g. induced by fire, grazing or climate change) within ecosystems.

3.3.3 Case study 2

In this case study, several of the approaches discussed above were applied to study SOC dynamics at the soil profile scale. These include: the space-for-time substitution (change from annual crops to the perennial bioenergy crop *Miscanthus*), SOC fractionation and the natural $\delta^{13}C$ abundance method, to study fraction and depth-specific SOC turnover (Poeplau and Don, 2014). *Miscanthus*, a perennial C4 grass from Eastern Asia, is considered a bioenergy crop in European agro-ecosystems, due to its high biomass production under low input conditions (Clifton-Brown et al., 2015). It was hypothesized that SOC sequestration could be an additional, climate-smart benefit of *Miscanthus* cultivation on soil that has been used as long-term cropland earlier. Soils of six *Miscanthus* plantations and their adjacent croplands across central Europe (Germany, the Netherlands, Switzerland and Denmark) were sampled to a depth of 80 cm to assess total SOC stock changes as well as C4-carbon accumulation and C3 depletion along the soil profile. For the topsoil (0–10, 10–20 and 20–30 cm increments), an SOC fractionation was conducted and the fraction of C4 carbon was determined for each of the five following fractions (Zimmermann et al., 2007): POM, dissolved organic C (DOC), sand and stable-aggregates-associated-C (S+A), silt and clay associated C that was (rSOC) or was not resistant to a chemical oxidation treatment (S+C-rSOC). Conceptually, the turnover rate of the different fractions follows the order POM > DOC > S+A > S+C−rSOC > rSOC. After 16 years of *Miscanthus* cultivation, SOC stocks increased at an annual rate of 0.4±0.2 Mg C ha^{-1} yr^{-1}. As expected, sequestration of *Miscanthus*-derived SOC was highest in the topsoil and decreased exponentially with depth (Fig. 6a). However, even in the deepest subsoil increment (50–80 cm), a clear *Miscanthus* signal was detected at all sites. The fractionation revealed that the POM fraction was most enriched in *Miscanthus*-derived SOC, while the lowest enrichment was found in the oxidation-resistant carbon of the silt and clay fraction (rSOC) (Fig. 6b). This case study represents a proof of concept for the applied fractionation method.

3.4 Other isotope approaches and alternatives to quantify soil organic carbon turnover

The use of C isotopes is not restricted to the natural abundance and stable isotopes ^{12}C and ^{13}C. The much lower natural abundance of the radioactive isotope ^{14}C is also used for studying SOC age and turnover, in the so-called radiocarbon dating method (Balesdent, 1987, Trumbore et al., 1989). This powerful method, which is widely used in age determination in many subject areas, is based on the radioactive decay of ^{14}C. It has fundamentally contributed to our understanding of SOC ages, which range from years to millennia (Paul

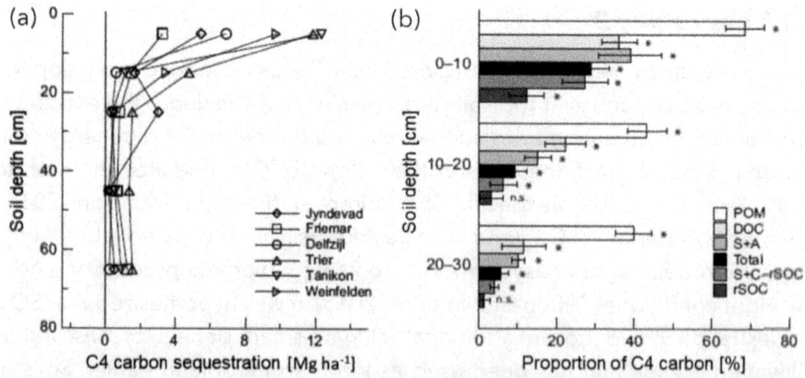

Figure 6 (a) *Miscanthus*-derived carbon sequestration as a function of depth at six central European sites; and (b) average proportion of *Miscanthus*-derived carbon in bulk SOC (total) and five different SOC fractions (POM = particulate organic matter, DOC = dissolved organic carbon, S + A = sand and stable aggregate-associated organic carbon, S + C − rSOC = silt and clay-associated organic carbon that is not resistant to oxidation with NaOCl, rSOC = NaOCL-resistant SOC). Adapted from Poeplau and Don (2014).

et al., 1997). Whole-profile SOC dynamics were extensively studied using radiocarbon dating, leading to improved understanding of soil-atmosphere carbon transfers (Balesdent et al., 2018). Paul et al. (2001) measured ^{14}C natural abundance in the profiles of agricultural soils in mid-west USA and determined the mean residence time (MRT) of SOC at different depths. The MRTs of the SOC of surface horizons ranged from modern to 1100 years (mean: 560 years). The MRT increased to an average of 1700 years in the 25–50-cm depth increment and 2757 years at 50–100 cm. One important feature of radiocarbon dating is the use of the so-called 'bomb peak,' which relates to the testing of nuclear weapons in the 1950s and 1960s, which increased the atmospheric ^{14}C content by as much as 70%. This peak helps to distinguish modern (post-bomb peak) from older (pre-bomb peak) carbon and can thus be used to estimate several parameters of SOC dynamics, such as annual input and turnover time (Goh, 1990). The steep depletion of atmospheric ^{14}C after the bomb peak can further be used as a calibration curve for exact dating of entirely modern (post-bomb peak) organic matter (Wild et al., 2000). Apart from natural abundance, ^{13}C and ^{14}C are also used as tracers in labeling studies, mainly to study shorter-term (days to years) SOC dynamics and related processes (Gregorich et al., 2015, Ziegler et al., 2005). When applied as pulses to living plants, carbon allocation within the plant and to the soil (microbes) can be directly quantified (Johnson et al., 2002, Epron et al., 2012, Swinnen et al., 1994).

Apart from the coupled use of isotopes and SOC fractionation, other methods, such as thermal analysis techniques, have been used (Plante et al., 2009). The main idea is to link SOC thermal stability to its biogeochemical stability, and in recent years, thermogravimetry, differential scanning calorimetry and temperature-controlled thermal analysis (e.g. Rock Eval pyrolysis) have been found to be promising high-throughput approaches (Peltre et al., 2013, Barré et al., 2016, Disnar et al., 2003). Several indices from temperature-controlled thermal analysis have been developed to successfully describe *in situ* biogeochemical stability (Cécillon et al., 2018, Poeplau et al., 2019). In combination with a machine-learning model, Rock Eval pyrolysis was found to be potentially applicable to estimate the proportion of centennially stable SOC in large sample sets, such as from soil inventories (Cécillon et al., 2018).

The ratio of SOC stocks and inputs or output (i.e. heterotrophic respiration) is often used as a measure of the MRT of SOC (Chen et al., 2013, Carvalhais et al., 2014). Due to several critical assumptions associated with using this approach, the absolute numbers should be interpreted with care, and the correct terminology is also important, as pointed out by Sierra et al. (2017). Moreover, it has recently been pointed out that the MRT of SOC is strongly underestimated by the ratio of total SOC divided by total annual C input or efflux, since a major part ~80% of fresh C inputs to the soil is decomposed rapidly and never appears in the pool that is usually defined as SOC (that is, organic carbon quantified in the < 2 mm soil fraction) (Luo et al., submitted, Poeplau et al., 2021). The stock/input or stock/output ratio might therefore define the MRT of organic carbon that freshly enters the soil rather than that of stabilized SOC, which should also explain the discrepancies usually detected when comparing estimates of MRT or SOC turnover based on $\delta^{13}C$ approaches (Sanderman et al., 2003).

4 Conclusion

This chapter gives a broad overview on advances in measuring SOC stocks and dynamics. Owing to its high relevance for scientists, land owners and policy makers, soil organic matter research has a long history in many parts of the globe. The focus of this research in the past has been on surface soils. However, in the past 2-3 decades, more attention has been given to SOC in the subsoil. Thus, there has been more effort to characterize SOC stocks and dynamics within the soil profile. The complexity and diversity of soils has led to the development of a wide range of different methods and approaches to sample soils, to design experiments and monitoring schemes, calculate SOC stocks, fractionate SOC and investigate its dynamics through the soil profile. This diversity can be problematic because one method, technique or concept may not be universally applicable or comparable to other methods. However, even in situations where no change in method and geographical position occurs, it remains a challenge to 'reproduce' SOC stock estimates, hampering the verification of SOC stock changes. Soil-sensing methods have become more widely used in recent years and more attention is currently devoted to developing proximal and remote-sensing-based methods for estimating SOC stocks. However, in the short term, it seems unlikely that soil-sensing methods will fully replace traditional soil sampling and analysis. It is therefore necessary to continue to work toward improving our techniques for measuring SOC stocks and dynamics at the profile scale.

5 Where to look for further information

A text book with a good overview on soil sampling and analysis methods

- Carter, M. R. and Gregorich, E. G. (Eds.). (2007). *Soil Sampling and Methods of Analysis*. CRC Press.

Further material on sensing methods, including mobile phone applications:

- http://blog.our-sci.net/2019/01/15/an-open-strategy-to-build-soil-carbon-part-2/.
- https://www.quickcarbon.org/.
- https://www.hutton.ac.uk/research/departments/information-and-computational-sciences/esmart.

A webpage on SOC fractionation methods, based on an international method comparison (Poeplau et al., 2018):

- https://www.somfractionation.org/.

Comprehensive guidelines for measuring and modeling SOC stocks and changes:

- http://www.fao.org/3/CA2934EN/ca2934en.pdf.

Project page of the European soil monitoring system LUCAS Soil, with links to data repositories:

- https://esdac.jrc.ec.europa.eu/projects/lucas.

Tea time! – An inspiring global citizen science project related to organic matter (litter) decomposition:

- http://www.teatime4science.org/.

6 References

Acton, P., Fox, J., Campbell, E., Rowe, H. and Wilkinson, M. 2013. Carbon isotopes for estimating soil decomposition and physical mixing in well-drained forest soils. *Journal of Geophysical Research: Biogeosciences* 118(4), 1532–1545.

Aitkenhead, M., Donnelly, D., Coull, M. and Gwatkin, R. 2016. Estimating soil properties with a mobile phone. In: *Digital Soil Morphometrics*. Cham: Springer.

Balesdent, J. 1987. The turnover of soil organic fractions estimated by radiocarbon dating. *Science of the Total Environment* 62, 405–408.

Balesdent, J., Basile-Doelsch, I., Chadoeuf, J., Cornu, S., Derrien, D., Fekiacova, Z. and Hatté, C. 2018. Atmosphere-soil carbon transfer as a function of soil depth. *Nature* 559(7715), 599–602.

Barré, P., Plante, A. F., Cécillon, L., Lutfalla, S., Baudin, F., Bernard, S., Christensen, B. T., Eglin, T., Fernandez, J. M., Houot, S., Kätterer, T., Le Guillou, C., Macdonald, A., Van Oort, F. and Chenu, C. 2016. The energetic and chemical signatures of persistent soil organic matter. *Biogeochemistry* 130(1–2), 1–12.

Batjes, N. H. 2014. Total carbon and nitrogen in the soils of the world. *European Journal of Soil Science* 65(1), 10–21.

Bellamy, P. H., Loveland, P. J., Bradley, R. I., Lark, R. M. and Kirk, G. J. 2005. Carbon losses from all soils across England and Wales 1978–2003. *Nature* 437(7056), 245–248.

Bolinder, M. A., Crotty, F., Elsen, A., Frac, M., Kismányoky, T., Lipiec, J., Tits, M., Tóth, Z., Kätterer, T. J. M. and Change, A. S. F. G. 2020. The effect of crop residues, cover crops, manures and nitrogen fertilization on soil organic carbon changes in agroecosystems: A synthesis of reviews. *Mitigation and Adaptation Strategies for Global Change* 25(6), 929–952.

Bolinder, M. A., Kätterer, T., Andrén, O., Ericson, L., Parent, L.-E. and Kirchmann, H. 2010. Long-term soil organic carbon and nitrogen dynamics in forage-based crop rotations in Northern Sweden (63–64°N). *Agriculture, Ecosystems and Environment* 138(3–4), 335–342.

Campbell, J. E., Fox, J. F., Davis, C. M., Rowe, H. D. and Thompson, N. 2009. Carbon and nitrogen isotopic measurements from southern Appalachian soils: Assessing soil carbon sequestration under climate and land-use variation. *Journal of Environmental Engineering* 135(6), 439–448.

Carter, M. R., Angers, D. A., Gregorich, E. G. and Bolinder, M. A. 2003. Characterizing organic matter retention for surface soils in eastern Canada using density and particle size fractions. *Canadian Journal of Soil Science* 83(1), 11–23.

Carvalhais, N., Forkel, M., Khomik, M., Bellarby, J., Jung, M., Migliavacca, M., Mu, M., Saatchi, S., Santoro, M., Thurner, M., Weber, U., Ahrens, B., Beer, C., Cescatti, A., Randerson, J. T. and Reichstein, M. 2014. Global covariation of carbon turnover times with climate in terrestrial ecosystems. *Nature* 514(7521), 213–217.

Cécillon, L., Barthès, B. G., Gomez, C., Ertlen, D., Genot, V., Hedde, M., Stevens, A. and Brun, J. J. 2009. Assessment and monitoring of soil quality using near-infrared reflectance spectroscopy (NIRS). *European Journal of Soil Science* 60(5), 770–784.

Cécillon, L., Baudin, F., Chenu, C., Houot, S., Jolivet, R., Kätterer, T., Lutfalla, S., Macdonald, A., Oort, F. V., Plante, A. F., Savignac, F., Soucémarianadin, I. N. and Barré, P. 2018. A model based on Rock-Eval thermal analysis to quantify the size of the centennially persistent organic carbon pool in temperate soils. *Biogeosciences* 15(9), 2835–2849.

Chalk, P. M., Balieiro, F. C. and Chen, D. 2020. Quantitative estimation of carbon dynamics in terrestrial ecosystems using natureal variation of the δ13C abundance of soils and biota. Advances in Agronomy 167, 63–104.

Chen, S., Huang, Y., Zou, J. and Shi, Y. 2013. Mean residence time of global topsoil organic carbon depends on temperature, precipitation and soil nitrogen. *Global and Planetary Change* 100, 99–108.

Clifton-Brown, J., Schwarz, K-U. and Hastings, A. 2015. History of the development of Miscanthus as a bioenergy crop: From small beginnings to potential realisation. *Biology and Environment: Proceedings of the Royal Irish Academy* 115B(1), 45-57.

Debreczeni, K. and Körschens, M. 2003. Long-term field experiments of the world. *Archives of Agronomy and Soil Science* 49(5), 465–483.

de Gruijter, J. J., Mcbratney, A. B., Minasny, B., Wheeler, I., Malone, B. P. and Stockmann, U. 2018. Farm-scale soil carbon auditing. In: Mcbratney, A. B., Minasny, B. and Stockmann, U. (Eds) *Pedometrics*. Cham: Springer International Publishing.

Derrien, D. and Amelung, W. 2011. Computing the mean residence time of soil carbon fractions using stable isotopes: Impacts of the model framework. *European Journal of Soil Science* 62(2), 237–252.

Disnar, J. R., Guillet, B., Kéravis, D., Di-Giovanni, C. and Sebag, D. 2003. Soil organic matter (SOM) characterization by Rock-Eval pyrolysis: Scope and limitations. *Organic Geochemistry* 34(3), 327–343.

Doetterl, S., Stevens, A., Van Oost, K. and Van Wesemael, B. 2013. Soil organic carbon assessment at high vertical resolution using closed-tube sampling and vis-NIR spectroscopy. *Soil Science Society of America Journal* 77(4), 1430–1435.

Don, A., Hagen, C., Grüneberg, E. and Vos, C. 2019. Simulated wild boar bioturbation increases the stability of forest soil carbon. *Biogeosciences* 16(21), 4145 –4155.

Dynarski, K. A., Bossio, D. A. and Scow, K. M. 2020. Dynamic stability of soil carbon: Reassessing the "Permanence" of soil carbon sequestration. *Frontiers in Environmental Science* 8, 218.

Ellert, B., Janzen, H., McConkey, B. and Lal, R. 2001. Measuring and comparing soil carbon storage. In: Kimble, J. M., Follett, R. F. and Stewart, B. A. (Eds) *Assessment Methods for Soil Carbon*, 131–146. Boca Raton, FL.

Ellert, B., Janzen, H., Vandenbygaart, A. and Bremer, E. 2008. *Measuring Change in Soil Organic Carbon Storage*. Boca Raton, FL: Taylor & Francis.

Ellert, B. H. and Bettany, J. R. 1995. Calculation of organic matter and nutrients stored in soils under contrasting management regimes. *Canadian Journal of Soil Science* 75(4), 529–538.

Elliott, E. T., Paustian, K. and Frey, S. D. 1996. Modelling the measurable or measuring the modelable: A hierarchical approach to isolating meaningful soil organic matter fractions. In: Smith, J. U. (Ed.) *Evaluation of Soil Organic Matter Models Using Existing Long-Term Datasets*. Berlin: Springer-Verlag.

England, J. R. and Viscarra Rossel, R. A. 2018. Proximal sensing for soil carbon accounting. *Soil* 4(2), 101–122.

Epron, D., Bahn, M., Derrien, D., Lattanzi, F. A., Pumpanen, J., Gessler, A., Högberg, P., Maillard, P., Dannoura, M., Gérant, D. and Buchmann, N. 2012. Pulse-labelling trees to study carbon allocation dynamics: A review of methods, current knowledge and future prospects. *Tree Physiology* 32(6), 776–798.

Fernández-Ugalde, O., Ballabio, C., Lugato, E., Scarpa, S. and Jones, A. 2020a. Assessment of changes in topsoil properties in LUCAS samples between 2009/2012 and 2015 surveys. *JRC Technical Research Reports*, Luxembourg.

Fernández-Ugalde, O., Jones, A. and Meuli, R. G. 2020b. Comparison of sampling with a spade and gouge auger for topsoil monitoring at the continental scale. *European Journal of Soil Science* 71(2), 137-150.

Freibauer, A., Rounsevell, M. D. A., Smith, P. and Verhagen, J. 2004. Carbon sequestration in the agricultural soils of Europe. *Geoderma* 122(1), 1-23.

Garten, C. T., Cooper, L. W., Post, W. M., III. and Hanson, P. J. M. 2000. Climate controls on forest soil C isotope ratios in the southern Appalachian mountains. *Ecology* 81(4), 1108-1119.

Goh, K. M. 1990. Bomb carbon. In: David Coleman, B. F., Melillo, J. and Eldor, P. (Eds) *Carbon Isotope Techniques*. San Diego, CA: Academic Press.

Goidts, E., Van Wesemael, B. and Crucifix, M. 2009. Magnitude and sources of uncertainties in soil organic carbon (SOC) stock assessments at various scales. *European Journal of Soil Science* 60(5), 723-739.

Gregorich, E. G., Gillespie, A. W., Beare, M. H., Curtin, D., Sanei, H. and Yanni, S. F. 2015. Evaluating biodegradability of soil organic matter by its thermal stability and chemical composition. *Soil Biology and Biochemistry* 91, 182-191.

Gregorich, E. G., Monreal, C. M. and Ellert, B. H. (1995). Turnover of soil organic matter and storage of corn residue carbon estimated from natural 13C abundance. Canadian Journal of Soil Science 75(2), 161-167.

Heikkinen, J., Keskinen, R., Regina, K., Honkanen, H. and Nuutinen, V. 2020. Estimation of carbon stocks in boreal cropland soils - methodological considerations. *European Journal of Soil Science* 72(2), 934-945.

Heikkinen, J., Ketoja, E., Nuutinen, V. and Regina, K. 2013. Declining trend of carbon in Finnish cropland soils in 1974-2009. *Global Change Biology* 19(5), 1456-1469.

Hermansen, C., Knadel, M., Moldrup, P., Greve, M. H., Karup, D. and De Jonge, L. W. 2017. Complete soil texture is accurately predicted by visible near-infrared spectroscopy. *Soil Science Society of America Journal* 81(4), 758-769.

Hicks Pries, C. E., Castanha, C., Porras, R. C. and Torn, M. S. 2017. The whole-soil carbon flux in response to warming. *Science* 355(6332), 1420-1423.

Hobley, E., Steffens, M., Bauke, S. L. and Kögel-Knabner, I. 2018. Hotspots of soil organic carbon storage revealed by laboratory hyperspectral imaging. *Scientific Reports* 8(1), 13900.

Jackson, R. B., Canadell, J., Ehleringer, J. R., Mooney, H. A., Sala, O. E. and Schulze, E. D. 1996. A global analysis of root distributions for terrestrial biomes. *Oecologia* 108(3), 389-411.

Jacobs, A., Flessa, H., Don, A., Heidkamp, A., Prietz, R., Dechow, R., Gensior, A., Poeplau, C., Riggers, C. and Schneider, F. 2018. Landwirtschaftlich genutzte Böden in Deutschland: Ergebnisse der Bodenzustandserhebung. *Thünen Report* 64, Braunschweig, Germany.

Jaconi, A., Don, A. and Freibauer, A. 2017. Prediction of soil organic carbon at the country scale: Stratification strategies for near-infrared data. *European Journal of Soil Science* 68(6), 919-929.

Jaconi, A., Poeplau, C., Ramirez-Lopez, L., Van Wesemael, B. and Don, A. 2019. Log-ratio transformation is the key to determining soil organic carbon fractions with near-infrared spectroscopy. *European Journal of Soil Science* 70(1), 127-139.

Jiang, Z.-D., Wang, Q.-B., Brye, K. R., Adhikari, K., Sun, F.-J., Sun, Z.-X., Chen, S. and Owens, P. R. 2021. Quantifying organic carbon stocks using a stereological profile imaging method to account for rock fragments in stony soils. *Geoderma* 385, 114837.

Johnson, D., Leake, J. R., Ostle, N., Ineson, P. and Read, D. J. 2002. In situ $13CO_2$ pulse-labelling of upland grassland demonstrates a rapid pathway of carbon flux from arbuscular mycorrhizal mycelia to the soil. *New Phytologist* 153(2), 327-334.

Kaur, R., Kumar, S. and Gurung, H. P. 2002. A pedo-transfer function (PTF) for estimating soil bulk density from basic soil data and its comparison with existing PTFs. *Soil Research* 40(5), 847-858.

Keeling, R. F., Graven, H. D., Welp, L. R., Resplandy, L., Bl, J., Piper, S. C., Sun, Y., Bollenbacher, A. and Meijer, H. A. J. 2017. Trend in 13C discrimination of land photosynthesis. Proceedings of the National Academy of Sciences 114, 10361-10366.

Kohl, L., Laganière, J., Edwards, K. A., Billings, S. A., Morrill, P. L., Van Biesen, G. and Ziegler, S. E. 2015. Distinct fungal and bacterial δ13C signatures as potential drivers of increasing δ13C of soil organic matter with depth. *Biogeochemistry* 124(1-3), 13-26.

Krull, E. S. and Retallack, G. J. 2000. δ13C depth profiles from paleosols across the Permian-Triassic boundary: Evidence for methane release. *GSA Bulletin* 112, 1459-1472.

Krull, E. S. and Skjemstad, J. O. 2003. δ13C and δ15N profiles in (14)C-dated Oxisol and Vertisols as a function of soil chemistry and mineralogy. *Geoderma* 112, 1-29.

Lavallee, J. M., Soong, J. L. and Cotrufo, M. F. 2020. Conceptualizing soil organic matter into particulate and mineral-associated forms to address global change in the 21st century. *Global Change Biology* 26(1), 261-273.

Luan, J., Liu, S., Chang, S. X., Wang, J., Zhu, X., Liu, K. and Wu, J. 2014. Different effects of warming and cooling on the decomposition of soil organic matter in warm-temperate oak forests: A reciprocal translocation experiment. *Biogeochemistry* 121(3), 551-564.

Luo, Z., Wang, E., Fillery, I. R. P., Macdonald, L. M., Huth, N. and Baldock, J. 2014. Modelling soil carbon and nitrogen dynamics using measurable and conceptual soil organic matter pools in APSIM. *Agriculture, Ecosystems and Environment* 186, 94-104.

Luo, Z., Wang, E. and Sun, O. J. 2010. Can no-tillage stimulate carbon sequestration in agricultural soils? A meta-analysis of paired experiments. *Agriculture, Ecosystems and Environment* 139(1-2), 224-231.

Luo, Z., Wang, G., Xiao, L., Mao, X., Guo, X. and Cowie, A. L., Zhang, S. and Wang, M. Submitted. *Most Root-Derived Carbon Inputs Do Not Contribute to the Global Bulk Soil Carbon Pool*. Research Square. DOI:10.21203/rs.3.rs-65178/v1.

Lutfalla, S., Chenu, C. and Barré, P. 2014. Are chemical oxidation methods relevant to isolate a soil pool of centennial carbon? *Biogeochemistry* 118(1-3), 135-139.

O'Gorman, E. J., Benstead, J. P., Cross, W. F., Friberg, N., Hood, J. M., Johnson, P. W., Sigurdsson, B. D. and Woodward, G. 2014. Climate change and geothermal ecosystems: Natural laboratories, sentinel systems, and future refugia. *Global Change Biology* 20(11), 3291-3299.

Orgiazzi, A., Ballabio, C., Panagos, P., Jones, A. and Fernández-Ugalde, O. 2018. LUCAS Soil, the largest expandable soil dataset for Europe: A review. *European Journal of Soil Science* 69(1), 140-153.

Paul, E. A., Collins, H. P. and Leavitt, S. W. 2001. Dynamics of resistant soil carbon of Midwestern agricultural soils measured by naturally occurring 14C abundance. *Geoderma* 104(3-4), 239-256.

Paul, E. A., Follett, R. F., Leavitt, S. W., Halvorson, A., Peterson, G. A. and Lyon, D. J. 1997. Radiocarbon dating for determination of soil organic matter pool sizes and dynamics. *Soil Science Society of America Journal* 61(4), 1058-1067.

Peltre, C., Fernández, J. M., Craine, J. M. and Plante, A. F. 2013. Relationships between biological and thermal indices of soil organic matter stability differ with soil organic carbon level. *Soil Science Society of America Journal* 77(6), 2020-2028.

Pickett, S. T. 1989. Space-for-time substitution as an alternative to long-term studies. In: Likens, G. E. (Eds) *Long-Term Studies in Ecology*. New York: Springer.

Plante, A. F., Fernández, J. M. and Leifeld, J. 2009. Application of thermal analysis techniques in soil science. *Geoderma* 153(1-2), 1-10.

Poeplau, C., Barré, P., Cécillon, L., Baudin, F. and Sigurdsson, B. D. 2019. Changes in the Rock-Eval signature of soil organic carbon upon extreme soil warming and chemical oxidation: A comparison. *Geoderma* 337, 181-190.

Poeplau, C., Bolinder, M. A. and Kätterer, T. 2016. Towards an unbiased method for quantifying treatment effects on soil carbon in long-term experiments considering initial within-field variation. *Geoderma* 267, 41-47.

Poeplau, C. and Don, A. 2014. Soil carbon changes under Miscanthus driven by C4 accumulation and C3 decompostion–toward a default sequestration function. *GCB Bioenergy* 6(4), 327-338.

Poeplau, C., Don, A. and Schneider, F. 2021. Roots are key to increasing the mean residence time of organic carbon entering temperate agricultural soils. *Global Change Biology* 27(19), 4921-4934.

Poeplau, C., Don, A., Six, J., Kaiser, M., Benbi, D., Chenu, C., Cotrufo, M. F., Derrien, D., Gioacchini, P., Grand, S., Gregorich, E., Griepentrog, M., Gunina, A., Haddix, M., Kuzyakov, Y., Kühnel, A., Macdonald, L. M., Soong, J., Trigalet, S., Vermeire, M., Rovira, P., Van Wesemael, B., Wiesmeier, M., Yeasmin, S., Yevdokimov, I. and Nieder, R. 2018. Isolating organic carbon fractions with varying turnover rates in temperate agricultural soils: A comprehensive method comparison. *Soil Biology and Biochemistry* 125, 10-26.

Poeplau, C., Jacobs, A., Don, A., Vos, C., Schneider, F., Wittnebel, M., Tiemeyer, B., Heidkamp, A., Prietz, R. and Flessa, H. 2020. Stocks of organic carbon in German agricultural soils–Key results of the first comprehensive inventory. *Journal of Plant Nutrition and Soil Science* 183(6), 665-681.

Poeplau, C., Vos, C. and Don, A. 2017. Soil organic carbon stocks are systematically overestimated by misuse of the parameters bulk density and rock fragment content. *Soil* 3(1), 61-66.

Powers, J. S. and Schlesinger, W. H. 2002. Geographic and vertical patterns of stable carbon isotopes in tropical rain forest soils of Costa Rica. *Geoderma* 109(1-2), 141-160.

Quezada, J. C., Etter, A., Ghazoul, J., Buttler, A. and Guillaume, T. 2019. Carbon neutral expansion of oil palm plantations in the Neotropics. *Science Advances* 5(11), eaaw4418.

Rovira, P., Sauras, T., Salgado, J. and Merino, A. 2015. Towards sound comparisons of soil carbon stocks: A proposal based on the cumulative coordinates approach. *Catena* 133, 420-431.

Saby, N. P. A., Bellamy, P. H., Morvan, X., Arrouays, D., Jones, R. J. A., Verheijen, F. G. A., Kibblewhite, M. G., Verdoodt, A., Üveges, J. B., Freudenschuss, A. and Simota, C. 2008. Will European soil-monitoring networks be able to detect changes in topsoil organic carbon content? *Global Change Biology* 14(10), 2432-2442.

Sanderman, J., Amundson, R. G. and Baldocchi, D. D. 2003. Application of eddy covariance measurements to the temperature dependence of soil organic matter mean residence time. *Global Biogeochemical Cycles* 17(2), 1-15.

Schneider, F., Amelung, W. and Don, A. 2020. Origin of carbon in agricultural soil profiles deduced from depth gradients of C:N ratios, carbon fractions, δ13C and δ15N values. *Plant and Soil* 460(1-2), 123-148.

Schöning, I., Totsche, K. U. and Kögel-Knabner, I. 2006. Small scale spatial variability of organic carbon stocks in litter and solum of a forested luvisol. *Geoderma* 136(3-4), 631-642.

Schrumpf, M., Schulze, E. D., Kaiser, K. and Schumacher, J. 2011. How accurately can soil organic carbon stocks and stock changes be quantified by soil inventories? *Biogeosciences* 8(5), 1193-1212.

Sierra, C. A., Müller, M., Metzler, H., Manzoni, S. and Trumbore, S. E. 2017. The muddle of ages, turnover, transit, and residence times in the carbon cycle. *Global Change Biology* 23(5), 1763-1773.

Sigurdsson, B. D., Leblans, N. I. W., Dauwe, S., Gudmundsdottir, E., Gundersen, P., Gunnarsdottir, G. E., Holmstrup, M., Ilieva-Makulec, K., Katterer, T., Marteinsdottir, B.-S., Maljanen, M., Oddsdóttir, E. S, Ostonen, I., Peñuelas, J., Poeplau, C., Richter, A., Sigurðsson, P., Van Bodegom, P., Wallander, H., Weedon, J. and Janssens, I. 2016. Geothermal ecosystems as natural climate change experiments: The ForHot research site in Iceland as a case study. *Icelandic Agricultural Sciences-Reykjavik, Akureyri* 29, 53-71.

Smith, P. 2004. How long before a change in soil organic carbon can be detected? *Global Change Biology* 10(11), 1878-1883.

Smith, P., Chapman, S. J., Scott, W. A., Black, H. I. J., Wattenbach, M., Milne, R., Campbell, C. D., Lilly, A., Ostle, N., Levy, P. E., Lumsdon, D. G., Millard, P., Towers, W., Zaehle, S. and Smith, J. U. 2007. Climate change cannot be entirely responsible for soil carbon loss observed in England and Wales, 1978-2003. *Global Change Biology* 13(12), 2605-2609.

Smith, P., Soussana, J. F., Angers, D., Schipper, L., Chenu, C., Rasse, D. P., Batjes, N. H., Van Egmond, F., Mcneill, S., Kuhnert, M., Arias-Navarro, C., Olesen, J. E., Chirinda, N., Fornara, D., Wollenberg, E., Álvaro-Fuentes, J., Sanz-Cobena, A. and Klumpp, K. 2020. How to measure, report and verify soil carbon change to realize the potential of soil carbon sequestration for atmospheric greenhouse gas removal. *Global Change Biology* 26(1), 219-241.

Sposito, G., Giraldez, J. V. and Reginato, R. J. 1976. The theoretical interpretation of field observations of soil swelling through a material coordinate transformation. *Soil Science Society of America Journal* 40(2), 208-211.

Steffens, M. and Buddenbaum, H. 2013. Laboratory imaging spectroscopy of a stagnic luvisol profile–High resolution soil characterisation, classification and mapping of elemental concentrations. *Geoderma* 195-196, 122-132.

Swinnen, J., Van Veen, J. A. and Merckx, R. 1994. 14C pulse-labelling of field-grown spring wheat: An evaluation of its use in rhizosphere carbon budget estimations. *Soil Biology and Biochemistry* 26(2), 161-170.

Tautges, N. E., Chiartas, J. L., Gaudin, A. C. M., O'Geen, A. T., Herrera, I. and Scow, K. M. 2019. Deep soil inventories reveal that impacts of cover crops and compost on soil carbon sequestration differ in surface and subsurface soils. *Global Change Biology* 25(11), 3753-3766.

Trumbore, S. E., Vogel, J. S. and Southon, J. R. 1989. AMS C-14 measurements of fractionated soil organic matter- an approach to deciphering the soil carbon cycle. *Radiocarbon* 31(3), 644–654.

Vadeboncoeur, M. A., Hamburg, S. P., Blum, J. D., Pennino, M. J., Yanai, R. D. and Johnson, C. E. 2012. The quantitative soil pit method for measuring belowground carbon and nitrogen stocks. *Soil Science Society of America Journal* 76(6), 2241–2255.

VandenBygaart, A. J. 2006. Monitoring soil organic carbon stock changes in agricultural landscapes: Issues and a proposed approach. *Canadian Journal of Soil Science* 86(3), 451–463.

VandenBygaart, A. J. and Angers, D. A. 2006. Towards accurate measurements of soil organic carbon stock change in agroecosystems. *Canadian Journal of Soil Science* 86(3), 465–471.

Vesterdal, L., Ritter, E. and Gundersen, P. 2002. Change in soil organic carbon following afforestation of former arable land. *Forest Ecology and Management* 169(1–2), 137–147.

Viscarra Rossel, R. A. and Brus, D. J. 2018. The cost-efficiency and reliability of two methods for soil organic C accounting. *Land Degradation and Development* 29(3), 506–520.

Viscarra Rossel, R. A., Brus, D. J., Lobsey, C., Shi, Z. and McLachlan, G. 2016. Baseline estimates of soil organic carbon by proximal sensing: Comparing design-based, model-assisted and model-based inference. *Geoderma* 265, 152–163.

Vohland, M., Ludwig, M., Thiele-Bruhn, S. and Ludwig, B. 2014. Determination of soil properties with visible to near- and mid-infrared spectroscopy: Effects of spectral variable selection. *Geoderma* 223-225, 88–96.

von Haden, A. C., Yang, W. H. and Delucia, E. H. 2020. Soils' dirty little secret: Depth-based comparisons can be inadequate for quantifying changes in soil organic carbon and other mineral soil properties. *Global Change Biology* 26(7), 3759–3770.

von Lützow, M., Kogel-Knabner, I., Ekschmittb, K., Flessa, H., Guggenberger, G., Matzner, E. and Marschner, B. 2007. SOM fractionation methods: Relevance to functional pools and to stabilization mechanisms. *Soil Biology and Biochemistry* 39(9), 2183–2207.

Waksman, S. A. 1925. What is humus? *Proceedings of the National Academy of Sciences of the United States of America* 11(8), 463–468.

Walter, K., Don, A., Tiemeyer, B. and Freibauer, A. 2016. Determining soil bulk density for carbon stock calculations: A systematic method comparison. *Soil Science Society of America Journal* 80(3), 579–591.

Wang, C., Houlton, B. Z., Liu, D., Hou, J., Cheng, W. and Bai, E. 2018. Stable isotopic constraints on global soil organic carbon turnover. *Biogeosciences* 15(4), 987–995.

Wendt, J. W. and Hauser, S. 2013. An equivalent soil mass procedure for monitoring soil organic carbon in multiple soil layers. *European Journal of Soil Science* 64(1), 58–65.

Wiese, L. D., Alcántara-Shivapatham, V. and Wollenberg, L. 2019. Enhancing Nationally Determined Contribution (NDC) ambition for soil organic carbon protection and sequestration. *CCAFS Info Notes*, Wageningen, NL. Available at: https://ccafs.cgiar.org/.

Wild, E. M., Arlamovsky, K. A., Golser, R., Kutschera, W., Priller, A., Puchegger, S., Rom, W., Steier, P. and Vycudilik, W. 2000. 14C dating with the bomb peak: An application to forensic medicine. *Nuclear Instruments and Methods in Physics Research Section B: Beam Interactions with Materials and Atoms* 172(1-4), 944–950.

Wills, S. A., Burras, C. L. and Sandor, J. A. 2007. Prediction of soil organic carbon content using field and laboratory measurements of soil color. *Soil Science Society of America Journal* 71(2), 380–388.

Ziegler, S. E., White, P. M., Wolf, D. C. and Thoma, G. J. 2005. Tracking the fate and recycling of 13C-labeled glucose in soil. *Soil Science* 170(10), 767–778.

Zimmermann, M., Leifeld, J., Schmidt, M. W. I., Smith, P. and Fuhrer, J. 2007. Measured soil organic matter fractions can be related to pools in the RothC model. *European Journal of Soil Science* 58(3), 658–667.

Chapter 12

Advances in digital soil mapping to assess baseline levels and carbon sequestration at the landscape scale

Amin Sharififar, University of Tehran, Iran; and Budiman Minasny, The University of Sydney, Australia

1 Introduction

Climate change is altering soil ecosystems and influences the storage and flux of carbon and nutrients. As soil varies spatially and temporally, a reliable soil organic carbon (SOC) stocks assessment is crucial to understanding the status of soil resources and their ecosystem services. Maps of soil carbon stocks can provide a basis for implementing soil carbon sequestration as a mitigation strategy for increased atmospheric carbon. Soils are the largest reservoir of terrestrial organic carbon (OC) and inorganic carbon (IC) and a source of greenhouse gases. Most soils of agroecosystems have been depleted of their OC, and the amount of the depletion depends on the severity of degradation (Lal, 2013). At the global scale, soils can retain two to three times more C than that present in the atmosphere or vegetation (Sanaullah et al., 2019). Thus, soil can be a source and a sink of greenhouse gases. For example, land use

http://dx.doi.org/10.19103/AS.2022.0106.11

and soil-related emissions can reach up to 350 Pg of CO_2 eq per annum, corresponding to approximately 21% of global C and N pools (Oertel et al., 2016). Generally, agriculture directly contributes 10% of all anthropogenic greenhouse gas (GHG) emissions. Considering agriculture-related industries, this ratio can sum up to 26–36% of all GHG emissions (Robertson, 2014). Nevertheless, increasing soil carbon by just a few per mil of its standing stock could compensate for these emissions (Minasny et al., 2017) and should be considered as nationally determined contributions required to meet the aims of the Paris Climate Agreement (www.4p1000.org). This chapter reviews and discusses the recent advances in digital mapping of soil carbon at landscape scales.

Digital soil mapping (DSM) technology has progressed rapidly in the past decade, making it operational for routine mapping over large areas. Mapping and knowing the spatial distribution of soil carbon is useful to provide a baseline carbon level. It will help implement trading schemes for soil carbon and provide incentives for adopting sustainable C sequestering practices to achieve better environmental management. Moreover, digital maps can identify priority areas for management action, i.e. better practices for enhancing carbon content and restoring degraded areas. Thus, understanding soil carbon stock distribution is necessary for atmospheric CO_2 management practices.

DSM can now deliver reliable quantitative soil information required by stakeholders, including scientists in related disciplines, landscape planners, and land managers. Scientists in ecology, hydrology, and agronomy who need critical soil information to inform their models can use these data to make predictions about biodiversity, water security, and food security.

There is a principle difference between mapping soil carbon and accounting of soil carbon. Mapping provides an image of the spatial distribution of soil carbon. In auditing, the variable of interest is the total carbon stock over an area for a particular depth at a specific time. There is no requirement to know the spatial distribution of soil carbon. The main difference is in sampling and inference. In DSM, the soil sampling strategy is usually a model-based design, and auditing uses statistically design-based sampling (Brus and De Gruijter, 2011; de Gruijter et al., 2018). In mapping, the total amount of carbon needs to be modeled, while in auditing, the carbon stock can be calculated directly using statistical inference. This chapter focuses on mapping soil carbon, which can be used for soil carbon baseline estimation and prediction of change.

2 Digital mapping of soil carbon

DSM follows the *scorpan* spatial prediction function approach (McBratney et al., 2003):

$$C_{x,y} = f(s,c,o,r,p,a,n) + e$$

where, soil carbon at spatial position x, y is a function of soil factors (s), climate (c), organisms which include land use, human effects, and management (o), relief (r), parent materials (p), age or time (a), spatial position (n), and e is the spatially correlated errors. Except for the 'time' or 'age' factor, most DSM examples have either explicitly or implicitly other factors.

Steps for creating digital maps of SOC can be summarized as follows:

- Define a soil sampling design and collect samples in the field, analyze soil samples for C content and bulk density.
- Compile the measured soil data and relevant environmental covariates.
- Harmonize the depth observed data using the equal-area spline (if collected at different depth intervals).
- For a particular depth interval, calculate soil C on a volume basis (C_v in Mg m^{-3}). If the measured bulk density is fine earth materials (excluding gravel or materials > 2 mm), C_v is calculated as (Poeplau et al., 2017):

  $$C_v = C \times \rho_{fe} \times (1 - CF_v)$$

  where, C is the gravimetric SOC content in kg C per kg of soil, ρ_{fe} is fine earth bulk density in Mg m^{-3}, and CF_v is the volume fraction of coarse materials (materials > 2 mm). If the measured bulk density is of whole earth materials (including materials > 2mm), C_v is calculated as (Hobley et al., 2018):

  $$C_v = C \times \rho_{we} \times (1 - CF_g)$$

  where, ρ_{fg} is whole earth bulk density in Mg m^{-3} and CF_g is the mass fraction of coarse materials (materials > 2 mm).
- Soil carbon density (C_d in Mg m^{-2}) can be calculated as integral of C_v from the upper to lower depth intervals z (in m):

$$C_d = \int_{z2}^{z1} C_v(z)\,dz$$

Soil carbon stock (in Mg) can be calculated by multiplying C_d by the area of the pixel size.

- Derive spatial prediction models from the calibration dataset to predict C_d.
- Use the calibrated spatial prediction model to map soil C density for the whole area.
- Calculate the uncertainty of the prediction.

After the review of digital mapping of soil carbon in 2013 (Minasny et al., 2013), there have been some advances in soil carbon mapping and evaluating its impact on climate regulation with regard to efflux of CO_2 to the atmosphere and potential of sequestration. Studies on the accuracy and uncertainty of digital soil maps and their spatiotemporal variations considering deeper soil depths have gained even more interest and attention. Studies also pay attention to subsoil storage and patterns of variations, which have vital roles in soil carbon explorations regionally and globally. For example, Minasny et al. (2017) reported that initial SOC stock as a baseline is correlated with its sequestration capacity.

Numerous local, regional, and national DSM studies have been carried out worldwide to map and quantify SOC storage. Most of these studies have focused on OC, while IC has been largely neglected so far. A review of 120 published papers on DSM of soil carbon carried out by Lamichhane et al. (2019) found that only very few studies have considered IC. Similarly, inorganic soil C has been largely ignored in the discussion on soil carbon sequestration and CO_2 emission. In addition, most studies reported 'soil carbon' and did not distinguish between OC and IC. Soil inorganic carbon (SIC) is as important as OC, at least in arid and semiarid areas where SIC constitutes the larger carbon pool (see Chapters 7 and 23 of this book). It has been estimated that more than 40% of the soil carbon stock is made up of IC (Lal, 2020).

Arid, semiarid, and hyperarid areas are estimated to be near 32.8% to 40% of the world's surface (Gaur and Squires, 2018; White and Nackoney, 2003). Also, Aridisols and Entisols, which are found in areas with low precipitation, make up about 12% and 16% of the world's ice-free land, respectively (Soil Survey Staff, 2010; https://www.nrcs.usda.gov/). Hence, a significant area of land is composed of soils with a large stock of IC. Some studies have reported a remarkable contribution of SIC efflux to atmospheric CO_2 (Ramnarine et al., 2012; Zamanian et al., 2018; Zamanian and Kuzyakov, 2019). Also, atmospheric CO_2 can be sequestered through biotic and abiotic calcification; although the rate of CO_2 exchange to IC needs further research, notable IC sequestration has been reported (Fa et al., 2016; Monger et al., 2015; Okyay and Rodrigues, 2015). IC can be dynamic in soil and plays an active role as a major stabilization agent of SOC (Guo et al., 2016; Wang et al., 2015; Yang et al., 2012). While many countries have soils with a low amount of IC and therefore do not require IC mapping, there is high uncertainty about SIC in vast areas of the world. Not enough attention has been paid to SIC, particularly in arid and semiarid regions. IC can contribute from 20% in topsoil to up to 60% in the subsoil of the whole soil CO_2 emission (Cardinael et al., 2020). Also, in agricultural soils, when nitrogen fertilization is implemented, remarkable amounts of IC-induced CO_2-C efflux can occur, contributing to greenhouse gas emissions (Raza et al., 2020; Chapter 23 of this book).

DSM has been a reliable and efficient tool for mapping and understanding soil carbon distribution and storage in different countries. According to recent studies, most of the DSM of soil carbon studies have been carried out with the help of machine learning algorithms. The use of soil spectroscopy technology to measure SOC and SIC has shown to be very promising. Especially proximal sensing in the field enables higher-resolution, cost-effective, rapid soil sampling and measurements, especially for profile investigations (Wijewardane et al., 2020). Furthermore, proximal sensing and soil spectroscopy have been shown to have a high correlation with OC, IC, and total carbon (e.g. Sepahvand et al., 2019; Sharififar et al., 2019b).

The use of remote sensing technology in mapping soil carbon and its vertical variation has not been as successful as proximal sensing. Gomez et al. (2008) compared soil spectra obtained via remotely sensed Hyperion data with laboratory spectra. They showed that predictions of SOC using the Hyperion spectra were less accurate than those of the proximally sensed spectroscopy data. Angelopoulou et al. (2019) noted that remote sensing estimation of SOC could be obtained via multi- and hyperspectral sensors mounted on satellites, airplanes, and unmanned aerial vehicles. However, external factors such as cloud cover, vegetation cover, soil moisture, and surface roughness can severely affect its prediction. Maynard and Johnson (2014) investigated the use of LiDAR satellite data as information of elevation for exploring soil total carbon and some other attributes. However, they reported that this method could be costly and has no significant improvement over standard modeling approaches. Also, satellite-driven images might not provide desirable resolution and accuracy (Venter et al., 2021).

3 Soil carbon spatial distribution and influencing factors

There have been efforts for quantifying and mapping SOC at the global scale within the GlobalSoilMap project (Arrouays et al., 2014) and later emulated by soilgrids (de Sousa et al., 2020) and the Global Soil Partnership consortium (https://www.fao.org/global-soil-partnership/en/). This map represents densities and distribution of SOC globally. Figure 1 shows the world map of SOC stocks. They vary from about less than 5 tonnes.ha^{-1} to about 750 tonnes.ha^{-1} in some areas of the world. Many countries and researchers are contributing to this project worldwide and it has been updated several times. However, such efforts have not been implemented on SIC, which was created by the United States Department of Agriculture in 2000. The world SIC distribution is discussed and shown in Chapter 23 of this book. Comparing Figure 1 (SOC map) with SIC map (in Chapter 23 of this book) shows that most areas with a low amount of SOC have a relatively high amount of SIC. Examples are areas located in Australia, northern Africa, and vast parts of the Middle East. As these maps are at a global scale, they can only inform roughly about variation due to climate and soil types.

It is important to know the influencing or driving factors on both lateral and vertical variations at the landscape scale to better understand soil carbon distribution and its quantification for management. This also helps for more certainty in the choice of covariates used in DSM of soil carbon, leading to more accurate and more certain maps. Drivers of soil carbon variation can vary with space, scale, and soil depth. Soil type, topography, and land use are among the most important factors affecting SOC stock (Mayer et al., 2019; Ramesh et al., 2019).

Environmental covariates can help describe soil carbon stock's spatial pattern only at proportionate scales. (Behrens et al., 2014). Considering the spatial structure and the range of variations of covariates with SOC will improve the map accuracy. In addition, the relevant spatial structure will present a more realistic spatial distribution pattern of SOC, which in turn leads to a more reliable and confident use of the map for soil management, planning, and decision-making. In cases where no statistically significant correlation between SOC (or any other soil property) and the set of covariates can be found, exploring the relevant range and scale of the variation, which could be due to intrinsic soil behavior, should be considered. Also, adding the pedologic knowledge in DSM and not mere machine learning and mapping procedure is needed to produce a reliable map (Ma et al., 2019). This can be achieved either using information from a variogram analysis or incorporating pedological expert knowledge within the framework of soil surveys and DSM (Behrens et al., 2019; Song et al., 2020). Wiesmeier et al. (2019) characterized key scale-dependent drivers and indicators of soil carbon variability based on a comprehensive literature review (Fig. 2).

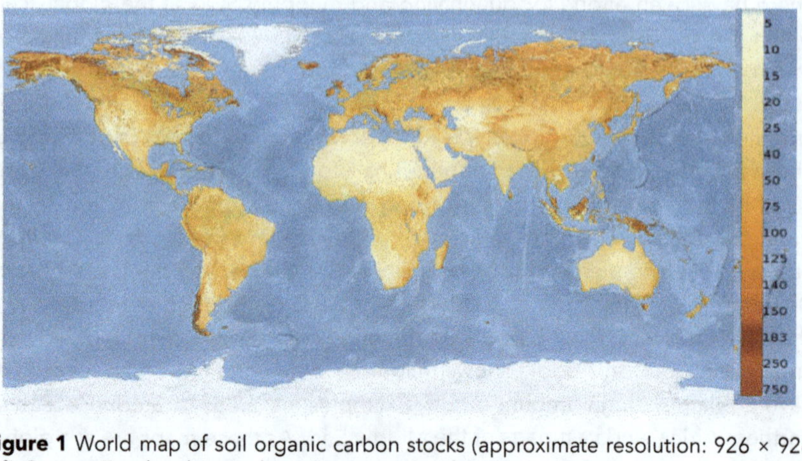

Figure 1 World map of soil organic carbon stocks (approximate resolution: 926 × 926 m^2). Source: Food and Agriculture Organization of the United Nations.

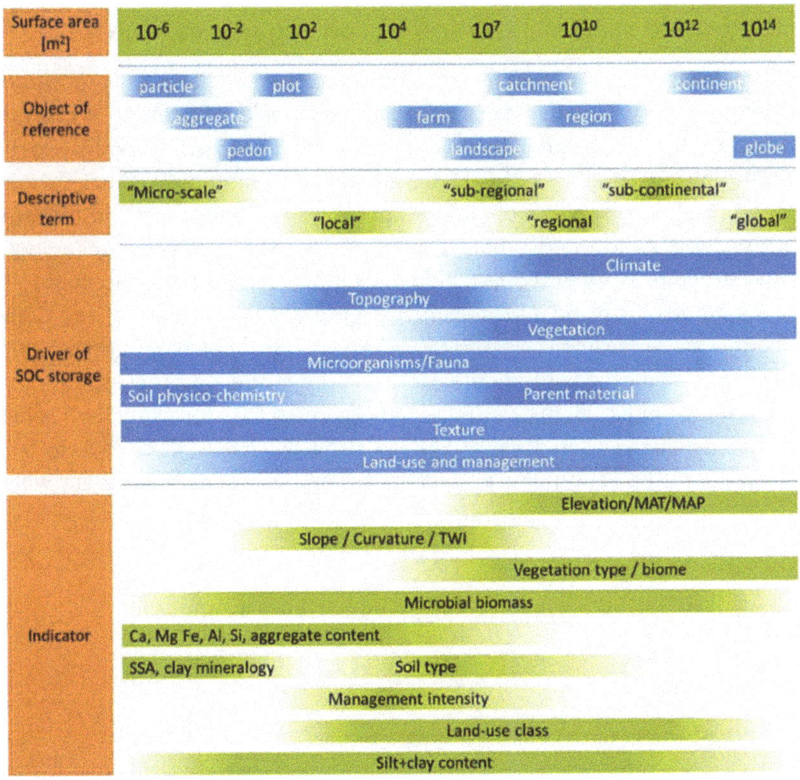

Figure 2 Overview of the scale-dependent hierarchy of drivers and indicators for soil organic carbon storage (MAT = mean annual temperature, MAP = Mean annual precipitation, TWI = topographic wetness index, SSA = specific surface area). Source: From Wiesmeier et al. (2019).

Figure 2 summarizes the drivers and indicators of SOC at different scales, which can be useful for indicative studies of soil carbon stock and potential storage at different extents. At the global scale, climate and vegetation are reported to be the main drivers of SOC variations, which may be addressed by using mean annual temperature and precipitation and vegetation type/ biome as proxy. However, these global factors may not be important for local-scale assessment. Among the explored factors, land use and management, soil texture, and microorganisms are drivers of OC at a wide range of scales from microscales in pedons to subcontinental. At the landscape scale, SOC can be influenced by land use, soil type, soil clay and silt content, topography and its related variables, vegetation and microbial biomass. These factors can be applied as useful covariates for SOC modeling. Surveying the influencing factors at different scales helps to better understand the variations of soil C, which could lead to more efficient sampling that results in time- and cost-effective efforts.

4 Field-scale spatial variation of soil organic carbon

Field-scale carbon explorations can be important for farmers and carbon marketing by surveying the status and spatial variability, and temporal changes of carbon for more precise management and decision-making. Spatial variography investigations of field-scale spatial variation of SOC have shown that the nugget-to-sill ratio of variograms varies with cropping and management systems (Singh and Whelan, 2020). The range in which a meaningful spatial structure is reflected can help with more certain sampling strategies and auditing soil carbon. In fact, surveying the spatial structure of soil C and its magnitude regarding scale and range can indicate spatial dependence and thus identify related environmental factors in the modeling process.

As shown in Fig. 3, the sill of the variograms differs for every farm, which indicates variation in the scale of SOC in farms, which could be due to farm management. The variograms also indicate that SOC varies significantly within

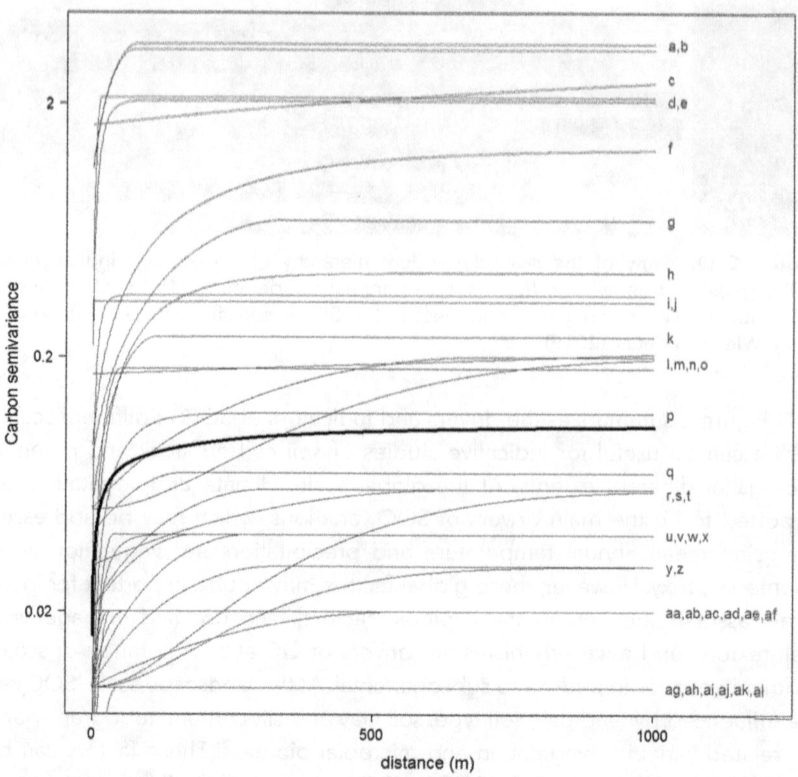

Figure 3 Compilation of field scale variograms for soil organic carbon. The bold black line represents the average variogram. The lines represent values obtained from various field-scale studies. Source: Based on Paterson et al. (2018).

200 m in the field. This analysis can be helpful for the choice of the sampling distance for survey purposes. Soil management practices are essential for decreasing soil CO_2 emissions and help climate change mitigation (Badgery et al., 2020; Wiesmeier et al., 2020). Sometimes, in the case of data scarcity and the availability of national grids, spatial downscaling can be a solution for mapping and auditing soil carbon at the farm scale. For example, Malone et al. (2017) used Gaussian random field simulation for downscaled mapping of a farm using the Australian national soil grid. They reported a significant improvement in accuracy when farm observations were incorporated as well. In addition, the use of proximal sensing by sensors such as electromagnetic induction and gamma radiometry could help in quantifying field-scale soil variation (Longo et al., 2020), which results in less costly, faster, and possibly more efficient detection of the relevant scale of variation, as higher resolution data are achievable.

Such investigations are required for SIC as well. However, SIC modeling can be challenging as it usually accumulates in deeper soil profile layers, and the data can be skewed. Filippi et al. (2020a) used a two-step mixture model to overcome the difficulty of mapping SIC in such situations by first predicting the presence or absence of IC and then quantifying it.

5 Soil carbon variation with depth

Initial DSM studies were concerned with mapping topsoil OC concentrations or stocks (de Gruijter et al., 2018). Studying soil carbon with depth is essential, given its remarkable amount and sequestration potential. For example, almost 50% of SOC stock was reported to be stored below 30 cm depth in Brazilian soils (Gomes et al., 2019). In addition, subsoil has a larger SOC sequestration potential than topsoil (0–30 cm) (Chen et al., 2018). Subsoils can contain a large amount of SOC, which is considered more stable and older when compared with topsoil (Gray et al., 2019; Hobley and Wilson, 2016; Rumpel and Kögel-Knabner, 2011).

SOC decreases with depth in most soils (Yost and Hartemink, 2019), while SIC usually increases with depth (Filippi et al., 2020a; Sharififar et al., 2019a). And thus, in DSM, the aim is to provide 3D maps of soil carbon. However, in most soil surveys, profile samples are not from the same depth increments, especially when the samples have been collected from genetic horizons from the perspective of soil classification and genesis studies. Also, soil depths and boundaries of specific variations and distribution vary within space due to soil management, soil type, land use, erosion, and other soil-forming and development factors. Hence, digital mapping should be done in 3D.

Harmonized depth layers are needed for mapping and comparable pedometric evaluations. The GlobalSoilMap project specified 6 depth intervals 0–5 cm, 5–15 cm, 15–30 cm, 30–60 cm, 60–100 cm, and 100–200 cm to better assess spatial profile variations (Arrouays et al., 2014). These standard depths are now adopted globally. However, in many cases, soil type changes within a defined study area, and as research has shown, this mere factor can influence the variations of soil carbon in topsoil and subsoil (Bai and Zhou, 2020; Sun et al., 2020). Therefore, harmonization and accounting for different influencing factors, particularly concerning different depths, should be considered.

6 2.5D and 3D mapping

The equal-area spline depth function approach has been applied to facilitate mapping soil carbon according to the GlobalSoilMap depth specification (Malone et al., 2011, 2009). Observations of soil layers are decomposed into a smooth function following a quadratic spline function (Bishop et al., 1999):

$$Y = \frac{1}{n}\sum\nolimits_{i=0}^{n}\left(y_i - \bar{f}_i\right)^2 + \lambda\int_{x_0}^{x_n}\left[f'(x)\right]^2 dx \tag{1}$$

where, n is the number of soil depth layers, y_i is observed soil data in layer i, $\bar{f}_i$ is mean of the n layers, x is the depth of the layers, and λ is the spline smoothing parameter which controls the trade-off between fidelity and roughness penalty.

Observed soil carbon at depth intervals can be harmonized using the spline function. Subsequently, soil carbon at the standard depth ranges can be predicted and mapped using the DSM method. A continuous vertical prediction for each location can be produced by the reconstruction of a continuous soil-depth function. This method is called 2.5D or the pseudo-3D method.

Some different approaches have been adopted for investigating soil carbon distribution and its mapping with depth. Some studies have considered depth data as point observations (i.e. the middle of a depth interval; Ma et al., 2021) which, in fact, is the data for the depth interval and not a point. This approach might cause some ambiguity or uncertainty within the modeling process, which considers the variation as fixed depths. In the mass preserving approach, such as the ones explored by integral functions (e.g. equal-area spline; Bishop et al., 1999; Malone et al., 2009), the upper and lower boundaries of a depth increment are accounted for in the computation. Some methodologies for the depth distribution of carbon and its so-called 3D mapping are discussed in the following paragraphs.

SOC is usually observed as an exponential depth function:

$$SOC(z) = A \times e^{\alpha.z} + SOC_{\infty} \tag{2}$$

where, $SOC(z)$ is SOC content at depth z, SOC_{∞} is SOC at the bottom of soil profile, α a constant determining the shape of the exponential, and A equals to the following equation:

$$A = SOC_0 - SOC_\infty \tag{3}$$

where, SOC_0 is topsoil OC content. According to soil formation processes and soil profile horizons pedologic view, Minasny et al. (2016) classified common soil depth variation into seven typologies, including uniform, gradational, exponential, wetting front, abrupt, peak, and minima-maxima variation types (Fig. 4). They reviewed these functions and concluded that depth variations can be soil-specific, and pedologic development should be considered. A few studies have attempted to compare different depth functions for SOC distribution. Bai et al. (2016) compared three functions, including allometric, exponential, and logistic, for modeling depth distribution of volumetric SOC and cumulative OC stock in different salt marshes. They reported better performance of exponential function for volumetric OC and higher accuracy of allometric function for OC stock depth distribution. Murphy et al. (2019) compared eight depth functions of exponential, power, and inverse types for modeling depth distribution of SOC in different soil types and land uses. They reported that exponential function with a constant is the most reliable one and that different processes were operating at different depths to control SOC distribution. Some case studies have shown that polynomial function is more accurate in 3D SOC mapping than exponential function (Jague et al., 2016; Rentschler et al., 2020). Minasny et al. (2006) spatially predicted and mapped parameters of the depth function, and thus SOC stock at any depth can be calculated. However, in this approach, SOC needs to follow a particular depth function.

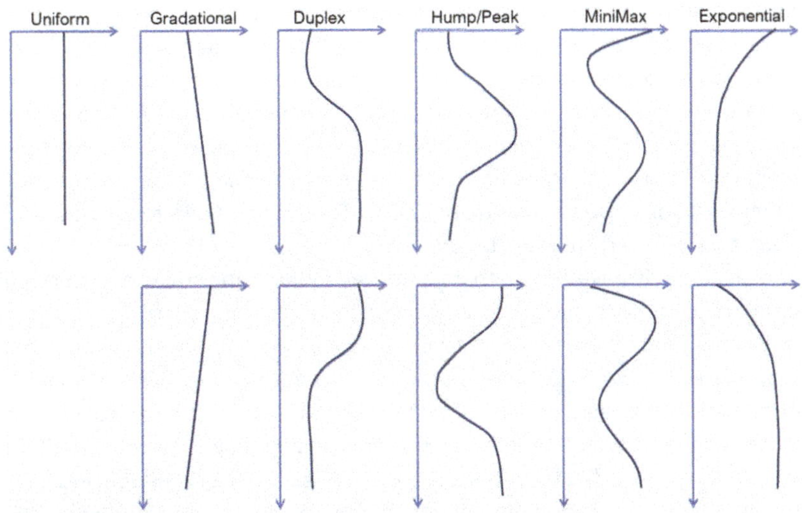

Figure 4 General typologies of depth functions.

In the so-called 3D modeling, depths are incorporated into covariates as one of the spatial predictors (Filippi et al., 2020b; Zhang et al., 2020). In this approach, harmonization is unnecessary, and values at any point at depth can be calculated. In such a modeling approach, observations from depth intervals need to be converted to point observations. Although the 3D approach seems more flexible and efficient for spatial prediction in any location and depth compared to the 2.5D approach, recent studies have reported similar or lower accuracy and higher uncertainty for this mapping approach (Brus et al., 2016; Ma et al., 2021; Nauman and Duniway, 2019). Three-dimensional modeling can outperform the multiple 2D approach when the residuals at two locations in different depths are correlated (Brus et al., 2016).

Apart from using depth functions and their related modeling, 3D geostatistics such as 3D kriging, 3D stochastic simulation (e.g. sequential Gaussian simulation), and Generalized Additive models have been attractive in modeling vertical soil C variations (Poggio and Gimona, 2017, 2014). In 3D geostatistical modeling, a 3D correlation space is used for variogram computation, and vertical and horizontal variations are considered simultaneously, which could show a volumetric variation of the soil property in 3D (Mariethoz and Caers, 2014). However, this approach has some constraints; for example, the estimated points usually have a much lower standard deviation compared to the observed points, which leads to the smoothing effect (Deutsch and Journel, 1998; Goovaerts, 1997). This effect results in a less realistic representation of spatial soil carbon variations. Also, considering points rather than intervals might cause uncertainties in interpretations. To overcome this issue, Orton et al. (2020) suggested increment-averaged kriging, considering non-stationarity and the interaction between depth and covariates.

Vertical soil carbon evaluation and mapping its pattern can be different locally. Soil carbon can vary significantly horizontally and vertically with land use and soil management strategies, parent material, climate, and surface phenomena such as erosion. Discontinuity and weathering can lead to uncertainties in modeling vertical patterns as sharp changes can be observed in some locations due to such phenomena. Therefore, in different landscapes, different modeling results can be expected.

Many studies have shown that lateral variations in the soil vary with depth, and influencing factors used as covariates for digital mapping of soil carbon can be different with depths (Mayer et al., 2019; Vaysse and Lagacherie, 2015; Vos et al., 2019; Zhou et al., 2016). The accuracy of 3D maps can be substantially decreased, especially for the subsoil layers, if the vertical variation is not considered with lateral variation concurrently (Pejović et al., 2018). For example, Chen et al. (2018) reported that land use is a major influencing factor for topsoil, while for subsoil, the critical factor was parent material. Hence, the essential covariates for topsoil subsoil carbon mapping should be different.

7 Soil carbon change and its mapping

Spatial and temporal changes quantification and mapping are crucial for understanding carbon emission, sequestration studies, and management strategies. In the literature, the soil carbon changes have been studied with two general approaches, including process-based models, such as CENTURY (Parton, 1996) and RothC model (Falloon et al., 2006), and statistical modeling using design-based modeling or DSM approach (Karunaratne et al., 2014; Sun et al., 2021). Usually, SOC of topsoil is more prone to changes with time and land use practices.

A common method for estimation of SOC change is based on DSM of SOC stock over two periods (time 1 and 2). Briefly, soil C is measured at two different times C_1 and C_2 where there is a detectable change, usually over a period of 5 years or longer. Digital maps of SOC at the two periods can be produced, and SOC change can be calculated as follows:

$$\hat{\Delta C} = \hat{C}_1 - \hat{C}_2 \tag{4}$$

where, $\hat{\Delta C}$ is the estimated change from digital maps predictions of $\hat{C}_1$ and $\hat{C}_2$.

Estimating the uncertainty of the change can be as important as estimating the change, as changes in soil carbon are small, and the variance is needed to make statistical inference about whether the estimated change is significantly different from zero. To test whether an observed change between two surveys is statistically significant, the following variance estimation is needed:

$$V\left(\hat{\Delta C}\right) = V\left(\hat{C}\right)_1 + V\left(\hat{C}\right)_2 - V\left(\hat{C}\right)_{1,2} \tag{5}$$

where, $V\left(\hat{\Delta C}\right)$ is the variance of the estimated change in SOC between two points in time, $V\left(\hat{C}\right)_1$ and $V\left(\hat{C}\right)_2$ are the variances in predicted SOC stock at the first and the second point in time, respectively, and $V\left(\hat{C}\right)_{1,2}$ is the covariance between the two time periods.

The covariance term needs to be included only when at least some observations are co-located in the two surveys. This suggests that if the second survey returns to the sites collected at the first survey, the uncertainty of estimated change would be reduced. However, for soil carbon auditing purposes, the second survey would be better to be collected independently as it would avoid land managers manipulating management only on surveyed sites.

Filippi et al. (2021) used a multivariate linear mixed model to assess the change in SOC at 0–10 cm and 30–50 cm depths between two time points (2002–2015) and reported a large difference map in their study area for topsoil

OC change, although only a small area presented a statistically significant change. While a small increase in topsoil OC in some areas, no significant change was found in subsoil. They related this outcome to a probable climatic factor of a change in rainfall. This approach is not a true space-time model as it does not model SOC as a function of time.

Another method of spatiotemporal change estimation is to include time explicitly in a hierarchical model called integrated nested Laplace approximation (INLA) as proposed by Cameletti et al. (2013). INLA was used for SOC change modeling by Sun et al. (2021), which is as follows:

$$y(s_i,t)=z(s_i,t)\beta+\xi(s_i,t)+\varepsilon(s_i,t),\varepsilon\sim N(0,\sigma_\varepsilon^2) \quad (6)$$

where, $y(s_i,t)$ denotes a response data at site s_i and time t, z is the vector of p covariates, β denotes the vector of corresponding coefficients of covariates, ξ denotes the realization of the so-called step process (the true unobserved process that changes soil), ε denotes a Gaussian white-noise process from measurement errors with a mean of 0 and a variance of σ_ε^2, and ξ is a spatiotemporal Gaussian field that changes with time (autoregressive model) with coefficients a and spatially correlated innovation x,

$$\xi(s_i,t)=a\xi(s_i,t-1)+x(s_i,t),x\sim N(0,\sigma_x^2) \quad (7)$$

In this method, SOC change is estimated as a function of time, while the linear mixed model only considers SOC change at two static observations.

In the literature, efforts have been made on quantification of SOC change concerning land use change and time with the help of DSM (e.g. Heuvelink et al., 2020; Luo et al., 2019; Song et al., 2018). At the regional to the global scale, attempts to predict SOC change using legacy soil data and time-series of optical imageries have been implemented. Using such an approach, Venter et al. (2021) mapped topsoil OC stocks of South Africa under natural vegetation between 1984 and 2019. They estimated that over that period, SOC had increased 0.3% relative to the long-term mean. While Stockmann et al. (2015) found that due to land use change, topsoil OC content in many parts of the tropical region falls below a critical level of 11 $g.kg^{-1}$. We note that such global analysis is not applicable at the landscape scale because of the fine spatial variability of SOC in the field.

8 Mapping soil functions

Soils have many functions: producing food, climate change regulation, biodiversity protection, filtering water, cycling of nutrients, and providing ecosystem service delivery (McBratney et al., 2014). However, most studies on soil carbon focused mostly on carbon sink and provision of food and biomass. Indeed, soil carbon is a key soil fertility factor and plays an important role in climate change regulation studies. However, soil carbon loss can also affect soil multifunctionality; the loss of carbon from soils causes erosion and loss of biodiversity and associated ecosystem services. Thus, linking soil carbon to soil functions would be the next step in DSM.

DSM can be used to support soil carbon sequestration initiatives, such as conservation and regenerative agriculture (Valkama et al., 2020). There are also regional and global initiatives such as agricultural land management practices within the soil carbon '4 per mille' initiative (https://www.4p1000.org). A strategy called 'perennialization' of the crops, which was presented by the Food and Agriculture Organization, emphasizes the role of perennial crops for mitigating climate change, increasing food security, and ecosystem services regulation (Glover et al., 2010). At the global scale, Ledo et al. (2020) quantified the effect of the change from annual to perennial crops on global SOC and its subsequent temporal changes and dynamics. They reported that, in general, perennialization leads to the accumulation of SOC through time. Digital mapping of soil carbon can further help assess the suitability and the reporting and verification of soil carbon stock and sequestration potential.

To account for potential soil greenhouse gas emission and carbon sequestration, we need to account for gas emissions and sequestration modeling along with SOC stock maps. Examples of such combined explorations are the work done by Farina et al. (2017) and Payen et al. (2021). Warner et al. (2019) used DSM for spatial quantification of soil CO_2 and CH_4 fluxes in forest lands and reported seasonal temperature and precipitation influence the gas fluxes. McDaniel et al. (2017) found a large spatial variation of soil CH_4 and N_2O fluxes which were governed by soil properties. Spatial variability in soil CH_4 fluxes was predicted best by electrical conductivity, while N_2O fluxes were predicted best by soil N and the gamma radiometric data. They concluded that field-scale 'hot spots' of soil CH_4 and N_2O fluxes need to be mapped to account for the uncertainty of the data. Therefore, spatiotemporal investigation of GHG fluxes and sequestration rates combined with DSM of soil C storage is required for agricultural lands carbon accounting.

9 Uncertainty quantification of soil carbon maps

Studies of soil carbon mapping and spatiotemporal distribution modeling have now considered spatial uncertainty quantification associated with their mapping outcomes. The review by Lamichhane et al. (2019) reported that 41% of the published papers on digital mapping of soil carbon have considered uncertainty quantification. There are several methods used in quantifying confidence limits of prediction, including bootstrapping for machine learning algorithms, spatial stochastic simulations (e.g. sequential Gaussian simulation), and the use of prediction variance (and subsequent confidence intervals) through Universal kriging or other kriging approaches (Liao et al., 2016; Nauman and Duniway, 2019; Szatmári and Pásztor, 2019). To validate the uncertainty of predictions, the prediction interval coverage probability assessment (accuracy plotting) (Goovaerts, 2001; Shrestha and Solomatine, 2006) is recommended.

As the variability in soil carbon in both horizontal and vertical directions is high due to pedogenic processes, spatial uncertainty evaluation and mapping the uncertainty can be particularly useful for risk and management planning. Most pedometricians have used the term 'prediction interval' for uncertainty evaluation, while other studies used 'confidence limits/intervals.' These two terms have different meanings. The prediction interval predicts the range of a future observation, while a confidence interval shows the likely range of values associated with a statistical parameter of the data (Altman et al., 2013; Wackerly et al., 2014; Weiss and Weiss, 2017).

Some minor but essential and effective points need to be noted when calculating the uncertainty of digital soil maps. Usually, the confidence limits of predicted map pixels are obtained by multiplying the standard error of predictions (e.g. from multiple bootstraps or multiple stochastic realizations) by a *Z*-value at a specified probability level:

$$CL = Z \times SE \tag{8}$$

where, *CL* is the confidence limit, *Z* denotes the *Z*-value (at a specified probability level), and *SE* is the standard error. Where:

$$SE = \frac{\sigma}{\sqrt{N}} \tag{9}$$

σ is the standard deviation of the mean, and *N* is the number of predictions (in bootstrapping and simulation methods, it equals the number of multiple maps or realizations).

$$\sigma = \sqrt{\frac{\Sigma(x_i - \mu)^2}{N}} \tag{10}$$

x_i is each of the multiple maps produced, μ is the mean map, and *N* is the number of all maps produced.

In some DSM uncertainty surveys, the confidence limit has been calculated by multiplying the *Z*-value by the standard deviation, which could be acceptable, but the obtained interval would be wider. This might be helpful, as a relatively wider interval might sometimes result in a more accurate validation concerning the interval coverage probability in the accuracy plot.

10 Conclusion

This chapter reviewed the advances and issues related to soil carbon and its digital mapping. This leads to the following conclusions:

- DSM can be a practical approach to obtain the baseline soil carbon stock for monitoring or management purposes.
- No single approach or function can be recommended to model 3D or depth distribution and mapping of soil carbon, as soil processes vary within space. Choosing the best covariates for DSM should be done carefully considering the landscape and soil processes.
- The scale of spatial surveys and sampling is an influencing factor in soil carbon mapping accuracy and uncertainty. Field-scale and landscape-scale variography can help to decide the number of samples and balance the cost with modeling uncertainty. Also, accounting for the scale and range of variations is essential for the drivers and influencing factors of soil carbon.
- In addition to digital mapping of soil carbon, mapping the distribution and amounts of soil greenhouse gas efflux and CO_2 sequestration rate is required to better understand the role of soil C in climate change regulation and ecosystem services.
- The multifunctionality of soils with its relation to soil carbon needs to be explored. Currently, most discussions focus on biomass production and climate change regulation. However, soil carbon controls soil microbial habitat and water flow as well. Thus, focusing on mapping soil carbon contribution to soil functions would be a future challenge.
- The role, importance, and contribution of SIC in CO_2 emission and sequestration and subsequent ecosystem services have been mostly ignored in most soil carbon DSM and spatiotemporal investigations. SIC can sometimes be as important as OC. In arid and semiarid areas, it can even be the main carbon source. Therefore, consideration of SIC needs attention both at local or regional and at the global scale. This conclusion has also been emphasized in Chapters 7 and 23 of this book.

11 Where to look for further information

Updated information regarding soil carbon data and national efforts for mapping can be found on Food and agriculture Organization initiative (https://www.fao.org/hand-in-hand/en/) and world soil information center (https://www.isric.org).

More information on soil carbon and digital soil mapping can be found in recent publications:

- Lamichhane, S., Kumar, L. and Wilson, B., 2019. Digital soil mapping algorithms and covariates for soil organic carbon mapping and their implications: A review. *Geoderma*, *352*, pp.395-413.
- Chen, S., Arrouays, D., Mulder, V.L., Poggio, L., Minasny, B., Roudier, P., Libohova, Z., Lagacherie, P., Shi, Z., Hannam, J. and Meersmans, J., 2022. Digital mapping of GlobalSoilMap soil properties at a broad scale: A review. *Geoderma*, *409*, p.115567.

12 References

Altman, D., Machin, D., Bryant, T. and Gardner, M. 2013. *Statistics with Confidence: Confidence Intervals and Statistical Guidelines*. British Medical Journal Publishing Group, United Kingdom.

Angelopoulou, T., Tziolas, N., Balafoutis, A., Zalidis, G. and Bochtis, D. 2019. Remote sensing techniques for soil organic carbon estimation: a review. *Remote Sens.* 11(6), 676.

Arrouays, D., McKenzie, N., de Forges, A. R., Hempel, J. and McBratney, A. B. 2014. *GlobalSoilMap: Basis of the Global Spatial Soil Information System*. CRC Press.

Badgery, W., Murphy, B., Cowie, A., Orgill, S., Rawson, A., Simmons, A. and Crean, J. 2020. Soil carbon market-based instrument pilot-the sequestration of soil organic carbon for the purpose of obtaining carbon credits. *Soil Res.* 59(1), 12–23

Bai, J., Zhang, G., Zhao, Q., Lu, Q., Jia, J., Cui, B. and Liu, X. 2016. Depth-distribution patterns and control of soil organic carbon in coastal salt marshes with different plant covers. *Sci. Rep.* 6, 34835.

Bai, Y. and Zhou, Y. 2020. The main factors controlling spatial variability of soil organic carbon in a small karst watershed, Guizhou Province, China. *Geoderma* 357, 113938.

Behrens, T., Rossel, R. A. V., Kerry, R., MacMillan, R., Schmidt, K., Lee, J., Scholten, T. and Zhu, A.-X. 2019. The relevant range of scales for multi-scale contextual spatial modelling. *Sci. Rep.* 9, 1.

Behrens, T., Schmidt, K., Ramirez-Lopez, L., Gallant, J., Zhu, A.-X. and Scholten, T. 2014. Hyper-scale digital soil mapping and soil formation analysis. *Geoderma* 213, 578–588.

Bishop, T. F. A., McBratney, A. B. and Laslett, G. M. 1999. Modelling soil attribute depth functions with equal-area quadratic smoothing splines. *Geoderma* 91(1-2), 27–45.

Brus, D. J. and De Gruijter, J. J. 2011. Design-based Generalized Least Squares estimation of status and trend of soil properties from monitoring data. *Geoderma* 164(3–4), 172–180.

Brus, D. J., Yang, R.-M. and Zhang, G.-L. 2016. Three-dimensional geostatistical modeling of soil organic carbon: a case study in the Qilian Mountains, China. *CATENA* 141, 46-55.

Cameletti, M., Lindgren, F., Simpson, D. and Rue, H. 2013. Spatio-temporal modeling of particulate matter concentration through the SPDE approach. *AStA Adv. Stat. Anal.* 97(2), 109–131.

Cardinael, R., Chevallier, T., Guenet, B., Girardin, C., Cozzi, T., Pouteau, V. and Chenu, C. 2020. Organic carbon decomposition rates with depth and contribution of inorganic carbon to CO_2 emissions under a Mediterranean agroforestry system. *Eur. J. Soil Sci.* 71, 909–923.

Chen, S., Martin, M. P., Saby, N. P. A., Walter, C., Angers, D. A. and Arrouays, D. 2018. Fine resolution map of top-and subsoil carbon sequestration potential in France. *Sci. Total Environ.* 630, 389–400.

de Gruijter, J. J., McBratney, A. B., Minasny, B., Wheeler, I., Malone, B. P. and Stockmann, U. 2018. Farm-scale soil carbon auditing. In: *Pedometrics*, Springer, Cham, pp. 693–720.

de Sousa, L. M., Poggio, L., Batjes, N. H., Heuvelink, G., Kempen, B., Riberio, E. and Rossiter, D. 2020. SoilGrids 2.0: producing quality-assessed soil information for the globe. *Soil Discuss.* 1, 1–37.

Deutsch, C. V. and Journel, A. G. 1998. *Geostatistical Software Library and User's Guide*. Oxford University Press, New York.

Fa, K., Liu, Z., Zhang, Y., Qin, S., Wu, B. and Liu, J. 2016. Abiotic carbonate dissolution traps carbon in a semiarid desert. *Sci. Rep.* 6, 23570.

Falloon, P., Smith, P., Bradley, R. I., Milne, R., Tomlinson, R., Viner, D., Livermore, M. and Brown, T. 2006. RothCUK–a dynamic modelling system for estimating changes in soil C from mineral soils at 1-km resolution in the UK. *Soil Use Manag.* 22(3), 274–288.

Farina, R., Marchetti, A., Francaviglia, R., Napoli, R. and Di Bene, C. D. 2017. Modeling regional soil C stocks and CO_2 emissions under Mediterranean cropping systems and soil types. *Agric. Ecosyst. Environ.* 238, 128–141.

Filippi, P., Cattle, S. R., Pringle, M. J. and Bishop, T. F. A. 2021. Space-time monitoring of soil organic carbon content across a semiarid region of Australia. *Geoderma Reg.* 24, e00367.

Filippi, P., Cattle, S. R., Pringle, M. J. and Bishop, T. F. A. 2020a. A two-step modelling approach to map the occurrence and quantity of soil inorganic carbon. *Geoderma* 371, 114382.

Filippi, P., Jones, E. J. and Bishop, T. F. A. 2020b. Catchment-scale 3D mapping of depth to soil sodicity constraints through combining public and on-farm soil databases–a potential tool for on-farm management. *Geoderma* 374, 114396.

Gaur, M. K. and Squires, V. R. 2018. Geographic extent and characteristics of the world's arid zones and their peoples. In: *Climate Variability Impacts on Land Use and Livelihoods in Drylands*. Springer, Cham, pp. 3–20.

Glover, J. D., Reganold, J. P., Bell, L. W., Borevitz, J., Brummer, E. C., Buckler, E. S., Cox, C. M., Cox, T. S., Crews, T. E., Culman, S. W., DeHaan, L. R., Eriksson, D., Gill, B. S., Holland, J., Hu, F., Hulke, B. S., Ibrahim, A. M., Jackson, W., Jones, S. S., Murray, S. C., Paterson, A. H., Ploschuk, E., Sacks, E. J., Snapp, S., Tao, D., Van Tassel, D. L., Wade, L. J., Wyse, D. L. and Xu, Y. 2010. Agriculture. Increased food and ecosystem security via perennial grains. *Science* 328(5986), 1638–1639.

Gomes, L. C., Faria, R. M., de Souza, E., Veloso, G. V., Schaefer, C. E. G. R. and Fernandes Filho, E. I. F. 2019. Modelling and mapping soil organic carbon stocks in Brazil. *Geoderma* 340, 337–350.

Gomez, C., Rossel, R. A. V. and McBratney, A. B. 2008. Soil organic carbon prediction by hyperspectral remote sensing and field vis-NIR spectroscopy: an Australian case study. *Geoderma* 146(3–4), 403–411.

Goovaerts, P. 2001. Geostatistical modelling of uncertainty in soil science. *Geoderma* 103(1–2), 3–26.

Goovaerts, P. 1997. Kriging vs stochastic simulation for risk analysis in soil contamination. In: *GeoENVI–Geostatistics for Environmental Applications* (Eds. A. Soares, J. Gomez-Hernandez and R. Froidevaux). Springer, pp. 247-258.

Gray, J., Karunaratne, S., Bishop, T., Wilson, B. and Veeragathipillai, M. 2019. Driving factors of soil organic carbon fractions over New South Wales, Australia. *Geoderma* 353, 213-226.

Guo, Y., Wang, X., Li, X., Wang, J., Xu, M. and Li, D. 2016. Dynamics of soil organic and inorganic carbon in the cropland of upper Yellow River Delta, China. *Sci. Rep.* 6, 36105.

Heuvelink, G. B. M., Angelini, M. E., Poggio, L., Bai, Z., Batjes, N. H., van den Bosch, R., Bossio, D., Estella, S., Lehmann, J., Olmedo, G. F. and Sanderman, J. 2020. Machine learning in space and time for modelling soil organic carbon change. *Eur. J. Soil Sci.* 72(4), 1607-1623.

Hobley, E. U. and Wilson, B. 2016. The depth distribution of organic carbon in the soils of eastern Australia. *Ecosphere* 7(1), e01214.

Hobley, E. U., Murphy, B. and Simmons, A. 2018. Comment on "Soil organic stocks are systematically overestimated by misuse of the parameters bulk density and rock fragment content" by Poeplau et al. (2017). *Soil* 4(2), 169-171.

Jague, E. A., Sommer, M., Saby, N. P. A., Cornelis, J.-T., Van Wesemael, B. and Van Oost, K. 2016. High resolution characterization of the soil organic carbon depth profile in a soil landscape affected by erosion. *Soil Till. Res.* 156, 185-193.

Karunaratne, S. B., Bishop, T. F. A., Odeh, I. O. A., Baldock, J. A. and Marchant, B. P. 2014. Estimating change in soil organic carbon using legacy data as the baseline: issues, approaches and lessons to learn. *Soil Res.* 52(4), 349-365.

Lal, R. 2020. Managing soils for negative feedback to climate change and positive impact on food and nutritional security. *Soil Sci. Plant Nutr.* 66(1), 1-9.

Lal, R. 2013. Soil carbon management and climate change. *Carbon Manag.* 4(4), 439-462.

Lamichhane, S., Kumar, L. and Wilson, B. 2019. Digital soil mapping algorithms and covariates for soil organic carbon mapping and their implications: a review. *Geoderma* 352, 395-413.

Ledo, A., Smith, P., Zerihun, A., Whitaker, J., Vicente-Vicente, J. L., Qin, Z., McNamara, N. P., Zinn, Y. L., Llorente, M., Liebig, M., Kuhnert, M., Dondini, M., Don, A., Diaz-Pines, E., Datta, A., Bakka, H., Aguilera, E. and Hillier, J. 2020. Changes in soil organic carbon under perennial crops. *Glob. Chang. Biol.* 26(7), 4158-4168.

Liao, K., Lai, X., Lv, L. and Zhu, Q. 2016. Uncertainty in predicting the spatial pattern of soil water temporal stability at the hillslope scale. *Soil Res.* 54(6), 739-748.

Longo, M., Piccoli, I., Minasny, B. and Morari, F. 2020. Soil apparent electrical conductivity-directed sampling design for advancing soil characterization in agricultural fields. *Vadose Zone J* 19(1), e20060.

Luo, Z., Eady, S., Sharma, B., Grant, T., Li Liu, D. L., Cowie, A., Farquharson, R., Simmons, A., Crawford, D., Searle, R. and Moore, A. 2019. Mapping future soil carbon change and its uncertainty in croplands using simple surrogates of a complex farming system model. *Geoderma* 337, 311-321.

Ma, Y., Minasny, B., McBratney, A., Poggio, L. and Fajardo, M. 2021. Predicting soil properties in 3D: should depth be a covariate? *Geoderma* 383, 114794.

Ma, Y. X., Minasny, B., Malone, B. P. and McBratney, A. B. 2019. Pedology and digital soil mapping (DSM). *Eur. J. Soil Sci.* 70(2), 216-235.

Malone, B. P., McBratney, A. B. and Minasny, B. 2011. Empirical estimates of uncertainty for mapping continuous depth functions of soil attributes. *Geoderma* 160(3-4), 614-626.

Malone, B. P., McBratney, A. B., Minasny, B. and Laslett, G. M. 2009. Mapping continuous depth functions of soil carbon storage and available water capacity. *Geoderma* 154(1-2), 138-152.

Malone, B. P., Styc, Q., Minasny, B. and McBratney, A. B. 2017. Digital soil mapping of soil carbon at the farm scale: a spatial downscaling approach in consideration of measured and uncertain data. *Geoderma* 290, 91-99.

Mariethoz, G. and Caers, J. 2014. *Multiple-Point Geostatistics: Stochastic Modeling with Training Images*. John Wiley & Sons.

Mayer, S., Kühnel, A., Burmeister, J., Kögel-Knabner, I. and Wiesmeier, M. 2019. Controlling factors of organic carbon stocks in agricultural topsoils and subsoils of Bavaria. *Soil Till. Res.* 192, 22-32.

Maynard, J. J. and Johnson, M. G. 2014. Scale-dependency of LiDAR derived terrain attributes in quantitative soil-landscape modeling: effects of grid resolution vs. neighborhood extent. *Geoderma* 230-231, 29-40.

McBratney, A., Field, D. J. and Koch, A. 2014. The dimensions of soil security. *Geoderma* 213, 203-213.

McBratney, A. B., Santos, M. L. M. and Minasny, B. 2003. On digital soil mapping. *Geoderma* 117(1-2), 3-52.

McDaniel, M. D., Simpson, R. R., Malone, B. P., McBratney, A. B., Minasny, B. and Adams, M. A. 2017. Quantifying and predicting spatio-temporal variability of soil CH_4 and N_2O fluxes from a seemingly homogeneous Australian agricultural field. *Agric. Ecosyst. Environ.* 240, 182-193.

Minasny, B., Malone, B. P., McBratney, A. B., Angers, D. A., Arrouays, D., Chambers, A., Chaplot, V., Chen, Z.-S., Cheng, K., Das, B. S., Field, D. J., Gimona, A., Hedley, C. B., Hong, S. Y., Mandal, B., Marchant, B. P., Martin, M., McConkey, B. G., Mulder, V. L., O'Rourke, S., Richer-de-Forges, A. C., Odeh, I., Padarian, J., Paustian, K., Pan, G., Poggio, L., Savin, I., Stolbovoy, V., Stockmann, U., Sulaeman, Y., Tsui, C., Vågen, T., van Wesemael, B. and Winowiecki, L. 2017. Soil carbon 4 per mille. *Geoderma* 292, 59-86.

Minasny, B., McBratney, A. B., Malone, B. P. and Wheeler, I. 2013. Digital mapping of soil carbon. *Adv. Agron.* 118, 1-47.

Minasny, B., McBratney, A. B., Mendonça-Santos, M. L., Odeh, I. O. A. and Guyon, B. 2006. Prediction and digital mapping of soil carbon storage in the Lower Namoi Valley. *Soil Res.* 44(3), 233-244.

Minasny, B., Stockmann, U., Hartemink, A. E. and McBratney, A. B. 2016. Measuring and modelling soil depth functions. In: *Digital Soil Morphometrics* (A. E. Hartemink and B. Minasny). Springer, Cham, pp. 225-240.

Monger, H. C., Kraimer, R. A., Khresat, S., Cole, D. R., Wang, X. and Wang, J. 2015. Sequestration of inorganic carbon in soil and groundwater. *Geology* 43(5), 375-378.

Murphy, B. W., Wilson, B. R. and Koen, T. 2019. Mathematical functions to model the depth distribution of soil organic carbon in a range of soils from New South Wales, Australia under different land uses. *Soil Syst.* 3(3), 46.

Nauman, T. W. and Duniway, M. C. 2019. Relative prediction intervals reveal larger uncertainty in 3D approaches to predictive digital soil mapping of soil properties with legacy data. *Geoderma* 347, 170-184.

Oertel, C., Matschullat, J., Zurba, K., Zimmermann, F. and Erasmi, S. 2016. Greenhouse gas emissions from soils–a review. *Geochemistry* 76(3), 327-352.

Okyay, T. O. and Rodrigues, D. F. 2015. Biotic and abiotic effects on CO_2 sequestration during microbially-induced calcium carbonate precipitation. *FEMS Microbiol. Ecol.* 91(3). https://doi.org/10.1093/femsec/fiv017.

Orton, T. G., Pringle, M. J., Bishop, T. F. A., Menzies, N. W. and Dang, Y. P. 2020. Increment-averaged kriging for 3-D modelling and mapping soil properties: combining machine learning and geostatistical methods. *Geoderma* 361, 114094.

Parton, W. J. 1996. The century model. In: *Evaluation of Soil Organic Matter Models*. Springer, , Berlin, Heidelberg, pp. 283-291.

Paterson, S., McBratney, A. B., Minasny, B. and Pringle, M. J. 2018. Variograms of soil properties for agricultural and environmental applications. *Pedometrics*, Springer, Cham, pp. 623-667.

Payen, F. T., Sykes, A., Aitkenhead, M., Alexander, P., Moran, D. and MacLeod, M. 2021. Predicting the abatement rates of soil organic carbon sequestration management in Western European vineyards using random forest regression. *Clean. Environ. Syst.* 2, 100024.

Pejović, M., Nikolić, M., Heuvelink, G. B. M., Hengl, T., Kilibarda, M. and Bajat, B. 2018. Sparse regression interaction models for spatial prediction of soil properties in 3D. *Comput. Geosci.* 118, 1-13.

Poggio, L. and Gimona, A. 2017. Assimilation of optical and radar remote sensing data in 3D mapping of soil properties over large areas. *Sci. Total Environ.* 579, 1094-1110.

Poggio, L. and Gimona, A. 2014. National scale 3D modelling of soil organic carbon stocks with uncertainty propagation–an example from Scotland. *Geoderma* 232-234, 284-299.

Poeplau, C., Vos, C. and Don, A. 2017. Soil organic carbon stocks are systematically overestimated by misuse of the parameters bulk density and rock fragment content. *SOIL* 3(1), 61-66, https://doi.org/10.5194/soil-3-61-2017.

Ramesh, T., Bolan, N. S., Kirkham, M. B., Wijesekara, H., Kanchikerimath, M., Rao, C. S., Sandeep, S., Rinklebe, J., Ok, Y. S., Choudhury, B. U., Wang, H., Tang, C., Wang, X., Song, Z. and Freeman, O. W. 2019. Soil organic carbon dynamics: impact of land use changes and management practices: a review. *Adv. Agron.* 156, 1-107.

Ramnarine, R., Wagner-Riddle, C., Dunfield, K. E. and Voroney, R. P. 2012. Contributions of carbonates to soil CO_2 emissions. *Can. J. Soil Sci.* 92(4), 599-607.

Raza, S., Miao, N., Wang, P., Ju, X., Chen, Z., Zhou, J. and Kuzyakov, Y. 2020. Dramatic loss of inorganic carbon by nitrogen-induced soil acidification in Chinese croplands. *Glob. Chang. Biol.* 26(6), 3738-3751.

Rentschler, T., Werban, U., Ahner, M., Behrens, T., Gries, P., Scholten, T., Teuber, S. and Schmidt, K. 2020. 3D mapping of soil organic carbon content and soil moisture with multiple geophysical sensors and machine learning. *Vadose Zone J.* 19(1), e20062.

Robertson, G. P. 2014. Soil greenhouse gas emissions and their mitigation. *Encycl. Agric. Food Syst.* 185-196.

Rumpel, C. and Kögel-Knabner, I. 2011. Deep soil organic matter–a key but poorly understood component of terrestrial C cycle. *Plant Soil* 338(1-2), 143-158.

Sanaullah, M., Afzal, T., Shahzad, T. and Wakeel, A. 2019. Carbon sequestration for sustainable agriculture. In: *Innovations in Sustainable Agriculture*. Springer, Cham, pp. 469-500.

Sepahvand, H., Mirzaeitalarposhti, R., Beiranvand, K., Feizian, M. and Müeller, T. 2019. Prediction of soil carbon levels in calcareous soils of Iran by mid-infrared reflectance spectroscopy. *Environ. Pollut. Bioavailab* 31(1), 9-17.

Sharififar, A., Sarmadian, F., Alikhani, H., Keshavarzi, A., Asghari, O. and Malone, B. P. 2019a. Lateral and vertical variations of soil organic and inorganic carbon content in Aridisols and Entisols of a rangeland. *Eurasian Soil Sci.* 52(9), 1051-1062. https://doi.org/10.1134/S1064229319090084.

Sharififar, A., Singh, K., Jones, E., Ginting, F. I. and Minasny, B. 2019b. Evaluating a low-cost portable NIR spectrometer for the prediction of soil organic and total carbon using different calibration models. *Soil Use Manag.* 35(4), 607-616. https://doi.org/10.1111/sum.12537.

Shrestha, D. L. and Solomatine, D. P. 2006. Machine learning approaches for estimation of prediction interval for the model output. *Neural Netw.* 19(2), 225-235.

Singh, K. and Whelan, B. 2020. Soil carbon change across ten New South Wales farms under different farm management regimes in Australia. *Soil Use Manag.* 36(4), 616-632.

Song, X.-D., Wu, H.-Y., Ju, B., Liu, F., Yang, F., Li, D.-C., Zhao, Y.-G., Yang, J.-L. and Zhang, G.-L. 2020. Pedoclimatic zone-based three-dimensional soil organic carbon mapping in China. *Geoderma* 363, 114145.

Song, X.-D., Yang, F., Ju, B., Li, D.-C., Zhao, Y.-G., Yang, J.-L. and Zhang, G.-L. 2018. The influence of the conversion of grassland to cropland on changes in soil organic carbon and total nitrogen stocks in the Songnen Plain of Northeast China. *CATENA* 171, 588-601.

Stockmann, U., Padarian, J., McBratney, A., Minasny, B., de Brogniez, D., Montanarella, L., Hong, S. Y., Rawlins, B. G. and Field, D. J. 2015. Global soil organic carbon assessment. *Glob. Food Sec.* 6, 9-16.

Sun, T., Wang, Y., Hui, D., Jing, X. and Feng, W. 2020. Soil properties rather than climate and ecosystem type control the vertical variations of soil organic carbon, microbial carbon, and microbial quotient. *Soil Biol. Biochem.* 148, 107905.

Sun, X.-L., Minasny, B., Wang, H.-L., Zhao, Y.-G., Zhang, G.-L. and Wu, Y.-J. 2021. Spatiotemporal modelling of soil organic matter changes in Jiangsu, China between 1980 and 2006 using INLA-SPDE. *Geoderma* 384, 114808.

Szatmári, G. and Pásztor, L. 2019. Comparison of various uncertainty modelling approaches based on geostatistics and machine learning algorithms. *Geoderma* 337, 1329-1340.

Soil Survey Staff. 2010. *Keys to Soil Taxonomy, 12th ed.* USDA-Natural Resources Conservation Service, Washington, DC.

Valkama, E., Kunypiyaeva, G., Zhapayev, R., Karabayev, M., Zhusupbekov, E., Perego, A., Schillaci, C., Sacco, D., Moretti, B., Grignani, C. and Acutis, M. 2020. Can conservation agriculture increase soil carbon sequestration? A modelling approach. *Geoderma* 369, 114298.

Vaysse, K. and Lagacherie, P. 2015. Evaluating digital soil mapping approaches for mapping GlobalSoilMap soil properties from legacy data in Languedoc-Roussillon (France). *Geoderma Reg.* 4, 20-30.

Venter, Z. S., Hawkins, H. J., Cramer, M. D. and Mills, A. J. 2021. Mapping soil organic carbon stocks and trends with satellite-driven high resolution maps over South Africa. *Sci. Total Environ.* 771, 145384.

Vos, C., Don, A., Hobley, E. U., Prietz, R., Heidkamp, A. and Freibauer, A. 2019. Factors controlling the variation in organic carbon stocks in agricultural soils of Germany. *Eur. J. Soil Sci.* 70(3), 550–564.

Wackerly, D., Mendenhall, W. and Scheaffer, R. L. 2014. *Mathematical Statistics with Applications*. Brooks/Cole [Cengage Learning], Belmont, CA.

Wang, X., Wang, J., Xu, M., Zhang, W., Fan, T. and Zhang, J. 2015. Carbon accumulation in arid croplands of northwest China: pedogenic carbonate exceeding organic carbon. *Sci. Rep.* 5, 11439.

Warner, D. L., Guevara, M., Inamdar, S. and Vargas, R. 2019. Upscaling soil-atmosphere CO_2 and CH4 fluxes across a topographically complex forested landscape. *Agric. For. Meteorol.* 264, 80–91.

Weiss, N. A. and Weiss, C. A. 2017. *Introductory Statistics*. Pearson.

White, R. P. and Nackoney, J. 2003. Drylands, people, and ecosystem goods and services: a web-based geospatial analysis (PDF version). *World Resour. Inst.* Available at: http//pdf. wri. org/drylands.Pdf (Accessed 30/01/2012).

Wiesmeier, M., Mayer, S., Burmeister, J., Hübner, R. and Kögel-Knabner, I. 2020. Feasibility of the 4 per 1000 initiative in Bavaria: a reality check of agricultural soil management and carbon sequestration scenarios. *Geoderma* 369, 114333.

Wiesmeier, M., Urbanski, L., Hobley, E., Lang, B., von Lützow, M., Marin-Spiotta, E., van Wesemael, B., Rabot, E., Ließ, M., Garcia-Franco, N., Wollschläger, U., Vogel, H. and Kögel-Knabner, I. 2019. Soil organic carbon storage as a key function of soils-a review of drivers and indicators at various scales. *Geoderma* 333, 149–162.

Wijewardane, N. K., Hetrick, S., Ackerson, J., Morgan, C. L. S. and Ge, Y. 2020. VisNIR integrated multi-sensing penetrometer for in situ high-resolution vertical soil sensing. *Soil Till. Res.* 199, 104604.

Yang, Y., Fang, J., Ji, C., Ma, W., Mohammat, A., Wang, S., Wang, S., Datta, A., Robinson, D. and Smith, P. 2012. Widespread decreases in topsoil inorganic carbon stocks across China's grasslands during 1980s-2000s. *Glob. Change Biol.* 18(12), 3672-3680.

Yost, J. L. and Hartemink, A. E. 2019. Soil organic carbon in sandy soils: a review. *Adv. Agron.* 158, 217-310.

Zamanian, K. and Kuzyakov, Y. 2019. Contribution of soil inorganic carbon to atmospheric CO_2: more important than previously thought. *Glob. Chang. Biol.* 25(1), e1-e3.

Zamanian, K., Zarebanadkouki, M. and Kuzyakov, Y. 2018. Nitrogen fertilization raises CO_2 efflux from inorganic carbon: a global assessment. *Glob. Chang. Biol.* 24(7), 2810–2817.

Zhang, Y., Ji, W., Saurette, D. D., Easher, T. H., Li, H., Shi, Z., Adamchuk, V. I. and Biswas, A. 2020. Three-dimensional digital soil mapping of multiple soil properties at a field-scale using regression kriging. *Geoderma* 366, 114253.

Zhou, Y., Biswas, A., Ma, Z., Lu, Y., Chen, Q. and Shi, Z. 2016. Revealing the scale-specific controls of soil organic matter at large scale in Northeast and North China Plain. *Geoderma* 271, 71–79.

Chapter 13

Modeling soil organic carbon dynamics, carbon sequestration and the climate benefit of sequestration

Carlos A. Sierra, Max Planck Institute for Biogeochemistry, Germany and Swedish University of Agricultural Sciences, Sweden; and Susan E. Crow, University of Hawai'i at Mānoa, USA

1 Introduction

Soil organic carbon (SOC) is a complex mix of organic molecules with different biological, chemical, and physical properties. This carbon (C) can be part of dead plant matter, live and dead microorganisms, or organic molecules dissolved in water or attached to the surface of minerals. Soil is an open system with respect to organic C, where it forms from the remains of plant parts or their root exudates, and it is lost as CO_2 from the respiratory activity of microorganisms, dissolved C in water leachates, or soil erosion (Swift et al., 1979; Sollins et al., 1996; Lützow et al., 2006). No single measurement or experiment can capture all this complexity in a comprehensive form that allows one to make predictions on the future dynamics of a particular soil. For a predictive understanding of soil C dynamics over time, the development of mathematical models is essential.

Mathematical models can integrate information that captures the heterogeneity of process rates, transformations of organic matter, and different input-output processes (Manzoni and Porporato, 2009; Sierra and Müller, 2015). Models, however, are imperfect tools that sometimes are a crude

http://dx.doi.org/10.19103/AS.2022.0106.12

approximation of reality and often make overly simplified assumptions in order to be useful. There has been a long debate in the soil science literature about the most appropriate models to represent C dynamics and their response to climate change and to different management practices. There are hundreds of soil C models proposed by many scientists, each developed for a specific purpose, to address a particular question, or to explain certain pattern in a dataset. There is no single model that can explain all observations, that can be used to explore every single question, or to analyze patterns at multiple scales.

We must accept and embrace the diversity of soil C models, including those that have been proposed already and those that will be proposed in the future. Each model has a particular advantage, either because it can predict well results from a certain experiment, address a scale or process of interest, or because it offers the best compromise between accuracy and parsimony.

In this chapter, instead of reviewing particular SOC models, we review families of models with common mathematical properties. We will also review a set of metrics that can be applied to any model in order to obtain a holistic understanding of the main factors that influence C sequestration and the climate benefit of C sequestration. With a set of examples, we will show how different predictions by different models can be used to assess C sequestration and the timescale at which C remains in soils.

2 Modeling soil organic carbon dynamics

2.1 Landscape of soil carbon models

In a comprehensive review of soil carbon and nitrogen models, Manzoni and Porporato (2009) identified 250 mathematical models developed over a period of 80 years. The models differed in their level of detail and spanned over a wide range of temporal and spatial scales of application. Up to that time (year 2009), the number of models were growing at an exponential rate of 6% per year, and it is likely that the number of proposed models continues growing exponentially. Given this diversity of models, it is important to develop a conceptual map of the landscape of available models, so they can be studied systematically.

For the purpose of classifying models according to fundamental mathematical structures, we will consider here the variable x as the mass of SOC in a specific soil. We are interested in studying how this variable changes over time, which may depend on four fundamental variables that define model structure

$$\frac{\partial x}{\partial t} = f(\psi, z, x, t) \tag{1}$$

where

1. Ψ: SOC heterogeneity with respect to process rates. This heterogeneity can be characterized with continuous functions, or with discrete compartments (Sierra et al., 2011). The continuous function can represent complex shapes of the distribution of SOC along a continuum of process rates using only a few parameters (Carpenter, 1981; Ågren and Bosatta, 1996). However, the solution of these systems is complex, and solutions are often derived for the average quality of SOC. In contrast, discrete pool models are very common, but require that total SOC is partitioned in discrete categories according to certain physical, chemical, or biological characteristic that do not necessarily represent the continuous distribution of rates. As the criteria to split SOC into discrete categories is somewhat arbitrary, there has been a long debate on whether these criteria should be based on the chemistry of the substrates, the mineralogical composition of the soil matrix, or the microbial communities that consume the C.
2. z: Vertical dimension, which can be characterized with continuous functions, or with discrete layers. Continuous functions of the vertical dimension lead to systems of partial differential equations, and the equations of advection and diffusion are commonly used as templates for SOC vertical transfer. For modeling discrete layers, the system of equations can be simplified to first-order differential equations (Koven et al., 2013; Metzler et al., 2020).
3. x: SOC mass. If the system of differential equations includes dependencies among the mass of multiple compartments, the system would be nonlinear. This is necessary for modeling the priming effect, or processes resembling Michaelis-Menten kinetics for substrate-enzyme interactions. In case the masses of multiple compartments do not interact, the system of differential equations would be linear.
4. t: Time dependencies. Environmental variables not explicitly modeled, and included as a driving factor, are usually modeled as time-dependent functions that modify process rates. Inputs of SOC to the soil can also be modeled as a time-dependent process, particularly if a data set of litter and root inputs is available. When time dependencies are explicitly included, the system of differential equations is classified as a nonautonomous dynamical system. If all coefficients are constant in the model and do not change over time, the system of differential equations is classified as an autonomous dynamical system. Additional details about the classification of models as autonomous and nonautonomous systems is provided below.

The presence or absence of any of these variables determines the particular structure of a model. For example, a model without vertical dimension $f(\Psi, x, t)$

describes the dynamics of SOC in different pools with possible nonlinear interactions, but without explicitly describing spatial patterns of C transfers along a soil profile. Although the structure of the equation above is useful to identify important components of models, it is too general to understand particular behaviors or mathematical properties of models. In the following section, we will describe a general equation for the classification of models into four groups with common mathematical properties, which in turn lead to different methods to study C sequestration in soils.

2.2 Basic ecological principles to classify and analyze SOC models

Despite the large variety of SOC models proposed in the literature, there is a set of six ecological principles that can be used to mathematically generalize all models. These principles are: mass balance, heterogeneity in process rates, substrate dependence of decomposition, internal transfers and transformations of SOC, environmental variability effects, and substrate interactions (Sierra and Müller, 2015). These principles lead to a general mathematical form expressed as a system of first-order differential equations

$$\frac{d\boldsymbol{C}}{dt} = \boldsymbol{I}(t) + \boldsymbol{T}(\boldsymbol{C},t) \cdot \boldsymbol{N}(\boldsymbol{C},t) \cdot \boldsymbol{C}(t) \tag{2}$$

where $\boldsymbol{C}$ is a vector of SOC contents in n discrete C pools; $\boldsymbol{I}$ is a vector of C inputs to the soil; $\boldsymbol{T}$ is a $n \times n$ matrix of transfers and transformations of organic C among the n pools; $\boldsymbol{N}$ is a $n \times n$ diagonal matrix of process rates that characterize how fast C is processed in each of the n pools. If any of the terms in this equation is time dependent, the system is called nonautonomous. If any of the terms of this equation has an interaction among different C pools, the entire system is nonlinear. This distinction between linearity and autonomy is fundamental to classify and analyze models. Some diagnostic metrics can only be obtained for certain groups of models. For example, only autonomous models may have a steady state, while nonautonomous models, by definition, have no steady-state solution because they have time dependencies that change C inputs and process rates all the time.

Almost all SOC models can be written in the form of equation (2), including models that represent vertical dynamics (Metzler et al., 2020). Most models with partial differential equations that include vertical C transfers cannot be solved analytically, and instead are solved numerically in computers for discrete vertical intervals, then equation (2) can also be used to generalize any continuous model that has been discretized (Sierra et al., 2011). This general mathematical form is also called a compartmental dynamical system, because due to mass constraints, the system of equations follows strict mathematical constraints (Anderson, 1983; Jacquez and Simon, 1993; Sierra et al., 2018a).

3 Families of models

3.1 Linear autonomous compartmental models

The simplest models are those that can be classified as linear autonomous models, i.e. those that have constant coefficients and no interactions among different types of SOC. For example, models with constant C inputs to the soil and constant decomposition rates that do not change over time can be classified into this category. These models are simple to analyze and can provide interesting insights, e.g. about the maximum C storage capacity of a soil or the average time it takes for C atoms to travel through the system. Models in this category converge to a steady-state solution, which can give an approximation of the maximum amount of C storage in a soil (assuming that C inputs are nearly constant and rates do not change significantly over time). Although these assumptions may not hold for very long time scales, the steady-state C storage concept is useful to consider as a potential upper limit of maximum soil C stocks as determined by the set input and loss rates. Linear autonomous models may help land managers to estimate potential changes in the maximum C stocks that could be obtained if significant changes in C inputs or rates are changed.

3.2 Nonlinear autonomous compartmental models

Models classified as nonlinear autonomous are useful to represent interactions among different forms of SOC. For example, it has been shown experimentally that the addition of fresh C promotes the decomposition of old C, a phenomenon called the priming effect (Kuzyakov, 2010). This type of interaction can be modeled as an interaction among different forms of SOC that leads to nonlinear systems of differential equations (Wutzler and Reichstein, 2008; Blagodatsky et al., 2010). This type of system may have one or multiple steady states. However, in some cases steady-state solutions may violate the basic principles mentioned above and lead to unrealistic solutions, e.g. negative SOC stocks. In most cases, the interactions among different SOC pools are represented with multiplicative terms or with Michaelis–Menten functions that express the saturation effect imposed by one type of substrate (e.g. enzymes) that limits decomposition (Schimel and Weintraub, 2003; Allison et al., 2010; Wang et al., 2012; Wieder et al., 2014). The predictions of these mathematical models often show oscillations among the different types of SOC (Wang et al., 2014); e.g. fluctuations of enzymes and substrates resembling Lotka–Volterra cycles, mostly because the equations in these models are in essence the same equations used to model predator–prey interactions.

3.3 Linear nonautonomous systems

Linear nonautonomous models usually take a linear autonomous model with constant coefficients as a template and add functions that change the values of these coefficients as other environmental variables change over time. These models are very common in many biogeochemical and land surface models. For example, it is common to use functions that depend on soil moisture and soil temperature to predict how changes in the environment change the values of decomposition rates over time (Sierra et al., 2015). It is also common to use a dataset of past C inputs to the soil derived from crop yields or model-based estimates of net primary productivity as a main driver in linear nonautonomous models. It is always necessary to hold knowledge of the temporal dynamics of the system to be able to run these models. Because inputs and rates are always changing, these models do not converge to a steady-state solution, but rather show the trajectory of SOC stocks imposed by the time-dependent inputs and rates (Sierra et al., 2018a). The linear nonautonomous system is the predominant paradigm in SOC models such as Century (Parton et al., 1987), RothC (Jenkinson and Rayner, 1977), MEMS (Robertson et al., 2019), Millennium (Abramoff et al., 2018), among many others. These models are very useful to predict how spatial patterns of SOC change over time, and to predict the effects of environmental change on SOC stocks. Most Earth system models include a soil C module built as a linear nonautonomous system.

3.4 Nonlinear nonautonomous systems

The most complex models have time dependencies in inputs and process rates, and include nonlinear interactions among SOC pools. These nonlinear nonautonomous models sometimes lead to complex, chaotic behaviors, and are difficult to generalize outside of the trajectories obtained from a particular simulation (Müller and Sierra, 2017). There are few examples of these models to represent SOC and assess C sequestration. One interesting example of a nonlinear nonautonomous model is the Bacwave model (Zelenev et al., 2000), used to represent the dynamics of soil bacterial populations in response to the time-dependent dynamics of the production of root exudates. The model reproduced well observations from a laboratory experiment, but to be able to run the model, a function that predicts exudate dynamics over time must be known. In general, it is challenging to use nonlinear nonautonomous models for large-scale applications because considerable knowledge of the system and the expected behavior over time must be known a priori.

4 Metrics to analyze models in the context of carbon sequestration for climate change mitigation

Despite the large variety of mathematical models used to represent the dynamics of SOC, it is possible to use metrics computed across diverse models and learn general system properties. In particular, we can use metrics to aggregate the combined effect of adding C to the soil with the amount of time this C will spend in the soil. The representation of models in the form of equation (2) is useful for this purpose because it allows us to use very general methods to compute aggregated metrics across all possible models. We will review two recently developed metrics (Sierra et al., 2021) that can be applied to any of the model categories mentioned above, these are carbon sequestration (CS) and the climate benefit of sequestration (CBS).

4.1 Carbon sequestration

Different perspectives on soil C sequestration exist. Carbon sequestration (CS) has been defined variously as the amount of C present in a soil at a given time, or the amount that enters the soil at a given time or during a particular period, or as a difference between the amount stored at one time versus the amount stored at a later time. Time is a particular aspect of utmost importance to consider when defining C sequestration. That is: Is new C entering the soil, or C already there, stored for a short period of time or at a longer timescale relevant for climate change mitigation? Researchers have struggle with this question before and pointed out that C sequestration in natural ecosystems is not permanent (Fearnside et al., 2000; Moura Costa and Wilson, 2000). Therefore, a meaningful definition of C sequestration in the context of climate change mitigation must include a consideration of the time C is stored.

Here, we use a definition of C sequestration that integrates the amount of C inputs entering the soil at a given time and the amount of time that those C inputs are stored in the soil. We assume the soil to be an open system, where all C that enters eventually leaves in solid, gas, or liquid form. We base this definition on new C entering the soil, tracking the fate of those C inputs, and quantifying how long this C remains. Because decomposition is a continuous process that gradually degrades a substrate, we need to use a model that predicts the dynamics of these inputs and select a time horizon of interest. More precisely,

> *Carbon sequestration is the storage of a certain amount of C input over a certain period of time within a system. It can be quantified as the area under the curve illustrating the fate of C input as it flows through an ecosystem over time.*

Graphically (Fig. 1), we can see that an amount of C that enters at a time t_0, decreases in an almost exponential form until eventually most of it is

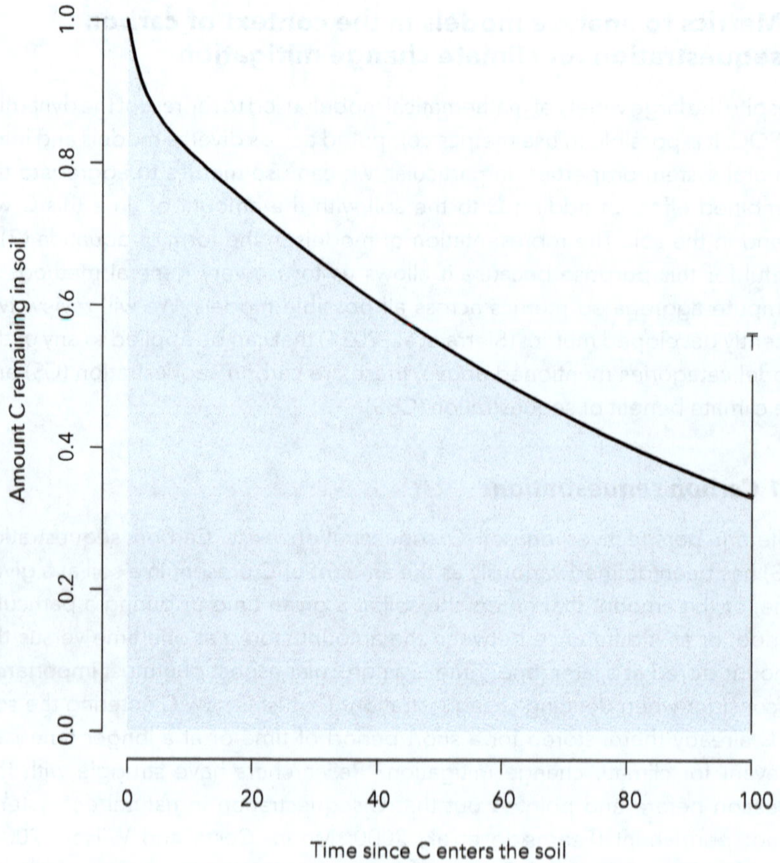

Figure 1 CS can be defined as the area under the curve of an amount of C inputs entering at t_0 observed for a time horizon *T*. In this example, 1 unit of C enters at $t_0 = 0$, and the area under the curve (shaded blue area) of the remaining C is computed until $T = 100$.

decomposed. If we fix a time horizon *T*, we can compute the area under the curve of the remaining amount of inputs and obtain a metric that tells us how much C is retained in the soil over a fixed period of time. The approach is similar as in pharmacology, where the area under the curve of a drug in the blood system tells researchers about the total drug exposure across time. It is also the same approach used to study the absolute global warming potential (AGWP) of an emission of a greenhouse gas (GHG) to the atmosphere: the area under the curve of an amount of emissions tell us how much the GHG is exposed to the absorption of solar radiation (Rodhe, 1990). We will see later that this approach is also useful to quantify the climate effect of sequestering C in soils.

CS is a metric that can be computed from any model and can be used to predict how much and for how long C will be stored in the soil. It can be computed for any model structure because its computational definition is independent of assumptions of linearity or the existence of a steady state.

As an example, let's consider two different ecosystems, a perennial grass system growing on a Mollisol in Hawai'i, and an arable Cambisol soil in Sweden. A linear autonomous model with three C pools for the perennial grass cultivation in Hawai'i was developed by Crow et al. (2018). Similarly, a linear autonomous model with two pools was developed for Swedish agricultural soils by Andren and Kätterer (1997) (the ICBM model). Both models can be written following the matrix form of equation (1), therefore we can compute CS from both models. For simplicity, we will assume that rates and inputs are constant, consistent with the family of linear autonomous models.

If we compare the fate of the same amount of inputs entering these soils (1 Mg C ha^{-1}), we can observe the remaining C over time (Fig. 2). The comparison between the amount of C remaining and the area under this curve, which we define as CS, depends strongly on the chosen time horizon for the comparison. We see that the 1 Mg C ha^{-1} entering the Swedish soil decays fast at first, but then the remaining C persists for a longer time in comparison to the Hawaiian Mollisol. On a 5-year time horizon, more C remains in the Hawaiian Mollisol (higher C sequestration) than on the Swedish Cambisol. However, on a 20-year time horizon, C sequestration is slightly larger in the Swedish soil. Therefore, according to this conceptual framework, the timescale in which soil C persistence is evaluated plays a large role in the assessment of CS in soils.

In addition to the timescale of persistence, it is also important to consider differences in the amount of inputs entering the soil. If we now consider the differences in plant productivity between the two sites, i.e. the amount of C inputs between the Hawaiian and the Swedish soil, we can see that the more productive Hawaiian soil with 3.5 Mg C ha^{-1} year^{-1} of inputs has a larger area under the curve for all time horizons in comparison to the less productive Swedish cropland with only 2.0 Mg C ha^{-1} year^{-1} of inputs (Fig. 3). The value of CS for the Hawaiian Mollisol is 11.88 Mg C ha^{-1} year on a 20-year time horizon, which is more C remaining (higher CS) than in the Swedish Cambisol on the same time horizon (6.93 Mg C ha^{-1} year). Therefore, both amount of inputs and the timescale at which soil C persistence is evaluated play a large role in quantifying soil CS in this conceptual framework.

This simple example shows that we can assess C sequestration from the perspective of new C inputs entering the soils, taking into account how long it will stay stored. Because the fate of the inputs can be obtained from different models that may include different mechanisms leading to different model structures, the CS metric helps to make comparisons independent of the type of model being used. The amount of C inputs can change over time, and it is possible that one is interested in assessing the fate of C entering the soil during a particular time window. In this case, it is a simple matter of summing the CS

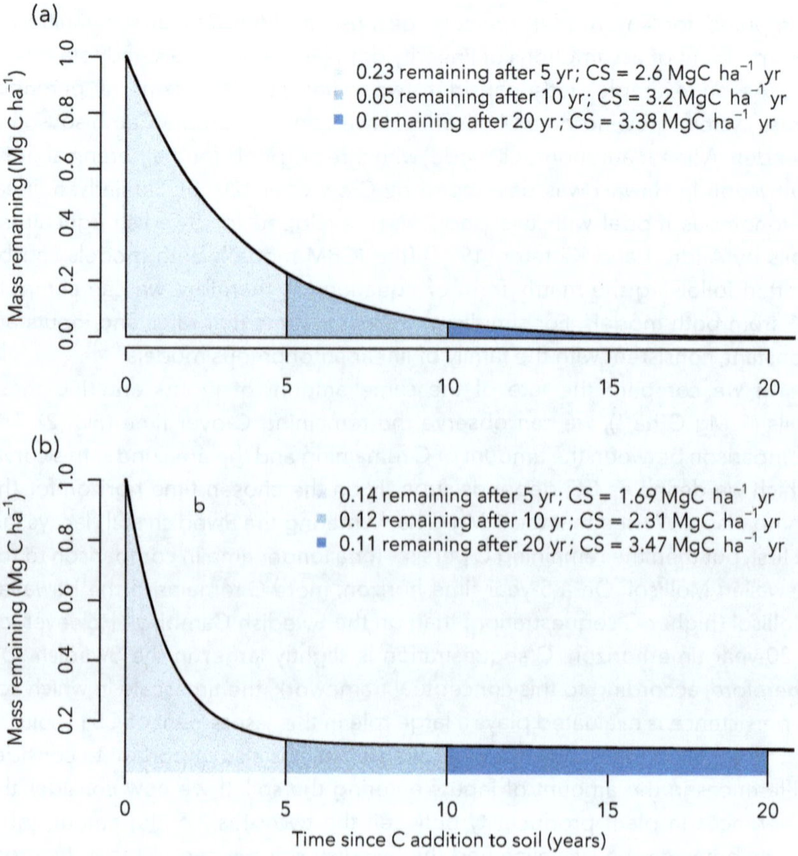

Figure 2 Mass remaining over time, and CS defined as the area under the curve for different time horizons, for two different soils: (a) a Mollisol under perennial cultivation in Hawai'i; (b) an arable Cambisol in Sweden. On a 5-year time horizon, more C remains from one Mg of C inputs in Hawai'i than in Sweden, therefore CS is higher for the tropical Mollisol. However, at a 20-year time horizon, the amount of remaining C is higher in the Swedish Cambisol.

values obtained for discrete time units of say one year each. The areas under the curves of discrete cohorts of inputs can be summed to obtain an overall CS value that integrates multiple cohorts. A detailed description of the procedure including all required mathematical formulas can be found in Sierra et al. (2021).

4.2 Climate benefit of sequestration

CS tells us about the amount of C stored over a period of time, but it does not inform us about the greenhouse effect this C avoids while it is stored in soil. To quantify this, we use a different metric, which we call the CBS. We define it as

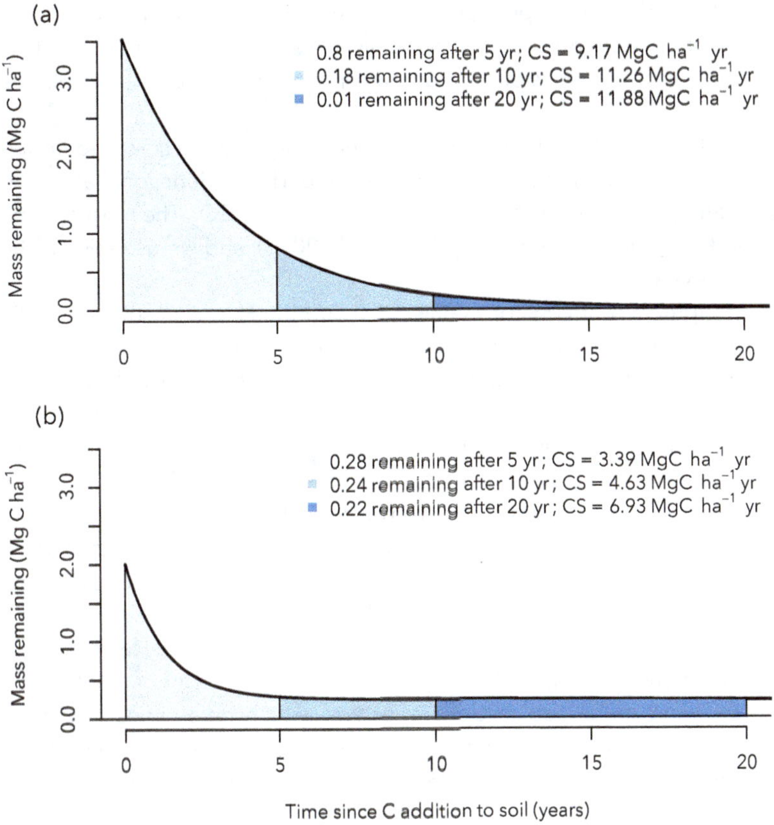

Figure 3 Mass remaining over time, and CS defined as the area under the curve for different time horizons, for two different soils with different productivity and annual C inputs: (a) a Mollisol under perennial cultivation in Hawai'i with inputs of 3.5 Mg C ha^{-1} year^{-1}; (b) an arable Cambisol in Sweden with inputs of 2.0 Mg C ha^{-1} year^{-1}.

The Climate Benefit of Sequestration is the radiative forcing (greenhouse) effect avoided by an amount of C inputs to the soil stored over a specific time horizon.

CBS is reported in units of Watts per square meter year (W m^{-2} year). It is very similar to the AGWP of CO_2, but instead of considering a C emission as an input to the atmosphere with a positive sign, it considers sequestration into a soil as a negative emission. Because most C that enters the soil returns to the atmosphere as heterotrophic respiration, CBS accounts for the temporary effect of storing a unit of C that enters at a particular time and returns to the atmosphere over a time horizon.

The concept of CBS can be used to compare different soils or different forms of management in terms of the avoided warming they can provide as an ecosystem service. For its computation, it is necessary to have a model that predicts the fate of the inputs of C to the soil and the timing of C return to the atmosphere from respiration. Practically any SOC model can provide these predictions, and it is possible to compare predictions of CBS from different models.

In addition to a SOC model, it is also necessary to use a so-called impulse response function (IRF) of CO_2 in the atmosphere. This is a function that predicts how the atmosphere reacts to additions or removals of CO_2. The most common IRF used today was developed by Joos et al. (2013), and we use this function for our analysis here.

A computation of CBS for the Swedish and the Hawaiian soils shows that the climate benefit of sequestration can be very different depending on the time horizon used for comparing these soils. For time horizons below 40 years, more warming is avoided by the Mollisol soil in Hawai'i than by the Cambisol in Sweden. This is because the more productive site in Hawai'i draws down more C from the atmosphere than the less productive Swedish soil. However, decomposition is also faster in the Hawaiian soil and C returns back to the atmosphere faster than the Swedish soil. Therefore, for time horizons longer than 40 years, CBS is larger for the Swedish Cambisol where a small proportion of the original inputs remains stored for a longer time (Fig. 4).

This example shows the importance of considering two key factors related to C sequestration in soils and its climate benefit: the amount of inputs added to the soil, and the speed of decomposition, which determines the time C persists. Although much research has been done in quantifying inputs and rates of decay of SOC, much less attention has been paid to the time C persists in soils.

From these examples, it is also possible to infer that management of soils for C sequestration purposes can be focused in balancing the amount of inputs and the time they will remain. This book contains many examples and methods to increase inputs or decrease decomposition of SOC. The conceptual framework provided by the concepts of CS and CBS, together with simulation models, can give important insights on how to best achieve climate goals through C management in soils.

In addition, CBS can be compared with atmospheric metrics that quantify the warming effect of an emission of a GHG (CH_4, N_2O, O_3, etc.), namely the AGWP, which is also expressed in units of W m^{-2} year. AGWP and CBS (with negative sign) can be added to obtain the net climate effect of simultaneous emissions and sequestration in a particular system, which can be useful to assess overall climate effects of multiple forms of ecosystem and soil management. In general, the AGWP of CO_2 is much larger for 1 unit of C emissions than for 1 unit of sequestration for any timescale because fossil fuel emissions stay for much longer in the atmosphere-ocean-biosphere system (Sierra et al., 2021). The emission of 1 Mg C to the atmosphere leads to a potential warming of 3.4

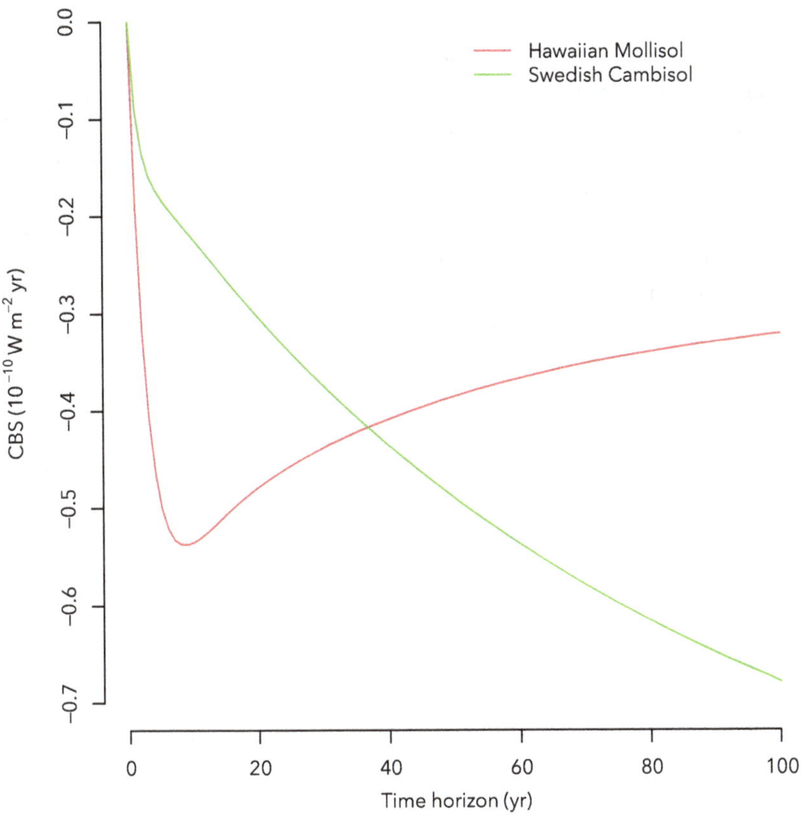

Figure 4 CBS for a Mollisol in Hawai'i with 3.5 Mg C ha^{-1} year^{-1} of inputs versus a Cambisol in Sweden with inputs of 2.0 Mg C ha^{-1} year^{-1}. The negative CBS values indicate the warming avoided by storing C in these soils for different time horizons. Because inputs are larger in the Mollisol, more warming is avoided for time horizons below 40 years, but because a small proportion of the original input stays in the Cambisol for a long time, CBS is larger in the Cambisol for time horizons longer than 40 years.

× 10^{-10} W m^{-2} year on a 100-year time horizon (Joos et al., 2013), one order of magnitude higher than the potential avoided warming in any of the two soils from our example.

5 Discussion

Models of SOC dynamics are important tools to quantify C sequestration and related climate benefits. There is a large variety of models that can be used for this purpose, and a systematic classification and analysis of these models helps to better integrate different models within one single conceptual framework. Mathematically, we can classify models according to the presence of nonlinearities (linear or nonlinear models) and time dependencies (autonomous and nonautonomous models). These classes help to define the type of analysis methods and metrics that can be used for each model. Two metrics to analyze models in the context of soil C sequestration are the concepts of CS and CBS we presented here. These two metrics take into consideration the amount of C entering the soil and for how long this C remains stored.

Decisions such as the location of a site, options for land use, or aspects of management can affect both the amount of C inputs and the time that C may be expected to remain in the system (Sierra et al., 2021). The amount of inputs are profoundly affected by land use and management changes such as deforestation, afforestation, conversion of managed pasture to conservation lands, removal of crop residues, among others (Lal, 2016; Chabbi et al., 2017; Minasny et al., 2017). The amount of time C inputs remain in the system is affected by a management choice that would affect SOC persistence; e.g. due to site selection based on climate or environmental factors (Sierra et al., 2018b; Shi et al., 2020), and/or soil mineralogy such as poorly or noncrystalline minerals (Torn et al., 1997), or selection of particular soil amendments such as biochar (Guo and Chen, 2014). Likewise, management decisions that consider both the organic C inputs and the persistence of those inputs can maximize the climate change mitigation potential through sequestration as a result of those choices.

Examples presented here were specifically for soil C sequestration, but the CS and CBS concepts may also be applied to any other system that can remove C from the atmosphere and store it for a particular amount of time. Examples include single trees or forest stands, aquatic ecosystems, or engineered C reservoirs (carbon capture and storage systems). The concepts can also be applied to 'cradle to grave' life cycle analyses that allow for systems-level understanding of the consequences of any form of C management. A system is defined here in a very general form (eq. 2) as a set of compartments or reservoirs that exchange mass among each other and interact with the environment. The definition of this system can explicitly consider the way human interventions affect its functioning by modifying inputs, transfer among components, and process rates. Soil is a classic example of a social-ecological system (Kibblewhite, 2018), where humans interact directly with environmental factors to influence the outcome, including soil C management (Amin et al., 2020). The climate benefit of soil C sequestration may only be one component of a system defined by the boundaries of a site, jurisdiction, or nation.

In system-level assessment, often there is a need to directly compare the benefit of sequestration to that of direct emissions avoidance elsewhere (Verkerk et al., 2020). For example, in the Hawaiian islands, undisturbed native forests with high productivity in volcanic ash-derived soils (that have very high capacity to stabilize soil C on poorly and noncrystalline mineral surfaces) are likely to have high CS (Giardina et al., 2014). In this case, the avoided warming quantified by CBS may provide the same or greater value than the warming predicted by AGWP from direct emissions avoidance, and argue for prioritizing the protection of these forests. Afforestation of rangelands is often assumed to promote climate change mitigation, however in some systems soil C loss during land preparation and planting is greater than the tree biomass produced during the same time frame (Crow et al., 2016). Furthermore, the time C inputs persist in the system as a whole is dependent on the fate of the final wood product, e.g. if the forest is managed for pulp/paper or pellets/biofuel (short timeframe), construction or furniture (medium timeframe), or conservation (long timeframe) (Sahoo et al., 2019; Schulze et al., 2020).

Finally, consider a food system, which is the social-ecological system encompassing all the drivers, activities, and resources that go into producing, distributing, and consuming food and includes waste stream and disposal (Miles et al., 2017). A third of anthropogenic GHG emissions originate from food systems globally (Crippa et al., 2021). Increasingly, incentives are offered to support producers to implement practices promoting soil health, for example, that also improve multiple co-benefits including water quality, human health, and climate change (Lehmann et al., 2020). Many practices that promote soil health will also build soil organic matter and C, but not predictably and unlikely in a form that would provide meaningful CS that persists over time (e.g. Gattinger et al., 2012; Leifeld et al., 2013). However, the CBS computation allows a more holistic quantification of improved sustainability of the food system that includes CS and avoided emissions associated with local production of food rather than importing. This is particularly important in island systems such as Hawai'i and other areas where regional accounting places system boundaries at geopolitical borders (Miles et al., 2017).

6 Conclusion

Mathematical models are essential for integrating different processes that control rates of soil C dynamics and for assessing C sequestration and related climate benefits. Many models have been proposed in the literature to predict C stocks and fluxes, with no overall consensus on the best model that can provide relevant insights at a large range of scales and for multiple questions. We reviewed general groups of models with their expected ranges of application. We also reviewed recent advances in using models of any level of detail to compute C sequestration, and the climate benefit of C sequestration. Using agricultural soils from Sweden and Hawai'i as examples, we show that new C inputs to the soil do not remain for long timescales, and only small proportions are stabilized. Although soils are a promising reservoir to store C and mediate emissions, long timescales are required to store amounts of C of relevance to mitigate climate change. The magnitude of climate benefit to mitigate warming through soil C sequestration is less than that of avoiding direct emissions, however, remains an important component of climate change mitigation and adaptation portfolios. Beyond the direct warming mitigation benefits, improved soil health through soil organic matter aggradation brings many co-benefits to the environment and local communities. Improved production practices and locally sourced food and energy feedstocks are associated directly with avoided emissions elsewhere in the food and energy system.

7 Where to look for further information

- A comprehensive review of soil C models can be found in Manzoni and Porporato (2009). A conceptual classification of soil C models based on ecological principles can be found in Sierra and Müller (2015), with a further classification scheme in Sierra et al. (2018a).
- The concepts of CS and CBS, including all required formulas for their computations for different types of models, can be found in Sierra et al. (2021). Background information on the concept of AGWP can be found in Rodhe (1990) and Joos et al. (2013).

8 References

Abramoff, R., Xu, X., Hartman, M., O'Brien, S., Feng, W., Davidson, E., Finzi, A., Moorhead, D., Schimel, J., Torn, M. and Mayes, M. A. 2018. The millennial model: in search of measurable pools and transformations for modeling soil carbon in the new century. *Biogeochemistry* 137(1–2): 51–71. Available at:https://doi.org/10.1007/s10533-017-0409-7.

Ågren, G. I. and Bosatta, E. 1996. *Theoretical Ecosystem Ecology: Understanding Element Cycles*. Cambridge: Cambridge University Press.

Allison, S. D., Wallenstein, M. D. and Bradford, M. A. 2010. Soil-carbon response to warming dependent on microbial physiology. *Nature Geoscience* 3(5): 336–40.

Amin, M. N., Hossain, M. S., Lobry de Bruyn, L. and Wilson, B. 2020 A systematic review of soil carbon management in Australia and the need for a social-ecological systems framework. *Science of the Total Environment* 719: 135182. Available at: https://doi.org/10.1016/j.scitotenv.2019.135182.

Anderson, D. H. 1983. *Compartmental Modeling and Tracer Kinetics* (vol. 50). Berlin: Springer Science & Business Media.

Andren, O. and Kätterer, T. 1997. ICBM: the introductory carbon balance model for exploration of soil carbon balances. *Ecological Applications* 7(4): 1226–36.

Blagodatsky, S., Blagodatskaya, E., Yuyukina, T. and Kuzyakov, Y. 2010. Model of apparent and real priming effects: linking microbial activity with soil organic matter decomposition. *Soil Biology and Biochemistry* 42(8): 1275–83. Available at: https://doi.org/10.1016/j.soilbio.2010.04.005.

Carpenter, S. R. 1981. Decay of heterogenous detritus: a general model. *Journal of Theoretical Biology* 89(4): 539–47.

Chabbi, A., Lehmann, J., Ciais, P., Loescher, H. W., Cotrufo, M. F., Don, A., SanClements, M., Schipper, L., Six, J., Smith, P. and Rumpel, C. 2017. Aligning agriculture and climate policy. *Nature Climate Change* 7(5): 307–9. Available at: https://doi.org/10.1038/nclimate3286.

Crippa, M., Solazzo, E., Guizzardi, D., Monforti-Ferrario, F., Tubiello, F. N. and Leip, A. 2021. Food systems are responsible for a third of global anthropogenic GHG emissions. *Nature Food* 2(3): 198–209. Available at: https://doi.org/10.1038/s43016-021-00225-9.

Crow, S. E., Deem, L. M., Sierra, C. A. and Wells, J. M. 2018. Belowground carbon dynamics in tropical perennial C4 grass agroecosystems. *Frontiers in Environmental Science* 6: 18. Available at: https://doi.org/10.3389/fenvs.2018.00018.

Crow, S. E., Reeves, M., Turn, S., Taniguchi, S., Schubert, O. S. and Koch, N. 2016. Carbon balance implications of land use change from pasture to managed Eucalyptus Forest in Hawaii. *Carbon Management* 7(3–4): 171–81. Available at: https://doi.org/10.1080/17583004.2016.1213140.

Fearnside, P. M., Lashof, D. A. and Moura-Costa, P. 2000. Accounting for time in mitigating global warming through land-use change and forestry. *Mitigation and Adaptation Strategies for Global Change* 5(3): 239–70. Available at: https://doi.org/10.1023/A:1009625122628.

Gattinger, A., Muller, A., Haeni, M., Skinner, C., Fliessbach, A., Buchmann, N., Mäder, P., Stolze, M., Smith, P., Scialabba, Nel-H. and Niggli, U. 2012. Enhanced top soil carbon stocks Under organic farming. *Proceedings of the National Academy of Sciences of the United States of America* 109(44): 18226–31. Available at: https://doi.org/10.1073/pnas.1209429109.

Giardina, C. P., Litton, C. M., Crow, S. E. and Asner, G. P. 2014. Warming-related increases in soil CO_2 efflux are explained by increased Below-ground carbon flux. *Nature Climate Change* 4(9): 822–27. Available at: https://doi.org/10.1038/nclimate2322.

Guo, J. and Chen, B. 2014. Insights on the molecular mechanism for the recalcitrance of biochars: interactive effects of carbon and silicon components. *Environmental Science and Technology* 48(16): 9103–12. Available at: https://doi.org/10.1021/es405647e.

Jacquez, J. A. and Simon, C. P. 1993. Qualitative theory of compartmental systems. *SIAM Review* 35(1): 43–79. Available at: https://doi.org/10.1137/1035003.

Jenkinson, D. S. and Rayner, J. H. 1977. The turnover of soil organic matter in some of the Rothamsted classical experiments. *Soil Science* 123(5): 298–305.

Joos, F., Roth, R., Fuglestvedt, J. S., Peters, G. P., Enting, I. G., von Bloh, W., Brovkin, V., Burke, E. J., Eby, M., Edwards, N. R., Friedrich, T., Frölicher, T. L., Halloran, P. R., Holden, P. B., Jones, C., Kleinen, T., Mackenzie, F. T., Matsumoto, K., Meinshausen, M., Plattner, G.-K., Reisinger, A., Segschneider, J., Shaffer, G., Steinacher, M., Strassmann, K., Tanaka, K., Timmermann, A. and Weaver, A. J. 2013. Carbon dioxide and climate impulse response functions for the computation of greenhouse gas metrics: a multi-model analysis. *Atmospheric Chemistry and Physics* 13(5): 2793–825. Available at: https://doi.org/10.5194/acp-13-2793-2013.

Kibblewhite, M. G. 2018. Soil and soil health: an overview. In: *Managing Soil Health for Sustainable Agriculture*. Reicosky, D. (Ed.), 3–15. Burleigh Dodds Science Publishing Limited. Available at: https://doi.org/10.1201/9781351114530.

Koven, C. D., Riley, W. J., Subin, Z. M., Tang, J. Y., Torn, M. S., Collins, W. D., Bonan, G. B., Lawrence, D. M. and Swenson, S. C. 2013. The effect of vertically resolved soil biogeochemistry and alternate soil C and N models on C dynamics of Clm4. *Biogeosciences* 10(11): 7109–31. https://doi.org/10.5194/bg-10-7109-2013.

Kuzyakov, Y. 2010. Priming effects: interactions Between living and dead organic matter. *Soil Biology and Biochemistry* 42(9): 1363-71. Available at: https://doi.org/10.1016/j.soilbio.2010.04.003.

Lal, R. 2016. "Beyond COP 21": potential and challenges of the "4 per thousand" initiative. *Journal of Soil and Water Conservation* 71(1): 20A-5A. Available at: https://doi.org/10.2489/jswc.71.1.20A.

Lehmann, J., Bossio, D. A., Kögel-Knabner, I. and Rillig, M. C. 2020. The concept and future prospects of soil health. *Nature Reviews. Earth and Environment* 1(10): 544-53. Available at: https://doi.org/10.1038/s43017-020-0080-8.

Leifeld, J., Angers, D. A., Chenu, C., Fuhrer, J., Kätterer, T. and Powlson, D. S. 2013. Organic farming gives no climate change benefit Through soil carbon sequestration. *Proceedings of the National Academy of Sciences of the United States of America* 110(11): E984. Available at: https://doi.org/10.1073/pnas.1220724110.

Lützow, Mv, Kögel-Knabner, I., Ekschmitt, K., Matzner, E., Guggenberger, G., Marschner, B. and Flessa, H. 2006 Stabilization of organic matter in temperate soils: mechanisms and their relevance under different soil conditions - a review. *European Journal of Soil Science* 57(4): 426-45.

Manzoni, S. and Porporato, A. 2009. Soil carbon and nitrogen mineralization: theory and models across scales. *Soil Biology and Biochemistry* 41(7): 1355-79. Available at: https://doi.org/10.1016/j.soilbio.2009.02.031.

Metzler, H., Zhu, Q., Riley, W., Hoyt, A., Müller, M. and Sierra, C. A. 2020. Mathematical reconstruction of land carbon models from their numerical output: computing soil radiocarbon from c dynamics. *Journal of Advances in Modeling Earth Systems* 12(1): e2019. Available at: https://doi.org/10.1029/2019MS001776.

Miles, A., DeLonge, M. S. and Carlisle, L. 2017. Triggering a positive research and policy feedback cycle to support a transition to agroecology and sustainable food systems. *Agroecology and Sustainable Food Systems* 41(7): 855-79. Available at: https://doi.org/10.1080/21683565.2017.1331179.

Minasny, B., Malone, B. P., McBratney, A. B., Angers, D. A., Arrouays, D., Chambers, A., Chaplot, V., Chen, Z., Cheng, K., Das, B. S., Field, D. J., Gimona, A., Hedley, C. B., Hong, S. Y., Mandal, B., Marchant, B. P., Martin, M., McConkey, B. G., Mulder, V. L., O'Rourke, S., Richer-de-Forges, A. C., Odeh, I., Padarian, J., Paustian, K., Pan, G., Poggio, L., Savin, I., Stolbovoy, V., Stockmann, U., Sulaeman, Y., Tsui, C., Vågen, T., van Wesemael, B. and Winowiecki, L. 2017. Soil carbon 4 per mille. *Geoderma* 292: 59-86. Available at: https://doi.org/10.1016/j.geoderma.2017.01.002.

Moura Costa, P. and Wilson, C. 2000. An equivalence factor Between CO2 avoided emissions and sequestration - description and applications in forestry. *Mitigation and Adaptation Strategies for Global Change* 5(1): 51-60. Available at: https://doi.org/10.1023/A:1009697625521.

Müller, M. and Sierra, C. A. 2017. Application of input to state stability to reservoir models. *Theoretical Ecology* 10(4): 451-75. Available at: https://doi.org/10.1007/s12080-017-0342-3.

Parton, W. J., Schimel, D. S., Cole, C. V. and Ojima, D. S. 1987. Analysis of factors controlling soil organic matter levels in great plains grasslands. *Soil Science Society of America Journal* 51(5): 1173-79. Available at: https://doi.org/10.2136/sssaj1987.03615995005100050015x.

Robertson, A. D., Paustian, K., Ogle, S., Wallenstein, M. D., Lugato, E. and Cotrufo, M. F. 2019. Unifying soil organic matter formation and persistence frameworks: the MEMS model. *Biogeosciences* 16(6): 1225–48. Available at: https://doi.org/10.5194/bg-16-1225-2019.

Rodhe, H. 1990. A comparison of the contribution of various gases to the greenhouse effect. *Science* 248(4960): 1217–19. Available at: https://doi.org/10.1126/science.248.4960.1217.

Sahoo, K., Bergman, R., Alanya-Rosenbaum, S., Gu, H. and Liang, S. 2019. Life cycle assessment of forest-based products: a review. *Sustainability* 11(17). https://doi.org/10.3390/su11174722.

Schimel, J. P. and Weintraub, M. N. 2003. The implications of exoenzyme activity on microbial carbon and nitrogen limitation in soil: a theoretical model. *Soil Biology and Biochemistry* 35(4): 549–63. Available at: https://doi.org/10.1016/S0038-0717(03)00015-4.

Schulze, E. D., Sierra, C. A., Egenolf, V., Woerdehoff, R., Irslinger, R., Baldamus, C., Stupak, I. and Spellmann, H. 2020. The climate change mitigation effect of bioenergy from sustainably managed forests in Central Europe. *GCB Bioenergy* 12(3): 186–97. Available at: https://doi.org/10.1111/gcbb.12672.

Shi, Z., Allison, S. D., He, Y., Levine, P. A., Hoyt, A. M., Beem-Miller, J., Zhu, Q., Wieder, W. R., Trumbore, S. and Randerson, J. T. 2020. The age distribution of global soil carbon inferred from radiocarbon measurements. *Nature Geoscience* 13(8): 555–59. Available at: https://doi.org/10.1038/s41561-020-0596-z.

Sierra, C. A., Crow, S. E., Heimann, M., Metzler, H. and Schulze, E.-D. 2021. The climate benefit of carbon sequestration. *Biogeosciences* 18(3): 1029–48. Available at: https://doi.org/10.5194/bg-18-1029-2021.

Sierra, C. A., Ceballos-Núñez, V., Metzler, H. and Müller, M. 2018a. Representing and understanding the carbon cycle using the theory of compartmental dynamical systems. *Journal of Advances in Modeling Earth Systems* 10(8): 1729–34. Available at: https://doi.org/10.1029/2018MS001360.

Sierra, C. A., Hoyt, A. M., He, Y. and Trumbore, S. E. 2018b. Soil organic matter persistence as a stochastic process: age and transit time distributions of carbon in soils. *Global Biogeochemical Cycles* 32(10): 1574–88. Available at: https://doi.org/10.1029/2018GB005950.

Sierra, C. A., Harmon, M. E. and Perakis, S. S. 2011. Decomposition of heterogeneous organic matter and its long-term stabilization in soils. *Ecological Monographs* 81(4): 619–34. Available at: https://doi.org/10.1890/11-0811.1.

Sierra, C. A. and Müller, M. 2015. A general mathematical framework for representing soil organic matter dynamics. *Ecological Monographs* 85(4): 505–24. Available at: https://doi.org/10.1890/15-0361.1.

Sierra, C. A., Trumbore, S. E., Davidson, E. A., Vicca, S. and Janssens, I. 2015. Sensitivity of decomposition rates of soil organic matter with respect to simultaneous changes in temperature and moisture. *Journal of Advances in Modeling Earth Systems* 7(1): 335–56. Available at: https://doi.org/10.1002/2014MS000358.

Sollins, P., Homann, P. and Caldwell, B. A. 1996. Stabilization and destabilization of soil organic matter: mechanisms and controls. *Geoderma* 74(1–2): 65–105.

Swift, M. J., Heal, O. W. and Anderson, J. M. 1979. *Decomposition in Terrestrial Ecosystems*. Berkeley: University of California Press.

Torn, M. S., Trumbore, S. E., Chadwick, O. A., Vitousek, P. M. and Hendricks, D. M. 1997. Mineral control of soil organic carbon storage and turnover. *Nature* 389(6647): 170–73. Available at: https://doi.org/10.1038/38260.

Verkerk, P. J., Costanza, R., Hetemaki, L., Kubiszewski, I., Leskinen, P., Nabuurs, G. J., Potočnik, J. and Palahi, M. 2020. Climate-smart forestry: the missing link. *Forest Policy and Economics* 115: 102164. Available at: https://doi.org/10.1016/j.forpol.2020.102164.

Wang, G., Gangsheng, W. M., Post, W. M. and Mayes, M. A. 2012. Development of microbial-enzyme-mediated decomposition model parameters Through steady-state and dynamic analyses. *Ecological Applications* 23(1): 255–72. Available at: https://doi.org/10.1890/12-0681.1.

Wang, Y. P., Chen, B. C., Wieder, W. R., Leite, M., Medlyn, B. E., Rasmussen, M., Smith, M. J., Agusto, F. B., Hoffman, F. and Luo, Y. Q. 2014. Oscillatory behavior of two nonlinear microbial models of soil carbon decomposition. *Biogeosciences* 11(7): 1817-31. Available at: https://doi.org/10.5194/bg-11-1817-2014.

Wieder, W. R., Grandy, A. S., Kallenbach, C. M. and Bonan, G. B. 2014. Integrating microbial physiology and physio-chemical principles in soils with the MIcrobial-MIneral carbon stabilization (MIMICS) model. *Biogeosciences* 11(14): 3899-917. Available at: https://doi.org/10.5194/bg-11-3899-2014.

Wutzler, T. and Reichstein, M. 2008. Colimitation of decomposition by substrate and decomposers: a comparison of model formulations. *Biogeosciences* 5(3): 749–59.

Zelenev, V. V., van Bruggen, A. H. C. and Semenov, A. M. 2000. 'BACWAVE', a spatial-temporal model for traveling waves of bacterial populations in response to a moving carbon source in soil. *Microbial Ecology* 40(3): 260–72. Available at: https://doi.org/10.1007/s002480000029.

Chapter 14

Digital tools for assessing soil organic carbon at farm and regional scale

M. J. Aitkenhead, The James Hutton Institute, UK

1 Introduction

The relationships and interactions between agriculture, society and policy are changing. Much of this change is being driven by climate change, and associated concerns that modern intensive agriculture is a net contributor to climate change but also vulnerable to its consequences. Nationally and internationally, almost all intensive agriculture is supported by subsidy payments which influence farmers' land management decision-making. In recent years, national climate change targets (in the United Kingdom and across the EU) have begun to change policy attitudes towards agricultural subsidies' alignment with environmental protection as well as food production.

One of the important aspects of this changing subsidy landscape revolves around the debate of 'payment for actions or payments for results?' (see Chapters 26 and 27 of this book). Under 'payment for results', which links payments to observed environmental improvements, farmers may not trust that previously untried activities will produce the desired outcomes, and so could

http://dx.doi.org/10.19103/AS.2022.0106.13

continue a risk-averse strategy of maintaining current activities. Under 'payment for actions', this risk would be removed. However, policymakers almost always strongly prefer to pay for results, transferring associated risk to the farmers. Payment for results will only work when farmers trust the science, which states that a certain action results in specific outcomes. The challenge here is that much of the science is still being carried out, or the findings from one experiment are difficult to translate to a different region or country.

One way to resolve this problem would be to provide farmers with direct evidence of the impacts of their actions. To do this in relation to soil organic carbon (SOC), we need a system that can provide an accurate estimate of soil carbon stocks and rates of change, rapidly, at very low cost and that operates in a way that can be verified and accredited. Currently, no such system exists but there is a strong demand for it.

Research into rapid, low-cost field-based soil assessment very definitely falls within current scientific, regulatory and policy priorities. Developments in this area would benefit from government-funded research (Aitkenhead, 2020) but could also be advanced by private investment. A risk here is that many commercial organisations are jumping on the bandwagon of SOC monitoring with a view to capitalising on likely agri-environment farm payment systems in the United Kingdom, Europe and elsewhere. This has created a 'Wild West' situation in which unfounded claims made about untested systems, with the intention of capturing market share, could damage the credibility of the wider soil carbon monitoring landscape.

Public acceptance of biological carbon sequestration and storage in soils (a nature-based solution) appears to be higher than for storage based on geological and technological solutions, although awareness also appears to be lower. These two observations are arguably linked as while nature-based solutions have a lower potential rate and capacity for carbon storage (at least per unit area), they also have less capacity for carbon release due to accidents. Alongside the relatively low-tech nature of many nature-based carbon capture solutions, makes them a lower risk but also less 'sexy' option and so less likely to be publicised or marketed as a solution to climate change. However, acceptance of a different kind is common to technology- and nature-based solutions: validation of effectiveness. On this side of things, technological solutions are more advanced as the monitoring equipment and approaches are relatively mature and quantifiable (e.g. Schloemer et al., 2013).

Things are moving rapidly, driven by current and likely future policy changes and supported on a wide base of public perception and concerns around climate change. One consequence of this rapid change is that this chapter is likely to be outdated quite soon! At the time of writing, the public, policy and academic debate is focused around two main areas that, once resolved, will provide greater focus for developments in this field:

First, there is significant disagreement about the capacity of nature-based solutions for climate change mitigation, adaptation and restoration. Within this debate and linked to perceived financial opportunities mentioned above, multiple solutions are being promoted, often with significant over-stating of the potential in each case. In the case of soil carbon, the reality of global carbon stocks of >3000 GT and an atmospheric carbon surplus of ~150 GT imply an apparently simple solution; that an increase in soil carbon stocks by a 'mere' 3–5% would correct the whole climate change problem. However, this theoretical potential is constrained by multiple factors to a lower technical potential that would require significant changes in agricultural management practices and rendering the realisation of this 'solution' very difficult.

Second, but strongly linked to the first point, is the uncertainty looming over farmers regarding future agricultural subsidies being linked to soil carbon storage and the environmental condition of their land. Signs so far indicate that in the 'payment for action' versus 'payment for results' debate, results are the preferred option of policymakers – meaning that any payments are going to be dependent on successful monitoring, reporting and verification frameworks.

For carbon trading or other market-based greenhouse gas (GHG) mitigation measures involving SOC, baseline soil carbon stock data is needed across different crop types and management practices (e.g. Suddick et al., 2013). Saby et al. (2008) cast doubt on the ability of European national soil monitoring networks to provide national carbon accounting frameworks due to their inability to detect change at a national level over timescales of less than a decade. Batjes and van Wesemael (2015) highlighted the need for soil monitoring networks with globally consistent protocols and measuring instruments, to allow national and local assessment of SOC changes linked to the effects of land management activities. Any global marketplace for soil carbon will also require international consistency and standardisation.

In this chapter, I explore the current scientific and technological state of the art and associated tool developments that may be employed for field and landscape scale evaluation of SOC stock and changes.

2 Monitoring, reporting and verification frameworks

In a review of GHG accounting methods for land-based approaches, Cowie et al. (2012) highlighted two priorities for emission offset/trading schemes, regardless of their nature, in GHG emission and carbon sequestration calculations: (1) a balance between credibility and cost-effectiveness; (2) flexibility to accommodate the effects of different site conditions and management actions. On the first point, Smith (2004) highlighted the range of effort and verifiability that are possible using different approaches to MRV, arguing that an approach between the two extremes (low effort, low verifiability and high effort, high verifiability) could provide a workable trade-off of cost and effectiveness.

Although there is a need for a global MRV framework for SOC change to support improvements in soil carbon management (Smith et al., 2020, Chapter 10 of this book), at present there may be an imbalance in monitoring and verification developments with a shift towards commercialisation and a lot of theoretical development (models and frameworks) but insufficient field experimentation providing data to support these (Jenkins, 2020).

The concept of global standardisation and harmonisation is important. Regional and national differences in soil sampling, analysis and mapping make it difficult to compare SOC stocks and rates of change around the world (Jandl et al., 2014), creating a need for agreed protocols for surveying and analysis. This inconsistency in approaches and lack of common standards (Bispo et al., 2017) means that it is also difficult to create global datasets of land management and SOC stock impacts, reducing any model calibration to effectiveness at local or regional level only.

Soil monitoring network harmonisation and inter-calibration across multiple sites, soil types and land management activities are required to achieve confidence in SOC stock and change assessments (Arrouays et al., 2014). A vital component of this is the sharing of data, requiring multiple issues of intellectual property, commercial sensitivity, metadata standards and data harmonisation to be addressed. Until these issues are resolved, any tools/methods for rapid SOC assessment will be harder to accommodate within international carbon trading, offset or farm payment schemes.

An interesting example of bottom-up development of standards for carbon credits is Trustable Credit (https://trustablecredit.com/), driven by Scotland's Rural College (SRUC) through a consortium of multiple partners. This framework takes a decentralised approach to the development and testing of open, internationally applicable standards for digital-based carbon trading. In the author's opinion, a bottom-up approach is more likely to provide benefit to land managers than top-down approaches and regulations. Multiple stakeholder partnerships including practitioners will be less likely to result in farmers being locked into long-term agreements with corporations looking to offset their emissions through land-based carbon storage schemes.

There appear to be barriers to the uptake of rapid assessment tools for SOC within monitoring systems. What are they – scientific, technological or practical in terms of implementation? Certainly, harmonisation of approaches is difficult when everyone wants to 'capture the market' with their own idea. Resolving this requires a top-down government investment approach at national and international level.

Finding ways to express the uncertainty in SOC stock assessments is vital for validation of monitoring approaches (Perie et al., 2014), and is also important for the evaluation of management change impacts. Baldock and Grundy (2017) demonstrated a system for monitoring the effects of agricultural management on SOC stocks, across a range of scales. An important part of this work is the estimation and presentation of uncertainty in SOC stock calculations. This provides a way of indicating risk in terms of the effectiveness of specific management practices.

Cost considerations are also important for MRV systems. Collection of physical soil samples by farmers provides SOC content information that is equivalent in quality to samples collected by researchers but is not suitable for MRV as it provides only topsoil information (Deluz et al., 2020); placing expectations on farmers to capture physical samples from the subsoil is impractical and unrealistic. So someone else must do it - either researchers interested in developing methods or practitioners paid by government or private companies.

Additionally, it is easier to measure stocks than to measure change, with even the best laboratory-based methods giving uncertainty ranges for single samples that correspond to 3–5 years of SOC change within agricultural soils. However, this could be resolved with improvements in sampling strategies to provide more statistically robust estimates of within-field and between-field variation.

3 Sampling strategies

Unless there is a hitherto-unexpected technological breakthrough, any low-cost rapid assessment tool for SOC estimation will need to make use of a library of data developed from in-field soil samples. This data is used to develop a calibration between SOC content and what the assessment tool is measuring. Significant cost and effort in collecting and analysing soil samples is required to develop calibration models for SOM and SOC content estimation (Gedminas and Martin, 2019). Models that can be applied as widely as possible with minimal sample collection are therefore desirable. A well-planned sampling strategy is therefore an important and cost-sensitive component of any SOC assessment tool development.

Collecting representative sample data and measurements is important for models to be able to estimate SOM within all features of a landscape (e.g. Huang et al., 2017) - the variability within not just the soil, but also within other landscape characteristics must be captured. Traditional approaches to capturing soil data involved physical sampling for laboratory analysis, but it is now possible for emplaced sensors to capture data on certain soil properties (e.g. temperature, moisture content, pH). Regardless of which approach is used, spatial distribution of sample or sensor locations needs to be optimised, to minimise the number of sample locations (and associated costs) while maximising the representation of site conditions.

Examples of sampling optimisation approaches include the use of genetic algorithms (Santacruz et al., 2014) and the Conditioned Latin Hypercube (Roudier, 2021). Other national-level sampling has incorporated grid-based sampling as the simplest, most efficient form of gathering nationally representative data (see Chapter 12 of this book). Conant et al. (2011) highlighted numerous challenges in measuring SOC stock changes and highlighted the effectiveness of different sampling, interpolation and modelling approaches.

For physical sampling, variation in selection and use of sampling tools for bulk density can result in different SOC stock values (Cihacek et al., 2015) with implications for verification of stock changes over time. Consistent, or at least harmonised, sampling approaches therefore need to be used if the sampling is carried out over a significant period of time.

The main challenge may not be technological but within the soil science community: given the number of local, regional and national soil surveys that have been carried out over the years, there is probably sufficient data available, but not enough has been shared using harmonised approaches to allow comparison. Getting soil scientists (or rather their institutional managers) to agree to work together for a common cause will probably be the most difficult challenge of all.

4 Integration of data sources

Plot- and regional-scale variability in soil conditions are a cause of uncertainty in SOC modelling (Hedley et al., 2012). The drivers of uncertainty are likely to differ across scales, with plot-scale variation driven by factors that can change over metres (e.g. hydrology, management) and regional-scale variation driven by changes in factors such as climate, geology or topography. The incorporation of environmental information into information from any sensor (more detail on sensor types is given in Sections 7 and 9) is therefore likely to improve SOC estimation accuracy, although the relative importance of each soil-forming factor will vary greatly from one location to another. Because of this, it is usually necessary to take a 'kitchen sink' approach and include a wide range of environmental factors into any modelling, particularly if the modelling is to be applied across a wide range of landscape types.

Similarly, integration of soil colour and spatial covariate data produces better SOC estimation than using soil colour alone or relying solely on spatial covariates (i.e. Digital Soil Mapping, Chapter 12 of this book). Integrating data from multiple spatial data sources is an effective approach for mapping SOC stocks (e.g. Paul et al., 2020). This also points towards potential improvement of existing field-based SOC sensors (i.e. spectroscopic or colour-based instruments) and land manager guidance. Information sources for land managers can provide location-specific model outputs for changing soil conditions, by integrating limited user data input, existing spatial datasets and near-real-time weather data (e.g. Li et al., 2018).

Swetha and Chakraborty (2021) demonstrated that adding texture information to colour from soil can significantly increase SOC estimation accuracy. If texture information can be extracted from digital image analysis of soil profiles, this could provide proxy soil texture and structure information. Alternatively, as texture is linked to a range of soil formation factors including parent material and topography (through hydrology), the integration of relevant spatial covariates may improve model performance (e.g. Aitkenhead et al., 2016a, 2018; Mallik et al., 2020).

This has been used in the development of a smartphone app (SOCIT) for soil organic matter content estimation in Scotland (Aitkenhead et al., 2016a,b). The principle of this app is that a shallow soil profile, suitably colour corrected, can be linked with user location to drive a machine learning model trained with SOM and colour data from across Scotland. Future work in this area aims to allow estimation of multiple soil properties including bulk density, thus allowing topsoil carbon stocks to be estimated. Figure 1 shows screenshots from the SOCIT app (Apple version).

Adhikari et al. (2020) found that topography has a stronger influence on SOC stock at finer spatial scales, while climate is more significant at coarser

Figure 1 SOCIT app screenshots.

scales. This points to the importance of considering minor topographic changes at field-scale SOC monitoring. It also highlights the need to use relatively high-resolution digital elevation maps. Such large datasets would be impossible to process on a smartphone or desktop-base app (without even considering the potential data licence issues with saving such datasets onto a smartphone), and so server-side processing of sensor data is likely to be important.

Remote sensing (RS) data is another example of a spatial covariate, but one that comes with an additional consideration: that the values for each location change over time. It can provide important information on small-scale variability on vegetative growth in situations where soil formation factors do not vary over short distances (or at least, where available information does not provide high-resolution variation, such as for uniform crop type on agricultural land where the only records may be what was planted, rather than within-field variability of crop yield) (e.g. Guo et al., 2020).

Inversely to the above point, RS can be effective for monitoring and verifying agricultural carbon sequestration through crop residue incorporation over large areas but is less effective for farm- and field-scale monitoring of this practice (Bricklemyer et al., 2007). So the application needs to be considered, and RS will not always provide an improvement. The accuracy of mapping SOC using RS data varies with the spatial resolution of modelled outputs, even using the same original sensor data (Zhou et al., 2021).

Temporal variation may be caused by changes on the ground, as well as updated imagery from repeated passes of the air- or space-borne sensors. Croft et al. (2012) identified opportunities for improving the use of RS for monitoring agricultural SOC dynamics. These included the integration of other spatial datasets with RS data, the development of spectral libraries for further, broader use and the integration of RS into SOC modelling work. In specific cases, multitemporal Sentinel-2 imagery can be used for mapping of topsoil SOC (Gholizadeh et al., 2018). And in semi-arid regions, the use of seasonal fractional cover information derived from RS data improves SOC stock estimates (Wang et al., 2018a). What this indicates is that RS is good for detection, but not quantification of change, at small scales; conversely, it can be used for quantification of SOC stock change at large scales when coupled with other datasets. As with all other sensor approaches, calibration against data from more accurate methods is vital.

Integration of RS and topographic indices derived from a Digital Elevation Map (DEM) can provide landscape-scale SOC mapping. Aitkenhead and Coull (2020) mapped SOC stocks at 100 m resolution for Scotland using an integrative approach incorporating survey data, spatial covariates and RS to train an artificial neural network model. They highlighted the amount of carbon held in non-peat, carbon-rich soils that are under grazing in Scotland. This kind of mapping can be repeated over time to detect the effects of changes to soil carbon, from climate change or land management.

In general, this data integration approach involves the harmonisation of multiple spatial datasets, usually acquired from different sources at different times, and each representing a specific factor important to soil development (e.g. multiple topographic indices derived from elevation, gridded climate data, land use/cover, geological maps). These datasets are of different ages, quality, spatial resolution and suitability in terms of the categories used for whatever they are mapping; as such, their integration and harmonisation is something of a black art that requires familiarity with both the datasets and the landscape they represent in order to be achieved effectively. The section on Model Calibration below gives further information on the types of statistical and machine learning models that are commonly used.

Spectroscopic data can also be used for rapid estimation of SOC content, although the effectiveness of this approach relies on finding the best spectral pre-processing approach for the combination of sensor, soils and environmental conditions (Srivastava et al., 2015; Singh et al., 2021). In a study evaluating the integration of data from multiple proximal sensors (soil electrical conductivity, visible-near-infrared (Vis-NIR) spectroscopy and insertion force probe) to estimate SOM, Wetterlind et al. (2015) found little improvement by combining different ground-based sensors. The most appropriate method therefore seems to be to select one type of sensor and optimise the data from it using pre-processing approaches, before combining this with multiple spatial covariates to develop a calibration model for determining soil carbon stocks at larger spatial scales (e.g. landscapes or regions).

Integration of new data with existing spatial datasets, including factors that influence soil carbon, can be used to produce high resolution national-scale maps of SOC values and uncertainty ranges (e.g. Rossel et al., 2014; Aitkenhead and Coull, 2020; see Chapter 12 of this book). These maps achieve a level of accuracy that allows for national-level SOC stock assessment and local-scale spatial variability but are not sufficiently accurate to act as the sole source of SOC stock at local levels. Local bias caused by imperfections in such 'global' national-scale models will still require local correction adjustments.

Yagasaki and Shirato (2014) demonstrated spatially explicit national-level SOC stock changes using a modelling approach that incorporated land management variation and changes, soil variability and climate. The inclusion of existing inventory data within this model was important for its calibration. Modelling rates of change of SOC in some manner, be it through sensor-driven process-based simulation or through correlation with direct observation of rapidly and cheaply measured soil properties, is needed if physical sampling and analysis is not a viable option (as in, for SOC stock assessments as explained earlier).

Topographic indices derived from digital elevation maps can provide landscape metrics that influence SOM content (e.g. Mallick et al., 2020). The selection and use of these indices for SOM estimation requires not just statistical evaluation but also awareness of the wide range of topographic indices that can be calculated, and an appreciation of how they relate to soil formation processes.

Integration of multiple datasets for calibration modelling requires sophisticated data mining or machine learning approaches. These approaches usually have 'hyperparameters' that control their architecture, learning rate or other operational factors, and optimisation of these hyperparameters can be time-consuming. Approaches that reduce this model optimisation effort can therefore improve performance within reasonable time constraints. An application of Particle Swarm Optimisation for estimating SOM from multiple RS data was demonstrated by Jin et al. (2017).

So if Digital Soil Mapping and RS methods cannot provide sufficient accuracy of SOC stock assessments, and field-based rapid assessment tools cannot provide this either, does that mean that the only available option is the expensive and time-consuming route of field sampling and laboratory analysis? From the findings mentioned above, perhaps the answer is no; an approach that integrates environmental spatial datasets and time-series RS with field-based sensors and limited physical sampling may be a suitable compromise, between accuracy, efficiency and costs.

However, any such system will need to be highly automated and rely on server-side processing of sensor readings with the minimum of user expertise and effort. Such a system will require significant effort to design, including engagement with likely users of the system to ensure that it is providing the kind of information that they need. It must also be able to deal with a range of conditions in the field that can influence the soil's appearance but that cannot be predicted in advance, including lighting conditions and moisture.

5 Effects of moisture on soil colour

Soil colour is strongly influenced by soil moisture content. This adds to the complexities of measuring SOC or SOM in the field, particularly if the intention is to use colour for rapid, low-cost sensing. Fu et al. (2020) evaluated the effect of soil moisture on the accuracy of SOM estimates based on smartphone imagery. They showed that above a threshold moisture value, models that included moisture content had better accuracy than those that did not. However, is it realistic to expect soil moisture readings to be taken alongside soil imaging? To do so would increase the cost and effort to the point where you might as well just measure the soil carbon directly using a lab-based approach.

Roberts et al. (2011) combined active (Vis-NIR modulated light source and reflectance) and passive (aerial photography) sensors using multiple approaches for estimating topsoil SOM. They showed that reducing surface residue increased mapping accuracy, as did accounting for soil moisture effects on sensor readings. Additionally, soil colour and depth are important for SOC estimation in both dry and moist soils (Stiglitz et al., 2017). Meanwhile, Raeesi et al. (2019) explored the use of soil colour information from digital camera imagery to estimate soil organic matter content. They achieved a level of reliability in arid, low-SOM soils that would allow change detection over a multi-year timescale.

Due to soil-specific variability in the way moisture influences soil colour, it is not possible to achieve dry soil colour accurately from moist soil using a grey colour correction card (Kirillova et al., 2021). However, dry soil colour may not be needed to estimate soil carbon content; it may be possible to determine SOC from moist soil colour if the effects of moisture on colour are taken into account in the calibration model.

This is a challenge that needs to be explored thoroughly, perhaps by using a digital soil mapping approach to provide the calibration model with an 'internal' calibration linking the effects of soil moisture level to colour change. The same may be true of soil depth, which has a different impact on soil properties and imagery due to the processes that vary as one goes down the profile.

6 Depth

Soils are not all the same depth. The first question to be asked when calculating the amount of carbon in a soil should therefore be 'to what depth'? Should the full profile always be considered? Are there depths beyond which the amount of carbon becomes negligible? Do human activities influence carbon stocks beyond a certain depth? Frustratingly, the answer to all of these questions is 'it depends'. There are practical considerations to sampling at depth, and as with much in life, the adopted practice almost always involves a trade-off: as deep as you can afford to go, and as shallow as will give useful information.

Commonly, soil carbon stocks are calculated to a fixed depth. This approach can introduce errors in detection of SOC change, as bulk density often changes in response to land management as well as SOC concentration. Solutions to this problem include the use of an equivalent mass procedure (Wendt and Hauser, 2013, Chapter 11 of this book), in which the same mass of soil is compared over time even if this means sampling to different depths.

van Wesemael et al. (2011) recommended fixed depth increment sampling for carbon content and bulk density, using stratified sampling of sites to capture variability in soil, land use, climate and other factors for national SOC stock assessments. This leads back to the earlier discussion about sampling strategies. However, accurately representing SOC stock variation with depth requires more sophistication than assuming mean SOC values for fixed depth intervals (Zhang et al., 2020).

Depth is an important predictor of SOC (Stiglitz et al., 2018), with significant variation down the profile in patterns that are common to specific soil types. Integration of depth information and colour data from low-cost imaging sensors can provide rapid assessment of SOC (Mikhailova et al., 2018). Heil et al. (2020) found that digital imagery can also be used to carry out mapping of soil properties across a soil profile, potentially allowing mapping of SOC at depth.

In Scottish topsoils, while the SOC content is higher in peat, the bulk density of organo-mineral soils is significantly higher. So while non-peat soils contain less carbon by weight, they can contain the same amount of carbon as peat soils in the upper layers (Aitkenhead and Coull, 2016).

What these above observations highlight is that even the most 'rapid' SOC estimates in the field will always require some time, attention to procedure and physical effort. There is no magic wand or Star Trek tricorder that can be waved at the ground to obtain an instant, fuss-free measurement.

7 Model calibration

Models calibrated on data from soils with a narrow range of SOC contents are more accurate within that range than models calibrated with data from all soils (Aitkenhead et al., 2015). Models used to estimate SOC for agricultural soils should therefore be calibrated with appropriate data. Field-specific calibration gives more accurate SOM estimation than global calibration models (Roberts et al., 2011).

Capturing the spatial heterogeneity of carbon flux measurements varying widely with weather conditions requires simultaneous measurement or modelling at a number of locations within a site/region (e.g. Wang et al., 2018b). In a local calibration study with high sampling density, Pouladi et al. (2019) found that kriging provided the best performance in comparison to models that integrated RS and topographic data for the estimation of SOM. However, this approach may not work so well for low sampling density or for larger areas.

Multiple linear regression is still commonly used for DSM approaches to SOC mapping, despite more advanced machine learning approaches giving better performance (Lamichhane et al., 2019). The same work showed that multiple factors are influential in estimating SOC content with variations in factor importance at different scales; this implies that sophisticated models capable of integrating multiple spatially variable factors are required for effective SOC mapping. Calibration against sites with similar pedoclimatic conditions and soil formation factors is also likely to be important for model accuracy.

Modelling SOC changes on extensive grassland is difficult due to uncertainties about grazing intensity and the impacts of drivers on soil carbon. In some countries, these extensive grasslands cover large areas and so direct monitoring of SOC is economically counterproductive (McConkey et al., 2017). If neither modelling nor monitoring can solve the SOC stock assessment challenge independently, then perhaps a hybrid approach can provide better results? Could direct estimates from a well-calibrated sensor provide readings to drive process-based or machine learning-based modelling?

Models need to be fed with site-specific data to be able to make estimates of SOC variability at the plot scale (Cerri et al., 2004). Estimates of SOC change in process-based modelling vary significantly depending on the processes and underlying assumptions implemented within the models (Dijkstra et al., 2009). For SOC monitoring, approaches that can integrate data from multiple sources are vital (i.e. sophisticated machine learning). Process-based modelling at small scales requires detailed local condition calibration and may not be cost-effective.

A better approach therefore might be a global model linking sensor data (e.g. colour or spectral data) and SOC, trained using machine learning and that can be adjusted using a small number of local measurements derived from field sampling. Once adjusted for local conditions, this model could provide rapid and accurate SOC assessment using local (proximal) sensor readings that can rapidly measure proxy indicators of SOC. 'Proxy indicators' are taken here to mean indicators that do not measure SOC directly but instead rely on a colorimetric response (see Section 9).

8 Proximal sensing

Even if their accuracy against a global dataset is insufficiently accurate for credible SOC stock estimates, low-cost sensing methods can be used to capture information on local spatial variation (e.g. Garcia-Tomillo et al., 2017). Model bias or error is likely to be relatively consistent over local sites where many factors are constant, so variation in estimates at individual locations within a site should reflect actual variation. Ground-based low-cost sensors in sufficient numbers can accurately represent the spatial variation in SOM captured using RS data (Nguyen and Kodagoda, 2016). The authors of this work did not specify the individual sensor types, stating only that various sensing features including temperature, humidity and light level were utilised.

Using soil colour alone, in-field SOM spatial variability can be accurately captured without knowing absolute SOM values at any location (Aimrun et al., 2014). In-field measurements of soil colour using a low-cost digital camera can provide local calibrations with high accuracy for SOC estimation (Gholizadeh et al., 2020). Wu et al. (2019) carried out a detailed analysis of the effects of adjustments to image capture using a digital camera, on correlation between colour and SOM. They highlighted the importance of using a standardised approach in a consistent manner to capture colour information. In addition, they showed that the standard deviation of colour bands provided important information for SOM estimation.

Regionally sourced image (i.e. digital photography of soil sample) data can provide accurate assessment of local SOC content if appropriate image analysis approaches are used (Morais et al., 2019). It is not that regional or national soil image data is not good enough for SOC assessment at a local scale, but that sufficiently good data analysis approaches are needed to create effective models.

Hyperspectral data (i.e. characterising the spectral response over many narrow wavelength bands) can be used for characterisation of SOC, although selection of appropriate sample density/location, model calibration and pre-calibration data transformation approaches has an important effect on model performance (Hou et al., 2014). Capturing hyperspectral data traditionally requires expensive equipment and much of the data generated is often redundant or unnecessary for SOM characterisation. This cost and applicability depends a lot on the specific sensor chosen (and also on the familiarity and skill of the user with its application for this purpose – this important point is often overlooked). Methods capable of using a smaller number of wavelength ranges from cheaper equipment (or freely available data sources) are therefore desirable, particularly if these data sources are standardised or already in operation (Hong et al., 2017).

Additionally, concerns about cost for this kind of approach are also reducing; it is possible to develop low-cost hyperspectral sensors with significantly greater spectral resolution than digital cameras (e.g. Song et al., 2020). Selection of the best modelling approach to translate the data into SOC stocks is something of a dark art and requires experience of both the sensor and the site from which samples are taken. For hyperspectral imaging, identification of a small number of spectral bands indicative of soil organic matter content can improve model performance (e.g. Yang et al., 2019). Identifying specific wavelengths (660 and 940 nm) that are important for SOM estimation has previously allowed a low-cost hand-held optical sensor to be developed (Kweon and Maxton, 2013). There has been a lot of development recently in reducing the cost and technical requirements of field-based hyperspectral sensing. However, a lot of solutions on the market do not provide access to the raw data, which is sent to the manufacturer and used to improve their calibrations. While this is arguably of benefit to the user, it is likely to be less beneficial than being able to develop their own, site-specific calibrations from the raw spectral data (particularly as the default calibration of the sensor manufacturer may not be optimal for the local conditions).

Vis-NIR spectroscopy and gamma-ray attenuation have strong potential for in-field assessment of SOC concentration and soil bulk density, respectively (England and Rossel, 2018). Knowledge of these two variables is vital for calculating SOC stocks. Machine learning approaches to SOC estimation with field-captured spectral data can provide high levels of accuracy, if the model hyperparameters are selected appropriately (Chen et al., 2020).

The use of spectral data of local samples to develop the calibration increases the model performance for SOM predictions (Shi et al., 2015). This points to the need to remove sample data from calibration datasets when the soil-forming factors that dominate those samples are not relevant to the region of interest; effectively, this reduces the risk of inter-class variability being confused with intra-class variability. Stevens et al. (2008) highlighted the potential of field-based spectroscopy to provide SOC estimates of similar accuracy to standard laboratory analytical techniques, provided the sampling and calibration models were sufficiently robust.

Integration of optical sensing with electrical conductivity measurements provides enhanced capability for SOM mapping, in terms of accuracy and spatial applicability (Kweon et al., 2013). Serrano et al. (2016) demonstrated how effectively soil electrical conductivity and Normalised Difference Vegetation Index from optimal imagery could be used to map SOC variability.

Standardised sensor technology is a problem, with users unable to make use of data from different instruments to develop common calibration models. There is little perceived incentive at the moment for manufacturers and researchers to work together to resolve this lack of standardisation; rather, there is a perception that whoever comes up with a working sensor first will corner the market.

To the author, this does not make sense; people should be aware that it is practically impossible to protect intellectual property under these conditions, and indeed the penalties for copying someone else's design are minor compared to the potential income. Effectively, the first mouse gets the trap (i.e. does all the work), and the second mouse gets the cheese. This may be another reason why there has been so little investment in the development of accurate, working platforms for the rapid assessment of SOC.

9 Tools and frameworks

There are multiple tools available for SOC estimation, but these are almost exclusively limited to online reporting systems that convert user information about land management practices into farm-business carbon emission budget estimates. These are different from tools for actual direct measurement or estimation of SOC stocks in the field, which are limited to research project output prototype systems. Online tools may be very useful for ex-ante assessment of the potential impact of management practices on soil carbon sequestration. However, they are unlikely to be useful for the characterisation of SOC stock or the properties that are used to calculate this (bulk density and SOC concentration at different depths) as this would require the land manager to carry out significant sampling effort using extremely specific protocols. There are two risks here: first that the results would be dependent on the efforts of people who while skilled are unlikely to be experienced in the specifics of soil sampling; second, it would open the system up to the possibility of deliberately misleading SOC stock calculations for the purposes of financial gain.

However, it is worth briefly describing some notable examples of these online systems, particularly those that are relevant for land managers in developing countries or where the costs of in-field surveys of SOC stocks is impractical to do so. For example, Vagen et al. (2017) demonstrated an operational framework for assessing soil health, including SOC mapping, for land management decision-making in Kenya.

The forecast tool EX-ACT from the FAO (http://www.fao.org/tc/exact/accueil-ex-act/fr/) provides various ways of estimating the GHG emissions of soils amongst other landscape components, under multiple policies and management activities. However, it does not include direct measurement of soil carbon, but rather makes use of evidence from previous projects and monitoring frameworks. A functionally similar example is the Carbon Benefits Project toolset (https://cbp.nrel.colostate.edu/), which allows users to estimate the impacts of various management activities on carbon stock and GHG emissions.

The EX-ACT tool provides a consistent approach to linking land use policies and investments to their impacts on GHG emissions and is contained within a standardised 'envelope' of training and administration. The Carbon Benefits Project tool aims to provide something similar but is targeted at exploring the GHG impacts of specific management actions. This is an important distinction as it will influence which of these two tools would be most important for specific stakeholders. Both of these tools can potentially be applied globally, and both rely on robust standardisation and characterisation of approaches; however, neither provide a mechanism for calculating baseline SOC stocks - they are both focused on emissions.

Fundamentally, there is a need for carbon sequestration and nature/biodiversity improvement projects to be able to make attested verification-based and creditable evaluations of their impacts. To be funded, ex-ante assessment of these project impacts may be necessary. These must be based on robust scientific methods proven by actual monitoring and measurement 'in the field', rather than being proxy/model based.

Currently, no systems exist that can provide this, although the Agricarbon system (https://agricarbon.co.uk) probably comes closest and has potential for being able to satisfy all of the requirements for a working system. This is achieved through a combination of field-specific sampling to maximise respresentativeness of the samples while minimising the number of samples actually required, and automation of sample analysis. The approach developed to achieve this is a synthesis of standardised soil science field sampling approaches with novel use of freely available spatial datasets.

The state of the art in assessing field-level SOC stocks is very clearly an evolving integration of fundamental soil science and geostatistical knowledge, and agricultural equipment/implementations that are being adapted to new field applications. Knowledge gaps still exist, although we are now in a position to identify the 'known unknowns' and actually start to tackle them. Much of the uncertainty around whether or not these systems are effective lies in a lack of large-scale field testing and validation of the approaches developed.

10 Knowledge gaps and uncertainties

Post et al. (2001) identified advances in technology (specifically for direct measurement) and mathematical methods (specifically for mapping) needed even then to enable sufficiently accurate SOC monitoring to be carried out. Arguably however, we have still not achieved these advances, particularly in relation to baseline mapping of SOC. Developments mentioned above have moved us a long way forward at the field scale, but these developments are still being piloted and it is unclear how they will achieve national-level, standardised access to tools that provide the level of detail required by land managers.

In this context, Pallegedara Dewage et al. (2020) discuss the challenges of disaggregating regional-extent digital soil maps for field-scale mapping of SOC, highlighting the uncertainty created by the disaggregation process. Moreover, Balkovic et al. (2020) reported that local inaccuracies in soil data prevent modelling of soil carbon changes to the level of accuracy required for verification of impacts from changing management practices. However, they also showed that agricultural land use statistics reported by land managers did provide suitable information for estimation of soil carbon dynamics. So while the existing maps are not sufficiently detailed for baseline mapping, once this barrier is overcome it should be possible to provide accreditation-level estimates of SOC change.

In modelling and mapping SOC stocks in Tanzania, Kempen et al. (2019) found that soil type and terrain information were important predictors of SOC, and that the main source of uncertainty in SOC stock came from uncertainty in SOC concentration. So, once again, the main issue concerns the baseline mapping. This issue can be particularly problematic where human influence has been greatest. For example, GHG emissions from urban soils can be significantly higher than those from nearby rural soils, due to imported carbon amendments (e.g. mulch, manure, compost) (Decina et al., 2016).

Many soil monitoring and modelling approaches assume that soils are currently at a steady state when setting out to evaluate the effects of management change on SOC stocks. However, this assumption is very often untrue. Soils are instead often trending towards a regional land use baseline (Goidts et al., 2009), a fact that could be used to make the distinction between current rate of SOC stock change due to ongoing land management and that driven by recent changes in management.

Overall then, better and more harmonised field data is needed to decrease the uncertainty of global models. Additionally, model calibrations typically lack information about historical land management. This means that it is difficult to tell how soils are different from their 'natural' state. So direct measurement in some form will always be needed. However, models may be able to calculate the regional land use baseline from which local variations can be detected by other means (e.g. proximal sensing or RS).

11 Where to look for further information

As progress is made towards integrating Natural Capital into agricultural payments systems, more robust frameworks for monitoring and assessing soil carbon are being developed. The IUCN UK Peatland Code (Introduction to the Peatland Code | IUCN UK Peatland Programme (iucn-uk-peatlandprogramme .org)) is focussed on best practice guidelines for peatland restoration practices, but also provides information on monitoring plans and carbon accounting. Across all soil types in the UK, the Soil Carbon Code (Soil Carbon Code | About the Code (sustainablesoils.org)) provides information on how a national-level framework is currently being developed across sequestration, monitoring and trading activities. Many other countries have, or are developing, similar frameworks. The 2021 European Commission working document on carbon farming (swd_2021_450_en_0.pdf (europa.eu)) contains a section on monitoring, reporting and verification of soil carbon although much of this is aspirational and identifies future research challenges.

Consistent and robust soil sampling is critical to soil carbon baseline and change monitoring using physical samples. Various countries have developed standards for soil sampling/survey methodologies in relation to carbon monitoring. Many of these have been developed from more general soil sampling guidelines or operating procedures, which contain a range of practical information. The US Environmental Protection Agency soil sampling guidance at Region 4 (epa.gov) contains details on acquiring and handling samples, for example. More soil carbon-specific guidance is given on a Government of Western Australia website at Measuring and reporting soil organic carbon | Agriculture and Food.

For further information on digital tools in this area, the reader is advised to pay more attention to ongoing research than current technology. Underlying research questions are still being tackled, particularly colorimetric or spectral approaches to rapid soil carbon assessment. The best methods currently available (and in the author's opinion, this situation will not change in the near to medium future) are hybrid approaches that combine targeted physical sampling with rapid physicochemical measurement and geostatistical methods; these approaches reduce but do not eliminate the number of physical samples required, and rely on cutting-edge statistical or modelling interpretation to provide field- to farm-scale carbon stock assessments. A good example of this kind of approach is Agricarbon (https://agricarbon.co.uk).

12 References

Adhikari, K., Mishra, U., Owens, P. R., Libohova, Z., Wills, S. A., Riley, W. J., Hoffman, F. M. and Smith, D. R. 2020. Importance and strength of environmental controllers of soil organic carbon changes with scale. *Geoderma* 375, AR 114472. DOI: 10.1016/j.geoderma.2020.114472.

Aimrun, W., Gholizadeh, A., Asminda, J. N. A. and Amin, M. S. M. 2014. Development of on-the-go soil organic matter sensor. *Indian Journal of Agricultural Sciences* 84(10), 1201–1204.

Aitkenhead, M. J. 2020. Climate change and soil organic matter in Scotland: time to turn over a new leaf? *Soil Research* 59(6). DOI: 10.1071/SR19351.

Aitkenhead, M. J., Cameron, C., Gaskin, G., Choisy, B., Coull, M. C. and Black, H. I. J. 2018. Digital RGB photography and visible-range spectroscopy for soil composition analysis. *Geoderma* 313, 265–275.

Aitkenhead, M. J. and Coull, M. C. 2016. Mapping soil carbon stocks across Scotland using a neural network model. *Geoderma* 262, 187. DOI: 10.1016/j.geoderma.2015.08.034.

Aitkenhead, M. J. and Coull, M. C. 2020. Mapping soil profile depth, bulk density and carbon stock in Scotland using remote sensing and spatial covariates. *European Journal of Soil Science* 71(4), 553. DOI: 10.1111/ejss.12916.

Aitkenhead, M. J., Coull, M. C., Gwatkin, R. and Donnelly, D. 2016a. Automated soil physical parameter assessment using smartphone and digital camera imagery. *Journal of Imaging* 2(4), 35. DOI: 10.3390/jimaging2040035.

Aitkenhead, M. J., Donnelly, D., Coull, M. C. and Gwatkin, R. 2016b. Chapter 7: estimating soil fertility indicators with a mobile phone. In: *"Digital Soil Morphometrics"; Book series "Progress in Soil Science"*. Springer. Available at: http://www.amazon.co.uk/Digital-Soil-Morphometrics-Progress-Science-ebook/dp/B01DXI295Y.

Aitkenhead, M. J., Donnelly, D., Sutherland, L., Miller, D. G., Coull, M. C. and Black, H. I. J. 2015. Predicting Scottish topsoil organic matter content from colour and environmental factors. *European Journal of Soil Science* 66(1), 112.

Aitkenhead, M. J., Poggio, L., Wardell-Johnson, D., Coull, M. C., Rivington, M., Black, H. I. J., Yacob, G., Boke, S. and Habte, M. 2020. Estimating soil properties from smartphone imagery in Ethiopia. *Computers and Electronics in Agriculture* 171. DOI: 10.1016/j.compag.2020.105322.

Arrouays, D., Marchant, B. P., Saby, N. P. A., Meersmans, J., Jolivet, C., Orton, T. G., Martin, M. P., Bellamy, P. H., Lark, R. M., Louis, B. P., Allard, D. and Kibblewhite, M. 2014. On soil carbon monitoring networks: soil carbon. *Progress in Soil Science*, 59–68. DOI: 10.1007/978-3-319-04084-4_6.

Baldock, J. A. and Grundy, M. 2017. Measuring and monitoring the impact of agricultural management on soil carbon stocks from point to continental scale in Australia. *Proceedings of the Global Symposium on Soil Organic Carbon 2017*, Rome, Italy, 21–23 March, 2017, 35–41.

Balkovic, J., Madaras, M., Skalsky, R., Folberth, C., Smatanova, M., Schmid, E., van der Velde, M., Kraxner, F. and Obersteiner, M. 2020. Verifiable soil organic carbon modelling to facilitate regional reporting of cropland carbon change: a test case in the Czech Republic. *Journal of Environmental Management* 274, 111206. DOI: 10.1016/j.jenvman.2020.111206.

Batjes, N. H. and van Wesemael, B. 2015. *Measuring and Monitoring Soil Carbon. Soil Carbon: Science, Management and Policy for Multiple Benefits*. SCOPE (series 71), 188–201.

Bispo, A., Andersen, L., Angers, D. A., Bernoux, M., Brossard, M., Cecillon, L., Comans, R. N. J., Harmsen, J., Jonassen, K., Lamé, F., Lhuillery, C., Maly, S., Martin, E., Mcelnea, A. E., Sakai, H., Watabe, Y. and Eglin, T. K. 2017. Accounting for carbon stocks in soils and measuring GHGs emission fluxes from soils: do we have the necessary standards? *Frontiers in Environmental Science* 5(41). DOI: 10.3389/fenvs.2017.00041.

Bricklemyer, R. S., Lawrence, R. L., Miller, P. R. and Battogtokh, N. 2007. Monitoring and verifying agricultural practices related to soil carbon sequestration with satellite imagery. *Agriculture, Ecosystems and Environment* 118(1–4). DOI: 10.1016/j.agee.2006.05.017.

Cerri, C. E. P., Cerri, C. C., Paustian, K., Bernoux, M. and Mellilo, J. M. 2004. Combining soil C and N spatial variability and modeling approaches for measuring and monitoring soil carbon sequestration. *Environmental Management* 33(S1). DOI: 10.1007/s00267-003-9137-y.

Chen, S., Xu, D., Li, S., Ji, W., Yang, M., Zhou, Y., Hu, B., Xu, H. and Shi, Z. 2020. Monitoring soil organic carbon in alpine soils using in situ vis-NIR spectroscopy and a multilayer Perceptron. *Land Degradation and Development* 31(8), 1026.

Cihacek, L. J., Foss, L. A. and Jacobson, K. A. 2015. Comparison of soil sampling devices for soil bulk density determination for carbon sequestration monitoring. *Communications in Soil Science and Plant Analysis* 46(2), 180.

Conant, R. T., Ogle, S. M., Paul, E. A. and Paustian, K. 2011. Measuring and monitoring soil organic carbon stocks in agricultural lands for climate mitigation. *Frontiers in Ecology and the Environment* 9(3), 169.

Cowie, A., Eckard, R. and Eady, S. 2012. Greenhouse gas accounting for inventory, emissions trading and life cycle assessment in the land-based sector: a review. *Crop and Pasture Science* 63(3). DOI: 10.1071/CP11188.

Croft, H., Kuhn, N. J. and Anderson, K. 2012. On the use of remote sensing techniques for monitoring spatio-temporal soil organic carbon dynamics in agricultural systems. *CATENA* 94, 64.

Decina, S. M., Hutyra, L. R., Gately, C. K., Getson, J. M., Reinmann, A. B., Gianotti, A. G. S. and Templer, P. H. 2016. Soil respiration contributes substantially to urban carbon fluxes in the greater Boston area. *Environmental Pollution* 212, 433–439. DOI: 10.1016/j.envpol.2016.01.012.

Deluz, C., Nussbaum, M., Sauzet, O., Gondret, K. and Boivin, P. 2020. Evaluation of the potential for soil organic carbon content monitoring With farmers. *Frontiers in Environmental Science* 8, AR 113. DOI: 10.3389/fenvs.2020.00113.

Dijkstra, J. P. M., Reinds, G. J., Kros, H., Berg, B. and de Vries, W. 2009. Modelling soil carbon sequestration of intensively monitored forest plots in Europe by three different approaches. *Forest Ecology and Management* 258(8), 1780.

England, J. R. and Viscarra Rossel, R. A. 2018. Proximal sensing for soil carbon accounting. *SOIL* 4(2), 101.

Fu, Y., Taneja, P., Lin, S., Ji, W., Adamchuk, V., Daggupati, P. and Biswas, A. 2020. Predicting soil organic matter from cellular phone images under varying soil moisture. *Geoderma* 361, AR 114020. DOI: 10.1016/j.geoderma.2019.114020.

Garcia-Tomillo, A., Miras-Avalos, J. M., Dafonte-Dafonte, J. and Paz-Gonzalez, A. 2017. Estimating soil organic matter using interpolation methods with a electromagnetic induction sensor and topographic parameters: a case study in a humid region. *Precision Agriculture* 18(5), 882.

Gedminas, L. and Martin, S. 2019. Soil organic matter mapping using hyperspectral imagery and elevation data. *IEEE Aerospace Conference*, March 02–09, 2019.

Gholizadeh, A., Saberioon, M., Rossel, R. A. V., Boruvka, L. and Klement, A. 2020. Spectroscopic measurements and imaging of soil colour for field scale estimation of soil organic carbon. *Geoderma* 357, AR 113972. DOI: 10.1016/j.geoderma.2019.113972.

Gholizadeh, A., Zizala, D., Saberioon, M. and Boruvka, L. 2018. Soil organic carbon content monitoring and mapping using airborne and Sentinel-2 spectral imaging. 6th International Conference on Remote Sensing and Geoinformation of the Environment (RSCy), March 26–29, 2018. AR UNSP 1077319. DOI: 10.1117/12.2323820.

Goidts, E., van Wesemael, B. and Van Oost, K. 2009. Driving forces of soil organic carbon evolution at the landscape and regional scale using data from a stratified soil monitoring. *Global Change Biology* 15(12), 2981–3000. DOI: 10.1111/j.1365-2486.2009.01940.x.

Guo, L., Fu, P., Shi, T. Z., Chen, Y. Y., Zhang, H. T., Meng, R. and Wang, S. Q. 2020. Mapping field-scale soil organic carbon with unmanned aircraft system-acquired time series multispectral images. *Soil and Tillage Research* 196, AR 104477. DOI: 10.1016/j.still.2019.104477.

Hedley, C. B., Payton, I. J., Lynn, I. H., Carrick, S. T., Webb, T. H. and McNeill, S. 2012. Random sampling of stony and non-stony soils for testing a national soil carbon monitoring system. *Soil Research* 50(1). DOI: 10.1071/SR11171.

Heil, J., Marschner, B. and Stumpe, B. 2020. Digital photography as a tool for microscale mapping of soil organic carbon and iron oxides. *CATENA* 193, AR 104610. DOI: 10.1016/j.catena.2020.104610.

Hong, Y.-S., Zhu, Y.-X., Su, X.-P., Zhu, Q., Zhou, Y. and Yu, L. 2017. Estimation of soil organic matter content using hyperspectral techniques combined with normalized difference spectral index. *Spectroscopy and Spectral Analysis* 37(11), 3537-3542. DOI: 10.3964/j.issn.1000-0593(2017)11-3537-06.

Hou, Y. J., Tashpolat, T., Mamat, S. and Zhang, F. 2014. Estimation model of desert soil organic matter content using hyperspectral data. *Transactions of the Chinese Society of Agricultural Engineering* 30(16), 113.

Huang, J., Pedrera-Parrilla, A., Vanderlinden, K., Taguas, E. V., Gomez, J. A. and Triantafilis, J. 2017. Potential to map depth-specific soil organic matter content across an olive grove using quasi-2d and quasi-3d inversion of DUALEM-21 data. *CATENA* 152. DOI: 10.1016/j.catena.2017.01.017.

Jandl, R., Rodeghiero, M., Martinez, C., Cotrufo, M. F., Bampa, F., van Wesemael, B., Harrison, R. B., Guerrini, I. A., Richter, D. D., Rustad, L., Lorenz, K., Chabbi, A. and Miglietta, F. 2014. Current status, uncertainty and future needs in soil organic carbon monitoring. *Science of the Total Environment* 468-469, 376-383. DOI: 10.1016/j.scitotenv.2013.08.026.

Jenkins, C. 2020. The State of the Art in Monitoring and Verification: an update five years on. *International Journal of Greenhouse Gas Control* 100, AR 103118. DOI: 10.1016/j.ijggc.2020.103118.

Jin, X., Song, K., Du, J., Liu, H. and Wen, Z. 2017. Comparison of different satellite bands and vegetation indices for estimation of soil organic matter based on simulated spectral configuration. *Agricultural and Forest Meteorology* 244-245, 57-71. DOI: 10.1016/j.agrformet.2017.05.018.

Kempen, B., Dalsgaard, S., Kaaya, A. K., Chamuya, N., Ruiperez-Gonzalez, M., Pekkarinen, A. and Walsh, M. G. 2019. Mapping topsoil organic carbon concentrations and stocks for Tanzania. *Geoderma* 337. DOI: 10.1016/j.geoderma.2018.09.011.

Kirillova, N. P., Zhang, Y., Hartemink, A. E., Zhulidova, D. A., Artemyeva, Z. S. and Khomyakov, D. M. 2021. Calibration methods for measuring the color of moist soils with digital cameras. *CATENA* 202, 164-180. DOI: 10.1016/j.geoderma.2018.09.011.

Kweon, G., Lund, E. and Maxton, C. 2013. Soil organic matter and cation-exchange capacity sensing with on-the-go electrical conductivity and optical sensors. *Geoderma* 199, 80.

Kweon, G. and Maxton, C. 2013. Soil organic matter sensing with an on-the-go optical sensor. *Biosystems Engineering* 115(1), 66.

Lamichhane, S., Kumar, L. and Wilson, B. 2019. Digital soil mapping algorithms and covariates for soil organic carbon mapping and their implications: a review. *Geoderma* 352, 395-413. DOI: 10.1016/j.geoderma.2019.05.031.

Li, H., Li, J., Shen, Y., Zhang, X. and Lei, Y. 2018. Web-based irrigation decision support system with limited inputs for farmers. *Agricultural Water Management* 210, 279-285. DOI: 10.1016/j.agwat.2018.08.025.

Mallick, J., Ahmed, M., Alqadhi, S. D., Falqi, I. I., Parayangat, M., Singh, C. K., Rahman, A. and Ijyas, T. 2020. Spatial stochastic model for predicting soil organic matter using remote sensing data. *Geocarto International* 37(2), 413.

Mallik, S., Bhowmik, T., Mishra, U. and Paul, N. 2020. Mapping and prediction of soil organic carbon by an advanced geostatistical technique using remote sensing and terrain data. *Geocarto International* 37(8), 2198.

McConkey, B. G., Angers, D., Cerkonwiak, D., Maillard, E. and Fan, J. 2017. Monitoring, reporting, and verification of soil carbon sequestration on the cropland and grassland of Canada. *Proceedings of the Global Symposium on Soil Organic Carbon 2017*, Rome, Italy, 21-23 March, 2017, 126-134.

Mikhailova, E. A., Stiglitz, R. Y., Post, C. J., Schlautman, M. A., Sharp, J. L. and Gerard, P. D. 2018. Predicting soil organic carbon and total nitrogen in the Russian chernozem from depth and wireless color sensor measurements. *Eurasian Soil Science* 50(12), 1414.

Morais, P. A. D., de Souza, D. M., Madari, B. E., Soares, AdS. and de Oliveira, A. E. 2019. Using image analysis to estimate the soil organic carbon content. *Microchemical Journal* 147, 775-781. DOI: 10.1016/j.microc.2019.03.070.

Nguyen, L. and Kodagoda, S. 2016. Soil organic matter estimation in precision agriculture using wireless sensor networks. *International Conference on Control Automation Robotics and Vision*. DOI: 10.1109/ICARCV.2016.7838635.

Pallegedara Dewage, S. N. S., Minasny, B. and Malone, B. 2020. Disaggregating a regional-extent digital soil map using Bayesian area-to-point regression kriging for farm-scale soil carbon assessment. *SOIL* 6(2), 359.

Paul, S. S., Dowell, L., Coops, N. C., Johnson, M. S., Krzic, M., Geesing, D. and Smukler, S. M. 2020. Tracking changes in soil organic carbon across the heterogeneous agricultural landscape of the Lower Fraser Valley of British Columbia. *Science of the Total Environment* 732, AR 138994. DOI: 10.1016/j.scitotenv.2020.13899.

Perie, E., Clothier, B. E. and McLaren, S. J. 2014. Soil-carbon in LCA: feasibility of measurement and monitoring systems. *Proceedings of the 9th International Conference on Life Cycle Assessment in the Agri-Food Sector (LCA Food 2014)*, San Francisco, CA, 8-10 October, 2014, 981-987.

Post, W. M., Izaurralde, R. C., Mann, L. K. and Bliss, N. 2001. Monitoring and verifying changes of organic carbon in soil. *Climatic Change* 51(1), 73.

Pouladi, N., Moller, A. B., Tabatabai, S. and Greve, M. H. 2019. Mapping soil organic matter contents at field level with Cubist, Random Forest and kriging. *Geoderma* 342, 85-92. DOI: 10.1016/j.geoderma.2019.02.019.

Raeesi, M., Zolfaghari, A. A., Yazdani, M. R., Gorji, M. and Sabetizade, M. 2019. Prediction of soil organic matter using an inexpensive colour sensor in arid and semiarid areas of Iran. *Soil Research* 57(3), 276-286. DOI: 10.1071/SR18323.

Roberts, D. F., Adamchuk, V. I., Shanahan, J. F., Ferguson, R. B. and Schepers, J. S. 2011. Estimation of surface soil organic matter using a ground-based active sensor and aerial imagery. *Precision Agriculture* 12(1), 82-102. DOI: 10.1007/s11119-010-9158-5.

Rossel, R. A. V., Webster, R., Bui, E. N. and Baldock, J. A. 2014. Baseline map of organic carbon in Australian soil to support national carbon accounting and monitoring under climate change. *Global Change Biology* 20(9), 2953-2970. DOI: 10.1111/gcb.12569.

Roudier, P. 2021. R package 'clhs', Conditioned Latin HyperCube Sampling. Available at: https://cran.r-project.org/web/packages/clhs/clhs.pdf.

Saby, N. P. A., Bellamy, P. H., Morvan, X., Arrouays, D., Jones, R. J. A., Verheijen, F. G. A., Kibblewhite, M. G., Verdoodt, A., Üveges, J. B., Freudenschuß, A. and Simota, C. 2008. Will European soil-monitoring networks be able to detect changes in topsoil organic carbon content? *Global Change Biology* 14(10), 2432-2442. DOI: 10.1111/j.1365-2486.2008.01658.x.

Santacruz, A., Rubiano, Y. and Melo, C. 2014. Evolutionary optimization of spatial sampling networks designed for the monitoring of soil organic carbon. Soil carbon. *Progress in Soil Science*, 77-84. DOI: 10.1007/978-3-319-04084-4_8.

Schloemer, S., Furche, M., Dumke, I., Poggenburg, J., Bahr, A., Seeger, C., Vidal, A. and Faber, E. 2013. A review of continuous soil gas monitoring related to CCS – technical advances and lessons learned. *Applied Geochemistry* 30, 148-160. DOI: 10.1016/j.apgeochem.2012.08.002.

Serrano, J. M., Shahidian, S., Marques da Silva, J. M. and Carvalho, M. 2016. Monitoring of soil organic carbon over 10 years in a Mediterranean silvo-pastoral system: potential evaluation for differential management. *Precision Agriculture* 17(3), 274–295. DOI: 10.1007/s11119-015-9419-4.

Shi, Z., Ji, W., Rossel, R. A. V., Chen, S. and Zhou, Y. 2015. Prediction of soil organic matter using a spatially constrained local partial least squares regression and the Chinese vis-NIR spectral library. *European Journal of Soil Science* 66(4), 679–687. DOI: 10.1111/ejss.12272.

Singh, K., Aitkenhead, M., Fidelis, C., Yinil, D., Sanderson, T., Snoeck, D. and Field, D. J. 2021. Optimisation of spectral preprocessing for estimating soil condition on small farms. *Soil Use and Management* 38(1), 150. DOI: 10.1111/SUM.12684.

Smith, P. 2004. Monitoring and verification of soil carbon changes under Article 3.4 of the Kyoto Protocol. *Soil Use and Management* 20, 264–270. DOI: 10.1079/SUM2004239.

Smith, P., Soussana, J. F., Angers, D., Schipper, L., Chenu, C., Rasse, D. P., Batjes, N. H., van Egmond, F., McNeill, S., Kuhnert, M., Arias-Navarro, C., Olesen, J. E., Chirinda, N., Fornara, D., Wollenberg, E., Alvaro-Fuentes, J., Sanz-Cobena, A. and Klumpp, K. 2020. How to measure, report and verify soil carbon change to realize the potential of soil carbon sequestration for atmospheric greenhouse gas removal. *Global Change Biology* 26(1), 219–241. DOI: 10.1111/gcb.14815.

Song, S., Gibson, D., Ahmadzadeh, S., Chu, H. O., Warden, B., Overend, R., Macfarlane, F., Murray, P., Marshall, S., Aitkenhead, M. J., Bienkowski, D. and Allison, R. 2020. Low-cost hyperspectral imaging system using a linear variable bandpass filter for agritech applications. *Applied Optics* 59(5), A167–A175.

Srivastava, R., Sarkar, D., Mukhopadhayay, S. S., Sood, A., Singh, M., Nasre, R. A. and Dhale, S. A. 2015. Development of hyperspectral model for rapid monitoring of soil organic carbon under precision farming in the Indo-Gangetic Plains of Punjab, India. *Journal of the Indian Society of Remote Sensing* 43(4), 751–759. DOI: 10.1007/s12524-015-0458-0.

Stevens, A., van Wesemael, B., Bartholomeus, H., Rosillon, D., Tychon, B. and Ben-Dor, E. 2008. Laboratory, field and airborne spectroscopy for monitoring organic carbon content in agricultural soils. *Geoderma* 144(1–2), 395–404. DOI: 10.1016/j.geoderma.2007.12.009.

Stiglitz, R., Mikhailova, E., Post, C., Schlautman, M. and Sharp, J. 2017. Using an inexpensive color sensor for rapid assessment of soil organic carbon. *Geoderma* 286, 98–103. DOI: 10.1016/j.geoderma.2016.10.027.

Stiglitz, R. Y., Mikhailova, E. A., Sharp, J. L., Post, C. J., Schlautman, M. A., Gerard, P. D. and Cope, M. P. 2018. Predicting soil organic carbon and total nitrogen at the farm scale using quantitative color sensor measurements. *Agronomy* 8(10). DOI: 10.3390/agronomy8100212.

Suddick, E. C., Ngugi, M. K., Paustian, K. and Six, J. 2013. Monitoring soil carbon will prepare growers for a carbon trading system. *California Agriculture* 67(3), 162–171. DOI: 10.3733/ca.v067n03p162.

Swetha, R. K. and Chakraborty, S. 2021. Combination of soil texture with Nix color sensor can improve soil organic carbon prediction. *Geoderma* 382, AR 114775. DOI: 10.1016/j.geoderma.2020.114775.

Vagen, T. G., Winowiecki, L. A., Neely, C., Chesterman, S. and Bourne, M. 2017. Spatial assessments for the mapping and monitoring of soil organic carbon: using stakeholder engagement processes. *Proceedings of the Global Symposium on Soil Organic Carbon 2017*, Rome, Italy, 21–23 March, 2017, 213–221.

van Wesemael, B., Paustian, K., Andren, O., Cerri, C. E. P., Dodd, M., Etchevers, J., Goidts, E., Grace, P., Katterer, T., McConkey, B. G., Ogle, S., Pan, G. and Siebner, C. 2011. How can soil monitoring networks be used to improve predictions of organic carbon pool dynamics and CO2 fluxes in agricultural soils? *Plant and Soil* 338(1-2), 247-259. DOI: 10.1007/s11104-010-0567-z .

Wang, B., Waters, C., Orgill, S., Gray, J., Cowie, A., Clark, A. and Liu, D. L. 2018a. High resolution mapping of soil organic carbon stocks using remote sensing variables in the semi-arid rangelands of eastern Australia. *Science of the Total Environment* 630, 367-378. DOI: 10.1016/j.scitotenv.2018.02.204.

Wang, G., Wu, X., Mo, L. and Zhao, J. 2018b. SCFSen: a sensor node for regional soil carbon flux monitoring. *Sensors* 18(11), AR 3986. DOI: 10.3390/s18113986.

Wendt, J. W. and Hauser, S. 2013. An equivalent soil mass procedure for monitoring soil organic carbon in multiple soil layers. *European Journal of Soil Science* 64(1), 58-65. DOI: 10.1111/ejss.12002.

Wetterlind, J., Piikki, K., Stenberg, B. and Soderstrom, M. 2015. Exploring the predictability of soil texture and organic matter content with a commercial integrated soil profiling tool. *European Journal of Soil Science* 66(4), 631-638. DOI: 10.1111/ejss.12228.

Wu, C. W., Yang, H., Xia, J., Chang, J., Yang, Y., Zhang, Y. and Cheng, F. 2019. Study on prediction of soil organic matter based on digital image color extraction. *Spectroscopy and Spectral Analysis* 39(4), 1222-1228. DOI: 10.3964/j.issn.1000-0593(2019)04-1222-07.

Yagasaki, Y. and Shirato, Y. 2014. Assessment on the rates and potentials of soil organic carbon sequestration in agricultural lands in Japan using a process-based model and spatially explicit land-use change inventories - Part 1: historical trend and validation based on nation-wide soil monitoring. *Biogeosciences* 11(16), 4429-4442. DOI: 10.5194/bg-11-4429-2014.

Yang, M., Xu, D., Chen, S., Li, H. and Shi, Z. 2019. Evaluation of machine learning approaches to predict soil organic matter and pH using vis-NIR spectra. *Sensors* 19(2), AR 263. DOI: 10.3390/s19020263.

Zhang, Z. Q., Li, J. Z., Tsui, C. C. and Chen, Z. S. 2020. The study of gaining more detailed variability information of soil organic carbon in surface soils and its significance to enriching the existing soil database. *Sustainability* 12(12), AR 4866. DOI: 10.3390/su12124866.

Zhou, T., Geng, Y. J., Ji, C., Xu, X. R., Wang, H., Pan, J. J., Bumberger, J., Haase, D. and Lausch, A. 2021. Prediction of soil organic carbon and the C:N ratio on a national scale using machine learning and satellite data: a comparison between Sentinel-2, Sentinel-3 and Landsat-8 images. *Science of the Total Environment* 755, AR 142661. DOI: 10.1016/j.scitotenv.2020.142661.

Part 3

Fostering carbon sequestration in soils

Chapter 15

Promoting carbon sequestration in soils: the importance of soil, region and context-specific interventions

Rattan Lal, CFAES Rattan Lal Center for Carbon Management and Sequestration, The Ohio State University, USA

1 Introduction

The global carbon (C) cycle (GCC), critical to the moderation of the Earth's climate, consists of diverse C pools (Fig. 1). Prominent among these are atmospheric C (gray), coastal C (blue), soil organic C or SOC (brown), soil inorganic C or SIC

http://dx.doi.org/10.19103/AS.2022.0106.14

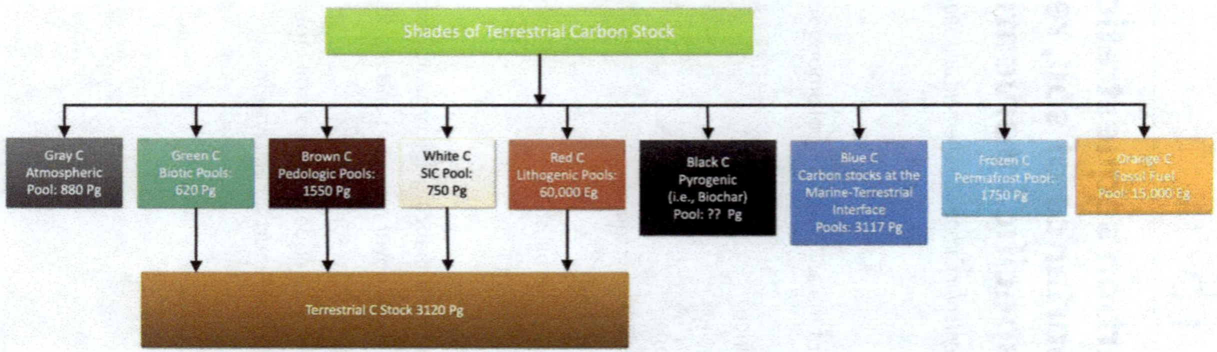

Figure 1 Different reservoirs of carbon, which interact with each other and are affected by anthropogenic activity. The data on Blue C from Atwood et al. (2020), terrestrial C and black C stock from Lal (2004), gray C from Lal (2018) and frozen C from Jungkunst et al. (2012).

(white), permafrost or Cryosol C (frozen), pyrogenic or fire-derived C (black), vegetation or biomass C (green), lithospheric C (red) and fossil fuel C (orange). Exchange of C among these diverse pools moderates the Earth's climate by influencing the gray or atmospheric C, which in turn strongly impacts the radiation balance of the Earth. Anthropogenic activities, especially deforestation and agriculture along with drainage of peatland and the irrigation of arid land (Aridisols), have drastically perturbed the GCC and set in motion the so-called Anthropocene, which began with the evolution of settled agriculture about 10 millennia ago (Ruddiman, 2003). Anthropogenic influence on the climate was accelerated by the onset of the Industrial Revolution around 1750 and has peaked during the twenty-first century (Crutzen and Stoermer, 2000).

The terrestrial C reservoir (brown, white, green, blue, black and frozen C pools combined) is the third largest pool, after the lithospheric (red) and oceanic reservoirs. In ice-free uplands, the terrestrial C pool comprises soil organic matter (SOM), SIC and vegetation pools. The SOM is composed of ~50% C, and most mineral soils have SOM contents of less than 5%. Yet, SOC has a strong effect on soil quality/functionality and agronomic productivity. The storage and dynamics of SOM, along with its different pools (e.g. labile, intermediate and passive) affect the soil structure and pore size distribution, the aeration and gaseous fluxes between the soil and the atmosphere, water retention and movement, nutrient retention and availability to plants, nutrient and energy availability for soil biota (micro, meso and macro). Indeed, SOM is the bloodstream, essence or elixir of all terrestrial life. Above all, as a major component of the terrestrial C pool, it is an important component of the GCC, which along with the coupled cycling of water (H_2O) and other elements (e.g. nitrogen, or N, and phosphorus, or P) forms the basis of numerous ecosystem services (ESs) essential to human well-being and nature conservancy. The decoupling of these critical cycles due to the impact of human activities has created some notable environmental problems including accelerated water runoff and soil erosion by water and wind, eutrophication of water along with the issues of algal bloom and anoxia, contamination of air and the alteration of atmospheric chemistry and changes in the radiative forcing of greenhouse gases (GHGs).

The objective of this chapter is to describe the distribution and dynamics of diverse components of the soil C or pedological reservoir (brown, white, blue, black, green and frozen C), discuss estimates of these pools, explain the impact of anthropogenic activities and outline management strategies to restore them through land use and management options. The literature on soil C reserves is voluminous. Therefore, rather than providing a comprehensive review, the focus of this chapter is on outlining critical features of this urgent topic and on discussing concepts of management that may be used to restore the depleted pedologic C reservoir, with a specific focus on SOC.

Table 1 Some estimates of SOC and SIC stocks for different ecosystems

Soil	Location	Country	Depth (cm)	Units	SOC	SIC	Reference
Saline	Yellow River Delta	China	100	Kg C/m²	2.3–11.7	13.3–24.7	Zhang et al. (2021)
Desert lands	Rajasthan	India	90	Mg C/ha	15.5	76.7	Moharana et al. (2021)
Loess Plateau	-	China	100	Kg C/m²	-	17.04	Tan et al. (2014)
Tropical forest succession	Kalimantan	Indonesia	40	Mg C/ha			
	(i) *Imperata*				36.2		
	(ii) Secondary forest				39.0		
	(iii) Primary forest				33.2		
Blue C	Pacific NW Coast	USA	300	Mg C/ha			Kauffman et al. (2020)
	(i) Seagrass				217 ± 60		
	(ii) Low marsh				417 ± 70		
	(iii) High marsh				551 ± 47		
	Total				1064 ± 38		
	Amazon (Mangroves)	Brazil	?	Mg C/ha			Kauffman et al. (2018)
	(i) Salt marshes				257		
	(ii) Mangroves				361–746		
	(iii) AGB				145		
	(iv) Sal				340		
	Florida Evergreens	USA		Mg C/ha			Reithmaier et al. (2021)
	(i) Mangroves				-	103 ± 61	
	(ii) Marshes				-	30 ± 30	
	(iii) Seagrass				-	39 ± 1	

Riverine marginal wetlands		China	100	Kg C/m^2	4.6–13.2	-	Bai et al. (2020)
Coniferous forest Taiga	Karelia Isthmus Bryanak Regions	Russia	50	Mg C/ha	47–116	-	Kuznetsova et al. (2020)
Tundra	Arctic	Greenland		Kg C/m^2	34.9–54.6	-	Bradley-Cook and Virginia (2018)
Histosols	Babia Gora National Park	Poland	?	Mg C/ha	50-905 (7550)	-	Reyna-Bowen et al. (2019)
Tropical rain forest	Amazon	Brazil		Mg C/ha	(Total stock 267.5)	-	Dantas et al. (2021)
	(i) AGB				35.2%		
	(ii) Soil				53.2%		
	(iii) Root				1.5%		
Grasslands	Pampas	Argentina		Mg C/ha	56–59	-	Alvarez et al. (2021)

SOC = Soil organic carbon, SIC = Soil inorganic carbon.

2 Global distribution of the soil carbon pool among soil types and ecoregions

The magnitude and dynamics of SOC differ widely depending on soil characteristics and region. SOC and SIC stock estimates are highly variable (Table 1), leading to uncertainty in erroneous interpretation of and contradictory conclusions from the data. Examples of the effects of the diverse variables affecting different SOC pools are outlined in this table. Major soil properties affecting SOC are texture (clay and silt content) (Arrouays et al., 2006; Chatterjee, 2020; Djagbletey et al., 2018; Larionova et al., 2008; Okpoho, 2018; Wattel-Koekkoek et al., 2003), mineralogy (Paul et al., 2008; Wattel-Koekkoek et al., 2003; Yu et al., 2013), soil depth (Tan et al., 2014), the soil wetness index (Mishra et al., 2021; Tomar and Baishya, 2020) and climate, comprising of mean annual precipitation (MAP) (Yu et al., 2013; Chatterjee, 2020), mean annual temperature (MAT) (Chen et al., 2013) and other climate variables (Yu et al., 2013). The SOC pool is also affected by interaction among these complex variables, and understanding the primary controls of SOC stocks and their dynamics is critically important.

Chatterjee (2020) measured the mineralizable SOC pool in six ecological regions of the United States and found that the pools varied with climate, vegetation and soil properties. The permanganate oxidizable C ($KMNO_4$-C) ranged from 7.9 mg/g in the state of Florida to 325 mg/g in Hawaii. Chatterjee observed that the MAP was an important factor for SOC and the resistant C pools, while the clay content was related to the mineralizable active C pools and bacterial abundance.

Gruba et al. (2014) assessed the influence of different parent materials on the spatial distribution of SOC stocks in some forest soils. The range of parent materials studied included Tertiary clay stones and sandstones and Quaternary sands of different origins. The <0.05 mm fraction, as a function of the parent material, was an important control for the distribution of C stocks in soils. The SOC content in the mineral fraction increased with an increase in the fines fraction and, in the O horizon, with an increase in the sand content. Furthermore, the SOC stock was also affected by the forest species and the biomass volume (Gruba et al., 2014). All other factors remaining the same, SOM content is strongly related to climate.

Both SOC and SIC stocks vary significantly with depth, as is exemplified by the data in Table 2. Whereas C stocks in subsoil (below 50 cm depth) do not change rapidly (except in soils prone to accelerated erosion and truncation of the topsoil), they must be accounted for to obtain accurate estimates of the C budget. In the Loess Plateau of China, Yu et al. (2020) assessed SOC and SIC stocks to 5 m depth for diverse land use systems (Table 2). Both SOC and SIC stocks were 1.6- and 2.1-fold higher within a 1–5 m depth range as compared

Table 2 Depth distribution of soil organic and inorganic C for diverse land uses in the Loess Plateau of China

	0-5 m depth (Kg C/m²)	
Land use	SOC	SIC
Shrubland	17.2	90.6
Grassland	16.3	104.1
Forest land	15.2	96.2
Cropland	14.1	82.4
Gully land	6.4	50.3

SOC = Soil organic carbon, SIC = Soil inorganic carbon.Source: Adapted from Yu et al. (2020).

with depths of 0–1 m, and both stocks decreased significantly from the 5th to the 15th year of cultivation in cropland. Management effects and rather rapid changes of deep C were observed in several studies. However, to date, the processes leading to the accumulation and loss of subsoil C have been poorly understood. Progress is prevented through these diverse and strongly interacting factors and different methodologies, leading to large uncertainties of global SOC stock estimates to 1 m depth and indicating the urgent need for an improved database by introducing a standardized methodology.

3 Soil carbon persistence

Persistence, as indicated by the mean residence time (MRT) of soil C stock, is a critical factor in the mitigation of climate change and is determined by a range of interacting biotic and abiotic factors. Indeed, there are numerous uncertainties in the MRT of SOC. Sierra et al. (2018) distinguished between persistence (age) and the transit time (MRT) and argued that C stored in soils is in general much older than that in the release flux, and that there is a large difference between 'mean ages' and the MRTs, with the former being of the order of centuries or millennia and the latter of the order of decades. They concluded that the age distribution may be a better metric to characterize persistence than the MRT (for more information, see Chapter 13 of this book).

In addition to the magnitude of SOC stock, its MRT may also depend on parameters inherent to the soil, for example, its texture. Based on a 40-year-long monoculture of corn grown on a Chernozem, Larionova et al. (2012) found that the mean duration of SOC renewal was 1271–1498 years, but it changed from 697 years in the upper horizon (0-40 cm) to 2742 years in the subsoil (40–60 cm). Furthermore, the MRT increased with decreasing particle size. In addition to particle size or the clay content (Leifeld et al., 2005), the MRT of recently incorporated organic matter (OM) also depends on the type of clay minerals present. Paul et al. (2008) reported that the MRT may be longer in soils dominated by smectite compared with non-crystalline minerals. Paul and colleagues also concluded that the presence of SOC stabilization mechanisms does not necessarily imply that recently incorporated OM is stabilized. Zeraatpishe and Khormali (2013) observed a significant relationship between SOC and the illite and chlorite contents. Wattel-Koekkoek et al. (2003) observed that SOC associated with kaolinite turned over faster (360 years) than that associated with bentonite, and the MRT of the whole clay size fraction can be as large as 1100 years.

There are several mechanisms of protection (e.g. chemical, physical, biological and ecological) (Lal, 2018; Paul et al., 2008; Six et al., 2002, Chapter 2 of this book) of SOC against microbial decay processes. Chemically, some SOC pools are more resistant to decomposition than others. Accordingly, the C:N ratio remains an important determinant of both the persistence and MRT of SOC and the decomposability of particulate OM or POM (Cotrufo et al., 2019). Soil pH is another important control factor for the MRT. For subalpine grasslands, Leifeld et al. (2013) observed that the MRT significantly increased with a decline in pH (5.9 to 3.9). However, differences in MRT along with the pH gradient may not be reflected in SOC stocks because the aboveground biomass (AGB) may be lower under acidic conditions, and relatively higher inputs from AGB may compensate for the faster turnover time at less acidic pH.

Physically, access of SOC to microbial processes may be a more important determinant than its chemical or molecular structure (Dungait et al., 2012). In addition to access, soil temperature is another primary physical (abiotic) determinant of the MRT (Sanderman et al., 2003). Chen et al. (2013) observed a significant declining trend in the global MRT between 1960 and 2008 in soils in high latitude regions characterized by large SOC stocks and a greater degree of climate change. For some French cultivated soils, Balesdent and Recous (1997) observed that 75% of the topsoil C had an MRT of 40 years and coarse fractions contained most of the younger SOC. Further, the MRT may also be related to the MAP (Parshotam et al., 2000).

The source of OM is another important control factor of the MRT. Rasse et al. (2005) argued that understanding the origin of OM stabilization in soils is essential to identifying and promoting management practices that can enhance the accrual of SOC. In this context, the importance of root versus shoot must never be overlooked. Based on the synthesis of the available literature, Rasse et al. observed that the MRT of root-derived C is only partly (~25%) attributable to a higher chemical resistance of the root tissues compared with the shoot tissues. Thus, they proposed that the higher MRT of root-derived C may be due to other factors, such as (a) physio-chemical protection in the deeper horizon, (b) physical protection through mycorrhiza and root hair activities and (c) chemical interactions with metal ions or the formation of so-called organo-mineral complexes. Thus, environmental conditions in the mineral soil, and especially below the A horizon (subsoil), may be an important determinant of the MRT. Krull et al. (2003) emphasized the importance of chemical recalcitrance of black C for SOM protection at very long timescales.

4 Temperature sensitivity

The temperature sensitivity of SOC stock and its MRT are major concerns regarding its vulnerability to enhanced decomposition under global warming (see Chapters 5, 8 and 13 of this book). By using the European-wide database, Lugato et al. (2021) estimated an SOC loss of 2.5 ± 1.2 Pg C by 2020 with climate change. Lugato and colleagues recommended coniferous forest management practices to increase plant inputs to soils and offset the vulnerability of the POM fraction. Huang et al. (2018) computed location- and scale-specific correlations between temperature and soil C sequestration at the global scale and observed a negative correlation between temperature and SOC at the regional scale between 52°N and 40°S and a positive one beyond this region. The authors hypothesized that large SOC stocks in low-temperature regions may increase with global warming while the low SOC stocks in high-temperature areas may decrease.

5 Soil inorganic carbon

The white C, or SIC, is an important C reservoir not only in arid and semi-arid uplands (Lal, 2019; Table 2 for the Loess Plateau of China) but also in tropical everglades and marshlands. The global magnitude of SIC is estimated at 750 Pg to 1 m and 2300 Pg to 2 m depth (Batjes, 1996, Table 3). In the northern temperate grassland of Canada, Bork et al. (2020) observed that SIC accounted for 34.6% of the total C stock and was particularly large below 30 cm depth. An assessment of a wheat-maize cropland in the North China Plains by Shi et al. (2017) indicated that the SOC content (g/kg) decreased from 5.5 in the 0–20 cm layer to 4.1 in the 20–40 cm layer, but that the SIC content increased from 8.1 in the 0–20 cm layer to 12.1 in the 80–100 cm layer. The SIC stock (g/kg) for horizons deeper than the 100 cm layer was 16.5 ± 4.7 and was more than twice the SOC stock (7.5 ± 1.4). Further, there was a significant positive correlation between SOC and SIC stocks ($R = 0.74$, $P < 0.001$) in the cropland of the North China Plains.

Soil management affects SIC in addition to SOC (see Chapter 23 of this book). Zamanian et al. (2021) estimated that, between 1970 and 2020, 0.41 Pg of SIC was released into the atmosphere and 0.72 Pg C will be released as CO_2 by 2050. The SIC stock is affected by irrigation, liming and fertilization (see

Table 3 A wide range of estimates of global soil organic carbon stock to 1 m depth

Reference	Stock	Range (Pg C)
Tifafi et al. (2018)	Total soil C	2500-3468
Köchy et al. (2015)	SOC	1230
	SOC (including peatland)	1325
	SOC tropical soils	421
	SOC tropical wetlands	40
	SOC with Histosol P_b value	1062
Cao and Woodward (1998)		1358
Batjes (1996)		1550
Eswaran et al. (1993)		1505
McNicol et al. (2019)		1000 to >3000
Scharlemann et al. (2014)		1415

P_b = soil bulk density, SOC = soil organic carbon.

Chapter 23 of this book), and the loss of SIC may create a positive, negative or neutral feedback to climate change (Sanderman, 2012), particularly in arid and semi-arid regions.

Similar to the loss of SOC with climate change, the decline of SIC stock may also occur with global warming. For example, based on a study across grasslands in China, Yang et al. (2012) reported an SIC stock decrease between 1980 and 2005 at a mean rate of 26.8 g/m^2 · yr in the top 10 cm layer. The dynamic nature of the SIC stock has also been reported in Chernozems (Mikhailova and Post, 2006) in which SIC is a large component of the total C stock below 1 m depth.

6 Blue carbon in coastal ecosystems

The global coastline is about 440000 km long and consists of coastal ecosystems that are highly diverse and include intertidal and subtidal areas on and above the continental shelf to a depth of 200 m and on immediately adjacent lands (Burke et al., 2001; Ouillon, 2018). Coastal ecosystems consist of coral reefs, mangroves, tidal wetlands, seagrass beds, barrier islands, estuaries, peatland swamps and so on (Burke et al., 2001). Blue C refers to the C stocks at the marine-terrestrial interface, which includes the variety of ecosystems listed above. The blue C stock includes C in AGB, BGB and sediment C. Atwood et al. (2020) estimated the global mean sedimentary C stocks at 3117 Pg C (3006-3209 PgC) in the top 1 m layer. The authors argued that marine sediments contain a large C reservoir (Table 4) that is vulnerable to climate change. Kauffman et al. (2020) evaluated the total blue C stock of the Pacific Northwest

Table 4 Examples of some estimates of the global Blue carbon stock

Reference	Units	Ecosystem carbon stock	Ecosystem
Jardine and Siikamäki (2014)	Mg C/ha	272 ± 49-703 ± 38	Mangroves
Atwood et al. (2020)	Pg C to 1 m depth	2239-2391	Marine sediment

Coast of the United States at 1064 ± 38 Mg C/ha, accounting for >98% of the total ecosystem C stock (TECS) in seagrass and marsh communities (low elevation, high salinity) and 78% in tidal forests (high elevation and low salinity). Kaufman and colleagues observed that measuring TECS to 100 cm depth, as is usually done, strongly underestimates both C stock and the emission of GHGs from tidal wetlands due to the impact of humans on the climate. In addition to containing large C stocks, coastal wetlands and their blue C provide numerous ESs, highlighting the need for the protection, conservation and restoration of blue C in diverse coastal ecosystems.

Stocks of 'Blue C' vary among different coastal ecosystems (Table 5). In the world's largest mangroves, located in Brazil's Amazon, Kauffman et al. (2018) observed that the mean TECS of the salt marshes was 257 Mg/ha and that of the mangroves ranged from 361 Mg C/ha to 746 Mg C/ha. The C stocks in the Amazon mangroves were over twice those of upland evergreen forests and almost ten-fold those of tropical dry forests (Kauffman et al., 2018). In China, Yang et al. (2020) observed that coastal salt marshes form important reservoirs for SOC, which reservoirs are highly heterogeneous and also system-specific. Seagrass ecosystems are considered among the most efficient C sinks globally. In this context, Ricart et al. (2015) observed that C stocks were ~20% higher inside seagrass patches than at seagrass–sand edges in a *Zostera muelleri* patchy seagrass landscape. Mangrove deforestation can lead to decline in ecosystem C stocks (Ouyang and Lee, 2020). Fu et al. (2021) estimated that

Table 5 Relative magnitude of SOC and biomass C in different coastal ecosystems

	C stock (Mg C/ha)	
Ecosystem	Soil organic carbon	Living biomass carbon
Seagrasses	137.1	3.0
Salt marsh	250.1	8.5
Estuarine mangroves	289.1	126.6
Oceanic mangroves	485.3	126.6
Tropical forest	54.3	162.9

Source: Herr and Landis (2016); IPCC (2013); IUCN (2021); Macreadie et al. (2019); The Blue Carbon Initiative (2019).

Chinese vegetated coastal habitat is a large C sink and stores C in the top 1 m of sediments at an annual rate (Gg C/yr) 44 ± 17 in mangroves, 159 ± 57 in salt marshes and 6 ± 45 in seagrass. Mangrove sediments are a large and critical C reservoir containing 2322 Pg C in the top 1 m (Atwood et al., 2020). Thus, 'Blue C' stored in coastal ecosystems can play an important role in climate change mitigation strategies and thus needs to be protected, conserved and restored (Nehren and Wicaksono, 2018).

7 Black or pyrogenic carbon

In this chapter, the term black carbon, or BC (Fig. 1), refers to any fire-derived (pyrogenic) C, including biochar. The latter is derived from the incomplete burning of biomass and has a long MRT (Nguyen et al., 2014). Fire, as a management tool, is widely used in tropical ecosystems, and it has strong pedologic, ecologic, economic and social impacts. Uncontrolled and indiscriminate use of fire can adversely impact the SOC stock and its dynamics, with implications for nutrient loss, volatilization, soil loss by accelerated erosion, loss of biodiversity and a shift in plant communities. A study on the effects of fire at the Bhoramdeo Wildlife Sanctuary of Chhattisgarh, India, by Jhariya and Singh (2021) indicated that SOM and N stocks, micronutrients and microbial biomass C (MBC) were higher in the 0–20 cm layer of no-fire zones than in those under fire treatment. Total soil C stock in the 0–20 cm layer was 69.5 Mg/ha under control compared with 66.5 Mg/ha under medium and 53.7 Mg/ha under low-fire intensity. The total N stock, equal to 4.1 Mg/ha, was also the highest under no-fire treatment (Jhariya and Singh, 2021). However, the MRT of biochar may be overestimated (Lutfalla et al., 2017) because the turnover of biochar may be affected both by properties of the biochar and those of the soil. Similar to pyrolysis for biochar production, the process of hydrothermal carbonization is used to convert biomass into hydrochar, which also leads to C sequestration after soil application. A microcosm study conducted by Gajić et al. (2012) documented that the application of hydrochar as a soil conditioner under field conditions has a moderate potential for SOC sequestration.

Burning of rice straw is practiced widely now in the Indo-Gangetic Plains. This practice has adverse effects on soil and the environment (Shyamsundar et al., 2019). Fire as a management tool was used for millennia in the Yangtze River Delta, China. Lehndorff et al. (2014) examined the contribution of BC derived from burning of rice straw by studying chronosequences of 0-2000 years of rice–wheat rotation compared with adjacent non-paddy systems (50-70 years) in the Bay of Hanzhou, China. The authors observed that, despite its regular long-term input, BC only accounted for 7-11% of total SOC in the surface soil and reached a steady state level in 297 years of 13 Mg of BC/ha, with an MRT of 303 years. They concluded that this low contribution of BC to the total SOC in paddy soils was similar to other global aerobic ecosystems. The fate of C originating from BC input thus remains an open question. Mechanisms leading to its removal from the site of deposition may include fragmentation and loss due to wind and/or water erosion (Rumpel et al., 2015). Effects of fire on SOC stocks under pine forests in the west of Russia (Karelia, Isthamus and Bryonsk regions) were reported to 50 cm depth by Kuznetsova et al. (2020), and it was observed that fire has an important effect on SOC stocks in addition to climate, parent materials, vegetation and agricultural activities.

8 Frozen carbon and permafrost soils

Permafrost soils (Cryosols or the frozen C) are a large reservoir of SOC stocks estimated at ~1750 Pg to 3 m depth (Jungkunst et al., 2012). In addition to their vulnerability to climate change and increase in Q_{10} factors (Bradley-Cook and Virginia, 2018; Huntzinger et al., 2020), SOC stocks in permafrost are also limited by a deficiency of nutrients such as P. Based on a study of the response of high Arctic dwarf shrub tundra to 20 years of low-level nutrient regimes, Street et al. (2018) reported that C stocks in vegetation and soil were halved over this period. In comparison, sites where P and N were added, C storage increased by more than 50%. Whereas the addition stimulated decomposition and decreased the C stock, P and N together co-stimulated moss productivity and increased OM accumulation. Street and colleagues hypothesized that if climate warming were to increase P availability, any increase in N enrichment from warming of the Artic may lead to increased C sequestration. If P is limited in the tundra region, increase in N would increase C loss. Temperature-induced melting of tundra soil, studied by Bradley-Cook and Virginia (2018), indicated that frozen soils are a 'hotspot' for soil C storage and CO_2 efflux. By analyzing datasets of >2700 profiles, Mishra et al. (2021) reported that the soil wetness index and the elevation are the dominant topographic control parameters (see Table 1), and the surface–air temperature and precipitation are important climate-related determinants of SOC stocks in permafrost soils. Huntzinger et al. (2020) assessed ecosystem C stocks in the Arctic/Boreal region of North America and indicated that errors exist in the secondary data generated by models.

In addition to permafrost, global organic (peat) soils are also an important reservoir of SOC and must be protected against land use change (e.g. drainage, deforestation or clearing, cultivation, fire) and mining. In Switzerland, Leifeld et al. (2005) observed that organic soils accounted for less than 3% of the total area but stored about 28% of the national SOC stock. Drainage and cultivation of organic soils, especially for the development of palm oil plantations in Southeast Asia, have created a large terrestrial C debt and have been a major source of GHG emissions. Moreover, climate change and increased fire frequency in boreal forests may lead to a loss of legacy C in organic soils and thus turn them from a carbon sink into a carbon source (Walker et al., 2019). Global loss of wetlands (Mitsch and Gosselink, 2015), and with them the SOC stored under anaerobic conditions, has been a major ecological issue since the 1950s. Thus, protection of existing wetlands and the rewetting of drained wetlands are high priorities. In China, Bai et al. (2020) evaluated changes in SOC stock in river marginal wetlands and reported that higher flooding frequencies could contribute to an accumulation of soil C and N due to favorable hydrological conditions. Thus, Bai and colleagues concluded that attention must be given to C and N stocks in deeper soils.

9 Land use and management

The magnitude of SOC stocks and their dynamics are governed by land use and management. Some researchers have argued that land use may have a greater impact on SOC stocks in temperate cultivated soils than that of climate change over the next 50 years (Balesdent and Recous, 1997). Land use may be an important factor in transforming the labile fraction into passive reservoir with a longer MRT (e.g. perennial land use). In addition to the effect of current land use, historic land use may also affect the SOC stock. However, a study of the impacts of historic land use and forest management over the past 200 years in Europe on SOC stocks by Wäldchen et al. (2013) indicated that there is no consistent trend in relation to the historical management of such lands.

A study of SOC stock in the Amazon Forest by Trumbore and de Camargo (2009) showed that mineral soils (0–30 cm) in the Amazon Basin account for 40 Pg C or 3% of the global SOC stock. When combined with the detrital C and SOC stocks to 1 m depth, the total C stock may be 4 times higher; however, in some cases, it may be extremely vulnerable to future land use and climate change. In Switzerland, Leifeld et al. (2005) reported that about 16% of the national SOC stock has been lost due to cultivation, urbanization and deforestation, and future changes in land use and management may not be sufficient to compensate for this historic loss. A global study by Huang et al. (2018) compared the SOC content of eroded, cultivated, forest and grassland soils. The lowest SOC stock of 55.6 Mg/ha was observed in eroded landscapes, and the highest SOC stock of 83.5 Mg/ha was under forest. To 1.05 m depth, forest soil contained 22.4 Mg C/ha more than cultivated soils. Based on a study on smallholder farms in Abuja, Nigeria, Okpoho (2018) showed that SOC stocks in the 0–20 cm layer amounted to 24.8 ± 0.2 Mg C/ha under cashew, 22.0 ± 0.3 Mg C/ha in maize-cowpea and 13.9 ± 0.1 Mg C/ha under maize monocultures. Given similar baselines at these sites, these results might indicate that the cultivation of cereals with extractive practices can deplete SOC stocks.

10 Landscape management

Limiting atmospheric CO_2 is a high global priority in the twenty-first century, and sustainable management of the landscape and afforestation are among the preferred options to achieve this. Management of the landscape, comprising both surface features and water resources, is based on a holistic concept. Garten and Ashwood (2002) observed that the effects of topography on soil quality indicators (SOC and SOM stocks and the C:N ratio) were secondary to those of the vegetation/land cover. The indicators were more favorable (e.g. higher stocks and lower ratio) with more land cover. A study conducted on the mountain soils of the Babia Gora National Park, Poland, by Reyna Bowen et al. (2019) showed that topographical factors influenced soil conditions and vegetation cover, which in turn affected C accumulation. In this landscape study, the highest SOC stock was recorded in Histosols situated at landscape positions that allow for high water levels. The SIC density in relation to land cover in the Loess Plateau of China was estimated in the order of grassland > forest land for all soil layers to 1 m depth. Tan et al. (2014) also observed that the highest SIC density at 100 cm depth was observed in alkaline soil and the lowest in subalpine meadow soil. Riparian flood plains, which receive sediments on a regular basis, have a high C sink capacity, for both SOC and SIC, depending on the specific site. Based on a study of a riparian zone in the Yellow River Basin, China, Hou et al. (2021) observed that SOC and SIC stocks to 1 m depth increased at average rates of 2.73 Mg C/ha and 5.54 Mg C/ha · yr and that the SOC and SIC stocks were closely correlated with one another and with the silt content. Jones et al. (2019) hypothesized that stabilizing the atmospheric CO_2 concentration can be facilitated by promoting secondary tropical regrowth on abandoned agricultural landscapes. The authors argued that the soil C of abandoned cropland may recover within 40 years of secondary forest growth, but the AGB-C stocks may continue to increase over 100 years.

11 Salinity management for enhancing soil organic carbon stocks

Soil salinity is a major issue in semi-arid and arid regions, including irrigated lands (Chapter 23 of this book). Saline soils cover 3.1% of the global land area, or approximately 397 M ha (Setia et al., 2013). Soil salinity affects SOC stock by limiting productivity through adverse changes in pedological processes. Setia and colleagues estimated that, globally, saline soils may have lost an average of 3.4 Mg SOC/ha since they became saline, and they predicted that the world's soils may further lose 6.8 Pg C due to salinity by 2100. Regardless of the salinity, projected global warming may also accelerate the decomposition of SOC stocks. The SOC stock in salinized soil is adversely affected by reduced biomass productivity and changes in biological processes. A study on cropland in the Yellow River Delta, China, by Zhang et al. (2021) showed a large range (kg C/m^2) of SOC (2.3–11.7) and SIC (13.3–24.7) over 1 m depth. Zhang and colleagues observed that SOC and SIC stocks were the lowest in alkaline soils with the highest soil pH (8.6–9.0). Further, SOC desorption was greater in saline soils with an alkaline pH, indicating low SOC stability in these soils. These results were corroborated by a study on the alkali-sodic soils of Northeast China, where the SOC stock was negatively correlated with soil pH and exchangeable sodium percentage (ESP) and positively with SIC content (Wang et al., 2020). The authors also observed a significant loss in total C stock in corn fields, although the stock was preserved at 120 Mg C/ha in paddies. They concluded that corn and rice cultivation reduced alkali-sodic conditions and favored SOC accumulation but reduced SIC stocks. Flooding of rice paddies facilitated the leaching of salts and transport of dissolved organic carbon (DOC) into the subsoil.

The DOC, being an easily decomposable, labile fraction, may be leached into the subsoil with seepage or percolating water. Sanderman and Amundson (2008, 2009) observed that DOC leaching is an important process of transporting labile C from surface layers and stabilizing it within the mineral soil in deeper layers, especially in soil types developed under high rainfall conditions. They reported that DOC transported into mineral soil accounts for 22% of the annual C inputs below 40 cm in coniferous forest compared with only 2% of C below 20 cm in prairie soil. The DOC transported may be subsequently absorbed on mineral soil and may have long MRTs of 90–150 years. Based on a study of agroforestry systems in Canada, Lim et al. (2018) observed that DOC concentrations under a silvopasture system were 22–24% higher than in other systems and that the same was also true of the SOC stock. Furthermore, the SOC stock in a 0–30 cm layer of mineral soil was more responsive to changes in land use type than to topography (Lim et al., 2018).

12 Conservation agriculture, cover cropping and agroforestry

The projected level of global warming may exacerbate the decomposition of SOC stocks and create a positive feedback to climate change. Jones et al. (2005) estimated that soil C stocks may decrease by 54 Pg C to 80 Pg C by 2100. Thus, there has been an increasing emphasis on the need to sequester atmospheric CO_2 into the SOM pool. Strategies to bring about this transformation include those that increase the input of biomass C into soil while reducing the risks of decomposition and erosion (Chapters 16, 20 and 22 of this book). Conservation agriculture, or the no-till system with cover cropping, is one such option (Machado, 2005). All other factors remaining the same, the conversion of plow tillage to a system-based conservation agriculture on a highly erodible landscape can enhance SOC stocks (Lal, 2015). This is especially true in tropical environments (Chapter 22 of this book). Based on a study in the Brazilian Cerrado, de Carvalho et al. (2014) observed that the use of a no-till system cover crop maintains or increases SOC and N stocks and enhances soil fertility. It also results in high concentrations of exchange bases, higher CEC and high base saturation in the surface layer. SOC stocks to 40 cm depth differed among cover crop species and the use of *Mucuna puriens* and *Canavalia brasiliensis* increased SOC and N stocks. *Imperata cylindrica* is a rhizomatic weed that occupies a large area of deforested land in Indonesia and West Africa. van der Kamp et al. (2009) observed that SOC storage in the A horizon increased by 14% from 14.5 g/kg under *Imperata* to 16.5 g/kg under secondary forest. *Imperata* can be controlled by growing an aggressive cover crop (e.g. *Mucuna utilis*) for several years followed by conservation agriculture. The latter, along with the use of cover cropping or pastured forages, can be used in conjunction with agroforestry to enhance SOC stocks. This can enhance the ecosystem C budget even in an arid climate. For example, a study in Rajasthan, India, by Tanwar et al. (2019) showed that the ecosystem C stock (Mg C/ha of biomass and soil) over a period of nine years was in the order of farm forestry (47.6) followed by *Ziziphus*-based systems (32.5-33.9). About 50–78% of the additional soil C stock was in the form of SIC. Tanwar and colleagues observed that total ecosystem C sequestered (Mg C/ ha biomass and soil) decreased in the order farm forestry (49.8) > silvo-arable systems (11.0-13.3) > hortipasture systems (8.3) > agri-horti (5.5), silvopasture (5.4), permanent pasture (5.3) and permanent cropping (1.0).

Site-specific adoption of a system-based conservation agriculture may also offset possible decline in the net primary productivity (NPP) due to projected climate change. In addition to the loss of global SOC stock by 54-80 Pg C because of the decomposition of SOM under a changing climate (Jones et al., 2005), Cao and Woodward (1998) predicted decline in NEP with climate change. Cao and Woodward estimated that global NPP is 57 Pg C/yr, that C stocks in vegetation and soil are 640 Pg and 1358 Pg, respectively, and that the NEP ranges from −0.5 Pg C in October to 1.6 Pg C in July. Cao and Woodward further predicted that, with the doubling of CO_2, NPP will increase to 69.6 Pg C/yr, the C stocks in vegetation and the world's soil may increase by 133 Pg (773 Pg) and 160 Pg (1518 Pg), respectively, and the seasonal amplitude in NPP will increase by 71%. Increase in NPP by 25% due to the CO_2 fertilization effect and the return of biomass C to soil under a conservation agriculture may accentuate the rate of SOC sequestration with a strong increase in global SOC stocks. Therefore, a global increase in land area under conservation agriculture from 180 M ha at present (Kassam et al., 2019) to a larger area may further enhance the SOC stocks in the world's cropland soils (Lal, 2015). Moreover, a less intensive system under conservation agriculture may also be better adapted to climate extremes and pests, which will most likely increase with climate change.

13 Grazing management

Livestock management and grazing are practiced on ~3.5 billion ha globally. Grazing management and the associated practices have strong impacts on soil C stocks, with implications for managing these ecosystems to optimize SOC storage (Smith et al., 2014, Chapter 18 of this book). Based on integrated crop/livestock systems (ICLS) in South Dakota, USA, Polat et al. (2020) observed that ICLS can improve SOC stock and soil quality. Their data showed that a low stocking rate in conjunction with ICLS enhanced SOC stock from 20.7 g/kg to 28.3 g/kg and total N from 2.06 g/kg to 2.60 g/kg for 0–5 cm depth. Similarly, Smith et al. (2014) studied the impact of 3 sheep stocking treatments (2.7 ewe/ha · yr, 0.9 ewe/ha · yr and no grazing) in the Scottish uplands. The SOC stock in *Molinia caerulea* swards were 3.83 Mg C/ha, 5.01 Mg C/ha and 6.85 Mg C/ha under high, low and no stocking, respectively. The authors concluded that no sheep and low-intensity sheep grazing are preferred upland management practices for enhancing soil and plant C stocks than commercial sheep grazing. In contrast to the general belief that grasslands are always a C sink, a study in the Pampas (Argentina) by Alvarez et al. (2021) showed that SOC stocks to 50 cm depth declined from 59 Mg C/ha in 2007 to 56.2 Mg C/ha in 2019. In northern temperate grasslands, Bork et al. (2020) observed that SOC stock increased from 24.7 Mg/ha to 57.4 Mg/ha to 0.6 m depth with an increase in the normalized stocking rate from 0.49 to 2.3 times the recommended rate because of the increased abundance of introduced vegetation, such as the rhizomatous grass *Poa pratensis*. Furthermore, SIC stock did not vary with the stocking rate, but SOC was associated with the long-term stocking rate. Grassland/savannas are widespread ecosystems in Sub-Saharan Africa. Based on a study in the Guinea Savanna region of Ghana, Djagbletey et al. (2018) found that SOC stocks at 0–10 cm depth (4.8–12.6 Mg C/ha) were associated with higher amounts with the silt and clay fraction than with micro- and small macro-aggregates.

14 Nutrient management for sequestration of soil organic carbon

Biomass, especially the residues of cereals, have larger C:N, C:P and C:S ratios than the SOC fractions (labile, intermediate and passive). Therefore, conversion of biomass into SOC pool, especially a stable pool with a larger MRT, requires inputs of additional nutrients such as N, P and S (Himes, 1998) and micronutrients. In general, a deficiency of N and P may limit the transformation of biomass C into SOC. Spohn (2020) observed that the storage of SOC in mineral soils leads to sequestration of large amounts of organic P. Spohn hypothesized that the formation of mineral-associated SOM is favorable for storing SOC in soil over decadal to millennial timescales and sequesters a large amount of organic P. For example, storage of 1 Mg of SOC in the clay fraction of the topsoil in croplands sequesters 13.1 kg P/ha (Spohn, 2020). P input is also important to enhancing SOC stock in tundra region permafrost soils (Street et al., 2018). Crowther et al. (2019) studied the sensitivity of global soil C stocks to combined nutrient enrichment and concluded that nitrogen and P enrichment in isolation had a minimal impact on soil C storage. When combined with potassium (K) and micronutrients, however, soil C stock increased on average by 0.04 kg C /m^2 · yr. Several studies have observed positive correlations of SOC stocks with the C:N ratio of the biomass (Devi and Sherpa, 2019). Jones et al. (2019) observed that soil C stocks in secondary tropical forest growth on abandoned cropland may be positively related to soil N reserves, and they concluded that soil C recovered within 40 years of forest rejuvenation. The studies show the important nutrient requirements for SOC storage in mineral soils. In contrast, the need for P and other nutrients for restoring SOC stocks in the organic soils of peatlands would be much less.

There is a growing trend in the adoption of organic farming and agriculture systems with reduced agrochemical input. The concept is based on restoring SOM content through recycling of biomass, using organic amendments and strengthening the mechanisms that couple the cycling of H_2O with those of C and essential elements (e.g. N, P, S). Leite et al. (2003) compared soil quality after 16 years of using organic vs. mineral fertilizers. While soils under organic farming systems had higher SOC and SON contents for the same soil type compared with unfertilized ones, they also contained more light (POM) or labile fractions. In general, management-induced changes in the POM and labile fractions are more evident than those in total SOC stocks. Although the labile and stable fractions both ensure different soil functions, their optimal proportions for soil function are as yet unknown.

15 Carbon sequestration under urban ecosystems

Urbanization is a phenomenon of the twenty-first century, and the rapid rate of urban encroachment is strongly affecting the landscape and pedological processes. However, the historic and current effects of urbanization on SOC stocks and dynamics are not well understood with regard to features such as surface sealing, functional zoning, green spaces, history and so on. Vasenev et al. (2013) evaluated the effects of urbanization in the Moscow (Russia) region on SOC for different land use types, soils, urban zones and the so-called 'cultural layers', the latter being the result of human residential activity and settlement history. Vasenev and colleagues observed that the

SOC stock was highly heterogeneous and differed with soil type, soil depth and land use factor and that the SOC content in urban soil has a 'unique character' that deserves an in-depth study. In a subsequent study, Vasenev and Kuzyakov (2018) reported that urban soils and cultural layers may sequester C for centuries, leading to large C stocks beneath cities. Data from a wide range of ecoregions showed that the total C content in urban soils was 1.5–3.0 times higher and the C accrual was much deeper compared with natural soils, resulting in 3 to 5 times larger C stocks. Furthermore, urban SOC stocks increase with latitude but SIC stocks are less affected by climate. Vasenev and colleagues hypothesized that with the long-term input of C from outside cities, urban soils are global hotspots of long-term C sequestration (see also Chapter 24 of this book).

16 Conclusion

Soil, or the pedological carbon (C) reservoir, comprising of the SOC and SIC pools, plays a critical role in the GCC and in moderating the Earth's climate. Both need to be considered when applying soil management practices intended to foster soil carbon. The SOC pool is a strong determinant of soil functionality, especially soil health, through which it affects quality and quantity of agronomic productivity (food and nutritional security) and generates numerous ESs essential to humans and nature. Global hotspots of SOC stocks, in terms of the magnitude and their vulnerability to climate change and anthropogenic perturbations, are in coastal ecosystems, permafrost soils, wetland and peat soils and tropical rainforests. Research priorities should include a strengthening of the database of carbon stocks differentiated by carbon type and their vulnerability in different global regions. This will allow for the identification of areas requiring particular attention and specific management strategies in terms of protecting existing stocks or increasing soil carbon when contents are low.

The literature review supports the following conclusions:

1 Terrestrial and soil C stocks are important components of the GCC and critical to moderating the Earth's climate.
2 Anthropogenic land use and soil/crop/livestock management have depleted soil C stock and caused emission of GHGs into the atmosphere. Agriculture contributes about 25% of global GHG emissions. Thus, most soils of agro-ecosystems are depleted of their SOC stocks.
3 Soil C stocks comprise of SOC and SIC components, and both are important to moderation of the climate. Land use and soil management affect both SOC and SIC stocks.

4 Estimates of the global C stocks are highly variable, obtained by diverse and unstandardized methods and often reported to 30 cm depth. Soil C stocks must be reported to at least 1 m depth and preferably to 2 m and 3 m depths through a standardized methodology. The development of simple tools, which can be used in the field to measure soil health and relate it to soil C stocks and other parameters, is a high priority (see Chapters 10–14 of this book).
5 The soil carbon stocks are temperature sensitive, and MRT also depends on the carbon types (brown, white, blue, black, green and frozen C) and their origins and locations in addition to soil inherent parameters such as texture and mineralogy.
6 Blue carbon, the ecosystem C stock in coastal ecosystems, is large, highly valuable and the most threatened by changes due to human activities. Yet, mangroves and other coastal ecosystems provide a range of ecosystem services, including C sequestration for the mitigation of climate change. The world's most C-rich mangroves contain 703 ± 38 Mg C/ha.
7 Global hot spots of C stocks are permafrost, tropical rainforest, coastal ecosystems, wetlands/peat lands and urban lands. Sustainable management strategies to maintain and foster the soil carbon pools in these environments need to be adopted urgently.
8 Technological options for the management of soil health through the restoration of SOC stocks are conservation agriculture, integrated nutrient management, agroforestry, landscape management and improved grazing systems, which are described in more detail in the following chapters.

17 Future trends in research

Additional research is needed, for site-specific conditions, with regards to the following:

1. Critical limits of soil organic carbon concentration in the root zone in relation to optimization of agronomic yield and maximization of the use efficiency of inputs (e.g. fertilizers, irrigation, improved varieties, pesticide use).
2. Incremental increase in agronomic yield with unit increase in soil organic carbon concentration or stock in the root zone.
3. Saving of inputs (e.g. fertilizer, irrigation, energy) with unit increase in soil organic carbon stock in the root zone.
4. In-field measurements of soil organic carbon concentration by cost-effective and easy-to-use hand-held devices.

5. Assessment of the rate of sequestration of soil inorganic carbon (e.g. secondary carbonates in soil of arid/semi-arid regions and in irrigated ecosystems.
6. Estimates of societal value of soil organic carbon for the purpose of payments to farmers and trading of carbon credits.
7. Identification and implementation of policies which promote and incentivize farmers and land manager for adoption of practices which lead to re-carbonization of soils and of the terrestrial biosphere.
8. Promoting curricula at school and levels which enhance awareness among students about the importance of soil are source or sink of atmosphere CO_2.

18 References

Alvarez, R., Berhongaray, G. and Gimenez, A. (2021). Are grassland soils of the pampas sequestering carbon? *Science of the Total Environment* 763: 142978. DOI: 10.1016/j.scitotenv.2020.142978.

Arrouays, D., Saby, N., Walter, C., Lemercier, B. and Schvartz, C. (2006). Relationships between particle-size distribution and organic carbon in French arable topsoils. *Soil Use and Management* 22(1): 48–51. DOI: 10.1111/j.1475-2743.2006.00020.x.

Atwood, T. B., Witt, A., Mayorga, J., Hammill, E. and Sala, E. (2020). Global patterns in marine sediment carbon stocks. *Frontiers in Marine Science* 7. DOI: 10.3389/fmars.2020.00165.

Bai, J., Yu, L., Du, S., Wei, Z., Liu, Y., Zhang, L., Zhang, G. and Wang, X. (2020). Effects of flooding frequencies on soil carbon and nitrogen stocks in river marginal wetlands in a ten-year period. *Journal of Environmental Management* 267: 110618. DOI: 10.1016/j.jenvman.2020.110618.

Balesdent, J. and Recous, S. (1997). Les temps de résidence du carbone et le potentiel de stockage de carbone dans quelques sols cultivés français. *Canadian Journal of Soil Science* 77(2): 187–193. DOI: 10.4141/S96-109.

Batjes, N. H. (1996). Total carbon and nitrogen in the soils of the world. *European Journal of Soil Science*. Blackwell Publishing Ltd 47(2): 151–163. DOI: 10.1111/j.1365-2389.1996.tb01386.x.

Bork, E. W., Raatz, L. L., Carlyle, C. N., Hewins, D. B. and Thompson, K. A. (2020). Soil carbon increases with long-term cattle stocking in northern temperate grasslands. *Soil Use and Management* 36(3): 387–399. DOI: 10.1111/sum.12580.

Bradley-Cook, J. I. and Virginia, R. A. (2018). Landscape variation in soil carbon stocks and respiration in an Arctic tundra ecosystem, west Greenland. *Arctic, Antarctic, and Alpine Research* 50(1). DOI: 10.1080/15230430.2017.1420283.

Burke, L., Kura, Y., Kassem, K., Revenga, C., Spalding, M. and McAllister, D. (2001). *Pilot Analysis of Global Ecosystems: Agroecosystems*. Washington, DC: World Resources Institute. Available at: http://pdf.wri.org/page_coastal.pdf.

Cao, M. and Woodward, F. I. (1998). Net primary and ecosystem production and carbon stocks of terrestrial ecosystems and their responses to climate change. *Global Change Biology* 4(2): 185–198. DOI: 10.1046/j.1365-2486.1998.00125.x.

Chatterjee, A. (2020). Soil carbon pools of six ecological regions of the United States. *Journal of Forestry Research* 31(5): 1933–1938. DOI: 10.1007/s11676-019-00976-z.

Chen, S., Huang, Y., Zou, J. and Shi, Y. (2013). Mean residence time of global topsoil organic carbon depends on temperature, precipitation and soil nitrogen. *Global and Planetary Change* 100: 99–108. DOI: 10.1016/j.gloplacha.2012.10.006.

Cotrufo, M. F., Ranalli, M. G., Haddix, M. L., Six, J. and Lugato, E. (2019). Soil carbon storage informed by particulate and mineral-associated organic matter. *Nature Geoscience* 12(12): 989–994. DOI: 10.1038/s41561-019-0484-6.

Crowther, T. W., Riggs, C., Lind, E. M., Borer, E. T., Seabloom, E. W., Hobbie, S. E., Wubs, J., Adler, P. B., Firn, J., Gherardi, L., Hagenah, N., Hofmockel, K. S., Knops, J. M. H., McCulley, R. L., MacDougall, A. S., Peri, P. L., Prober, S. M., Stevens, C. J. and Routh, D. (2019). Sensitivity of global soil carbon stocks to combined nutrient enrichment. *Ecology Letters* 22(6): 936–945. DOI: 10.1111/ele.13258.

Crutzen, P. J. and Stoermer, E. F. (2000). The anthropocene, global change. *International Geosphere-Biosphere Programme (IGBP)* 41: 17–18.

Dantas, D., Terra, MdC. N. S. de, Pinto, L. O. R., Calegario, N. and Maciel, S. M. (2021). Above and belowground carbon stock in a tropical forest in Brazil. *Acta Scientiarum. Agronomy* 43: e48276. DOI: 10.4025/actasciagron.v43i1.48276.

de Carvalho, A. Md, Marchão, R. L., Souza, K. W. and Bustamante, M. MdC. (2014). Soil fertility status, carbon and nitrogen stocks under cover crops and tillage regimes. *Revista Ciencia Agronomica* 45(5spe): 914–921. DOI: 10.1590/S1806-66902014000500007.

Devi, S. B. and Sherpa, S. S. S. S. (2019). Soil carbon and nitrogen stocks along the altitudinal gradient of the Darjeeling Himalayas, India. *Environmental Monitoring and Assessment* 191(6): 361. DOI: 10.1007/s10661-019-7470-8.

Djagbletey, E. D., Logah, V., Ewusi-Mensah, N. and Tuffour, H. O. (2018). Carbon stocks in the Guinea savanna of Ghana: estimates from three protected areas. *Biotropica* 50(2): 225–233. DOI: 10.1111/btp.12529.

Dungait, J. A. J., Hopkins, D. W., Gregory, A. S. and Whitmore, A. P. (2012). Soil organic matter turnover is governed by accessibility not recalcitrance. *Global Change Biology*. John Wiley & Sons, Ltd 18(6): 1781–1796. DOI: 10.1111/j.1365-2486.2012.02665.x.

Eswaran, H., Van Den Berg, E. and Reich, P. (1993). Organic carbon in soils of the world. *Soil Science Society of America Journal* 57(1): 192–194. DOI: 10.2136/sssaj1993.03615995005700010034x.

Fu, C., Li, Y., Zeng, L., Zhang, H., Tu, C., Zhou, Q., Xiong, K., Wu, J., Duarte, C. M., Christie, P. and Luo, Y. (2021). Stocks and losses of soil organic carbon from Chinese vegetated coastal habitats. *Global Change Biology* 27(1): 202–214. DOI: 10.1111/gcb.15348.

Gajić, A., Ramke, H. G., Hendricks, A. and Koch, H. (2012). Microcosm study on the decomposability of hydrochars in a cambisol. *Biomass and Bioenergy* 47: 250–259. DOI: 10.1016/j.biombioe.2012.09.036.

Garten, C. T. and Ashwood, T. L. (2002). Landscape level differences in soil carbon and nitrogen: implications for soil carbon sequestration. *Global Biogeochemical Cycles: 61-1-61-14* 16(4): 61-61. DOI: 10.1029/2002GB001918.

Gruba, P., Socha, J., Błońska, E., Lasota, J., Suchanek, A. and Gołąb, P. (2014). Influence of parent material on the spatial distribution of organic carbon stock in the forest soils. *Sylwan* 158(6): 443–452.

Herr, D. and Landis, E. (2016). *Coastal Blue Carbon Ecosystems. Opportunities for Nationally Determined Contributions. Policy Brief.* Gland, Switzerland and Washington, DC: IUCN and TNC.

Himes, F. L. (1998). Nitrogen, sulfur, and phosphorus and the sequestering of carbon. In: Lal, R., Kimble, J. M., Follett, R. F., et al. (Eds) *Soil Processes and the Carbon Cycle*. Boca Raton, FL: CRC Press, pp. 315–319. DOI: 10.1201/9780203739273.

Hou, C., Li, Y., Huang, Y., Zhu, H., Ma, J., Yu, F. and Zhang, X. (2021). Reclamation substantially increases soil organic and inorganic carbon stock in riparian floodplains. *Journal of Soils and Sediments* 21(2): 957–966. DOI: 10.1007/s11368-020-02836-4.

Huang, J., Minasny, B., McBratney, A. B., Padarian, J. and Triantafilis, J. (2018). The location- and scale- specific correlation between temperature and soil carbon sequestration across the globe. *Science of the Total Environment* 615: 540–548. DOI: 10.1016/j.scitotenv.2017.09.136.

Huntzinger, D. N., Schaefer, K., Schwalm, C., Fisher, J. B., Hayes, D., Stofferahn, E., Carey, J., Michalak, A. M., Wei, Y., Jain, A. K., Kolus, H., Mao, J., Poulter, B., Shi, X., Tang, J. and Tian, H. (2020). Evaluation of simulated soil carbon dynamics in Arctic-Boreal ecosystems. *Environmental Research Letters* 15(2). DOI: 10.1088/1748-9326/ab6784.

IPCC (2013). 2013: the physical science basis. Contribution of working group I to the fifth assessment report of the Intergovernmental Panel on Climate Change. In Stocker, T. F., Qin, D., Plattner, G.-K., et al. (Eds). *Climate Change*. Cambridge and New York: CNY. Cambridge University Press.

IUCN (2021). *Blue Carbon*. Available at: https://www.iucn.org/resources/issues-briefs/blue-carbon.

Jardine, S. L. and Siikamäki, J. V. (2014). A global predictive model of carbon in mangrove soils. *Environmental Research Letters* 9(10). DOI: 10.1088/1748-9326/9/10/104013.

Jhariya, M. K. and Singh, L. (2021). Effect of fire severity on soil properties in a seasonally dry forest ecosystem of Central India. *International Journal of Environmental Science and Technology* 18(12): 3967–3978. DOI: 10.1007/s13762-020-03062-8.

Jones, C., McConnell, C., Coleman, K., Cox, P., Falloon, P., Jenkinson, D. and Powlson, D. (2005). Global climate change and soil carbon stocks; predictions from two contrasting models for the turnover of organic carbon in soil. *Global Change Biology* 11(1): 154–166. DOI: 10.1111/j.1365-2486.2004.00885.x.

Jones, I. L., DeWalt, S. J., Lopez, O. R., Bunnefeld, L., Pattison, Z. and Dent, D. H. (2019). Above- and belowground carbon stocks are decoupled in secondary tropical forests and are positively related to forest age and soil nutrients respectively. *Science of the Total Environment* 697: 133987. DOI: 10.1016/j.scitotenv.2019.133987.

Jungkunst, H. F., Krüger, J. P., Heitkamp, F., et al. (2012). Accounting more precisely for peat and other soil carbon resources. In: Lal, R., Lorenz, K., Hüttl, R. F., et al. (Eds) *Recarbonization of the Biosphere: Ecosystem and the Global Carbon Cycle*. Dordrecht, Holland: Springer, pp. 127–157.

Kassam, A., Friedrich, T. and Derpsch, R. (2019). Global spread of Conservation Agriculture. *International Journal of Environmental Studies*. Routledge 76(1): 29–51. DOI: 10.1080/00207233.2018.1494927.

Kauffman, J. B., Bernardino, A. F., Ferreira, T. O., Giovannoni, L. R., de O. Gomes, L. E., Romero, D. J., Jimenez, L. C. Z. and Ruiz, F. (2018). Carbon stocks of mangroves and salt marshes of the Amazon region, Brazil. *Biology Letters* 14(9). DOI: 10.1098/rsbl.2018.0208.

Kauffman, J. B., Giovanonni, L., Kelly, J., Dunstan, N., Borde, A., Diefenderfer, H., Cornu, C., Janousek, C., Apple, J. and Brophy, L. (2020). Total ecosystem carbon stocks at the marine-terrestrial interface: blue carbon of the Pacific Northwest Coast, United States. *Global Change Biology* 26(10): 5679–5692. DOI: 10.1111/gcb.15248.

Köchy, M., Hiederer, R. and Freibauer, A. (2015). Global distribution of soil organic carbon - Part 1: masses and frequency distributions of SOC stocks for the tropics, permafrost regions, wetlands, and the world. *SOIL* 1(1): 351-365. DOI: 10.5194/soil-1-351-2015.

Krull, E. S., Baldock, J. A. and Skjemstad, J. O. (2003). Importance of mechanisms and processes of the stabilisation of soil organic matter for modelling carbon turnover. *Functional Plant Biology* 30(2): 207-222. DOI: 10.1071/FP02085.

Kuznetsova, A. I., Lukina, N. V., Gornov, A. V., Gornova, M. V., Tikhonova, E. V., Smirnov, V. E., Danilova, M. A., Tebenkova, D. N., Braslavskaya, T. Y., Kuznetsov, V. A., Tkachenko, Y. N. and Genikova, N. V. (2020). Carbon stock in sandy soils of pine forests in the west of Russia. *Eurasian Soil Science* 53(8): 1056-1065. DOI: 10.1134/S1064229320080104.

Lal, R. (2004). Soil carbon sequestration impacts on global climate change and food security. *Science* 304(5677): 1623-1627. DOI: 10.1126/science.1097396.

Lal, R. (2015). A system approach to conservation agriculture. *Journal of Soil and Water Conservation* 70(4): 82A-88A. DOI: 10.2489/jswc.70.4.82A.

Lal, R. (2018). Digging deeper: a holistic perspective of factors affecting soil organic carbon sequestration in agroecosystems. *Global Change Biology*. Blackwell Publishing Ltd 24(8): 3285-3301. DOI: 10.1111/gcb.14054.

Lal, R. (2019). Carbon cycling in global drylands. *Current Climate Change Reports*. Springer 5(3): 221-232. DOI: 10.1007/s40641-019-00132-z.

Larionova, A. A., Zolotareva, B. N., Yevdokimov, I. V., Sapronov, D. V., Kuzyakov, Y. V. and Buegger, F. (2008). The rates of organic matter renewal in gray forest soils and chernozems. *Eurasian Soil Science* 41(13): 1378-1386. DOI: 10.1134/S106422930813005X.

Larionova, A. A., Stulin, A. F., Zanina, O. G., Yevdokimov, I. V., Khokhlova, O. S., Buegger, F., Schloter, M. and Kudeyarov, V. N. (2012). Distribution of stable carbon isotopes in an agrochernozem during the transition from C3 vegetation to a corn monoculture. *Eurasian Soil Science* 45(8): 768-778. DOI: 10.1134/S1064229312060063.

Lehndorff, E., Roth, P. J., Cao, Z. H. and Amelung, W. (2014). Black carbon accrual during 2000 years of paddy-rice and non-paddy cropping in the Yangtze River Delta, China. *Global Change Biology* 20(6): 1968-1978. DOI: 10.1111/gcb.12468.

Leifeld, J., Bassin, S. and Fuhrer, J. (2005). Carbon stocks in Swiss agricultural soils predicted by land-use, soil characteristics, and altitude. *Agriculture, Ecosystems and Environment* 105(1-2): 255-266. DOI: 10.1016/j.agee.2004.03.006.

Leifeld, J., Bassin, S., Conen, F., Hajdas, I., Egli, M. and Fuhrer, J. (2013). Control of soil pH on turnover of belowground organic matter in subalpine grassland. *Biogeochemistry* 112(1-3): 59-69. DOI: 10.1007/s10533-011-9689-5.

Leite, L. F. C., Mendonça, E. S., Neves, J. C. L., Machado, P. L. O. A. and Galvão, J. C. C. (2003). Estoques totais de carbono orgânico e seus compartimentos em argissolo sob floresta e sob milho cultivado com adubação mineral e orgânica. *Revista Brasileira de Ciência do Solo* 27(5): 821-832. DOI: 10.1590/S0100-06832003000500006.

Lim, S. S., Baah-Acheamfour, M., Choi, W. J., Arshad, M. A., Fatemi, F., Banerjee, S., Carlyle, C. N., Bork, E. W., Park, H. and Chang, S. X. (2018). Soil organic carbon stocks in three Canadian agroforestry systems: from surface organic to deeper mineral soils. *Forest Ecology and Management* 417: 103-109. DOI: 10.1016/j.foreco.2018.02.050.

Lugato, E., Lavallee, J. M., Haddix, M. L., Panagos, P. and Cotrufo, M. F. (2021). Different climate sensitivity of particulate and mineral-associated soil organic matter. *Nature Geoscience* 14(5): 295-300. DOI: 10.1038/s41561-021-00744-x.

Lutfalla, S., Abiven, S., Barré, P., Wiedemeier, D. B., Christensen, B. T., Houot, S., Kätterer, T., Macdonald, A. J., van Oort, F. and Chenu, C. (2017). Pyrogenic carbon lacks long-term persistence in temperate arable soils. *Frontiers in Earth Science* 5. DOI: 10.3389/feart.2017.00096.

Machado, P. L. OdA. (2005). Soil carbon and the mitigation of global climate change. *Quimica Nova* 28(2): 329–334. DOI: 10.1590/S0100-40422005000200026.

Macreadie, P. I., Anton, A., Raven, J. A., Beaumont, N., Connolly, R. M., Friess, D. A., Kelleway, J. J., Kennedy, H., Kuwae, T., Lavery, P. S., Lovelock, C. E., Smale, D. A., Apostolaki, E. T., Atwood, T. B., Baldock, J., Bianchi, T. S., Chmura, G. L., Eyre, B. D., Fourqurean, J. W., Hall-Spencer, J. M., Huxham, M., Hendriks, I. E., Krause-Jensen, D., Laffoley, D., Luisetti, T., Marbà, N., Masque, P., McGlathery, K. J., Megonigal, J. P., Murdiyarso, D., Russell, B. D., Santos, R., Serrano, O., Silliman, B. R., Watanabe, K. and Duarte, C. M. (2019). The future of blue carbon science. *Nature Communications* 10(1): 3998. DOI: 10.1038/s41467-019-11693-w.

McNicol, G., Bulmer, C., D'Amore, D., Sanborn, P., Saunders, S., Giesbrecht, I., Arriola, S. G., Bidlack, A., Butman, D. and Buma, B. (2019). Large, climate-sensitive soil carbon stocks mapped with pedology-informed machine learning in the North Pacific coastal temperate rainforest. *Environmental Research Letters* 14(1). DOI: 10.1088/1748-9326/aaed52.

Mikhailova, E. A. and Post, C. J. (2006). Effects of land use on soil inorganic carbon stocks in the Russian chernozem. *Journal of Environmental Quality* 35(4): 1384–1388. DOI: 10.2134/jeq2005.0151.

Mishra, U., Hugelius, G., Shelef, E., Yang, Y., Strauss, J., Lupachev, A., Harden, J. W., Jastrow, J. D., Ping, C. L., Riley, W. J., Schuur, E. A. G., Matamala, R., Siewert, M., Nave, L. E., Koven, C. D., Fuchs, M., Palmtag, J., Kuhry, P., Treat, C. C., Zubrzycki, S., Hoffman, F. M., Elberling, B., Camill, P., Veremeeva, A. and Orr, A. (2021). Spatial heterogeneity and environmental predictors of permafrost region soil organic carbon stocks. *Science Advances* 7(9): eaaz5236. DOI: 10.1126/sciadv.aaz5236.

Mitsch, W. J. and Gosselink, J. G. (2015). Wetlands. 5th ed. Hoboken, NJ.

Moharana, P. C., Jena, R. K., Kumar, N., Singh, R. S. and Rao, S. S. (2021). Assessment of soil organic and inorganic carbon stock at different soil depths after conversion of desert into arable land in the hot arid regions of India. *Carbon Management* 12(2): 153–165. DOI: 10.1080/17583004.2021.1893128.

Nehren, U. and Wicaksono, P. (2018). Mapping soil carbon stocks in an oceanic mangrove ecosystem in Karimunjawa Islands, Indonesia. *Estuarine, Coastal and Shelf Science* 214: 185–193. DOI: 10.1016/j.ecss.2018.09.022.

Nguyen, B. T., Koide, R. T., Dell, C., Drohan, P., Skinner, H., Adler, P. R. and Nord, A. (2014). Turnover of soil carbon following addition of switchgrass-derived biochar to four soils. *Soil Science Society of America Journal* 78(2): 531–537. DOI: 10.2136/sssaj2013.07.0258.

Okpoho, N. A. (2018). Smallholder agriculture land use impact on soil organic carbon stock in Federal Capital Territory of Nigeria. *Journal of Agriculture and Environment for International Development* 112(1): 109–119. DOI: 10.12895/jaeid.20181.702.

Ouillon, S. (2018). Why and how do we study sediment transport? Focus on coastal zones and ongoing methods. *Water* 10(4): 390. DOI: 10.3390/w10040390.

Ouyang, X. and Lee, S. Y. (2020). Improved estimates on global carbon stock and carbon pools in tidal wetlands. *Nature Communications* 11(1): 317. DOI: 10.1038/s41467-019-14120-2.

Parshotam, A., Saggar, S., Searle, P. L., Daly, B. K., Sparling, G. P. and Parfitt, R. L. (2000). Carbon residence times obtained from labelled ryegrass decomposition in soils under contrasting environmental conditions. *Soil Biology and Biochemistry* 32(1): 75–83. DOI: 10.1016/S0038-0717(99)00131-5.

Paul, S., Flessa, H., Veldkamp, E. and López-Ulloa, M. (2008). Stabilization of recent soil carbon in the humid tropics following land use changes: evidence from aggregate fractionation and stable isotope analyses. *Biogeochemistry* 87(3): 247–263. DOI: 10.1007/s10533-008-9182-y.

Polat, A., Chim, B., Kumar, S. and Osborne, S. (2020). On-farm assessment of soil quality in low and high grazing under integrated crop-livestock system in south Dakota. *Tarim Bilimleri Dergisi* 26(4): 434–441. DOI: 10.15832/ankutbd.557832.

Rasse, D. P., Rumpel, C. and Dignac, M. F. (2005). Is soil carbon mostly root carbon? Mechanisms for a specific stabilisation. *Plant and Soil* 269(1–2): 341–356. DOI: 10.1007/s11104-004-0907-y.

Reithmaier, G. M. S., Johnston, S. G., Junginger, T., Goddard, M. M., Sanders, C. J., Hutley, L. B., Ho, D. T. and Maher, D. T. (2021). Alkalinity production coupled to pyrite formation represents an unaccounted blue carbon sink. *Global Biogeochemical Cycles* 35(4). DOI: 10.1029/2020GB006785.

Reyna-Bowen, L., Lasota, J., Vera-Montenegro, L., Vera-Montenegro, B. and Błońska, E. (2019). Distribution and factors influencing organic carbon stock in mountain soils in Babia Góra National Park, Poland. *Applied Sciences (Switzerland)* 9(15). DOI: 10.3390/app9153070.

Ricart, A. M., York, P. H., Rasheed, M. A., Pérez, M., Romero, J., Bryant, C. V. and Macreadie, P. I. (2015). Variability of sedimentary organic carbon in patchy seagrass landscapes. *Marine Pollution Bulletin* 100(1): 476–482. DOI: 10.1016/j.marpolbul.2015.09.032.

Ruddiman, W. F. (2003). The anthropogenic greenhouse era began thousands of years ago. *Climatic Change* 61(3): 261–293. DOI: 10.1023/B:CLIM.0000004577.17928.

Rumpel, C., Dignac, M. F., Remuscat, L. and Barré, P. (2015). Nanoscale evidence of contrasted process for root-derived organic matter stabilization by mineral interactions depending on soil depth. *Soil Biology and Biochemistry*. https://doi.org/10.1016/j.soilbio.2015.02.017.

Sanderman, J. (2012). Can management induced changes in the carbonate system drive soil carbon sequestration? A review with particular focus on Australia. *Agriculture, Ecosystems and Environment* 155: 70–77. DOI: 10.1016/j.agee.2012.04.015.

Sanderman, J. and Amundson, R. (2008). A comparative study of dissolved organic carbon transport and stabilization in California forest and grassland soils. *Biogeochemistry* 89(3): 309–327. DOI: 10.1007/s10533-008-9221-8.

Sanderman, J. and Amundson, R. (2009). A comparative study of dissolved organic carbon transport and stabilization in California forest and grassland soils. *Biogeochemistry* 92(1–2): 41–59. DOI: 10.1007/s10533-008-9249-9.

Sanderman, J., Amundson, R. G. and Baldocchi, D. D. (2003). Application of eddy covariance measurements to the temperature dependence of soil organic matter mean residence time. *Global Biogeochemical Cycles* 17(2). DOI: 10.1029/2001GB001833.

Scharlemann, J. P. W., Tanner, E. V. J., Hiederer, R. and Kapos, V. (2014). Global soil carbon: understanding and managing the largest terrestrial carbon pool. *Carbon Management*. Taylor & Francis 5(1): 81–91. DOI: 10.4155/cmt.13.77.

Setia, R., Gottschalk, P., Smith, P., Marschner, P., Baldock, J., Setia, D. and Smith, J. (2013). Soil salinity decreases global soil organic carbon stocks. *Science of the Total Environment* 465: 267-272. DOI: 10.1016/j.scitotenv.2012.08.028.

Shi, H. J., Wang, X. J., Zhao, Y. J., Xu, M. G., Li, D. W. and Guo, Y. (2017). Relationship between soil inorganic carbon and organic carbon in the wheat-maize cropland of the North China Plain. *Plant and Soil* 418(1-2): 423-436. DOI: 10.1007/s11104-017-3310-1.

Shyamsundar, P., Springer, N. P., Tallis, H., Polasky, S., Jat, M. L., Sidhu, H. S., Krishnapriya, P. P., Skiba, N., Ginn, W., Ahuja, V., Cummins, J., Datta, I., Dholakia, H. H., Dixon, J., Gerard, B., Gupta, R., Hellmann, J., Jadhav, A., Jat, H. S., Keil, A., Ladha, J. K., Lopez-Ridaura, S., Nandrajog, S. P., Paul, S., Ritter, A., Sharma, P. C., Singh, R., Singh, D. and Somanathan, R. (2019). Fields on fire: alternatives to crop residue burning in India. *Science* 365(6453): 536-538. DOI: 10.1126/science.aaw4085.

Sierra, C. A., Hoyt, A. M., He, Y. and Trumbore, S. E. (2018). Soil organic matter persistence as a stochastic process: age and transit time distributions of carbon in soils. *Global Biogeochemical Cycles* 32(10): 1574-1588. DOI: 10.1029/2018GB005950.

Six, J., Conant, R. T., Paul, E. A. and Paustian, K. (2002). Stabilization mechanisms of soil organic matter: implications for C-saturation of soils. *Plant and Soil* 241(2): 155-176. DOI: 10.1023/A:1016125726789.

Smith, S. W., Vandenberghe, C., Hastings, A., Johnson, D., Pakeman, R. J., van der Wal, R. and Woodin, S. J. (2014). Optimizing carbon storage Within a spatially heterogeneous upland grassland Through sheep grazing management. *Ecosystems* 17(3): 418-429. DOI: 10.1007/s10021-013-9731-7.

Spohn, M. (2020). Increasing the organic carbon stocks in mineral soils sequesters large amounts of phosphorus. *Global Change Biology* 26(8): 4169-4177. DOI: 10.1111/gcb.15154.

Street, L. E., Mielke, N. and Woodin, S. J. (2018). Phosphorus availability determines the response of tundra ecosystem carbon stocks to nitrogen enrichment. *Ecosystems* 21(6): 1155-1167. DOI: 10.1007/s10021-017-0209-x.

Tan, W. F., Zhang, R., Cao, H., Huang, C., Yang, Q., Wang, M. and Koopal, L. K. (2014). Soil inorganic carbon stock under different soil types and land uses on the Loess Plateau region of China. *CATENA* 121: 22-30. DOI: 10.1016/j.catena.2014.04.014.

Tanwar, S. P. S., Kumar, P., Verma, A., Bhatt, R. K., Singh, A., Lal, K., Patidar, M. and Mathur, B. K. (2019). Carbon sequestration potential of agroforestry systems in the Indian arid zone. *Current Science* 117(12): 2014-2022. DOI: 10.18520/cs/v117/i12/2014-2022.

The Blue Carbon Initiative (2019). *Mitigating Climate Change through Coastal Ecosystem Management*. Conservation International, IUCN, Intergovernmental Oceanographic Cooperation. Available at: https://www.thebluecarboninitiative.org/.

Tifafi, M., Guenet, B. and Hatté, C. (2018). Large differences in global and regional total soil carbon stock estimates based on SoilGrids, HWSD, and NCSCD: intercomparison and evaluation based on field data From USA, England, wales, and France. *Global Biogeochemical Cycles*. John Wiley & Sons, Ltd 32(1): 42-56. DOI: 10.1002/2017GB005678.

Tomar, U. and Baishya, R. (2020). Moisture regime influence on soil carbon stock and carbon sequestration rates in semi-arid forests of the National Capital Region, India. *Journal of Forestry Research* 31(6): 2323-2332. DOI: 10.1007/s11676-019-01032-6.

Trumbore, S. and de Camargo, P. B. (2009). Soil carbon dynamics. In: *Geophysical Monograph Series*. Keller, M., Bustamante, M. and Gash, J. (Eds) Geophysical Monograph Series, pp. 451-462. DOI: 10.1029/2008GM000741.

van der Kamp, J., Yassir, I. and Buurman, P. (2009). Soil carbon changes upon secondary succession in Imperata grasslands (east Kalimantan, Indonesia). *Geoderma* 149(1-2): 76–83. DOI: 10.1016/j.geoderma.2008.11.033.

Vasenev, V. and Kuzyakov, Y. (2018). Urban soils as hot spots of anthropogenic carbon accumulation: review of stocks, mechanisms and driving factors. *Land Degradation and Development* 29(6): 1607–1622. DOI: 10.1002/ldr.2944.

Vasenev, V. I., Stoorvogel, J. J. and Vasenev, I. I. (2013). Urban soil organic carbon and its spatial heterogeneity in comparison with natural and agricultural areas in the Moscow region. *CATENA* 107: 96–102. DOI: 10.1016/j.catena.2013.02.009.

Wäldchen, J., Schulze, E. D., Schöning, I., Schrumpf, M. and Sierra, C. (2013). The influence of changes in forest management over the past 200years on present soil organic carbon stocks. *Forest Ecology and Management* 289: 243–254. DOI: 10.1016/j.foreco.2012.10.014.

Walker, X. J., Baltzer, J. L., Cumming, S. G., Day, N. J., Ebert, C., Goetz, S., Johnstone, J. F., Potter, S., Rogers, B. M., Schuur, E. A. G., Turetsky, M. R. and Mack, M. C. (2019) Increasing wildfires threaten historic carbon sink of boreal forest soils. *Nature* 572(7770): 520–523. doi: 10.1038/s41586-019-1474-y. Epub 2019 Aug 21. PMID: 31435055.

Wang, J., Tang, J., Li, Z., Yang, W., Yang, P. and Qu, Y. (2020). Corn and rice cultivation affect soil organic and inorganic carbon storage through altering soil properties in alkali sodic soils, Northeast of China. *Sustainability* 12(4). DOI: 10.3390/su12041627.

Wattel-Koekkoek, E. J. W., Buurman, P., Van Der Plicht, J., Wattel, E. and Van Breemen, N. (2003). Mean residence time of soil organic matter associated with kaolinite and smectite. *European Journal of Soil Science* 54(2): 269–278. DOI: 10.1046/j.1365-2389.2003.00512.x.

Yang, D., Miao, X. Y., Wang, B., Jiang, R. P., Wen, T., Liu, M. S., Huang, C. and Xu, C. (2020). System-specific complex interactions shape soil organic carbon distribution in coastal salt marshes. *International Journal of Environmental Research and Public Health* 17(6). DOI: 10.3390/ijerph17062037.

Yang, Y., Fang, J., Ji, C., Ma, W., Mohammat, A., Wang, S., Wang, S., Datta, A., Robinson, D. and Smith, P. (2012). Widespread decreases in topsoil inorganic carbon stocks across China's grasslands during 1980s–2000s. *Global Change Biology* 18(12): 3672–3680. DOI: 10.1111/gcb.12025.

Yu, W., Ding, X., Xue, S., Li, S., Liao, X. and Wang, R. (2013). Effects of organic-matter application on phosphorus adsorption of three soil parent materials. *Journal of Soil Science and Plant Nutrition* 13(ahead): 0. DOI: 10.4067/S0718-95162013005000079.

Yu, X., Zhou, W., Wang, Y., Cheng, P., Hou, Y., Xiong, X., Du, H., Yang, L. and Wang, Y. (2020). Effects of land use and cultivation time on soil organic and inorganic carbon storage in deep soils. *Journal of Geographical Sciences* 30(6): 921–934. DOI: 10.1007/s11442-020-1762-3.

Zamanian, K., Zhou, J. and Kuzyakov, Y. (2021). Soil carbonates: the unaccounted, irrecoverable carbon source. *Geoderma* 384. DOI: 10.1016/j.geoderma.2020.114817.

Zeraatpisheh, M. and Khormali, F. (2013). Carbon stock and mineral factors controlling soil organic carbon in a climatic gradient, Golestan province. *Journal of Soil Science and Plant Nutrition* 12: 637–654. https://doi.org/10.4067/S0718-95162012005000022.

Zhang, K., Wang, X., Wu, L., Lu, T., Guo, Y. and Ding, X. (2021). Impacts of salinity on the stability of soil organic carbon in the croplands of the Yellow River Delta. *Land Degradation and Development* 32(4): 1873–1882. DOI: 10.1002/ldr.3840.

Chapter 16

Agriculture practices to improve soil carbon storage in upland soil

Thomas Kätterer and Martin A. Bolinder, Swedish University of Agricultural Sciences (SLU), Sweden

1 Introduction

To meet the climate goal of the Paris Agreement, signed in 2015 by 196 parties at the COP 21 meeting, the mean global air temperature increase must be kept well below 2°C and efforts to limit the increase to 1.5°C above pre-industrial levels should be encouraged. In addition to drastic reductions in the use of fossil carbon and emissions of non-CO_2 greenhouse gases (GHG), negative carbon emissions technologies are needed to achieve this goal. Promising technologies include bioenergy with carbon capture and storage, direct air carbon capture and storage, afforestation and reforestation, enhanced weathering, ocean fertilization, biochar, and soil carbon sequestration (Minx et al., 2018). It is unlikely that one single technology will achieve the carbon uptake required to meet the climate target in a sustainable way and not all negative emissions technologies are available at a reasonable cost (Fuss et al., 2018; Bednar et al., 2019). The role of soils in the global carbon cycle and

http://dx.doi.org/10.19103/AS.2022.0106.15

the importance of reducing GHG emissions from agriculture are increasingly being acknowledged (IPCC, 2018). Increased carbon storage in plants and soils through afforestation, reforestation, and soil carbon sequestration (SCS) are generally considered to be more cost-effective than the other methods mentioned above, with high potential of SCS in particular (Fuss et al., 2018; Minx et al., 2018). SCS has other advantages because agricultural techniques enhancing soil organic carbon (SOC) are known and well tested and do not require land-use changes.

The technical potential for SCS in all soils is high, 1.2–3.1 Pg C $year^{-1}$ according to Lal (2013). Conversion of biomass into biochar, which may be used for soil amelioration, increases this potential further, by about 0.5 Pg C $year^{-1}$, and according to Woolf et al. (2010), it can be achieved sustainably. Global estimates of SCS in cropland soils due to improved agricultural management practices, excluding biochar, range between 0.069 and 1.85 Pg C $year^{-1}$ according to a recent review (Roe et al., 2019). Estimates at the higher end of this range refer solely to technical potential, while values up to about 0.5 Pg C $year^{-1}$ are regarded as economically feasible. On considering some limitations to implementation, Paustian et al. (2016) estimated that 0.08–0.4 Pg C $year^{-1}$ can be sequestered in upland agricultural soils, but that further technological developments in coming decades, such as plant breeding for enhanced root phenotypes of crops, could add an additional 0.3 Pg C $year^{-1}$. According to the Food and Agriculture Organization of the United Nations (FAO), the total area of harvested cropland in 2017 was 1.424 Gha, of which 0.165 Gha was paddy rice systems. Dividing the estimated annual SCS potential at the higher end of the range reported in several publications (0.3 Pg C $year^{-1}$) by the area of upland cropland soils (1.259 Gha) results in an average annual SCS potential of 0.24 Mg C ha^{-1} $year^{-1}$, which corresponds to 0.4% of average SOC stocks to 20 cm depth in European cropland (Lugato et al., 2021). This high-end estimate meets the target set by the '4 per mille' initiative to increase SOC by 0.4% per year (Minasny et al., 2017). On considering only low-cost and cost-effective measures to increase SOC in upland soil, that is, cover crops and more trees in agricultural landscapes, the estimated SCS potential decreases to less than half, 0.14 Pg C, corresponding to 0.11 Mg C ha^{-1} $year^{-1}$ (Bossio et al., 2020).

In this chapter, we synthesize the current state of knowledge on agricultural measures that could be implemented on cropland to increase SCS in upland mineral soils. Following a brief overview of the principles of SCS, we synthesize the results from reviews and meta-analyses quantifying the effect of different agricultural management practices at field scale and illustrate these with examples familiar to us. We also discuss uncertainties associated with the effect size of these measures and problems with upscaling data obtained in field studies to regional or global scale.

Published by Burleigh Dodds Science Publishing Limited, 2023.

2 Basic principles of soil organic carbon storage and sequestration

The SOC balance is controlled by inputs of organic carbon and outputs resulting from decomposition (Fig. 1). Inputs derive from rhizodeposition during the growing season, post-harvest residues, and ex situ-derived inputs of other organic materials, such as manures and recycled organic wastes originating from agricultural production and/or food processing and consumption that may have been transformed through composting, fermentation, or pyrolysis. Carbon outputs from the soil are mainly controlled by pedoclimatic conditions and are therefore more difficult to manage than inputs, which are determined by farming practices (Kätterer et al., 2012). Losses in form of dissolved organic carbon (DOC) are often neglected but may be important to consider (Kindler et al., 2011). Soil erosion leads to loss of carbon at the plot scale, but the effect on the carbon balance at larger scales depends on where the eroded material is deposited (Lal, 2004). In some cases, erosion may even create a carbon sink at a larger scale (Doetterl et al., 2016). Net primary production (NPP), that is, the fraction of carbon fixed by photosynthesis that is converted to biomass, is the first step in capturing carbon from the atmosphere. The proportion of this fixed carbon that enters the soil depends on the crop grown, how it is managed, and how much carbon is retained in field, exported in products, or recycled through manure or through waste from society.

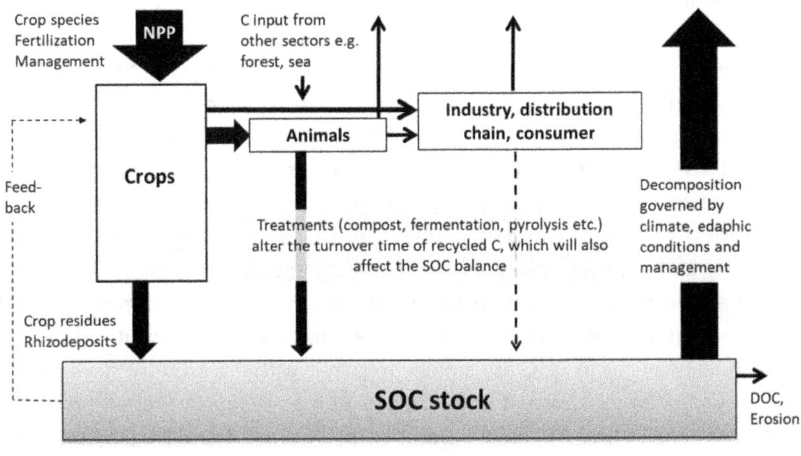

Figure 1 The soil organic carbon (SOC) balance of agricultural fields is driven by net primary production (NPP) and management decisions taken by farmers, that is, choice of crop, residue handling, input of organic amendments, etc. Losses of carbon, through decomposition, erosion, and dissolved organic carbon (DOC), are mainly under pedoclimatic control and less controllable by management (diagram modified after Bolinder et al., 2020).

The organic material that enters the soil through rhizodeposition, harvest residues, and recycled waste decomposes rapidly, but a fraction of it is stabilized through physical, chemical, and biological processes, depending on its characteristics and pedoclimatic conditions (Basile-Doelsch et al., 2020; see Chapter 3 of this book). These include microbial anabolism related to substrate quality and nutrient availability (Liang et al., 2017; Poeplau et al., 2019). The turnover time differs between different sources of organic inputs. Root-derived carbon is particularly important since it has a longer turnover time than aboveground plant material (Hénin and Dupuis, 1945; Bolinder et al., 1999; Wilhelm et al., 2004; Rasse et al., 2005; Kätterer et al., 2011). Plant breeding for cultivars with deep and extended root systems could be a way to increase SOC stocks (Eckersten et al., 2017). Organic material that has been digested in the gut of animals or through composting, fermentation, or pyrolysis prior to soil application generally has longer turnover times than the fresh plant material (e.g. Kätterer et al., 2011; Dechow et al., 2019; Levavasseur et al., 2020).

In this chapter, we review measures that lead to increased carbon storage compared with a reference state. This does not necessarily mean that carbon is sequestered (Chenu et al., 2019). According to Olson et al. (2014), SCS is defined as 'the process of transferring CO_2 from the atmosphere into the soil of a land unit, through plants, plant residues and other organic solids which are stored or retained in the unit as part of the soil organic matter (humus). The retention time of sequestered carbon in the soil (terrestrial pool) can range from short-term (not immediately released back to the atmosphere) to long-term (millennia) storage'.

Whether increased carbon storage will lead to SCS, and thereby to negative emissions, or only to mitigation of carbon losses compared with the status quo depends on historical land use and management at a specific site (Fig. 2). The carbon inputs required to keep the soil at a steady-state are proportional to the initial SOC stock. Increased carbon storage after a change in management contributes to climate change mitigation either by real SCS or by decreasing carbon losses compared with the status quo. The advantage of focusing on changes in carbon storage rather than changes in carbon sequestration is that the former is independent of site history, at least in theory, which makes it much easier to compare the impact of agricultural measures between sites (Fig. 2).

Annual changes in SOC are small compared with the amount stored in the soil. Due to spatial variation within fields, it takes many years until changes in SOC are measurable. Long-term field experiments (LTEs) are therefore indispensable for quantifying the effect of agricultural management practices on soil carbon stocks. Primary results from LTEs have been compiled in several reviews and meta-analyses, in which changes in SOC are presented as either stock change rate (SCR, Mg C ha^{-1} $year^{-1}$) or response ratio (RR; %), that is, relative changes in SOC concentrations or stocks caused by a specific experimental treatment

Published by Burleigh Dodds Science Publishing Limited, 2023.

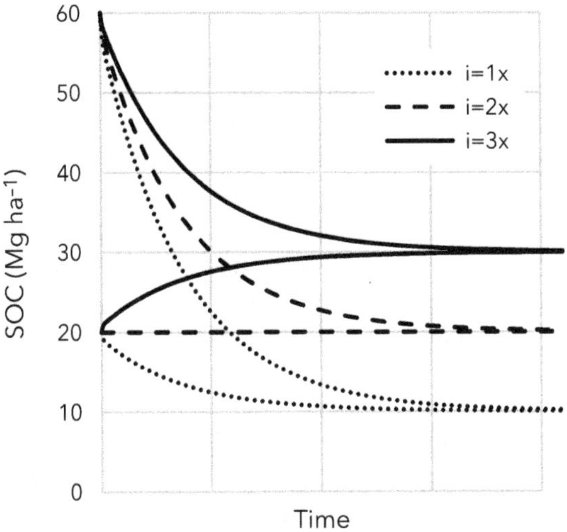

Figure 2 Soil organic carbon (SOC) dynamics in two hypothetical soils with different land use and management history, both receiving carbon inputs, *i*, at three different rates, 1×, 2×, and 3× (highest rate). In this example, we used a two-pool SOC model (ICBM; Andrén and Kätterer, 1997), but any other commonly used carbon model would produce similar curves. The soil with high initial SOC loses much more carbon over time than the soil starting with low initial SOC, but absolute differences in SOC between input rates are identical at any point in time, irrespective of initial conditions. Response ratio, i.e. the ratio of SOC stocks at different input rates, is dependent on both time and initial conditions. Only the low SOC soil at the highest input rate leads to net soil carbon sequestration, but carbon storage increases in both soils when input rates increase and both soils approach the same steady-state SOC stock when carbon inputs equal outputs.

compared with a control. Response ratio is a relative measure and is therefore scale-dependent, that is, a 10% change corresponds to twice as much in absolute terms in a soil with 2% SOC than in a soil with 1% SOC. However, since both SOC concentration and bulk density measurements are required to calculate SCR, RR based on concentration measurements is more frequently reported, although recent studies suggest that, when feasible, it is important to discuss SCR and RR in order to avoid misleading conclusions (e.g. Bolinder et al., 2020; Gross and Glaser, 2021). Therefore, we considered both these effect size indices in the present synthesis. Changes in SOC can also be expressed using absolute differences in SOC content (mg g^{-1}; see, e.g. Land et al., 2021). For scaling-up purposes, SCR is the most common and useful measure.

3 Crop residue retention, cover crops, recycling of organic materials and nitrogen fertilization

In this section we report the effect of four management practices (crop residue retention, cover crops, recycling of organic materials, and nitrogen fertilization) on SOC, based on the main outcomes of synthesis by Bolinder et al. (2020) of 20 reviews and meta-analyses, some covering several management practices. A majority of the publications were derived from the Systematic Map (Haddaway et al., 2015) and applied according to a systematic review protocol that includes a detailed and transparent description of the objectives and methods (Söderström et al., 2014). All publications used paired comparisons, involving a total of 2388 data entries on RR and 855 observations on SCR obtained from LTEs, some of which permitted assessment of their interactions with pedo-climatic conditions and cropping system characteristics. The mean duration of LTEs on which the publications were based generally ranged from 10 to 25 years, and they generally addressed changes in SOC in the topsoil layer (0–30 cm), with mean sampling depth typically between 20 cm and 30 cm. The reviews and meta-analyses rarely included subsoil samples (>30 cm), so the proportion in the databases was very small (for details, see Bolinder et al. 2020). In our discussion below on the four management practices, we also consider the main conclusions from other studies, in particular some recent meta-analyses not included in Bolinder et al. (2020).

3.1 Aboveground crop residue retention

Nine of the reviews covered by Bolinder et al. (2020) addressed issues relating to aboveground crop residue retention (Liu et al., 2014; Lemke et al., 2010; Powlson et al., 2011; Lehtinen et al., 2014; Smith et al., 1997; VandenBygaart et al., 2003; Wang et al., 2015; Lu, 2015; Luo et al., 2010a). Compared with the reference treatment (crop residue removal), calculated across reviews, retaining aboveground crop residues in the field resulted in a mean RR value of 10.3% (range of mean values in all reviews = 2.7–18.2%, number of paired observations (N = 995), and an SCR value of 0.12 Mg C ha^{-1} year^{-1} (0.053–0.59 Mg C ha^{-1} year^{-1}, N=279)). There was a large variation in mean values between the reviews. A few extreme treatments, such as the inclusion of bare fallow as a reference treatment or application of large amounts of straw in the residue retention treatments, can partly explain this variation. The reviews had different degrees of coverage (i.e. country-specific, continental, and global), further contributing to the variation. Most importantly, there were also variations in average soil depth considered and in the duration of the LTEs on which the different meta-databases were based.

The positive effects on SOC of leaving crop residues in the field are mainly a consequence of increasing inputs of carbon to the soil, although reduced losses of carbon from the topsoil via erosion is another major mechanism. According to two of the reviews, crop residue retention can also affect crop yields, thereby increasing NPP and leading to higher residue carbon inputs of the subsequent crop to the soil, which may in turn increase SOC and positively influence SCR (Lu, 2015; Wang et al., 2015). All reviews mainly considered small-grain cereal-based agroecosystems, although the relative importance of retaining aboveground residues increases with the amount of aboveground biomass produced. For example, grain maize produces around twice as many aboveground crop residues as small-grain cereals (Wilhelm et al., 2004). As shown for maize cropping systems in North American studies, SCR can then be as high as 0.3–0.8 Mg C ha^{-1} $year^{-1}$ (Anderson-Texeira et al., 2009). Indeed, one of the reviews found that the effect on RR increased (from 9% to 14%) with the proportion of maize in the rotation and that RR was higher for chopped (13%) compared with unchopped (9%) aboveground crop residues (Lu, 2015).

Reviews with global (Liu et al., 2014) or European coverage (Lehtinen et al., 2014) were unable to establish any interactions between climate and changes in RR or SCR with crop residue retention. However, some potential interactions were identified in country-scale reviews for China (Lu, 2015) and Australia (Luo et al., 2010a), with, for example, the latter showing greater increases in RR in areas with low precipitation. Among the reviews assessing interactions between soil texture and RR or SCR, two reported insignificant interactions for China (Lu, 2015; Wang et al., 2015). Interactions were significant in three of the reviews (Luo et al., 2010a; Liu et al., 2014; Lehtinen et al., 2014), but with contrasting results (i.e., either lower or higher RR with lower or higher clay content).

3.2 Cover crops

Five reviews on the effect of cover crops, using a reference treatment without cover crops, reported mean RR of 9% (7.8–13.1%, N = 129) and SCR of 0.33 Mg C ha^{-1} $year^{-1}$ (0.27–0.43 Mg C ha^{-1} $year^{-1}$, N = 176) (Poeplau and Don, 2015; McDaniel et al., 2014; Aguilera et al., 2013; Blanco-Canqui, 2013; Poeplau et al., 2015). Compared with the effect of aboveground crop residue retention, the variation in mean RR and SCR calculated across reviews was less variable for cover crops. Reviews on cover crops at a global scale showed similar values of RR (McDaniel et al., 2014) and SCR (Poeplau and Don, 2015) to the mean values calculated across all reviews.

However, studies and reviews on the effect of cover crops are complex, involving different species of main crops and cover crops, mixture of specifies, or reference treatments. Poeplau and Don (2015), for example, were comparing a winter cover crop with a reference cropland with winter fallow, and Poeplau et al. (2015) examined only the effect of perennial ryegrass (mostly undersown) as a cover crop. In the review by McDaniel et al. (2014), almost all paired comparisons (97%) involved leguminous cover crops. Aguilera et al. (2013) specifically examined scenarios where cover crops were substituting bare soils. Blanco-Canqui (2013) also included a specific discussion on the effect of multispecies (i.e., mixtures) of cover crops. The mechanism behind the positive effect of cover crops in agroecosystems is primarily increased carbon inputs into the soil since cover crops provide an additional source of aboveground and belowground crop residue carbon entering the soil. However, cover crops can also play an important role in reducing losses of carbon via soil erosion, particularly in permanent woody cropping systems, such as olive groves and vineyards, where RR can be as high as 27–55% (Palese et al., 2014; Gonzalez-Sanchez et al., 2012; Favretto et al., 1992). As seen for retention of aboveground crop residues, cover crops can stimulate NPP of the main crop, and thereby increase carbon inputs to the soil. Some studies specifically addressing the effect of cover crops on the yield of the main crop indicate grain yield increases in small-grain cereals of up to 5% with undersown legume crops, while non-legume cover crops may have no or even negative effects on yield (Valkama et al., 2015). Daryanto et al. (2018) reviewed the effects of cover crops on both RR and yield and reported a mean RR of 9% from 28 studies. This is similar to the mean values calculated across all reviews by Bolinder et al. (2020). Daryanto et al. (2018) reported a relative yield increase of the cash crop of up to 27% following leguminous cover crops (N = 1005) and 6% with non-leguminous cover crops (N = 1282). The 'no cover crop' reference treatment in that study was a fallow period and the analysis only involved annual cash crops following this cover crop period, which may explain the high yield increases. In our current synthesis of the reviews, we noted that only Poeplau et al. (2015) assessed the effect on yield of the main crop under Swedish conditions (i.e., cover crops undersown in the main crop consisting mostly of cereals) and found that the relationship was not significant. A global review by Poeplau and Don (2015) also found no influence on SCR when testing the effect of plant functional types such as non-legume versus legume cover crops. These reviews were the only two examining the interactions of soil texture and climate with RR and SCR, and both concluded that such interactions were not significant.

Higher SOC changes were recently found in a global meta-analysis (1195 paired comparisons, 60% from North America) by Jian et al. (2020), which showed that including cover crops in rotations resulted in an overall RR of 15.5% and SCR of 0.56 Mg C ha^{-1} year^{-1}. That meta-analysis also reported higher RR for fine-textured soils (39.5%) than for medium- and coarse-textured soils (10.3–11.4%), an interaction with climate, with RR being twice as high in temperate (18.7%) compared with tropical (7.2%) climates and no significant effect of grass cover crop species on SOC changes. That was the only review on cover crops to include a relatively large number of observations including both topsoil and subsoil >30 cm (20% of the database). The authors stressed that, although SOC changes are generally reported to be negligible in deeper layers, this may be a consequence of limited subsoil observations.

Published by Burleigh Dodds Science Publishing Limited, 2023.

3.3 Recycling of organic materials

The effect of applying organic materials (mainly manures) was a mean RR of 30% (23.5-43.4%, $N = 418$) and SCR of 0.41 Mg C ha^{-1} year^{-1} (0.20-1.3 Mg C ha^{-1} year^{-1}, $N = 217$), based on seven reviews reporting effect size indices using a mineral-fertilized treatment as the reference (Maillard and Angers, 2014; Ladha et al., 2011; Kopittke et al., 2017; Körschens et al., 2013; Smith et al., 1997; Aguilera et al., 2013; VandenBygaart et al., 2003). However, some reviews derived effect size indices using either a mineral-fertilized or an unfertilized treatment as the reference, with RR and SCR being much higher in the latter case, for example, RR of 46% instead of 33% (Körschens et al., 2013) and SCR of 0.52 Mg C ha^{-1} year^{-1} instead of 0.31 Mg C ha^{-1} year^{-1} (Maillard and Angers, 2014).

The effect of recycling organic materials to soil on SOC obviously depends on the quantity applied and also on the quality, which determines the relative proportions transformed into more stable SOC components. The organic materials applied in the studies compiled by the reviews presented in Bolinder et al. (2020) were broadly defined as manures. In addition to varying amounts applied, some of these manures may have been more or less decomposed, fermented, or composted. This lack of exact information contributed to the variation in the RR and SCR values. The review by Maillard and Angers (2014) was the only one to assess the influence of soil texture on RR and SCR after manure addition and found that it was not significant. The review by Ladha et al. (2011) concluded that RR is not climate-dependent, while Maillard and Angers (2014) reported a trend of higher SCR only for temperate compared with tropical climates.

In another review, Han et al. (2016) conducted a global meta-analysis with 652 paired observations comparing manure and chemical fertilizer treatments against an unfertilized reference treatment and found a mean RR of 36.2%. In a more recent worldwide meta-analysis by Gross and Glaser (2021) using 529 paired comparisons, the mean RR was 35.4%, with a variation in RR similar to that reported in the synthesis by Bolinder et al. (2020). Gross and Glaser (2021) was attributing this variation to factors such as management, site properties, and manure characteristics. Their study considered a wide variety of manure types and found that, for example, pig, cattle, and farmyard manures had the highest RR values (50%, 32%, and 41%, respectively). They also found that more clayey soils showed higher SCR than sandy soils, but that sandy soils had higher RR than clayey soils, partly because of lower initial SOC contents. Similar to Bolinder et al. (2020), Gross and Glaser (2021) stressed the importance of considering changes in both RR and SCR to avoid misleading conclusions. They also found a trend for lower SCR in tropical soils compared with soils in cool and humid climates although, due to small numbers of observations for tropical soils, the trend was not significant.

In comparison with the reviews summarized by Bolinder et al. (2020), the database analyzed by Gross and Glaser (2021) included a relatively large proportion of data for sampling depth >30 cm (N = 103). The results showed that recycling organic materials to soil affected both RR (24%) and SCR (4.6 Mg C ha^{-1}) in the subsoil.

Although some reviews presented data for liquid manures, it proved challenging to make realistic comparisons between solid and liquid manures because of a lack of data (Maillard and Angers, 2014), and the effects were often not significant (e.g. Aguilera et al., 2013). Manure management systems have changed drastically over time toward liquid forms, for example, in Sweden more than 90% of all dairy and pig manures are now handled in liquid form with a low dry matter content (about 5%). Other changes in the treatment and use of manures are also occurring, for example, in Sweden about 4% of manures were used as substrate in biogas plants in 2015 (Anonymous, 2017). The result is a byproduct commonly named biofertilizer, which is returned to agricultural land almost entirely in liquid form (i.e. dry matter content typically only 1-5%). There is a lack of research on the effect of this biofertilizer on changes in SOC.

Application of recycled organic materials such as sewage sludge and municipal solid waste can induce very strong effects on SOC, as shown by two reviews with RR ranging from 98% to 117% and SCR from 1.65 to 5.29 Mg C ha^{-1} year^{-1}, respectively (Smith et al., 1997; Aguilera et al., 2013). However, sewage sludge application is subject to statutory restrictions and, as shown by Kirchmann et al. (2017), this limits the SOC sequestration rate to around 0.08 Mg C ha^{-1} year^{-1} for sewage sludge in Sweden.

3.4 Nitrogen fertilization

The effect of nitrogen fertilization was a mean RR of 6.2% (3.5-10.0%, N = 846) and SCS of 0.23 Mg C ha^{-1} year^{-1} (0.20-0.48 Mg C ha^{-1} year^{-1}, N = 183). The six reviews assessing this used an unfertilized or control plot as the reference (Ladha et al., 2011; Alvarez, 2005; Lu et al., 2011; Körschens et al., 2013; Aguilera et al., 2013; VandenBygaart et al., 2003). With the exception of one study with a high SCR value (Aguilera et al., 2013), similar to that of cover crops, the nitrogen fertilization effect was fairly consistent between studies. Two global reviews on RR (Ladha et al., 2011) and SCR (Alvarez, 2005) showed the closest agreement with the mean calculated from all reviews. The review by Ladha et al. (2011) found that RR was highest for tropical conditions and lowest for temperate climates, while that by Alvarez (2005) found that SCR was higher under humid and temperate conditions than under dry conditions or in tropical climates. Only Alvarez (2005) assessed the effect of soil texture and found a greater effect of nitrogen fertilization on SCR for sandy soils than for fine-textured soils.

In contrast to the control used for other management practices, the reference treatment with no application of nitrogen fertilizer is not common agronomic practice. In situations where no mineral nitrogen is applied, farmers usually compensate for this using nitrogen from other sources, such as manures. However, nitrogen may sometimes not be applied in regions where there is a lack of alternative sources. Further, the nutrient supply affects SOC dynamics in a complex manner through its simultaneous effects on NPP and on the rate of heterotrophic respiration through nitrogen mining and mineralization of organic matter (Poeplau et al., 2016). Nonetheless, it is generally recognized for agroecosystems that the effect of nitrogen fertilization on SOC is positive because of increasing NPP (and yields), resulting in higher carbon inputs to the soil from aboveground and belowground post-harvest crop residues (Christopher and Lal, 2007; Kätterer et al., 2012). The review by Alvarez (2005) demonstrated this by establishing a relationship indicating that SOC storage increased by 2 kg C ha^{-1} for each additional 1 kg N ha^{-1} applied. On analyzing LTEs under Nordic conditions, Kätterer et al. (2012) obtained similar results, with SOC in the topsoil (0–20 cm) increasing by 1-2 kg C ha^{-1} $year^{-1}$ for each extra kg of nitrogen applied. According to VandenByggart et al. (2003), using an unfertilized treatment as the reference may overestimate the effect of nitrogen fertilization, since both SOC and yield responses are lower at higher nitrogen application rates. However, even when zero-nitrogen treatments are excluded, by considering only the range of values receiving nitrogen, such positive relationships between nitrogen fertilization and SOC are still present.

4 Crop rotations

Crop rotations that include perennial forage crops, often termed leys, have components of cropland and temporary grassland. Although leys occupy a significant proportion of arable land at high latitudes, for example, almost 50% in Sweden, their impact on SOC has not been as thoroughly documented as that of the four management practices discussed in the previous section. Since SOC stocks in grassland are generally, but not always, higher than those in cropland (Poeplau et al., 2011; Mukumbuta and Hatano, 2019), rotations dominated by leys can be expected to approach SOC stocks intermediate to those in cropland and grassland (Crème et al., 2020). Kätterer et al. (2013) compiled data for 15 sites in Canada, Estonia, Norway, Sweden, and the UK, taken from eight publications (Uhlen, 1991; Viiralt, 1998; Yang and Kay, 2001; VandenBygaart et al., 2003; Quenum et al., 2004; Reintam, 2007; Johnston et al., 2009; Bolinder et al., 2010) and four unpublished datasets. They calculated mean annual SOC stock changes in pairs of rotations for sites including leys with those consisting of only annual crops. The duration of the LTEs in those cases varied between 10 and 58 years and the soil depth to which SOC was measured varied between

20 cm and 70 cm. The studies differed in ley species composition and frequency of leys in the rotation. According to the analysis, on average 0.52 Mg C ha^{-1} year^{-1} (range 0.3–1.1 Mg C ha^{-1} year^{-1}) more carbon was retained in soils in ley-arable systems (including pure grass and legume leys as well as grass/legume mixtures) than in exclusively annual cropping systems (Kätterer et al., 2013).

Recently, a comprehensive review on the effect of selected crop rotations on SOC was conducted following the strict protocol for systematic reviews (Land et al., 2017). The outcomes from this work, including a meta-analysis, were recently published in Swedish (Land et al., 2021), but with supplementary material including the database in English, which can be freely accessed. Among the different crop rotations investigated, the effect of perennial forage crops was strongest and highly significant. It was found that SOC concentrations were on average 3.49 mg g^{-1} higher in rotations with perennial forage crops than in rotations with different annual crops. The difference between the final SOC concentration in 39 individual treatment pairs divided by the duration of the LTE (on average 24.5 years) was 0.16 mg g^{-1} year^{-1} soil (Land et al., 2017). Assuming a bulk density of 1.3 Mg m^{-3}, data on which were only provided in a few studies, this equates to an SCR of 0.51 Mg C ha^{-1} year^{-1} to 25 cm depth.

The study by Land et al. (2021) also analyzed the effect of various rotations and monocultures on SOC changes. Selected major findings from these meta-analyses are highlighted here. Effects of diverse rotations in general compared with repeated monocultures (203 pairs) were not significant. However, diverse rotations including legumes led to significantly higher SOC content than repeated monocultures without legumes (113 pairs, mean effect 0.351 mg g^{-1}), which roughly equates to stock changes of about 0.05 Mg C ha^{-1} year^{-1}, based on the same assumptions for bulk density presented above. Subsets from this dataset showed that the effect of legumes in diverse rotations compared with repeated monoculture without mineral fertilization was highly positive (18 pairs, 1.425 mg g^{-1}, corresponding to about 0.2 Mg C ha^{-1} year^{-1}). Due to the higher production of legumes under nitrogen-limited conditions, resulting in higher carbon inputs to soil, this response was expected. However, this comparison only applied to practical situations where no mineral fertilizer was applied when growing non-legume crops. Compared with repeated monocultures receiving fertilizer, the inclusion of legumes in diverse rotations still had a slightly positive mean effect on SOC, but this effect was not significantly different from zero (102 pairs). Comparing diverse rotations with and without legumes, the inclusion of legumes had a significantly positive effect on SOC (22 pairs, 0.611 mg g^{-1}, corresponding to about 0.09 Mg C ha^{-1} year^{-1}). In general, the inclusion of perennial forage crops in rotations was far more effective in increasing SOC than the inclusion of legumes (Land et al., 2017).

The positive impact of leys in arable rotations on SOC described above was seen in studies with different proportions of leys and other crops in rotations. Thus, the pure impact of leys compared with annual crops was probably diluted, since rotations entirely dominated by leys would probably have a stronger effect. This is supported by results from three LTEs in northern Sweden, where leys are grown with different frequencies in arable rotations (5, 3, 2, and 1 years in 6-year rotations A, B, C, and D, respectively) (Bolinder et al., 2010). Using annual SOC stock estimates according to Bolinder et al. (2010), which were based on equivalent mass principles, the calculated mean annual SOC mass increase compared with rotation D was 0.55 Mg C ha^{-1} year^{-1} for rotation A and 0.30 Mg C ha^{-1} year^{-1} for rotations B and C across the three sites. However, these SOC changes did not represent a pure ley effect, since crops and amendments differed between the four rotations. Nevertheless, ley frequency was probably the main driver according to a modeling study for one of the sites (Bolinder et al., 2012) and the estimated changes were within the range reported in the reviews cited earlier.

5 Soil tillage

Reducing the intensity of soil disturbance through tillage, particularly no-till practices, is desirable in many respects, for example, for saving labor and fuel and for preserving soil moisture and fauna in soil. Early assessments of no-till in the USA (Paustian et al., 1997) led to optimistic extrapolations of potential carbon stock increases at regional and global scale (Kern and Johnson, 1993; Freibauer et al., 2004). During recent years, several meta-analyses have revealed both positive effects (West and Post, 2002; Angers and Eriksen-Hamel, 2008; Gonzalez-Sanchez et al., 2012; Virto et al., 2012; Aguilera et al., 2013) and null effects on SOC (Dimassi et al., 2014; Powlson et al., 2014). Global meta-analyses have shown increases in SOC concentrations in the upper topsoil layers and decreases in the layers below, compared with conventional tillage (Luo et al., 2010b; Ogle et al., 2019). Only a few studies have included subsoil layers (i.e., below 30 cm), and these found that the positive effect of no-till was highly variable (Govaerts et al., 2009), reduced (Angers and Eriksen-Hamel, 2008), or even absent for some regions (VandenByggart et al., 2003; Luo et al., 2010b). Due to the high variability of its effects on SOC sequestration, the value of reduced tillage practices for climate change mitigation is now being questioned (Powlson et al., 2014; VandenBygaart, 2016).

The most comprehensive recent systematic review on this subject, including 351 LTEs, reported rather moderate effects (4.61 Mg C ha^{-1} on average) of no-till over conventional tillage on SOC stocks in the topsoil (0–30 cm), while no significant tillage effect was detected in the full soil profile (Haddaway et al., 2017). In the majority of studies, effect size indices were calculated considering a fixed soil depth but, as bulk density is often affected by tillage, this may introduce considerable bias (Ellert et al., 2002). Meurer et al. (2018) selected a subset of 101 LTEs from the database provided by Haddaway et al. (2017), which reported both SOC and bulk densities for calculating SOC stocks with the equivalent soil mass approach (Ellert and Bettany, 1995). According to their analysis, the carbon gain in no-till over conventional tillage decreased with depth and even became non-significant in studies where SOC stocks could be calculated to an equivalent depth of 60 cm, that is, 1.15 Mg ha^{-1}, which divided by the mean duration of the LTEs gave 0.06 Mg ha^{-1} year^{-1} (Meurer et al., 2018). The effect size of no-till compared with conventional tillage to an equivalent soil depth of 30 cm (0.2 Mg C ha^{-1}) was lower in their study than that presented by Haddaway et al. (2017) (4.2 Mg C ha^{-1} year^{-1}), which may partly be due to slightly higher bulk densities in no-till treatments. Thus, the sequestration potential of no-till seems indeed to be over-estimated when neglecting deeper soil depths. However, only 11 LTEs permitted calculation of SOC stocks to 60 cm depth, so there is a need for additional assessments in ongoing long-term tillage studies including deeper depth layers to confirm these findings.

Moving to the other extreme of tillage operations, deep plowing to 1 m depth to improve soil structure and/or break up hard pans in podzols has been shown to increase SOC stocks significantly at some sites (Alcántara et al., 2016; Schneider and Don, 2019). However, this practice is irreversible and can affect soil ecosystem services in several ways, an issue that needs to be further assessed (Schiedung et al., 2019). Studies from New Zealand show that full inversion tillage of pastures to 30 cm depth resulted in increased SOC stocks and that the additional costs associated with this can be offset by the increase in productivity already in the first year after pasture renewal (Beare et al., 2020; Lawrence-Smith et al., 2021). The generality of these results also needs further consideration.

6 Subsoil

The amount of SOC stored in subsoil (>30 to 100 cm depth) is similar to that in the topsoil arable layer at 0–30 cm depth (Morari et al., 2019). However, LTEs examining common management practices for SCS often only consider the arable layer, where most SOC changes are assumed to occur because carbon cycling is more dynamic in topsoil than in deeper soil layers (Bolinder et al., 2020). In fact, the mean carbon age of SOC decreases sharply with depth (Balesdent et al., 2018), but this does not imply that subsoil SOC is insensitive to agricultural management practices. There is evidence that common management practices affect SOC stocks in the upper part of the subsoil or in deeper layers at decadal time scales (e.g., Börjesson et al., 2018; Dal Ferro et al., 2020; Kätterer et al., 2014; Kirchmann et al., 2013; Menichetti et al., 2015). However, this is not the case at all sites (Jarvis et al., 2017; Börjesson et al., 2018). For instance, Börjesson et al. (2018) presented SCRs calculated according to equivalent mass principles to

50 cm depth for two Swedish LTEs, on loam and sand, with an identical experimental set-up comparing a cereal monoculture with a ley-dominated rotation. During the 35 years of the trials, SCR was positive in the fully fertilized ley-dominated rotation at both sites (0.47 and 0.17 Mg ha^{-1} year^{-1} in the loam and clay, respectively) and negative in cereal monoculture (−0.38 and −0.15 M ha^{-1} year^{-1}, respectively). Around 27% of the crop rotation effect occurred below 20 cm depth in the loam, but no effect at all on subsoil C was detected in the clay soil. This illustrates that site-specific conditions such as soil texture play an important role in both carbon inputs and retention in soil profiles (Poeplau and Kätterer, 2017).

The rationale for including deeper soil layers in any analysis of SOC stock changes is obvious when considering tillage practices. Including deeper soil layers then significantly influences the overall conclusions drawn on total SOC stock changes, since tillage affects the depth distribution of SOC (Angers and Eriksen-Hamel, 2008; Luo et al., 2010b; Meurer et al., 2018). Including subsoil is also important when evaluating and comparing annual and perennial crops with more or less well-developed root systems or other deep-rooting species (Carter and Gregorich, 2010; Collins et al., 2010; VandenBygaart et al., 2011; Guan et al., 2016). It has been shown that major land-use changes, such as cropland to grassland or pasture or cropland to forest, and vice versa, may in some cases also induce changes in subsoil carbon (Guo and Gifford, 2002; Poeplau and Don, 2013). A recent analysis on the conversion of former cropland on environmentally sensitive soils in the USA into grassland within the Conservation Reserve Program showed that gains in topsoil carbon may have been negated by net losses at depth (Yang et al., 2021). There is thus strong reason to include SOC changes in the subsoil in carbon accounting systems (Rumpel and Kögel-Knabner, 2011). However, the responses are site-specific and the mechanisms governing the effects on subsoil SOC have been insufficiently studied, which has been identified as a major knowledge gap (Lorenz and Lal, 2005; Chenu et al., 2019).

Since carbon changes are small relative to total SOC stocks, LTEs are indispensable when quantifying the net effect of different management practices on subsoil carbon. To our knowledge, reviews focusing on agricultural management impacts on subsoil SCS are currently lacking. Future efforts are urgently needed in data compilation and analysis of existing data, and further research is needed on the mechanisms controlling the movement of organic carbon into subsoils and its stabilization (Balesdent et al., 2018; Kögel-Knabner and Amelung, 2021; Simo et al., 2019; Torres-Sallan et al., 2017).

7 Case study: soil carbon and fertility

Rebuilding SOC in agricultural soils is frequently mentioned as a key component of sustainable intensification, due to its positive feedback on soil fertility and food security (e.g. Lal, 2004, 2013; Foley et al., 2011; Henryson et al., 2018). Although many studies show synergies between SOC and crop productivity (Lal, 2013; Oldfield et al., 2019), the responses seem not to be universal (Edmeades, 2003; Hijbeek et al., 2017) and may even be reversed in data analyses at regional or continental scale (Oelofse et al., 2015; Vonk et al., 2020). Confounding correlations, such as negative correlations between SOC and pH, further complicate this type of analysis (Kirchmann et al., 2020).

In the following, we present a case study of a Swedish LTE comparing two treatments, with and without the addition of cereal straw corresponding to 1.8 Mg C ha^{-1} year $^{-1}$ (Kätterer et al., 2011). The most frequent confounding factors, such as treatment differences in pH and crop nutrient supply, are most likely negligible in this comparison. Both treatments are fertilized annually with nitrogen, phosphorus, and potassium, and these elements are probably not limiting for crop growth. Differences in crop yield between the two treatments have increased over time since the start of the LTE in 1956. We regressed the significant linear slope of a regression model fitted to the time series of annual yield ratios in the straw-amended compared with control treatments on measured SOC concentration changes in the straw treatment compared with the control. The results clearly showed increasing yield with increasing SOC concentration, with a slope of 4.4% (Fig. 3).

This case study was performed on clay soil in Sweden, and extrapolation of the results is not straightforward. However, the results show that the synergy between SOC storage and yield potential can be considerable, at least in some regions. They also support findings in previous analyses of LTEs, for example, by Körschens et al. (2013), who reported a yield benefit of 6% in manured treatments compared with mineral fertilization alone. We interpreted the strong crop yield response to SOC increase in the Ultuna LTE as being mainly governed by soil structural changes related to SOC, through increasing the amount of plant-available water (Meurer et al., 2020) and creating a better environment for root growth, thereby increasing the uptake efficiency of water and nutrients. This complies with the interpretation of manure-induced yield increases in European LTEs (Zavatarro et al., 2017). The dynamic interactions between SOC and soil structure have recently been formalized in a modeling approach that represents the observed soil structural changes in this trial very well (Meurer et al., 2020). An important difference between the Ultuna experiment and most other LTEs, which may exaggerate the effect of SOC on soil structure to some extent, is that Ultuna is a small-plot trial (2 m × 2 m) that is managed entirely by hand, thus preventing soil compaction. In large plot experiments using full-scale machinery, the situation may differ, and under real-life field conditions, it is widely recognized by farmers that soil compaction results in yield reductions through reduced aeration and water infiltration, which increases the risk of waterlogging.

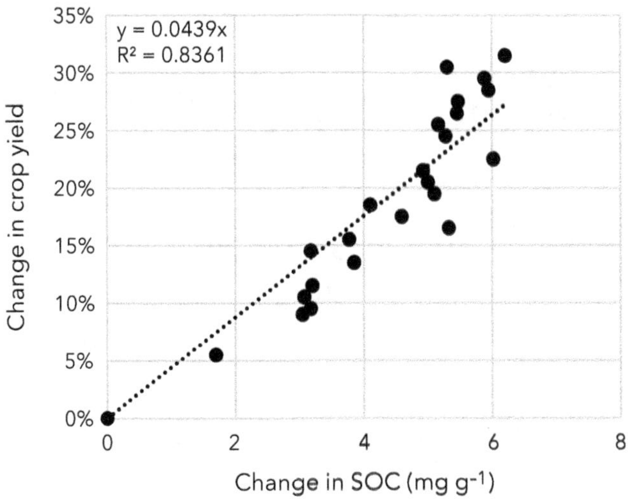

Figure 3 Crop yield response to increasing soil organic carbon (SOC) concentration in the topsoil during 63 years in the Ultuna long-term experiment (Kätterer et al., 2011). Yield changes correspond to linear regression (significant slope) of crop yield over time in a treatment amended with straw compared with treatment without straw. Changes in SOC between the two treatments were calculated from direct measurements recorded for several years and differences steadily increased over time. In 2019, 63 years after the start of the experiment, topsoil SOC concentration was 6.2 mg g^{-1} higher in the straw-amended treatment compared with the control. Both treatments received mineral fertilizer, so the yield response is probably mainly due to changes in soil physical properties.

8 Conclusion and future trends

Since annual changes in SOC are small compared with total SOC stocks, which vary greatly even at small scales, data from LTEs are essential when quantifying the impact of agricultural management on SOC stocks. The responses of different management practices on SCR synthesized in this review are summarized in Fig. 4. Studies reporting SCR were less common than those reporting RR because bulk density, which is laborious to measure, is not required to calculate RR based on SOC concentration measures. Among the practices considered, crop residue retention compared with removal was that most commonly investigated in LTEs, followed by manure addition, nitrogen fertilization, cover crops, and rotations. Changes in SOC stocks were highest when comparing perennial with annual crops and lowest, though still statistically significant when comparing diverse rotations including legumes with monocultures. LTEs investigating tillage effects were frequently described, but only a limited number of those reported bulk density values, which is a prerequisite for calculating SOC stock changes. Carbon stocks increased in no-till treatments compared with conventional

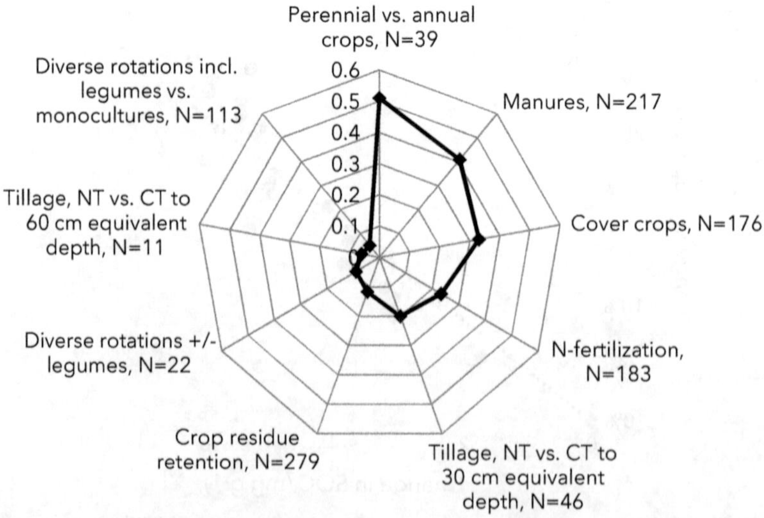

Figure 4 The overall average effect of different management practices on mean stock change rate (Mg C ha^{-1} year^{-1}), weighted by the number of paired comparisons (*N*), in the reviews presented in this chapter (Meurer et al., 2018; Bolinder et al., 2020; Land et al., 2021). NT, no-till; CT, conventional tillage.

tillage to 30 cm depth, but in studies reporting SOC to 60 cm depth, the SOC gain was much lower, and not significantly different from zero.

8.1 Carbon stock changes and long-term field experiments

The majority of the field-based estimates presented in Fig. 4 were found to be statistically significant in meta-analyses. Hence, they are valid for inclusion in large-scale analyses and planning. Most estimates of global SCS potential were directly derived from LTEs or indirectly from models based on LTEs. Nevertheless, extrapolation from LTEs to larger scales is not straightforward. For example, LTEs are more abundant in temperate than in tropical climate zones, and publications from certain regions, for example, Russia, are under-represented due to lack of cataloging in digital scientific databases or the use of native languages, making them difficult to find (Haddaway et al., 2015). There is a lack of LTEs in some regions, especially developing countries and the distribution of LTEs in reviews and meta-analyses is frequently also biased toward regions where certain management practices are common, as illustrated by the over-representation of tillage studies from North America (Haddaway et al., 2015, 2017).

Potential biases, especially in long time series, should also be considered, since management practices in LTEs, such as cultivars, fertilization, crop protection, or liming, do not always match common agricultural practices. Consequently, the yields are not necessarily optimized and the relative difference in yields, which is the main driver of annual crop residue carbon inputs to soil, may be lower in LTEs than on real farms. This could also cause some difficulties in the interpretation of effect size indices when combining data from shorter- and medium-term LTEs with data from much longer-term experiments. The RR and SCR values are relative assessments, most commonly using only the most recent measurements of SOC in LTEs, but it is possible that SOC content may have declined over time in both the management practice treatment and the control treatment, depending on the initial conditions at the start of the experiment (Bolinder et al., 2020). Using the time series of SOC measurements in LTEs could improve the estimates of RR and SCR (Haddaway et al., 2016).

Differences in tillage depth over time or in drainage conditions may also be prevalent and may create biases (Warren et al., 2020). Although the exact time when the changes take place is difficult to establish, some studies indicate that plowing depth may have increased by as much as 7.5–10 cm between around 1950–1960 and 1970–1990 (e.g. Van Meirvenne et al., 1996; Riley and Bakkegard, 2006). This can create a distortion in the time series of data used for SOC measurement through dilution of SOC by the inclusion of poor subsoil material into the topsoil. Hence, conclusions on absolute trends in SOC may be biased when comparing management practices from different databases, such as observations from LTEs with large plot sizes using modern full-scale machinery or soil monitoring programs. This dilution can also affect the calibration of SOC models and needs to be accounted for when modeling SOC stocks, as shown by, for example, van Wesemael et al. (2005). The movement of soil from one plot to another through tillage operations is another problem in small-plot experiments (Sibbesen, 1986). Furthermore, some LTEs include extreme treatments that are not scalable due to legal restrictions, as we pointed out with the example of application rates of sewage sludge in Sweden (Kirchmann et al., 2017). Another difficulty relating to the effects of recycled organic materials on SOC changes arises in LTEs with manures, since the amounts applied in these experiments do not necessarily reflect current agronomic practices and the SCR estimates are probably too optimistic in some cases. In fact, during recent decades there has been a lowering of application rates because of the introduction of nutrient-related regulations in many countries.

8.2 Other management practices

Several important agricultural management practices are under-represented in LTEs. Kirchmann et al. (2020) found that only 9 out of 735 LTEs that they reviewed investigated liming interventions according to the systematic map presented by Haddaway et al. (2015), despite soil acidity being a major yield-limiting factor in large agricultural areas. Lime application has been shown to decrease nitrous oxide emissions from soils (Scheer et al., 2008; Wang et al., 2018) and increase methane oxidation (Kunhikrishnan et al., 2016), but its effect on the carbon balance is understudied. It is well-known that soil acidity affects the composition and diversity of microbial communities, with associated feedbacks on carbon and nitrogen cycling in soil (Bahram et al., 2018; Malik et al., 2018), including stabilization mechanisms mediated, for example, by changes in bacteria/fungi ratio (Rousk et al., 2009). In their review, Paradelo et al. (2015) identified this topic as a knowledge gap and encouraged the soil science community to synthesize unpublished data from existing LTEs.

Field application of biochar is a measure that can increase SOC to the same order of magnitude as SCS (Paustian et al., 2016; Minx et al., 2018). The thermal decomposition of biomass under low oxygen conditions during pyrolysis makes biochar highly resistant to heterotrophic organisms in soil (Lehmann et al., 2006). As most of the biochar is not transformed to soil organic matter, the SOC increase after its application does not comply with the strict definition of SCS according to Olson et al. (2014). Nevertheless, due to its positive effects on soil fertility, biochar application generally leads to higher crop yields, especially in sub-tropical and tropical climates, as shown in several meta-analyses (Liu et al., 2013; Jeffery et al., 2017; Ye et al., 2020). Yield responses appear to persist for at least 20 growing seasons, as shown for LTEs in Kenya (Kätterer et al., 2019). The total climate change mitigation benefit of biochar technologies depends on the origin of the biomass and how efficiently the energetic gases evolving during pyrolysis are used, including substation effects (Tisserant and Cherubini, 2019; Sundberg et al., 2020). LTEs quantifying carbon stock changes due to biochar application is still scarce. Furthermore, biochar-induced nitrogen mining through priming of soil organic matter is a potential risk that should be evaluated before scaling-up this technology under nutrient-poor conditions (Wardle et al., 2008; Wang et al., 2016).

Irrigation is another topic that is insufficiently studied in relation to SOC. Despite its significance for food production and use on more than 0.3 billion ha, Trost et al. (2013) found only 14 LTEs reporting SOC changes due to irrigation practices. The results from these studies were highly variable, with up to 500% increases in SOC in irrigated desert soils, moderate increases in semi-arid regions, and no consistent effects in humid regions. Consideration of potential tradeoffs relating to water scarcity elsewhere, groundwater depletion, and nitrous oxide emissions are likely to be particularly relevant for irrigation practices.

Planting more trees or bushes in agricultural landscapes could also boost aboveground carbon storage and SCS (Menichetti et al., 2020). However, few LTEs on agroforestry cover more than a decade (Cardinael et al., 2018; Corbeels et al., 2018). Agroforestry systems are also highly diverse and region-specific, and their contribution as a carbon sink is often not accounted for in national inventories (Rosenstock et al., 2019). As seen for agroforestry, the introduction of perennial energy crops on cropland will often lead to SCR, as shown in a meta-analysis for woody species based mainly on European trials (Don et al., 2012). In addition, energy grasses have the potential to sequester carbon, as shown by Poeplau and Don (2012). Rough estimates of SCR provided in these two reviews lie around 0.4 Mg C ha^{-1} $year^{-1}$, which is similar to that leys in crop rotation trials. Much higher SCR values, that is, 1.56 Mg C ha^{-1} $year^{-1}$ and 0.68 Mg C ha^{-1} $year^{-1}$ on average for herbaceous and woody perennials, respectively, were reported in the meta-analysis by Agostini et al. (2015). However, if energy production is restricted to abandoned and marginal land in order to avoid competition for land with food productions, these estimates for SCR are probably not scalable.

Avoiding soil compaction through controlled traffic, i.e. constraining field traffic to the smallest possible area of permanent traffic lines, may also have the potential to increase crop productivity and SOC stocks, and mitigate nitrous oxide emissions (Antille et al., 2015; Hernandez-Ramirez et al., 2021). However, this technology is rather new and LTEs quantifying its effects on SOC are still lacking. Amelioration of subsoil conditions to improve soil fertility is another topic that should be investigated further. This may involve soil drainage, a measure for increasing both productivity and decomposition, and tillage operations aimed at loosening compact subsoil layers. Besides deep plowing operations, which have been shown to result in SOC sequestration in the few studies available, loosening of upper subsoil layers combined with injection of organic materials may be another way to increase soil fertility and SOC stocks (Getahun et al., 2018).

Organic farming practices are promoted in many countries, especially in the European Union, based in part on the argument that it leads to SCS (Freibauer et al., 2004). Higher average SOC stocks in organic than in conventional systems may be interpreted as an advantage of organic systems at farm scale (Smith, 2004; Goh, 2011; Gattinger et al., 2012). However, comparing conventional stockless farms with low external carbon inputs with organic mixed farms with inputs from manures, or even off-farm carbon sources transferred from conventional systems, requires system boundaries to be considered (Goulding et al., 2009; Kirchmann et al, 2016). Higher SOC stocks in organic farming can be due to high and often disproportionate application of organic fertilizers (Leifeld and Fuhrer, 2010; Leifeld et al., 2013) and transfer of nutrients from conventional to organic systems (Nowak et al., 2013). Moreover, organic farming often involves a combination of several of the management practices covered in this chapter, such as cover crops and leaving crop residues in the field, based on the belief that there is a synergetic effect. This may indeed be the case but needs further investigation, especially as these practices are also used in conventional farming systems. More importantly, all of the above-mentioned considerations make it very difficult to perform appropriate and equitable assessments on the effect of organic farming systems, both at farm scale and in LTEs, because these systems involve too many confounding factors acting simultaneously (Kirchmann et al., 2016). The lower yields in organic compared with conventional systems implies a SOC stock decrease in situ and a requirement for more cropland to produce an equivalent amount of food (Kirchmann et al., 2007). This may induce indirect land-use change, including deforestation, with accompanying losses of SOC at larger spatial scales (Kirchmann et al., 2016; Searchinger et al., 2018).

8.3 Uncertainty and scaling-up in space and time

At the 21st Conference of the Parties meeting in Paris, the French government launched the 4 per 1000 initiative (www.4p1000.org) to promote SCS, based on the rationale that an increase of 0.4% in global SOC stocks could offset all anthropogenic carbon dioxide emissions. Although the intention of the initiative was to set this as a normative target (Rumpel et al., 2020), it has created lively discussions in the scientific literature about the feasibility of SCS in practical implementation (Minasny et al., 2017; VandenBygaart, 2018; Poulton et al., 2018; Amelung et al., 2020). Among other issues, the sources of nitrogen and other nutrients required for building soil organic matter are a crucial climate-related issue as long as fertilizers are produced using fossil energy (van Groenigen et al., 2017). The carbon inputs required to reach the target would have to increase by 43%, or 0.66 Mg ha^{-1} $year^{-1}$, according to a recent study (Bruni et al., 2021). No single measure among those presented in Fig. 4 would be sufficient to reach this target.

Despite the uncertainties mentioned above and the incomplete understanding of how SCS is influenced by management and pedo-climatic factors (Chenu et al., 2019; Smith et al., 2020), the evidence from LTEs on several major management practices affecting SOC stocks is quite solid and can be used to design more climate-smart production systems and applied in bottom-up approaches to estimate biophysical SCS potential at larger scales. Several of these practices, like cover crops, more perennial crops, and surface mulch due to reduced tillage, will also directly lead to global cooling through associated increases in surface albedo (Lugato et al., 2020). However, some practices may lead to increased nitrous oxide emissions, which may lower the mitigation potential (Guenet et al., 2021). Moreover, when extrapolating the effects obtained from field measurements in one region to another with different soils, climate conditions, and agricultural practices, the estimates may become highly uncertain or even false. Proper boundary conditions must be set and tradeoffs with food production and other ecosystem services, leakage through indirect land-use change, non-CO_2 GHG, and the United Nations

Sustainable Development Goals must be considered when upscaling SCS measures in space.

It must be emphasized that not all measures are scalable in space. For example, manures that increase SOC stocks at field or farm scale (Fig. 4) will not contribute to SCS at larger scales as long as the number of animals or manure handling/treatment methods in a given region does not change. The same applies to other limited organic resources, such as sewage sludge, composts, or other wastes from the food chain. To optimize the use of biomass, alternative pathways for any organic resource have to be evaluated. Although crop residue retention in the field will most likely improve soil health and lead to SOC gains, it may not always be the best option. Instead, part of the residues could be pyrolyzed and the evolving gases could be upgraded to fuels, replacing fuels of fossil origin, while the biochar produced could first be used for water purification or as a feed additive to cows to reduce methane emissions, and then applied to soil, where it will improve soil fertility and lead to negative carbon emissions. This is just one example of a win-win-win strategy that can result in more SOC with several other co-benefits. We view SCS as one component in the sustainable development of emerging bioeconomies. Transdisciplinary multi-actor approaches will be necessary to identify and develop region-specific pathways for more sustainable societies encompassing all dimensions of sustainable development.

9 Where to look for further information

Many of the reviews cited in this chapter provide excellent overviews of the research area.

Dr. Rattan Lal, Professor of Soil Science and Director of the Carbon Management and Sequestration Center at Ohio State University, has edited many textbooks that include articles from many leading scientists. He is easy to find on the internet. The following book is just an example: Lal, R. and Stewart, B.A. (2018), *Soil and Climate*. CRC Press, Boca Raton, doi:10.1201/b21225

Key research in this area can be found on the websites of the following organizations:

- United States Department of Agriculture (USDA; www.usda.gov).
- Joint European Research Program to develop knowledge, tools and an integrated research community to foster climate-smart sustainable agricultural soil (https://ejpsoil.eu/).
- The Global Research Alliance on Agricultural Greenhouse Gases https://globalresearchalliance.org/.
- All major soil science departments at universities and research institutes.

10 Acknowledgement

Financial support was provided by the Swedish Farmers' Foundation for Agricultural Research, grant number O-18-23-141.

11 References

Agostini, F., Gregory, A. S. and Richter, G. M. (2015). Carbon sequestration by perennial energy crops: is the jury still out? *Bioenergy Res.* 8, 1057-1080.

Aguilera, E., Lassaletta, L., Gattinger, A. and Gimeno, B. S. (2013). Managing soil carbon for climate change mitigation and adaptation in Mediterranean cropping systems: a meta-analysis, *Agric. Ecosyst. Environ.* 168, 25-36.

Alcántara, V., Don, A., Well, R. and Nieder, R. (2016). Deep ploughing increases agricultural soil organic matter stocks, *Glob. Change Biol.* 22(8), 2939-2956.

Alvarez, R. (2005). A review of nitrogen fertilizer and conservation tillage effects on soil organic carbon storage, *Soil Use Manag.* 21(1), 38-52.

Amelung, W., Bossio, D., de Vries, W., Kögel-Knabner, I., Lehmann, J., Amundson, R., Bol, R., Collins, C., Lal, R., Leifeld, J., Minasny, B., Pan, G., Paustian, K., Rumpel, C., Sanderman, J., van Groenigen, J. W., Mooney, S., van Wesemael, B., Wander, M. and Chabbi, A. (2020). Towards a global-scale soil climate mitigation strategy, *Nat. Commun.* 11(1), 5427.

Anderson-Texeira, K. J., Davis, S. C., Masters, M. D. and Delucia, E. H. (2009). Changes in soil organic carbon under biofuel crops, *GCB Bioenergy* 1(1), 75-96.

Andrén, O. and Kätterer, T. (1997). ICBM - the Introductory Carbon Balance Model for exploration of soil carbon balances, *Ecol. Appl.* 7(4), 1226-1236.

Angers, D. A. and Eriksen-Hamel, N. S. (2008). Full-inversion tillage and organic carbon distribution in soil profiles: a meta-analysis, *Soil Sci. Soc. Am. J.* 72(5), 1370-1374.

Anonymous (2017). Statistiska meddelanden MI 30 SM 1702, Gödselmedel i jordbruket 2015/16, Mineral- och stallgödsel till olika grödor samt hantering och lagring av stallgödsel [Use of fertilisers and animal manure in agriculture in 2015/16], Statistiska centralbyrån, Stockholm, Sweden.

Antille, D. L., Chamen, W. C. T., Tullberg, J. N. and Lal, R. (2015). The potential of controlled traffic farming to mitigate greenhouse gas emissions and enhance carbon sequestration in arable land, *Trans. ASABE* 58, 707-731.

Bahram, M., Hildebrand, F., Forslund, S. K., Anderson, J. L., Soudzilovskaia, N. A., Bodegom, P. M., Bengtsson-Palme, J., Anslan, S., Coelho, L. P., Harend, H., Huerta-Cepas, J., Medema, M. H., Maltz, M. R., Mundra, S., Olsson, P. A., Pent, M., Põlme, S., Sunagawa, S., Ryberg, M., Tedersoo, L. and Bork, P. (2018). Structure and function of the global topsoil microbiome, *Nature* 560(7717), 233-237.

Balesdent, J., Basile-Doelsch, I., Chadoeuf, J., Cornu, S., Derrien, D., Fekiacova, Z. and Hatté, C. (2018). Atmosphere-soil carbon transfer as a function of soil depth, *Nature* 559(7715), 599-602.

Basile-Doelsch, I., Balesdent, J. and Pellerin, S. (2020). Reviews and syntheses: the mechanisms underlying carbon storage in soil, *Biogeosciences* 17(21), 5223-5242.

Beare, M. H., McNally, S. R., Calvelo Pereira, R., Tregurtha, C. S., Gillespie, R. N., Van der Klei, G. R. and Hedley, M. J. (2020). The agronomic and Environmental benefits and risks of autumn pasture renewal with full inversion tillage. In: Christensen, C. L., Horne, D. J.

and Singh, R. (Eds) *Nutrient Management in Farmed Landscapes. Occasional Report No. 33*. Farmed Landscapes Research Centre, Massey University, Palmerston North, New Zealand. 6 pages. Available at: http://flrc.massey.ac.nz/publications.html.

Bednar, J., Obersteiner, M. and Wagner, F. (2019). On the financial viability of negative emissions, *Nat. Commun.* 10(1), 1783.

Blanco-Canqui, H. (2013). Crop residue removal for bioenergy reduces soil carbon pools: how can we offset carbon losses?, *Bioenerg. Res.* 6(1), 358–371.

Bolinder, M. A., Angers, D. A., Giroux, M. and Laverdière, M. R. (1999). Estimating C inputs retained as soil organic matter from corn (*Zea mays* L.), *Plant Soil* 215(1), 85–91.

Bolinder, M. A., Kätterer, T., Andrén, O., Ericson, L., Parent, L.-E. and Kirchmann, H. (2010). Long-term soil organic carbon and nitrogen dynamics in forage-based crop rotations in Northern Sweden (63–64 °N), *Agric. Ecosyst. Environ.* 138(3–4), 335–342.

Bolinder, M. A., Kätterer, T., Andrén, O. and Parent, L.-E. (2012). Estimating carbon inputs to soil in forage-based crop rotations and modeling the effects on soil carbon dynamics in a Swedish long-term field experiment, *Can. J. Soil Sci.* 92(6), 821–833.

Bolinder, M. A., Crotty, F., Elsen, A., Frac, M., Kismanyoky, T., Lipiec, J., Tits, M., Toth, Z. and Kätterer, T., (2020). The effect of crop residues, cover crops, manures and nitrogen fertilization on soil organic carbon changes: a synthesis of reviews, *Mitig Adapt. Strat. Gl* 25, 929–952.

Börjesson, G., Bolinder, M. A., Kirchmann, H. and Kätterer, T. (2018). Organic carbon stocks in topsoil and subsoil in long-term ley and cereal monoculture rotations, *Biol. Fertil. Soils* 54(4), 549–558.

Bossio, D. A., Cook-Patton, S. C., Ellis, P. W., Fargione, J., Sanderman, J., Smith, P., Wood, S., Zomer, R. J., von Unger, M., Emmer, I. M. and Griscom, B. W. (2020). The role of soil carbon in natural climate solutions, *Nat. Sustain.* 3(5), 391–398.

Bruni, E., Guenet, B., Huang, Y., Clivot, H., Virto, I., Farina, R., Kätterer, T., Ciais, P., Martin, M., and Chenu, C. (2021). Additional carbon inputs to reach a 4 per 1000 objective in Europe: feasibility and projected impacts of climate change based on Century simulations of long-term arable experiments, *Biogeosciences* 18, 3981–4004. https://doi.org/10.5194/bg-18-3981-2021.

Cardinael, R., Umulisa, V., Toudert, A., Olivier, A., Bockel, L. and Bernoux, M. (2018). Revisiting IPCC Tier 1 coefficients for soil organic and biomass carbon storage in agroforestry systems, *Environ. Res. Lett.* 13(12), 124020.

Carter, M. R. and Gregorich, E. G. (2010). Carbon and nitrogen storage by deep-rooted tall fescue (*Lolium arundinaceum*) in the surface and subsurface soil of a fine sandy loam in eastern Canada, *Agric. Ecosyst. Environ.* 136(1–2), 125–132.

Chenu, C., Angers, D. A., Barré, P., Derrien, D., Arrouays, D. and Balesdent, J. (2019). Increasing organic stocks in agricultural soils: knowledge gaps and potential innovations, *Soil Till. Res.* 188, 41–52.

Christopher, S. F. and Lal, R. (2007). Nitrogen management affects carbon sequestration in North American cropland soils, *Crit. Rev. Plant Sci.* 26(1), 45–64.

Collins, H. P., Smith, J. L., Fransen, S., Alva, A. K., Kruger, C. E. and Granatstein, D. M. (2010). Carbon sequestration under irrigated switchgrass (*Panicum virgatum* L.) production, *Soil Sci. Soc. Am. J.* 74(6), 2049–2058.

Corbeels, M., Cardinael, R., Naudin, K., Guibert, H. and Torquebiau, E. (2018). The 4 per 1000 goal and soil carbon storage under agroforestry and conservation agriculture systems in sub-Saharan Africa, *Soil Till. Res.* 188, 16–26.

Crème, A., Rumpel, C., Malone, S. L., Saby, N. P. A., Vaudour, E., Decau, M. and Chabbi, A. (2020). Monitoring grassland management effects on soil organic carbon–a matter of scale, *Agronomy* 10(12).

Dal Ferro, N., Piccoli, I., Berti, A., Polese, R. and Morari, F. (2020). Organic carbon storage potential in deep agricultural soil layers: evidence from long-term experiments in northeast Italy, *Agr. Ecosyst. Environ.* 300, 106967.

Daryanto, S., Fu, B., Wang, L., Jacinthe, P. and Zhao, W. (2018). Quantitative synthesis on the ecosystem services of cover crops, *Earth Sci. Rev.* 185, 357-373.

Dechow, R., Franko, U., Kätterer, T. and Kolbe, H. (2019). Evaluation of the RothC model as a prognostic tool for the prediction of SOC trends in response to management practices on arable land, *Geoderma* 337, 463-478.

Dimassi, B., Mary, B., Wylleman, R., Labreuche, J., Couture, D., Piraux, F. and Cohan, J. (2014). Long-term effect of contrasted tillage and crop management on soil carbon dynamics during 41 years, *Agr. Ecosyst. Environ.* 188, 134-146.

Don, A., Osborne, B., Hastings, A., Skiba, U., Carter, M. S., Drewer, J., Flessa, H., Freibauer, A., Hyvönen, N., Jones, M. B., Lanigan, G. J., Mander, Ü, Monti, A., Djomo, S. N., Valentine, J., Walter, K., Zegada-Lizarazu, W. and Zenone, T. (2012). Land-use change to bioenergy production in Europe: implications for the greenhouse gas balance and soil carbon, *GCB Bioenergy* 4(4), 372-391.

Doetterl, S., Berhe, A. A., Nadeu, E., Wang, Z., Sommer, M. and Fiener, P. (2016). Erosion, deposition and soil carbon: a review of process-level controls, experimental tools and models to address C cycling in dynamic landscapes, *Earth Sci. Rev.* 154, 102-122.

Eckersten, H., Bolinder, M., Johnsson, H., Kätterer, T., Mårtensson, K., Collentine, D., Tidåker, P. and Marstorp, H. (2017). Nitrogen leaching and soil organic carbon sequestration of a barley crop with improved N use efficiency - A regional case study, *Acta. Agr. Scand. B. - S. P.* 67(7), 615-627.

Edmeades, D. C. (2003). The long-term effects of manures and fertilisers on soil productivity and quality: a review, *Nutr. Cycl. Agroecosys* 66(2), 165-180.

Ellert, B. H. and Bettany, J. R. (1995). Calculation of organic matter and nutrients stored in soils under contrasting management regimes, *Can. J. Soil Sci.* 75(4), 529-538.

Ellert, B. H., Janzen, H. H. and Entz, T. (2002). Assessment of a method to measure temporal change in soil carbon storage, *Soil Sci. Soc. Am. J.* 66(5), 1687-1695.

Favretto, M. R., Paoletti, M. G., Caporali, F., Nannipieri, P., Onnis, A. and Tomei, P. E. (1992). Invertebrates and nutrients in a Mediterranean vineyard mulched with subterranean clover (*Trifolium subterraneum* L.), *Biol. Fertil. Soils* 14(3), 151-158.

Foley, J. A., Ramankutty, N., Brauman, K. A., Cassidy, E. S., Gerber, J. S., Johnston, M., Mueller, N. D., O'Connell, C., Ray, D. K., West, P. C., Balzer, C., Bennett, E. M., Carpenter, S. R., Hill, J., Monfreda, C., Polasky, S., Rockström, J., Sheehan, J., Siebert, S., Tilman, D. and Zaks, D. P. (2011). Solutions for a cultivated planet, *Nature* 478(7369), 337-342.

Freibauer, A., Rounsevell, M. D. A., Smith, P. and Verhagen, J. (2004). Carbon sequestration in the agricultural soils of Europe, *Geoderma* 122(1), 1-23.

Fuss, S., Lamb, W. F., Callaghan, M. W., Hilaire, J., Creutzig, F., Amann, T., Beringer, T., de Oliveira Garcia, W., Hartmann, J., Khanna, T., Luderer, G., Nemet, G. F., Rogelj, J., Smith, P., Vicente, J. L. V., Wilcox, J., del Mar Zamora Dominguez, M. and Minx, J. C. (2018). Negative emissions–Part 2: costs, potentials and side effects, *Environ. Res. Lett.* 13(6), 063002.

Gattinger, A., Muller, A., Haeni, M., Skinner, C., Fliessbach, A., Buchmann, N., Mäder, P., Stolze, M., Smith, P., Scialabba, Nel-H. and Niggli, U. (2012). Enhanced top soil carbon stocks under organic farming, *Proc. Natl Acad. Sci. U. S. A.* 109(44), 18226-18231.

Getahun, G. T., Kätterer, T., Munkholm, L. J., Parvage, M. M., Keller, T., Rychel, K. and Kirchmann, H. (2018). Short-term effects of loosening and incorporation of straw slurry into the upper subsoil on soil physical properties and crop yield, *Soil Till. Res.* 184, 62-67.

Goh, K. M. (2011). Greater mitigation of climate change by organic than conventional agriculture: a review, *Biol. Agric. Hortic.* 27(2), 205–229.

Gonzalez-Sanchez, E. J., Ordonez-Fernandez, R., Carbonell-Bojollo, R., Veroz-González, O. and Gil-Ribes, J. A. (2012). Meta-analysis on atmospheric carbon capture in Spain through the use of conservation agriculture, *Soil Till. Res.* 122, 52–60.

Goulding, K. W. T., Trewavas, A. J. and Giller, K. (2009). Can organic farming feed the world? A contribution to the debate on the ability of organic farming systems to provide sustainable supplies of food, International Fertilizer Society, Proceedings 663, 1-27.

Govaerts, B., Verhulst, N., Castellanos-Navarrete, A., Sayre, K. D., Dixon, J. and Dendooven, L. (2009). Conservation agriculture and soil carbon sequestration: between myth and farmer reality, *Crit. Rev. Plant Sci.* 28(3), 97–122.

Gross, A. and Glaser, B. (2021). Meta-analysis on how manure application changes soil organic storage, *Sci. Rep.* 11, 5516.

Guan, X.-K., Turner, N. C., Song, L., Gu, Y.-J., Wang, T.-C. and Li, F.-M. (2016). Soil carbon sequestration by three perennial legume pastures is greater in deeper soil layers than in the surface soil, *Biogeosciences* 13(2), 527–534.

Guenet, B., Gabrielle, B., Chenu, C., Arrouays, D., Balesdent, J., Bernoux, M., Bruni, E., Caliman, J., Cardinael, R., Chen, S., Ciais, P., Desbois, D., Fouche, J., Frank, S., Henault, C., Lugato, E., Naipal, V., Nesme, T., Obersteiner, M., Pellerin, S., Powlson, D. S., Rasse, D. P., Rees, F., Soussana, J., Su, Y., Tian, H., Valin, H. and Zhou, F. (2021). Can N_2O emissions offset the benefits from soil organic carbon storage?, *Glob. Change Biol.* 27(2), 237–256.

Guo, L. B. and Gifford, R. M. (2002). Soil carbon stocks and land use change: a meta analysis, *Glob. Change Biol.* 8(4), 345–360.

Haddaway, N. R., Hedlund, K., Jackson, L. E., Kätterer, T., Lugato, E., Thomsen, I. K., Jørgensen, H. B. and Söderström, B. (2015). What are the effects of agricultural management on soil organic carbon in boreo-temperate systems?, *Environ. Evid.* 4(1), 23.

Haddaway, N. R., Hedlund, K., Jackson, L. E., Kätterer, T., Lugato, E., Thomsen, I. K., Jørgensen, H. B. and Isberg, P. (2016). Which agricultural management interventions are most influential on soil organic carbon (using time series data)?, *Environ. Evid.* 5(1), 2.

Haddaway, N. R., Hedlund, K., Jackson, L. E., Kätterer, T., Lugato, E., Thomsen, I. K., Jørgensen, H. B. and Isberg, P. (2017). How does tillage intensity affect soil organic carbon? A systematic review, *Environ. Evid.* 6(30), 48.

Han, P., Zhang, W., Wang, G., Sun, W. and Huang, Y. (2016). Changes in soil organic carbon in croplands subjected to fertilizer management: a global meta-analysis, *Sci. Rep.* 6, 27199.

Hénin, S. and Dupuis, M. (1945). Essai de bilan de la matière organique du sol. *Ann. Agron.* 15, 17–29.

Henryson, K., Sundberg, C., Kätterer, T. and Hansson, P.-A. (2018). Accounting for long-term soil fertility effects when assessing the climate impact of crop cultivation, *Agr. Syst.* 164, 185–192.

Hernandez-Ramirez, G., Ruser, R. and Kim, D.-G. (2021). How does soil compaction alter nitrous oxide fluxes? A meta-analysis, *Soil Till. Res.* 211, 105036.

Hijbeek, R., van Ittersum, M. K., ten Berge, H. F. M., Gort, G., Spiegel, H. and Whitmore, A. P. (2017). Do organic inputs matter – a meta-analysis of additional yield effects for arable crops in Europe, *Plant Soil* 411(1-2), 293-303.

IPCC (2018). Global warming of 1.5°C. An IPCC Special Report on the impacts of global warming of 1.5°C above pre-industrial levels and related global greenhouse gas emission pathways, in the context of strengthening the global response to the threat of climate change, sustainable development, and efforts to eradicate poverty. Intergovernmental Panel on Climate Change. Available at: http://www.ipcc.ch/report/sr15.

Jarvis, N., Forkman, J., Koestel, J., Kätterer, T., Larsbo, M. and Taylor, A. (2017). Long-term effects of grass-clover leys on the structure of a silt loam soil in a cold climate, *Agr. Ecosyst. Environ.* 247, 319–328.

Jeffery, S., Abalos, D., Prodana, M., Bastos, A. C., van Groenigen, J. W., Hungate, B. A. and Verheijen, F. (2017). Biochar boosts tropical but not temperate crop yields, *Environ. Res. Lett.* 12(5), 053001.

Jian, J., Du, X., Reiter, M. S. and Stewart, R. D. (2020). A meta-analysis of global cropland soil carbon changes due to cover cropping, *Soil Biol. Biochem.* 143, 107735.

Johnston, A. E., Poulton, P. R. and Coleman, K. (2009). Soil organic matter: its importance in sustainable agriculture and carbon dioxide fluxes, *Adv. Agron.* 101, 1-57.

Kätterer, T., Bolinder, M. A., Andrén, O., Kirchmann, H. and Menichetti, L. (2011). Roots contribute more to refractory soil organic matter than above-ground crop residues, as revealed by a long-term field experiment, *Agr. Ecosys Environ.* 141(1-2), 184–192.

Kätterer, T., Bolinder, M. A., Berglund, K. and Kirchmann, H. (2012). Strategies for carbon sequestration in agricultural soils in northern Europe, *Acta. Agr. Scand. A* 62(4), 181–198.

Kätterer, T., Bolinder, M. A., Thorvaldsson, G. and Kirchmann, H. (2013). Influence of ley-arable systems on soil carbon stocks in Northern Europe and Eastern Canada. In Helgadóttir, A. and Hopkins, A. (Eds), The Role of Grasslands in a Green Future - Threats and Perspectives in Less Favoured Areas. Proceedings of the 17th Symposium of the European Grassland Federation, Akureyri, Iceland, 23-26 June 2013, Grassland Science in Europe, (vol. 18), 47-56.

Kätterer, T., Börjesson, G. and Kirchmann, H. (2014). Changes in organic carbon in topsoil and subsoil and microbial community composition caused by repeated additions of organic amendments and N fertilisation in a long-term field experiment in Sweden, *Agric. Ecosyst. Environ.* 189, 110–118.

Kätterer, T., Roobroeck, D., Andrén, O., Kimutai, G., Karltun, E., Kirchmann, H., Nyberg, G., Vanlauwe, B. and Röing de Nowina, K. (2019). Biochar addition persistently increased soil fertility and yields in maize-soybean rotations over 10 years in sub-humid regions of Kenya, *Field Crops Res.* 235, 18–26.

Kern, J. S. and Johnson, M. G. (1993). Conservation tillage impacts on national soil and atmospheric carbon levels, *Soil Sci. Soc. Am. J.* 57(1), 200–210.

Kindler, R., Siemens, J. A. N., Kaiser, K., Walmsley, D. C., Bernhofer, C., Buchmann, N., Cellier, P., Eugster, W., Gleixner, G., Grũnwald, T., Heim, A., Ibrom, A., Jones, S. K., Jones, M., Klumpp, K., Kutsch, W., Larsen, K. S., Lehuger, S., Loubet, B., Mckenzie, R., Moors, E., Osborne, B., Pilegaard, K., Rebmann, C., Saunders, M., Schmidt, M. W. I.,

Schrumpf, M., Seyfferth, J., Skiba, U., Soussana, J., Sutton, M. A., Tefs, C., Vowinckel, B., Zeeman, M. J. and Kaupenjohann, M. (2011). Dissolved carbon leaching from soil is a crucial component of the net ecosystem carbon balance, *Glob. Change Biol.* 17(2), 1167–1185.

Kirchmann, H., Bergström, L., Kätterer, T., Mattsson, L. and Gesslein, S. (2007). Comparison of Swedish long-term organic and conventional crop-livestock systems on a previously nutrient-depleted soil in Sweden, *Agron. J.* 99(4), 960-972.

Kirchmann, H., Kätterer, T., Schön, M., Börjesson, G. and Hamnér, K. (2013). Properties of soils in the Swedish long-term fertility experiments: VII. Changes in topsoil and upper subsoil at Örja and Fors after 50 years of nitrogen fertilisation and manure application, *Acta. Agr. Scand. B. - S. P.* 63, 25-36.

Kirchmann, H., Kätterer, T., Bergström, L., Börjesson, G. and Bolinder, M. A. (2016). Flaws and criteria for design and evaluation of comparative organic and conventional cropping systems, *Field Crops Res.* 186, 99-106.

Kirchmann, H., Börjesson, G., Kätterer, T. and Cohen, Y. (2017). From agricultural use of sewage sludge to nutrient extraction: a soil science outlook, *Ambio* 46(2), 143-154.

Kirchmann, H., Börjesson, G., Bolinder, M. A., Kätterer, T. and Djodjic, F. (2020). Soil properties currently limiting crop yields in Swedish agriculture - an analysis of 90 yield survey districts and 10 long-term field experiments, *Eur. J. Agron.* 120, 126132.

Kögel-Knabner, I. and Amelung, W. (2021). Soil organic matter in major pedogenic soil groups, *Geoderma* 384, 114785.

Kopittke, P. M., Dalal, R. C., Finn, D. and Menzies, N. W. (2017). Global changes in soil stocks of carbon, nitrogen, phosphorus, and sulphur as influenced by long-term agricultural production, *Glob. Change Biol.* 23(6), 2509–2519.

Körschens, M., Albert, E., Armbruster, M., Barkusky, D., Baumecker, M., Behle-Schalk, L., Bischoff, R., Čergan, Z., Ellmer, F., Herbst, F., Hoffmann, S., Hofmann, B., Kismanyoky, T., Kubat, J., Kunzova, E., Lopez-Fando, C., Merbach, I., Merbach, W., Pardor, M. T., Rogasik, J., Rühlmann, J., Spiegel, H., Schulz, E., Tajnsek, A., Toth, Z., Wegener, H. and Zorn, W. (2013). Effect of mineral and organic fertilization on crop yield, nitrogen uptake, carbon and nitrogen balances, as well as soil organic carbon content and dynamics: results from 20 European long-term field experiments of the twenty-first century, *Arch. Agron. Soil Sci.* 59(8), 1017–1040.

Kunhikrishnan, A., Thangarajan, R., Bolan, N. S., Xu, Y., Mandal, S., Gleeson, D. B., Seshadri, B., Zaman, M., Barton, L., Tang, C., Luo, J., Dalal, R., Ding, W., Kirkham, M. B. and Naidu, R. (2016), Chapter One.Functional relationships of soil acidification, liming, and greenhouse gas flux, *Adv. Agron.* 139, 1–71.

Ladha, J. K., Reddy, C. K., Padre, A. T. and van Kessel, C. (2011). Role of nitrogen fertilization in sustaining organic matter in cultivated soils, *J. Environ. Qual.* 40(6), 1756–1766.

Lal, R. (2004). Soil carbon sequestration impacts on global climate change and food security, *Science* 304(5677), 1623-1627.

Lal, R. (2013). Intensive agriculture and the soil carbon pool, *J. Crop Improve* 27(6), 735–751.

Land, M., Haddaway, N. R., Hedlund, K., Jørgensen, H. B., Kätterer, T. and Isberg, P. (2017). How do selected crop rotations affect soil organic carbon in boreo-temperate systems? A systematic review protocol, *Environ. Evid.* 6(1), 9.

Land, M., Hedlund, K., Kätterer, T., Bracht Jörgensen, H. and Haddaway, N. (2021). Del 1: Övergripande evidensgraderade slutsatser av den systematiska översikten.

In M. Land och H. Scharin (Eds), *Växtföljders påverkan på inlagring av organiskt kol i jordbruksmark*. Formas Rapport F1:2021 (in Swedish with English summary; https://formas.se).

Lawrence-Smith, E. J., Curtin, D., Beare, M. H., McNally, S. R., Kelliher, F. M., Calvelo Pereira, R. and Hedley, M. J. (2021). Full inversion tillage during pasture renewal to increase soil carbon storage: New Zealand as a case study, *Glob. Change Biol.* 27(10), 1998-2010.

Leifeld, J. and Fuhrer, J. (2010). Organic farming and soil carbon sequestration: what do we really know about the benefits?, *Ambio* 39(8), 585-599.

Leifeld, J., Angers, D. A., Chenu, C., Fuhrer, J., Kätterer, T. and Powlson, D. S. (2013). Organic farming gives no climate change benefit through soil carbon sequestration, *Proc. Natl Acad. Sci. U. S. A.* 110(11), E984.

Lehmann, J., Gaunt, J. and Rondon, M. (2006). Bio-char sequestration in terrestrial Ecosystems: a review, *Mitig Adapt. Strat. Gl* 11(2), 403-427.

Lehtinen, T., Schlatter, N., Baumgarten, A., Bechini, L., Krüger, J., Grignani, C., Zavattaro, L., Costamagna, C. and Spiegel, H. (2014). Effect of crop residue incorpororation on soil organic carbon and greenhouse gas emissions in European agricultural soils, *Soil Use Manag.* 30(4), 524-538.

Lemke, R. L., VandenBygaart, A. J., Campbell, C. A., Lafond, G. P. and Grant, B. (2010). Crop residue removal and fertilizer N: effects on soil organic carbon in a long-term crop rotation experiment on a Udic Boroll, *Agric. Ecosyst. Environ.* 135(1-2), 42-51.

Levavasseur, F., Mary, B., Christensen, B. T./ Duparque, A., Ferchaud, F., Kätterer, T., Lagrange, H., Montenach, D., Resseguier, C. and Houot, S. (2020). The simple AMG model accurately simulates organic carbon storage in soils after repeated application of exogenous organic matter, *Nutr. Cycl. Agroecosyst.* 117(2), 215-229.

Liang, C., Schimel, J. P. and Jastrow, J. D. (2017). The importance of anabolism in microbial control over soil carbon storage, *Nat. Microbiol.* 2, 17105.

Liu, X. Y., Zhang, A. F., Ji, C. Y., Joseph, S., Bian, R., Li, L., Pan, G. and Paz-Ferreiro, J. (2013). Biochar's effect on crop productivity and the dependence on experimental conditions–a meta-analysis of literature data, *Plant Soil* 373(1-2), 583-594.

Liu, C., Lu, M., Cui, J., Li, B. and Fang, C. (2014). Effects of straw carbon input on carbon dynamics in agricultural soils: a meta-analysis, *Glob. Change Biol.* 20(5), 1366-1381.

Lorenz, K. and Lal, R. (2005). The depth distribution of soil organic carbon in relation to land use and management and the potential of carbon sequestration in subsoil horizons, *Adv. Agron.* 88, 35-66.

Lu, M., Zhou, X., Luo, Y., Yang, Y., Fang, C., Chen, J. and Li, B. (2011). Minor stimulation of soil carbon storage by nitrogen addition: a meta-analysis, *Agric. Ecosyst. Environ.* 140(1-2), 234-244.

Lu, F. (2015). How can straw incorporation management impact on soil carbon storage? A meta-analysis, *Mitig Adapt. Strateg. Glob. Change* 20(8), 1545-1568.

Lugato, E., Cescatti, A., Jones, A., Ceccherini, G. and Duveiller, G. (2020). Maximising climate mitigation potential by carbon and radiative agricultural land management with cover crops, *Environ. Res. Lett.* 15(9), 094075.

Lugato, E., Lavallee, J. M., Haddix, M. L., Panagos, P. and Cotrufo, M. F. (2021). Different climate sensitivity of particulate and mineral-associated soil organic matter, *Nat. Geosci.* 14(5), 295-300.

Luo, Z., Wang, E. and Sun, O. J. (2010a). Soil carbon change and its responses to agricultural practices in Australian agro-ecosystems: a review and synthesis, *Geoderma* 155(3-4), 211-223.

Luo, Z., Wang, E. and Sun, O. J. (2010b). Can no-tillage stimulate carbon sequestration in agricultural soils? A meta-analysis of paired experiments, *Agric. Ecosyst. Environ.* 139(1-2), 224-231.

Maillard, É and Angers, D. A. (2014). Animal manure application and soil organic carbon stocks: a meta-analysis, *Glob. Change Biol.* 20(2), 666-679.

Malik, A. A., Puissant, J., Buckeridge, K. M., Goodall, T., Jehmlich, N., Chowdhury, S., Gweon, H. S., Peyton, J. M, Masson, K. E., van Agtmaal, M., Blaud, A., Clark, I. M., Whitaker, J., Pywell, R. F., Ostle, N., Gleixner, G. and Griffiths, R. I. (2018). Land use driven change in soil pH affects microbial carbon cycling processes, *Nat. Comun.* 9, 3591 (2018). https://doi.org/10.1038/s41467-018-05980-1.

McDaniel, M. D., Tiemann, L. K. and Grandy, A. S. (2014). Does agricultural crop diversity enhance soil microbial biomass and organic matter dynamics? A meta-analysis, *Ecol. Appl.* 24(3), 560-570.

Menichetti, L., Ekblad, A. and Kätterer, T. (2015). Contribution of roots and amendments to soil carbon accumulation within the soil profile in a long-term field experiment in Sweden, *Agric. Ecosyst. Environ.* 200, 79-87.

Menichetti, L., Bolinder, M. A. and Kätterer, T. (2020). A Bayesian modeling framework for estimating equilibrium soil organic C sequestration in agroforestry systems, *Agric. Ecosyst. Environ.* 303, 107118.

Meurer, K. H. E., Haddaway, N. R., Bolinder, M. A. and Kätterer, T. (2018). Tillage intensity affects total SOC stocks in boreo-temperate regions only in the topsoil–A systematic review using an ESM approach, *Earth Sci. Rev.* 177, 613-622.

Meurer, K. H. E., Chenu, C., Coucheney, E., Herrmann, A. M., Keller, T., Kätterer, T., Nimblad Svensson, D. and Jarvis, N. (2020). Modelling dynamic interactions between soil structure and the storage and turnover of soil organic matter, *Biogeosciences* 17(20), 5025-5042.

Minasny, B., Malone, B. P., McBratney, A. B., Angers, D. A., Arrouays, D., Chambers, A., Chaplot, V., Chen, Z., Cheng, K., Das, B. S., Field, D. J., Gimona, A., Hedley, C. B., Hong, S. Y., Mandal, B., Marchant, B. P., Martin, M., McConkey, B. G., Mulder, V. L., O'Rourke, S., Richer-de-Forges, A. C., Odeh, I., Padarian, J., Paustian, K., Pan, G., Poggio, L., Savin, I., Stolbovoy, V., Stockmann, U., Sulaeman, Y., Tsui, C., Vågen, T., van Wesemael, B. and Winowiecki, L. (2017). Soil carbon 4 per mille, *Geoderma* 292, 59-86.

Minx, J. C., Lamb, W. F., Callaghan, M. W., Fuss, S., Hilaire, J., Creutzig, F., Amann, T., Beringer, T., de Oliveira Garcia, W., Hartmann, J., Khanna, T., Lenzi, D., Luderer, G., Nemet, G. F., Rogelj, J., Smith, P., Vicente Vicente, J. L., Wilcox, J. and del Mar Zamora Dominguez, M. (2018). Negative emissions - part 1: research landscape and synthesis, *Environ. Res. Lett.* 13(6), 063001.

Morari, F., Berti, A., Dal Ferro, N. and Piccoli, I. (2019). Deep carbon sequestration in cropping systems. In: *Sustainable Agriculture Reviews* Lal, R., Francaviglia, R. (Eds) 29, 33-65. doi:10.1007/978-3-030-26265-5_2.

Mukumbuta, I. and Hatano, R. (2019). Do tillage and conversion of grassland to cropland always deplete soil organic carbon?, *Soil Sci. Plant Nutr.* 66(1), 76-83.

Nowak, B., Nesme, T., David, C. and Pellerin, S. (2013). To what extent does organic farming rely on nutrient inflows from conventional farming?, *Environ. Res. Lett.* 8(4), 044045.

Oelofse, M., Markussen, B., Knudsen, L., Schelde, K., Olesen, J. E., Jensen, L. S. and Bruun, S. (2015). Do soil organic carbon levels affect potential yields and nitrogen use efficiency? An analysis of winter wheat and spring barley field trials, *Eur. J. Agron.* 66, 62–73.

Ogle, S. M., Alsaker, C., Baldock, J., Bernoux, M., Breidt, F. J., McConkey, B., Regina, K. and Vazquez-Amabile, G. G. (2019). Climate and soil characteristics determine where no-till management can store carbon in soils and mitigate greenhouse gas emissions, *Sci. Rep.* 9(1), 11665.

Oldfield, E. E., Bradford, M. A. and Wood, S. A. (2019). Global meta-analysis of the relationship between soil organic matter and crop yields, *SOIL* 5(1), 15-32.

Olson, K. R., Al-Kaisi, M. M., Lal, R. and Lowery, B. (2014). Experimental consideration, treatments, and methods in determining soil organic carbon sequestration rates, *Soil Sci. Soc. Am. J.* 78(2), 348-360.

Palese, A. M., Vignozzi, N., Celano, G., Agnelli, A. E., Pagliai, M. and Xiloyannis, C. (2014). Influence of soil management on soil physical characteristics and water storage in a mature rainfed olive orchard, *Soil Till. Res.* 144, 96–109.

Paradelo, R., Virto, I. and Chenu, C. (2015). Net effect of liming on soil organic carbon stocks: a review, *Agric. Ecosyst. Environ.* 202, 98–107.

Paustian, K., Andrén, O., Janzen, H. H., Lal, R., Smith, P., Tian, G., Tiessen, H., Noordwijk, M. and Woomer, P. L. (1997). Agricultural soils as a sink to mitigate CO_2 emissions, *Soil Use Manag.* 13(s4), 230-244.

Paustian, K., Lehmann, J., Ogle, S., Reay, D., Robertson, G. P. and Smith, P. (2016). Climate-smart soils, *Nature* 532(7597), 49-57.

Poeplau, C. and Don, A. (2012). Soil carbon changes under *Miscanthus* driven by C4 accumulation and C3 decompostion – toward a default sequestration function, *GCB Bioenergy* 6, 327-338.

Poeplau, C. and Don, A. (2013). Sensitivity of soil organic carbon stocks and fractions to different land.use changes across Europe, *Geoderma* 192, 189-201.

Poeplau, C. and Don, A. (2015). Carbon sequestration in agricultural soils via cultivation of cover crops – A meta-analysis, *Agric. Ecosyst. Environ.* 200, 33–41.

Poeplau, C., Don, A., Vesterdal, L., Leifeld, J., Van Wesemael, B., Schumacher, J. and Gensior, A. (2011). Temporal dynamics of soil organic carbon after land-use change in the temperate zone – carbon response functions as a model approach, *Glob. Change Biol.* 17(7), 2415-2427.

Poeplau, C., Aronsson, H., Myrbeck, Å. and Kätterer, T. (2015). Effect of perennial ryegrass cover crop on soil organic carbon stocks in southern Sweden, *Geoderma Reg.* 4, 126-133.

Poeplau, C., Bolinder, M. A., Kirchmann, H. and Kätterer, T. (2016). Phosphorus fertilisation under nitrogen limitation can deplete soil carbon stocks: evidence from Swedish meta-replicated long-term field experiments, *Biogeosciences* 13(4), 1119-1127.

Poeplau, C. and Kätterer, T. (2017). Is soil texture a major controlling factor of root:shoot ratio in cereals?, *Eur. J. Soil Sci.* 68(6), 964-970.

Poeplau, C., Helfrich, M., Dechow, R., Szoboszlay, M., Tebbe, C. C., Don, A., Greiner, B., Zopf, D., Thumm, U., Korevaar, H. and Geerts, R. (2019). Increased microbial anabolism contributes to soil carbon sequestration by mineral fertilization in temperate grasslands, *Soil Biol. Biochem.* 130, 167–176.

Poulton, P., Johnston, J., Macdonald, A., White, R. and Powlson, D. (2018). Major limitations to achieving "4 per 1000" increases in soil organic carbon stock in temperate regions: evidence from long-term experiments at Rothamsted Research, United Kingdom, *Glob. Change Biol.* 24(6), 2563–2584.

Published by Burleigh Dodds Science Publishing Limited, 2023.

Powlson, D. S., Glendining, M. J., Coleman, K. and Whitmore, A. P. (2011). Implications for soil properties of removing cereal straw: results from long-term studies 1, *Agron. J.* 103(1), 279-287.

Powlson, D. S., Stirling, C. M., Jat, M. L., Gerard, B. G., Palm, C. A., Sanchez, P. A. and Cassman, K. G. (2014). Limited potential of no-till agriculture for climate change mitigation, *Nat. Clim. Change* 4(8), 678-683.

Quenum, M., Giroux, M. and Royer, R. (2004). Étude sur le bilan humique des sols dans des systèmes culturaux sous prairies et sous cultures commerciales selon les modes de fertilisation, *Agrosol* 15, 57-71. [In French with English abstract].

Rasse, D. P., Rumpel, C. and Dignac, M.-F. (2005). Is soil carbon mostly root carbon? Mechanisms for a specific stabilization, *Plant Soil* 269(1-2), 341-356.

Reintam, L. (2007). Soil formation on reddish-brown calcareous till under herbaceous vegetation during forty years, *Est. J. Earth Sci.* 56, 65-84.

Riley, H. and Bakkegard, M. (2006). Declines of soil organic matter content under arable cropping in southeast Norway, *Acta. Agr. Scand. B. - S. P.* 56(3), 217-223.

Roe, S., Streck, C., Obersteiner, M., Frank, S., Griscom, B., Drouet, L., Fricko, O., Gusti, M., Harris, N., Hasegawa, T., Hausfather, Z., Havlík, P., House, J., Nabuurs, G., Popp, A., Sánchez, M. J. S., Sanderman, J., Smith, P., Stehfest, E. and Lawrence, D. (2019). Contribution of the land sector to a 1.5°C world, *Nat. Clim. Change* 9(11), 817-828.

Rosenstock, T. S., Wilke, A., Jallo, C. Namoi, N., Bulusu, M., Suber, M., Mboi, D., Mulia, R., Simelton, E., Richards, M. and Gurwick, N. (2019). Making trees count: measurement and reporting of agroforestry in UNFCCC national communications of non-Annex I countries, *Agric. Ecosyst. Environ.* 284, 106569.

Rousk, J., Brookes, P. C. and Bååth, E. (2009). Contrasting soil pH effects on fungal and bacterial growth suggest functional redundancy in carbon mineralization, *Appl. Environ. Microbiol.* 75(6), 1589-1596.

Rumpel, C. and Kögel-Knabner, I. (2011). Deep soil organic matter - a key but poorly understood component of terrestrial C cycle, *Plant Soil* 338(1-2), 143-158.

Rumpel, C., Amiraslani, F., Chenu, C., Garcia Cardenas, M., Kaonga, M., Koutika, L. S., Ladha, J., Madari, B., Shirato, Y., Smith, P., Soudi, B., Soussana, J. F., Whitehead, D. and Wollenberg, E. (2020). The 4p1000 initiative: opportunities, limitations and challenges for implementing soil organic carbon sequestration as a sustainable development strategy, *Ambio* 49(1), 350-360.

Scheer, C., Wassmann, R., Kienzler, K., Ibragimov, N., Lamers, J. P. A. and Martius, C. (2008). Methane and nitrous oxide fluxes in annual and perennial land-use systems of the irrigated areas in the Aral Sea Basin, *Glob. Change Biol.* 14(10), 2454- 2468.

Schiedung, M., Tregurtha, C. S., Beare, M. H., Thomas, S. M. and Don, A. (2019). Deep soil flipping increases carbon stocks of New Zealand grasslands, *Glob. Change Biol.* 25(7), 2296-2309.

Schneider, F. and Don, A. (2019). Root-restricting layers in German agricultural soils. Part II: adaptation and melioration strategies, *Plant Soil* 442(1-2), 419-432.

Searchinger, T. D., Wirsenius, S., Beringer, T. and Dumas, P. (2018)Assessing the efficiency of changes in land use for mitigating climate change, *Nature* 564(7735), 249-253.

Sibbesen, E. (1986). Soil movement in long-term field experiments, *Plant Soil* 91(1), 73-85.

Simo, I., Schulte, R., O'Sullivan, L. and Creamer, R. (2019). Digging deeper: understanding the contribution of subsoil carbon for climate mitigation, a case study of Ireland, *Environ. Sci. Policy* 98, 61-69.

Smith, P., Powlson, D. S., Glendining, M. J. and Smith, J. U. (1997). Potential for carbon sequestration in European soils: preliminary estimates for five scenarios using results from long-term experiments, *Glob. Change Biol.* 3(1), 67–79.

Smith, P. (2004). Carbon sequestration in croplands: the potential in Europe and the global context, *Eur. J. Agron.* 20(3), 229–236.

Smith, P., Soussana, J. F., Angers, D., Schipper, L., Chenu, C., Rasse, D. P., Batjes, N. H., van Egmond, F., McNeill, S., Kuhnert, M., Arias-Navarro, C., Olesen, J. E., Chirinda, N., Fornara, D., Wollenberg, E., Álvaro-Fuentes, J., Sanz-Cobena, A. and Klumpp, K. (2020). How to measure, report and verify soil carbon change to realize the potential of soil carbon sequestration for atmospheric greenhouse gas removal, *Glob. Change Biol.* 26(1), 219–241.

Söderström, B., Hedlund, K., Jackson, L. E., Kätterer, T., Lugato, E., Thomsen, I. K. and Bracht Jørgensen, H. (2014). What are the effects of agricultural management on soil organic carbon (SOC) stocks? protocol, *Environ. Evid.* 3(1), 1–8.

Sundberg, C., Karltun, E., Gitau, J. K., Kätterer, T., Kimutai, G. M., Mahmoud, Y., Njenga, M., Nyberg, G., Roing de Nowina, K., Roobroeck, D. and Sieber, P. (2020). Biochar from cookstoves reduces greenhouse gas emissions from smallholder farms in Africa, *Mitig. Adapt. Strateg. Glob. Change* 25(6), 953-967.

Tisserant, A. and Cherubini, F. (2019). Potentials, limitations, co-benefits, and trade-offs of biochar applications to soils for climate change mitigation, *Land* 8(12), 179.

Torres-Sallan, G., Schulte, R., Lanigan, G. J., Byrne, K. A., Reidy, B., Simó, I., Six, J. and Creamer, R. E. (2017). Clay illuviation provides a longterm sink for C sequestration in subsoils, *Sci Rep.*, 7, 45635.

Trost, B., Prochnow, A., Drastig, K., Meyer-Aurich, A., Ellmer, F. and Baumecker, M. (2013). Irrigation, soil organic carbon and N_2O emissions. A review, *Agron. Sustain. Dev.* 33(4), 733–749.

Uhlen, G. (1991). Long-term effects of fertilizers, manure, straw and crop rotation on total-N and total-C in soil, *Acta. Agr. Scand. B. - S. P.* 41(2), 119-127.

Valkama, E., Lemola, R., Känkänen, H. and Turtola, E. (2015). Meta-analysis of the effects of undersown catch crops on nitrogen leaching loss and grain yields in the Nordic countries, *Agric. Ecosyst. Environ.* 203, 93–101.

VandenBygaart, A. J. (2016). The myth that no-till can mitigate global climate change, *Agric. Ecosyst. Environ.* 216, 98–99.

VandenBygaart, A. J. (2018). Comments on soil carbon 4 per mille by Minasny et al. 2017, *Geoderma*, 309, 113–114.

VandenByggart, A. J., Gregorich, E. G. and Angers, D. A. (2003). Influence of agricultural management on soil organic carbon: a compendium and assessment of Canadian studies, *Can. J. Soil Sci.* 83(4), 363–380.

VandenBygaart, A. J., Bremer, E., McConkey, B. G., Ellert, B. H., Janzen, H. H., Angers, D. A., Carter, M. R., Drury, C. F., Lafond, G. P. and McKenzie, R. H. (2011). Impact of sampling depth on differences in soil carbon stocks in long-term agroecosystem experiments, *Soil Sci. Soc. Am. J.* 75(1), 226–234.

van Groenigen, J. W., van Kessel, C., Hungate, B. A., Oenema, O., Powlson, D. S. and van Groenigen, K. J. (2017). Sequestering soil organic carbon: a nitrogen dilemma, *Environ. Sci. Technol.* 51(9), 4738–4739.

Van Meirvenne, M., Pannier, J., Hofman, G. and Louwagie, G. (1996). Regional characterization of the long-term change in soil organic carbon under intensive agriculture, *Soil Use Manag.* 12(2), 86–94.

Viiralt, R. (1998). Productivity of perennial herbage on carbonated moraine in soil formation model experiment, *Grassl. Sci. Eur.* 3, 153–156.

Virto, I., Barré, P., Burlot, A. and Chenu, C. (2012). Carbon input differences as the main factor explaining the variability in soil organic C storage in no-tilled compared to inversion tilled agroecosystems, *Biogeochemistry* 108(1–3), 17–26.

Vonk, W. J., van Ittersum, M. K., Reidsma, P., Zavattaro, L., Bechini, L., Guzmán, G., Pronk, A., Spiegel, H., Steinmann, H. H., Ruysschaert, G. and Hijbeek, R. (2020). European survey shows poor association between soil organic matter and crop yields, *Nutr. Cycl. Agroecosyst.* 118(3), 325–334.

Wang, J., Wang, X., Xu, M., Feng, G., Zhang, W. and Lu, C. (2015). Crop yield and soil organic matter after long-term straw return to soil in China, *Nutr. Cycl. Agroecosyst.* 102(3), 371–381.

Wang, J., Xiong, Z. and Kuzyakov, Y. (2016). Biochar stability in soil: meta-analysis of decomposition and priming effects, *GCB Bioenergy* 8(3), 512–523.

Wang, Y. J., Guo, J. H., Vogt, R. D., Mulder, J., Wang, J. and Zhang, X. (2018). Soil pH as the chief modifier for regional nitrous oxide emissions: new evidence and implications for global estimates and mitigation, *Glob. Change Biol.* 24(2), e617–e626.

Wardle, D. A., Nilsson, M. C.. and Zackrisson, O. (2008). Fire-derived charcoal causes loss of forest humus, *Science* 320(5876), 629.

Warren, C. J., Saurette, D. D.. and Gillespie, A. W. (2020). Soil organic carbon content: decreases partly attributed to dilution by increased depth of cultivation in southern Ontario, *Can. J. Soil Sci.* 101(2), 335–338.

Wesemael, B. v., Caroline Roelandt, S. L. and Orshoven, J. V. (2005). Modelling the evolution of regional carbon stocks in Belgian cropland soils, *Can. J. Soil Sci.* 85(Special Issue), 511–521.

West, T. O. and Post, W. M. (2002). Soil organic carbon sequestration rates by tillage and crop rotation, *Soil Sci. Soc. Am. J.* 66(6), 1930–1946.

Wilhelm, W. W., Johnson, J. M. F., Hatfield, J. L., Voorhees, W. B. and Linden, D. R. (2004). Crop and soil productivity response to corn residue removal: a literature review, *Agron. J.* 96(1), 1–17.

Woolf, D., Amonette, J. E., Street-Perrott, F. A., Lehmann, J. and Joseph, S. (2010). Sustainable biochar to mitigate global climate change, *Nat. Commun.* 1, 56.

Yang, X. M. and Kay, B. D. (2001). Rotation and tillage effects on soil organic carbon sequestration in a typic Hapludalf in Southern Ontario, *Soil Till. Res.* 59(3–4), 107–114.

Yang, Y., Loecke, T. and Knops, J. (2021). Surface soil organic carbon sequestration under post agricultural grasslands offset by net loss at depth, *Biogeochemistry* (preprint doi:10.21203/rs.3.rs-449031/v1).

Ye, L., Camps-Arbestain, M., Shen, Q., Lehmann, J., Singh, B. and Sabir, M. (2020). Biochar effects on crop yields with and without fertilizer: a meta-analysis of field studies using separate controls, *Soil Use Manag.* 36(1), 2–18.

Zavatarro, L., Bechini, L., Grignani, C., van Evert, F. K., Mallast, J., Spiegel, H., Sandén, T., Pecio, A., Giráldez Cervera, J. V., Guzmán, G., Vanderlinden, K., D'Hose, T., Ruysschaert, G. and ten Berge, H. F. M. (2017). Agronomic effects of bovine manure: a review of long-term European field experiments, *Eur. J. Agron.* 90, 127–138.

Chapter 17

Agricultural practices to improve soil carbon sequestration in rice paddy soils

Hyeon Ji Song and Pil Joo Kim, Gyeongsang National University, South Korea

1 Introduction

Rice is a nutritious staple crop for over half of the global population, even though rice fields cover only 9% of global cropland (Maclean et al., 2002). Rice paddies are widely distributed from tropical to temperate climatic regions on all continents (except Antarctica), with high concentrations in South-East Asia. Global rice demand is predicted to grow by around 28% by

http://dx.doi.org/10.19103/AS.2022.0106.16

2050 compared to the early 2000s (Alexandratos and Bruinsma, 2012). Paddy rice soils are important carbon (C) sinks, compared with upland soils, since the anaerobic conditions in flooded rice paddies slow down organic matter (OM) decomposition, consequently favoring soil organic carbon (SOC) stock increases (Yan et al., 2013; Chen et al., 2021).

However, due to cultivation under submerged conditions, rice cropping fields act as major greenhouse gas (GHG) emission sources in the form of methane (CH_4) and nitrous oxide (N_2O). Globally, 8.2% of anthropogenic CH_4 (which has 28 times higher global warming potential (GWP) than carbon dioxide (CO_2) over a 100-year period; IPCC, 2013) is emitted from rice cultivation (Saunois et al., 2020). IPCC estimated an average of 0.3% of applied nitrogen (N) to be converted into N_2O which has 298 times higher GWP than CO_2 over a 100-year period (Klein et al., 2006; IPCC, 2013). However, total N_2O flux from rice paddies has not been properly accounted for.

SOC stock is significantly affected by agricultural management practices (i.e. crop rotation, fertilizer and organic amendment applications, irrigation, tillage management, etc.) (Windeatt et al., 2014; Vicente-Vicente et al., 2016; Paustian et al., 2019; see also Chapter 16 of this book). The periodic application of organic amendments (i.e. crop residues, farmyard and livestock manure, compost, biochar, etc.) is accepted as the most beneficial practice to increase SOC stock in upland soils (Goyal et al., 1999; Diacono and Montemurro, 2011; Huang et al., 2014). In paddy rice soils, cover cropping and straw recycling were historically recommended to increase SOC stock and improve soil quality.

However, as noted, the anaerobic conditions of paddy rice systems facilitate microbial methanogenesis and convert rice paddies into a main CH_4 emission source (Chen et al., 2013), particularly after OM applications under flooded conditions (Hwang et al., 2017; Jiang et al., 2019; Song et al., 2021). Rice has the highest GHG intensity (GHGI) which indicates net GWP per grain yield among major cereal crops (Linquist et al., 2012; Carlson et al., 2017). In particular, management practices intended to increase SOC stocks, such as organic and/or inorganic fertilizer use can provide energy sources and/or substrates for methanogens and denitrification in anaerobic soils, and thus amplify GHG emissions during rice cultivation (Neue et al., 1996; Le Mer and Roger, 2001; Lee et al., 2020a,b). This should be considered for mitigation of global warming through enhancing soil C sequestration (Lee et al., 2020a,b; Song et al., 2021). In rice paddies, the overall changes of SOC stock and GHG fluxes should be estimated together when evaluating the impact of agricultural management practices on environmental sustainability.

In this chapter, we present the specific conditions leading to high SOC stocks and low CH_4 and N_2O emissions of paddy rice soils. We indicate how to evaluate their GWP considering both SOC stock and GHG emissions under different management practices.

2 Carbon sequestration potential of rice paddy soils

Rice paddy soils have a significant potential to sequester atmospheric CO_2 through increasing SOC storage (Lal, 2004; Pan et al., 2004; Yan et al., 2013; Chen et al., 2021). According to a recent global meta-analysis (Liu et al., 2021), rice paddy harbors over 14% of land SOC stocks, proportionally more than any other crop type (Table 1), mainly due to higher biomass productivity and reduced OM decomposition under anaerobic conditions. The SOC sequestration potential of rice is, at 401 kg C ha^{1} year^{1}, much higher than that of barley, corn, and wheat residues, which amounts to 247 kg C ha^{1} year^{1}, 292 kg C ha^{1} year^{1}, and 272 kg C ha^{1} year^{1}, respectively (Fig. 1) (Jarecki and Lal, 2003).

Rice paddies are flooded before transplanting, and water is drained several days before harvesting. In traditional rice cropping systems, the soil is managed under submerged conditions for at least 85–90% of the cropping period. Under these soil conditions, both the mineralization of OM inputs and decomposition of native SOC are slower than those under upland soil conditions (Witt et al., 2000). Indeed, under anaerobic conditions, OM decomposition partially affects labile compounds whilst recalcitrant aromatic and aliphatic compounds accumulate (Herndon et al., 2015). This leads to changes in SOC quality and may slow down soil organic matter (SOM) decomposition in flooded paddy soils. In addition, the formation of recalcitrant complexes with OM can make them less available for microbial decomposition (Six et al., 2002). Moreover, biological N fixation, coupled with decreased overall humification and higher primary productivity, could lead to a net accumulation of SOM (Liu et al., 2018). On the other hand, the availability of biologically fixed N may also provide substrates for N_2O emissions under reduced conditions.

In rice paddy fields, hydrologic regimes that rotate periodical flooding and drying may drive iron (Fe) cycling (Sahrawat, 2004) and thereby impact the formation of organo-mineral interactions, which were shown to be stable SOM forms in upland soils. Under the dried soil condition, Fe species may be oxidized, and during the flooding period, Fe species may be reduced. Fe cycling has the potential to stabilize SOC and reduce C oxidation and CH_4

Table 1 Estimates of global land use and SOC storage

Ecosystem	Global area (ha)	Contribution of rice paddy (%)	Mean SOC stock (Mg ha^{1})		Global SOC storage (Pg C)		
			0–30 cm	0–100 cm	0–30 cm	0–100 cm	Contribution of rice paddy (%)
All soils	1.49×10^{10}	1.1	45	98	710	1456	1.2
Cropland	1.87×10^{9}	8.9	41	89	58.4	127	14.2
Rice paddy	1.67×10^{8}	-	51	108	8.5	18.0	-

Note: Data was modified from Liu et al. (2021).

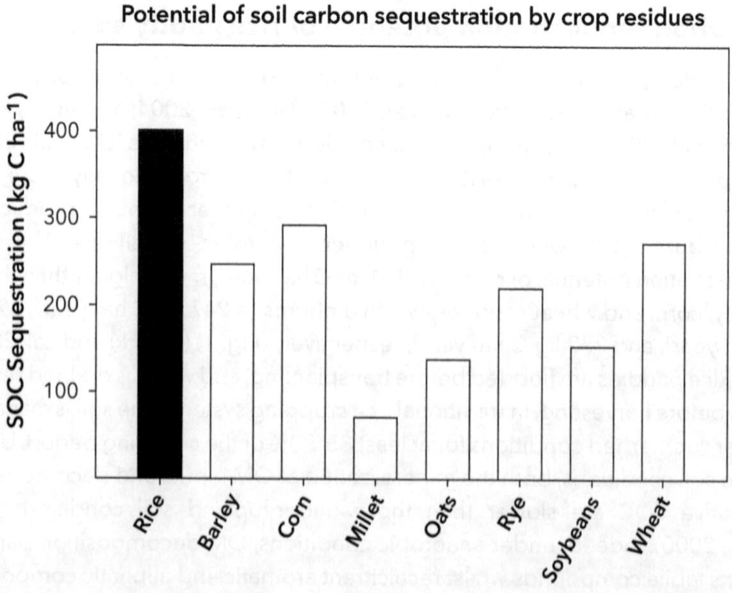

Figure 1 Potential of SOC sequestration by crop residues. (Data was modified from Jarecki and Lal, 2003.)

formation by reacting as an oxidizing agent and an electron acceptor for microbial respiration. In contrast, the formation of SOM can be maximized in partially oxidizing environments. High mineralization rates occur under completely aerobic conditions, but oxidative polymerization is favored by lack of oxygen (O_2). Therefore, repeated wetting and drying prevent the stagnation that appears in either oxidizing or reducing environments and promotes the oxidative polymerization that stabilizes OC compounds in soils (Post et al., 2004). These processes may have led to the net retention of OM and plant debris in most wetland soils (Mitsch and Gosselink, 2007).

3 Methane production, oxidation and emissions in rice paddy soils

Flooded rice paddy fields are an important anthropogenic CH_4 emission source (Newton, 2016; Nisbet et al., 2016; Saunois et al., 2016; Schaefer et al., 2016). Global CH_4 emissions from rice fields were estimated at an average of 30 Tg CH_4 year^{-1}, accounting for around 8.2% of total anthropogenic CH_4 emissions (average 364 Tg CH_4 year^{-1}) (Saunois et al., 2020). Rice production has been estimated as increasing from 97 million tons in 1990 to a predicted 145 million tons by 2025, leading to an estimated proportionate CH_4 flux increase of 1.1% per year from paddy rice systems (Anastasi et al., 1992).

CH_4 flux in rice fields is the result of the difference between CH_4 production and consumption (oxidation). Soil flooding prevents gas exchange between the soil and the atmosphere. Methane is formed by methanogens under extremely anaerobic conditions (less than −200 mV of E_h value), and this CH_4 can also be oxidized by methanotrophs at conditions over −200 mV of soil E_h value (Garcia et al., 2000; Conrad, 2007). Rice fields are generally flooded during the cropping season while being kept dry during a succeeding fallow season. In addition, a small amount of CH_4 is oxidized during the dry fallow season, but this flux is not comparable with the big CH_4 fluxes during the flooded rice cultivation period (Fig. 2). At the early rice growth stage, CH_4 is emitted at a lower rate, but its flux steadily increases with plant growth and development of anaerobic soil conditions (Ali et al., 2009a; Kim et al., 2015; Haque et al., 2016; Jeong et al., 2018; Song et al., 2019; Lee et al., 2020a,b; Song et al., 2021).

Methanogenesis is strongly affected by the availability of C substrates, and therefore OM addition strongly stimulates CH_4 formation. Under anaerobic conditions where nitrate (NO_3^-) and sulfate (SO_4^{2-}) concentrations are low, OM [$(CH_2O)_n$] can be completely mineralized into CH_4 and CO_2 via methanogenic fermentation (Le Mer and Roger, 2001). This change needs successive reactions from four populations of microorganisms that degrade complex organic molecules into simpler compounds:

- hydrolysis of polymeric organic compounds into monomers by a hydrolytic microflora;
- acidogenesis by a fermentative microflora;
- acetogenesis by a homoacetogenic or syntrophic microflora;
- CH_4 formation by methanogens.

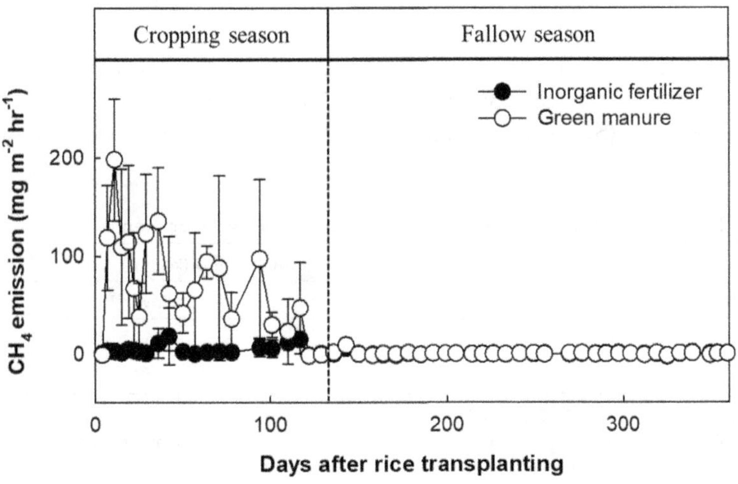

Figure 2 Annual changes of CH_4 emission rate in a mono-rice paddy.

Published by Burleigh Dodds Science Publishing Limited, 2023.

Methanogens belonging to the Archaea domain (Woese et al., 1978) include strictly anaerobic, specialized microflora that can develop in synergy with other anaerobic bacteria. The two major pathways of CH_4 formation are acetotrophic and hydrogenotrophic (Schütz et al., 1989). In paddy rice soils, acetotrophy is believed to be responsible for approximately two-thirds of the CH_4 produced (Le Mer and Roger, 2001).

Methane is consumed in soils by microbial oxidation. This reaction takes place in the aerobic zone of dried soils and methanogenic soils. In dried rice paddies, small amounts of CH_4 can be oxidized mainly during the fallow season. Methanotrophy is most highly developed in soils that are often water saturated or submerged and where methanogenic activity develops (Nesbit and Breitenbeck, 1992). Paddy soils for rice cultivation often have a very high potential for methanotrophic activities. However, rice paddies generally have a positive balance between CH_4 production and oxidation and consequently react like a huge CH_4 source, not a sink (Table 2).

In rice fields, variations in CH_4 flux are mainly caused by differences in methanotrophic bacteria activity (Schütz et al., 1989; Sass et al., 1990). The contribution of CH_4 oxidation to the net CH_4 budget varies, depending on agricultural practices like the management of irrigation water and organic amendments. Methanogens and methanotrophs are ubiquitous, and both population densities have a highly positive correlation with each other. It is known that cultivable methanotroph densities and potential methanotrophic activities are greater than cultivable methanogen densities and potential methanogenic activities (Joulian et al., 1997).

Methane is oxidized by two different methanotrophs (high and low CH_4 affinity). In general, O_2 availability is the main factor controlling methanotrophs' activity. The high-affinity CH_4 oxidation occurs at a low CH_4 concentration (<12 ppm) close to that of the atmosphere (Topp and Hanson, 1991). Approximately 10% of total CH_4 consumption is contributed by methanotrophs (Topp and Pattey, 1997). In comparison, low-affinity CH_4 oxidation appears at high CH_4 concentrations (>40 ppm). In flooded rice cropping soils, over 90% of the

Table 2 Annual CH_4 fluxes in a typical mono-rice paddy field under different fertilization

Treatment	Total CH_4 flux (kg ha^{1})		
	Rice cropping	Fallow	Total
Chemical fertilizer (NPK)[a]	220	−1.6	218
Green manure[b]	904	22.7	881

[a] Chemical fertilization: N-P_2O_5-K_2O = 90-45-57 kg ha^{1}.
[b] The mixture of barley and hairy vetch was cultivated during the cold fallow season, and its whole biomass (8.4 Mg ha^{1} on dry weight) was applied for rice cultivation.

CH_4 formed may be reoxidized by methanotrophs in aerobic soil areas (Sass et al., 1990; Frenzel et al., 1992; Oremland and Culbertson, 1992). Methane oxidation in the rhizosphere most strongly influences net CH_4 flux but varies according to rice growth stage (Van der Gon and Neue, 1996). Methanotrophs' activity and root oxidation activity show a highly positive correlation (King et al., 1990).

Most of CH_4 is emitted through the rice aerenchyma channel to the atmosphere, but a small portion of CH_4 is emitted as bubbles from soils via ebullition and diffusion. Rice plants' aerenchyma acts as a chimney, allowing for gaseous CH_4 exchange from soil to the atmosphere (Neue and Roger, 1994). Paddy fields with rice plants generally emit more CH_4 than flooded paddy fields without plants, due to easier transfer to the atmosphere and higher organic substrate availability for methanogenesis (Schütz et al., 1989).

Methane passively transfers from soils to the atmosphere (Nouchi et al., 1994). The CH_4 fluxes vary with rice varieties, mainly due to differences in root exudate production, oxidation potential, and morphological characteristics (Adhya et al., 1994; Butterbach-Bahl et al., 1997; Gutierrez et al., 2013). At the early rice-growing stage, CH_4 is emitted through a vertical movement of gas bubbles in the soil, but the diffusion through the aerenchyma becomes the dominant CH_4 transport pathway as plants grow. At the rice reproductive stage, over 90% of CH_4 is transported via the aerenchyma channel (Cicerone and Shetter, 1981; Schütz et al., 1989; Tyler et al., 1997).

4 Nitrous oxide emissions in rice paddy soils

The impact of N_2O emission from rice paddies on net GWP was found to be negligible. IPCC (Klein et al., 2006) proposed 0.003 kg N_2O-N kg^1 N as the emission factor of N_2O for flooded rice fields. Nitrous oxide formation is influenced by many factors including soil redox, available N and OM, but N_2O is principally produced via microbial denitrification and nitrification (Butterbach-Bahl et al., 2013; Shakoor et al., 2021; Thilakarathna and Hernandez-Ramirez, 2021).

In flooded rice paddies, soil redox conditions are not ideal for biological N_2O formation. Water drainage for mitigating CH_4 formation does not lead to high amounts of N_2O formation (Hou et al., 2000; Johnson-Beebout et al., 2009). More intermittent flooding or midseason drainage can lead to higher nitrification and denitrification rates and consequently increase N_2O emissions (Harrison-Kirk et al., 2013; Lagomarsino et al., 2016; Liang et al., 2016). However, intermittent irrigation over five cycles increased seasonal N_2O fluxes by only three times over continuous flooding of rice fields (Kritee et al., 2018), but the GWP of these fluxes is much lower than CH_4 fluxes.

5 Evaluation of the net global warming potential and greenhouse gas intensity of rice paddy soils

Since paddy rice soils can be a GHG source and sink, the effect of agricultural management practices on global warming should be measured by net GWP. This can be done through the integration of two GHG fluxes (CH_4 and N_2O) and SOC stock changes measured in CO_2 equivalents (Mosier et al., 2006). The closed static chamber method is broadly utilized to estimate CH_4 and N_2O fluxes in paddy rice soil (Fig. 3) (Schütz and Seiler, 1989). Diffusive GHG fluxes are directly quantified using the change of gas concentrations over short time intervals multiplied by the chamber volume (m^3) per area (m^2) ratio (Rolston, 1986). Because rice plants significantly influence CH_4 production, oxidation, and emission dynamics, the rice seedling inside the chamber should be properly planted to reflect the way that they experience representative growth conditions in rice fields. To calculate net GWP (Eq. 1), SOC stock change (ΔSOC) should be properly estimated along with CH_4 and N_2O fluxes. The yield scaled GHG emission impact can be compared using GHGI (kg CO_2-eq. Mg^{1} grain) that indicates the net GWP (kg CO_2-eq. ha^{1}) per grain yield (Mg ha^{1}) (Li et al., 2006; Weller et al., 2016):

$$\text{Net GWP} = CH_4 \text{ flux} \times 25 + N_2O \text{ flux} \times 298 - \Delta SOC \times 44/12 \quad (1)$$

SOC stock changes are conventionally used to estimate CO_2 exchange in arable lands. However, this practice is not precise enough to represent short-term SOC stock changes, mainly because of the big background pool and large spatial variation of SOC contents (Ciais et al., 2010; Conant et al., 2011; Smith et al., 2020). Theoretically, SOC change can be estimated using net ecosystem C budget (NECB) representing the balance between C input and output under

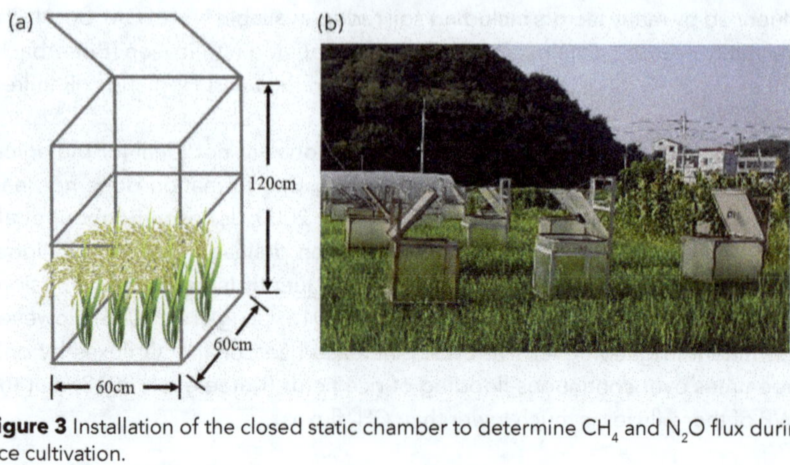

Figure 3 Installation of the closed static chamber to determine CH_4 and N_2O flux during rice cultivation.

the limited system boundary (Smith et al., 2010; Jia et al., 2012; Ma et al., 2013). For example, in rice paddies, the C input source includes the net primary production (NPP) of rice and weeds, organic amendments, fertilizers, etc. In comparison, the C output source includes harvest C removal and heterotrophic respiration (CO_2-C and CH_4-C fluxes) (Eq. 2). NPP implies total C uptake by plant biomass via photosynthesis. Organic amendments lead to the C addition via organic materials. Harvest C removal means the removed C via crop plant harvesting. Respired C loss indicates heterotrophically respired C losses (CO_2 and CH_4 fluxes) from soils:

$$\text{NECB} = \sum \text{C input} - \sum \text{C output} \tag{2}$$

The NPP (kg C ha^{1}) of rice and weeds can be quantified by their C uptakes (Smith et al., 2010) (Eq. 3).

$$\text{NPP} = \text{NPP}_{\text{above ground}} + \text{NPP}_{\text{root}} + \text{NPP}_{\text{litter}} + \text{NPP}_{\text{rhizo deposit}} \tag{3}$$

The NPP of aboveground biomass can be calculated using biomass productivity and its C content. The NPP of root biomass is calculated as 10% of the NPP of aboveground biomass (Huang et al., 2013). The NPP of litter accounts for the average 5% of whole biomass's NPP (Kimura et al., 2004). Rhizosphere-deposited NPP is estimated to range within 7–15% of total biomass's NPP (Mandal et al., 2008).

6 Effect of water management on global warming potential of rice paddy soils

Since CH_4 is biologically produced by methanogens under strongly anaerobic soil conditions, water-management techniques (i.e. midseason drainage and intermittent irrigation) are expected to be the most promising measure to suppress CH_4 fluxes in flooded rice cropping fields (Yagi et al., 1997). For instance, midseason drainage decreased total CH_4 flux by around 50% in a Japanese rice field (Kimura et al., 1992). To minimize the negative effects of soil flooding and reduction in rice growth and development (i.e. sulfide toxicity and excess tillering, Kanno et al., 1997), short-term floodwater drainage is commonly used in rice cropping regions. However, water drainage during rice cultivation also increases microbial activity, which can increase N_2O emission and microbial respiration (CO_2 emission) (Miyata et al., 2000).

Similar results were obtained in our 2-year field study in South Korea (Haque et al., 2016), where single midseason drainage practice for 3 weeks during the high CH_4 emission period considerably increased soil redox (E_h value) (Table 3). During paddy rice cultivation under continuous flooding conditions, the soil C stock (NECB) was increased by around 3.0 Mg ha^{1} year^{1},

Table 3 Changes of GHG fluxes, soil C stock, and rice productivity by midseason drainage under chemical fertilization (NPK) during rice cultivation

Irrigation system	Seasonal flux (kg CO_2-eq. ha[1])				Grain yield (kg ha[1])	GHGI (kg CO_2-eq. kg[1] grain)
	CH_4	N_2O	ΔSOC	Net GWP		
Continuous flooding	6438	145	3012	3570	6650	0.54
Midseason drainage	3313	185	1775	1723	6650	0.26
Statistical analysis	***	***	***	***	ns	***

Note: Data was modified from Haque et al. (2016).
ns and *** denote not significant and significance at 0.1% level, respectively.

but one single midseason drainage decreased this C stock by around 60%. Midseason drainage also increased seasonal N_2O flux by approximately 30% over that in the continuous flooding conditions (Table 3).

However, to evaluate the net effect of water management on global warming, the trade-off between GHG (CH_4 and N_2O) emissions and soil C stock change must be assessed. A trade-off between GHG emissions and soil C stock changes is frequently reported in rice cropping systems changing from continuous flooding to midseason drainage or intermittent irrigation (Cai et al., 1997, 1999; Yagi et al., 1997; Zheng et al., 2000; Zou et al., 2005). Despite increased SOC loss and N_2O emissions, our single midseason drainage considerably decreased the net GWP value by around 50% over that of continuous flooding, mostly due to a big suppression of CH_4 fluxes.

In addition, the midseason drainage did not influence rice growth and yield properties (Table 3). As a result, the single midseason drainage for 3 weeks can reduce GHGI by approximately 50% over continuous flooding. Such practice may be beneficial for the GWP of rice systems, because rice plants require plenty of water during their root development stage, whereas flooding may not be required at the other growth stages (Minamikawa and Sakai, 2005). Therefore, midseason drainage might be a very useful water management practice to mitigate the impact of GHG emissions without rice productivity changes.

7 Effect of green manure management on global warming potential of rice paddy soils

In rice paddy fields, cover crop cultivation during the fallow season and its biomass recycling as green manure (GM) for rice cultivation is broadly recommended to improve soil quality via SOC stock increase and to replace chemical fertilizers (Garcia-Franco et al., 2015; Yao et al., 2019). In temperate mono-rice cropping paddies like Korea and Japan, winter cover crops (i.e. hairy vetch and Chinese milk vetch as a leguminous crop, rye, barley, and wheat as nonleguminous crops) are cultivated during the cold fallow season, and their biomass as GM is incorporated before rice transplanting.

In general, cover crop biomass productivity determines nutrient accumulation under the same management (i.e. cultivar selection, growing stage, fertilization, water management, etc.) and then directly affects subsequent crop (rice) productivity and its NPP as the main C sink (Hwang et al., 2015). Recently, the mixing cultivation of leguminous and nonleguminous cover crops has been recommended to improve biomass and nutrient accumulation. For example, the mixed seeding of barley and hairy vetch significantly increased cover crop biomass and their nutrient contents compared to those of hairy vetch or barley monocultures (Table 4) (Hwang et al., 2015). The increased biomass productivity accumulated enough nutrients to fulfill the recommended fertilization level (N-P_2O_5-K_2O = 90-45-57 kg ha^{1}, RDA, 2017) for rice cultivation, and significantly increased rice productivity without chemical fertilization.

However, fresh biomass application also provides large amounts of readily available organic C substrates for methanogens and thus strongly increases CH_4 emission in flooded rice fields (Conrad, 2007; Yang et al., 2010; Li et al., 2013; Kim et al., 2015; Hwang et al., 2017; Lee et al., 2020b; Song et al., 2021). Cover crop biomass incorporation increased the annual CH_4 flux by 2.6–4.2 times over that of chemical fertilization within the rice cropping boundary (Table 5).

Compared with CH_4 flux in a rice paddy, very small N_2O fluxes were detected during the rice cultivation period. Compared with chemical fertilization (NPK), seasonal N_2O flux was not increased by GM application, even though its flux was significantly affected by cover crop species choice (Table 5). Low N_2O emissions were observed when barley with a high C/N ratio was used as a cover crop, whereas seasonal N_2O flux was significantly increased when hairy vetch

Table 4 Effect of mixed seeding of hairy vetch and barley on biomass and nutrient productivities

Seeding ratio (%)		Biomass yield (Mg ha^{1}, dry weight)			Nutrient accumulation (kg ha^{1})			
Hairy vetch	Barley	Hairy vetch	Barley	Sum	C	N	P_2O_5	K_2O
100	0	4.80	0	4.80	1983	110	53	101
75	25	5.79	1.27	7.06	2933	143	72	142
50	50	4.92	2.32	7.24	3020	132	68	140
25	**75**	4.56	3.30	7.86	3287	132	70	149
0	100	0	5.89	5.89	2501	50	36	94
Statistical analysis		***	***	***	***	***	***	***

Note: Data was modified from Hwang et al. (2015).
*** denotes significance at the 0.1% level.

Table 5 Changes in GHG fluxes and rice productivity under different fertilization regimes

Fertilization	Annual flux (Mg CO_2-eq. ha[1])				Grain yield (Mg ha[1])	GHGI (Mg CO_2-eq. Mg[1] grain)
	CH_4	N_2O	ΔSOC	Net GWP		
Chemical fertilizer	5.5	3.0	−2.1	10.6	6.7	1.6
Green manure						
Barley	14.4	1.5	5.3	10.6	6.2	1.7
Hairy vetch	15.8	3.5	−0.2	19.5	7.1	2.7
Mixture	23.2	3.4	8.8	17.8	8.1	2.2
Statistical analysis	***	**	**	**	**	***

Note: Data was modified from Hwang et al. (2017).
** and *** denote significance at the 1% and 0.1% levels, respectively.

with a low C/N ratio was used as a cover crop. Whilst chemical fertilization alone slightly decreased SOC stock, biomass application with a high C/N ratio (barley, and a mixture of barley and hairy vetch) significantly increased SOC stock. As far as GWP is considered, cover cropping and its biomass application as GM significantly increased net GWP by around 60–80% compared to chemical fertilization (NPK).

Interestingly, short-term aerobic predigestion of biomass applied to soils before irrigation strongly reduced CH_4 emission during rice cultivation (Table 6) (Lee et al., 2020b; Song et al., 2021). Since the same biomass was applied at different time intervals before irrigation for rice cultivation, aerobic decomposition of labile organic C substrates under dried soil conditions might effectively mitigate CH_4 emission during flooded rice cultivation. Only 10 days of aerobic predigestion of cover crop biomass incorporated in soils decreased seasonal CH_4 flux by 55–60% as compared to flooding without aerobic predigestion (0 days). This mitigation effect was significantly increased with longer aerobic periods. Moreover, short-term aerobic predigestion of biomass applied soils before irrigation did not influence seasonal N_2O fluxes.

Short-term aerobic predigestion of GM-amended soils did not influence seasonal SOC stock changes estimated by NECB analysis (Table 6). The same volumes of cover crop biomass were amended as the C input source, and the rice plant's NPP did not change. As a C output source, rice harvest C removal was not affected by aerobic predigestion. The respired C loss thus worked as the main determinant of the NECB scale. In aerobic predigestion plots, high C output through CO_2 emission was observed, but the CH_4-C flux was decreased, resulting in no statistical differences of total respired-C loss among different aerobic predigestion treatments.

In addition, short-term aerobic predigestion in soils with cover crop biomass effectively decreased net GWP, but was not significantly affected by

Table 6 Effect of aerobic predigestion of biomass amended soils on GHG fluxes and rice productivity

Days before flooding	Seasonal flux (Mg CO_2-eq. ha^{1})				Grain yield (Mg ha^{1})	GHGI (Mg CO_2-eq. Mg^{1} grain)
	CH_4	N_2O	ΔSOC	Net GWP		
30	8.6	2.8	10.9	0.4	5.7	0.1
20	13.6	3.0	10.9	5.6	5.8	1.0
10	18.1	3.4	10.4	11.1	5.8	1.9
0	43.4	2.4	10.1	35.7	5.9	6.1
Statistical analysis	***	ns	ns	***	ns	***

Note: Data was modified from Song et al. (2021).
ns and *** denote not significant and significance at the 0.1% level, respectively.

the duration of aerobic predigestion (10–30 days) before flooding. We found that aerobic predigestion of cover crop biomass over 10 days considerably decreased GHGI by 65–98% as compared to the cover crop treatment without aerobic predigestion (Song et al., 2021). As a result, in GM-amended rice paddies, aerobic predigestion before flooding may be essential to reduce net GWP and GHGI.

8 Effect of straw management on global warming potential of rice paddy soils

Straw recycling is regarded as the most effective management practice to increase SOC stock in rice fields. However, most straw has been removed for livestock feeding and burnt for site preparation, thereby depleting SOC stocks (Wang et al., 2015a; Xia et al., 2018). In Korean rice paddies, straw application level officially decreased from 3.7 Mg ha^{-1} in 1992 to 1.1 Mg ha^{-1} in 1998 (Kim et al., 2003), which consequently resulted in depletion of SOM contents in nationwide rice paddies from the average of 26 g kg^{-1} in the 1960s to 24 g kg^{-1} in the 2000s (RDA, 2010).

In soil C balance (NECB) analysis in a typical rice paddy (Lee et al., 2020a), only chemical fertilization (NPK) without straw return decreased SOC stock by 0.24–1.12 Mg C ha^{-1} year^{-1}. This was caused by harvest C removal (65–73% of total C output) and the respired C loss (27–35% of total C output). However, straw retention as an organic amendment considerably increased SOC stocks by 1.48–2.82 Mg C ha^{-1} year^{-1}. Straw spreading over the soil surface was more efficient than straw incorporation to increase SOC stocks, due to favorable microbial decomposition of incorporated rice straw during the dried fallow season. Moreover, straw incorporation significantly increased CH_4 fluxes during flooded rice cropping. Its addition also increased N_2O emissions, but this

Table 7 Characteristics of GHG emissions and rice productivity under different rice straw management strategies

Fertilizer management	Straw application	Annual flux (Mg CO_2-eq. ha[1]) CH_4	N_2O	ΔSOC	Net GWP	Grain yield (Mg ha[1])	GHGI (Mg CO_2-eq. Mg[1] grain)
NPK		6.4	0.9	−2.5	9.8	5.1	1.9
NPK + straw	Mixing	10.2	1.6	7.1	4.8	6.0	0.8
	Spreading	17.8	1.3	8.8	10.4	5.7	1.8
Statistical analysis		***	**	***	**	**	*

Note: Data was modified from Lee et al. (2020a).
*, ** and *** denote significance at the 5%, 1%, and 0.1% levels, respectively.

enhanced flux was still too small to influence the net GWP. SOC stock increase via straw recycling may conflict with CH_4 flux increase in a rice paddy (Hsu et al., 2009), and both fluxes must be considered to evaluate net GWP in rice paddies (see above; Table 7).

In the conventional rice cropping system in which straw is removed and chemical fertilizer is applied, CH_4 flux and SOC stock change represent approximately 65% and 25% of the annual net GWP (average 9.8 Mg CO_2-eq. ha^{-1}), respectively. However, straw retention did not only increase CH_4 flux but also SOC accumulation. In contrast to the general assumption that organic amendments in rice fields significantly increase GHG fluxes and global warming impact (Le Mer and Roger, 2001), rice straw incorporation into soil and aerobic decomposition during the dried fallow season were very effective in decreasing net GWP by around 50% over NPK, primarily due to SOC stock increase. In addition, rice straw addition stimulated rice growth and increased rice grain productivity by 10–20% as compared to paddies managed with straw removal. In contrast, although straw spreading over the surface layer and aerobic decomposition during the long fallow season was effective as a way to increase SOC stocks, it also strongly increased CH_4 emissions. As a result, this kind of straw management was not effective in decreasing net GWP.

In conclusion, straw recycling may be essential to improve soil productivity through SOC stock increase without an equivalent increase in GHG emission impact. However, straw should be incorporated into the mineral soil, not spread over soil surface, to be aerobically decomposed during the long fallow season.

9 Effect of fertilizer management on global warming potential of rice paddy soils

Intensive cropping systems may increase crop productivity but have a negative environmental impact. Of all rice management practices, appropriate fertilization is one of the most effective ways to improve crop productivity and quality. Proper fertilizer management can improve rice plants' photosynthetic capacity, resistance to biotic stress, nutrient uptake, and productivity. For example, in a long-term fertilized rice paddy in Korea, N fertilizer application enhanced rice grain productivity by an average of 45%, which was much higher than 9.8% and 5.1% of grain yield increase by phosphorus (P) and potassium (K) fertilization, respectively (Lee et al., 2008).

However, high N fertilization exceeding plant requirements can decrease N use efficiency (NUE) (Peng et al., 2006; Liu et al., 2015) as well as cause resource loss and environmental pollution. Appropriate N fertilization is important to improve crop productivity and sustain high environmental quality (Tilman et al., 2002; Yousaf et al., 2014). Furthermore, N fertilizer application may increase GHG fluxes from soils (Schimel, 2000; Ma et al., 2007; Liu et al., 2015; Yang et al., 2015). Generally, the N fertilization level shows a highly positive relationship with N_2O emission levels (Ma et al., 2013), because it provides the substrate necessary for biological N_2O formation through nitrification and denitrification (Paul et al., 1993).

There were many contrasting findings related to the influence of N fertilization on CH_4 fluxes in rice paddies (Bodelier et al., 2000; Shrestha et al., 2010; Kim et al., 2019). In several studies (Xie et al., 2010; Dong et al., 2011; Yao et al., 2012), N fertilization considerably decreased CH_4 fluxes, mostly because of enhanced CH_4 oxidation in the rice rhizosphere. However, in many other cases, N fertilization strongly increased CH_4 formation and emissions, due to stimulation of methanogen activity (Schimel, 2000; Cai et al., 2007). In our 2-year field study (Kim et al., 2019), N fertilization significantly increased seasonal N_2O fluxes (Table 8). Moreover, seasonal CH_4 fluxes also increased with N fertilization increase, peaking at around 130 kg N ha^{-1} of urea addition and thereafter decreased. Consequently, N fertilization has a considerable

Table 8 Characteristics of GHG emissions and rice productivity under different N application levels

N application (kg N ha^{1})	Annual flux (Mg CO_2-eq. ha^{1})				Grain yield (Mg ha^{1})	GHGI (Mg CO_2-eq. Mg^{1} grain)
	CH_4	N_2O	ΔSOC	Net GWP		
0	6.2	0.17	−5.0	11.3	4.6	2.5
45	6.6	0.23	−4.8	11.7	5.6	2.1
90	7.4	0.27	−4.6	12.2	6.8	1.8
180	7.1	0.33	−4.7	12.2	5.9	2.1
Statistical analysis	*	***	***	**	**	*

Note: Data was modified from Kim et al. (2019).
*, ** and *** denote significance at the 5%, 1%, and 0.1% levels, respectively.

effect on GWP, but in contrast to what may be expected, net GWP was mostly influenced by the CH_4 flux, which represents more than 90% of the total GWP.

Similar to changes in GWP, GHGI is increased by N addition (Kim et al., 2019). In a typical Korean rice paddy, the lowest GHGI was observed at 104-112 kg N ha^{-1} of urea application, and thereafter, the GHGI was increased with N fertilization increase, mainly due to a decrease in grain productivity at high N fertilization. Theoretically, any soil management practice that can decrease N fertilization levels without changing crop productivity will result in the lowest GHGI (Mosier et al., 2006). Consequently, N fertilization should be carefully managed to achieve sustainable rice cultivation systems with high rice productivity and low GWP and GHGI.

10 Effect of soil amendments on global warming potential of rice paddy soils

Methane is produced through the anaerobic decomposition of organic substrates when CO_2 is used as an electron acceptor. Soil microorganisms, which can reduce energetically more favorable electron acceptors (i.e. O_2, NO_3^-, Mn^{4+}, Fe^{3+}, SO_4^{2-}) may outcompete methanogens using the less favorable electron acceptor (i.e. CO_2) (Lovley et al., 2004). Methanogenesis may thus be suppressed by adding alternative electron acceptors to extremely reduced soils. This suppression might result in a combination of inhibition and competition effects for common electron donors (Achtnich et al., 1995; Jakobsen et al., 1981).

Soil amendments that contain electron acceptors (i.e. ammonium nitrate, ammonium sulfate, iron slag-based amendments, etc.) can be utilized to mitigate CH_4 production in flooded rice fields. For example, iron oxide (Fe_2O_3) can react as an important oxidizing agent and control the formation of organic acids (Asami and Takai, 1970) and CH_4 (Watanabe and Kimura, 1999) in anaerobic soils.

For example, blast furnace slag (BFS, also known as iron slag) based silicate fertilizer contains around 4.8-5.4% and 0.3-1.1% of Fe and Mn oxides, respectively (Ali et al., 2009a; Lim et al., 2021). In Korean and Japanese rice paddies, silicate fertilizers have been used as alkaline amendments to improve acidic soil pH and provide valuable elements for rice over 50-100 years (Datnoff et al., 1997). During rice cultivation, silicate fertilization strongly increases dissolved Fe contents in soils (Ali et al., 2008a; Wang et al., 2015b), which suppresses methanogens` activity and CH_4 emissions.

Moreover, the addition of silicate fertilizer stimulates rice root activities (Zhang et al., 2020). The solubility of Fe and Mn oxides can be expected as the anaerobic condition continues to develop (Gotoh and Patrick Jr, 1974; Miao et al., 2006). Rice roots emit a high amount of O_2 to reduce Fe and Mn toxicity, thus decreasing their solubility (Mei et al., 2012), as the solubility of oxidized

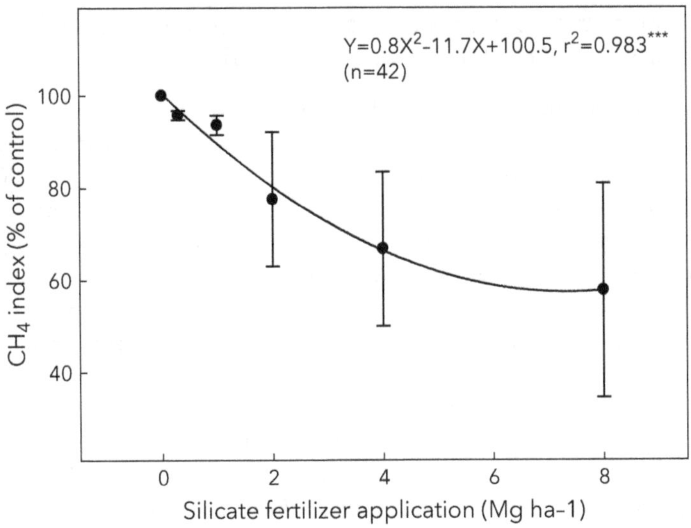

Figure 4 Changes of CH_4 flux index (% of control) under different levels of silicate fertilizer (blast furnace slag) application as soil amendment in rice paddies (Lim et al., 2021).

Fe and Mn is much lower than those of the reduced Fe and Mn compounds (Schwertmann, 1991). Hence, CH_4 inside the rice rhizosphere can be indirectly oxidized by methanotrophic bacteria. Therefore, silicate fertilizer application was effective in suppressing CH_4 emission during rice cultivation, probably due to the added Fe^{3+} and Mn^{4+} as an electron acceptor and the improved oxidation potential in rice roots (Liang et al., 2007; Ali et al., 2008a,b; Ali et al., 2009a,b; Lee et al., 2012; Das et al., 2020). By analyzing data from several studies, Lim et al. (2021) showed that the CH_4 flux index (%), which is used to compare the total CH_4 flux from blast furnace slag treated plots to that of control plots, considerably decreased as the fertilization increased (Fig. 4).

Silicate fertilizers have alkaline pH and contain high amounts of calcium oxide (CaO) and silicate (SiO_2) as their main components (Proctor et al., 2000; Gwon et al., 2018; Lim et al., 2021). In irrigated paddy fields, rice accumulates SiO_2 by around 10% of the total biomass on dry weight (Savant et al., 1996; Ma et al., 2006). Silicate fertilizer enhances rice plant erectness and reduces lodging damage (Idris et al., 1975; Hossain et al., 2002). This improvement in rice plants' physical properties increased rice productivity and improved rice quality. Rice grain productivity may be maximized at around 4 Mg ha^{-1} of silicate fertilizer application with an average of 18% increase (Lim et al., 2021). Therefore, application of soil amendments containing electron acceptors and silicate can be an effective soil management strategy to reduce CH_4 emission in rice fields and concomitantly increase rice productivity via improvement of soil properties.

11 Effect of biochar and compost addition on global warming potential of rice paddy soils

To simultaneously increase SOC stocks and decrease CH_4 emission in rice fields, the utilization of stabilized organic amendments (i.e. biochar and compost) has been recommended (Woolf et al., 2010; Feng et al., 2012; Weng et al., 2017: Qi et al., 2018; Jeong et al., 2019). However, in most studies, the effect of stabilized organic amendment applications on GHG fluxes and SOC stock changes was evaluated only within the cropping system but did not consider additional GHG fluxes from industrial processes when producing biochar and compost (Zhong et al., 2013; Jeong et al., 2018). It is largely unknown if biochar and compost amendments in rice soils can fully reduce GHG emission impacts within the whole process from their production to land utilization.

To assess the impact of organic amendment applications on GHG emissions (taking into account their production and application) (Canatoy et al., 2022), three different types of swine manure (fresh, compost, and biochar) were selected for comparison. Air-dried swine manure was mixed with sawdust (50% of dried weight) and used to make compost and biochar. To make biochar, the manure mixture was pyrolyzed at 400°C for 4 h. To investigate GHG fluxes in rice cropping systems, the three different types of organic amendments (fresh, compost, and biochar manure) treatments were applied with the same rate (12 Mg ha^1 on dry weight) under the same fertilization and managements. The whole system analysis was divided into industrial and cropping processes. During the industrial process, to produce compost and biochar, direct GHG fluxes were evaluated by the closed chamber method. Additional GHG fluxes from transportation and electricity consumption were taken into consideration by using European Emission Standards 3 and 4 (European Parliament Council, 2000). In the rice cropping process, biogenic GHG (CH_4 and N_2O) fluxes were directly monitored using the closed chamber method, and SOC stock changes were evaluated by analyzing the NECB which means the difference between C input and output (Smith et al., 2010).

In comparison with fresh manure, compost showed a high potential to increase SOC stocks and reduce CH_4 fluxes after soil application. Although compost application reduced net GWP by around 43% over fresh manure, it also showed huge GHG emissions, amounting to 25.8 Mg CO_2-eq ha^1 of net GWP. Biochar was much more effective than compost in decreasing net GWP in a rice paddy, via strong reduction in CH_4 emissions and a high increase in SOC stocks (Agegnehu et al., 2016; Phuong et al., 2020). Biochar amendment resulted in favorable environmental conditions for methanotrophic proteobacteria to proliferate, which resulted in an enhanced rate of CH_4 oxidation (Feng et al., 2012). Moreover, biochar functionally reduced N_2O as it aged in the soil due to an enhanced sorption of nitrate and ammonium via oxidative reactions on its

Table 9 Life cycle analysis to evaluate GWP and GHGI under different types of manure amendments

System boundary	Fresh manure	Compost manure	Biochar manure: Without syngas recycling	Biochar manure: With syngas recycling	Statistical analysis
Industrial process (Mg CO_2-eq. ha^{1})					
Net GWP	0.07	34.5	32.6	13.5	
Cropping process (Mg CO_2-eq. ha^{1})					
CH_4 flux	54.9	40.9	17.9	17.9	***
N_2O flux	2.2	1.7	1.1	1.1	**
ΔSOC	12.1	16.8	19.0	19.0	**
Net GWP	45.0	25.8	−0.12	−0.12	**
Whole process					
Net GWP (Mg CO_2-eq. ha^{1})	45.1	60.3	32.5	13.4	**
Grain yield (Mg ha^{1})	7.8	8.4	8.2	8.2	**
GHGI (Mg CO_2-eq. Mg^{1} grain)	5.8	7.1	3.9	1.6	***

Note: Data was modified from Canatoy et al. (2022) (under revision in *Science of the Total Environment*).
** and *** denote significance at the 1% and 0.1% levels, respectively.

surfaces (Singh et al., 2010). SOC stocks increased under biochar addition by around 57% and 13% over fresh manure and compost, respectively. As a result, biochar application converted rice paddy into a GHG sink with minus 119 kg CO_2-eq. ha^{1} of net GWP. Similar results were reported in several rice cropping studies (Liu et al., 2016; Mohammadi et al., 2020).

In the industrial process (compost and biochar production and transport processes), large amounts of GHGs were emitted (Table 9). Composting takes place under aerobic conditions, which lead to decomposition of labile organic substrates, resulting in their transformation into a more stabilized form via enzymatic biochemical degradation, releasing mainly CO_2 during the conversion process (Fukumoto et al., 2003; Mehta et al., 2014). In our study, a total of 34.3 Mg CO_2-eq. of biogenic GHGs were released to make 12 Mg of compost from 24.2 Mg of manure mixture. This net GWP consisted of nearly 55%, 37%, and 8% of CO_2, N_2O, and CH_4 fluxes, respectively. Although CO_2 is an intrinsic byproduct of composting not accounted for by IPCC calculations, it is still important for implementing abatement technologies to reduce global warming (Sánchez et al., 2015). Without gas capturing facilities during the compost production process, these levels of GHGs could markedly influence global warming and, therefore, overall GHG emissions should be carefully considered in evaluating the effect of compost utilization on GHG emissions in paddy fields.

In our study, without a syngas recycling system during the pyrolysis process, a total of 19.2 Mg CO_2-eq. syngas GHGs were emitted to make 12 Mg of biochar from 24.8 Mg of manure mixture (Table 9). This net GWP consisted of around 66% and 34% of CO_2 and CH_4 fluxes, respectively. However, the use of a syngas recycling system was reported to impede CO_2 and CH_4 formation and further convert them into non-GHGs like CO, H_2, and water (Shen et al., 2017; You et al., 2018; Schmidt et al., 2019). Conversely, N_2O emission during pyrolysis negligibly influenced net GWP of the industrial process (Pennise et al., 2001; Clark et al., 2017). As a result, pyrolysis with syngas recycling can reduce this net GWP by 59% over a no syngas recycling system.

Considering the whole life cycle from industrial to rice cropping processes, compost utilization as organic amendment significantly increased net GWP compared to fresh manure and biochar treatment, due to huge amounts of biogenic GHG emissions during the composting process. In comparison with fresh manure, compost application was very effective in reducing GHG emission impact during the rice cropping process (Islam et al., 2020), but the additional GHG fluxes from composting process markedly increased net GWP within the life cycle analysis (Table 9). In the biochar treatment, although similar levels (32.6 Mg CO_2-eq.) of GHGs were released during pyrolysis with the composting process, the longevity of biochar after soil application and the use of syngas recycling system may have positive effects on decreasing GWP. Hence, when the whole life cycle is considered, biochar application significantly reduced net GWP by a minimum of 28% and 46% and a maximum of 70% and 78% over fresh manure and compost, respectively. The minimum and maximum reductions were estimated from industrial processes without and with a syngas recycling system, respectively. Irrespective of the syngas recycling system, lifecycle analysis showed that biochar application to rice paddies may convert them into a huge GHG sink.

Compost manure application considerably increased rice grain productivity over fresh manure which might be due to higher available nutrients and improved soil qualities (Sadegh-Zadeh et al., 2018). No statistical difference was noted among the three selected manure treatments, only a trend to higher rice productivity in compost treatments (Table 9). In comparison with fresh manure, compost application significantly increased GHGI, while biochar highly decreased GHGI by 33% and 72% over fresh manure without and with a syngas recycling system, respectively. Therefore, biochar utilization rather than compost as organic amendment could be a promising option to maintain soil quality and suppress GHG emissions in rice cultivation systems.

12 Conclusion

Anaerobic conditions in paddy soils during flooding lead to SOC accumulation but also significantly increase CH_4 emissions, which has a much greater impact than changes in SOC stock or N_2O emissions on GWP of rice cropping systems.

In these systems, agricultural management practices must consider trade-offs in the form of CH_4 emissions, which could outweigh SOC sequestration.

The combined management of aerobic decomposition of rice straw or predigestion of organic amendments before flooding and irrigation of water drainage during rice cropping can significantly decrease net GWP over conventional soil management. In addition, careful management of fertilizers and utilization of soil amendments that have electron acceptors can reduce CH_4 emissions and net GWP, in addition to improving soil productivity. As a stable organic amendment, biochar (rather than compost) is effective to mitigate GHG emission impact and improve soil quality. In conclusion, more stable organic amendments like biochar may be useful to decrease the global warming impact and improve soil quality in irrigated rice fields. The combined management of organic amendments with water and soil amendments, which impede methanogenesis, can also be an important option to mitigate GHG emissions of rice paddies.

13 Acknowledgements

This work was supported by Basis Science Research Program through the National Research Foundation, Ministry of Education, Korea (NRF-2015R1A6A1A03031413). Hyeon Ji Song was supported by scholarships from the BK21+program of the Ministry of Education and Human Resources Development, Korea.

14 References

Achtnich, C., Bak, F. and Conrad, R. 1995. Competition for electron donors among nitrate reducers, ferric iron reducers, sulfate reducers, and methanogens in anoxic paddy soil. *Biology and Fertility of Soils* 19(1), 65–72.

Adhya, T. K., Rath, A. K., Gupta, P. K., Rao, V. R., Das, S. N., Parida, K. M., Parashar, D. C. and Sethunathan, N. 1994. Methane emission from flooded rice fields under irrigated conditions. *Biology and Fertility of Soils* 18(3), 245–248.

Agegnehu, G., Bass, A. M., Nelson, P. N. and Bird, M. I. 2016. Benefits of biochar, compost and biochar-compost for soil quality, maize yield and greenhouse gas emissions in a tropical agricultural soil. *Science of the Total Environment* 543(A), 295–306.

Alexandratos, N. and Bruinsma, J. 2012. World agriculture towards 2030/2050: the 2012 revision. ESA Working paper No. 12-03. Food and Agriculture Organization, Rome.

Ali, M. A., Lee, C. H. and Kim, P. J. 2008a. Effect of silicate fertilizer on reducing methane emission during rice cultivation. *Biology and Fertility of Soils* 44(4), 597–604.

Ali, M. A., Oh, J. H. and Kim, P. J. 2008b. Evaluation of silicate iron slag amendment on reducing methane emission from flood water rice farming. *Agriculture, Ecosystems and Environment* 128(1-2), 21-26.

Ali, M. A., Lee, C. H., Kim, S. Y. and Kim, P. J. 2009a. Effect of industrial by-products containing electron acceptors on mitigating methane emission during rice cultivation. *Waste Management* 29(10), 2759-2764.

Ali, M. A., Lee, C. H., Lee, Y. B. and Kim, P. J. 2009b. Silicate fertilization in no-tillage rice farming for mitigation of methane emission and increasing rice productivity. *Agriculture, Ecosystems and Environment* 132(1-2), 16-22.

Anastasi, C., Dowding, M. and Simpson, V. J. 1992. Future CH4 emissions from rice production. *Journal of Geophysical Research* 97(D7), 7521-7525.

Asami, T. and Takai, Y. 1970. Behaviour of free iron oxide in paddy soils. 4. Relationship between reduction of free iron oxide and formation of gases in paddy soils. *Journal of the Science of Soil and Manure, Japan* 41(1), 48-55.

Bodelier, P. L. E., Hahn, A. P., Arth, I. R. and Frenzel, P. 2000. Effects of ammonium-based fertilisation on microbialprocesses involved in methane emission from soilsplanted with rice. *Biogeochemistry* 51(3), 225-257.

Butterbach-Bahl, K., Papen, H. and Rennenberg, H. 1997. Impact of gas transport through rice cultivars on methane emission from rice paddy fields. *Plant, Cell and Environment* 20(9), 1175-1183.

Butterbach-Bahl, K., Baggs, E. M., Dannenmann, M., Kiese, R. and Zechmeister-Boltenstern, S. 2013. Nitrous oxide emissions from soils: how well do we understand the processes and their controls? *Philosophical Transactions of the Royal Society of London. Series B, Biological Sciences* 368(1621), 20130122.

Cai, Z., Xing, G., Yan, X., Xu, H., Tsuruta, H., Yagi, K. and Minami, K. 1997. Methane and nitrous oxide emissions from rice paddy fields as affected by nitrogen fertilisers and water management. *Plant and Soil* 196(1), 7-14.

Cai, Z. C., Xing, G. X., Shen, G. Y., Xu, H., Yan, X. Y., Tsuruta, H., Yagi, K. and Minami, K. 1999. Measurements of CH4 and N20 emissions from rice paddies in Fengqiu, China. *Soil Science and Plant Nutrition* 45(1), 1-13.

Cai, Z., Shan, Y. and Xu, H. 2007. Effects of nitrogen fertilization on CH4 emissions from rice fields. *Soil Science and Plant Nutrition* 53(4), 353-361.

Canatoy, R. C., Cho, S. R., Ok, Y. S., Jeong, S. T. and Kim, P. J. 2022. Critical evaluation of biochar utilization effect on mitigating global warming in whole rice cropping boundary. *Science of the Total Environment* 827, 154344.

Carlson, K. M., Gerber, J. S., Mueller, N. D., Herrero, M., MacDonald, G. K., Brauman, K. A., Havlik, P., O'Connell, C. S., Johnson, J. A., Saatchi, S. and West, P. C. 2017. Greenhouse gas emissions intensity of global croplands. *Nature Climate Change* 7(1), 63-68.

Chen, H., Zhu, Q. A., Peng, C., Wu, N., Wang, Y., Fang, X., Jiang, H., Xiang, W., Chang, J., Deng, X. and Yu, G. 2013. Methane emissions from rice paddies natural wetlands, lakes in China: synthesis new estimate. *Global Change Biology* 19(1), 19-32.

Chen, X., Hu, Y., Xia, Y., Zheng, S., Ma, C., Rui, Y., He, H., Huang, D., Zhang, Z., Ge, T., Wu, J., Guggenberger, G., Kuzyakov, Y. and Su, Y. 2021. Contrasting pathways of carbon sequestration in paddy and upland soils. *Global Change Biology* 27(11), 2478-2490.

Ciais, P., Wattenbach, M., Vuichard, N., Smith, P., Piao, S. L., Don, A., Luyssaert, S., Janssens, I. A., Bondeau, A., Dechow, R., Leip, A., Smith, P., Beer, C., Van Der Werf, G. R., Gervois, S., Van Oost, K., Tomelleri, E., Freibauer, A. and Schulze, E. D. 2010. The European greenhouse gas balance revisited: Part 2: Croplands. *Global Change Biology* 16(5), 1409-1428.

Cicerone, R. J. and Shetter, J. D. 1981. Sources of atmospheric methane: measurements in rice paddies and a discussion. *Journal of Geophysical Research* 86(C8), 7203-7209.

Clark, S. C., Ryals, R., Miller, D. J., Mullen, C. A., Pan, D., Zondlo, M. A., Boateng, A. A. and Hastings, M. G. 2017. Effluent gas flux characterization during pyrolysis of chicken manure. *ACS Sustainable Chemistry and Engineering* 5(9), 7568-7575.

Conant, R. T., Ogle, S. M., Paul, E. A. and Paustian, K. 2011. Measuring and monitoring soil organic carbon stocks in agricultural lands for climate mitigation. *Frontiers in Ecology and the Environment* 9(3), 169-173.

Conrad, R. 2007. Microbial ecology of methanogens and methanotrophs. *Advances in Agronomy* 96, 1-63.

Das, S., Galgo, S. J., Alam, M. A., Lee, J. G., Hwang, H. Y., Lee, C. H. and Kim, P. J. 2020. Recycling of ferrous slag in agriculture: potentials and challenges. *Critical Reviews in Environmental Science and Technology* 52(8), 1247-1281.

Datnoff, L. E., Deren, C. W. and Snyder, G. H. 1997. Silicon fertilization for disease management of rice in Florida. *Crop Protection* 16(6), 525-531.

Denier van der Gon, H. A. C. and Neue, H.-U. 1996. Oxidation of methane in the rhizosphere of rice plants. *Biology and Fertility of Soils* 22(4), 359-366.

Diacono, M. and Montemurro, F. 2011. Long-term effects of organic amendments on soil fertility. *Sustainable Agriculture* 2, 761-786.

Dong, H., Yao, Z., Zheng, X., Mei, B., Xie, B., Wang, R., Deng, J., Cui, F. and Zhu, J. 2011. Effect of ammonium-based, non-sulfate fertilizers on CH4 emissions from a paddy field with a typical Chinese water management regime. *Atmospheric Environment* 45(5), 1095-1101.

European Parliament 2000. Directive 98/69/EC relating to measures to be taken against air pollution by emissions from motor vehicles and amending council directive 70/220/EEC. https://eurlex.europa.eu/resource.html?uri¼cellar:9ee5d16b-1a4a-4a72-ac90-5e3a0bb1d745.0008.02/DOC_1&format¼PDF.

Feng, Y., Xu, Y., Yu, Y., Xie, Z. and Lin, X. 2012. Mechanisms of biochar decreasing methane emission from Chinese paddy soils. *Soil Biology and Biochemistry* 46, 80-88.

Frenzel, P., Rothfuss, F. and Conrad, R. 1992. Oxygen profiles and methane turnover in a flooded rice microcosm. *Biology and Fertility of Soils* 14(2), 84-89.

Fukumoto, Y., Osada, T., Hanajima, D. and Haga, K. 2003. Patterns and quantities of NH3, N2O and CH4 emissions during swine manure composting without forced aeration--effect of compost pile scale. *Bioresource Technology* 89(2), 109-114.

Garcia, J. L., Patel, B. K. and Ollivier, B. 2000. Taxonomic, phylogenetic, and ecological diversity of methanogenic Archaea. *Anaerobe* 6(4), 205-226.

Garcia-Franco, N., Albaladejo, J., Almagro, M. and Martínez-Mena, M. 2015. Beneficial effects of reduced tillage and green manure on soil aggregation and stabilization of organic carbon in a Mediterranean agroecosystem. *Soil and Tillage Research* 153, 66-75.

Gotoh, S. and Patrick Jr., W. H. 1974. Transformation of iron in a waterlogged soil as influenced by redox potential and pH. *Soil Science Society of America Journal* 38(1), 66-71.

Goyal, S., Chander, K., Mundra, M. C. and Kapoor, K. K. 1999. Influence of inorganic fertilizers and organic amendments on soil organic matter and soil microbial properties under tropical conditions. *Biology and Fertility of Soils* 29(2), 196-200.

Gutierrez, J., Kim, S. Y. and Kim, P. J. 2013. Effect of rice cultivar on CH4 emissions and productivity in Korean paddy soil. *Field Crops Research* 146, 16–24.

Gwon, H. S., Khan, M. I., Alam, M. A., Das, S. and Kim, P. J. 2018. Environmental risk assessment of steel-making slags and the potential use of LD slag in mitigating methane emissions and the grain arsenic level in rice (Oryza sativa L.). *Journal of Hazardous Materials* 353, 236–243.

Haque, M. M., Kim, G. W., Kim, P. J. and Kim, S. Y. 2016. Comparison of net global warming potential between continuous flooding and midseason drainage in monsoon region paddy during rice cropping. *Field Crops Research* 193, 133–142.

Harrison-Kirk, T., Beare, M. H., Meenken, E. D. and Condron, L. M. 2013. Soil organic matter and texture affect responses to dry/wet cycles: effects on carbon dioxide and nitrous oxide emissions. *Soil Biology and Biochemistry* 57, 43-55.

Herndon, E. M., Mann, B. F., Roy Chowdhury, T., Yang, Z., Wullschleger, S. D., Graham, D., Liang, L. and Gu, B. 2015. Pathways of anaerobic organic matter decomposition in tundra soils from Barrow, Alaska. *Journal of Geophysical Research: Biogeosciences* 120(11), 2345-2359.

Hossain, M. T., Mori, R., Soga, K., Wakabayashi, K., Kamisaka, S., Fujii, S., Yamamoto, R. and Hoson, T. 2002. Growth promotion and an increase in cell wall extensibility by silicon in rice and some other Poaceae seedlings. *Journal of Plant Research* 115(1117), 23-27.

Hou, A. X., Chen, G. X., Wang, Z. P., Van Cleemput, O. and Patrick Jr., W. H. 2000. Methane and nitrous oxide emissions from a rice field in relation to soil redox and microbiological processes. *Soil Science Society of America Journal* 64(6), 2180–2186.

Hsu, Y., Singh, S. K., Chiang, M., Wu, Y. and Chang, I. 2009. Strategies to lower greenhouse gas level by rice agriculture. *African Journal of Biotechnology* 8(2), 126–132.

Huang, S., Zeng, Y., Wu, J., Shi, Q. and Pan, X. 2013. Effect of crop residue retention on rice yield in China: a meta-analysis. *Field Crops Research* 154, 188–194.

Huang, Q. H., Li, D. M., Liu, K. L., Yu, X. C., Ye, H. C., Hu, H. W., Xu, X. L., Wang, S. L., Zhou, L. J., Duan, Y. H. and Zhang, W. J. 2014. Effects of long-term organic amendments on soil organic carbon in a paddy field: a case study on red soil. *Journal of Integrative Agriculture* 13(3), 570–576.

Hwang, H. Y., Kim, G. W., Lee, Y. B., Kim, P. J. and Kim, S. Y. 2015. Improvement of the value of green manure via mixed hairy vetch and barley cultivation in temperate paddy soil. *Field Crops Research* 183, 138–146.

Hwang, H. Y., Kim, G. W., Kim, S. Y., Haque, M. M., Khan, M. I. and Kim, P. J. 2017. Effect of cover cropping on the net global warming potential of rice paddy soil. *Geoderma* 292, 49–58.

Idris, M. D., Hossain, M. M. and Choudhury, F. A. 1975. The effect of silicon on lodging of rice in presence of added nitrogen. *Plant and Soil* 43(1–3), 691–695.

IPCC 2013. Annex V: contributors to the IPCC WGI fifth assessment report. In: Stocker, T. F., Qin, D., Plattner, G.-K., Tignor, M., Allen, S. K., Boschung, J., Nauels, A., Xia, Y., Bex, V. and Midgley, P. M. (Eds), *Climate Change 2013: The PhysicalScience Basis. Contribution of Working Group I to the Fifth Assessment Reportof the Intergovernmental Panel on Climate Change*. Cambridge UniversityPress, Cambridge, UK and New York, NY, pp. 1477–1496.

Islam, S. F. U., de Neergaard, A., Sander, B. O., Jensen, L. S., Wassmann, R. and van Groenigen, J. W. 2020. Reducing greenhouse gas emissions and grain arsenic and lead levels without compromising yield in organically produced rice. *Agriculture, Ecosystems and Environment* 295, 106922.

Jakobsen, P., Patrick Jr., W. H. and Williams, B. G. 1981. Sulfide and methane formation in soils and sediments. *Soil Science* 132(4), 279-287.

Jarecki, M. K. and Lal, R. 2003. Crop management for soil carbon sequestration. *Critical Reviews in Plant Sciences* 22(6), 471-502.

Jeong, S. T., Kim, G. W., Hwang, H. Y., Kim, P. J. and Kim, S. Y. 2018. Beneficial effect of compost utilization on reducing greenhouse gas emissions in a rice cultivation system through the overall management chain. *Science of the Total Environment* 613-614, 115-122.

Jeong, S. T., Cho, S. R., Lee, J. G., Kim, P. J. and Kim, G. W. 2019. Composting and compost application: trade-off between greenhouse gas emission and soil carbon sequestration in whole rice cropping system. *Journal of Cleaner Production* 212, 1132-1142.

Jia, J. X., Ma, Y. C. and Xiong, Z. Q. 2012. Net ecosystem carbon budget, net global warming potential and greenhouse gas intensity in intensive vegetable ecosystems in China. *Agriculture, Ecosystems and Environment* 150, 27-37.

Jiang, Y., Qian, H., Huang, S., Zhang, X., Wang, L., Zhang, L., Shen, M., Xiao, X., Chen, F., Zhang, H., Lu, C., Li, C., Zhang, J., Deng, A., van Groenigen, K. J. and Zhang, W. 2019. Acclimation of methane emissions from rice paddy fields to straw addition. *Science Advances* 5(1), eaau9038.

Johnson-Beebout, S. E., Angeles, O. R., Alberto, M. C. R. and Buresh, R. J. 2009. Simultaneous minimization of nitrous oxide and methane emission from rice paddy soils is improbable due to redox potential changes with depth in a greenhouse experiment without plants. *Geoderma* 149(1-2), 45-53.

Joulian, C., Escoffier, S., Le Mer, J., Neue, H. U. and Roger, P. A. 1997. Populations and potential activities of methanogens and methanotrophs in rice fields: relations with soil properties. *European Journal of Soil Biology* 33(2), 105-116.

Kanno, T., Miura, Y., Tsuruta, H. and Minami, K. 1997. Methane emission from rice paddy fields in all of Japanese prefecture. *Nutrient Cycling in Agroecosystems* 49(1), 147-151.

Kim, S. C., Park, Y. H., Lee, Y., Lee, J. Y., Kim, C. S. and Kim, P. J. 2003. Comparison of farm based fertilizer usage in 1992 and 1999. *Korean Journal of Soil Science and Fertilizer* 36(5), 344-355.

Kim, S. Y., Park, C. K., Gwon, H. S., Khan, M. I. and Kim, P. J. 2015. Optimizing the harvesting stage of rye as a green manure to maximize nutrient production and to minimize methane production in mono-rice paddies. *Science of the Total Environment* 537, 441-446.

Kim, G. W., Gutierrez-Suson, J. and Kim, P. J. 2019. Optimum N rate for grain yield coincides with minimum greenhouse gas intensity in flooded rice fields. *Field Crops Research* 237, 23-31.

Kimura, M., Miura, Y., Watanabe, A., Murase, J. and Kuwatsuka, S. 1992. Methane production and its fate in paddy fields: I. Effects of rice straw application and percolation rate on the leaching of methane and other soil components into the subsoil. *Soil Science and Plant Nutrition* 38(4), 665-672.

Kimura, M., Murase, J. and Lu, Y. 2004. Carbon cycling in rice field ecosystems in the context of input, decomposition and translocation of organic materials and the fates of their end products (CO2 and CH4). *Soil Biology and Biochemistry* 36(9), 1399-1416.

King, G. M., Roslev, P. and Skovgaard, H. 1990. Distribution and rate of methane oxidation in sediments of the Florida Everglades. *Applied and Environmental Microbiology* 56(9), 2902-2911.

Klein, C. D., Novoa, R. S. A., Ogle, S., Smith, K. A., Rochette, P., Wirth, T. C., McConkey, B. G., Moiser, A., Rypdal, K., Walsh, M. and Williams, S. 2006. N2O emissions from managed soils, and CO2 emissions from lime and urea application. In: *IPCC Guidelines for National Greenhouse Gas Inventories* (vol. 2006). Cambridge University Press, Cambridge, UK and New York, NY.

Kritee, K., Nair, D., Zavala-Araiza, D., Proville, J., Rudek, J., Adhya, T. K., Loecke, T., Esteves, T., Balireddygari, S., Dava, O. and Ram, K. 2018. High nitrous oxide fluxes from rice indicate the need to manage water for both long-and short-term climate impacts. *Proceedings of the National Academy of Sciences of the United States of America* 115(39), 9720-9725.s.

Lagomarsino, A., Agnelli, A. E., Pastorelli, R., Pallara, G., Rasse, D. P. and Silvennoinen, H. 2016. Past water management affected GHG production and microbial community pattern in Italian rice paddy soils. *Soil Biology and Biochemistry* 93, 17-27.

Lal, R. 2004. Soil carbon sequestration impacts on global climate change and food security. *Science* 304(5677), 1623-1627.

Le Mer, J. and Roger, P. 2001. Production, oxidation, emission and consumption of methane by soils: a review. *European Journal of Soil Biology* 37(1), 25-50.

Lee, C. H., Kang, U. G., Park, K. D., Lee, D. K. and Kim, P. J. 2008. Long-term fertilization effects on rice productivity and nutrient efficiency in Korean paddy. *Journal of Plant Nutrition* 31(8), 1496-1506.

Lee, S. B., Joo, J. H., Shin, J. D., Kim, C. G. and Jung, Y. S. 2012. Effect of water-soluble silicate fertilizers on stem strength and yield of paddy rice. *Korean Journal of Soil Science and Fertilizer* 45(6), 1017-1021.

Lee, J. H., Lee, J. G., Jeong, S. T., Gwon, H. S., Kim, P. J. and Kim, G. W. 2020a. Straw recycling in rice paddy: trade-off between greenhouse gas emission and soil carbon stock increase. *Soil and Tillage Research* 199, 104598.

Lee, J. H., Park, M. H., Song, H. J. and Kim, P. J. 2020b. Unexpected high reduction of methane emission via short-term aerobic pre-digestion of green manured soils before flooding in rice paddy. *Science of the Total Environment* 711, 134641.

Li, C., Salas,W., DeAngelo, B. and Rose, S. 2006. Assessing alternative for mitigating net greenhouse gas emissions and increasing yields from rice production in China over the next twenty years. *Journal of Environmental Quality* 35(4), 1554-1565. https://doi.org/10.2134/jeq2005.0208.

Li, Y., Zhang, R., Liu, G., Chen, C., He, Y. and Liu, X. 2013. Comparison of methane production potential, biodegradability, and kinetics of different organic substrates. *Bioresource Technology* 149, 565-569.

Liang, Y., Sun, W., Zhu, Y. G. and Christie, P. 2007. Mechanisms of silicon-mediated alleviation of abiotic stresses in higher plants: a review. *Environmental Pollution* 147(2), 422-428.

Liang, L. L., Grantz, D. A. and Jenerette, G. D. 2016. Multivariate regulation of soil CO2 and N2O pulse emissions from agricultural soils. *Global Change Biology* 22(3), 1286-1298.

Lim, J. Y., Kang, Y. G., Sohn, K. M., Kim, P. J. and Galgo, S. J. C. 2021. Creating new value of blast furnace slag as soil amendment to mitigate methane emission and improve rice cropping environments. *Science of the Total Environment* 806(4), 150961.

Linquist, B., Groenigen, K. J., Adviento-Borbe, M. A., Pittelkow, C. and Kessel, C. 2012An agronomic assessment of greenhouse gas emissions from major cereal crops. *Global Change Biology* 18(1), 194-209.

Liu, Y., Zhou, Z., Zhang, X., Xu, X., Chen, H. and Xiong, Z. 2015. Net global warming potential and greenhouse gas intensity from the double rice system with integrated soil-crop system management: A three-year field study. *Atmospheric Environment* 116, 92-101.

Liu, Q., Liu, B., Ambus, P., Zhang, Y., Hansen, V., Lin, Z., Shen, D., Liu, G., Bei, Q., Zhu, J., Wang, X., Ma, J., Lin, X., Yu, Y., Zhu, C. and Xie, Z. 2016. Carbon footprint of rice production under biochar amendment-a case study in a Chinese rice cropping system. *GCB Bioenergy* 8(1), 148-159.

Liu, Q., Zhuang, L., Ni, X., You, C., Yang, W., Wu, F., Tan, B., Yue, K., Liu, Y., Zhang, L. and Xu, Z. 2018. Nitrogen additions stimulate litter humification in a subtropical forest, southwestern China. *Scientific Reports* 8(1), 17525.

Liu, Y., Ge, T., van Groenigen, K. J., Yang, Y., Wang, P., Cheng, K., Zhu, Z., Wang, J., Li, Y., Guggenberger, G. and Sardans, J. 2021. Rice paddy soils are a quantitatively important carbon store according to a global synthesis. *Communications Earth & Environment* 2(1), 1-9.

Lovley, D. R., Holmes, D. E. and Nevin, K. P. 2004. Dissimilatory fe (iii) and mn (iv). *Advances in Microbial Physiology* 49(2), 219-286.

Ma, J. F., Tamai, K., Yamaji, N., Mitani, N., Konishi, S., Katsuhara, M., Ishiguro, M., Murata, Y. and Yano, M. 2006. A silicon transporter in rice. *Nature* 440(7084), 688-691.

Ma, J., Li, X. L., Xu, H., Han, Y., Cai, Z. C. and Yagi, K. 2007. Effects of nitrogen fertiliser and wheat straw application on CH4 and N2O emissions from a paddy rice field. *Soil Research* 45(5), 359-367.

Ma, Y. C., Kong, X. W., Yang, B., Zhang, X. L., Yan, X. Y., Yang, J. C. and Xiong, Z. Q. 2013. Net global warming potential and greenhouse gas intensity of annual rice-wheat rotations with integrated soil crop system management. *Agriculture, Ecosystems and Environment* 164, 209-219. https://doi.org/10.1016/j.agee.2012.11.003.

Maclean, J. L., Dawe, D. C. and Hettel, G. P. (Eds.) 2002. *Rice Almanac: Source Book for the Most Important Economic Activity on Earth*. CABI Publishing, Wallingford.

Mandal, U. K., Warrington, D. N., Bhardwaj, A. K., Bar-Tal, A., Kautsky, L., Minz, D. and Levy, G. J. 2008. Evaluating impact of irrigation water quality on a calcareous clay soil using principal component analysis. *Geoderma* 144(1-2), 189-197.

Mehta, C. M., Palni, U., Franke-Whittle, I. H. and Sharma, A. K. 2014. Compost: its role, mechanism and impact on reducing soil-borne plant diseases. *Waste Management* 34(3), 607-622.

Mei, X. Q., Wong, M. H., Yang, Y., Dong, H. Y., Qiu, R. L. and Ye, Z. H. 2012. The effects of radial oxygen loss on arsenic tolerance and uptake in rice and on its rhizosphere. *Environmental Pollution* 165, 109-117.

Miao, S., DeLaune, R. D. and Jugsujinda, A. 2006. Influence of sediment redox conditions on release/solubility of metals and nutrients in a Louisiana Mississippi River deltaic plain freshwater lake. *Science of the Total Environment* 371(1-3), 334-343.

Minamikawa, K. and Sakai, N. 2005. The effect of water management based on soil redox potential on methane emission from two kinds of paddy soils in Japan. *Agriculture, Ecosystems and Environment* 107(4), 397-407.

Mitsch, W. J. and Gosselink, J. G. 2007. *Wetlands* (4th edn.). Wiley, New York, pp. 11–24.

Miyata, A., Leuning, R., Denmead, O. T., Kim, J. and Harazono, Y. 2000. Carbon dioxide and methane fluxes from an intermittently flooded paddy field. *Agricultural and Forest Meteorology* 102(4), 287–303.

Mohammadi, A., Khoshnevisan, B., Venkatesh, G. and Eskandari, S. 2020. A critical review on advancement and challenges of biochar application in Paddy fields: environmental and life cycle cost analysis. *Processes* 8(10), 1275.

Mosier, A. R., Halvorson, A. D., Reule, C. A. and Liu, X. J. 2006. Net global warming potential and greenhouse gas intensity in irrigated cropping systems in northeastern Colorado. *Journal of Environmental Quality* 35(4), 1584–1598.

Nesbit, S. P. and Breitenbeck, G. A. 1992. A laboratory study of factors influencing methane uptake by soils. *Agriculture, Ecosystems and Environment* 41(1), 39–54.

Neue, H. U. and Roger, P. A. 1994. *Potential of Methane Emission in Major Rice Ecologies*.

Neue, H. U., Wassmann, R., Lantin, R. S., Alberto, M. C. R., Aduna, J. B. and Javellana, A. M. 1996. Factors affecting methane emission from rice fields. *Atmospheric Environment* 30(10–11), 1751–1754.

Newton, A. 2016. Shifting sources. *Nature Geoscience* 9(5), 346–346.

Nisbet, E. G., Dlugokencky, E. J., Manning, M. R., Lowry, D., Fisher, R. E., France, J. L., Michel, S. E., Miller, J. B., White, J. W. C., Vaughn, B., Bousquet, P., Pyle, J. A., Warwick, N. J., Cain, M., Brownlow, R., Zazzeri, G., Lanoisellé, M., Manning, A. C., Gloor, E., Worthy, D. E. J., Brunke, E. -G., Labuschagne, C., Wolff, E. W. and Ganesan, A. L. 2016. Rising atmospheric methane: 2007-2014 growth and isotopic shift. *Global Biogeochemical Cycles* 30(9), 1356–1370.

Nouchi, I., Hosono, T., Aoki, K. and Minami, K. 1994. Seasonal variation in methane flux from rice paddies associated with methane concentration in soil water, rice biomass and temperature, and its modelling. *Plant and Soil* 161(2), 195–208.

Oremland, R. S. and Culbertson, C. W. 1992. Importance of methane-oxidizing bacteria in the methane budget as revealed by the use of a specific inhibitor. *Nature* 356(6368), 421–423.

Pan, G., Li, L., Wu, L. and Zhang, X. 2004. Storage and sequestration potential of topsoil organic carbon in China's paddy soils. *Global Change Biology* 10(1), 79–92.

Paul, J. W., Beauchamp, E. G. and Zhang, X. 1993. Nitrous and nitric oxide emissions during nitrification and denitrification from manure-amended soil in the laboratory. *Canadian Journal of Soil Science* 73(4), 539–553.

Paustian, K., Larson, E., Kent, J., Marx, E. and Swan, A. 2019. Soil C sequestration as a biological negative emission strategy. *Frontiers in Climate* 1, 8.

Peng, S., Buresh, R. J., Huang, J., Yang, J., Zou, Y., Zhong, X., Wang, G. and Zhang, F. 2006. Strategies for overcoming low agronomic nitrogen use efficiency in irrigated rice systems in China. *Field Crops Research* 96(1), 37–47.

Pennise, D. M., Smith, K. R., Kithinji, J. P., Rezende, M. E., Raad, T. J., Zhang, J. and Fan, C. 2001. Emissions of greenhouse gases and other airborne pollutants from charcoal making in Kenya and Brazil. *Journal of Geophysical Research: Atmospheres* 106(D20), 24143–24155.

Phuong, N. T. K., Khoi, C. M., Ritz, K., Sinh, N. V., Tarao, M. and Toyota, K. 2020. Potential use of rice husk biochar and compost to improve P availability and reduce GHG emissions in acid sulfate soil. *Agronomy* 10(5), 685.

Post, W. M., Izaurralde, R. C., Jastrow, J. D., McCarl, B. A., Amonette, J. E., Bailey, V. L., Jardine, P. M., West, T. O. and Zhou, J. 2004. Enhancement of carbon sequestration in US soils. *BioScience* 54(10), 895-908.

Proctor, D. M., Fehling, K. A., Shay, E. C., Wittenborn, J. L., Green, J. J., Avent, C., Bigham, R. D., Connolly, M., Lee, B., Shepker, T. O. and Zak, M. A. 2000. Physical and chemical characteristics of blast furnace, basic oxygen furnace, and electric arc furnace steel industry slags. *Environmental Science and Technology* 34(8), 1576-1582.

Qi, L., Niu, H. D., Zhou, P., Jia, R. J. and Gao, M. 2018. Effects of biochar on the net greenhouse gas emissions under continuous flooding and water-saving irrigation conditions in paddy soils. *Sustainability* 10(5), 1403.

RDA (Rural Development Administration, Korea) 2010. Low-carbon & natural cycling agriculture. National Institute of Agricultural Science and Technology, Suwon.

RDA (Rural Development Administration) 2017. Fertilization standard of crop plants. National Institute of Agricultural Science and Technology, Jeonju, Republic of Korea (In Korean).

Rolston, D. E. 1986. Gas flux. In: Klute, A. (Ed.), *Methods of Soil Analysis, Part 1*, 2nd ed., ASA and SSSA, Madison, WI, pp. 1103-1119 (Agron. Monogr. 9).

Sadegh-Zadeh, F., Tolekolai, S. F., Bahmanyar, M. A. and Emadi, M. 2018. Application of biochar and compost for enhancement of rice (Oryza sativa L.) grain yield in calcareous sandy soil. *Communications in Soil Science and Plant Analysis* 49(5), 552-566.

Sahrawat, K. L. 2004. Terminal electron acceptors for controlling methane emissions from submerged rice soils. *Communications in Soil Science and Plant Analysis* 35(9-10), 1401-1413.

Sánchez, A., Artola, A., Font, X., Gea, T., Barrena, R., Gabriel, D., Sánchez-Monedero, M. Á., Roig, A., Cayuela, M. L. and Mondini, C. 2015. Greenhouse gas emissions from organic waste composting. *Environmental Chemistry Letters* 13(3), 223-238.

Sass, R. L., Fisher, F. M., Harcombe, P. A. and Turner, F. T. 1990. Methane production and emission in a Texas rice field. *Global Biogeochemical Cycles* 4(1), 47-68.

Saunois, M., Bousquet, P., Poulter, B., Peregon, A., Ciais, P., Canadell, J. G., Dlugokencky, E. J., Etiope, G., Bastviken, D., Houweling, S., Janssens-Maenhout, G., Tubiello, F. N., Castaldi, S., Jackson, R. B., Alexe, M., Arora, V. K., Beerling, D. J., Bergamaschi, P., Blake, D. R., Brailsford, G., Brovkin, V., Bruhwiler, L., Crevoisier, C., Crill, P., Covey, K., Curry, C., Frankenberg, C., Gedney, N., Höglund-Isaksson, L., Ishizawa, M., Ito, A., Joos, F., Kim, H., Kleinen, T., Krummel, P., Lamarque, J., Langenfelds, R., Locatelli, R., Machida, T., Maksyutov, S., McDonald, K. C., Marshall, J., Melton, J. R., Morino, I., Naik, V., O'Doherty, S., Parmentier, F. W., Patra, P. K., Peng, C., Peng, S., Peters, G. P., Pison, I., Prigent, C., Prinn, R., Ramonet, M., Riley, W. J., Saito, M., Santini, M., Schroeder, R., Simpson, I. J., Spahni, R., Steele, P., Takizawa, A., Thornton, B. F., Tian, H., Tohjima, Y., Viovy, N., Voulgarakis, A., van Weele, M., van der Werf, G. R., Weiss, R., Wiedinmyer, C., Wilton, D. J., Wiltshire, A., Worthy, D., Wunch, D., Xu, X., Yoshida, Y., Zhang, B., Zhang, Z. and Zhu, Q. 2016. The global methane budget 2000-2012. *Earth System Science Data* 8(2), 697-751.

Saunois, M., Stavert, A. R., Poulter, B., Bousquet, P., Canadell, J. G., Jackson, R. B., Raymond, P. A., Dlugokencky, E. J., Houweling, S., Patra, P. K., Ciais, P., Arora, V. K., Bastviken, D., Bergamaschi, P., Blake, D. R., Brailsford, G., Bruhwiler, L., Carlson, K. M., Carrol, M., Castaldi, S., Chandra, N., Crevoisier, C., Crill, P. M., Covey, K., Curry, C. L., Etiope, G., Frankenberg, C., Gedney, N., Hegglin, M. I., Höglund-Isaksson, L., Hugelius, G., Ishizawa, M., Ito, A., Janssens-Maenhout, G., Jensen, K. M., Joos, F.,

Kleinen, T., Krummel, P. B., Langenfelds, R. L., Laruelle, G. G., Liu, L., Machida, T., Maksyutov, S., McDonald, K. C., McNorton, J., Miller, P. A., Melton, J. R., Morino, I., Müller, J., Murguia-Flores, F., Naik, V., Niwa, Y., Noce, S., O'Doherty, S., Parker, R. J., Peng, C., Peng, S., Peters, G. P., Prigent, C., Prinn, R., Ramonet, M., Regnier, P., Riley, W. J., Rosentreter, J. A., Segers, A., Simpson, I. J., Shi, H., Smith, S. J., Steele, L. P., Thornton, B. F., Tian, H., Tohjima, Y., Tubiello, F. N., Tsuruta, A., Viovy, N., Voulgarakis, A., Weber, T. S., van Weele, M., van der Werf, G. R., Weiss, R. F., Worthy, D., Wunch, D., Yin, Y., Yoshida, Y., Zhang, W., Zhang, Z., Zhao, Y., Zheng, B., Zhu, Q., Zhu, Q. and Zhuang, Q. 2020. The global methane budget 2000-2017. *Earth System Science Data* 12(3), 1561-1623.

Savant, N. K., Snyder, G. H. and Datnoff, L. E. 1996. Silicon management and sustainable rice production. *Advances in Agronomy* 58, 151-199.

Schaefer, H., Fletcher, S. E. M., Veidt, C., Lassey, K. R., Brailsford, G. W., Bromley, T. M., Dlugokencky, E. J., Michel, S. E., Miller, J. B., Levin, I., Lowe, D. C., Martin, R. J., Vaughn, B. H. and White, J. W. 2016. A 21st-century shift from fossil-fuel to biogenic methane emissions indicated by 13CH4. *Science* 352(6281), 80-84.

Schimel, J. 2000. Rice, microbes and methane. *Nature* 403(6768), 375-377.

Schmidt, H. P., Anca-Couce, A., Hagemann, N., Werner, C., Gerten, D., Lucht, W. and Kammann, C. 2019. Pyrogenic carbon capture and storage. *GCB Bioenergy* 11(4), 573-591.

Schütz, H. and Seiler, W. 1989. *Methane Flux Measurements: Methods and Results. Exchange of Trace Gases between Terrestrial Ecosystems and the Atmosphere*, pp. 209-228.

Schütz, H., Seiler, W. and Conrad, R. 1989. Processes involved in formation and emission of methane in rice paddies. *Biogeochemistry* 7(1), 33-53.

Schwertmann, U. 1991. Solubility and dissolution of iron oxides. *Plant and Soil* 130(1-2), 1-25.

Shakoor, A., Shahzad, S. M., Chatterjee, N., Arif, M. S., Farooq, T. H., Altaf, M. M., Tufail, M. A., Dar, A. A. and Mehmood, T. 2021. Nitrous oxide emission from agricultural soils: application of animal manure or biochar? A global meta-analysis. *Journal of Environmental Management* 285, 112170.

Shen, Y., Ma, D. and Ge, X. 2017. CO2-looping in biomass pyrolysis or gasification. *Sustainable Energy and Fuels* 1(8), 1700-1729.

Shrestha, M., Shrestha, P. M., Frenzel, P. and Conrad, R. 2010. Effect of nitrogen fertilization on methane oxidation, abundance, community structure, and gene expression of methanotrophs in the rice rhizosphere. *The ISME Journal* 4(12), 1545-1556.

Singh, B. P., Hatton, B. J., Singh, B. and Cowie, A. L. 2010. The role of biochar in reducing nitrous oxide emissions and nitrogen leaching from soil. In: 19th World Congress of soil science, soil solutions for a changing World, Brisbane, Australia, pp. 1-6.

Six, J., Conant, R. T., Paul, E. A. and Paustian, K. 2002. Stabilization mechanisms of soil organic matter: implications for C-saturation of soils. *Plant and Soil* 241(2), 155-176.

Smith, P., Lanigan, G., Kutsch, W. L., Buchmann, N., Eugster, W., Aubinet, M., Ceschia, E., Béziat, P., Yeluripati, J. B., Osborne, B., Moors, E. J., Brut, A., Wattenbach, M., Saunders, M. and Jones, M. 2010. Measurements necessary for assessing the net ecosystem carbon budget of croplands. *Agriculture, Ecosystems and Environment* 139(3), 302-315.

Smith, P., Soussana, J. F., Angers, D., Schipper, L., Chenu, C., Rasse, D. P., Batjes, N. H., van Egmond, F., McNeill, S., Kuhnert, M., Arias-Navarro, C., Olesen, J. E., Chirinda, N., Fornara, D., Wollenberg, E., Álvaro-Fuentes, J., Sanz-Cobena, A. and Klumpp, K. 2020. How to measure, report and verify soil carbon change to realize the potential of soil carbon sequestration for atmospheric greenhouse gas removal. *Global Change Biology* 26(1), 219–241.

Song, H. J., Lee, J. H., Jeong, H. C., Choi, E. J., Oh, T. K., Hong, C. O. and Kim, P. J. 2019. Effect of straw incorporation on methane emission in rice paddy: conversion factor and smart straw management. *Applied Biological Chemistry* 62(1), 1–13.

Song, H. J., Lee, J. H., Canatoy, R. C., Lee, J. G. and Kim, P. J. 2021. Strong mitigation of greenhouse gas emission impact via aerobic short pre-digestion of green manure amended soils during rice cropping. *Science of the Total Environment* 761, 143193.

Thilakarathna, S. K. and Hernandez-Ramirez, G. 2021. Primings of soil organic matter and denitrification mediate the effects of moisture on nitrous oxide production. *Soil Biology and Biochemistry* 155, 108166.

Tilman, D., Cassman, K. G., Matson, P. A., Naylor, R. and Polasky, S. 2002. Agricultural sustainability and intensive production practices. *Nature* 418(6898), 671–677.

Topp, E. and Hanson, R. S. 1991. *Methane Oxidising Bacteria. Microbial Production and Consumption of Greenhouse Gases: Methane, Nitrogen Oxides, and Halomethanes.*

Topp, E. and Pattey, E. 1997. Soils as sources and sinks for atmospheric methane. *Canadian Journal of Soil Science* 77(2), 167–177.

Tyler, S. C., Bilek, R. S., Sass, R. L. and Fisher, F. M. 1997. Methane oxidation and pathways of production in a Texas paddy field deduced from measurements of flux, δl3C, and δD of CH4. *Global Biogeochemical Cycles* 11(3), 323–348.

Vicente-Vicente, J. L., García-Ruiz, R., Francaviglia, R., Aguilera, E. and Smith, P. 2016. Soil carbon sequestration rates under Mediterranean woody crops using recommended management practices: A meta-analysis. *Agriculture, Ecosystems and Environment* 235, 204–214.

Wang, J., Zhang, X., Xiong, Z., Khalil, M. A. K., Zhao, X., Xie, Y. and Xing, G. 2012. Methane emissions from a rice agroecosystem in South China: effects of water regime, straw incorporation and nitrogen fertilizer. *Nutrient Cycling in Agroecosystems* 93(1), 103–112.

Wang, W., Lai, D. Y. F., Wang, C., Pan, T. and Zeng, C. 2015a. Effects of rice straw incorporation on active soil organic carbon pools in a subtropical paddy field. *Soil and Tillage Research* 152, 8–16.

Wang, P., Liu, Y., Li, L., Cheng, K., Zheng, J., Zhang, X., Zheng, J., Joseph, S. and Pan, G. 2015b. Long-term rice cultivation stabilizes soil organic carbon and promotes soil microbial activity in a salt marsh derived soil chronosequence. *Scientific Reports* 5(1), 15704.

Watanabe, A. and Kimura, M. 1999. Influence of chemical properties of soils on methane emission from rice paddies. *Communications in Soil Science and Plant Analysis* 30(17–18), 2449–2463.

Weller, S., Janz, B., Jörg, L., Kraus, D., Racela, H. S., Wassmann, R., Butterbach-Bahl, K. and Kiese, R. 2016. Greenhouse gas emissions and global warming potential of traditional and diversified tropical rice rotation systems. *Global Change Biology* 22(1), 432–448.

Weng, Z. H., Van Zwieten, L., Singh, B. P., Tavakkoli, E., Joseph, S., Macdonald, L. M., Rose, T. J., Rose, M. T., Kimber, S. W. L., Morris, S., Cozzolino, D., Araujo, J. R., Archanjo,

B. S. and Cowie, A. 2017. Biochar built soil carbon over a decade by stabilizing rhizodeposits. *Nature Climate Change* 7(5), 371–376.

Windeatt, J. H., Ross, A. B., Williams, P. T., Forster, P. M., Nahil, M. A. and Singh, S. 2014. Characteristics of biochars from crop residues: potential for carbon sequestration and soil amendment. *Journal of Environmental Management* 146, 189–197.

Witt, C., Cassman, K. G., Olk, D. C., Biker, U., Liboon, S. P., Samson, M. I. and Ottow, J. C. G. 2000. Crop rotation and residue management effects on carbon sequestration, nitrogen cycling and productivity of irrigated rice systems. *Plant and Soil* 225(1), 263–278.

Woese, C. R., Magrum, L. J. and Fox, G. E. 1978. Archaebacteria. *Journal of Molecular Evolution* 11(3), 245–251.

Woolf, D., Amonette, J. E., Street-Perrott, F. A., Lehmann, J. and Joseph, S. 2010. Sustainable biochar to mitigate global climate change. *Nature Communications* 1(1), 56.

Xia, L., Lam, S. K., Wolf, B., Kiese, R., Chen, D. and Butterbach-Bahl, K. 2018. Trade-offs between soil carbon sequestration and reactive nitrogen losses under straw return in global agroecosystems. *Global Change Biology* 24(12), 5919–5932.

Xie, B., Zheng, X., Zhou, Z., Gu, J., Zhu, B., Chen, X., Shi, Y., Wang, Y., Zhao, Z., Liu, C. and Yao, Z. 2010. Effects of nitrogen fertilizer on CH 4 emission from rice fields: multi-site field observations. *Plant and Soil* 326(1), 393–401.

Yagi, K., Tsuruta, H. and Minami, K. 1997. Possible options for mitigating methane emission from rice cultivation. *Nutrient Cycling in Agroecosystems* 49(1), 213–220.

Yan, X., Zhou, H., Zhu, Q. H., Wang, X. F., Zhang, Y. Z., Yu, X. C. and Peng, X. 2013. Carbon sequestration efficiency in paddy soil and upland soil under long-term fertilization in southern China. *Soil and Tillage Research* 130, 42–51.

Yang, B., Xiong, Z., Wang, J., Xu, X., Huang, Q. and Shen, Q. 2015. Mitigating net global warming potential and greenhouse gas intensities by substituting chemical nitrogen fertilizers with organic fertilization strategies in rice–wheat annual rotation systems in China: a 3-year field experiment. *Ecological Engineering* 81, 289–297.

Yang, X., Shang, Q., Wu, P., Liu, J., Shen, Q., Guo, S. and Xiong, Z. 2010. Methane emissions from double rice agriculture under long-term fertilizing systems in Hunan, China. *Agriculture, Ecosystems and Environment* 137(3–4), 308–316.

Yao, Z., Zheng, X., Dong, H., Wang, R., Mei, B. and Zhu, J. 2012. A 3-year record of N2O and CH4 emissions from a sandy loam paddy during rice seasons as affected by different nitrogen application rates. *Agriculture, Ecosystems and Environment* 152, 1–9.

Yao, Z., Zhang, D., Liu, N., Yao, P., Zhao, N., Li, Y., Zhang, S., Zhai, B., Huang, D., Wang, Z., Cao, W., Adl, S. and Gao, Y. 2019. Dynamics and sequestration potential of soil organic carbon and total nitrogen stocks of leguminous green manure-based cropping systems on the Loess Plateau of China. *Soil and Tillage Research* 191, 108–116.

You, S., Ok, Y. S., Tsang, D. C. W., Kwon, E. E. and Wang, C. H. 2018. Towards practical application of gasification: a critical review from syngas and biochar perspectives. *Critical Reviews in Environmental Science and Technology* 48(22–24), 1165–1213.

Yousaf, M., Fahad, S., Shah, A. N., Shaaban, M., Khan, M. J., Sabiel, S. A. I., Ali, S. A. I., Wang, Y. and Osman, K. A. 2014. The effect of nitrogen application rates and timings of first irrigation on wheat growth and yield. *International Journal of Agricultural Research* 2(4), 645–665.

Zhang, X., Chen, J., Jiang, J., Li, J., Tyagi, R. D. and Surampalli, R. Y. 2020. The potential utilization of slag generated from iron-and steelmaking industries: a review. *Environmental Geochemistry and Health* 42(5), 1321–1334.

Zheng, X., Wang, M., Wang, Y., Shen, R., Gou, J., Li, J., Jin, J. and Li, L. 2000. Impacts of soil moisture on nitrous oxide emission from croplands: a case study on the rice-based agro-ecosystem in Southeast China. *Chemosphere - Global Change Science* 2(2), 207–224.

Zhong, J., Wei, Y., Wan, H., Wu, Y., Zheng, J., Han, S. and Zheng, B. 2013. Greenhouse gas emission from the total process of swine manure composting and land application of compost. *Atmospheric Environment* 81, 348–355.

Zou, J., Huang, Y., Jiang, J., Zheng, X. and Sass, R. L. 2005. A 3-year field measurement of methane and nitrous oxide emissions from rice paddies in China: effects of water regime, crop residue, and fertilizer application. *Global Biogeochemical Cycles* 19(2), doi: 10.1029/2004GB002401.

Chapter 18

Managing grasslands to optimize soil carbon sequestration

A. Chabbi, Institute National de Recherche Agronomique et Environnement (INRAE) – Unité de Recherche Pluridisciplinaire Prairies et Plantes Fourragères (UR P3F), France; C. Rumpel, CNRS, Sorbonne University, Institute of Ecology and Environmental Sciences Paris, France; K. Klumpp, INRAE – VetAgro Sup, UMR 874 Ecosystème Prairial, France; and A. J. Franzluebbers, USDA-ARS, USA

1 Introduction

Grasslands have large soil carbon (C) sequestration potential due to permanent vegetation cover and prolific belowground biomass inputs, making their implementation a negative emission technology (IPCC, 2018). However, the extent of carbon sequestration in grassland soils will be dependent on a variety of management approaches and environmental conditions that affect the balance between productivity (i.e. C fixation) and decomposition (i.e. soil respiration). Grasslands are typically used for grazing livestock or to grow hay to feed livestock elsewhere. Both practices may ultimately affect soil microbial activity and influence soil organic carbon (SOC) storage (Gilmullina

http://dx.doi.org/10.19103/AS.2022.0106.17

et al., 2020), SOC release and emissions of CO_2 and other potent greenhouse gases. While both systems are intended for feeding ruminants, the choice of management will have an impact on a variety of ecosystem services, including SOC sequestration, water quality and above- and belowground biodiversity.

Grassland management may be intensive or extensive involving contrasting animal types, stocking rates, mineral fertilizer input and/or irrigation (Box 1). It is generally region-specific with typically extensive management with naturalized botanical composition in semi-arid and arid regions and more intensive management in sub-humid and humid regions. Intensive management in the latter regions occurs with inputs of fertilizer, reseeding with annual and/or preferred perennial species and variations in stocking method and season of grazing.

Box 1 Grassland management systems

Permanent grasslands

- Prairies — cut for hay and typically fed to livestock off-site
- Pastures — grazed directly by livestock

Temporary grasslands

- Leys — Short-term (typically <5 years), grazed or hayed, in rotation with cropping
- Integrated systems — seasonal forages (<1 year) composed of cover crops and/or crop residues grazed by livestock, in rotation with cropping

---

Grasslands, in rotation with cultivated croplands (temporary grasslands or leys) can be considered non-permanent as their renewal is carried out with a variety of frequency intervals suited to the nature of the agricultural systems. Short-term leys are sometimes used for soil fertility restoration of otherwise crop-dominated systems, whereas longer-term pastures rotated with crops may be a part of a diverse agricultural operation with a greater focus on livestock production.

At grassland installation, the choice of botanical composition often determines various other management decisions. Forage legumes for high-quality hay production have lower N fertilizer requirements and may be more resistant to climatic hazards such as drought (Sanaullah et al., 2012). Similarly, planting mixed-species annual forages may have greater nutritive value for pasture-based finishing operations but must be renewed seasonally to keep land productive. Robust perennial forages are often planted with the intent to maintain pasture production on the same parcel of land for a decade or more. The choice of warm-season or cool-season forage species will determine when the forage is best utilized and, depending on environmental conditions, whether opportunities exist to overseed these perennial forages with annual forages to further enhance productivity. These management choices have an impact on greenhouse gas emission and SOC sequestration (Franzluebbers et al., 2000), as they impact the biogeochemical cycling of elements and may lead to disruption of C, N and P cycles (Rumpel et al., 2015; Vertes et al., 2019), resulting in element losses from the system (Fig. 1).

When these losses occur in gaseous form, such as CO_2, CH_4 and N_2O, they may contribute to greenhouse gas accumulation in the atmosphere. In terms of global warming potential, N_2O is ~300 times and CH_4 ~40 times more potent than CO_2. N_2O emission rates from grassland systems are fortunately often orders of magnitude lower than that of CO_2. Agriculture is globally responsible for nearly a quarter of all greenhouse gas emissions (IPCC, 2014), with livestock production systems being major contributors.

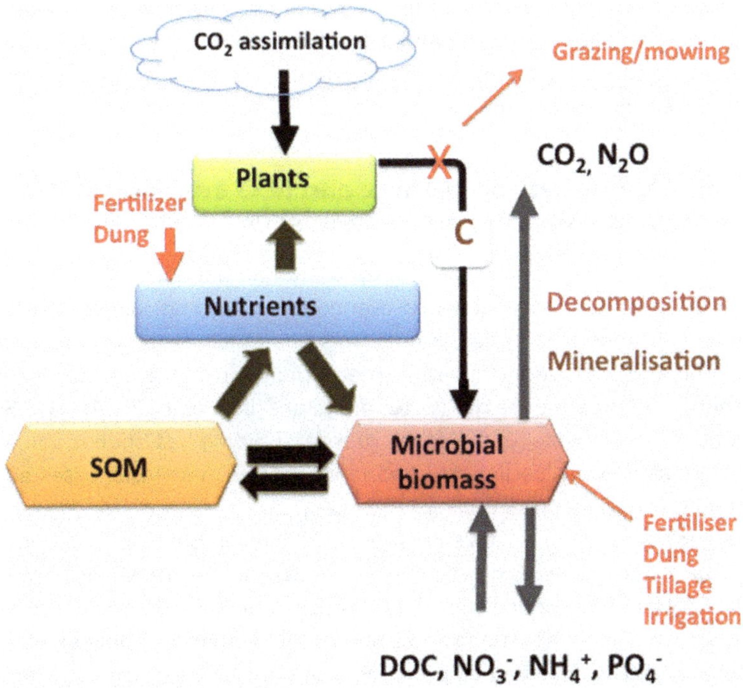

Figure 1 Disruption of biogeochemical cycles leading to greenhouse gas emissions of pastures and prairies.

Published by Burleigh Dodds Science Publishing Limited, 2023.

Prominent greenhouse gases lost from grassland soils are CO_2 during ubiquitous soil respiration across systems and N_2O, particularly following pulses of N availability either from N fertilization or disruption of perennial sods with tillage.

Greenhouse gas production in pasture-based livestock systems occurs during the digestive process via CH_4 production in the rumen, and following excretion with N_2O production in urine patches and during decomposition of fecal pats. The feedlot phase of cattle production can exacerbate N_2O losses (Aguilar et al., 2014) but typically reduces CH_4 emissions due to higher quality dietary intake (Beauchemin and McGinn, 2006; Cole et al., 2020). Monogastric livestock production on pastures, such as swine and poultry, will likely have reduced greenhouse gas emissions than when produced in confinement due to the spatial distribution of excreta that reduces the concentration of C and N resources. Robust forage production and its effects on SOC sequestration in grassland soils may counterbalance part of these emissions when well managed with few inputs, appropriate stocking rates and site-specific vegetation composition.

In this chapter, we will discuss the effect of different grassland management practices on greenhouse gas emissions and SOC sequestration. This includes a comparison of grasslands with arable croplands, the role of N fertilization and grazing strategies. Special emphasis will be given to grasslands in rotation with cropping systems and their integration with cropping or timber systems to improve sustainable land management and SOC sequestration.

2 Soil organic carbon storage and N_2O emissions from grassland soils versus cropland soils

Grasslands and arable croplands are both integral parts of agriculture. Industrial agriculture since the mid-twentieth century has largely led to the separation of these land use systems. This has exacerbated the decoupling of elemental cycles leading to widespread pollution from N and P leaching and runoff and greenhouse gas (GHG) emissions (Lam et al., 2021; IPCC Smith et al., 2014). Worldwide, ~60% of the grassland biome has been modified by human interventions, leading to important changes in soil processes and, in particular, biogeochemical cycling.

In general, grassland soils store more SOC than arable croplands, and they tend to have more closed biogeochemical cycles resulting in fewer opportunities to lose C, N and P to surface and ground waters, but they may emit more N_2O if intensively managed (Soussana and Lemaire, 2014; Franzluebbers et al., 2014; Crème et al., 2018). Due to their permanent cover, they also positively affect the energy balance as compared to continuous cropping systems, which include periods with bare soil (Martin et al., 2020).

Differences in surface SOC stock between grasslands and arable croplands vary considerably, depending upon geographical setting, soil type, landscape position and the various management approaches deployed in each system. Paired land use comparisons offer an excellent opportunity to separate the environmental and edaphic factors from important management factors. This approach also sets the stage for better characterizing the relative influence of environmental, edaphic, and management factors on the magnitude of difference in SOC between grassland and cropland systems. With sufficient observations, these data can be assembled into a robust dataset. Some examples of this approach are described in the following sections.

From a comparison of broad land use categories across 28 studies in the southeastern United States, which is defined as a warm-humid region with relatively weathered and coarse-textured soils, SOC stock (surface layer of 24 ± 6 cm) was estimated as 31.1 Mg C ha^{-1} under cropland <47.4 Mg C ha^{-1} under grassland = 49.9 Mg C ha^{-1} under forestland (Franzluebbers, 2005). In the same region, a total of 92 tall fescue [*Schedonorus arundinaceus* (Schreb.) Dumort.] pastures were sampled at 0–10 cm depth and characterized with regard to two environmental factors (elevation and soil texture) and two management factors (forage utilization and pasture age) (Franzluebbers and Poore, 2020b). Using a novel approach to estimate the mass of SOC enriched within the upper rooting zone (0–30 cm depth) in the southeastern United States, SOC enrichment was estimated as 16-26 Mg C ha^{-1} (interquartile range) under grassland management (n = 29), while it was estimated as 4–9 Mg C ha^{-1} under conventional-till cropland (n = 45) and 14–30 Mg C ha^{-1} under no-till cropland (n = 97) (Franzluebbers, 2021).

With this same novel approach used across an additional 52 studies from the literature in other regions of the world, SOC enrichment was estimated as 6-16 Mg C ha^{-1} (interquartile range) under conventional-till cropland (n = 210), 9–22 Mg C ha^{-1} under no-till cropland (n = 270), 13–26 Mg C ha^{-1} under grassland (n = 88), and 9–40 Mg C ha^{-1} under forestland (n = 23) (A. J. Franzluebbers, unpublished data).

Using a multiple regression analysis to separate potentially confounding factors, SOC concentration:

- was 26% greater with improved grazing management (i.e. rotational stocking) than with conventional management (i.e. haying and/or continuous stocking);
- increased in a non-linear manner with pasture age (effective SOC sequestration rate of 0.9 Mg C ha^{-1} yr^{-1} during the first 20 years);
- increased with increasing elevation gradient (effective increase of 1.3 g kg^{-1} for every 100 m of elevation);
- was greater in fine-textured than coarse-textured soils.

These data illustrate the high SOC storage potential of grasslands with regards to croplands and the importance of pedoclimatic conditions and grassland management practices. It is highly likely that, with more detailed environmental, edaphic and management conditions going beyond determination of SOC stocks and/or enrichment, a robust assessment could be made of how management factors influence SOC sequestration in specific pedoclimatic situations. Assembling historical information on previous management practices may be challenging in some situations, but essential to understanding the cumulative effects of practices on SOC sequestration. Environmental and edaphic information may be less difficult to obtain due to readily available online databases for historical weather and relatively static soil conditions, at least in many developed countries.

Similar to croplands, harvest of grasslands for hay removes nutrients from the site of production. And, unless manure from livestock fed elsewhere is returned to these fields, this practice results in a potential need for nutrient replacement to maintain fertility so that productive forage swards can continue to provide their ecosystem services (fix atmospheric CO_2 via photosynthesis, produce high-quality forage and enhance SOC storage, e.g. Poeplau et al., 2018). However, nitrogen fertilization can lead to soil acidification and increase in nitrifier and denitrifier activity, both of which can lead to N_2O losses (Hijbeek et al., 2019). Fertilization as a strategy to enhance SOC storage may thus generate trade-offs, in the form of reduced soil quality and other greenhouse gas (GHG) emissions (see Chapter 6 of this book).

Despite the large global warming potential of N_2O and its importance as a GHG from grassland soils, adequate characterization of the controls of these emissions is lacking in many cases, and the processes are still poorly understood (Chapter 6 of this book). While it is recommended to apply mineral N fertilization carefully and synchronized with plant needs, microbial acquisition of N is necessary to store SOC when its levels are low, and there is large potential for additional SOC sequestration. Mineral N fertilization may be necessary in some instances simply due to the close coupling of C and N in soil so that microbial transformation of plant and animal residues into soil organic matter can be effective.

Grassland management strategies that minimize GHG emissions and enhance SOC sequestration must be developed to maximize coupling between the cycling of C and N avoiding leakage of both elements. To do this, GHG fluxes of CO_2, CH_4 and N_2O should be evaluated simultaneously. When N is deficient in soil, fertilization leads to greater biomass production and C input, which in turn stimulates microbial activity (Lu et al., 2011) (Fig. 2). Enhanced production of microbial sugars leads to greater aggregate formation, and deposition of microbial necromass, which is beneficial for enhancing aggregation and SOC storage. In some pedoclimatic environments, these processes may be stimulated under grassland by P input in addition to N (Poblete et al., 2020).

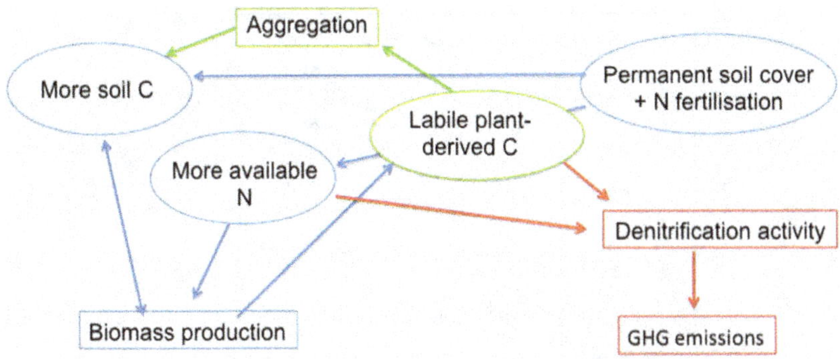

Figure 2 Processes leading to SOC storage and greenhouse gas emissions in productive grasslands.

On the other hand, mineral N fertilization and labile plant litter input stimulate denitrifier activity and N_2O emissions (Fig. 2) (Szukics et al., 2019). The evaluation of SOC sequestration capacity of grassland soils has to take into account these competing processes (Lugato et al., 2018; Guenet et al., 2020; Frank et al., 2015). A recent study suggested that grassland and cropping systems could have similar emissions, when SOC stock changes along with all emission types including those from livestock are considered (Meier et al., 2020).

As with croplands, the accurate determination of N fertilizer requirements to optimize productivity of grasslands and minimize trade-offs in form of associated N_2O emissions has been complicated by the variable nature of N mineralization processes, which are strongly affected by antecedent temperature and moisture conditions, as well as the type of management deployed and depth distribution and quality of soil organic matter. These complications have befuddled researchers in making specific N fertilizer recommendations for the numerous soil, landscape and management conditions within a particular region.

Recently, a short-term measure of soil biological activity that is strongly associated with potential N mineralization was shown to adequately predict the N fertilizer requirements in a variety of soil types, landscape settings and management conditions in the southeastern United States. On fall-stockpiled tall fescue, the level of soil-test biological activity was inversely related to the economically optimum N fertilizer rate (Franzluebbers and Poore, 2020). Young pastures and those that typically had biomass removal with hay cutting were the most responsive to N fertilizer while older pastures that were grazed were typically unresponsive to N fertilizer. This same predictive capability was shown for a gradient of conservation cropping systems with maize (Zea mays) production in the region (Franzluebbers, 2020). More refined N fertilizer recommendations will allow grassland management systems to achieve high productivity when soil cannot supply enough N and still likely limit the extent of

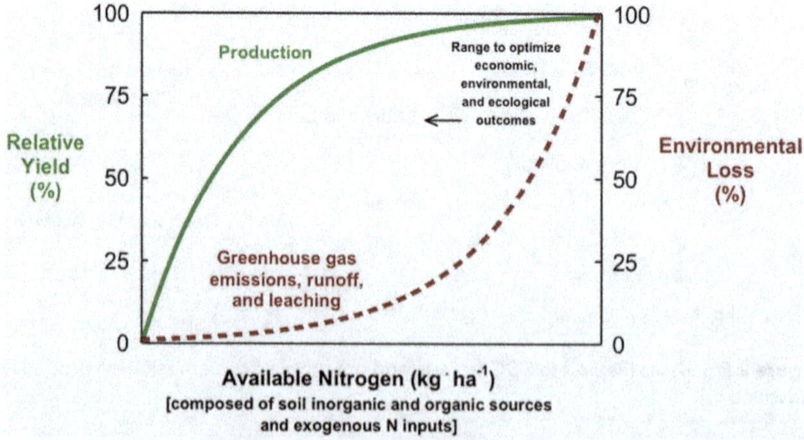

Figure 3 Conceptual diagram of how soil N availability can lead to trade-offs between grassland productivity and environmental degradation.

N_2O emission (Fig. 3). Moreover, high grassland productivity while minimizing loss of N from runoff, leaching and N_2O emissions can be equally achieved if N fertilizer inputs can be reduced or eliminated when soil is capable of supplying sufficient N. Soil analytical tools like the simple, rapid and reliable approach of determining pre-season soil-test biological activity can help to determine these gradations in soil N supplying capacity.

Grassland management needs to recognize the positive and negative effects of fertilization on SOC sequestration and GHG emissions. This is particularly important considering the functions of temporary grasslands in arable crop rotations, since this ley-cropping rotation scheme has been suggested as a solution to increase SOC stocks and to reduce GHG emissions and other environmental externalities from agriculture (Lemaire et al., 2015; Launay et al., 2021).

3 Temporary (ley) grasslands

Grass-crop rotations can experience high fluctuations of SOC stock – large increases in SOC stock during early years of temporary grassland development followed by large declines in SOC stock with tillage termination of these grasslands (Studdert et al., 1997). However, when temporary grasslands are terminated without tillage and by chemical desiccation only, then there may be preservation of surface SOC concentration (Follett et al., 2009; Franzluebbers and Stuedemann, 2009).

The location in the soil profile, where SOC changes as a result of grass-crop rotations occur, is of particular interest in understanding how to manage these transitions to avoid loss of SOC and reduce the risk of enhanced N_2O emissions (Franzluebbers et al., 2014). Some data suggest that changes in SOC may be relegated to the superficial zone of soil within the plow layer (Franzluebbers and Stuedemann, 2010), while other data suggest there may be changes below this zone (Carter and Gregorich, 2010). Other studies indicate little detectable change in SOC over short periods of time following pasture termination (Pierce et al., 1994; Grandy and Robertson, 2006). A decade of differently managed temporary grasslands seemed to maintain SOC stocks under temperate climate conditions in contrast to cropped soils, which lost SOC and continuous grasslands, which gained SOC (Crème et al., 2020).

Recently, deep soil flipping (1–3 m full inversion of the soil profile) of podzolized soils in New Zealand has been suggested as a valuable grassland conversion practice to increase SOC storage due to burial of SOC-rich A horizons and exposure of SOC-poor B horizons at the soil surface (Schiedung et al., 2019). Randomized and replicated plot studies or paired land use studies with this treatment of ultra-deep plowing are needed in different types of soils to know the full magnitude of such a large effect on profile SOC stocks and other soil properties.

Although positive effects of the introduction of temporary grasslands on SOC storage in arable croplands occur, various soil properties and, in particular, SOC stocks decline more rapidly when grasslands are terminated than when they are being established (Sanderman and Baldock, 2010; Poeplau et al., 2011). Recent meta-analyses have shown that SOC stocks with an average of 14 years of grassland management were 18% greater than under arable cropland (Kaempf et al., 2016). Even after 40 years since perennial grassland establishment, SOC stock may still accumulate at a substantial rate, e.g. 0.62 Mg C ha^{-1} yr^{-1} (McLauchlan et al., 2006). However, the rate of SOC storage may vary with pasture age (Franzluebbers et al., 2000; Johnston et al., 2017; Franzluebbers and Poore, 2021). SOC stock increases as well as decreases upon grassland establishment or conversion may not be linear as indicated by some studies. However, at the moment, there are not enough data to confirm that the trends observed are generalizable and applicable to contrasting management systems and/or pedoclimatic regions. In particular, most data were recorded in temperate regions with ruminant grazing, whereas relatively little is known about tropical systems, non-ruminant grazing animals and cut and carry systems, which may have contrasting effects on soil microbial functioning and biogeochemical cycling (Moinet et al., 2019; Gilmullina et al., 2020, 2021).

Temporary (ley) grasslands have numerous (positive) effects on ecosystem services and may be one solution for the diversification of tomorrows cropping systems (Martin et al., 2020). However, the effects of contrasting grassland management practices and their legacy effect on organic matter dynamics, and SOC sequestration (including their quantitative effects on trade-offs and other ecosystem services) are only just being elucidated. For example, legacy effects of temporary grasslands include soil structure maintenance and biodiversity conservation (Hoeffner et al., 2021). However, these effects may depend on grassland age and may be quantitatively less important than those of permanent grasslands, as even 91 years after the abandonment of arable cropland, certain prairies recovered only half of their productivity and 75% of their biodiversity (Isbell et al., 2019).

It is clear that temporary grasslands have the potential for SOC accrual when rotated with arable cropland, (Johnston et al., 2017), thereby supporting soil fertility restoration and SOC sequestration (Minasny et al., 2017). Temporary grasslands can also carry legacy effects into the cropping phase in terms of GHG emissions, SOC storage and organic matter composition (Crème et al., 2018; Rumpel and Chabbi, 2010). Although there are positive effects of short-term grasslands for SOC accumulation (Johnston et al., 2017), conservation of permanent grasslands should not be neglected as these lands often hold greater quantities of SOC in larger land areas that are not suitable for arable cropping (Minasny et al., 2017).

4 Grassland management options: species choice, fertilization and irrigation

Since grassland botanical diversity and SOC storage are intimately linked (Cong et al., 2014; Lange et al., 2015), focusing on grassland biodiversity may be the best way to improve ecological functioning of restored grasslands. Greater plant diversity increases the type and location of rhizosphere C inputs, as well as soil microbial utilization and processing, resulting in greater microbial activity and greater SOC storage of recently fixed C (see Chapter 3 of this book). Although enormous biodiversity can be achieved in permanent grasslands, temporary prairies in intensive agricultural systems usually have relatively low biodiversity because they are composed of only a few intentional forage species. It takes time for seed recruitment and development of more functional biodiversity in newly planted swards. Nonetheless, a balance should be sought among forage species diversity, grassland productivity, SOC storage and grassland age (Soussana and Lemaire, 2014; Kohler et al., 2020).

There are two reasons for applying N fertilization to productive grasslands (Crème et al., 2018):

(1) High nutritive value of harvested hay may only be possible, depending on the type of soil and its concentration and depth of soil organic matter, with intermittent application of N fertilizer during the two to three harvests expected to be taken each year.
(2) N-addition may be necessary to store SOC

Indeed, recent meta-analyses confirmed that a significant increase of SOC stocks (0.6 Mg C ha^{-1} yr^{-1}) can occur with N fertilization (Conant et al., 2017; Eze et al., 2018). However, there is a non-linear relationship between the amount of N fertilizer added and SOC accrual, depending intrinsically on grassland management. Poeplau et al. (2018) showed that 1.15 kg N ha^{-1} was needed to sequester 1 kg SOC ha^{-1}. However, since the C cost embedded within N fertilizer is of the same magnitude (West and Marland, 2002), there is a need to make good evaluations of the need for enhancing forage nutritive value and quantity of additional forage biomass harvested. There may be situations in which the quantity of N fertilizer applied to productive prairie soils exceeds that applied to cropland soil due to greater biomass export, thereby increasing the potential for greater N_2O emissions (Guenet et al., 2020; Hauggaard-Nielsen et al., 2016). In grass-crop rotations, duration of the grassland phase can have an impact on plant available mineral N in the soil solution and can reduce leaching loss (Kunrath et al., 2015).

Dominant forage species within a sward can determine forage productivity, SOC storage and the extent of GHG emissions (Conant et al., 2017; Sebastia et al., 2018). For instance, long-term biodiversity restoration increased SOC and total soil N storage, especially when forage treatments were combined with the legume, *Trifolium pratense* (De Deyn et al., 2022). Moreover, the use of legumes in hayed grasslands can be a viable option to reduce mineral N fertilization requirements (IPCC, 2013), to increase protein concentration of forage and to increase resilience to short-term drought (Sanaullah et al., 2012).

Relying on biological N fixation (BNF) with incorporation of legume species in temporary grasslands may require additional mineral P fertilizer inputs, as P availability is known to limit the development and growth of many legume species (Graham and Vance, 2003). The high demand of legume species for soil P may be attenuated when grown in a mixture with gramineous plants, depending on the species (Crème et al., 2016), partly because various grass-legume combinations can influence the biogeochemical composition of soil organic matter (Crème et al., 2017). Therefore in grass-legume mixtures, optimization of P fertilization to maximize N inputs may not be so straightforward but requires site-specific strategies (Mendoza et al., 2016; Stiles et al., 2017).

While grass-legume mixtures often enhance forage N content and digestibility of herbage (de Wit et al., 1966), the impacts of such mixtures on N_2O emission from soil have only been recently studied (Fuchs et al., 2018). Although some BNF from legumes in a mixed sward may increase grass N nutrition through direct N transfer, the majority of benefit is more likely through slow release of mineral N following decomposition of legume plant litter (Fustec et al., 2010). Direct N transfer was found to be species-specific (Marty et al., 2009), while indirect transfer is a general phenomenon and depends on soil microbial activity.

Several recent studies have shown that replacement of mineral N fertilization with BNF from legumes in the sward can reduce N_2O emissions while maintaining or even improving grassland productivity (Fuchs et al., 2018; Barneze et al., 2020). Use of legumes in temporary grasslands may therefore be highly beneficial, especially on N-poor soil with low organic matter following many years of arable cropping with inversion tillage.

In some regions, irrigation is used to overcome water limitations and to increase herbage production and subsequent SOC sequestration (Hunt et al., 2016). Despite greater forage biomass production with irrigation, change in SOC stock may be limited due to greater soil microbial activity that consumes additional C resources (Mudge et al., 2017). The net effect of irrigation on C input via production and C output via soil respiration needs to be considered (Condron et al., 2014), along with its effect on N_2O emissions after fertilization due to changing pH and redox conditions (Yu and Patrick, 2004, Chapter 6 of this book).

5 Harvesting strategies: grazing versus mowing and grazing regimes

Grazing of grasslands by ruminant livestock is a fundamental change in how forage is harvested and utilized as compared with cutting for hay. Instead of two or three abrupt and complete biomass removals per year with export of harvested nutrients, grazing tends to gradually consume biomass with a continuous stocking method (i.e. long duration, high frequency of return).

The implications of grazing style on SOC storage and GHG emissions have received more attention recently and more assessments are needed under different environments to understand the impact of grazing style on production and environmental outcomes (e.g. SOC sequestration, GHG emissions, water quality and soil health), as well as on livestock herd health, farm economics, farmers' quality of life and rural community development.

A comprehensive review of the literature on the impacts of prescribed grazing on key ecological outcomes was conducted in the United States (Sollenberger et al., 2012). This literature review illustrated some key findings of stocking rate on SOC sequestration and other soil conditions that will be covered in the following but also spurred subsequent research on the effects of stocking method.

Livestock grazing on pasture leads to consumption of forage while moving around the pasture with nutrients and C return in the form of dung and urine. Placement of these nutrients depends on how pastures are managed, i.e. continuous or rotational stocking. With 50–70% of C embedded in forage considered digestible, a large portion of the C fixed in forage is returned directly back to the land with dung and urine deposition. Dung is the primary source of returned C, and urine is the primary source of returned N. Biomass

Published by Burleigh Dodds Science Publishing Limited, 2023.

carbon is utilized for maintenance and growth of livestock, and at least half of the C can be considered either stored in animal tissue or respired as CO_2, CH_4 and other volatile organic compounds. Biomass nitrogen is utilized for the production of animal proteins, and the vast majority of N (70%) is returned to the land via urine and dung.

The fate of this N is dependent on soil and environmental conditions, but a portion of it can be (1) nitrified and denitrified leading to N_2O emissions, (2) volatilized as NH_3, (3) utilized by soil microorganisms and (4) transformed into stable organic structures. Organic N may also be mineralized into inorganic N, and this inorganic N can be leached or lost via runoff if not taken up by plants or microorganisms. In which of these compartments C and N are found and at what rate and intensity they are transformed are complex questions and require detailed analyses of pasture ecosystems and thus have received attention but require greater scrutiny of the factors explaining these fluxes under various soil, environmental and management conditions.

Well-managed pastures with appropriate stocking rates to maintain and improve botanical sward structure can lead to significant SOC sequestration with increasing pasture age (Franzluebbers et al., 2000, 2010, 2012a;). At specific pasture ages with grazing management, more SOC may be sequestered relative to hayed management (Franzluebbers et al., 2001, 2012b). Net GHG emissions can also be reduced when extensively managed pastures are grazed by livestock as compared with cutting forage for hay and feeding livestock elsewhere (Koncz et al., 2017). Greater SOC stock has been observed in grazed native grassland as compared to an ungrazed treatment (Orgill et al., 2018), and in planted pastures relative to a conservation reserve without harvest (Franzluebers and Studdemann, 2009). Positive grazing impacts on SOC storage are often attributed to enhanced plant productivity (Franzluebbers et al., 2004) and greater allocation of C to a more robust root system. It has been suggested that plants increase root exudation and thus C input to soil as a response to grazing (Hamilton et al., 2008), thus leading to greater soil microbial activity and OC use (Klumpp et al., 2009). Greater SOC accumulation with grazing of pastures might also be explained by decreased soil organic matter mineralization due to shifts in the microbial community composition (Gilmullina et al., 2020) and reduced abundance of fungi (Shahzad et al., 2012).

To achieve greater SOC storage and productivity, grazing systems with moderate stocking density have been suggested (Briske et al., 2008; Derner and Schuman, 2007). In fact, the effect of stocking rate on optimization of SOC storage was demonstrated in a warm-season pasture in the southeastern United States (Fig. 4). However, the impacts of stocking rate and grazing style will be dependent on soil and environmental conditions, such that region-specific grazing strategies need to be developed (McSherry and Ritchie, 2013; Abdalla et al., 2018).

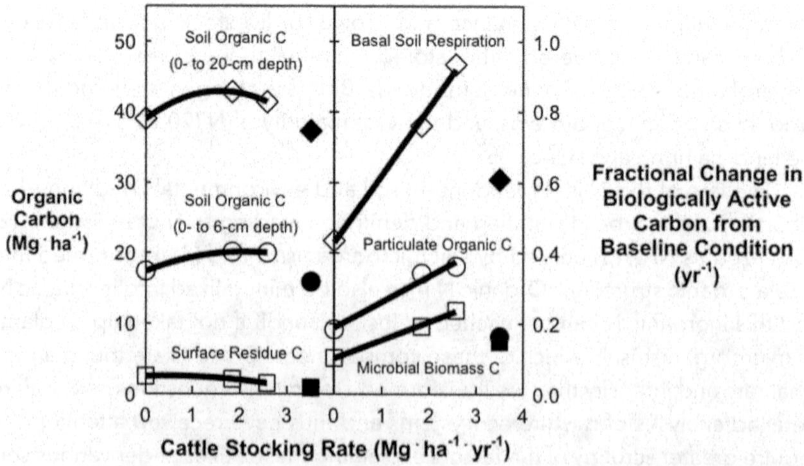

Figure 4 Effect of stocking rate on soil organic C accumulation during the first five years of bermudagrass (Cynodon dactylon) management in the southeastern United States (data from Franzluebbers et al., 2001).

When livestock graze pastures, they are not typically homogenously distributed unless restricted to small paddocks for short durations. Selective grazing may impact pasture production and therefore organic C input to soils. In particular, in continuous grazing systems, when livestock graze large pastures, they tend to repeatedly graze preferred plants and in areas close to water and shade (Bloor and Pottier, 2014; Bloor et al., 2020). Such animal behavior leads to greater than average stocking rate in preferred areas and lower than average stocking rate or total avoidance of other spaces (Wallisdevries et al., 1999; Teague et al., 2011). Pastures affected by selective grazing can lead to a change in species composition toward less palatable grasses, weedy forbs and woody plants. The resulting grasslands with undesirable species are more vulnerable to drought (Teague et al., 2011, 2013), may have greater proportions of bare ground, reduced aboveground and belowground biomass productivity and essentially enter into a state of degraded condition with numerous undesirable traits.

These processes also affect SOC storage due to reduced C inputs (Knopf, 1994; Frank et al., 1998; Teague et al., 2013; Park et al., 2017). On degraded grasslands, regulating the frequency and intensity of grazing may be the only low-cost solution to prevent further spiraling into eventual soil degradation (i.e. loss of SOC) by maintaining aboveground and root biomass, which act as primary C input sources to soil (Liu et al., 2016; Sun et al., 2014; McSherry and Ritchie, 2013). If existing botanical composition is sufficient, then using the livestock themselves to renovate a pasture is possible by subdividing the pasture so that high-density stocking disrupts the sward with heavy trampling

Published by Burleigh Dodds Science Publishing Limited, 2023.

followed by a resting period of one to several months so that forage species can fully recover. This short-duration, high-density impact moves to all other parcels of the pasture in sequence so that natural regeneration of desirable grasses and forbs can emerge following these disturbances that may occur one to several times a year depending on the environmental conditions of a region (Teague et al., 2011, 2013; Sanderman et al., 2015).

Another practice, known as rotational stocking, shifts grazing to another extreme of forage utilization (i.e. short duration, low frequency of return). This system is typically based on the axiom 'graze half, leave half'. Of course, there can be gradations in how much actual forage is consumed and how much is returned to the soil to feed soil fauna and microorganisms. It can be as simple as a two-paddock system with one paddock grazed for a period of time while allowing the other paddock to rest and accumulate biomass. Livestock are moved to the other paddock when the farm manager decides, thereby allowing the previously grazed paddock to accumulate biomass until the next round of grazing. The duration of grazing and length of time for pasture rest and recovery can be set with different values depending on paddock size and number on a farm, or it can vary based on weather conditions, forage availability, livestock needs and farm management timelines. Rotational stocking has been found to be beneficial for SOC storage in different environments (Wang et al., 2015; Teague, 2018;), but not all (Ingram et al., 2008; Derner et al., 2019). An increasing number of farmers are adopting rotational stocking to increase profitability and improve ecosystem health (Ferguson et al., 2013).

An extreme case of rotational stocking is sometimes called 'mob grazing' or Adaptive Multi-Paddock (AMP) grazing with 10–1000 Mg body weight ha^{-1}, but only for a limited time (Photo 1). Livestock are typically restricted with temporary, electrified polywire fencing and forced to consume all forage species and trample some fraction of forage into the ground without removing apical meristems of desirable forage species for more rapid regrowth compared with hay cutting. The animals are moved frequently to new areas of a pasture. New paddocks may be offered a few times a day to once every few days. There can be gradients in between several movements of livestock to fresh forage per day to a move once every few weeks, depending on the type of forage and soil, landscape, and climatic conditions. This practice is becoming popular among some producers due to its perceived benefits on SOC sequestration and maintaining forage diversity (Wang et al., 2021). While few studies reported extremely high rates of SOC sequestration with a shift to mob or AMP grazing (Russell et al., 2013) and an increase of persistent SOC and N retention (Mosier and Cotrufo, 2022), the practice has also been found beneficial for controlling woody plant encroachment in rangelands (Mesléard et al., 2017).

However, scientific evidence for the benefits and trade-offs of this practice and its effect on SOC sequestration is weak as studies have been conducted

Photo 1 Mob grazing (photo: Tom Chapman).

on a limited number of grazing systems with a chronosequence approach to estimate change in SOC stock, which makes some assumptions of uniform conditions that may not always be verifiable. In addition, unreplicated field experiments are often used, due to the large space required for completely randomized block design with pasture management.

Moreover, different grazing techniques are compounded by the inherent variability within and between paddocks, making detection of a small real change difficult within a short time period (Crème et al., 2020). Similarly, comparison of two fixed categories (i.e. rotational vs. continuous stocking) may not always be the most appropriate approach for comparing diverse management styles, given (a) the large spatial variability within a large farm with multiple paddocks and (b) a diversity of specific management approaches among farmers using what could be considered the same approach.

In the last decades, studies have become more common describing the effects of forages (grass, legumes or mixture) on animal N excretion and resulting N_2O emissions and C fluxes. Pasture-finished livestock approaches have also gained interest, thus bringing together pasture ecosystem services and livestock production strategies. Critically, forage digestibility and intake has become recognized as a major factor affecting rumen CH_4 emissions (Hristov et al., 2013). There has been increased research on forage nutritive value to develop appropriate mitigation practices. High nutritive value grass silage can reduce CH_4 emissions when fed to ruminant livestock, depending on stage of forage maturity that controls fiber and protein concentrations. These variables are dependent on harvest date, climate conditions, and botanical composition of the sward (Elgersma and Søegaard, 2018).

Published by Burleigh Dodds Science Publishing Limited, 2023.

6 Integrated crop-livestock systems and silvopastures

Since the mid-twentieth century, greater specialization of agriculture has led to greater separation of crop and livestock production (Russelle et al., 2007). Agricultural expansion and its specialization has led to reduced diversity, greater agrochemical use and more damaging externalities from declines in water quality/availability, greenhouse gas emissions, rural community stresses, reliance on global markets, disruption of local supply chains and so on.

To counteract these negative effects of specialized agriculture, integrated crop-livestock systems and silvopasture systems offer great opportunities to enhance diversity from field to landscape scale, enhance agricultural resilience and reduce environmental pollution (Lemaire et al., 2014). These systems have been suggested as sustainable alternatives for diversified agriculture while, at the same time, leading to sustainable intensification without bringing additional land into cultivation (Kumar et al., 2019; Peterson et al., 2020). Such systems could diversify farm income sources, but they may also require new skills and investment.

Regarding nutrient cycling, integrated crop-livestock systems have an advantage over specialized cropping systems due to animal excreta deposited directly onto the land in a distributed manner as organic fertilizer. Organic fertilization may be more beneficial to SOC sequestration than mineral fertilization (Calabi-Floody et al., 2018). Moreover, re-use of animal waste materials may contribute to a circular economy. In contrast, on-farm storage and transformation of manure into organic amendments can generate N_2O and CH_4 emissions. Such emissions may be reduced by improving management at the farm level (Ditzler et al., 2018) and with improved waste management practices, e.g. composting and/or co-composting in the presence of minerals (Barthod et al., 2018) or biochar (Agegnehu et al., 2017).

Silvopastures and integrated crop-livestock-forestry systems (Photo 2) offer potential system-level production improvements, better animal welfare with thermal regulation of the microclimate in the dispersed shade of trees, greater water infiltration and nutrient cycling, and enhanced scenic beauty of rural landscapes (Carpinelli et al., 2020; Da Pontes et al., 2016). Randomized landscape distribution of trees in pastures has been a historical feature of many regions around the world. In fact, vast areas of semi-arid rangelands supporting the grazing of cattle, sheep and goats are common yet today, including the dehesa on the Iberian Peninsula and dispersed oak woodlands in California.

Since the mid-twentieth century, new silvopasture systems have also been introduced using more geometric arrangement of trees with an understory of planted and desirable forages, such as in Florida (Nair et al., 2008). In the southern part of the United States, tree-based pastures were shown to increase the soils' capacity to accumulate stabilized SOC and therefore increase SOC

Photo 2 Silvopastural system in Argentina (photo: F. Montagnini).

sequestration (Haile et al., 2008). Such regular arrangements of trees lining alleys of managed pastures are gaining popularity in countries with a large livestock industry, such as in Brazil (Alves et al., 2017).

From on-farm sampling in Alberta, Canada, SOC stock within the surface (i.e. 30 cm) was greater under silvopasture systems than under trees of hedgerow and shelterbelts (Lim et al., 2018). From on-farm sampling in Chiapas, Mexico, SOC stock at 0–15 cm depth was 23% greater under 5–50-year-old silvopastures than under open pastures dominated by forage species of *Cynodon dactylon*, *Cynodon plectostachium*, *Hiparrhenia rufa* and *Brachiaria dictyoneura* (Aryal et al., 2019). In Michoacán, Mexico, SOC stock at 0–30 cm depth was 54% greater under a silvopasture of *Leucaena leucocephala* undersown with *Panicum maximum* than under open pasture of *C. plectostachyus* (Lopez-Santiago et al., 2019). N_2O emissions were found to be reduced, while SOC stocks were increased, underneath the trees of silvopastoral systems (Amadi et al., 2016; Franzluebbers et al., 2017). Indeed, in a 4-year-old Eucalyptus silvopasture in Minas Gerais Brazil, SOC stock at 0–30 cm depth was up to 19% greater under nearest Eucalyptus trees than in the open pasture between tree lines (de Abreu et al., 2020). Consequently, it was calculated that 17–44 trees ha^{-1} are needed to offset the greenhouse gas emissions from livestock grazing pastures with trees in Brazilian silvopastoral systems. This tree density is easily exceeded in most tropical regions (Torres et al., 2017). Carbon storage in tree biomass is highly significant in silvopastoral systems, but significant SOC stock change has also been frequently reported in these and other regions of the world (Lorenz and Lal, 2014).

Published by Burleigh Dodds Science Publishing Limited, 2023.

Silvopastures have been shown to effectively limit net greenhouse gas emissions by increasing C storage in woody biomass and soil, as well as potentially mitigating CH_4 emissions from enteric methane (Resende et al., 2020). They are promoted as one of the top land use activities to mitigate against greenhouse gas emissions due to the large plant biomass accumulation on existing agricultural land (Jameel et al., 2022). However, due to the scarcity of investigations, their impact on the processes leading to SOC sequestration remains largely unknown.

7 Life-cycle assessment of agricultural production systems with grasslands

Table 1 summarizes the effect of the various grassland management practices on SOC storage and GHG emissions. In order to account for the benefits and trade-offs of prairies and pastures in terms of the GHG emission balance, a whole system modeling approach is necessary (Stewart et al., 2009; Chang et al., 2015) to account for lateral C and N fluxes. Managed grasslands are in general used to produce forage for animal feed in livestock or dairy production systems. This means that mineral fertilizers and/or manure are often needed to improve forage nutritive value and overall production. Therefore, in terms of GHG emissions, CO_2 respiration by soils and animals, CH_4 production by ruminants, and N_2O and CH_4 emissions from dung, urine and stored manure must be taken into account, as well as SOC sequestration that can mitigate some or all of these emissions (Fig. 5).

Generally, dairy and beef production systems have large enteric CH_4 emissions (about 40% of all methane emitted, FAO, 2013). This is a significant ecological footprint (Vermeulen et al., 2012). However, many life-cycle analyses

Table 1 Potential effects of management practices on SOC storage and GHG emissions

Grassland type	Management practice	Impact on SOC storage	Impact on GHG emissions
Hayed	Mineral fertilization	+++	+++
	Legume use	+++	-
	Grassland renewal	-	+
Grazed	Continuous stocking	+	++
	Rotational stocking	+++	++
	Short-duration, high-density stocking	++	+
	Irrigation	?	+

SOC, soil organic carbon; GHG, greenhouse gas.
+++ = high impact; ++ = medium impact; + = low impact; - = negative impact; ? = unknown impact.

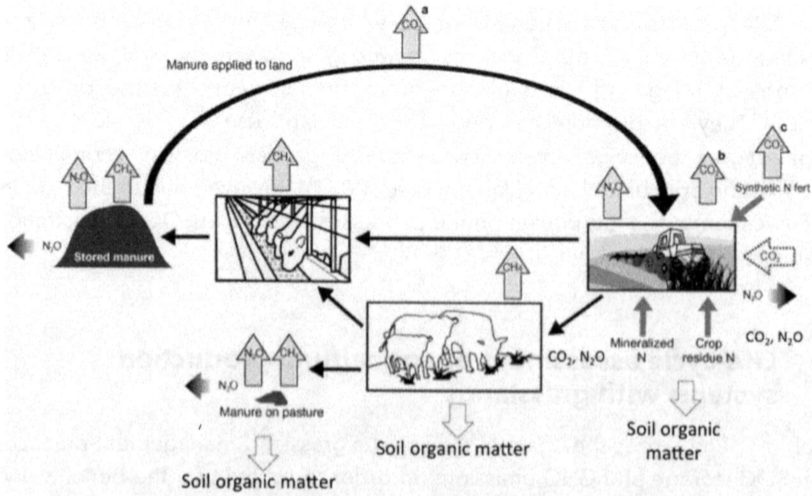

Figure 5 Greenhouse gas emissions and storage in animal production systems to be included in lifecyle analyses (Beauchemin et al., 2010, modified).

of pasture-based livestock systems (de Vries et al., 2015) do not include C and N changes in the form of soil organic matter. When SOC sequestration potential was included into life-cycle analyses of beef production systems on pastures in the Midwest United States, grazing strategies with low and high stocking densities were found to be an overall CO_2-equivalent sink (Rowntree et al., 2020). Recently, other life-cycle analyses have also suggested that SOC sequestration in well-managed grasslands can partially or fully offset GHG emissions (Stanley et al., 2018; Frank et al., 2015; Guenet et al., 2020). For this reason, it is necessary to include SOC sequestration as an integral part of any life-cycle analysis to evaluate the GHG emission contribution of agricultural systems including grasslands.

8 Conclusion

Managed grasslands, such as prairies and pastures, are favorable systems for increasing SOC storage. With ruminant livestock as the primary consumer of these forages, enteric CH_4 emission is the primary driver of greenhouse gas emissions from either hay-harvested prairies or grazed pastures. Nitrous oxide is a potent greenhouse gas and its emission is linked to the periodic concentration of available N in soil. Therefore, mineral fertilization of prairies and pastures can contribute substantially to global warming potential from N_2O emissions. Manure management of confined animal feeding operations can be a source of CH_4 and N_2O emissions. Accumulation of SOC with grassland development can be a strong sink for atmospheric CO_2, thus potentially negating other greenhouse gas emissions.

Grassland systems can be managed to reduce net greenhouse gas emissions by understanding the benefits and trade-offs of different management approaches that ultimately affect plant and animal productivity, environmental quality, biodiversity and socioeconomic outcomes. The net greenhouse gas emission potential of managed grasslands should be based on life-cycle analyses to include all components of the agricultural system.

In hayed systems particularly, N fertilization becomes a required input to increase forage production and nutritive value but can also lead to N_2O emissions. Such emissions may be limited when using organic fertilizers and leguminous species that reduce the high concentration of mineral N in soil at one time. Grassland management for ruminant livestock production varies by animal species and class, stocking rate, stocking method, season of grazing and livestock distribution on the landscape. Due to the selective grazing behavior of livestock, rotational stocking can be an effective strategy to maintain high biomass production and contribute to SOC accumulation over the entire pasture and not just isolated portions of the pasture due to uneven distribution of grazing, loafing and treading.

Achieving a positive balance between SOC sequestration and CH_4 and N_2O emissions from livestock, management strategies may need to acknowledge trade-offs in ecosystem service outcomes, which will be dependent on soil and climatic conditions of a particular region. Innovative systems to maximize the benefits of SOC sequestration and minimize trade-offs in form of greenhouse gas emissions may be derived with integrated crop-livestock systems or silvopastures that are more ecologically diverse. Co-benefits of these more complex agroecosystems include income stability for producers and increased resilience of agricultural systems against climate change.

9 Where to look for further information

- EIP AGRI Focus Group (2018) Grazing for carbon. Final Report. (https://ec.europa.eu/eip/agriculture/en/publications/eip-agri-focus-group-grazing-carbon-final-report).
- Lemaire, G., Carvalho, P. C., Krohnberg, S. and Recous, S. 2018. Agroecosystem Diversity, Édition 1 Reconciling Contemporary Agriculture and Environmental Quality. Elsevier.

10 References

Abdalla, M., Hastings, A., Chadwick, D. R., Jones, D. L., Evans, C. D., Jones, M. B., Rees, R. M. and Smith, P. 2018. Critical review of the impacts of grazing intensity on soil organic carbon storage and other soil quality indicators in extensively managed grasslands. *Agriculture, Ecosystems and Environment* 253, 62–81.

Agegnehu, G., Srivastava, A. K. and Bird, M. I. 2017. The role of biochar and biochar-compost in improving soil quality and crop performance: a review. *Applied Soil Ecology* 119, 156–170.

Aguilar, O. A., Maghirang, R., Rice, C. W., Trabue, S. L. and Erickson, L. E. 2014. Nitrous oxide fluxes from a commercial beef cattle feedlot in Kansas. *Air, Soil and Water Research* 7, 35–45.

Alves, B. J. R., Madari, B. E. and Boddey, R. M. 2017. Integrated crop-livestock-forestry systems: prospects for a sustainable agricultural intensification. *Nutrient Cycling in Agroecosystems* 108(1), 1–4.

Amadi, C. C., van Rees, K. C. J. and Farrell, R. E. 2016. Soil-atmosphere exchange of carbon dioxide, methane and nitrous oxide in shelterbelts compared with adjacent cropped fields. *Agriculture, Ecosystems and Environment* 223, 123–134.

Aryal, D. R., Gomez-Gonzalez, R. R., Hernandez-Nuriasmu, R. and Morales-Ruiz, D. E. 2019. Carbon stocks and tree diversity in scattered tree silvopastoral systems in Chiapas, Mexico. *Agroforestry Systems* 93(1), 213–227.

Barneze, A. S., Whitaker, J., McNamara, N. P. and Ostle, N. J. 2020. Legumes increase grassland productivity with no effect on nitrous oxide emissions. *Plant and Soil* 446(1–2), 163–177.

Barthod, J., Rumpel, C. and Dignac, M.-F. 2018. Composting with additives to improve organic amendments. A review. *Agronomy for Sustainable Development* 38(2), 17.

Beauchemin, K. A., Janzen, H. H., Little, S. M., McAllister, T. A. and McGinn, S. M. 2010. Life cycle assessment of greenhouse gas emissions from beef production in western Canada: a case study. *Agricultural Systems* 103(6), 371–379.

Beauchemin, K. A. and McGinn, S. M. 2006. Enteric methane emissions from growing beef cattle as affected by diet and level of intake. *Canadian Journal of Animal Science* 86(3), 401–408.

Bloor, J. M. G. and Pottier, J. 2014. Grazing and spatial heterogeneity: implications for grassland structure and function. In: Mariotte, P. and Kardol, P. (Eds) *Grassland Biodiversity and Conservation in a Changing World. Nova Science Publishers, Inc.* Hauppage, New York, pp 135–162.

Bloor, J. M. G., Tardi, A. and Pottier, J. 2020. Spatial heterogeneity of vegetation structure, plant N pools and soil N Content in relation to grassland management. *Agronomy* 10(5), 716; 10.3390/agronomy10050716.

Briske, D. D., Derner, J. D., Brown, J. R., Fuhlendorf, S. D., Teague, W. R., Havstad, K. M., Gillen, R. L., Ash, A. J. and Willms, W. D. 2008. Rotational grazing on rangelands: reconciliation of perception and experimental evidence. *Rangeland Ecology and Management* 61(1), 3–17.

Calabi-Floody, M., Medina, J., Rumpel, C., Condron, L. M., Hernandez, M., Dumont, M. and Mora, MdlL. 2018. Smart fertilizers as a strategy for sustainable agriculture. *Advances in Agronomy* 147, 119–157.

Carpinelli, S., da Fonseca, A. F., Weirich Neto, P. H., Dias, S. H. B. and Pontes, LdS. 2020. Spatial and temporal distribution of cattle dung and nutrient cycling in integrated crop-livestock systems. *Agronomy* 10(5), 672; 10.3390/agronomy10050672.

Carter, M. R. and Gregorich, E. G. 2010. Carbon and nitrogen storage by deep-rooted tall fescue (Lolium arundinaceum) in the surface and subsurface soil of a fine sandy loam in eastern Canada. *Agriculture, Ecosystems and Environment* 136(1–2), 125–132.

Chang, J., Viovy, N., Vuichard, N., Ciais, P., Campioli, M., Klumpp, K., Martin, R., Leip, A. and Soussana, J. F. 2015. Modeled changes in potential grassland productivity and in grass-fed ruminant livestock density in Europe over 1961-2010. *PLoS ONE* 10(5), e0127554. 10.1371/journal.pone.0127554.

Cole, N. A., Meyer, B. E., Parker, D. B., Neel, J., Turner, K. E., Northrup, B. K., Jennings, T. and Jennings, J. S. 2020. Effects of diet quality on energy metabolism and methane production by beef steers fed a warm-season grass-based hay diet. *Applied Animal Science* 36(5), 652-667.

Conant, R. T., Cerri, C. E. P., Osborne, B. B. and Paustian, K. 2017. Grassland management impacts on soil carbon stocks: a new synthesis. *Ecological Applications* 27(2), 662-668.

Condron, L. M., Hopkins, D. W., Gregorich, E. G., Black, A. and Wakelin, S. A. 2014. Long-term irrigation effects on soil organic matter under temperate grazed pasture. *European Journal of Soil Science* 65(5), 741-750.

Cong, W. F., van Ruijven, J., Mommer, L., De Deyn, G. B., Berendse, F. and Hoffland, E. 2014. Plant species richness promotes soil carbon and nitrogen stocks in grasslands without legumes. *Journal of Ecology* 102(5), 1163-1170. 10.1111/1365-2745.12280.

Crème, A., Chabbi, A., Gastal, F. and Rumpel, C. 2017. Biogeochemical nature of grassland soil organic matter under plant communities with two nitrogen sources. *Plant and Soil* 415(1-2), 189-201.

Crème, A., Rumpel, C., Gastal, F., Mora, M.-L. and Chabbi, A. 2016. Effect of grasses and a legume grown in monoculture or mixture on soil organic matter and phosphorus forms. *Plant and Soil* 402(1-2), 117-128.

Crème, A., Rumpel, C., Le Roux, X., Romian, A., Lan, T. and Chabbi, A. 2018. Fertilised ley grassland has a legacy effect on soil organic matter quantity and quality and microbial activities favourable to C sequestration. *Soil Biology and Biochemistry* 122, 203-210.

Crème, A., Rumpel, C., Malone, S. L., Saby, N. P. A., Vaudour, E., Decau, M. L. and Chabbi, A. 2020. Monitoring grassland management effects on soil organic carbon - a matter of scale. *Agronomy* 10(12), 2016; 10.3390/agronomy10122016.

Da Pontes, L. S., Giostri, A. F., Baldissera, T. C., Barro, R. S., Stafin, G., Porfírio-da-Silva, V., Moletta, J. L. and Césarde Faccio Carvalho, P. 2016. Interactive effects of trees and nitrogen supply on the agronomic characteristics of warm-climate grasses. *Agronomy Journal* 108(4), 1531-1541.

de Abreu, L. H. G., de Freitas, I. C., Santana, P. H. L., de Almeida Barbosa, D. L., Santos, L. D. T., Santos, M. V., Sanglard, D. A. and Frazão, L. A. 2020. Variation in soil carbon, nitrogen and microbial attributes within a silvopastoral system in the Brazilian cerrado. *Agroforestry Systems* 94(6), 2343-2353.

De Deyn, G. B., Shiel, R. S., Ostle, N. F., McNamara, N. P., Oakley, S., Young, I., Freeman, C., Fenner, N., Quirk, H. and Bardgett, R. D. 2022. Additional carbon sequestration benefits of grassland diversity restoration. *Journal of Applied Ecology* 48, 600-608.

Derner, J. D., Augustine, D. J. and Frank, D. A. 2019. Does grazing matter for soil organic carbon sequestration in the western North American Great Plains? *Ecosystems* 22(5), 1088-1094.

Derner, J. D. and Schuman, G. E. 2007. Carbon sequestration and rangelands: a synthesis of land management and precipitation. *Journal of Soil and Water Conservation* 62, 77-85.

de Vries, M., van Middelaar, C. E. and de Boer, I. J. M. 2015. Comparing environmental impacts of beef production systems: a review of life cycle assessments. *Livestock Science* 178, 279-288.

Ditzler, L., Breland, T. A., Francis, C., Chakraborty, M., Singh, D. K., Srivastava, A., Eyhorn, F., Groot, J. C. J., Six, J. and Decoc, C. 2018. Identifying viable nutrient management interventions at the farm level: the case of smallholder organic Basmati rice production in Uttarakhand, India. *Agricultural Systems* 161, 61–71.

Elgersma, A. and Soegaard, K. 2018. Changes in nutritive value and herbage yield during extended growth intervals in grass-legume mixtures: effects of species, maturity at harvest, and relationships between productivity and components of feed quality. *Grass and Forage Science* 73(1), 78–93.

Eze, S., Palmer, S. M. and Chapman, P. J. 2018. Soil organic carbon stock in grasslands: effects of inorganic fertilizers, liming and grazing in different climate settings. *Journal of Environmental Management* 223, 74–84. 10.1016/j.jenvman.2018.06.013.

FAO 2013. *Greenhouse Gas Emissions from Ruminant Supply Chains, a Global Life Cycle Assessment*. Food and Agriculture Organization, Rome.

Ferguson, B. G., Diemont, S. A. W., Alfaro-Arguello, R., Martin, J. F., Nahed-Toral, J., Alvarez-Solís, D. and Pinto-Ruíz, R. 2013. Sustainability of holistic and conventional cattle ranching in the seasonally dry tropics of Chiapas, Mexico. *Agricultural Systems* 120, 38–48.

Follett, R. F., Kimble, J. M., Pruessner, E. G., Samson-Liebig, S. and Waltman, S. (2009). Soil organic carbon stocks with depth and land use at various US sites. In: Lal, R. and Follett, R. F. (Eds) *Soil Carbon Sequestration and the Greenhouse Effect, vol 57* (2nd edn.), SSSA Special Publications, pp. 29–46. 410p. Madison, WI.

Frank, D. A., McNaughton, S. J. and Tracy, B. F. 1998. The ecology of the earth's grazingecosystems. *BioScience* 48(7), 513-521.

Frank, S., Schmid, E., Havlík, P., Schneider, U. A., Böttcher, H., Balkovič, J. and Obersteiner, M. 2015. 2015 the dynamic of soil organic carbon mitigation potential of European cropland. *Global Environmental Change* 35, 269-278. 10.1016/j.gloenvcha.2015.08.004.

Franzluebbers, A. J. 2005. Soil organic carbon sequestration and agricultural greenhouse gas emissions in the southeastern USA. *Soil and Tillage Research* 83(1), 120-147.

Franzluebbers, A. J. 2020. Soil-test biological activity with the flush of CO2: V. Validation of nitrogen prediction for corn production. *Agronomy Journal* 112(3), 2188-2204.

Franzluebbers, A. J. 2021. Soil organic carbon sequestration calculated from depth distribution. *Soil Science Society of America Journal* 85(1), 158-171.

Franzluebbers, A. J., Chappell, J. C., Shi, W. and Cubbage, F. W. 2017. Greenhouse gas emissions in an agroforestry system of the southeastern USA. *Nutrient Cycling in Agroecosystems* 108(1), 85–100.

Franzluebbers, A. J., Endale, D. M., Buyer, J. S. and Stuedemann, J. A. 2012a. Tall fescue management in the Piedmont: sequestration of soil organic carbon and total nitrogen. *Soil Science Society of America Journal* 76(3), 1016-1026.

Franzluebbers, A. J., Paine, L. K., Winsten, J. R., Krome, M., Sanderson, M. A., Ogles, K. and Thompson, D. 2012b. Well-managed grazing systems: a forgotten hero of conservation. *Journal of Soil and Water Conservation* 67(4), 100A–104A.

Franzluebbers, A. J. and Poore, M. H. 2020. Soil-test biological activity with the flush of CO2: VII. Validating nitrogen needs for fall-stockpiled forage. *Agronomy Journal* 112(3), 2240-2255.

Franzluebbers, A. J. and Poore, M. H. 2021. Tall fescue management and environmental influences on soil, surface residue, and forage properties. *Agronomy Journal* 113(2), 2029–2043.

Published by Burleigh Dodds Science Publishing Limited, 2023.

Franzluebbers, A. J., Sawchik, J. and Taboada, M. A. 2014. Agronomic and environmental impacts of pasture-crop rotations in temperate North and South America. *Agriculture, Ecosystems and Environment* 190, 18–26.

Franzluebbers, A. J. and Stuedemann, J. A. 2009. Soil-profile organic carbon and total nitrogen during 12 years of pasture management in the Southern Piedmont USA. *Agriculture, Ecosystems and Environment* 129(1-3), 28–36.

Franzluebbers, A. J. and Stuedemann, J. A. 2010. Surface soil changes during twelve years of pasture management in the Southern Piedmont USA. *Soil Science Society of America Journal* 74(6), 2131-2141.

Franzluebbers, A. J., Stuedemann, J. A., Schomberg, H. H. and Wilkinson, S. R. 2000. Soil organic C and N pools under long-term pasture management in the southern Piedmont USA. *Soil Biology and Biochemistry* 32(4), 469–478.

Franzluebbers, A. J., Stuedemann, J. A. and Wilkinson, S. R. 2001. Bermudagrass management in the Southern Piedmont USA: I. Soil and surface residue carbon and sulfur. *Soil Science Society of America Journal* 65(3), 834–841.

Franzluebbers, A. J., Wilkinson, S. R. and Stuedemann, J. A. 2004. Bermudagrass management in the Southern Piedmont USA: X. Coastal productivity and persistence in response to fertilization and defoliation regimes. *Agronomy Journal* 96(5), 1400–1411.

Fuchs, K., Hortnagl, L., Buchmann, N., Eugster, W., Snow, V. and Merbold, L. 2018. Management matters: testing a mitigation strategy for nitrous oxide emissions using legumes on intensively managed grassland. *Biogeosciences* 15(18), 5519-5543.

Fustec, J., Lesuffleur, F., Mahieu, S. and Cliquet, J.-B. 2010. Nitrogen rhizodeposition of legumes. A review. *Agronomy for Sustainable Development* 30(1), 57–66.

Gilmullina, A., Rumpel, C., Blagodatskaya, E. and Chabbi, A. 2020. Management of grasslands by mowing versus grazing - impacts on soil organic matter quality and microbial functioning. *Applied Soil Ecology* 156, 103701.

Gilmullina, A., Rumpel, C., Klumpp, K. and Chabbi, A. 2021. Do grassland management practices affect soil lignin chemistry by changing the composition of plant-derived organic matter input? *Plant and Soil* 469(1–2), 443–455, 10.1007/s11104-021-05174-7.

Graham, P. H. and Vance, C. P. 2003. Legumes: importance and constraints to greater use. *Plant Physiology* 131(3), 872-877.

Grandy, A. S. and Robertson, G. P. 2006. Aggregation and organic matter protection following tillage of a previously uncultivated soil. *Soil Science Society of America Journal* 70(4), 1398–1406.

Guenet, B., Gabrielle, B., Chenu, C., Arrouays, D., Balesdent, J., Bernoux, M., Bruni, E., Caliman, J. P., Cardinael, R., Chen, S., Ciais, P., Desbois, D., Fouche, J., Frank, S., Henault, C., Lugato, E., Naipal, V., Nesme, T., Obersteiner, M., Pellerin, S., Powlson, D. S., Rasse, D. P., Rees, F., Soussana, J. F., Su, Y., Tian, H., Valin, H. and Zhou, F. 2020. Can N2O emissions offset the benefits from soil organic carbon storage? *Global Change Biology* 27(2), 237-256. 10.1111/gcb.15342.

Haile, S. G., Nair, P. K. R. and Nair, V.D., V. D. 2008. Carbon storage in different soil size fractions in Florida silvopastoral systems. *Journal of Environmental Quality* 37, 1789–1797.

Hamilton, E. W., Frank, D. A., Hinchey, P. M. and Murray, T. R. 2008. Defoliation induces root exudation and triggers positive rhizospheric feedbacks in a temperate grassland. *Soil Biology and Biochemistry* 40(11), 2865-2873.

Hauggaard-Nielsen, H., Lachouani, P., Knudsen, M. T., Ambus, P., Boelt, B. and Gislum, R. 2016. Productivity and carbon footprint of perennial grass-forage legume intercropping strategies with high or low nitrogen fertilizer input. *Science of the Total Environment* 541, 1339–1347. 10.1016/j.scitotenv.2015.10.013.

Hijbeek, R., van Loon, M. P. and van Ittersum, M. K. 2019. Fertiliser use and soil carbon sequestration: opportunities and trade-offs. CCAFS Working Paper no. 264. Wageningen, the Netherlands: CGIAR Research Program on Climate Change, Agriculture and Food Security (CCAFS). Available online at: www.ccafs.cgiar.org.

Hoeffner, K., Beylich, A., Chabbi, A., Cluzeau, D., Dascalu, D., Graefe, U., Guzmán, G., Hallaire, V., Hanisch, J., Landa, B. B., Linsler, D., Menasseri, S., Öpik, M., Potthoff, M., Sandor, M., Scheu, S., Schmelz, R. M., Engell, I., Schrader, S., Vahter, T., Banse, M., Nicolaï, A., Plaas, E., Runge, T., Roslin, T., Decau, M. L., Sepp, S. K., Arias-Giraldo, L. F., Busnot, S., Roucaute, M. and Pérès, G. 2021. Legacy effects of temporary grassland in annual crop rotation on soil ecosystem services. *Science of the Total Environment* 780, 146140.

Hristov, A. N., Oh, J., Firkins, J. L., Dijkstra, J., Kebreab, E., Waghorn, G., Makkar, H. P. S., Adesogan, A. T., Yang, W., Lee, C., Gerber, P. J., Henderson, B. and Tricarico, J. M. 2013. SPECIAL TOPICS – mitigation of methane and nitrous oxide emissions from animal operations: I. A review of enteric methane mitigation options. *Journal of Animal Science* 91(11, November), 5045–5069.

Hunt, J. E., Laubach, J., Barthel, M., Fraser, A. and Phillips, R. L. 2016. Carbon budgets for an irrigated intensively grazed dairy pasture and an unirrigated winter-grazed pasture. *Biogeosciences* 13(10), 2927–2944. 10.5194/bg-13-2927-2016.

Ingram, L. J., Stahl, P. D., Schuman, G. E., Buyer, J. S., Vance, G. F., Ganjegunte, G. K., Welker, J. M. and Derner, J. D. 2008. Grazing impacts on soil carbon and microbial communities in a mixed-grass ecosystem. *Soil Science Society of America Journal* 72(4), 939–948.

IPCC 2013. 2013: the physical science basis. Contribution of working group I to the fifth assessment report of the intergovernmental panel on climate change. *Climate Change*. Cambridge University Press, Cambridge, United Kingdom and New York, NY.

IPCC 2014. *Contribution of Working Group III to the Fifth Assessment Report of the Intergovernmental Panel on Climate Change*. Cambridge University Press, Cambridge, United Kingdom and New York, NY.

IPCC 2018. Index. In: Global Warming of 1(5)°C. An IPCC Special Report on the impacts of global warming of 1.5°C above preindustrial levels and related global greenhouse gas emission pathways, in the context of strengthening the global response to the threatof climate change, sustainable development, and efforts to eradicate poverty [Masson-Delmotte, V., P. Zhai, H.-O. Pörtner, D. Roberts,J. Skea, P.R. Shukla, A. Pirani, W. Moufouma-Okia, C. Péan, R. Pidcock, S. Connors, J.B.R. Matthews, Y. Chen, X. Zhou, M.I. Gomis, E. Lonnoy, T. Maycock, M. Tignor, and T. Waterfield (eds.)].

Isbell, F., Tilman, D., Reich, P. B. and Clark, A. T. 2019. Deficits of biodiversity and productivity linger a century after agricultural abandonment. *Nature Ecology and Evolution* 3(11), 1533–1538.

Jameel, Y., Patrone, C. M., Patterson, K. P. and West, P. C. (2022). Climate-poverty connections: opportunities for synergistic solutions at the intersection of planetary and human well-being. Project Drawdown. https://doi.org/10.55789/y2c0k2p2.

Johnston, A. E., Poulton, P. R., Coleman, K., Macdonald, A. J. and White, R. P. 2017. Changes in soil organic matter over 70 years in continuous arable and ley-arable rotations on a sandy loam soil in England. *European Journal of Soil Science* 68(3), 305–316.

Kaempf, I., Hoelzel, N., Stoerrle, M., Broll, G. and Kiehl, K. 2016. Potential of temperate agricultural soils for carbon sequestration: a meta-analysis of land-use effects. *Science of the Total Environment* 566–567, 428–435.

Klumpp, K., Fontaine, S., Attard, E., Le Roux, X., Gleixner, G. and Soussana, J. 2009. Grazing triggers soil carbon loss by altering plant roots and their contol on soil microbial community. *Journal of Ecology* 97(5), 876–885.

Knopf, F. L. 1994. Avian assemblages on altered grasslands. *Studies in Avian Biology* 15, 247–257.

Kohler, C., Morvan-Bertrand, A., Cliquet, J., Klumpp, K. and Lemauviel-Lavenant, S. 2020. Tradeoff between the conservation of soil C Stocks and vegetation productivity in temperate grasslands. *Agronomy* 10(7), 1024; 10.3390/agronomy10071024.

Koncz, P., Pintér, K., Balogh, J., Papp, M., Hidy, D., Csintalan, Z., Molnár, E., Szaniszló, A., Kampfl, G., Horváth, L. and Nagy, Z. 2017. Extensive grazing in contrast to mowing is climate-friendly based on the farm-scale greenhouse gas balance. *Agriculture, Ecosystems and Environment* 240, 121–134. 10.1016/j.agee.2017.02.022.

Kumar, S., Sieverding, H., Lai, L., Thandiwe, N., Wienhold, B., Redfearn, D., Archer, D., Ussiri, D., Faust, D., Landblom, D., Grings, E., Stone, J. J., Jacquet, J., Pokharel, K., Liebig, M., Schmer, M., Sexton, P., Mitchell, R., Smalley, S., Osborne, S., Ali, S., Şentürklü, S., Sehgal, S., Owens, V. and Jin, V. 2019. Facilitating crop-livestock reintegration in the northern Great Plains. *Agronomy Journal* 111(5), 2141–2156, 10.2134/agronj2018.07.0441.

Kunrath, T. R., de Berranger, C., Charrier, X., Gastal, F., de Faccio Carvalho, P. C., Lemaire, G., Emile, J. and Durand, J. 2015. How much do sod-based rotations reduce nitrate leaching in a cereal cropping system? *Agricultural Water Management* 150, 46–56.

Lam, W. Y., Chatterton, J., Sim, S., Kulak, M., Mendoza Beltran, A. and Huijbregts, M. A. J. 2021. Estimating greenhouse gas emissions from direct land use change due to crop production in multiple countries. *Science of the Total Environment* 755(2), 143338. 10.1016/j.scitotenv.2020.143338.

Lange, M., Eisenhauer, N., Sierra, C. A., Bessler, H., Engels, C., Griffiths, R. I., Mellado-Vázquez, P. G., Malik, A. A., Roy, J., Scheu, S., Steinbeiss, S., Thomson, B. C., Trumbore, S. E. and Gleixner, G. 2015. Plant diversity increases soil microbial activity and soil carbon storage. *Nature Communications* 6, 6707.

Launay, C., Constantin, J., Chlebowski, F., Houot, S., Graux, A. I., Klumpp, K., Martin, R., Mary, B., Pellerin, S. and Therond, O. 2021. Estimating the carbon storage potential and greenhouse gas emissions of French arable cropland using high-resolution modeling. *Global Change Biology* 27(8), 1645–1661. 10.1111/GCB.15512.

Lemaire, G., Franzluebbers, A., Carvalho, P. CdF. and Dedieu, B. 2014. Integrated crop-livestock systems: strategies to achieve synergy between agricultural production and environmental quality. *Agriculture, Ecosystems and Environment* 190, 4–8.

Lemaire, G., Gastal, F., Franzluebbers, A., Chabbi, A., 2015. Grassland-cropping rotations: An avenue for agricultural diversification to reconcile high production with environmental quality. *Environ. Manage.* 56, 1065–1077.

Lim, S.-S., Baah-Acheamfour, M., Choi, W.-J., Arshad, M. A., Fatemi, F., Barnerjee, S., Carlyle, C. N., Bork, E. W., Park, H.-J. and Chang, S. X. 2018. Soil organic carbon stocks in three Canadian agroforestry systems: From surface organic to deeper mineral soils. *Forest Ecology and Management* 417, 103-109.

Liu, C., Song, X., Wang, L., Wang, D., Zhou, X., Liu, J., Zhao, X., Li, J. and Lin, H. 2016. Effects of grazing on soil nitrogen spatial heterogeneity depend on herbivore assemblage and pre-grazing plant diversity. *Journal of Applied Ecology* 53(1), 242-250.

Lopez-Santiago, J. G., Casanova-Lugo, F., Villanueva-Lopez, G., Diaz-Echeverria, V. F., Solorio-Sanchez, F. J., Martinez-Zurimendi, P., Aryal, D. R. and Chay-Canul, A. J. 2019. Carbon storage in a silvopastoral system compared to that in a deciduous dry forest in Michoacan, Mexico. *Agroforestry Systems* 93(1), 199-211.

Lorenz, K. and Lal, R. 2014. Soil organic carbon sequestration in agroforestry systems: a review. *Agronomy for Sustainable Development* 34(2), 443-454.

Lu, M., Zhou, X., Luo, Y., Yang, Y., Fang, C., Chen, J. and Li, B. 2011. Minor stimulation of soil carbon storage by nitrogen addition: a meta-analysis. *Agriculture, Ecosystems and Environment* 140(1-2), 234-244.

Lugato, E., Leip, A. and Jones, A. 2018. Mitigation potential of soil carbon management overestimated by neglecting N2O emissions. *Nature Climate Change* 8(3), 219-223.

Martin, G., Durand, J.-L., Duru, M., Gastal, F., Julier, B., Litrico, I., Louarn, G., Médiène, S., Moreau, D., Valentin-Morison, M., Novak, S., Parnaudeau, V., Paschalidou, F., Vertès, F., Voisin, A., Cellier, P. and Jeuffroy, M. 2020. Role of ley pastures in tomorrow's cropping systems: a review. *Agronomy for Sustainable Development* 40(3), 17.

Marty, C., Pornon, A., Escaravage, N., Winterton, P. and Lamaze, T. 2009. Complex interactions between a legume and two grasses in a subalpine meadow. *American Journal of Botany* 96(10), 1814-1820.

McLauchlan, K. K., Hobbie, S. E. and Post, W. M. 2006. Conversion from agriculture to grassland builds soil organic matter on decadal timescales. *Ecological Applications* 16(1), 143-153.

McSherry, M. E. and Ritchie, M. E. 2013. Effects of grazing on grassland soil carbon: a global review. *Global Change Biology* 19(5), 1347-1357.

Meier, E. A., Thorburn, P. J., Bell, L. W., Harrison, M. T. and Biggs, J. S. 2020. Greenhouse gas emissions from cropping and grazed pastures are similar: a simulation analysis in Australia. *Frontiers in Sustainable Food Systems* 3, UNSP 121.

Mendoza, R., Bailleres, M., García, I. and Ruiz, O. 2016. Phosphorus fertilization of a grass-legume mixture: effect on plant growth, nutrients acquisition and symbiotic associations with soil microorganisms. *Journal of Plant Nutrition* 39(5), 691-701.

Mesléard, F., Yavercovski, N., Lefebvre, G., Willm, L. and Bonis, A. 2017. High stocking density controls Phillyrea angustifolia in mediterranean grasslands. *Environmental Management* 59(3), 455-463.

Minasny, B., Malone, B. P., McBratney, A. B., Angers, D. A., Arrouays, D., Chambers, A., Chaplot, V., Chen, Z. S., Cheng, K., Das, B. S., Field, D. J., Gimona, A., Hedley, C. B., Hong, S. Y., Mandal, B., Marchant, B. P., Martin, M., McConkey, B. G., Mulder, V. L., O'Rourke, S., Richer-de-Forges, A. C., Odeh, I., Padarian, J., Paustian, K., Pan, G., Poggio, L., Savin, I., Stolbovoy, V., Stockmann, U., Sulaeman, Y., Tsui, C., Vågen, T., van Wesemael, B. and Winowiecki, L. 2017. Soil carbon 4 per mille. *Geoderma* 292, 59-86.

Moinet, G. Y. K., Midwood, A. J., Hunt, J. E., Rumpel, C., Millard, P. and Chabbi, A. 2019. Grassland management influences the response of soil respiration to drought. *Agronomy* 9(3), 124; 10.3390/agronomy9030124.

Mosier, S. and Cotrufo, F. 2022. Adaptive multi-paddock grazing enhances soil carbon and nitrogen stocks and stabilization through mineral association in southeastern U.S. grazing lands. *Journal of Environmental Management* 288, 112409.

Mudge, P. L., Kelliher, F. M., Knight, T. L., O'Connell, D., Fraser, S. and Schipper, L. A. 2017. Irrigating grazed pasture decreases soil carbon and nitrogen stocks. *Global Change Biology* 23(2), 945–954, 10.1111/gcb.13448.

Nair, P. K. R., Gordon, A. M. and Mosquera-Losada, M. R. 2008. Agroforestry. In: Jorgensen, S. E. and Fath, B. D. (Eds) *Encyclopedia of Ecology*. Elsevier, Oxford, pp. 101–110.

Orgill, S. E., Condon, J. R., Conyers, M. K., Morris, S. G., Alcock, D. J., Murphy, B. W. and Greene, R. S. B. 2018. Removing grazing pressure from a native pasture decreases soil organic carbon in Southern New South Wales, Australia. *Land Degradation and Development* 29(2), 274–283.

Park, J. Y., Ale, S. and Teague, W. R. 2017. Simulated water quality effects of alternate grazing management practices at the ranch and watershed scales. *Ecological Modelling* 360, 1–13.

Peterson, C. A., Deiss, L. and Gaudin, A. C. M. 2020. Commercial integrated crop-livestock systems achieve comparable crop yields to specialized production systems: a meta-analysis. *PLoS ONE* 15(5), e0231840. 10.1371/journal.pone.0231840.

Pierce, F. J., Fortin, M.-C. and Staton, M. J. 1994. Periodic plowing effects on soil properties in a no-till farming system. *Soil Science Society of America Journal* 58(6), 1782–1787.

Poblete-Grant, P., Suazo-Hernández, J., Condron, L., Rumpel, C., Demanet, R., Malone, S. L. and Mora, MdL. L. 2020. Soil available P, soil organic carbon and aggregation as affected by longterm poultry manure application to andisols under pastures in Southern Chile. *Geoderma Regional* 21, e00271.

Poeplau, C., Don, A., Vesterdal, L., Leifeld, J., Van Wesemael, B., Schuhmacher, J. and Gensior, A. 2011. Temporal dynamics of soil organic carbon after land-use change in the temperate zone - carbon response functions as a model approach. *Global Change Biology* 17(7), 2415–2427.

Poeplau, C., Zopf, D., Greiner, B., Geerts, R., Korvaar, H., Thumm, U., Don, A., Heidkamp, A. and Flessa, H. 2018. Why does mineral fertilization increase soil carbon stocks in temperate grasslands? *Agriculture, Ecosystems and Environment* 265, 144–155.

Resende, LdO., Müller, M. D., Kohmann, M. M., Pinto, L. F. G., Cullen Junior, L., de Zen, S. and Rego, L. F. G. 2020. Silvopastoral management of beef cattle production for neutralizing the environmental impact of enteric methane emission. *Agroforestry Systems* 94(3), 893-903.

Rowntree, J. E., Ryals, R., DeLonge, M., Teague, W. R., Chiavegato, M., Byck, P., Wang, T. and Xu, S. 2016. Potential mitigation of midwest grass-finished beef production emissions with soil carbon sequestration in the United States of America. *Future Food* 4, 31-38.

Rowntree, J. E., Stanley, P. L., Maciel, I. C. F., Thorbecke, M., Rosenzweig, S. T., Hancock, D. W., Guzman, A. and Raven, M. R. 2020. Ecosystem impacts and productive capacity of a multi-species pastured livestock system. *Frontiers in Sustainable Food Systems* 4, 544984.

Rumpel, C. and Chabbi, A. 2010. Response of bulk chemical composition, lignin and carbohydrate signature to grassland conversion in a ley-arable cropping system. *Nutrient Cycling in Agroecosystems* 88(2), 173–182.

Rumpel, C., Crème, A., Ngo, P. T., Valesquez, G., Mora, M. L. and Chabbi, A. 2015. The impact of grassland management on biogeochemical cycles involving carbon, nitrogen and phosphorus. *Journal of Soil Science and Plant Nutrition* 15(ahead), 0.

Russell, J. R., Barnhart, S. K., Morrical, D. G. and Sellers, H. J. 2013. *Use of mob grazing to improve cattle production, enhance legume establishment and increase carbon sequestration in iowa pastures*. Leopold Center Completed Grant Reports.

Russelle, M. P., Entz, M. H. and Franzluebbers, A. J. 2007. Reconsidering integrated crop-livestock Systems in North America. *Agronomy Journal* 99(2), 325–334.

Sanaullah, M., Chabbi, A., Rumpel, C. and Kuzyakov, Y. 2012. Carbon allocation in grassland communities under drought stress followed by 14C pulse labelling. *Soil Biology and Biochemistry* 55, 132–139.

Sanderman, J. and Baldock, J.A. 2010. Accounting for soil carbon sequestration in national inventories: a soil scientist's perspective. *Environmental Research Letters* 5, 3.

Sanderman, J., Reseigh, J., Wurst, M., Young, M.-A., Austin, J. 2015. Impacts of rotational grazing on soil carbon in native grass-based pastures in Southern Australia. *PLoS ONE* 10(8), e0136157.

Schiedung, M., Tregurtha, C. S., Beare, M. H., Thomas, S. M. and Don, A. 2019. Deep soil flipping increases carbon stocks of New Zealand grasslands. *Global Change Biology* 25(7), 2296–2309.

Sebastia, M.-T., Chabbi, A., Cordovil, C. M.d.S., De Vliegher, A., Hennessy, D., Kramberger, B., Newell Price, P., Richmond, R. and van Rijn, C. 2018. EIP-AGRI Focus Group - Grazing for Carbon Mini-Paper – Mixtures of species, Final report, 32p.

Senapati, N., Chabbi, A., Gastal, F., Smith, P., Mascher, N., Loubet, B., Cellier, P. and Naisse, C. 2014. Net carbon storage measured in a mowed and grazed temperate sown grassland shows potential for carbon sequestration under grazed system. *Carbon Management* 5(2), 131–144.

Shahzad, T., Chenu, C., Repincay, C., Mougin, C., Ollier, J. L. and Fontaine, S. 2012. Plant clipping decelerates the mineralization of recalcitrant soil organic matter under multiple grassland species. *Soil Biology and Biochemistry* 51, 73–80.

Sollenberger, L. E., Agouridis, C. T., Vanzant, E. S., Franzluebbers, A. J. and Owens, L. B. 2012. Prescribed grazing on pasturelands. In: Nelson, C. J. (Ed.). *Conservation Outcomes from Pastureland and Hayland Practices: Assessment, Recommendations, and Knowledge Gaps*. United States Department of Agriculture, Natural Resources Conservation Service, pp 111–204.

Smith, P. C., Heinrich, G., Suklitsch, M., Gobiet, A., Stoffel, M. and Fuhrer, J. 2014. Station-scale bias correction and uncertainty analysis for the estimation of irrigation water requirements in the Swiss Rhone catchment under climate change. *Climatic Change* 127(3–4), 521–534.

Soussana, J. F. and Lemaire, G. 2014. Coupling carbon and nitrogen cycles for environmentally sustainable intensification of grasslands and crop-livestock systems. *Agriculture, Ecosystems and Environment* 190, 9–17.

Stanley, P. L., Rowntree, J. E., Beede, D. K., DeLonge, M. S. and Hamm, M. W. 2018. Impacts of soil carbon sequestration on life cycle greenhouse gas emissions in midwestern USA beef finishing systems. *Agricultural Systems* 162, 249–258.

Stewart, A. A., Little, S. M., Ominski, K. H., Wittenberg, K. M. and Janzen, H. H. 2009. Evaluating greenhouse gas mitigation practices in livestock systems: an illustration of a whole-farm approach. *Journal of Agricultural Science* 147(4), 367–382.

Stiles, W. A. V., Rowe, E. C. and Dennis, P. 2017. Long-term nitrogen and phosphorus enrichment alters vegetation species composition and reduces carbon storage in upland soil. *Science of the Total Environment* 593-594, 688–694. 10.1016/j.scitotenv.2017.03.136.

Studdert, G. A., Echeverria, H. E. and Casanovas, E. M. 1997. Crop-pasture rotation for sustaining the quality and productivity of a typic Argiudoll. *Soil Science Society of America Journal* 61(5), 1466–1472.

Sun, J., Wang, X., Cheng, G., Wu, J., Hong, J. and Niu, S. 2014. Effects of grazing regimes on plant traits and soil nutrients in an alpine steppe, Northern Tibetan Plateau. *PLoS ONE* 9(9), e108821.

Szukics, U., Grigulis, K., Legay, N., Kastl, E. M., Baxendale, C., Bardgett, R. D., Clément, J. C., Lavorel, S., Schloter, M. and Bahn, M. 2019. Management versus site effects on the abundance of nitrifiers and denitrifiers in European mountain grasslands. *Science of the Total Environment* 648, 745–753. 10.1016/j.scitotenv.2018.08.039.

Teague, W. R. 2018. Forages and pastures symposium: cover crops in livestock production: whole-system approach: managing grazing to restore soil health and farm livelihoods. *Journal of Animal Science* 96(4), 1519–1530.

Teague, W. R., Dowhower, S. L., Baker, S. A., Haile, N., DeLaune, P. B. and Conover, D. M. 2011. Grazing management impacts on vegetation, soil biota and soil chemical, physical and hydrological properties in tall grass prairie. *Agriculture, Ecosystems and Environment* 141(3–4), 310–322.

Teague, W. R., Provenza, F., Kreuter, U. P., Steffens, T. and Barnes, M. 2013. Multi-paddock grazing on rangelands: why the perceptual dichotomy between research results andrancher experience? *Journal of Environment Management* 128, 699–717.

Torres, C. M. M. E., Jacovine, L. A. G., Nolasco de Olivera Neto, S., Fraisse, C. W., Soares, C. P. B., de Castro Neto, F., Ferreira, L. R., Zanuncio, J. C. and Lemes, P. G. 2017. Greenhouse gas emissions and carbon sequestration by agroforestry systems in southeastern Brazil. *Scientific Reports* 7(1), 16738.

Vermeulen, S. J., Campell, B. M. and Ingram, J. S. I. 2012. Climate change and food systems. *Annual Review of Environment and Resources* 37(1), 195–222.

Vertes, F., Delaby, L., Klumpp, K. and Bloor, J. 2019. C-N-P uncoupling in grazed grasslands and environmental implications of management intensification. In: Lemaire, G., Carvalho, P. C. D. F., Kronberg, S. and Recous, S. (Eds), *Agroecosystem Diversity: Reconciling Contemporary Agriculture and Environmental Quality*. Elsevier, Academic Press, pp. 15–34.

Wallisdevries, M. F., Laca, E. A. and Demment, M. W. 1999. The importance of scale of patchiness for selectivity in grazing herbivores. *Oecologia* 121(3), 355–363.

Wang, F., Apfelbaum, S. I., Thompson, R. L., Teague, R. and Byck, P. 2021. Effects of adaptive multiple paddock and continuous grazing on fine-scale spatial patterns of vegetation species and biomass in commercial ranches. *Landscape Ecology* 36(9), 2725–2741.

Wang, T., Teague, W., Park, S. and Bevers, S. 2015. GHG mitigation potential of different grazing strategies in the United States Southern Great Plains. *Sustainability* 7(10), 13500–13521.

West, T. O. and Marland, G. 2002. A synthesis of carbon sequestration, carbon emissions, and net carbon flux in agriculture: comparing tillage practices in the United States. *Agriculture, Ecosystems and Environment* 91(1–3), 217–232.

Wit, C. T. de, Tow, P. G. and Ennick, G. C. 1966. Competition between legumes and grasses. *Verslagen van Landbouwkundige Onderzoekingen* 687, 1–30.

Yu, K. and Patrick, W. H. 2004. Redox window with minimum global warming potential contribution from rice soils. *Soil Science Society of America Journal* 68(6), 2086–2091.

Chapter 19

Optimizing forest management for soil carbon sequestration

Andreas Schindlbacher, Federal Research and Training Centre for Forests, Natural Hazards and Landscape (BFW), Austria; Mathias Mayer, Swiss Federal Institute for Forest, Snow and Landscape Research (WSL), Switzerland and University of Natural Resources and Life Sciences (BOKU), Austria; Robert Jandl, Federal Research and Training Centre for Forests, Natural Hazards and Landscape (BFW), Austria; and Stephan Zimmermann and Frank Hagedorn, Swiss Federal Institute for Forest, Snow and Landscape Research (WSL), Switzerland

1 Introduction

Forest carbon (C) cycling is a key component of the terrestrial C sink (Friedlingstein et al., 2020; Harris et al., 2021). C capture and storage in forest biomass and soils, therefore, has been identified as a keystone 'natural climate solution' (Bossio et al., 2020; Griscom et al., 2017) and forests have been assigned important roles in national and international CO_2 mitigation frameworks/strategies, such as the IPCC, the Bonn challenge, and the 4 per mill initiative (Minasny et al., 2017; Smith et al., 2014; Stanturf et al., 2019).

Forests store more than 90% of biomass C across terrestrial ecosystems and thus forest management can contribute to C sequestration (Bastin et al., 2019; Pan et al., 2011; Smith et al., 2016). However, practices have primarily been optimized to fulfill ecosystems services such as timber production, biodiversity, recreation, and natural hazard protection (e.g. Führer, 2000), while C storage in forest soils has received less attention, although it represents an important C pool, storing greater amounts of C than biomass in most eco-regions (Scharlemann et al., 2014).

http://dx.doi.org/10.19103/AS.2022.0106.18

Repeated soil inventories in Germany and France, for example, observed that forest soils have sequestered about 0.4 Mt C/ha/yr during the last few decades (Grüneberg et al., 2014; Jonard et al., 2017); this increase in C meets the targets of the 4 per mill initiative and increases the expectation that forest soils have high potential for C sequestration. On the other hand, climate change already affects forests, driving forest management towards adaptive measures, leading to the risk of sequestered C being re-released by accelerated soil organic matter (SOM) decomposition (Melillo et al., 2017; Prietzel et al., 2016; Schindlbacher et al., 2015) and/or by disturbances such as forest fires (Seidl et al., 2014).

The capacity for sustainable storage of soil organic SOC, therefore, is a moving target. Concepts for defining reference levels or benchmarks for stocks of organic C of specific forest soils have been proposed, but are difficult to validate or apply due to a lack of required soil characterization data (e.g. texture, clay content, aggregate size distribution, etc.) (De Vos et al., 2015; Rabot et al., 2018; Wiesmeier et al., 2019) and the high diversity of forest soil types in general (Fig. 1). Principally, SOC stocks are assumed to be in dynamic equilibrium under given environmental conditions (Böttcher and Springob, 2001; Schlesinger, 1990). This concept implies that if forest management

Figure 1 Forest soils are highly variable in soil types and organic carbon stocks. (a) Skeletic Stagnic Fluvisol, 73 tC ha^{-1} (18% in organic-layer); (b) Albic Podzol, 140 tC ha^{-1} (23% in organic-layer); (c) Folic Gleysol, 126 tC ha^{-1} (56% in organic-layer); (d) Ferralic Cambisol, 170 tC ha^{-1} (22% in organic-layer); (e) Calcaric Skeletic Cambisol, 140 tC ha^{-1} (5% in organic-layer). The displayed soils are examples from Central Europe, Austria. Photo credits: Rainer Reiter - BFW.

increases C inputs, SOC stocks increase until a new steady state is reached (Jandl et al., 2007). However, soils have only limited capacity to store SOC as the availability of reactive mineral surfaces for long-term C stabilization is limited and once sorption sites are occupied with organic matter, only a small fraction of C input residues may remain in the soil for longer time periods (Kaiser and Guggenberger, 2003). Consequently, C turnover increases with increasing SOC concentrations and C inputs are balanced out by SOC outputs, which leads to so-called C saturation (Stewart et al., 2007), with the maximal SOC stocks depending on site conditions (texture, mineralogy, and climate). In comparison to agricultural soils, forest soils with inherently high SOC stocks appear close to C saturation, but quantitative evidence is limited.

Forest management practices are primarily dedicated to optimizing wood and biomass production, but these practices also impact soils by altering the quantity and quality of C inputs, changing microclimatic conditions by modifying the light and water regime, and by physically disturbing soils during management operations (Jandl et al., 2007; Mayer et al., 2020). In the following sections, we present information on the effects of forest management on soil C sequestration and discuss whether management of forests (e.g. choice of tree species, harvest, density regulation and thinning, fertilization, etc.) can be optimized under expected climate change conditions. We also discuss afforestation and reforestation options and highlight the role of disturbance-management. Furthermore, we present results from a case study in Central Europe investigating forest soil C storage in mountain regions of Switzerland and Austria.

2 Forest management and soil carbon sequestration

2.1 Afforestation and reforestation

Afforestation of formerly non-forested land and reforestation of deforested land have been highlighted as important strategies to reduce atmospheric CO_2 (Bastin et al., 2019; Griscom et al., 2017). Although the majority (>90%) of the CO_2 is thought to be sequestered in the new forest biomass, a significant CO_2 sink has been attributed to C sequestration in soils (Bossio et al., 2020; Nave et al., 2018). Whether this soil C sequestration potential is realistic remains uncertain since most case studies demonstrate that both afforestation and reforestation do not necessarily result in increased SOC stocks (Mayer et al., 2020).

Cropland SOC pools are typically depleted by persistent agricultural biomass removal. After re- or afforestation, organic matter accumulates until the soil reaches a new equilibrium between C input (litter fall, rhizodeposition) and C output (respiration, leaching). Accordingly, afforested cropland soils have been shown to typically sequester C across various climatic zones (Ashwood et al., 2019; Don et al., 2011; Poeplau and Don, 2013). Soil C sequestration is generally faster in warmer (topical and sub-tropical) climatic regions (Don et al., 2011; Silver et al., 2000), but reforested croplands in temperate and hemi-boreal regions can also recover SOC stocks close to those of managed or natural forest soils within 100 years or less (Ashwood et al., 2019). Cropland afforestation in temperate and cold biomes first leads to an accumulation of litter and forest floor, particularly if conifers are planted. After a period of years to decades, the sequestered C enters the mineral soil C pool (Bárcena et al., 2014a,b). Faster increases in mineral soil C were observed in reforested tropical cropland (Silver et al., 2000). As croplands are a limited land resource, global population growth and food security do, however, constrain their use for reforestation (Griscom et al., 2017).

Grasslands have been proposed as favorable afforestation targets due to their large spatial distribution. In tropical and sub-tropical regions, reforested C-depleted pastures proved to sequester C in soil, but to a lesser extent as reforested croplands (Don et al., 2011; Silver et al., 2000). In temperate and colder biomes, the situation is different. Temperate grassland soils typically store similar or even higher amounts of C than temperate forest soils. This already indicates that the capacity to sequester additional C following reforestation is limited. Reforestation of European pastures had only modest effects on SOC and often led to, at least transient, soil C losses (Hiltbrunner et al., 2013; Poeplau and Don, 2013). A further common observation is that reforestation of temperate grasslands reshuffles the C density within the soil profile. Carbon content often decreases in deeper mineral soil layers and increases in the organic layers and forest floor (Hiltbrunner et al., 2013; Mobley et al., 2015). The SOC that accumulates at the soil surface is more labile and less protected than mineral-bound SOC in deeper soil layers (Poeplau and Don, 2013) and therefore could decompose faster if environmental conditions become more favorable for microbial activity (e.g. under climate warming) or if soil is disturbed (e.g. after harvest). Afforestation of native grasslands can cause considerable habitat and biodiversity loss (Temperton et al., 2019; Veldman et al., 2019) and has recently been shown to decrease soil carbon (Santos et al., 2020).

In Northern Europe, many upland sites with thick organic soil layers have been afforested with conifers. Lowering the water table through drainage and site preparation prior to tree planting typically leads to substantial C losses from the aerated peat layer (Simola et al., 2012). However, it has been suggested that initial C losses following afforestation can be compensated by C input throughout several tree rotations, and that afforestation will sequester C in afforested organic soil in the long term (Vanguelova et al., 2019). Recent studies again highlight the risks associated with tree planting on organic soils.

Planting of native tree species on organic moorland at four sites in Scotland, for example, resulted in soil C losses on a decadal timescale (Friggens et al., 2020). Similarly, afforested peatland soils at eight different locations across Ireland have been shown to act as net C sources to the atmosphere (Jovani-Sancho et al., 2021; Friggens et al., 2020), suggesting that, even if there is no change in peatland water tables, tree roots shift the fungal/mycorrhizal community composition in the subsoil and thereby could trigger an increased decomposition of organic matter.

An extensive study in Chinese forests, comparing more than 600 control and afforested plot pairs, clearly confirmed that afforestation effects depend on SOC stocks of the former land (Hong et al., 2020). At low SOC stocks, planting forests leads to C gains in the soil, while at high SOC stocks (>100 t SOC/ha), such as in grasslands and organic soils, afforestation caused SOC losses. Overall, C sequestration in soils was much lower than expected when extrapolated to the total investigation area (Northern China, 120 000 km^2), suggesting that the global soil C sequestration potential of re- and afforestation may be smaller than in the above-mentioned optimistic scenarios. It further highlights the importance for forest managers to consider soil properties/processes when implementing re- or afforestation measures. Carbon gains by newly forming forest biomass can be partially or even totally offset if trees are not planted at appropriate sites. It should also be considered that tree plantations experience higher water use than non-forest vegetation. This can significantly affect site-, catchment-, and landscape-hydrology and potentially increase soil salinization and acidification (Jackson et al., 2005).

2.2 Management of tree species

Planting tree species is an active measure to enhance forest productivity, produce specific wood, promote biodiversity and enhance soil fertility. Tree species have distinct impacts on soil C cycling by altering the quantity and quality of above- and below-ground inputs and by affecting below-ground communities associated with tree species (Peng et al., 2020; Reich et al., 2005; Vesterdal et al., 2013). These processes are frequently interwoven and although tree species might have strong effects on individual soil processes, observed net effects on SOC stocks are often relatively small (Grüneberg et al., 2019; Wiesmeier et al., 2013).

In general, broadleaf trees have deeper rooting systems and litter that is more easily decomposed, thereby favoring biological activity. These species provide greater C inputs into mineral soils as compared to most conifer trees, which show litter accumulation at the soil surface. Below-ground C inputs decompose more slowly than above-ground litter (Hansson et al., 2010), and as they interact also more strongly with mineral surfaces, root-derived C is assumed to be of primary importance for the formation of stable SOM (Rasse et al., 2005; Schmidt et al., 2011). Tree species may also influence SOC storage by their association with distinct mycorrhizal fungi. SOC stocks and vertical SOC distribution were found to be greater under trees associated with arbuscular mycorrhiza than under those associated with ectomycorrhiza (ECM) (Craig et al., 2018; Peng et al., 2020). The negative effect of ectomycorrhizal fungi on SOC stocks is attributed to the secretion of powerful oxidizing enzymes which allow the acquisition of nutrients bound to SOM but promote SOM decomposition (Zak et al., 2019).

Due to the greater rooting depth, increasing the uptake of base cations from subsoils, and to the annually renewed foliage, litter from broadleaf trees generally has a higher quality with a greater nutrient content than litter from conifer trees (e.g. Berger et al., 2006; Reich et al., 2005). Litter quality is associated with a distinct microbial community composition and functioning that may in turn affect SOC storage (Heděnec et al., 2020). Moreover, earthworms profit from calcium-rich litter of broadleaf species (Reich et al., 2005; De Wandeler et al., 2018). As earthworms transfer litter C into the mineral soil, broadleaf trees species are associated with smaller C stocks in the forest floor (Desie et al., 2020), but greater C stocks in the mineral soil and, thus, potentially sustainable C sequestration (Jandl et al., 2007; Mayer et al., 2020).

The effects of tree species frequently overlap with other site conditions (Grüneberg et al., 2019). For instance, spruce naturally grows in a colder and more humid climate with inherently greater SOC stocks than broadleaf trees, which impedes the direct identification of tree species effects. However, common garden experiments and soil C inventories observed pronounced tree species effects in forest floors with greater C stocks under coniferous trees due to reduced litter decomposition as compared to high-quality litter from broadleaf trees (Reich et al., 2005; Vesterdal et al., 2013; Wiesmeier et al., 2013). The opposite pattern exists for mineral soils with greater SOC stocks under broadleaf forests often associated with arbuscular mycorrhiza (Peng et al., 2020; Wiesmeier et al., 2013). However, although stabilization of SOC is generally enhanced by the interaction with reactive mineral surfaces, indicating that soils from broadleaf forests have greater potential to sequester C in the long-term, observed differences in total SOC stocks (organic layer + mineral soil) are rather small (Grüneberg et al., 2019; Wiesmeier et al., 2013). One reason for these relatively small tree species effects could be 'priming' by rhizodeposition where fresh C inputs in the mineral soils stimulate microbial activity, which in turn accelerates decomposition of old SOC (Dijkstra et al., 2021).

Management options to promote tree species have to consider the adaptation of tree species composition to ongoing climatic changes. For instance, as spruce responds sensitively to drought and becomes vulnerable to bark beetle infestations, promotion of spruce in low elevation forests in Central Europe will be inappropriate in a future climate with more frequent and severe drought (Hlásny et al., 2021; Jandl, 2020). Silver fir and oak, by contrast, form deeper rooting systems (Vitasse et al., 2019), which potentially provide a greater C input into the mineral soil (Reich et al., 2005). Planting Douglas fir, another drought-resistant conifer originating from North America, may conflict with the paradigm of basing forests on native tree species.

2.3 Harvest and harvest residues

The harvesting of timber is one of the most important management activities in forests worldwide. Harvesting or logging operations commonly focus on the stem section of a tree, as it yields the highest merchantable value. Harvest residues such as leaves/needles, twigs, branches, or stumps, have a comparable low economic value and are often left on-site following harvesting operations. However, growing demand for biofuels as a renewable energy source has also led to an interest in utilizing harvest residues (Popp et al., 2014). Harvesting and the utilization of harvest residues strongly affects the quantity and quality of organic matter input to soil. Moreover, decomposition processes may be altered by more favorable microclimatic conditions (e.g. warmer and moister), shifts in microbial community composition, and/or soil disturbance (e.g. compaction, erosion). These changes may result in a misbalance of C input and C output, which ultimately impacts forest soil C stocks (Jandl et al., 2007; Mayer et al., 2020). To maintain the long-term C sequestration potential of forest soils (e.g. over several rotation periods) it is therefore crucial to select harvesting practices that minimize net soil C losses.

Global meta-analyzes investigating the effects of forest harvesting operations on soil C storage (mainly by clearcut logging of the entire tree layer) revealed diverging results; soil C stocks following harvest have been shown to decrease when compared to unharvested control stands (Achat et al., 2015; James and Harrison, 2016; Nave et al., 2010) or to be unaffected (Hume et al., 2018; Johnson and Curtis, 2001). Comprehensive meta-analysis (James and Harrison, 2016) (112 studies), however, reports harvest-related C losses of −30%, −3%, and −18% in organic, mineral topsoil (0–15 cm depth), and deep soil horizons (60-100+ cm), respectively. Greater negative harvesting effects on C stocks in organic horizons than in mineral soil horizons were also shown (Achat et al., 2015; Nave et al., 2010). Carbon losses from organic horizons were generally larger in hardwood forests (−36%) than in coniferous or mixed forests (−20%) (Nave et al., 2010). Following a period of net C losses, soil C stocks in organic and mineral soil horizons begin to recover after 10 to 50 years following forest harvest (Achat et al., 2015; James and Harrison, 2016; Nave et al., 2010; Sun et al., 2004; Tang et al., 2009).

The impact of harvesting operations on soil C stocks has been shown to depend on the amounts of harvest residues left on site. Meta-analyzes showed soil C stocks were unchanged following stem-only (i.e. bole fraction) harvest (Hume et al., 2018; James et al., 2021) and increased (+18%) when compared to control stands (Johnson and Curtis, 2001). By contrast, whole-tree harvest (including branches, leaves/needles) decreased soil C stocks by −15% (James et al., 2021). Achat et al. (2015) showed larger C losses in the organic horizon (10–45%) and in deeper mineral soil horizons (10%) after harvest when residues were removed. This meta-analysis also found a larger effect of harvest residue removal on soil C stocks in temperate than in boreal forests (Achat et al., 2015). Other studies, however, did not find clear evidence that whole-tree harvest would decrease soil C stocks (Hume et al., 2018; James and Harrison, 2016). In a Sitka spruce forest in the UK, C stocks of apeaty gley soil were even higher after whole-tree harvest when compared to stem-only harvest (Vanguelova et al., 2010). This finding was related to higher decomposition rates underneath residues. Stump harvesting has been shown to strongly decrease C stocks in organic horizons across a range of forests in Scandinavia, while mineral top- and sub-soil horizons were unaffected (Clarke et al., 2021). In a temperate forest in Washington, USA, stump harvesting resulted in 24% lower soil C stocks (Zabowski et al., 2008). However, others found either no reduction in soil C stocks (Jurevics et al., 2016; Strömgren et al., 2013) or only a small decrease following stump harvest (Hyvönen et al., 2016; Kaarakka et al., 2016).

Whether partial harvesting practices that retain living trees on site would reduce logging-related soil C losses remains inconclusive. Shelterwood harvest in a hardwood forest in New England, USA, for example, resulted in a decrease in mineral soil C stocks, while C stocks in the litter layer were unaffected (Warren and Ashton, 2014). In a Chilean Lenga forest, soil C stocks decreased for 3 years after shelterwood harvest, but recovered 8 years after logging (Klein et al., 2008). Other studies report little or no differences in soil C stocks when comparing partial to conventional clearcutting practices (Christophel et al., 2015; Hoover, 2011; Puhlick et al., 2016).

The effects of harvesting operations on soil C stocks have been suggested to be ameliorated if harvest rotation periods are extended (Law and Waring, 2015; Noormets and Nouvellon, 2015) or soil disturbance from, for example, machinery movements is reduced (Achat et al., 2015). Machinery-related soil disturbance in particular is supposed to strongly enhance harvesting effects on soil C stocks (James and Harrison, 2016; Mayer et al., 2021). These negative effects may be reduced by selecting harvesting systems that alleviate soil disturbance (e.g. harvesting during winter when soils are frozen or snow-covered, cable-yard or helicopter logging) (Mayer et al., 2021).

2.4 Density regulation and thinning

Density regulation already starts with tree planting if natural tree regeneration is lacking or undesirable. Figure 2 shows the results of a density regulation trial in an Austrian spruce forest in which saplings were planted in various densities on an arable field. No differences were found in total soil C stocks among the different density treatments 30 years after planting (Fig. 2). Na et al. (2021), however, found a significant increase in total SOC stocks with increasing planting densities in 8-year-old larch plantations, concurrent with significantly increasing fine root biomass and above-ground litter input.

Once a forest stand is growing, thinning (the removal of designated trees to increase space for the remainder) is the traditional silvicultural practice used to maximize merchantable timber volume and financial value (Gonçalves, 2020).

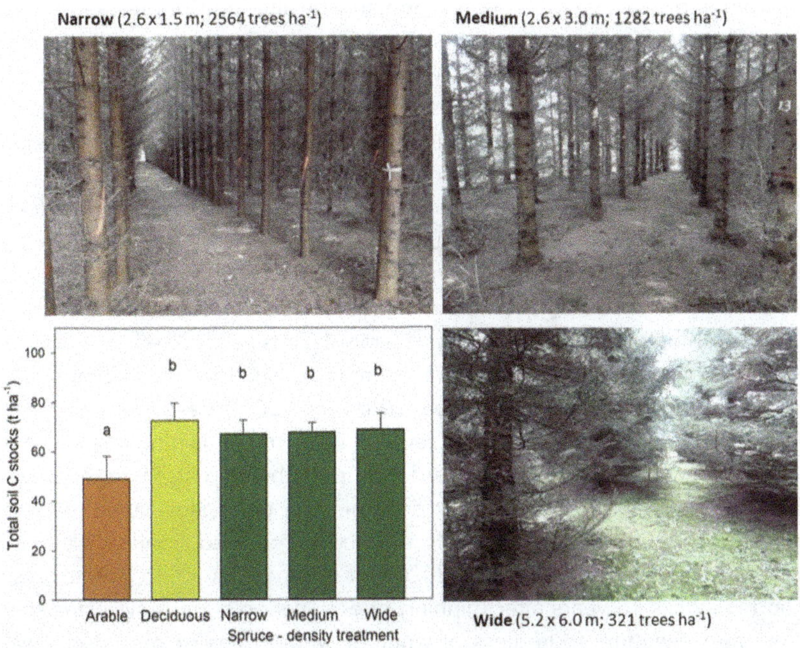

Figure 2 Soil carbon stocks in a density regulation trial by the Austrian Forest Research Centre - Institute of Forest Growth. Norway spruce (*Picea abies*) was planted in 1992 in narrow, medium and wide spacing. Deciduous trees (*Acer pseudoplatanus* and *Carpinus betulus*) were planted with a single spacing of 2.6 × 1.5 m. All trees were planted on an arable field. Total soil C stocks (including forest floor and litter) down to 40 cm soil depth were assessed in autumn 2020. Different letters above the individual bars indicate statistically significant differences (ANOVA, $p < 0.05$, n = 3-4 plots; 50 × 50 m each). While the soil C stocks of all forest plots were significantly higher than those of the adjacent arable field, there was no difference between deciduous and spruce stands, nor between the different density treatments.

Thinning is applied at various intensities, standages, intervals, and by different approaches (from above, below, selective, etc.). Thinning is also a management measure to reduce vulnerability to climate changes such as drought (Sohn et al., 2016), fire (Finkral and Evans, 2008; Fulé et al., 2001), and to facilitate structural and species diversity (D'Amato et al., 2013). Thinning always leads to a temporary reduction in stand density and therefore to a temporary reduction in above-ground biomass C stocks. After thinning, increased growth of the remaining trees can partly or fully compensate for the thinning biomass C loss – depending on thinning intensity and time elapsed thereafter. Thus above-ground biomass C stocks of thinned stands were often found to be lower than those of unmanaged stands (Bravo-Oviedo et al., 2015; Nilsen and Strand, 2008; Powers et al., 2011; Ruiz-Peinado et al., 2016) but they were also observed to fully regain above-ground C stocks of unmanaged stands in the longer run (Powers et al., 2012).

Thinning impacts soil C cycling primarily by altering C inputs and the soil microclimate. Thinning prompts a significant initial C and nutrient input from onsite harvest residues (except for whole trees harvest) and dead roots. Above-ground litterfall typically decreases after thinning (Zhang et al., 2018). Soil temperatures increase due to the increased solar penetration of the ground surface (Zhang et al., 2018). The magnitude of the temperature increase depends on the thinning intensity and to what extent the remaining understory or forest floor vegetation contributes to ground shading (Li et al., 2021). Soil moisture is mostly unaffected by thinning since increased transpiration is balanced by overall reduced water uptake. Under an improved soil microclimate, litter and SOM decomposition can become accelerated. Thus soil CO_2 effluxes increase in general during the first few years after thinning (summarized in Zhang et al., 2018) and forest floor C stocks decrease (Kim et al., 2018; Vesterdal et al., 1995).

Initial thinning effects, however, level off when the remaining trees increase their growth and the canopy closes during post-thinning stand regeneration. Most studies encompassing longer post-thinning timespans, stand-rotations, or even whole tree lifespans, therefore have not reported significant thinning effects on soil C stocks. Lim et al. (2020), for instance, found no effects on soil C and N stocks 18 years after thinning of a 35-year-old *Pinus sylvestris* stand in northern Sweden, regardless of whether whole trees or only stems were harvested (sampling depth 20 cm). Nilsen and Strand (2008) did not detect any significant effects on humus and mineral soil (sampling depth 100 cm) C storage 33 years after thinning of a Norway spruce stand in southern Norway either. No significant effects of various intensity repeated-thinning treatments over 50+ years were detected in the forest floor and mineral soil (sampling depth 30 cm) of a red Pine stand in Minnesota, and a northern hardwood forest in Wisconsin, USA (Jurgensen et al., 2012; Powers et al., 2011). However, in a pine stand in the Great Lakes region, Canada, almost 90 years of repeated intensive thinning not only substantially reduced above-ground biomass C but also reduced soil C stocks (sampling depth 30 cm) as well (Dávila Reátegui et al., 2021). In a global meta-analysis, Zhang et al. (2018) did not observe an overall significant effect of thinning on soil C stocks, though heavy thinning tended to decrease total soil C, whereas light thinning increased total soil C stocks. A similar trend toward soil C losses under heavy thinning and soil C gain under light and moderate thinning was observed by Gong et al. (2021) when meta-analyzing thinning effects in Chinese forest plantations.

Overall, the rather indifferent effects on forest soil C stocks do not indicate a certain way of adapting specific thinning strategies to optimize soil C sequestration. The role of thinning in future forest soil C sequestration should not, however, be underestimated. Thinning typically increases the resistance of the remaining trees to disturbances such as windthrow, snow pressure and drought, and allows for adaptive adjustments in tree species composition and vertical stand structure. Therefore, thinning can be a proper silvicultural measure to protect forest (soils) from disturbance-related C loss (see below). That such disturbances are mostly absent in the relatively rare published longer-term thinning trials might be an indication of a publishing bias toward experiments conducted at 'safe sites'. Accordingly, the role of thinning on the landscape scale might be somewhat underrated. As an example, Gong et al. (2021) reported an overall positive effect of thinning on the soil C stocks of more than 200 Chinese forest plantations – with the most pronounced effects in drought-prone areas.

2.5 Fertilization, liming and wood-ash application

Fertilization represents a potential measure to overcome nutrient deficiency, thereby promoting forest productivity. Fertilization impacts have most intensively been studied for N, and mechanisms include increased litter input through higher forest productivity and reduced SOM decomposition (Mayer et al., 2020). The latter can be attributed (1) to N-induced shifts in the decomposer community and reduced extracellular enzyme activities, (2) to a preferential substrate use shifting from recalcitrant SOM with high N content to fresh substrate with low N content, and/or (3) by adding N required for forming microbial necromass (e.g. Fog, 1988; Hagedorn et al., 2003). The review by Nave et al. (2009) indicated that N fertilization increases total soil C stocks (combined forest floor and mineral soil) by 7.7%, but the magnitude likely depends on the inherent N status of the soil. These benefits must be weighed against the associated environmental costs, such as nitrate leaching, soil acidification, enhanced soil N_2O emissions, and loss of biodiversity. Moreover, the production, transport, and application of synthetic fertilizers all entail fossil fuel combustion and emission of CO_2.

Liming aims at reversing soil acidification and improving soil fertility (Hildebrand, 1996). In principle, liming influences SOC storage by potentially increasing C inputs into soils through improved tree growth and a deeper rooting system (Schäffer et al., 2001). However, it also accelerates SOM decomposition (particularly in the forest floor) by enhancing C solubility and stimulating biological activity associated with an increased soil pH and a higher litter quality with higher Ca-contents (Kreutzer, 1995). Liming enhances the amount and activity of earthworms (Schäffer et al., 2001), which leads to a translocation of litter-derived C into the mineral soil and hence to a C loss in the forest floor but a potential C gain in the mineral soil. This C transfer is promoted by enhanced leaching of dissolved organic C through the rising pH (Feger et al., 2000).

Soil surveys and liming experiments show that liming effectively transforms thick forest floors into more 'active' humus forms (e.g. Court et al., 2018), which leads to SOC losses in the forest floor (e.g. Bauhus et al., 2004; Table 1). However, observed effects on SOC stocks in the mineral soil under the forest floor are inconsistent and it remains uncertain whether the C losses in the forest floor are outbalanced by C gains in the mineral soil (Table 1). The small effect sizes in the mineral soil could be related to stimulated SOC mineralization by liming, the C stocks in the mineral soil being already close to saturation, or the difficulty in detecting changes in the large C reservoir of mineral soils. While liming experiments tended to result in negative effects on total SOC stocks, SOC changes between repeated surveys of 385 limed sites in German forests showed slightly positive (+0.18 tC/ha/yr) liming effects (Grüneberg et al., 2019). The authors explained their findings by potentially deeper rooting trees and stabilization of SOM by Ca^{2+} added with the lime. Potentially, the enhanced C transfer into the mineral soil leads to a long-term stabilization of soil C.

The magnitude of liming responses appears to depend on site and soil conditions with greater SOC losses to be expected in soils with a thick forest floor. Potential C gains in soil C storage have to be outweighed against the release of CO_2 from carbonate in the lime (~12% of its mass), which corresponds to a release of 0.036 t CO_2–C/ha/yr at a standard addition of 3 t lime/ha every 10 years under temperate climate conditions. Moreover, liming alters microbial community composition (e.g. of ectomycorrhizal fungi (Kjøller and Clemmensen, 2008; van der Linde et al., 2018)) and bears the risk of increased NO_3^- leaching due to stimulated SOM and N mineralization in combination with enhanced nitrification (transformation of NH_4^+ to NO_3^-) at higher pH values. Potential negative effects can be minimized by the application of lime on small areas only, following shelterwood cutting interventions and by application of lime with low solubility.

Wood-ash application aims at returning base cations removed with harvesting back to the soil. Mechanistically, impacts of wood ash on soil

Table 1 Effect of liming on SOC storage in European forest soils

			Years after liming	Forest floor	Forest floor	Mineral soil (0–20 cm)	Mineral soil (0–20 cm)	Total soil	
		Mode of liming	y	tC/ha	%change	tC/ha	%change	tC/ha/y	Comments and limitations
Experiments									
Bauhus et al. (2004)	Beech (Solling) Stand	Dolomite (3t/ha)	8	−0.5	−3%	−14.4	−22%	−1.86	
	Beech (Solling) Gap	Dolomite (3t/ha)	8	−9.0	−60%	20.6	38%	1.45	
Kreutzer (1995)	Spruce (Höglwald)	Dolomite (4t/ha)	7	−7.2	−23%	0.8	4%	−0.91	0–5 cm depth; no data on deeper soil
Court et al. (2018)	Beech (N-France)	CaCO3 (2.5 t/ha)	20-40	−2.1	−0.1	n. sign.	n. sign.		Average of five sites
Marschner and Wilczynski (1991)	Spruce (Berlin)		3	−6.9	−23%	−2.9	−5%	−3.27	Heterogenous sandy site, liming was combined with K fertilization
Matzner et al. (1985)	Spruce (Solling)	CaCO3 (5 t/ha)	10	−7.1	−15%	−12.5	−19%	−1.96	Combined with N-fertilizer
	Beech (Solling)	CaCO3 (5 t/ha)	10	4.4	18%	20.6	37%	2.50	Combined with N-fertilizer
Persson et al. (1995)	Spruce (Sweden)	CaCO3 (9 t/ha)	42	−6.0	−40%	3.0	4%	−0.07	Includes a stand rotation
(values 0–50 cm)	Spruce (Sweden)	CaCO3 (12 t/ha)	37	−10.3	−30%	−10.7	−12%	−0.57	Planted former heathland
	Spruce (Sweden)	CaCO3 (10 t/ha)	38	−7.5	−94%	−8.0	−8%	−0.41	Includes a stand rotation
	Beech (Sweden)	CaCO3 (10 t/ha)	38	−20.0	−80%	0.0	0%	−0.53	
Soil survey									Large data set
Grüneberg et al. (2019)	German forests	mostly 3 t/ha	variable	−0.25 tC/ha/y		0.43 tC/ha/y		0.18 tC/ha/y	Comparability of sites uncertain

processes are similar to those of lime, with an increased soil pH and an improved supply with base cations (review by Huotari et al., 2015). lthough 'net effects on soil C balance have not been thoroughly evaluated' (see Huotari et al., 2015), in agreement with liming studies, C and N mineralization rates were found to be stimulated by wood-ash addition (Rosenberg et al., 2010). In a field study in Switzerland, Zimmermann and Frey (2002) observed long-lasting increases in soil CO_2 effluxes and a decrease in SOC content in the uppermost 5 cm by approximately 20% during the first 3 years after wood-ash application. Whether the C losses in the forest floor and upper mineral soils are compensated for by C gains in the deeper mineral soil through a C transfer mediated by soil fauna or by the development of a deeper rooting system remains unstudied.

2.6 Disturbance control

Forests around the world are increasingly being damaged by natural disturbance agents such as wildfires, droughts, pest outbreaks, or windthrows (Allen et al., 2010; Seidl et al., 2017; Senf et al., 2018; Williams et al., 2016). Climate change is considered to reinforce disturbance frequency and severity, and model predictions suggest that natural forest disturbance will be even more pronounced in the next few decades (Seidl et al., 2014). Natural disturbance can strongly impact soil C stocks of forest stands. Global meta-analyzes indicated that soil C concentrations and storage generally decrease following disturbance by fire, insects, and wind (Nave et al., 2011; Thom and Seidl, 2016; Wang et al., 2012; Zhang et al., 2015). Severe wildfires can cause a rapid loss in forest floor C by direct combustion of organic matter (Bowd et al., 2019; Nave et al., 2011; Pérez-Izquierdo et al., 2021). Forest ecosystems with thick organic layers (e.g. boreal forests) may thus be more vulnerable to C losses after fire than ecosystems with less thick organic layers (e.g. tropical, Mediterranean forests). On the other hand, forest fires lead to the formation of pyrogenic C, which is biogeochemically stable (Schmidt et al., 2011). Fire-prone systems may thus have particularly large SOC stocks although they lose C in the forest floor during single fires (Eckmeier et al., 2010).

Large reductions in soil C stocks have also been evidenced after stand-replacing windthrows in mountainous forests (Christophel et al., 2015; Mayer et al., 2017b). Post-disturbance C losses were suggested to be caused by decreased organic matter input and increased decomposition rates due to warmer soil conditions following canopy removal. Canopy loss following disturbance can make soils prone to accelerated surface erosion, which may contribute to a reduction in soil C stocks (Berhe et al., 2018; Gerber et al., 2002). Blown-over trees and an associated soil disturbance (e.g. pit-and-mound topography) may also cause soil C losses from deeper subsoil horizons. However, studies also found neutral or even positive effects on soil C stocks following natural forest disturbance (Don et al., 2012; Kobler et al., 2015; Kosunen et al., 2020; Santos et al., 2016). In the following section, selected management activities and disturbance control measures are presented that may mitigate and/or prevent potential negative effects on soil C stocks after natural disturbance.

The intensity of wildfires is often related to the amount of flammable organic material (i.e. branches, twigs, dry plants) on the soil surface. Prescribed burning represents a management activity that reduces fuel loads by planned low-intensity fires to prevent uncontrolled severe wildfires (Alcañiz et al., 2018). Although prescribed burning has been shown to reduce forest floor C stocks, the reductions were smaller (−46%) than after wildfires (−67%) (Nave et al., 2011). The same study showed that prescribed burning did not affect mineral soil C stocks and that the presence of hardwood tree species with less flammable leaves and litter reduced fire impacts on soil C storage. Another strategy to control fire severity may be the establishment of tree-free corridors to reduce fire spread across a forest and to decrease the continuity of fuel loads. Fire suppression often delays but does not prevent forest fires in many cases, as it favors an accumulation of fuel loads that increases the risk of severe wildfires in the long term (Jandl et al., 2007). Following windthrows, insect attacks, and fires, damaged trees are commonly salvage-logged to utilize the timber's merchantable value and to reduce the risk of further pest outbreaks (Leverkus et al., 2018). Although salvage logging is meant to remove dead and dying trees, surviving trees are often also logged, primarily due to more cost-efficient harvesting operations. As additional forest disturbance, salvage logging has been shown to amplify forest floor C losses after windthrow and fire in coniferous forests in North America (Bradford et al., 2012; Kishchuk et al., 2015). Similarly, Hotta et al. (2020) reported lower forest floor C stocks when comparing salvage-logged to non-salvage-logged sites six decades after windthrow in coniferous stands in Japan. Other studies, however, did not find additional effects of salvage logging on soil C stocks following fire and bark-beetle attacks (Avera et al., 2020; Poirier et al., 2014). In some cases, positive effects on soil C storage, either due to higher organic matter input rates from surviving tree- and deadwood litter, decreased decomposition rates due to cooler soil conditions from shading, or lower soil erosion rates were achieved when salvage logging was omitted and dead wood was left in place. As shown in harvesting operations (see *harvesting and harvest residues*), retaining surviving trees and deadwood could be a potential strategy to reduce soil C losses following natural disturbances in certain cases (James et al., 2021; Wan et al., 2018). Deadwood retention, however, may require additional measures (e.g. debarking) to prevent further pest outbreaks in neighboring stands.

Potential negative disturbance effects on soil C stocks may also be mitigated by proactive silvicultural activities that increase stand stability and stand recovery following disturbance. Mixed forests, for example, are less susceptible to damage from bark beetles (Faccoli and Bernardinelli, 2014). A study from North America showed also that the probability of individual tree attack in pine stands was lower at smaller tree diameters and at lower stand densities (Negrón et al., 2008). An uneven-aged stand structure in a Swiss forest was further hypothesized to decrease forest damage by wind (Hanewinkel et al., 2014). Thus, thinned and uneven-aged stands may be more resilient against bark-beetle infestations and windthrows, thereby mitigating net soil C losses (see also *thinning* and *stand density*). Moreover, the promotion of a fast tree regeneration has been suggested to slow post-disturbance soil C losses (Mayer et al., 2017a). Net soil C losses following disturbance may also be reduced if a fast recovery of regenerating trees (either naturally or planted) is fostered.

3 Case study: forest soil carbon storage in Central Europe mountain regions

Mountainous terrains show strong gradients in environmental conditions (climate, parent material, tree species) within a small area, which provide an excellent natural laboratory to identify the relative importance of these factors for SOC storage (Hagedorn et al., 2019). Here, we evaluate the effects of forest types on SOC stocks in Austria and Switzerland with similar topography but different intensity of forest management.

In Austria, the dominating tree species is Norway spruce, both due to its natural dominance in mountain regions and its cultivation at lower elevations. Analyzing the data of the Austrian Forest Soil Survey (Englisch et al., 1992) yielded very small strata for individual deciduous species. Therefore all broadleaf-dominated forests were compared to coniferous forests (Jandl et al., 2021). Overall, the soils of coniferous forests had higher stocks of organic carbon down to 50 cm depth (max. sampling depth). The accumulation of organic C from slowly decomposing tree needles in the forest floor has often been described yet, even in the mineral soil, the SOC stock is higher under conifers than under deciduous trees. The differences in Fig. 3 are statistically insignificant because the tree species effect is overlain by the effects of geology, elevation, and a climatic gradient from a Pannonian climate in the East to a more Atlantic climate in the west of the country.

In Switzerland, forests have primarily been managed as plenter forests with single-tree harvest and promotion of adapted tree species, leading to a near-natural structure of forests. Due to the high costs of wood harvesting, management intensity has declined during the last few decades (Brändli et al., 2020). Consequently, Swiss forests are characterized by a high mean tree age (more than a quarter of Swiss accessible forest area is older than 120 years, see Brändli et al., 2020) and by the greatest stand biomass in Europe (Liski et al.,

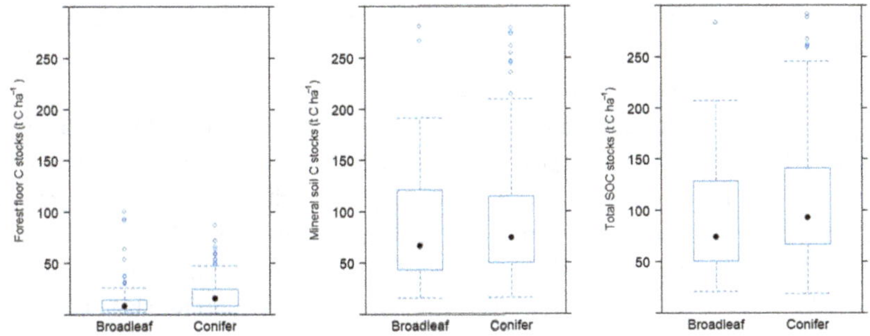

Figure 3 Organic carbon in the soils of Austrian forests. Values are reported to a depth of 50 cm.

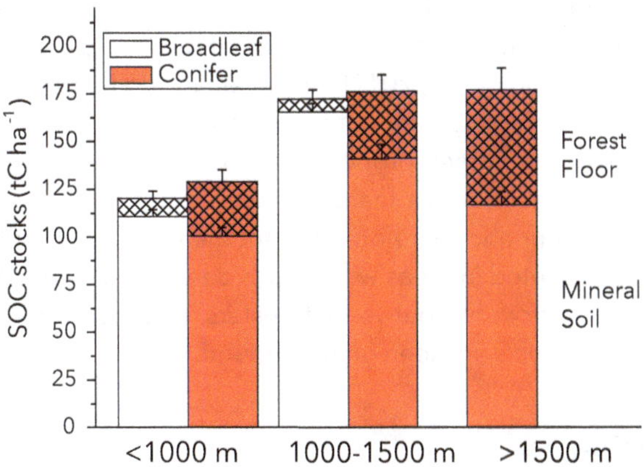

Figure 4 Smaller SOC stocks in the forest floor but greater stocks in the mineral soils in broadleaf than in coniferous forests of Switzerland. Total SOC stocks do not differ between forest types but depend upon elevation (based on approx. 1000 soil profiles).

2002). Swiss forests show a decreasing share of broadleaf trees with increasing elevation due to decreasing temperatures. In their adequate growth region, broadleaf forests dominate (e.g. 68% broadleaf forest in the Swiss plateau).

In Swiss forests, SOC stocks show a general increase with increasing elevation due to less favorable climatic conditions for decomposition (Fig. 4). At a given elevation, total SOC stocks do not differ significantly between the coniferous and broadleaf forests. Nonetheless, the forest types show different vertical SOC distribution with smaller SOC stocks in the forest floor but greater ones in the mineral soil in broadleaf than coniferous forests. This statistical analysis across Swiss forests indicates that forest type is not significantly related to total SOC stocks, while soil chemistry and mineralogy (pH, Fe-, Al-oxides and Ca^{2+}) and precipitation are the most important drivers (Gosheva et al., 2017; Solly et al., 2020). Overall, these findings strongly suggest that, despite the strong effects of tree species on biogeochemical fluxes, vertical SOC distribution and below-ground communities and their functioning, promotion of tree species seems to have only limited potential to influence carbon sequestration.

Favoring deciduous trees over more vulnerable coniferous tree species is expected to increase the stability of forests against emerging climate change-related pressures in Central Europe. Although broadleaf forests have greater SOC stocks in the mineral soil than conifer forests, there is little evidence that total SOC stocks are substantially increased (Grüneberg et al., 2019; Wiesmeier et al., 2013). Forestry follows refined and proven concepts to reach sustainable high-productive forests. High biomass production and optimized forest management have gradually enriched the forest SOC stock during the last few decades. An important factor in Central Europe has been the discontinuation of litter removal for use as animal bedding material (i.e. litter raking) (Führer, 2000). Resorting to other tree species combinations could lead to lower SOC stocks (Prietzel and Bachmann, 2012). However, a wide range of evidence has been provided and the effects are not fully congruent.

Moving toward a higher proportion of deciduous trees also requires auxiliary measures within forest management. Deciduous trees are browsed selectively by roe deer and therefore efficient measures of population control need to be implemented in areas with a high deer population. This argument reflects an old controversy between foresters and hunters. Solutions are discussed but few effects are seen on the ground.

4 Conclusion

Afforestation and reforestation of arable land with low SOC stocks can significantly enhance C sequestration in soils (and biomass) over decades. Yet, it is questionable whether the need to provide food and fuel from agricultural soils for a growing world population is reconcilable with maximized C sequestration in terrestrial ecosystems. A less controversial concept is optimized forest management to maximize stocks of organic carbon sustainably. In our assessment of available data and studies (predominantly temperate forests) we have found several options to increase SOC stocks, but potential effect sizes seem rather limited. Table 2 summarizes the effects of different forest management measures on soil C sequestration and highlights the associated uncertainties and research needs.

Tree species selection can influence C sequestration along the soil profile. Coniferous forests tend to store more C in the topsoil whereas broadleaf forests store more C in the subsoil. However, on the landscape scale (see case study), total soil C stocks are often similar when comparing coniferous and broadleaf forests. Adapting present forests to climate change will entail intermixing of more heat-, drought- and pest-resistant tree species. To what extent such species changes

Table 2 Forest management practices and how they affect forest soil C stocks[a]

Management practice	Effect on soil C	Gaps in knowledge	Research needs
Afforestation and reforestation	Positive on low-C soil (croplands); neutral to negative on high-C soils (grasslands)	Effects on deeper soil C; sequestration pathways (litter input vs. labile C from roots and mycorrhiza)	Long-term studies encompassing several forest rotations; processes involved in soil C buildup (and loss)
Management of tree species	Neutral to positive	Interactions with site preconditions; legacy effects of past tree species; climate change interactions	Moving from common garden experiments to larger scale and integration of legacy effects
Harvest and harvest residues	Negative to neutral	Many soil types are not covered in available studies; nutrient benchmarks for residue removal; long-term effects	Detailed studies on specific harvest techniques and machinery setups; long-term studies in different ecosystem types
Density regulation and thinning	Neutral	Limited information on landscape-scale effects; long-term studies are rare, primarily focusing on temperate and boreal forests	Chronosequence and landscape gradient studies; long-term experiments; inclusion of indirect effects such as disturbance resistance
Fertilization, liming and wood-ash application	Positive for fertilization; unclear for liming and wood-ash application	Long-term effects of liming and wood ash on soil microbiome; long-term effects on deeper soil	Offset emissions from fertilizer production and application; CO_2 emissions from lime; effects on soil biology
Disturbance management	Mostly positive (through avoidance of C loss)	Interaction of climate change and disturbance effects; disturbance-related species shifts	Effects of herbivory and browsing on resilience; integration of remote sensing; storage in charcoal after fire

For each practice, existing knowledge gaps and suggested research to address them are given.

affect soil C sequestration is as yet unclear, since C sequestration not only depends on litter quality and quantity change but also on C-flow below-ground, or the composition of mycorrhizal- and decomposer communities. The potential effects on SOM decomposition (and C sequestration) are correspondingly complex, depending on site-specific pedoclimatic conditions, and the overall effects of a tree species change on soil C sequestration are difficult to predict.

Forests in many eco-regions are already under pressure due to increasing drought, heat waves, and other disturbances. Current forest management therefore often focuses on adaptation (rather than CO_2 mitigation) and the promotion of increased resistance and resilience. Management measures that reduce disturbance risks also reduce the risks of associated soil C losses. For instance, silvicultural practices such as thinning can be utilized to improve stand stability and to adapt tree species composition. Although thinning intensity *per se* seems to have limited effects on soil C sequestration, thinning-related stand stabilization, and structural adaptations may lead to stable SOC stocks in the longer term. Stand-replacing disturbance can impose significant soil C losses, especially if tree regeneration is hampered or delayed. Controlling the density of browsing game populations can be such a management measure in regions where natural predators are extinct.

As a man-made disturbance, tree harvesting operations can lead to transient soil C losses. The magnitude of soil C losses depends on the severity of the harvest intervention, whereby large-scale clear cut with heavy machinery and, for example, whole tree plus stump removal can cause soil C losses that require several years to decades to be balanced out by new C input during stand regeneration. Moreover, it is necessary to align harvesting operations to the prevalent site and soil conditions. On nutrient-rich soils with low C stocks in the forest floor, even whole tree harvest might not affect soil C sequestration, whereas on nutrient-poor soils with generally thick organic horizons, as well as at erosion-prone sites, removal of biomass, and/or harvest residues can substantially reduce soil C sequestration. Physical disturbance, such as soil compaction, can also negatively affect soil C sequestration.

Presently, monetary incentives to optimize tree harvest and forest operations toward soil preservation are rare, and soil C sequestration is usually not the highest priority of forest managers. Harvesting techniques that minimize soil disturbance are available but cost and the high demand for wood and other biomass components make its extraction more attractive than leaving it behind for increased soil C storage. Forest (soil) C accounting and C offset-credit systems might be a way forward, but their implementation will require scientific support since soil C cycling is complex and actual soil C sequestration rates are difficult to quantify.

5 Future trends in research

- What is the carbon sequestration potential of an individual site/soil? It is as yet unclear to what extent soils are already C saturated and thus how strongly current forest soil C stocks can be increased. Specific soil traits (e.g. clay content, Fe and Al-oxides) have been identified as predictors of C sequestration potentials. However, forest management needs more practical guidance; e.g. which soil type at which location and under which forest cover is a promising candidate for C sequestration. A special case is forest soils on carbonate bedrock. The mechanisms in C stabilization in carbonate soils are less well understood than those in acidic soils, and their potential for further C sequestration (or loss) remains uncertain.

- What happens to deep soil C? Effects on soil C storage in deeper subsoil horizons have not been studied in depth, although deep soil C has the longest residence time. Tree species selection can affect rooting depth and thereby alter C inputs into deeper soil. These new C inputs may enhance the decomposition of old existing SOC by priming, leading to few net changes. The rates of C downward leaching can be altered by silvicultural practices (e.g. thinning intensity, harvest method etc.). This could not only trigger deep soil C storage but its decomposition as well. The overall effects are as yet unclear.
- How to ensure sustainable soil C sequestration? How sequestered C is stabilized by the soil matrix will influence its residence time. As an example, afforestation or tree species change from broadleaf to conifer can significantly increase C storage in the forest floor. Forest floor represents a labile C pool that can rapidly decompose again after, for example, a clear-cut harvest or be combusted during fire (slow in fast out). Long-term studies and modeling, including SOC storage in distinct SOC pools, can help determine how current C sequestration measures play out in the longer term.
- Going from case studies to areas: Findings on the sequestration of organic C are often based on case studies with an unclear representativity for regions. There is already an ongoing trend toward regional assessments and landscape gradient approaches. Better integration of remote sensing could provide further opportunities.
- Confining the effects of specific silvicultural activities: while the effects of the top-tier management options have been dealt with (e.g. tree species selection, clear cut or permanent cover, etc.), many specific silvicultural practices are not yet, or only poorly, studied. For instance, it remains unclear if and how different harvesting systems (e.g. cable yarding vs. forwarder vs. tractor winch) affect soil C stocks and sequestration.
- Socioeconomic implications and forest management: there exists only poor understanding of the perception of soil C sequestration within forest management. Specific funding regimes, soil C accounting strategies, as well as monetary incentives for forest soil C sequestration are currently lacking.

6 Where to look for further information

- Jandl, R., Lindner, M., Vesterdal, L., Bauwens, B., Baritz, R., Hagedorn, F., Johnson, D. W., Minkkinen, K. and Byrne, K. A. (2007). How strongly can forest management influence soil carbon sequestration? *Geoderma*, 137(3-4), 253-268.

- Mayer, M., Prescott, C. E., Abaker, W. E. A., Augusto, L., Cécillon, L., Ferreira, G. W. D., James, J., Jandl, R., Katzensteiner, K., Laclau, J.-P., Laganière, J., Nouvellon, Y., Paré, D., Stanturf, J. A., Vanguelova, E. I. and Vesterdal, L. (2020). Influence of forest management activities on soil organic carbon stocks: A knowledge synthesis. *Forest Ecology and Management*, 466, 118127. https://doi.org/10.1016/j.foreco.2020.118127.

7 References

Achat, D. L., Fortin, M., Landmann, G., Ringeval, B. and Augusto, L. 2015. Forest soil carbon is threatened by intensive biomass harvesting. *Scientific Reports* 5, 15991.

Alcañiz, M., Outeiro, L., Francos, M. and Úbeda, X. 2018. Effects of prescribed fires on soil properties: a review. *Science of the Total Environment* 613-614, 944-957. https://doi.org/10.1016/j.scitotenv.2017.09.144.

Allen, C. D., Macalady, A. K., Chenchouni, H., Bachelet, D., McDowell, N., Vennetier, M., Kitzberger, T., Rigling, A., Breshears, D. D., Hogg, E. T., Gonzalez, P., Fensham, R., Zhang, Z., Castro, J., Demidova, N., Lim, J., Allard, G., Running, S. W., Semerci, A. and Cobb, N. 2010. A global overview of drought and heat-induced tree mortality reveals emerging climate change risks for forests. *Forest Ecology and Management* 259(4), 660-684.

Ashwood, F., Watts, K., Park, K., Fuentes-Montemayor, E., Benham, S. and Vanguelova, E. I. 2019. Woodland restoration on agricultural land: long-term impacts on soil quality. *Restoration Ecology* 27(6), 1381-1392. https://doi.org/10.1111/rec.13003.

Avera, B. N., Rhoades, C. C., Calderón, F. and Cotrufo, M. F. 2020. Soil C storage following salvage logging and residue management in bark beetle-infested lodgepole pine forests. *Forest Ecology and Management* 472, 118251.

Bárcena, T. G., Gundersen, P. and Vesterdal, L. 2014a. Afforestation effects on SOC in former cropland: oak and spruce chronosequences resampled after 13 years. *Global Change Biology* 20(9), 2938-2952. https://doi.org/10.1111/gcb.12608.

Bárcena, T. G., Kiær, L. P., Vesterdal, L., Stefánsdóttir, H. M., Gundersen, P. and Sigurdsson, B. D. 2014b. Soil carbon stock change following afforestation in northern Europe: a meta-analysis. *Global Change Biology* 20(8), 2393-2405. https://doi.org/10.1111/gcb.12576.

Bastin, J. F., Finegold, Y., Garcia, C., Mollicone, D., Rezende, M., Routh, D., Zohner, C. M. and Crowther, T. W. 2019. The global tree restoration potential. *Science* 365(6448), 76-79.

Bauhus, J., Vor, T., Bartsch, N. and Cowling, A. 2004. The effects of gaps and liming on forest floor decomposition and soil C and N dynamics in a Fagus sylvatica forest. *Canadian Journal of Forest Research* 34(3), 509-518.

Berger, T. W., Swoboda, S., Prohaska, T. and Glatzel, G. 2006. The role of calcium uptake from deep soils for spruce (Picea abies) and beech (Fagus sylvatica). *Forest Ecology and Management* 229(1-3), 234-246.

Berhe, A. A., Barnes, R. T., Six, J. and Marín-Spiotta, E. 2018. Role of soil erosion in biogeochemical cycling of essential elements: carbon, nitrogen, and phosphorus. *Annual Review of Earth and Planetary Sciences* 46(1), 521-548.

Bossio, D. A., Cook-Patton, S. C., Ellis, P. W., Fargione, J., Sanderman, J., Smith, P., Wood, S., Zomer, R. J., von Unger, M., Emmer, I. M. and Griscom, B. W. 2020. The role of soil carbon in natural climate solutions. *Nature Sustainability* 3(5), 391–398. https://doi.org/10.1038/s41893-020-0491-z.

Böttcher, J. and Springob, G. 2001. A carbon balance model for organic layers of acid forest soils. *Journal of Plant Nutrition and Soil Science* 164(4), 399–405.

Bowd, E. J., Banks, S. C., Strong, C. L. and Lindenmayer, D. B. 2019. Long-term impacts of wildfire and logging on forest soils. *Nature Geoscience* 12(2), 113–118.

Bradford, J. B., Fraver, S., Milo, A. M., D'Amato, A. W., Palik, B. and Shinneman, D. J. 2012. Effects of multiple interacting disturbances and salvage logging on forest carbon stocks. *Forest Ecology and Management* 267, 209-214.

Brändli, U., Abegg, M. and Allgaier Leuch, B. 2020. Schweizerisches Landesforstinventar: Ergebnisse der vierten Erhebung 2009-2017 (results of the fourth Swiss National Forest Inventory 2009–2017). Swiss Federal Research Institute for Forest, Snow and Landscape Research, Birmensdorf (ZH) and Federal Office for the Environment (FOEN). Bern.

Bravo-Oviedo, A., Ruiz-Peinado, R., Modrego, P., Alonso, R. and Montero, G. 2015. Forest thinning impact on carbon stock and soil condition in southern European populations of P. sylvestris L. *Forest Ecology and Management* 357, 259–267. https://doi.org/10.1016/j.foreco.2015.08.005.

Christophel, D., Höllerl, S., Prietzel, J. and Steffens, M. 2015. Long-term development of soil organic carbon and nitrogen stocks after shelterwood- and clear-cutting in a mountain forest in the Bavarian Limestone Alps. *European Journal of Forest Research* 134(4), 623–640. https://doi.org/10.1007/s10342-015-0877-z.

Clarke, N., Kiær, L. P., Janne Kjønaas, O., Bárcena, T. G., Vesterdal, L., Stupak, I., Finér, L., Jacobson, S., Armolaitis, K., Lazdina, D., Stefánsdóttir, H. M. and Sigurdsson, B. D. 2021. Effects of intensive biomass harvesting on forest soils in the Nordic countries and the UK: a meta-analysis. *Forest Ecology and Management* 482, 118877. https://doi.org/10.1016/j.foreco.2020.118877.

Court, M., van der Heijden, G., Didier, S., Nys, C., Richter, C., Pousse, N., Saint-André, L. and Legout, A. 2018. Long-term effects of forest liming on mineral soil, organic layer and foliage chemistry: insights from multiple beech experimental sites in Northern France. *Forest Ecology and Management* 409, 872–889.

Craig, M. E., Turner, B. L., Liang, C., Clay, K., Johnson, D. J. and Phillips, R. P. 2018. Tree mycorrhizal type predicts within-site variability in the storage and distribution of soil organic matter. *Global Change Biology* 24(8), 3317–3330.

D'Amato, A. W., Bradford, J. B., Fraver, S. and Palik, B. J. 2013. Effects of thinning on drought vulnerability and climate response in north temperate forest ecosystems. *Ecological Applications* 23(8), 1735–1742. https://doi.org/10.1890/13-0677.1.

Dávila Reátegui, H., Poirier, V., Coyea, M. R. and Munson, A. D. 2021. Repeated thinning treatments reduce long-term soil carbon and nitrogen storage: an 87-year study at the Petawawa Research Forest, Canada. *Canadian Journal of Forest Research* 51(2), 190–197.

De Vos, B., Cools, N., Ilvesniemi, H., Vesterdal, L., Vanguelova, E. and Carnicelli, S. 2015. Benchmark values for forest soil carbon stocks in Europe: results from a large scale forest soil survey. *Geoderma* 251-252, 33–46.

De Wandeler, H., Bruelheide, H., Dawud, S. M., Dănilă, G., Domisch, T., Finér, L., Hermy, M., Jaroszewicz, B., Joly, F.-X., Müller, S., Ratcliffe, S., Raulund-Rasmussen, K., Rota, E., Van Meerbeek, K., Vesterdal, L. and Muys, B. 2018. Tree identity rather than tree diversity drives earthworm communities in European forests. *Pedobiologia* 67, 16–25.

Desie, E., Van Meerbeek, K., De Wandeler, H., Bruelheide, H., Domisch, T., Jaroszewicz, B., Joly, F., Vancampenhout, K., Vesterdal, L. and Muys, B. 2020. Positive feedback loop between earthworms, humus form and soil pH reinforces earthworm abundance in European forests. *Functional Ecology* 34(12), 2598–2610.

Dijkstra, F. A., Zhu, B. and Cheng, W. 2021. Root effects on soil organic carbon: a double-edged sword. *New Phytologist* 230(1), 60–65.

Don, A., Bärwolff, M., Kalbitz, K., Andruschkewitsch, R., Jungkunst, H. F. and Schulze, E.-D. 2012. No rapid soil carbon loss after a windthrow event in the High Tatra. *Forest Ecology and Management* 276, 239–246. https://doi.org/10.1016/j.foreco.2012.04.010.

Don, A., Schumacher, J. and Freibauer, A. 2011. Impact of tropical land-use change on soil organic carbon stocks–a meta-analysis. *Global Change Biology* 17(4), 1658–1670.

Eckmeier, E., Egli, M., Schmidt, M. W. I., Schlumpf, N., Nötzli, M., Minikus-Stary, N. and Hagedorn, F. 2010. Preservation of fire-derived carbon compounds and sorptive stabilisation promote the accumulation of organic matter in black soils of the Southern Alps. *Geoderma* 159(1-2), 147–155. https://doi.org/10.1016/j.geoderma.2010.07.006.

Englisch, M., Karrer, G. and Mutsch, F. 1992. Methodische grundlagen. *Mitteilungen der Forstlichen Bundesversuchsanstalt* 168, 5–22.

Faccoli, M. and Bernardinelli, I. 2014. Composition and elevation of spruce forests affect susceptibility to bark beetle attacks: implications for forest management. *Forests* 5(1), 88–102.

Feger, K., Lorenz, K., Raspe, S. and Armbruster, M. 2000. *Mittel-bis langfristige Auswirkungen von Kompensations-bzw. Bodenschutzkalkungen auf die Pedo-und Hydrosphäre. Schlussbericht.-Freiburg*.

Finkral, A. J. and Evans, A. M. 2008. The effects of a thinning treatment on carbon stocks in a northern Arizona ponderosa pine forest. *Forest Ecology and Management* 255(7), 2743–2750. https://doi.org/10.1016/j.foreco.2008.01.041.

Fog, K. 1988. The effect of added nitrogen on the rate of decomposition of organic matter. *Biological Reviews* 63(3), 433–462.

Friedlingstein, P., O'Sullivan, M., Jones, M. W., Andrew, R. M., Hauck, J., Olsen, A., Peters, G. P., Peters, W., Pongratz, J., Sitch, S., Le Quéré, C., Canadell, J. G., Ciais, P., Jackson, R. B., Alin, S., Aragão, L. E. O. C., Arneth, A., Arora, V., Bates, N. R., Becker, M., Benoit-Cattin, A., Bittig, H. C., Bopp, L., Bultan, S., Chandra, N., Chevallier, F., Chini, L. P., Evans, W., Florentie, L., Forster, P. M., Gasser, T., Gehlen, M., Gilfillan, D., Gkritzalis, T., Gregor, L., Gruber, N., Harris, I., Hartung, K., Haverd, V., Houghton, R. A., Ilyina, T., Jain, A. K., Joetzjer, E., Kadono, K., Kato, E., Kitidis, V., Korsbakken, J. I., Landschützer, P., Lefèvre, N., Lenton, A., Lienert, S., Liu, Z., Lombardozzi, D., Marland, G., Metzl, N., Munro, D. R., Nabel, J. E. M. S., Nakaoka, S.-I., Niwa, Y., O'Brien, K., Ono, T., Palmer, P. I., Pierrot, D., Poulter, B., Resplandy, L., Robertson, E., Rödenbeck, C., Schwinger, J., Séférian, R., Skjelvan, I., Smith, A. J. P., Sutton, A. J., Tanhua, T., Tans, P. P., Tian, H., Tilbrook, B., van der Werf, G., Vuichard, N., Walker, A. P., Wanninkhof, R., Watson, A. J., Willis, D., Wiltshire, A. J., Yuan, W., Yue, X. and Zaehle, S. 2020. Global carbon budget 2020. *Earth System Science Data* 12(4), 3269–3340. https://doi.org/10.5194/essd-12-3269-2020.

Friggens, N. L., Hester, A. J., Mitchell, R. J., Parker, T. C., Subke, J. A. and Wookey, P. A. 2020. Tree planting in organic soils does not result in net carbon sequestration on decadal timescales. *Global Change Biology* 26(9), 5178–5188. https://doi.org/10.1111/gcb.15229.

Führer, E. 2000. Forest functions, ecosystem stability and management. *Forest Ecology and Management* 132(1), 29–38.

Fulé, P. Z., Waltz, A. E. M., Covington, W. W. and Heinlein, T. A. 2001. Measuring forest restoration effectiveness in reducing hazardous fuels. *Journal of Forestry* 99, 24–29. https://doi.org/10.1093/jof/99.11.24.

Gerber, W., Rickli, C. and Graf, F. 2002. Surface erosion in cleared and uncleared mountain windthrow sites. *Forest Snow and Landscape Research* 77, 109–116.

Gonçalves, A. C. 2020. Thinning: an overview. In: *Silviculture*. IntechOpen. https://www.intechopen.com/chapters/73015.

Gong, C., Tan, Q., Liu, G. and Xu, M. 2021. Forest thinning increases soil carbon stocks in China. *Forest Ecology and Management* 482, 118812. https://doi.org/10.1016/j.foreco.2020.118812.

Gosheva, S., Walthert, L., Niklaus, P. A., Zimmermann, S., Gimmi, U. and Hagedorn, F. 2017. Reconstruction of historic forest cover changes indicates minor effects on carbon stocks in Swiss forest soils. *Ecosystems* 20(8), 1512–1528.

Griscom, B. W., Adams, J., Ellis, P. W., Houghton, R. A., Lomax, G., Miteva, D. A., Schlesinger, W. H., Shoch, D., Siikamäki, J. V., Smith, P., Woodbury, P., Zganjar, C., Blackman, A., Campari, J., Conant, R. T., Delgado, C., Elias, P., Gopalakrishna, T., Hamsik, M. R., Herrero, M., Kiesecker, J., Landis, E., Laestadius, L., Leavitt, S. M., Minnemeyer, S., Polasky, S., Potapov, P., Putz, F. E., Sanderman, J., Silvius, M., Wollenberg, E. and Fargione, J. 2017. Natural climate solutions. *Proceedings of the National Academy of Sciences of the United States of America* 114(44), 11645–11650.

Grüneberg, E., Schöning, I., Riek, W., Ziche, D. and Evers, J. 2019. Carbon stocks and carbon stock changes in German Forest Soils. In: N. Wellbrock and A. Bolte (eds) *Status and Dynamics of Forests in Germany. Ecological Studies, vol 237*. Springer, Cham, 167–198. https://doi.org/10.1007/978-3-030-15734-0_6.

Grüneberg, E., Ziche, D. and Wellbrock, N. 2014. Organic carbon stocks and sequestration rates of forest soils in Germany. *Global Change Biology* 20(8), 2644–2662.

Hagedorn, F., Gavazov, K. and Alexander, J. M. 2019. Above-and belowground linkages shape responses of mountain vegetation to climate change. *Science* 365(6458), 1119–1123.

Hagedorn, F., Spinnler, D. and Siegwolf, R. 2003. Increased N deposition retards mineralization of old soil organic matter. *Soil Biology and Biochemistry* 35(12), 1683–1692.

Hanewinkel, M., Kuhn, T., Bugmann, H., Lanz, A. and Brang, P. 2014. Vulnerability of uneven-aged forests to storm damage. *Forestry* 87(4), 525–534.

Hansson, K., Kleja, D. B., Kalbitz, K. and Larsson, H. 2010. Amounts of carbon mineralised and leached as DOC during decomposition of Norway spruce needles and fine roots. *Soil Biology and Biochemistry* 42(2), 178–185. https://doi.org/10.1016/j.soilbio.2009.10.013.

Harris, N. L., Gibbs, D. A., Baccini, A., Birdsey, R. A., de Bruin, S., Farina, M., Fatoyinbo, L., Hansen, M. C., Herold, M., Houghton, R. A., Potapov, P. V., Suarez, D. R., Roman-Cuesta, R. M., Saatchi, S. S., Slay, C. M., Turubanova, S. A. and Tyukavina, A. 2021. Global maps of twenty-first century forest carbon fluxes. *Nature Climate Change* 11(3), 234–240. https://doi.org/10.1038/s41558-020-00976-6.

Heděnec, P., Nilsson, L. O., Zheng, H., Gundersen, P., Schmidt, I. K., Rousk, J. and Vesterdal, L. 2020. Mycorrhizal association of common European tree species shapes biomass and metabolic activity of bacterial and fungal communities in soil. *Soil Biology and Biochemistry* 149, 107933. https://doi.org/10.1016/j.soilbio.2020.107933.

Hildebrand, E. 1996. Warum müssen wir Waldböden kalken. *Agrarforschung in Baden-Württemberg* 26, 53–65.

Hiltbrunner, D., Zimmermann, S. and Hagedorn, F. 2013. Afforestation with Norway spruce on a subalpine pasture alters carbon dynamics but only moderately affects soil carbon storage. *Biogeochemistry* 115(1–3), 251–266.

Hlásny, T., Zimová, S., Merganičová, K., Štěpánek, P., Modlinger, R. and Turčáni, M. 2021. Devastating outbreak of bark beetles in the Czech Republic: drivers, impacts, and management implications. *Forest Ecology and Management* 490, 119075.

Hong, S., Yin, G., Piao, S., Dybzinski, R., Cong, N., Li, X., Wang, K., Peñuelas, J., Zeng, H. and Chen, A. 2020. Divergent responses of soil organic carbon to afforestation. *Nature Sustainability* 3(9), 694–700.

Hoover, C. M. 2011. Management impacts on forest floor and soil organic carbon in northern temperate forests of the US. *Carbon Balance and Management* 6, 1–8.

Hotta, W., Morimoto, J., Inoue, T., Suzuki, S. N., Umebayashi, T., Owari, T., Shibata, H., Ishibashi, S., Hara, T. and Nakamura, F. 2020. Recovery and allocation of carbon stocks in boreal forests 64 years after catastrophic windthrow and salvage logging in northern Japan. *Forest Ecology and Management* 468, 118169.

Hume, A. M., Chen, H. Y. H. and Taylor, A. R. 2018. Intensive forest harvesting increases susceptibility of northern forest soils to carbon, nitrogen and phosphorus loss. *Journal of Applied Ecology* 55(1), 246–255.

Huotari, N., Tillman-Sutela, E., Moilanen, M. and Laiho, R. 2015. Recycling of ash–For the good of the environment? *Forest Ecology and Management* 348, 226–240.

Hyvönen, R., Kaarakka, L., Leppälammi-Kujansuu, J., Olsson, B. A., Palviainen, M., Vegerfors-Persson, B. and Helmisaari, H.-S. 2016. Effects of stump harvesting on soil C and N stocks and vegetation 8-13 years after clear-cutting. *Forest Ecology and Management* 371, 23–32.

Jackson, R. B., Jobbágy, E. G., Avissar, R., Roy, S. B., Barrett, D. J., Cook, C. W., Farley, K. A., Maitre, D. C. le, McCarl, B. A. and Murray, B. C. 2005. Trading water for carbon with biological carbon sequestration. *Science* 310(5756), 1944–1947. https://doi.org/10.1126/science.1119282.

James, J. and Harrison, R. 2016. The effect of harvest on forest soil carbon: a meta-analysis. *Forests* 7(12), 308. https://doi.org/10.3390/f7120308.

James, J., Page-Dumroese, D., Busse, M., Palik, B., Zhang, J., Eaton, B., Slesak, R., Tirocke, J. and Kwon, H. 2021. Effects of forest harvesting and biomass removal on soil carbon and nitrogen: two complementary meta-analyses. *Forest Ecology and Management* 485, 118935.

Jandl, R. 2020. Climate-induced challenges of Norway spruce in Northern Austria. *Trees, Forests and People* 1, 100008. https://doi.org/10.1016/j.tfp.2020.100008.

Jandl, R., Ledermann, T., Kindermann, G. and Weiss, P. 2021. Soil organic carbon stocks in mixed-deciduous and coniferous forests in Austria. *Frontiers in Forests and Global Change* 4, 69.

Jandl, R., Lindner, M., Vesterdal, L., Bauwens, B., Baritz, R., Hagedorn, F., Johnson, D. W., Minkkinen, K. and Byrne, K. A. 2007. How strongly can forest management influence soil carbon sequestration? *Geoderma* 137(3–4), 253–268.

Johnson, D. W. and Curtis, P. S. 2001. Effects of forest management on soil C and N storage: meta analysis. *Forest Ecology and Management* 140(2-3), 227-238.

Jonard, M., Nicolas, M., Coomes, D. A., Caignet, I., Saenger, A. and Ponette, Q. 2017. Forest soils in France are sequestering substantial amounts of carbon. *Science of the Total Environment* 574, 616-628.

Jovani-Sancho, A. J., Cummins, T. and Byrne, K. A. 2021. Soil carbon balance of afforested peatlands in the maritime temperate climatic zone. *Global Change Biology* 27(15), 3681-3698. https://doi.org/10.1111/gcb.15654.

Jurevics, A., Peichl, M., Olsson, B. A., Strömgren, M. and Egnell, G. 2016. Slash and stump harvest have no general impact on soil and tree biomass C pools after 32-39 years. *Forest Ecology and Management* 371, 33-41.

Jurgensen, M., Tarpey, R., Pickens, J., Kolka, R. and Palik, B. 2012. Long-term effect of silvicultural thinnings on soil carbon and nitrogen pools. *Soil Science Society of America Journal* 76(4), 1418-1425.

Kaarakka, L., Hyvönen, R., Strömgren, M., Palviainen, M., Persson, T., Olsson, B. A., Launonen, E., Vegerfors, B. and Helmisaari, H.-S. 2016. Carbon and nitrogen pools and mineralization rates in boreal forest soil after stump harvesting. *Forest Ecology and Management* 377, 61-70.

Kaiser, K. and Guggenberger, G. 2003. Mineral surfaces and soil organic matter. *European Journal of Soil Science* 54(2), 219-236.

Kim, S., Kim, C., Han, S. H., Lee, S.-T. and Son, Y. 2018. A multi-site approach toward assessing the effect of thinning on soil carbon contents across temperate pine, oak, and larch forests. *Forest Ecology and Management* 424, 62-70. https://doi.org/10.1016/j.foreco.2018.04.040.

Kishchuk, B. E., Thiffault, E., Lorente, M., Quideau, S., Keddy, T. and Sidders, D. 2015. Decadal soil and stand response to fire, harvest, and salvage-logging disturbances in the western boreal mixedwood forest of Alberta, Canada. *Canadian Journal of Forest Research* 45(2), 141-152.

Kjøller, R. and Clemmensen, K. E. 2008. *The Impact of Liming on Ectomycorrhizal Fungal Communities in Coniferous Forests in Southern Sweden*. Skogsstyrelsens förlag.

Klein, D., Fuentes, J. P., Schmidt, A., Schmidt, H. and Schulte, A. 2008. Soil organic C as affected by silvicultural and exploitative interventions in Nothofagus pumilio forests of the Chilean Patagonia. *Forest Ecology and Management* 255(10), 3549-3555.

Kobler, J., Jandl, R., Dirnböck, T., Mirtl, M. and Schindlbacher, A. 2015. Effects of stand patchiness due to windthrow and bark beetle abatement measures on soil CO_2 efflux and net ecosystem productivity of a managed temperate mountain forest. *European Journal of Forest Research* 134(4), 683-692. https://doi.org/10.1007/s10342-015-0882-2.

Kosunen, M., Peltoniemi, K., Pennanen, T., Lyytikäinen-Saarenmaa, P., Adamczyk, B., Fritze, H., Zhou, X. and Starr, M. 2020. Storm and Ips typographus disturbance effects on carbon stocks, humus layer carbon fractions and microbial community composition in boreal Picea abies stands. *Soil Biology and Biochemistry* 148, 107853.

Kreutzer, K. 1995. Effects of forest liming on soil processes. *Plant and Soil* 168-169(1), 447-470.

Law, B. E. and Waring, R. H. 2015. Carbon implications of current and future effects of drought, fire and management on Pacific Northwest forests. *Forest Ecology and Management* 355, 4-14. https://doi.org/10.1016/j.foreco.2014.11.023.

Leverkus, A. B., Rey Benayas, J. M., Castro, J., Boucher, D., Brewer, S., Collins, B. M., Donato, D., Fraver, S., Kishchuk, B. E., Lee, E.-J., Lindenmayer, D. B., Lingua, E., Macdonald, E., Marzano, R., Rhoades, C. C., Royo, A., Thorn, S., Wagenbrenner, J. W., Waldron, K., Wohlgemuth, T. and Gustafsson, L. 2018. Salvage logging effects on regulating and supporting ecosystem services–a systematic map. *Canadian Journal of Forest Research* 48(9), 983–1000.

Li, R., Guan, X., Han, J., Zhang, Y., Zhang, W., Wang, J., Huang, Y., Xu, M., Chen, L., Wang, S. and Yang, Q. 2021. Litter decomposition was retarded by understory removal but was unaffected by thinning in a Chinese fir [Cunninghamia lanceolata (Lamb.) Hook] plantation. *Applied Soil Ecology* 163, 103968. https://doi.org/10.1016/j.apsoil.2021.103968.

Lim, H., Olsson, B. A., Lundmark, T., Dahl, J. and Nordin, A. 2020. Effects of whole-tree harvesting at thinning and subsequent compensatory nutrient additions on carbon sequestration and soil acidification in a boreal forest. *GCB Bioenergy* 12(11), 992–1001.

Liski, J., Perruchoud, D. and Karjalainen, T. 2002. Increasing carbon stocks in the forest soils of Western Europe. *Forest Ecology and Management* 169(1–2), 159–175.

Marschner, B. and Wilczynski, A. W. 1991. The effect of liming on quantity and chemical composition of soil organic matter in a pine forest in Berlin, Germany. *Plant and Soil* 137(2), 229–236.

Matzner, E., Khanna, P. K., Meiwes, K. J. and Ulrich, B. 1985. Effects of fertilization and liming on the chemical soil conditions and element distribution in forest soils. *Plant and Soil* 87(3), 405–415. https://doi.org/10.1007/BF02181907.

Mayer, M., Matthews, B., Rosinger, C., Sandén, H., Godbold, D. L. and Katzensteiner, K. 2017a. Tree regeneration retards decomposition in a temperate mountain soil after forest gap disturbance. *Soil Biology and Biochemistry* 115, 490–498.

Mayer, M., Sandén, H., Rewald, B., Godbold, D. L. and Katzensteiner, K. 2017b. Increase in heterotrophic soil respiration by temperature drives decline in soil organic carbon stocks after forest windthrow in a mountainous ecosystem. *Functional Ecology* 31(5), 1163–1172.

Mayer, M., Pousse, N. and James, J. 2021. Harvest systems that limit soil disturbance and reduced impact logging. In: *Recarbonizing Global Soils: A Technical Manual of Recommended Management Practices*. Food and Agriculture Organization of the United Nations, Rome, Italy.

Mayer, M., Prescott, C. E., Abaker, W. E. A., Augusto, L., Cécillon, L., Ferreira, G. W. D., James, J., Jandl, R., Katzensteiner, K., Laclau, J.-P., Laganière, J., Nouvellon, Y., Paré, D., Stanturf, J. A., Vanguelova, E. I. and Vesterdal, L. 2020. Tamm review: Influence of forest management activities on soil organic carbon stocks: a knowledge synthesis. *Forest Ecology and Management* 466, 118127. https://doi.org/10.1016/j.foreco.2020.118127.

Melillo, J. M., Frey, S. D., DeAngelis, K. M., Werner, W. J., Bernard, M. J., Bowles, F. P., Pold, G., Knorr, M. A. and Grandy, A. S. 2017. Long-term pattern and magnitude of soil carbon feedback to the climate system in a warming world. *Science* 358(6359), 101–105.

Minasny, B., Malone, B. P., McBratney, A. B., Angers, D. A., Arrouays, D., Chambers, A., Chaplot, V., Chen, Z.-S., Cheng, K., Das, B. S., Field, D. J., Gimona, A., Hedley, C. B., Hong, S. Y., Mandal, B., Marchant, B. P., Martin, M., McConkey, B. G., Mulder, V. L., O'Rourke, S., Richer-de-Forges, A. C., Odeh, I., Padarian, J., Paustian, K., Pan, G.,

Poggio, L., Savin, I., Stolbovoy, V., Stockmann, U., Sulaeman, Y., Tsui, C., Vågen, T., van Wesemael, B. and Winowiecki, L. 2017. Soil carbon 4 per mille. *Geoderma* 292, 59–86.

Mobley, M. L., Lajtha, K., Kramer, M. G., Bacon, A. R., Heine, P. R. and Richter, D. D. 2015. Surficial gains and subsoil losses of soil carbon and nitrogen during secondary forest development. *Global Change Biology* 21(2), 986–996. https://doi.org/10.1111/gcb.12715.

Na, M., Sun, X., Zhang, Y., Sun, Z. and Rousk, J. 2021. Higher stand densities can promote soil carbon storage after conversion of temperate mixed natural forests to larch plantations. *European Journal of Forest Research* 140(2), 373–386. https://doi.org/10.1007/s10342-020-01346-9.

Nave, L. E., Vance, E. D., Swanston, C. W. and Curtis, P. S. 2009. Impacts of elevated N inputs on north temperate forest soil C storage, C/N, and net N-mineralization. *Geoderma* 153(1–2), 231–240.

Nave, L. E., Domke, G. M., Hofmeister, K. L., Mishra, U., Perry, C. H., Walters, B. F. and Swanston, C. W. 2018. Reforestation can sequester two petagrams of carbon in US topsoils in a century. *Proceedings of the National Academy of Sciences of the United States of America* 115(11), 2776–2781.

Nave, L. E., Vance, E. D., Swanston, C. W. and Curtis, P. S. 2011. Fire effects on temperate forest soil C and N storage. *Ecological Applications* 21(4), 1189–1201.

Nave, L. E., Vance, E. D., Swanston, C. W. and Curtis, P. S. 2010. Harvest impacts on soil carbon storage in temperate forests. *Forest Ecology and Management* 259(5), 857–866. https://doi.org/10.1016/j.foreco.2009.12.009.

Negrón, J. F., Allen, K., Cook, B. and Withrow Jr., J. R. 2008. Susceptibility of ponderosa pine, Pinus ponderosa (Dougl. ex Laws.), to mountain pine beetle, Dendroctonus ponderosae Hopkins, attack in uneven-aged stands in the Black Hills of South Dakota and Wyoming USA. *Forest Ecology and Management* 254(2), 327–334.

Nilsen, P. and Strand, L. T. 2008. Thinning intensity effects on carbon and nitrogen stores and fluxes in a Norway spruce (Picea abies (L.) Karst.) stand after 33 years. *Forest Ecology and Management* 256(3), 201–208.

Noormets, A. and Nouvellon, Y. 2015. Introduction for special issue: carbon, water and nutrient cycling in managed forests. *Forest Ecology and Management* 355, 1–3. https://doi.org/10.1016/j.foreco.2015.08.022.

Pan, Y., Birdsey, R. A., Fang, J., Houghton, R., Kauppi, P. E., Kurz, W. A., Phillips, O. L., Shvidenko, A., Lewis, S. L., Canadell, J. G., Ciais, P., Jackson, R. B., Pacala, S. W., McGuire, A. D., Piao, S., Rautiainen, A., Sitch, S. and Hayes, D. 2011. A large and persistent carbon sink in the world's forests. *Science* 333(6045), 988–993. https://doi.org/10.1126/science.1201609.

Peng, Y., Schmidt, I. K., Zheng, H., Heděnec, P., Bachega, L. R., Yue, K., Wu, F. and Vesterdal, L. 2020. Tree species effects on topsoil carbon stock and concentration are mediated by tree species type, mycorrhizal association, and N-fixing ability at the global scale. *Forest Ecology and Management* 478, 118510.

Pérez-Izquierdo, L., Clemmensen, K. E., Strengbom, J., Granath, G., Wardle, D. A., Nilsson, M. and Lindahl, B. D. 2021. Crown-fire severity is more important than ground-fire severity in determining soil fungal community development in the boreal forest. *Journal of Ecology* 109(1), 504–518.

Persson, T., Rudebeck, A. and Wirén, A. 1995. Pools and fluxes of carbon and nitrogen in 40-year-old forest liming experiments in southern Sweden. *Water, Air, and Soil Pollution* 85(2), 901–906.

Poeplau, C. and Don, A. 2013. Sensitivity of soil organic carbon stocks and fractions to different land-use changes across Europe. *Geoderma* 192, 189–201. https://doi.org/10.1016/j.geoderma.2012.08.003.

Poirier, V., Paré, D., Boiffin, J. and Munson, A. D. 2014. Combined influence of fire and salvage logging on carbon and nitrogen storage in boreal forest soil profiles. *Forest Ecology and Management* 326, 133–141.

Popp, J., Lakner, Z., Harangi-Rakos, M. and Fari, M. 2014. The effect of bioenergy expansion: food, energy, and environment. *Renewable and Sustainable Energy Reviews* 32, 559–578.

Powers, M., Kolka, R., Palik, B., McDonald, R. and Jurgensen, M. 2011. Long-term management impacts on carbon storage in Lake States forests. *Forest Ecology and Management* 262(3), 424–431. https://doi.org/10.1016/j.foreco.2011.04.008.

Powers, M. D., Kolka, R. K., Bradford, J. B., Palik, B. J., Fraver, S. and Jurgensen, M. F. 2012. Carbon stocks across a chronosequence of thinned and unmanaged red pine (Pinus resinosa) stands. *Ecological Applications* 22(4), 1297–1307. https://doi.org/10.1890/11-0411.1.

Prietzel, J. and Bachmann, S. 2012. Changes in soil organic C and N stocks after forest transformation from Norway spruce and Scots pine into Douglas fir, Douglas fir/spruce, or European beech stands at different sites in Southern Germany. *Forest Ecology and Management* 269, 134–148. https://doi.org/10.1016/j.foreco.2011.12.034.

Prietzel, J., Zimmermann, L., Schubert, A. and Christophel, D. 2016. Organic matter losses in German Alps forest soils since the 1970s most likely caused by warming. *Nature Geoscience* 9(7), 543–548.

Puhlick, J. J., Fernandez, I. J. and Weiskittel, A. R. 2016. Evaluation of forest management effects on the mineral soil carbon pool of a lowland, mixed-species forest in Maine, USA. *Canadian Journal of Soil Science* 96(2), 207–218.

Rabot, E., Wiesmeier, M., Schlüter, S. and Vogel, H.-J. 2018. Soil structure as an indicator of soil functions: a review. *Geoderma* 314, 122–137. https://doi.org/10.1016/j.geoderma.2017.11.009.

Rasse, D. P., Rumpel, C. and Dignac, M.-F. 2005. Is soil carbon mostly root carbon? Mechanisms for a specific stabilisation. *Plant and Soil* 269(1–2), 341–356. https://doi.org/10.1007/s11104-004-0907-y.

Reich, P. B., Oleksyn, J., Modrzynski, J., Mrozinski, P., Hobbie, S. E., Eissenstat, D. M., Chorover, J., Chadwick, O. A., Hale, C. M. and Tjoelker, M. G. 2005. Linking litter calcium, earthworms and soil properties: a common garden test with 14 tree species. *Ecology Letters* 8(8), 811–818.

Rosenberg, O., Persson, T., Högbom, L. and Jacobson, S. 2010. Effects of wood-ash application on potential carbon and nitrogen mineralisation at two forest sites with different tree species, climate and N status. *Forest Ecology and Management* 260(4), 511–518.

Ruiz-Peinado, R., Bravo-Oviedo, A., Montero, G. and Del Río, M. 2016. Carbon stocks in a Scots pine afforestation under different thinning intensities management. *Mitigation and Adaptation Strategies for Global Change* 21, 1059–1072.

Santos, L. T. dos, Magnabosco Marra, D., Trumbore, S., Camargo, P. B. de, Negrón-Juárez, R. I., Lima, A. J. N., Ribeiro, G. H. P. M., Santos, J. dos and Higuchi, N. 2016. Windthrows increase soil carbon stocks in a central Amazon forest. *Biogeosciences* 13(4), 1299–1308.

Santos, R. S., Oliveira, F. C. C., Ferreira, G. W. D., Ferreira, M. A., Araujo, E. F. and Silva, I. R. 2020. Carbon and nitrogen dynamics in soil organic matter fractions following eucalypt afforestation in southern Brazilian grasslands (Pampas). *Agriculture, Ecosystems and Environment* 301, 106979.

Schäffer, J., Geißen, V., Hoch, R. and von Wilpert, K. 2001. Waldkalkung belebt Böden wieder. *AFZ-Der Wald* Bd 21, 1106-1109.

Scharlemann, J. P., Tanner, E. V., Hiederer, R. and Kapos, V. 2014. Global soil carbon: understanding and managing the largest terrestrial carbon pool. *Carbon Management* 5(1), 81-91. https://doi.org/10.4155/cmt.13.77.

Schindlbacher, A., Schnecker, J., Takriti, M., Borken, W. and Wanek, W. 2015. Microbial physiology and soil CO_2 efflux after 9 years of soil warming in a temperate forest - no indications for thermal adaptations. *Global Change Biology* 21(11), 4265-4277. https://doi.org/10.1111/gcb.12996.

Schlesinger, W. H. 1990. Evidence from chronosequence studies for a low carbon-storage potential of soils. *Nature* 348(6298), 232-234. https://doi.org/10.1038/348232a0.

Schmidt, M. W. I., Torn, M. S., Abiven, S., Dittmar, T., Guggenberger, G., Janssens, I. A., Kleber, M., Kögel-Knabner, I., Lehmann, J., Manning, D. A. C., Nannipieri, P., Rasse, D. P., Weiner, S. and Trumbore, S. E. 2011. Persistence of soil organic matter as an ecosystem property. *Nature* 478(7367), 49-56. https://doi.org/10.1038/nature10386.

Seidl, R., Schelhaas, M. J., Rammer, W. and Verkerk, P. J. 2014. Increasing forest disturbances in Europe and their impact on carbon storage. *Nature Climate Change* 4(9), 806-810.

Seidl, R., Thom, D., Kautz, M., Martin-Benito, D., Peltoniemi, M., Vacchiano, G., Wild, J., Ascoli, D., Petr, M., Honkaniemi, J., Lexer, M. J., Trotsiuk, V., Mairota, P., Svoboda, M., Fabrika, M., Nagel, T. A. and Reyer, C. P. O. 2017. Forest disturbances under climate change. *Nature Climate Change* 7, 395-402. https://doi.org/10.1038/nclimate3303.

Senf, C., Pflugmacher, D., Zhiqiang, Y., Sebald, J., Knorn, J., Neumann, M., Hostert, P. and Seidl, R. 2018. Canopy mortality has doubled in Europe's temperate forests over the last three decades. *Nature Communications* 9(1), 4978. https://doi.org/10.1038/s41467-018-07539-6.

Silver, W. L., Ostertag, R. and Lugo, A. E. 2000. The potential for carbon sequestration Through reforestation of abandoned tropical agricultural and pasture lands. *Restoration Ecology* 8(4), 394-407. https://doi.org/10.1046/j.1526-100x.2000.80054.x.

Simola, H., Pitkänen, A. and Turunen, J. 2012. Carbon loss in drained forestry peatlands in Finland, estimated by re-sampling peatlands surveyed in the 1980s. *European Journal of Soil Science* 63(6), 798-807. https://doi.org/10.1111/j.1365-2389.2012.01499.x.

Smith, P, Clark, H., Dong, H., Elsiddig, E. A., Haberl, H., Harper, R., House, J., Jafari, M., et al. 2014. Chapter 11 - Agriculture, forestry and other land use (AFOLU). In: *Climate Change 2014: Mitigation of Climate Change. IPCC Working Group III Contribution to AR5*. Cambridge University Press. https://pure.iiasa.ac.at/id/eprint/11115/.

Smith, P., Davis, S. J., Creutzig, F., Fuss, S., Minx, J., Gabrielle, B., Kato, E., Jackson, R. B., Cowie, A., Kriegler, E., van Vuuren, D. P., Rogelj, J., Ciais, P., Milne, J., Canadell, J. G., McCollum, D., Peters, G., Andrew, R., Krey, V., Shrestha, G., Friedlingstein, P., Gasser, T., Grübler, A., Heidug, W. K., Jonas, M., Jones, C. D., Kraxner, F., Littleton, E., Lowe, J., Moreira, J. R., Nakicenovic, N., Obersteiner, M., Patwardhan, A., Rogner, M., Rubin, E.,

Sharifi, A., Torvanger, A., Yamagata, Y., Edmonds, J. and Yongsung, C. 2016. Biophysical and economic limits to negative CO2 emissions. *Nature Climate Change* 6(1), 42–50.

Sohn, J. A., Saha, S. and Bauhus, J. 2016. Potential of forest thinning to mitigate drought stress: a meta-analysis. *Forest Ecology and Management* 380, 261–273. https://doi.org/10.1016/j.foreco.2016.07.046.

Solly, E. F., Weber, V., Zimmermann, S., Walthert, L., Hagedorn, F. and Schmidt, M. W. I. 2020. A critical evaluation of the relationship between the effective cation exchange capacity and soil organic carbon content in Swiss forest soils. *Frontiers in Forests and Global Change* 3, 98.

Stanturf, J. A., Kleine, M., Mansourian, S., Parrotta, J., Madsen, P., Kant, P., Burns, J. and Bolte, A. 2019. Implementing forest landscape restoration under the Bonn Challenge: a systematic approach. *Annals of Forest Science* 76(2), 50. https://doi.org/10.1007/s13595-019-0833-z.

Stewart, C. E., Paustian, K., Conant, R. T., Plante, A. F. and Six, J. 2007. Soil carbon saturation: concept, evidence and evaluation. *Biogeochemistry* 86(1), 19–31.

Strömgren, M., Egnell, G. and Olsson, B. A. 2013. Carbon stocks in four forest stands in Sweden 25 years after harvesting of slash and stumps. *Forest Ecology and Management* 290, 59–66.

Sun, O. J., Campbell, J., Law, B. E. and Wolf, V. 2004. Dynamics of carbon stocks in soils and detritus across chronosequences of different forest types in the Pacific Northwest, USA. *Global Change Biology* 10(9), 1470–1481.

Tang, J., Bolstad, P. V. and Martin, J. G. 2009. Soil carbon fluxes and stocks in a Great Lakes forest chronosequence. *Global Change Biology* 15(1), 145–155.

Temperton, V. M., Buchmann, N., Buisson, E., Durigan, G., Kazmierczak, L., Perring, M. P., de Sá Dechoum, M., Veldman, J. W. and Overbeck, G. E. 2019. Step back from the forest and step up to the Bonn Challenge: how a broad ecological perspective can promote successful landscape restoration. *Restoration Ecology* 27, 705–719.

Thom, D. and Seidl, R. 2016. Natural disturbance impacts on ecosystem services and biodiversity in temperate and boreal forests. *Biological Reviews of the Cambridge Philosophical Society* 91(3), 760–781. https://doi.org/10.1111/brv.12193.

van der Linde, S., Suz, L. M., Orme, C. D. L., Cox, F., Andreae, H., Asi, E., Atkinson, B., Benham, S., Carroll, C., Cools, N., De Vos, B., Dietrich, H. P., Eichhorn, J., Gehrmann, J., Grebenc, T., Gweon, H. S., Hansen, K., Jacob, F., Kristöfel, F., Lech, P., Manninger, M., Martin, J., Meesenburg, H., Merilä, P., Nicolas, M., Pavlenda, P., Rautio, P., Schaub, M., Schröck, H. W., Seidling, W., Šrámek, V., Thimonier, A., Thomsen, I. M., Titeux, H., Vanguelova, E., Verstraeten, A., Vesterdal, L., Waldner, P., Wijk, S., Zhang, Y., Žlindra, D. and Bidartondo, M. I. 2018. Environment and host as large-scale controls of ectomycorrhizal fungi. *Nature* 558(7709), 243–248. https://doi.org/10.1038/s41586-018-0189-9.

Vanguelova, E., Pitman, R., Luiro, J. and Helmisaari, H.-S. 2010. Long term effects of whole tree harvesting on soil carbon and nutrient sustainability in the UK. *Biogeochemistry* 101(1–3), 43–59.

Vanguelova, E. I., Crow, P., Benham, S., Pitman, R., Forster, J., Eaton, E. L. and Morison, J. I. L. 2019. Impact of Sitka spruce (Picea sitchensis (Bong.) Carr.) afforestation on the carbon stocks of peaty gley soils – a chronosequence study in the north of England. *Forestry* 92(3), 242–252. https://doi.org/10.1093/forestry/cpz013.

Veldman, J. W., Aleman, J. C., Alvarado, S. T., Anderson, T. M., Archibald, S., Bond, W. J., Boutton, T. W., Buchmann, N., Buisson, E., Canadell, J. G., Dechoum, M. S.,

Diaz-Toribio, M. H., Durigan, G., Ewel, J. J., Fernandes, G. W., Fidelis, A., Fleischman, F., Good, S. P., Griffith, D. M., Hermann, J. M., Hoffmann, W. A., Le Stradic, S., Lehmann, C. E. R., Mahy, G., Nerlekar, A. N., Nippert, J. B., Noss, R. F., Osborne, C. P., Overbeck, G. E., Parr, C. L., Pausas, J. G., Pennington, R. T., Perring, M. P., Putz, F. E., Ratnam, J., Sankaran, M., Schmidt, I. B., Schmitt, C. B., Silveira, F. A. O., Staver, A. C., Stevens, N., Still, C. J., Strömberg, C. A. E., Temperton, V. M., Varner, J. M. and Zaloumis, N. P. 2019. Comment on "The global tree restoration potential." *Science* 366(6463). https://doi.org/10.1126/science.aay7976.

Vesterdal, L., Clarke, N., Sigurdsson, B. D. and Gundersen, P. 2013. Do tree species influence soil carbon stocks in temperate and boreal forests? *Forest Ecology and Management* 309, 4-18.

Vesterdal, L., Dalsgaard, M., Felby, C., Raulund-Rasmussen, K. and Jørgensen, B. B. 1995. Effects of thinning and soil properties on accumulation of carbon, nitrogen and phosphorus in the forest floor of Norway spruce stands. *Forest Ecology and Management* 77(1-3), 1-10. https://doi.org/10.1016/0378-1127(95)03579-Y.

Vitasse, Y., Bottero, A., Rebetez, M., Conedera, M., Augustin, S., Brang, P. and Tinner, W. 2019. What is the potential of silver fir to thrive under warmer and drier climate? *European Journal of Forest Research* 138(4), 547-560.

Wan, X., Xiao, L., Vadeboncoeur, M. A., Johnson, C. E. and Huang, Z. 2018. Response of mineral soil carbon storage to harvest residue retention depends on soil texture: a meta-analysis. *Forest Ecology and Management* 408, 9-15. https://doi.org/10.1016/j.foreco.2017.10.028.

Wang, Q., Zhong, M. and Wang, S. 2012. A meta-analysis on the response of microbial biomass, dissolved organic matter, respiration, and N mineralization in mineral soil to fire in forest ecosystems. *Forest Ecology and Management* 271, 91-97. https://doi.org/10.1016/j.foreco.2012.02.006.

Warren, K. L. and Ashton, M. S. 2014. Change in soil and forest floor carbon after shelterwood harvests in a New England Oak-Hardwood Forest, USA. *International Journal of Forestry Research* 2014, 1-9.

Wiesmeier, M., Prietzel, J., Barthold, F., Spörlein, P., Geuß, U., Hangen, E., Reischl, A., Schilling, B., von Lützow, M. and Kögel-Knabner, I. 2013. Storage and drivers of organic carbon in forest soils of southeast Germany (Bavaria)-Implications for carbon sequestration. *Forest Ecology and Management* 295, 162-172.

Wiesmeier, M., Urbanski, L., Hobley, E., Lang, B., von Lützow, M., Marin-Spiotta, E., van Wesemael, B., Rabot, E., Ließ, M., Garcia-Franco, N., Wollschläger, U., Vogel, H. and Kögel-Knabner, I. 2019. Soil organic carbon storage as a key function of soils-A review of drivers and indicators at various scales. *Geoderma* 333, 149-162.

Williams, C. A., Gu, H., MacLean, R., Masek, J. G. and Collatz, G. J. 2016. Disturbance and the carbon balance of US forests: a quantitative review of impacts from harvests, fires, insects, and droughts. *Global and Planetary Change* 143, 66-80. https://doi.org/10.1016/j.gloplacha.2016.06.002.

Zabowski, D., Chambreau, D., Rotramel, N. and Thies, W. G. 2008. Long-term effects of stump removal to control root rot on forest soil bulk density, soil carbon and nitrogen content. *Forest Ecology and Management* 255(3-4), 720-727.

Zak, D. R., Pellitier, P. T., Argiroff, W., Castillo, B., James, T. Y., Nave, L. E., Averill, C., Beidler, K. V., Bhatnagar, J., Blesh, J., Classen, A. T., Craig, M., Fernandez, C. W., Gundersen, P., Johansen, R., Koide, R. T., Lilleskov, E. A., Lindahl, B. D., Nadelhoffer, K. J., Phillips,

R. P. and Tunlid, A. 2019. Exploring the role of ectomycorrhizal fungi in soil carbon dynamics. *New Phytologist* 223(1), 33–39.

Zhang, B., Zhou, X., Zhou, L. and Ju, R. 2015. A global synthesis of below-ground carbon responses to biotic disturbance: a meta-analysis. *Global Ecology and Biogeography* 24(2), 126–138. https://doi.org/10.1111/geb.12235.

Zhang, X., Guan, D., Li, W., Sun, D., Jin, C., Yuan, F., Wang, A. and Wu, J. 2018. The effects of forest thinning on soil carbon stocks and dynamics: a meta-analysis. *Forest Ecology and Management* 429, 36–43.

Zimmermann, S. and Frey, B. 2002. Soil respiration and microbial properties in an acid forest soil: effects of wood ash. *Soil Biology and Biochemistry* 34(11), 1727–1737. https://doi.org/10.1016/S0038-0717(02)00160-8.

Chapter 20

The contribution of agroforestry systems to improving soil carbon sequestration

Lydie-Stella Koutika, Research Centre on the Durability and the Productivity of Industrial Plantations (CRDPI), Republic of the Congo; Nicolas Marron, UMR 1434 Silva, INRAE Grand-Est Nancy, Université de Lorraine, AgroParisTech 54000 Nancy, France; and Rémi Cardinael, AIDA, University of Montpellier, CIRAD, Montpellier, France, CIRAD, UPR AIDA, Harare and University of Zimbabwe, Zimbabwe

1 Introduction

Considered as a climate-smart production system by the FAO (FAO, 2010), agroforestry systems (AFSs) are defined as complex agro-ecosystems combining trees and crops, or trees and pastures, within the same field. Agroforestry is a generic term that includes a wide variety of systems differing in tree and crop species arrangements depending on climate zone and region, making AFS classification challenging (Somarriba, 1992; Torquebiau, 2000). AFSs in the tropics often include livestock (FAO, 2010). AFSs are sometimes defined as 'trees outside forests'. Under this definition, AFSs may be present on about 46% of all agricultural lands supporting 30% of global rural populations (FAO, 2010; Zomer et al., 2016). AFSs cover approximately 1 billion hectares of land and contribute to the livelihoods of over 900 million people (Zomer et al., 2014). A review of a total of 72 scientific, peer-reviewed articles linked to biomass carbon (C) and soil organic C

http://dx.doi.org/10.19103/AS.2022.0106.19

(SOC) storage identified eight main types of AFS comprising alley cropping fallows, hedgerows, multistrata systems, parklands, shaded perennial crops, and silvoarable and silvopastoral systems (Cardinael et al., 2018a), but also much more complex classifications exist (Nair, 1985). Cardinael et al. (2018a) defined alley cropping as intercropping with small shrubs, often N_2-fixing trees, which are regularly pruned or coppiced, to improve soil fertility; this system is mainly found in tropical areas. In contrast, silvoarable systems are defined as arable crops associated with mainly large trees, used either for fuel/firewood production (poplar, for instance) or for timber wood (walnut trees, for instance); these systems are often found in temperate ecosystems, but also exist in subtropical India and China. In temperate regions, hedgerows, windbreaks and silvopastoral systems are the most widespread AFS. In the dry tropics, parklands (e.g. *Faidherbia albida* (Delile) A. Chev.) and fallows are common, while in humid tropics, complex agroforests and shaded perennial-crop systems are widespread.

Farming using AFS has been accepted as a C sequestration activity under the afforestation and reforestation approach (Albrecht and Kandji, 2003; Nair et al., 2009a,b) since the United Nations Framework Convention on Climate Change (UNFCCC) includes the use of C sequestration through afforestation and reforestation in greenhouse-gas offsetting (http://unfccc.int/essential_background/glossary/items/). AFSs have also been included as a greenhouse gas (GHG) mitigation strategy under the Kyoto Protocol since 2007 (Stavi and Lal, 2013; Baah-Acheamfour et al., 2017). The Kyoto Protocol was adopted on 11 December 1997 and entered into force on 16 February 2005, and has 192 signatory Parties (https://unfccc.int/kyoto_protocol). The Protocol initiated by the UNFCCC was designed to accompany industrialized countries, recognized as largely responsible for the current high levels of GHG emissions in the atmosphere, in their transition to regulate and lower GHG emissions in accordance with agreed individual targets.

Indeed, farming using AFSs is a promising C sequestration strategy according to the Intergovernmental Panel for Climate Change (IPCC, 2000, 2019). The C sequestration potential of AFS was estimated to be around 600×10^6 Mg C year^{-1} on an area of 630 million ha of infertile croplands by 2040, while the potential of other nature based solutions was much lower (i.e. 20×10^6 Mg C year^1 for wetland restoration, 50 Mg C year^1 for restoration of degraded land, 250 Mg C year^1 for forest management, 375 Mg C year^1 for grazing management, 20 Mg C year^1 for rice management and 150 Mg C year^1 for cropland management (IPCC, 2000; Jose, 2009; Abbas et al., 2017; Chapman et al., 2020). Between 2000 and 2010, tree cover on agricultural lands increased by 3.7% worldwide, resulting in an increase of more than 2×10^9 Mg C (2 Pg C) of stored biomass carbon (Zomer et al., 2016). AFS can also contribute to reducing the pressure on natural forests (Asaah et al., 2011; Getnet, 2020).

Direct and indirect pathways for C sequestration in soils can be distinguished (Nair et al., 2009b). The conversion of CO_2 into soil inorganic C compounds (calcium and magnesium carbonates) during inorganic chemical reactions is a direct process (see Chapters 7 and 23 of this book). Organic C accumulation by plants during photosynthesis and the subsequent transfer of C to the soil during the decomposition of plant material is an indirect pathway (Nair et al., 2009b; Chapter 3 of this book). AFSs improve SOC sequestration (Feliciano et al., 2018; Cardinael et al., 2018a) through indirect pathways, and also enhance soil faunal and microbial diversity, abundance and activity (Sileshi and Mafongoya, 2006; Marsden et al., 2019; Cardinael et al., 2019). When N_2-fixing species (NFS) are used in AFSs, SOC sequestration may be boosted by enhanced microbial activity in the litter and soil, thus favouring litter mineralisation (Bini et al., 2012; Santos et al., 2017a,b; Ortaş et al., 2017; Wang et al., 2017) and improving C cycling and SOC sequestration efficiency (Bini et al., 2012, 2013; Pereira et al., 2017, 2018). AFS have great potential to (1) increase SOC accumulation and ease adaptation and resilience to climate change (IPCC, 2000; Albrecht and Kandji, 2003; Verchot et al., 2007; Nair et al., 2009a; Nguyen et al., 2013; Mbow et al., 2014; van Noordwijk et al., 2021; Cardinael et al., 2021); (2) provide more food and income diversification for the growing populations, mainly in developing countries such as sub-Saharan Africa or Asia (Sileshi et al., 2008; Mbow et al., 2014; Prasad et al., 2016); (3) restore degraded lands (Sileshi et al., 2007; Verchot et al., 2007); and (4) conserve and boost ecosystem biodiversity while simultaneously improving human welfare (Sileshi et al., 2007, 2008; Asaah et al., 2011; Chowdhury et al., 2020).

Reviewing data from 86 published and peer-reviewed studies, Feliciano et al. (2018) argued that time since conversion to AFS was the major factor affecting biomass C sequestration; SOC sequestration, on the other hand, was mostly influenced by climate. Ma et al. (2020) found that previous land use and tree age were the most important drivers of SOC sequestration in AFSs. In a review of 15 publications, Kim et al. showed that the SOC sequestration rate was divided by four by the end of the first year after AFS establishment and then gradually diminished with time while SOC stocks reach a new equilibrium (Kim et al., 2016). Furthermore, AFSs have many advantages linked their ability to improve SOC sequestration, for example, enhancing the overall productivity and efficiency of a land-use system (Nair et al., 2009a).

The '4 per 1000' Initiative encourages the implementation of SOC sequestration practices considering the specificities of the targeted region, such as climate, social and economic factors, but also the barriers, risks and trade-offs (Rumpel et al., 2019). In this context, it should be considered that AFSs are diverse and their SOC sequestration potential is controlled by several factors like topography, climate, geographical area and socioeconomic context (Nair et al., 2009a; Soto-Pinto et al., 2010; Stainback et al., 2012; Feliciano et al., 2018; Corbeels et al., 2019). SOC sequestration in AFSs also strongly depends on species composition and soil characteristics (Asaah et al., 2011; Abbas et al., 2017; Ma et al., 2020).

This chapter shows to which extent climate, geographical regions, soil characteristics, type of AFS and technical factors such as the choice of planted species, tree density and tree management control SOC sequestration, both in terms of the amount of SOC stored and in terms of the SOC form. The chapter also compares SOC sequestration in AFSs and other agroecosystems in different geographical and climatic areas. The main objectives are (1) to highlight how AFSs may improve SOC sequestration relative to other practices or approaches; and (2) to disentangle the main factors determining SOC sequestration processes in AFSs, (3) to identify barriers and (4) provide recommendations on how to improve SOC sequestration in AFS.

2 Improved soil carbon sequestration in agroforestry relative to other systems

SOC sequestration potential in AFSs may be evaluated in comparison to other practices such as afforestation, reforestation or agriculture without trees (Nair et al., 2009a). Studies have shown that AFSs may sequester C in both soil and biomass and that they improve soil quality more than agricultural systems without trees (Feliciano et al., 2018; Chowdhury et al., 2020; Gusli et al., 2020; Ma et al., 2020). Abbas et al. (2017) argued that AFS maintain greater amounts of C in both above- and belowground biomass compared to grazing and pure cropping systems. AFSs do indeed boost SOC sequestration while improving soil quality (Isaac et al., 2012; Feliciano et al., 2018). Chowdhury et al. (2020) reported higher concentrations of soil organic matter (SOM), available phosphorus (P) and exchangeable potassium (K) in AFSs (4.75%, 12.17 μg g^{1}, 0.39 mg kg^{1}, respectively) than after reforestation (3.18%, 6.50 μg g^{1} and 0.21 mg kg^{1}, respectively) or in slash-and-burn plots (1.83%, 5.90 μg g^{1} and 0.03 mg kg^{1}, respectively) in the hilly regions of south-eastern Bangladesh. Globally, SOC stocks increase following land-use changes from less complex systems such as pure cropping systems to more complex ones like AFSs. The greater increase in SOC stocks found in AFSs converted from previous croplands compared to AFSs converted from previous grasslands may be due to the lower initial SOC stock found in croplands. Increasing SOC stocks is a slow process; it usually takes several years, or even decades, after a land-use change before a measurable change in SOC stocks can be observed (Poeplau et al., 2011; Ma et al., 2020).

Several recent meta-analyses have quantified the carbon sequestration potential of the AFS as compared to other land-use systems. de Stefano and Jacobson (2018) compared the C sequestration potential of transitioning from diverse land uses toward an AFS. A decrease in SOC stocks was reported for a conversion from forest to AFS (−26% and 24% at 0–15 cm and 0–30 cm soil depths, respectively), while SOC stocks after conversion from agriculture to AFS increased (+26%, 40% and 34% at 0–15 cm, 0–30 cm and 0–100 cm soil depths, respectively). Even though several meta-analyses have shown no difference in SOC for grasslands converted to silvopasture or even a loss of SOC in temperate regions (Mayer et al., 2022), an increase of 9% and 10% at 0–10 cm and 0–30 cm soil depths has been reported globally after grassland/pasture conversion to AFS, but also after conversion from uncultivated or other land uses to AFS (+25% at 0–30 cm with, however, a decrease of −23% at the 0–60 cm soil depth) (de Stefano and Jacobson, 2018). Indeed, silvopastoral AFSs generally do have the highest SOC stocks. Another global meta-analysis reported the highest increase in SOC stocks for silvopastoral AFSs compared to croplands or pastures (Shi et al., 2018). However, exceptions exist; for instance, under lowland humid tropical and subtropical conditions, multistrata AFSs may have the potential to sequester 12.6% more SOC than forests (Chatterjee et al., 2018, Fig. 1).

AFSs can store large amounts of C in both above- and belowground compartments (Cardinael et al., 2018a). Nevertheless, it is a well-documented fact that natural forests have a greater potential to sequester C than AFS. Indeed, they lose SOC and biomass C when converted to an AFS (Cardinael et al., 2018a; Chatterjee et al., 2018; Feliciano et al., 2018; de Stefano and Jacobson, 2018). The following land-use ranking in terms of average SOC stocks can be found in the literature: old forests (288 Mg C ha^{-1}) > mixed-species reforestation (181 Mg C ha^{-1}) > tree plantations (142 Mg C ha^{-1}) > grasslands (110 Mg C ha^{-1}) > AFS (99 Mg C ha^{-1}) > cropland (40 Mg C ha^{-1}) (Pandey, 2002; Albrecht and Kandji, 2003). Based on data reported in the literature for all climates and AFS types combined, Shi et al. (2018) also calculated a mean SOC stock in the AFS (at a 1-m soil depth) of 126 Mg C ha^{-1}, which is 19% more than in treeless agricultural systems. Comparing SOC sequestration among systems has also shown that sequestration potential depends on geographical zone. Chatterjee et al. (2018) showed that, compared to treeless agricultural systems, SOC stocks under AFS were higher by 27% in arid and semi-arid regions, by 26% in lowland humid tropics and by 5.8% in Mediterranean regions, and that they were lower by 5.3% in temperate conditions in the 0–100 cm soil layer (Fig. 1).

Tree root systems may induce C storage in very deep soil layers (Nair et al., 2009a; Ma et al., 2020; Cardinael et al. (2015b, 2018b). This deep rooting of agroforestry trees may be due to the competition with associated crops and to soil tillage (Cardinael et al., 2015a). SOC in AFSs may be sequestered in

Figure 1 Percentages of variation in SOC stocks (to 40 cm in depth) when AFSs are compared to agriculture, pasture, forest or uncultivated land in different agro-ecological regions, according to the meta-analysis by Chatterjee et al. (2018). Positive values indicate higher SOC stocks under AFS.

deeper layers since tree root systems can grow very deep, and this, in turn, results in a higher input of organic matter to the deeper soil layers (Nair et al., 2010; Germon et al., 2016). At the same time, high surface input of organic matter (e.g. leaf litter) from trees favours the production of dissolved organic C that can be transported to deeper soil layers, thereby contributing to subsoil C storage (Lorenz and Lal, 2005; Ma et al., 2020). In their meta-analysis, Shi et al. (2018) showed that SOC stocks were higher in AFSs (averaged over all AFS types) than in croplands in all soil layers, though the difference was only significant for the 60–80 soil layer. Nygren et al. (2012) argued that root nodule dynamics and N rhizo-deposition from N_2-fixing species in AFSs are sources of readily available N to crops and indirectly affect the dynamics of SOC sequestration (Van Groningen et al., 2017). However, the authors reported that these processes have been poorly investigated to date, even though they highlight the interactions between tree N rhizo-deposition to soil via exudates and root turnover.

Other soil factors play a role in the more efficient SCS in AFSs compared to more conventional agricultural systems. Several studies have shown an increase in soil aggregate stability in AFSs (Udawatta et al., 2008), and this should allow better stabilization and protection of soil C. Moreover, the abundance, biomass and diversity of earthworms are also higher in AFSs than in agricultural plots (Hauser, 1993; Price and Gordon, 1998; Cardinael et al., 2019), and this macrofauna plays a crucial role in nutrient cycling, litter decomposition and SOC stabilization in AFSs (Sileshi and Mafongoya, 2006; Marsden et al., 2019).

3 Factors driving soil carbon sequestration in agroforestry systems

SOC sequestration in AFSs is linked to various factors (Nair et al., 2009b). Four main groups of strongly interconnected factors can be identified: (1) climatic and topographic conditions (altitude, climate); (2) plant characteristics (tree species, diversity and age of planted material) and stand characteristics (tree density); (3) management factors (tillage, fertilization, residue management, pruning, thinning, harvesting regimes and holding size); and (4) soil characteristics (texture, structure, nutrient status, and physical, chemical and biological conditions). In addition, the effect of an AFS on SOC stocks increases with tree age and development (Ma et al., 2020).

3.1 Climatic and topographic conditions

The impact of climate is decisive for the SOC sequestering efficiency of an AFS. Some meta-analyses have reported more effective SOC sequestration in AFSs in the tropics than in temperate areas (Feliciano et al., 2018; Shi et al., 2018; Ma et al., 2020). Higher SOC sequestration has also been reported in temperate and tropical zones than in boreal or arid regions (Nair et al., 2009b; Ma et al., 2020), while SOC sequestration in the subtropical regions tends to be greater than in their tropical counterparts (Shi et al., 2018). The potential of sequestering C aboveground is also estimated to be greater in the tropics than in temperate AFSs (Abbas et al., 2017; Feliciano et al., 2018), while the contrary is observed for SOC sequestration (Fig. 2). Aboveground C sequestration in both temperate and tropical zones is higher than in boreal and arid areas (Ma et al., 2020). It takes about 2.5 times longer to sequester C in arid and semi-arid regions (intercropped tree systems) than in humid tropical regions (multistrata shaded perennial and silvopastoral systems), with the latter sequestering up to 20 times more C than the former (Nair et al., 2009b). Moreover, differences in SOC sequestration potential within agro-ecological regions in the tropics have been found (Nair et al., 2009b). SOC sequestration potential (down to 1 m in soil depth) after about 25 years ranges from 5 Mg C ha^{1} to 10 Mg C ha^{1} (intercropped systems with trees) in

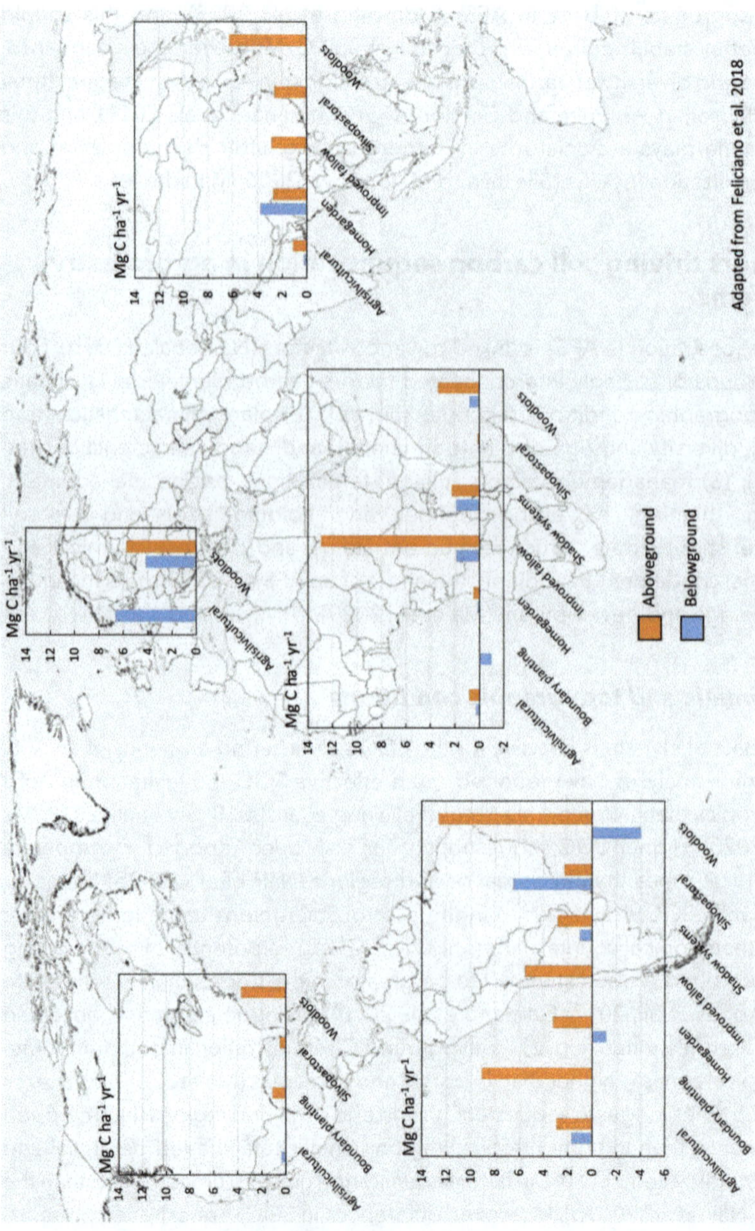

Figure 2 Mean above- and belowground biomass C sequestration (Mg C ha^{-1} year^{-1}) by AFS type and world region according to the meta-analysis by Feliciano et al. (2018).

the arid and semi-arid lands while in humid tropics, the potential ranges from 100 Mg C ha^{1} to 250 Mg C ha^{1} (multistrata shaded perennial systems and home gardens) after about 10 years (Nair et al., 2009b). In their meta-analysis, Feliciano et al. (2018) showed that the benefits of the AFS are greater in tropical climates than in other climates, both in terms of SOC and aboveground C stocks. AFSs in tropical zones have the ability to increase SOC quickly, whereas in temperate zones, SOC usually increases at a slower rate, though it peaks at a higher overall level than in other climates (Ma et al., 2020). Primary productivity, plant biomass and soil C input are typically greater in wet than in dry climates, whereas higher temperatures are associated with greater rates of decomposition and reduced SOC storage. Globally, C accumulation in both the biomass and the soil components in AFSs tends to differ among climatic zones, with overall positive effects from a high mean annual precipitation (MAP) and negative effects from a high mean annual temperature (MAT) (Ma et al., 2020).

3.2 Plant and stand characteristics

In conjunction with pedoclimatic factors, the ability of an AFS to store SOC also depends on the type of agroforestry implemented (choice of tree species, species diversity, etc. - Cardinael et al., 2018a; Feliciano et al., 2018; Hübner et al., 2021), although there may often be an interaction between agroforestry type and region or climate. Eight main AFSs were identified in a systematic review on above- and belowground C storage and SOC in AFSs (Table 1, Cardinael et al., 2018a).

Table 1 Mean aboveground (ABG) and belowground (BLG) biomass carbon accumulation rates (Mg C ha^{1} year^{1}, except for hedgerows: Mg C km^{1} year^{1}), with standard deviations, for all regions combined

	Number of studies	Tree density per hectare	ABG C accumulation rate	BLG C accumulation rate
Alley cropping systems	90	8568 ± 8403	2.37 ± 1.45	0.55 ± 0.34
Fallows	69	6074 ± 4529	4.42 ± 2.86	1.49 ± 1.56
Hedgerows	15	979 ± 824	0.79 ± 0.69	0.20 ± 0.18
Multistrata systems	51	929 ± 901	3.25 ± 2.54	0.80 ± 0.60
Parklands	7	152 ± 102	0.59 ± 0.46	0.21 ± 0.11
Shaded perennial-crop system	28	4236 ± 2347	2.40 ± 1.54	0.55 ± 0.36
Silvoarable systems	36	574 ± 440	2.56 ± 2.66	0.63 ± 0.67
Silvopastoral systems	28	1329 ± 981	2.70 ± 3.33	0.76 ± 0.89

Source: Adapted from Cardinael et al. (2018a).

(1) Alley cropping systems (or intercropping systems) are usually found in tropical regions and comprise fast-growing, usually leguminous, woody species (mainly shrubs) grown in crop fields, usually at high densities. The woody species are regularly pruned and the pruning residues are applied as mulch on the alleys as a source of organic matter and nutrients. Multistrata systems are multistorey combinations of a large number of various trees at high densities and perennial and annual crops. They include (2) home gardens and (3) agroforests and are mainly found in the humid tropics. In these regions, shaded perennial-crop systems are also common for shade-tolerant species such as cacao and coffee, grown under or between overstory shade trees that can be used for timber or other commercial products. In dry tropics like the Sahel, (4) parklands are widespread. In these systems, agricultural crops or grazing areas occur under scattered mature trees (e.g. *Faidherbia albida*). (5) Fallows are a type of sequential AFS including both natural and improved fallows.

In temperate regions, (6) hedgerows and (7) silvopastures are the most common AFSs. Hedgerows consist of linear plantations around the fields and also include shelterbelts, windbreaks and live fences. Silvopastures are systems where woody species are planted on permanent, often grazed, grasslands. Hedgerows and silvopastures are also commonly found in tropical regions. (8) Silvoarable systems comprise woody species planted in parallel rows at low densities, alternating with an annual crop. This system allows the use of mechanised equipment and, the trees can be harvested for timber (e.g. *Juglans* spp.), but also for fuel (e.g. *Populus* spp.). According to the meta-analysis by Feliciano et al. (2018), silvopastoral AFSs have high SOC stocks whereas in improved fallows, C sequestration in the aboveground biomass was preponderant. The SOC sequestration gained by planting trees in pastures is almost null due to high initial SOC stocks, some authors even reported a SOC loss following conversion of grasslands to silvopastures in temperate regions (Mayer et al., 2022).

In their review of the literature, Pellerin et al. (2020) calculated a mean SOC storage rate of 0.25 Mg C ha^{1} year^{1} for silvoarable systems under temperate conditions in North American and Europe (between −0.23 Mg C ha^{1} year^{1} and +0.73 Mg C ha^{1} year^{1}, for a mean soil depth of 36.4 cm, a mean tree density of 182 trees per ha and a mean tree age of 15.8 years; 25 study cases in 11 publications). This value is close to the one found by Cardinael et al. (2017) for five silvoarable systems in France (a mean of 0.24 Mg C ha^{1} year^{1}, with a range of 0.09 Mg C ha^{1} year^{1} to 0.46 Mg C ha^{1} year^{1}, for the 0–30 cm soil layer, a mean tree density of 75 trees par ha, and a mean tree age of 17.8 years). However, significant additional SOC sequestration in silvoarable systems has also been observed in the subsoil down to 1 m in depth (Cardinael et al., 2015b; Shi et al., 2018). For silvopastoral systems located in Europe and North America, Pellerin et al. (2020) found no significant effect of tree plantation on SOC stocks in perennial grasslands, with a mean value of 0.05 Mg C ha^{1} year^{1} (between −0.20 Mg C ha^{1} year^{1} and 0.29 Mg C ha^{1} year^{1}, for a mean soil depth of 42.9 cm, a mean tree density of 272 trees per ha and a mean tree age of 25.3 years; 12 study cases in 7 publications). Finally, for hedgerows, the mean SOC storage rate was 0.75 Mg C ha^{1} hedgerow year^{1} (between 0.49 Mg C ha^{1} hedgerow year^{1} and 1.02 Mg C ha^{1} hedgerow year^{1}, for a mean soil depth of 38.8 cm, a mean tree density of 739 trees per km and a mean tree age of 26.6 years). The same ranking was found for temperate AFS by Mayer et al. (2022) in their meta-analysis: hedgerows (0.32 ± 0.26 and 0.28 ± 0.15 Mg C ha^{1} year^{1} for topsoils and subsoils, respectively) > alley cropping systems (0.26 ± 1.15 and 0.23 ± 0.25 Mg C ha^{1} year^{1}) > silvopastoral systems (−0.17 ± 0.50 and 0.03 ± 0.26 Mg C ha^{1} year^{1}).

Tree diversity also has an impact on C sequestration. AFSs with multiple tree species contain greater biomass C stocks and accumulate biomass C faster than systems with only one tree species. Including multiple tree species in an AFS may promote a more efficient use of resources compared with a single-species AFS, thereby leading to greater net primary production and, consequently, greater C sequestration in above- and belowground biomass (Ma and Chen, 2017). Given that the positive effect of tree species diversity on biomass production increases over time as a result of increasing interspecific complementarity (Reich et al., 2012), the general pattern of biomass C accumulation as a function of tree age is expected to differ between AFSs with multiple tree species and those with a single tree species (Ma et al., 2020). Under temperate conditions, it has also been shown that SOC sequestration tended to be higher for AFS with broadleaf tree species compared to coniferous species (Mayer et al., 2022).

3.3 Management factors

Trees generally grow faster in AFSs than in forests due to weaker intra-specific competition and to fertilising inputs (Balandier and Dupraz, 1998). Thus, each individual tree's contribution to SOC storage is much greater (Pellerin et al., 2020). In the AFS, agricultural inputs concern fertilisers (chemical or organic amendments: manure, wastewater, sewage sludge, etc.) and/or animal dung and urine which also stimulate tree growth and their subsequent ability to store C. Fertilisation can also occur through symbiotic N_2 fixation, particularly when either herbaceous or woody species of the *Fabaceae* family (legumes) are a component of the AFS.

Introducing N_2-fixing species (NFS) in an AFS can lead to a more N self-sufficient system, as argued by Prasad et al. (2016) in a review on AFSs linked to food production and security in India, a country with a very long history of agroforestry practices. The importance of N in the SOC sequestration process is well known; it results from the interaction among the C, N and P cycles as well as constraints linked to stoichiometry (Van Groningen et al., 2017). In one study, the total amount of N_2 biologically fixed from the atmosphere by the inoculated *Leucaena leucocephala* (Lam) de Wit was estimated to be more than 500 kg of N ha[1] year[1] for the subsequent maize crop (Sanginga et al., 1986). Almost the same range was reported in another study: up to 472 kg N ha[1] year[1] for *L. leucocephala*, *Gliricidia sepium* and *Calliandra calothyrsus*, though lower values were found for *Acacia melanoxylon* and *Acacia holoserica* (<50 kg N ha[1] year[1]) (Giller, 2001).

Integrating NFS in AFSs not only improves soil N status, but also enhances the AFS's potential to sequester C and improve soil health and quality (Nair et al., 2009a; Kaonga, 2005; Lorenz and Lal, 2014; Sebukyu and Mosango, 2012; Dubiez et al., 2019). Enhanced SOC sequestration in NFS-based agroforestry may be attributed to the decomposition of large quantities of high-quality organic matter from N_2-fixing trees (von Haden et al., 2019). NFS may also transfer the atmospheric N_2 they fix to non-NFS species (Epron et al., 2013; Paula et al., 2015). This was demonstrated for pot culture on red Mediterranean soils where legume tree species facilitated N transfer to interspersed crops stimulated by high P conditions, thus evidencing the effective nutrient economy of a properly managed NFS-based AFS (Isaac et al., 2012). Under field conditions in north-eastern France, a succession of alfalfa and clover crops grown together with poplars significantly enhanced soil N, thus increasing growth and growing season length for the trees, compared to poplars in a monoculture (Thomas et al., 2020).

In most rural populations, the concept of increasing SOC sequestration to improve soil health and quality is unknown; most farmers verify and evaluate the benefits of practices through crop yields and income (ASB, 2000; Schroth and Sinclair, 2003; Degrande et al., 2007). To stimulate interest in adopting new methodologies, a project in north and northwest Cameroon promoted an agroforestry approach with N_2-fixing trees and selected cultivars of local fruit and nut trees with local and regional markets as using 'fertilizer trees' to improve soil quality and obtain higher incomes (Asaah et al., 2011).

3.4 Soil characteristics

The success of implementing an AFS obviously depends on the soil characteristics of the target area (Nair et al., 2009a,b; Hübner et al., 2021). The benefits of an AFS may be more perceptible in unproductive or degraded lands than on richer or rehabilitated parcels, especially when N_2-fixing trees are used (Bisiaux et al., 2009; Chowdhury et al., 2020). An agroforestry project of 8000 hectares including *Acacia auriculiformis* A. Cunn. ex Benth. established from 1987 to 1992 on the infertile sandy soils of the Batéké plateau in the Democratic Republic of Congo led to both improved soil quality (increased C and N concentrations and increased soil acidity) and an increase in crop yields (cassava, maize) and honey production (Bisiaux et al., 2009; Nsombo, 2016). The higher SOM turnover than commonly observed in similar ecosystems were linked to soil intrinsic characteristics in addition to the higher quality and quantity of aboveground biomass production. There may also be higher SOM decomposition or mineralisation in tropical ecosystems due to high temperatures and mean annual rainfall (Hassink, 1997; Koutika et al., 1999). These processes may be enhanced in ecosystems with fragile soils with a limited ability to store and protect added SOM due, for instance, to their coarse texture or low nutrient availability (Derrien et al., 2014). The ability of the silt and clay particles in the soil to protect organic material, obviously depends on intrinsic soil characteristics (Hassink, 1997; Koutika et al., 1999) and is strongly linked to physical soil characteristics (Nair et al., 2009a,b; Chowdhury et al., 2020). Coarse textured and/or disturbed soils contain few organic materials and have a poor structure and aggregation (Elliott and Coleman, 1988; Six et al., 2000). In addition to the above-mentioned benefits on soil attributes observed in AFSs, especially when NFS are introduced, AFS have a positive impact on the diversity and functioning of soil invertebrates (Sileshi and Mafongoya, 2006; Marsden et al., 2019). The combined soil benefits provided by implementing an AFS usually lead to enhanced faunal and microbial activity, more robust arbuscular mycorrhizal fungi and better nutrient cycling (Bini et al., 2012, 2013; Pereira et al., 2017, 2018, 2020; Ortaş et al., 2017; Wang et al., 2017; Dubiez et al., 2019).

4 Other co-benefits of sequestering soil carbon in agroforestry systems

Food availability and security are two of the expected co-benefits of improving SOC sequestration through AFSs, which could benefit poorer populations in some regions of Africa or Asia (Mbow et al., 2014; Prasad et al., 2016). For certain types of AFSs, increases in crop yields can be expected. Food security occurs through increased crop yields following an improvement in soil quality (Sileshi et al., 2007, 2008). Yields for maize, a staple crop in many African countries, increased from 1.3 Mg ha^{1} year^{1} to 1.6 Mg ha^{1} year^{1} in AFSs in sub-Saharan Africa as compared to pure cropping systems (Sileshi et al., 2008). In the case mentioned above (Sileshi et al., 2008), improved fallows and coppicing woody legumes were key as there was a temporal association with no competition. This also occurs in parklands with *Faidherbia albida* where no competition occurs since the trees bear leaves only during the dry season when there are no crops; in addition, soil N status is improved through N_2 fixation by the NFS. However, in most cases, one can expect competition to be higher than facilitation and crop yields to decrease in an AFS. In addition to the above-mentioned benefits of SOC sequestration in terms of improved soil fertility and

Table 2 Improved SOC sequestration in agroforestry systems and co-benefits in different parts of the world according to the literature

		Benefits of improved SOC sequestration			Other co-benefits of improved C sequestration			
Region and climate	Type of paper	Soil fertility and land restoration	Food availability and security	Adaptation to climate change and resilience	Fuel wood energy	Forest products	C payment markets	References
Overall	Review	+	+	+	+			Abbas et al. (2017)
Cameroon	Review	+	+	+	+	+		Asaah et al. (2011)
Tropics	Review	+	+	+	+			Albrecht and Kandji (2003)
Overall	Review	+					+	Cardinael et al. (2018a)
Bangladesh	Research	+		+				Chowdhury et al. (2020)
DR Congo	Research	+	+		+	+		Dubiez et al. (2019)
Overall	Meta-analysis	+	+	+	+	+		Feliciano et al. (2018)
Ethiopia	Review			+			+	Getnet (2020)
Indonesia	Research	-	+	+				Gusli et al. (2020)
Central Africa	Research						-	Lescuyer et al. (2009)
Overall	Review	+	+	+	+	+		Lorenz and Lal (2014)
Africa	Review	+	+	+	+	+		Mbow et al. (2014)
Overall	Review	+	+	+			+	Nair et al. (2009a,b)
India	Review	+	+	+	+	+		Prasad et al. (2016)
Uganda	Research	+	+	+	+	+		Sebukyu and Mosango (2012)
Eastern and Southern Africa	Review	+	+	+	+	+	+	Sileshi et al. (2007)
Mexico	Research	+	+	+		+		Soto-Pinto et al. (2010)
Rwanda	Review	+	+			+	+	Stainback et al. (2012)
Overall	Review			+				Verchot et al. (2007)

biodiversity, AFSs can also provide tree products, i.e. fodder, fuelwood, food and building materials (Sileshi et al., 2007, 2008; Nair et al., 2009a; Lescuyer et al., 2009; Sebukyu and Mosango, 2012; Gusli et al., 2020, Table 2). Other ecosystem services such as land restoration and biodiversity protection are also provided by AFSs (Lescuyer et al., 2009; Sebukyu and Mosango, 2012; Mbow et al., 2014). In some tropical areas, AFSs help to decrease pressure on forests for fuel wood energy (Sileshi et al., 2007; Prasad et al., 2016). C sequestration could also be an economic opportunity for subsistence farmers in developing countries if trading C sequestered through agroforestry activities becomes more widespread (Lescuyer et al., 2009; Nair et al., 2009b; Feliciano et al., 2018).

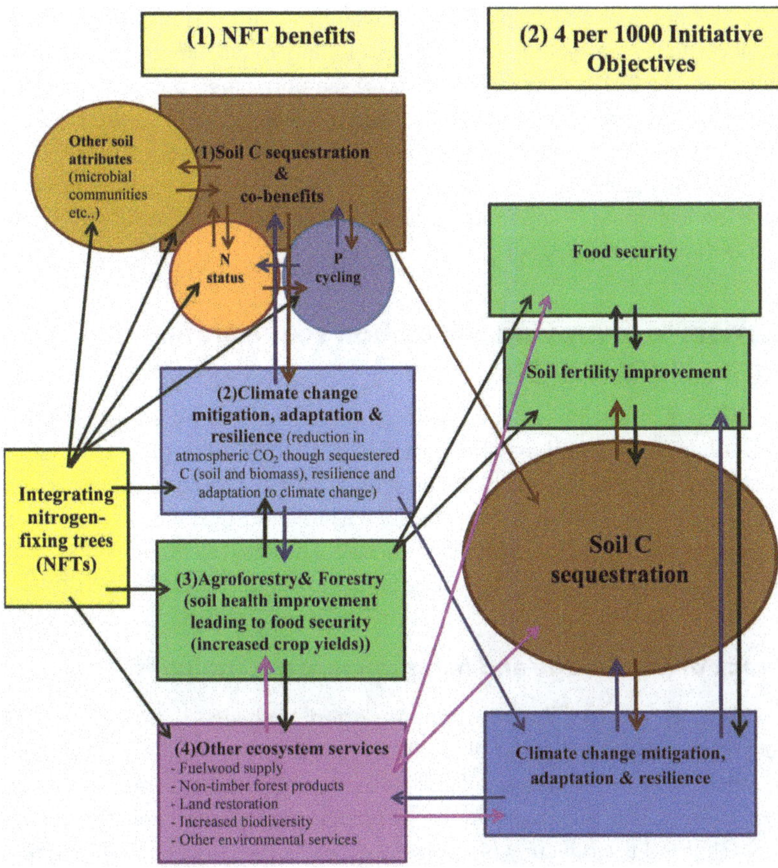

Figure 3 Conceptual diagram showing (1) N-fixing tree (NFT) benefits and links to (2) the '4 per 1000' Initiative objectives: how introducing NFTs in agroforestry and forestry leads to C sequestration and co-benefits in other ecosystem services, and promotes the '4 per 1000' Initiative in the Congo Basin (DR Congo and Rep. Congo). Source: Adapted from Koutika et al. (2021).

The co-benefits of sequestering C in AFSs in Central Africa (mainly in DR Congo and the Republic of the Congo) have been summarized in a review (Koutika et al., 2021). The authors developed a conceptual scheme of the benefits of N_2-fixing trees related to the main objectives of the '4 per 1000' 'Soils for food security and climate' Initiative (Fig. 3). C sequestration, with its links to N and P cycling, microbial communities and other soil attributes, helps to control not only the quantity but also the quality of sequestered SOC (Bini et al., 2012, 2013; Pereira et al., 2018, 2020; Koutika et al., 2019, 2020a,b). Not only climate change mitigation, by reducing atmospheric CO_2 through SOC sequestration, but also improved resilience and adaptation to climate change are also benefits of planting N_2-fixing trees and implementing agroforestry practices (Verchot et al., 2007; Stavi and Lal, 2013; Baah-Acheamfour et al., 2017). Well-managed AFSs play an important role in boosting resilience and adaptation to a changeable climate by creating microclimates, for instance (Nguyen et al., 2013; Mbow et al., 2014), among other benefits related to adaptation and resilience to climate change (Arenas-Corraliza et al., 2018; Blanchet et al., 2021; van Noordwijk et al., 2021; Cardinael et al., 2021). In the AFOLU (Agriculture, Forestry and Other Land Use) sector, mitigation and adaptation to climate change have been identified as a necessity in developing countries (Mbow et al., 2014).

5 Barriers to improving soil carbon sequestration in agroforestry systems

Some barriers lessening, or even halting, the ability of an AFS to sequester C have been identified and may be classified in two main categories: (1) lack of knowledge and measurement uncertainties, and (2) socioeconomic factors (linked to social and political contexts such as relationships with policy makers or governments, lack of funds or infrastructure. This may lead to absence of AFS projects in regions with the greatest potential to sequester C (Table 3).

5.1 Lack of knowledge and measurement uncertainties

Even though studies have often reported enhanced biological SOC sequestration in AFSs (Nair et al., 2009a), estimating or evaluating biological SOC stocks in AFSs may be a difficult task due to limited accuracy (Nair et al., 2009a) or technical constraints in the poorest regions. Cardinael et al. (2018a) also highlighted the difficulty of correctly estimating changes in SOC stocks in AFSs due to the lack of rigorous long-term measurements, as already reported by Nair et al. (2009a) and Nair (2012). Even when SOC is measured in AFSs, it is rarely compared to an adequate reference system (Lorenz and Lal, 2014; Cardinael et al., 2018a; Feliciano et al., 2018).

Table 3 Main barriers to improving SOC sequestration in agroforestry systems

Types of barriers	References
(i) Lack of knowledge, technical barriers Incorrect measurements of sequestered C Lack of ability to measure sequestered C Inappropriate AFS implementation (wrong choice of tree species, animal damage, etc.) Lack of access to plant seedlings, irrigation or soil drainage systems, mechanization, inappropriate field size, etc.	Balandier and Dupraz (1998), Cardinael et al. (2018a), Chatterjee et al. (2018), Feliciano et al. (2018), Lorenz and Lal (2014), Nair et al. (2009a), Shi et al. (2018)
(ii) Socioeconomic barriers Social and political contexts Policymakers (lack of policies) Governments (insufficient support of infrastructure, etc.) Lack of training or extension services Lack of funds, infrastructure or AFS-enabling projects in the regions with the greatest potential to sequester C Land tenure issues	Nair et al. (2009a), Mbow et al. (2014), Stainback et al. (2012), Bucagu et al. (2013), Foundjem-Tita et al. (2013)

Many studies on SOC sequestration in AFSs are based on biomass and soil sampling undertaken at a specific period, others are based on field measurements and others use simulation models based on various assumptions. Almost all the studies to date report the amount of SOC stocks for specific systems and only at the time of study. While these data are extremely valuable for understanding existing conditions, they represent only one point in the continuum of land-use changes. They are not sufficient when the goal is to predict the rate of change in SOC when land-management practices change from agriculture or forest to an AFS. Long-term monitoring information is needed to properly assess the benefits, not only in terms of SOC but also in terms of the economic advantages associated with the adoption of AFS for local farmers (Chatterjee et al., 2018).

Shi et al. (2018) identified the following important knowledge gaps regarding SOC sequestration in AFS: (1) how to quantitatively assess C allocation and stocks by various AFS components, with a special focus on deep soil C and its dependence on tree species and age; (2) how to optimise the relative area allocated to trees and crops within each AFS type to maximise C sequestration, yield, ecosystem services and to improve environmental conditions; and (3) how to develop new remote sensing approaches to distinguish AFSs from the background of forests, plantations and gardens, and to quantify the AFS area. An investigation of several forestry and agroforestry plantations in the Mediterranean area, in dry central temperate plains, in cold wet central uplands and in mild oceanic areas revealed that a wrong choice of species for the prevailing local conditions may cause some AFSs to fail (Balandier and Dupraz, 1998). Finally, a comprehensive understanding of what happens belowground in the AFS is also lacking, for all its dimensions (temporal, spatial, soil depth). Studies integrating AFS functioning as a whole, embracing the very diverse forms of AFS and considering the multifunctionality of these systems are urgently needed (Cardinael et al., 2020).

5.2 Socioeconomic barriers

There are also socioeconomic barriers that must be overcome, mainly in developing or poor countries in sub-Saharan Africa, to enable SOC sequestration improvement (Demenois et al., 2020) through AFS introduction (Nair et al., 2009a; Stainback et al., 2012; Bucagu et al., 2013; Foundjem-Tita et al., 2013). Success in implementing an AFS and its ability to improve soil C sequestration strongly relies on different factors such as land tenure, policies, technical advice and assistance, government support, etc. Stainback et al. (2012) linked AFS success for smallholders in Rwanda to favourable support from the government and the efficient technical capacity to design, implement and monitor carbon projects included in environmental service markets. In order to overcome barriers to adoption, a participatory approach allowing smallholders to share their knowledge and identify key opportunities, problems and constraints seems appropriate (Bucagu et al., 2013). Other barriers can be a lack of funding, infrastructure or specific AFS projects in regions with the greatest potential to sequester C (Mbow et al., 2014; Stainback et al., 2012; Bucagu et al., 2013; Foundjem-Tita et al., 2013). Rosenstock et al. (2019) formulated four recommendations in order to operationalize broad-scale measurement, reporting and verification (MRV) systems for AFSs under the UNFCCC: (1) develop accessible approaches for representation of lands with AFS; (2) create guidelines for AFS reporting to improve relevance to national policy and transparency; (3) develop carbon-stock change and emission databases relevant for reporting requirements; and (4) navigate the institutional arrangements needed to include agroforestry in MNV.

6 Recommendations

To improve the positive impact of AFSs on SOC sequestration, not only the type of AFS but also the environmental, climatic, socioeconomic and political contexts must be considered prior to adoption and implementation of such systems. SOC sequestration is related to soil characteristics (content of other nutrients such as N and P, soil texture, etc.) and strongly linked to the factors mentioned above. The way the AFS is managed will have a crucial impact on the quantities of C stored in the soil. For example, because of the low density of the trees in agroforestry, there is no natural pruning as there is in the forest. Trees must be pruned, not only to produce wood, but also to limit their competition with crops and facilitate the passage of agricultural machinery.

The management of pruning residues can be a lever to increase SOC storage, especially if they are crushed and brought to the ground in the form of ramial chipped wood (Fungo et al., 2021). Moreover, alignments of trees induce, both directly (input of organic matter from the trees) and indirectly (sub-tree herbaceous vegetation), a heterogeneous distribution of soil carbon. The largest SOC stocks are usually observed near trees rather than in the middle of the cultivated inter-row (Bambrick et al., 2010; Cardinael et al., 2015b). Therefore, managing the sub-tree vegetation in silvoarable systems (sowing or spontaneous cover, mowing or not, etc.) seems crucial for SOC storage. Finally, AFSs should also be managed to increase the SOC reservoir by avoiding burning and by minimizing soil disturbance through limited or zero tillage practices and appropriate erosion control measures (Soto-Pinto et al., 2010).

7 Conclusion

In most situations, AFSs store C (especially in deeper soil layers) more efficiently than croplands but less efficiently than forests. Diversified AFSs in terms of tree species are more efficient for SOC sequestration due to the complementarity among species for the use of resources. Regional climate also has a strong impact, with tropical AFSs being, on average, more efficient than temperate AFSs thanks to the higher biomass production and greater soil C sequestration occurring under wetter conditions. SOC sequestration also depends on the type of AFS: in temperate latitudes, literature reviews have shown that silvoarable systems are, on average, more efficient than silvopastoral systems in storing additional C in soils because pastures already have high SOC stocks. The use of woody or herbaceous N_2-fixing species (NFS) can also be a way to enhance SOC sequestration by stimulating the growth of the non-fixing species in the association. Recommendations to improve the probability of successful SOC sequestration in AFSs include (1) carefully matching the selected tree species to local pedoclimatic conditions in order to maximize plant growth and subsequent C storage, and (2) applying good plantation management practices by, for instance, reducing soil disturbance as much as possible, mulching with fragmented tree pruning residues, and adequately managing sub-tree vegetation (the area where the largest SOC stocks are found).

8 Where to look for further information

8.1 Some guidelines for future research

- More studies including socio-economic aspects of agroforestry systems and linked to sustainable development goals of Agenda 2030 of the United Nations.

- Estimate appropriate agroforestry system according to specific agro ecological zone taking into account climate warming.

8.2 Organisations involved with respect to this topic

- CRDPI (Congo), INRAE (France) and CIRAD (France), University of Zimbabwe (Zimbabwe).

9 References

Abbas, F., Hammad, H. M., Fahad, S., Cerdà, A., Rizwan, M., Farhad, W., Ehsan, S. and Bakhat, H. F. (2017). Agroforestry: a sustainable environmental practice for carbon sequestration under the climate change scenarios - a review, *Environ. Sci. Pollut. Res. Int.* 24(12), 11177–11191. DOI: 10.1007/s11356-017-8687-0.

Albrecht, A. and Kandji, S. T. (2003). Carbon sequestration in tropical agroforestry systems, *Agric. Ecosyst. Environ.* 99(1–3), 15–27.

Arenas-Corraliza, M. G., López-Díaz, M. L. and Moreno, G. (2018). Winter cereal production in a Mediterranean silvoarable walnut system in the face of climate change, *Agric. Ecosyst. Environ.* 264, 111–118. DOI: 10.1016/j.agee.2018.05.024.

Asaah, E. K., Tchoundjeu, Z., Leakey, R. R. B., Takousting, B., Njong, J. and Edang, I. (2011). Trees, agroforestry and multifunctional agriculture in Cameroon, *Int. J. Agric. Sustain.* 9(1), 110–119. DOI: 10.3763/ijas.2010.0553.

ASB (2000). Summary report and synthesis of phase II in Cameroon. Alternatives-to-Slash-and-Burn, Nairobi, Kenya, pp. 2–7.

Baah-Acheamfour, M., Chang, S. X., Bork, E. W. and Carlyle, C. N. (2017). The potential of agroforestry to reduce atmospheric greenhouse gases in Canada: insight from pairwise comparisons with traditional agriculture, data gaps and future research, *The Forestry Chronicle* 93(2), 180–189.

Balandier, P. and Dupraz, C. (1998). Growth of widely spaced trees: a case study from young agroforestry plantations in France, *Agroforest. Syst.* 43(1/3), 151–167.

Bambrick, A. D., Whalen, J. K., Bradley, R. L., Cogliastro, A., Gordon, A. M., Olivier, A. and Thevathasan, N. V. (2010). Spatial heterogeneity of soil organic carbon in tree-based intercropping systems in Quebec and Ontario, Canada, *Agroforest. Syst.* 79(3), 343–353.

Bini, D., Figueiredo, A. F., da Silva, M. C. P., de Figueiredo Vasconcellos, R. L. and Cardoso, E. J. B. N. (2012). Microbial biomass and activity in litter during the initial development of pure and mixed plantations of *Eucalyptus grandis* and *Acacia mangium*, *Rev. Bras. Cienc. Solo* 37, 76–85.

Bini, D., Santos, C. Ad, Bouillet, J., Gonçalves, J. LdM. and Cardoso, E. J. B. N. (2013). *Eucalyptus grandis* and *Acacia mangium* in monoculture and intercropped plantations: evolution of soil and litter microbial and chemical attributes during early stages of plant development, *Appl. Soil Ecol.* 63, 57–66.

Bisiaux, F., Peltier, R. and Muliele, J. C. (2009). Plantations industrielles et agroforesterie au service des populations des plateaux Batéké, Mampu, en République Démocratique du Congo, *Bois for trop* 301(3), 21–32.

Blanchet, G., Barkaoui, K., Bradley, M., Dupraz, C. and Gosme, M. (2021). Interactions between drought and shade on the productivity of winter pea grown in a 25-year-old walnut-based alley cropping system, *J. Agron. Crop Sci.*, 1–16. DOI: 10.1111/jac.12488.

Bucagu, C., Vanlauwe, B., Van Wijk, M. T. and Giller, K. E. (2013). Assessing farmers' interest in agroforestry in two contrasting agro-ecological zones of Rwanda, *Agroforest. Syst.* 87(1), 141–158. DOI: 10.1007/s10457-012-9531-7.

Cardinael, R., Mao, Z., Prieto, I., Stokes, A., Dupraz, C., Kim, J. H. and Jourdan, C. (2015a). Competition with winter crops induces deeper rooting of walnut trees in a Mediterranean alley cropping agroforestry system, *Plant Soil* 391(1–2), 219–235. DOI: 10.1007/s11104-015-2422-8.

Cardinael, R., Chevallier, T., Barthès, B. G., Saby, N. P. A., Parent, T., Dupraz, C., Bernoux, M. and Chenu, C. (2015b). Impact of alley cropping agroforestry on stocks, forms and spatial distribution of soil organic carbon - a case study in a Mediterranean context, *Geoderma* 259–260, 288–299. DOI: 10.1016/j.geoderma.2015.06.015.

Cardinael, R., Chevallier, T., Cambou, A., Béral, C., Barthès, B. G., Dupraz, C., Durand, C., Kouakoua, E. and Chenu, C. (2017). Increased soil organic carbon stocks under agroforestry: a survey of six different sites in France, *Agric. Ecosyst. Environ.* 236, 243–255. DOI: 10.1016/j.agee.2016.12.011.

Cardinael, R., Umulisa, V., Toudert, A., Olivier, A., Bockel, L. and Bernoux, M. (2018a). Revisiting IPCC Tier 1 coefficients for soil organic and biomass carbon storage in agroforestry systems, *Environ. Res. Lett.* 13(12), 124020. DOI: 10.1088/1748-9326/aaeb5f.

Cardinael, R., Guenet, B., Chevallier, T., Dupraz, C., Cozzi, T. and Chenu, C. (2018b). High organic inputs explain shallow and deep SOC storage in a long-term agroforestry system - combining experimental and modeling approaches, *Biogeosciences* 15(1), 297–317. DOI: 10.5194/bg-15-297-2018.

Cardinael, R., Hoeffner, K., Chenu, C., Chevallier, T., Béral, C., Dewisme, A. and Cluzeau, D. (2019). Spatial variation of earthworm communities and soil organic carbon in temperate agroforestry, *Biol. Fertil. Soils* 55(2), 171–183. DOI: 10.1007/s00374-018-1332-3.

Cardinael, R., Mao, Z., Chenu, C. and Hinsinger, P. (2020). Belowground functioning of agroforestry systems: recent advances and perspectives, *Plant Soil* 453(1–2), 1–13. DOI: 10.1007/s11104-020-04633-x.

Cardinael, R., Cadis, G., Gosme, M., Oelbermann, M. and van Noordwijk, M. (2021). Climate change mitigation and adaptation in agriculture: why agroforestry should be part of the solution, *Agric. Ecosyst. Environ.* 319, 107555. DOI: 10.1016/j.agee.2021.107555.

Chapman, M., Walker, W. S., Cook-Patton, S. C., Ellis, P. W., Farina, M., Griscom, B. W. and Baccini, A. (2020). Large climate mitigation potential from adding trees to agricultural lands, *Glob. Chang. Biol.* 26(8), 4357–4365. DOI: 10.1111/gcb.15121.

Chatterjee, N., Nair, P. K. R., Chakraborty, S. and Nair, V. D. (2018). Changes in soil carbon stocks across the forest-agroforest-agriculture/pasture continuum in various agroecological regions: a meta-analysis, *Agr. Eco. Environ.* 266, 55–67.

Chowdhury, F. I., Barua, I., Chowdhury, A. I., Resco de Dios, V. and Alam, M. S. (2020). Agroforestry shows higher potential than reforestation for soil restoration after slash-and-burn: a case study from Bangladesh, *Geology Ecology and Landscapes* 6(1), 48–54. DOI: 10.1080/24749508.2020.1743528.

Corbeels, M., Cardinael, R., Naudin, K., Guibert, H. and Torquebiau, E. (2019). The 4 per 1000 goal and soil carbon storage under agroforestry and conservation agriculture systems in sub-Saharan Africa, *Soil Till. Res.* 188, 16–26. DOI: 10.1016/j.still.2018.02.015.

Degrande, A., Asaah, E., Tchoundjeu, Z., Kanmegne, J., Duguma, B. and Franzel, S. (2007). Opportunities for and constraints to adoption of improved fallows: ICRAF's experience in the humid tropics of Cameroon. In: *Advances in Integrated Soil Fertility Management in Sub-Saharan Africa: Challenges and Opportunities* Bationo, A.,Waswa, B., Kihara, J. and Kimetu, J. (Eds), Springer, The Netherlands, pp. 901–909.

Demenois, J., Torquebiau, E., Arnoult, M. H., Eglin, T., Masse, D., Assouma, M. H., Blanfort, V., Chenu, C., Chapuis-Lardy, L., Medoc, J. and Sall, S. N. (2020). Barriers and strategies to boost soil carbon sequestration in agriculture, *Front. Sustain. Food Syst.* 4, 1–14. DOI: 10.3389/fsufs.2020.00037.

Derrien, D., Plain, C., Courty, P. E., Gelhaye, L., Moerdijk-Poortvliet, T. C. W., Thomas, F., Versini, A., Zeller, B., Koutika, L. S., Boschker, H. T. S. and Epron, D. (2014). Does the addition of labile substrate destabilise old soil organic matter?, *Soil Biol. Biochem.* 76, 149–160.

de Stefano, A. and Jacobson, M. G. (2018). Soil carbon sequestration in agroforestry systems: a meta-analysis, *Agroforest. Syst.* 92, 285–299. DOI: 10.1007/s10457-017-0147-9.

Dubiez, E., Freycon, V., Marien, J. M., Peltier, R. and Harmand, J. M. (2019). Long term impact of *Acacia auriculiformis* woodlots growing in rotation with cassava and maize on the carbon and nutrient contents of savannah sandy soils in the humid tropics (Democratic Republic of Congo), *Agroforest. Syst.* 93(3), 1167–1178. 10.1007/s10457-018-0222-x.

Elliott, E. T. and Coleman, D. C. (1988). Let the soil work for us, *Ecol. Bull.* 39, 23–32.

Epron, D., Nouvellon, Y., Mareschal, L., Moreira, R. Me, Koutika, L.-S., Geneste, B., Delgado-Rojas, J. S., Laclau, J. P., Sola, G., Gonçalves, J. LdM. and Bouillet, J. P. (2013). Partitioning of net primary production in *Eucalyptus* and *Acacia* stands and in mixed-species plantations: two case-studies in contrasting tropical environments, *Forest Ecol. Manag.* 301, 102–111.

FAO (Food and Agriculture Organization of the United Nations) (2010). Climate-Smart Agriculture policies, practices and financing for food security, adaptation and mitigation. Available at: http://www.fao.org/docrep/013/i1881e/i1881e00.pdf.

Feliciano, D., Ledo, A., Hillier, J. and Nayak, D. R. (2018). Which agroforestry options give the greatest soil and above ground carbon benefits in different world regions?, *Agric. Ecosyst. Environ.* 254, 117–129. DOI: 10.1016/j.agee.2017.11.032.

Foundjem-Tita, D., Tchoundjeu, Z., Speelman, S., D'Haese, M., Degrande, A., Asaah, E., van Huylenbroeck, G., van Damme, P. andNdoye, O. (2013). Policy and legal frameworks governing trees: incentives or disincentives for smallholder tree planting decisions in Cameroon?, *Small-Scale For.* 12(3), 489–505. DOI: 10.1007/s11842-012-9225-z.

Fungo, B., Wiesmeier, M. and Cardinael, R. (2021). Agroforestry 1: agrisilvicultural systems. In: *Recarbonizing Global Soils: A Technical Manual of Recommended Management Practices (vol. 3), Cropland, Grassland, Integrated Systems and Farming Approaches - Practices Overview*, Food and Agriculture Organization and ITPS, Rome, pp. 474–486. DOI: 10.4060/cb6595en.

Germon, A., Cardinael, R., Prieto, I., Mao, Z., Kim, J., Stokes, A., Dupraz, C., Laclau, J. P. and Jourdan, C. (2016). Unexpected phenology and lifespan of shallow and deep fine roots of walnut trees grown in a silvoarable Mediterranean agroforestry system, *Plant Soil* 401(1–2), 409–426. DOI: 10.1007/s11104-015-2753-5.

Getnet, D. (2020). GHGs reduction capacity of agroforestry systems in tropical Africa: a review, *Curr. Agri. Res. Jour.* 8(3), 166–177. DOI: 10.12944/CARJ.8.3.02.

Giller, K. E. (2001). *Nitrogen Fixation in Tropical Cropping Systems* (2nd edn.), CAB International Publishing, Wallingford.

Gusli, S., Sumeni, S., Sabodin, R., Muqfi, I. H., Nur, M., Hairiah, K., Useng, D. and van Noordwijk, M. (2020). Soil organic matter, mitigation of and adaptation to climate change in cocoa-based agroforestry systems, *Land* 9(9), 323. DOI: 10.3390/land9090323.

Hassink, J. (1997).The capacity of soils to preserve organic C and N by their association with clay and silt particles, *Plant Soil* 191(1), 77–87.

Hauser, S. (1993). Distribution and activity of earthworms and contribution to nutrient recycling in alley cropping, *Biol. Fertil. Soils* 15(1), 16–20.

Hübner, R., Kühnel, A., Lu, J., Dettmann, H., Wan, W. and Wiesmeier, M. (2021). Soil carbon sequestration by agroforestry systems in China: a meta-analysis, *Agric. Ecosyst. Environ.* 315, 107437. DOI: 10.1016/j.agee.2021.107437.

IPCC (2019). Summary for policymakers. In: *Climate Change and Land: an IPCC Special Report on Climate Change, Desertification, Land Degradation, Sustainable Land Management, Food Security, and Greenhouse Gas Fluxes in Terrestrial Ecosystems'* Shukla, P. R., Skea, J., Calvo Buendia, E., Masson-Delmotte, V., Pörtner, H.- O., Roberts, D. C., Zhai, P., Slade, R., Connors, S., van Diemen, R., et al. (Eds). Cite Report – Special Report on Climate Change and Land (ipcc.ch).

IPCC (2000). Land-Use, Land-Use Change and Forestry. Special Report of the Intergovernmental Panel on Climate Change, Cambridge University Press, Cambridge, p. 375.

Isaac, M. E., Hinsinger, P. and Harmand, J. M. (2012). Nitrogen and phosphorus economy of a legume tree-cereal intercropping system under controlled conditions, *Sci. Total Environ.* 434, 71–78. DOI: 10.1016/j.scitotenv.2011.12.071.

Jose, S. (2009). Agroforestry for ecosystem services and environmental benefits: an overview, *Agroforest. Syst.* 76(1), 1–10.

Kaonga, M. L. (2005). Understanding carbon dynamics in agroforestry systems in eastern Zambia. Ph.D. Dissertation, Fitzwilliam College, University of Cambridge, UK, p. 431.

Kim, D. G., Kirschbaum, M. U. F. and Beedy, T. L. (2016). Carbon sequestration and net emissions of CH_4 and N_2O under agroforestry: synthesizing available data and suggestions for future studies', *Agric. Ecosyst. Environ.* 226, 65–78. DOI: 10.1016/j.agee.2016.04.011.

Koutika, L.-S., Chone, T., Andreux, F., Burtin, G. and Cerri, C. C. (1999). Factors influencing carbon decomposition of topsoils from the Brazilian Amazon Basin, *Biol. Fertil. Soils* 28(4), 436–438.

Koutika, L. S., Ngoyi, S., Cafiero, L. and Bevivino, A. (2019). Soil organic matter quality along rotations in acacia and eucalyptus plantations in the Congolese coastal plains, *Forest Ecosyst.* 6, 39. DOI: 10.1186/s40663-019-0197-8.

Koutika, L. S., Cafiero, L., Bevivino, A. and Merino, A. (2020a). Organic matter quality of forest floor as a driver of C and P dynamics in acacia and eucalypt plantations established on a ferralic arenosols, Congo, *Forest Ecosyst.* 7(1), 40. DOI: 10.1186/s40663-020-00249-w.

Koutika, L. S., Fiore, A., Tabacchioni, S., Aprea, G., Pereira, A. PdA. and Bevivino, A. (2020b). Influence of *Acacia mangium* on soil fertility and bacterial community in *Eucalyptus* Plantations in the Congolese Coastal Plains, *Sustainability* 12(21), 8763. DOI: 10.3390/su12218763.

Koutika, L.-S., Taba, K., Ndongo, M. and Kaonga, M. (2021). Nitrogen-fixing trees increase organic carbon sequestration in forest and agroforestry ecosystems in the Congo basin, *Reg. Environ. Change* 21(4). DOI: 10.1007/s10113-021-01816-9.

Lescuyer, G., Karsenty, A. and Eba'a Atyi, R. (2009) A new tool for sustainable forest management in Central Africa: payments for environmental services. Chap 8, 127–139.

Lorenz, K. and Lal, R. (2005). The depth distribution of soil organic carbon in relation to land use and management and the potential of carbon sequestration in subsoil horizons, Adv. Agron. 88, 35–66.

Lorenz, K. and Lal, R. (2014). Soil organic carbon sequestration in agroforestry systems: a review, *Agron. Sustain. Dev.* 34(2), 443–454. DOI: 10.1007/s13593-014-0212-y.

Ma, Z. and Chen, H. Y. H. (2017). Effects of species diversity on fine root productivity increase with stand development and associated mechanisms in a boreal forest, *J. Ecol.* 105(1), 237–245. DOI: 10.1111/1365-2745.12667.

Ma, Z. L., Chen, H. Y. H., Bork, E. W., Carlyle, C. N. and Chang, S. X. (2020). Carbon accumulation in agroforestry systems is affected by tree species diversity, age and regional climate: a global meta-analysis, *Glob. Ecol. Biogeogr.* 29(10), 1817–1828. DOI: 10.1111/geb.13145.

Marsden, C., Martin-Chave, A., Cortet, J., Hedde, M. and Capowiez, Y. (2019). How agroforestry systems influence soil fauna and their functions - a review, *Plant Soil* 453(1–2), 29–44. DOI: 10.1007/s11104-019-04322-4.

Mayer, S., Wiesmeier, M., Sakamoto, E., Hübner, R., Cardinael, R., Kühnel, A. and Kögel-Knabner, I. (2022). Soil organic carbon sequestration in temperate agroforestry systems - a meta-analysis, *Agric. Ecosyst. Environ.* 323, 107689. DOI: 10.1016/j.agee.2021.107689.

Mbow, C., Smith, P., Skole, D., Duguma, L. and Bustamante, M. (2014). Achieving mitigation and adaptation to climate change through sustainable agroforestry practices in Africa, *Current Opinion in Environmental Sustainability* 6, 8–14. DOI: 10.1016/j.cosust.2013.09.002.

Nair, P. K. R. (1985). Classification of agroforestry systems, *Agroforest. Syst.* 3(2), 97–128. DOI: 10.1007/BF00122638.

Nair, P. K. R., Kumar, B. M. and Nair, V. D. (2009a). Agroforestry as a strategy for carbon sequestration, *J. Plant Nutr. Soil Sci.* 172, 10–23. DOI: 10.1002/jpln.200800030.

Nair, P. K. R., Nair, V. D., Kumar, B. M. and Haile, S. G. (2009b). Soil carbon sequestration in tropical agroforestry systems: a feasibility appraisal, *Environ. Sci. Policy* 12(8), 1099–1111. DOI: 10.1016/j.envsci.2009.01.010.

Nair, P. K. R., Nair, V. D., Mohan Kumar, B. and Showalter, J. M. (2010). Carbon sequestration in agroforestry systems, *Adv. Agron.* 108, 237-307. DOI: 10.1016/S0065-2113(10)08005-3.

Nair, P. K. R. (2012). Carbon sequestration studies in agroforestry systems: a reality-check, *Agroforest. Syst.* 86(2), 243–253.

Nguyen, Q., Hoang, M. H., Oborn, I. and Noordwijk, M. V. (2013). Multipurpose agroforestry as a climate change resiliency option for farmers: an example of local adaptation in Vietnam, *Clim. Change* 117(1–2), 241–257.

Nsombo, B. M. (2016). Evaluation des nutriments et du carbone organique du sol dans le système agroforestier du plateau des Batéké en République Démocratique du Congo. Thèse de Doctorat. Ecole Régionale Post-Universitaire d'Aménagement et de Gestion Intégrés des forêts et Territoires Tropicaux (ERAIFT), pp. 79–80.

Nygren, P., Fernandez, M. P., Harmand, J. M. and Leblanc, H. A. (2012). Symbiotic dinitrogen fixation by trees: an underestimated resource in agroforestry systems?, *Nutr. Cycl. Agroecosyst.* 94(2-3), 123-160. DOI: 10.1007/s10705-012-9542-9.

Ortaş, İ, Lal, R. and Kapur, S. (2017). Carbon sequestration and mycorrhizae in Turkish soils, *The Anthropocene: Politik–Economics–Society–Science*, 139-149. DOI: 10.1007/978-3-319-45035-3_10.

Pandey, D. N. (2002). Carbon sequestration in agroforestry systems, *Clim. Policy* 2, 367-377.

Paula, R. R., Bouillet, J., Ocheuze Trivelin, P. C., Zeller, B., Leonardo de Moraes Gonçalves, J., Nouvellon, Y., Bouvet, J., Plassard, C. and Laclau, J. (2015). Evidence of short-term belowground transfer of nitrogen from *Acacia mangium* to *Eucalyptus grandis* trees in a tropical planted forest, *Soil Biol. Biochem.* 91, 99-108.

Pellerin, S., Bamière, L., Launay, C., Martin, R., Schiavo, M., Angers, D., Augusto, L., Balesdent, J., Basile-Doelsch, I., Bellassen, V., Cardinael, R., Cécillon, L., Ceschia, E., Chenu, C., Constantin, J., Darroussin, J., Delacote, P., Delame, N., Gastal, F., Gilbert, D., Graux, A. I., Guenet, B., Houot, S., Klumpp, K., Letort, E., Litrico, I., Martin, M., Menasseri, S., Mézière, D., Morvan, T., Mosnier, C., Roger-Estrade, J., Saint-André, L., Sierra, J., Thérond, O., Viaud, V., Grateau, R., Le Perchec, S., Savini, I. and Réchauchère, O. (2020). Stocker du carbone dans les sols français, Quel potentiel au regard de l'objectif 4 pour 1000 et à quel coût ? *Rapport scientifique de l'étude*, INRA, France, 540 p.

Pereira, A. P. A., de Andrade, P. A. M., Bini, D., Durrer, A., Robin, A., Bouillet, J. P., Andreote, F. D. and Cardoso, E. J. B. N. (2017). Shifts in the bacterial community composition along deep soil profiles in monospecific and mixed stands of *Eucalyptus grandis* and *Acacia mangium*, *PLoS ONE* 12(7), e0180371.

Pereira, A. P. A., Zagatto, M. R. G., Brandani, C. B., Mescolotti, D. L., Cotta, S. R., Gonçalves, J. L. M. and Cardoso, E. J. B. N. (2018). Acacia changes microbial indicators and increases C and N in soil organic fractions in intercropped Eucalyptus plantations, *Front. Microbiol.* 9, 655. DOI: 10.3389/fmicb.2018.00655.

Pereira, A. P. A., Santana, M. C., Zagatto, M. R. G., Brandani, C. B., Wang, J. T., Verma, J. P., Singh, B. K. and Cardoso, E. J. B. N. (2020). Nitrogen-fixing trees in mixed forest systems regulate the ecology of fungal community and phosphorus cycling, *Sci. Total Environ.* 758, 143711 DOI: 10.1016/j.scitotenv.2020.143711.

Poeplau, C., Don, A., Vesterdal, L., Leifeld, J., Van Wesemael, B., Schumacher, J. and Gensior, A. (2011). Temporal dynamics of soil organic carbon after land-use change in the temperate zone - carbon response functions as a model approach, *Glob. Change Biol.* 17(7), 2415-2427.

Prasad, R., Dhyani, S. K., Newaj, R., Kumar, S. and Tripathi, D. V. (2016). Contribution of advanced agroforestry research in sustaining soil quality for increased food production and food security, *J. Soil Water Conserv.* 15(1), 31-39.

Price, G. W. and Gordon, A. M. (1998). Spatial and temporal distribution of earthworms in a temperate intercropping system in southern Ontario, Canada, *Agroforest. Syst.* 44(2/3), 141-149.

Reich, P., Tilman, D., Forest, I., Mueller, K., Hobbie, S. E., Flynn, D. F. B. and Eisenhauer, N. (2012). Impacts of biodiversity loss escalate through time as redundancy fades, *Science* 336(6081) 589-592. DOI: 10.1126/science.1217909.

Rosenstock, T. S., Wilkes, A., Jallo, C., Namoi, N., Bulusu, M., Suber, M., Mboi, D., Mulia, R., Simelton, E., Richards, M., Gurwick, N. and Wollenberg, E. (2019). Making trees count: measurement and reporting of agroforestry in UNFCCC national communications of non-Annex I countries, *Agric. Ecosyst. Environ.* 284, 106569. DOI: 10.1016/j.agee.2019.106569.

Rumpel, C., Amiraslani, F., Chenu, C., Garcia Cardenas, M., Kaonga, M., Koutika, L. S., Ladha, J., Madari, B., Shirato, Y., Smith, P., Soudi, B., Soussana, J. F., Whitehead, D. and Wollenberg, E. (2019). The 4p1000 Initiative: opportunities, limitations and challenges for implementing soil organic carbon sequestration as a sustainable development strategy, *Ambio* 49(1), 350–360.

Sanginga, N., Mulungoy, K. and Ayanaba, A. (1986). Inoculation of *Leucaena leucocephala* Lam de Witt with Rhizobium and its nitrogen contribution to a subsequent maize crop, *Biol. Agric. Hortic.* 3, 341–352.

Santos, F. M., Chaer, G. M., Diniz, A. R. and Balieiro, FdC. (2017a). Nutrient cycling over five years of mixed-species plantations of Eucalyptus and Acacia on a sandy tropical soil, *Forest Ecol. Manag.* 384, 110–121. DOI: 10.1016/j.foreco.2016.10.041.

Santos, F. M., Balieiro, FdC., Fontes, M. A. and Chaer, G. M. (2017b). Understanding the enhanced litter decomposition of mixed-species plantations of *Eucalyptus* and *Acacia mangium*, *Plant Soil* 423(1–2), 141–155. DOI: 10.1007/s11104-017-3491-7.

Schroth, G. and Sinclair, F. L. (2003). Chapter 1: impacts of trees on the fertility of agricultural soils. In: *Trees, Crops and Soil Fertility. Concepts and Research Methods* Schroth, G. and Sinclair, F. L. (Eds), CAB International; Oxford University Press, UK, pp. 1–13.

Sebukyu, V. B. and Mosango, M. (2012). Adoption of agroforestry systems by farmers in Masaka District of Uganda, *Ethnobot. Res. Appl.* 10, 059–068. Available at: ethnobotanyjournal.org/vol10/i1547-3465-10-059.pdf.

Shi, L. L., Feng, W. T., Xu, J. C. and Kuzyakov, Y. (2018). Agroforestry systems: meta-analysis of soil carbon stocks, sequestration processes, and future potentials, *Land Degrad. Dev.* 29(11), 3886–3897. DOI: 10.1002/ldr.3136.

Sileshi, G. and Mafongoya, P. L. (2006). Long-term effects of improved legume fallows on soil invertebrate macrofauna and maize yield in eastern Zambia, *Agric. Ecosyst. Environ.* 115(1–4), 69–78. DOI: 10.1016/j.agee.2005.12.010.

Sileshi, G., Akinnifesi, F. K., Ajayi, O. C., Chakeredza, S., Kaonga, M. and Matakala, P. W. (2007). Contributions of agroforestry to ecosystem services in the Miombo eco-region of eastern and southern Africa, *Afr. J. Environ. Sci. Technol.* 1(4), 68–80.

Sileshi, G., Akinnifesi, F. K., Ajayi, O. C. and Place, F. (2008). Meta-analysis of maize yield response to woody and herbaceous legumes in the sub-Saharan Africa, *Plant Soil* 307(1–2), 1–19.

Six, J., Elliott, E. T. and Paustian, K. (2000). Soil macroaggregate turnover and microaggregate formation: a mechanism for C sequestration under no-tillage agriculture, *Soil Biol. Biochem.* 32(14), 2099–2103.

Somarriba, E. (1992). Revisiting the past - an essay on agroforestry definition, *Agroforest. Syst.* 19(3), 233–240.

Soto-Pinto, L., Anzueto, M., Mendoza, J., Ferrer, G. J. and de Jong, B. (2010). Carbon sequestration through agroforestry in indigenous communities of Chiapas, Mexico, *Agroforest. Syst.* 78(1), 39–51.

Stainback, G. A., Masozera, M., Mukuralinda, A. and Dwivedi, P. (2012). Smallholder agroforestry in Rwanda: a SWOT-AHP analysis, *Small-Scale For.* 11(3), 285–300. DOI: 10.1007/s11842-011-9184-9.

Stavi, I. and Lal, R. (2013). Agroforestry and biochar to offset climate change: a review, *Agron. Sustain. Dev.* 33(1), 81–96. DOI: 10.1007/s13593-012-0081-1.

Thomas, A. L., Kallenbach, R., Sauer, T. J., Brauer, D. K., Burner, D. M., Coggeshall, M. V., Dold, C., Rogers, W., Bardhan, S. and Jose, S. (2020). Carbon and nitrogen accumulation within four black walnut alley cropping sites across Missouri and Arkansas, USA. *Agrofor. Syst.* 94(5), 1625–1638.

Torquebiau, E. F. (2000). A renewed perspective on agroforestry concepts and classification, *C. R. Acad. Sci. III* 323(11), 1009–1017.

Udawatta, R. P., Kremer, R. J., Adamson, B. W. and Anderson, S. H. (2008). Variations in soil aggregate stability and enzyme activities in a temperate agroforestry practice, *Appl. Soil Ecol.* 39(2), 153–160.

Van Groningen, J. W., van Kessel, C., Hungate, B. A., Oenema, O., Powlson, D. S. and van Groenigen, K. J. (2017). Sequestering soil organic carbon: a nitrogen Dilemma, *Environ. Sci. Technol.* 51(9), 4738–4739. DOI: 10.1021/acs.est.7b01427.

van Noordwijk, M., Coe, R., Sinclair, F. L., Luedeling, E., Bayala, J., Muthuri, C. W., Cooper, P., Kindt, R., Duguma, L., Lamanna, C. and Minang, P. A. (2021). Climate change adaptation in and through agroforestry: four decades of research initiated by Peter Huxley, *Mitig Adapt. Strateg. Glob. Chang* 26(5), 1–33. DOI: 10.1007/s11027-021-09954-5.

Verchot, L. V., Van Noordwijk, M., Kandji, S., Tomich, T., Ong, C., Albrecht, A., Mackensen, J., Bantilan, C., Anupama, K. V. and Palm, C. (2007). Climate change: linking adaptation and mitigation through agroforestry, *Mitig Adapt. Strat. Glob. Change* 12(5), 901–918. DOI: 10.1007/s11027-007-9105-6.

von Haden, A. C., Kucharik, C. J., Jackson, R. D. and Marin-Spiotta, E. (2019). Litter quantity, litter chemistry, and soil texture control changes in soil organic carbon fractions under bioenergy cropping systems of the North Central US, *Biogeochemistry* 143(3), 313–326. DOI: 10.1007/s10533-019-00564-7.

Wang, Z. G., Bi, Y. L., Jiang, B., Zhakypbek, Y., Peng, S. P., Liu, W. W. and Liu, H. (2017). Arbuscular mycorrhizal fungi enhance soil carbon sequestration in the coalfields, northwest China, Nature Scient. Rep. 6, 34336. DOI: 10.1038/srep34336.

Zomer, R., Trabucco , C. R., Place, F., van Noordwijk, M. and Xu, J. (2014). Trees on farms: an update and reanalysis of agroforestry's global extent and socio-ecological characteristics, *Working Paper 179. Bogor, Indonesia: World Agroforestry Centre (ICRAF) Southeast Asia Regional Program*. DOI: 10.5716/WP14064.PDF.

Zomer, R. J., Neufeldt, H., Xu, J., Ahrends, A., Bossio, D., Trabucco, A., Van Noordwijk, M. and Wang, M. (2016). Global tree cover and biomass carbon on agricultural land: the contribution of agroforestry to global and national carbon budgets, *Sci. Rep.* 6, 29987. DOI: 10.1038/srep29987.

Chapter 21

Management of organic soils to reduce soil organic carbon losses

Sonja Paul and Jens Leifeld, Agroscope, Switzerland

1 Introduction

Peatlands are ecosystems where soil organic matter (SOM) forms under water-saturated conditions, leading to anoxic conditions, which slow down microbial decomposition processes. The corresponding organic soils, Histosols, are important soil organic carbon (SOC) stores that accumulated 600 Pg SOC during the Holocene (Yu et al., 2010). Organic soils spread over almost 500 Mha worldwide, of which the majority, around 400 Mha, is situated in boreal and temperate regions, and 40–70 Mha is located in the tropics (Frolking et al., 2011; Leifeld and Menichetti, 2018; Page et al., 2011). Correspondingly, these soils have high SOC densities of often 1000–2000 t SOC ha^{-1} (Leifeld and Menichetti, 2018; Page et al., 2011). The quasi-continuous SOC accumulation rate of intact peatlands is estimated at 0.22 (northern) and 0.45 (tropical) t C ha^{-1} yr^{-1} (Leifeld et al., 2019; Loisel et al., 2014), resulting in a net accumulation of annually 0.41-0.51 Pg CO_2-eq, mostly as SOC (Gallego-Sala et al., 2018; Leifeld et al., 2019). As a result of differences in accumulation rates and attributed areas, about 75% of total SOC in Histosols is stored in boreal and temperate peatlands, and 25% in tropical regions.

SOM accumulation under anoxic conditions has trade-offs, as it implies redox processes that lead to the formation of methane, a greenhouse gas, which is, over a 100 years' time scale, around 28 times more potent per kg than CO_2 (Myhre et al., 2013). Indeed, current methane emissions from peatland ecosystems are estimated at 0.03 Pg CH_4 yr^{-1} (0.84 Pg CO_2-eq. yr^{-1}; Frolking et al., 2011). This is similar to the fluxes attributed to global emissions from paddy rice production (Chapter 17 of this book) and corresponds to 20% of

http://dx.doi.org/10.19103/AS.2022.0106.20

the CH_4 flux from all natural wetlands (Saunois et al., 2020). Despite of being a substantial methane source, the long-term GHG effect of intact peatlands is negative with a global cooling of −0.2 W m^{-2} to −0.5 W m^{-2}, mostly owing to the long-lasting C sink and the much longer atmospheric lifetime of CO_2 over CH_4 (Frolking and Roulet, 2007). Emissions of nitrous oxide, the third important biogenic GHG, are very small in intact peatlands but raise after drainage.

Glossary of peatland terms.

Peat	Fibric organic sedentarily accumulated material with virtually all of the organic matter allowing the identification of plant forms; consists of at least 30% (dry weight) of dead organic material (Parish et al., 2008).
Peatland	An area with or without vegetation with a naturally accumulated peat layer at the surface of at least 30 cm depth (Parish et al., 2008).
Mire	Synonymous with any peat-accumulating wetland. A peatland where peat is currently forming and accumulating (Parish et al., 2008).
Bog	Mire raised above the surrounding landscape and only fed by precipitation (Parish et al., 2008).
Fen	Peatland receiving inflow of water and nutrients from the mineral soil. Distinguished from swamp forest by a lack of tree cover or with only a sparse crown cover (Parish et al., 2008).
Swamp	Usually forested minerotrophic peatland (minerotrophic peatland receiving nutrients through an inflow of water that has filtered through mineral soil). Separate from wooded fens due to a denser tree canopy. Also called peat swamp forest (Parish et al., 2008).
Acrotelm	Upper peat producing layer of mire with a distinct hydraulic conductivity gradient in which water level fluctuations and most of horizontal water flow occur (Parish et al., 2008).
Catotelm	The lower permanently water-saturated layer in a peatland, with relatively low hydraulic conductivity and rate of decay (Parish et al., 2008).
Mesotelm	The intermediate zone, through which carbon is transferred from acrotelm to catotelm (Clymo and Bryant, 2008).
Blanket Bog	Bog in a very humid climate, which forms a blanket-like layer over the underlying mineral soil (Parish et al. 2008).
Cutover	Peatland area where peat was extracted, but where still economically useful peat is left (Price and Ketcheson, 2009).

Peat extraction	The excavation and drying of wet peat and the collection, transport, and storage of the dried product (Parish et al., 2008).
Wetland	Areas of marsh, fen, peatland, or water, whether natural or artificial, permanent or temporary, with water that is static, flowing, fresh, brackish, or salt, including areas of marine water, the depth of which at low tide does not exceed 6 m (definition according to Ramsar convention).
Histosol	Histosol is a soil type that comprise soils formed in organic material accumulating as groundwater peat (fen), rainwater peat (raised bog) or mangroves (WRB, 2015).

Climatic factors, particularly temperature and moisture, mostly controlled the rates of peat accumulation since the last glacial maximum (Charman et al., 2015). With ongoing climate change, it is discussed controversially whether intact peatlands will continue being a C sink throughout the twenty-first century. Gallego-Sala et al. (2018) argued that the net CO_2 uptake of peatlands is going to slightly increase during the next decades, while Huang et al. (2021) found that, owing to lowered water tables with a warming climate, peatlands will release more CO_2 by 2100 through microbial decomposition. In addition, fire-induced losses may increase when peatlands are drained (Turetsky et al., 2015). An expert assessment indicated that ongoing and expected future changes in the peatlands' carbon balance will be more strongly driven by land-use change, fire and permafrost thaw than in the past (Loisel et al., 2021). Land-use change of peatlands towards agriculture, forestry, or peat extraction usually implies drainage of the peat deposit. Drainage for management induces a variety of system changes, of which biodiversity loss, loss of water storing capacity, soil subsidence and compaction, leaching and SOM decomposition owing to peat aeration with subsequent release of CO_2 and N_2O, but reduced CH_4 emissions, are the most important ones.

Losses of SOC and, more generally, changes in the GHG balance of organic soils after drainage are pronounced and well documented (Hooijer et al., 2012; Kasimir-Klemedtsson et al., 1997; Tiemeyer et al., 2020). The area of peatlands drained for management has been estimated at 43–51 Mha, representing 10% of its total area (Joosten, 2010; Leifeld and Menichetti, 2018), of which almost half is situated in the tropics. Because tropical peatlands contribute less than 20% to the total peatland biome, this number indicates that they are exploited disproportionately. In consequence, the global peatland biome, representing managed and unmanaged organic soils, has shifted from a persistent GHG sink to a strong GHG source since 1850. The annual global GHG sink of intact peatlands contrasts with a GHG source from drained and managed organic soils in the order of 0.8–1.91 Pg CO_2-eq. yr^{-1} (Evans et al., 2021; Humpenöder et al., 2020; Joosten, 2010; Leifeld et al., 2019; Leifeld and Menichetti, 2018).

According to the same authors, tropical organic soils managed by agriculture contribute 59-82% to these emissions. By source, CO_2 is the most important GHG emitted from organic soils drained for agriculture, contributing between 75–78% (boreal and temperate) and 88–90% (tropical) to total GHG emissions (i.e. sum of CO_2, CH_4, N_2O; IPCC, 2014). These annual SOC losses from a relatively small area of only 50 Mha are higher as compared to the total SOC loss, which occurred in the last century from 5000 Mha mineral soil under crop- and grassland (estimated at 0.48 Pg CO_2-eq yr^{-1}; Sanderman et al., 2017). Addressing these losses is thus of major importance in the context of the soils' role in climate change.

In this chapter, we give an overview on the assessment of C losses and processes occurring after peatland drainage in different environments. We will also present management strategies to reduce SOC loss and GHG emissions from managed peatlands.

2 Soil properties and carbon dynamic following peatland drainage

2.1 Methods for estimating C loss after drainage

The SOC loss from drained peatlands can be determined by two different approaches: (i) inventory methods and (ii) direct flux measurements (Simola et al., 2012). By integrating the SOC loss over long periods (several years to decades, up to several hundred years), inventory studies deliver robust estimates of long-term changes in SOC stocks (e.g. Krüger et al., 2016; Minkkinen et al., 1999). They incorporate gaseous and aqueous carbon losses as well as erosion. Detailed information on the drainage history is necessary to calculate annual carbon loss rates. Inventory studies can be broadly classified into (a) repeated inventories that allow to compare historic SOC stocks with present day or (b) profile methods, where peat profiles of drained and undrained (reference) sites are compared, following the space-by-time approach. As natural reference sites are often unavailable, the subsoil is - as an approximation - taken as a reference for peat properties before drainage. The assumptions for this approximation are that the subsoil is not affected by drainage and that changes in carbon stock in the topsoil derivate from peat decomposition. Recent annual changes can only be investigated by tracing the gas exchange of the soil or the whole ecosystem. Flux measurements provide insights into processes and drivers of the GHG exchange. Mechanistic understanding of the underlying carbon cycling that is needed for modelling can be gained as well. However, these methods are time and cost-intensive.

2.1.1 Soil inventory methods

2.1.1.1 Repeated soil sampling

Repeated inventories of SOC stocks can detect carbon losses over time. As the volume of peat is decreasing upon drainage, it is crucial to consider the entire carbon stock down to the mineral subsoil or to compare the carbon stock above a layer of known radiocarbon age. The challenge of this method is to detect a relatively small annual carbon loss within a large carbon stock. Moreover, the small-scale spatial heterogeneity of carbon stock adds uncertainty to this method. Yet, it has been successfully applied in a repeated inventory in Finland where at 37 drained forest sites an average C loss of 7.4 kg C m^{-2} was observed over a 30-year period (Simola et al., 2012).

2.1.1.2 Ash method

The ash method is based on the simplified assumption that the carbon to ash content is constant during peat accumulation. Changes in the ratio are attributed to the decomposition of peat and subsequent loss of SOC. Accordingly, the occurrence of mineral layers – as often found in fens – restricted this approach to organic peatlands without variable mineral input, in the absence of a natural reference site (e.g. Krüger et al., 2016). The method can provide reasonable estimates of past drainage-induced C losses in bogs (e.g. Rogiers et al., 2008; Wüst-Galley et al., 2016) but has been criticized because of large errors due to the fact that conditions for its application are often not matched (Laiho and Pearson, 2016).

2.1.1.3 Bulk density method

The peat surface subsides over time due to physical compaction as well as peat decomposition (Section 2.3.1). As only decomposition leads to a carbon loss, the contribution of this process to the overall soil subsidence has to be determined (Schothorst, 1977). For this method, depth profiles of peat bulk density and carbon concentrations before and after drainage must be available. An example of this approach by Kluge et al. (2008) indicated average losses of 0.7 kg C m^{-2} yr^{-1} for drained organic soils in Germany. Soil subsidence together with changes in soil bulk density were used by Hooijer et al. (2012) to estimate C losses from drained tropical peatlands. More simplified approaches that derive drainage-induced losses just from the difference in actual soil bulk density and carbon content in degraded topsoils and intact deeper peat layers have also been applied (e.g. Leifeld et al., 2011; Egli et al., 2021). Yet, the latter methodology suffers from limitations similar to the ash method (see above). New approaches to monitor soil subsidence with satellite data and modelling offer the opportunity to assess peat carbon loss at a large scale (e.g. Hoyt et al., 2020).

2.1.1.4 Radiocarbon method

Because drainage-induced peat decomposition proceeds from the surface to deeper layers and is accompanied by continuous loss of surface peat, it comes along with an increasing peat age at the surface as compared to the native state because younger material previously situated at the surface of the profile is lost first (Section 2.2). Whereas in an intact peatland the uppermost peat layers are of recent age, strongly degraded peatlands that were drained for many decades reveal topsoil radiocarbon ages of centuries to millennia (Bader et al. 2018b; Leifeld, 2018; Mrotzek et al., 2020), providing strong evidence for the loss of young peat. The age difference of degraded versus intact topsoil has been used to estimate carbon losses from drainage *a posteriori* (Krüger et al., 2016; Leifeld et al., 2018). There are two principal approaches to use ^{14}C for estimating previous C losses. First, the SOC stock above a reference layer of same radiocarbon age can be measured for the drained situation and compared with the stock of a nearby natural peatland to infer SOC losses. This might be hampered by fact that peatland drainage often concerns larger areas and intact reference sites are not available. A second approach uses the age-depth gradient of intact peat below the drained horizon to infer the site-specific SOC accumulation rates of the peatland and to calculate expected SOC stocks that are compared to the actual ones. With this approach, Leifeld et al. (2018) showed that drained European peatlands lost on average 50% of their former C stock, corresponding to 56 kg C m^{-2}. A recent application by Wang et al. (2021) extended the approach to managed organic soils with mineral cover. However, limitations of this approach are due to the dilution of the older topsoil peat by recent assimilates (Leifeld et al., 2021), which leads to an overestimation of the remaining peat if not quantitatively considered, and to varying peat accumulation rates over time.

2.1.2 Net ecosystem carbon balance approach

The net ecosystem carbon balance (NECB) approach is determined by direct measurement of all carbon fluxes entering and leaving the investigated compartment (soil-vegetation to atmosphere). Generally, the data are reported as positive and negative fluxes with negative values indicating a net downward flux (i.e. a gain of carbon from the atmosphere = carbon sink). Positive values indicate a loss (e.g. harvested biomass or through oxidation/decomposition = carbon source). Fluxes include the exchange of gaseous (CO_2, CH_4) and, ideally, waterborne fractions (dissolved organic carbon (DOC),

Published by Burleigh Dodds Science Publishing Limited, 2023.

particulate organic carbon (POC), and dissolved inorganic carbon (DIC)) and, in case of a managed system, any carbon import (e.g. organic fertilizer) or export (harvest). The NECB approach is equivalent to the change in SOC storage. However, in systems where substantial biomass is accumulating during a year, such as tree growth in young plantations, it is essential to account for this storage change; otherwise, the SOC loss is underestimated (e.g. Hommeltenberg et al., 2014; Minkkinen et al., 2018). Long-term time series for peatlands generally include not more than several years. In addition, for a complete NECB approach, fluviatile carbon losses have to be quantified, which is complicated and time consuming. As the major part of carbon is released as CO_2, numerous studies do not quantify waterborne carbon losses (e.g. Tiemeyer et al., 2016). Thus, in many peatlands, carbon losses may have been underestimated by this approach. The direct gas exchange between surface and atmosphere can be measured by two different techniques: Eddy-Covariance (EC) and chambers.

2.1.2.1 Eddy-Covariance

EC measures the undisturbed gas exchange of the net ecosystem exchange (NEE), that is, the sum of CO_2 uptake by plants (gross primary production, GPP) and ecosystem respiration (ER) over an area of about 0.5-5 ha. The ER consists of autotrophic and heterotrophic respiration. NEE is partitioned into GPP and ER by a modelling approach. The advantage of EC is that the CO_2, and also CH_4, exchange can be measured continuously in the free atmosphere without disturbing the investigated ecosystem. As there is no limitation of vegetation height, EC can be used for forests as well. However, NEE is integrated over an area, which is variable in its extension and direction. Thus, capturing the small-scale heterogeneity of mosaic structures, as often found in intact peatlands, is impossible. Limitations of the method under specific micrometeorology conditions generate gaps within the continuous data set, which have to be filled. Furthermore, the equipment is expensive, and expert knowledge is needed to perform flux measurements (e.g. Minkkinen et al., 2018; Peacock et al., 2019; Paul et al., 2021).

2.1.2.2 Chambers

The most widely used method to measure gas exchange is the chamber method. Briefly, this method consists of putting a closed chamber on the soil and measuring the gas, which is accumulating over a specific time, often covering an area of 0.01-1 m^2. Gas samples are taken to detect the change in gas concentration *in situ* or offline. The most common chamber systems operate manually, which limits the frequency of flux measurements. Automatic chamber systems can increase the temporal resolution, especially covering the whole day and night-time measurements (Wang et al., 2022). However, these systems are expensive and electricity is needed. Using a transparent chamber, NEE can be measured, and opaque chambers, excluding the light, are used for ER. Advantages are that specific vegetation communities and the small-scale variability can be captured; however, the use is restricted to small vegetation fitting within the chamber. Artefacts due to the creation of closed spaces (changes in temperature, humidity, or pressure) may influence the measurements (e.g. Alm et al., 2007; Leiber-Sauheitl et al., 2014).

2.2 Peat profile changes following drainage

Drainage is a prerequisite for the agricultural use of peatlands, which became widespread in the twentieth century, particularly in Europe and North America (Davidson, 2014). Agricultural practices on drained peatlands range from extensively used pastures and highly productive agriculture for food production to forestry. Drainage-induced mineralization transforms the original fibric to hemic structure of the undrained peat to a new material called mursh, which is characterized by a moderate to strongly granular or blocky structure, or the structure is even pulverized (WRB, 2015). Mineralization of peat decreases the organic carbon content and C/N ratio while bulk density increases (Fig. 1). The most severe shift in the pore structure of peat occurs within the first 20 years of drainage (Liu et al., 2020). The decline in total porosity is accompanied by a shift in pore size distribution (Wallor et al., 2018), and the altered pore structure reduces saturated hydraulic conductivity (Kechavarzi et al., 2010). The infiltration rate may decrease while the runoff from drained peatlands may increase, thereby enhancing soil erosion (Li et al., 2018; van Seters and Price, 2002).

Surface desiccation can lead to an increased hydrophobicity of peat (Eggelsmann et al., 1993). A dry surface crust is often developed, which is susceptible to shrinkage and cracking and thus may enhance superficial flooding. Dry peat becomes flammable. Peat fires generally are dominated by smouldering combustion, and these flameless fires can burn for long periods. In natural peatlands, fires occur as well, but deeper peat layers are protected from burning due to a high water table (Turetsky et al., 2015). Drainage and also climate change increase the frequency and extent of peat fires thereby also affecting the mineralization of old carbon (Konecny et al., 2016; Turetsky et al., 2015; Wiggins et al., 2018). In natural peatlands, the age of SOC linearly increases with depth (Fig. 1). Drainage and fire increase the mineralization of the peat, and the age of the carbon at the soil surface increases successively (Fig. 1). The contribution of old peat-derived carbon to total SOM decreases, while the contribution of the fresh litter input from current non-peat vegetation increases over time (Fig. 1).

Published by Burleigh Dodds Science Publishing Limited, 2023.

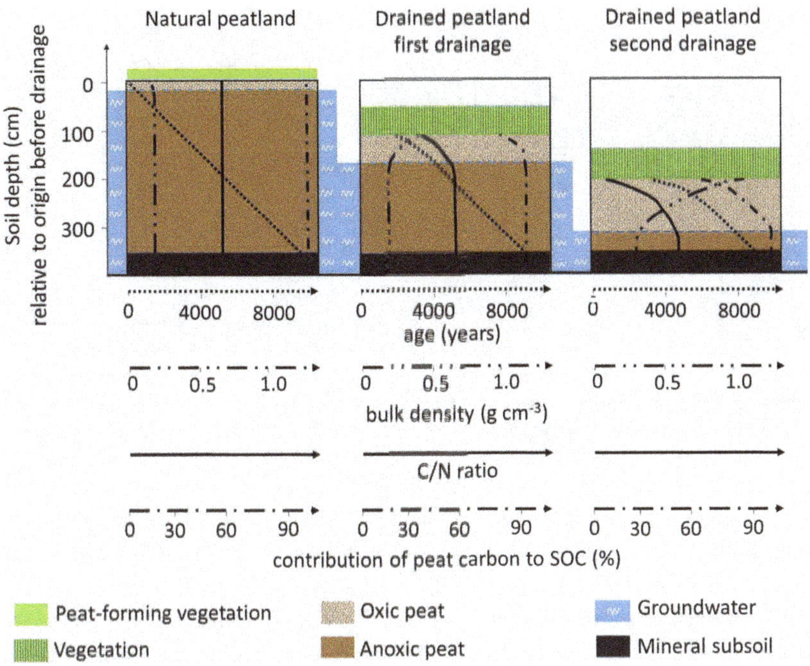

Figure 1 Changes of soil properties through drainage over 100 years in a temperate peatland.

2.3 Soil subsidence and carbon loss from drained peatlands

2.3.1 Soil subsidence

In drained peatlands, SOC loss will continue as long as decomposable organic material is available in the aerated soil horizons. The loss of peat by wind and water erosion, burning, and peat extraction also contributes to the loss of organic material and thus soil subsidence. After 40 years of drainage, Schipper and McLeod (2006) found average subsidence rates of 3.4 cm yr^{-1} in a New Zealand under grassland. The authors attributed about 63% of the soil subsidence to consolidation and shrinkage, with the remainder (37%) attributed to losses of OM by peat mineralization. For longer periods (1000 years), the estimated contribution of decomposition to soil subsidence was 85% (Schothorst, 1977).

After strong initial soil subsidence due to rapid consolidation and shrinkage, the rate of subsidence decreases because peat decomposition processes occur on longer timescales (Fig. 2). For the first 10 years after drainage, average soil subsidence rates of -1.6 ± 1.0 cm yr^{-1} to -6.2 ± 2.2 cm yr^{-1} were found for subarctic/boreal, temperate, and tropical peatlands (Evans et al., 2019; Liu et al., 2020). Subsidence was the strongest in the first 5 years in tropical systems, with rates up to 28 cm yr^{-1} (Hooijer et al., 2012). The primary thickness

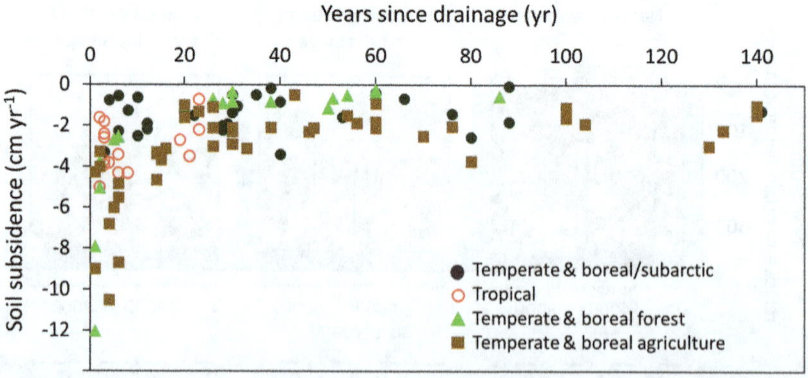

Figure 2 Soil subsidence rates over time (data of temperate and boreal forest and temperate and boreal agriculture from Liu et al., 2020: data of temperate and boreal/subarctic and of tropical peatlands from Evans et al., 2019 and references therein).

of peat, drainage intensity, and physical properties of the peat are important factors influencing soil subsidence rates in the initial phase (Oleszczuk et al., 2020). Thereafter, soil subsidence rates decreased with lower subsidence rates in boreal and temperate forest (-0.67 ± 0.3 cm yr^{-1}) compared to boreal and temperate croplands (-2.1 ± 0.9 cm yr^{-1}, Liu et al., 2020) and to tropical forest (-2.3 ± 1.0 cm yr^{-1}; Evans et al., 2019). The considerably shallower water table of the forest sites compared to the arable sites (35 cm vs. 87 cm) might explain the low subsidence rates for forests (Liu et al., 2020).

Besides time since drainage, the rate of subsidence varies strongly with drainage depth. As the groundwater table determines the volume of exposed peat material, the drainage depth is a key factor: A linear increase in subsidence rate with decreasing ground water table was found for high-latitude and tropical peatlands (Evans et al., 2019). A decrease in drainage depth from 40 cm to 80 cm resulted in a threefold increase in subsidence rates from -0.42 cm yr^{-1} to -1.27 cm yr^{-1} in temperate and boreal/subarctic peatlands. In tropical peatlands, a similar decrease in water table seems to result in higher subsidence rates as an increase from -2.8 to -4.5 cm yr^{-1} was recorded. However, this might also have resulted from the shorter drainage time of 8.7 years in the tropical system, while the temperate peatlands had been drained for 37 years.

Peat properties such as density and thickness add to the observed variation in subsidence rates. For the Sacramento Delta, a strong relationship was found between the content of organic matter (OM) and subsidence rates. After 100 years of initial drainage, subsidence rates were less than 1 cm yr^{-1} in areas where organic matter content in the topsoil was below 10% but considerably higher (3.5 cm yr^{-1}) in areas with high organic matter content (60%; Deverel and Leighton, 2010). In agreement with this finding, the spatial variation in

SOM content, ranging from 4% to 60%, explained over 55% of the variation in average subsidence rates from 1978 to 2006 in the same area (Rojstaczer and Deverel, 1995; Deverel and Leigthon, 2010). Similarly, thicker peat deposits showed higher subsidence rates than shallow peat deposits over a 38-year observation period in Poland's peatlands (Grzywna, 2017).

With ongoing subsidence, the soil surface (relative to the ground water table) decreases and the peatland becomes wet again when subsidence depth approaches the groundwater table. To maintain an aerated zone in the organic soils for agriculture, repeated drainage is necessary, thus becoming a self-perpetuating process. In temperate regions, the drainage system has to be renewed every 40–50 years (Ferré et al., 2019). Long-term drainage will finally lead to a loss of the whole peat deposit - converting the organic into a mineral soil. The subsequent long-term agricultural suitability depends strongly on the properties of the underlying mineral material. Coastal regions with substantial soil subsidence over the last century of up to 8 m lead to levee instability or failure, island inundation, and increased seepage penetration into the groundwater (Deverel and Leigthon, 2010; Erkens et al., 2016). In addition, high mineralization rates of the peat cause enhanced nutrient mobilization and thus the risk of water eutrophication (Kløve et al., 2017, 2010; Prévost et al., 1999).

2.3.2 Pathways and forms of C loss in drained peatlands

During the aerobic decomposition processes of peat in the aerated zone, most of the carbon is lost as CO_2 (Fig. 3; IPCC, 2014). Only a minor part is released as CH_4 via the ditch waters. Besides gaseous carbon loss, carbon may be transported in the liquid phase. Losses of particulate organic carbon (POC) are often connected to water erosion processes. Moreover, POC can also be eroded by wind, especially when the peat is not covered by vegetation. Peat fires represent another substantial source of carbon release. Occasionally they are man-made fires for management purposes, but fires may also occur naturally in drained peatlands after prolonged drought periods. In addition to CO_2, CO, and CH_4 emissions from these fires, smoke is released reducing light and growth conditions and consequently CO_2 uptake by plants (Turetsky et al., 2015). The most direct carbon loss in peatlands is peat extraction for horticulture and energy use – still practised in several northeastern European countries. It is estimated that global emissions from peat mining activities amount to 0.066 Pg CO_2-eq yr^{-1}, in contrast to about 1 Pg CO_2-eq yr^{-1} released by peat fires and 1.5-2 Pg CO_2-eq. yr^{-1} by microbial decomposition of peat (Leifeld et al., 2019).

For estimation of SOC losses, default values as compiled by IPCC (2014) can be used if no detailed country-specific information is available (see Chapter 9 of this book). These default values were derived from studies of CO_2 and CH_4 emissions from land and ditches waters, DOC loss and carbon loss from peat fires. Emission

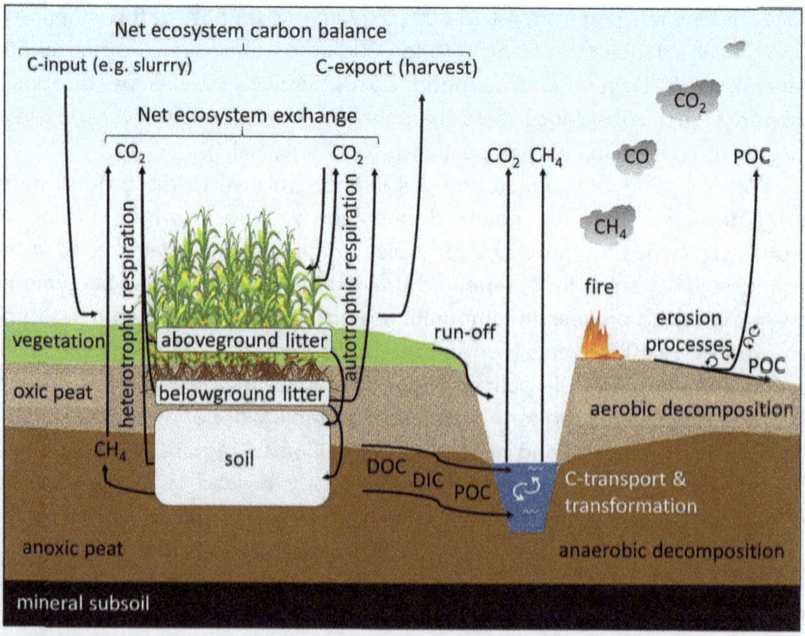

Figure 3 Summary of main carbon fluxes from drained organic soils contributing to SOC stock changes and additional carbon exchange fluxes due to management of agricultural systems on organic soils (see Section 2.1.1; modified after IPCC, 2014 and Jauhiainen et al., 2019).

factors were assigned to different climate zones and land use (Fig. 4). Depending on the available number of studies, more specific subclasses such as nutrient status or drainage depth are given for some categories. Integration of new data will be used to update and improve the estimated emission factors. Generally, total carbon losses from drained peatlands are dominated by CO_2 emissions, contributing to the total C loss by 93% in the tropics, 94% in the temperate zone, and 96% in the boreal zone, followed by DOC losses (IPCC, 2014). CH_4 emissions amount for <0.1% of the total SOC losses, but are more relevant when fluxes are expressed as CO_2-equivalents. However, the contribution of these fluxes may differ considerably across peatland types: In drained blanket peat in the UK, DOC fluxes were the dominating pathway, followed by DIC, POC, CH_4, while the net CO_2 balance was positive (Rowson et al., 2010).

2.3.3 Drivers of CO_2 emissions in drained peatlands

Following Fig. 4, temperature is a key driver for CO_2 emissions from drained peatlands, causing particularly high emissions in the tropics. In addition, in all climate zones, the groundwater level strongly determines the CO_2 emission (Evans et al., 2021; Huang et al., 2021). Independent of the climatic region,

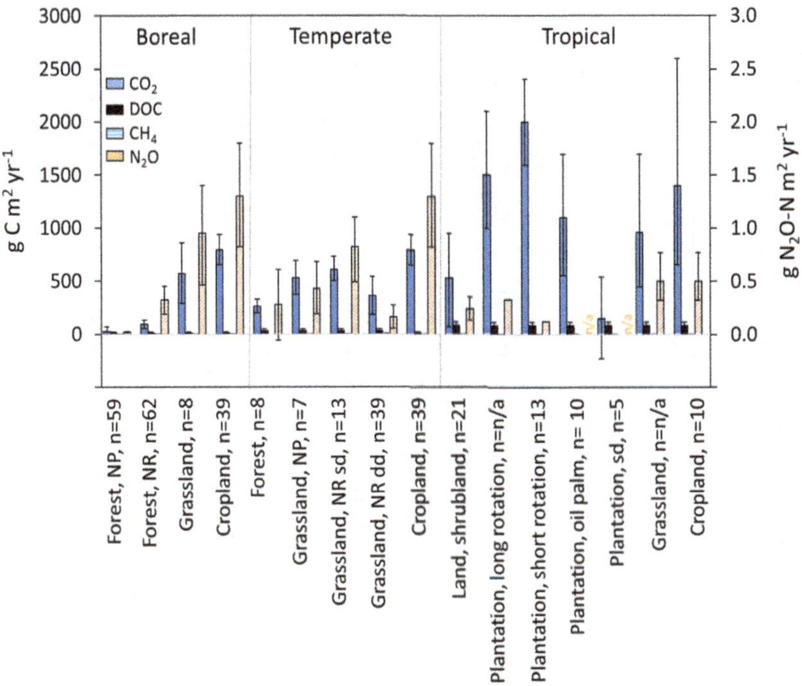

Figure 4 Summary of IPCC default values of drained peatlands (IPCC, 2014; 95% confidence interval; NR, nutrient-rich; NP, nutrient-poor; sh, shallow-drained (<30 cm); dd, deep-drained (>30 cm); n, number of studies for emissions factor of CO_2). No differentiation was made for the emission factor of temperate and boreal cropland. CH_4 includes direct emissions from land and from ditch waters. CH_4 fluxes are all <7 g C m^{-2} yr^{-1} and consequently too small for visualization.

considerably higher emissions are found at a drainage depth of −80 cm compared with −40 cm (Table 1). Generally, CO_2 emissions increase linearly with falling water table: per 10 cm water table drawdown by 188–267 g C m^{-2} yr^{-1} in tropical forests, by 134 in temperate peatlands in the UK and Ireland and by 16–33 g C m^{-2} yr^{-1} in boreal forests (Evans et al., 2021; Hooijer et al., 2012; Ojanen and Minkkinen, 2019; Table 1). In contrast, German peatlands are more sensitive and showed high site-to-site variability in CO_2 emissions at the same water table (Tiemeyer et al., 2020). Studies with water tables below 60 cm are scarce (Ojanen and Minkkinen (2019).

In some cases, the relationship between water table depth and CO_2 emissions is not linear (e.g. Tiemeyer et al., 2020; Couwenberg et al., 2010). In dry summers, the mineralization of peat in the upper layer may be limited by soil moisture (Glatzel et al., 2006; Hahn-Schöfl et al., 2011; Tiemeyer et al., 2016). Moreover, in shallow peat, no additional carbon may be mineralized with further lowering of the water table (Paul et al., 2021). While groundwater levels

Table 1 Change of carbon emissions per 10 cm water table drawdown (ΔWT_{10cm}) and resulting carbon emissions at a water table of 80 cm (WT_{80}) and 40 cm (WT_{40}) estimated from the relationship of water table and emission derived from studies of different peatlands, climatic region and land use.

Peatland	Land use	ΔWT_{10cm} (g C m^{-2} yr^{-1})	Drainage depth (cm)	Emission WT_{80} (g C $m^{-2}yr^{-1}$)	Emission WT_{40} (g C m^{-2} yr^{-1})	Method	*N*	Reference
Boreal NR, Finland	Forest	33	1-91		100	EC (NECB) SR + L	37	Ojanen and Minkkinen (2019)
Boreal NP, Finland	Forest	16	0-60		−5	EC (NECB) SR + L	33	Ojanen and Minkkinen (2019)
Temperate + boreal NR + NP	Natural, rewetted cropland, grassland peat extraction	92	3-100	570	200	EC (NECB)	65	Evans et al. (2021)
Temperate NR + NP, UK, Ireland	Natural, rewetted cropland, grassland	134	1-91	900	360	EC (NECB)	16	Evans et al. (2021)
Temperate NR + NP, Germany	Rewetted, semi-natural, sphagnum farm, grassland	140-486	0-20[a]		543 WT_{20}	CH (NECB)	44	Tiemeyer et al. (2020)
Temperate NR + NP, Germany	Rewetted, grassland cropland	360-111	20-40[a]		960	CH (NECB)	34	Tiemeyer et al. (2020)
Temperate NR + NP, Germany	Grassland, cropland peat extraction	32-9	40-60[a]			CH (NECB)	19	Tiemeyer et al. (2020)
Temperate NR + NP, Germany	Grassland, cropland	0-2.5	60-120[a]	1000		CH (NECB)	15	Tiemeyer et al. (2020)
Tropical	Forest	189	3-64	1190	435	EC (NEE)	6	Evans et al. (2021)

Tropical	Forest Indonesia	267	0–110	2138	1069	SS	51[b]	Hooijer et al. (2012)
Tropical	Plantation Indonesia	188	30–120	2078	1325	SS	167[b]	Hooijer et al. (2012)

NR = nutrient rich peatland, NP = nutrient poor peatland, Methods: EC, Eddy-covariance; CH, chamber; NECB, net ecosystem carbon balance; SR+L, soil respiration and litter production; SS, soil subsidence; EC (NEE), only NEE was measured, the tree stand biomass increment was not considered, therefore, NECB may be higher than NEE, all approaches of NECB did not include fluvial carbon losses, *N* = number of site. References Ojanen and Minkkinen (2019), Evans et al. (2021), and Hooijer et al. (2012) used linear models.

[a]Tiemeyer et al. (2020) fitted non-linear models, therefore, an average of emissions for a change in drainage depth was calculated to different 20-cm depth intervals.

[b]Number of measurement points, which were distributed over 16 transects with a longitude of 0.5–12 km.

can have a high predictive power for the CO_2 emissions in consecutive years for specific sites (Fortuniak et al., 2017; Hirano et al., 2012), it is unclear whether this is a general pattern that also holds for deeply drained peatlands.

In a newly established palm oil plantation, McCalmont et al. (2021) revealed that respiration from the upper soil horizons is more sensitive to water table changes than that from deeper layers. A 30-cm water table decrease from −50 to −80 cm resulted in 7% higher soil respiration compared to 37% when the water table fell from −10 cm to −40 cm (McCalmont et al., 2021). In line with this observation, Hirano et al. (2012) found a significantly stronger response of a tropical peat swamp upon water table lowering during an El Nino event, compared to a drained peatland. Recent EC measurements in an oil palm plantation revealed a very high carbon loss rate of −3.6 kg C m^{-2} in the first year following plantation establishment, which was reduced to −2.9 kg C m^{-2} in the second and third year and even further in the mature phase (−1.3 kg C m^{-2} yr^{-1}, McCalmont et al., 2021). One part of this considerable loss was explained by a higher water table in mature plantations compared to newly established ones (−0.25 cm vs. −0.60 cm). In addition, different peat qualities, originating from the antecedent decomposition of peat may be one factor explaining differences in mineralization rates with depth (Section 2.3). Also differences in vegetation induce different peat qualities which was found to affect decomposability (e.g. Duval and Radu, 2018; Maie et al., 2019; Moore and Dalva, 1997; Scanlon and Moore, 2000). Whether peatland type also affects CO_2 release after drainage is not yet clear. Whereas two studies found higher CO_2 mineralization rates for bogs than for fens (e.g. Alm et al., 1999a; Deppe et al., 2010), other studies reported only small differences in CO_2 emission between different peatland types (e.g. Moore and Dalva, 1993; Updegraff et al., 2001). Moreover, a meta-analyses by Huang et al. (2021) found a considerably stronger increase in CO_2 emissions with increasing drainage depth for fens than for bogs.

It is well established that peat decomposition increases with soil temperature in all climatic regions (Carter et al., 2012; Wilson et al., 2016a). Laboratory incubations indicated that Q10 values for CO_2 release under aerobic conditions are in the range of 1.4–2.5 for tropical peatlands (Girkin et al., 2020; Jauhiainen et al., 2019; Maie et al., 2019; Inglett et al., 2012; Ali et al., 2006), 1.8–4.2 for temperate peatlands (Bader et al., 2017; Hardie et al., 2011; Lund et al., 2007; Moore and Dalva, 1993), and 1.2–5.0 for boreal peatlands (Bubier et al., 1998; Yavitt et al., 2000). Noticeably, the boreal peatlands showed the highest variability in temperature sensitivity. The variability for CH_4 production rates was even more pronounced in boreal and temperate peatlands, which could be related to vegetation composition. CH_4 emissions from Sphagnum-dominated peatlands were more temperature sensitive with an average Q10 of 8.0 (at single sites Q10 values of up to 30) than wetter minerotrophic sedge-dominated peatlands (Q10 = 4.3, Lupascu et al., 2012). Q10 values of 1.9–5.8 were measured for an arctic permafrost layer in Sweden (Lupascu et al., 2012).

Determining the temperature dependency of CO_2 emission rates using field measurements is very challenging as land-use history, time since drainage, water table depth, management intensity, and other confounding factors differ between peatlands and regions. While in the tropics major drainage systems are relatively young, drainage is ongoing since centuries in the boreal and particularly in the temperate zone.

As the emitted CO_2 results from the combined effect of various controlling factors, including the composition of peat, climatic conditions (soil moisture and temperature), nutrient availability, land-use history, the influence of a single parameter on the CO_2 emissions is site-specific. It is up to now impossible to predict the emissions for a specific peatland and concurrently estimate the reduction potential of GHG emissions of management strategies for a particular site. However, on a broader scale, it can be concluded that the deeper the water table, the lower the latitude, and the nutrient richer the peatland is, the higher the CO_2 emissions and consequently the SOC loss.

2.3.4 Waterborne carbon losses in form of dissolved organic carbon (DOC) and particulate organic carbon (POC) in drained peatlands

Waterborne C losses of peatlands comprise different carbon species of various origins. DOC is released through biological activities during decomposition processes and plant growth. It is transported horizontally and vertically and can be exported from the peatland (Fig. 4) with DOC fluxes ranging from 5 g C m^{-2} yr^{-1} in subarctic and boreal peatlands to 100 g C m^{-2} yr^{-1} in tropical peat swamps (Evans et al., 2016; Moore et al., 2013). Temperature was found to be a major driver for DOC fluxes. Also within single sites, peat temperature may control the baseline flux of DOC (Rosset et al., 2020). In addition, drainage is usually considered as a driver for DOC production as it aerates the peat volume and consequently accelerates the decomposition rate and induces a shift in vegetation (e.g. Billett et al., 2006; Ritson et al., 2017; Strack et al., 2008). For example, Evans et al. (2016) found an overall higher DOC mobilization of 67% in drained as compared to natural peatlands. A positive correlation of water table depth with DOC fluxes was observed in drained oil palm plantations (Cook et al., 2018). Drainage depth was found to influence the age of DOC with deeper drainage leading to mobilization of older peat-derived carbon than shallow-drained peatland. Dean et al. (2019) analysed ^{14}C data from studies that measured DOC export to aquatic systems from drained peatland catchments. As a general feature, DOC from arctic, boreal, temperate, and tropical undisturbed and only slightly disturbed peatlands (n = 179) often had modern ages (87–90%), while 70% of disturbed peatlands (n = 30) had DOC, which showed older ages (Dean et al., 2019). This implies that drainage may trigger mobilization of older peat in form of DOC.

The contribution of DOC to the total carbon balance depends on the ecosystem type, disturbance degree, and climatic region and may be a quantitatively important component of the carbon balance. DOC fluxes of about 60 g C m^{-2} yr^{-1} in an undrained and about 100 g C m^{-2} yr^{-1} in a drained peat swamp were reported by Moore et al. (2013). In the latter case, DOC contributed 16–24% to the total carbon loss, when compared to CO_2 fluxes from the previous years (Hirano et al., 2012). In contrast, DOC fluxes of only 0.53 g C m^{-2} yr^{-1} contributed less than 2% to the total carbon loss of a temperate drained peat under grassland (Tiemeyer and Kahle, 2014), and in intensively used peatland in northern Germany, DOC fluxes of 4.28 g C m^{-2} yr^{-1} contributed about 8% to total carbon loss (Tiemeyer et al., 2016; Frank, 2016). In contrast to temperate or tropical systems, DOC is more relevant for the overall carbon budget in northern peatlands of the cool temperate or boreal zone with moderate disturbance (IPCC, 2014). For example, adding the DOC fluxes of 13 g C m^{-2} yr^{-1} in a peatland in northern Sweden to the gaseous exchange reduced the total carbon sink considerably from −37.5 g C m^{-2} yr^{-1} to −23.3 g C m^{-2} yr^{-1} (Nilsson et al., 2008). In a raised bog in Southern Scotland, waterborne fluxes of 30 ± 6.2 g C m^{-2} yr^{-1} even equalled CO_2 fluxes (28 ± 2.5 g C m^{-2} yr^{-1}, (Billett et al., 2004).

In comparison to DOC, POC contributes significantly only in disturbed peatlands to waterborne carbon fluxes (Billett et al., 2010; Cook et al., 2018; Dawson et al., 2002; Shuttleworth et al., 2015; Yupi et al., 2016). As particles are primarily detached and transported during erosion, relevant POC fluxes occur for bare peat during runoff situations (Billett et al., 2010; Holden, 2006; Pawson et al., 2012). In strongly disturbed blanket peatlands in the UK with eroding sites, POC fluxes of 13–79 g C m^{-2} yr^{-1} were recorded (Pawson et al., 2012). Evans et al. (2019) calculated that POC fluxes in these systems increased by around 4 g C m^{-2} yr^{-1} per 1% of exposed bare peat area. For tropical peat swamps, Moore et al. (2013) also found higher POC fluxes in disturbed (10 g C m^{-2} yr^{-1}) than in intact peat swamp forests (1 g m^{-2} yr^{-1}), and a high ratio DOC/POC was found to indicate erosion processes (Moore et al., 2013).

2.4 OM composition in intact and drained peatlands

The composition of OM in peat profiles is mostly related to (i) vegetation changes which occur following variations in climate and/or hydrology during millennial peat formation and (ii) ongoing decomposition of the formed peat

(Moore, 2002). The latter is of particular relevance at it provides insight into mechanisms of OM transformation in the absence of minerals. Although intact peatlands almost steadily accumulate new carbon (Yu et al., 2010), peat formation is coupled to ongoing decomposition both in the acrotelm and in deeper peat layers (Clymo, 1984), albeit transformation rates in the permanently water-saturated zone are thermodynamically strongly constrained (Worrall et al., 2018).

Accordingly, deeper and older peat should reveal a stronger degree of transformation and a more pounced microbial signature at the expense of degradable plant remains. Further, lignins as vascular plant markers might be better preserved owing to the anoxic conditions below the acrotelm. This was confirmed by a range of studies from different peatland ecosystems. Zaccone et al. (2008) studied a cool temperate bog in Switzerland and not only showed that phenols were relatively enriched in the catotelm as compared to the mesotelm but also indicated that lignin decomposition, as revealed from a significant demethoxylation and higher degree of lignin oxidation, continues with depth, albeit at low rates. In general, anoxic conditions prevent lignin decomposition, requiring high amounts of oxygen. Indeed, preferential loss of carbohydrates and selective preservation of alkyl-C and aromatic C with increasing depth, including lignin, was found for tropical and temperate peatlands in many studies (Preston et al., 1987; Upton et al., 2018; Leifeld et al., 2012; Moody et al., 2018; Schellekens and Buurman, 2011; Broder et al., 2012).

In contrast, peat C/N ratios as a possible indicator for the microbial imprint show no clear pattern. Declining peat C/N ratios with depth have been interpreted as to represent the declining rate of peat decay over time with a concurrent preservation of nitrogen in the deeper peat layers (Kuhry and Vitt, 1996), but this pattern cannot be considered universal. For example, C/N ratios decreased from 50 to 70 in the upper 3.5 m of the profile to around 30 below 5.5 m (Schellekens and Buurman, 2011). Yet, within the upper 3.5 m, C/N ratios increased with depth. Peat bogs studied by Broder et al. (2012) showed no clear trend of the C/N ratio with depth. A compilation of data from dozens of northern peatlands by Loisel et al. (2014) confirms that peat C/N ratios do not follow a systematic trend with increasing peat age (Fig. 5). The diverse sources of N inputs from fixation, atmospheric deposition, and nitrogen inflow, outputs via denitrification and runoff and their variation over longer timescales makes the N content of peat highly variable within a profile (Limpens et al., 2006).

2.4.1 Changes in OM composition of peat after drainage

In drained peatlands, new vegetation is established following agricultural practices, which provides new litter input. For understanding the change in

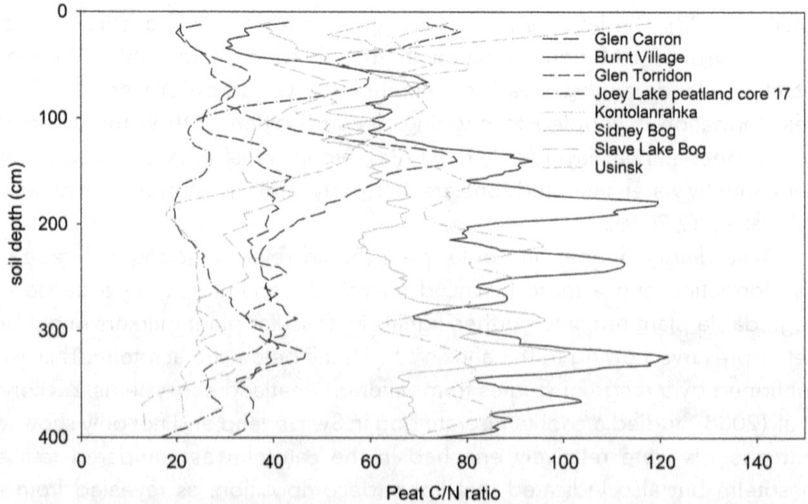

Figure 5 Depth trends of peat C/N ratios for selected sites from Loisel et al. (2014). Data are 20 cm running averages. Site description follows Loisel et al. (2014).

OM composition, peat inherent decomposition processes and incorporation of new inputs are of particular relevance. The contribution of new inputs to topsoil organic carbon (0–30 cm) in degrading peatlands was reported to be in the range of 14–24% after 2 decades, based on results from natural ^{13}C labelling when C3 vegetation was changed to C4 vegetation (Leifeld et al., 2021). The C4 accumulation rate in these topsoils was not significantly different to that of mineral topsoils, suggesting an overall similar dynamics of incorporation of new C. The new assimilates are relatively labile and contribute more to the CO_2 as the peat OM was from managed organic soils (e.g. Bader et al., 2017; Fig. 1).

After drainage, microbial transformation of the peat deposit accelerates, leading to an increase in the N content of OM. This increase can be substantial and is a major distinction between intact and drained peatland soils (Leifeld et al., 2020). A study from North-western Germany indicated that decadal drainage and grassland management reduced C/N ratios from 40–60 to around 20 in the topsoil (0–20 cm). Importantly, the trend was similar at fertilized and unfertilized sites, indicating that C/N in drained peatlands is related to OM mineralization and is a relative enrichment of nitrogen (Krüger et al., 2015).

With ongoing decomposition, exposure of older peat changes the overall peat composition towards a higher contribution of aryl-C and alkyl-C at the expense of O-alkyl originating from polysaccharides in boreal and temperate peatlands (Leifeld et al., 2018). In the same study, a relative enrichment of pyrogenic C with ongoing degradation was observed. A higher contribution of aromatics and other recalcitrant moieties was also measured for drained tropical peatlands (Könönen et al., 2016; Gandois et al., 2013; Cooper et al., 2019).

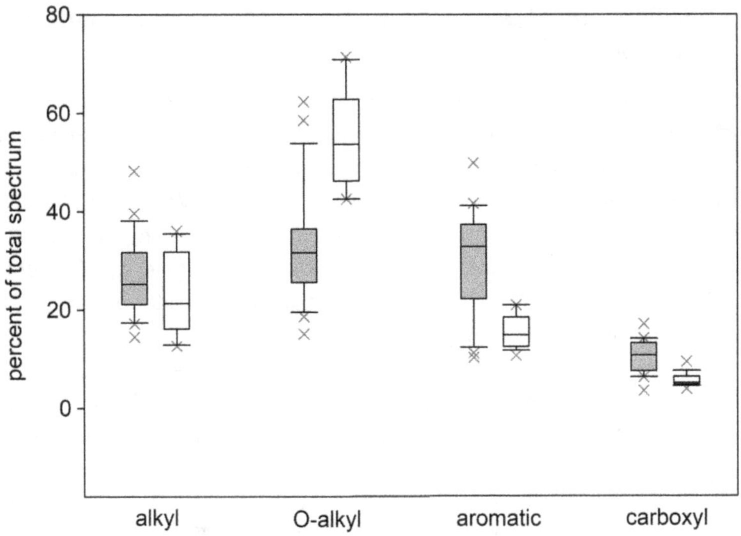

Figure 6 Chemical shift regions of ^{13}C-NMR spectra from tropical (grey) and northern (white) peat samples (alkyl-C: 0–45 ppm or 0–50 ppm; O-alkyl-C: 45–110 ppm or 50–110 ppm, aromatic 110–160 ppm, carboxyl 160–210 ppm or 160–220 ppm). Tropical data are from Purwanto et al. (2005), Sangok et al. (2017, 2020), Wright et al. (2011), northern are from Leifeld et al. (2018). All samples were derived from intact peatlands and were taken in the upper 1 m of the profile. Tropical $n = 22$, northern $n = 17$.

An increase in alkyl-C and a corresponding decline in O-alkyl-C is generally related to microbial decomposition of organic material, which is characterized by the accumulation of recalcitrant aliphatic plant compounds and/or microbial material (Baldock et al., 1997).

2.4.2 Latitudinal gradients in peat composition

Lowland tropical peats often form in peat swamp forests, where not only broadleaf trees but also palm trees (palm forest swamps) and sedges can be peat building. In addition, the vegetation in temperate and boreal peatlands is highly variable, including sphagnum spp sedges, broadleaf, and coniferous species, providing organic material for peat formation. The different conditions regarding vegetation, hydrology, and climate result in different peat properties. Tropical peatlands show lower contribution of polysaccharides and higher aromaticity than northern peatlands (Fig. 6; Hodgkins et al., 2018) because they contain higher contribution of woody species than temperate and boreal peatlands. In addition, tropical peatlands show on average lower C/N ratios (35.0 ± 17.8 SD; Leifeld and Menichetti, 2018) than northern peatlands (average 42.9 ± 18.8; Loisel et al., 2014). For the latter, sphagnum bogs contain the least nitrogen, resulting in wide C/N ratios of 81.0 ± 49.2.

2.4.3 Implications of peat composition for C loss of managed peatlands

The accumulation of peat is strongly related to the anoxic conditions of the peat deposit. The low abundance of oxygen directs microbial transformation towards other terminal electron acceptors, which implies a lower energy return and, finally, lower rates of decomposition (LaRowe and Van Cappellen, 2011). In addition, even reactions with a negative Gibbs energy might not occur owing to the energy requirement for ATP production and an efficiency of reaction of below 100% (Worrall et al., 2018). With drainage, these limitations are released and peat decomposition by microbes occurs at high rates.

In line with the thermodynamic constraints of water-logged peatlands, CO_2 fluxes from drained peatlands very much depend on climate and water table depth and thus oxygen availability as discussed above. Yet, these factors leave a large unexplained variability, and it has been hypothesized that the chemical composition of different peats might be a determining factor for CO_2 emissions from managed organic soils (Glatzel et al., 2004; Leifeld et al., 2012). In addition, also the abundance or absence of key controlling molecules such as polyphenols or tannins and their effect on OM decomposition might guide peat decomposition under drier conditions (Fenner and Freeman, 2020).

In many studies, peat samples from different soil depths were incubated in the laboratory and their CO_2, CH_4, or N release was monitored. Such an approach might be useful also to disentangle the role of peat composition on decomposability. Almost all of these studies found higher CO_2 release rates under aerobic conditions (Brake et al., 1999; Hardie et al., 2011; Hogg, 1993) and also under anaerobic conditions (Glatzel et al., 2004; Sjögersten et al., 2016) for surface samples, although depth effects were not always pronounced (Bader et al., 2018a). The van Post degree of humification, a widely used but unspecific field indicator of peat degradation (Malterer et al., 1992), was found to be positively related to the CO_2 release, that is, more decomposed peat, typically situated in the aerated topsoil of managed organic soils, was more labile (Glatzel et al., 2004; Säurich et al., 2019). However, simple indicators such as nutrient content or peat stoichiometry failed to explain peat decomposability (Bader et al., 2018a; Säurich et al., 2019). A higher degradability of surface peat might not be related to peat composition but to changes in microbial community composition and higher soil microbial activity in aerated topsoil peats (Brake et al., 1999), and to a high contribution of labile OM that derives from the current vegetation on former peatlands which may also induce priming effects (Bader et al., 2018b).

More specific approaches confirmed a regulatory role of OM composition for peat decomposability. An incubation of various tropical peats revealed a reduced nitrogen mineralization rate with increasing alkyl/O-alkyl-C ratio, that is,

Published by Burleigh Dodds Science Publishing Limited, 2023.

with increasing microbial transformation of the OM, as measured by solid-state ^{13}C-NMR spectroscopy (Purwanto et al., 2005). More specifically and also for tropical peat, Hoyos-Santillan et al. (2016) showed that both in anoxic and oxic environments, lignin moieties, longer-chain fatty acids, and polysaccharides, all analysed by TMAH-pyrolysis-GC-MS, were positively related to CH_4 and CO_2 release. For the anaerobic incubation of temperate peat, Reiche et al. (2010) showed that the peats' thermal stability, depending on the amount of carbohydrates in peat, explained 71% (CH_4) and 49% (CO_2) of the release of these two GHGs. Importantly, the relationship was non-linear. Leifeld et al. (2012), Sjögersten et al. (2016), and Normand et al. (2021) found a positive relationship between the O-alkyl-C content, as measured by ^{13}C NMR, and CO_2 release from various peatland soils. Together, these findings indicate that (i) peat composition varies steadily within the peat profile, (ii) peat quality changes systematically after drainage, and (iii) peat quality is one relevant factor for the release of CO_2 after drainage.

3 Management strategies to reduce soil organic carbon loss from drained peatlands

The most efficient ways to prevent further SOC loss from peatlands is to protect existing intact peatlands and to raise the water table of managed (drained) ones. Rewetting is then accompanied by either restoration of the peatland with the loss of the agricultural production function or with cropping of water-tolerant crops. In addition, engineering measures with continued management at deeper water tables were suggested.

3.1 Restoration and rewetting of peatlands without use

Rewetting of peatlands means restoring the water table or hydrological regime towards a condition where the new groundwater level is close to the surface of the peatland, with the aim of a partial or total reversal of drainage. However, the original ecosystem functions may be only partly restored. Restoration of peatlands aims at revitalizing the peat accumulation process. This includes the rehabilitation of original fauna and flora and thus the original ecosystem functions including carbon sequestration. In addition to rewetting measures, active restoration techniques that reintroduce peatland plant species are important (Landry and Rochefort, 2012), while the term 'rehabilitation' refers to natural succession after the abandonment of active management.

The continuous loss of SOC from degraded and drained peatlands can be reversed by rewetting as the main trigger for carbon release - the aerobic mineralization of the peat - is prevented (Evans et al., 2021; Huang et al., 2021). A common observation is that while CO_2 and N_2O fluxes decrease CH_4

emissions increase (Figs 4 and 7; Wilson et al., 2016a). After the successful recovery of the original vegetation, the carbon fluxes match those of undisturbed peatlands (IPCC, 2014; Nugent et al., 2019; Wilson et al., 2016b). Natural peatlands are characterized by a mosaic of habitat structure, including small-scale differences in water tables resulting in specific plant communities and GHG emission patterns (e.g. Lai et al., 2014; Laine et al., 2007; Maanavilja et al., 2011). Small-scale heterogeneity is often restored by rewetting (Wilson et al., 2013; Strack et al., 2014). However, carbon fluxes may differ in the initial phase after rewetting. Observed responses include (i) a stronger carbon sink due to higher carbon uptake during vegetation establishment, (ii) a consistent large carbon source due to high CO_2 emissions, or (iii) high CH_4 emissions (Valach et al., 2021; Waddington et al., 2010; Wilson et al., 2013). These short-term rewetting effects are often attributed to a transition period before a new equilibrium is established. Restoring the original ecosystem functions may be difficult and is a long-term objective (Vanselow-Algan et al., 2015; Renou-Wilson et al., 2019; Samaritani et al., 2011; Wilson et al., 2016b). The restoration success also depends on the state of degradation of the peatland and restoration techniques. In some cases, restoration may not be possible due to hydrological restriction or altered peat properties (Kløve et al., 2017). In some cases, restored wetlands may shift to stable states that differ from their original condition (Moreno-Mateos et al., 2012).

3.1.1 CO_2 flux changes after rewetting of drained peatlands

Strong reductions of CO_2 emissions upon rewetting was observed for temperate and boreal peatlands (IPCC, 2014; Wilson et al., 2016a). Most

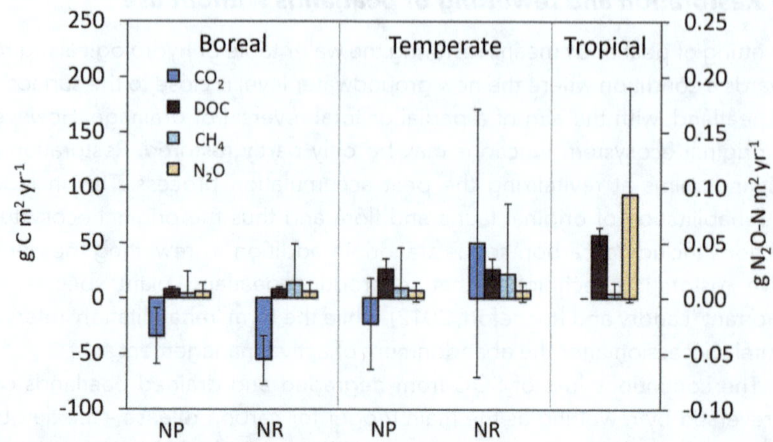

Figure 7 Carbon fluxes of rewetted and natural peatlands for boreal, temperate, and tropical climate zone. Data from Wilson et al. (2016a). NP, nutrient-poor; NR, nutrient-rich.

rewetted peatland types are, on average, carbon sinks. The strongest annual CO_2 sink is found in nutrient-rich boreal peatlands (−53 g C m^{-2} yr^{-1}), followed by nutrient-poor peatlands (−41 g C m^{-2} yr^{-1} for boreal and −33 g C m^{-2} yr^{-1} for temperate peatlands, Wilson et al., 2016a, Fig. 7). As an exception, nutrient-rich temperate peatlands showed lower CO_2 emissions of 26 g C m^{-2} yr^{-1} (Wilson et al., 2016a). Nutrient-rich peatlands display a wider range of fluxes than nutrient-poor sites in temperate and boreal peatlands, which can be explained by their high hydrological, biogeochemical, and biological diversity (Wilson et al., 2016a). Different functional plant types, including mosses, grasses, and sedges, are growing in fens (Joosten and Clarke, 2002). Especially boreal fen sedges are known to be a highly productive system, explaining the greatest carbon uptake in this system (IPCC, 2014). The elevated diversity of previous land use of temperate fens might add to the high site-to-site variability, as these fens were more intensively used than nutrient-poor bogs (Joosten and Clarke, 2002). In addition, a higher disturbance degree or the incorporation of studies on incompletely rewetted sites, where the natural peat-forming vegetation has not re-established, or recently rewetted sites that are still undergoing transitional changes, may explain the observed CO_2 emissions from rewetted nutrient-rich temperate peatlands (Wilson et al., 2016a). Also severe droughts might shift peatlands into a temporary carbon source (Alm et al., 1999b; Lund et al., 2012; Rinne et al., 2020).

Like in drained peatlands, an influence of mean water table on the CO_2 balance was observed for rewetted peatlands, with lower emission or higher sequestration for elevated water tables. The decrease in CO_2 emissions per cm change in the water table is more pronounced in temperate climate (44 g C m^{-2} yr^{-1} per 10 cm water table increase) than in boreal climate (17 g C m^{-2} yr^{-1} per 10 cm water table increase, IPCC, 2014). A water table above −10 cm depth is necessary to achieve a CO_2 sequestration in temperate peatlands, while boreal peatlands showed a net CO_2 uptake down to a water table depth of −30 cm. Up to date, no data of rewetted tropical peatlands are available; therefore, Wilson et al. (2016a) suggested a default value of 0 based on the observation that soil subsidence stopped when rewetting tropical organic soils. Due to the low number of observation in tropical peatlands, these emission factors have a high uncertainty (Fig. 4).

3.1.2 CH_4 emissions after rewetting of peatlands

As methane is produced under anaerobic conditions, increased CH_4 emissions after restoration of anaerobic conditions is a logical consequence. Up to date, the database for CH_4 emissions of rewetted sites is limited. As Wilson et al. (2016a) found no significant differences in CH_4 emissions between natural (undrained) and rewetted sites, data of natural sites are useful to explain the driving factors of CH_4 emission. Several factors, like water table height, plant species composition, and nutrient status, control CH_4 emissions of intact peatlands. Similar to the CO_2 emissions, the water table is considered the key driver for CH_4 emissions. A water table depth of <−0.25 m is sufficient to markedly reduce CH_4 emissions as CH_4 produced in the anoxic zone is oxidized by methanotrophs in the oxic zone (Turetsky et al., 2015; Evans et al., 2021; Tiemeyer et al., 2020). Annual CH_4 fluxes exponentially increase with raising the water table but with very high site-to-site variability. Comparing 108 peatlands from the boreal and temperate zone, Abdalla et al. (2016) reported higher CH_4 emissions for nutrient-rich environments, such as fens (15.4 g C m^{-2} yr^{-1}), compared to naturally nutrient-poor bogs (7.1 g C m^{-2} yr^{-1}). This pattern was confirmed for temperate peatlands in Germany (Tiemeyer et al., 2016). However, CH_4 emissions were considerably higher for shallow-drained nutrient-rich peatlands (79 ± 104 g C m^{-2} yr^{-1}) and for nutrient-poor peatlands (10.3 ± 24 g C m^{-2} yr^{-1}), indicating overall higher CH_4 emission upon rewetting in temperate peatlands (Abdalla et al., 2016; Tiemeyer et al., 2016).

In addition to hydrology, peatland vegetation is a key factor to explain the magnitude of CH_4 fluxes. Some graminoid plant species have an aerenchyma tissue, facilitating the direct gas exchange between roots and the atmosphere, thereby allowing methane to bypass the upper oxic peat layers (Schimel, 1995). Accordingly, wet temperate peatlands without this specialized plant species showed considerably lower CH_4 emissions (mean: 5 (range: −0.02 to 25 g C m^{-2} yr^{-1})) than peatlands supporting a vegetation with an aerenchyma tissue (mean: 17 (range: 0–76 g C m^{-2} yr^{-1})) (Couwenberg and Fritz, 2012; Turetsky et al., 2014).

Temperature is another crucial driver for CH_4 emissions. Plant productivity and input of labile carbon from plant exudates and mineralization processes depend on temperature (Conant et al., 2011). CH_4 fluxes decreased with increasing latitude from temperate to subarctic in most fens (Turetsky et al., 2014). In contrast, tropical peatlands showed an opposite trend. Wet tropical peatlands released on average only 6.1 g C m^{-2} yr^{-1} (Wilson et al., 2016a). Recent CH_4 fluxes from EC indicate considerably higher CH_4 emissions of 7.5–22 g C m^{-2} yr^{-1} in tropical peat swamps (Deshmukh et al., 2020; Griffis et al., 2020; Wong et al., 2018). Higher CH_4 emissions from these recent EC measurement suggested that plant-mediated CH_4 transport may play a role in the overall CH_4 emissions to the atmosphere, and this process is usually not captured by chamber measurement on which earlier observations were predominantly based (IPCC, 2014; Pangala et al., 2017; Welch et al., 2019). However, compared to boreal peatlands, CH_4 emissions of tropical sites are low despite the higher temperatures. This may be explained by high mineralization rates under warm temperatures in combination with a high contribution of lignin from tropical forests, which result in recalcitrant peat (Section 2.3) with a lower availability of labile substrate for methanogenesis.

In most cases, rewetting drained peatlands reduces GHG emissions (Section 3.4), however, high CH_4 emissions after rewetting can counteract mitigation strategies of peatlands that aspire an improved GHG balance (Waddington and Day, 2007; Wilson et al., 2009). High annual emissions up to 370 g CH_4-C m^{-2} yr^{-1} were measured in a nutrient-rich fen in the first years after rewetting (Augustin and Chojnicki, 2008). Rapid flooding of the site led to a plant die-off, with the decomposing litter serving as an easily degradable carbon source fuelling microbial decomposition and subsequent development of anaerobic conditions leading to the high CH_4 emissions (Hahn-Schöfl et al., 2011). In addition, high nutrient availability and lateral carbon import from floating biomass supported high methanogenesis rates (Augustin and Chojnicki, 2008). However, 8 years later considerably lower annual CH_4 emissions of about 53 g CH_4-C m^{-2} yr^{-1} from open water surfaces and 13 g CH_4-C m^{-2} yr^{-1} from emerging littoral vegetation were measured (Franz et al., 2016). Similarly, CH_4 emission after rewetting of a highly degraded coastal fen decreased in the first 2 years after rewetting (Hahn et al., 2015; Koebsch et al., 2015). These studies suggest that rapid flooding of nutrient-rich fens, where the vegetation was not adapted to flooding conditions, may induce a CH_4 peak and seems to represent a transition state with decreasing CH_4 emissions over time. However, other studies indicated high CH_4 emission rates depend on the vegetation types even after 30 years of rewetting (Vanselow-Algan et al., 2015). These emissions were explained by a strongly fluctuating water table, including inundation periods with the anaerobic decomposition of easy decomposable plants and shunt species (Vanselow-Algan et al., 2015). Similarly, mulching a rewetted peatland where the grass cut was left on the side, with temporary inundation, resulted in relatively high CH_4 emissions (Kandel et al., 2019b). Thus, it appears that not the time since rewetting is important, but that the hydrological regime and plant communities should be controlled to prevent high CH_4 emissions.

CH_4 emissions may additionally be affected by nutrient availability: removal of the upper 30 cm of degraded topsoil prior to rewetting of a former intensively used bog reduced CH_4 emissions considerably in the first year (Huth et al., 2020). Alongside the CH_4 reduction following topsoil removal, reduced fluxes of DOC (60%) and cations (80–90%) from a rewetted former agricultural peatland were recorded (Harpenslager et al., 2015). Former land use and hydrologic condition should be considered before rewetting. In particular, additional input of easily decomposable carbon and nutrients and prolonged flooding conditions, especially when vegetation is not adapted to inundation, should be avoided.

3.1.3 DOC changes following rewetting

DOC fluxes contribute essentially to total carbon fluxes in some peatlands systems (Section 2.2.2). Drainage usually increases DOC fluxes (Section 2.2.4). It is assumed that rewetting of drained organic soils generally decreases the DOC fluxes from peatland (IPCC, 2014; Evans et al., 2016; Wilson et al., 2016a). Limited information is available on the influence of land use and nutrient status on DOC fluxes from rewetted peatlands. Accordingly, only aggregated values are given for the climate zones (51, 24, and 8 g C m^{-2} yr^{-1} for tropical, temperate, and boreal zones, respectively (IPCC, 2014, Fig. 7). Several studies report higher DOC fluxes in the initial phase after rewetting (Glatzel et al., 2003; Rowson et al., 2010; Worrall et al., 2007). Zak and Gelbrecht (2007) explained the increased mobilization of DOC with increased amounts of redox-sensitive substances and enhanced availability of decomposable OM in the upper highly decomposed peat of a nutrient-rich fen. In 24 Finnish peatlands drained for forestry, DOC concentration in pore water increased following the first years of rewetting, thereafter DOC decreased (Menberu et al., 2017). However, 6 years post-rewetting, DOC was still higher in rewetted sites than in pristine peatlands. The increase of DOC together with P-concentration in recently rewetted peatlands under forest was explained by the leaching from forest residues and soil disturbances following clear-felling (Howson et al., 2021).

Differences in the vegetation of rewetted and natural sites were attributed to different DOC concentrations in a cutover bog in Canada: Ten years post-restoration, DOC in a rewetted site remained intermediate between natural and unrestored peatlands (Strack et al., 2015). In addition, DOC was characterized by smaller and less-aromatic structures at the restoration site, which indicated an input of DOC from the fresh litter of growing vegetation at the restored peatland (Strack et al., 2015). In line with this, Herzsprung et al. (2017) found higher DOC concentrations with a higher polyphenol contribution in summer time on sites rewetted for 9 years compared to a natural bog in the Harz Mountain in Germany. The authors attributed this to the higher degree and biomass of vascular plants at the rewetted site.

Given the limited data set, the increased DOC fluxes following drainage generally seem reversible by rewetting. However, as DOC is derived from turnover processes of peat and the living vegetation, the temporal evolution of vegetation after rewetting may be accompanied by temporal changes in DOC as well.

3.2 Agricultural management of peatlands without drainage: paludiculture

The term 'Paludiculture' refers to farming and agroforestry systems, which grow a commercial plant under wetland conditions in a sustainable way. The idea is that the SOC stocks of peatlands are preserved under a high water table and that a cultivated plant species - adapted to wet conditions - generates an economic income to the farmers (Wichtmann et al., 2016). Co-benefits of paludiculture include biodiversity and water purification (Giannini et al., 2017; Vroom et al., 2018; Wichtmann et al., 2016). From a conceptual viewpoint, paludiculture would - under optimal conditions - restore a peat-forming system, although the aboveground part of plant biomass is removed and can consequently not contribute to carbon input. To date, much research on this practice is conducted on pilot studies, which have been recently emerging. The research in northern peatlands focuses on the climate mitigation potential, growth conditions, practical implementation barriers, and political barriers for the cultivation of paludiculture plants. Many of the plants suggested for paludiculture are wetland species that tolerate a wide range of environmental conditions (Abel et al., 2013; Joosten and Clarke, 2002; Melts et al., 2019). As paludiculture is a relatively recent technique, the number of studies providing empirical data concerning field measurement of GHG exchange is limited. Particularly, studies covering multiple years of the complete carbon balance of paludicultures with a whole or several production cycles are missing. As paludiculture plants are colonizing natural habitats as well, few studies exist of these unmanaged systems. It is speculated that these unmanaged systems can serve as a proxy for the carbon balance of managed system and that the effect of biomass removal will not significantly alter the complete carbon balance (Tanneberger et al., 2020).

Recently, the transfer of the paludiculture concept to the tropics has been attempted (Tan et al., 2021). After severe peatland fires in 2015 in Indonesia, there was a need to rewet degraded and burned peatland and simultaneously provide benefit for livelihood, which was supported by the Indonesian Government new peatland restoration strategy (Tata, 2019). In this context, paludiculture research focuses on short-term socio-economic and environmental aspects of existing agroforestry systems on peatlands (Budiman et al., 2020; Tata, 2019). Those socio-economic aspects receive most attention that attempt to find alternatives to land clearance by fire for agriculture (Budiman et al., 2020). Tropical paludiculture systems are even less studied than temperate or boreal ones.

3.2.1 *Sphagnum farming on nutrient-poor temperate peatlands*

Sphagnum is a typical paludiculture plant for northern bogs, which can substitute natural peat for horticulture and thus simultaneously reduce peat mining activities (Gaudig et al., 2017; Wichtmann et al., 2016). A 6-year-old sphagnum farm, established on a rewetted cutover bog without topsoil removal, was characterized at the 2 last years of the production cycle by CH_4 emissions of 2 ± 0.6 g C m^{-2} yr^{-1} and a net carbon uptake amounting to −98 ± 20.1 g C m^{-2} yr^{-1} (Beyer and Höper, 2015). Optimization of the sphagnum farm in terms of availability of labile C and water management might lead to even lower CH_4 emissions (Günther et al., 2017). Particular attention should be paid to CO_2 emissions from ditch water and from constructed dams, which have the potential for further improvements in sphagnum farming (Günther et al., 2017). One option would be to use subsoil irrigation to avoid high emissions from ditch waters (Brown et al., 2017). A constant water table was shown to be essential in providing optimal conditions for growth and carbon balances in a North American sphagnum farm (Brown et al., 2017). However, all experimental plots were CO_2 sources in the second year of establishment, which was partly caused by the decomposition of straw, which was introduced for establishing the sphagnum cultivation (Brown et al., 2017). In addition, a low water table (around 10–25 cm below peat surface) may have added to the higher CO_2 emissions from this farm.

3.2.2 Cultivation of bioenergy plants on nutrient-rich temperate peatlands

In contrast to sphagnum farming on nutrient-poor bogs, reed (Phragmites) and Thypa are common wetland plants in fens. Several studies measured GHG emissions from naturally occurring reed and Thypha sites without harvesting (Franz et al., 2016; Mander et al., 2012; Minke et al., 2016; Shurpali et al., 2009; van den Berg et al., 2016). However, no GHG measurement took place at the harvested sites (Mander et al., 2012; Minke et al., 2016; Shurpali et al., 2009). Like in other peatlands, a high water table lowered carbon emissions from peat decomposition, while on the other hand, CH_4 emissions increased up to around 100 g C m^{-2} yr^{-1} under flooding (Franz et al., 2016; Mander et al., 2012; Minke et al., 2016; Shurpali et al., 2009; van den Berg et al., 2016). To the best of our knowledge, the only study that explicitly compares the effect of biomass harvest on the carbon balance was conducted in temperate rewetted fen by comparing gas fluxes of harvested and non-harvested plots of Phragmites, Thypa, and Carex vegetation (Günther et al., 2015). Cutting above ground biomass changed the net CO_2 uptake depending on the year and vegetation type. When including the biomass export from the harvested plots, harvested paludiculture plots became in most cases a stronger carbon source compared to the non-harvest plots. However, carbon losses of these 15-year-old rewetted sites were lower compared to their drained counterpart and thus would present a mitigation option (Günther et al., 2015). Günther et al. (2015) concluded that multiple-year studies are needed to evaluate the long-term effect of harvest on the carbon balance of managed peatlands. In view of increasing the yield for bioenergy plants and intensifying cultivation in future, practices including more frequent biomass harvest and fertilization have to be carefully evaluated (Mander et al., 2012; Kandel et al., 2019a).

3.2.3 Rice as potential paludiculture crop in subtropics and tropics

Rice is a staple crop with high water tolerance. Whereas the GHG balance and CH_4 release of paddy rice on mineral soil is in general well studied, few field experiments evaluated these parameters for paddy rice cropping on organic soils. Studies in Indonesia showed that rice crops had favourable GHG balances compared to other plant species and that even if rice cultivation let to high CH_4 emissions, compared with other corps, these were overcompensated by reduced CO_2 release from peat degradation (Hadi et al., 2005; Furukawa et al., 2005). In a review of other data from SE Asia, Hergoualc'h and Verchot (2014) compared data sets from measurements of various land-use and management types on intact and drained tropical peat. Paddy rice emitted less CO_2 than croplands or acadia plantations but did not reveal an improved GHG balance over most of the land-use types when CH_4 was taken into account. Water levels were the highest with rice (−4.5 cm), whereas water levels of other land-use types were between −18 and −78 cm, resulting in partially high CO_2 emissions. Hatala et al. (2012) measured CO_2 and CH_4 fluxes on degraded peatlands managed as paddy rice or pasture using EC over 2 years in California. The water table of the pasture was −50 to −80 cm throughout the year, whereas the paddy rice was inundated most of the year but drained 45 days before the start of the growing season for cultivation and planting and for about 55 days at the end of the growing season for harvest. Their study showed that paddy rice was a net GHG sink in both years owing to a substantial CO_2 uptake, whereas the pasture released large amounts of CO_2, in addition to CH_4 from cattle, and was overall a strong GHG source. With harvest exports taken into account, paddy rice management turned into an overall GHG source which was, however, substantially smaller than the one from the pasture system.

Together these few available GHG measurements with paddy rice on organic soil reveal that CO_2 emissions from peat decomposition can be largely reduced whereas CH_4 increases. The latter has long been recognized as an issue of paddy rice cropping (Seiler et al., 1984), and CH_4 release can be reduced by measures such as intermittent flooding to keep periods with preferable conditions for methanogens short (Minamikawa et al., 2018). Paddy rice on organic soils might be a more suitable agricultural crop than many others, but optimization of its GHG balance is still an understudied topic.

3.2.4 Potential of paludiculture to reduce carbon loss

Based on the available data and mechanistic understanding, a high water table reduces aerobic peat decomposition and thus CO_2 emissions of paludiculture systems compared to management after drained peatland. However, the empirical database is too small to say how the full carbon balance changes when harvest is included and hence, if paludiculture systems can be carbon neutral or even carbon sinks. More field studies investigating the complete carbon balance of paludicultures are needed to give the scientific background towards reducing GHG emissions from drained and rewetted peatlands. In addition, site-specific management options, including technical requirements to stabilize water tables throughout the year, and the importance of topsoil removal have to be taken into account.

3.3 Engineering measures to reduce greenhouse gas emissions under sustained agricultural management

3.3.1 Water management

Due to the strong dependence of the GHG emissions on the water table, the idea of finding the optimal water table - as high as possible but at the same time maintaining crop productivity and reducing CH_4 emissions - is obvious. Evans et al. (2021) estimated that raising average water tables of organic soils from currently −90 for cropland and −60 cm for grassland by half could save 3.3 Mt CO_2 yr^{-1} (2.7-4.0 Mt CO_2 yr^{-1}) with a negligible increase in CH_4 emissions in UK peatlands. Globally, a reduction potential of 65% corresponding to 508 Mt CO_2 yr^{-1} was approximated by reducing drainage depth by half in all agriculturally used peatlands (Evans et al., 2021).

However, this technique has received little attention so far, resulting in few experimental field studies quantifying the potential of reducing carbon losses through active water management. Short-term mesocosm experiments revealed that raising the water table from the typical drainage depth of −70 cm to an optimum of −30 cm would reduce total net emissions by about 42% in grassland systems (Regina et al., 2015). Musarika et al. (2017) tried to find the best balance between peat preservation and yield of radish. Raising the water table from −50 cm to −30 cm in mesocosms not only reduced the peat mineralization by half but also improved the productivity of radish. In contrast, mineralization was reduced by 31% but also decreased the yield of celery by 19% (Matysek et al., 2019). To avoid high economic loss, the authors suggested to raise the water table only in the non-growing season. Wen et al. (2020) compared precisely this setting. While raising the water table from −50 cm to −30 cm for both, fallow period and growing season, CO_2 emissions and yield of lettuce decreased by 36% and by 37%, respectively. In contrast, a higher water table restricted to the fallow period reduced annual CO_2 emissions by 30% without affecting lettuce yield. To balance the reduction of soil subsidence and yield in sugarcane fields in the Everglades, Florida, Glaz and Morris (2010) investigated the influence of water table on yield. A constant water table of −20 cm resulted in a lower sucrose yield compared to a water table of −45 cm. However, periodic flooding of up to 14 days did not reduce the yield but showed potential to reduce soil subsidence (Glaz and Morris, 2010). In contrast, short flooding cycles could also induce higher CO_2 emissions as compared to a constant drainage depth of −30 cm (Rodriguez et al., 2021). As the major part of emission with constant drainage occurs in summertime, constant flooding restricted to the summer period reduced CO_2 emissions. The authors, therefore, suggested to implement crop rotations that allow flooding during summer to reduce peat loss. Although most studies agree that raising the water table reduces carbon emissions, Berglund and Berglund (2011) observed that raising the water table from −80 to −40 cm resulted in slightly higher CO_2 emissions. These results demonstrate that the direction and size of effects following water table manipulations depend on crop and peatland type.

One technical approach to reach the targeted water table depth is subsurface drainage, also called subsoil irrigation. The basic principle is that irrigation pipes are placed at about −70 cm in agriculturally used peatlands to drain the sites at high water tables in autumn, winter, and spring and to raise the water table in summer to reduce peat mineralization. A relative dense network of the sub-irrigation pipes as narrow as 10 m was suggested to maintain a stable water table in the whole field for a lowland UK peat soil (Kechavarzi et al., 2007). Subsoil irrigation was suggested to reduce soil subsidence from drained pastures in the Netherlands (Querner et al., 2012). The effect of this technique on carbon loss was tested at four dairy farms by Weideveld et al. (2021). However, neither substantial higher yield nor reduced peat mineralization was found in 2 consecutive years (Weideveld et al., 2021). This could be related to a relatively small increase in the water table by about 6-18 cm in the summer. Indeed, only at periods with a water table difference of >20 cm, a reduction in soil respiration was measured.

In the tropics, a reduction of drainage depth in palm oil plantations to 40-60 cm below surface was proposed as a best management practice in order to reduce soil subsidence and CO_2 emissions and thus to extend the economic life span of a drained peat area (Lim et al., 2012). Moreover, the best plant growth of oil palms was found at a drainage depth of −55 cm compared to a drainage depth of −85 and −25 cm (Hashim et al., 2019). However, the average drainage depth is often considerably lower than −55 cm. Uda et al. (2020) evaluated the potential of low-drainage food crops to not only reduce CO_2 emissions but also include profitability, scalability of market, and acceptability to farmers. While sago palm (Metroxylon sagu) and illipe nut/tengkawang (Shorea spp.)

provided the most sustainable crops, as they grew at near-surface water levels, bananas (Musa paradisiaca), pineapples (Ananas comosus), and sweet melons (Cucumis melo) yielded the highest scores in terms of scalability of market and acceptability to farmers but needed a drainage depth of at least −30 cm (Uda et al., 2020). Yet, experimental data on the carbon balance of these systems are missing.

3.3.2 Soil management and cropping

To improve trafficability and increase the yield on drained peatlands under agriculture, different strategies are applied (Blankenburg, 2015; Sognnes et al., 2006). In principle, peat is covered with mineral material - mainly sand - with a thickness of about 20–40 cm. With shallow ploughing without mixing a deeper peat layer with mineral soil, the sand remains as a covering layer on top of the peat. Another strategy is to mix the mineral material with peat in a ratio of 1:1 (Blankenburg, 2015). These strategies were often applied at sites with underlying sand, which could be directly taken as cover material. Another way of mixing the peat is deep ploughing, where soils with underlying sand layers are deeply ploughed at a maximum of 2 m to shift and turn the soil layer at 130–150°, thereby creating alternating tilted bars of sand and peat (Schindler and Müller, 1999). However, sometimes other material than sand is used as covering material, including moraine material, material from construction sites, or clayey material in coastal regions (Blankenburg, 2015; Sognnes et al., 2006; Ferré et al., 2019). Addition of mineral material consistently alters the physical properties of drained peat. Hydraulic properties change and result in increased drainage and aeration of the upper soil layers, attenuation of spring frost, increased warming in spring, reduced fire risk, increased trafficability with heavy machinery, and ultimately increased yield (Bambalov, 1999; Blankenburg, 2015; Schindler and Müller, 1999; Sognnes et al., 2006). Yet the effects of these soil treatments on soil subsidence and finally carbon loss are poorly studied. Long-term soil subsidence measurements in the Netherlands revealed that peat soils with a naturally occurring clay cover subsided less as compared to peat soils without mineral covering (van den Akker et al., 2008). Moreover, Schindler and Müller (1999) investigated the soil subsidence of a sand-mixed site during the first 10 years after establishment in Germany. After initial subsidence of 1.5 cm yr^{-1}, rates decreased to 0.3 cm yr^{-1}, which was interpreted as a reduction in SOC loss. In peatlands of north-east Germany, a mineral top layer reduced soil subsidence from 0.5 cm yr^{-1} to 0.27 cm yr^{-1} when the initial underlying peat had a thickness of 30–70 cm but increased subsidence to 1.33 cm yr^{-1} at sites with an underlying peat layer of more than 100 cm (Fell et al., 2016). In contrast, in a peat soil with sand mixing and sand coverage, both treatments showed higher carbon mineralization rates, which was explained by higher soil temperatures

(Zaidelman and Shvarov, 2000). Some recent studies suggest that the addition of sand does not principally alter the CO_2 fluxes. Beyer (2014) measured an average SOC loss of 370–650 g C m^{-2} yr^{-1} on three peatland sites covered with sand. Furthermore, organic soils mixed with sand, resulting in a carbon content of about 10%, lost SOC in a similar range as 'true' peat soils with high carbon content (Leiber-Sauheitl et al., 2014). However, Wang et al. (2022) showed significantly reduced N_2O emissions after addition of mineral soil material on a drained organic soil.

As for sand additions, field experiments in the Netherlands showed that GHG emissions from peat soils with a naturally occurring clay layer were not systematically different as compared to peat soils without clay (Weideveld et al., 2021). Based on the current knowledge, addition of mineral material does not seem to reduce carbon mineralization. Interaction of the organic material with the mineral phase does not seem to play a role, in contrast to mineral soils. However, addition of mineral soils might lead to SOC protection if it allows to keep higher water levels which are needed to protect the underlying peat.

Based on the current research, the choice of crop or soil tillage practice seems to have a minor effect on SOC loss. Norberg et al. (2016a,b) concluded from 11 field experiments on different organic soil types that differences between years and sites were in general much larger than differences between crops (cereals and row crops) grown on the same field. Comparisons of barley with grassland revealed no clear effect of crop type on CO_2 emissions (Elsgaard et al., 2012; Maljanen et al., 2001; Lohila, 2004; Lohila et al., 2003) and also Tiemeyer et al. (2020) found no clear difference between the response of CO_2 emissions to water table differences between crop- and grasslands. In addition, Taft et al. (2018) concluded from the incubation of intact soil columns of horticultural peatland that zero or minimum tillage could not reduce SOC loss. Even after cessation of cultivation practices, soil carbon loss remains high in northern peatlands when cultivation was abandoned (Kløve et al., 2010; Maljanen et al., 2010). Similarly, bare peatlands, where a vegetation cover is missing, showed slightly increased carbon loss in three northern peatlands (Maljanen et al., 2010).

3.3.3 Can afforestation reduce soil carbon loss from drained peatlands?

Under specific conditions, afforestation of shallow-drained Fennoscandia peatlands may lead to an increase in SOC stocks. Measuring the components of NECB with the EC methods showed that a naturally forested pine bog with drainage and afforestation lead to an increase on average of about 60 g C m^{-2} yr^{-1} SOC in 5 consecutive years (Lohila et al., 2011; Minkkinen et al., 2018).

Even in a severe drought year, the drained peatland remained a carbon sink (Minkkinen et al., 2018). This is in line with earlier research, based on SOC stock measurement (Minkkinen and Laine, 1998; Minkkinen et al., 1999). These results were explained with the relatively shallow drainage depth of about −30 to −40 cm in Finnish peatlands, a similar plant structure compared to undrained sites and a significantly increased tree stand growth and litter production (Minkkinen et al., 2018). In addition, in drained afforested peatland, a new secondary humus layer formed on the top of peat (Minkkinen et al., 2008). Drainage induced a decrease in pH due to oxidation processes, enhanced nutrient uptake by trees, and decreased enzyme activity. In addition, soil temperature decreased in the long-term, caused by the decrease of thermal conductivity in the drier surface peat and increasing shading by growing trees (Minkkinen et al., 2008). It was shown that drier surface conditions at the drained peatlands lead to retarded decomposition of pine needles and fine roots compared to undisturbed peatlands (Laiho et al., 2004). However, this SOC increase is restricted to nutrient-poor sites, whereas nutrient-rich forests were net carbon sources (Ojanen et al., 2013). Also Meyer et al. (2013) observed that a spruce-dominated fertile organic soil on a former agricultural land in southwestern Sweden represents a small carbon sink during peak forest production but would become a carbon source in the long run. Laine et al. (1996) evaluated the climatic impact of drainage in nutrient-poor and -rich southern Finnish peatland by modelling the change in radiative forcing. Drainage in nutrient-rich sedge fen lead to a cooling effect due to the low CH_4 emission, although the peat layer turned from a C sink to a C source, but increased CO_2 emissions would have generated a warming effect over longer time scales. A similar, but smaller, effect was predicted for bogs (Laine et al., 1996).

The assessment of temporal dynamics through the full forest production cycle is important, as GHG fluxes change with stand age. Forest clearance comes with soil preparation and drastic change in site conditions. Directly after clear-cutting, high CO_2 emissions of around 840 ± 40 g C m^{-2} yr^{-1} were measured for the first and 565 ± 34 g C m^{-2} yr^{-1} for the second year in a nutrient-rich boreal peatland in southern Finland (Korkiakoski et al., 2019). These high CO_2 emissions were explained with a high rate of decomposition of peat and logging residues, which were left on site. It was estimated that about 49% of total ecosystem respiration was attributed to the decomposition of logging residues. In addition, GPP was sharply reduced because of the removal of assimilating trees and a declining understory. Reduction of evaporation led to a rise in the water table of about 20 cm, which reduced the volume of aerated peat and likely reduced decomposition rates of peat. However, increased soil temperatures may have an inverse effect. Recovery of ground vegetation in the second year and decreased decomposition of logging residues caused a reduction in CO_2 emissions by about 41%.

Unlike the Fennoscandia-afforested peatlands, considerably fewer data exist for other global regions, and it was questioned if results from northern peatlands can be transferred to afforested peatlands under different climate and management conditions in UK and Ireland (Sloan et al., 2018). In contrast to relative shallow-drained Fennoscandia afforestation, closely spaced plough furrows between deeper drainage is a common soil preparation in the UK, which may also influence carbon mineralization (Sloan et al., 2018). Mature forests in eight sites in Ireland represented soil carbon sources of about 60–300 g C (Jovani-Sancho et al., 2021). Using a snap-shot measurement and modelling, Hargreaves et al. (2003) estimated that the ecosystem of a Scottish peatland conifers acts as a net source of about 200–400 g C m^{-2} yr^{-1} in the first years after ploughing. After 4–8 years when trees started to grow it became a sink of about −300 g C m^{-2} yr^{-1} that was further increased up to −500 g C m^{-2} yr^{-1} in the following years when tree growth dominated. After accounting for the carbon stored in the biomass, the soil appeared to lose carbon at a rate of about 100 g C m^{-2} yr^{-1}. Similarly, a temperate 44-years-old spruce afforestation on a German bog was a carbon sink of about −157 ± 36 g C m^{-2} yr^{-1} (Hommeltenberg et al., 2014). However, when accounting for SOC loss since plantation and average rotation length of 60 years, the site was found to be a strong SOC source. Quantifying the carbon balance over a whole production cycle for forests is challenging but unavoidable for the robust estimate of the effect of peatland afforestation on the carbon budget. Up to date, available data suggest that - apart from nutrient-poor bogs in Fennoscandia - many afforested peatlands are still considerable carbon sources, especially in the tropics.

3.4 Metrics to evaluate reduced CO_2 emissions versus increased CH_4 emissions after rewetting

The balance between CO_2 sinks and CH_4 emissions strongly determines the net climatic impact of rewetted peatlands, and these two GHGs show a strong and inverse dependence on water table depth. To assess the climatic impact of rewetting activities through ecosystem GHG exchange, the differing radiative properties and atmospheric lifetimes of GHGs need to be accounted for. Compared to CO_2, CH_4 has a much larger radiative efficiency but a much shorter lifetime of 9-12 years (Myhre et al., 2013). The huge difference in lifetime leads to strongly time-dependent climatic effects. A steady emission of CH_4 at constant rates lead to rapid increase of atmospheric CH_4 that stabilizes after a few decades when natural atmospheric removal of CH_4 balances out ongoing emissions (Lynch et al., 2020). Correspondingly, the radiative forcing by CH_4 emissions rapidly increases due to the high radiative efficiency of CH_4 but then largely stabilizes. In contrast, continuous emissions of CO_2 lead to steady increases of atmospheric CO_2 concentrations and a steady increase of

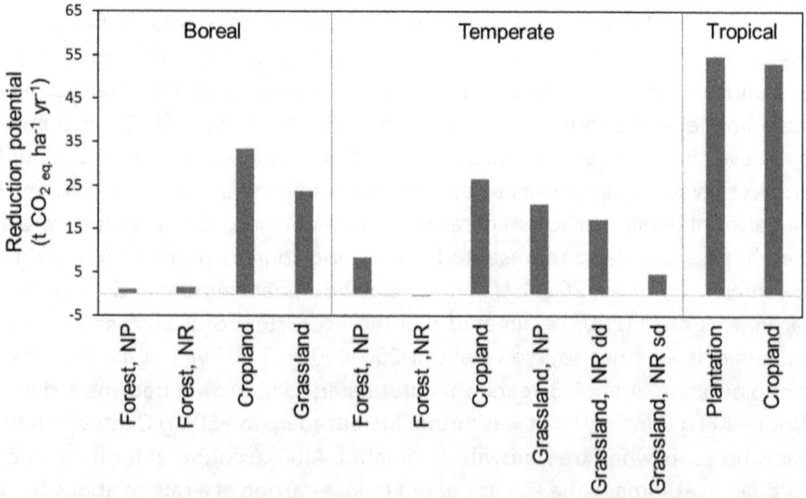

Figure 8 Emission reduction through rewetting for different land use and climatic zones (data from Wilson et al., 2016a). NR, nutrient-rich; NP, nutrient-poor; dd, deep-drained (>30 cm); sd, shallow-drained (<30 cm).

radiative forcing as long as the emissions sustain (Lynch et al., 2020) because of the long atmospheric lifetime of CO_2. In consequence, CO_2-induced radiative forcing is strongly depending on the cumulative emissions, while CH_4 emissions contribute predominately as a function their emission rate (Allen et al., 2016).

The most commonly used metric global warming potential (GWP) integrates the radiative forcing of a GHG emission pulse over a prescribed period, often 20 or 100 years. Using the GWP, non-CO_2 GHG fluxes are converted to a common unit called CO_2 equivalent (i.e. the cumulative radiative forcing of that gas relative to that of CO_2). Owing to the different lifetimes of the GHGs', the GWP is time-dependent. While this measure is convenient and widely adopted, it does not properly reflect the radiative forcing of quasi-continuous emissions and uptakes of GHGs' from peatlands. To overcome this shortcoming, two approaches were suggested. First, the GWP of sustained emissions and removals, also called sustained flux GWP, captures more precisely the time-dependent radiative forcing of short-lived GHGs (Allen et al., 2018; Lynch et al., 2020; Neubauer and Megonigal, 2015). It is able to relate the change in the emission rate of a short-lived GHG such as CH_4 rather than the emission rate itself to the radiative effect of CO_2. Yet, Lynch et al. (2020) encouraged to report the radiative forcing effect of each GHG separately as the most transparent way for GHG reporting and communication. Despite these recent advantages in metrics, the most straightforward approach for describing the temperature effect of GHG emissions and removals is to report the radiative forcing over time either as instantaneous or cumulative forcing.

This is particularly meaningful because emission and removal rates may change during, for example, rewetting, which limits the suitability of static approaches. Calculating the change in radiative forcing of peatlands over time, based on GHG fluxes, has been first done by Frolking et al. (2006) who showed that 'the current radiative forcing of intact peatlands is determined primarily by a trade-off between the total C sequestered since the peatland's formation and the recent methane fluxes' and that the net CO_2 sink of an intact peatland outweighs its CH_4 emissions after some centuries or millennia. The time after which a net cooling (cumulative radiative forcing becomes negative) is achieved can be up to 14 000 years for intact wetlands as shown by Neubauer (2014). This point in time, termed radiative forcing switchover time, is strongly and non-linearly related to the ratio of CO_2 uptake to CH_4 release (Frolking et al., 2006; Neubauer, 2014). Another way of approaching the warming versus cooling effect of peatlands is to look at the instantaneous, not the cumulative, forcing. Ojanen and Minkkinen (2020) showed that, by calculating instantaneous radiative forcing, rewetting offers climate benefits for tropical and agricultural peatlands but not for peatlands previously drained for forestry because of the different balance of CO_2 emission savings versus rewetting-induced CH_4. Despite long switchover times, rewetting peatlands reduces their instantaneous radiative forcing already after a few decades (Günther et al., 2020).

In this chapter, we have followed the conventional GWP notion as it is still the most used metric to compare the climatic impact of different GHGs. Based on the difference of IPCC default values of drained peatlands and an updated default value for rewetted peatlands, Wilson et al. (2016a) calculated the emission reduction through rewetting for different climate regions and land uses, following the IPCC classification (IPCC, 2014; Fig. 8). The magnitude of the reduction potentials is displayed by climate zone, land-use intensity, and drainage depth. The highest reduction potential is located in tropical peatlands, followed by boreal and temperate peatlands used as croplands, and boreal and temperate peatlands under grass. In almost all cases, rewetted peatlands turn into CO_2 sinks whereas CH_4 emissions increase and N_2O becomes negligible. Based on the 100-years GWP applied by Wilson et al. (2016a), the increased CH_4 is the dominant factor for the overall GWP after rewetting for all climate and land-use combinations. Yet, in all combinations except under nutrient-rich temperate forests, rewetting provides a climate benefit. Because CH_4 plays a relatively important role when GWP100 is used, the climate benefit will likely increase over longer time scales.

3.5 Outlook

In the course of the discussion about carbon sequestration potentials in soils, scientific attempts of how to restore the C sink or to reduce the climatic impact of organic soils and finally to implement them in climatic mitigation strategies are rapidly emerging (e.g. Günther et al., 2020; Humpenöder et al., 2020; Leifeld and Menichetti, 2018; Paustian et al., 2016; Tanneberger et al., 2020). Rewetting and restoration of natural peatlands is the main option to avoid SOC losses, while at the same time increasing CH_4 losses and economic losses as main trade offs. The first major rewetting activities occurred in the temperate and boreal zone at sites, which were not intensively used before. Therefore, research up to now has focused on little or comparably moderately degraded peatlands, which were rewetted with the aim of re-establishing the origin habitat conditions or of restoring peat extraction sites (Wilson et al., 2016b). Owing to their high emissions, the demand for mitigation and rewetting activities shifted to more intensively used peatlands as well (Harpenslager et al., 2015; Tiemeyer et al., 2020), emphasizing the need to investigate the GHG dynamics following rewetting activities in these formerly intensively used sites with a high potential of CH_4 and N_2O emissions. Rewetting such sites has a high conflict potential between economic revenue and GHG reduction potential (Ferré et al., 2019). Further investigations are needed to develop sustainable management strategies for the agricultural use of peatlands where restoration is not an option due to hydrological restrictions or the socio-economic constraints. Paludiculture was recently introduced as an agricultural practice on rewetted organic soils to achieve biomass production with water tolerant species such as sphagnum mosses, reed, and cattail.

Reducing high carbon loss from peat decomposition and fires in tropical drained peatlands by rewetting offers a substantial mitigation potential (Ojanen and Minkkinen, 2020). Yet, implementation is challenging because of the high pressure to clear land by fire for agriculture or plantations (Budiman et al., 2020), as well as owing to the very high costs of re-establishing ecosystem services (Hansson and Dargusch, 2017). To overcome this land-use conflict, more research about tropical paludiculture species and implementation strategies is urgently needed (Giesen, 2021), particularly in the light of the tremendous and tragic failure of converting peat swamp forests to rice production in Indonesia in the 1990es (Goldstein, 2015).

The need for paludiculture research likewise applies to temperate agriculture, as rewetting offers also rapid climatic benefits (Ojanen and Minkkinen, 2020). In addition, technical measures such as combining soil coverage with active water management enhancing SOC protection and reducing GHG emissions are worth to be investigated further. Finally, for the mechanistic understanding and modelling of carbon dynamics in drained and rewetted peatlands, more long-term field studies are needed that integrate different phases of rewetting and production cycles for various peatland types and historical uses. These data, together with drainage depth and peat quality, are necessary to further develop existing peatland models (e.g. Bona et al., 2020; Frolking et al., 2010; Premrov et al., 2021) as tools for evaluating mitigation options.

4 Where to look for further information

There are several societies, non-profit organizations and networks aiming at the conservation and restoration of peatlands, their sustainable management, interdisciplinary research on the functioning of these ecosystems, and knowledge transfer to policy makers and the society. Below is a selection of the most relevant webpages (all webpages were accessed on 25/4/2022).

- The International Peatland Society (IPS, https://peatlands.org) was founded in Canada in 1968. Its mission is to serve all those involved in peatlands and peat through the promotion, gathering, exchange and communication of knowledge and experience, by means of events and projects, which address key issues, including climate change, biodiversity, the need for responsible use and restoration.
- The International Mire Conservation Group (IMCG, http://www.imcg.net/) is an international network of specialists promoting conservation and exchange knowledge of mires and related ecosystems. Together with the IPS it publishes the peer-reviewed journal Mires & Peat (http://www.mires-and-peat.net/), covering all aspects of peatland science, technology and wise use.
- The aim of the Global Peatland Initiative (https://www.globalpeatlands.org/) is to save the carbon stored in peatlands by improving the conservation, restoration and sustainable management of peatlands.
- Ramsar Convention on Wetlands of International Importance especially as Waterfowl Habitat (https://www.ramsar.org/) is an international convention for the conservation and sustainable use of wetlands.
- Wetlands International https://www.wetlands.org/ is a global not-for-profit organization dedicated to the conservation and restoration of wetlands.
- The IUCN UK Peatlands Programme (https://www.iucn-uk-peatlandprogramme.org) is a national program to promote peatland restoration in the UK. The Program advocates the multiple benefits of peatlands through partnerships, science, policy and practice. The aspect of climate change mitigation is a central theme.
- The Greifswald Mire Center (https://www.greifswaldmoor.de/home.html) performs interdisciplinary research, provides policy makers and society with knowledge and imparts theoretical and practical knowledge on climate protection, biodiversity and sustainable use of peatlands. As one special research focus, they study paludiculture.

Published by Burleigh Dodds Science Publishing Limited, 2023.

5 References

Abdalla, M., Hastings, A., Truu, J., Espenberg, M., Mander, Ü. and Smith, P. 2016. Emissions of methane from northern peatlands: a review of management impacts and implications for future management options. *Ecology and Evolution* 6(19), 7080–7102. https://doi.org/10.1002/ece3.2469.

Abel, S., Couwenberg, J., Dahms, T. and Joosten, H. 2013. The Database of Potential Paludiculture Plants (DPPP) and results for western Pomerania. *Plant Diversity and Evolution* 130(3–4), 219–228. https://doi.org/10.1127/1869-6155/2013/0130-0070.

Ali, M., Taylor, D. and Inubushi, K. 2006. Effects of environmental variations on CO2 Efflux from a tropical peatland in eastern Sumatra. *Wetlands* 26(2), 612–618. https://doi.org/10.1672/0277-5212(2006)26[612:EOEVOC]2.0.CO;2.

Allen, M. R., Fuglestvedt, J. S., Shine, K. P., Reisinger, A., Pierrehumbert, R. T. and Forster, P. M. 2016. New use of global warming potentials to compare cumulative and short-lived climate pollutants. *Nature Climate Change* 6(8), 773–776. https://doi.org/10.1038/nclimate2998.

Allen, M. R., Shine, K. P., Fuglestvedt, J. S., Millar, R. J., Cain, M., Frame, D. J. and Macey, A. H. 2018. A solution to the misrepresentations of CO2-equivalent emissions of short-lived climate pollutants under ambitious mitigation. *NPJ Climate and Atmospheric Science* 1(1), 16. https://doi.org/10.1038/s41612-018-0026-8.

Alm, J., Saarnio, S., Nykänen, H., Silvola, J. and Martikainen, P. 1999a. Winter CO_2, CH_4 and N_2O fluxes on some natural and drained boreal peatlands. *Biogeochemistry* 44(2), 163–186. https://doi.org/10.1007/BF00992977.

Alm, J., Schulman, L., Walden, J., Nykänen, H., Martikainen, P. J. and Silvola, J. 1999b. Carbon balance of a boreal bog during a year with an exceptionally dry summer. *Ecology* 80(1), 161–174.

Alm, J., Shurpali, N. J., Tuittila, E.-S., Laurila, T., Maljanen, M., Saarnio, S. and Minkkinen, K. 2007. Methods for determining emission factors for the use of peat and peatlands–flux measurements and modelling. *Boreal Environment Research* 12, 16.

Augustin, J. and Chojnicki, B. 2008. Austausch von klimarelevanten Spurengasen, Klimawirkung und Kohlenstoffdynamik in den ersten Jahren nach der Wiedervernässung von degradiertem Niedermoorgrünland. In: Gelbrecht, J., Zak, D. and Augustin, J. (Eds), *Phosphor- und Kohlenstoff-Dynamikund Vegetationsentwicklung in wiedervernässten Mooren des Peenetals in Mecklenburg-Vorpommern -Status, Steuergrößen und Handlungsmöglichkeiten*, Berichte des IGB Heft 26. IGB, Berlin. pp. 50–67.

Bader, C., Müller, M., Schulin, R. and Leifeld, J. 2018a. Peat decomposability in managed organic soils in relation to land use, organic matter composition and temperature. *Biogeosciences* 15(3), 703–719. https://doi.org/10.5194/bg-15-703-2018.

Bader, C., Müller, M., Szidat, S., Schulin, R. and Leifeld, J. 2018b. Response of peat decomposition to corn straw addition in managed organic soils. *Geoderma* 309, 75–83. https://doi.org/10.1016/j.geoderma.2017.09.001.

Bader, C., Müller, M., Schulin, R. and Leifeld, J. 2017. Amount and stability of recent and aged plant residues in degrading peatland soils. *Soil Biology and Biochemistry* 109, 167–175. https://doi.org/10.1016/j.soilbio.2017.01.029.

Baldock, J. A., Oades, J. M., Nelson, P. N., Skene, T. M., Golchin, A. and Clarke, P. 1997. Assessing the extent of decomposition of natural organic materials using solid-state 13C NMR spectroscopy. *Soil Research* 35(5), 1061–1083.

Bambalov, N. 1999. Dynamic of organic matter in peat soil and under the conditions of sand-mix culture during 15 years. *International Agrophysics* 13, 269-272.

Berglund, Ö. and Berglund, K. 2011. Influence of water table level and soil properties on emissions of greenhouse gases from cultivated peat soil. *Soil Biology and Biochemistry* 43(5), 923-931. https://doi.org/10.1016/j.soilbio.2011.01.002.

Beyer, C. 2014. Greenhouse gas exchange of organic soils in northwest Germany: effects of organic soil cultivation, agricultural land use and restoration. PhD-thesis, Universität Bremen. Available at: http://nbn-resolving.de/urn:nbn:de:gbv:46-00103939-13.

Beyer, C. and Höper, H. 2015. Greenhouse gas exchange of rewetted bog peat extraction sites and a Sphagnum cultivation site in northwest Germany. *Biogeosciences* 12(7), 2101-2117. https://doi.org/10.5194/bg-12-2101-2015.

Billett, M., Charman, D., Clark, J., Evans, C., Evans, M., Ostle, N., Worrall, F., Burden, A., Dinsmore, K., Jones, T., McNamara, N., Parry, L., Rowson, J. and Rose, R. 2010. Carbon balance of UK peatlands: current state of knowledge and future research challenges. *Climate Research* 45, 13-29. https://doi.org/10.3354/cr00903.

Billett, M. F., Deacon, C. M., Palmer, S. M., Dawson, J. J. C. and Hope, D. 2006. Connecting organic carbon in stream water and soils in a peatland catchment: organic carbon in stream water and soils. *Journal of Geophysical Research: Biogeosciences* 111(G2), n/a-n/a. https://doi.org/10.1029/2005JG000065.

Billett, M. F., Palmer, S. M., Hope, D., Deacon, C., Storeton-West, R., Hargreaves, K. J., Flechard, C. and Fowler, D. 2004. Linking land-atmosphere-stream carbon fluxes in a lowland peatland system: land-atmosphere-stream carbon fluxes. *Global Biogeochemical Cycles* 18(1), n/a-n/a. https://doi.org/10.1029/2003GB002058.

Blankenburg, J. 2015. Die landwirtschaftliche Nutzung von Mooren in Nordwestdeutschland. *Telma* 5, 29-58.

Bona, K. A., Shaw, C., Thompson, D. K., Hararuk, O., Webster, K., Zhang, G., Voicu, M. and Kurz, W. A. 2020. The Canadian Model for Peatlands (CaMP): a peatland carbon model for national greenhouse gas reporting. *Ecological Modelling* 431, 109164. https://doi.org/10.1016/j.ecolmodel.2020.109164.

Brake, M., Höper, H. and Joergensen, R. G. 1999. Land use-induced changes in activity and biomass of microorganisms in raised bog peats at different depths. *Soil Biology and Biochemistry* 31(11), 1489-1497.

Broder, T., Blodau, C., Biester, H. and Knorr, K. H. 2012. Peat decomposition records in three pristine ombrotrophic bogs in southern Patagonia. *Biogeosciences* 9(4), 1479-1491. https://doi.org/10.5194/bg-9-1479-2012.

Brown, C. M., Strack, M. and Price, J. S. 2017. The effects of water management on the CO2 uptake of Sphagnum moss in a reclaimed peatland. *Mires and Peat*, 1-15. https://doi.org/10.19189/MaP.2016.OMB.258.

Bubier, J. L., Crill, P. M., Moore, T. R., Savage, K. and Varner, R. K. 1998. Seasonal patterns and controls on net ecosystem CO2 exchange in a boreal peatland complex. *Global Biogeochemical Cycles* 12(4), 703-714. https://doi.org/10.1029/98GB02426.

Budiman, I., Bastoni, Sari, E. N., Hadi, E. E., Asmaliyah, Siahaan, H., Januar, R. and Hapsari, R. D. 2020 Progress of paludiculture projects in supporting peatland ecosystem restoration in Indonesia. *Global Ecology and Conservation* 23, e01084. https://doi.org/10.1016/j.gecco.2020.e01084.

Carter, M. S., Larsen, K. S., Emmett, B., Estiarte, M., Field, C., Leith, I. D., Lund, M., Meijide, A., Mills, R. T. E., Niinemets, Ü., Peñuelas, J., Portillo-Estrada, M., Schmidt, I. K., Selsted, M. B., Sheppard, L. J., Sowerby, A., Tietema, A. and Beier, C. 2012. Synthesizing greenhouse gas fluxes across nine European peatlands and shrublands - responses to climatic and environmental changes. *Biogeosciences* 9(10), 3739-3755. https://doi.org/10.5194/bg-9-3739-2012.

Clymo, R. S. 1984. The limits to peat bog growth. *Philosophical Transactions of the Royal Society of London. B, Biological Sciences* 303(1117), 605-654. https://doi.org/10.1098/rstb.1984.0002.

Clymo, R. S. and Bryant, C. L. 2008. Diffusion and mass flow of dissolved carbon dioxide, methane, and dissolved organic carbon in a 7-m deep raised peat bog. *Geochimica et Cosmochimica Acta* 72(8), 2048-2066. https://doi.org/10.1016/j.gca.2008.01.032.

Charman, D. J., Amesbury, M. J., Hinchliffe, W., Hughes, P. D. M., Mallon, G., Blake, W. H., Daley, T. J., Gallego-Sala, A. V. and Mauquoy, D. 2015. Drivers of Holocene peatland carbon accumulation across a climate gradient in northeastern North America. *Quaternary Science Reviews* 121, 110-119.

Conant, R. T., Ogle, S. M., Paul, E. A. and Paustian, K. 2011. Measuring and monitoring soil organic carbon stocks in agricultural lands for climate mitigation. *Frontiers in Ecology and the Environment* 9(3), 169-173. https://doi.org/10.1890/090153.

Cook, S., Whelan, M. J., Evans, C. D., Gauci, V., Peacock, M., Garnett, M. H., Kho, L. K., Teh, Y. A. and Page, S. E. 2018. Fluvial organic carbon fluxes from oil palm plantations on tropical peatland (preprint). *Biogeosciences* 15(24), 7435-7450. https://doi.org/10.5194/bg-15-7435-2018.

Cooper, H. V., Vane, C. H., Evers, S., Aplin, P., Girkin, N. T. and Sjögersten, S. 2019. From peat swamp forest to oil palm plantations: the stability of tropical peatland carbon. *Geoderma* 342, 109-117. https://doi.org/10.1016/j.geoderma.2019.02.021.

Couwenberg, J., Dommain, R. and Joosten, H. 2010. Greenhouse gas fluxes from tropical peatlands in south-east Asia. *Global Change Biology* 16(6), 1715-1732. https://doi.org/10.1111/j.1365-2486.2009.02016.x.

Couwenberg, J. and Fritz, C. 2012. Towards developing IPCC methane 'emission factors' for peatlands (organic soils). *Mires and Peat* 10(03), 1-17. Available at: http://mires-and-peat.net/pages/volumes/map10/map1003.php.

Davidson, N. C. 2014. How much wetland has the world lost? Long-term and recent trends in global wetland area. *Marine and Freshwater Research* 65(10), 934. https://doi.org/10.1071/MF14173.

Dawson, J. J. C., Billett, M. F., Neal, C. and Hill, S. 2002. A comparison of particulate, dissolved and gaseous carbon in two contrasting upland streams in the UK. *Journal of Hydrology* 257(1-4), 226-246. https://doi.org/10.1016/S0022-1694(01)00545-5.

Dean, J. F., Garnett, M. H., Spyrakos, E. and Billett, M. F. 2019. The potential hidden age of dissolved organic carbon exported by peatland streams. *Journal of Geophysical Research: Biogeosciences* 124(2), 328-341. https://doi.org/10.1029/2018JG004650.

Deppe, M., Knorr, K.-H., McKnight, D. M. and Blodau, C. 2010. Effects of short-term drying and irrigation on CO_2 and CH_4 production and emission from mesocosms of a northern bog and an alpine fen. *Biogeochemistry* 100(1-3), 89-103. https://doi.org/10.1007/s10533-010-9406-9.

Deshmukh, C. S., Julius, D., Evans, C. D., Nardi, Susanto, A. P., Page, S. E., Gauci, V., Laurén, A., Sabiham, S., Agus, F., Asyhari, A., Kurnianto, S., Suardiwerianto, Y. and Desai, A. R. 2020. Impact of forest plantation on methane emissions from tropical peatland. *Global Change Biology* 26, 2477-2495. https://doi.org/10.1111/gcb.15019.

Deverel, S. J. and Leigthon, D. A. 2010. Historic, recent, and future subsidence, Sacramento-San Joaquin Delta, California, USA. *San Francisco Estuary and Watershed Science* 8(2), 2. https://doi.org/10.15447/sfews.2010v8iss2art1.

Duval, T. P. and Radu, D. D. 2018. Effect of temperature and soil organic matter quality on greenhouse-gas production from temperate poor and rich fen soils. *Ecological Engineering* 114, 66–75. https://doi.org/10.1016/j.ecoleng.2017.05.011.

Eggelsmann, R., Heathwaite, A. L., Gross-Braukmann, G., Kuster, E., Naucke, W., Schich, M. and Schweikle, V. 1993. Physical processes and properties of mires. In: Heathwaite, A. L. and Gottlich, K. H. (Eds), *Mires, Process, Exploration and Conservation*. John Wiley & Sons, Chichester, UK. pp. 171-262.

Egli, M., Wiesenberg, G., Leifeld, J., Gärtner, H., Seibert, J., Röösli, C., Wingate, V., Dollenmeier, W., Griffel, P., Suremann, J., Weber, J., Zyberaj, M. and Musso, A. 2021. Formation and decay of peat bogs in the vegetable belt of Switzerland. *Swiss Journal of Geosciences* 114(1), 2. https://doi.org/10.1186/s00015-020-00376-0.

Elsgaard, L., Görres, C.-M., Hoffmann, C. C., Blicher-Mathiesen, G., Schelde, K. and Petersen, S. O. 2012. Net ecosystem exchange of CO2 and carbon balance for eight temperate organic soils under agricultural management. *Agriculture, Ecosystems and Environment* 162, 52-67. https://doi.org/10.1016/j.agee.2012.09.001.

Erkens, G., van der Meulen, M. J. and Middelkoop, H. 2016. Double trouble: subsidence and CO2 respiration due to 1,000 years of Dutch coastal peatlands cultivation. *Hydrogeology Journal* 24(3), 551–568. https://doi.org/10.1007/s10040-016-1380-4.

Evans, C. D., Peacock, M., Baird, A. J., Artz, R. R. E., Burden, A., Callaghan, N., Chapman, P. J., Cooper, H. M., Coyle, M., Craig, E., Cumming, A., Dixon, S., Gauci, V., Grayson, R. P., Helfter, C., Heppell, C. M., Holden, J., Jones, D. L., Kaduk, J., Levy, P., Matthews, R., McNamara, N. P., Misselbrook, T., Oakley, S., Page, S. E., Rayment, M., Ridley, L. M., Stanley, K. M., Williamson, J. L., Worrall, F. and Morrison, R. 2021. Overriding water table control on managed peatland greenhouse gas emissions. *Nature* 593(7860), 548–552. https://doi.org/10.1038/s41586-021-03523-1.

Evans, C. D., Renou-Wilson, F. and Strack, M. 2016. The role of waterborne carbon in the greenhouse gas balance of drained and re-wetted peatlands. *Aquatic Sciences* 78(3), 573–590. https://doi.org/10.1007/s00027-015-0447-y.

Evans, C. D., Williamson, J. M., Kacaribu, F., Irawan, D., Suardiwerianto, Y., Hidayat, M. F., Laurén, A. and Page, S. E. 2019. Rates and spatial variability of peat subsidence in Acacia plantation and forest landscapes in Sumatra, Indonesia. *Geoderma* 338, 410–421. https://doi.org/10.1016/j.geoderma.2018.12.028.

Fell, H., Roßkopf, N., Bauriegel, A. and Zeitz, J. 2016. Estimating vulnerability of agriculturally used peatlands in north-east Germany to carbon loss based on multi-temporal subsidence data analysis. *CATENA* 137, 61–69. https://doi.org/10.1016/j.catena.2015.08.010.

Fenner, N. and Freeman, C. 2020. Woody litter protects peat carbon stocks during drought. *Nature Climate Change* 10(4), 363–369. https://doi.org/10.1038/s41558-020-0727-y.

Ferré, M., Muller, A., Leifeld, J., Bader, C., Müller, M., Engel, S. and Wichmann, S. 2019. Sustainable management of cultivated peatlands in Switzerland: insights, challenges, and opportunities. *Land Use Policy* 87. https://doi.org/10.1016/j.landusepol.2019.05.038, 104019.

Fortuniak, K., Pawlak, W., Bednorz, L., Grygoruk, M., Siedlecki, M. and Zieliński, M. 2017. Methane and carbon dioxide fluxes of a temperate mire in Central Europe.

Agricultural and Forest Meteorology 232, 306–318. https://doi.org/10.1016/j.agrformet.2016.08.023.

Frank, S. 2016. Factors controlling concentrations and losses of dissolved carbon and nitrogen from disturbed bogs in lower saxony. PhD-thesis. Gottfried Wilhelm Leibniz Universität Hannover. https://doi.org/10.15488/8674.

Franz, D., Koebsch, F., Larmanou, E., Augustin, J. and Sachs, T. 2016. High net CO_2 and CH_4 release at a eutrophic shallow lake on a formerly drained fen. *Biogeosciences* 13(10), 3051–3070. https://doi.org/10.5194/bg-13-3051-2016.

Frolking, S., Roulet, N. and Fuglestvedt, J. 2006. How northern peatlands influence the Earth's radiative budget: sustained methane emission versus sustained carbon sequestration. *Journal of Geophysical Research* 111(G1), G01008. https://doi.org/10.1029/2005JG000091.

Frolking, S. and Roulet, N. T. 2007. Holocene radiative forcing impact of northern peatland carbon accumulation and methane emissions. *Global Change Biology* 13(5), 1079–1088. https://doi.org/10.1111/j.1365-2486.2007.01339.x.

Frolking, S., Roulet, N. T., Tuittila, E., Bubier, J. L., Quillet, A., Talbot, J. and Richard, P. J. H. 2010. A new model of Holocene peatland net primary production, decomposition, water balance, and peat accumulation. *Earth System Dynamics* 1(1), 1–21. https://doi.org/10.5194/esd-1-1-2010.

Frolking, S., Talbot, J., Jones, M. C., Treat, C. C., Kauffman, J. B., Tuittila, E.-S. and Roulet, N. 2011. Peatlands in the Earth's 21st century climate system. *Environmental Reviews* 19, 371–396. https://doi.org/10.1139/a11-014.

Furukawa, Y., Inubushi, K., Ali, M., Itang, A. M. and Tsuruta, H. 2005. Effect of changing groundwater levels caused by land-use changes on greenhouse gas fluxes from tropical peat lands. *Nutrient Cycling in Agroecosystems* 71(1), 81–91. https://doi.org/10.1007/s10705-004-5286-5.

Gallego-Sala, A. V., Charman, D. J., Brewer, S., Page, S. E., Prentice, I. C., Friedlingstein, P., Moreton, S., Amesbury, M. J., Beilman, D. W., Björck, S., Blyakharchuk, T., Bochicchio, C., Booth, R. K., Bunbury, J., Camill, P., Carless, D., Chimner, R. A., Clifford, M., Cressey, E., Courtney-Mustaphi, C., De Vleeschouwer, F., de Jong, R., Fialkiewicz-Koziel, B., Finkelstein, S. A., Garneau, M., Githumbi, E., Hribjlan, J., Holmquist, J., Hughes, P. D. M., Jones, C., Jones, M. C., Karofeld, E., Klein, E. S., Kokfelt, U., Korhola, A., Lacourse, T., Le Roux, G., Lamentowicz, M., Large, D., Lavoie, M., Loisel, J., Mackay, H., MacDonald, G. M., Makila, M., Magnan, G., Marchant, R., Marcisz, K., Martínez Cortizas, A., Massa, C., Mathijssen, P., Mauquoy, D., Mighall, T., Mitchell, F. J. G., Moss, P., Nichols, J., Oksanen, P. O., Orme, L., Packalen, M. S., Robinson, S., Roland, T. P., Sanderson, N. K., Sannel, A. B. K., Silva-Sánchez, N., Steinberg, N., Swindles, G. T., Turner, T. E., Uglow, J., Väliranta, M., van Bellen, S., van der Linden, M., van Geel, B., Wang, G., Yu, Z., Zaragoza-Castells, J. and Zhao, Y. 2018. Latitudinal limits to the predicted increase of the peatland carbon sink with warming. *Nature Climate Change* 8(10), 907–913. https://doi.org/10.1038/s41558-018-0271-1.

Gandois, L., Cobb, A. R., Hei, I. C., Lim, L. B. L., Salim, K. A. and Harvey, C. F. 2013. Impact of deforestation on solid and dissolved organic matter characteristics of tropical peat forests: implications for carbon release. *Biogeochemistry* 114(1–3), 183–199. https://doi.org/10.1007/s10533-012-9799-8.

Gaudig, G., Krebs, M. and Joosten, H. 2017. Sphagnum farming on cut-over bog in NW Germany: long-term studies on Sphagnum growth. *Mires and Peat*, 1–19. https://doi.org/10.19189/MaP.2016.OMB.238.

Giannini, V., Silvestri, N., Dragoni, F., Pistocchi, C., Sabbatini, T. and Bonari, E. 2017. Growth and nutrient uptake of perennial crops in a paludicultural approach in a drained Mediterranean peatland. *Ecological Engineering* 103, 478–487. https://doi.org/10.1016/j.ecoleng.2015.11.049.

Giesen, W. 2021. Tropical peatland restoration in Indonesia by replanting with useful indigenous peat swamp species: paludiculture. In: Osaki, M., Tsuji, N., Foead, N. and Rieley, J. (Eds), *Tropical Peatland Eco-Management*. Springer, Singapore. https://doi.org/10.1007/978-981-33-4654-3_14.

Girkin, N. T., Dhandapani, S., Evers, S., Ostle, N., Turner, B. L. and Sjögersten, S. 2020. Interactions between labile carbon, temperature and land use regulate carbon dioxide and methane production in tropical peat. *Biogeochemistry* 147(1), 87–97. https://doi.org/10.1007/s10533-019-00632-y.

Glatzel, S., Basiliko, N. and Moore, T. 2004. Carbon dioxide and methane production potentials of peats from natural, harvested and restored sites, eastern Québec, Canada. *Wetlands* 24(2), 261–267. https://doi.org/10.1672/0277-5212(2004)024[0261:CDAMPP]2.0.CO;2.

Glatzel, S., Kalbitz, K., Dalva, M. and Moore, T. 2003. Dissolved organic matter properties and their relationship to carbon dioxide efflux from restored peat bogs. *Geoderma* 113(3–4), 397–411. https://doi.org/10.1016/S0016-7061(02)00372-5.

Glatzel, S., Lemke, S. and Gerold, G. 2006. Short-term effects of an exceptionally hot and dry summer on decomposition of surface peat in a restored temperate bog. *European Journal of Soil Biology* 42(4), 219–229. https://doi.org/10.1016/j.ejsobi.2006.03.003.

Glaz, B. and Morris, D. R. 2010. Sugarcane responses to water-table depth and periodic flood. *Agronomy Journal* 102(2), 372–380. https://doi.org/10.2134/agronj2009.0262.

Goldstein, J. 2015. Mega-development, scientific expertise, and the remaking of Indonesia's degraded peatlands. dissertation. University of California. Available at: https://escholarship.org/uc/item/0671h97t.

Griffis, T. J., Roman, D. T., Wood, J. D., Deventer, J., Fachin, L., Rengifo, J., Del Castillo, D., Lilleskov, E., Kolka, R., Chimner, R. A., del Aguila-Pasquel, J., Wayson, C., Hergoualc'h, K., Baker, J. M., Cadillo-Quiroz, H. and Ricciuto, D. M. 2020. Hydrometeorological sensitivities of net ecosystem carbon dioxide and methane exchange of an Amazonian palm swamp peatland. *Agricultural and Forest Meteorology* 295, 108167. https://doi.org/10.1016/j.agrformet.2020.108167.

Grzywna, A. 2017. The degree of peatland subsidence resulting from drainage of land. *Environmental Earth Sciences* 76(16), 559. https://doi.org/10.1007/s12665-017-6869-1.

Günther, A., Barthelmes, A., Huth, V., Joosten, H., Jurasinski, G., Koebsch, F. and Couwenberg, J. 2020. Prompt rewetting of drained peatlands reduces climate warming despite methane emissions. *Nature Communications* 11(1), 1644. https://doi.org/10.1038/s41467-020-15499-z.

Günther, A., Huth, V., Jurasinski, G. and Glatzel, S. 2015. The effect of biomass harvesting on greenhouse gas emissions from a rewetted temperate fen. *GCB Bioenergy* 7(5), 1092–1106. https://doi.org/10.1111/gcbb.12214.

Günther, A., Jurasinski, G., Albrecht, K., Gaudig, G., Krebs, M. and Glatzel, S. 2017. Greenhouse gas balance of an establishing Sphagnum culture on a former bog grassland in Germany. *Mires and Peat*, 1–16. https://doi.org/10.19189/MaP.2015.OMB.210.

Hadi, A., Inubushi, K., Furukawa, Y., Purnomo, E., Rasmadi, M. and Tsuruta, H. 2005. Greenhouse gas emissions from tropical peatlands of Kalimantan,Indonesia. *Nutrient Cycling in Agroecosystems* 71(1), 73–80. https://doi.org/10.1007/s10705-004-0380-2.

Hahn, J., Köhler, S., Glatzel, S. and Jurasinski, G. 2015. Methane exchange in a coastal fen in the first year after flooding - A systems shift. *PLoS ONE* 10(10), e0140657. https://doi.org/10.1371/journal.pone.0140657.

Hahn-Schöfl, M., Zak, D., Minke, M., Gelbrecht, J., Augustin, J. and Freibauer, A. 2011. Organic sediment formed during inundation of a degraded fen grassland emits large fluxes of CH_4 and CO_2. *Biogeosciences* 8(6), 1539–1550. https://doi.org/10.5194/bg-8-1539-2011.

Hansson, A. and Dargusch, P. 2017. An estimate of the financial cost of peatland restoration in Indonesia. *Case Studies in the Environment* 2(1), 1–8. https://doi.org/10.1525/cse.2017.000695.

Hardie, S. M. L., Garnett, M. H., Fallick, A. E., Rowland, A. P., Ostle, N. J. and Flowers, T. H. 2011. Abiotic drivers and their interactive effect on the flux and carbon isotope (14C and δ13C) composition of peat-respired CO2. *Soil Biology and Biochemistry* 43(12), 2432–2440. https://doi.org/10.1016/j.soilbio.2011.08.010.

Hargreaves, K. J., Milne, R. and Cannell, M. G. R. 2003. Carbon balance of afforested peatland in Scotland. *Forestry* 76(3), 299–317. https://doi.org/10.1093/forestry/76.3.299.

Harpenslager, S. F., van den Elzen, E., Kox, M. A. R., Smolders, A. J. P., Ettwig, K. F. and Lamers, L. P. M. 2015. Rewetting former agricultural peatlands: topsoil removal as a prerequisite to avoid strong nutrient and greenhouse gas emissions. *Ecological Engineering* 84, 159–168. https://doi.org/10.1016/j.ecoleng.2015.08.002.

Hashim, S. A., Teh, C. B. S. and Ahmed, O. H. 2019. Influence of water table depths, nutrients leaching losses, subsidence of tropical peat soil and oil palm (Elaeis guineensis Jacq.) seedling growth. *Malaysian Journal of Soil Science* 23, 13–30.

Hatala, J. A., Detto, M., Sonnentag, O., Deverel, S. J., Verfaillie, J. and Baldocchi, D. D. 2012. Greenhouse gas (CO_2, CH_4, H_2O) fluxes from drained and flooded agricultural peatlands in the Sacramento-San Joaquin Delta. *Agriculture, Ecosystems and Environment* 150, 1–18. https://doi.org/10.1016/j.agee.2012.01.009.

Hergoualc'h, K. and Verchot, L. V. 2014. Greenhouse gas emission factors for land use and land-use change in Southeast Asian peatlands. *Mitigation and Adaptation Strategies for Global Change* 19(6), 789–807. https://doi.org/10.1007/s11027-013-9511-x.

Herzsprung, P., Osterloh, K., von Tümpling, W., Harir, M., Hertkorn, N., Schmitt-Kopplin, P., Meissner, R., Bernsdorf, S. and Friese, K. 2017. Differences in DOM of rewetted and natural peatlands - results from high-field FT-ICR-MS and bulk optical parameters. *Science of the Total Environment* 586, 770–781. https://doi.org/10.1016/j.scitotenv.2017.02.054.

Hirano, T., Segah, H., Kusin, K., Limin, S., Takahashi, H. and Osaki, M. 2012. Effects of disturbances on the carbon balance of tropical peat swamp forests. *Global Change Biology* 18(11), 3410–3422. https://doi.org/10.1111/j.1365-2486.2012.02793.x.

Hodgkins, S. B., Richardson, C. J., Dommain, R., Wang, H., Glaser, P. H., Verbeke, B., Winkler, B. R., Cobb, A. R., Rich, V. I., Missilmani, M., Flanagan, N., Ho, M., Hoyt, A. M., Harvey, C. F., Vining, S. R., Hough, M. A., Moore, T. R., Richard, P. J. H., De La Cruz, F. B., Toufaily, J., Hamdan, R., Cooper, W. T. and Chanton, J. P. 2018. Tropical peatland carbon storage linked to global latitudinal trends in peat recalcitrance. *Nature Communications* 9(1), 3640. https://doi.org/10.1038/s41467-018-06050-2.

Hogg, E. H. 1993. Decay potential of hummock and hollow Sphagnum peats at different depths in a Swedish raised bog. *Oikos* 66(2), 269. https://doi.org/10.2307/3544814.

Holden, J. 2006. Sediment and particulate carbon removal by pipe erosion increase over time in blanket peatlands as a consequence of land drainage. *Journal of Geophysical Research* 111(F2), F02010. https://doi.org/10.1029/2005JF000386.

Hommeltenberg, J., Schmid, H. P., Drösler, M. and Werle, P. 2014. Can a bog drained for forestry be a stronger carbon sink than a natural bog forest? *Biogeosciences* 11(13), 3477–3493. https://doi.org/10.5194/bg-11-3477-2014.

Hooijer, A., Page, S., Jauhiainen, J., Lee, W. A., Lu, X. X., Idris, A. and Anshari, G. 2012. Subsidence and carbon loss in drained tropical peatlands. *Biogeosciences* 9(3), 1053–1071. https://doi.org/10.5194/bg-9-1053-2012.

Howson, T., Chapman, P. J., Shah, N., Anderson, R. and Holden, J. 2021. A comparison of porewater chemistry between intact, afforested and restored raised and blanket bogs. *Science of the Total Environment* 766, 144496. https://doi.org/10.1016/j.scitotenv.2020.144496.

Hoyos-Santillan, J., Lomax, B. H., Large, D., Turner, B. L., Boom, A., Lopez, O. R. and Sjögersten, S. 2016. Quality not quantity: organic matter composition controls of CO_2 and CH_4 fluxes in neotropical peat profiles. *Soil Biology and Biochemistry* 103, 86–96. https://doi.org/10.1016/j.soilbio.2016.08.017.

Hoyt, A. M., Chaussard, E., Seppalainen, S. S. and Harvey, C. F. 2020. Widespread subsidence and carbon emissions across Southeast Asian peatlands. *Nature Geoscience* 13(6), 435–440. https://doi.org/10.1038/s41561-020-0575-4.

Huang, Y., Ciais, P., Luo, Y., Zhu, D., Wang, Y., Qiu, C., Goll, D. S., Guenet, B., Makowski, D., De Graaf, I., Leifeld, J., Kwon, M. J., Hu, J. and Qu, L. 2021. Tradeoff of CO_2 and CH_4 emissions from global peatlands under water-table drawdown. *Nature Climate Change* 11(7), 618–622. https://doi.org/10.1038/s41558-021-01059-w.

Humpenöder, F., Karstens, K., Lotze-Campen, H., Leifeld, J., Menichetti, L., Barthelmes, A. and Popp, A. 2020. Peatland protection and restoration are key for climate change mitigation. *Environmental Research Letters* 15(10). https://doi.org/10.1088/1748-9326/abae2a, 104093.

Huth, V., Günther, A., Bartel, A., Hofer, B., Jacobs, O., Jantz, N., Meister, M., Rosinski, E., Urich, T., Weil, M., Zak, D. and Jurasinski, G. 2020. Topsoil removal reduced in-situ methane emissions in a temperate rewetted bog grassland by a hundredfold. *Science of the Total Environment* 721, 137763. https://doi.org/10.1016/j.scitotenv.2020.137763.

Inglett, K. S., Inglett, P. W., Reddy, K. R. and Osborne, T. Z. 2012. Temperature sensitivity of greenhouse gas production in wetland soils of different vegetation. *Biogeochemistry* 108(1–3), 77–90. https://doi.org/10.1007/s10533-011-9573-3.

IPCC. 2014. In: Hiraishi, T., Krug, T., Tanabe, K., Srivastava, N., Baasansuren, J., Fukuda, M. and Troxler, T.G. (Eds), *2013 Supplement to the 2006 IPCC Guidelines for National Greenhouse Gas Inventories: Wetlands*. 354 p. IPCC, Switzerland.

Jauhiainen, J., Alm, J., Bjarnadottir, B., Callesen, I., Christiansen, J. R., Clarke, N., Dalsgaard, L., He, H., Jordan, S., Kazanavičiūtė, V., Klemedtsson, L., Lauren, A., Lazdins, A., Lehtonen, A., Lohila, A., Lupikis, A., Mander, Ü., Minkkinen, K., Kasimir, Å., Olsson, M., Ojanen, P., Óskarsson, H., Sigurdsson, B. D., Søgaard, G., Soosaar, K., Vesterdal, L. and Laiho, R. 2019. Reviews and syntheses: greenhouse gas exchange data from drained organic forest soils – a review of current approaches and recommendations for future research. *Biogeosciences* 16(23), 4687–4703. https://doi.org/10.5194/bg-16-4687-2019.

Joosten, H. and Clarke, D. 2002. *Wise Use of Mires and peatlands: background and principles including a framework for decision-making*. International Peat Society; International Mire Conservation Group, Saarijärvi [Greifswald].

Joosten, H. (2010). *The Global Peatland CO_2 Picture: Peatland Status and Drainage Related Emissions in All Countries of the World*. Wetland International, Ede, The Netherlands. https://www.wetlands.org/download/6672.

Jovani-Sancho, A. J., Cummins, T. and Byrne, K. A. 2021. Soil carbon balance of afforested peatlands in the maritime temperate climatic zone. *Global Change Biology* 27(15), 3681–3698. https://doi.org/10.1111/gcb.15654.

Kandel, T. P., Karki, S., Elsgaard, L. and Lærke, P. E. 2019a. Fertilizer-induced fluxes dominate annual N_2O emissions from a nitrogen-rich temperate fen rewetted for paludiculture. *Nutrient Cycling in Agroecosystems* 115(1), 57–67. https://doi.org/10.1007/s10705-019-10012-5.

Kandel, T. P., Lærke, P. E., Hoffmann, C. C. and Elsgaard, L. 2019b. Complete annual CO_2, CH_4, and N_2O balance of a temperate riparian wetland 12 years after rewetting. *Ecological Engineering* 127, 527–535. https://doi.org/10.1016/j.ecoleng.2017.12.019.

Kasimir-Klemedtsson, Å., Klemedtsson, L., Berglund, K., Martikainen, P., Silvola, J. and Oenema, O. 1997. Greenhouse gas emissions from farmed organic soils: a review. *Soil Use and Management* 13(s4), 245–250. https://doi.org/10.1111/j.1475-2743.1997.tb00595.x.

Kechavarzi, C., Dawson, Q. and Leeds-Harrison, P. B. 2010. Physical properties of low-lying agricultural peat soils in England. *Geoderma* 154(3–4), 196–202. https://doi.org/10.1016/j.geoderma.2009.08.018.

Kechavarzi, C., Dawson, Q., Leeds-Harrison, P. B., Szatyłowicz, J. and Gnatowski, T. 2007. Water-table management in lowland UK peat soils and its potential impact on CO2 emission. *Soil Use and Management* 23(4), 359–367. https://doi.org/10.1111/j.1475-2743.2007.00125.x.

Kløve, B., Berglund, K., Berglund, Ö., Weldon, S. and Maljanen, M. 2017. Future options for cultivated Nordic peat soils: can land management and rewetting control greenhouse gas emissions? *Environmental Science and Policy* 69, 85–93. https://doi.org/10.1016/j.envsci.2016.12.017.

Kløve, B., Sveistrup, T. E. and Hauge, A. 2010. Leaching of nutrients and emission of greenhouse gases from peatland cultivation at Bodin, Northern Norway. *Geoderma* 154(3–4), 219–232. https://doi.org/10.1016/j.geoderma.2009.08.022.

Kluge, B., Wessolek, G., Facklam, M., Lorenz, M. and Schwärzel, K. 2008. Long-term carbon loss and CO2-C release of drained peatland soils in northeast Germany. *European Journal of Soil Science* 59(6), 1076–1086. https://doi.org/10.1111/j.1365-2389.2008.01079.x.

Koebsch, F., Jurasinski, G., Koch, M., Hofmann, J. and Glatzel, S. 2015. Controls for multi-scale temporal variation in ecosystem methane exchange during the growing season of a permanently inundated fen. *Agricultural and Forest Meteorology* 204, 94–105. https://doi.org/10.1016/j.agrformet.2015.02.002.

Konecny, K., Ballhorn, U., Navratil, P., Jubanski, J., Page, S. E., Tansey, K., Hooijer, A., Vernimmen, R. and Siegert, F. 2016. Variable carbon losses from recurrent fires in drained tropical peatlands. *Global Change Biology* 22(4), 1469–1480. https://doi.org/10.1111/gcb.13186.

Könönen, M., Jauhiainen, J., Laiho, R., Spetz, P., Kusin, K., Limin, S. and Vasander, H. 2016. Land use increases the recalcitrance of tropical peat. *Wetlands Ecology and Management* 24(6), 717–731. https://doi.org/10.1007/s11273-016-9498-7.

Korkiakoski, M., Tuovinen, J.-P., Penttilä, T., Sarkkola, S., Ojanen, P., Minkkinen, K., Rainne, J., Laurila, T. and Lohila, A. 2019. Greenhouse gas and energy fluxes in a boreal peatland forest after clear-cutting. *Biogeosciences* 16(19), 3703–3723. https://doi.org/10.5194/bg-16-3703-2019.

Krüger, J. P., Alewell, C., Minkkinen, K., Szidat, S. and Leifeld, J. 2016. Calculating carbon changes in peat soils drained for forestry with four different profile-based methods. *Forest Ecology and Management* 381, 29–36. https://doi.org/10.1016/j.foreco.2016.09.006.

Krüger, J. P., Leifeld, J., Glatzel, S., Szidat, S. and Alewell, C. 2015. Biogeochemical indicators of peatland degradation – a case study of a temperate bog in northern Germany. *Biogeosciences* 12(10), 2861–2871. https://doi.org/10.5194/bg-12-2861-2015.

Kuhry, P. and Vitt, D. H. 1996. Fossil carbon/nitrogen ratios as a measure of peat decomposition. *Ecology* 77(1), 271–275. https://doi.org/10.2307/2265676.

Lai, D. Y. F., Roulet, N. T. and Moore, T. R. 2014. The spatial and temporal relationships between CO_2 and CH_4 exchange in a temperate ombrotrophic bog. *Atmospheric Environment* 89, 249–259. https://doi.org/10.1016/j.atmosenv.2014.02.034.

Laiho, R., Laine, J., Trettin, C. C. and Finér, L. 2004. Scots pine litter decomposition along drainage succession and soil nutrient gradients in peatland forests, and the effects of inter-annual weather variation. *Soil Biology and Biochemistry* 36(7), 1095–1109. https://doi.org/10.1016/j.soilbio.2004.02.020.

Laiho, R. and Pearson, M. 2016. Surface peat and its dynamics following drainage - do they facilitate estimation of carbon losses with the C/ash method? *Mires and Peat*, 1–19. https://doi.org/10.19189/MaP.2016.OMB.247.

Laine, A., Byrne, K. A., Kiely, G. and Tuittila, E.-S. 2007. Patterns in vegetation and CO2 dynamics along a water level gradient in a lowland blanket bog. *Ecosystems* 10(6), 890–905. https://doi.org/10.1007/s10021-007-9067-2.

Laine, J., Silvola, J., Tolonen, K., Alm, J., Nykänen, H., Vasander, H., Sallantaus, T., Savolainen, I., Sinisalo, J. and Martikainen, P. J. 1996. Effect of water-level drawdown on global climatic warming: northern peatlands. *Ambio – A Journal of the Human Environment* 25, 179–184.

Landry, J. and Rochefort, L. 2012. The drainage of peatlands. Impacts and rewetting techniques. Département de phytologie, Université Laval. Département de phytologie, Université Laval, Quebec. Available online: https://www.gret-perg.ulaval.ca/uploads/tx_centrerecherche/Drainage_guide_Web.pdf (Accessed 16 July 2021).

LaRowe, D. E. and Van Cappellen, P. 2011. Degradation of natural organic matter: a thermodynamic analysis. *Geochimica et Cosmochimica Acta* 75(8), 2030–2042. https://doi.org/10.1016/j.gca.2011.01.020.

Leiber-Sauheitl, K., Fuß, R., Voigt, C. and Freibauer, A. 2014. High CO2 fluxes from grassland on histic gleysol along soil carbon and drainage gradients. *Biogeosciences* 11(3), 749–761. https://doi.org/10.5194/bg-11-749-2014.

Leifeld, J. 2018. Distribution of nitrous oxide emissions from managed organic soils under different land uses estimated by the peat C/N ratio to improve national GHG inventories. *Science of the Total Environment* 631–632, 23–26. https://doi.org/10.1016/j.scitotenv.2018.02.328.

Leifeld, J., Alewell, C., Bader, C., Krüger, J. P., Mueller, C. W., Sommer, M., Steffens, M. and Szidat, S. 2018. Pyrogenic carbon contributes substantially to carbon storage in intact and degraded northern peatlands: pyrogenic carbon in peatlands. *Land Degradation and Development* 29(7), 2082–2091. https://doi.org/10.1002/ldr.2812.

Leifeld, J., Alewell, C. and Paul, S. M. 2021. Accumulation of C4-carbon from *Miscanthus* in organic-matter-rich soils. *GCB Bioenergy* 13(8), 1319–1328. https://doi.org/10.1111/gcbb.12861.

Leifeld, J., Klein, K. and Wüst-Galley, C. 2020. Soil organic matter stoichiometry as indicator for peatland degradation. *Scientific Reports* 10(1), 7634. https://doi.org/10.1038/s41598-020-64275-y.

Leifeld, J. and Menichetti, L. 2018. The underappreciated potential of peatlands in global climate change mitigation strategies. *Nature Communications* 9(1), 1071. https://doi.org/10.1038/s41467-018-03406-6.

Leifeld, J., Müller, M. and Fuhrer, J. 2011. Peatland subsidence and carbon loss from drained temperate fens. *Soil Use and Management* 27(2), 170–176.

Leifeld, J., Steffens, M. and Galego-Sala, A. 2012. Sensitivity of peatland carbon loss to organic matter quality: peat quality and carbon loss. *Geophysical Research Leters* 39(14), n/a–n/a. https://doi.org/10.1029/2012GL051856.

Leifeld, J., Wüst-Galley, C. and Page, S. 2019. Intact and managed peatland soils as a source and sink of GHGs from 1850 to 2100. *Nature Climate Change* 9(12), 945–947. https://doi.org/10.1038/s41558-019-0615-5.

Li, C., Grayson, R., Holden, J. and Li, P. 2018. Erosion in peatlands: recent research progress and future directions. *Earth-Science Reviews* 185, 870–886. https://doi.org/10.1016/j.earscirev.2018.08.005.

Lim, K. H., Lim, S. S., Parish, F. and Suharto, R. (Eds) 2012. *Summary: RSPO Manual on Best Management Practices (BMPs) for Existing Oil Palm Cultivation on Peat*. RSPO, Kuala Lumpur.

Limpens, J., Heijmans, M. M. P. D. and Berendse, F. 2006. The nitrogen cycle in boreal peatlands. In: Wieder, R. K. and Vitt, D. H. (Eds), *Boreal Peatland Ecosystems, Ecological Studies*. Springer, Berlin Heidelberg. pp. 195–230. https://doi.org/10.1007/978-3-540-31913-9_10.

Liu, H., Price, J., Rezanezhad, F. and Lennartz, B. 2020. Centennial-scale shifts in hydrophysical properties of peat induced by drainage. *Water Resources Research* 56(10). https://doi.org/10.1029/2020WR027538.

Lohila, A. 2004. Annual CO2 exchange of a peat field growing spring barley or perennial forage grass. *Journal of Geophysical Research* 109(D18), D18116. https://doi.org/10.1029/2004JD004715.

Lohila, A., Aurela, M., Regina, K. and Laurila, T. 2003. Soil and total ecosystem respiration in agricultural fields: effect of soil and crop type. *Plant and Soil* 251(2), 303–317. https://doi.org/10.1023/A:1023004205844.

Lohila, A., Minkkinen, K., Aurela, M., Tuovinen, J.-P., Penttilä, T., Ojanen, P. and Laurila, T. 2011. Greenhouse gas flux measurements in a forestry-drained peatland indicate a large carbon sink. *Biogeosciences* 8(11), 3203–3218. https://doi.org/10.5194/bg-8-3203-2011.

Loisel, J., Gallego-Sala, A. V., Amesbury, M. J., Magnan, G., Anshari, G., Beilman, D. W., Benavides, J. C., Blewett, J., Camill, P., Charman, D. J., Chawchai, S., Hedgpeth, A., Kleinen, T., Korhola, A., Large, D., Mansilla, C. A., Müller, J., van Bellen, S., West, J. B., Yu, Z., Bubier, J. L., Garneau, M., Moore, T., Sannel, A. B. K., Page, S., Väliranta, M.,

Bechtold, M., Brovkin, V., Cole, L. E. S., Chanton, J. P., Christensen, T. R., Davies, M. A., De Vleeschouwer, F., Finkelstein, S. A., Frolking, S., Gałka, M., Gandois, L., Girkin, N., Harris, L. I., Heinemeyer, A., Hoyt, A. M., Jones, M. C., Joos, F., Juutinen, S., Kaiser, K., Lacourse, T., Lamentowicz, M., Larmola, T., Leifeld, J., Lohila, A., Milner, A. M., Minkkinen, K., Moss, P., Naafs, B. D. A., Nichols, J., O'Donnell, J., Payne, R., Philben, M., Piilo, S., Quillet, A., Ratnayake, A. S., Roland, T. P., Sjögersten, S., Sonnentag, O., Swindles, G. T., Swinnen, W., Talbot, J., Treat, C., Valach, A. C. and Wu, J. 2021. Expert assessment of future vulnerability of the global peatland carbon sink. *Nature Climate Change* 11(1), 70–77. https://doi.org/10.1038/s41558-020-00944-0.

Loisel, J., Yu, Z., Beilman, D. W., Camill, P., Alm, J., Amesbury, M. J., Anderson, D., Andersson, S., Bochicchio, C., Barber, K., Belyea, L. R., Bunbury, J., Chambers, F. M., Charman, D. J., De Vleeschouwer, F., Fiałkiewicz-Kozieł, B., Finkelstein, S. A., Gałka, M., Garneau, M., Hammarlund, D., Hinchcliffe, W., Holmquist, J., Hughes, P., Jones, M. C., Klein, E. S., Kokfelt, U., Korhola, A., Kuhry, P., Lamarre, A., Lamentowicz, M., Large, D., Lavoie, M., MacDonald, G., Magnan, G., Mäkilä, M., Mallon, G., Mathijssen, P., Mauquoy, D., McCarroll, J., Moore, T. R., Nichols, J., O'Reilly, B., Oksanen, P., Packalen, M., Peteet, D., Richard, P. J., Robinson, S., Ronkainen, T., Rundgren, M., Sannel, A. B. K., Tarnocai, C., Thom, T., Tuittila, E.-S., Turetsky, M., Väliranta, M., van der Linden, M., van Geel, B., van Bellen, S., Vitt, D., Zhao, Y. and Zhou, W. 2014. A database and synthesis of northern peatland soil properties and Holocene carbon and nitrogen accumulation. *The Holocene* 24(9), 1028–1042. https://doi.org/10.1177/0959683614538073.

Lund, M., Christensen, T. R., Lindroth, A. and Schubert, P. 2012. Effects of drought conditions on the carbon dioxide dynamics in a temperate peatland. *Environmental Research Letters* 7(4). https://doi.org/10.1088/1748-9326/7/4/045704, 045704.

Lund, M., Lindroth, A., Christensen, T. R. and Strö, L. 2007. Annual CO2 balance of a temperate bog. *Tellus B* 59(5), 804–811. https://doi.org/10.1111/j.1600-0889.2007.00303.x.

Lupascu, M., Wadham, J. L., Hornibrook, E. R. C. and Pancost, R. D. 2012. Temperature sensitivity of methane production in the permafrost active layer at Stordalen, Sweden: a comparison with non-permafrost northern wetlands. *Arctic, Antarctic, and Alpine Research* 44(4), 469–482. https://doi.org/10.1657/1938-4246-44.4.469.

Lynch, J., Cain, M., Pierrehumbert, R. and Allen, M. 2020. Demonstrating GWP*: a means of reporting warming-equivalent emissions that captures the contrasting impacts of short- and long-lived climate pollutants. *Environmental Research Letters : ERL [Web Site]* 15(4), 044023. https://doi.org/10.1088/1748-9326/ab6d7e.

Maanavilja, L., Riutta, T., Aurela, M., Pulkkinen, M., Laurila, T. and Tuittila, E.-S. 2011. Spatial variation in CO2 exchange at a northern aapa mire. *Biogeochemistry* 104(1–3), 325–345. https://doi.org/10.1007/s10533-010-9505-7.

Maie, N., Maeda, M., Murouchi, A., Melling, L., Takamatsu, R., Sangok, F., Kakino, W., Tanji, H. and Watanabe, A. 2019. Influence of temperature and water conditions on the mineralization rate of tropical peat. *Humic Substances Research* 15, 8.

Maljanen, M., Martikainen, P. J., Walden, J. and Silvola, J. 2001. CO2 exchange in an organic field growing barley or grass in eastern Finland. *Global Change Biology* 7, 679–692. https://doi.org/10.1111/j.1365-2486.2001.00437.x.

Maljanen, M., Sigurdsson, B. D., Guðmundsson, J., Óskarsson, H., Huttunen, J. T. and Martikainen, P. J. 2010. Greenhouse gas balances of managed peatlands in the Nordic countries – present knowledge and gaps. *Biogeosciences* 7(9), 2711–2738. https://doi.org/10.5194/bg-7-2711-2010.

Malterer, T. J., Verry, E. S. and Erjavec, J. 1992. Fiber content and degree of decomposition in peats: review of national methods. *Soil Science Society of America Journal* 56(4), 1200-1211.

Mander, Ü., Järveoja, J., Maddison, M., Soosaar, K., Aavola, R., Ostonen, I. and Salm, J.-O. 2012. Reed canary grass cultivation mitigates greenhouse gas emissions from abandoned peat extraction areas. *GCB Bioenergy* 4(4), 462-474. https://doi.org/10.1111/j.1757-1707.2011.01138.x.

Matysek, M., Leake, J., Banwart, S., Johnson, I., Page, S., Kaduk, J., Smalley, A., Cumming, A. and Zona, D. 2019. Impact of fertiliser, water table, and warming on celery yield and CO_2 and CH_4 emissions from fenland agricultural peat. *Science of the Total Environment* 667, 179-190. https://doi.org/10.1016/j.scitotenv.2019.02.360.

McCalmont, J., Kho, L. K., Teh, Y. A., Lewis, K., Chocholek, M., Rumpang, E. and Hill, T. 2021. Short- and long-term carbon emissions from oil palm plantations converted from logged tropical peat swamp forest. *Global Change Biology* 27(11), 2361-2376. https://doi.org/10.1111/gcb.15544.

Melts, I., Ivask, M., Geetha, M., Takeuchi, K. and Heinsoo, K. 2019. Combining bioenergy and nature conservation: an example in wetlands. *Renewable and Sustainable Energy Reviews* 111, 293-302. https://doi.org/10.1016/j.rser.2019.05.028.

Menberu, M. W., Marttila, H., Tahvanainen, T., Kotiaho, J. S., Hokkanen, R., Kløve, B. and Ronkanen, A.-K. 2017. Changes in pore water quality After peatland restoration: assessment of a large-scale, replicated before-after-control-impact study in Finland: restoration and pore water quality. *Water Resources Research* 53(10), 8327-8343. https://doi.org/10.1002/2017WR020630.

Meyer, A., Tarvainen, L., Nousratpour, A., Björk, R. G., Ernfors, M., Grelle, A., Kasimir Klemedtsson, Å., Lindroth, A., Räntfors, M., Rütting, T., Wallin, G., Weslien, P. and Klemedtsson, L. 2013. A fertile peatland forest does not constitute a major greenhouse gas sink. *Biogeosciences* 10(11), 7739-7758. https://doi.org/10.5194/bg-10-7739-2013.

Minamikawa, K., Yamaguchi, T., Tokida, T., Sudo, S. and Yagi, K. 2018. Handbook of monitoring, reporting, and verification for a greenhouse gas mitigation project with water management in irrigated rice paddies. *Institute for Agro-Environmental Sciences*. NARO, Japan, Tsukuba, 42.

Minke, M., Augustin, J., Burlo, A., Yarmashuk, T., Chuvashova, H., Thiele, A., Freibauer, A., Tikhonov, V. and Hoffmann, M. 2016. Water level, vegetation composition, and plant productivity explain greenhouse gas fluxes in temperate cutover fens after inundation. *Biogeosciences* 13(13), 3945-3970. https://doi.org/10.5194/bg-13-3945-2016.

Minkkinen, K., Byrne, K. A. and Trettin, C. 2008. Climate impacts of peatland forestry. In: Strack, M. (Ed.), *Peatlands and Climate Change*. International Peatland Society, Jyväskylä. pp. 98-122. ISBN 978-952-99401-1-0.

Minkkinen, K., Ojanen, P., Penttilä, T., Aurela, M., Laurila, T., Tuovinen, J.-P. and Lohila, A. 2018. Persistent carbon sink at a boreal drained bog forest. *Biogeosciences* 15(11), 3603-3624. https://doi.org/10.5194/bg-15-3603-2018.

Minkkinen, K., Vasander, H., Jauhiainen, S., Karsisto, M. and Laine, J. 1999. Post-drainage changes in vegetation composition and carbon balance in Lakkasuo mire, Central Finland. *Plant and Soil* 207, 107-120.

Minkkinen, K. and Laine, J. 1998. Long-term effect of forest drainage on the peat carbon stores of pine mires in Finland. *Canadian Journal of Forest Research* 28(9), 1267-1275. https://doi.org/10.1139/x98-104.

Moody, C. S., Worrall, F., Clay, G. D., Burt, T. P., Apperley, D. C. and Rose, R. 2018. A molecular budget for a peatland based Upon 13 C solid-state nuclear magnetic resonance. *Journal of Geophysical Research: Biogeosciences* 123(2), 547–560. https://doi.org/10.1002/2017JG004312.

Moore, P. D. 2002. The future of cool temperate bogs. *Environmental Conservation* 29(1), 3–20. https://doi.org/10.1017/S0376892902000024.

Moore, S., Evans, C. D., Page, S. E., Garnett, M. H., Jones, T. G., Freeman, C., Hooijer, A., Wiltshire, A. J., Limin, S. H. and Gauci, V. 2013. Deep instability of deforested tropical peatlands revealed by fluvial organic carbon fluxes. *Nature* 493(7434), 660–663. https://doi.org/10.1038/nature11818.

Moore, T. R. and Dalva, M. 1997. Methane and carbon dioxide exchange potentials of peat soils in aerobic and anaerobic laboratory incubations. *Soil Biology and Biochemistry* 29(8), 1157–1164. https://doi.org/10.1016/S0038-0717(97)00037-0.

Moore, T. R. and Dalva, M. 1993. The influence of temperature and water table position on carbon dioxide and methane emissions from laboratory columns of peatland soils. *Journal of Soil Science* 44(4), 651–664. https://doi.org/10.1111/j.1365-2389.1993.tb02330.x.

Moreno-Mateos, D., Power, M. E., Comín, F. A. and Yockteng, R. 2012. Structural and functional loss in restored wetland ecosystems. *PLoS Biology* 10(1), e1001247. https://doi.org/10.1371/journal.pbio.1001247.

Mrotzek, A., Michaelis, D., Günther, A., Wrage-Mönnig, N. and Couwenberg, J. 2020. Mass balances of a drained and a rewetted peatland: on former losses and recent gains. *Soil Systems* 4(1), 16. https://doi.org/10.3390/soilsystems4010016.

Musarika, S., Atherton, C. E., Gomersall, T., Wells, M. J., Kaduk, J., Cumming, A. M. J., Page, S. E., Oechel, W. C. and Zona, D. 2017. Effect of water table management and elevated CO2 on radish productivity and on CH_4 and CO_2 fluxes from peatlands converted to agriculture. *Science of The Total Environment* 584–585, 665–672. https://doi.org/10.1016/j.scitotenv.2017.01.094.

Myhre, G., Shindell, D., Breon, F.-M., Collins, W., Fuglestvedt, J., Huang, J., Koch, D., Lamarque, J.-F., Lee, D., Mendoza, B., Nakajima, T., Robock, A., Stephens, G., Takemura, T. and Zhang, H. 2013. Climate change 2013: the physical science basis. Contribution of working group I to the fifth assessment report of the Intergovernmental panel on climate change. In: Stocker, T. F., Qin, D., Plattner, G.-K., Tignor, M., Allen, S. K., Boschung, J., Nauels, A., Xia, Y., Bex, V. and Midgley, P. M. (Eds), *Anthropogenic and Natural Radiative Forcing*. Cambridge University Press, Cambridge, UK and New York. pp. 659–740.

Neubauer, S. C. 2014. On the challenges of modeling the net radiative forcing of wetlands: reconsidering Mitsch et al. 2013. *Landscape Ecology* 29(4), 571–577. https://doi.org/10.1007/s10980-014-9986-1.

Neubauer, S. C. and Megonigal, J. P. 2015. Moving beyond global warming potentials to quantify the climatic role of ecosystems. *Ecosystems* 18(6), 1000–1013. https://doi.org/10.1007/s10021-015-9879-4.

Nilsson, M., Sagerfors, J., Buffam, I., Laudon, H., Eriksson, T., Grelle, A., Klemedtsson, L., Weslien, P. and Lindroth, A. 2008. Contemporary carbon accumulation in a boreal oligotrophic minerogenic mire - a significant sink after accounting for all C-fluxes: contemporary carbon accumulation. *Global Change Biology* 14(10), 2317–2332. https://doi.org/10.1111/j.1365-2486.2008.01654.x.

Norberg, L., Berglund, Ö. and Berglund, K. 2016a. Seasonal CO2 emission under different cropping systems on histosols in southern Sweden. *Geoderma Regional* 7(3), 338–345. https://doi.org/10.1016/j.geodrs.2016.06.005.

Norberg, L., Berglund, Ö. and Berglund, K. 2016b. Nitrous oxide and methane fluxes during the growing season from cultivated peat soils, peaty marl and gyttja clay under different cropping systems. *Acta Agriculturae Scandinavica, Section B – Soil and Plant Science* 66(7), 602–612. https://doi.org/10.1080/09064710.2016.1205126.

Normand, A. E., Turner, B. L., Lamit, L. J., Smith, A. N., Baiser, B., Clark, M. W., Hazlett, C., Kane, E. S., Lilleskov, E., Long, J. R., Grover, S. P. and Reddy, K. R. 2021. Organic matter chemistry drives carbon dioxide production of peatlands. *Geophysical Research Letters* 48(18), e2021.

Nugent, K. A., Strachan, I. B., Roulet, N. T., Strack, M., Frolking, S. and Helbig, M. 2019. Prompt active restoration of peatlands substantially reduces climate impact. *Environmental Research Letters* 14(12). https://doi.org/10.1088/1748-9326/ab56e6, 124030.

Ojanen, P. and Minkkinen, K. 2019. The dependence of net soil CO2 emissions on water table depth in boreal peatlands drained for forestry. *Mires and Peat*, 1–8. https://doi.org/10.19189/MaP.2019.OMB.StA.1751.

Ojanen, P. and Minkkinen, K. 2020. Rewetting offers rapid climate benefits for tropical and agricultural peatlands but not for forestry-drained peatlands. *Global Biogeochemical Cycles* 34(7). https://doi.org/10.1029/2019GB006503.

Ojanen, P., Minkkinen, K. and Penttilä, T. 2013. The current greenhouse gas impact of forestry-drained boreal peatlands. *Forest Ecology and Management* 289, 201–208. https://doi.org/10.1016/j.foreco.2012.10.008.

Oleszczuk, R., Zając, E. and Urbański, J. 2020. Verification of empirical equations describing subsidence rate of peatland in Central Poland. *Wetlands Ecology and Management* 28(3), 495–507. https://doi.org/10.1007/s11273-020-09727-y.

Page, S. E., Rieley, J. O. and Banks, C. J. 2011. Global and regional importance of the tropical peatland carbon pool: tropical peatland carbon pool. *Global Change Biology* 17(2), 798–818. https://doi.org/10.1111/j.1365-2486.2010.02279.x.

Pangala, S. R., Enrich-Prast, A., Basso, L. S., Peixoto, R. B., Bastviken, D., Hornibrook, E. R. C., Gatti, L. V., Marotta, H., Calazans, L. S. B., Sakuragui, C. M., Bastos, W. R., Malm, O., Gloor, E., Miller, J. B. and Gauci, V. 2017. Large emissions from floodplain trees close the Amazon methane budget. *Nature* 552(7684), 230–234. https://doi.org/10.1038/nature24639.

Parish, F., Sirin, A., Charman, D., Joosten, H., Minaeva, T. and Silvius, M. (Eds) 2008. *Assessment on Peatlands, Biodiversity and Climate Change*. Global Environment Centre, Kuala Lumpur and Wetlands International Wageningen, 179 p. Available at: http://www.imcg.net/media/download_gallery/books/assessment_peatland.pdf (accessed 25 April 2022).

Paul, S., Ammann, C., Alewell, C. and Leifeld, J. 2021. Carbon budget response of an agriculturally used fen to different soil moisture conditions. *Agricultural and Forest Meteorology* 300. https://doi.org/10.1016/j.agrformet.2021.108319, 108319.

Paustian, K., Lehmann, J., Ogle, S., Reay, D., Robertson, G. P. and Smith, P. 2016. Climate-smart soils. *Nature* 532(7597), 49–57. https://doi.org/10.1038/nature17174.

Pawson, R. R., Evans, M. G. and Allott, T. E. H. A. 2012. Fluvial carbon flux from headwater peatland streams: significance of particulate carbon flux: fluvial carbon flux from eroding peatlands. *Earth Surface Process and Landforms* 37(11), 1203–1212. https://doi.org/10.1002/esp.3257.

Peacock, M., Gauci, V., Baird, A. J., Burden, A., Chapman, P. J., Cumming, A., Evans, J. G., Grayson, R. P., Holden, J., Kaduk, J., Morrison, R., Page, S., Pan, G., Ridley, L. M., Williamson, J., Worrall, F. and Evans, C. D. 2019. The full carbon balance of a rewetted cropland fen and a conservation-managed fen. *Agriculture, Ecosystems and Environment* 269, 1-12. https://doi.org/10.1016/j.agee.2018.09.020.

Premrov, A., Wilson, D., Saunders, M., Yeluripati, J. and Renou-Wilson, F. 2021. CO_2 fluxes from drained and rewetted peatlands using a new ECOSSE model water table simulation approach. *Science of the Total Environment* 754, 142433. https://doi.org/10.1016/j.scitotenv.2020.142433.

Preston, C. M., Shipitalo, S. E., Dudley, R. L., Fyfe, C. A., Mathur, S. P. and Levesque, M. 1987. Comparison of ^{13}C CPMAS NMR and chemical techniques for measuring the degree of decomposition in virgin and cultivated peat profiles. Canadian Journal of Soil Science 67(1), 187–198.

Prévost, M., Plamondon, A. P. and Belleau, P. 1999. Effects of drainage of a forested peatland on water quality and quantity. *Journal of Hydrology* 214(1-4), 130-143. https://doi.org/10.1016/S0022-1694(98)00281-9.

Price, J. S. and Ketcheson, S. J. 2009. Water relations in cutover peatlands. *Geophysical Monograph Series* 184, 277–287.

Purwanto, B. H., Watanabe, A., Shoon, J. F., Kakuda, K. and Ando, H. 2005. Kinetic parameters of gross N mineralization of peat soils as related to the composition of soil organic matter. *Soil Science and Plant Nutrition* 51(1), 109–115. https://doi.org/10.1111/j.1747-0765.2005.tb00013.x.

Querner, E. P., Jansen, P. C., van den Akker, J. J. H. and Kwakernaak, C. 2012. Analysing water level strategies to reduce soil subsidence in Dutch peat meadows. *Journal of Hydrology* 446–447, 59–69. https://doi.org/10.1016/j.jhydrol.2012.04.029.

Regina, K., Sheehy, J. and Myllys, M. 2015. Mitigating greenhouse gas fluxes from cultivated organic soils with raised water table. *Mitigation and Adaptation Strategies for Global Change* 20(8), 1529–1544. https://doi.org/10.1007/s11027-014-9559-2.

Reiche, M., Gleixner, G. and Küssel, K. 2010. Effect of peat quality on microbial greenhouse gas formation in an acidic fen. *Biogeosciences* 7(1), 187–198.

Renou-Wilson, F., Moser, G., Fallon, D., Farrell, C. A., Müller, C. and Wilson, D. 2019. Rewetting degraded peatlands for climate and biodiversity benefits: results from two raised bogs. *Ecological Engineering* 127, 547-560. https://doi.org/10.1016/j.ecoleng.2018.02.014.

Rinne, J., Tuovinen, J. P., Klemedtsson, L., Aurela, M., Holst, J., Lohila, A., Weslien, P., Vestin, P., Łakomiec, P., Peichl, M., Tuittila, E. S., Heiskanen, L., Laurila, T., Li, X., Alekseychik, P., Mammarella, I., Ström, L., Crill, P. and Nilsson, M. B. 2020. Effect of the 2018 European drought on methane and carbon dioxide exchange of northern mire ecosystems. *Philosophical Transactions of the Royal Society of London. Series B, Biological Sciences* 375(1810), 20190517. https://doi.org/10.1098/rstb.2019.0517.

Ritson, J. P., Brazier, R. E., Graham, N. J. D., Freeman, C., Templeton, M. R. and Clark, J. M. 2017. The effect of drought on dissolved organic carbon (DOC) release from peatland soil and vegetation sources. *Biogeosciences* 14(11), 2891–2902. https://doi.org/10.5194/bg-14-2891-2017.

Rodriguez, A. F., Daroub, S. H., Gerber, S., Jennewein, S. P. and Singh, M. P. 2021. Water management effect on soil oxidation, greenhouse gas emissions, and nitrogen leaching in drained peat soils. *Soil Science Society of America Journal* 85(3), 814–828. https://doi.org/10.1002/saj2.20247.

Rogiers, N., Conen, F., Furger, M., Stöckli, R. and Eugster, W. 2008. Impact of past and present land-management on the C-balance of a grassland in the Swiss Alps: impact of land-management on the c-balance of a grassland. *Global Change Biology* 14(11), 2613–2625. https://doi.org/10.1111/j.1365-2486.2008.01680.x.

Rojstaczer, S. and Deverel, S. J. 1995. Land subsidence in drained histosols and highly organic mineral soils of California. *Soil Science Society of America Journal* 59(4), 1162–1167.

Rosset, T., Binet, S., Antoine, J.-M., Lerigoleur, E., Rigal, F. and Gandois, L. 2020. Drivers of seasonal- and event-scale DOC dynamics at the outlet of mountainous peatlands revealed by high-frequency monitoring. *Biogeosciences* 17(13), 3705–3722. https://doi.org/10.5194/bg-17-3705-2020.

Rowson, J. G., Gibson, H. S., Worrall, F., Ostle, N., Burt, T. P. and Adamson, J. K. 2010. The complete carbon budget of a drained peat catchment: carbon budget of drained peat catchment. *Soil Use and Management* 26(3), 261–273. https://doi.org/10.1111/j.1475-2743.2010.00274.x.

Samaritani, E., Siegenthaler, A., Yli-Petäys, M., Buttler, A., Christin, P.-A. and Mitchell, E. A. D. 2011. Seasonal net ecosystem carbon exchange of a regenerating cutaway bog: how long does it take to restore the C-sequestration function? *Restoration Ecology* 19(4), 480–489. https://doi.org/10.1111/j.1526-100X.2010.00662.x.

Sanderman, J., Hengl, T. and Fiske, G. J. 2017. Soil carbon debt of 12,000 years of human land use. *Proceedings of the National Academy of Sciences of the United States of America* 114(36), 9575–9580. https://doi.org/10.1073/pnas.1706103114.

Sangok, F. E., Maie, N., Melling, L. and Watanabe, A. 2017. Evaluation on the decomposability of tropical forest peat soils after conversion to an oil palm plantation. *Science of the Total Environment* 587–588, 381–388. https://doi.org/10.1016/j.scitotenv.2017.02.165.

Sangok, F. E., Sugiura, Y., Maie, N., Melling, L., Nakamura, T., Ikeya, K. and Watanabe, A. 2020. Variations in the rate of accumulation and chemical structure of soil organic matter in a coastal peatland in Sarawak, Malaysia. *CATENA* 184. https://doi.org/10.1016/j.catena.2019.104244, 104244.

Saunois, M., Stavert, A. R., Poulter, B., Bousquet, P., Canadell, J. G., Jackson, R. B., Raymond, P. A., Dlugokencky, E. J., Houweling, S., Patra, P. K., Ciais, P., Arora, V. K., Bastviken, D., Bergamaschi, P., Blake, D. R., Brailsford, G., Bruhwiler, L., Carlson, K. M., Carrol, M., Castaldi, S., Chandra, N., Crevoisier, C., Crill, P. M., Covey, K., Curry, C. L., Etiope, G., Frankenberg, C., Gedney, N., Hegglin, M. I., Höglund-Isaksson, L., Hugelius, G., Ishizawa, M., Ito, A., Janssens-Maenhout, G., Jensen, K. M., Joos, F., Kleinen, T., Krummel, P. B., Langenfelds, R. L., Laruelle, G. G., Liu, L., Machida, T., Maksyutov, S., McDonald, K. C., McNorton, J., Miller, P. A., Melton, J. R., Morino, I., Müller, J., Murguia-Flores, F., Naik, V., Niwa, Y., Noce, S., O'Doherty, S., Parker, R. J., Peng, C., Peng, S., Peters, G. P., Prigent, C., Prinn, R., Ramonet, M., Regnier, P., Riley, W. J., Rosentreter, J. A., Segers, A., Simpson, I. J., Shi, H., Smith, S. J., Steele, L. P., Thornton, B. F., Tian, H., Tohjima, Y., Tubiello, F. N., Tsuruta, A., Viovy, N., Voulgarakis, A., Weber, T. S., van Weele, M., van der Werf, G. R., Weiss, R. F., Worthy, D., Wunch, D., Yin, Y., Yoshida, Y., Zhang, W., Zhang, Z., Zhao, Y., Zheng, B., Zhu, Q., Zhu, Q. and Zhuang, Q. 2020. The global methane budget 2000-2017. *Earth System Science Data* 12(3), 1561–1623. https://doi.org/10.5194/essd-12-1561-2020.

Säurich, A., Tiemeyer, B., Dettmann, U. and Don, A. 2019. How do sand addition, soil moisture and nutrient status influence greenhouse gas fluxes from drained organic soils? *Soil Biology and Biochemistry* 135, 71–84. https://doi.org/10.1016/j.soilbio.2019.04.013.

Scanlon, D. and Moore, T. 2000. Carbon dioxide production from peatland soil profiles: the influence of temperature, oxic/anoxic conditions and substrate. *Soil Science* 165(2), 153–160. https://doi.org/10.1097/00010694-200002000-00006.

Schellekens, J. and Buurman, P. 2011. n-Alkane distributions as palaeoclimatic proxies in ombrotrophic peat: the role of decomposition and dominant vegetation. *Geoderma* 164(3–4), 112–121. https://doi.org/10.1016/j.geoderma.2011.05.012.

Schimel, J. P. 1995. Plant transport and methane production as controls on methane flux from arctic wet meadow tundra. *Biogeochemistry* 28(3), 183–200. https://doi.org/10.1007/BF02186458.

Schindler, U. and Müller, L. 1999. Rehabilitation of the soil quality of a degraded peat site. In: Stott, D. E., Mohtar, R. H. and Steinhardt, G. C. (Eds) 2001. Sustaining the global farm. Selected papers from the 10th International Soil Conservation Organization Meeting held 24–29, 1999, Purdue University and the USDA-ARS National Soil Erosion Research Laboratory.

Schipper, L. A. and McLeod, M. 2006. Subsidence rates and carbon loss in peat soils following conversion to pasture in the Waikato Region, New Zealand. *Soil Use and Management* 18(2), 91–93. https://doi.org/10.1111/j.1475-2743.2002.tb00225.x.

Schothorst, C. J. 1977. Subsidence of low moor peat soils in the western Netherlands. *Geoderma* 17(4), 265–291. https://doi.org/10.1016/0016-7061(77)90089-1.

Seiler, W., Holzapfelpschorn, A., Conrad, R. and Scharffe, D. 1984. Methane emission from rice paddies. *Journal of Atmospheric Chemistry* 1(3), 241-268.

Shurpali, N. J., Hyvönen, N. P., Huttunen, J. T., Clement, R. J., Reichstein, M., Nykänen, H., Biasi, C. and Martikainen, P. J. 2009. Cultivation of a perennial grass for bioenergy on a boreal organic soil - carbon sink or source? *GCB Bioenergy* 1(1), 35–50. https://doi.org/10.1111/j.1757-1707.2009.01003.x.

Shuttleworth, E. L., Evans, M. G., Hutchinson, S. M. and Rothwell, J. J. 2015. Peatland restoration: controls on sediment production and reductions in carbon and pollutant export: peatland restoration: controls on sediment production. *Earth Surface Process and Landforms* 40(4), 459–472. https://doi.org/10.1002/esp.3645.

Simola, H., Pitkänen, A. and Turunen, J. 2012. Carbon loss in drained forestry peatlands in Finland, estimated by re-sampling peatlands surveyed in the 1980s. *European Journal of Soil Science* 63(6), 798–807. https://doi.org/10.1111/j.1365-2389.2012.01499.x.

Sjögersten, S., Caul, S., Daniell, T. J., Jurd, A. P. S., O'Sullivan, O. S., Stapleton, C. S. and Titman, J. J. 2016. Organic matter chemistry controls greenhouse gas emissions from permafrost peatlands. *Soil Biology and Biochemistry* 98, 42–53. https://doi.org/10.1016/j.soilbio.2016.03.016.

Sloan, T. J., Payne, R. J., Anderson, A. R., Bain, C., Chapman, S., Cowie, N., Gilbert, P., Lindsay, R., Mauquoy, D., Newton, A. J. and Andersen, R. 2018. Peatland afforestation in the UK and consequences for carbon storage. *Mires and Peat*, 1–17. https://doi.org/10.19189/MaP.2017.OMB.315.

Sognnes, L. S., Fystro, G., Øpstad, S. L., Arstein, A. and Børresen, T. 2006. Effects of adding moraine soil or shell sand into peat soil on physical properties and grass yield in western Norway. *Acta Agriculturae Scandinavica, Section B - Plant Soil Science* 56(3), 161–170. https://doi.org/10.1080/09064710500218845.

Strack, M., Keith, A. M. and Xu, B. 2014. Growing season carbon dioxide and methane exchange at a restored peatland on the Western Boreal Plain. *Ecological Engineering* 64, 231–239. https://doi.org/10.1016/j.ecoleng.2013.12.013.

Strack, M., Waddington, J. M., Bourbonniere, R. A., Buckton, E. L., Shaw, K., Whittington, P. and Price, J. S. 2008. Effect of water table drawdown on peatland dissolved organic carbon export and dynamics. *Hydrological Processes* 22(17), 3373–3385. https://doi.org/10.1002/hyp.6931.

Strack, M., Zuback, Y., McCarter, C. and Price, J. 2015. Changes in dissolved organic carbon quality in soils and discharge 10years after peatland restoration. *Journal of Hydrology* 527, 345–354. https://doi.org/10.1016/j.jhydrol.2015.04.061.

Taft, H. E., Cross, P. A. and Jones, D. L. 2018. Efficacy of mitigation measures for reducing greenhouse gas emissions from intensively cultivated peatlands. *Soil Biology and Biochemistry* 127, 10–21. https://doi.org/10.1016/j.soilbio.2018.08.020.

Tan, Z. D., Lupascu, M. and Wijedasa, L. S. 2021. Paludiculture as a sustainable land use alternative for tropical peatlands: a review. *Science of the Total Environment* 753, 142111. https://doi.org/10.1016/j.scitotenv.2020.142111.

Tanneberger, F., Schröder, C., Hohlbein, M., Lenschow, †U., Permien, T., Wichmann, S. and Wichtmann, W. 2020. Climate change mitigation through land use on rewetted peatlands – cross-sectoral spatial planning for Paludiculture in Northeast Germany. *Wetlands* 40(6), 2309–2320. https://doi.org/10.1007/s13157-020-01310-8.

Tata, H. L. 2019. Mixed farming systems on peatlands in Jambi and Central Kalimantan provinces, Indonesia: should they be described as paludiculture? *Mires and Peat*, 1–17. https://doi.org/10.19189/MaP.2018.KHR.360.

Tiemeyer, B., Albiac Borraz, E., Augustin, J., Bechtold, M., Beetz, S., Beyer, C., Drösler, M., Ebli, M., Eickenscheidt, T., Fiedler, S., Förster, C., Freibauer, A., Giebels, M., Glatzel, S., Heinichen, J., Hoffmann, M., Höper, H., Jurasinski, G., Leiber-Sauheitl, K., Peichl-Brak, M., Roßkopf, N., Sommer, M. and Zeitz, J. 2016. High emissions of greenhouse gases from grasslands on peat and other organic soils. *Global Change Biology* 22(12), 4134–4149. https://doi.org/10.1111/gcb.13303.

Tiemeyer, B., Freibauer, A., Borraz, E. A., Augustin, J., Bechtold, M., Beetz, S., Beyer, C., Ebli, M., Eickenscheidt, T., Fiedler, S., Förster, C., Gensior, A., Giebels, M., Glatzel, S., Heinichen, J., Hoffmann, M., Höper, H., Jurasinski, G., Laggner, A., Leiber-Sauheitl, K., Peichl-Brak, M. and Drösler, M. 2020. A new methodology for organic soils in national greenhouse gas inventories: data synthesis, derivation and application. *Ecological Indicators* 109. https://doi.org/10.1016/j.ecolind.2019.105838, 105838.

Tiemeyer, B. and Kahle, P. 2014. Nitrogen and dissolved organic carbon (DOC) losses from an artificially drained grassland on organic soils. *Biogeosciences* 11(15), 4123–4137. https://doi.org/10.5194/bg-11-4123-2014.

Turetsky, M. R., Benscoter, B., Page, S., Rein, G., van der Werf, G. R. and Watts, A. 2015. Global vulnerability of peatlands to fire and carbon loss. *Nature Geoscience* 8(1), 11–14. https://doi.org/10.1038/ngeo2325.

Turetsky, M. R., Kotowska, A., Bubier, J., Dise, N. B., Crill, P., Hornibrook, E. R. C., Minkkinen, K., Moore, T. R., Myers-Smith, I. H., Nykänen, H., Olefeldt, D., Rinne, J., Saarnio, S., Shurpali, N., Tuittila, E. S., Waddington, J. M., White, J. R., Wickland, K. P. and Wilmking, M. 2014. A synthesis of methane emissions from 71 northern, temperate, and subtropical wetlands. *Global Change Biology* 20(7), 2183–2197. https://doi.org/10.1111/gcb.12580.

Uda, S. K., Hein, L. and Adventa, A. 2020. Towards better use of Indonesian peatlands with paludiculture and low-drainage food crops. *Wetlands Ecology and Management* 28(3), 509-526. https://doi.org/10.1007/s11273-020-09728-x.

Updegraff, K., Bridgham, S. D., Pastor, J., Weishampel, P. and Harth, C. 2001. Response of CO2 and CH $_4$ emissions FROM peatlands to warming and water table manipulation. *Ecological Applications* 11(2), 311-326. https://doi.org/10.1890/1051-0761(2001)011[0311:ROCACE]2.0.CO;2.

Upton, A., Vane, C. H., Girkin, N., Turner, B. L. and Sjögersten, S. 2018. Does litter input determine carbon storage and peat organic chemistry in tropical peatlands? *Geoderma* 326, 76-87. https://doi.org/10.1016/j.geoderma.2018.03.030.

Valach, A. C., Kasak, K., Hemes, K. S., Szutu, D., Verfaillie, J. and Baldocchi, D. D. 2021. Carbon flux trajectories and site conditions from restored impounded marshes in the Sacramento-San Joaquin Delta. *Geophysical Monograph Series* 267, 247-271.

van den Akker, J. J. H., Jansen, P. C., Hendriks, R. F. A., Hoving, I. and Pleijter, M. 2008. Submerged infiltration to halve subsidence and ghg emissions of agricultural peat soils. 14th International Peat Congress.

van der Berg, M. V. D., Ingwersen, J., Lamers, M. and Streck, T. 2016. The role of Phragmites in the CH_4 and CO_2 fluxes in a minerotrophic peatland in southwest Germany. *Biogeosciences* 13(21), 6107-6119.

van Seters, T. E. and Price, J. S. 2002. Towards a conceptual model of hydrological change on an abandoned cutover bog, Quebec. *Hydrological Processes* 16(10), 1965-1981. https://doi.org/10.1002/hyp.396.

Vanselow-Algan, M., Schmidt, S. R., Greven, M., Fiencke, C., Kutzbach, L. and Pfeiffer, E.-M. 2015. High methane emissions dominated annual greenhouse gas balances 30 years after bog rewetting. *Biogeosciences* 12(14), 4361-4371. https://doi.org/10.5194/bg-12-4361-2015.

Vroom, R. J. E., Xie, F., Geurts, J. J. M., Chojnowska, A., Smolders, A. J. P., Lamers, L. P. M. and Fritz, C. 2018. Typha latifolia paludiculture effectively improves water quality and reduces greenhouse gas emissions in rewetted peatlands. *Ecological Engineering* 124, 88-98. https://doi.org/10.1016/j.ecoleng.2018.09.008.

Waddington, J. M. and Day, S. M. 2007. Methane emissions from a peatland following restoration: methane and peatland restoration. *Journal of Geophysical Research: Biogeosciences* 112(G3), n/a-n/a. https://doi.org/10.1029/2007JG000400.

Waddington, J. M., Strack, M. and Greenwood, M. J. 2010. Toward restoring the net carbon sink function of degraded peatlands: short-term response in CO_2 exchange to ecosystem-scale restoration. *Journal of Geophysical Research* 115(G1), G01008. https://doi.org/10.1029/2009JG001090.

Wallor, E., Rosskopf, N. and Zeitz, J. 2018. Hydraulic properties of drained and cultivated fen soils part I - Horizon-based evaluation of van Genuchten parameters considering the state of moorsh-forming process. *Geoderma* 313, 69-81. https://doi.org/10.1016/j.geoderma.2017.10.026.

Wang, Y., Paul, S. M., Jocher, M., Alewell, C. and Leifeld, J. 2022. Reduced nitrous oxide emissions from drained temperate agricultural peatland after coverage with mineral soil. *Frontiers in Environmental Sciences* 157. https://doi.org/10.3389/fenvs.2022.856599.

Wang, Y., Paul, S. M., Jocher, M., Espic, C., Alewell, C., Szidat, S. and Leifeld, J. 2021. Soil carbon loss from drained agricultural peatland after coverage with mineral soil. *Science of the Total Environment* 800, 149498.

Weideveld, S. T. J., Liu, W., van den Berg, M., Lamers, L. P. M. and Fritz, C. 2021. Conventional subsoil irrigation techniques do not lower carbon emissions from drained peat meadows. *Biogeosciences* 18(12), 3881–3902. https://doi.org/10.5194/bg-18-3881-2021.

Welch, B., Gauci, V. and Sayer, E. J. 2019. Tree stem bases are sources of CH_4 and N_2O in a tropical forest on upland soil during the dry to wet season transition. *Global Change Biology* 25(1), 361–372. https://doi.org/10.1111/gcb.14498.

Wen, Y., Freeman, B., Ma, Q., Evans, C. D., Chadwick, D. R., Zang, H. and Jones, D. L. 2020. Raising the groundwater table in the non-growing season can reduce greenhouse gas emissions and maintain crop productivity in cultivated fen peats. *Journal of Cleaner Production* 262. https://doi.org/10.1016/j.jclepro.2020.121179, 121179.

Wichtmann, W., Schröder, C. and Joosten, H. (Eds) 2016. *Paludiculture - Productive Use of Wet Peatlands* (vol. VIII). Schweizerbart Science Publishers, Stuttgart. 272 p.

Wiggins, E. B., Czimczik, C. I., Santos, G. M., Chen, Y., Xu, X., Holden, S. R., Randerson, J. T., Harvey, C. F., Kai, F. M. and Yu, L. E. 2018. Smoke radiocarbon measurements from Indonesian fires provide evidence for burning of millennia-aged peat. *Proceedings of the National Academy of Sciences of the United States of America* 115(49), 12419–12424. https://doi.org/10.1073/pnas.1806003115.

Wilson, D., Alm, J., Laine, J., Byrne, K. A., Farrell, E. P. and Tuittila, E.-S. 2009. Rewetting of cutaway peatlands: are we Re-creating hot spots of methane emissions? *Restoration Ecology* 17(6), 796–806. https://doi.org/10.1111/j.1526-100X.2008.00416.x.

Wilson, D., Blain, D. and Couwenberg, J. 2016a. Greenhouse gas emission factors associated with rewetting of organic soils. *Mires and Peat*, 1–28. https://doi.org/10.19189/MaP.2016.OMB.222.

Wilson, D., Farrell, C. A., Fallon, D., Moser, G., Müller, C. and Renou-Wilson, F. 2016b. Multiyear greenhouse gas balances at a rewetted temperate peatland. *Global Change Biology* 22(12), 4080–4095. https://doi.org/10.1111/gcb.13325.

Wilson, D., Farrell, C., Mueller, C., Hepp, S., and Renou-Wilson, F. 2013. Rewetted industrial cutaway peatlands in western Ireland: a prime location for climate change mitigation? *Mires and Peat* 11(01), 1–22.

Wong, G. X., Hirata, R., Hirano, T., Kiew, F., Aeries, E. B., Musin, K. K., Waili, J. W., Lo, K. S. and Melling, L. 2018. Micrometeorological measurement of methane flux above a tropical peat swamp forest. *Agricultural and Forest Meteorology* 256-257, 353–361. https://doi.org/10.1016/j.agrformet.2018.03.025.

Worrall, F., Armstrong, A. and Holden, J. 2007. Short-term impact of peat drain-blocking on water colour, dissolved organic carbon concentration, and water table depth. *Journal of Hydrology* 337(3–4), 315–325. https://doi.org/10.1016/j.jhydrol.2007.01.046.

Worrall, F., Moody, C. S., Clay, G. D., Burt, T. P., Kettridge, N. and Rose, R. 2018. Thermodynamic control of the carbon budget of a peatland. *Journal of Geophysical Research: Biogeosciences* 123(6), 1863–1878. https://doi.org/10.1029/2017JG003996.

WRB. 2015. USS Working Group WRB. World reference base for soil resources 2014, update 2015. International soil classification system for naming soils and creating legends for soil maps. World Soil Resources Reports No. 106. FAO, Rome. Available at: https://www.fao.org/3/i3794en/I3794en.pdf (Accessed 25 April 2022).

Wright, E. L., Black, C. R., Cheesman, A. W., Drage, T., Large, D., Turner, B. L. and Sjögersten, S. 2011. Contribution of subsurface peat to CO_2 and CH_4 fluxes in a neotropical peatland: carbon fluxes in a neotropical peatland. *Global Change Biology* 17(9), 2867-2881. https://doi.org/10.1111/j.1365-2486.2011.02448.x.

Wüst-Galley, C., Mössinger, E. and Leifeld, J. 2016. Loss of the soil carbon storage function of drained forested peatlands. *Mires and Peat*, 1-22. https://doi.org/10.19189/MaP.2015.OMB.189.

Yavitt, J. B.,Williams, C. J. and Wieder, R. K. 2000. Controls on microbial production of methane and carbon dioxide in three *Sphagnum*-dominated peatland ecosystems as revealed by a reciprocal field peat transplant experiment. *Geomicrobiology Journal* 17(1), 61-88. https://doi.org/10.1080/014904500270503.

Yu, Z., Loisel, J., Brosseau, D. P., Beilman, D. W. and Hunt, S. J. 2010. Global peatland dynamics since the last glacial maximum: global peatlands since the LGM. *Geophysical Research Letters* 37(13), n/a-n/a. https://doi.org/10.1029/2010GL043584.

Yupi, H. M., Inoue, T. and Bathgate, J. 2016. Concentrations, loads and yields of organic carbon from two tropical peat swamp forest streams in Riau Province, Sumatra, Indonesia. *Mires and Peat*, 1-15. https://doi.org/10.19189/MaP.2015.OMB.181.

Zaccone, C., Said-Pullicino, D., Gigliotti, G. and Miano, T. M. 2008. Diagenetic trends in the phenolic constituents of Sphagnum-dominated peat and its corresponding humic acid fraction. *Organic Geochemistry* 39(7), 830-838. https://doi.org/10.1016/j.orggeochem.2008.04.018.

Zaidelman, F. R. and Shvarov, A. P. 2000. Hydrothermic regime, dynamics of organic matter and nitrogen in drained peaty soils at different sanding modes. *Archives of Agronomy and Soil Science* 45(2), 123-142. https://doi.org/10.1080/03650340009366117.

Zak, D. and Gelbrecht, J. 2007. The mobilisation of phosphorus, organic carbon and ammonium in the initial stage of fen rewetting (a case study from NE Germany). *Biogeochemistry* 85(2), 141-151. https://doi.org/10.1007/s10533-007-9122-2.

Chapter 22

Fostering carbon sequestration in humid tropical and subtropical soils

Deborah Pinheiro Dick and Cimélio Bayer, Federal University of Rio Grande do Sul, Brazil; and Jeferson Dieckow, Federal University of Paraná, Brazil

1 Introduction

Humid tropical and subtropical regions are characterised by intense solar radiation and high rainfall volume and erosivity (Table 1), receiving an average of 400 calories (cal) cm^{-2} day^{-1} of solar energy and 1500 mm of annual precipitation, which is twice as high as that of temperate regions at 200 cal cm^{-2} day^{-1} and 750 mm $year^{-1}$ (Sanchez, 1976; Greenland et al., 1992; van Wambeke, 1992). Rainfall erosivity varies from less than 400 megajoules millimeter per hectare per hour per year (MJ mm ha^{-1} h^{-1} $year^{-1}$) in temperate regions to more than 5000 MJ mm ha^{-1} h^{-1} $year^{-1}$ in tropical/subtropical regions (Panagos et al., 2017).

Due to heat and humidity, the soils of tropical regions are generally deep, highly weathered and have a physical structure that allows free drainage; Ferralsols and Acrisols are typical soil types of this region. Owing to the high degree of weathering, their chemical, physical and biological attributes are dependent on the organic matter (OM) content of the soil, which plays a vital role in soil function (Sanchez, 1976; van Wambeke, 1992).

These highly weathered soils are commonly regarded as poor in OM relative to temperate soils. However, the average C stock down to 1 m depth in tropical

http://dx.doi.org/10.19103/AS.2022.0106.21

Table 1 General averages of climate and soil conditions in humid tropical and subtropical regions and in temperate regions

Characteristic	Tropical and subtropical	Temperate
Climate		
Solar radiation, cal/cm^2/day	400	200
Rainfall volume, mm/year	1500	750
Rainfall erosivity, MJ mm ha^{-1} h^{-1} year^{-1}	>5000	200–400
Soils		
Main classes, % area		
Ferralsol	23	-
Acrisol/Luvisol	35	20
Variable charge soils, % area	60	10
Main minerals in clay fraction	kaolinite, iron and aluminium oxides	2:1 minerals

Sources: McGregor and Nieuwolt (1998); Panagos et al. (2017); Sanchez and Logan (1992); Uehara and Gillman (1981).

soils is approximately 102 Mg ha^{-1} (Eswaran et al., 1993), which does not differ greatly from the global mean of 116 Mg ha^{-1} (Eswaran et al., 1993) and supports the conclusion of Greenland et al. (1992) that tropical soils are not C poor. Research has revealed that due to the greater net primary production (NPP) and SOC stabilisation in humid tropical and subtropical regions, the range of soil OM content is variable and comparable to temperate regions under native vegetation. However, due to the higher turnover of OM, and the potentially higher activity of microorganisms involved in soil decomposition, maintaining OM levels in cultivated soils in humid tropical and subtropical ecosystems is very difficult (Greenland et al., 1992). This challenge must be addressed by greater attention to soil management than is required in temperate regions.

The exceptional biological activity in humid tropical and subtropical ecosystems, including higher NPP of vegetation and greater activity of soil fauna and microorganisms (van Wambeke, 1992; Mielniczuk et al., 2003), represents an ecological situation that is extremely favourable agronomically in terms of food, fibre and energy production. In some regions it is possible to grow 2 or more high-yielding crops per year providing the natural chemical limitations of soils are overcome (Mielniczuk et al., 2003). However, it also represents an ecological condition that requires the application of environmental conservation practices, as heat and humidity can also be highly damaging in terms of land degradation via OM depletion and soil erosion if effective management and conservation practices are not adopted (Ogle et al., 2005; Blanco-Canqui and Lal, 2010).

Maintaining soil OM in tropical and subtropical regions was promoted by the development and adoption of sustainable conservation agriculture based on three principles: (1) no or minimum mechanical soil disturbance, (2) permanent soil cover and (3) crop rotation/succession. The global area under conservation agriculture has grown from less than 20 million hectare (Mha) in the 1990s to a current area close to 160 Mha, showing an exponential growth of 140 Mha in the last 20–30 years, occupying approximately 60% of the cropland in South America, 35.9% in Australia and New Zealand and 24% in North America (Kassam et al., 2015). These 3 continents combined utilise conservation agriculture on more than 88% of land worldwide (Kassam et al., 2015).

The main principle of conservation agriculture is the no-tillage system, which should be combined with cropping systems with high crop residue addition aiming to keep permanent soil cover, and maintain or ameliorate soil quality and soil OM content. As a general reference, adding 10 Mg crop residue ha^{-1} yr^{-1} is necessary to maintain OM content in tropical and subtropical soils under no-tillage (Bayer and Dieckow, 2020), which is achieved by including high input crops, such as maize or sorghum, or cover crops in cropping schemes. The best management of soil cover, soil structure and soil OM is via crop rotation and cover crops, in addition to adopting beneficial conservation practices such as contouring and terracing, which minimise the risk of soil degradation in the tropics and subtropics, even in no-tillage systems (Mielniczuk et al., 2003; Blanco-Canqui and Lal, 2010; Merten and Minella, 2013).

2 Evolution of conservation agriculture and its impact on soil carbon

2.1 *Early conventional management and the decline of soil organic carbon stocks*

The original steady state of SOC stocks in tropical and subtropical soils was created naturally by native forests, savannas or grasslands under precipitation and temperatures that are comparatively higher than in temperate regions, thus leading to much higher biomass production and residue input. The conversion of natural ecosystems to managed agricultural systems and, in particular, the introduction of conventional methods of soil tillage based on ploughing and disking/harrowing operations have historically been the main cause of soil OM losses, significantly contributing to CO_2 emissions into the atmosphere and problems related to climate change (Lal, 2004b).

The mechanisation of agriculture following World War II led to the degree of soil disturbance increasing considerably. In South America, the soil management model based on mechanisation came largely from countries with temperate climate and was characterised by intensive tillage, which due to high rainfall erosivity and high temperatures inevitably degraded the tropical and subtropical soils much more rapidly than in temperate regions. The removal of soil cover, the disaggregation of soil structure and the depletion of OM by intensive tillage triggered physical soil degradation, mainly erosion, but also compaction. Such processes were further aggravated by the reduced natural fertility of these soils, which severely limited the NPP of crops and, consequently, the contribution of plant residue and C input to the soil.

The SOC stocks of the first 20 cm of Brazilian subtropical soils, under native forest or native grassland, drastically reduced by half in the first 15 years after conversion to annual croplands managed under intensive conventional tillage, as shown in the study by Pöttker (1977) (Fig. 1a). Other studies in this region showed a less drastic scenario, but still significant losses of 20–30% in the first 30–40 years (Tornquist et al., 2009; Dieckow et al., 2009). For sandy loam Ferralsols from the Brazilian Cerrado, Silva et al. (1994) reported that such SOC losses were >50% in relation to the initial stock in the first 5 years of conventional soybean cultivation (Fig. 1b). These results clearly demonstrate

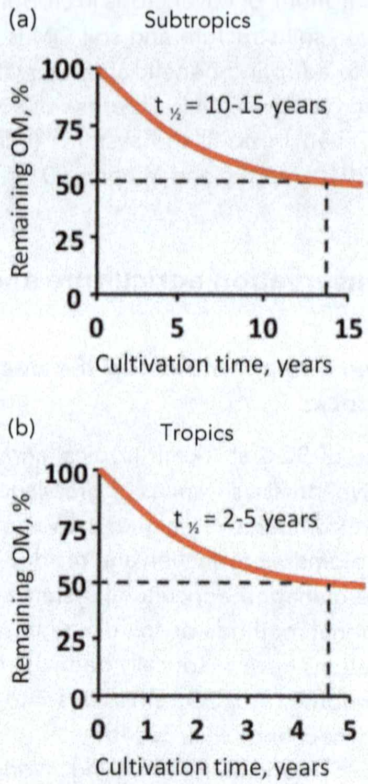

Figure 1 Soil organic matter losses (0–20 cm) of subtropical (a) and tropical (b) Ferralsols, respectively, in southern and centre-west regions of Brazil, after conversion of the natural vegetation into annual croplands managed under intensive conventional tillage. Adapted from Pöttker (1977) (a) and Silva et al. (1994) (b).

how susceptible the SOC stocks of tropical and subtropical soils are to losses caused by conventional management, which is incompatible with a warmer and wetter climate.

In terms of soil OM composition, land use changes such as the conversion of Cerrado or native grassland into annual cropland with conventional tillage significantly altered the original OM composition of a sandy clay Ferralsol in the Brazilian Cerrado, in Campo Grande-Mato Grosso do Sul State (MS), and of a sandy clay loam Acrisol in the Brazilian subtropics, in Eldorado do Sul-Rio Grande do Sul State (RS) (Table 1). This conversion decreased the concentration of O-alkyl C and alkyl C and increased aromatic C and carbonyl C, resulting in a higher aromatic/O-alkyl ratio (Table 1). Furthermore, the land use change also increased the H_{LIF} (humification index) and the concentration of semiquinone free radicals (Table 1). The results suggest that disturbance of the soil originally under native vegetation favoured preferential mineralisation of constituents such as polysaccharides (O-alkyl) and lipids (alkyl), thus increasing the relative concentration of more recalcitrant structures, such as aromatic, but not its absolute quantity. This outcome indicates the loss of OM in soils under conventional tillage compared with soils under native vegetation.

2.2 Soil carbon accumulation promoted by conservation tillage

Some of the first studies on conservation tillage in subtropical/tropical environments started in Brazil in the 1970s. At that time the level of soil degradation caused by conventional tillage practices reached an unsustainable level, characterised by average annual soil losses via erosion of 30–40 Mg ha^{-1} year^{-1} (Derpsch et al., 1991; Casão-Junior et al., 2012). As a first step towards conservation tillage, ploughing was initially replaced by chiselling, which reduced both the disturbance of the soil structure and the incorporation of litter deposited on the soil surface (Derpsch et al., 1991). Later, after the development of adapted machines, no-tillage was consolidated as the routine soil management system, with total maintenance of soil cover and disturbance of the structure confined to row planting (Casão-Junior et al., 2012).

With minimal disturbance of the soil structure, the annual rate of soil OM mineralisation under no-tillage dropped to approximately half compared with that under conventional tillage, as demonstrated in a subtropical Acrisol where the mineralisation rate of 0.040 year^{-1} under conventional tillage reduced to 0.019 year^{-1} under no-tillage (Bayer et al., 2006a). Thus, the mean residence time of C increased from 25 years with conventional tillage to 50 years with no-tillage. The lower mineralisation rate with no-tillage is associated with what Jastrow (1996) defined as the reciprocity of aggregate turnover and OM turnover. The physical protection of OM by occlusion in aggregates hinders the action of microorganisms and their enzymes on the organic substrate, as it acts as a physical barrier and reduces the diffusion of O_2 required for the oxidative processes of decomposition (Balesdent et al., 2000; von Lützow et al., 2006).

The idea of aggregate 'turnover' and OM, used by Six et al. (2000), proposed a model to mechanistically explain the accumulation of SOC in soils under no-tillage. According to this model, the plant residue with either no-tillage or conventional tillage is initially incorporated in macroaggregates, in the form of coarse (>250 μm) particulate OM (POM). Over time, the coarse POM decomposes to fine POM and is incorporated into microaggregates (<250 μm) within the macroaggregates, providing a greater degree of physical protection. Upon decomposition of coarse POM, the macroaggregates eventually become destabilised and release the microaggregates; however, upon the addition of new crop residues, the microaggregates form macroaggregates, restarting the 'cycle of aggregation'. In the case of soil managed under conventional tillage, the disturbance breaks down the macroaggregates and the accelerated decomposition of coarse POM begins, without having an 'opportunity' to be incorporated and protected in the form of fine POM inside microaggregates. On the other hand, under no-tillage, physical protection in macroaggregates can provide longer OM residence in soil, which aids interaction with specific adsorption sites on the surface of minerals, as proposed by Balabane and Plante (2004) regarding the role of occlusion.

In nearly 5 decades of no-tillage in Brazil, a great deal has been discovered regarding the environmental and economic benefits of this conservation system. In terms of SOC accumulation, soils under no-tillage may sequester 0.48–0.68 Mg ha^{-1} yr^{-1} and 0.35–0.65 Mg ha^{-1} yr^{-1} in the 0–20 cm layer in Brazilian subtropical and tropical regions, respectively (Bernoux et al., 2006; Bayer et al., 2006b). However, counting stocks down to 1 m depth, Boddey et al. (2010) estimated that C sequestration rates in subtropical Ferralsols under no-tillage could be increased by nearly 60%, or even higher, in the case of legume cover crop-based cropping systems. Measuring SOC stocks down to 1 m depth of a subtropical Acrisol, Veloso et al. (2018) found that no-tillage combined with cover crop-based cropping systems may reach mean SOC accumulation rates as high as 1.15 Mg C ha^{-1} $year^{-1}$ over 30 years, with about half of this accumulation occurring in the deeper 30–100 cm layer.

Although there is clear evidence that no-tillage leads to accumulation of SOC in tropical and subtropical Brazilian soils (Amado et al., 2006; Bayer et al., 2006b; Sá et al., 2001; Bernoux et al., 2006; Boddey et al., 2010; Veloso et al., 2018), some studies from temperate regions question the role of no-tillage in accumulating OM (Baker et al., 2007; Angers and Eriksen-Hamel, 2008; Blanco-Canqui and Lal, 2008; Powlson et al., 2014; VandenBygaart, 2016). The study of Angers and Eriksen-Hamel (2008) showed that SOC stocks are equivalent under no-tillage or conventional tillage, with the argument that SOC accumulation in the upper layer of no-tillage soil is offset by SOC accumulation in the bottom of the 'arable' layer in conventional tillage, which is due to layer inversion and the incorporation of surface residue with ploughing. However, Brazilian studies indicate that SOC accumulation under no-tillage can extend well beyond 20–25 cm depth (Dieckow et al., 2005b; Sisti et al., 2004; Boddey et al., 2010; Veloso et al., 2018).

In South America, well-managed no-tillage combined with cropping systems of high NPP, and consequently high input of plant biomass, presents a promising avenue in terms of SOC sequestration. However, it is reasonable to question how no-tillage is being practiced and whether crop rotation and management of soil cover and structure are being effectively implemented in accordance with the principles of conservation agriculture. In most grain production regions in Brazil, the challenge of converting from conventional tillage to no-tillage has been overcome, or at least is closer to being overcome, thanks to new machinery and weed-control technologies. This is of utmost importance as, to date, no-tillage required high amounts of pesticides, which accumulate in soil (da Silva et al., 2021) and may be harmful to soil health, ultimately also affecting SOC sequestration (Nathan et al., 2020; Chapter 6 of this book). Now the challenge is to improve the cropping system with the aim of high NPP and input of crop residue and C into the soil, while reducing external input.

As for the soil OM composition in the conservation agriculture system, the adoption of no-tillage has shown distinct results, ranging from the absence of effect to a partial recovery of the original composition of OM, closer to that of the soil under native vegetation. In studies conducted in medium/sandy textured soils in the Brazilian Cerrado (Campo Grande-MS) and subtropics (Eldorado do Sul-RS), there was an increase in the concentration of O-alkyl C and a decrease in the proportion of aromatic C groups associated with the increased SOC stock in soil after conversion from conventional tillage to no-tillage (Table 2), suggesting a recovery of the original OM characteristics under native vegetation. These results also suggest that selective preservation (recalcitrance) was not the main mechanism responsible for stabilising C in soil under no-tillage, reinforcing the hypothesis that physical protection by occlusion, enhanced by more stable aggregates, and surface interaction play the most relevant roles in the stabilisation of C in soil under no-tillage (Six et al., 2000).

On the other hand, the adoption of no-tillage for 17 years in a clayey Ferralsol in the Brazilian Cerrado (Dourados-MS) did not significantly change the composition of OM compared with conventional tillage (Table 2), possibly because it did not also change the SOC stock (data not shown). With regards to this clayey Ferralsol, more than 17 years are probably necessary for no-tillage to significantly change the composition and stocks of OM. The same was observed in another clayey Ferralsol in the subtropics (Santo Ângelo-RS, Table 2), although there was a (non-significant) trend of increasing concentrations of O-alkyl and

Table 2 Relative distribution of functional groups of SOC (carbonyl, aromatic, O-alkyl and alkyl) determined by ^{13}C-nuclear magnetic resonance spectroscopy, aromatic/O-alkyl ratio, H_{LIF} humification index obtained by laser induced fluorescence and free radical concentration obtained by electron spin resonance in the 0–5 cm layer of soils under native vegetation (NV), conventional tillage (CT) and no-tillage (NT), in 2 locations in the tropics and 2 locations in the subtropics of Brazil.

	Carbonyl C (%)	Aromatic C (%)	O-alkyl C (%)	Alkyl C (%)	Aromatic/O-alkyl ratio	H_{LIF} (a.u.)	Free radicals ($\times 10^{17}$ spins g^{-1} C)
Tropics – sandy clay Ferralsol (Campo Grande-MS)							
NV-Cerrado	9.9 b	18.5 b	45.9 b	25.7 a	0.40 b	87 b	12.2 b
CT	12.7 a	22.2 a	42.7 c	22.4 b	0.52 a	137 a	17.0 a
NT	11.0 ab	18.5 b	47.6 a	22.9 b	0.39 b	107 ab	11.9 b
Tropics – clayey Ferralsol (Dourados-MS)							
NV-Cerrado	8.8 b	14.5 a	53.7 a	23.0 a	0.27 b	97 b	10.9 a
CT	10.1 ab	15.6 a	49.5 b	24.8 a	0.31 a	148 a	4.1 b
NT	10.5 a	15.7 a	49.2 b	24.6 a	0.32 a	132 ab	3.1 b
Subtropics – clayey Ferralsol (Santo Ângelo-RS)							
NV-Grassland	9.7 a	16.1 a	50.8 a	23.4 a	0.32 b	69 b	5.6 b
CT	12.5 a	19.0 a	47.6 a	20.9 a	0.40 a	110 a	9.5 a
NT	10.0 a	17.1 a	49.5 a	23.4 a	0.35 ab	95 ab	7.8 ab
Subtropics – sandy clay loam Acrisol (Eldorado do Sul-RS)							
NV-Grassland	9.3 b	11.7 c	52.1 a	26.9 a	0.23 b	225 b	8.1 b
CT	10.4 a	16.1 a	49.1 b	24.3 b	0.33 a	408 a	12.3 a
NT	9.7 ab	14.4 b	52.1 a	23.8 b	0.28 ab	262 ab	10.2 ab

Letters in the same column compare the effect of management at each location – Tukey Test $p < 0.10$. Sources: Adapted from Boeni (2007) and Dieckow et al. (2009).

alkyl C groups and decreasing concentrations of aromatic and carbonyl groups with the adoption of no-tillage.

2.3 Carbon accumulation promoted by cropping systems

The adoption of cropping systems with a high input of residues to increase organic C input in soils under no-tillage is as important as adopting conservation tillage systems to reduce SOC losses due to mineralisation (Conceição et al., 2013; Leal et al., 2020). Ecologically, the cropping system, with its temporal and spatial arrangement of cultivated species, is a key factor in taking advantage of the higher temperature and precipitation in the tropics/subtropics. Optimisation is needed to increase the rate of photosynthesis and NPP in agroecosystems, with positive effects on the incorporation of organic C into soil OM.

The NPP of a cropping system depends on biotic factors, such as the plant species that make up the system, and on abiotic factors such as climate, nitrogen fertilisation, soil type, etc. While crops such as maize, soybeans, rice, wheat, etc. are grown for direct economic income, cover crops are also grown to maintain soil cover, promote green manure, improve the degree of soil aggregation, etc. Crops of economic importance, such as maize, and cover crops with high biomass production play an essential role in cropping systems when it is intended to increase the addition of plant residue in order to increase SOC stocks in no-tillage soils.

In a study conducted in a subtropical Acrisol, plots kept as bare soil and without any tillage disturbance (weed plants uprooted as early as possible but not removed from the plot) had an annual addition of 0.6 Mg C ha^{-1} (Bayer et al., 2000). It was estimated for this bare soil that over 100 years the initial SOC stock in the 0–17.5 cm layer would be reduced by 70%, which highlights that not disturbing the soil is insufficient to prevent SOC losses and C input is required. In a legume cover crop-based cropping system of pigeon pea (*Cajanus cajan* (L.) Millsp.) plus maize, where the addition was 8.9 Mg C ha^{-1} $year^{-1}$, the steady-state SOC stock in the 0–17.5 cm layer after 100 years was estimated to be almost twice that of the initial SOC stock (61.3 compared with 32.2 Mg C ha^{-1}). On the other hand, a black oat (*Avena strigosa* Schreb.)-maize succession in the same study, with an intermediate addition of 4.4 Mg C ha^{-1} $year^{-1}$, tended to only maintain the initial SOC stock of Acrisol.

The amount of crop residue added to the soil depends on plant species and other management aspects such as fertilisation. In cropping systems in a subtropical Acrisol, the application of 180 kg N ha^{-1} to maize increased the amount of C added to the soil by 92% when maize was in an oat-maize system and, on average, 38% when maize was intercropped with legumes (vetch-maize or oat+vetch-maize+cowpea) (Lovato, 2001; Lovato et al., 2004). These results show the importance of nitrogen fertilisation, especially in grass-based systems, in increasing the addition of residues to the soil and therefore the C and N stocks. Several other studies under temperate climate conditions reached similar conclusions regarding the role of nutrients on SOC sequestration (Kirkby et al., 2014; Wang et al., 2015; Campbell et al., 1991).

In addition to the amount of crop biomass added to the soil, the portion that will actually be converted to OM stocks also depends on the conversion coefficient k_1 (Henin and Dupuis, 1945), which may vary mainly as a function of two factors (Balesdent and Balabane, 1996; Bolinder et al., 1999). The first factor refers to the intrinsic characteristics of the residue, such as the lignin/N ratio, where a higher concentration of lignin increases the material's resistance to microbial decomposition and thus increases the k_1 value. The second factor refers to the accessibility of the microbiota to the residue, where it is important to highlight the differences between residue from aboveground and residue from roots. Residue from roots has a higher k_1 than residue from aboveground, possibly due to being deposited below ground and directly into the soil and is physically protected inside stable aggregates (Bolinder et al., 1999). Additionally, the higher lignin/N ratio of roots compared with shoots may contribute to greater preservation of organic material derived from roots (Chapter 3 of this book).

Regarding the potential of legume cover crops to increase the stock of OM in tropical and subtropical soils, it is important to highlight their role in biological nitrogen fixation and the consequent benefit that this nitrogen will bring directly or indirectly to the accumulation of soil OM. In subtropical conditions, legume cover crops include a range of species such as common vetch (*Vicia sativa* L.), hairy vetch (*Vicia villosa* Roth), blue lupine (*Lupinus angustifolius* L.), chicharo (*Lathyrus sativus* L.), among others. The oil radish (*Raphanus sativus* L. var. *oleiferus* Metzg.), despite not being a legume, has a prominent role as a cover crop to recycle and supply N in the cool/cold season. Legumes cultivated under a tropical climate include cowpea (*Vigna unguiculata* (L.) Walp.), sunn hemp (*Crotalaria* sp.), pigeon pea (*C. cajan* (L.) Millsp.), lablab (*Lablab purpureus* (L.) Sweet) and velvet bean (*Mucuna pruriens* var. *utilis*). The importance of legume cover crops in a tropical climate, in addition to nitrogen fixation, is their high phytomass production, directly contributing to C input into the soil. Soybean, although a very important legume cash crop in the tropics and subtropics, cannot be considered a contributor of N to the soil, as almost all of its fixed nitrogen is exported via harvesting (Alves et al., 2002).

In a subtropical Acrisol under no-tillage, Veloso et al. (2018) reported SOC sequestration rates as high as 1.0 t ha^{-1} year^{-1} at the ninth year of a legume-based cropping system comprised of oat+vetch/maize+cowpea, while in the cereal system without legumes (oat/maize) the sequestration was only 0.2 t ha^{-1} year^{-1}. In this same study it was also shown that N fixed by legume cover crops was more efficient at inducing C sequestration than

N applied as mineral fertiliser in the cereal systems, which agrees with the hypothesis that SOC originating from organic residue with a low C:N ratio might be efficiently stabilised in organo-mineral interactions on clay surfaces due to highly efficient C use by soil microorganisms (Cotrufo et al., 2013, 2015).

In integrated crop-livestock systems, with phases of temporary pasture rotating with phases of temporary cropland (Moraes et al., 2014), interesting results have been obtained in terms of improving SOC stocks. In this integrated system, the animal component was a key factor, as grazing stimulates regrowth aboveground and by the root system and, consequently, stimulates the addition of C via roots (Chapter 18 of this book). In a clayey Ferralsol of a tropical Cerrado, the contribution of integrated systems was clearly evidenced by Salton et al. (2014). In their study, 9 years after adopting an integrated crop-livestock system with *Brachiaria decumbens* pasture for 2 years and wheat/soybean cropland for 2 more years, the average SOC sequestration rate in the 0–20 cm layer was 0.44 Mg ha^{-1} $year^{-1}$; whereas in the continuous cropping system including oat, wheat, soybean and maize, no SOC sequestration occurred.

In terms of the chemical characteristics of soil OM in different cropping systems, there is currently less information available compared with quantitative information regarding SOC stocks. However, from the available results, it is inferred that although the characteristics of the original plant tissue are different, the qualitative characteristics of OM most associated with minerals do not appear to be changed by different plant materials. In work investigating a subtropical soil with three cropping systems under no-tillage, including oat/maize, pigeon pea+maize and lablab+maize, Dieckow et al. (2005a) used ^{13}C nuclear magnetic resonance NMR spectroscopy and observed that the composition of the aboveground organic material varied considerably among species comprising the cropping systems. Pigeon pea plants had the highest proportion of alkyl C (17–18%) compared with oat plants (11%) and maize (~7%), probably as pigeon peas have a higher cutin content. Regarding aromatic C, pigeon pea and maize had a similar proportion, which was about 11% higher than the 7% observed for oat plants, probably due to the higher lignin content in pigeon pea and maize. The proportion of carbonyl C among species followed a similar trend to that of alkyl C, with pigeon pea presenting the highest proportion (8%), followed by oat (6%) and finally, maize (4%). In the top 2.5 cm of soil, the POM chemical composition was affected by the cropping systems, reflecting the trend observed for the composition of the original aboveground residue. However, in the silt and clay fractions, these differences practically disappeared, in agreement with other studies suggesting little influence of plant material type on the OM composition in the smaller fractions, where the characteristics of products of microbial origin probably have a more relevant role (Grandy and Neff, 2008; Boeni et al., 2014).

3 Carbon stabilisation by organo-mineral interactions in highly weathered soils

In tropical and subtropical soils, organo-mineral interactions play a key role in OM stabilisation. Considering that OM encompasses a wide variety of functional groups and presents a complex and dynamic molecular structure (Kleber et al., 2021), and highly weathered soils have a predominance of variable charge minerals, many mechanisms and factors may act simultaneously in the OM interaction and thus, affect C stabilisation in soil.

The clay fraction of these soils is usually composed of kaolinite clay (1:1 clay) followed by the Fe oxides, goethite (α-FeOOH), haematite (α-Fe_2O_3) and gibbsite ($Al(OH)_3$), occurring in different proportions (Kämpf and Curi, 2000, 2003; Oliveira et al., 2020). To a lesser extent, pedogenic *maghaemite* (γ-Fe_2O_3) is found in some Ferralsols developed from mafic and ultramafic rocks, as is the case for soils in southeastern Brazil (Carvalho et al., 2015; Camêlo et al., 2018) and/or in Ferralsols subjected to fire (Cornell and Schwertmann, 2003). In addition to the high specific area of these minerals, their hydroxylated surface groups are highly reactive with ionic and molecular moieties in these soils, and among them, the OM.

The organo-mineral association is therefore considered the key mechanism on the molecular scale for OM stabilisation in tropical and subtropical soils, and the following interactions are theoretically foreseen: surface coordination (anion ligand exchange), hydrogen bonding, cation bridging, hydrophobic interaction and electrostatic interaction (von Lützow et al., 2006; Dick et al., 2010; Dignac et al., 2017). However, a direct observation of the above-mentioned mechanisms in natural soil samples is difficult even when using current state-of-the-art instrumental techniques, hence most of the information on this subject, to date, has been obtained indirectly through correlations or by comparison approaches.

In a group of six Ferralsols under native vegetation of a toposequence in southern Brazil, the relation of the clay fraction, with C content of the Aw horizon, confirmed the importance of the fine fraction in stabilising soil OM (Dalmolin et al., 2006). Regarding the mineralogy, it has been recognised that Fe oxides are the main effective drivers of OM stabilisation in tropical and subtropical soils due to strong adsorption on the oxide surface (Dick et al., 2005). For instance, in a study with 8 representative subtropical altitude soils (4 Leptosols, 3 Cambissols and 1 Ferralsol) under native vegetation, an inverse correlation was obtained between C content recovered after sample demineralisation (C_R) and pedogenic Fe oxide content (Fed) for the 0–10 cm layer (Fig. 2) (Dick et al., 2008). This result suggests that dissolution of Fe oxides, via treatment with a 10% hydrofluoric acid (HF) solution, results in the hydrophilic and low molecular weight OM directly associated with the Fe oxides being solubilised

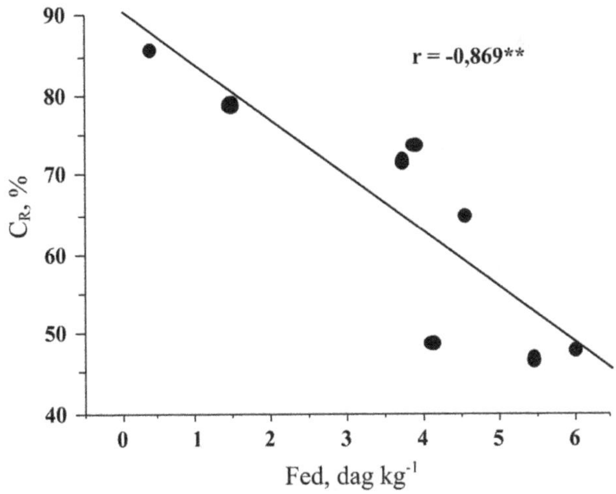

Figure 2 Carbon recovery (C_R) after 10% HF treatment and Fed content for a group of 8 high-altitude subtropical soils from southern Brazil (Dick et al., 2008).

and removed from the sample along with the minerals. The more relevant role of goethite (Gt) in comparison to haematite (Hm) in this process was observed using the same approach, where an inverse correlation between C_R and Gt/Gt+Hm was obtained for the BA and Bw horizons of a group of subtropical Ferralsols (Fig. 3) (Dalmolin et al., 2006).

Studies dealing with highly weathered soils have indicated that carboxylic C and O-alkyl C are the most important OM groups participating in the surface coordination reactions. In profile samples of four subtropical soils derived from basalt (three Ferralsols and one Chernozem), the mean crystallite dimension (MCD) of kaolinite, heamatite and goethite correlated inversely with –COOH and O-alkyl C proportions, suggesting that a continuous dissolution/reprecipitation process on the mineral surface promoted by adsorption of these two groups hindered further crystallisation of the mineral (Hanke et al., 2015). The association of Fe oxide fractions of low crystallinity with OM, was also observed in a previous study with subtropical high-altitude soils (Leptosols) where the ratio of Feo/Fed correlated with (O-alkyl + N-alkyl) groups (determined by ^{13}C NMR) (Potes et al., 2010) (Fig. 4).

Interactions with Fe oxides by carbohydrate-like structures present in soil OM were also identified by Schoening et al. (2005), when comparing the solid state ^{13}C NMR NMR spectra of German soils before and after demineralisation with a 10% HF solution. The authors observed that the N/O-alkyl groups were less 'visible' in the untreated sample, which was assigned to the strong interaction with Fe oxides. After the removal of Fe oxides, the signal intensity of this functional group in the NMR spectrum increased due to disruption of the Fe-O-C interaction. Furthermore, advanced techniques, such as NanoSims and Near

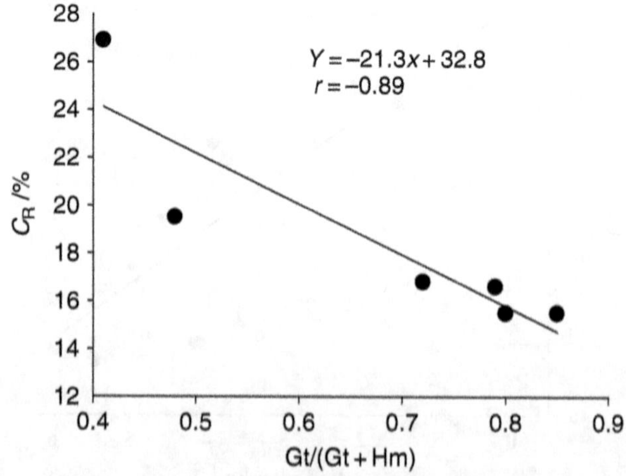

Figure 3 Carbon recovery (C_R) after 10% HF treatment and the Gt/Gt+Hm ratio determined for the B horizons of subtropical Ferralsols under native vegetation (Dalmolin et al., 2006).

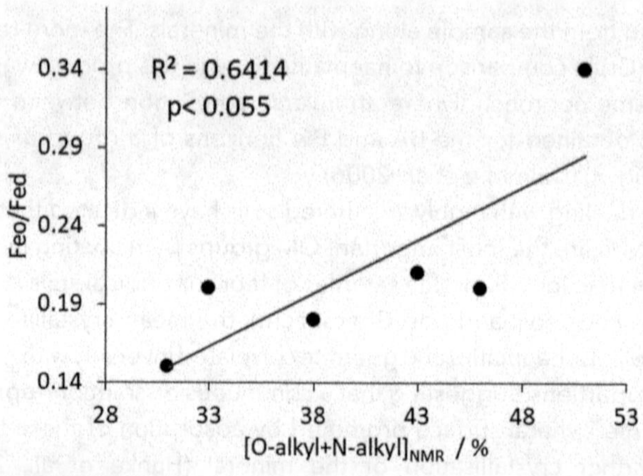

Figure 4 Correlation between the Feo/Fed ratio and the [O-alkyl+ N-alkyl] ratio determined by NMR for a group of Leptosols in southern Brazil (Potes et al., 2010).

Edge X-ray Absorption Fine Structure (NEXAFS) when applied to pure OM and mineral systems or to soil samples evidenced that oxygen-containing groups are prone to coordinating with the Fe oxide surfaces (Liu et al., 2013; Rumpel et al., 2015; Chen et al., 2016; Tamrat et al., 2019). Recently, a 3D conceptual model of a short-range order mineral associated with OM, named 'nanoCLICS', has been proposed (Tamrat et al., 2019), where Fe, Al and Si are involved. The possibility of such a structure occurring in microsites in tropical and subtropical soils should not be excluded and its contribution to OM stabilisation requires investigation.

The role of Fe oxides in protecting O-alkyl C groups in tropical and subtropical agriculture soils was revealed by investigations focusing on conservation soil management systems, as already exemplified by the data in Table 2 and discussed in Section 2.2 of this chapter. The results of Dieckow et al. (2009) were later corroborated by Briedis et al. (2021) with a group of four subtropical soils under no-tillage. The positive correlation between O-alkyl-C and silt + clay contents (0.42, p <.01) indicates that the accumulation of SOC in these soils is mainly associated with labile C compounds, which are presumably protected by interactions. Furthermore, the coarse fraction (>50 μm) presented a greater O-alkyl proportion than the fine fraction (<50 μm), indicating that biochemically labile OM was protected via surface interactions with clay-size minerals, which in turn build up structures greater than 50 μm.

Hydrophobic interaction is another mechanism of OM stabilisation in soils that requires further investigation in highly weathered soils, which has been evidenced by the work of Reis et al. (2014) in two clayey subtropical Ferralsols. The authors showed that in addition to the organo-mineral interaction, a mechanism of OM self-assemblage, enhanced by long periods of non-disturbance, contributed to its stabilisation in both soils, as indicated by the greater C retention in the silt compared with the clay fraction. This assumption is in line with the findings of Vogel et al. (2014) who showed that new OM preferentially binds to already existing patches of old OM adsorbed on mineral surfaces. Therefore, the capacity of soil minerals to stabilise OM would also depend on their ability to adsorb OM in a large number of patches (Dignac et al., 2017). In the study carried out by Reis et al. (2014), the specificity of the mineralogy interacting with SOM was also shown: the goethite+gibbsite-rich Ferralsol showed a greater C sequestration capacity than the heamatite+kaolinite-rich soil.

4 Greenhouse gas balance at the soil-atmosphere interface

The sink function of soils for C dioxide (CO_2) from the atmosphere may benefit from conservation agriculture in tropical and subtropical ecosystems (Bayer et al., 2006b, 2016; Veloso et al., 2018). However, this environmental benefit should be considered along with trade-offs from the potential impact of soil management practices (Chapter 6 of this book) on the emission of GHG other than CO_2, namely methane (CH_4) and nitrous oxide (N_2O), which have a global warming potential 25 and 298 times higher than CO_2, respectively (IPCC, 2013). Additionally, the CO_2 equivalent costs from inputs and farm operations must also be accurately accounted for (Lal, 2004a; Zanatta et al., 2007).

4.1 Soil nitrous oxide and methane fluxes are affected by soil management

The production of N_2O and CH_4 in soils is mediated by microbial processes, which are affected by several soil and climatic factors, including those influenced by soil management. The main soil processes related to N_2O production are nitrification, denitrification and nitrifier denitrification, which are mainly controlled by substrate (NH_4^+, NO_3^- and labile C) and O_2 availability. On the other hand, soil CH_4 fluxes are derived from the balance of methanotrophic and methanogenic activity. In general, soil disturbance by conventional tillage creates a hostile environment for methanotrophic organisms (Hütsch, 1998), while no-tillage is usually suggested as an alternative for restoring the CH_4 uptake capacity of soils (Bayer et al., 2012, 2013).

Early GHG emission results obtained mainly in temperate regions showed an increase in soil N_2O emissions under no-tillage, attributed mainly to soil compaction, which would partially offset the environmental benefit of SOC sequestration (Gregorich et al., 2005; Baggs et al., 2003; Ball et al., 1999). However, as GHG research advanced in tropical and subtropical regions, those early results from temperate regions have not been confirmed. In the tropics and subtropics, N_2O emissions in no-tillage soils are usually lower than those of conventionally tilled soils, possibly due to the lower mineralisation rates of OM and plant residues under no-tillage, and thus lower availability of NH_4^+ and NO_3^- for N_2O production via soil microbial processes (Piva et al., 2012; Escobar et al., 2010; Jantalia et al., 2008; Bayer et al., 2015). In relation to CH_4 fluxes in soil, despite the potential impact of conservation agriculture on the capacity of CH_4 uptake in soils, these effects are considered to be extremely slow, with several decades necessary for the soil to become a significant CH_4 sink (Bayer et al., 2012, 2013). In addition, the CH_4 emission rates in aerated soils are usually very low and thus do not significantly impact the GHG balance of tropical and subtropical soils.

4.2 Global warming potential and yield-scaled greenhouse gases emissions

Figure 5 depicts the impact of conservation agriculture on CO_2, N_2O and CH_4 emissions from subtropical soils in southern Brazil, in comparison to conventional cultivation. In addition, the balance of GHG (here named global warming potential) was calculated taking in account the respective fluxes and global warming potential of N_2O (265) and CH_4 (25), in comparison to CO_2 (1), being expressed in kg CO_2 equivalent ha^{-1}.

In spite of N_2O emission mitigation, SOC accumulation in soil is the main determining factor for the overall mitigation of GHG emissions in no-tillage

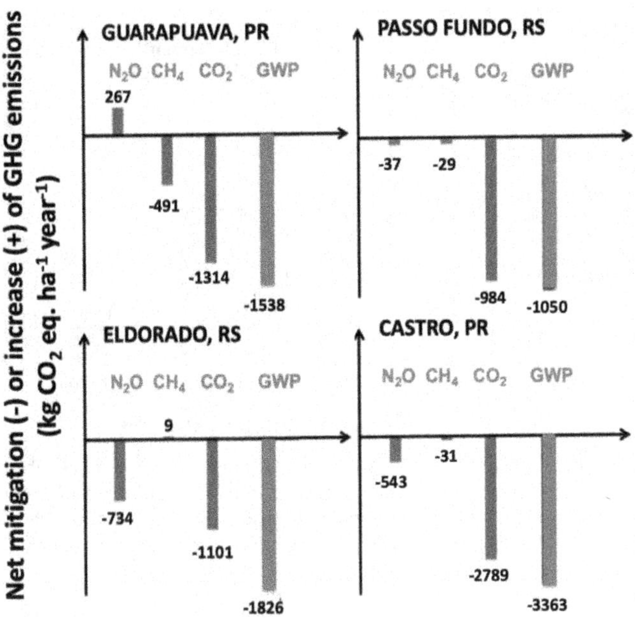

Figure 5 Net greenhouse gas emissions (N_2O, CH_4 and CO_2) and mitigation of global warming potential in no-tillage soils in subtropical environment in southern Brazil [3 Ferralsols (Guarapuava, Passo Fundo and Castro) and 1 Acrisol (Eldorado)]. The data are expressed in kg CO_2 eq ha^{-1} year^{-1} taking the soil under conventional tillage as reference. Source: Bayer, C. (data not published).

soil compared with conventional tillage. Figure 5 shows that 75% of the overall mitigation of GHG emissions of approximately 2000 kg CO_2 eq ha^{-1} year^{-1} in no-tillage soils was due to SOC accumulation. However, the GHG balance of no-tillage systems is highly dependent on the adopted cropping system and its effect on SOC stocks; in general, the higher the annual biomass input, the greater the environmental benefits (Piva et al., 2012; Bayer et al., 2016).

Thus, considering that conservation management systems have a positive impact on crop productivity (Bayer and Dieckow, 2020) and at the same time reduce GHG emissions, the normalised effect on the intensity of GHG emissions (emission of GHG per unit of food or fibre produced) is highly beneficial. Figure 6 illustrates the positive impact of no-tillage and cover crop-based cropping systems on the reduction of yield-scaled GHG emissions in subtropical agriculture. The figure shows that to produce 1 Mg of maize grain, GHG emission decreased from more than 300 kg CO_2 equivalent under a conventional tillage system to less than 50 kg CO_2 equivalent when no-tillage was combined with a legume cover crop-based cropping system.

Thus, conservation agriculture simultaneously allows for food and fibre production while contributing to SOC sequestration and mitigating non-CO_2

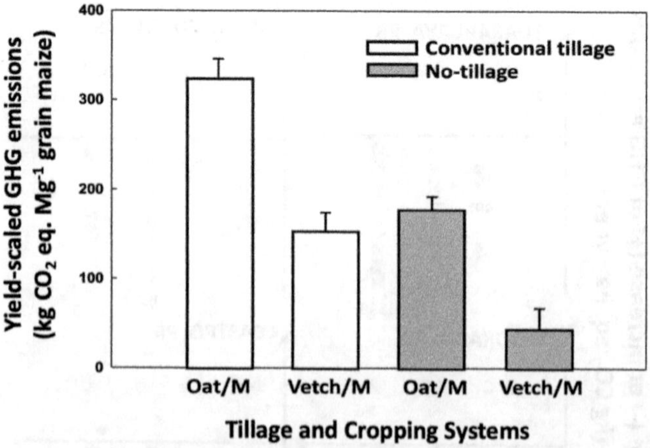

Figure 6 Intensity of greenhouse gas emissions (yield-scaled emissions, GHG emissions per unit of maize grain produced, kg CO_2 eq Mg^{-1} grain) in a subtropical Acrisol subjected to different tillage and cropping systems in southern Brazil. Oat and vetch are winter cover crops, and maize (M) the cash crop grown in summer. Source: Bayer, C. (data not published).

gases, thus having a crucial role in the development of sustainable production systems in tropical and subtropical environments.

5 Conclusion

The conservation agriculture practised in the tropics and subtropics over the past 3–4 decades has played a significant role in curbing soil degradation and establishing a more sustainable agricultural scenario based on no-tillage and improved cropping systems. However, several points still need to be addressed in terms of further improvements in soil management, C sequestration and the mitigation of GHG, either via practical farming systems or scientific research. We believe the current cropping system models based on soybean commodity production can be improved into more sophisticated crop rotation systems based on crops with high NPP, such as maize or sorghum, that eventually increase the C input into the soil. This would involve more efficient utilisation of natural resources such as light, water and soil, which are abundant in most tropical and subtropical environments. Integrated crop-livestock systems are another option for improving NPP and C input into soil, in addition to diversifying the farming system. A wider adoption of cover crops, especially legumes, combined in time and space with cash crops is also a valuable strategy, especially for those regions where fallow periods are current practice. In terms of research, effort should be made to better understand the mechanisms or processes involved in SOC accumulation in deeper soil layers, below 0.3 m, which is responsible for half of the SOC stabilisation in the 1 m soil profile, and where the role of roots and soil fauna is unquestionable. In tropical and subtropical regions, the GHG mitigation potential through SOC accumulation and decrease of other GHG emissions from agricultural soils under no-tillage may be quite different from those in temperate regions. In these humid and warm climates there is considerable potential for conservation agriculture to mitigate GHG emissions. The processes of SOC accumulation in stable aggregates and on mineral surfaces through organo-mineral interactions and the role of the soil microbial biomass must also be elucidated for tropical and subtropical soils, and hopefully new analytical techniques will provide a clearer understanding of these mechanisms. The effect of external input on these processes also requires clarification. The majority of increasing food and fibre demand over the coming years is to be supplied by tropical and subtropical agriculture, and this will certainly require more actions towards fostering sustainable SOC sequestration in humid tropical and subtropical soils under conservation agriculture.

6 Where to look for further information

There are several international initiatives designed to promote soil carbon sequestration. Here we present two of them.

In the framework of the Global Soil Partnership/FAO, the RECSOIL initiative (Recarbonization of global agricultural soils, https://www.fao.org/global-soil-partnership/areas-of-work/recarbonization-of-global-soils/en/) aims to support and improve national and regional GHG mitigation and carbon sequestration initiatives.

The international '4 per 1000' Initiative (https://4p1000.org/?lang=en) launched by the French government has an aspiration to increase global soil organic matter stocks by 4 per 1000 (or 0.4%) per year as a way to cut global emissions of greenhouse gases.

A recent and comprehensive review about soil conservation management systems and their impact on C sequestration can also be found in 'Intensification of no-till agricultural systems: An opportunity for carbon sequestration' from Nicoloso, R. S. and Rice, C. W. (*Soil Sci Soc Am J. 85:*1395–1409. 2021. https://doi.org/10.1002/saj2.20260).

7 References

Alves, B. J. R., Zotarelli, L., Lara-Cabezas, W. A. R., Torres, E., Hungria, M., Urquiaga, S. and Boddey, R. M. (2002). Benefit of legume-fixed N in crop rotations under zero-tillage. *Current Plant Science and Biotechnology in Agriculture* 38: 533–534.

Amado, T. J., Bayer, C., Conceição, P. C., Spagnollo, E., de Campos, B. H. and da Veiga, M. (2006). Potential of carbon accumulation in no-till soils with intensive use and cover crops in southern Brazil. *Journal of Environmental Quality* 35(4): 1599–1607.

Angers, D. A. and Eriksen-Hamel, N. S. (2008). Full-inversion tillage and organic carbon distribution in soil profiles: A meta-analysis. *Soil Science Society of America Journal* 72(5): 1370-1374.

Baggs, E. M., Stevenson, M., Pihlatie, M., Regar, A., Cook, H. and Cadisch, G. (2003). Nitrous oxide emissions following application of residues and fertiliser under zero and conventional tillage. *Plant and Soil* 254(2): 361-370.

Baker, J. M., Ochsner, T. E., Venterea, R. T. and Griffis, T. J. (2007). Tillage and soil carbon sequestration - What do we really know? *Agriculture, Ecosystems and Environment* 118(1-4): 1-5.

Balabane, M. and Plante, A. F. (2004). Aggregation and carbon storage in silty soil using physical fractionation techniques. *European Journal of Soil Science* 55(2): 415-427.

Balesdent, J. and Balabane, M. (1996). Major contribution of roots to soil carbon storage inferred from maize cultivated soils. *Soil Biology and Biochemistry* 28(9): 1261-1263.

Balesdent, J., Chenu, C. and Balabane, M. (2000). Relationship of soil organic matter dynamics to physical protection and tillage. *Soil and Tillage Research* 53(3-4): 215-230.

Ball, B. C., Scott, A. and Parker, J. P. (1999). Field N_2O, CO_2 and CH_4 fluxes in relation to tillage, compaction and soil quality in Scotland. *Soil and Tillage Research* 53(1): 29-39.

Bayer, C. and Dieckow, J. (2020). Lessons learnt from long-term no-till systems regarding soil management in humid tropical and subtropical regions. In: Dang, Y. P., Dalal, R. C. and Menzies, N. W. (Eds) *No-Till Farming Systems for Sustainable Agriculture: Challenges and Opportunities*. Cham: Springer International Publishing, 437-457.

Bayer, C., Gomes, J., Beber Vieira, F. C. B., Zanatta, J. A., de Cássia Piccolo, M. and Dieckow, J. (2012). Methane emission from soil under long-term no-till cropping systems. *Soil and Tillage Research* 124: 1-7.

Bayer, C., Gomes, J., Beber Vieira, F. C. B., Zanatta, J. A., Piccolo, M. D. C. and Dieckow, J. (2013). Soil methane oxidation in a long-term no-tillage system in Southern Brazil. *Semina: Ciencias Agrarias* 34(4): 1695-1706.

Bayer, C., Gomes, J., Zanatta, J. A., Vieira, F. C. B. and Dieckow, J. (2016). Mitigating greenhouse gas emissions from a subtropical ultisol by using long-term no-tillage in combination with legume cover crops. *Soil and Tillage Research* 161: 86-94.

Bayer, C., Gomes, J., Zanatta, J. A., Vieira, F. C. B., Piccolo, MdC., Dieckow, J. and Six, J. (2015). Soil nitrous oxide emissions as affected by long-term tillage, cropping systems and nitrogen fertilization in Southern Brazil. *Soil and Tillage Research* 146: 213-222.

Bayer, C., Martin-Neto, L., Mielniczuk, J. and Ceretta, C. A. (2000). Effect of no-till cropping systems on soil organic matter in a sandy clay loam acrisol from Southern Brazil monitored by electron spin resonance and nuclear magnetic resonance. *Soil and Tillage Research* 53(2): 95-104.

Bayer, C., Lovato, T., Dieckow, J., Zanatta, J. A. and Mielniczuk, J. (2006a). A method for estimating coefficients of soil organic matter dynamics based on long-term experiments. *Soil and Tillage Research* 91(1-2): 217-226.

Bayer, C., Martin-Neto, L., Mielniczuk, J., Pavinato, A. and Dieckow, J. (2006b). Carbon sequestration in two Brazilian cerrado soils under no-till. *Soil and Tillage Research* 86(2): 237-245.

Bernoux, M., Cerri, C. C., Cerri, C. E. P., Neto, M. S., Metay, A., Perrin, A., Scopel, E., Razafimbelo, T., Blavet, D., Piccolo, MdC., Pavei, M. and Milne, E. (2006). Cropping systems, carbon sequestration and erosion in Brazil, a review. *Agronomy for Sustainable Development* 26(1): 1–8.

Blanco-Canqui, H. and Lal, R. (2008). No-tillage and soil-profile carbon sequestration: an on-farm assessment. *Soil Science Society of America Journal* 72(3): 693–701.

Blanco-Canqui, H. and Lal, R. (2010). *Principles of Soil Conservation and Management*. Netherlands: Springer.

Boddey, R. M., Jantalia, C. P., Conceição, P. C., Zanatta, J. A., Bayer, C., Mielniczuk, J., Dieckow, J., Dos Santos, H. P., Denardin, J. E., Aita, C., Giacomini, S. J., Alves, B. J. R. and Urquiaga, S. (2010). Carbon accumulation at depth in Ferralsols under zero-till subtropical agriculture. *Global Change Biology* 16(2): 784–795.

Boeni, M. (2007). *Proteção física da matéria orgânica em Latossolos sob sistemas com pastagens na região do Cerrado Brasileiro*. Porto Alegre: Programa de Pós Graduação em Ciência do Solo. Universidade Federal do Rio Grande do Sul.

Boeni, M., Bayer, C., Dieckow, J., Conceição, P. C., Dick, D. P., Knicker, H., Salton, J. C. and Macedo, M. C. M. (2014). Organic matter composition in density fractions of cerrado Ferralsols as revealed by CPMAS 13C NMR: influence of pastureland, cropland and integrated crop-livestock. *Agriculture, Ecosystems and Environment* 190: 80–86.

Bolinder, M. A., Angers, D. A., Giroux, M. and Laverdière, M. R. (1999). Estimating C inputs retained as soil organic matter from corn (*Zea mays* L.). *Plant and Soil* 215(1): 85–91.

Briedis, C., Baldock, J., Sa, J. C. D., Santos, J. B., McGowan, J. and Milori, D. M. B. P. (2021). Organic carbon pools and organic matter chemical composition in response to different land uses in southern Brazil. *European Journal of Soil Science* 72(3): 1083–1100.

Camêlo, DdL., Ker, J. C., Fontes, M. P. F., Costa, A. C. Sd, Corrêa, M. M. and Leopold, M. (2018). Mineralogy, magnetic susceptibility and geochemistry of Fe-rich oxisols developed from several parent materials. *Scientia Agricola* 75(5): 410–419.

Campbell, C. A., Zentner, R. P., Bowren, K. E., Townley-Smith, L. and Schnitzer, M. (1991). Effect of crop rotations and fertilization on soil organic matter and some biochemical properties of a thick Black chernozem. *Canadian Journal of Soil Science* 71(3): 377–387.

Carvalho, Ad, Inda, A. V., Fink, J. R. and Curi, N. (2015). Iron oxides in soils of different lithological origins in Ferriferous Quadrilateral (Minas Gerais, Brazil). *Applied Clay Science* 118: 1–7.

Casão-Junior, R., Araújo, A. Gd. and Llanillo, R. F. (2012). *No-Till Agriculture in Southern Brazil*. Londrina: Food and Agriculture Organization/IAPAR.

Chen, K. Y., Chen, T. Y., Chan, Y. T., Cheng, C. Y., Tzou, Y. M., Liu, Y. T. and Teah, H. Y. (2016). Stabilization of natural organic matter by short-range-order iron hydroxides. *Environmental Science and Technology* 50(23): 12612–12620.

Conceição, P. C., Dieckow, J. and Bayer, C. (2013). Combined role of no-tillage and cropping systems in soil carbon stocks and stabilization. *Soil and Tillage Research* 129: 40–47.

Cornell, R. M. and Schwertmann, U. (2003). *The Iron Oxides: Structure, Properties, Reactions, Occurrences and Uses*. John Wiley & Sons; Weinheim, Germany.

Cotrufo, M. F., Soong, J. L., Horton, A. J., Campbell, E. E., Haddix, M., Wall, D. H. and Parton, W. J. (2015). Formation of soil organic matter via biochemical and physical pathways of litter mass loss. *Nature Geoscience* 8(10): 776–779.

Cotrufo, M. F., Wallenstein, M. D., Boot, C. M., Denef, K. and Paul, E. (2013). The Microbial Efficiency-Matrix Stabilization (MEMS) framework integrates plant litter decomposition with soil organic matter stabilization: do labile plant inputs form stable soil organic matter? *Global Change Biology* 19(4): 988-995.

da Silva, K. A., Nicola, V. B., Dudas, R. T., et al. (2021). Pesticides in a case study on no-tillage farming systems and surrounding forest patches in Brazil. *Scientific Reports* 11(1): 9839. https://doi.org/10.1038/s41598-021-88779-.

Dalmolin, R. S. D., Gonçalves, C. N., Dick, D. P., Knicker, H., Klamt, E. and Kögel-Knabner, I. (2006). Organic matter characteristics and distribution in Ferralsols profiles of a climosequence in southern Brazil. *European Journal of Soil Science* 57(5): 644–654.

Derpsch, R., Roth, C. H., Sidiras, N., et al. (1991). *Controle da Erosão no Paraná, Brasil: sistemas de cobertura do solo, plantio direto e preparo conservacionista do solo*. Eschborn: Deutsche Gesellschaft fur Technische Zusammenarbeit.

Dick, D. P., Gonçalves, C. N., Dalmolin, R. S. D., Knicker, H., Klamt, E., Kögel-Knabner, I., Simões, M. L. and Martin-Neto, L. (2005). Characteristics of soil organic matter of different Brazilian Ferralsols under native vegetation as a function of soil depth. *Geoderma* 124(3–4): 319–333.

Dick, D. P., Martinazzo, R., Dalmolin, R. S. D., Jacques, A. V. Á, Mielniczuk, J. and Rosa, A. S. (2008). Impacto da queima nos atributos químicos e na composição química da matéria orgânica do solo e na vegetação. *Pesquisa Agropecuaria Brasileira* 43(5): 633–640.

Dick, D. P., Martinazzo, R., Knicker, H. and Almeida, P. S. G. (2010). Matéria orgânica em quatro tipos de solos brasileiros: composição química e sorção de atrazina. *Química Nova* 33(1): 14–19.

Dieckow, J., Bayer, C., Conceição, P. C., Zanatta, J. A., Martin-Neto, L., Milori, D. B. M., Salton, J. C., Macedo, M. M., Mielniczuk, J. and Hernani, L. C. (2009). Land use, tillage, texture and organic matter stock and composition in tropical and subtropical Brazilian soils. *European Journal of Soil Science* 60(2): 240–249.

Dieckow, J., Mielniczuk, J., Knicker, H., Bayer, C., Dick, D. P. and Kogel-Knabner, I. (2005a). Composition of organic matter in a subtropical acrisol as influenced by land use, cropping and N fertilization, assessed by CPMAS ^{13}C NMR spectroscopy. *European Journal of Soil Science* 56: 705–715.

Dieckow, J., Mielniczuk, J., Knicker, H., Bayer, C., Dick, D. P. and Kögel-Knabner, I. (2005b). Soil C and N stocks as affected by cropping systems and nitrogen fertilisation in a southern Brazil acrisol managed under no-tillage for 17 years. *Soil and Tillage Research* 81(1): 87–95.

Dignac, M.-F., Derrien, D., Barré, P., Barot, S., Cécillon, L., Chenu, C., Chevallier, T., Freschet, G. T., Garnier, P., Guenet, B., Hedde, M., Klumpp, K., Lashermes, G., Maron, P., Nunan, N., Roumet, C. and Basile-Doelsch, I. (2017). Increasing soil carbon storage: mechanisms, effects of agricultural practices and proxies. A review. *Agronomy for Sustainable Development* 37(2): 14.

Escobar, L. F., Amado, T. J. C., Bayer, C., Chavez, L. F., Zanatta, J. A. and Fiorin, J. E. (2010). Postharvest nitrous oxide emissions from a subtropical oxisol as influenced by summer crop residues and their management. *Revista Brasileira de Ciencia do Solo* 34(2): 507–516.

Eswaran, H., van den Berg, E. and Reich, P. (1993). Organic carbon in soils of the World. *Soil Science Society of America Journal* 57(1): 192–194.

Grandy, A. S. and Neff, J. C. (2008). Molecular C dynamics downstream: the biochemical decomposition sequence and its impact on soil organic matter structure and function. *Science of the Total Environment* 404(2–3): 297–307.

Greenland, D. J., Wild, A. and Adams, D. (1992). Organic matter dynamics in soils of the tropics—From myth to complex reality. In: Lal, R. and Sanchez, P. A. (Eds) *Myths and Science of Soils of the Tropics*. Madison, WI: Soil Science Society of America and American Society of Agronomy, 17-33.

Gregorich, E. G., Rochette, P., van den Bygaart, A. J. and Angers, D. (2005). Greenhouse gas contributions of agricultural soils and potential mitigation practices in Eastern Canada. *Soil and Tillage Research* 83(1): 53-72.

Hanke, D., Melo Vd, VdF., Dieckow, J., Dick, D. P. and Bognola, I. A. (2015). Influência da matéria orgânica no diâmetro médio de minerais da fração argila de solos desenvolvidos de basalto no Sul do Brasil. *Revista Brasileira de Ciencia do Solo* 39(6): 1611-1622.

Henin, S. and Dupuis, M. (1945). Essai de bilan de la matiere organique du sol. *Annales Agronomiques* 3: 17-29.

Hütsch, B. W. (1998). Tillage and land use effects on methane oxidation rates and their vertical profiles in soil. *Biology and Fertility of Soils* 27(3): 284-292.

IPCC, Intergovermental Panel on Climate Change (2013). *Climate Change 2013: The Physical Science Basis. Contribution of Working Group I to the Fifth Assessment Report of the Intergovernmental Panel on Climate Change*. Stocker, T. F., Qin, D., Plattner, G.-K., Tignor, M., Allen, S. K., Boschung, J., Nauels, A., Xia, Y., Bex, V. and Midgley, P. M. (Eds). *Technical Summary*. Cambridge: Cambridge University Press.

Jantalia, C. P., Santos, H. P., Urquiaga, S., Boddey, R. M. and Alves, B. J. R. (2008). Fluxes of nitrous oxide from soil under different crop rotations and tillage systems in the South of Brazil. *Nutrient Cycling in Agroecosystems* 82(2): 161-173.

Jastrow, J. D. (1996). Soil aggregate formation and the accrual of particulate and mineral-associated organic matter. *Soil Biology and Biochemistry* 28(4-5): 665-676.

Kämpf, N. and Curi, N. (2000). Óxidos de Ferro: indicadores de ambientes pedogênicos. *Tópicos em Ciência do Solo. Sociedade Brasileira de Ciência do Solo*: 107-138.

Kämpf, N. and Curi, N. (2003). Argilominerais em solos brasileiros. *Tópicos em Ciência do Solo. Sociedade Brasileira de Ciência do Solo* 3: 1-54.

Kassam, A., Friedrich, T., Derpsch, R., and Kienzle, J. (2015). Overview of the worldwide spread of conservation agriculture. *Field Actions Science Reports* 8: 1-11.

Kirkby, C. A., Richardson, A. E., Wade, L. J., Passioura, J. B., Batten, G. D., Blanchard, C. and Kirkegaard, J. A. (2014). Nutrient availability limits carbon sequestration in arable soils. *Soil Biology and Biochemistry* 68: 402-409.

Kleber, M., Bourg, I. C., Coward, E. K., Hansel, C. M., Myneni, S. C. B. and Nunan, N. (2021). Dynamic interactions at the mineral-organic matter interface. *Nature Reviews Earth and Environment* 2(6): 402-421.

Lal, R. (2004a). Carbon emission from farm operations. *Environment International* 30(7): 981-990.

Lal, R. (2004b). Soil carbon sequestration to mitigate climate change. *Geoderma* 123(1-2): 1-22.

Leal, O. A., Amado, T. J. C., Fiorin, J. E., Keller, C., Reimche, G. B., Rice, C. W., Nicoloso, R. S., Bortolotto, R. P. and Schwalbert, R. (2020). Linking cover crop residue quality and tillage system to CO_2-C emission, soil C and N stocks and crop yield based on a long-term experiment. *Agronomy* 10(12): 1848.

Liu, X. R., Eusterhues, K., Thieme, J., Ciobota, V., Höschen, C., Mueller, C. W., Küsel, K., Kögel-Knabner, I., Rösch, P., Popp, J. and Totsche, K. U. (2013). STXM and

NanoSIMS investigations on EPS fractions before and after adsorption to goethite. *Environmental Science and Technology* 47(7): 3158–3166.

Lovato, T. (2001). *Dinâmica do carbono e do nitrogênio do solo afetada por preparos do solo, sistemas de cultura e adubo nitrogenado (Soil Carbon and Nitrogen Dynamics as Affected by Tillage, Cropping System and Nitrogen Fertilizer).* Porto Alegre: Programa de Pós Graduação em Ciência do Solo. Universidade Federal do Rio Grande do Sul, 132.

Lovato, T., Mielniczuk, J., Bayer, C. and Vezzani, F. (2004). Adição de carbono e nitrogênio e sua relação com os estoques no solo e com o rendimento do milho em sistemas de preparo (Carbon and nitrogen addition related to stocks of these elements in soil and corn yield under management systems). *Revista Brasileira de Ciencia do Solo* 28(1): 175–187.

McGregor, G. R. and Nieuwolt, S. (1998). *Tropical Climatology: an Introduction to the Climates of the Low Latitudes*. West Sussex: Wiley.

Merten, G. H. and Minella, J. P. G. (2013). The expansion of Brazilian agriculture: soil erosion scenarios. *International Soil and Water Conservation Research* 1(3): 37–48.

Mielniczuk, J., Bayer, C., Vezzani, F., et al. (2003). Soil and crop management and their relationship to soil carbon and nitrogen stocks (in Portuguese). *Tópicos em Ciência do Solo* 3: 165–208.

Moraes, A., Carvalho, P. CdF., Anghinoni, I., Lustosa, S. B. C., Costa, S. E. V. GdA. and Kunrath, T. R. (2014). Integrated crop-livestock systems in the Brazilian subtropics. *European Journal of Agronomy* 57: 4–9.

Nathan, V. K., Jasna, V. and Parvathi, A. (2020). Pesticide application inhibit the microbial carbonic anhydrase-mediated carbon sequestration in a soil microcosm. *Environmental Science and Pollution Research International* 27(4): 4468–4477. https://doi.org/10.1007/s11356-019-06503-1.

Ogle, S. M., Breidt, F. J. and Paustian, K. (2005). Agricultural management impacts on soil organic carbon storage under moist and dry climatic conditions of temperate and tropical regions. *Biogeochemistry* 72(1): 87–121.

Oliveira, J. S., Inda, A. V., Barron, V., Torrent, J., Tiecher, T. and de Oliveira Camargo, F. A. (2020). Soil properties governing phosphorus adsorption in soils of Southern Brazil. *Geoderma Regional* 22.

Panagos, P., Borrelli, P., Meusburger, K., Yu, B., Klik, A., Jae Lim, K., Yang, J. E., Ni, J., Miao, C., Chattopadhyay, N., Sadeghi, S. H., Hazbavi, Z., Zabihi, M., Larionov, G. A., Krasnov, S. F., Gorobets, A. V., Levi, Y., Erpul, G., Birkel, C., Hoyos, N., Naipal, V., Oliveira, P. T. S., Bonilla, C. A., Meddi, M., Nel, W., Al Dashti, H., Boni, M., Diodato, N., Van Oost, K., Nearing, M. and Ballabio, C. (2017). Global rainfall erosivity assessment based on high-temporal resolution rainfall records. *Scientific Reports* 7(1): 4175.

Piva, J. T., Dieckow, J., Bayer, C., Zanatta, J. A., de Moraes, A., Pauletti, V., Tomazi, M. and Pergher, M. (2012). No-till reduces global warming potential in a subtropical ferralsol. *Plant and Soil* 361(1-2): 359-373.

Potes, MdL., Dick, D. P., Dalmolin, R. S. D., Knicker, H. and Rosa, A. Sd (2010). Matéria orgânica em Neossolo de altitude: influência do manejo da pastagem na sua composição e teor. *Revista Brasileira de Ciencia do Solo* 34(1): 23–32.

Pöttker, D. (1977). *Efeito do tipo de solo, tempo de cultivo e da calagem sobre a mineralização da matéria orgânica em solos do Rio Grande do Sul*. Porto Alegre: Programa de Pós-Graduação em Agronomia. Universidade Federal do Rio Grande do Sul, 128.

Powlson, D. S., Stirling, C. M., Jat, M. L., Gerard, B. G., Palm, C. A., Sanchez, P. A. and Cassman, K. G. (2014). Limited potential of no-till agriculture for climate change mitigation. *Nature Climate Change* 4(8): 678–683.

Reis, C. E. Sd, Dick, D. P., Caldas, JdS. and Bayer, C. (2014). Carbon sequestration in clay and silt fractions of Brazilian soils under conventional and no-tillage systems. *Scientia Agricola* 71(4): 292–301.

Rumpel, C., Baumann, K., Remusat, L., Dignac, M., Barré, P., Deldicque, D., Glasser, G., Lieberwirth, I. and Chabbi, A. (2015). Nanoscale evidence of contrasted processes for root-derived organic matter stabilization by mineral interactions depending on soil depth. *Soil Biology and Biochemistry* 85: 82–88.

Sá, J. C. M., Cerri, C. C., Dick, W. A., Lal, R., Filho, S. P. V., Piccolo, M. C. and Feigl, B. E. (2001). Organic matter dynamics and carbon sequestration rates for a tillage chronosequence in a Brazilian oxisol. *Soil Science Society of America Journal* 65(5): 1486–1499.

Salton, J. C., Mercante, F. M., Tomazi, M., Zanatta, J. A., Concenço, G., Silva, W. M. and Retore, M. (2014). Integrated crop-livestock system in tropical Brazil: Toward a sustainable production system. *Agriculture, Ecosystems and Environment* 190: 70–79.

Sanchez, P. A. (1976). *Properties and Management of Soils in the Tropics*. New York: John Wiley.

Sanchez, P. A. and Logan, T. J. (1992). Myths and science about the chemistry and fertility of soils in the tropics. In: Lal, R. and Sanchez, P. A. (Eds) *Myths and Science of Soils of the Tropics*. Madison, WI: Soil Science Society of America and American Society of Agronomy, 35–46.

Schoening, I., Knicker, H. and Kogel-Knabner, I. (2005). Intimate association between O/N-alkyl carbon and iron oxides in clay fractions of forest soils. *Organic Geochemistry* 36(10): 1378–1390.

Silva, J. E., Lemainski, J. and Resck, D. V. S. (1994). Perdas de matéria orgânica e suas relações com a capacidade de troca catiônica em solos da região de Cerrados do oeste baiano. *Revista Brasileira de Ciencia do Solo* 18: 541–547.

Sisti, C. P. J., Santos, H. P., Kohhann, R., Alves, B. J. R., Urquiaga, S. and Boddey, R. M. (2004). Change in carbon and nitrogen stocks in soil under 13 years of conventional or zero tillage in southern Brazil. *Soil and Tillage Research* 76(1): 39–58.

Six, J., Elliott, E. T. and Paustian, K. (2000). Soil macroaggregate turnover and microaggregate formation: a mechanism for C sequestration under no-tillage agriculture. *Soil Biology and Biochemistry* 32(14): 2099–2103.

Tamrat, W. Z., Rose, J., Grauby, O., Doelsch, E., Levard, C., Chaurand, P. and Basile-Doelsch, I. (2019). Soil organo-mineral associations formed by co-precipitation of Fe, Si and Al in presence of organic ligands. *Geochimica Et Cosmochimica Acta* 260: 15–28.

Tornquist, C. G., Gassman, P. W., Mielniczuk, J., Giasson, E. and Campbell, T. (2009). Spatially explicit simulations of soil C dynamics in Southern Brazil: integrating century and GIS with i_Century. *Geoderma* 150(3–4): 404–414.

Uehara, G. and Gillman, G. P. (1981). *The Mineralogy, Chemistry and Physics of Tropical Soils with Variable Charge Clays*. Boulder, CO: West View Press.

van Wambeke, A. (1992). *Soils of the Tropics: Properties and Appraisal*. New York: McGraw Hill.

VandenBygaart, A. J. (2016). The myth that no-till can mitigate global climate change. *Agriculture, Ecosystems and Environment* 216: 98–99.

Veloso, M. G., Angers, D. A., Tiecher, T., Giacomini, S., Dieckow, J. and Bayer, C. (2018). High carbon storage in a previously degraded subtropical soil under no-tillage with legume cover crops. *Agriculture, Ecosystems and Environment* 268: 15-23.

Vogel, C., Mueller, C. W., Hoschen, C., Buegger, F., Heister, K., Schulz, S., Schloter, M. and Kögel-Knabner, I. (2014). Submicron structures provide preferential spots for carbon and nitrogen sequestration in soils. *Nature Communications* 5: 2947.

von Lützow, Mv, Kögel-Knabner, I., Ekschmitt, K., Matzner, E., Guggenberger, G., Marschner, B. and Flessa, H. (2006). Stabilization of organic matter in temperate soils: mechanisms and their relevance under different soil conditions - a review. *European Journal of Soil Science* 57(4): 426-445.

Wang, Y. D., Hu, N., Xu, M. G., Li, Z., Lou, Y., Chen, Y., Wu, C. and Wang, Z. (2015). 23-year manure and fertilizer application increases soil organic carbon sequestration of a rice-barley cropping system. *Biology and Fertility of Soils* 51(5): 583-591.

Zanatta, J. A., Bayer, C., Dieckow, J., Vieira, F. and Mielniczuk, J. (2007). Soil organic carbon accumulation and carbon costs related to tillage, cropping systems and nitrogen fertilization in a subtropical acrisol. *Soil and Tillage Research* 94(2): 510-519.

Chapter 23

Management of carbonate-rich soils and trade-offs with soil inorganic carbon cycling

Iñigo Virto, Isabel de Soto and Rodrigo Antón, Universidad Pública de Navarra, Spain; and Rosa M. Poch, Universitat de Lleida, Spain

1 Introduction

Calcareous soils are present in many arid and semi-arid areas worldwide. Calcisols, the soil group defined in the World Reference Base comprising those soils with a substantial secondary accumulation of calcium carbonate, cover up to 1 billion ha (IUSS Working Group WRB, 2015; Fig. 1).

In these soils, lithogenic and secondary carbonates can account for different proportions of the soil matrix and play a significant role in it. The estimated amount of inorganic carbon (C) stored in soils (SIC) in the form of carbonates worldwide is indeed relevant - between 695 Pg and 748 Pg in the first meter of soil, compared to 1500 Pg of organic C, or 950 Pg down to 2 meters deep (Zamanian et al., 2016). In many arid and semiarid regions, SIC stored in carbonate minerals (mainly calcite) can exceed that in the organic fraction of the soil (Suarez, 2000), with observations of up to 20 times more soil carbonates-C than soil organic C (SOC) in some regions (Wang et al., 2015). Chapter 7 in this book explains in detail the typology and dynamics of SIC in calcareous soils, as well as the relationship between SIC and soil functioning, including SOC turnover and stabilization, soil biology, and aggregation.

In reality, the organic and inorganic compartments of soil C cannot be considered completely independent: when the C contained in pedogenic carbonates comes from microbial or root respiration in soil, there will be a

http://dx.doi.org/10.19103/AS.2022.0106.22

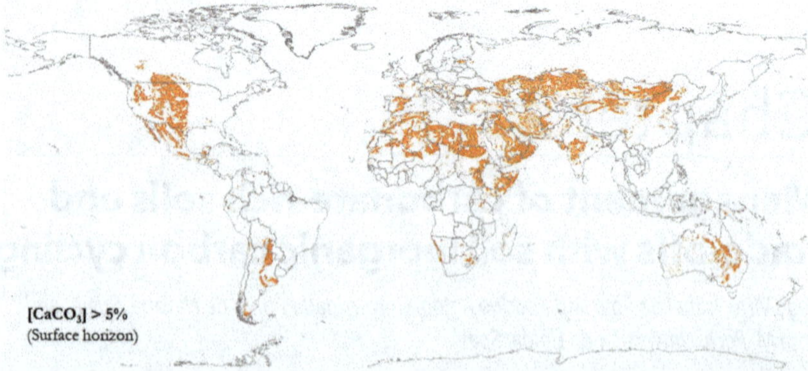

Figure 1 Distribution of soils with concentrations of calcium carbonate higher than 50 mg/g soil in the upper horizon. Source: Own elaboration from data at FAO Soil Portal (http://www.fao.org/soils-portal/soil-survey/soil-maps-and-databases/other-global-soil-maps-and-databases/en/).

'transfer' of C from the organic fraction to the inorganic fraction, as summarized by the chain of reactions described by Zamanian et al., (2016) for the case of calcite:

$$\begin{aligned} CaC^GO_3 + C^RO_2 + H_2O \leftrightarrow Ca\left(HC^GO_3\right)^+ + HC^RO_3^- \leftrightarrow Ca\left(HC^RO_3\right)^+ \\ + HC^GO_3^- \leftrightarrow CaC^RO_3 \downarrow + C^GO_2 + H_2O \end{aligned} \tag{1}$$

where, C^G is the C from 'geogenic' or primary calcite, and C^R is the C from respiration.

In addition, as presented in Chapter 7 of this book, soil carbonates can indirectly affect soil organic matter (SOM) stabilization and destabilization processes at different scales by influencing the formation of organo-mineral interactions and aggregates of contrasting size.

This dual role of carbonates in the soil carbon cycle is especially relevant considering the interest that the stabilization and dynamics of SOC and, in particular, its interaction with agricultural management, have gained in recent decades (Rumpel et al., 2020).

As introduced in Chapter 7 of this book, there is solid evidence that some agricultural management activities can generate changes in the dissolution and precipitation of soil carbonates (Plaza-Bonilla et al., 2015), which can, for instance, result in the formation of new precipitates in relatively short periods (Magaritz and Amiel, 1981; Wu et al., 2008; Bughio et al., 2016).

Considering that Ca^{2+} is superabundant in calcareous soils, several reports indicate that the main factors that determine the dissolution and precipitation cycle of calcite shown in Equation 1 are the soil water content, the partial pressure of CO_2 in the soil atmosphere, and the concentration of HCO_3^- in the soil solution (de Soto et al., 2017). Other factors influencing calcite precipitation in soils are the composition of the soil solution, in particular, the availability of calcium and/or water and temperature (Hasinger et al., 2015). Agricultural management normally implies changes in these driving factors. In addition to the respiration of roots and soil biota, which affects the partial pressure of CO_2, the chemistry of the soil solution can be affected by fertilization, by the incorporation of amendments such as liming, by irrigation with waters of different quality, or by contributions of exogenous elements such as heavy metals, among others. Finally, the availability of water can also be affected by practices such as irrigation or drainage.

This suggests that the management of carbonate-rich soils can result in changes in the inorganic fraction of soil C, which, in turn, may result in indirect alterations of the properties and condition of these soils, in addition to the carbon sequestration potential. However, the attention devoted to the dynamics of SIC has therefore historically been very small compared to that centered on SOC and mainly focused on genesis (ex. Francis and Poch, 2019) or paleoenvironmental studies (Zamanian et al., 2016). The general inverse relationship between soil carbonate and organic matter contents, both in the landscape and the soil profile, adds to this lower attention.

In this chapter, we summarize the major observed effects in this sense, describe a case study in a newly irrigated area, and collect the most relevant questions still pending to be answered, as well as future trends in research for the study of the consequences of the agricultural management of carbonate-rich soils.

2 Consequences of agricultural management on soil inorganic carbon

Although it has generally been assumed that agricultural practices, with maybe the exception of liming (Paradelo et al., 2015), do not affect the mineral fraction of soils, different authors have been interested in the potential consequences of agricultural management on SIC. These consequences are dependent on the modifications that soil management may induce in carbonates' cycling. They may result in gains or losses of SIC and in changes in the type (pedogenic vs. lithogenic), size, and distribution of soil carbonates.

The reactions leading to these changes include those altering the natural Ca^{2+} and/or HCO_3^- concentrations in the soil solution and changes in its acidity. As described in the early works of C. H. Monger (Monger and Gallegos, 2000; Monger and Martínez-Ríos, 2001; Monger et al., 2015), these alterations in the dissolution-precipitation cycle of carbonates can lead to net atmospheric CO_2 sequestration, CO_2 emissions, or have a neutral effect on atmospheric CO_2. This will depend on their net effect on carbonates dissolution vs. precipitation (see Chapter 7 of this book for a complete description of the factors and processes) and on the origin of Ca^{2+}.

From the simplified dissolution reaction

$$CaCO_3 + H_2O + CO_2 \leftrightarrow 2HCO_3^- + Ca^{2+} \quad (2)$$

it can be deduced that, as explained by Sanderman (2012), Ca^{2+} supplied in the cationic form will be carbon neutral and the supply of HCO_3^- will determine whether the reaction is a net source or sink of carbon. If the HCO_3^- is supplied from the dissolution of biologically derived soil CO_2 (Equation 1), then the reaction will be a carbon sink regardless of the origin of Ca^{2+} supply. If HCO_3^- is supplied from irrigation water, for example, then 1 mol of formerly stable C (residing in a long-term reservoir such as groundwater) will be released, so the supply of Ca^{2+} will dictate whether the reaction is a carbon source (cationic Ca^{2+}), sink (silicate weathering or other non-carbonate or 'exogenous' sources of Ca^{2+}), or neutral (carbonate dissolution).

Exogenous sources of Ca^{2+} would result in the precipitation of the so-called *pedo-atmogenic* carbonates (Monger and Martínez-Ríos, 2001) and include, in addition, natural processes such as weathering of Ca^{2+}-bearing silicates, atmospheric and aeolian additions of non-carbonate Ca^{2+}, dissolution of soluble salts and gypsum, anthropogenic sources related to the agricultural use of SIC-rich soils: irrigation water, the applications of biosolids and of mineral fertilizers (Sanderman, 2012; Bughio et al., 2016).

The reaction can also be shifted toward carbonate dissolution through the addition of protons, as a consequence of the neutralization of this acidity by $CaCO_3$. The fate of HCO_3^- generated in this dissolution will determine the net source or sink character of the process. Only if HCO_3^- leaches from the soil to long-lived reservoirs (groundwater or the ocean, Sanderman, 2012), a net carbon sink would exist. Otherwise, acidification will lead to CO_2 release, as explained in Chapter 7 in this book, and by the chain of reactions described, for example, by Zamanian et al. (2018) for the case of acidification from N-fertilizers:

$$2\left[CaCO_3 + H^+ + NO_3^- \rightarrow Ca^{2+} + HCO_3^- + NO_3^-\right] \quad (3)$$

$$2\times\left[HCO_3^- + H^+ \leftrightarrow CO_2^- + H_2O\right] \quad (4)$$

From this, it can be deduced that the net consequences of agricultural management on SIC will be dependent on different parameters, mostly associated with the use of fertilization, irrigation, and amendments of different origins (both organic and inorganic). There are, however, two aspects that merit attention in this respect.

One is that, in drylands, agricultural activity itself can result in increased soil biological activity, leading to elevated CO_2 levels in soils, which would foster carbonate dissolution-precipitation reactions, compared to non-cultivated soils in these regions. As introduced in Chapter 7 of this book, the formation of pedogenic carbonates as a result of increased partial pressure of CO_2 from the respiration of crop roots in arid and semi-arid conditions has been observed in many studies (Gocke et al., 2012; Wang et al., 2015). This can explain the greater SIC stocks observed in cultivated soils compared to native grasslands in the U.S. Great Plains or some Mollisols in Russia. In this sense, primary mineral weathering is expected to play an important role in supplying Ca^{2+}in areas of intensive agriculture where biological activity is greater than that of undisturbed soils (Bughio et al., 2016).

Other authors, however, have observed a loss of carbonates associated with agricultural management, such as Moreno et al. (2006) in the tilled horizon of a fluventic soil in xeric Mediterranean conditions of SW Spain after more than 10 years of wheat-sunflower rotation, with and without tillage.

The second issue that needs attention is the fact that, regardless of the net effect on SIC gains or losses, and/or the source or sink effect of these processes, the activation of carbonate dissolution-precipitation reactions can result in differences in the proportion of pedogenic carbonates vs. lithogenic carbonates (e.g. de Soto et al., 2019) and on the size distribution of these carbonates (e.g. Magaritz and Amiel, 1981). Considering the indirect role of pedogenic carbonates in soil functioning, and in particular, in SOC cycling (see Chapter 7 of this book), these changes can also have consequences in cultivated soils when they are the result of agricultural management.

In summary, there is increasing evidence of the need of accounting for SIC when assessing the relationship between agricultural soil management, greenhouse gas emissions, and the global C cycle. An example of this is the recent recommendation of the members of the Scientific Committee of the 4per1000 initiative (Rumpel et al., 2020), who indicate the need of considering SIC dynamics in climate change models since changes in temperature and microbial activities in cultivated soils could liberate relevant amounts of CO_2 from SIC.

2.1 Nitrogen fertilization

The effect of fertilization on SIC has been the object of attention of several studies in the last decades. Among macronutrients, N is the one most affecting soil pH, being ammonium-based products the most acidifying ones. Soil acidification due to the use of phosphorus fertilizers is small compared to N, except for phosphoric acid. Indeed, the addition of N-fertilizers to soils can produce changes in the chemical, physical, and microbial properties of the soil (Matocha et al., 2016), some of which can have a direct effect on SIC. The most relevant one is that the use of N-fertilizers can induce the dissolution of $CaCO_3$, resulting in CO_2 emissions, and producing a decrease in soil pH and the leaching of Ca_2^+ in the long term, once the buffering capacity of carbonates is overcome (Equations 3 and 4).

As described earlier, this is associated with the incorporation of protons into the soil solution both in the nitrification process oxidizing ammonium to nitrate in aerated soils and in the uptake of NH_4^+ by crop roots in exchange for H^+ in less aerobic conditions (Raza et al., 2021). These processes would increase carbonate dissolution rates by up to one order of magnitude after N fertilization, especially with mineral-N fertilizers (Zamanian et al., 2018).

The acidification potential of fertilizers, in particular N-fertilizers, can be assessed through different methods, as the calcium carbonate equivalent (kg CCE/ton of fertilizer), which refers to the pure calcium carbonate required to neutralize the acidity resulting from the application of the fertilizer, while for basic reaction fertilizers, the CCE would be the equivalent calcium carbonate precipitated in the soil after their application (Pierre, 1933).

As a result of these processes, SIC can be lost in agricultural soils at different paces. One of the few long-term fertilization trials (22 years, Zhang et al., 2016) measuring $CaCO_3$ contents stated that the lowest decreases in soil $CaCO_3$ occurred in the combined organic+inorganic fertilization treatment, compared with N alone (approximately 4% and 1% decrease, respectively) while the inorganic NP treatment did not result in significant decreases, which is explained by the authors by $Ca_3(PO_4)_2$ precipitation on the surface of carbonate particles.

The loss of SIC through acidification of agricultural soils has consequences at different levels. At the catchment scale, the relevance of this process was already noticed and quantified by Perrin et al. (2008) by observing the continuous reduction of the alkalinity of water over 30 years in a stream draining an agricultural catchment in France, while the nitrate and calcium plus magnesium contents increased. From these observations, these authors concluded that the deficit of CO_2 uptake by riverine alkalinity can be estimated as up to 12-29 $\times10^{12}$ g CO_2 yr^{-1} at a global scale, representing 1.6-3.8% of the total CO_2 flux naturally consumed by carbonate dissolution. For comparison, present models estimate annual net CO_2 emissions from land-use changes in the period 2007-2016 at 5.2 ± 2.6 $\times10^{15}$ g CO_2 yr^{-1} (Jia et al., 2019). Other publications have related NO_3^- contents coming from fertilizers in waters with carbonate dissolution in small watersheds in France (Brunet et al., 2011) and Switzerland (Müller et al., 2020).

Concerning greenhouse gases emission balances, Zamanian et al. (2018) made a global estimation of the CO_2 release from pedogenic and geogenic carbonates in the upper 1 m soil depth as a consequence of the neutralization of N fertilization-induced acidity by carbonate dissolution, based on a global N-fertilization map and the worldwide distribution of soils containing $CaCO_3$. This would result in an annual release of 27.4$\times10^{12}$ g CO_2 (corresponding to

$7.48x10^{12}$ g SIC). The actual effect of N-fertilization can in reality be modulated locally by the buffering capacity of SOM and clays, acidity localization at the fertilization point and SIC localization in the subsoil (Datta and Mandal, 2018), and might have been underestimated (Zamanian and Kuzyakov, 2019). For comparison, global annual net soil emissions of greenhouse gases have been estimated to be $\geq 350 \times 10^{15}$ g eq CO_2 (CO_2 equivalents; Oertel et al. (2016).

More recent calculations have shown that at least $0.41x10^{15}$ g C have been released from SIC as CO_2 to the atmosphere from agricultural soils, and an additional 0.72×10^{15} g C will be released by 2050 (Zamanian et al., 2021). In China, the loss of SIC stocks in response to N-induced soil acidification was estimated at 0.15×10^{15}g C, corresponding to 1.1 Mg C/ha in the period between 1980 and 2020, which corresponds to 7% of the SIC stored in 1980 (Raza et al., 2020). From the point of view of atmospheric C, these processes result in a relevant and unaccounted source of CO_2 from soil C stocks (Zamanian et al., 2021).

Finally, the potential consequences of this continuous loss of SIC can spread beyond net agricultural soil CO_2 emissions. Considering the role of soil carbonates in organic matter stabilization, an indirect consequence of this process would be the progressive loss of organic matter stability and protection (Rasmussen et al., 2018; Rowley et al., 2018). The intensity of this effect will greatly depend on the initial SIC content in the soil layers most affected by agricultural management. In soils with low SIC contents, the process can lead to complete SIC depletion. In China, Raza et al. (2020) have accounted for 7 million ha of cropland becoming carbonate-free in the last 40 years and predicted another 30 million under the same condition by 2100 if N fertilization follows the present scenario.

From these observations, an urgent need emerges to consider measurements to increase N-use efficiency and the rational use of N fertilization so that excess N is limited as much as possible in croplands worldwide.

2.2 Irrigation

The introduction and expansion of irrigation in arid and semi-arid regions is leading to an important change in soil management of these regions. In addition, it is estimated that the amount of water needed for irrigated crops in semi-arid areas such as Mediterranean Europe will increase in the coming years (European Environmental Agency, 2012), which makes it necessary to study the possible consequences of irrigation in soil C in these areas. However, the impact of irrigated agriculture on soil C storage is not completely known (Wu et al., 2008; Trost et al., 2013) and still little studied compared to that of other agricultural management strategies (Apesteguía et al., 2015). In particular, its effect on the lithogenic and pedogenic carbonate content of soils has not been studied in depth (Magaritz and Amiel, 1981; Lal and Kimble, 2000; Wu et al., 2009; Bughio et al., 2016).

Irrigation implies an important change in the water balance of agricultural soils, which, according to the reaction shown in Equations 1 and 2, can produce alterations in the precipitation/dissolution cycles of carbonates and can also accelerate these processes by favoring the addition of biomass into the soil and therefore stimulating the biological activity of soils and the generation of CO_2. At the same time, the accumulation and distribution of soil carbonates is greatly affected by the quality of irrigation water (Sanderman, 2012) as it can be an external source of Ca^{2+}, Mg^{2+}, and HCO_3^- (Lal and Kimble, 2000) and promote the formation of pedogenic carbonates (Bughio et al., 2016). For example, Mohammed et al. (2017) explained that irrigation water used in agricultural soils in Provence (SE France) was supersaturated with respect to calcite. The resulting increase in carbonates with irrigation water can accelerate the precipitation of pedogenic carbonates (Hannam et al., 2016; Cox et al., 2018).

Early observations (Magaritz and Amiel, 1981) already found an effect of long-term irrigation in the Jordan Valley in Israel, resulting in 7% of the total amount of $CaCO_3$ being leached (about 500 metric Mg ha^{-1}) and an even larger effect on the amount of recrystallized carbonate in the soil (about 11% of total carbonates). They also found that most of this SIC loss, as well as accumulation of most of the pedogenic carbonates, occurred in the finest soil fractions due to preferential precipitation of calcite in these fractions.

More recent studies have shown that the introduction of irrigation modified the carbonate distribution in the tilled layer of Mediterranean soil in a very short period since there was a reduction in the concentration of the total soil carbonate content in the finest fractions (de Soto et al., 2017), and at the same time, a preferential accumulation of pedogenic carbonates in these fractions (de Soto et al., 2019, see Section 3 Case study). Similar results have been obtained by Lu et al. (2020), who observed a reduction in the SIC content associated with irrigation and fertilization in the finest fractions of soils in North China. On the contrary, Zornoza et al. (2016) did not detect any important effect of irrigation on the SIC content in soils of the Mediterranean area, and Wu et al. (2008) observed opposite trends of the content of SIC in the first meter after decades of irrigation of two soils in California. These authors found that carbonate-rich soil profiles irrigated with water rich in Ca^{2+} in the Imperial Valley increased their SIC stocks by 10% after 85 years of irrigation, while those carbonate-poor ones irrigated with water of low ionic concentration lost almost completely their SIC stocks. The composition of irrigation water stands out as a key driver of SIC gains or losses.

Many of these studies incorporated the use of isotopic tools to determine the origin of soil carbonates, as the isotopic signature of soil carbonates-C

($\delta^{13}C$) allows to differentiate the source of C (lithogenic or pedogenic carbonates), if that of lithogenic or primary and pedogenic carbonates are known. Although this is not possible in most cases, different methodological approaches assuming pedogenic carbonates as formed from respired soil CO_2 have been developed (Salomons and Mook, 1976; Mermut et al., 2000), which can be used to determine the relative effect of agricultural practices. For instance, in the study by Lu et al. (2020), it was found that pedogenic carbonates had an opposed trend to total SIC when comparing two irrigation districts in China, which they attributed to different levels of SOC and Ca^{2+} and Mg^{2+} in the soils of each district. They also found an interesting correlation between the concentration of pedogenic carbonates and SOC stocks, although no significant relationship between SOC and total SIC was observed.

Other examples can be found in soils in which the effect of irrigation was combined with that of different fertilization or management strategies. This is a relevant issue when assessing the effect of irrigation, as in most cases, it cannot be separated from that of other management strategies accompanying it (Apesteguía et al., 2015). As an example, Bughio et al. (2016) analyzed the formation of pedogenic carbonates in an irrigated Fluvic Cambisol in a semi-humid region of China under different fertilization systems (organic and mineral) at medium and low doses, over 39 years. They observed annual accumulation rates between 0.23 Mg ha^{-1} and 0.12 Mg ha^{-1} of C in the form of pedogenic carbonates in the different treatments, compared to the 0.12 Mg ha^{-1} observed in an unfertilized control, where they assumed that this relatively high rate was possible as a result of primary mineral weathering under intensive agriculture, where biological activity is greater than in undisturbed soils in the study area. This and other examples using the isotopic signature of soil carbonates and the mix equation proposed by Salomons and Mook (1976) are summarized in Table 1.

Enhanced SIC accumulation by an increase of SOC was also observed by Shi et al. (2017) in the irrigated systems in the North China Plain. Like Lu et al. (2020) (Table 1), these authors determined indeed that, for the conditions of their study (an irrigated winter wheat-summer maize rotation under semi-arid continental temperate monsoon climate), the maximum potential for neoformation of pedogenic carbonates depended on the availability of Ca^{2+} and Mg^{2+} in irrigation water. This supports the idea that the source of Ca^{2+} and HCO_3^- is the key factor determining if the observed increased rate of soil carbonates precipitation with irrigation can be an atmospheric C sink, source, or C neutral (Sanderman, 2012). On the same soil and management conditions, Bughio et al. (2017) studied the implication of some agricultural practices such as tillage and the incorporation of straw (Table 1). In this case, neoformation of silicatic pedogenic carbonates sequestered between 0.29 Mg C ha^{-1} and 0.49 Mg C ha^{-1} annually, with tillage and the incorporation of crop residues

Table 1 Information about pedogenic carbonates in different agricultural soils under different managements.

Study	Soil type	Climate	Managements	Pedogenic carbonates (PC)
Bughio et al. (2016)	Fluvic Cambisol	Warm, sub-humid climate with a severely dry spring, wet summer, and dry-cold winter. Average annual temperature of 13.2°C, ranging from a maximum of 26.8°C in July to a minimum of −2.9°C in January. Mean precipitation of 543 mm.	Natural fallow; control without fertilization; mineral fertilizers; organic fertilizers at low rate; organic fertilizers at high rate.	4.8 Mg C ha^{-1} to 8.9 Mg C ha^{-1} of PC-C at 0-20 cm. Treatments with organic fertilizers at high and low rates, mineral fertilizers, and a control site without fertilizers showed an increase in PC compared to a natural fallow plot.
Bughio et al. (2017)	Fluvic Cambisol	As in Bughio et al. (2016).	Tillage with wheat and maize straw return, tillage with wheat straw return, wheat and maize straw return with no-tillage and native plantation plot.	1.7- 4.3 Mg PC-C ha^{-1} at 0-20 cm. There was an increase of PCs compared to a native plantation plot.
de Soto, et al. (2019)	Xeric Haplocalcid	Semi-arid Mediterranean, with mean annual rainfall of 440 mm and mean annual temperature of 13.7°C.	Irrigated corn - Non-irrigated wheat.	87-92% PC contribution to total SIC with irrigation for 61% to 74% without irrigation in the 0-20 cm layer.
Lu et al. (2020)	Calcaric Cambisol and Fluvo-Aquic in Hebei Plain (HBP) and Salic Fluvisols, Calcaric Fluvisols and Gleyic Solonchaks in Yellow River Delta (YRD)	Semi-humid region with a cold and dry winter, hot and rainy summer, and severely dry spring. Annual mean temperature is 12.5°C and 13.4°C, annual mean precipitation 550 mm and 600 mm, and annual mean evaporation 1900 mm and 2500 mm in the HBP and YRD, respectively.	Typical cropland with a maize-wheat rotation. The Hebei Plain (HBP) under intensive cropping and the Yellow River Delta (YRD) under soil salinization.	Slightly higher values for PC (3.9-14.5 kg PC-C m^{-2}) in the top 1 m. PC stock was smaller in the YRD (8.67 kg PC-C m^{-2}) relative to the HBP (9.41 kg PC-C m^{-2}).

Table 1 (*Continued*)

Study	Soil type	Climate	Managements	Pedogenic carbonates (PC)
Wang et al. (2015)		Arid (Urumqi) and semi-arid (Yangling and Zhengzhou) climatic conditions in China. Annual average temperature and precipitation are lower in Urumqi (7.7°C and 299 mm) than in Yangling and Zhengzhou (13–14.5°C and 550–632 mm). Annual evaporation varies from 993 mm to 2570 mm. Irrigation is usually applied during the crop's growing season only.	Two sets (65 each) of archived soil samples collected in the early and late 2000s from three long-term experiment sites under wheat-corn cropping with various fertilization treatments.	Accumulation of PCs under fertilization varied from 60 g PC-C m^{-2} y^{-1} to 179 g PC-C m^{-2} y^{-1} in the 0–100 cm layer.
Wang et al. (2015)		Arid and semi-arid region in China.	100 soil samples from the Yanqi Basin from various land uses (conversion of shrub land to cropland). In addition, other 18 archived soil samples from a long-term experiment (LTE) in Pingliang, Gansu, which was established in 1979 with a mono-cropping system dominated by maize-wheat rotations.	The data from the Yanqi Basin showed that converting shrub land to cropland increased PC-C stock by 5.2 kg PC-C m^{-2}. The data from the LTE showed greater accumulation of PC-C (21–49 g PC-C m^{-2} $year^{-1}$) than SOC in 0–20 cm.

from two crops (maize and wheat), resulting in higher rates than no-till and the incorporation of residues from one single crop.

Still, the determination of the function of SIC for sink or source of atmospheric C by using these data is not simple. As explained by Sanderman (2012), stable isotope approaches cannot distinguish between newly precipitated and dissolved/reprecipitated $CaCO_3$, which would be crucial in determining if the net process is a sink or a source of atmospheric CO_2. Therefore, an accurate analysis, in addition to identifying the sources of Ca^{2+} and HCO_3^-, needs to consider the final fate of HCO_3^- originating from carbonates dissolution. In some cases, the apparent loss of SIC in the tilled layers may correspond to a simple redistribution in the soil profile. If this is not the case, the process can be considered a C sink only if HCO_3^- is lost through leaching or drainage and enters a long-term reservoir. When it ends up in a surface water system, precipitation within the new system can counteract the C sequestration occurring during initial dissolution.

Beyond these consequences, the dissolution and recrystallization of carbonates resulting from repeated irrigation cycles can also be noticed as changes in the soil fabric. Kharitonova et al. (2020) have recently provided evidence of the consequences of the formation of pedogenic carbonates in microaggregation in the upper horizon (0–20 cm) of a Vertisol and a Terra Rosa irrigated with fresh and wastewater in Israel. An average increase in the size of microaggregates (from 350 to 500 μm in the Vertisol) was observed to be associated with the formation of micro- and meso-calcite crystals. These crystals resulted in microaggregates sealing or cementation. Similarly, in the Terra Rossa, calcite-smectite fabrics were observed on the surface of microaggregates, suggesting that the changes induced in SIC dynamics by irrigation can have consequences on the physical condition of the soil and, indirectly, in organic matter stabilization within aggregates.

In summary, these observations support the idea of irrigation inducing major changes in SIC dynamics, which may have consequences for total C cycling in soils, including C emission or sequestration, and other soil properties. As the number of studies directly addressing these potential changes is still very limited, the evaluation of pedogenic carbonates' dynamics under irrigation and its consequences on soil functioning CO_2 emissions merit further research (Wang et al., 2015).

2.3 Reclamation of calcareous sodic soils

Sodic soils are one of the two types of salt-affected soils, characterized by high exchangeable sodium contents, which cause colloid dispersion and, therefore, structure collapse. The effects are mainly physical (crusting, erosion, piping, compaction) but may also be biological due to sodium toxicity in Na-sensitive crops. According to the FAO/UNESCO Soil Map of the World (FAO, 2008), the total salt-affected area is higher than 1100 Mha, out of which more than a half are sodic soils. When the source of Na is sodium carbonate, the pH of these soils is buffered around 9; in these cases, we are dealing with alkaline soils.

The objective when reclaiming sodic soils is to reduce the proportion of exchangeable sodium; therefore, the main management practices consist in supplying high amounts of soluble calcium to the soil solution and flocculating the colloids by decreasing the proportion of exchangeable Na (ESP), mainly through gypsum amendments. Other cations, such as K^+, Mg^{2+}, or Fe^{2+} have also been used to displace sodium (Artiola et al., 2000; Chun et al., 2007; Day et al., 2019). Although the mechanism may seem rather simple, the main problem is that the adverse physical properties of these soils prevent a thorough leaching of the profile. To counteract this effect, gypsum is normally applied together with organic amendments to increase aggregation, infiltration, and permeability (Qadir et al., 2006). Organic matter is also causing a higher cation exchange capacity, so its application also exerts a dilution effect on the existing sodium in the soil, lowering the ESP.

The exchange reaction that reduces sodification by gypsum application is as follows:

$$CaSO_4 \cdot 2H_2O \rightarrow Ca^{2+} + SO_4^{2-} + 2H_2O \quad (5)$$

$$X - 2Na^+ + Ca^{2+} \leftrightarrow X - Ca^{2+} + 2Na^+ \quad (6)$$

In alkaline (Na_2CO_3-bearing) soils, dissolution of gypsum provides a source of Ca^{2+}, which precipitates as $CaCO_3$ and $Ca(HCO_3)_2$, resulting in a decrease in soil pH (Wong et al., 2009).

Other amendments that proved to be efficient in the reclamation of sodic soils, like sulfur or sulfuric acid, also affect soil carbonates, since the desired effect is acidification, dissolution of carbonates, and release of calcium to the solution, which is made available for the exchange reaction with sodium. While direct application of sulfuric acid (H_2SO_4) is effective (Amezketa et al., 2005; Mahmoodabadi et al., 2013), it cannot be widely adopted because it requires special equipment and is considered impractical for large-scale applications. The use of elemental S, in pelletized form, is more common. The reactions involved in this case are the microbial oxidation of S to sulfate, which reacts with soil water to form sulfuric acid, which solubilizes $CaCO_3$, releasing Ca^{2+} which displaces exchangeable Na^+:

$$S^0 + \frac{3}{2}O_2 + H_2O \rightarrow SO_4^{2-} + 2H^+ \quad (7)$$

$$CaCO_3 + H^+ \rightarrow HCO_3^- + Ca^{2+} \quad (8)$$

It is evident that all these reactions are indeed affecting the amount and dynamics of calcium carbonate in soils, but the diverse experiments carried out both in the field and in the laboratory on calcareous sodic soils to evaluate these amendments have mostly measured only soil pH, ESP, Sodium Adsorption Ratio, and cation concentrations in depth, but up to date, none of the studies did measure their impact on calcium carbonate contents. The pH decreases may range from 0.5 (Sadiq et al., 2007) to more than 2 units (Wong et al., 2009), depending on the treatments, which indicate the extent of carbonate mobilization due to these management practices; but a better-focused research on quantifying their effect on soil carbonates and CO_2 emissions is highly needed.

2.4 Organic matter management and soil biota

As the organic and inorganic fractions of soil C maintain a dynamic relationship, at shorter rates and higher magnitudes than previously thought, SOM management can be thought to have consequences on SIC. It offers therefore an additional opportunity for carbon sequestration if the increases in SIC derived from this management are accounted for, besides those of SOC.

From the point of view of agriculture, these benefits are indirect: they probably will not result in clear increases in soil productivity by themselves, when compared with the benefits of SOC increases in soil quality. In any case, the effect of many SOM management practices on SIC is not known - because simply it has not been the object of attention, and no accurate measurements exist. In these circumstances, there is a clear lack of knowledge and the body of literature on the state of research on the relationships between SIC and SOC in organically-amended soils. This only allows us to give some clues on the consequences of organic and biogenic amendments and possible practices related to them.

As detailed in previous sections, adequate tools are needed for an accurate assessment of changes induced in soil carbonates by agricultural management. In the case of the activation of the dissolution and precipitation cycles of carbonates resulting from changes in the organic fraction, soil micromorphology has been and can be a tool of paramount relevance because it is the technique that gives more information on the types and origin of secondary calcium carbonate accumulations in soils. For instance, the study of calcite biomineralizations or precipitation of calcite due to the metabolism of living organisms is possible in soil thin sections (Durand et al., 2018). As described in Chapter 7 of this book, calcite biomineralizations or biocalcifications are in fact well known and take very different forms.

Some of these metabolic processes include the synthesis/transformation of carbonates from calcium oxalates, which are believed to be an intermediate stage before their transformation to calcite by oxalotrophic bacteria (Klappa, 1979; Philips et al., 1987; Uren, 2018) and some types of fungi (Hervé et al., 2021). The oxalate-carbonate transformation is possible by the generation of a localized rise in pH around these microorganisms. Syed et al. (2020) have recently reviewed the role of oxalotrophic bacterial communities existing in agricultural and forest soils and their potential role in carbonates neoformation from plant-derived oxalates oxidation in the so-called oxalate-carbonate pathway. This pathway can be affected by many factors and represents a high potential for indirect agricultural SIC sequestration. According to these authors, for it to be effective, not only oxalogenic plants are needed but also a soil biological community able to degrade their oxalogenic biomass and release and maintain the oxalate pool in the soil, which would be used by the oxalotrophic bacteria. In this framework, oxalogenic plants of agricultural interest appear as a promising tool for promoting this pathway in long-term SIC storage in cultivated land. Some species of Cactaceae, for instance, have been identified as efficient in the accumulation of oxalate crystals such as weddellite in their tissues (Bárcenas-Argüello et al., 2015), which can be used as an alternative source of CO_2 for photosynthesis under drought conditions (Tooulakou et al., 2016). Nopal (*Opuntia ficus-indica* L.), a widely cultivated and consumed plant in Central America, has been identified as one of these plants (Contreras-Padilla et al., 2011; Zúñiga et al., 2018). An optimization of nopal management or the use of other species of the kind as cover crops are worth to be considered as strategies to store SIC in cultivated soils of arid regions.

In addition to these biologically driven processes, even if some other calcitic pedofeatures observable in soil thin sections (such as hypocoatings or calcite nodules) are believed to form through a physicochemical process of precipitation, they will always be affected by the soil biological activity, since it affects the carbonate-bicarbonate equilibrium in the soil solution, for instance, by root metabolism (Wieder and Yaalon, 1982; Zamanian et al., 2016) or by the mere respiration of microorganisms (Monger et al., 1991; Rehman et al., 2021). In relation to this, some references indicate that soils under land covers promoting higher organic matter contents and biological activity contain higher carbonate contents than neighboring soils under more intensive land uses (Wang et al., 2016; Rehman et al., 2021).

Finally, it has been observed that precipitation of $CaCO_3$ in experimental columns of soil fed with microorganisms can proceed in periods from days to weeks (Monger et al., 1991). This precipitation seems especially effective in the case of urea-hydrolyzing bacteria producing CO_2 by ureolytic hydrolysis and causing the precipitation of $CaCO_3$ (Okwadha and Li, 2010; Sarmast et al., 2014). Experiences with some algal biofertilizers also indicate gains in $CaCO_3$, attributed to CO_2 production through heterotrophic metabolism (Marks et al., 2019). Other studies have however reported that soil microorganisms can retard carbonate recrystallization in alkaline soils, theoretically by growing on carbonate surfaces like coatings (Zhao et al., 2020).

These evidences suggest that not only organic matter management *per se* has a potential to stimulate changes in SIC storage and/or its composition but also that the use of certain types of microorganisms may have a potential to influence these parameters. Yet, the interactions within the soil biota and between them and the soil inorganic components are very complex and far from being understood (Francis and Poch, 2019). More research is needed to quantify these interactions before it will be possible to design management strategies to sequester carbon in inorganic form, in synergy with the rest of soil ecosystem services.

2.5 Best management practices

Some clues on the agricultural management strategies and practices that can contribute to preserving SIC in agricultural soils while maintaining agricultural productivity can be deduced from the previous sections.

Managing fertilization strategies, especially N fertilization, to avoid acidification is needed to reduce SIC losses and prevent SIC-derived CO_2 emissions. In addition to promoting the use of N-fertilizers with the lowest possible acidification potential, optimizing N-use efficiency, which has also other environmental benefits (Zhang et al., 2021), seems also needed for SIC loss control. This can be achieved, for instance, by adequate N scheduling N additions by adjusting them to the needs of the crops and/or by the use of nitrification and urease inhibitors, which decrease H^+ release by retarding the nitrification process (Raza et al., 2019).

The spreading of adequate management strategies of cover crops and crops management contributing to SIC accumulation, such as those related to *Cactaceae* and organic inputs to soil can also be evaluated in relation to their consequences on the total soil C stock.

Details on other strategies such as adequate land planning, including the preservation of marginal land in calcareous environments and those related to liming management are provided in Chapter 7 in this book.

In any case, the trade-offs of these practices with SOC storage and/or other aspects of soil functioning should be considered from a holistic perspective.

3 Case study: effect of irrigation on soil inorganic carbon dynamics in the tilled layer (0–20 cm) of Mediterranean agricultural soils

In these studies, the effect of irrigation on carbonates dynamics in the tilled layer of Mediterranean agricultural soils was assessed (de Soto et al., 2017, 2019). Three sites were selected at three different locations (Miranda, Funes, Valtierra) in Navarre, NE Spain, corresponding to three major irrigation districts. In each of these sites, two contiguous or neighboring agricultural plots were selected, one under rainfed wheat cropping since at least 50 years, and the other with irrigated corn since irrigation was implemented in each area (5, 12 and 19 consecutive years in Miranda, Funes, and Valtierra, respectively). In each of these areas, one composite soil sample from 25 subsamples and one intact soil core for bulk density were collected in the tilled layer (0–20 cm) at 0–10 cm and 10–20 cm.

The soils were *Xeric Haplocalcids* in Valtierra and Funes, and a *Typic Calcixerept* in Miranda ((S.S.S.) Soil Survey Staff, 2014) Climate in the three irrigation districts is semi-arid Mediterranean. The mean annual rainfall and mean annual temperatures are 429 mm and 13.9°C for Miranda, 437 mm and 13.7°C for Funes, and 380 mm and 13.9°C for Valtierra. Irrigated corn gets around 700 extra mm of water annually in these areas.

Soils were managed using the conventional routines for wheat and corn in the region. This includes inversion tillage at 0–20 cm, and fertilization with mineral N. Carbonates' storage and their size distribution were quantified at the three selected sites. In addition, a numerical simulation of the geochemical interactions between the soil, the soil solution, and irrigation water was developed with the actual data of Funes district using PHREEQC (Parkhurst and Appelo, 1999). A sensitive analysis was also conducted to investigate the potential impact of water quality and crop types as sources of variability on the model outputs. Finally, the proportion and distribution of pedogenic carbonates in different size fractions in Funes, was estimated using the isotopic signature of carbonates-C, and the mix equation proposed by (Salomons and Mook, 1976).

The results of this study showed that lower concentrations of carbonates in the fine fraction (silt and clay) were systematically observed with irrigation overall (Fig. 2; de Soto et al., 2017). In particular, the concentration of carbonates in the silt+clay size fraction was systematically smaller with irrigation in Valtierra

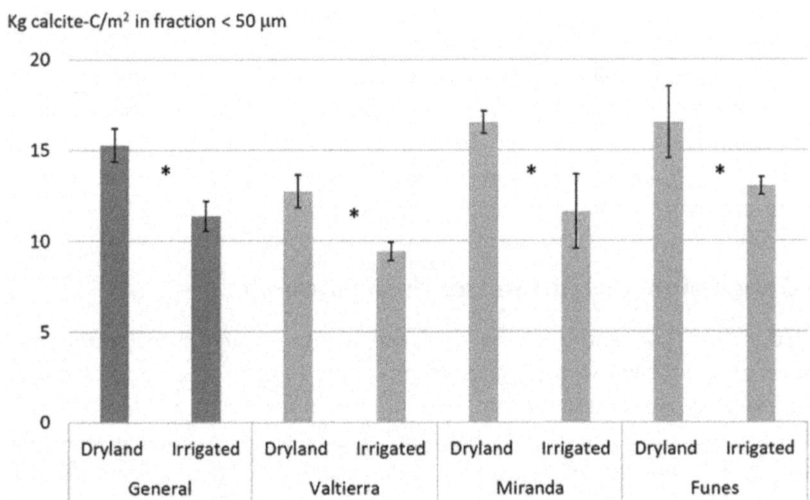

Figure 2 Concentration of carbonates in the silt + clay size fractions (<50 µm) in the irrigated and non-irrigated fields overall (left), and in each irrigation district (Valtierra, Miranda, and Funes, right) in the topsoil (0–30 cm). * Significant differences at $P < 0.05$ ($n = 3$). Source: Adapted from own data in de Soto et al. (2017).

(22.2 ± 1.4 g carbonates 100 g soil fraction^{-1} without irrigation and 16.1 ± 0.9 with irrigation for 19 years), Miranda (26.7 ± 1.1 and 19.1 ± 3.8 g carbonates 100 g soil fraction^{-1}, after 5 years of irrigation) and Funes (27.8 ± 2.9 and 22.7 ± 1.5, g carbonates 100 g soil fraction^{-1}, after 12 years).

The modeling results indicated that the introduction of irrigation for corn resulted in a reduction in the concentration of soil carbonates, which amounted to annual carbonates-C loss between 13.52 g m^{-2} year^{-1} and 12.06 g m^{-2} year^{-1} in 0–30 cm depth under irrigation, depending on the quality of irrigation water, and 0.46 gm^{-2} without irrigation. These field results were qualitatively in line with the simulation results. However, the net balances of carbonates-C in the fine fraction in Valtierra, Funes, and Miranda, were one order of magnitude greater than in the model, which confirms the complexity of the biogeochemical processes taking place in agricultural soils under dryland and irrigation conditions (de Soto et al., 2017).

Finally, the calculated proportion of pedogenic carbonates was estimated from the isotopic ^{13}C signature of soil carbonates, soil organic C, and common values of δ^{13}C found for lithogenic carbonates (between −2 and 2‰; Ryskov et al., 2000). It was observed to be higher with irrigation, both for bulk soil samples and for all the studied size-factions (Fig. 3). From these observations, it was concluded that in the study region, irrigation can alter the distribution of soil carbonates and produce a preferential accumulation of pedogenic carbonates (87% to 92% with irrigation for 61% to 74% without irrigation), in the finer soil fractions (de Soto et al., 2019).

As a conclusion, these results suggest that the introduction of irrigation modified the distribution of carbonates in the tilled layer in a very short period of time. In terms of SIC balance, this was expressed as a general loss in the finest soil fractions. In addition, the greater proportion of pedogenic carbonates estimated due to irrigation can be understood as the result of this alteration: pedogenic carbonate formation can be attributed to the precipitation process in the short term, occurring mainly at the microscale.

4 Conclusion and future trends in research

Despite the clear and growing evidence of agricultural management as a driver of SIC cycling with consequences on the total soil C budget, there are large knowledge gaps in terms of quantitative understanding of the processes affecting SIC and its interaction with SOC. If 'climate-smart' agriculture is to be promoted, the design of agricultural systems should account not only for their net consequences on emissions or as a C sink but also for other aspects such as their resilience and adaptation capacity to climate change. This includes, in the case of carbonate-rich soils, usually poor in SOC, sustained ability for SOC stabilization, and water storage, for instance.

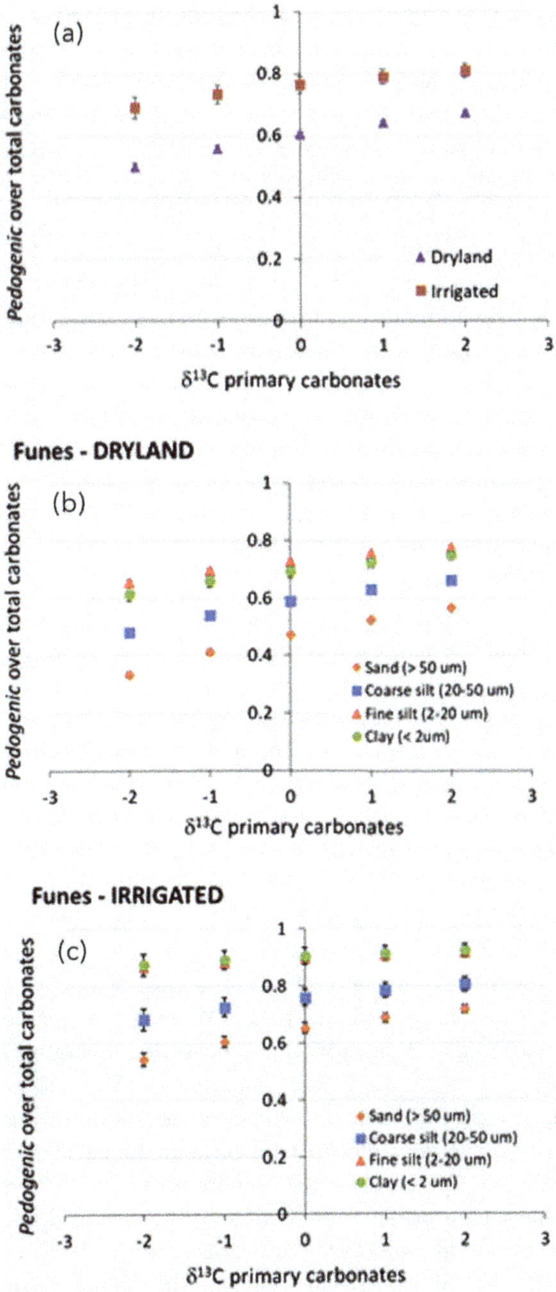

Figure 3 Estimated proportion of pedogenic carbonates over total soil carbonates for the common range of $\delta^{13}C$ in primary or lithogenic carbonates between −2 and 2‰ (Ryskov et al., 2000), for total carbonates (a) and particle size-fractions without (b) and with (c) irrigation. Source: Adapted from de Soto et al. (2019).

Understanding the interaction of changes in SIC dynamics related to management needs an accurate assessment not only of the amounts of SIC gained or lost but of the changes in the typology and size distribution of soil carbonates, and their different roles in SOC cycling and soil structure stabilization. In this sense, also there is increasing interest in the role of carbonates as abiotic agents in aggregates formation and SOC stabilization in calcareous soils (Fernández-Ugalde et al., 2011; Garcia-Franco et al., 2020), the relationship between the stabilization of organic matter and the formation of pedogenic carbonates, as well as the relationship between the abundance and nature of organic compounds and that of carbonates at different levels of soil structure (from organo-mineral complexes to macroaggregates), is still rather unknown.

Another field not completely developed is the inclusion of SIC dynamics in landscape-scale biogeochemical modeling (Hirmas et al., 2010) beyond natural landscapes, i.e. together with agricultural management and land uses that can affect it. An example of this would be the work of Perrin et al. (2008) at the watershed level in France, which could be transferred to other regions and locations, where land-use and water quality data are available. In the case of transformations of dryland to irrigation, for instance, this type of exercise could help us understand the actual effect of these transformations on the regional C budget.

Part of the lack of knowledge is also the result of some methodological barriers hindering our capacity to completely understand the interaction of soil management and carbonates. There is a need for finding techniques or analytical strategies able to quantify and determine the typology of SIC. As stated by Apesteguía et al. (2018), in carbonate-rich soils, the quantification of SOC and SIC remains challenging: The measurement of SOC by dry combustion cannot be directly obtained due to the presence of SIC, and the determination of SIC characteristics, such as its stable carbon isotope signature, requires the removal of SOC. The pros and cons of traditional approaches followed to remove SIC and SOC from soil samples have been debated for decades, and all procedures have demonstrated experimental limitations.

Recent approaches have tried to overcome these limitations. Ramped combustion, which takes advantage of the large difference in the temperatures at which SOC and SIC combust or decompose, was tested in samples containing different proportions of SOC and SIC, and it appears to be an adequate single-sample method for the simultaneous quantification of SOC and SIC in calcareous soils by Apesteguía et al. (2018). Mid-infrared reflectance spectroscopy using spectral libraries has also been shown to be effective in quantifying SIC without the need for sample pretreatments (Barthès et al., 2020; Gomez et al., 2020).

The use of these techniques, and other emerging ones such as Rock-Eval pyrolysis, coupled to different strategies of soil and organic matter fractionation, is a promising perspective for a better understanding of the relationship between SOC and SIC cycling and the possible consequences of agricultural management on it. In particular, Rock-Eval pyrolysis allows using the pyrograms obtained to identify components that can be associated with different constituents of the organic fraction of the soil (Sebag et al., 2006, 2016) and their 'energy signature' (Barré et al., 2016). It can also be used to study the level of interaction of organic compounds with the mineral fraction (Cécillon et al., 2018; Poeplau et al., 2019). This technique has been also used by Pillot et al. (2014) in the identification and quantification of carbonates, which suggests that it can be used for advancing in the development of quantitative indicators of the typology of carbonates associated with soil and organic matter fractions, simultaneously with the study of the characteristics of organic components.

In terms of the quantification of net CO_2 emissions from calcareous soils, and the consequences of agricultural management on them, a present challenge is the quantification of the partition of the emitted CO_2 issued from SOC decomposition from that issued from SIC dissolution. Oren and Steinberger (2008) already identified that the addition of substrates of different nature can lead to an overestimation or underestimation of actual respiration rates, resulting from the interaction of the substrates and SIC. Isotopic analysis of CO_2 has been used successfully to demonstrate the temperature-dependence of SIC dissolution (Chevallier et al., 2016) and to understand the actual consequences of agricultural management (such as agroforestry) on SOC mineralization in calcareous soils (Cardinael et al., 2020). These authors found that, in a silty calcareous Fluvisol in SE France, SIC-derived CO_2 represented about 20% of the total emitted CO_2 in the topsoil and 60% in the subsoil.

Finally, the use of other known techniques in minerals and soil studies, such as X-ray diffraction analysis, scanning electron microscopy, or soil micromorphology, can be expanded for understanding the consequences of soil management on soil carbonate dynamics. For instance, a shift in the d-spacing calcite peak in X-ray diffractograms, from 3.03 to 3.01 Å was observed by Landi et al. (2003) and Khalaf and Al-Zamel (2016), which suggests the incorporation of Mg in the calcite structure of these carbonates. These authors also reported a change in the morphology of calcite crystals when observing the samples by SEM. These techniques can therefore be of great help for the study of pedogenic carbonates changes in cultivated soils. They can complement the already mentioned potential of soil micromorphology for a deeper understanding of the consequences of the agricultural management of calcareous soils, carbon sequestration, and their benefits and trade-offs in terms of SIC cycling and carbonate dynamics.

5 Where to look for further information

The following articles provide a good overview of the subject:

- Sanderman, J. (2012). Can management-induced changes in the carbonate system drive soil carbon sequestration? A review with particular focus on Australia, *Agriculture, Ecosystems and Environment* 155, 70–77.
- Rowley M. C., Grand, S. and Verrecchia, É. P. (2018). Calcium-mediated stabilisation of soil organic carbon, *Biogeochemistry* 137, 27–49.
- Cardinael, R., Chevallier, T., Guenet, B. et al. (2020). Organic carbon decomposition rates with depth and contribution of inorganic carbon to CO_2 emissions under a Mediterranean agroforestry system, *European Journal of Soil Science* 71, 909-923.
- Zamanian, Z., Zhou, J. and Kuzyakov, Y. (2021). Soil carbonates: the unaccounted, irrecoverable carbon source, *Geoderma* 384, 114817.

6 References

Amezketa, E., Aragüés, R. and Gazol, R. (2005). Efficiency of sulfuric acid, mined gypsum, and two gypsum by-products in soil crusting prevention and sodic soil reclamation, *Agronomy Journal* 97(3), 983–989. doi: 10.2134/agronj2004.0236.

Apesteguía, M., Plante, A. F. and Virto, I. (2018). Methods assessment for organic and inorganic carbon quantification in calcareous soils of the Mediterranean region, *Geoderma Regional*. Elsevier 12, 39–48. doi: 10.1016/j.geodrs.2017.12.001.

Apesteguía, M., Virto, I., Orcaray, L., Enrique, A. and Bescansa, P. (2015). Effect of the conversion to irrigation of semiarid Mediterranean dryland agrosecoystems on soil carbon dynamics and soil aggregation, *Arid Land Research and Management* 29(4), 399–414. doi: 10.1080/15324982.2015.1016245.

Artiola, J. F., Gebrekidan, H. and Carty, D. J. (2000). Use of langbeinite to reclaim sodic and saline sodic soils, *Communications in Soil Science and Plant Analysis* 31(17–18), 2829–2842. doi: 10.1080/00103620009370631.

Bárcenas-Argüello, M. L., Gutiérrez-Castorena, M. C. and Terrazas, T. (2015) The polymorphic weddellite crystals in three species of Cephalocereus (Cactaceae), *Micron*. Elsevier Ltd 77, 1–8. doi: 10.1016/j.micron.2015.05.014.

Barré, P., Plante, A. F., Cécilllon, L., Lutfalla, S., Baudin, F., Bernard, S., Christensen, B. T., Eglin, T., Fernandez, J. M., Houot, S., Kätterer, T., Le Guillou, C., Macdonald, A., van Oort, F. and Chenu, C. (2016). *The Energetic and Chemical Signatures of Persistent Soil Organic Matter, Biogeochemistry*. Springer International Publishing. doi: 10.1007/s10533-016-0246-0.

Barthès, B. G., Kouakoua, E., Coll, P., Clairotte, M., Moulin, P., Saby, N. P. A., Le Cadre, E., Etayo, A. and Chevallier, T. (2020). Improvement in spectral library-based quantification of soil properties using representative spiking and local calibration – the case of soil inorganic carbon prediction by mid-infrared spectroscopy, *Geoderma*. Elsevier 369(February), 114272. doi: 10.1016/j.geoderma.2020.114272.

Brunet, F., Potot, C., Probst, A. and Probst, J. L. (2011). Stable carbon isotope evidence for nitrogenous fertilizer impact on carbonate weathering in a small agricultural watershed, *Rapid Communications in Mass Spectrometry* 25(19), 2682–2690. doi: 10.1002/rcm.5050.

Bughio, M. A., Wang, P., Meng, F., Chen, Q., Li, J. and Shaikh, T. A. (2017). Neoformation of pedogenic carbonate and conservation of lithogenic carbonate by farming practices and their contribution to carbon sequestration in soil, *Journal of Plant Nutrition and Soil Science* 180(4), 454–463. doi: 10.1002/jpln.201500650.

Bughio, M. A., Wang, P., Meng, F., Qing, C., Kuzyakov, Y., Wang, X. and Junejo, S. A. (2016). Neoformation of pedogenic carbonates by irrigation and fertilization and their contribution to carbon sequestration in soil, *Geoderma* 262, 12–19. doi: 10.1016/j.geoderma.2015.08.003.

Cardinael, R., Chevallier, T., Guenet, B., Girardin, C., Cozzi, T., Pouteau, V. and Chenu, C. (2020). Organic carbon decomposition rates with depth and contribution of inorganic carbon to CO_2 emissions under a Mediterranean agroforestry system, *European Journal of Soil Science* 71(5), 909–923. doi: 10.1111/ejss.12908.

Cécillon, L., Baudin, F., Chenu, C., Houot, S., Jolivet, R., Kätterer, T., Lutfalla, S., Macdonald, A., van Oort, F., Plante, A. F., Savignac, F., Soucémarianadin, L. N. and Barré, P. (2018). A model based on Rock-Eval thermal analysis to quantify the size of the centennially persistent organic carbon pool in temperate soils, *Biogeosciences* 15(9), 2835–2849. doi: 10.5194/bg-15-2835-2018.

Chevallier, T., Cournac, L., Hamdi, S., Gallali, T. and Bernoux, M. (2016). Temperature dependence of CO_2 emissions rates and isotopic signature from a calcareous soil, *Journal of Arid Environments*. Elsevier Ltd 135, 132–139. doi: 10.1016/j.jaridenv.2016.08.002.

Chun, S., Rai, H., Nishiyama, M. and Matsumoto, S. (2007). Using organic matter with chemical amendments to improve calcareous sodic soil, *Communications in Soil Science and Plant Analysis* 38(1), 205–216. doi: 10.1080/00103620601094163.

Contreras-Padilla, M., Pérez-Torrero, E., Hernández-Urbiola, M. I., Hernández-Quevedo, G., del Real, A., Rivera-Muñoz, E. M. and Rodríguez-García, M. E. (2011). Evaluation of oxalates and calcium in nopal pads (*Opuntia ficus-indica* var. redonda) at different maturity stages, *Journal of Food Composition and Analysis* 24(1), 38–43. doi: 10.1016/j.jfca.2010.03.028.

Cox, C., Jin, L., Ganjegunte, G., Borrok, D., Lougheed, V. and Ma, L. (2018). Soil quality changes due to flood irrigation in agricultural fields along the Rio Grande in western Texas, *Applied Geochemistry*. Elsevier 90, 87–100. doi: 10.1016/j.apgeochem.2017.12.019.

Datta, A. and Mandal, B. (2018). Production (via N-fertilization) and correction (by liming) of acidity in soils contribute a huge efflux of CO2 to atmosphere: real or arbitrary ?, *Global Change Biology* 24(8), (May), 3280–3281. doi: 10.1111/gcb.14318.

Day, S. J., Norton, J. B., Strom, C. F., Kelleners, T. J. and Aboukila, E. F. (2019). Gypsum, langbeinite, sulfur, and compost for reclamation of drastically disturbed calcareous saline-sodic soils, *International Journal of Environmental Science and Technology*. Berlin Heidelberg: Springer 16(1), 295–304. doi: 10.1007/s13762-018-1671-5.

de Soto, I. S., Virto, I., Barré, P. and Enrique, A. (2019). Effect of irrigation on carbonate dynamics in a calcareous soil using isotopic determinations, *Spanish Journal of Soil Science* 9(2), 142–147. doi: 10.3232/SJSS.2019.V9.N2.06.

de Soto, I. S., Virto, I., Barré, P., Fernández-Ugalde, O., Antón, R., Martínez, I., Chaduteau, C., Enrique, A. and Bescansa, P. (2017). A model for field-based evidences of the impact of irrigation on carbonates in the tilled layer of semi-arid Mediterranean soils, *Geoderma*. Elsevier B.V. 297, 48–60. doi: 10.1016/j.geoderma.2017.03.005.

Durand, N., Monger, H. C. and Canti, M. G. (2018). Calcium carbonate features. In: Stoops, G., Marcelino, V. and Mees, F. (Eds) *Interpretation of Micromorphological Features of Soils and Regoliths*. Amsterdam, The Netherlands: Elsevier, pp. 205–258.

European Environmental Agency (2012). Climate Change, impacts and vulnerability in Europe 2012. An Indicator-Based Report. EEA Report No 12/2012. Luxembourg.

FAO, Food and Agriculture Organization of the United Nations. (2008). *Harmonized World Soil Database* (version 1.0). Rome, Italy.

Fernández-Ugalde, O., Virto, I., Barré, P., Gartzia-Bengoetxea, N., Enrique, A., Imaz, M. J. and Bescansa, P. (2011). Effect of carbonates on the hierarchical model of aggregation in calcareous semi-arid Mediterranean soils, *Geoderma*. Elsevier B.V. 164(3–4), 203–214. doi: 10.1016/j.geoderma.2011.06.008.

Francis, M. L. and Poch, R. M. (2019). Calcite accumulation in a South African heuweltjie: role of the termite Microhodotermes viator and oribatid mites, *Journal of Arid Environments* 170(May). doi: 10.1016/j.jaridenv.2019.05.009.

Garcia-Franco, N., Walter, R., Wiesmeier, M., Hurtarte, L. C. C., Berauer, B. J., Buness, V., Zistl-Schlingmann, M., Kiese, R., Dannenmann, M. and Kögel-Knabner, I. (2020). Biotic and abiotic controls on carbon storage in aggregates in calcareous alpine and prealpine grassland soils, *Biology and Fertility of Soils* 57(2), 203–218. doi: 10.1007/s00374-020-01518-0.

Gocke, M., Pustovoytov, K. and Kuzyakov, Y. (2012). Pedogenic carbonate formation: recrystallization versus migration process rates and periods assessed by ^{14}C labeling, *Global Biogeochemical Cycles* 26(1), n/a–n/a. doi: 10.1029/2010GB003871.

Gomez, C., Chevallier, T., Moulin, P., Bouferra, I., Hmaidi, K., Arrouays, D., Jolivet, C. and Barthès, B. G. (2020). Prediction of soil organic and inorganic carbon concentrations in Tunisian samples by mid-infrared reflectance spectroscopy using a French national library, *Geoderma*. Elsevier 375, 114469. doi: 10.1016/j.geoderma.2020.114469.

Hannam, K. D., Kehila, D., Millard, P., Midwood, A. J., Neilsen, D., Neilsen, G. H., Forge, T. A., Nichol, C. and Jones, M. D. (2016). Bicarbonates in irrigation water contribute to carbonate formation and CO_2 production in orchard soils under drip irrigation, *Geoderma*. Elsevier B.V. 266, 120–126. doi: 10.1016/j.geoderma.2015.12.015.

Hasinger, O., Spangenberg, J. E., Millière, L., Bindschedler, S., Cailleau, G. and Verrecchia, E. P. (2015). Carbon dioxide in scree slope deposits: a pathway from atmosphere to pedogenic carbonate, *Geoderma* 247–248, 129–139. doi: 10.1016/j.geoderma.2015.02.012.

Hervé, V., Simon, A., Randevoson, F., Cailleau, G., Rajoelison, G., Razakamanarivo, H., Bindschedler, S., Verrecchia, E. and Junier, P. (2021). Functional diversity of the litter-associated fungi from an oxalate-carbonate pathway ecosystem in Madagascar, *Microorganisms* 9(5), 1–12. doi: 10.3390/microorganisms9050985.

Hirmas, D. R., Amrhein, C. and Graham, R. C. (2010). Spatial and process-based modeling of soil inorganic carbon storage in an arid piedmont, *Geoderma*. Elsevier B.V. 154(3–4), 486–494. doi: 10.1016/j.geoderma.2009.05.005.

IUSS Working Group (2015). World reference base for soil resources 2014. International soil classification system for naming soils and creating legends for soil maps. WRB, Update. Rome: Italy. Available at: http://www.fao.org/3/i3794en/I3794en.pdf.

Jia, G., et al. (2019). Climate change and land. Chapter 2: land-climate interactions, *IPCC SPECIAL REPORT Global Warming of 1.5°C*. Available at: https://www.ipcc.ch/site/assets/uploads/2019/08/2c.-Chapter-2_FINAL.pdf, pp. 1–186.

Khalaf, F. I. and Al-Zamel, A. (2016). Petrography, micromorphology and genesis of Holocene pedogenic calcrete in Al-Jabal Al-Akhdar, Sultanate of Oman, *CATENA*. Elsevier B.V. 147, 496–510. doi: 10.1016/j.catena.2016.07.044.

Kharitonova, G. V., Kot, F. S. and Krutikova, V. O. (2020). Carbonate and concomitant microaggregation in irrigated Mediterranean soils of Israel, *Irrigation Science*. Berlin Heidelberg: Springer 38(4), 431–447. doi: 10.1007/s00271-020-00685-0.

Klappa, C. F. (1979). Calcified filaments in Quaternary calcretes; organo-mineral interactions in the subaerial vadose environment, *Journal of Sedimentary Research* 49(3), 955–968.

Lal, R. and Kimble, J. M. (2000). Pedogenic carbonates and the global C cycle. In: Lal, R., et al. (Eds) *Global Change and Pedogenic Carbonates*. Boca Raton, FL: Lewis Publishers, pp. 43–64.

Landi, A., Mermut, A. R. and Anderson, D. W. (2003). Origin and rate of pedogenic carbonate accumulation in Saskatchewan soils, Canada, *Geoderma* 117(1–2), 143–156. doi: 10.1016/S0016-7061(03)00161-7.

Lu, T., Wang, X., Xu, M., Yu, Z., Luo, Y. and Smith, P. (2020). Dynamics of pedogenic carbonate in the cropland of the North China Plain: influences of intensive cropping and salinization, *Agriculture, Ecosystems and Environment*. Elsevier 292(January), 106820. doi: 10.1016/j.agee.2020.106820.

Magaritz, M. and Amiel, A. J. (1981). Influence of intensive cultivation and irrigation on soil properties in the Jordan Valley, Israel: recrystallization of carbonate minerals, *Soil Science Society of America Journal* 45(6), 1201–1205.

Mahmoodabadi, M., Yazdanpanah, N., Sinobas, L. R., Pazira, E. and Neshat, A. (2013). Reclamation of calcareous saline sodic soil with different amendments (I): redistribution of soluble cations within the soil profile, *Agricultural Water Management*. Elsevier B.V. 120, 30–38. doi: 10.1016/j.agwat.2012.08.018.

Marks, E. A. N., Montero, O. and Rad, C. (2019). The biostimulating effects of viable microalgal cells applied to a calcareous soil: increases in bacterial biomass, phosphorus scavenging, and precipitation of carbonates, *Science of the Total Environment*. Elsevier B.V. 692, 784–790. doi: 10.1016/j.scitotenv.2019.07.289.

Matocha, C. J., Grove, J. H., Karathanasis, T. D. and Vandiviere, M. (2016). Changes in soil mineralogy due to nitrogen fertilization in an agroecosystem, *Geoderma*. Elsevier B.V. 263, 176–184. doi: 10.1016/j.geoderma.2015.09.002.

Mermut, A. R., Amudson, R. and Cerling, T. E. (2000). The use of stable isotopes in studying carbonate dynamics in soils. In: Lal, R., et al. (Eds) *Global Climate Change and Pedogenic Carbonates*. Boca Raton, FL: CRC Press- Lewis Publisher, pp. 65–86.

Mohammed, G., Trolard, F., Gillon, M., Cognard-Plancq, A. L., Chanzy, A. and Bourrié, G. (2017). Combination of a crop model and a geochemical model as a new approach to evaluate the sustainability of an intensive agriculture system, *Science of the Total Environment*. Elsevier B.V. 595, 119–131. doi: 10.1016/j.scitotenv.2017.03.146.

Monger, C. and Gallegos, R. A. (2000). Biotic and abiotic processes and rates of pedogenic carbonate accumulation in the Southwestern United States-Relationships to atompsheric CO_2 sequestration. In: Lal, R., et al. (Eds) *Global*

Climate Change and Pedogenic Carbonates. Boca Raton, FL: CRC Lewis Publishers, pp. 273-290.

Monger, C. H. and Martínez-Ríos, J. J. (2001). Inorganic C sequestration in grazing lands. In: Follett, R. F., Kimble, J. M. and Lal, R. (Eds) *The Potential of US Grazing Lands to Sequester Carbon and Mitigate the Greenhouse Effect*. Boca Raton, FL: Taylor & Francis, pp. 87-117.

Monger, H. C., Daugherty, L. A., Lindemann, W. C. and Liddell, C. M. (1991). Microbial precipitation of pedogenic calcite, *Geology* 19(10), 997-1000.

Monger, H. C., Kraimer, R. A., Khresat, S., Cole, D. R., Wang, X. and Wang, J. (2015). Sequestration of inorganic C in soil and groundwater, *Geology* 43(5), 375-378.

Moreno, F., Murillo, J. M., Pelegrín, F. and Girón, I. F. (2006). Long-term impact of conservation tillage on stratification ratio of soil organic carbon and loss of total and active $CaCO_3$, *Soil and Tillage Research* 85(1-2), 86-93. doi: 10.1016/j.still.2004.12.001.

Müller, B., Meyer, J. S. and Gächter, R. (2020). Nitrogen fertilization of soils fuels carbonate weathering and translocation in calcareous watersheds, *Aquatic Sciences*. Springer International Publishing, 82(2), 1-44. doi: 10.1007/s00027-020-0712-6.

Oertel, C., Matschullat, J., Zurba, K., Zimmermann, F. and Erasmi, S. (2016). Greenhouse gas emissions from soils–a review, *Geochemistry*. Elsevier GmbH 76(3), 327-352. doi: 10.1016/j.chemer.2016.04.002.

Okwadha, G. D. O. and Li, J. (2010). Optimum conditions for microbial carbonate precipitation, *Chemosphere*. Elsevier Ltd 81(9), 1143-1148. doi: 10.1016/j.chemosphere.2010.09.066.

Oren, A. and Steinberger, Y. (2008). Coping with artifacts induced by $CaCO_3$-CO_2-H_2O equilibria in substrate utilization profiling of calcareous soils, *Soil Biology and Biochemistry* 40(10), 2569-2577. doi: 10.1016/j.soilbio.2008.06.020.

Paradelo, R., Virto, I. and Chenu, C. (2015). Net effect of liming on soil organic carbon stocks: a review, *Agriculture, Ecosystems and Environment* 202, 98-107. doi: 10.1016/j.agee.2015.01.005.

Parkhurst, D. and Appelo, C. A. J. (1999). User's Guide to PHREEQC (2nd version) - a computer program for speciation, batch-reaction, one-dimensional transport, and inverse geochemical calculations, Water-Resources Investigations Report 99-4259. Denver, CO: U.S. Geological Survey, p. 4259.

Perrin, A. S., Probst, A. and Probst, J. L. (2008). Impact of nitrogenous fertilizers on carbonate dissolution in small agricultural catchments: implications for weathering CO_2 uptake at regional and global scales, *Geochimica et Cosmochimica Acta* 72(13), 3105-3123. doi: 10.1016/j.gca.2008.04.011.

Philips, S. E., Milnes, A. R. and Foster, R. C. (1987). Calcified filaments-an example of biological influences in the formation of calcrete in South Australia, *Soil Research* 25(4), 405-428.

Pierre, W. H. (1933). Determination of equivalent acidity and basicity of fertilizers, *Industrial and Engineering Chemistry Analytical Edition* 5(4), 229-234.

Pillot, D., Deville, E. and Prinzhofer, A. (2014). Détermination et quantification des carbonates par pyrolyse Rock-Eval, *Oil and Gas Science and Technology - Revue d'IFP Energies Nouvelles* 69(2), 341-349. doi: 10.2516/ogst/2012036.

Plaza-Bonilla, D., Arrúe, J. L., Cantero-Martínez, C., Fanlo, R., Iglesias, A. and Álvaro-Fuentes, J. (2015). Carbon management in dryland agricultural systems. A review, *Agronomy for Sustainable Development* 35(4), 1319-1334. doi: 10.1007/s13593-015-0326-x.

Poeplau, C., Barré, P., Cécillon, L., Baudin, F. and Sigurdsson, B. D. (2019). Changes in the Rock-Eval signature of soil organic carbon upon extreme soil warming and chemical oxidation - a comparison, *Geoderma*. Elsevier 337, 181–190. doi: 10.1016/j.geoderma.2018.09.025.

Qadir, M., Noble, A. D., Schubert, S., Thomas, R. J. and Arslan, A. (2006). Sodicity-induced land degradation and its sustainable management: problems and prospects, *Land Degradation and Development* 17(6), 661–676. doi: 10.1002/ldr.751.

Rasmussen, C., Heckman, K., Wieder, W. R., Keiluweit, M., Lawrence, C. R., Berhe, A. A., Blankinship, J. C., Crow, S. E., Druhan, J. L., Hicks Pries, C. E., Marin-Spiotta, E., Plante, A. F., Schädel, C., Schimel, J. P., Sierra, C. A., Thompson, A. and Wagai, R. (2018). Beyond clay: towards an improved set of variables for predicting soil organic matter content, *Biogeochemistry* 137(3), 297–306. doi: 10.1007/s10533-018-0424-3.

Raza, S., Li, X., Miao, N., Ahmed, M., Liu, Z. and Zhou, J. (2019). Dicyandiamide increased ammonia volatilisation and decreased carbon dioxide emission from calcareous soil during wheat-maize rotation on the Loess Plateau, *Soil Research* 57(7), 767–778. doi: 10.1071/SR18088.

Raza, S., Miao, N., Wang, P., Ju, X., Chen, Z., Zhou, J. and Kuzyakov, Y. (2020). Dramatic loss of inorganic carbon by nitrogen-induced soil acidification in Chinese croplands, *Global Change Biology* 26(6), 3738–3751. doi: 10.1111/gcb.15101.

Raza, S., Kuzyakov, Y. and Zhou, J. (2021). Facts to acidification-induced carbonate losses from Chinese croplands, *Global Change Biology* 27(5), e7–e10. doi: 10.1111/gcb.15478.

Rehman, H. U., Poch, R. M., Scarciglia, F. and Francis, M. L. (2021). A carbon-sink in a sacred forest: Biologically-driven calcite formation in highly weathered soils in Northern Togo (West Africa). *Catena*, 198, 105027. doi: 10.1016/j.catena.2020.105027.

Rowley, M. C., Grand, S. and Verrecchia, É. P. (2018). Calcium-mediated stabilisation of soil organic carbon, *Biogeochemistry* 137(1–2), 27–49. doi: 10.1007/s10533-017-0410-1.

Rumpel, C., Amiraslani, F., Chenu, C., Garcia Cardenas, M., Kaonga, M., Koutika, L. S., Ladha, J., Madari, B., Shirato, Y., Smith, P., Soudi, B., Soussana, J. F., Whitehead, D. and Wollenberg, E. (2020). The 4p1000 initiative: opportunities, limitations and challenges for implementing soil organic carbon sequestration as a sustainable development strategy, *Ambio* 49(1), 350–360. doi: 10.1007/s13280-019-01165-2.

Ryskov, Y. A., Borisov, A. V., Oleinik, S. A., Ryskova, E. A. and Demkin, V. A. (2000). The relationship between lithogenic and pedogenic carbonate fluxes in steppe soils, and regularities of their profile dynamics for the last four millennia. In: Lal, R., et al. (Eds) *Global Change and Pedogenic Carbonates*. Boca Raton, FL: Lewis Publishers, pp. 121–133.

Sadiq, M., Hassan, G., Mehdi, S. M., Hussain, N. and Jamil, M. (2007). Amelioration of saline-sodic soils with tillage implements and sulfuric acid application, *Pedosphere* 17(2), 182–190. doi: 10.1016/S1002-0160(07)60024-1.

Salomons, W. and Mook, W. G. (1976). Isotope geochemistry of carbonate dissolution and reprecipitation in soils, *Soil Science* 1(1), 15–24.

Sanderman, J. (2012). Can management induced changes in the carbonate system drive soil carbon sequestration? A review with particular focus on Australia, *Agriculture, Ecosystems and Environment*. Elsevier B.V. 155, 70–77. doi: 10.1016/j.agee.2012.04.015.

Sarmast, M., Farpoor, M. H., Sarcheshmehpoor, M. and Eghbal, M. K. (2014). Micromorphological and biocalcification effects of *Sporosarcina pasteurii* and

Sporosarcina ureae in sandy soil columns, *Journal of Agricultural Science and Technology* 16(3), 681–693.

Sebag, D., Disnar, J. R., Guillet, B., Di Giovanni, C., Verrecchia, E. P. and Durand, A. (2006). Monitoring organic matter dynamics in soil profiles by "Rock-Eval pyrolysis": bulk characterization and quantification of degradation, *European Journal of Soil Science* 57(3), 344–355. doi: 10.1111/j.1365-2389.2005.00745.x.

Sebag, D., Verrecchia, E. P., Cécillon, L., Adatte, T., Albrecht, R., Aubert, M., Bureau, F., Cailleau, G., Copard, Y., Decaens, T., Disnar, J. -R., Hetényi, M., Nyilas, T. and Trombino, L. (2016). Dynamics of soil organic matter based on new Rock-Eval indices, *Geoderma*. Elsevier B.V. 284, 185–203. doi: 10.1016/j.geoderma.2016.08.025.

Shi, H. J., Wang, X. J., Zhao, Y. J., Xu, M. G., Li, D. W. and Guo, Y. (2017). Relationship between soil inorganic carbon and organic carbon in the wheat-maize cropland of the North China Plain, *Plant and Soil* 418(1–2), 423–436. doi: 10.1007/s11104-017-3310-1.

Soil Survey Staff (S.S.S.) (2014) *Keys to Soil Taxonomy, USDA*. doi: 10.1063/1.1698257.

Suarez, D. L. (2000). Impact of agriculture on CO_2 as affected by changes in inorganic carbon. In: Lal, R., et al. (Eds) *Global Change and Pedogenic Carbonates*. Boca Raton, FL: Lewis Publishers, pp. 257–272.

Syed, S., Buddolla, V. and Lian, B. (2020). Oxalate carbonate pathway–conversion and fixation of soil carbon–a potential scenario for sustainability, *Frontiers in Plant Science* 11(December), 1–7. doi: 10.3389/fpls.2020.591297.

Tooulakou, G., Giannopoulos, A., Nikolopoulos, D., Bresta, P., Dotsika, E., Orkoula, M. G., Kontoyannis, C. G., Fasseas, C., Liakopoulos, G., Klapa, M. I. and Karabourniotis, G. (2016). Alarm photosynthesis: calcium oxalate crystals as an internal CO_2 source in plants, *Plant Physiology* 171(4), 2577–2585. doi: 10.1104/pp.16.00111.

Trost, B., Prochnow, A., Drastig, K., Meyer-Aurich, A., Ellmer, F. and Baumecker, M. (2013). Irrigation, soil organic carbon and N_2O emissions: a review, *Agronomy for Sustainable Development* 33(4), 733–749. doi: 10.1007/s13593-013-0134-0.

Uren, N. C. (2018). Calcium oxalate in soils, its origins and fate-a review, *Soil Research* 56(5), 443–450. doi: 10.1071/SR17244.

Wang, J., Monger, C., Wang, X., Serena, M. and Leinauer, B. (2016). Carbon sequestration in response to grassland-shrubland-turfgrass conversions and a test for carbonate biomineralization in desert soils, New Mexico, USA, *Soil Science Society of America Journal* 80(6), 1591–1603. doi: 10.2136/sssaj2016.03.0061.

Wang, X., Wang, J., Xu, M., Zhang, W., Fan, T. and Zhang, J. (2015). Carbon accumulation in arid croplands of northwest China: pedogenic carbonate exceeding organic carbon, *Scientific Reports*. Nature Publishing Group 5(May), 11439. doi: 10.1038/srep11439.

Wieder, M. and Yaalon, D. H. (1982). Micromorphological fabrics and developmental stages of carbonate nodular forms related to soil characteristics, *Geoderma* 28(3–4), 203–220.

Wong, V. N. L., Dalal, R. C. and Greene, R. S. B. (2009). Carbon dynamics of sodic and saline soils following gypsum and organic material additions: a laboratory incubation, *Applied Soil Ecology* 41(1), 29–40. doi: 10.1016/j.apsoil.2008.08.006.

Wu, H., Guo, Z., Gao, Q. and Peng, C. (2009). Distribution of soil inorganic carbon storage and its changes due to agricultural land use activity in China, *Agriculture, Ecosystems and Environment* 129(4), 413–421. doi: 10.1016/j.agee.2008.10.020.

Wu, L., Wood, Y., Jiang, P., Li, L., Pan, G., Lu, J., Chang, A. C. and Enloe, H. A. (2008). Carbon sequestration and dynamics of two irrigated agricultural soils in California, *Soil Science Society of America Journal* 72(3), 808–814. doi: 10.2136/sssaj2007.0074.

Zamanian, K. and Kuzyakov, Y. (2019). Contribution of soil inorganic carbon to atmospheric CO_2: more important than previously thought, *Global Change Biology* 25(1), e1–e3. doi: 10.1111/gcb.14463.

Zamanian, K., Pustovoytov, K. and Kuzyakov, Y. (2016). Pedogenic carbonates: forms and formation processes, *Earth-Science Reviews*. Elsevier B.V. 157, 1–17. doi: 10.1016/j.earscirev.2016.03.003.

Zamanian, K., Zarebanadkouki, M. and Kuzyakov, Y. (2018). Nitrogen fertilization raises CO_2 efflux from inorganic carbon: a global assessment, *Global Change Biology* 24(7), 2810–2817. doi: 10.1111/gcb.14148.

Zamanian, K., Zhou, J. and Kuzyakov, Y. (2021). Soil carbonates: the unaccounted, irrecoverable carbon source, *Geoderma*. Elsevier B.V. 384, 114817. doi: 10.1016/j.geoderma.2020.114817.

Zhang, L., Liang, Z., Hu, Y., Schmidhalter, U., Zhang, W., Ruan, S. and Chen, X. (2021). Integrated assessment of agronomic, environmental and ecosystem economic benefits of blending use of controlled-release and common urea in wheat production, *Journal of Cleaner Production*. Elsevier Ltd 287, 125572. doi: 10.1016/j.jclepro.2020.125572.

Zhang, Y., Zhang, S., Wang, R., Cai, J., Zhang, Y., Li, H., Huang, S. and Jiang, Y. (2016). Impacts of fertilization practices on pH and the pH buffering capacity of calcareous soil, *Soil Science and Plant Nutrition*. Taylor & Francis 62(5–6), 432–439. doi: 10.1080/00380768.2016.1226685.

Zhao, X., Zhao, C., Stahr, K., Kuzyakov, Y. and Wei, X. (2020). The effect of microorganisms on soil carbonate recrystallization and abiotic CO_2 uptake of soil, *CATENA*. Elsevier 192, 104592. doi: 10.1016/j.catena.2020.104592.

Zornoza, R., Rosales, R. M., Acosta, J. A., de la Rosa, J. M., Arcenegui, V., Faz, Á and Pérez-Pastor, A. (2016). Efficient irrigation management can contribute to reduce soil CO_2emissions in agriculture, *Geoderma*. Elsevier B.V. 263, 70–77. doi: 10.1016/j.geoderma.2015.09.003.

Zúñiga, R., Gutiérrez Castorena, E. V., Gutiérrez Castorena, M. D. C., Olivares Sáenz, E., Méndez Gallegos, SdJ., Carranza de la Rosa, R. and Vázquez Alvarado, R. E. (2018). Characterization $CaCO_3$ and CaC_2O_4 with microphotographic analysis in *Opuntia ficus-indica* (L.) Miller, *Revista Mexicana de Ciencias Agrícolas* 9(7), 1524–1531.

Chapter 24

Management of soil carbon sequestration in urban areas

C. Rumpel, CNRS, Sorbonne University, Institute of Ecology and Environmental Sciences Paris, France; F. Amiraslani, Ulster University, UK; J.-C. Lata, Sorbonne University, Institute of Ecology and Environmental Sciences Paris, France; C. Marques-dos-Santos Cordovil, Universidade de Lisboa, Portugal; E. Nartey, University of Ghana, Ghana; C. Staudhammer, The University of Alabama, USA; and E. Yeboah, CSIR – Soil Research Institute, Ghana

1 Introduction

Urban areas are characterised by a high population density and built infrastructure. According to a global, people-based definition, urban centres are places with at least 1500 inhabitants per square kilometre and a total population of at least 50 000 people (EC, 2022). For the first time in history, in 2007, more people lived in urban than in rural areas with the urban population expected to reach 60.4% of the global population by 2030 (UN-Habitat, 2020). Urban soils are vital components of urban ecosystems (Li et al., 2018), although urbanisation also leads to soil loss following the construction of buildings and (grey) infrastructures.

Given continued urban growth, urban soils are becoming increasingly important for the ecosystem services they can deliver (O'Riordan et al., 2021). These range from potential roles in improving air quality, water quality and management, recreational services, through to organic carbon sequestration to combat climate change (Jansson, 2013). Promoting ecosystem services includes a diverse range of practices to improve soil structure and organic

http://dx.doi.org/10.19103/AS.2022.0106.23

matter content, for example, through bioremediation of degraded soils, adding compost or biosolids or planting vegetation (Logsdon et al., 2017). Enriching urban soils with organic matter requires putting the focus on recycling waste materials within urban areas under the framework of a circular economy rather than relying on importing organic matter from outside. Indeed, it has recently been suggested that locally available resources should be used to restore soils and to revitalize metropolitan areas for improving the overall quality of life for a rapidly growing urban population (Kumar and Hundal, 2016).

Urban soils may have substantial soil organic carbon (SOC) storage potential per unit area (Pouyat et al., 2006). Data from 116 cities worldwide showed that the total carbon (C) content of urban soils was 1.5–3 times higher and that SOC storage occurred at greater depth than in some rural mineral soils, particularly in residential areas (Mazurek et al., 2016; Vasenev and Kuzyakov, 2018). Urban soils also contain substantial amounts of inorganic C and have been identified as global hotspots in terms of soil C storage, partly because they may have the potential to gain or lose C rapidly (Riddle et al., 2021). Some types of urban structures and carbon types, which may be found in urban environments, are shown in Fig. 1.

This chapter aims to present the urban soil and carbon types and the management of SOC sequestration in different urban infrastructures. We will analyse SOC sequestration in man-made and semi-natural (infra-) structures. In the context of a circular economy, this chapter also addresses the use of recycling strategies of urban waste materials to foster SOC sequestration.

Urban infrastructures

Grey infrastructures

- Water retention systems
- Roads
- Piplines

Green infrastructures

- Parks, gardens, lawns
- Urban forests
- Community gardens
- Green roofs
- Bioretention systems

Carbon types

- Inorganic C (SIC)
- Organic C (SOC)
- Plant litter
- Compost
- Black C
- Plastic
- Organic contaminants

Figure 1 Infrastructures and carbon types, which may be found in an urban context.

2 Characteristics of soils in urban environments

Urban soils created as part of the process of urbanisation are strongly affected by human activities and are typically heterogeneous in structure, composition and properties (Rossiter, 2007; Lorenz and Lal, 2015). They were defined as a continuum depending on the degree of anthropogenic alterations (Pouyart and Treammell, 2019) and are included in soil classification systems as Technosols and Anthrosols (WRB, 2014). Human activity influences urban soils through material movement and redistribution, transport and deposition, introduction of man-made materials and accumulation of toxic compounds (Burghardt et al., 2015). Urban soils include a wide range of different soil types. Relict 'native' soil is presently protected from urban development, in green, semi-natural spaces such as parks and urban forests. These areas may contain a mix of local and imported soils designed to support trees and other plants. Additionally, urban soils may be composed of highly heterogeneous combinations of materials resulting from building or industrial activity. These may be characterized by a high content of synthetic materials such as glass, metal, plastic and so on, frequently with high levels of chemical contamination (Craul, 1985; Li et al., 2018; Sager, 2020; O'Riordan et al., 2021). The latter types of soil are usually characterised by disturbed structure, high compaction, reduced pore space, limited moisture, higher average temperature and pH, high levels of toxic pollutants, low organic matter content and limited vegetation cover (Craul, 1985). In general, urban soilscape is complex and influenced by past and current human interventions, pre-urban geomorphology and hydrography (Delbecque et al., 2022). A recent study concluded that urban soils are global hotspots of soil C sequestration due to high stocks and accumulation rates (Vasenev and Kuzyakov, 2018). Different forms of C with contrasting stability and functions are present in urban soils. They include SOC derived from plant litter, soil inorganic carbon (SIC), black carbon derived from combustion processes and xenobiotic carbon - such as plastics - polycyclic aromatic hydrocarbons and other organic pollutants (Fig. 1). Management of soil C in these systems is thus complex but relates in many cases to best management practices recommended for soils in rural areas (refer to Chapters 15–23 of this book).

3 Fostering soil organic carbon sequestration in urban infrastructures

In urban areas, two types of infrastructures are present (Fig. 1): (1) grey infrastructures, which are engineered structures that use concrete and steel and generally seal floors; and (2) green infrastructures, which comprise natural, semi-natural and artificial areas for vegetation with and without trees (such as parks, gardens and road-side verges) and also man-made systems such as bioretention systems and green roofs (Tzoulas et al., 2007). Management activities to foster SOC sequestration need to be specific in each of these different infrastructures.

3.1 Grey infrastructures

Construction of buildings and infrastructure such as roads often leads to sealing of the soil surface with impervious materials such as concrete or tarmac (Scalenghe and Marsan, 2009). Such sealing largely influences biogeochemical cycling and soils under impervious surfaces are often depleted in SOC compared to open soils (Raciti et al., 2012; Wei et al., 2014). Soil organic carbon losses may occur following decomposition of organic matter beneath the impervious surface or through topsoil removal and erosion during the construction process (Wei et al., 2014). Sealing disrupts the complex interactions between the atmosphere, plants and soil, reducing the capacity of the soil to deliver ecosystem services such as reducing runoff, filtering pollutants and sequestering SOC.

Therefore, soil sealing should be avoided and wherever possible, semi-pervious materials such as gravel, stone aggregates, porous asphalt or wood chip should be used to allow for water infiltration and potential plant growth (Scalenghe and Marsan, 2009). It has been shown that de-sealing can favour SOC storage and other soil ecosystem services by rapid recovery of physical, chemical and biological properties following the establishment of pioneer vegetation (Renella, 2020).

3.2 Gardens, parks, street trees and lawns

Urban parks, gardens and lawns are often established on natural soils, which had been present before urbanisation or had been imported from rural areas. These green infrastructures are usually managed by regular cutting (and disposal of cuttings), weed removal (often using herbicides), irrigation and application of synthetic fertilisers. These intensive management practices have environmental costs and trade-offs for SOC sequestration (Chapter 6 of this book). For example, an evaluation of SOC sequestration and greenhouse gas (GHG) emissions from turfgrass in athletic fields and ornamental lawns in urban parks found that GHG emissions through fuel use from machinery, fertilisation and irrigation outweighed SOC sequestration (Townsend-Small and Czimzik, 2010). In addition, taking organic matter out of the system, following management activities such as mowing and removal of grass clippings and/or cleaning measures such as tree leave removal in autumn, was also found to reduce SOC contents (Yoon et al., 2016).

In order to manage SOC in parks, gardens and lawns sustainably, semi-natural gardens and lawns requiring less intensive management should be established. High tree density has been found to be related to SOC storage (Mexia et al., 2018). More sustainable practices include retention of plant litter such as tree leaves and grass cuttings to provide continuous soil cover, improving moisture retention and nutrient cycling which likely also reduces the need for irrigation and fertilisation. Soil organic carbon storage in park soils may be enhanced by appropriate management of plant litter. Where grass clippings and tree leaves are not contaminated (e.g. by pollutants from traffic emissions), organic waste treatments such as composting can be used. Increased SOC storage was, for example, achieved in the Seoul Forest Park, where litter was added back to soils after composting (Bae and Ryu, 2015).

3.3 Urban forests

Urban forests provide a variety of essential ecosystem services, including decreasing air, water and noise pollution, mitigating flood risk and providing recreational areas (Escobedo et al., 2011; Roy et al., 2012). Organic carbon sequestration is a climate change mitigation service of urban forests in addition to their function of reducing the urban heat island effect (Kleerekoper et al., 2012). While some studies have shown that total SOC storage in urban forests is similar to those of rural ones (Pouyat et al., 2002), other studies found that SOC sequestration is depending on tree density (Lv et al., 2016). Moreover, more significant proportions of recalcitrant C have been found in urban forest soils than in rural ones due to poorer quality leaf litter, enhanced mineralisation of readily available SOC due to non-native earthworms and higher soil temperatures (Groffman et al., 1995). According to Pouyat et al. (2002), urban forests could potentially sequester and store more SOC than rural forests.

While relatively few studies exist that specifically address soil C management in urban forests, their higher SOC stocks compared to those of other urban sites may partly be explained by their age and a management legacy of wastewater irrigation, liming and charcoal application (Foti et al., 2021). Long-term studies of forest management strategies of rural forests recommend increasing productivity, for example, through afforestation and planting of fast-growing tree species, which have immediate effects on SOC sequestration (Jandl et al., 2007, Chapter 19 of this book); however, in an urban context, environmental and political constraints may limit such practices. For example, the presence of construction materials leading to high carbonate contents, limited water storage capacity and/or presence of pollutants in urban soils may limit the species able to grow under such conditions. Sustainable forest management after their establishment should include minimal disturbance of stand structure and soils to promote SOC retention (Jandl et al., 2007, Chapter 19 of this book).

Application of compost and/or mulching could increase urban forest SOC storage (Brown et al., 2012; Beesley, 2012), as could biochar; however, it is not known how management practices might impact inorganic C from carbonate reactions (Lorenz and Lal, 2015).

3.4 Green roofs

Green roofs, green walls and vertical gardens have been introduced into building environments because of their positive effects such as insulation, air quality, soundproofing and as a way of improving the ecological footprint of cities (Getter et al., 2009).

They can also be used to sequester SOC. It has been found that green roofs can potentially store about 375 g m^2 SOC with substrate contributing to about one-third of the storage (Getter et al., 2009). This highlights the importance of the type and depth of the growth substrate as well as the proportion of inorganic and organic materials used when generating the substrate (Lata et al., 2018). A 10–20% organic component seems optimal for plant growth (Ondono et al., 2016). Local waste materials such as compost and municipal sewage sludge have been used for green roof construction in the UK (Molineux et al., 2009) and China (Luo et al., 2015). The addition of sewage sludge to the substrate increased SOC sequestration up to 13 kg SOC per m^2 (Luo et al., 2015).

It is important to take the impact of management intensity into account. Extensive green roofs are light weighted and typically support herbal vegetation on shallow substrates requiring minimal maintenance. Intensive green roofs are heavier with a deeper layer of growing substrate to support a wider variety of plant types, including shrubs and trees (Besir and Cuce, 2017). Soil organic carbon storage may thus be higher in intensively managed green roofs but may also require more external inputs in terms of irrigation, fertilisation and overall maintenance, thus also increasing the trade offs of SOC sequestration (Chapter 6 of this book).

Whilst green roofs have been widely used in temperate regions such as Northern Europe, their use in arid regions (such as the Mediterranean) is more challenging without potentially intensive irrigation (with its associated environmental costs). One potential solution is the use biocrusts, desert-based communities composed of a complex mosaic of cyanobacteria, green algae, lichens, mosses, microfungi and other bacteria. These are well adapted to survive drought conditions with minimal substrates and may be suited for SOC sequestration on green roofs in drier climates (Paço et al., 2014).

3.5 Bioretention systems

Bioretention systems are used to channel, retain and purify rainwater. They use vegetation to moderate water flow and act as a filter to improve water quality. Cities such as New York have invested significantly in these systems

(The City of New York, 2010; Joyner et al., 2019). The type and amount of organic materials used to construct bioretention systems such as bioswales affect their water infiltration, filtration properties and also SOC sequestration potential. Biochar, for example, has a high adsorption potential for contaminants and may sequester SOC in the long term (Biswal et al., 2022). A potential problem to avoid in their construction and maintenance is nitrate leaching from use of nitrogen-rich vegetation supplemented by fertilisers (Shetty et al., 2018). Despite the role bioretention systems could play in SOC sequestration, their SOC sequestration potentials have not yet been studied in detail (Gill et al., 2017).

4 Organic waste recycling to foster soil organic carbon storage in urban soils in the context of a circular urban economy

Organic matter is employed in many urban soils for restoration, construction purposes and also green infrastructure development. In particular, Technosols are man-made engineered structures with technic material. They contain artefacts or extracted unconsolidated material and bedrocks, and may have a sealing layer or a geo-membrane (Burghardt et al., 2015). Construction and demolition waste may be re-used together with organic waste for their construction. Technosols are an alternative to importing soil for restoring degraded urban areas (Deeb et al., 2017; Barredo et al., 2020; Fabbri et al., 2021). They have been used, for example, in urban community gardens in New York City (Egendorf et al., 2018). Moreover, Technosols are used for green infrastructure construction (Deeb et al., 2020) and also urban farming is often based on specific engineered systems, such as lasagne gardening (Lanza, 1998), consisting of different layers of materials. In many cities, rooftop farming is nowadays practised, and management of organic matter in soils or engineered growth substrates used for food production is thus of primary importance in urban agriculture. Organic materials used as growth substrates in urban agriculture may include bark, composted materials including green (yard) wastes, municipal solid wastes and even sewage sludge (Carlile et al., 2015).

The type of organic material determines the stability of SOC as well as other soil properties. High SOC concentrations lead to soil decompaction as well as increased drought resistance through greater water retention and rooting depth (Robin et al., 2018) but may also lead to groundwater contamination due to N and P leaching and production of greenhouse gases. In order to optimise the properties of Technosols, the presence of worms and plants is required (Deeb et al., 2017). It has been shown that the earthworm species *Dendrobaena veneta* was able to improve properties such as plant available water content in Technosols (Ulrich et al., 2021). Earthworm activity can also promote SOC stabilisation (Le Mer et al., 2020). Optimising SOC sequestration in Technosols requires the right combination of organic, inorganic and biological components. Generally, the addition of more than 30% of organic matter should be avoided (Deeb et al., 2020). It is also important to consider the SIC content of construction material as active dissolution and redistribution processes of calcium carbonate are frequently observed in Technosols under humid climate conditions (Prokof'eva et al., 2021).

Materials for Technosol construction should ideally originate from inside the urban setting. Organic materials should be transformed by recycling organic wastes from urban areas into bio-fertilisers and soil amendments to create a circular local economy with minimal environmental impact (Tedesco et al., 2017; Moinard et al., 2021, Fig. 2).

This is all the more important given the scale of organic urban waste and the limited access of urban farmers to resources such as organic matter, fertilisers and water. Therefore, food production as well as soil restoration and SOC sequestration in urban agriculture may be favoured by recycling of bio-solids and bio-liquids produced by human settlements (Lal, 2017). Policy

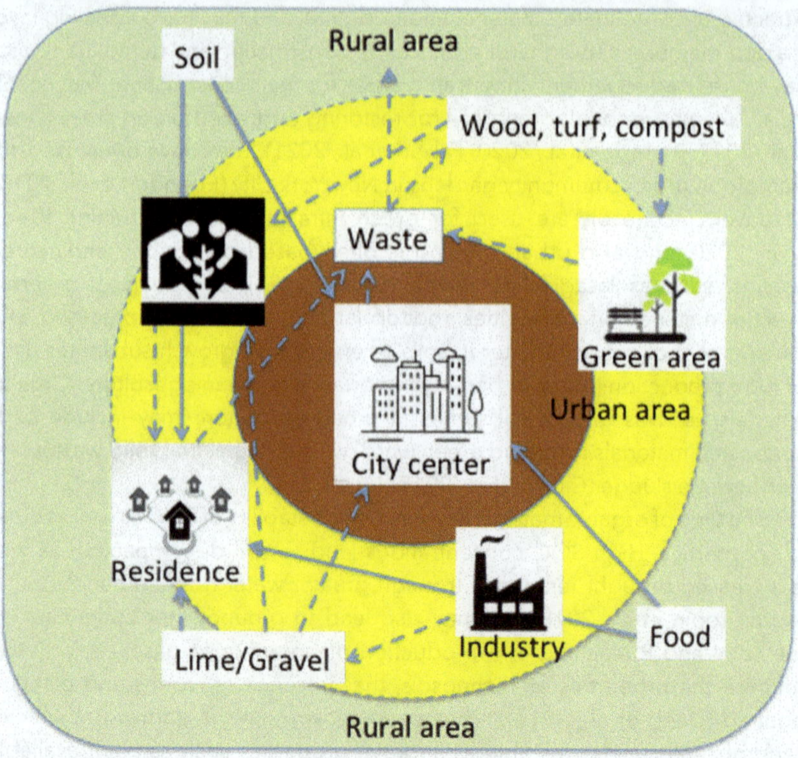

Figure 2 Material flows in an urban context (inspired by Vasenev and Kuzyakov, 2018).

support and strategies for urban agricultural development are often missing (Crush et al., 2012) although they are extremely important in urban areas, where SOC sequestration should be integrated to reduce their C-footprint.

Organic urban waste can be categorised in various ways such as food waste, garden waste and animal waste. The volume and type of such waste vary widely. Organic waste forms more than 50% of the total municipal solid waste generated in low and middle-income countries of West Africa (Henry et al., 2006; Couth and Trois, 2010) and 30% of municipal solid waste in developed countries (Hoornweg et al., 2013). The European Union, for example, generates around 88 million tonnes of food waste each year, coming mainly from homes and food retail outlets (Stenmarck et al., 2016), while around 17% of total global food production amounting to 931 million tonnes may be wasted (UNEP, 2021) and thus available for recycling.

Recycling this waste is a major challenge for municipal authorities (Pollans et al., 2017; Chen et al., 2020). Some European countries have schemes to separately collect and sort organic municipal waste, for example, Germany, Austria, Belgium and France. Some municipalities go further and encourage citizens to recycle organic waste themselves. The City of Paris, for example, distributes composting units and provides guidance on their use. However, in some parts of the world, such as Asia and the Pacific, a large amount of municipal waste still goes to landfills (Horrocks et al., 2016; Jara-Samaniego et al., 2017). This has significant negative environmental impacts in terms of odours, methane emissions, soil and groundwater contamination (Xiaoli et al., 2007; Prechthai et al., 2008) and also is a missed opportunity to increase SOC storage.

In the US, for example, recycling of food waste has doubled since 2010 but is still at a very low level (5%). A pioneer in establishing a progressive food waste policy was the State of California, which introduced the Integrated Waste Management Act in 1989. This law has been put into practice by the City of San Francisco, which successfully reduced waste going to landfills by 50% by 2000 and is currently aiming to achieve zero food waste (AzCentral, 2017). Other cities such as New York have announced new rules requiring restaurants and grocery stores to recycle food waste (Recycling Today, 2018). Another successful example is the green exchange programme (Cambio Verde) developed by the City of Curitiba (Brazil) in 1989, in which organic wastes are collected by people living in slums and exchanged for surplus food produced on smallholder farms. Korea has developed a zero food waste system, which allows onsite treatment of organic wastes in apartment complexes (Oh and Lee, 2018).

There are a range of techniques to transform organic waste into organic fertilisers (Chapter 9 of this book). Briefly, these include anaerobic digestion, composting, vermicomposting (using earthworms) and thermal treatment to create biochar. Composting has been widely used in many low-income countries in West Africa, Asia and South America to process organic waste

(Zurbrügg et al., 2005; Hoornweg and Bhada-Tata, 2012; Jara-Samaniego et al., 2017). Advanced composting techniques can now transform organic waste into bio-fertiliser in as little as 12 days (Wei et al., 2021). A key challenge is removing chemical and biological contaminants from mixed wastes (Farrell and Jones, 2009). Suitable composting strategies include separating the waste into pools with different compositions and co-composting with bulking agents, mineral, organic or microbiological compounds (Barthod et al., 2018). Vermicomposting has also been found to reduce contaminant levels in organic urban waste (Singh et al., 2011). Co-composting and co-vermicomposting also shorten processing times, help to stabilise mineral content (especially available N) and promote SOC sequestration (Vidal et al., 2020). Soil organic carbon residence time in the soil can be increased by mixing with recalcitrant compounds such as biochar before field application (Ngo et al., 2016).

5 Conclusion

In urban areas, the nature and distribution of soil C are highly heterogeneous, ranging from organic matter in soils of parks or lawns to carbon-rich artefacts and contaminants in man-made structures. Similar to rural areas, SOC sequestration in urban soils is influenced by land-use and management practices. Extensive management of urban infrastructures may counteract trade offs of SOC sequestration in form of greenhouse gas emissions, water use and pollution. Although many urban soils may contain substantial amounts of C, their SOC sequestration potential has been poorly quantified. Improving SOC sequestration may be necessary to secure the broad range of ecosystem services that urban soils can deliver. More generally, the challenge of fostering SOC sequestration in urban soils needs to be seen in the wider context of creating a more sustainable circular urban economy in which organic materials are recycled to reduce waste and at the same time improve the quality of urban soils. This requires skills in planning, governance, communication and education to develop and implement management practices and technologies aiming to protect and foster SOC storage without generating trade-offs.

6 Where to look for further information

FAO and ITPS. 2021. Recarbonizing global soils: A technical manual of recommended management practices. Volume 5 and 6. Rome, FAO. https://doi.org/10.4060/cb6606en.

Levin/Kim/Morel/Burghardt/Charzynski/Shaw (Eds.) 2017. Soils within cities. Global approaches to their sustainable management - composition, properties, and functions of soils of the urban environment. Schweizerbart Science Publishers, 253 pages.

7 References

AzCentral. (2017). How San Francisco's mandatory composting laws turn food waste into profit. Available at: https://eu.azcentral.com/story/entertainment/dining/food-waste/2017/08/03/san-francisco-mandatory-composting-law-turns-food-waste-money/440879001/ (accessed 25 October 2021).

Bae, J. and Ryu, Y. (2015). Land use and land cover changes explain spatial and temporal variations of the soil organic carbon stocks in a constructed urban park. *Landscape and Urban Planning* 136, 57-67.

Barredo, O., Vilela, J., Garbisu, C., Besga, G., Alkorta, I. and Epelde, L. (2020). Technosols made from urban and industrial wastes are a good option for the reclamation of abandoned city plots. *Geoderma* 377, 114563.

Barthod, J., Rumpel, C. and Dignac, M.-F. (2018). Composting with additives to improve organic amendments: A review. *Agronomy for Sustainable Development* 38(2), 17.

Beesley, L. (2012). Carbon storage and fluxes in existing and newly created urban soils. *Journal of Environmental Management* 104, 158–165.

Besir, A. B. and Cuce, E. (2017). Green roofs and facades: A comprehensive review. *Renewable and Sustainable Energy Reviews* 82, 915–939.

Biswal, B. K., Vijayaraghavan, K., Tsen-Tieng, D. L. and Balasubramanian, R. (2022). Biochar-based bioretention systems for removal of chemical and microbial pollutants from stormwater: A critical review. *Journal of Hazardous Materials* 422, 126886.

Brown, S., Miltner, E. and Cogger, C. (2012). Carbon sequestration potential in urban soils. In: Lal, R. and Augustin, B. (Eds). *Carbon Sequestration in Urban Ecosystems.* Springer: New York, 173–196.

Burghardt, W., More, J. L. and Zhang, G.-L. (2015). Development of the soil research about urban, industrial, traffic, mining and military areas (SUITMA). *Soil Science and Plant Nutrition* 61(Suppl 1), 3–21.

Carlile, W. R., Cattivello, C. and Zaccheo, P. (2015). Organic growing media: Constituents and properties. *Vadoze Zone Journal*. doi: 10.2136/v zj2014.09.0125.

Chen, T., Zhang, S. and Yua, Z. (2020). Adoption of solid organic waste composting products: A critical review. *Journal of Cleaner Production* 272, 122712.

Couth, R. and Trois, C. (2010). Carbon emissions reduction strategies in Africa from improved waste management: A review. *Waste Management* 30(11), 2336–2346.

Craul, P. J. (1985). A description of urban soils and their desired. *Journal of Arboriculture* 11(11), 330–339.

Crush, J., Frayne, B. and Pendletn, W. (2012). The crisis of food insecurity in African cities. *Journal of Hunger and Environmental Nutrition* 7(2–3), 271–292.

Deeb, M., Groffman, P. M., Blouin, M., Egendorf, S. P., Vergnes, A., Vasenev, V., Cao, D. L., Walsh, D., Morin, T. and Séré, G. (2020). Using constructed soils for green infrastructure – Challenges and limitations. *Soil* 6, 413–434.

Deeb, M., Dejardins, T., Podwojewski, P., Pando, A., Blouin, M. and Lerch, T. Z. (2017). Interactive effects of compost, plants and earthworms on the aggregations of constructed Technosols. *Geoderma* 305, 305–313.

Delbecque, N., Dondeyne, S., Gelaude, F., Mouazen, A. M., Vermeir, P. and Verdoodt, A. (2022). Urban soil properties, land use, time since urbanization and pre-urban geomorphology. *Geoderma* 413, 115719.

EC. (2022). European Commission urban data platform plus. Available at: https://urban.jrc.ec.europa.eu/thefutureofcities/what-is-a-city#the-chapter (accessed 14 April 2022).

Egendorf, S. P., Cheng, Z., Deeb, M., Flores, V., Paltseva, M., Walsh, C., Groffman, P.and Mielke, H. W. (2018). Constructed soils for mitigating lead (Pb) exposure and promoting urban community gardening: The New York City Clean Soil Bank pilot study. *Landscapt and Urban Planning* 175, 164–194.

Escobedo, F. J., Kroeger, T. and Wagner, J. E. (2011). Urban forests and pollution mitigation: Analyzing ecosystem services and disservices. *Environmental Pollution* 159(8–9), 2078–2087.

Fabbri, D., Pizzol, R., Calza, P., Malandrino, M., Gaggero, E., Padoan, E. and Ajmone-Marsan, F. (2021). Constructed technosols: A strategy toward a circular economy. *Applied Sciences (Switzerland)* 11(8), 3432. doi: 10.3390/app11083432.

Farrell, M. and Jones, D. L. (2009). Critical evaluation of municipal solid waste composting and potential compost markets. *Bioresource Technology* 100(19), 4301–4310.

Foti, L., Barot, S., Ginoux, J., Grimaldi, M., Lata, J.-C., Lerch, T. Z., Nold, F., Nunan, N., Raynaud, X., Abbadie, L. and Dubs, F. (2021). Topsoil characteristics of forests and lawns along an urban-rural gradient in the Paris region (France). *Soil Use and Management* 37(4), 749–761.

Getter, K. L., Rowe, D. B., Robertson, G. P., Cregg, B. M. and Andresen, J. A. (2009). Carbon sequestration potential of extensive green roofs. *Environmental Science and Technology* 43(19), 7564–7570.

Gill, A. S., Lee, A.and McGuire, K. L. (2017). Phylogenetic and functional diversity of total (DNA) and expressed (RNA) bacterial communities in urban green infrastructure bioswale soils. *Applied and Environmental Microbiology* 83(16), e00287-17.

Groffman, P. M., Pouyat, P. V., McDonnell, M. J., Pickett, S. T. A., Zipperer, W. C. (1995), Carbon pools and trace gas fluxes in urban forest soils. In: Lat, R. (Ed.). *Advances in Soil Science: Soil Management and the Greenhouse Effect*. Boca Raton, FL: CRC Press.

Henry, R. K., Yongsheng, Z. and Jun, D. (2006). Municipal solid waste management challenges in developing countries - Kenyan case study. *Waste Management* 26(1), 92–100.

Hoornweg, D. and Bhada-Tata, P. (2012). *What a Waste: A Global Review of Solid Waste Management*. Urban Development Series, Knowledge Papers 15.

Hoornweg, D., Bhada-Tata, P. and Kennedy, C. (2013). Environment: Waste production must peak this century. *Nature* 502(7473), 615–617.

Horrocks, A., Curtin, D., Tregurtha, C. and Meenken, E. (2016). Municipal compost as a nutrient source for organic crop production in New Zealand. *Agronomy* 6(2), 35.

Jandl, R., Lindner, M., Vesterdal, L., Bauwens, B., Baritz, R., Hagedorn, F., Johnson, D. W., Minkkinen, K. and Byrne, K. A. (2007). How strongly can forest management influence soil carbon sequestration? *Geoderma* 137(3–4), 253–268.

Jansson, Å. (2013). Reaching for a sustainable, resilient urban future using the lens of ecosystem services. *Ecological Economics* 86, 285–291.

Jara-Samaniego, J., Pérez-Murcia, M. D., Bustamante, M. A., Paredes, C., Pérez-Espinosa, A., Gavilanes-Terán, I., López, M., Marhuenda-Egea, F. C., Brito, H. and Moral, R. (2017). Development of organic fertilizers from food market waste and urban gardening by composting in Ecuador. *PLoS ONE* 12(7), e0181621. doi: 10.1371/journal.pone.0181621.

Joyner, J. L., Kerwin, J., Deeb, M., Lozefski, G., Prithiviraj, B., Paltseva, A., McLaughlin, J., Groffman, P., Cheng, Z. and Muth, T. R. (2019). Green infrastructure design influences communities of urban soil bacteria. *Frontiers in Microbiology* 10, 982.

Kleerekoper, L., van Esch, M. and Salcedo, T. B. (2012). How to make a city climate-proof, addressing the urban heat island effect. *Resources, Conservation and Recycling* 64, 30–38.

Kumar, R. and Hundal, L. (2016). Soil in the city: Sustainably improving urban soils. *Journal of Environment Quality* 45(1), 208.

Lal, R. (2017). Urban agriculture and food security. In: Levin, M. J. (Ed.). *Soils Within Cities, Catena Soil Sciences* (vol. 177), Catena-Schweizerbart, Stuttgart.

Lanza, P. (1998). *Lasagna Gardening*. Emmaus: Rodale Press Inc.

Lata, J.-C., Dusza, Y., Abbadie, L., Barot, S., Carmignac, D., Gendreau, E., Kraepiel, Y., Mériguet, J., Motard, E. and Raynaud, X. (2018). Role of substrate properties in the provision of multifunctional green roof ecosystem services. *Applied Soil Ecology* 123, 464–468.

Le Mer, G., Barthod, J., Dignac, M.-F., Barré, P., Baudin, F. and Rumpel, C. (2020). Inferring earthworms' impact on the stability of organo-mineral associations by Rock-Eval pyrolysis and 13C NMR spectroscopy. *Organic Geochemistry* 144, 104016.

Li, G., Sun, G. -X., Ren, Y., Luo, X. -S. and Zhu, Y. -G. (2018). Urban soil and human health: A review. *European Journal of Soil Science* 69(1), 196–215. doi: 10.1111/ejss.12518.

Logsdon, S. D., Sauer, P. A. and Shipitalo, M. J. (2017). Compost improves urban soil and water quality. *Journal of Water Resource and Protection* 9(4), 345–357. doi: 10.4236/jwarp.2017.94023.

Lorenz, K. and Lal, R. (2015). Managing soil carbon stocks to enhance the resilience of urban ecosystems. *Carbon Management* 6(1–2), 35–50.

Luo, H., Liu, X., Anderson, B. C., Zhang, K., Li, X., Huang, B., Li, M., Mo, Y., Fan, L., Shen, Q., Chen, F. and Jiang, M. (2015). Carbon sequestration potential of green roofs using mixed-sewage-sludge substrate in Chengdu World Modern Garden City. *Ecological Indicators* 49, 247–259.

Lv, H., Wang, W., He, X., Xiao, L., Zhou, W. and Zhang, B. (2016). Quantifying tree and soil carbon stocks in a temperate urban forest in Northeast China. *Forests* 7(12), 200–218.

Mazurek, R., Kowalska, J., Gasiorek, M. and Setlak, M. (2016). Micromorphological and physico-chemical analyses of cultural layers in the urban soil of a medieval city–A case study from Krakow, Poland. *Catena* 141, 73–84.

Mexia, T., Vieira, J., Príncipe, A., Anjos, A., Silva, P., Lopes, N., Freitas, C., Santos-Reis, M., Correia, O., Branquinho, C. and Pinho, P. (2018). Ecosystem services: Urban parks under a magnifying glass. *Environmental Research* 160, 469–478.

Moinard, V., Lavavasseur, F. and Houot, S. (2021). Current and potential recycling of exogenous organic matter as fertilizers and amendments in a French peri-urban territory. *Resources, Conservation and Recycling* 169, 105523.

Molineux, C. J., Fentiman, C. H. and Gange, A. C. (2009). Characterising alternative recycled waste materials for use as green roof growing media in the U.K. *Ecological Engineering* 35(10), 1507–1513.

New York City. (2010). *Green Infrastructure Program*. http://www.nyc.gov/html/dep/html/stormwater/using_green_infra_to_manage_stormwater.shtml.

Ngo, P. T., Rumpel, C., Janeaud, J. L., Dang, D.-K., Doan, T. T. and Jouquet, P. (2016). Mixing of biochar with organic amendments reduces carbon removal after field exposure under tropical conditions. *Ecological Engineering* 91, 378–380.

Oh, J. and Lee, H. (2018). Exploring a zero food waste system for sustainable residential buildings in urban areas. *Environmental Engineering Research* 23(1), 46–53.

Ondono, S., Martinez-Sanchez, J. J. and Moreno, J. L. (2016). The composition and depth of green roof substrates affect the growth of Silene vulgaris and Lagurus ovatus species and the C and N sequestration under two irrigation conditions. *Journal of Environmental Management* 166, 330–340.

O'Riordan, R., Davies, J., Stevens, C., Quinton, J. N. and Boyko, C. (2021). The ecosystem services of urban soils: A review. *Geoderma* 395, 115076.

Paço, T. A., Cameira, M. R., Branquinho, C., Cruz de Carvalho, R., Luís, L., Espírito-Santo, D., Valente, F., Brandão, C., Soares, A. L., Anico, A., Abreu, F. and Pereira, L. S. (2014). Innovative green roofs for southern Europe: Biocrusts and native species with low water use. In: *Proceedings of the 40th IAHS World Congress on Housing–"Sustainable Housing Construction,"* Funchal, Portugal, 16–19 December 2014.

Pollans, L. B., Krones, J. S. and Ben-Joseph, E. (2017). Patterns in municipal food scrap programming in mid-sized U.S. cities. *Resources, Conservation and Recycling* 125, 308–314.

Pouyat, R., Groffman, P., Yesilonis, I. and Hernandez, L. (2002). Soil carbon pools and fluxes in urban ecosystems. *Environmental Pollution* 116 (Suppl 1), S107–S118.

Pouyat, R. V. and Trammell, T. L. E. (2019). Climate change and urban forest soils. *Developments in Soil Science* 36, 189–211.

Pouyat, R. V., Yesilonis, I. D. and Nowak, D. J. (2006). Carbon storage by urban soils in the United States. *Journal of Environmental Quality* 35(4), 1566–1575.

Prechthai, T., Parkpian, P. and Visvanathan, C. (2008). Assessment of heavy metal contamination and its mobilization from municipal solid waste open dumping site. *Journal of Hazardous Materials* 156(1–3), 86–94.

Prokof'eva, T., Shishkov, V. and Kiriushin, A. (2021). Calcium carbonate accumulations in Technosols of Moscow city. *Journal of Soils and Sediments* 21(5), 2049–2058.

Raciti, S. M., Hutyra, L. R., Rao, P. and Finzi, A. C. (2012). Inconsistent definitions of "urban" result in different conclusions about the size of urban carbon and nitrogen stocks. *Ecological Applications* 22(3), 1015–1035.

Recycling Today. (2018). *New York Businesses Now Required to Put Food Waste to Beneficial Use*. http://www.recyclingtoday.com/article/maryland-introduces-slow-down-to-get-around-law/.

Renella, G. (2020). Evolution of physico-chemical properties, microbial biomass and microbial activity of an urban soil after de-sealing. *Agriculture* 10(12), 596.

Riddle, R. L., Shaw, R. K., Cheng, Z., Tucker, R., Ryan, K. A., Vargas, O. M. and Rumpel, C. (2021). Technosols and urban soils. In: FAO and ITPS (Eds). *Recarbonizing Global Soils: A Technical Manual of Recommended Management Practices* (vols. 5 and 6). Rome: Food and Agriculture Organization.

Robin, P., Morel, C., Vial, F., Landrain, B., Toudic, A., Li, Y. and Akkal-Corfini, N. (2018). Effect of three types of exogenous organic carbon on soil organic matter and physical properties of a sandy Technosol. *Sustainability* 10(4), 1146.

Rossiter, D. G. (2007). Classification of urban and industrial soils in the World Reference Base for Soil Resources. *Journal of Soils and Sediments* 7(2), 96–100.

Roy, S., Byrne, J. and Pickering, C. (2012). A systematic quantitative review of urban tree benefits, costs, and assessment methods across cities in different climatic zones. *Urban Forestry and Urban Greening* 11(4), 351-363.

Sager, M. (2020). Urban soils and road dust–Civilization effects and metal pollution–A review. *Environments* 7(11), 1-65. doi: 10.3390/environments7110098.

Scalenghe, R. and Marsan, F. A. (2009). The anthropogenic sealing of soils in urban areas. *Landscape and Urban Planning* 90(1-2), 1-10.

Shetty, N., Hu, R., Hoch, J., Mailloux, B., Palmer, M., Menge, D. N. L., McGuire, K., McGillis, W. and Culligan, P. (2018). Quantifying urban bioswale nitrogen cyclinG in the soil, gas, and plant phases. *Water* 10(11), 1627.

Singh, R. P., Singh, P., Araujo, A. S. F., Ibrahim, M. H. and Sulaiman, O. (2011). Management of urban solid waste: Vermicomposting a sustainable option. *Resources, Conservation and Recycling* 55(7), 719-729.

Stenmarck, A., Jensen, C., Quested, T. and Moates, G. (2016). Estimates of European food waste levels. Available at: https://ec.europa.eu/food/safety/food-waste_en (accessed 15 February 2022).

Tedesco, C., Petit, C., Billen, G., Granier, J. and Personne, E. (2017). Potential for recoupling production and consumption in peri-urban territories: The case-study of the Saclay plateau near Paris, France. *Food Policy* 69, 35-45.

Townsend-Small, A. and Czimczik, C. I. (2010). Carbon sequestration and greenhouse gas emissions in urban turf. *Geophysical Research Letters* 37(2). doi: 10.1029/2009GL041675.

Tzoulas, K., Korpela, K., Venn, S., Yli-Pelkonen, V., Kazmierczak, A., Niemela, J. and James, P. (2007). Promoting ecosystem and human health in urban areas using Green Infrastructure: A literature review. *Landscape and Urban Planning* 81(3), 167-178.

Ulrich, S., Willaredt, M., Nehls, T. and van Schaik, L. (2021). Do earthworms (D. veneta) influence plant-available water in technogenic soil-like substrate from bricks and compost? (*D. veneta*). *Journal of Soils and Sediments* 21(5), 2013-2024.

UNEP. (2021). *United Nations Environment Programme, Food Waste Index Report 2021*. Nairobi: UNEP.

UN-Habitat. (2020). World cities report 2020: the value of sustainable urbanization, p. 418.

Vasenev, V. and Kuzyakov, Y. (2018). Urban soils as hot spots of anthropogenic carbon accumulation: Review of stocks, mechanisms and driving factors. *Land Degradation and Development* 29(6), 1607-1622.

Vidal, A., Lenhart, T., Dignac, M. F., Biron, P., Barthod, J., Vedere, C., Vaury, V., Bariac, T. and Rumpel, C. (2020). Promoting plant growth and carbon transfer to soil with composts and vermicomposts produced with mineral additives. *Geoderma* 174, 114454.

Wei, Y., Wang, N., Lin, Y., Zhan, Y., Ding, X., Liu, Y., Zhang, A., Ding, G., Xu, T. and Li, J. (2021). Recycling of nutrients from organic waste by advanced compost technology - A case study. *Bioresource Technology* 337, 125411.

Wei, Z., Wu, S., Yan, X. and Zhou, S. (2014). Density and stability of soil organic carbon beneath impervious surfaces in urban areas. *PLoS ONE* 9(10), e109380. doi: 10.1371/journal.pone.0109380.

WRB. (2014). *World Soil Resourcesreports, 106*. Rome, Italy: FAO, UNESCO.

Xiaoli, C., Shimaoka, T., Xianyan, C., Qiang, G. and Youcai, Z. (2007). Characteristics and mobility of heavy metals in an MSW landfill: Implications in risk assessment and reclamation. *Journal of Hazardous Materials* 144(1-2), 485–491.

Yoon, T. K., Seo, K. W., Park, G. S., Son, Y. M. and Son, Y. (2016). Surface soil carbon storage in urban green spaces in three major South Korean cities. *Forests* 7(12), 115.

Zurbrügg, C., Drescher, S., Rytz, I., Sinha, A. H. and Enayetullah, I. (2005). Decentralised composting in Bangladesh, a win-win situation for all stakeholders. *Resources, Conservation and Recycling* 43(3), 281–292.

Part 4

Socioeconomic, legal and policy issues

Chapter 25

Soil organic carbon on the political agenda

Luca Montanarella, European Commission, Joint Research Centre (JRC), Italy

1 Introduction

In recent years, we have observed a steady increase in interest in soil organic carbon (SOC) from policymakers and the public. This attention has been mainly driven by increased recognition of the importance of SOC within the global greenhouse gas (GHG) balance. Soils hold the second-largest carbon pool on earth, after the oceans. They constitute the largest terrestrial carbon pool: an estimated total of 2344 Gt C which is more than the sum of carbon contained in the atmosphere and vegetation (Stockmann et al., 2013). Soils perform crucial functions in the global carbon balance. As a result, recognition of the importance of soils and their sustainable management in addressing climate change adaptation and mitigation is increasing. Soils play a role as a major carbon sink. However, poor soil management could also be a threat since it results in more SOC being mineralized and adding CO_2 and other GHGss to the global atmospheric GHG balance. Soil organic carbon is therefore very relevant to the United Nations Framework Convention for Climate Change (UNFCCC).

Soil organic carbon is also relevant to policy beyond the climate change debate. Soil organic carbon is of central importance for maintaining soil fertility and therefore key to debate about food security and agricultural policy.

http://dx.doi.org/10.19103/AS.2022.0106.24

Soil organic carbon also plays a major role in water retention. It is therefore important to have discussions about drought mitigation and has a role within the desertification policy arena, especially in relation to the implementation of the United Nation Convention to Combat Desertification (UNCCD). Soil organic carbon is one of the three metrics defining land degradation according to the UNCCD land degradation neutrality target setting programmme (Orr et al., 2017).

Soil organic carbon provides a key energy source supporting soil biodiversity and is, therefore, also highly relevant to biodiversity protection. It therefore has a role in supporting the Convention on Biological Diversity (CBD), as recently documented by the new global assessment of soil biodiversity (FAO et al., 2020). Finally, SOC increases the capacity of soils to buffer and store contaminants, making it relevant for policies relating to pollution control and environmental health.

Given the relevance of SOC to a wide range of policy areas, there has been a growing interest in SOC from very different policy perspectives including agricultural policies, environmental policies, climate change policies, biodiversity and nature protection policies. Unfortunately, existing multilateral environmental agreements were largely negotiated well before the relevance of SOC was fully recognized. The UNFCCC has largely ignored the importance of SOC for more than 20 years. Only at COP 21 in Paris in 2015, thanks to the initiative of the French government, has SOC moved to the heart of the climate change debate. The launch of the '4 per 1000' initiative after COP 21 has attracted the attention of the climate change community to this neglected aspect of organic carbon on the planet.

Given the lack of an overarching policy framework on a global scale, relevant policies have been developed over many years at a national level in many countries of the world (see http://www.fao.org/soils-portal/soilex/en/). Probably the best-known national soil protection policy is the US Soil Conservation Act of 1935 which is now considered a textbook example of a successful soil protection policy. The US Soil Conservation Act succeeded in reversing a catastrophic soil degradation process known as the 'dust bowl' affecting the Midwest of the USA in the 1930s. Extensive soil erosion caused large areas of the Midwest to be seriously degraded and ultimately abandoned by farmers and local communities. The introduction of effective soil management measures by the US Department of Agriculture and the creation of an effective Soil Conservation Service reversed the situation with erosion rates rapidly declining within less than 20 years.

The only transnational SOC policy framework established so far has been developed at a regional scale by the European Union and is known as the EU Soil Thematic Strategy. In this chapter, we will focus on this example of a policy framework addressing SOC.

2 Soil organic carbon within the European Union Soil Thematic Strategy

The EU Soil Thematic Strategy was formally presented by the European Commission in September 2006. It includes four main pillars of action:

1. binding legislation.
2. integration in existing and new EU legislation.
3. research.
4. awareness raising.

While the proposed binding legislation, known as the EU Soil Framework Directive, was not finally adopted, the three other pillars of action have been fully implemented and have resulted in important improvements in the sustainable management of EU soil resources. At the core of the strategy is the recognition of a range of vital soil functions delivering ecosystem services for everyone, beyond the limited boundaries of actual land ownership. The strategy states that the common ecosystem services and benefits provided by soils need to be protected by appropriate legislation and policy measures. Soil organic carbon is at the core of the strategy. The role of soils as a major carbon pool has been formally recognized in the strategy as one of the main soil functions to be protected. Soil organic carbon delivers key benefits to all of us by acting directly on the GHG balance and ultimately influencing climate change.

The strongest impact of the strategy has been in the pillar of action in relation to research. Several large EU research projects have been financed since 2006, contributing towards a much-improved understanding of SOC dynamics and functions in European soils. The series of research projects has culminated with the European Joint Programming initiative on agricultural soil management. European Union research efforts have also had a significant impact outside the EU, including the initiative to establish an International Research Consortium (IRC) on SOC.

The other two pillars of the strategy have also had a significant impact on soil management in the EU. In particular, the integration pillar of the EU Soil Thematic Strategy has helped to ensure that soils are properly accounted for in all relevant EU legislation. This has been particularly important in developing a new EU Common Agricultural Policy and in EU Land Use/Land Use Change and Forestry (LULUCF) regulation.

More recently, the President of the European Commission, Ursula von der Leyen, has outlined her priorities 2019–2024, including her vision for a greener Europe: 'Climate change, biodiversity, food security, deforestation and land degradation go together. We need to change the way we produce, consume and trade. Preserving and restoring our ecosystem needs to guide all of our work' (Von der Leyen, 2019). This vision underpins the proposal for a European Green Deal (EGD), striving to be the first climate-neutral continent. Becoming the world's first climate-neutral continent by 2050 is one of the greatest challenges and opportunities of our times.

The EGD represents an ambitious package of measures to enable a sustainable green transition. An initial roadmap of key policies ranges from ambitiously cutting emissions, investing in cutting-edge research and innovation, to preserving Europe's natural environment. Achieving sustainable soil management (SSD) in the EU will be crucial to several of the planned actions within the EGD. Principles and voluntary guidelines for SSD were established by the FAO in 2016 (FAO, 2017). All FAO member countries, including EU member states, have endorsed these guidelines. A key challenge is now how to translate these guidelines into actions within the EGD.

The proposed EGD has several elements (Fig. 1) designed to lead to the final goal of a climate-neutral continent. Soils are explicitly mentioned in the EU Farm to Fork Strategy and in the zero-pollution action plan (European Commission, 2019). They are also indirectly relevant to achieving the overall goal of climate neutrality by 2050 by preserving and restoring ecosystems and biodiversity. The important role that land degradation and restoration play in preserving terrestrial ecosystems has been well documented in the recent Intergovernmental Science Policy Platform on Biodiversity and Ecosystem Services (IPBES) Land Degradation and Restoration Assessment (IPBES, 2018). Achieving the SDG target 15.3 of a land degradation neutral world by 2030 will substantially contribute to the preservation of terrestrial biodiversity. Soils will

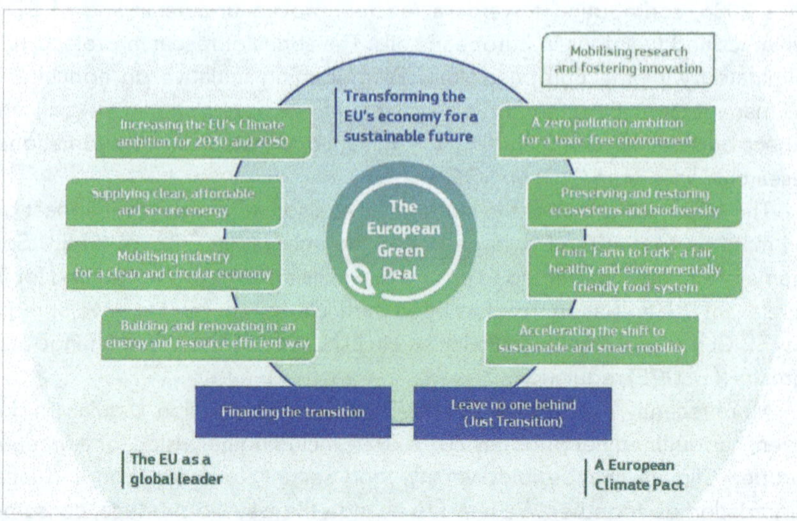

Figure 1 The various elements of the European Green Deal (COM (2019) 640 final).

therefore play an important role in future EU agricultural policy (Farm to Fork Strategy), environmental protection (biodiversity strategy) and climate change (climate legislation).

3 Soils and agriculture

European food must remain safe, nutritious and of high quality. It must also be produced with minimum impact on nature. To achieve this important goal in the EGD, there is a need to implement measures to preserve soil quality and limit soil contamination. Healthy food from healthy soils needs to be one of the slogans of the EGD.

Introducing innovative agricultural practices and technology is key, as envisaged by the focus on soil health and food within the Horizon Europe research framework programme beginning in 2021. A focus on soil health and food will be a powerful way to raise awareness of the importance of soils, engage with citizens, create knowledge and develop solutions for restoring soil health and soil functions.

It will be important to address the main soil degradation processes that affect agricultural soils in the EU (EEA, 2019). Soil erosion, soil contamination, soil compaction, soil sealing and the loss of soil organic matter and biodiversity continue to be major threats to soil health in Europe. The EU Farm to Fork Strategy will need to consistently address these issues and implement an efficient monitoring system in order to measure progress. An effective monitoring, verification and reporting (MRV) system for SOC in agricultural soils will be necessary to account for SOC stocks relevant to climate change mitigation. The proposed components of a suitable MRV system for SOC in agricultural soils (Smith et al., 2019) need to be translated into an operational MRV system for the EU. This needs to include elements such as:

- benchmark sites based on long-term experiments, such as Rothamsted Research (UK) for calibration;
- remote sensing tools based on the most recent satellite platforms (EU Copernicus program);
- ground observations and measurements, like in the LUCAS soil monitoring system;
- advanced modelling, including spatial data for final estimates and assessments.

This integrated monitoring system can also provide necessary indicators for implementing the new Common Agricultural Policy (CAP). The necessary innovation and research to fully develop such an integrated system will rely on the establishment of an International Research Consortium on SOC in agricultural soils. Such an IRC should be open to all interested global players and stakeholders in order to achieve full consensus on the MRV system to be adopted on a global scale.

4 Soils and biodiversity

Soils host one-fourth of the world's biodiversity and provide ecosystem services such as nutritious food, clean drinking water and carbon sequestration that are essential to meet societal challenges like climate change, food security, biodiversity loss and improved human health. Unfortunately, soil degradation is a pervasive, systemic phenomenon that occurs in all parts of the world and which can take many forms in the EU and globally. Combating land degradation and restoring degraded land is an urgent priority to protect the biodiversity and ecosystem services vital to all life on earth and to ensure human well-being. Land cover change and management intensity significantly affect soil condition and soil biodiversity in the EU. Progress in the remediation of polluted soils is slow. Despite recent reductions in soil sealing, soil fertility and biodiversity continue to decline. Soil biodiversity is endangered and soil carbon depleted on intensively managed land. Soil loss from erosion is still significant and results in economic costs for European citizens (Panagos et al., 2018).

In order to these trends in soil degradation, there is the need to devise a comprehensive land restoration program as recommended by IPBES (2018). This should include a commitment by the European Union:

- to protect soil functions, particularly fertility, and achieve land degradation neutrality in the EU, addressing specific drivers that reduce soil biodiversity, carbon storage and fertility and implementing extensively sustainable soil management practices;
- substantially increase efforts to reduce soil erosion and increase soil organic matter, and to enhance the integration of land use aspects into decision-making at all levels of government, supported by the adoption of targets on soil and on land;
- [for] EU [to] policies take into account their direct and indirect impact on land use in the EU and globally, and the rate of land take is well on track with an aim to achieve no net land take by 2050;
- to progress considerably in identification and remediation of polluted sites.

All of these actions were clearly outlined in the EU Soil Thematic Strategy adopted by the European Commission in 2006 (European Commission, 2006). Fourteen years after the adoption of the strategy, there is the need for a clear and objective assessment of the reasons for the failure of the proposed strategy in achieving full soil protection in the EU. A new strategy is needed based on new scientific evidence and including fewer, more measurable targets with a clear timeline. A new EU Soil Observatory should be established to collect relevant data and indicators for regular assessment of progress.

5 Soils and climate change

In order to achieve climate neutrality by 2050, there is a need to exploit the full potential of European soils in contributing to mitigation and adaptation strategies. Increasing the SOC pool in agricultural soils by implementing SSD practices will contribute to the final achievement of this ambitious goal. Halting land degradation and restoring degraded lands, as recommended by IPBES (2018) and IPCC (2019), should be at the core of such a strategy. Achieving land degradation neutrality by 2030 (SDG target 15.3) should be a pre-condition for the later achievement of a climate-neutral continent in 2050. Plenty of land restoration initiatives and programmes are possible in the EU. There will be the need for a coherent action plan on how to achieve land degradation neutrality, including the financial mechanisms that will allow the massive investments in land restoration that are needed. As already pointed out by the recent evaluation by the European Court of Auditors (ECA, 2018), the EU is still lacking a coherent approach for combating land degradation and desertification, with only one EU member state (Italy) that has presented a programme establishing land degradation neutrality targets as recommended by the United Nations Convention to Combat Desertification (UNCCD).

Land restoration results in a measurable increase of SOC. Healthy soils store large quantities of carbon (C) in the form of SOC. For this reason, SOC is included as a metric in the regular assessment of land degradation in reporting for SDG target 15.3. Recent discussions within UNFCCC under the Koronivia Joint Work on Agriculture (KJWA) initiative have highlighted the positive role of soils in climate change adaptation and mitigation, sustainable agriculture and improved food security.

There exists substantial scientific and practical evidence on how SSM can provide multiple benefits for the environment, people and livelihoods. Sustainable soil management preserves and increases SOM, a key element of soil health, which regulates many soil functions, including carbon storage in the form of SOC. In this way, SSM supports the retention and enhancement of carbon stocks in soils. It mitigates climate change and helps to adapt to it while generating benefits for agriculture, food security and nutrition, provision of ecosystem services, and advancing multiple sustainable development goals (SDGs). Investing in SSM constitutes a cost-effective and feasible climate change mitigation option which, at the same time, enhances soil health and climate resilience.

One hundred twenty-eight countries include the Agriculture, Forestry and Land Use sector in their pledges for the Paris Agreement. However, only a few refer to soils or SOC (see Wiese et al., 2019). Limiting warming to 1.5°C will require the use of 'negative emissions technologies' - methods that remove CO_2 from the atmosphere. These methods include SOC sequestration (Soussana et al., 2019). Soil organic carbon sequestration (SCS) is a major climate mitigation option. Up to 1.4 Gt C could be stored annually in agricultural soils (after IPCC, 2007, 2014). About 20% of the mitigation from SCS is achieved at negative cost and 80% below US$100/t$CO_2$, making SCS a low-cost mitigation option (Amelung et al., 2020). SCS includes measures such as:

- storing carbon by adding soil organic matter to agricultural soils;
- adding biomass through agroforestry;
- restoring drained peatlands and wetlands so that they store rather than release CO_2;
- better recycling of organic carbon through recycling urban and agri-food industry organic wastes back into agriculture and other industries, contributing to more circular systems and a more sustainable bioeconomy.

Soil organic carbon conservation and sequestration have multiple co-benefits for food security, climate change adaptation, land degradation neutrality (an objective agreed by the UNCCD), desertification, biodiversity and water resources as shown by the latest IPCC Special Report on Climate Change and Land. Soil organic carbon content in soils is also a target (15.3) of the life on land Sustainable Development Goal 15.

Unfortunately, the SDGs and existing multilateral environmental agreements (UNFCCC, CBD, UNCCD) are not binding. Parties ratifying these agreements have no legal obligation to comply, implementing measures on a voluntary basis. Reporting actions implemented by parties therefore becomes of crucial importance, since it is the only mean to monitor the implementation rate of proposed measures. In the case of UNCCD and for UNFCCC, countries are requested to report on their SOC status. However, only a few countries have the capabilities and methods to monitor agricultural soil carbon at a national level.

While there is considerable private and public interest in soil carbon and health, adoption of soil-enhancing agricultural practices appears to be slow. Farmers across the globe identify the main barriers to adoption as mainly socio-economic (e.g. additional costs are too high; lack of funds to access technology or machinery; farm extension services do not have knowledge and capacity). Overcoming these barriers requires a strengthened knowledge base and advisory services, improved awareness among the public, increased availability of indicators and tools, as well as financial support for agricultural transition and payments for soil carbon and other ecosystem services (CIRCASA, 2021).

6 The soil and water nexus

Any soil-related strategy cannot be considered in isolation from the close interdependence between soil and water management. Although it seems intuitive that soil and water need to be considered jointly in research and policy addressing agricultural production and sustainable livelihoods, there is still plenty of evidence that these two topics are often discussed separately. In hydrology, research soil is often reduced to a transfer zone which ignores the significance of soil in areas such as water flow, composition and purity. In soil science, on the other hand, hydrological processes are often reduced to boundary conditions which ignore the importance of water dynamics within the unsaturated zone for all kinds of biogeochemical processes. Beyond issues affecting the science, the separation of water and soil issues can lead to poorly integrated and counter-productive rural development policies. It is critical both to develop integrated research programmes and ensure that policies are integrated and informed by the results of such research.

A further problem is that responsibility for soil and water resources usually falls under different national ministries. At the international level, many organizations are either water or soil (land)-focussed bodies and few address the two resources jointly. To achieve sustainable livelihoods in both rural and indirectly in urban areas, we need to place more emphasis on tackling soil and water issues simultaneously both for agricultural development and for mitigating environmental hazards. This perspective is at the core of the only global multilateral environmental agreement addressing both, soil and water, UNCCD. This convention addresses land degradation and is the main tool for implementing a land degradation neutral world by 2030, as set out within SDG target 15.3.

Achieving such a world by 2030 requires a major shift in overall human behaviour, starting with consumption habits. As highlighted in the IPBES Land Degradation and Restoration Assessment (IPBES, 2018), we need to address land degradation and engage in major land restoration efforts if we want to reverse current degradation trends. Ten strategies have been proposed to overcome the policy barriers to achieve land degradation neutrality, and thus transform the effectiveness of land protection and restoration (Willemen et al., 2020):

1 Recognize that the benefits generated by healthy and productive land area global good.

2 Set clear, quantifiable, legally binding and ambitious targets to ensure that policies to halt and reverse land degradation match the scale and urgency of the problem.
3 Routinely collect and evaluate information on the state of land.
4 Promote local action to tackle land degradation based on local contexts and needs.
5 Build on all pertinent knowledge sources, not exclusively on conventional science.
6 Take into account all the substantive costs and benefits when making decisions that impact land.
7 Reduce the human demands for services delivered by land to match the capacity of the land to supply those services sustainably.
8 Encourage responsible trade and consumption.
9 Strengthen judicial institutions for environmental action by citizens.
10 Re-evaluate what it means to live well.

The recent global pandemic outbreak of SARS-COVID2019 has been an eye-opener in recognizing that a successful life is not synonymous with increasing purchasing power and levels of consumption. Alternative views exist, based on values such as solidarity and respect for nature. They can provide a foundation for more sustainable relationships between humans and the land we rely on.

7 Conclusion

Soils will play a central role in achieving the SDGs by 2030 (Bouma et al., 2019). Soil organic carbon is key to mitigating climate change, preserving soil biodiversity and protecting land from desertification. It therefore needs to become central within the related political frameworks of the United Nations Framework Convention for Climate Change (UNFCCC), the Convention on Biodiversity (CBD) and thUNCCD. Unfortunately, soils still have a marginal role in these Conventions. We are still far away from the full recognition of the central role of SOC in achieving the goals of the Rio+20 Sustainable Development Agenda. There is an urgent need to include soils in the global debate about sustainable development and climate change.

For Europe, this means that soils will have to be included as a key element in the proposed EGD. In particular, soils play a key role in achieving the ambitious European target of a climate-neutral EU by 2050. As a major carbon sink, they play an important role in mitigating GHG emissions and therefore should be an important element of the new EU Climate Law. In addition, soils are a major reservoir of biodiversity (Jefferey et al., 2010) and therefore need to be included in the new EU Biodiversity Strategy. Finally, soils are the foundation of agriculture, and therefore will have to play an important role in the EU Farm to Fork Strategy. Incorporating a coherent SSD framework within all three of these strategies will be challenging, given the necessary trade-offs between sometimes contradicting goals and targets. A coherent framework could be a revised EU Soil Thematic Strategy which takes into account the goals and ambitions of theEGD.

8 References

Amelung, W., Bossio, D., de Vries, W., Kögel-Knabner, I., Lehmann, J., Amundson, R., Bol, R., Collins, C., Lal, R., Leifeld, J., Minasny, B., Pan, G., Paustian, K., Rumpel, C., Sanderman, J., van Groenigen, J. W., Mooney, S., van Wesemael, B., Wander, M. and Chabbi, A. 2020. Towards a global-scale soil climate mitigation strategy. *Nat. Commun.* 11(1), 5427. doi: 10.1038/s41467-020-18887-7.

Bouma, J., Montanarella, L. and Evanylo, G. 2019. The challenge for the soil science community to contribute to the implementation of the UN Sustainable Development Goals. *Soil Use Manage.* 35(4), 538–546. doi: 10.1111/sum.12518.

CIRCASA 2021. Deliverable D3.2. *IRC Work Programme Report Presenting Recommendations for the set- up of an IRC on Soil Carbon,* Arias-Navarro, C., Soussana, J.-F., Trouillet, M., and Reynders, S. (Eds). European Union's Horizon 2020 research and innovation programme grant agreement No 774378 -Coordination of International Research Cooperation on soil CArbon Sequestration. doi: 10.15454/Q4IWTP.

ECA 2018. *Combating Desertification in the EU: A Growing Threat in Need of More Action*, Special Report No 33. European Court of Auditors, Luxembourg. https://www.eca.europa.eu/Lists/ECADocuments/SR18_33/SR_DESERTIFICATION_EN.pdf.

European Commission 2006. *Communication from the Commission to the Council, the European Parliament, the European Economic and Social Committee and the Committee of the Regions – Thematic Strategy for Soil Protection*. COM(2006) 231 Final.

European Commission 2019. *The European Green Deal*. Com(2019) 640 final.

European Environment Agency 2019. *The European Environment – State and Outlook 2020*. ISBN: 978-92-9480-090-9.

FAO 2017. *Voluntary Guidelines for Sustainable Soil Management*. Food and Agriculture Organization of the United Nations, Rome, Italy. http://www.fao.org/3/a-bl813e.pdf.

FAO, ITPS, GSBI, SCBD, and EC 2020. *State of Knowledge of Soil Biodiversity - Status, Challenges and Potentialities*, Report 2020. Food and Agriculture Organization, Rome, Italy.

IPBES 2018. *Land Degradation and Restoration Assessment*. IPBES Secretariat, Bonn, Germany.

IPCC 2019. *Climate Change and Land: an IPCC Special Report on Climate Change, Desertification, Land Degradation, Sustainable Land Management, Food Security, and Greenhouse Gas Fluxes in Terrestrial Ecosystems*, Shukla, P. R., Skea, J., Calvo Buendia, E., Masson-Delmotte, V., Pörtner, H.-O., Roberts, D. C., Zhai, P., Slade, R., Connors, S., van Diemen, R., Ferrat, M., Haughey, E., Luz, S., Neogi, S., Pathak, M., Petzold, J., Portugal Pereira, J., Vyas, P., Huntley, E., Kissick, K., Belkacemi, M. and Malley, J. (Eds).

Jefferey, S., Gardi, C., Jones, A., Montanarella, L., Marmo, L., Miko, L., Ritz, K., Peres, G., Römbke, J. and van der Putten, W. H. (Eds.) 2010. *European Atlas of Soil Biodiversity*. European Commission, Publications Office of the European Union, Luxembourg. EUR 24375 EN, 128pp.

Orr, B. J., Cowie, A. L., Castillo Sanchez, V. M., Chasek, P., Crossman, N. D., Erlewein, A., Louwagie, G., Maron, M., Metternicht, G. I., Minelli, S., Tengberg, A. E., Walter, S. and Welton, S. 2017. *Scientific Conceptual Framework for Land Degradation Neutrality. A Report of the Science-Policy Interface*. U.N. Convention to Combat Desertification (UNCCD), Bonn, Germany.

Panagos, P., Standardi, G., Borrelli, P., Lugato, E., Montanarella, L. and Bosello, F. 2018. Cost of agricultural productivity loss due to soil erosion in the European Union: From direct cost evaluation approaches to the use of macroeconomic models. *Land Degrad. Develop.* 29(3), 471–484. doi: 10.1002/ldr.2879.

Smith, P., Soussana, J.-F., Angers, D., Schipper, L., Chenu, C., Rasse, D. P., Batjes, N. H., van Egmond, F., McNeill, S., Kuhnert, M., Arias-Navarro, C., Olese, J. E., Chirinda, N., Firnara, D., Wollenberg, E., Álvaro-Fuentes, J., Sanz-Cobena, A. and Klumpp, K. 2019. How to measure, report and verify soil carbon change to realize the potential of soil carbon sequestration for atmospheric greenhouse gas removal. *Glob. Change Biol.* 26(1), 219–241. doi: 10.1111/gcb.14815.

Soussana, J. F., Lutfalla, S., Ehrhardt, F., Rosenstock, T., Lamanna, C., Havlík, P., Richards, M., Wollenberg, E., Chotte, J.-L., Torquebiau, E., Ciais, P., Smith, P., and Lal, R. 2019. Matching policy and science: Rationale for the '4 per 1000-soils for food security and climate' initiative. *Soil Tillage Res.* 188, 3–15.

Stockmann, U., Adams, M. A., Crawford, J. W., Field, D. J., Henakaarchchi, N., Jenkins, M., Minasny, B., McBratney, A. B., Courcelles, VdRd, Singh, K., Wheeler, I., Abbott, L., Angers, D. A., Baldock, J., Bird, M., Brookes, P. C., Chenu, C., Jastrow, J. D., Lal, R., Lehmann, J., O'Donnell, A. G., Parton, W. J., Whitehead, D. and Zimmermann, M. 2013. The knowns, known unknowns and unknowns of sequestration of soil organic carbon. *Agric. Ecosyst. Environ.* 164, 80–99.

Von der Leyen, U. 2019. *A Union that Strives for More. My Agenda for Europe*. https://ec.europa.eu/commission/sites/beta-political/files/political-guidelines-next-commission_en.pdf.

Wiese, L. D., Alcántara-Shivapatham, V. and Wollenberg, L. 2019. *Enhancing Nationally Determined Contribution (NDC) Ambition for Soil Organic Carbon Protection and Sequestration*. Info Note, CCAFS.

Willemen, L., Barger, N. N., Brink, B. t., Cantele, M., Erasmus, B. F. N., Fisher, J. L., Gardner, T., Holland, T. G., Kohler, F., Kotiaho, J. S., von Maltitz, G. P., Nangendo, G., Pandit, R., Parrotta, J. A., Potts, M. D., Pince, S. D., Sankaran, M., Brainich, A., Montanarella, L. and Scholes, R. 2020. How to halt the global decline of lands. *Nat. Sustain.* 3(3), 164–166. doi: 10.1038/s41893-020-0477-x.

Chapter 26

Creating frameworks to foster soil carbon sequestration

Beverley Henry, Queensland University of Technology, Australia; Ram Dalal, The University of Queensland, Australia; Matthew Tom Harrison, University of Tasmania, Australia; and Brian Keating, The University of Queensland, Australia

1 Introduction

It is now widely acknowledged that if the world is to avoid the worst of the projected impacts of climate change, then, in addition to urgent actions to reduce emissions of greenhouse gases (GHGs) to the atmosphere, removals are also needed (IPCC 2019, Rockström et al. 2017). Of the removal or 'negative emissions' technologies, natural climate solutions are an attractive option (Griscom et al. 2017). Soil organic carbon (SOC) sequestration, the process of removal of atmospheric carbon dioxide (CO_2) and storage as organic matter in soils, has immediate and broad applicability, offers potential for substantial climate change mitigation (Minasny et al. 2017, Paustian et al. 1997, Sanderman et al. 2010, Sanderman et al. 2017), and can contribute co-benefits for food security, ecosystem services and achievement of multiple Sustainable Development Goals (SDGs) (Keesstra et al. 2016, Keesstra et al. 2018, Paustian et al. 2016).

A systematic review of the literature has indicated that sequestration of soil carbon may contribute to removals of as much as 5 Gt CO_2 year^{-1} (Fuss et al. 2018). However, sequestration is variable across agro-ecological areas (Bossio et al. 2020) and will be highly dependent on levels of adoption of SOC-positive practices by landholders at scale. Past studies indicate that barriers to uptake can be substantial, and may include insufficient clarity on the impacts

http://dx.doi.org/10.19103/AS.2022.0106.25

of recommended practices on agricultural productivity, yields and profitability; poor or unknown compatibility of actions with farm operations; and uncertainty in whether any significant increase in SOC stocks will occur and, whether it is practical to measure the change (Box 1). Even where there is strong scientific evidence for positive productivity and environmental outcomes of carbon-sequestering practices, uptake among land managers has generally remained low (Paustian et al. 2019).

Policy frameworks that can provide information, training and financial or other incentives to adopt more sustainable practices represent an opportunity for governments or private organisations to invest in increasing soil carbon sequestration for climate change and other objectives. In this chapter, we explore how creation of national- or regional-scale frameworks appropriately linked to broader farm and forestry management, and supported by research, governance and rigorous practical measurement, reporting and verification (MRV) systems, can foster adoption of soil carbon sequestration practices. Frameworks can take a range of forms, but in this context, they would be expected to set out objectives, procedures, and incentives to guide soil management decisions for climate change, food security and sustainability benefits.

Box 1 Evolution of soil carbon research informing future action

Some of the earliest scientific reports of the eighteenth and nineteenth centuries focused on the role of soils and atmosphere in enabling plant growth. The early history was reported by Sir John Russell of Rothamsted Research Station in his landmark book 'Soil Conditions and Plant Growth' (Russell 1912). A mix of speculation, observation and experimentation over the prior two centuries helped resolve controversies on how soil, water, nutrients and atmospheric carbon interacted to control plant growth (see Keating and Thorburn (2018) for a more detailed summary). Over the past 70 years, research on soil organic matter and measurement of soil carbon content have evolved in a way that provides a context for current understanding of how good practice for soil carbon sequestration may be fostered more effectively (Fig. 1).

The founders of Rothamsted Experimental Station (JH Gilbert and JB Lawes) were pioneers in what we know now as soil fertility and crop agronomy science, and the long-term trials they established in the 1840s are still continuing to the present day. Some of these trials have been important for building and testing soil carbon models (Jenkinson 1990).

Building on these centuries of studies on soil fertility and crop yields, intensive research on SOC and climate change mitigation began in earnest during the 1990s (Barnwell et al. 1992, Paustian et al. 1997).

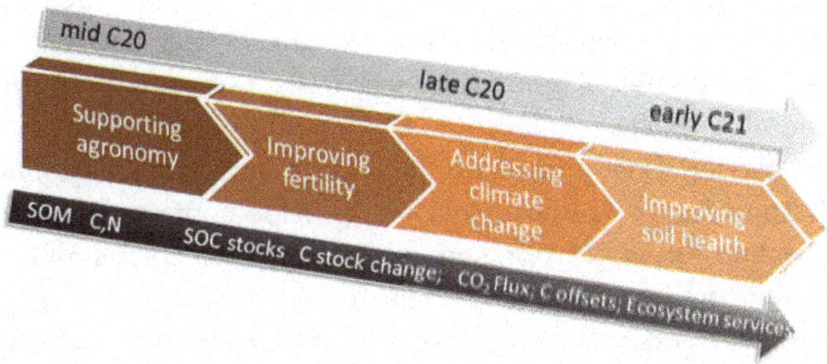

Figure 1 Illustrative timeline of research and monitoring of organic matter and carbon in soils, showing the evolution of interest driving research and development of metrics.

The initial focus was on understanding the contribution of land conversion and land use to increasing atmospheric CO_2 (e.g. Dalal et al. 2021a, Guo and Gifford 2002, Henry et al. 2018, Murphy 2020). By the late 1990s, research on rates of SOC sequestration moved towards assessing the value of management practices that increase soil organic matter (SOM) in diverse land uses and eco-regions as a climate change mitigation strategy. A variety of practices in croplands (Paustian et al. 1997, Lal 2004, Johnson et al. 2005, Franzluebbers 2005, Ogle et al. 2005, Causarano et al. 2006, Balkcom et al. 2013, Martens et al. 2005), such as the inclusion of cover crops (Causarano et al. 2006, Lal 2015a,b,c, Poeplau and Don 2015, Sainju et al. 2008), and in grazing lands (Conant et al. 2001, Schuman et al. 2002, Franzluebbers and Stuedemann 2009) and forestry and biofuel planting (Liebig et al. 2008, Follett et al. 2012) have been evaluated. As noted in the review by Stockmann et al. (2013), the results of field trials extending over periods of decades, and preferably for at least 100 years (referred to in this article as the 'permanence' period for storage) are more valuable than the outcomes of short-term experiments (Johnston et al. 2009, Powlson et al. 2011). Long-term monitoring enables the comparison of soil carbon under different practices over the time scales that are needed to understand the dynamics of SOC stock changes and their contribution to climate change mitigation.

2 Designing a framework to foster soil organic carbon sequestration

The key considerations in a framework for incentivising soil carbon sequestering actions by land managers include:

1 Sound scientific basis linking a management practice to soil carbon stock change, and real and achievable sequestration with permanence (long-term stability). It is essential that the evidence accounts for the

full system-wide consequences in terms of changes in SOC stocks or greenhouse gas emissions elsewhere, that is, avoids leakage, limits other environmental trade-offs and does not have negative socio-economic impacts;

2 Policy and legal instruments to enable payments or other incentivisation;
3 MRV and auditing protocols to provide appropriately rigorous accounting for soil carbon sequestration claims, credits and incentive payments; and
4 Consultation on socio-economic and cultural barriers to adoption of new practices, provision for training and education on the synergies or antagonisms with existing farm operations, and co-benefits of increasing SOC stocks.

This list is not exhaustive but is intended to facilitate discussion on how frameworks can be designed to improve opportunities for soil carbon sequestration in managed lands (Fig. 2).

2.1 Evidence linking practices to soil carbon outcomes

The evolving focus of soil science research for multiple benefits is expanding the body of evidence on the pivotal role of SOM and SOC in ecosystem

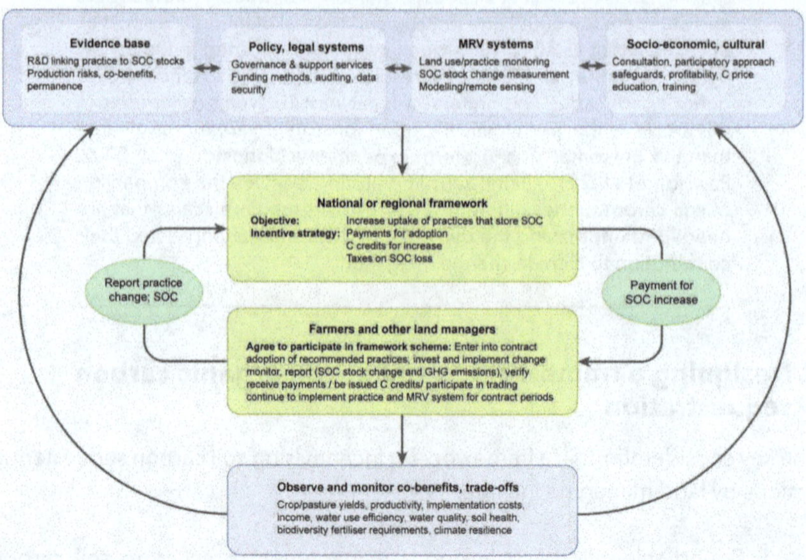

Figure 2 Schematic of a framework to foster adoption of management practices that increase sequestration of SOC in agricultural and other lands at the national or regional scale.

Published by Burleigh Dodds Science Publishing Limited, 2023.

processes and their potential role in the societal response to climate change and towards meeting the SDGs (Keesstra et al. 2016, 2018). In general, government programmes require sound scientific evidence in order to have the high level of integrity that is expected for public funding (Macintosh et al. 2019, Schneider and La Hoz Theuer 2019). Similarly, private investment in markets for carbon offsets requires confidence that the activities and practices being implemented will result in genuine climate abatement.

An incentivisation framework based on sound evidence for recommended practices will help manage the risk to participants in terms of non-achievement of SOC increase and a resultant loss of the confidence needed to continue implementing the practice, but it cannot eliminate the uncertainty entirely. Importantly, the range of prospective practices identified in experimental trials (Sanderman and Baldock 2010, Dalal et al. 2021b) cannot be assumed to apply uniformly, due to the heterogeneity in natural and human systems across global or national climatic, geographical, socio-economic and cultural landscapes (Johnston et al. 2009, Rumpel et al. 2020). In addition, there remain gaps in data and process understanding of how current and future climate, land use, management and edaphic factors will interact to influence the stocks and stability of SOC (Stockmann et al. 2013, Sun et al. 2020). Hence, systems and procedures are needed within the framework design to manage the risks in changing management practices, including the uncertainties around achieving soil carbon storage that is above the business-as-usual baseline, and the relationships with agricultural or forestry yields. An approach that incorporates flexibility in enabling instruments to allow incorporation of improved data, nascent knowledge and/or emerging technologies is recommended due to evolving knowledge and data (Arrouays et al. 2014, de Gruijter et al. 2018).

Chenu et al. (2019) discussed the potential to increase carbon stocks in agricultural soils and the prospective practices shown to influence organic matter dynamics, such as conservation agriculture (including no-tillage), deep-rooting crops, organic manures, irrigation (where available) and balanced fertilisation. They highlighted the existence of knowledge gaps and summarised key research questions relevant to the design of a framework to equip farmers with the confidence and accounting capacity needed for investment in fostering good practices for SOC increase:

- What practices can increase soil carbon stocks, what is the rate of increase and for how long would the rate of increase continue?
- Where should soil carbon stock increase be prioritised?
- How should potential gain in carbon be estimated?

There is now an improved understanding of the biophysical processes that follow implementation of practices such as conversion to reduced tillage, increasing irrigation, applying chemical fertiliser or organic amendments (including biochar), and rotational grazing (Paustian et al. 2019, Dalal et al. 2021b). Access to local data to inform farm management decisions is also expanding, and these improvements flow through to refinement of farm-scale simulations of SOC sequestration for crediting schemes. Process models now allow for predictions of longer-term outcomes under rising atmospheric carbon dioxide levels and climate changes such as warming soils and more severe droughts (Dalal et al. 2021a), but further research is needed to resolve outstanding questions (Amundson et al. 2015, Dalal et al. 2021a). The areas of active research include the role of microorganisms in stabilising organic matter applied to soil and the risk of 'leakage.' Organic amendments *per se* add carbon that was produced elsewhere; but its constituent nutrients such as nitrogen (N) and phosphorus (P) may contribute to enhanced carbon sequestration via increased biomass production as well as via a C:N:P stoichiometry approach (Chenu et al. 2019). These complex dynamics present accounting challenges (e.g. Australian Government 2018), as also do uncertainties in the extent of change in SOC storage in deeper soil layers (Murphy et al. 2019). Some differences in reported outcomes for management implementation likely reflect the lack of consistent definitions of practices and the absence of standardised measurement and verification protocols at the spatial and temporal scales required for quantifying change.

The most valuable insights come from long-term experimental trials, but they are costly to maintain and monitor (see Box 1). Globally, there are few continuous multi-decadal trials, and most are in temperate agricultural regions (Poulton et al. 2018, Powlson et al. 1998). For other regions, many of the assumptions on the dynamics of SOC have necessarily been derived from local short-term experiments. The risk of extrapolating rates of sequestration from short-term monitoring and across climatic zones can be illustrated by data from studies over different durations in eastern Australia. A five-year pilot study in the central-west region of New South Wales examining the potential for farmers to earn carbon credits from implementation of new practices, measured rates of increase in SOC as high as 1.2 Mg C ha^{-1} $year^{-1}$ in carbon-depleted land converted from cropping to permanent pasture (Badgery et al. 2020). In contrast, three longer-term trials (13–25 years) in permanent pastures near Wagga Wagga, a somewhat more arid area of New South Wales, also with low initial SOC concentrations, suggested that improved soil nutrient and grazing management may increase sequestration by only 0.5–0.7 Mg C ha^{-1} $year^{-1}$ (Chan et al. 2011). Such comparisons suggest that caution is needed in assuming that the initially high rates of SOC storage following management change can be maintained. While this may be due to sequestration slowing as a new equilibrium SOC level is approached, the results also point to the risk of extrapolation to other locations or practices.

The evidence base required to establish soil carbon sequestration frameworks should also extend beyond the site-specific soil biophysical processes to socio-economic impacts (Poulton et al. 2018; see Section 2.4) and systemic consequences, including for greenhouse gas emissions associated with practice changes such as irrigation (Kong et al. 2009, Trost et al. 2013), fertilisation (Rumpel et al. 2020, Van Groenigen et al. 2017) or livestock management (Chang et al. 2021, Harrison et al. 2021). Soil carbon sequestration activities that involve changes in land use or management, such as shifting from intensive crop production to reduced cropping intensity and pasture or 'set-aside' transitions can have wider land-use consequences (Gil et al. 2018). Further, the framework development phase should seek to understand any limitations to landholder participation due to legal (e.g. land rights) and policy issues.

2.2 Policy and governance requirements

2.2.1 The policy context for incentive frameworks

A primary objective of SOC sequestration policy frameworks and programmes is to incentivise increases in SOC stocks above the business-as-usual baseline while ensuring an improvement in the net greenhouse gas balance. Consideration is generally given to the global context by seeking consistency with international climate change agreements as well as domestic policy priorities. The United Nations Framework Convention on Climate Change (UNFCCC) and its Kyoto Protocol established a system of national communications and greenhouse gas inventory reporting, to be compiled by parties and published by the UNFCCC. Under the 2015 Paris Agreement, 194 countries and the European Union (UNFCCC 2021) have signed up to taking action to limit global warming to less than 2°C, and preferably no more than 1.5°C, and to report progress against a Nationally Determined Contribution (NDC). For countries (120 at the time or writing) who have committed to achieving net-zero emissions before the second half of this century, carbon sequestration in vegetation and soil is needed to offset 'hard-to-abate' emissions. This effectively embeds frameworks to encourage SOC storage in policy and carbon market contexts (Table 1). At the global scale, the multi-stakeholder '4 per 1000' initiative (https://www.4p1000.org/) which was launched at UNFCCC COP21 in 2015 seeks to galvanise action to improve soil carbon storage and soil health as set out in the *Soils for Food Security and Climate* vision (Rumpel et al. 2020).

The effectiveness of these policies, programmes and initiatives to foster adoption of practices that increase SOC storage will help to determine how much of the biophysical potential sequestration will be achieved in practice (Sanderman et al. 2017).

Table 1 Examples of policies and programmes which credit soil organic carbon offsets

Policy or Programme	Credit name	Scope	Reference
Australian Emissions Reduction Fund (ERF)	Australian Carbon Credit Unit (ACCU)	Australia	Macintosh et al. (2019)
Alberta Emissions Offset Program (AEOP)	Alberta Emissions Offset Credit (AEOC)	Alberta, Canada	Goddard (2008)
The Gold Standard	Verified Emissions Reduction (VER)	International (non-government)	Gold Standard (2020)
Verified Carbon Standard	Verified Carbon Unit (VCU)	International (non-government)	Verra (2012)

2.2.2 Insights from existing frameworks and programmes

Frameworks or programmes that incentivise SOC sequestration through carbon credit mechanisms commonly include functions for: (i) developing and approving/legislating methods or standards to quantify credits; (ii) reviewing, registering and auditing projects against the method or standard requirements; and (iii) issuing, transferring and operating a registry of offset credits (Broekhoff et al. 2019). It is crucial that policy frameworks aiming to incentivise new practices for credible increases in soil carbon consider:

- *permanence* (considered to be ≥ 100 years) of the SOC sequestration, including the risk of reversal through changes in land management or natural events such as drought;
- *additionality* beyond business-as-usual management and baseline SOC stocks when credits are to be counted as offsets;
- *leakage* risk, that is, whether eligible SOC practices lead to increased emissions elsewhere; and
- *accuracy and uncertainty* in the SOC stock change, allowing for assessment of bias or errors, data limitations, modelling assumptions or projected future values.

The design of effective frameworks also requires appropriate consideration of the risks and complexities for landholders and jurisdictions associated with practice change. Risks to jurisdictions depend on context-specific factors including priorities and budgets (COWI 2021, Peng et al. 2021), and the nature of individual and industry behavioural resistance to existing options available for soil carbon management (Alexander et al. 2015). Experience shows that diverse economic, social, cultural and psychological barriers will need to be overcome for uptake of practices to increase carbon storage – whether primarily for climate mitigation or to rehabilitate degraded, SOM-depleted lands (Alexander et al. 2015, Amundson et al. 2015, Lal et al. 2003, Lal 2015a, Rumpel et al. 2018). For example, policy design must consider whether change in land management to increase sequestration could potentially have social and environmental consequences such as reduced community access to productive agricultural land for food and fibre crops or could encroach on land valued for natural ecosystems and conservation reserves. The design challenge is exacerbated because research undertaken to inform policy frameworks may identify these issues poorly, if at all, in projections focussed on mitigation potential (Dooley et al. 2018). Nevertheless, these authors suggest that cross-disciplinary consultation and research can lead to a better understanding of the links between land use and climate change mitigation and support analysis of multiple objectives and interests to clarify how limited land resources can optimally be managed.

A significant design question for SOC frameworks is how to provide incentives which ensure that the adoption of new practices results in a long-term commitment to achieving genuine and sustained increases in SOC storage and minimises the risks of reversal of gains. This is only possible if the framework includes provisions for measurement and verification of changes in SOC stocks. Key technical considerations in the development of an MRV strategy include: (i) the purpose of measurement and level of confidence required; (ii) how to balance trade-offs between accuracy and measurement costs; and (iii) the extent to which the intensity and frequency of measurement and reporting may act as a barrier to voluntary participation. There is an expanding number of carbon-offset programmes (Table 1), each having its own rules for crediting and level of quality assurance, and this provides a sometimes confusing array of complex information that land managers need to weigh up before committing to practice change. The situation is exacerbated for frameworks or schemes that seek to create credits with value across jurisdictions or markets. From a policy perspective, more rigorous MRV systems will be required for recognition of offset credits as fungible commodities certified as representing a reduction of one metric tonne of carbon dioxide-equivalent emissions (t CO_2-e) (Schneider et al. 2019). The requirements and challenges for farm- or field-scale MRV to quantify SOC stock change are discussed in more detail below.

2.2.3 Common policy features in soil carbon sequestration programmes

The structure of frameworks varies between international, centralised schemes such as the Clean Development Mechanism (CDM), schemes that are private and independent (including some that issue credits that can be used in public or private compliance or market mechanisms) and domestic programmes with responsibilities to government. Here, a short overview is provided of the policy elements common to most programmes.

Governance, administration and strategic direction: A framework requires some form of governing body and administrative support, whose responsibilities could include developing the rules or methods and guidelines and approving, registering and overseeing implementation of projects and credit issuance. Some of these roles could be delegated, for example, in the not-for-profit organisation Verra, which is guided by a board of directors while private advisory committees develop standards and methodologies (Verra 2012). In all cases, a formal process is needed to establish eligibility of activities and projects and to maintain an account and registry of projects and credits to aid credibility.

Integrity: A process is needed to explicitly ensure that the GHG abatement (increased SOC sequestration less any associated emissions) is genuine and additional to what would occur under business-as-usual management practices, that is, in the absence of the incentive offered by the programme. Concepts related to ensuring environmental integrity include baseline setting, additionality, permanence of abatement/sequestration, avoidance of double counting and not causing other environmental harm such as negative impacts on biodiversity or loss of agricultural productivity leading to food insecurity.

MRV systems, tools and protocols: To ensure that policies support transparency and accountability in the issuance of carbon credits, offsets must be quantified in a way that is conservative and verifiable. The aim is to provide confidence that each offset corresponds to a real increase in SOC sequestration for the nominated 'permanence' period and represents a fungible unit (t CO_2-e) for offset markets. However, balancing cost-effective practical monitoring and reporting at farm scale with accuracy is challenging, as discussed below.

Stakeholder engagement: Actively engaging stakeholders (including farmers and advisers) in the scheme design process and regularly consulting during implementation can be effective in increasing uptake and commitment. Economic incentives are an important attractor for farmers but consulting on simplifying design or minimising initial baseline measurement costs have also been shown to influence participation (Macintosh et al. 2019). Integrating training and advisory opportunities into the scheme from the outset facilitates farmer learning and encourages adoption, especially if combined with raising awareness of the co-benefits of SOC sequestration for the farming business and as a societal approach to climate change mitigation.

Carbon pricing instruments and market mechanisms: Carbon market instruments may be voluntary or legislative. Policy issues for national frameworks to foster soil carbon sequestration extend across scales from transnational (e.g. market distortions) to farmer behaviour (Sampson and Sedjo 1997). Further, the policy design may include the capacity to generate demand for credits through linkage with other pricing or market instruments. Industries are increasingly adopting action on climate change as part of business risk management, which is creating a demand for eligible offset credits (e.g. TFCD 2021). Emissions trading systems, at the national or regional scale, are enabling trade in emission permits and offset credits issued for SOC sequestration. Alternatively, jurisdictions may link their policy instruments to allow permits or credits to be traded across borders. The scope of trades may expand through the provision for international transfers to achieve country targets under Article 6 of the Paris Agreement (Müller and Michaelowa 2019, Schneider and La Hoz Theuer 2019).

Co-benefits and trade-offs: Recognising the co-benefits such as biodiversity enhancement, water retention capability and reduced soil erosion, through ecosystem services credits (Lal et al. 2015, DES 2020, Flores-Rios et al. 2020) enhances farmers' ability to see value in improving their management practices to increase carbon sequestration (COWI 2021). There may also be trade-offs which should be identified. However, in general, few holistic assessments account for both benefits and trade-offs between agricultural, environmental, economic and social dimensions in land-use policies. While model-based holistic assessments can be challenging to formulate and fraught with uncertainties (Bishop and Welsh 1992, Harrison et al. 2012a,b), they can provide some cross-discipline evaluation for informing decision-makers about potential trade-offs and co-benefits and inform assessment of the possible opportunity costs of achieving environmental goals. The economic estimates of the minimum value of carbon sequestration credits and other environmental goods can assist a policy framework to achieve a net gain in social welfare, and opportunities for payments for ecosystem services credits (Baumber et al. 2019, DES 2020) may provide additional incentives for practice change by farmers (Lal et al. 2015, Kragt et al. 2016). Across these elements, policy and governance requirements for frameworks must take into consideration the objectives and the geographical and temporal scope of the desired carbon stock increase, and include explicit measures for ensuring environmental integrity and long-term maintenance of management changes.

2.3 Measurement, reporting and verification (MRV) systems for frameworks to incentivise practice change

While measurements of the carbon or organic matter content of soils (percentage SOC or SOM) have informed agronomic decisions over many decades (Box 1, Paustian et al. 2019), recent imperatives for greenhouse-gas-reporting and project-scale-monitoring to enable crediting of soil carbon offsets necessitate more rigorous quantification of SOC stocks and stock change. Policy frameworks or market mechanisms specify appropriately accurate estimation at field or farm scale, which together with the reporting and verification requirements aim to ensure consistency, reliability and integrity. If implemented properly, the framework will issue high-quality SOC sequestration offsets that have credibility in market mechanisms, demonstrate genuine long-term removal of atmospheric CO_2 (Chenu et al. 2019), and ideally, be consistent with national targets and inventory reporting. To achieve this, an MRV system for offsets must also include protocols to account for CO_2, nitrous oxide (N_2O) and methane (CH_4) emissions arising from new land management activities.

2.3.1 Overview of MRV approaches for soil organic carbon sequestration

There are challenges in designing MRV systems that balance accuracy, cost and practicality in a way that enables accounting that is fit-for-purpose (Paustian et al. 2019). Recognizing that access to resources and capacity for MRV can limit participation in incentive schemes, some have rewarded the adoption of SOC sequestering practices by action-based payments calculated on the expected result, rather than on measures of ex-post changes in SOC, as required in a result-based scheme such as Australia's ERF (Baldock and Burgess 2017) or Canada's Alberta Offset System (Goddard 2008). A framework providing government grants or payments for SOC management that is based on overall conservation benefits and could include improving soil health, ecosystem function and farm productivity (Govaerts et al. 2009) would normally require less rigorous quantification and lower MRV costs. However, the European experience has shown that action-oriented schemes not only have higher uncertainty (Burton and Schwarz 2013) but can be less successful in building confidence in the value of good practice over the long term (COWI 2021).

Traditional in-field measurements using destructive sampling and laboratory analysis are still the most widely accepted methods to accurately quantify SOC sequestration, but high costs are a barrier to participation (Commonwealth of Australia 2020). Prospects for lower-cost alternative technologies for SOC quantification are being explored, but issues remain to be overcome to enable their routine use in SOC crediting schemes (Costa Jr et al. 2020, Paustian et al. 2019, Smith et al. 2020, Jackson Hammond et al. 2021). Promising approaches include spectroscopy (Australian Government 2018, Viscarra Rossel et al. 2016), flux measurements (FAO 2019, Mudge et al. 2020), indicator or proxy measures (Wiesmeier et al. 2019, Terrer et al. 2021) and modelling (Powlson et al. 1998, Smith et al. 1997). Modelling and hybrid methods that use a combination of direct measurement, modelling and remote sensing are being trialled as alternatives to direct measurement for practical quantification options with appropriate accuracy (Zhang et al. 2019, Jackson Hammond et al. 2021). Systems for SOC MRV are presented in detail elsewhere in this volume, while here we discuss issues for MRV relating specifically to national frameworks for fostering SOC-sequestering practices.

Quantifying SOC storage attributed to adoption of new management practices is challenging because it necessitates measuring a small percentage change in a much larger, dynamic pool (Batjes 2014) largely driven by climate and soil characteristics (Allen et al. 2013, Rabbi et al. 2015, Sun et al. 2020). As described by Paustian et al. (2019), targeting a potential annual stock change of 1% or less means that measurement intervals of 5 years or more are generally required to detect statistically significant cumulative SOC stock changes for a moderate sampling density. To encourage participation in frameworks, strategies may be needed to enable early payments before confidence in detected SOC stock change is established (Commonwealth of Australia 2020).

Table 2 summarises the main components of an MRV system relevant to carbon crediting frameworks and provides examples of how emerging measurement and modelling capabilities may, in future, be integrated into carbon-crediting frameworks.

While a key strategic need is for accessible, harmonised, globally reconciled SOC databases that include management history, improvements in MRV could be readily achieved at low cost with better integration of current sources of data and commitment to maintain and consistently monitor long-term and short-term field trials, investment in advancing modelling capability, analysis of uncertainty and bias in simulations, and evaluation of how to use the increasing spatially-explicit activity data and the higher resolution remote-sensing platforms now available (Angelopoulou et al. 2019). Together these would enable the development of more accurate spatial data for model inputs and validation data for land use and management. Smith et al. (2020) summarised the gaps and needs to achieve better SOC monitoring and MRV practice that would support frameworks to foster practice for SOC sequestration:

- the provision of long-term continuity and consistency of monitoring in MRV systems under changing conditions;
- scientifically and politically appropriate spatial and temporal resolution for measurements;
- quality assurance at all stages of measurement and monitoring;
- documentation and measurement of all potential drivers of change in SOC stocks and greenhouse gas emissions; and
- geo-referenced samples archived with associated (harmonised) data made accessible.

Table 2 Major components of MRV systems and their possible integration in an SOC incentivisation framework

Purpose	Component activity	Applications for SOC data (examples)
Measurement/ monitoring	Long-term field trials/ experiments	Impact of management/practices on yield; Soil health; SOM and SOC content
	Short-term experiments; On-farm soil tests	Process studies; Model calibration; Agronomic decisions
Measurement, Reporting	SOC/GHG models Calibration - short-, long-term data Validation - long-term data Verification - survey data, remote sensing	Deriving Tier 2 Emission Factors Tier 3 methodologies Process/predictive data for policy and planning
	Spatial data	Model inputs (e.g. climate, soils, land cover layers)
	Activity data	Management data at farm/paddock scale
Measurement, Verification	Spatial soil re-sampling survey Site/paddock re-sampling	Ground-truthing SOC change; activity data for models; ground-truthing remote sensing Project monitoring; SOC change/ verification
Measurement, Reporting, Verification (Current and Emerging)	Remote sensing Flux tower measurements Spectroscopy Process-based modelling Other new technologies	• Soils, vegetation data; Inputs to run or verify models • Verifying project activity; Non-destructive sampling • Higher-density, spatial sensor and satellite data layers

2.4 Economic, social and cultural considerations in incentivisation frameworks

Review of the literature indicates a greater focus on biophysical and technological aspects of frameworks to incentivise climate change mitigation actions by landholders than on economic aspects (Harrison et al. 2021), and still less evaluation of social and cultural barriers. This observation likely reflects the broader traditional pathway in which scientific research is conducted, from an initial focus on technological questions, followed by biophysical (productivity etc.), then economic, and lastly, social and cultural aspects (Alcock et al. 2015, Ho et al. 2014). However, it can be argued that the policies and design for frameworks intended to influence behavioural and management decisions by farmers over extended periods, such as for adoption of SOC sequestering practices, would benefit from more, and earlier, economic and social science research (Sykes et al. 2020).

2.4.1 Economic considerations

Proponents of frameworks to foster SOC increase argue that positive incentives to change land management can result in substantial carbon sequestration in agricultural soils at lower cost than many alternative abatement activities (Kragt et al. 2012). However, considerable differences currently exist between economic viability at the farm level and financial support in national programmes and policies (Sampson and Sedjo 1997). The economic potential of agricultural carbon storage activities is likely to be lower than the technical potential (Bangsund and Leistritz 2008, Sanderman et al. 2017).

Landholders are more likely to adopt carbon sequestration practices if they see a clear financial advantage (Morgan et al. 2015), but the economic benefit of participating in a soil carbon sequestration payment framework depends on site-specific opportunity costs of changing production practices, the price of each tonne of carbon stored (as t CO_2-e) and rates of sequestration and hence the carbon credits issued over time. For example, changing rotations in a crop-livestock farm in the Western Australian wheat belt to increase SOC stocks could mean foregoing more than A$80 in profit for every additional t CO_2-e stored, depending on the crop residue retention practices adopted (Kragt et al. 2012), a value three- to four-fold the average carbon price under Australia's ERF scheme (at the time of writing in mid-2021, one Australian dollar (A$1) was equivalent to approximately US$0.75). Economic incentives to encourage landholder adoption of SOC sequestration practices may take the form of payments or subsidies for increases in SOC stocks or penalties, such as taxes for SOC losses. A study of land use in France sought to compare the cost-effectiveness of approaches based on either payments to compensate for costs of implementing better practice, or taxes (Bamière et al. 2021). The study, which examined three SOC sequestration measures - no-till, extension of temporary grasslands and hedgerows - found a disparity between net SOC sequestration and costs of implementation. Hedgerows and extended pasture phases in mixed crop/pasture systems resulted in sequestration of, respectively, 14.5 and 1.6 Mt CO_2-e, but very different average costs of €75/t CO_2-e sequestered for the hedgerow option but - €259/t CO_2-e in the case of extended pasture phases. Bamière et al. (2021) found that increasing the carbon price from €50 to €100 per t CO_2-e shifted the viability of actions and the dominance of farming enterprise participation. At the higher carbon price, the area of fallow land entering the scheme increased at the expense of the least-cost carbon storage option of temporary grasslands. In a similar large-area land-use study, Nelson et al. (2008) examined trade-offs in land-use policies in the Willamette Basin (Oregon, USA) under three levels of economic stimulus using efficiency frontiers. In all cases, increasing payment amounts to private landholders to restore natural land cover simultaneously increased ecosystems services, biodiversity conservation and carbon sequestration (Nelson et al. 2008).

From a landholder perspective, the costs of participation and income from carbon are often the key determinants of entry into voluntary SOC sequestration schemes. Burton and Schwarz (2013) contend that result-oriented schemes with payments based on validated changes in measured SOC are likely to be more cost-effective than action-oriented schemes (payment for adopting sequestration practices), and that using competitive reverse-auction-based pricing further established cost-effectiveness in incentivisation payments. Directly linking payments to outcomes and offering contracts for bids that deliver higher sequestration benefits at lower cost have been identified as effective design options for incentivisation frameworks in other analyses (e.g. Latacz-Lohmann and Schilizzi 2007, Claassen et al. 2008, Matzdorf and Lorenz 2010). This approach has been adopted in the Australian ERF policy framework (Commonwealth of Australia 2020, Macintosh et al. 2019) and is under active consideration for the EU (COWI 2021) at the time of writing.

2.4.2 Socio-cultural considerations

Social and cultural factors are the least explored areas with respect to frameworks to foster adoption of practices for SOC sequestration, but available studies indicate the relevant factors that include:

1. *Knowledge and trust:* Many farmers maintain conservative value systems, place importance on personal independence and have high regard for autonomy (as opposed to egalitarian views of authority and shared decision-making). Amundson and Biardeau (2018) suggest that farmers, even those practising innovative methods, may be wary of information from non-farming experts and a perceived government intervention through environmental schemes. They propose that farm advisors, who are local community members with relevant practical experience, may help lower cultural barriers to adoption, indicating that understanding people and how to motivate practice change is likely to be more important in terms of fostering farmer uptake than soil expertise.
2. *Human and social capital:* In his landmark work on the diffusion of innovations, Rogers (2004) proposed that every innovation relies on

human and social capital for adoption, depending on the innovation (here, national frameworks to foster uptake of SOC sequestering practices), communication channels, time and the social system and, further, that every innovation has some degree of status conferral. The provision of environmental capital through increased SOC stocks can be used to better establish social relationships that are beneficial over the long term (Bourdieu 1986, Burton and Schwarz 2013).

3 *Cultural assets and values of Indigenous peoples:* Important considerations in some regions are: (i) the need to obtain free and prior consent before developing a 'carbon farming' project; (ii) whether local Indigenous people have control over benefit-sharing; and (iii) whether traditional land owners will have increased quality of life as a result of projects (Robinson et al. 2016). It is plausible that carbon credit trading may present risks in commoditising indigenous cultural practices and landscapes (Gerrard 2008) and, therefore, the design of national frameworks for fostering SOC sequestration needs to define institutional and regulatory mechanisms to ensure appropriate incentives that respect indigenous values and achieve agreed, sustainable outcomes (Heckbert et al. 2008).

4 *Long-term behavioural change:* To be successful, any policy framework should foster not only shorter-term adoption of good practice for SOC increase but also beneficial long-term behavioural change (Matzdorf and Lorenz 2010). The design challenge is how to predict the level of [financial] incentive sufficient to elicit a desired behavioural change, in addition to what would occur in the ordinary course of events, due, for instance, to yield increase, but not so high as to attract a level of participation that is either not affordable under the scheme or, in a carbon market system, results in supply that exceeds demand, causing a drop in prices of carbon credits (Sampson and Sedjo 1997).

5 *Social and societal benefits of fostering SOC sequestration practices*: While landholders bear much of the direct costs of increasing SOC stocks, the associated ecosystem services are largely public goods (Brady et al. 2019), through contributions to water quality and climate regulation and providing resilience to future disturbances. There is a risk of trade-offs between private benefits, for example, higher yields, and wider societal benefits such as biodiversity and aesthetic services. A Swedish case study showed that the perceived value of conserving SOC stocks in agricultural land diverged between farmers and society in their decision-making, indicating that optimal conservation of soil natural capital from a societal perspective would benefit from innovative information systems and governance institutions as well as market mechanisms (Brady et al. 2019).

2.4.3 Multi-dimensional aspects in framework design

Incentives to foster SOC sequestration currently operating through government-sponsored frameworks and private market mechanisms usually focus on carbon stock change metrics aligned with IPCC accounting (Section 2.2). They may incorporate premium pricing for associated ecosystem services or social/economic credits (DES 2020, Verra 2012), and there is growing interest in considering more holistic approaches and interdisciplinary metrics in frameworks supporting sequestration (Chang-Fung-Martel et al. 2017, Rawnsley et al. 2016). Appropriately contextualised socio-economic and environmental 'bundling' of incentives for SOC-positive practices in policy frameworks may enhance their adoption. However, uptake may not increase if farmers and other stakeholders do not understand how multiple policy instruments that are operating concurrently can interact to affect their interests (Börner et al. 2017).

3 An Australian case study

This case study examines the theory and practice of a policy framework for government and industry action and provides an example of the design of a national approach to incentivising voluntary adoption of practice to increase soil carbon in agricultural lands. The historical context for the framework is provided in Box 2.

Box 2 Developing a framework to foster increase in SOC in agricultural land: Historical context 2000–2020

Climate change policy in Australia has been highly contested from a political perspective over the last two decades. There has been a litany of start, stop, revise, re-name and re-start of different policies, including a broad-based emissions trading system, a carbon tax, a direct-action program, a national energy guarantee (NEG) and many variations on these themes. Beneath the surface, however, there has been a broadly consistent program and policy development of national carbon accounts and emissions registers, and policy interventions to reduce the emissions intensity of different industry sectors and to incentivise carbon sequestration where possible. This consistent effort has been supported by technical capacity in policy departments and science agencies as well as a well-informed 'community of practice' in industry, particularly for agriculture and land management, and in the broader population.

Industry interest in carbon offsets was growing rapidly in the late 2000s and various voluntary carbon offset schemes were emerging internationally (e.g. Gold Standard https://www.goldstandard.org/; Verra https://verra.org/). There was also angst in land-based industries over the potential for emissions trading or carbon tax schemes to threaten industry viability in circumstances where the industry had few options for abatement action (e.g. technical options for CH_4 emissions reduction in the livestock sector seemed few and far between at that time). At the same time, some suggested government incentivisation was needed to encourage carbon sequestration in trees and soils. The term 'bio-sequestration' gained currency to the extent that the Department of Prime Minister and Cabinet (PMC) established an Inter-Departmental Committee (IDC) in 2009 to explore associated policy options.

3.1 Policy settings for Australian sequestration projects

The main policy and implementation elements underpinning Australia's carbon offset scheme are summarised below, followed by an overview of the ERF scheme legislated to incentivise actions to reduce GHG emissions or increase carbon sequestration across all sectors of the economy. The following sections explain how the ERF operates to incentivise adoption of soil carbon sequestering practices by landholders and highlights some of the lessons learnt in the six years that it has been in place.

3.1.1 Overview of policy, governance and implementation elements

3.1.1.1 National Greenhouse Accounts

The National Greenhouse Accounts are a series of databases and comprehensive reports (National Inventory Reports, NIR) that estimate and account for Australia's GHG emissions. The annual NIRs, which have fulfilled reporting requirements under the UNFCCC, its Kyoto Protocol and the Paris Agreement, are supported by a national spatial land sector program that in 1998 established a national carbon accounting system (NCAS) for land sector GHG emissions and carbon sequestration. The accounting system set up modelling tools and datasets centred on the Full Carbon Accounting Model (FullCAM), in which SOC is simulated in agricultural (including grazed native vegetation) and forest lands across Australia's total 769 million hectare land area through a sub-model based on the Roth C model (Jenkinson 1990).

3.1.1.2 NGER - National Greenhouse and Energy Reporting

The NGER scheme was established in 2007 as a national framework for reporting and disseminating company information, including GHG emissions, energy production and energy consumption.

3.1.1.3 CFI - Carbon Farming Initiative

The CFI was established through the *Carbon Credits (Carbon Farming Initiative) Act 2011* as a legislated project-based, baseline-and-credit offset scheme. Registered offset projects could generate certified offsets, Australian Carbon Credit Units (ACCUs), from land use, land-use change and forestry (LULUCF), agriculture and waste sectors. The CFI Act and its amendments are administered by the responsible Minister and Department.

3.1.1.4 CER - Clean Energy Regulator

The Clean Energy Regulator (CER) is an independent statutory authority that, since 2007, has administered schemes legislated by the Australian Government for measuring, managing, reducing or offsetting Australia's carbon emissions, including NGER, ERF and the Australian National Registry of Emissions Units which supports the CFI Act.

3.1.1.5 Climate Active

In 2019, Climate Active was formed as an update of the National Carbon Offsets Standard, which came into effect in 2010 to support the voluntary carbon offset market. This industry partnership, supported by government, aims to drive voluntary action towards carbon neutrality, based on the Climate Active Carbon Neutral Standard.

3.1.1.6 ERF - The Emissions Reduction Fund

The ERF was legislated in 2014 through the CFI Amendment Act to the *Carbon Credits (Carbon Farming Initiative) Act 2011*, to enable the CFI to extend voluntary abatement arrangements to all sectors of the economy (Box 3). The ERF was allocated A$2.55 billion in 2014 to purchase emissions reductions and extended by an additional A$2 billion in 2019 through the Climate Solutions Fund (CSF) to support international GHG emissions reduction targets.

3.1.1.7 ERAC - Emissions Reduction Assurance Committee

The 2011 *Carbon Farming Initiative (CFI Act)* established an independent expert committee to oversee the integrity of emissions abatement methodologies credited under the Act. Initially called the Domestic Offsets Integrity Committee (DOIC), and, after legislation of the ERF in 2014, the Emissions Reduction Assurance Committee (ERAC), this Committee assesses whether draft methods comply with the set of legislated offsets integrity standards (OIS; Box 4), designed to ensure that only genuine, additional abatement and sequestration is credited. Only if determined by ERAC as meeting all OIS can a drafted ERF method be made into legislation.

3.1.2 How the Emissions Reduction Fund (ERF) incentivises abatement

The ERF is a carbon crediting and pricing mechanism aimed at providing incentives for organisations and individuals to adopt new practices and technologies that reduce their emissions and/or increase sequestration (Fig. 4). Participation is voluntary. ERF methods set out the rules to enable projects to register under the scheme and to be issued with ACCUs. The ACCUs may be sold either through government reverse auctions or in private secondary markets as certified high-quality carbon offsets, providing income for participants. Activities that are eligible under ERF methods may provide public good through environmental, economic and social services as well as private good, for example through increased farm and forestry productivity, and social and financial benefits for Indigenous communities, and may attract additional value through other schemes (e.g. DES 2020). The ERF crediting, purchasing and compliance roles are summarised in Box 3.

Box 3 Summary: Emissions Reduction Fund (ERF) policy context

The Carbon Farming Initiative (CFI) Act 2011 was established to supply offsets to Australia's carbon pricing mechanism (Fig. 3).

In 2014, the CFI transitioned to the ERF through amendment to the Act, covering all sectors of the economy.

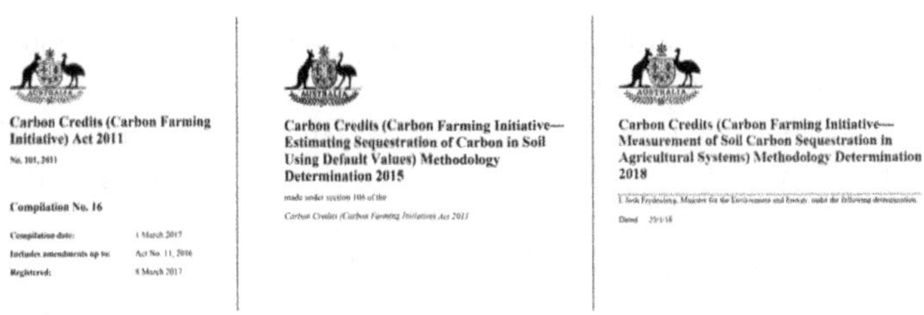

Carbon Credits (Carbon Farming Initiative) Act 2011

Compilation No. 16

Carbon Credits (Carbon Farming Initiative—Estimating Sequestration of Carbon in Soil Using Default Values) Methodology Determination 2015

Carbon Credits (Carbon Farming Initiative—Measurement of Soil Carbon Sequestration in Agricultural Systems) Methodology Determination 2018

Figure 3 ERF legislative instruments directly relevant to soil carbon sequestration projects are (left to right) the CFI Act 2011, the 2015 model-based soil carbon method and the 2018 measurement-based method for soil carbon sequestration in agricultural lands.

The three components of the ERF are:

- A *crediting scheme* for eligible offset projects, with one ACCU earned for each t CO_2-e abatement.
- A *purchasing scheme* enabling purchase of ACCUs through periodic reverse auctions from project owners who have voluntarily entered into a carbon abatement contract with the government.
- A *safeguard mechanism* that places legal emission obligations on major emitters (facilities emitting >100 000 t CO_2-e per year), which can be met through relinquishment of ACCUs (potentially supporting a market demand).

ERF methods set out the eligibility, legal and procedural requirements for activities and projects to be issued with credits. Two soil carbon sequestration methods are available at the time of writing (Fig. 3)

By March 2021, about 205 million tons of abatement had been contracted under the ERF over a total of 962 projects registered under the ERF, with about 80% from agriculture- and forestry- related activities.

Governance responsibilities for the ERF scheme are shared by the Minister, the relevant government department, the CER and the ERAC. The Minister is responsible for developing new methods, varying and revoking existing methods and, with consideration of Departmental advice, determining method-development priorities. The CER is responsible for developing new methods and method variations and administering the scheme, including registering and crediting projects, conducting auctions, managing the Safeguard Mechanism and providing the Secretariat role for the ERAC. An independent body, the Climate Change Authority (CCA), which was set up to provide expert advice on climate change policy, undertakes a review of the ERF at three-year intervals.

3.1.3 Integrity and practicality in the ERF

The set of integrity standards (the OIS) has remained at the core of the CFI legislation and its derivatives, which have broadened its scope, since inception in 2011 to the present day. These standards sit behind and guide the development of the approved methods and the governance structure that seeks to ensure practical arrangements to encourage participation. In this way, the ERF aims to provide opportunities for uptake of methods and income from credits that have high credibility in primary and secondary market mechanisms.

The OIS, legislated in Section 133 of the CFI Act, are intended to provide confidence that the offsets credited are: (i) additional; (ii) measurable and verifiable; (iii) able to provide abatement eligible to be counted towards

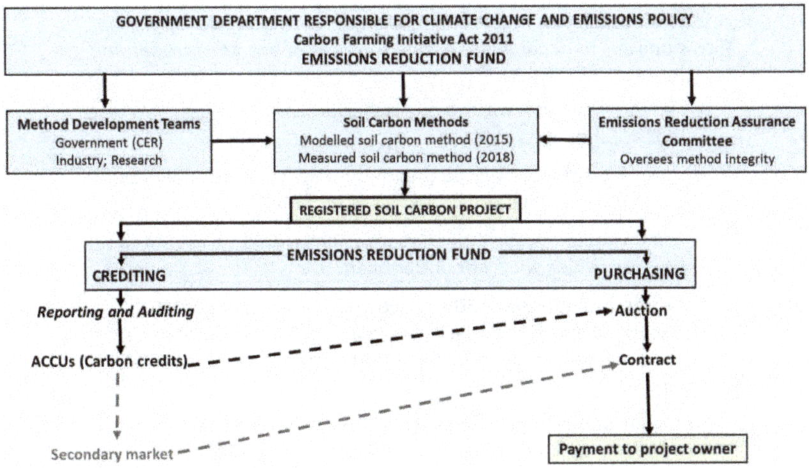

Figure 4 Schematic illustrating the steps for implementing an agricultural soil carbon project in Australia's ERF framework (Based on: http://www.cleanenergyregulator.gov.au/ERF).

Australia's international emissions reduction obligations; (iv) evidence based; (v) deducted for GHG emissions that result from project activities; and (vi) conservative. In interpreting the OIS to assess compliance of new and varied methods, the ERAC seeks consistency in its approach but considers requirements in individual methods against each standard (Box 4, ERAC 2021). The ERAC is also required to adopt an interpretation of the OIS that would best achieve the objects of the CFI Act:

- to remove GHGs from the atmosphere and avoid emissions of GHGs, in order to meet Australia's international mitigation obligations;
- to create incentives for people to undertake offsets projects; and
- to increase carbon abatement in a manner that is consistent with the protection of the natural environment and improves resilience to the effects of climate change; and to authorise the purchase by the government of units that represent carbon abatement.

Box 4 How the ERAC interprets the Offset Integrity Standards

The ERAC seeks to interpret the integrity standards to ensure projects registered under the scheme are issued with credits only for emissions reduction or sequestration that is real, additional and able to be measured accurately. A method must be assessed as complying with all six standards to be approved (ERAC 2021).

1. ***Additionality:*** *A method should result in carbon abatement that is unlikely to occur in the ordinary course of events (disregarding the effect of the Act).*

- At the method level, the general approach is to apply two tests for additionality:
    - Project test - Would the method activities occur without the incentive provided by the scheme?
    - Baseline test - What net emissions would be likely if the activities were not undertaken?
- Three project-level additionality requirements set out in the Act must also be satisfied (Newness, regulatory additionality, and possible funding under another government program or grant).

2. ***Measurable and verifiable:*** *Where a method results in emissions removal, reduction in emissions or emissions of greenhouse gases these should be measurable and capable of being verified.*

- The method must provide a rigorous and reliable way to measure or estimate and verify removals, emissions reductions and project emissions. Each method specifies how verification is achieved, e.g. by audits, models, satellite imagery. Interaction with other OIS, particularly 'evidence-based' and 'conservative' is considered.

3. ***Eligible carbon abatement:*** *A method should provide abatement that is able to be used to meet Australia's international mitigation obligations.*

- Interpretation of this standard requires the abatement to be from sources and sinks that are accounted for to meet Australia's mitigation targets under international commitments, notably the Paris Agreement.

4. ***Evidence-based:*** *A method should be supported by clear and convincing evidence.*

- This standard is interpreted as requiring clear and convincing evidence on the impact of a method activity on emissions or sequestration, their measurement and verification, and for the approach to ensure no leakage.

5. ***Project emissions:*** *Material greenhouse gas emissions directly resulting from a project should be deducted.*

- The project emissions standard requires methods to include appropriate deductions for all material project emissions, and their estimation to be consistent with the measurable and verifiable standard and evidence-based standard. Estimates of material project emissions (those that cumulatively exceed 5% of net project abatement) also must result in estimated net abatement being conservative.

6. ***Conservative:*** *Where a method involves an estimate, projection or assumption, it should be conservative.*

- Estimates, projections and assumptions that influence the calculation of net abatement for eligible projects are required to be conservative, that is, unlikely to be an over-estimation. However, requiring every estimate, projection and assumption to have a high probability of being an under-estimate, would be highly conservative and would be a disincentive for uptake of projects.
- In applying the conservative standard, the Committee also generally considers:
  - the potential for direct and indirect leakage to arise as a consequence of project activities;
  - the risk of non-permanence for sequestration projects and whether the permanence period discount is sufficiently conservative.

3.2 Performance of the Australian Emissions Reduction Fund (ERF) scheme

3.2.1 ERF status and soil carbon projects

The following section reflects publicly available data as at April 2021, noting that ERF project and ACCU issuance data should be interpreted in the context of the scheme and method procedures and are subject to updating by government.

As at April 2021, there were 966 projects registered under the CFI Act (2011) including initial CFI projects and ERF projects (CER 2021a). Contracts were in place to deliver 205 billion t CO_2-e abatement with 92.7 million issued by the CER up to April 2021. Of the 966 registered projects, more than half (532) use vegetation and savannah ERF methods, that is, they relate to the LULUCF sector of the national greenhouse accounts. Of the remainder, 155 projects use agriculture methods, the majority for avoidance of emissions relating to livestock manure management and ruminant enteric fermentation. There has been a rapidly growing interest in project registrations to increase carbon sequestration in soil, with more than 80% of all agriculture sector projects using methods for soil carbon sequestration at April 2021.

Under the ERF, there are (as at mid-2021) two methods under which farmers can be issued with carbon credits for adopting new practices that increase soil carbon sequestration:

1 A model-based soil carbon method (2015) for grazing land soils - land managers who adopt one of three eligible activities can earn carbon credits using regionally specified model-based default values that

give conservative estimates of sequestration. This method has had no uptake.

2 A measurement method for soil carbon sequestration in agricultural systems (2018) developed to improve the 2014 measurement method. As at March 2021, there were approximately 130 projects registered under measurement methods.

The number of soil carbon ACCUs delivered up to mid-2021 is low - approximately 1900. This reflects the expected delay between a project start and first measurement and reporting of SOC stock change which, under the ERF method, is permitted to be up to five years. It may also be a function of the complexity of the method and expense of sampling and analysis. Early experience has led to discussions on modifications and additional incentives to reduce costs and improve usability, including forward payments for baseline soil carbon measurements. Nevertheless, registration of new projects has accelerated in 2020 and 2021 (Fig. 5), and stakeholder interest in participating in soil carbon sequestration projects is high. Development of a new method with reduced measurement costs and improved flexibility is a departmental priority for release in 2021.

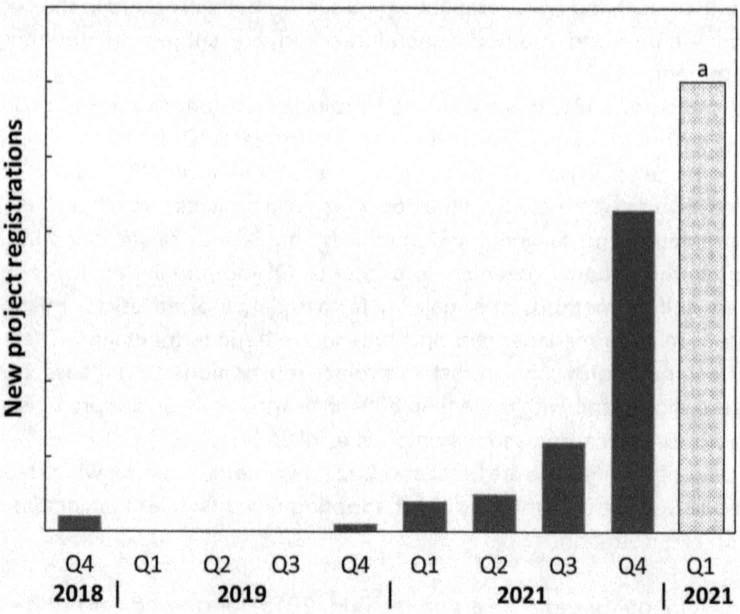

Figure 5 New registrations of ERF soil carbon sequestration projects in agricultural systems. Data for 2018 to 2020 from CER (2021b); a - Quarter 1, 2021 estimated from registrations http://www.cleanenergyregulator.gov.au/ERF; Accessed May 2021).

3.2.3 Practical issues limiting uptake of sequestration projects

The experience gained through implementation of the Australian ERF policy as an incentive framework for abatement actions highlights some of the practical issues for SOC practice adoption by land managers. It demonstrated the challenges in balancing the costs and rigorous integrity and MRV accuracy under the scheme, and also farmers' difficulty in understanding and applying technically complex methods and schemes without the help of specialist [sometimes costly] service providers. Several sequestration methods legislated in the ERF, including the modelled method for SOC, have had no uptake, with reasons including that they are not financially positive or that eligible activities are not compatible with managing a real-life farm enterprise as a profitable business. A review of uptake of ERF vegetation and soil sequestration projects by farmers (Macintosh et al. 2019) identified eight barriers to uptake: low carbon prices; perceived risk of rule changes affecting participation and crediting; uncertainty about future carbon prices; difficulties in getting third-party consents (land rights); permanence requirements; lack of awareness amongst farmers of carbon market opportunities; the scope of methods; and a lack of trust on behalf of farmers in parties offering carbon market information. Another key learning was that uptake could be improved if co-benefits to carbon sequestering activities were optimised and their value communicated to land managers.

The major constraints on participation in the ERF SOC model-based method (Australian Government 2015) reported by stakeholders concern the conservative nature of the default values and the method's narrow eligibility requirements. These problems reflect the state of knowledge in SOC sequestration rates for practices adopted on agricultural lands at the time the method was developed, and the requirements for integrity under the policy, including being evidence-based and using conservative assumptions to ensure high-quality carbon credits. In contrast, the main constraint on uptake of the less conservative measurement method has been the high costs of sampling and analysis. Stakeholders also cited the limited eligible practices and discounts applied to maintain conservativeness in the revised measurement method legislated in 2018 (Australian Government 2018) as barriers. More recently, greater confidence and higher prices for carbon offsets have resulted in accelerated participation in ERF projects.

This acceleration in participation is likely to continue as emerging techniques, such as spectrometry, remote sensing and flux measurements and new hybrid approaches using modelling capacity verified by accessible datasets, enable more cost-effective MRV for SOC sequestration methods in policy frameworks (Smith et al. 2020). Increased flexibility in MRV and in the range of eligible practices are being evaluated for an updated ERF SOC method, which is expected to be more attractive to stakeholders while maintaining the focus on integrity.

4 Conclusion and future trends in research

4.1 Opportunities and challenges to foster adoption of soil organic carbon practices

Lal et al. (2003) proposed that, with the implementation of suitable policy initiatives, substantial SOC sequestration potential could be realised for up to 30 years, or when the soil C sink capacity is filled. This would provide a valuable near-term contribution to restricting global warming to no more than 2°C, as specified in the Paris Agreement and supporting the Global Agenda for Sustainable Development set out in the SDGs. The challenges for design of national or regional frameworks relate not only to understanding the technical and biophysical potential for SOC sequestration but also to understanding the nature of behavioural resistance to practices and incentives already available for adoption. There is an urgent need for frameworks that enable investment in soil carbon mitigation to be practical and effective for achieving climate change mitigation, food security and ecosystem health benefits (Alexander et al. 2015).

Investment in research on biophysical and technological aspects of soil carbon sequestration and its climate change mitigation potential over more than a decade has provided strong evidence as a basis for action. More recently, research has accelerated as countries increase their ambition under the Paris Agreement, and as awareness is raised through initiatives such as '4 per 1000.' To support action, studies are being undertaken into the impacts of increasing SOC storage on agricultural yields and productivity, but economic, social and cultural opportunities and barriers to uptake of positive practices have generally received less attention (Harrison et al. 2021). Consideration of all four areas - biophysical, technological, economic and socio-cultural - will be essential for successful implementation of policy frameworks fostering adoption of good practice for long-term SOC management.

Experience with soil carbon measurement methods in the Australian ERF policy has also highlighted the practical challenges of implementing outcome-based soil carbon projects, including the barrier imposed by lack of early income from credits to cover participation expenses such as baseline sampling and analysis as part of a financial incentive model. Not unexpectedly, there is a tension between the timing of reporting for earlier financial returns and integrity of carbon credits issued. Early reporting (i.e. within 1-2 years) of [often small] changes in SOC stocks challenges the legislated requirement for verifiable and conservative crediting. Scientists caution that a rapid rise in measured

SOC stocks following implementation of new practices, especially for soil with initially low SOC, may reflect an increase in the labile carbon pool that improves soil health but may not be sustained over a 25-year or 100-year period. Sequestration rates are also affected by climatic and soil nutrient variations over time. The rate of increase in SOC stocks generally decreases as a new equilibrium level is approached. Managing the risk of over-crediting through provisions for discounted or delayed payments to ensure total crediting under an SOC method is conservative and is unlikely to align with farmers' need to manage costs. Sampling and analysis for project reporting incur significant up-front cost and there may also be disappointment if crediting is lower than expected, due, for example, to climate factors such as drought.

There can also be a need to resolve knowledge gaps relating to policy design. For example, a policy requirement that estimates of sequestration are conservative is difficult until there is more scientific certainty about what a 'conservative' estimate means for practices. Research and long-term trials will improve confidence in the expected relationship between practice change and soil carbon stocks in various climates and soil types (e.g. Schlesinger and Amundson 2019). The limited period of operation of incentivisation frameworks, including the Australian ERF, highlights the need for more experience to better understand and overcome tensions that act to constrain adoption of SOC sequestering practices on the ground.

4.2 Future study in soil carbon research

Soil carbon research will continue to build the underpinning evidence base for investment in frameworks and promote sequestration for climate change mitigation and soil health. As experience better identifies knowledge and data gaps, research is evolving to meet biophysical process and data needs for recognition of high-quality soil carbon offsets in market mechanisms. At the same time, multi-dimensional collaborations are contributing to overcoming socio-economic and cultural barriers to adoption by land managers. Here we provide illustrative examples, specific to the needs for developing frameworks for fostering adoption of good practice. More detail on biophysical research questions is found in a number of review papers, including Chenu et al. (2019), Paustian et al. (2019) and Smith et al. (2020), and socio-economic and cultural issues are identified in Burton and Schwarz (2013) and COWI (2021).

SOC stabilisation: From a policy framework perspective, key questions relate to the processes of stabilisation of carbon in soils. This understanding is needed to foster those practices that are consistent with the long-term SOC sequestration essential for genuine removals to offset emissions of GHGs. The role of processes in aggregate stabilization must be elucidated to identify effective sequestration practices for storing carbon in soil.

Research on stabilising soil aggregates will help the understanding of how to increase the pool of protected carbon that is relatively inaccessible to microbes and hydrolysing enzymes, and yet allows sufficient biogeochemical cycling for production (Cotrufo et al. 2013, Wilpiszeski et al. 2019). Next, advances in soil carbon modelling are needed to incorporate practices such as soil microbiome-aggregate processes affecting the stocks and dynamics of organic carbon.

Co-benefits and trade-offs: A transdisciplinary research effort is needed to understand the links between SOC sequestration, ecosystem services, and agricultural productivity and food security, to avoid trade-offs and optimise co-benefits of SOC sequestration. For example, selecting crops on their partitioning of biomass between aboveground and belowground carbon is a strategy that may increase soil carbon storage. Perennial crops such as perennial rice and intermediate wheat grass, (Cox et al. 2006), which invest more carbon into the roots, are likely to sequester more carbon than annual crops. However, research shows this practice may also lead to low grain productivity. Therefore, they may not prove economical to grow and the practice would not be feasible for adoption by farmers.

MRV: Another priority for research investment is improved capacity for rigorous but practical measurement, modelling and verification systems for SOC stock change. The target areas for investment include:

- Commitment to maintain and monitor short- and long-term field trials to quantify SOC sequestration across scales and management practices;
- More harmonious integration and quality control of data sources; and
- Advances in modelling capability linked to strategic measurement of validation sites and high-resolution remote sensing (Angelopoulou et al. 2019) and accurate spatial data for model initialisation and validation for diverse land use and management options (Rumpel et al. 2020).

Multi-disciplinary aspects: To clarify and define SOC sequestering practices for agriculture and forestry, including agroforestry and conservation forestry, research targeting sociological and economic considerations is needed. This research is of specific importance in assistance-deserving communities but is generally relevant to the design of all frameworks intended to influence landholders' long-term management decisions for adoption of practices to increase SOC (Sykes et al. 2020). The potential for private and public co-benefits and trade-offs of such decisions are not well understood (Brady et al. 2019) across the diverse range of management systems, landscapes and climates. The lack of confidence in environmental, economic and social impacts is a major deterrent to participation by farmers in outcome-based frameworks to foster uptake of better practice. Multidisciplinary collaboration based on cooperative research to understand realistic expectations (Rumpel et al. 2020) is needed to achieve improvements (e.g. Vermeulen et al. 2019).

4.3 Lessons learned on the role of government as the catalyst for action

Reflecting on ten years of experience with Australian policy and outcomes of a national framework to promote emissions abatement and carbon sequestration, it is clear that there is a consistent dynamic at play between 'government action and control' and 'industry action and initiative.' The notion that, at least in many nations, government can only ever be a 'catalyst' for industry initiative and action is evidenced in distinct ways in the Australian case study. Method development, a key element of the framework, was initially open to all in order to stimulate industry engagement. Later, method development moved more 'in-house' to the government, to reduce the duplication and work associated with the numerous industry-initiated methods being proposed. More recently, modular methods are being developed which set the overall framework and integrity structures but allow industry and/or others from within the science community to proposed 'models' as sub-components of methods. The umbrella method, however, sets the validation standard and process for such sub-models. There has been a consistent effort at consultation and review from the potential 'user community' (www.cleanenergyregulator.gov.au).

In parallel with these dynamics of method development, the Australia experience provides some insights into the evolution in the carbon offsets marketplace. Pre-2010, the markets were small and limited to voluntary schemes being offered to individuals and businesses. The advent of the CFI and later ERF policies introduced a government-funded marketplace that has invested more than A$575 million since 2014, with another A$1.4 billion committed by government, in the purchase of ACCUs that are eligible to be counted in the national greenhouse inventory accounts (CER 2021a). More recently, the private sector marketplace has been strengthening, which is something the government scheme seeks to encourage. For example, owners of ACCUs contracted to the government scheme under an 'optional' contract can choose to sell them onto the private marketplace rather than deliver them to the government at a fixed contracted price. Increasingly, there is a discussion of the development of a 'Carbon Exchange' that can facilitate public and private buying and selling of carbon offsets (www.cleanenergyregulator.gov.au/ 28 April 2021).

Irrespective of all these developments in methods and markets, our overarching conclusion is that a focus on the integrity of the carbon offsets must remain central at all stages of any policy framework if it is to continue to meet its objectives.

5 Where to look for further information

Recent publications provide informative overviews of the practical potential for SOC sequestration (Rumpel et al.), the challenges and outlook for accurately quantifying SOC change (Smith et al.) and the status and gaps in research (Chenu et al.), while websites discuss incentive mechanisms and illustrate a national policy framework to foster SOC sequestration.

5.1 Key publications

- Chenu, C., Angers, D. A., Barré, P., Derrien, D., Arrouays, D. and Balesdent, J. (2019). Increasing organic stocks in agricultural soils: Knowledge gaps and potential innovations. *Soil and Tillage Research* 188, 41-52.
- Rumpel, C., Amiraslani, F., Chenu, C., Cardenas, M. G., Kaonga, M., Koutika, L. S., Ladha, J., Madari, B., Shirato, Y., Smith, P. and Soudi, B. (2020). The 4p1000 initiative: Opportunities, limitations and challenges for implementing soil organic carbon sequestration as a sustainable development strategy. *Ambio* 49(1), 350–360.
- Smith, P., Soussana, J. F., Angers, D., Schipper, L., Chenu, C., Rasse, D. P., Batjes, N. H., Van Egmond, F., McNeill, S., Kuhnert, M. and Arias-Navarro, C. (2020). How to measure, report and verify soil carbon change to realize the potential of soil carbon sequestration for atmospheric greenhouse gas removal. *Global Change Biology* 26(1), 219–241.

5.2 Web resources

- Costa Jr, C., Dittmer, K., Shelton, S., Bossio, D., Zinyengere, N., Luu, P., Heinz, S., Egenolf, K., Rowland, B., Zuluaga, A., Klemme, J., Mealey, T., Smith, M. and Wollenberg, E. (2020). *How Soil Carbon Accounting Can Improve to Support Investment-Oriented Actions Promoting Soil Carbon Storage*. CCAFS Info Note. Wageningen, The Netherlands: CGIAR Research Program on Climate Change, Agriculture and Food Security (CCAFS). Available at: https://ccafs.cgiar.org/resources/publications/how-soil-carbon-accounting-can-improve-support-investment-oriented.
- The Australian Emissions Reduction Fund – How it works. http://www.cleanenergyregulator.gov.au/ERF/About-the-Emissions-Reduction-Fund/emissions-reduction-fund-schematic.
- The Australian ERF Measurement of soil carbon sequestration in agricultural systems method. http://www.cleanenergyregulator.gov.au/ERF/Pages/Choosing%20a%20project%20type/Opportunities%20for%20the%20land%20sector/Agricultural%20methods/The-measurement-of-soil-carbon-sequestration-in-agricultural-systems-method.aspx.

6 References

Alcock, D. J., Harrison, M. T., Rawnsley, R. P. and Eckard, R. J. (2015). Can animal genetics and flock management be used to reduce greenhouse gas emissions but also maintain productivity of wool-producing enterprises? *Agricultural Systems* 132, 25-34.

Alexander, P., Paustian, K., Smith, P. and Moran, D. (2015). The economics of soil C sequestration and agricultural emissions abatement. *SOIL* 1(1), 331-339.

Allen, D. E., Pringle, M. J., Bray, S., Hall, T. J., O'Reagain, P. O., Phelps, D., Cobon, D. H., Bloesch, P. M. and Dalal, R. C. (2013). What determines soil organic carbon stocks in the grazing lands of north-eastern Australia? *Soil Research* 51(8), 695-706.

Amundson, R. and Biardeau, L. (2018). Opinion: Soil carbon sequestration is an elusive climate mitigation tool. *Proceedings of the National Academy of Sciences of the United States of America* 115(46), 11652-11656.

Amundson, R., Berhe, A. A., Hopmans, J. W., Olson, C., Sztein, A. E. and Sparks, D. L. (2015). Soil science: Soil and human security in the 21st century. *Science* 348(6235), 1261071.

Angelopoulou, T., Tziolas, N., Balafoutis, A., Zalidis, G. and Bochtis, D. (2019). Remote sensing techniques for soil organic carbon estimation: A review. *Remote Sensing* 11(6), 676.

Arrouays, D., McBratney, A. B., Minasny, B., Hempel, J. W., Heuvelink, G. B. M., MacMillan, R. A., Hartemink, A. E., Lagacherie, P. and McKenzie, N. J. (2014). The GlobalSoilMap project specifications. *GlobalSoilMap: Basis of the Global Spatial Soil Information System*. CRC Press, Taylor and Francis Group, London, 494.

Australian Government (2015). Carbon credits (carbon farming initiative–estimating sequestration of carbon in soil using default values) methodology determination 2015. Available at: https://www.legislation.gov.au/Details/F2018C00126 [Accessed 6 December 2021].

Australian Government (2018). Carbon credits (carbon farming initiative–measurement of soil carbon sequestration in agricultural systems) methodology determination 2018. Available at: https://www.legislation.gov.au/Details/F2018L00089 [Accessed 6 December 2021].

Badgery, W., Murphy, B., Cowie, A., Orgill, S., Rawson, A., Simmons, A. and Crean, J. (2020). Soil carbon market-based instrument pilot-the sequestration of soil organic carbon for the purpose of obtaining carbon credits. *Soil Research* 59(1), 12-23.

Baldock, J. and Burgess, R. (2017). Soil Carbon Accounting-the Australian example. In: Workshop on "Soil Organic Carbon Measurement for Management and Management for Measurement". Available at: https://jahnresearchgroup.cals.wisc.edu/wp-content/uploads/sites/200/2017/08/3.-Soil-carbon-accounting-in-Australia-1.pdf [Accessed March 2021].

Balkcom, K. S., Arriaga, F. J. and Van Santen, E. (2013). Conservation systems to enhance soil carbon sequestration in the Southeast US Coastal Plain. *Soil Science Society of America Journal* 77(5), 1774-1783.

Bamière, L., Jayet, P. A., Kahindo, S. and Martin, E. (2021). Carbon sequestration in French agricultural soils: A spatial economic evaluation. *Agricultural Economics* 52(2), 301-316.

Bangsund, D. A. and Leistritz, F. L. (2008). Review of literature on economics and policy of carbon sequestration in agricultural soils. *Management of Environmental Quality: An International Journal* 19(1), 85–99. https://doi.org/10.1108/14777830810840381.

Barnwell, T. O., Jackson, R. B., Elliott, E. T., Burke, I. C., Cole, C. V., Paustian, K., Paul, E. A., Donigian, A. S., Patwardhan, A. S., Rowell, A. and Weinrich, K. (1992). An approach to assessment of management impacts on agricultural soil carbon. *Water, Air, and Soil Pollution* 64(1-2), 423–435.

Batjes, N. H. (2014). Total carbon and nitrogen in the soils of the world. *European Journal of Soil Science* 65(1), 10–21.

Baumber, A., Metternicht, G., Cross, R., Ruoso, L. E., Cowie, A. L. and Waters, C. (2019). Promoting co-benefits of carbon farming in Oceania: Applying and adapting approaches and metrics from existing market-based schemes. *Ecosystem Services* 39, 100982.

Bishop, R. C. and Welsh, M. P. (1992). Existence values in benefit-cost analysis and damage assessment. *Land Economics* 68(4), 405–417.

Börner, J., Baylis, K., Corbera, E., Ezzine-de-Blas, D., Honey-Rosés, J., Persson, U. M. and Wunder, S. (2017). The effectiveness of payments for environmental services. *World Development* 96, 359-374.

Bossio, D. A., Cook-Patton, S. C., Ellis, P. W., Fargione, J., Sanderman, J., Smith, P., Wood, S., Zomer, R. J., von Unger, M., Emmer, I. M. and Griscom, B. W. (2020). The role of soil carbon in natural climate solutions. *Nature Sustainability* 3(5), 391-398.

Bourdieu, P. (1986). The forms of capital. In: Richardson, J. (Ed.), *Handbook of Theory of Research for the Sociology of Education*. Greenwood, New York, 241–258.

Brady, M. V., Hristov, J., Wilhelmsson, F. and Hedlund, K. (2019). Evaluating farmers' provisioning of soil ecosystem services to inform agri-environmental policy. Brussels, Belgium 289822. European Association of Agricultural Economists, 172nd EAAE Seminar, May 28–29, 2019.

Broekhoff, D., Gillenwater, M., Colbert-Sangreeb, T. and Cage, P. (2019). *Securing Climate Benefit: A Guide to Using Carbon Offsets*. Offsetguide.org/pdf-download/. Stockholm Environment Institute & Greenhouse Gas Management Institute.

Burton, R. J. F. and Schwarz, G. (2013). Result-oriented agri-environmental schemes in Europe and their potential for promoting behavioural change. *Land Use Policy* 30(1), 628–641.

Causarano, H. J., Franzluebbers, A. J., Reeves, D. W. and Shaw, J. N. (2006). Soil organic carbon sequestration in cotton production systems of the southeastern United States: A review. *Journal of Environmental Quality* 35(4), 1374–1383.

CER (2021a). Clean energy regulator. *Emissions Position as* [Accessed at April 2021]. Available at: http://www.cleanenergyregulator.gov.au/Infohub/Media-Centre/Pages/Resources/ERF%20media%20resources/Emissions-Position-as-at-April-2021.aspx [Accessed 30 May 2021].

CER (2021b). Clean energy regulator. Quarterly Carbon Market Report. *Quarter* 4, 2020. Available at: http://www.cleanenergyregulator.gov.au/csf/market-information/Pages/quarterly-Market-report.aspx [Accessed 30 March 2021].

Chan, K. Y., Conyers, M. K., Li, G. D., Helyar, K. R., Poile, G., Oates, A. and Barchia, I. M. (2011). Soil carbon dynamics under different cropping and pasture management in temperate Australia: Results of three long-term experiments. *Soil Research* 49(4), 320–328.

Chang, J., Ciais, P., Gasser, T., Smith, P., Herrero, M., Havlík, P., Obersteiner, M., Guenet, B., Goll, D. S., Li, W., Naipal, V., Peng, S., Qiu, C., Tian, H., Viovy, N., Yue, C. and Zhu, D. (2021). Climate warming from managed grasslands cancels the cooling effect of carbon sinks in sparsely grazed and natural grasslands. *Nature Communications* 12(1), 118.

Chang-Fung-Martel, J., Harrison, M. T., Rawnsley, R., Smith, A. P. and Meinke, H. (2017). The impact of extreme climatic events on pasture-based dairy systems: A review. *Crop and Pasture Science* 68(12), 1158–1169.

Chenu, C., Angers, D. A., Barré, P., Derrien, D., Arrouays, D. and Balesdent, J. (2019). Increasing organic stocks in agricultural soils: Knowledge gaps and potential innovations. *Soil and Tillage Research* 188, 41–52.

Claassen, R., Cattaneo, A. and Johansson, R. (2008). Cost-effective design of agri-environmental payment programs: US experience in theory and practice. *Ecological Economics* 65(4), 737–752.

Commonwealth of Australia (2020). Report of the Expert Panel examining additional sources of low cost abatement. *Department of Industry, Science, Energy and Resources*. Available at: https://www.industry.gov.au/sites/default/files/2020-05/expert-panel-report-examining-additional-sources-of-low-cost-abatement.pdf [Accessed May 2021].

Conant, R. T., Paustian, K. and Elliott, E. T. (2001). Grassland management and conversion into grassland: Effects on soil carbon. *Ecological Applications* 11(2), 343–355.

Costa Jr., C., Dittmer, K., Shelton, S., Bossio, D., Zinyengere, N., Luu, P., Heinz, S., Egenolf, K., Rowland, B., Zuluaga, A., Klemme, J., Mealey, T., Smith, M. and Wollenberg, E. (2020). How soil carbon accounting can improve to support investment-oriented actions promoting soil carbon storage. *CCAFS Info Note*. CGIAR Research Program on Climate Change, Agriculture and Food Security (CCAFS), Wageningen, The Netherlands.

Cotrufo, M. F., Wallenstein, M. D., Boot, C. M., Denef, K. and Paul, E. (2013). The Microbial Efficiency Matrix Stabilization (MEMS) framework integrates plant litter decomposition with soil organic matter stabilization: Do labile plant inputs form stable soil organic matter? *Global Change Biology* 19(4), 988–995.

COWI, Ecologic Institute and IEEP (2021). *Technical Guidance Handbook - Setting up and Implementing Result-Based Carbon Farming Mechanisms in the EU Report to the European Commission*, DG Climate Action, under Contract No. CLIMA/C.3/ETU/2018/007. COWI, Kongens Lyngby.

Cox, T. S., Glover, J. D., Van Tassel, D. L., Cox, C. M. and DeHaan, L. R. (2006). Prospects for developing perennial grain crops. *BioScience* 56(8), 649–659.

Dalal, R. C., Thornton, C. M., Allen, D. E. and Kopittke, P. M. (2021a). A study over 33 years shows that carbon and nitrogen stocks in a subtropical soil are increasing under native vegetation in a changing climate. *Science of the Total Environment* 772, 145019.

Dalal, R. C., Thornton, C. M., Allen, D. E., Owens, J. S. and Kopittke, P. M. (2021b). Long-term land use change in Australia from native forest decreases all fractions of soil organic carbon, including resistant organic carbon, for cropping but not sown pasture. *Agriculture, Ecosystems and Environment* 311, 107326.

de Gruijter, J. J., McBratney, A. B., Minasny, B., Wheeler, I., Malone, B. P. and Stockmann, U. (2018). Farm-scale soil carbon auditing. In: McBratney, A., Minasny, B. and Stockmann, U. (Eds) *Pedometrics. Progress in Soil Science*. Springer, Cham, 693–720. https://doi.org/10.1007/978-3-319-63439-5_23.

DES (2020). Land restoration fund: Priority investment plan. Department of Environment and Science Government. Available at: https://www.qld.gov.au/__data/assets/pdf_file/0024/116547/lrf-priority-investment-plan.pdf [Accessed May 2021].

Dooley, K., Christoff, P. and Nicholas, K. A. (2018). Co-producing climate policy and negative emissions: Trade-offs for sustainable land-use. *Global Sustainability* 1, 1-10.

ERAC (2021). Emissions Reduction Assurance Committee Information Paper: Committee considerations for interpreting the Emissions Reduction Fund's offsets integrity standards Version 2.0 [Accessed March 2021]. Available at: http://www.cleanenergyregulator.gov.au/DocumentAssets/Documents/Information%20Paper%20on%20the%20Offsets%20Integrity%20Standards.pdf [Accessed May 2021].

FAO (2019). Measuring and Modelling Soil Carbon Stocks and Stock Changes in Livestock Production Systems: Guidelines for Assessment (1st version). Livestock Environmental Assessment and Performance (LEAP) Partnership. Food and Agriculture Organization, Rome. 170 pp. Licence: CC BY-NC-SA 3.0 IGO.

Flores-Rios, A., Thomas, E., Peri, P. P., Amelung, W., Duarte-Guardia, S., Borchard, N., Lizárraga-Travaglini, A., Vélez-Azañero, A., Sheil, D., Tscharntke, T., Steffan-Dewenter, I. and Ladd, B. (2020). Co-benefits of soil carbon protection for invertebrate conservation. *Biological Conservation* 252, 108859.

Follett, R. F., Vogel, K. P., Varvel, G. E., Mitchell, R. B. and Kimble, J. (2012). Soil carbon sequestration by switchgrass and no-till maize grown for bioenergy. *Bioenergy Research* 5(4), 866-875.

Franzluebbers, A. J. (2005). Soil organic carbon sequestration and agricultural greenhouse gas emissions in the southeastern USA. *Soil and Tillage Research* 83(1), 120-147.

Franzluebbers, A. J. and Stuedemann, J. A. (2009). Soil-profile organic carbon and total nitrogen during 12 years of pasture management in the Southern Piedmont USA. *Agriculture, Ecosystems and Environment* 129(1-3), 28-36.

Fuss, S., Lamb, W. F., Callaghan, M. W., Hilaire, J., Creutzig, F., Amann, T., Beringer, T., de Oliveira Garcia, W., Hartmann, J., Khanna, T., Luderer, G., Nemet, G. F., Rogelj, J., Smith, P., Vicente, J. L. V., Wilcox, J., del Mar Zamora Dominguez, M. and Minx, J. C. (2018). Negative emissions–Part 2: Costs, potentials and side effects. *Environmental Research Letters* 13(6), 063002.

Gerrard, E. (2008). Impacts and opportunities of climate change: Indigenous participation in environmental markets. *Land, Rights, Laws: Issues of Native Title* 3(13), 1-14.

Gil, J. D. B., Garrett, R. D., Rotz, A., Daioglou, V., Valentim, J., Pires, G. F., Costa, M. H., Lopes, L. and Reis, J. C. (2018). Tradeoffs in the quest for climate smart agricultural intensification in Mato Grosso, Brazil. *Environmental Research Letters* 13(6), 064025.

Goddard, T. (2008). The Alberta carbon offset system. *Proceedings of the Agronomy Update* 16, 90-93.

Gold Standard (2020). *Soil Organic Carbon Methodologies*. Available at: https://globalgoals.goldstandard.org/400-sdg-impact-quantification/ [Accessed June 2021].

Govaerts, B., Verhulst, N., Castellanos-Navarrete, A., Sayre, K. D., Dixon, J. and Dendooven, L. (2009). Conservation agriculture and soil carbon sequestration: Between myth and farmer reality. *Critical Reviews in Plant Sciences* 28(3), 97-122.

Griscom, B. W., Adams, J., Ellis, P. W., Houghton, R. A., Lomax, G., Miteva, D. A., Schlesinger, W. H., Shoch, D., Siikamäki, J. V., Smith, P., Woodbury, P., Zganjar, C., Blackman, A., Campari, J., Conant, R. T., Delgado, C., Elias, P., Gopalakrishna, T., Hamsik, M. R., Herrero, M., Kiesecker, J., Landis, E., Laestadius, L., Leavitt, S. M., Minnemeyer, S.,

Polasky, S., Potapov, P., Putz, F. E., Sanderman, J., Silvius, M., Wollenberg, E. and Fargione, J. (2017). Natural climate solutions. *Proceedings of the National Academy of Sciences of the United States of America* 114(44), 11645-11650.

Guo, L. B. and Gifford, R. M. (2002). Soil carbon stocks and land use change: A meta analysis. *Global Change Biology* 8(4), 345-360.

Harrison, M. T., Cullen, B. R., Mayberry, D. E., Cowie, A. L., Bilotto, F., Badgery, W. B., Liu, K., Davison, T., Christie, K. M., Muleke, A. and Eckard, R. J. (2021). Carbon myopia: The urgent need for integrated social, economic and environmental action in the livestock sector. *Global Change Biology* 27(22), 5726-5761. Available: matthew.harrison@utas.edu.au.

Harrison, M. T., Evans, J. R. and Moore, A. D. (2012a). Using a mathematical framework to examine physiological changes in winter wheat after livestock grazing: 1. *Field Crops Research* 136, 116-126.

Harrison, M. T., Evans, J. R. and Moore, A. D. (2012b). Using a mathematical framework to examine physiological changes in winter wheat after livestock grazing: 2. *Field Crops Research* 136, 127-137.

Heckbert, S., Davies, J., Cook, G., McIvor, J., Bastin, G. and Liedloff, A. (2008). *Land Management for Emissions Offsets on Indigenous Lands*. CSIRO Ecosystem Sciences, Townsville, 62 pp.

Henry, B., Murphy, B. and Cowie, A. (2018). Sustainable land management for environmental benefits and food security. *A Synthesis Report for the GEF*. Washington, DC.

Ho, C. K. M., Jackson, T., Harrison, M. T. and Eckard, R. J. (2014). Increasing ewe genetic fecundity improves whole-farm production and reduces greenhouse gas emissions intensities: 2. Economic performance. *Animal Production Science* 54(9), 1248-1253.

IPCC (2019). In: Shukla, P. R., Skea, J., Calvo Buendia, E., Masson-Delmotte, V., Pörtner, H.-O., Roberts, D. C., Zhai, P., Slade, R., Connors, S., van Diemen, R., Ferrat, M., Haugheym, E., Luz, S., Neogi, S., Pathak, M., Petzold, J., Portugal Pereira, J., Vyas, P., Huntley, E., Kissick, K., Belkacemi, M. and Malley, J. (Eds) Climate change and land: An IPCC special report on climate change, desertification, land degradation, sustainable land management, food security, and greenhouse gas fluxes in terrestrial ecosystems. Intergovernmental Panel on Climate Change, 1-41.

Jackson Hammond, A. A., Motew, M., Brummitt III, C. D., DuBuisson, M. L., Pinjuv, G., Harburg, D. V., Campbell, E. E. and Kumar, A. A. (2021). Implementing the soil enrichment protocol at scale: Opportunities for an agricultural carbon market. *Frontiers in Climate* 3, 64-71.

Jenkinson, D. S. (1990). The turnover of organic carbon and nitrogen in soil. *Philosophical Transactions of the Royal Society of London. Series B: Biological Sciences* 329(1255), 361-368.

Johnson, J. M. F., Reicosky, D. C., Allmaras, R. R., Sauer, T. J., Venterea, R. T. and Dell, C. J. (2005). Greenhouse gas contributions and mitigation potential of agriculture in the central USA. *Soil and Tillage Research* 83(1), 73-94.

Johnston, A. E., Poulton, P. R. and Coleman, K. (2009). Soil organic matter: Its importance in sustainable agriculture and carbon dioxide fluxes. *Advances in Agronomy* 101, 1-57.

Keating, B. A. and Thorburn, P. J. (2018). Modelling crops and cropping systems–evolving purpose, practice and prospects. *European Journal of Agronomy* 100, 163-176.

Keesstra, S. D., Bouma, J., Wallinga, J., Tittonell, P., Smith, P., Cerdà, A., Montanarella, L., Quinton, J. N., Pachepsky, Y., van der Putten, W. H., Bardgett, R. D., Moolenaar, S.,

Mol, G., Jansen, B. and Fresco, L. O. (2016). The significance of soils and soil science towards realization of the United Nations Sustainable Development Goals. *SOIL* 2(2), 111-128.

Keesstra, S., Nunes, J., Novara, A., Finger, D., Avelar, D., Kalantari, Z. and Cerdà, A. (2018). The superior effect of nature based solutions in land management for enhancing ecosystem services. *Science of the Total Environment* 610-611, 997-1009.

Kong, A. Y. Y., Fonte, S. J., van Kessel, C. and Six, J. (2009). Transitioning from standard to minimum tillage: Trade-offs between soil organic matter stabilization, nitrous oxide emissions, and N availability in irrigated cropping systems. *Soil and Tillage Research* 104(2), 256-262.

Kragt, M. E., Gibson, F. L., Maseyk, F. and Wilson, K. A. (2016). Public willingness to pay for carbon farming and its co-benefits. *Ecological Economics* 126, 125-131.

Kragt, M. E., Pannell, D. J., Robertson, M. J. and Thamo, T. (2012). Assessing costs of soil carbon sequestration by crop-livestock farmers in Western Australia. *Agricultural Systems* 112, 27-37.

Lal, R. (2004). Soil carbon sequestration impacts on global climate change and food security. *Science* 304(5677), 1623-1627.

Lal, R. (2015a). A system approach to conservation agriculture. *Journal of Soil and Water Conservation* 70(4), 82A-88A.

Lal, R. (2015b). Cover cropping and the "4 per Thousand" proposal. *Journal of Soil and Water Conservation* 70(6), 141A-141A.

Lal, R. (2015c). Sequestering carbon and increasing productivity by conservation agriculture. *Journal of Soil and Water Conservation* 70(3), 55A-62A.

Lal, R., Follett, R. F. and Kimble, J. M. (2003). Achieving soil carbon sequestration in the United States: A challenge to the policy makers. *Soil Science* 168(12), 827-845.

Lal, R., Negassa, W. and Lorenz, K. (2015). Carbon sequestration in soil. *Current Opinion in Environmental Sustainability* 15, 79-86.

Latacz-Lohmann, U. and Schilizzi, S. (2007). Quantifying the benefits of conservation auctions. *EuroChoices* 6(3), 32-39.

Liebig, M. A., Schmer, M. R., Vogel, K. P. and Mitchell, R. B. (2008). Soil carbon storage by switchgrass grown for bioenergy. *Bioenergy Research* 1(3-4), 215-222.

Macintosh, A., Roberts, G. and Buchan, S. (2019). Improving carbon markets to increase farmer participation. AgriFutures Australia Publication No. 19-026. Commonwealth of Australia. [Accessed July 2019]. Available at: www.agrifutures.com.au [Accessed March 2020].

Martens, D. A., Emmerich, W., McLain, J. E. and Johnsen, T. N. (2005). Atmospheric carbon mitigation potential of agricultural management in the southwestern USA. *Soil and Tillage Research* 83(1), 95-119.

Matzdorf, B. and Lorenz, J. (2010). How cost-effective are result-oriented agri-environmental measures? An empirical analysis in Germany. *Land Use Policy* 27(2), 535-544.

Minasny, B., Malone, B. P., McBratney, A. B., Angers, D. A., Arrouays, D., Chambers, A., Chaplot, V., Chen, Z. S., Cheng, K., Das, B. S., Field, D. J., Gimona, A., Hedley, C. B., Hong, S. Y., Mandal, B., Marchant, B. P., Martin, M., McConkey, B. G., Mulder, V. L., O'Rourke, S., Richer-de-Forges, A. C., Odeh, I., Padarian, J., Paustian, K., Pan, G., Poggio, L., Savin, I., Stolbovoy, V., Stockmann, U., Sulaeman, Y., Tsui, C., Vågen, T., van Wesemael, B. and Winowiecki, L. (2017). Soil carbon 4 per mille. *Geoderma* 292, 59-86.

Morgan, M. I., Hine, D. W., Bhullar, N. and Loi, N. M. (2015). Landholder adoption of low emission agricultural practices: A profiling approach. *Journal of Environmental Psychology* 41, 35–44.

Mudge, P., McNeill, S., Hedley, C., Roudier, P., Poggio, M. and Whenua, M. (2020). Design of an on-farm soil carbon benchmarking and monitoring approach for individual pastoral farms. MPI Technical Paper No: 2020/02. Ministry for Primary Industries, Wellington, NZ Available at: http://www.mpi.govt.nz/news-resources/publications.aspx.

Müller, B. and Michaelowa, A. (2019). How to operationalize accounting under Article 6 market mechanisms of the Paris Agreement. *Climate Policy* 19(7), 812–819, Article 6.

Murphy, B. (2020). Chapter 20: Soil carbon sequestration as an elusive climate mitigation tool. In: Dang Y., Dalal R. and Menzies, N. (Eds) *No-Till Farming Systems for Sustainable Agriculture*. Springer, Cham, 337–353. https://doi.org/10.1007/978-3-030-46409-7_20.

Murphy, B. W., Wilson, B. R. and Koen, T. (2019). Mathematical functions to model the depth distribution of soil organic carbon in a range of soils from New South Wales, Australia under different land uses. *Soil Systems* 3(3), 46.

Nelson, E., Polasky, S., Lewis, D. J., Plantinga, A. J., Lonsdorf, E., White, D., Bael, D. and Lawler, J. J. (2008). Efficiency of incentives to jointly increase carbon sequestration and species conservation on a landscape. *Proceedings of the National Academy of Sciences of the United States of America* 105(28), 9471–9476.

Ogle, S. M., Breidt, F. J. and Paustian, K. (2005). Agricultural management impacts on soil organic carbon storage under moist and dry climatic conditions of temperate and tropical regions. *Biogeochemistry* 72(1), 87–121.

Paustian, K. A. O. J. H., Andren, O., Janzen, H. H., Lal, R., Smith, P., Tian, G., Tiessen, H., Van Noordwijk, M. and Woomer, P. L. (1997). Agricultural soils as a sink to mitigate CO_2 emissions. *Soil Use and Management* 13(s4), 230–244.

Paustian, K., Collier, S., Baldock, J., Burgess, R., Creque, J., DeLonge, M., Dungait, J., Ellert, B., Frank, S., Goddard, T., Govaerts, B., Grundy, M., Henning, M., Izaurralde, R. C., Madaras, M., McConkey, B., Porzig, E., Rice, C., Searle, R., Seavy, N., Skalsky, R., Mulhern, W. and Jahn, M. (2019). Quantifying carbon for agricultural soil management: From the current status toward a global soil information system. *Carbon Management* 10(6), 567–587.

Paustian, K., Lehmann, J., Ogle, S., Reay, D., Robertson, G. P. and Smith, P. (2016). Climate-smart soils. *Nature* 532(7597), 49–57.

Peng, W., Iyer, G., Bosetti, V., Chaturvedi, V., Edmonds, J., Fawcett, A. A., Hallegatte, S., Victor, D. G., van Vuuren, D. and Weyant, J. (2021). Climate policy models need to get real about people—here's how. *Nature* 594(7862), 174–176.

Poeplau, C. and Don, A. (2015). Carbon sequestration in agricultural soils via cultivation of cover crops-A meta-analysis. *Agriculture, Ecosystems and Environment* 200, 33–41.

Poulton, P., Johnston, J., Macdonald, A., White, R. and Powlson, D. (2018). Major limitations to achieving "4 per 1000" increases in soil organic carbon stock in temperate regions: Evidence from long-term experiments at Rothamsted Research, United Kingdom. *Global Change Biology* 24(6), 2563–2584.

Powlson, D. S., Smith, P., Coleman, K., Smith, J. U., Glendining, M. J., Körschens, M. and Franko, U. (1998). A European network of long-term sites for studies on soil organic matter. *Soil and Tillage Research* 47(3–4), 263–274.

Powlson, D. S., Whitmore, A. P. and Goulding, K. W. T. (2011). Soil carbon sequestration to mitigate climate change: a critical re-examination to identify the true and the false. *European Journal of Soil Science* 62(1), 42–55.

Rabbi, S. M. F., Tighe, M., Delgado-Baquerizo, M., Cowie, A., Robertson, F., Dalal, R., Page, K., Crawford, D., Wilson, B. R., Schwenke, G., Mcleod, M., Badgery, W., Dang, Y. P., Bell, M., O'Leary, G., Liu, D. L. and Baldock, J. (2015). Climate and soil properties limit the positive effects of land use reversion on carbon storage in Eastern Australia. *Scientific Reports* 5(1), 17866.

Rawnsley, R., Dynes, R. A., Christie, K. M., Harrison, M. T., Doran-Browne, N. A., Vibart, R. and Eckard, R. (2016). A review of whole farm-system analysis in evaluating greenhouse-gas mitigation strategies from livestock production systems. *Animal Production Science* 58(6), 980–989.

Robinson, C. J., Renwick, A. R., May, T., Gerrard, E., Foley, R., Battaglia, M., Possingham, H., Griggs, D. and Walker, D. (2016). Indigenous benefits and carbon offset schemes: An Australian case study. *Environmental Science and Policy* 56, 129–134.

Rockström, J., Gaffney, O., Rogelj, J., Meinshausen, M., Nakicenovic, N. and Schellnhuber, H. J. (2017). A roadmap for rapid decarbonization. *Science* 355(6331), 1269–1271.

Rogers, E. M. (2004). A prospective and retrospective look at the diffusion model. *Journal of Health Communication* 9 (Suppl. 1), 13–19.

Rumpel, C., Amiraslani, F., Koutika, L. S., Smith, P., Whitehead, D. and Wollenberg, E. (2018). Put more carbon in soils to meet Paris climate pledges. *Nature* 564(7734), 32–34.

Rumpel, C., Amiraslani, F., Chenu, C., Cardenas, M. G., Kaonga, M., Koutika, L. S., Ladha, J., Madari, B., Shirato, Y., Smith, P., Soudi, B., Soussana, J. F., Whitehead, D. and Wollenberg, E. (2020). The 4p1000 initiative: Opportunities, limitations and challenges for implementing soil organic carbon sequestration as a sustainable development strategy. *Ambio* 49(1), 350–360.

Russell, E. J. (1912). *Soil Conditions and Plant Growth*. 1st edn. Longmans, London.

Sainju, U. M., Jabro, J. D. and Stevens, W. B. (2008). Soil carbon dioxide emissions as affected by irrigation, tillage, cropping system and nitrogen fertilization. *Journal of Environmental Quality* 37(1), 98–106.

Sampson, R. N. and Sedjo, R. A. (1997). Economics of carbon sequestration in forestry: An overview. *Critical Reviews in Environmental Science and Technology* 27(Suppl.1), 1–8.

Sanderman, J. and Baldock, J. A. (2010). Accounting for soil carbon sequestration in national inventories: A soil scientist's perspective. *Environmental Research Letters* 5(3), 034003.

Sanderman, J., Farquharson, R. and Baldock, J. (2010). Soil carbon sequestration potential: A review for Australian agriculture. Report to the Australian Government Department of Climate Change and Energy Efficiency. CSIRO, Canberra, Australia.

Sanderman, J., Hengl, T. and Fiske, G. J. (2017). Soil carbon debt of 12,000 years of human land use. *Proceedings of the National Academy of Sciences of the United States of America* 114(36), 9575–9580.

Schlesinger, W. H. and Amundson, R. (2019). Managing for soil carbon sequestration: Let's get realistic. *Global Change Biology* 25(2), 386–389.

Schneider, L., Duan, M., Stavins, R., Kizzier, K., Broekhoff, D., Jotzo, F., Winkler, H., Lazarus, M., Howard, A. and Hood, C. (2019). Double counting and the Paris Agreement rulebook. *Science* 366(6462), 180–183.

Schneider, L. and La Hoz Theuer, S. (2019). Environmental integrity of international carbon market mechanisms under the Paris Agreement. *Climate Policy* 19(3), 386-400.

Schuman, G. E., Janzen, H. H. and Herrick, J. E. (2002). Soil carbon dynamics and potential carbon sequestration by rangelands. *Environmental Pollution* 116(3), 391-396.

Smith, P., Smith, J. U., Powlson, D. S., McGill, W. B., Arah, J. R. M., Chertov, O. G., Coleman, K., Franko, U., Frolking, S., Jenkinson, D. S., Jensen, L. S., Kelly, R. H., Klein-Gunnewiek, H., Komarov, A. S., Li, C., Molina, J. A. E., Mueller, T., Parton, W. J., Thornley, J. H. M. and Whitmore, A. P. (1997). A comparison of the performance of nine soil organic matter models using datasets from seven long-term experiments. *Geoderma* 81(1-2), 153-225.

Smith, P., Soussana, J. F., Angers, D., Schipper, L., Chenu, C., Rasse, D. P., Batjes, N. H., Van Egmond, F., McNeill, S., Kuhnert, M., Arias-Navarro, C., Olesen, J. E., Chirinda, N., Fornara, D., Wollenberg, E., Álvaro-Fuentes, J., Sanz-Cobena, A. and Klumpp, K. (2020). How to measure, report and verify soil carbon change to realize the potential of soil carbon sequestration for atmospheric greenhouse gas removal. *Global Change Biology* 26(1), 219-241.

Stockmann, U., Adams, M. A., Crawford, J. W., Field, D. J., Henakaarchchi, N., Jenkins, M., Minasny, B., McBratney, A. B., De Courcelles, VdRd, Singh, K., Wheeler, I., Abbott, L., Angers, D. A., Baldock, J., Bird, M., Brookes, P. C., Chenu, C., Jastrow, J. D., Lal, R., Lehmann, J., O'Donnell, A. G., Parton, W. J., Whitehead, D. and Zimmermann, M. (2013). The knowns, known unknowns and unknowns of sequestration of soil organic carbon. *Agriculture, Ecosystems and Environment* 164, 80-99.

Sun, W., Canadell, J. G., Yu, L., Yu, L., Zhang, W., Smith, P., Fischer, T. and Huang, Y. (2020). Climate drives global soil carbon sequestration and crop yield changes under conservation agriculture. *Global Change Biology* 26(6), 3325-3335.

Sykes, A. J., Macleod, M., Eory, V., Rees, R. M., Payen, F., Myrgiotis, V., Williams, M., Sohi, S., Hillier, J., Moran, D., Manning, D. A. C., Goglio, P., Seghetta, M., Williams, A., Harris, J., Dondini, M., Walton, J., House, J. and Smith, P. (2020). Characterising the biophysical, economic and social impacts of soil carbon sequestration as a greenhouse gas removal technology. *Global Change Biology* 26(3), 1085-1108.

Terrer, C., Phillips, R. P., Hungate, B. A., Rosende, J., Pett-Ridge, J., Craig, M. E., van Groenigen, K. J., Keenan, T. F., Sulman, B. N., Stocker, B. D., Reich, P. B., Pellegrini, A. F. A., Pendall, E., Zhang, H., Evans, R. D., Carrillo, Y., Fisher, J. B., Van Sundert, K., Vicca, S. and Jackson, R. B. (2021). A trade-off between plant and soil carbon storage under elevated CO_2. *Nature* 591(7851), 599-603.

TFCD (2021). Task Force on Climate-related Financial Disclosures. Available at: https://www.fsb-tcfd.org/ [Accessed June 2021].

Trost, B., Prochnow, A., Drastig, K., Meyer-Aurich, A., Ellmer, F. and Baumecker, M. (2013). Irrigation, soil organic carbon and N_2O emissions. A review. *Agronomy for Sustainable Development* 33(4), 733-749.

UNFCCC (2021). Paris agreement - Status of ratification. Available at: https://unfccc.int/process/the-paris-agreement/status-of-ratification [Accessed 24 June 2021].

Van Groenigen, J. W., Van Kessel, C., Hungate, B. A., Oenema, O., Powlson, D. S. and Van Groenigen, K. J. (2017). Sequestering soil organic carbon: A nitrogen dilemma. *Environmental Science and Technology* 51(9), 4738-4739.

Vermeulen, S., Bossio, D., Lehmann, J., Luu, P., Paustian, K., Webb, C., Augé, F., Bacudo, I., Baedeker, T., Havemann, T., Jones, C., King, R., Reddy, M., Sunga, I., Von Unger, M. and Warnken, M. (2019). A global agenda for collective action on soil carbon. *Nature Sustainability* 2(1), 2-4.

Verra (2012). Verified carbon standard, VCS. Available at: https://verra.org/wp-content/uploads/2018/03/VM0021-Soil-Carbon-Quantification-Methodology-v1.0.pdf [Accessed May 2021].

Viscarra Rossel, R. A., Behrens, T., Ben-Dor, E., Brown, D. J., Demattê, J. A. M., Shepherd, K. D., Shi, Z., Stenberg, B., Stevens, A., Adamchuk, V., Aïchi, H., Barthès, B. G., Bartholomeus, H. M., Bayer, A. D., Bernoux, M., Böttcher, K., Brodský, L., Du, C. W., Chappell, A., Fouad, Y., Genot, V., Gomez, C., Grunwald, S., Gubler, A., Guerrero, C., Hedley, C. B., Knadel, M., Morrás, H. J. M., Nocita, M., Ramirez-Lopez, L., Roudier, P., Campos, E. M. R., Sanborn, P., Sellitto, V. M., Sudduth, K. A., Rawlins, B. G., Walter, C., Winowiecki, L. A., Hong, S. Y. and Ji, W. (2016). A global spectral library to characterize the world's soil. *Earth-Science Reviews* 155, 198-230.

Wiesmeier, M., Urbanski, L., Hobley, E., Lang, B., von Lützow, M., Marin-Spiotta, E., van Wesemael, B., Rabot, E., Ließ, M., Garcia-Franco, N., Wollschläger, U., Vogel, H. and Kögel-Knabner, I. (2019). Soil organic carbon storage as a key function of soils-A review of drivers and indicators at various scales. *Geoderma* 333, 149-162.

Wilpiszeski, R. L., Aufrecht, J. A., Retterer, S. T., Sullivan, M. B., Graham, D. E., Pierce, E. M., Zablocki, O. D., Palumbo, A. V. and Elias, D. A. (2019). Soil aggregate microbial communities: Towards understanding microbiome interactions at biologically relevant scales. *Applied and Environmental Microbiology* 85(14), e00324-19.

Zhang, Y., Guo, L., Chen, Y., Shi, T., Luo, M., Ju, Q., Zhang, H. and Wang, S. (2019). Prediction of soil organic carbon based on landsat 8 monthly NDVI data for the Jianghan Plain in Hubei Province, China. *Remote Sensing* 11(14), 1683.

Chapter 27

Economic considerations for the development of a carbon farming scheme

Siân Mooney and Kathryn Janoski, O'Neill School of Public and Environmental Affairs, Indiana University, USA

1 Introduction

A conversation around reducing atmospheric concentrations of greenhouse gases (GHGs) through additional sequestration of soil carbon (C) has been ongoing for more than 20 years. During that time, significant work has been undertaken to support the creation of market opportunities that will enable the trade of C-credits from projects that increase sequestration of C[1] in agricultural soils. Despite this, there still exist very few soil C projects globally (Amelung et al., 2020; Bossio et al., 2020).

To meet the goal of the Paris Agreement and limit global warming to 1.5°C, emissions of GHGs must fall to 45% below 2010 levels by 2030 (IPCC, 2018). Estimates suggest that C prices must reach \$50–100/tCO_2 by 2030 to cost-effectively reduce emissions in line with these temperature goals (High-Level Commission on Carbon Prices 2017). Emissions from industrial, transportation and other sector sources of GHGs have been included in national GHG inventories and C markets for about 20 years. The Paris Agreement explicitly recognizes the potential contributions of nature-based

1 This chapter focuses on soil organic carbon sequestration.

http://dx.doi.org/10.19103/AS.2022.0106.26

activities as a means for nations to reduce their GHGs, renewing interest in the potential to develop GHG mitigation activities from the agricultural sector by avoiding CO_2 and non-CO_2 emissions and increased accumulation of C in soils (IPCC, 2018).

Although the technical potential to sequester additional C in soils is large, the actual economic and social potential is smaller (Antle et al., 2003; Amelung et al., 2020; Fuss et al., 2018; Griscom et al., 2017) and is constrained by factors such as lack of markets or financial supports for soil carbon sequestration (SCS), producer risk aversion, social constraints and existing policy structures among other factors (Amelung et al., 2020; Bossio et al., 2020). These challenges to developing robust mechanisms that support C farming[2] can be roughly divided into two types, those facing individual agricultural producers who want to receive payments for sequestering additional C and those associated with creating transparent and credible mechanisms to support broader market activity. This chapter presents the current state of C markets and some of the challenges faced by individual producers, private companies and governments interested in C farming. We identify the two primary payment design mechanisms proposed for C payments and the complexities faced by individual producers navigating the requirements of the current credit markets including lack of uniformity across markets (compliance and voluntary) and low acceptance of credits from SCS that affect finance. We also address design elements that affect transaction costs such as measurement, reporting and verification; the spatial scale of projects; the need for producer technical assistance and opportunities; and questions surrounding compensation for other bundled attributes such as ecosystem services.

2 Global potential to sequester additional soil carbon

Soil C can be increased by changing current agricultural production and management practices to those that either increase C inputs to the soil from plants and/or store a larger proportion of plant C within the soil for longer term and/or slow decomposition (Paustian et al., 1997). Management and production choices that can increase soil C sequestration include shifting from annual to perennial plants, increasing the duration of vegetation cover (e.g. through planting cover crops and reducing summer fallow), increasing the vegetative productivity of existing crops, reducing soil disturbance (such as moving to no-till systems) and flood management/restoration of wetland conditions (Alexander et al., 2015). A very large scientific literature identifies the

2 The term carbon (C) farming is sometimes used more broadly to include management of all GHGs on a farm. This chapter focuses on soil C, but many of the discussions and considerations are relevant to managing other GHGs.

soil C sequestration potential of numerous management changes globally (e.g. Eagle et al. 2012, Franzlubbers 2010, Ogle et al. 2005, among others).

2.1 Estimates of global technical potential

Fuss et al. (2018) synthesize the literature containing estimates of the technical potential for SCS and provide an excellent discussion of many studies. They conclude that the mean global technical potential for SCS ranges between 2.91 and 5.65 $GtCO_{2e}yr^{-1}$. Global estimates of C sequestration from the use of cover crops only are 0.44 ± 0.11 $PgCO_{2e}yr^{-1}$ (Poeplau and Don, 2015). There is significant heterogeneity in the potential to sequester additional soil C regionally and many different agricultural production and land management practices that could be adopted (Amelung et al., 2020).

2.2 Estimates of global carbon supply and regional participation at different incentive levels

Opportunities to farm C vary regionally and are dependent on the suite of practices suitable for each region, historical production practices, climate, soils, capital availability, policy environment, market opportunities, economics and other factors (Amelung et al., 2020). These factors constrain the SCS potential globally to less than the technical potential. Incentives/market prices for soil C will have a significant influence on producers' participation in C farming.

There is significant heterogeneity in producer costs regionally, which will affect the global pattern of farmer participation. Sixty-two economic analyses that address producer participation at different incentive levels are summarized in Tang et al. (2016). Results reveal that producers experience very different costs associated with sequestering additional soil C, ranging between US\$3/tCO_{2e} to US\$130/tCO_{2e}.[3] These are the minimum payments necessary to encourage producers in these regions to sequester additional soil C using the practices listed and highlight the considerable heterogeneity of opportunity globally. Table 1 summarizes results from studies that estimate the costs of SCS across different regions of the world.

The global soil carbon sequestration potential at US\$100/tCO_{2e} is estimated between 3.56 $GtCO_{2e}yr^{-1}$ and 3.83 $GtCO_{2e}yr^{-1}$, at US\$50/tCO_{2e} between 2.23 $GtCO_{2e}yr^{-1}$ and 2.44 $GtCO_{2e}$ yr^{-1} and at US\$20/tCO_{2e} between 1.34 $GtCO_{2e}yr^{-1}$ and 1.42 $GtCO_{2e}yr^{-1}$ (Fuss et al., 2018[4]).

There are some concerns that practices adopted to increase SCS may displace food production or destroy wildlife habitats. Henderson et al. (2021)

3 Tang et al. (2016) convert all estimates to 2012 US\$.
4 These estimates draw on data from Smith et al. (2008) and Smith (2012, 2016).

Table 1 Example cost estimates for soil carbon sequestration

Management strategy	Study	Study region	Target farm system	Cost (US$/ tCO_{2e})
Conservation tillage	Antle et al. (2007)	Central USA	Dryland grain production system	19
Conservation tillage	Grace et al. (2010)	South-eastern region of Australia	Dryland grain production system	11
Conservation tillage	Grace et al. (2012)	India's Indo-Gangetic Plain	Rice-wheat agricultural system	25
Continuous cropping	Antle et al. (2001, 2002)	Montana, USA	Dryland grain production system	20
Crop rotation[a]	Tschakert (2004)	Old Peanut Basin, Senegal	Semi-arid mixed agriculture	129
Crop rotation	Kragt et al. (2012)	Wheatbelt, Western Australia	Dryland grain production system	52

[a] Tschakert (2004) examined 15 different management options. This estimate is for poor households adopting a millet-groundnut rotation with compost on millet. Additional inputs required to increase SCS are very costly to poor households and the increase in C is relatively small per hectare leading to high cost per additional unit of C sequestered. Medium and rich households can sequester C using this practice at costs estimated at $26/$tCO_{2e}$. Over half of the household in the study are identified as poor (Tschakert, 2004).

use the Global Biosphere Management Model (GLOBIOM) to examine the potential for changes in agricultural practices to contribute to global GHG efforts. Their models suggest that at price levels of up to $70/$tCO_{2e}$ an additional 0.83 $GtCO_{2e}yr^{-1}$ could be sequestered in agricultural soils globally by 2050, relative to a baseline in the same year, with very little change in agricultural production and food consumption (Henderson et al., 2021).

3 Incentives to change agricultural practices: payments for carbon

Across the globe, most producers are focused on producing commodities for private benefits obtained from their commercial sale or for subsistence consumption. It is reasonable to assume that producers selected their current production and management systems because these provide the best economic returns and possibly achieve other goals important to the farmer (Antle et al., 2003), given capital, social, bio-physical and other constraints. To increase SCS, an individual landowner will need to make changes to their existing management and production practices. Many producers will require financial incentives and possibly additional technical assistance to manage risk and other concerns as they make changes to current practices and adopt those that sequester additional soil C.

3.1 Payment/incentive designs

Two commonly proposed financial incentive mechanisms are payments for each unit of C produced and payments for adopting a practice likely to increase soil C (Antle et al., 2003; Antle and Mooney, 2002; COWI, Ecological Institute and IEEP, 2021). Hybrid approaches that include elements of each incentive type can also be created. Each payment design rewards farmers for slightly different outcomes. Payment for each unit of additional C sequestered directly rewards specific outcomes attained after changing managing practices. This design treats C as a commodity, and, like other farm products, producers are compensated for the total number of units sold. The ability to measure, verify and agree on the specific quantity of C transacted is an important aspect underlying this payment design. Both buyers and sellers require this information to support financial exchange.

A second design offers payments for adopting a management practice or undertaking an action that will increase SCS. This design is similar in concept to agricultural support payments already used by many governments.[5] Under this design, it is not necessary to directly measure the amount of C sequestered because payments are tied to observed changes in production practices. Instead, it is important to confirm in some way that the agreed-upon practices were adopted.

3.2 Decision-making under each payment design

A farmer is more likely to change practices and management if they are made no worse off than they were before. This means that payments for C must at least cover any reductions in profit occurring as a result of altering production. Under an incentive design that links payments to specific performance, e.g. payment per unit of additional CO_{2e} sequestered, the minimum compensation required to induce a farmer to supply additional soil C can be calculated as in (1).

$$\text{Minimum payment per unit } CO_{2e} \geq \frac{\text{Change in profit}}{\text{Change in } CO_{2e}} \quad (1)$$

This shows us that the market price offered per unit sequestered (payment per unit) would need to meet or exceed the minimum payment required/unit CO_{2e} before a producer would change practices to sequester additional C.

5 For example, the U.S. Department of Agriculture supported Environmental Quality Incentives Program https://www.nrcs.usda.gov/wps/portal/nrcs/main/national/programs/financial/eqip/ and Conservation Reserve Programs https://www.fsa.usda.gov/programs-and-services/conservation-programs/conservation-reserve-program/.

Incentives designed to pay producers for adopting practices that increase soil C also need to cover any decrease in profit. The minimum payment required for a producer to participate is calculated as in (2).

$$\text{Minimum payment for practice adoption}(\text{ha}^{-1}) \geq \text{Change in profit}(\text{ha}^{-1}) \quad (2)$$

This design is commonly used in government programs that incentivize producers to adopt beneficial environmental practices. Producers are more likely to enroll land into the program if they are compensated for any reduction in profit experienced because of changing management practices.

3.3 Incentive size, participation and total carbon supply

The size of payments/prices needed to encourage farmers to change practices will vary spatially, reflecting individual site-specific baseline conditions, local markets, geographic and other factors (Antle et al., 2003). The number of producers willing to change practices at a given price/incentive amount also drives the total additional soil C that is sequestered. At low prices, a small number of producers will find it economically attractive to adopt new management practices. As prices increase, more and more producers will adopt practices that sequester additional C, increasing the total quantity of sequestered C (see example Box 1). The price offered for each unit of C and the suite of potential management options that are available to increase SCS together play an important role in producer participation and total C sequestered within a given region or specific farm. The optimal practices to adopt will differ by region (Amelung et al., 2020; Antle et al., 2003).

Although financial incentives are important, there are also additional considerations that affect behavior, e.g. risk aversion, availability of inputs, transactions costs, social capital and cultural environment (Antle and Stoorvogel, 2008; Fleming et al., 2019; Roesch-McNally et al., 2018). Reducing these and other barriers to adoption would increase the speed and scale of system transitions and the feasibility of limiting warming (deConick et al., 2018).

Box 1: Example of how price/incentive size influences the participation of producers and affects the total amount of additional soil C sequestered

Consider three hypothetical farmers, each with one hectare of land but facing different agricultural markets, soils and climates that affect their potential profits and ability to adopt practices that sequester additional C.

Farmer identifier	Current profit under existing management ($)	Profit under management practice that sequesters additional CO_{2e} ($)	Change in profit ha^{-1} ($)	Unit change in CO_{2e} from management change	Minimum payment required per unit CO_{2e}
Farmer A	150	135	15	1	15
Farmer B	100	90	10	1	10
Farmer C	100	70	30	2	15

Under a payment for performance program, prices less than $10/unit CO_{2e} would not result in additional soil C sequestered; if prices rise to $10/unit CO_{2e}, Farmer B could join the market and provide 1 unit of CO_{2e}; once prices reach $15/unit CO_{2e}, Farmers A and C enter the market and a total of four units of CO_{2e} would be sequestered. This simple example illustrates that the overall quantity of CO_{2e} sequestered will vary depending on market prices for C-credits or offsets. At higher prices, the activities associated with sequestering additional C become financially viable for a larger number of producers. A similar outcome is expected when payments are provided to farmers for adopting practices. Payments less than $10 ha^{-1} would not induce any of our three farmers to adopt practices that sequester additional soil C. When payments increase per unit land area placed in approved practices, more farmers participate and more soil C is sequestered. Farmer B participates at payments at or greater than $10 ha^{-1}, Farmer A participates with payments at or greater than $15 ha^{-1} and Farmer C participates when payments are at or above $30 ha^{-1}.

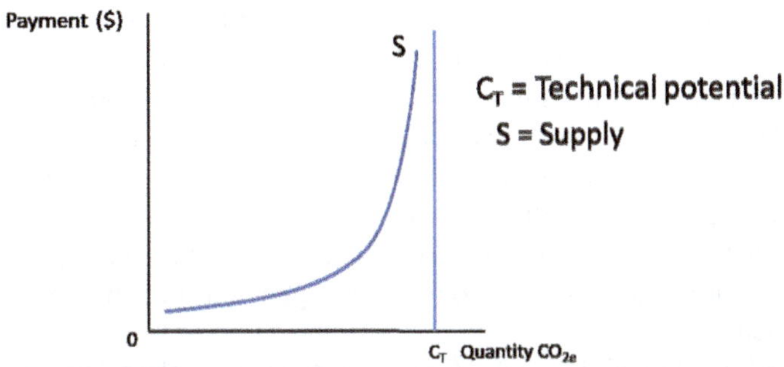

Figure 1 Carbon supply curve. Antle et al. (2003) provide more detail. Williams et al. (2005) provide an introductory discussion of factors affecting the C supply curve.

As financial incentives increase, more producers are likely to participate in C markets by creating C-credits or joining programs that reward the adoption of practices that sequester additional Soil C (Antle et al., 2003; Antle and Stoorvogel, 2008) and the amount of SOC sequestered will rise toward the technical potential (Fig. 1).

4 Financial incentives and market opportunities

Although there have been significant efforts to develop pilot schemes or proof-of-concept marketplaces, finance and ease of broad market, acceptance of SCS remains a significant barrier to C farming. Existing markets for GHG emissions reductions have diverged into two main types, compliance markets and voluntary markets, each with different opportunities to include SCS. Market opportunities or other incentives to engage in SCS are vitally important to encourage C farming.

4.1 Compliance markets

Almost 75% of the four billion tCO_{2e} of C-credits have been developed under the Clean Development Mechanism and Joint Implementation provisions of the Kyoto Protocol (World Bank, 2020). Most credits in these compliance markets are generated from projects that focus on industrial gases, renewable energy and fugitive emissions. Market activity reached a peak of almost one million $ktonCO_{2e}$ in 2012 and has since remained close to 200 000 $ktonCO_{2e}yr^{-1}$ (World Bank, 2020). Mitigation and GHG reduction options from agriculture have played a negligible role in compliance markets because agriculture was largely excluded from the Kyoto framework due to factors such as the complexity associated with measurement and monitoring and debates surrounding permanence among other factors.

The 2015 Paris Agreement provided more flexibility for countries to set their own Nationally Determined Contributions and focus on 'bottom-up' solutions. Concepts such as climate-smart agriculture[6] and the 4 per 1000 initiative[7] (soils for food security and climate) are of increasing interest, with more awareness of the potential co-benefits generated from SCS, such as increased yields and reduced erosion and runoff. Since 2015, there has been a change in market preference from industrial to land use-based sources of credits. Approximately 42% of credits are now issued from projects generated by the forestry sector (World Bank, 2020).

Globally, there are over 60 carbon pricing initiatives almost equally divided between emission trading schemes and carbon taxes (World Bank, 2020). The Australian emissions reduction fund (ERF)[8] is one market that incorporates projects from agricultural activities including practices that increase soil C sequestration,[9] such as retaining stubble after a crop is harvested and

6 http://www.fao.org/climate-smart-agriculture/en/.
7 https://www.4p1000.org/ (accessed June 8, 2021).
8 Chapter 25 of this book provides additional detail about the ERF.
9 http://www.cleanenergyregulator.gov.au/ERF/Choosing-a-project-type (accessed June 30, 2021).

conversion to no-till[10]. Projects are registered with the fund, and credits are sold at auction to the Australian government to meet the country's commitments for national emissions reductions. ERF credits are also traded privately. Despite more interest in land-based credits, soil C projects represent a tiny part of compliance markets.

Very recently, the European Commission (EC) engaged in a significant study focused on how to design and develop mechanisms to support a C farming market, slated to launch in 2021 (COWI, Ecological Institute and IEEP, 2021). Interest in C farming coincides with the EC Farm to Fork activities and Sustainable Finance Taxonomy and could provide additional incentives to engage in farming that generates increased environmental benefits.

4.2 Voluntary markets

Voluntary markets and voluntary crediting systems have been established outside of the Kyoto Framework and include project types and credit sources that are broader than those present in compliance markets. Credits bought through the voluntary markets are used by sectors, businesses and individuals to voluntarily offset emissions; rarely are they transferable into compliance markets (World Bank, 2020). Since 2019, activity in voluntary markets has increased significantly, now representing over 60% of new credits created annually, primarily in the form of C-offsets (World Bank, 2020; Forest Trends Ecosystem Marketplace, 2020b).

The United States is home to three of the largest credit providers: the Climate Action Reserve,[11] the American Carbon Registry[12] and Verra.[13] Each provider supports credits generated through land use, land use change and forestry projects; between 20% and 42% of credits are generated from forestry sources (World Bank, 2020). Credits from agricultural sources have accounted for less than 2% of credit volume (World Bank, 2020). A fourth provider, the Gold Standard[14] (headquartered in Switzerland), supplies C-credits and tracks additional project co-benefits including water protection and employment and health improvements for local communities (World Bank, 2020 and goldstandard.org). There are also several regional, national and subregional credit systems across the United States, Canada, China and Australia, among other locations (World Bank, 2020).

10 http://www.cleanenergyregulator.gov.au/ERF/Pages/Choosing%20a%20project%20type/Opportunities%20for%20the%20land%20sector/Agricultural%20methods/The-measurement-of-soil-carbon-sequestration-in-agricultural-systems-method.aspx (accessed June 30, 2021).
11 https://www.climateactionreserve.org (accessed June 23, 2021).
12 https://americancarbonregistry.org (accessed June 16, 2021).
13 https://verra.org/ (accessed June 20, 2021).
14 https://www.goldstandard.org (accessed June 23, 2021).

Voluntary markets and standardized exchanges have been in operation for many years, and in addition to being important facilitators of trades in C-credits, many have been early leaders in creating 'proof-of-concept' project design and early experimentation to support the creation of robust markets.[15] Voluntary markets have some design elements that resemble commodity markets such as standardization. Many more closely resemble real-estate markets, where commodities are not standardized and instead are characterized by a bundle of attributes that are site specific such as C sequestered, biodiversity and conservation co-benefits, contributions to sustainable development goals, impacts and opportunities for indigenous peoples and other factors (Forest Trends Ecosystem Marketplace, 2020b).

4.2.1 Global demand for nature-based carbon-credits

A recent survey of voluntary market participants conducted by Forest Trends' Ecosystem Marketplace (2021a)[16] found increased demand for credits generated from nature, with significant buyer interest in the added value of co-benefits that could be used to demonstrate progress toward Sustainable Development Goals. Almost two-thirds of the voluntary offsets purchased in 2019 had certified co-benefits and these credits received a price premium (Forest Trends' Ecosystem Marketplace, 2020a, 2021a). Most credit buyers are located within the United States, France, Germany and the United Kingdom. U.S. buyers primarily purchase domestically produced credits, while European buyers have focused on purchasing credits from Peru, Brazil and India (Forest Trends' Ecosystem Marketplace, 2020b). Survey respondents indicate that almost all demand for C-offsets in 2019 was driven by for-profit/private sector purchasers, with greater interest from non-profits within North America.

Buyers represent many different sectors and include large multinationals (such as Microsoft and Shell Oil); sporting events (Olympics, NASCAR); industries such as energy, transportation, hospitality and recreation; educational establishments; and individuals among many other purchasers. A primary reason to purchase offsets is to make activities carbon neutral and reduce the GHG impact of business practices and personal choices on the environment and atmosphere. There are also other reasons for purchases, such as replacing older retiring credits, a desire to help local communities, UN sustainable development goals and supporting proof-of-concept markets or projects.[17]

15 Influential early actors include the Chicago Climate Exchange, Winrock International and The Nature Conservancy.

16 A list of respondents can be found at https://www.ecosystemmarketplace.com/2020-em-carbon-survey-respondents/ (accessed July 18, 2021).

17 The multiple marketplaces each post buyer information, e.g. https://www.climateactionreserve.org/how/voluntary-offset-program/ (accessed September 28, 2021). Forest Trends Ecosystem Market Place (2021b) presents survey results by buyer sector and identifies some of the factors motivating buyers. The is currently no single database of global buyers.

4.2.2 Valuing co-benefits

Management practices that increase SCS can provide private and public co-benefits including improved water management, decreased erosion and improved soil fertility that may improve crop yields (Bryan et al., 2014; Griscom et al., 2017; Fleming et al., 2019). Co-benefits will differ depending on the geographic region, management changes and other factors. Private co-benefits are those that accrue directly to the landowner or producer. Public co-benefits are those positive externalities that accrue to society. Feng et al. (2007) estimated the co-benefits that could accrue from targeting payments of US$500 million at either increasing SCS, reducing erosion, nitrogen runoff or leaching (Table 2). They found that targeting payments to increasing SCS in their study area (mid-western United States) would also yield additional co-benefits of 3.2 million tonnes of erosion reduction, 7.4 thousand tonnes of reduced nitrogen runoff and 10 thousand tonnes of reduced nitrogen leaching. Interestingly, they also show that SCS can be a co-benefit of payments targeted to obtain other environmental services (Table 2). Jones et al. (2013) estimate that the U.S. Conservation Reserve Program (primarily focused on erosion reduction) also resulted in 16 $Tgyr^{-1}CO_{2e}$ sequestered in grassland soil and an addition of 15 $Tgyr^{-1}CO_{2e}$ of C sequestered through afforestation.

Co-benefits from many ecosystem services are difficult to value, but there are significant unrealized opportunities to link additional payments to the positive co-benefits created through changing management to increase soil C (the societal values of several co-benefits are summarized within Mooney and Williams (2007)). Although there is not an explicit 'price' for the additional co-benefits, the ability to sell C-credits faster and evidence of premium prices

Table 2 Co-benefits from targeting payments at different environmental services

	Target increases in soil C	Target erosion reduction	Target reduction in nitrogen runoff	Target reduction in nitrogen leaching
Area (million ha)	1.5	1.7	1.3	1.6
Carbon increase (million tonnes)	3.2	0.8	0.6	1
Erosion reduction (million tonnes)	7.4	40.5	14.1	9.7
Nitrogen runoff reduction (thousand tonnes)	2.8	5.1	11.7	2.8
Nitrogen leaching reduction (thousand tonnes)	10	6.4	5.6	30.6

Source: Feng et al. (2007).

suggest that markets are beginning to factor in these additional values. From a producer perspective, the provision of additional ecosystem services because of generating C-credits may motivate farmers to enter the market because the overall impact has social and environmental meaning to them personally (Fleming et al., 2019).

5 Measurement, reporting and verification

Measurement, reporting and verification (MRV) of soil C has been a significant challenge to the forward momentum and progress of C farming (Amelung et al., 2020; Paustian et al., 2019; Costa et al., 2020; Antle et al., 2003; COWI, Ecological Institute and IEEP, 2021). A lot of progress has been achieved over the last 20 years; however, results from a recent poll of personnel working in environmental organizations, businesses, academic researchers and government entities still identify MRV as one of the largest challenges still faced by entities developing C farming schemes (European Commission, 2021).

Confidence and assurance that C sequestration occurred is crucial to support trades and soil C accounting. It is also important that measurement protocols achieve accuracy and replicability at a low cost (Costa et al., 2020; Antle et al., 2003; Mooney et al, 2004). The development of measurement protocols is further complicated by the spatial scale over which measurement is needed and the payment mechanism and incentives used to support farmers (Costa et al., 2021; Mooney et al., 2004). MRV for a landscape or country scale assessment will look very different from those to support project-level exchange and payment for C-credits (Costa et al., 2021; Paustian et al., 2019). Even at the project level, measurement protocols will differ dependent on the payment design (Mooney et al., 2004).

Options for MRV[18] include a direct sampling of the soil on individual farms, aggregate sampling across many farms, modeling approaches and remote sensing, among others (Amelung et al., 2020; COWI, Ecological Institute and IEEP, 2021; Paustian et al., 2019; Costa et al., 2021). The approach taken will significantly influence MRV costs as well as the degree of uncertainty associated with the outcomes. The MRV design to support producer payments will vary depending on the design of the policy or payment scheme (Mooney et al., 2004). For example, a payment for performance is dependent on measuring changes in soil C, while payments for practice adoption require verification that the practice itself was adopted and are not as reliant on detailed measurement of soil C change. The ability to document and measure success will increase support for these programs with policymakers, private buyers and other entities.

18 Chapters 9-13 of this book provide more detail and insight into different techniques for measurement, reporting and verification.

Mooney et al. (2004, 2007) work through many of the factors that influence the costs of measurement using direct soil sampling under payment for performance and payment for practice incentive designs. They show that the magnitude of change in soil C, spatial heterogeneity of C within the study area; tolerance for sampling error, confidence intervals, number of producers covered (e.g. a single producer or aggregation over a large region) and frequency of measurement are all factors that drive MRV costs from direct sampling (Mooney et al., 2004, 2007).

Models to estimate changes in soil C have been developed in recent years. One example is the CarbOn Management Evaluation Tool (COMET)[19] that is used to assess changes in soil C and on-farm GHGs by several markets and government programs in the United States. Modeled estimates are likely to carry with them greater uncertainty than field sampling; however, MRV is achieved at a much-reduced cost. Remote sensing is another technique employed to track management changes, e.g. OpTIS.[20] Paustian et al. (2019) describe direct sampling and modeled approaches and empirical case studies.

The availability of different MRV approaches and the acceptance of buyers (and possibly governments) for measurement error and uncertainty will influence the ability of producers to trade in soil C. Amelung et al. (2020) Paustian et al. (2019) and Olesen et al. (2021) suggest that efforts to create and fund a global soil information system could provide data to further lower the costs of soil C MRV and support C-trades. Outside of compliance markets, the choice of MRV technique is at the discretion of market participants, who need to balance the trade-offs between costs, uncertainty and transparency trust related to transactions.[21]

6 Producer technical assistance

Currently, most soil C-credits or offsets are sold in bilateral transactions with little standardization (Forest Trends Ecosystem Marketplace, 2020b). The 'boutique' nature of many projects creates a barrier to producer participation because it is difficult to gain market experience through repeated transactions or clearly identify the economic payoffs from C farming. Suppliers and those demanding C-offsets may need considerable assistance to find suitable trading partners and create MRV protocols. Adopting a new enterprise can also increase production risk. Producers are looking to external consultants to navigate the management, compliance and production changes needed to support payments for soil C and identify the available market opportunities.

19 https://comet-farm.com/ (accessed February 10, 2021).
20 Operational Tillage Information System https://ctic.org/optis (accessed August 22, 2021).
21 A good introduction to some of these considerations if found within Costa et al. (2020).

Table 3 Examples of entities supporting trade in C-credits by providing client services to producers

Company	Summary information
Indigo Ag[a]	Private company that assists farmers with the transition from traditional farming to incorporating regenerative agriculture techniques. Support provided for cover crops, reducing tillage, diversifying crop rotation and reducing fertilizer. Farm must be located within Arkansas, Colorado, Georgia, Illinois, Indiana, Iowa, Kansas, Kentucky, Louisiana, Minnesota, Mississippi, Missouri, Nebraska, North Carolina, North Dakota, Ohio, Oklahoma, South Carolina, South Dakota, Tennessee and Texas. Payments are tied to producers adopting or improving at minimum one farming practice that sequesters additional C for up to 5 years.
Trimble[b]	A private, national program that aids farmers in quantifying their farming practices by providing their own technological system. Their consulting-based approach aids in different products such as crop planning, yield monitoring, land preparation, etc. They are a subscription-based program where the user pays $199 per year to gain access to their technology. With the help of their technology, farmers can plan their crops more efficiently and keep track of crop yields and management. Practices covered include tillage and amendment applications, among others.
National Indian Carbon Coalition[c]	A nonprofit organization dedicated to tribal nations and land that provides consulting on carbon farming practices and aids them in entering the carbon market. Their website lists benefits for farmers to start carbon farming as 'additional revenue-generation opportunities from land, preservation of tribal land ownership, promotion of land stewardship, greenhouse gas emissions reductions, and promotion of soil health, ecological diversity, and water and air quality.'
Truterra[d]	A private company that provides a marketplace for C. They evaluate sellers' farming practices and make recommendations to increase the amount of carbon sequestered from their agricultural lands. Practices must be implemented within the past 5 years, and reporting is required for 20 years.

[a] https://www.indigoag.com/ (accessed May 15, 2021).
[b] https://agriculture.trimble.com/product/carbon-credits/ (accessed May 19, 2021).
[c] https://www.indiancarbon.org/ (accessed May 15, 2021).
[d] https://www.truterraag.com/Carbon (accessed August 22, 2021).

New businesses provide technical expertise to producers interested in supplying C-credits. Carbon farmers of Australia have a central hub that connects interested farmers with a business that can support them through project creation and credit generation and provide market information in addition to connecting them with buyers.[22] Table 3 shows examples of similar business developed for technical assistance that are located within the United States.

22 https://carbonfarmersofaustralia.com.au/marketplace-hub/ (accessed August 21, 2021).

Services range from general technical and production support to project aggregation and brokerage services. The complexity and lack of market standardization leads to reliance on these third parties who charge for services. This further increases producer costs associated with participating in C farming.

7 Conclusion

For over 20 years, there has been interest in the opportunities to offset GHG emissions by sequestering C in soils. Additional scientific enquiry and advances in efforts for measuring soil C, as well as pilot projects that put science into practice (and suggest new areas of scientific enquiry), have all moved us closer to supporting trades in soil C. The 2015 Paris Accord helped to increase acceptance by acknowledging a role for nature-based solutions. However, many barriers to C farming remain: agreement on what constitutes acceptable methodologies for MRV and the availability of sufficient financial incentives and technical support for producers. There is likely a higher bar to support performance-based payments for soil C because these require the development of transparent and well-functioning markets to support trades of a commodity (C) that can, by its nature, vary over time. The science, policy and business communities are already actively engaged in trying to remove these barriers.

In the near term, governments could build upon their experiences with policies that provide payments for practice adoption and explore how to quickly increase the provision of SCS (and other environmental co-benefits). Effective decision support will be critical to guide producer decisions and avoid unintended consequences such as increasing emissions of other GHGs.

8 Where to look for further information

- A comprehensive overview of current markets and pricing globally is contained within World Bank. 2020. State and Trends of Carbon Pricing 2020 (May), World Bank, Washington, D.C. Doi: 10.1596/978-1-4648-1586-7.
- Detail of activity within Europe to develop carbon farming can be found in COWI, Ecologic Institute and IEEP (2021) *Technical Guidance Handbook - setting up and implementing result-based carbon farming mechanisms in the EU* Report to the European Commission, DG Climate Action, under Contract No. CLIMA/C.3/ETU/2018/007. COWI, Kongens Lyngby. https://op.europa.eu/en/publication-detail/-/publication/10acfd66-a740-11eb-9585-01aa75ed71a1/language-en. The report also discusses many of the same market barriers touched on in this chapter.
- Many papers, discussions and evaluation tools to aid in assessing ecosystem services and values from nature can be found on the Natural Capital Project web page https://naturalcapitalproject.stanford.edu/.

- Antle et al. (2003) and Mooney et al. (2004) provide detailed discussions of the relative efficiency of different payment designs and considerations for creating MRV schemes.

9 References

Alexander, P. K., Paustian, P. S. and Moran, D. 2015. The economics of soil C sequestration and agricultural emissions abatement. *Soil* 1, 331–339.

Amelung, W., Bossio, D., de Vries, W., Kogel-Knabner, I., Lehman, J., Amundson, R., Bol, R., Collins, C., Lal, R., Leifeld, J., Minasny, B., Pan, G., Paustian, K., Rumpel, C., Sanderman, J., van Groenigen, J. W., Mooney, S., van Wesemael, B., Wander, M. and Chabbi, A. 2020. Towards a global-scale soil climate mitigation strategy. *Nature Communications* 11(1), 5427 https://doi.org/10.1038/s41467-020-18887-7.

Antle, J. M., Capalbo, S., Mooney, S., Elliott, E. T. and Paustian, K. H. 2001. Economic analysis of agricultural soil carbon sequestration: an integrated assessment approach. *Journal of Agricultural and Resource Economics* 26(2), 344–367.

Antle, J. M., Capalbo, S., Mooney, S., Elliott, E. T. and Paustian, K. H. 2002. Sensitivity of carbon sequestration costs to soil carbon rates. *Environmental Pollution* 116(3), 413–422.

Antle, J. M., Capalbo, S., Mooney, S., Elliott, E. T. and Paustian, K. H. 2003. Spatial heterogeneity, contract design, and the efficiency of carbon sequestration policies for agriculture. *Journal of Environmental Economics and Management* 46(2), 231–250.

Antle, J. M., Capalbo, S. M., Paustian, K. H. and Ali, M. K. 2007. Estimating the economic potential for agricultural soil carbon sequestration in the Central United States using an aggregate econometric-process simulation model. *Climatic Change* 80(1–2), 145–171.

Antle, J. M. and Mooney, S. 2002. Chapter 31: designing efficient policies for agricultural soil carbon sequestration. In: Kimble, J., Lal, R. and Follett, R. (Eds) *Agriculture Practices and Policies for Carbon Sequestration in Soil*. CRC Press LLC, pp. 323–336.

Antle, J. M. and Stoorvogel, J. J. 2008. Agricultural carbon sequestration, poverty, and sustainability. *Environment and Development Economics* 13(3), 327–352.

Bossio, D. A., Cook-Patton, S. C., Ellis, P. W., Fargione, J., Sanderman, J., Smith, P., Wood, S., Zomer, R. J., Von Unger, M., Emmer, I. M. and Griscom, B. W. 2020. The role of soil carbon in natural climate solutions. *Nature Sustainability* 3(5), 391–398. https://doi.org/10.1038/s41893-020-0491-z.

Bryan, B. A., Nolan, M., Harwood, T. D., Connor, J. D., Navarro-Garcia, J., King, D., Summers, D. M., Newth, D., Cai, Y., Grigg, N., Harman, I., Crossman, N. D., Grundy, M. J., Finnigan, J. J., Ferrier, S., Williams, K. J., Wilson, K. A., Law, E. A. and Hatfield-Dodds, S. 2014. Supply of carbon sequestration and biodiversity services from Australia's agricultural land under global change. *Global Environmental Change* 28, 166–181. http://doi.org/10.1016/j.gloenvcha.2014.06.013.

Costa Jr., C., Dittmer, K., Shelton, S., Bossio, D., Zinyengere, N., Luu, P., Heinz, S., Egenolf, K., Rowland, B., Zuluaga, A., Klemme, J., Mealey, T., Smith, M. and Wollenberg, E. 2020. How soil carbon accounting can improve to support investment- oriented actions promoting soil carbon storage. CCAFS Info Note. Wageningen, The

Netherlands: CGIAR Research Program on Climate Change, Agriculture and Food Security (CCAFS). Available at: https://hdl.handle.net/10568/110284.

Costa, C. Jr., Seabaur, M., Schwarz, M., Dittmer, K. and Wollenberg, E. 2021. *Scaling Soil Organic Carbon Sequestration for Climate Change Mitigation*. Wageningen, the Netherlands: CGIAR Research Program on Climate Change, Agriculture and Food Security (CCAFS).

COWI, Ecologic Institute and IEEP 2021. *Technical Guidance Handbook - Setting up and Implementing Result-Based Carbon Farming Mechanisms in the EU Report to the European Commission, DG Climate Action, under Contract No. CLIMA/C.3/ ETU/2018/007*. Kongens Lyngby: COWI.

de Coninck, H., Revi, A., Babiker, M., Bertoldi, P., Buckeridge, M., Cartwright, A., Dong, W., Ford, J., Fuss, S., Hourcade, J.-C., Ley, D., Mechler, R., Newman, P., Revokatova, A., Schultz, S., Steg, L. and Sugiyama, T. 2018. Strengthening and implementing the global response. In: MassonDelmotte, V., Zhai, P., Pörtner, H.-O., Roberts, D., Skea, J., Shukla, P. R., Pirani, A., Moufouma-Okia, W., Péan, C., Pidcock, R., Connors, S., Matthews, J. B. R., Chen, Y., Zhou, X., Gomis, M. I., Lonnoy, E., Maycock, T., Tignor, M. and Waterfield, T. (Eds.) *Global Warming of 1.5°C*. An IPCC Special Report on the impacts of global warming of 1.5°C above pre-industrial levels and related global greenhouse gas emission pathways, in the context of strengthening the global response to the threat of climate change, sustainable development, and efforts to eradicate poverty.

Eagle, A., Olander, L., Henry, L. R., Haugen-Kozyra, K., Millar, N. and Robertson, G. P. 2012. *Greenhouse Gas Mitigation Potential of Agricultural Land Management in the United States: A Synthesis of the Literature*. Report NIR 10-04 (3rd edn.). Durham, NC: Nicholas Institute for Environmental Policy Solutions, Duke University.

European Commission 2021. Reviewing the contribution of the land use, land-use change and forestry sector to the green deal. Workshop IV Carbon Farming in the CAP Strategic Plans. Available at: https://ec.europa.eu/clima/sites/default/files/forests/docs/20210525_workshop_iv_report_en.pdf.

Feng, H., Kurkalova, L. A., Kling, C. L. and Gassman, P. W. 2007. Transfers and environmental co-benefits of C sequestration in agricultural soils: retiring agricultural land in the Upper Mississippi River Basin. *Climatic Change* 80(1–2), 91–107.

Fleming, A., Stitzlein, C., Jakku, E. and Fielke, S. 2019. Missed opportunity? Framing actions around co-benefits for carbon mitigation in Australian agriculture. *Land Use Policy* 85, 230–238. https://doi.org/10.1016/j.landusepol.2019.03.050.

Forest Trends' Ecosystem Marketplace 2020a. *Voluntary Carbon and the Post-pandemic Recovery, State of the Voluntary Carbon Markets Report, Special Climate Week NYC Installment 21 September 2020*. Washington, DC: Forest Trends Association.

Forest Trends' Ecosystem Marketplace 2020b. *The Only Constant is Change. State of the Voluntary Carbon Markets 2020. Second Installment Featuring Core Carbon & Additional Attributes Offset Prices, Volumes and Insights*. Washington, DC: Forest Trends Association, December 2020.

Forest Trends' Ecosystem Marketplace 2021a. *Buyers of Voluntary Carbon Offsets, A Regional Analysis. State of the Voluntary Carbon Markets 2020. Third Installment. Featuring European Buyers Offset Prices, Volumes and Insights*. Washington, DC: Forest Trends Association.

Forest Trends' Ecosystem Marketplace 2021b. *Markets in Motion State of the Voluntary Carbon Markets 2021 Installment 1*. Washington, DC: Forest Trends Association.

Franzluebbers, A. J. 2010. Achieving soil organic carbon sequestration with conservation agricultural systems in the southeastern United States. *Soil Science Society of America Journal* 74(2), 347–357.

Fuss, S., Lamb, W. F., Callaghan, M. W., Hilaire, J., Creutzig, F., Amann, T., Beringer, T., de Oliveira Garcia, W., Hartmann, J., Khanna, T., Luderer, G., Nemet, G. F., Rogelj, J., Smith, P., Vicente, J. L. V., Wilcox, J., del Mar Zamora Dominguez, M. and Minx, J. C. 2018. Negative emissions—Part 2: costs, potentials and side effects. *Environmental Research Letters* 13(6) https://doi.org/10.1088/1748-9326/aabf9f.

Grace, P. R., Antle, J., Aggarwal, P. K., Ogle, S., Paustian, K. and Basso, B. 2012. Soil carbon sequestration and associated economic costs for farming systems of the Indo-Gangetic plain: a meta-analysis. *Agriculture, Ecosystems and Environment* 146(1), 137–146.

Grace, P. R., Antle, J., Ogle, S., Paustian, K. and Basso, B. 2010. Soil carbon sequestration rates and associated economic costs for farming systems of south-eastern Australia. *Australian Journal of Soil Research* 48, 1–10.

Griscom, B. W., Adams, J., Ellis, P. W., Houghton, R. A., Lomax, G., Miteva, D. A., Schlesinger, W. H., Shoch, D., Siikamäki, J. V., Smith, P., Woodbury, P., Zganjar, C., Blackman, A., Campari, J., Conant, R. T., Delgado, C., Elias, P., Gopalakrishna, T., Hamsik, M. R., Herrero, M., Kiesecker, J., Landis, E., Laestadius, L., Leavitt, S. M., Minnemeyer, S., Polasky, S., Potapov, P., Putz, F. E., Sanderman, J., Silvius, M., Wollenberg, E. and Fargione, J. 2017. Natural climate solutions. *Proceedings of the National Academy of Sciences of the United States of America* 114(44), 11645–11650.

Henderson, B., Frank, S., Havlik, P. and Valin, H. 2021. *Policy Strategies and Challenges for Climate Change Mitigation in the Agriculture, Forestry and Other Land Use (AFOLU) Sector*. OECD Food, Agriculture and Fisheries Papers No. 149. https://doi.org/10.1787/47b3493b-en.

High-Level Commission on Carbon Prices 2017. *Report of the High-Level Commission on Carbon Prices*. Washington, DC: World Bank. License: Creative Commons Attribution CC BY 3.0 IGO.

IPCC 2018. Global warming of 1.5°C. In: *An IPCC Special Report on the Impacts of Global Warming of 1.5°C above Pre-industrial Levels and Related Global Greenhouse Gas Emission Pathways, in the Context of Strengthening the Global Response to the Threat of Climate Change, Sustainable Development, and Efforts to Eradicate Poverty*. Masson-Delmotte, V., Zhai, P., Pörtner, H.-O., Roberts, D., Skea, J., Shukla, P. R., Pirani, A., Moufouma-Okia, W., Péan, C., Pidcock, R., et al. (Eds).

Jones, C. A., Nickerson, C. J. and Heisey, P. W. 2013. New uses of old tools? Greenhouse gas mitigation with agricultural sector policies. *Applied Economic Perspectives and Policy* 35(3), 398–434.

Kragt, M. E., Pannell, D. J., Robertson, M. J. and Thamo, T. 2012. Assessing costs of soil carbon sequestration by crop-livestock farmers in Western Australia. *Agricultural Systems* 112, 27–37.

Mooney, S., Antle, J. M., Capalbo, S. M. and Paustian, K. 2004. Design and costs of a measurement protocol for trades in soil carbon credits. *Canadian Journal of Agricultural Economics/Revue Canadienne d'Agroeconomie* 52(3), 257–287.

Mooney, S., Gerow, K., Antle, J. M., Capalbo, S. M. and Paustian, K. 2007. Reducing standard errors by incorporating spatial autocorrelation into a measurement scheme for soil carbon credits. *Climatic Change* 80(1–2), 55–72.

Mooney, S. and Williams, J. 2007. Chapter 4: private and public values of soil carbon management. In Kimble Reed, D., Mooney, S., Follett, R. and Lal, R. (Eds) *Soil Carbon Management: Economic, Environmental and Societal Benefits*. Rice: Jackson College. Taylor and Francis Group, LLC., pp. 67-96.

Ogle, S. M., Breidt, F. J. and Paustian, K. 2005. Agricultural management impacts on soil organic carbon storage under moist and dry climatic conditions of temperate and tropical regions. *Biogeochemistry* 72(1), 87-121.

Olesen, A., Kowalczewski, T., Kennedy, K., Bellassen, V., Bird, N., Von Unger, M., Eaton, D., Leistner, S., Steinert, M. and Gionfra, S. 2021. *Reviewing the Contribution of the Land-Use Change and Forestry Sector to the Green Deal*. Final Study. European Commission (July).

Paustian, K., Andrén, O., Janzen, H. H., Lal, R., Smith, P., Tian, G., Tiessen, H., van Noordwijk, M. and Woomer, P. L. 1997. Agricultural soils as a sink to mitigate CO2 emissions. *Soil Use and Management* 13, 229-244.

Paustian, K., Collier, S., Baldock, J., Burgess, R., Creque, J., DeLonge, M., Dungait, J., Ellert, B., Frank, S., Goddard, T., Govaerts, B., Grundy, M., Henning, M., Izaurralde, R. C., Madaras, M., McConkey, B., Porzig, E., Rice, C., Searle, R., Seavy, N., Skalsky, R., Mulhern, W. and Jahn, M. 2019. Quantifying carbon for agricultural soil management: from the current status toward a global soil information system. *Carbon Management* 10(6), 567-587, https://doi.org/10.1080/17583004.2019.1633231.

Poeplau, C. and Don, A. 2015. Carbon sequestration in agricultural soils via cultivation of cover crops - a 1158 meta-analysis. *Agriculture, Ecosystems and Environment* 200, 33-41.

Roesch-McNally, G., Arbuckle, J. G. and Tyndall, J. C. 2018. Soil as social-ecological feedback: examining the "ethic" of soil stewardship among Corn Belt farmers. *Rural Sociology* 83(1), 145-173.

Smith, P. 2012. Agricultural greenhouse gas mitigation potential globally, in Europe and in the UK: what have we learnt in the last 20 years? *Global Change Biology* 18(1), 35-43.

Smith, P. 2016. Soil carbon sequestration and biochar as negative emissions technologies. *Global Change Biology* 22(3), 1315-1324. https://doi.org/10.1111/gcb.13178.

Smith, P., Martino, D., Cai, Z., Gwary, D., Janzen, H., Kumar, P., McCarl, B., Ogle, S., O'Mara, F., Rice, C., Scholes, B., Sirotenko, O., Howden, M., McAllister, T., Pan, G., Romanenkov, V., Schneider, U., Towprayoon, S., Wattenbach, M. and Smith, J. 2008. Greenhouse gas mitigation in agriculture. *Philosophical Transactions of the Royal Society. Series B, Biological Sciences* 363(1492), 789-813 https://doi.org/10.1098/rstb.2007.2184.

Tang, K., Kragt, M. E., Hailu, A. and Ma, C. 2016. Carbon farming economics: what have we learned? *Journal of Environmental Management* 172, 49-57.

Tschakert, P. 2004. The costs of soil carbon sequestration: an economic analysis for small-scale farming systems in Senegal. *Agricultural Systems* 81(3), 227-253.

Williams, J. R., Peterson, J. M. and Mooney, S. 2005. The value of carbon credits: is there a final answer? *Journal of Soil and Water Conservation* 60(2), 36A-40A.

World, B. 2020. *State and Trends of Carbon Pricing 2020 (May)*, Washington, DC: World Bank. https://doi.org/10.1596/978-1-4648-1586-7.

Chapter 28

Understanding the value of and reasoning behind farmer adoption of carbon centric practices

Michelle M. Wander and Carmen M. Ugarte, University of Illinois at Urbana-Champaign, USA

1 Introduction

Recognition of the need to reduce atmospheric greenhouse gas concentrations and appreciation of the role that agricultural soils can immediately play in climate change mitigation have created a demand for policies and programmes that promote C sequestration in soil. Public and private support for efforts to increase soil organic carbon (SOC) stocks are justified by the fact that we possess both the knowledge and technologies needed to make rapid progress towards sequestration goals to mitigate climate change while enhancing soil's provisioning, buffering, and regulating services (Amelung et al., 2020; Stockmann et al., 2013). While this is nothing new (Wander and Nissen, 2004), interest has grown rapidly due to governmental and private willingness to invest in mechanisms to promote C sequestration in arable soils (Honeycutt et al., 2020; Rumpel et al., 2020; Keenor et al., 2021). Twenty-eight countries currently include SOC in their Nationally Determined Contributions to meet pledges to the Paris Agreement (Wiese et al., 2021). The positive correlation between SOC and ecosystem services has made soil organic matter and related indicators of soil quality proxies for soil stewardship that are valued separately or in addition to climate offsetting (Paustian et al., 2019; Rumpel et al., 2020). The association between SOC and other ecosystem services beyond C storage (Amelung et al., 2020; Weil and Magdoff, 2004) explains why soil organic matter features

http://dx.doi.org/10.19103/AS.2022.0106.27

prominently in the battle for public perceptions. Different farming sectors are vying for legitimacy even as farmers' standing as moral actors has been eroded due to perceptions of agriculture's environmental and social harms (Williams and Martin, 2011).

Adoption of land stewardship practices tied to C sequestration by farmers depends on a variety of factors including the agroecological region, scale of production, incentives from public or private institutions, and cultural and the economic context (Piñeiro et al., 2020). Recognition of these variables is important as organizations acting at different levels of society develop outreach programmes and work with stakeholders to shape policy and implement practices for C stewardship. Typologies should not only consider individual but also the household and community services in addition to access to infrastructure (Ng'ang'a et al., 2019). This chapter explores factors including perceptions of risk or gaps in knowledge, technology, and resources, along with economic or social inducements that influence farmers' commitments to SOC stewardship. To learn about both the voluntary and regulatory government programmes as well as business-backed efforts influencing farmers' decision-making, we engaged farmers using conservation, conventional, or organic grain-producing practices in the US Midwest. We considered socio-economic factors determining farmers' willingness or ability to adopt conservation behaviours associated with soil building through C sequestration to understand how to adapt knowledge systems to promote the use of C-centric practices. We also explore how efficacy of information or inducements vary among groups.

2 Farmers' rationale, mechanisms and tactics

2.1 Farmers' eye view: between a rock and a hard place

Many mechanisms or/and strategies have been used to promote soil stewardship of arable lands by influencing individual behaviours or societal norms and actions through economic costs or opportunities. To improve these tactics it is crucial to understand what factors most determine farmers' behaviours assuming that they, and private land owners, determine land-use practices based on agronomic, economic, and social factors where lands are privately held. Even though collective land ownership remains important to subsistence agriculture, collective rights give way to individual decision-making when there is competition for land and opportunity for profit (Debolini et al., 2015; Wily, 2018). Through history, colonial lands and collectivized farming operations have been converted to private enterprises due to the belief that group rights hinder the adoption of farm improvements and, that private property rights provide the best way to reduce poverty because they allow individual operators to invest in, and profit from, improved productivity

Published by Burleigh Dodds Science Publishing Limited, 2023.

(Debolini et al., 2015). The farmers' actions are predicated by a tangle of issues and constraints that influence their interest and ability to act as stewards of the land. Tauger (2010) argues that this farmers' eye view must be seen as one that incurs a dual subordination to nature, with climate and land hazards, and to society, including its markets, taxes, rents, and institutions. Farmers are asked to be stewards of the land even as they face these pressures. Societal trust is placed because agrarian ideals commonly cast farmers as morally superior to industry agents (Dixon and Hapke, 2003; Freyfogle, 2007). This perspective is commonly challenged where agriculture is perceived to have harmed public health and the environment and failed to reduce but rather increased rural poverty (de Janvry and Sadoulet, 2000; Williams and Martin, 2011; Kannuri and Jadhav, 2018).

2.2 Extensive production systems

Historically SOC in agricultural systems has been equated with soil fertility and thus sought through the use of practices that increase organic matter return to the soil and/or prevent loss of C through soil erosion or depletion. Awareness of the value of SOC is reflected by early agricultural traditions manifested across the globe through the use of practices like terracing, intercropping, and manure management to build soil and maintain productivity (Mafongoya et al., 2006). Techniques commonly used by extensive systems vary depending upon the factor (typically nutrients, water, or pests) that is limiting productivity. In pre-industrialized societies and currently where limited-resource farmers do not rely upon intensive use of external inputs to maintain fertility, producers must avoid land overuse that results in land degradation (see references in Aref and Wander (1997) and King (2004)) by implementing a variety of practices. Low input or extensive production systems that rely on rest or rotation, strategic stocking or grazing, slash and burn, fallowing, green manuring or diversification become unsustainable if the area required to support farmers is inadequate, or when economic conditions prevent the purchase of supplements needed to overcome limits (Demir et al., 2015; Tauger, 2010). While these traditional systems are often seen as static, some of the earliest producer organizations were created to protect their rights and help them gain access to new ideas and practices. Conservative motives and commitment to self-reliance are commonly associated with grower groups and social movements pursuing sustainability (Monk, 2011). This is because practices used to improve and sustain productivity within extensive or low input production systems that include methods to conserve, concentrate, and recycle resources also tend to promote C sequestration. Current examples of commercially viable farming systems engaged with stewardship that explicitly identify C sequestration as an aspirational goal include organic, biodynamic, conservation and regenerative agriculture, and permaculture (Balfour, 1976; Brouder and Gomez-Macpherson, 2014; Conford, 2001; Heckman, 2006; Kirchmann et al., 2008).

The mechanisms and inducements that support and encourage the success of C-centric farming systems often differ from commodity-focused production in that they decommodify crops through certifications or product claims. Organic certification is an example of a voluntary market-based programme in which farmers may engage if there is enough knowledge of the opportunity costs to compensate for costs of practice implementation (Piñeiro et al., 2020). Farmers opting for organic certification must adopt practices that enhance soil biological processes through soil organic matter management. These standards have been effective due to their offer of clearly articulated guidelines and independent auditing and verification (Monk, 2011). The National Organic Standards (USDA-AMS, 2002) and The International Federation of Organic Agriculture Movements (IFOAM, 2005) emphasize that organic production fosters the natural cycling of nutrients by using practices that 'promote ecological balance and conserve biodiversity.' To maintain certification, farmers must comply with a generalized set of practices and restrictions. There is no verification of indicators of soil functioning. Expansion of certified organic production, or other similar systems, will depend upon consumer demand. The burden of record keeping, system complexity, and costs of certification may limit adoption by some (Piñeiro et al., 2020). Critiques of conventionalization associated with organic commodities have fostered interest in local food movements (Guthman, 2004; Fonte, 2008).

2.3 Intensive production systems

A shift from traditional farming systems to industrial production reliant on the use of modern technologies occurred following the green revolution. Increased reliance on inorganic fertilizers and pesticides decoupled the link farmers previously made between soil organic matter and productivity (Aref and Wander, 1997; Tauger, 2010). Industrial or commodity based agriculture has expanded due to the support of national-scale agricultural practices developed countries designed to promote commodities and trade (Lapola et al., 2014). Despite societal recognition of soils' importance to national economic well-being (Robinson et al., 2012), the national policies that have promoted the use of modern agricultural practices including mechanization, intensification of production, shortened rotations, and use of inorganic fertilizers and synthetic pest control measures, have frequently reduced SOC levels and associated services (Al-Kaisi and Kwaw-Mensah, 2020; Guo and Gifford, 2002; Lal et al., 2003; Lapola et al., 2014). Pressure to compete globally has prompted agricultural intensification and has compromised industrial-scale agriculture's social license to operate. In an attempt to reverse this trend, food-security-focused agricultural policies in both the US and EU were reformed in the 1980s and 1990s to require participants in governmental programmes to adopt conservation measures. Conservation compliance provisions in the US that included protections of highly erodible lands and wetlands or riparian habitat required minimal measures to participate in direct payments (Doering and Smith, 2012). Questions about the efficacy of conservation compliance and the underlying perception that these programmes were not truly voluntary undermined their support (Claassen et al., 2008; Cox, 2007). In the US, fixed-price support approaches have been replaced by crop revenue insurance coverage based on expected market prices (Glauber, 2013), and working lands provisions that reward the use of conservation practices are assumed to provide ecosystem services (Coppess, 2018). Both cost-share and technical service programmes encourage farmer adoption of C-centric practices to improve soil health (Bowman and Lynch, 2019). However, the practice-based enrolment or ranking tools that the government relies on do not measure actual delivery of services including C sequestration (Ugarte et al., 2014, 2018; Zilberman and Segerson, 2012).

Similar reforms to the EU Common Agricultural Policy (CAP) were made to reduce large surpluses of agricultural commodities, high transaction costs, distortion of markets, and concerns from consumers and taxpayers over environmental harms caused by intensive agriculture practices (Pe'er et al., 2019; Williams and Martin, 2011). In the most recent version (2023-2027), the CAP requires farmers to provide environmental services through compulsory cross-compliance of practice implementation (Buitenhuis et al., 2020). Cross-compliance provides a basis for payments when farmers undertake environmental commitments that exceed mandatory requirements. The CAP seeks compliance with environmental rules, including new requirements on public, animal and plant health, animal welfare, and the maintenance of all productive land in good agricultural and environmental conditions. These practices are implemented based on the standards of Good Agricultural and Environmental Conditions (GAEC) (Borrelli et al., 2016; FAO, 2021). The GAEC includes standards for protection against soil erosion, maintenance of soil organic matter and structure, avoidance of the deterioration of habitats, and water management (MARS, 2014). Programmatic approaches that are heavily focused on productivity have challenged the goals outlined in the CAP (Moschitz and Home, 2014; Pe'er et al., 2019, 2020). To address this, participatory efforts and systems approaches have been recommended to improve policy success (Moschitz and Home, 2014).

2.4 Markets and valorization

Efforts like the Global Business Compact, undertaken by multilateral organizations to spawn public-private partnerships and, calls for sustainable intensification have dramatically increased interest and support for market-backed trading schemes. The EU efforts built upon the Thematic Strategy for Soil Protection, which recognized farmers' central role in protecting soils, and focused on market mechanisms as a primary tactic for encouraging the use of sequestration and other conservation measures (European Commission, COM (2006)231). The '4 per mille' initiative launched by the French government embraced the SOC sequestration goals realized in different countries with the aspirational target for sequestration of 0.4% of existing SOC stocks per year (le Foll, 2015; Rumpel et al., 2020). Parallel efforts in the US and Australia have leveraged public-private partnerships to promote SOC sequestration in association with soil health and established C trading markets. These efforts have spurred innovation and captured large amounts of speculative conservation investment that is being marketed to mainstream agricultural sectors (Honeycutt et al., 2020; Sullivan, 2013; Chapters 26 and 27 of this book).

At present government-backed conservation programmes fill a market gap to secure environmental benefits to society that would otherwise go unrealized (Zilberman and Segerson, 2012). To replace or supplant the government's role, commercial markets must verify services rendered. Individualized outputs, however, may be difficult to verify. Association between best practices and outcomes like no-tillage, which is widely promoted as a conservation practice and important deterrent of erosion, do not consistently result in increased C sequestration. In the US Midwest there is little to no difference in the SOC contents of soils in conventional row crop systems under conventional or conservation management (Al-Kaisi and Kwaw-Mensah, 2020; Ugarte et al., 2018). In other regions, however, the combined implementation of reduced or no-tillage with cover crops commonly promotes SOC accumulation (Snapp et al., 2022), particularly in drylands (Thapa et al., 2019).

In the US, voluntary initiatives like the Climate Action Reserve aim to register projects in agricultural lands seeking C credits that would be assessed using protocols devised by a commercial entity (Oldfield et al., 2021). Importantly, sampling requirements for verification purposes are left up to the verifier (Climate Action Reserve, 2020; Jackson Hammond et al., 2021). Given the fluidity and hype in this arena, farmers are rightly asking about transaction costs and the veracity of claims as they wonder how to engage with markets and weigh in on policy. The resulting partnerships between corporations and civic organizations are undergoing 'paradigm struggle' as they negotiate assumptive worldviews, vocabularies, goals, and loyalties as they struggle to capture and hopefully deserve credibility (Botan, 1993). Whether these market-based approaches can deliver their promise is a source of concern (Baveye et al., 2016; Bracking and Leffel, 2021; Ghosh and Wolf, 2021). Many fear markets will compound problems of industrial agriculture and have detrimental outcomes (Ainscough et al., 2019; Clapp and Isakson, 2018).

3 Sequestration strategies and motivations of Midwest grain farmers

To understand why and how farmers in the US undertake stewardship behaviours, we first sought to identify practices that perform well enough to compete for C trading and conservation programmes in Illinois. This is located in the Midwest, where grain cropping systems vary in rotation length, quantity and quality of inputs, and frequency and intensity of tillage. We engaged farmers classified as conventional or conservation that typically manage 2-year rotations in this region and include a corn phase that is followed by soybean. Nitrogen and other fertilizers are applied in the years when corn is grown. The conservation group uses a mixture of 'best tillage practices' that include reduced or strip tillage frequently made in spring to maintain soil surface coverage and habitat for wildlife during the winter months. In this group, they frequently use split N applications and different formulations of N that typically have nitrification inhibitors. We also engaged organic farmers certified by the USDA National Organic Standard who use longer rotations that typically include the 2-year corn and soybean sequence followed by a year of small grains and use organic amendments and green manures to satisfy crop nutrient requirements. Soil sampling was carried out in spring to explore how information could be used at the individual level (e.g. conservation compliance, verification of C credits), or at the group level, with pooled data being used to assess differences at the practice-based level. Samples were collected from 21 groups (blocks) containing the comparisons of the above-described management systems and geographically distributed to account for differences in soil inherent characteristics as detailed in Ugarte et al. (2018). Within each field, 16 samples (0–30 cm divided into 15 cm increments) were collected using a hydraulic probe (4.06 cm of internal diameter) in a grid sampling pattern to account for field variability. In the field, sampling points were defined within a 4-ha area and based on the dominant soil series in a given group of fields and separated at least 30 m apart. Samples were collected in the spring before the start of field operations from 2011 to 2013. Each farm operator also provided detailed information on management practices leading to field preparation, nutrient management, and average yields. Samples were processed to determine SOC concentrations and stocks as well as other fractions of soil organic matter known to respond to management.

3.1 Farmers' individual and collective use of data

Field specific data were shared with individual farmers in a format that was similar to a soil testing report in that it included targets and threshold values (Fig. 1). While farmers knew how to use information resulting from standard soil tests, results describing stocks of SOC or labile fractions of SOC that are useful indicators of soil quality needed to be supplemented with other inputs to be useful or informative. Fortunately, recent improvements in scoring curves for SOC that draw on large national datasets can provide regional curves that allow farmers to compare their SOC stock values with a reference to determine where their soil ranks on a logit regression curve (Nunes et al., 2021).

We used the data to explore whether it was possible for farmers to use their individual data to assess change over time and to qualify for C credits. To detect changes at the plot or field scale, farmers need to sample using a sampling density that overcomes spatial variability. Based on our sampling from the 4-ha area within a field, we found it may not be logistically feasible to sample fields to detect small changes in C that are in line with C trading protocols (Table 1). Using the aspirational goal of 0.4% C per year as an example, we note that over a 10-year period, sampling needs are reduced fourfold but are still quite demanding. Sampling demands would pose a challenge for verification as institutions and practitioners rely on limited or no budgets for this purpose (Amelung et al., 2020). We considered change within the 30 cm of soil profile as protocols for C sequestration verification require taking samples to at least 30 cm where the greatest proportional change occurs (VandenBygaart et al., 2011). While the 4 per mille recommends sampling at 30 cm, many protocols

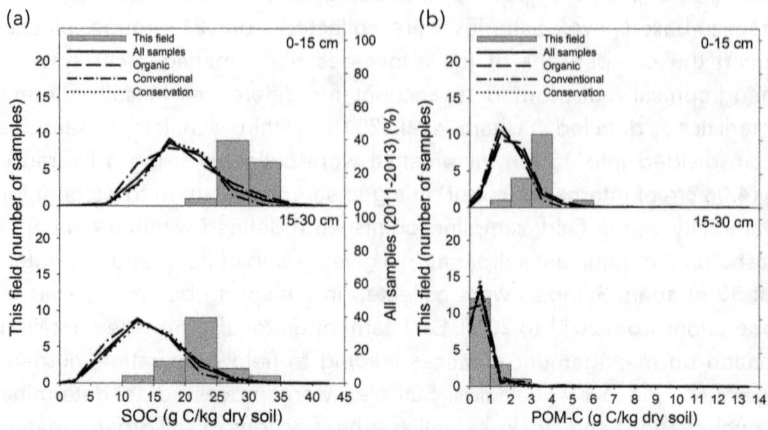

Figure 1 Example of data shared with farmers comparing specific fields with the project-wide results for SOC (a) and particulate organic C (b).

Published by Burleigh Dodds Science Publishing Limited, 2023.

recommend sampling to 1 m (FAO, 2020; The Earth Partners, 2012). This may be fiscally impossible or at least unreasonable as sampling to deeper depths would greatly increase sampling density due to increased variability and reduced statistical power (<80%) (Kravchenko and Robertson, 2011; Necpálová et al., 2014).

One might have expected sampling demands to be lower and statistical power to be higher for the evaluation of changes in labile fractions of organic matter like particulate organic C (POM-C), which is known to reflect changes due to management in a shorter time frame than total soil organic matter stocks (Magdoff and Weil, 2004; Wander, 2004); but, we did not find this to be the case. Only by pooling data, from 21 locations could we successfully separate management treatments with a minimum detectable difference of 10% (Table 1). It must be noted that 10% is an extremely unrealistic rate of change unless dramatic changes in land use occurred.

Sampling efforts and costs might be lowered if the information gained from samples and site history is augmented through the use of process-based models or indirect ranking tools. For example, the Soil Carbon Protocol (Climate Action Reserve, 2020) relies on a combination of soil measures and biogeochemical modelling in the delineation of a baseline and projection of C gains. Aside from costs related to sampling, the costs associated with laboratory analysis should also be considered. In this region, only a few commercial laboratories are equipped to conduct SOC testing using combustion analysis, and typical costs per sample remain around US$10. Current verification of C sequestration is assumed based on surveys of practice adoption. While physical documentation of C sequestration at the field level is currently not required, its costs could limit enrolment in C sequestration programmes.

Table 1 Plot requirement and within plot sample density needed to account for differences in soil organic carbon (SOC) and particulate organic carbon (POM-C) with a Power = 0.8 and $\alpha = 0.05$

		Samples needed to detect differences at the 4-ha plot scale		
	Plots (fields) needed to detect differences of 10% change*	0.4% change per year	0.4% change per year in a 10-year period	10% change
Mg C ha^{-1} in ESM 0–30 cm	21	15 226	153	25
Mg POM-C ha^{-1} in ESM 0–30 cm	78	81 744	818	131

*Considering regional variability, these represent the number of fields that would need to be sampled using the 16 samples per 4-ha area used in this study.

Data analysis using concentrations and soil equivalent mass basis (ESM) better accounts for differences in bulk density that can contribute to inaccurate comparisons of SOC and POM-C stocks (Ellert and Bettany, 1995). Comparisons of surface layers (0–30 cm) showed that conservation and organic systems had significantly higher SOC than their conventional counterparts (Table 2, $P < 0.05$). A similar trend was observed for concentrations of POM-C with stocks in the conservation category being intermediate ($P < 0.05$). Even with data-pooling a tremendous amount of research support is needed to account for regional variability and supply sufficient experimental power (>80%) Table 2.

Farm interviews allowed us to score stewardship using the Conservation Measurement Tool (CMT), which was previously used to determine enrolment in the Conservation Stewardship Program (CSP). Practice-based tools like this that are low cost compared to sampling are used to prioritize applications for government programs that reward the use of practices that increase C sequestration and soil health (Bowman and Lynch, 2019; Wallander et al., 2021). Based on actual field samples we were able to pool results to compare farm practices and farm sectors. This analysis determined that the CMT scores overestimated the benefits of reduced tillage and use of manure on SOC (Ugarte et al., 2018). Improvements in the Conservation Assessment Ranking Tool (CART), which replaces CMT and inventories management and conservation practices to compute baseline levels of soil stewardship at the farm operation scale (NRCS, 2021) are more regionalized. Pooled data from on-farm studies like this have tremendous potential to validate and refine tools like CART that assess soil quality and erosion, air quality and water quality and quantity. This kind of participatory work is a valuable tool for policy and programme design. However, we

Table 2 Mean separation for the effect of management practices on SOC and particulate organic C (POM-C) expressed in concentration and equivalent soil mass (ESM)

	SOC (g C Kg^{-1} soil 0–30 cm)	SOC (Mg C ha^{-1} in ESM 0–30 cm)	POM-C (g POM-C kg^{-1} soil 0–30 cm)	POM-C (Mg POM-C ha^{-1} in ESM 0–30 cm)
Conservation	20.28 ± 0.89 a	86.25 ± 4.64 a	1.62 ± 0.08 ab	10.34 ± 0.64 ab
Conventional	18.45 ± 0.85 b	78.77 ± 4.50 b	1.47 ± 0.07 b	9.21 ± 0.58 b
Organic	20.29 ± 0.85 a	86.67 ± 4.48 a	1.73 ± 0.07 a	10.71 ± 0.57 a
Power	0.94	0.84	0.94	0.91

Note: Means within columns not followed by the same letter differ at the 0.05 percent probability level.

must remember that researcher-driven explorations like ours that carry out between-group comparisons for farm-sector and treatment-based analyses are unfortunately not immediately useful to farmers as decision support tools for on-farm management.

3.2 Farmer behaviours

3.2.1 Attitudes, norms, and constraints

To understand factors influencing their use of practices likely to sequester C, we used mixed methods, surveys, and interviews by exploring the theory of planned behaviour (Ajzen, 1991). This was implemented in three stages (Fig. 2) that included generative, evaluative and, verification phases. The generative phase started with the framework as modified by Morrison et al. (2012), who replaced perceived constraints (time and capitol constraints) to include business orientation, access to information, and prior action. Participants (>200 farmers and farm industry professionals) at a grower field day were surveyed with electronic clickers to identify attitudes and norms influencing their use of practices thought to increase or conserve SOC. This was followed by three focus groups with farmer cohorts representing conservation, conventional, and organic farmer categories in which we trialled questions for individual interviews. The generative phase included farmer interviews carried out in association with the 21 general locations we sampled (Fig. 2). Farmers were asked open-ended questions about attitudes and norms (Table 3; evaluative phase) before being asked about (1) threats to sustainability where they lived (i.e. weather, economic trends, pressure to produce, farm-size expansion, lack of knowledge about best practices, and others); (2) specific issues (i.e. changing weather, drought, flooding, loss of wildlife habitat, water pollution, spray drift, community health, jobs, poverty, and others); next, they were asked what (3) practices they associated with good stewardship; (4) attitudes about technology and innovation; and (5) where or how they marketed their crops (e.g. the local elevator with no on-farm storage, on a contract basis with on-farm storage, with a contract based on certification or an identity-preserved trait). Finally, they were then asked to identify any added value tactics (i.e. certifications or eco-labels including USDA Organic, animal welfare, American Grassfed Beef, Global Good Ag Practices, GMO free, Fair Trade) and to list any government programmes promoting soil conservation in which they participated. In the verification phase, summary information was refined through engagement with conservation experts at a workshop, reviewed by participating farmers in small group webinars and at a 'Soil and Water Conservation Society' meeting.

3.2.2 Realized actions as individuals or groups

Measured results, interviews, and surveys suggest business orientation and social connectedness most influence farmers' practice choices and propensity to participate in voluntary stewardship efforts. This information can be used to segment producers into groups with distinct information and programmatic needs.

Farmers' individual choices are mostly levied through purchasing, participation, and actions that extend beyond practice or land use choices that are the typical focus of most research on practice adoption. Commercial interventions that target information use by individual operators attempt to relate information or product use to economic gain achieved principally through increasing productivity (yield) and/or crop quality or by reducing risk. This explains why the use of standard soil test information and crop insurance are far more common than participation in C trading markets (Table 4). As an individual operator engaging in a C trading market, a farmer might be expected to verify changes in organic matter stocks using approaches similar to soil testing performed to inform soil fertilization routines. Based on our analysis, the expected rates of SOC change needed to secure payments (US$10–15 per C ton) (Keenor et al., 2021) are not practical for grain cropping systems in this region. Additionally, little is known about costs associated with the adoption of C-centric practices. Further, fine-scale indicators of change in SOC that are measured on-farm may not adequately reflect change over time or predict benefits provided at the watershed or regional climatological scale. This creates scale mismatch wherein individual farmer decision-making about tillage or fertilization practices with stewardship and climate abatement inadequately address or overcome differences in the spatial or temporal scale over which decision-making and ecological processes occur (Guerrero et al., 2013). Farmers did not express concern over the lack of evidence directly relating stewardship to ecosystem services and provided

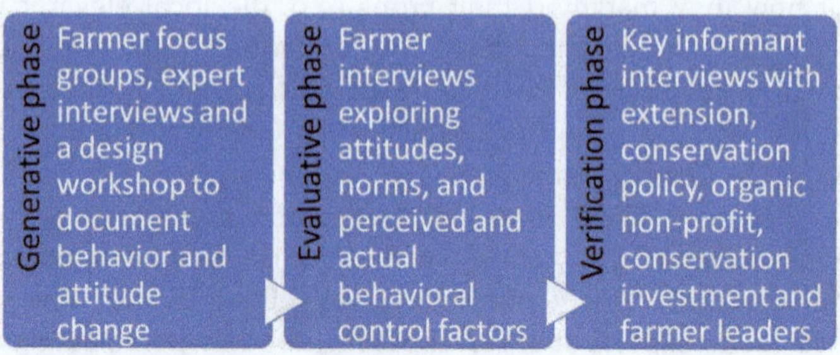

Figure 2 Phases of engagement with stakeholders are various levels to identify barriers preventing farmers from implementing stewardship practices using the theory of planned behaviour.

Published by Burleigh Dodds Science Publishing Limited, 2023.

Table 3 Evolution of the theory of planned behaviour during the three stages of the study

	Generative phase	Evaluative phase
Attitudes towards behaviour	Economic motivations Trust Innovativeness Satisfaction	Economic Trust Goodness Success Innovation Privacy (information)
Subjective norms	Environmental responsibility	Environmental responsibility **Family responsibility** **Group responsibility**
Perceived behavioural control	Business orientation Information seeking behaviour Past environmental behaviour Connectedness	Time constraints Capitol constraints **Competition** **Government regulations/programmes**

Source: A generative list of attitudes, norms, and constraints was taken from Morrison et al. (2012). Items in bold during the evaluative phase were identified during the generative phase.

no indication that lack of knowledge influenced their decision-making about farming practices.

Voluntary participation by individuals in public or market mechanisms helped to distinguish farm sectors. While farmers selling into conventional markets consistently took advantage of crop insurance and precision technologies, farmers in the other two categories varied much more in their attitudes about independence from government and new technologies. Individuals within the organic and conservation groups represented the most and least intensive users of government programmes and precision tillage, guidance, and vision systems. In general, the farmers in the region were receptive to technologies with smaller operations being more judicious in their use. Larger scale operations may have benefited more from tax incentives that encourage purchase of new equipment. Lease agreements and access to land were points of concern for all but large operators more frequently complained about outsiders renting ground 'out from under them'. Key factors influencing conventional farmers were pressure to produce and expand the size of farm operations. Organic price premiums being received at the time allowed organic farmers to secure and retain leases. Even though farmers complained about transaction costs associated with paperwork needed for government programmes and certification, these opportunities enabled conservation behaviours. Farmers engaged in alternative markets relied more heavily upon social networks (Table 4). Cooperative coping strategies offer risk avoidance tactics that are not necessarily economically rationale but are socially wise where social capitol can be traded (Mustafa et al., 2019; Okonya et al., 2013;

Table 4 Summary of results from the implementation of the theory of planned behaviour to identify barriers preventing farmers from implementing stewardship practices or using information to direct management practices

Programmes and mechanisms promoting conservation	Organic	Conservation	Conventional
Soil testing	Many	All	All
Certification (USDA National Organic Program)	All	None	None
USDA Conservation Reserve	Many	Many	Some
USDA Conservation Stewardship or Environmental Quality Incentives Programs	Some	Many	Some
Crop insurance	Many	Many	All
Attitudes and interests in approaches	**Organic**	**Conservation**	**Conventional**
Independent (anti-governmental programs)	Several	Few	None
Interest in C sequestration contracts	Some	Many	Some
Interest in sustainability certification	Some	Many	Some
Concerns	**Organic**	**Conservation**	**Conventional**
Agricultural sustainability	Many	Some	Some
Changing weather (drought and flooding)	Many	Many	Many
Loss of wildlife habitat	Many	Many	Few
Water pollution	Many	Many	Few
Spray drift	Many	Many	Few
Input costs	Some	Few	Many
Land rental by community outsiders	Few	Some	Many
Land price	Some	Many	Few
Important values	**Organic**	**Conservation**	**Conventional**
Profitability	Many	Many	All
Environment	Many	All	Some
Social/religious network	All	Many	Few
Family	Many	Many	All
Goodness	Many	Many	Some

All = 100%, Many > 75%, Some > 25%, Few < 10%.

Sima et al., 2015). Conventional farmers' responses suggested they value economic success and family quite highly, often farming at a larger scale and selling into commodity markets through their local elevators. Because farmers live where they work, there is a natural tension between competitive business and communal motives. A dominant concern for all was depopulation and loss of jobs within their communities. Despite this, larger operators reported benefiting financially from lower input prices sourced in bulk outside their communities. Farmers in all groups expressed interest in reversal of the trend in farm-size expansion and increased reliance on crop insurance that promote resource degradation and discourage conservation.

Results showed resources and policies that align with the different attitudes, norms, and behavioural controls determine farmers' conservation behaviours. Farming styles, which include norms and practices used by more or less connected groups of farmers, evolve over time (Vanclay et al., 2006). Differences among these groups provide a useful way to target efforts and assess outcomes. Government-backed conservation programmes that reflect cultural and social goals, and not just economics, achieve community goals (Lago et al., 2019). Evolving private markets that aim to market ecosystem services associated with C sequestration will need to verify services in an affordable way. Pooling of on-farm data has already been used to demonstrate how remote sensing can be used to document the use of C-centric practices (Xia et al., 2021) and how rapid hand-held sensors can successfully and cheaply quantify C based on colour (Ewing et al., 2021).

We found that a focus only on mechanisms influencing the use of in-field farm practices would limit our understanding of farmers' decision-making by blinding us to important social and economic factors. Farmers not only influence C sequestration directly by practice adoption but also through engagement with farm or community organizations, service providers, the research community, governmental agencies and, consumers to influence whether or how they are rewarded for their efforts (Ledingham and Bruning, 1998). Again, we show that pooling farmer data and farmer networks can empower cross-scale efforts to effectively address individual and institutional concerns about management as was suggested by Cash et al. (2006).

4 Conclusion

1. Field-based verification by direct soil sampling may be unrealistic for producers wanting to be compensated during the years immediately following the adoption of practices that sequester C.
2. Existing tools and datasets are not adequate to support site- and system-based assessments of ecosystem services derived by implementing C-centric practices. Scale mismatch and improper allocation of

responsibility to individual operators charged with delivering services through market mechanisms exacerbate this problem.

3 Blended approaches that leverage reference datasets that verify benefits for site-specific farming systems and use models or rapid soil testing methods (e.g. apps that sample colour and remote sensing that documents residue cover, presence of cover crops) will reduce costs and make this more practical.
4 Farmers' engagement in groups increases both the development and use of effective sequestration practices by embracing social norms, spurring innovation, orienting research, and justifying policy, premiums, or product claims that promote the use of C-centric practices.
5 Use of C sequestration as a proxy for ecosystem services likely undervalues the use of conservation practices and will not help farmers overcome trends in farm size that are generally seen as threats to sustainability.
6 Clear differences in farming styles, values, and norms argue strongly for targeting efforts to promote C sequestration among different farm sectors.

5 Acknowledgements

We thank the farmers who participated in the research project 'Organic systems and climate change' and the USDA Organic Transitions program Grant No. 2010 51106-21824 (Project No.: ILLU-875-634).

6 Where to look for further information

- The Food and Agriculture Organization is a primary source for world agriculture that considers all types of farmers and farming systems: https://www.fao.org/home/en.
- The Food and Agriculture Organization's Global Soil Organic Carbon Map is a key resource: https://www.fao.org/soils-portal/data-hub/soil-maps-and-databases/global-soil-organic-carbon-map-gsocmap/en/.
- The Food and Agriculture Organization's Map of Soil Carbon Sequestration is also of interest: https://www.fao.org/soils-portal/data-hub/soil-maps-and-databases/global-soil-organic-carbon-sequestration-potential-map-gsocseq/en/.

7 References

Ainscough, J., de Vries Lentsch, A., Metzger, M., Rounsevell, M., Schröter, M., Delbaere, B., de Groot, R. and Staes, J. (2019). Navigating pluralism: understanding perceptions of the ecosystem services concept. *Ecosystem Services* 36. DOI: 10.1016/j.ecoser.2019.01.004.

Ajzen, I. (1991). The theory of planned behavior. *Organizational Behavior and Human Decision Processes* 50(2): 179–211.

Al-Kaisi, M. M. and Kwaw-Mensah, D. (2020). Quantifying soil carbon change in a long-term tillage and crop rotation study across Iowa landscapes. *Soil Science Society of America Journal* 84(1): 182–202. DOI: 10.1002/saj2.20003.

Amelung, W., Bossio, D., de Vries, W., Kögel-Knabner, I., Lehmann, J., Amundson, R., Bol, R., Collins, C., Lal, R., Leifeld, J., Minasny, B., Pan, G., Paustian, K., Rumpel, C., Sanderman, J., van Groenigen, J. W., Mooney, S., van Wesemael, B., Wander, M. and Chabbi, A. (2020). Towards a global-scale soil climate mitigation strategy. *Nature Communications* 11(1): 5427. DOI: 10.1038/s41467-020-18887-7.

Aref, S. and Wander, M. M. (1997). Long-term trends of corn yield and soil organic matter in different crop sequences and soil fertility treatments on the morrow plots. *Advances in Agronomy* 62(C): 1704–1711. DOI: 10.1016/S0065-2113(08)60568-4.

Balfour, E. B. (1976). *The Living Soil and the Haughley Experiment*. New York: Universe Books.

Baveye, P. C., Baveye, J. and Gowdy, J. (2016). Soil "ecosystem" services and natural capital: critical appraisal of research on uncertain ground. *Frontiers in Environmental Science* 4. DOI: 10.3389/fenvs.2016.00041.

Borrelli, P., Paustian, K., Panagos, P., Jones, A., Schütt, B. and Lugato, E. (2016). Effect of good agricultural and environmental conditions on erosion and soil organic carbon balance: a national case study. *Land Use Policy* 50: 408–421. DOI: 10.1016/j.landusepol.2015.09.033.

Botan, C. (1993). Introduction to the paradigm struggle in public relations. *Public Relations Review* 19(2): 107–110. DOI: 10.1016/0363-8111(93)90001-S.

Bowman, M. and Lynch, L. (2019). Government programs that support farmer adoption of soil health practices: a focus on Maryland's agricultural water quality cost-share program. *Choices* 34(2): 1–8.

Bracking, S. and Leffel, B. (2021). Climate finance governance: fit for purpose? *WIREs Climate Change* 12(4). DOI: 10.1002/wcc.709.

Brouder, S. M. and Gomez-Macpherson, H. (2014). The impact of conservation agriculture on smallholder agricultural yields: a scoping review of the evidence. *Agriculture, Ecosystems and Environment* 187: 11–32. DOI: 10.1016/j.agee.2013.08.010.

Buitenhuis, Y., Candel, J. J. L., Termeer, K. J. A. M. and Feindt, P. H. (2020). Does the common agricultural policy enhance farming systems' resilience? applying the Resilience Assessment Tool (ResAT) to a farming system case study in the Netherlands. *Journal of Rural Studies* 80: 314-327. DOI: 10.1016/j.jrurstud.2020.10.004.

Cash, D. W., Adger, W. N., Berkes, F., Garden, P., Lebel, L., Olsson, P., Pritchard, L. and Young, O. (2006). Scale and cross-scale dynamics: governance and information in a multilevel world. *Ecology and Society* 11(2). DOI: 10.5751/ES-01759-110208.

Claassen, R., Cattaneo, A. and Johansson, R. (2008). Cost-effective design of agri-environmental payment programs: U.S. experience in theory and practice. *Ecological Economics* 65(4): 737–752. DOI: 10.1016/j.ecolecon.2007.07.032.

Clapp, J. and Isakson, S. R. (2018). Risky returns: the implications of financialization in the food system. *Development and Change* 49(2): 437–460. DOI: 10.1111/dech.12376.

Climate Action Reserve (2020). *Climate Action Reserve Soil Enrichment Protocol Version 1.0*. Available at: https://www.climateactionreserve.org/wp-content/uploads/2020/10/Soil-Enrichment-Protocol-V1.0.pdf (accessed 12 April 2022).

Conford, P. (2001). *The Origins of the Organic Movement*. Glasgow, UK: Floris Books.

Coppess, J. (2018). *The Fault Lines of Farm Policy: A Legislative and Political History of the Farm Bill*. Lincoln: University of Nebraska Press.

Cox, C. (2007). U.S. agriculture conservation policy & programs: history, trends, and implications. In: Arha, K., Josling, T., Sumner, D. A., et al. (Eds), *U.S. Agricultural Polivy and the 2007 Farm Bill*. Stanford, CA: Woods Institute for the Environment, pp. 113–146.

de Janvry, A. and Sadoulet, E. (2000). Growth, poverty, and inequality in Latin America: a causal analysis, 1970–94. *Review of Income and Wealth* 46(3): 267-287. DOI: 10.1111/j.1475-4991.2000.tb00843.x.

Debolini, M., Valette, E., François, M. and Chéry, J. (2015). Mapping land use competition in the rural-urban fringe and future perspectives on land policies: a case study of Meknès (Morocco). *Land Use Policy* 47: 373–381. DOI: 10.1016/j.landusepol.2015.01.035.

Demir, O., Inan, H. I., Biyik, C. and Uzun, B. (2015). Land management for erosion prevention: a case study for a Turkish nature reserve. *Land Use Policy* 47: 394–400. DOI: 10.1016/j.landusepol.2015.01.036.

Dixon, D. P. and Hapke, H. M. (2003). Cultivating discourse: the social construction of agricultural legislation. *Annals of the Association of American Geographers* 93(1): 142-164. DOI: 10.1111/1467-8306.93110.

Doering, O. and Smith, K. (2012). *Examining the Relationship of Conservation Compliance & Farm Program Incentives*. C-FARE Reports 156624, Council on Food, Agricultural, and Resource Economics (C-FARE).

Ellert, B. H. and Bettany, J. R. (1995). Calculation of organic matter and nutrients stored in soils under contrasting management regimes. *Canadian Journal of Soil Science* 75(4): 529–538. DOI: 10.4141/cjss95-075.

Ewing, P. M., TerAvest, D., Tu, X. and Snapp, S. S. (2021). Accessible, affordable, fine-scale estimates of soil carbon for sustainable management in sub-Saharan Africa. *Soil Science Society of America Journal* 85(5): 1814–1826. DOI: 10.1002/saj2.20263.

FAO (2020). *A Protocol for Measurement, Monitoring, Reporting and Verification of Soil Organic Carbon in Agricultural Landscapes*. DOI: 10.4060/cb0509en.

FAO (2021). *Standards on Good Agricultural and Environmental Conditions of Land: Introductory Handbook*. Budapest, Hungary.

Fonte, M. (2008). Knowledge, food and place: a way of producing, a way of knowing. *Sociologia Ruralis* 48(3): 200–222. DOI: 10.1111/j.1467-9523.2008.00462.x.

Freyfogle, E. (2007). *Agrarianism and the Good Society: Land, Culture, Conflict, and Hope*. Lexington: University Press of Kentucky.

Ghosh, R. and Wolf, S. (2021). Hybrid governance and performances of environmental accounting. *Journal of Environmental Management* 284: 111995. DOI: 10.1016/j.jenvman.2021.111995.

Glauber, J. W. (2013). The growth of the federal crop insurance program, 1990–2011. *American Journal of Agricultural Economics* 95(2): 482–488. DOI: 10.1093/ajae/aas091.

Guerrero, A., Mcallister, R., Corcoran, J., and Wilson, K. (2013). Scale mismatches, conservation planning, and the value of social-network analyses. *Conservation Biology* 27: 35–44. DOI: 10.1111/j.1523-1739.2012.01964.x.

Guo, L. B. and Gifford, R. M. (2002). Soil carbon stocks and land use change: a meta analysis. *Global Change Biology* 8(4): 345–360. DOI: 10.1046/j.1354-1013.2002.00486.x.

Guthman, J. (2004). The trouble with "organic lite" in California: a rejoinder to the "conventionalisation" debate. *Sociologia Ruralis* 44(3): 301–316. DOI: 10.1111/j.1467-9523.2004.00277.x.

Heckman, J. (2006). A history of organic farming: transitions from Sir Albert Howard's war in the soil to USDA national organic program. *Renewable Agriculture and Food Systems* 21(3): 143–150. DOI: 10.1079/RAF2005126.

Honeycutt, C. W., Morgan, C. L. S., Elias, P., Doane, M., Mesko, J., Myers, R., Odom, L., Moebius-clune, B. and Nichols, R. (2020). Soil health: model programs in the USA. *Frontiers of Agricultural Science and Engineering* 7(3). DOI: 10.15302/J-FASE-2020340.

IFOAM (2005). Principles of organic agriculture. Available at: https://www.ifoam.bio/sites/default/files/2020-03/poa_english_web.pdf (accessed 22 August 2021).

Jackson Hammond, A. A., Motew, M., Brummitt, C. D., DuBuisson, M. L., Pinjuv, G., Harburg, D. V., Campbell, E. E. and Kumar, A. A. (2021). Implementing the soil enrichment protocol at scale: opportunities for an agricultural carbon market. *Frontiers in Climate* 3. DOI: 10.3389/fclim.2021.686440.

Kannuri, N. K. and Jadhav, S. (2018). Generating toxic landscapes: impact on well-being of cotton farmers in Telangana, India. *Anthropology and Medicine* 25(2): 121–140. DOI: 10.1080/13648470.2017.1317398.

Keenor, S. G., Rodrigues, A. F., Mao, L., Latawiec, A. E., Harwood, A. R. and Reid, B. J. (2021). Capturing a soil carbon economy. *Royal Society Open Science* 8(4): 202305. DOI: 10.1098/rsos.202305.

King, F. H. (2004). *Farmers of Forty Centuries: Organic Farming in China, Korea, and Japan*. Mineola, NY: Dover Publications, Inc.

Kirchmann, H., Thorvaldsson, G., Bergström, L., Gerzabek, G., Andrén, O., Eriksson, L.-O. and Winninge, M. (2008). Fundamentals of organic agriculture - past and present. In: *Organic Crop Production - Ambitions and Limitations*. DOI: 10.1007/978-1-4020-9316-6_2.

Kravchenko, A. N. and Robertson, G. P. (2011). Whole-profile soil carbon stocks: the danger of assuming too much from analyses of too little. *Soil Science Society of America Journal* 75(1): 235–240. DOI: 10.2136/sssaj2010.0076.

Lago, M.G., Plant, R. and Jacobs, B. (2019). Re-politicising soils: what is the role of soil framings in setting the agenda? *Geoderma* 349: 97–106.

Lal, R., Follett, R. F. and Kimble, J. M. (2003). Achieving soil carbon sequestration in the United States: a challenge to the policy makers. *Soil Science* 168(12): 827–845. DOI: 10.1097/01.ss.0000106407.84926.6b.

Lapola, D. M., Martinelli, L. A., Peres, C. A., Ometto, J. P. H. B., Ferreira, M. E., Nobre, C. A., Aguiar, A. P. D., Bustamante, M. M. C., Cardoso, M. F., Costa, M. H., Joly, C. A., Leite, C. C., Moutinho, P., Sampaio, G., Strassburg, B. B. N. and Vieira, I. C. G. (2014). Pervasive transition of the Brazilian land-use system. *Nature Climate Change* 4(1): 27–35. DOI: 10.1038/nclimate2056.

le Foll, S. (2015). *4 Per 1000: A New Program for Carbon Sequestration in Agriculture*. Paris, France: French Minister of Agriculture, Agrifood and Forestry.

Ledingham, J. A. and Bruning, S. D. (1998). Relationship management in public relations: dimensions of an organization-public relationship. *Public Relations Review* 24(1): 55–65. DOI: 10.1016/S0363-8111(98)80020-9.

Mafongoya, P. L., Bationo, A., Kihara, J. and Waswa, B. S. (2006). Appropriate technologies to replenish soil fertility in southern Africa. *Nutrient Cycling in Agroecosystems* 76(2–3): 137–151. DOI: 10.1007/s10705-006-9049-3.

Magdoff, F. and Weil, R. (2004). Soil organic matter management strategies. *Advances in Agroecology*. DOI: 10.1201/9780203496374.ch2.

MARS (2014). *Good Agricultural and Environmental Conditions (GAEC) Database*. Available at: https://www.eea.europa.eu/data-and-maps/data/external/good-agricultural-environmental-conditions-gaec-2010 (accessed 12 April 2022).

Monk, A. (2011). Organic poetic licence: consumer moral norms driving farming systems. In: Martin, P. and Williams, J. (Eds), *Defending the Social License of Farming: Issues, Challenges and New Directions for Agriculture*, Collingwood, VIC: CSIRO Publishing. pp. 39-46.

Morrison, M., Durante, J., Greig, J., Ward, J. and Oczkowski, E. (2012). Segmenting landholders for improving the targeting of natural resource management expenditures. *Journal of Environmental Planning and Management* 55(1): 17-37. DOI: 10.1080/09640568.2011.575630.

Moschitz, H. and Home, R. (2014). The challenges of innovation for sustainable agriculture and rural development: integrating local actions into European policies with the reflective learning methodology. *Action Research* 12(4): 392–409. DOI: 10.1177/1476750314539356.

Mustafa, G., Latif, I. A., Bashir, M. K., Shamsudin, M. N. and Daud, W. M. N. W. (2019). Determinants of farmers' awareness of climate change. *Applied Environmental Education and Communication* 18(3): 219–233. DOI: 10.1080/1533015X.2018.1454358.

Necpálová, M., Anex, R. P., KxravchenkoProf, A. N., Abendroth, L. J., Del Grosso, S. J., Dick, W. A., Helmers, M. J., Herzmann, D., Lauer, J. G., Nafziger, E. D., Sawyer, J. E., Scharf, P. C., Strock, J. S. and Villamil, M. B. (2014). What does it take to detect a change in soil carbon stock? a regional comparison of minimum detectable difference and experiment duration in the north central united states. *Journal of Soil and Water Conservation* 69(6): 517-531. DOI: 10.2489/jswc.69.6.517.

Ng'ang'a, S. K., Jalang'o, D. A. and Girvetz, E. H. (2019). Soil carbon enhancing practices: a systematic review of barriers and enablers of adoption. *SN Applied Sciences* 1(12). DOI: 10.1007/s42452-019-1747-y.

NRCS (2021). *Conservation Assessment and Ranking (CART)*. Available at: https://www.nrcs.usda.gov/wps/portal/nrcs/detail/ca/programs/easements/acep/?cid=nrcseprd1554436 (accessed on 12 April 2022).

Nunes, M. R., Veum, K. S., Parker, P. A., Holan, S. H., Karlen, D. L., Amsili, J. P., Es, H. M., Wills, S. A., Seybold, C. A. and Moorman, T. B. (2021). The soil health assessment protocol and evaluation applied to soil organic carbon. *Soil Science Society of America Journal* 85(4): 1196–1213. DOI: 10.1002/saj2.20244.

Okonya, J. S., Syndikus, K. and Kroschel, J. (2013). Farmers' perception of and coping strategies to climate change: evidence from six agro-ecological zones of Uganda. *Journal of Agricultural Science* 5(8). DOI: 10.5539/jas.v5n8p252.

Oldfield, E. E., Eagle, A. J., Rubin, R. L., Rudek, J., Sanderman, D. R. and Gordon, D. R. (2021). *Agricultural Soil Carbon Credits: Making Sense of Protocols for Carbon Sequestration and Net Greenhouse Gas Removals*. Available at: https://www.edf.org

/sites/default/files/content/agricultural-soil-carbon-credits-protocol-synthesis.pdf (accessed 12 April 2022).

Paustian, K., Larson, E., Kent, J., Marx, E. and Swan, A. (2019). Soil C sequestration as a biological negative emission strategy. *Frontiers in Climate* 1. DOI: 10.3389/fclim.2019.00008.

Pe'er, G., Bonn, A., Bruelheide, H., Dieker, P., Eisenhauer, N., Feindt, P. H., Hagedorn, G., Hansjürgens, B., Herzon, I., Lomba, Â, Marquard, E., Moreira, F., Nitsch, H., Oppermann, R., Perino, A., Röder, N., Schleyer, C., Schindler, S., Wolf, C., Zinngrebe, Y. and Lakner, S. (2020). Action needed for the EU common agricultural policy to address sustainability challenges. *People and Nature* 2(2): 305–316. DOI: 10.1002/pan3.10080.

Pe'er, G., Zinngrebe, Y., Moreira, F., Sirami, C., Schindler, S., Müller, R., Bontzorlos, V., Clough, D., Bezák, P., Bonn, A., Hansjürgens, B., Lomba, A., Möckel, S., Passoni, G., Schleyer, C., Schmidt, J. and Lakner, S. (2019). A greener path for the EU common agricultural policy. *Science* 365(6452): 449–451. DOI: 10.1126/science.aax3146.

Piñeiro, V., Arias, J., Dürr, J., Elverdin, P., Ibáñez, A. M., Kinengyere, A., Opazo, C. M., Owoo, N., Page, J. R., Prager, S. D. and Torero, M. (2020). A scoping review on incentives for adoption of sustainable agricultural practices and their outcomes. *Nature Sustainability* 3(10): 809–820. DOI: 10.1038/s41893-020-00617-y.

Robinson, D. A., Hockley, N., Dominati, E., Lebron, I., Scow, K. M., Reynolds, B., Emmett, B. A., Keith, A. M., de Jonge, L. W., Schjønning, P., Moldrup, P., Jones, S. B. and Tuller, M. (2012). Natural capital, ecosystem services, and soil change: why soil science must embrace an ecosystems approach. *Vadose Zone Journal* 11(1). DOI: 10.2136/vzj2011.0051.

Rumpel, C., Amiraslani, F., Chenu, C., Garcia Cardenas, M., Kaonga, M., Koutika, L. S., Ladha, J., Madari, B., Shirato, Y., Smith, P., Soudi, B., Soussana, J. F., Whitehead, D. and Wollenberg, E. (2020). The 4p1000 initiative: opportunities, limitations and challenges for implementing soil organic carbon sequestration as a sustainable development strategy. *Ambio* 49(1): 350–360. DOI: 10.1007/s13280-019-01165-2.

Sima, M., Popovici, E.-A., Bălteanu, D., Micu, D. M., Kucsicsa, G., Dragotă, C. and Grigorescu, I. (2015). A farmer-based analysis of climate change adaptation options of agriculture in the Bărăgan Plain, Romania. *Earth Perspectives* 2(1). DOI: 10.1186/s40322-015-0031-6.

Snapp, S. S., Ugarte, C. M., Hunter, D. W., and Wander, M. M. (2022). Assessing the effects of cover crops on soil health. In: Horwath, W. (Ed.), *Improving Soil Health*. Cambridge, UK: Burleigh Dodds Science Publishing.

Stockmann, U., Adams, M. A., Crawford, J. W., Field, D. J., Henakaarchchi, N., Jenkins, M., Minasny, B., McBratney, A. B., Courcelles, VdRd, Singh, K., Wheeler, I., Abbott, L., Angers, D. A., Baldock, J., Bird, M., Brookes, P. C., Chenu, C., Jastrow, J. D., Lal, R., Lehmann, J., O'Donnell, A. G., Parton, W. J., Whitehead, D. and Zimmermann, M. (2013). The knowns, known unknowns and unknowns of sequestration of soil organic carbon. *Agriculture, Ecosystems and Environment* 164: 80–99. DOI: 10.1016/j.agee.2012.10.001.

Sullivan, S. (2013). Banking nature? The spectacular financialisation of environmental conservation. *Antipode* 45(1): 198–217. DOI: 10.1111/j.1467-8330.2012.00989.x.

Tauger, M. B. (2010). *Agriculture in World History*. DOI: 10.4324/9780203847480.

Thapa, V. R., Ghimire, R., Duval, B. D. and Marsalis, M. A. (2019). Conservation systems for positive net ecosystem carbon balance in semiarid drylands. *Agrosystems, Geosciences and Environment* 2(1): 1-8. DOI: 10.2134/age2019.03.0022.

The Earth Partners (2012). *VMD0021 Estimation of Stocks in the Soil Carbon Pool, v1.0*. Available at: https://verra.org/wp-content/uploads/2018/03/VMD0021-Estimation-of-Stocks-in-the-Soil-Carbon-Pool-v1.0.pdf (acessed 12 April 2022).

Ugarte, C. M., Kwon, H., Andrews, S. S. and Wander, M. M. (2014). A meta-analysis of soil organic matter response to soil management practices: an approach to evaluate conservation indicators. *Journal of Soil and Water Conservation* 69(5): 422-430. DOI: 10.2489/jswc.69.5.422.

Ugarte, C. M., Kwon, H. and Wander, M. M. (2018). Conservation management and ecosystem services in midwestern United States agricultural systems. *Journal of Soil and Water Conservation* 73(4): 422-433. DOI: 10.2489/jswc.73.4.422.

USDA-AMS (2002). *National Organic Program; Final Rule: 7CFR Part 205*. US Department of Agriculture, Agriculture Marketing Service. 7 CFR Part 205. Washington, DC. Available at: https://www.ams.usda.gov/about-ams/programs-offices/national-organic-program (accessed 12 August 2021).

Vanclay, F., Howden, P., Mesiti, L. and Glyde, S. (2006). The social and intellectual construction of farming styles: testing Dutch ideas in Australian agriculture. *Sociologia Ruralis* 46: 61-82. DOI:10.1111/j.1467-9523.2006.00404.x.

VandenBygaart, A. J., Bremer, E., McConkey, B. G., Ellert, B. H., Janzen, H. H., Angers, D. A., Carter, M. R., Drury, C. F., Lafond, G. P. and McKenzie, R. H. (2011). Impact of sampling depth on differences in soil carbon stocks in long-term agroecosystem experiments. *Soil Science Society of America Journal* 75(1): 226-234. DOI: 10.2136/sssaj2010.0099.

Wallander, S., Smith, D., Bowman, M. and Claassen, R. (2021). *Cover Crop Trends, Programs, and Practices in the United States*. United States Department of Agriculture, Economic Research Service (EIB 222).

Wander, M. (2004). Soil organic matter fractions and their relevance to soil function. *Advances in Agroecology*. DOI: 10.1201/9780203496374.ch3.

Wander, M. and Nissen, T. (2004). Value of soil organic carbon in agricultural lands. *Mitigation and Adaptation Strategies for Global Change* 9(4): 417-431.

Weil, R. and Magdoff, F. (2004). Significance of soil organic matter to soil quality and health. *Advances in Agroecology*. DOI: 10.1201/9780203496374.ch1.

Wiese, L., Wollenberg, E., Alcántara-Shivapatham, V., Richards, M., Shelton, S., Hönle, S. E., Heidecke, C., Madari, B. E. and Chenu, C. (2021). Countries' commitments to soil organic carbon in nationally determined contributions. *Climate Policy* 21(8): 1005-1019. DOI: 10.1080/14693062.2021.1969883.

Williams, J. and Martin, P. (2011). *Defending the social licence of farming: issues, challenges and new directions for agriculture*. Collingwood VIC: CSIRO Publishing.

Wily, L. A. (2018). Collective land ownership in the 21st century: overview of global trends. *Land* 7(2). DOI: 10.3390/land7020068.

Xia, Y., Guan, K., Copenhaver, K. and Wander, M. (2021). Estimating cover crop biomass nitrogen credits with Sentinel-2 imagery and sites covariates. *Agronomy Journal* 113(2): 1084-1101. DOI: 10.1002/agj2.20525.

Zilberman, D. and Segerson, K. (2012). *Top Ten Design Elements to Achieve More Efficient Conservation Programs*. C-FARE Reports 156623. Council on Food, Agricultural, and Resource Economics (C-FARE).

Chapter 29

Legal issues of implementing agricultural soil organic carbon sequestration as negative emission technology[1]

Alexandra Langlais, CNRS-Université de Rennes, France

1 Introduction

In approving the Paris Agreement in 2015, the international community set itself the goal of constraining the rise in the average temperature of the planet to below 2°C compared to pre-industrial levels or even to limit it to 1.5°C. Meeting these thresholds for global warming means that to achieve this ambitious goal by mid-century, it will be necessary to remove CO_2 from the atmosphere, in addition to reducing global CO_2 emissions to zero. Achieving 'carbon neutrality' by 2050, i.e. reaching a balance between the amount of greenhouse gases (GHGs) emitted worldwide and the earth's capacity to capture and store CO_2, requires drastic GHG emission reductions. In its 2018 report (IPCC, 2018), the IPCC considers other complementary options for achieving carbon neutrality.

In particular, the IPCC envisages CO_2 removal from the atmosphere by negative emission technologies (NETs or carbon dioxide removal (CDR) technologies). The purpose of NETs is to remove carbon dioxide from the earth's atmosphere in a sustainable manner (Fuss et al., 2018; Minx et al.,

1 This work was carried out under two research contracts: the national MITI-CNRS-France Neutraclim project, coordinated by M. Lemoine-Schonne, "Interrogating the role of carbon capture and storage to meet the challenge of climate neutrality" (translated title) and the project H2020 CONSOLE *"COntract solutions for effective and lasting delivery of agri-environmental climate public goods by EU agriculture and forestry"*, coordinated by D. Viaggi, University of Bologna, Italy, call H2020 RUR-O3-2018 for Contract for effective and lasting delivery of agri-environmental public goods.

http://dx.doi.org/10.19103/AS.2022.0106.28

2018; Nemet et al., 2018). However, NETs do not represent a homogeneous set of techniques or a strong definition (Tanzer and Ramirez, 2019; Langlais et Lemoine-Schonne, 2022). In simple terms, they will adopt a biological approach or use industrial technologies to extract CO_2 from the atmosphere and then store it permanently. More precisely, CO_2 'absorbed by biomass (photosynthesis) or captured by chemical processes (air filtration or fixation in mineral materials), or the carbon (C) it contains, will be stored in surface biomass (e.g. wood), in soil, in geological subsoil settings, in mineral materials or in the seabed' (OFEV, no year). A non-exhaustive list of these techniques has been detailed by the IPCC: 'Existing and potential CDR measures include afforestation and reforestation, land restoration and soil carbon sequestration, BECCS, direct air carbon capture and storage (DACCS), enhanced weathering and ocean alkalinisation. These techniques differ widely in terms of maturity, potentials, costs, risks, co-benefits and trade-offs (high confidence). To date, only a few published pathways include CDR measures other than afforestation and BECCS' (IPCC, 2018, pt C. 3.1).

Because of its close relationship with soils, agriculture has a definite role to play as an option for mitigating GHG emissions, in addition to reducing its emissions as a priority. This role in mitigating emissions was recognised in particular by the Kyoto Protocol established in December 1997, an international agreement aimed at reducing GHG emissions and which supplements the United Nations Framework Convention on Climate Change (Article 3.4). More specifically, this protocol aimed to reduce, through legally binding commitments, emissions of six GHGs (carbon dioxide, methane, nitrous oxide and three chlorofluorocarbons substitutes) by at least 5% compared to 1990 levels. The Paris Agreement, whose measures concern the post-2020 period, follows on from the Kyoto Protocol. The contribution of the Kyoto Protocol is to have recognised the capacity of the agricultural sector to be a source and sink of carbon (through sequestration in the biomass and in the soil). To underline the potential of soils, it is also regularly pointed out that soils are the second largest natural reservoir of carbon after the oceans; soils alone contain between 1500 and 2400 billion tonnes of carbon stored in the form of organic matter (Bispo et al., 2016) and only small changes to this large reservoir have the potential to influence atmospheric CO_2 concentrations (Balesdent and Arrouays, 1999).

While forest soils generally store more carbon than agricultural soils, the latter have an important role to play, as they are depleted in carbon and strongly affected by management. Their crucial role in climate change mitigation, adaptation and food security was emphasised in particular by the 4/1000 initiative,[2] which increased the carbon stock of agricultural soils by 0.4%

2 https://4p1000.org/fr.

(or 4 per 1000) each year in the top 40 cm of the soil would, in theory, be equivalent to a significant proportion of annual carbon emissions caused by human activities (Rumpel et al., 2020). This initiative launched by the French government at the 21st Conference of the Parties of the United Nations Framework Convention on Climate Change under the Lima-Paris Action Plan is seen as an additional opportunity to counteract the increase in atmospheric CO_2 concentrations. This initiative concerns both forest and agricultural soils. However, the initiative, at least in the discourse, seems to give more importance to sustainable agricultural practices that sequester carbon in the soil. This is expressed in the presentation of the initiative, which is only aimed at agriculture and its soils: 'The initiative aims to show that agriculture, and in particular agricultural soils, can play a crucial role in food security and climate change'.

The potential of carbon sequestration in agricultural land is now fully recognised and encouraged (e.g. National Academies of Sciences, Engineering, and Medicine, 2019) and thus its contribution to climate change mitigation. However, when it comes to the practical implementation, and even more so the legal implementation, of carbon sequestration in soils as NET, many questions arise. In particular, the relevance of carbon sequestration as NET has been questioned, due to the necessity to implement sustainable practices at scale to have a global impact. It is therefore essential to examine these legal issues related to carbon sequestration in agricultural soils insofar as, once they have been resolved, they can contribute to the development and widespread use of carbon sequestration in soils by securing the conditions for its implementation.

To be fully realised, carbon sequestration in soils requires, first of all, a legal prerequisite that has not yet been achieved. It is a question of protecting soils in both their quantitative and qualitative dimensions to guarantee their capacity to store and sequester carbon. In addition, to avoid leakage and to sustain soil carbon accumulation, the long-term nature of soil protection must be considered. These soil-based preconditions are essential for thinking of carbon sequestration as NET.

In addition, the implementation of soil carbon sequestration as NET as such and the legal support it requires must be addressed. The analysis of the implementation of this technique reveals the many obstacles that the law faces to achieve scalable results.

2 Legal protection of the capacity of agricultural soils to sequester carbon as a legal precondition

The climate issue associated with carbon sequestration in soils, including agricultural soils, should also serve as a powerful reminder of the urgent need to protect soils. Indeed, for soils to play their role, it is necessary to protect their capacity to sequester carbon from both a quantitative and qualitative point of view. However, it has to be said that legal action in favour of soils, and *a fortiori* agricultural soils, remains largely declaratory and therefore not very binding, although many calls have been made to raise awareness and take action (Section 2.1). The growing interest in soil organic carbon (SOC) sequestration as an ecosystem service (ES) also provides an opportunity to strengthen the legal protection of soils to recognise them as an ecosystem. It is, therefore, a question of recognising and protecting the qualities of this ecosystem and not just considering it as a support for activities (Section 2.2).

2.1 A discrepancy between calls for soil protection and the legal protection granted

Land, including soils, has a special role in law insofar as it is one of the components of the state. It is indeed the material support of the territory where the sovereignty of the States is exercised. Land is also the material basis of all activities carried out and the place where property or use rights are claimed. It is essentially through this prism that land as such is understood by the law and that specific legal rules are attached to it. For example, the distribution of land uses on the national territory can be achieved by reserving portions of the territory for agricultural activity. Land and soil are thus closely linked in terms of governance, and this link is important to achieve sustainability, as a lack of land rights can lead to poor management and soil degradation (Juergens and Hansjürgens, 2016).

Soil may benefit from protection through binding legal provisions, which do not mention it specifically. This is the case, for example, for water resource protection mechanisms that may consider the soil. However, the objective of the applicable legal texts is to satisfy the needs of water resource protection and not those of soil. This difference in legal treatment between water and land is explained by the fact that water has been legally considered as a natural resource to meet human needs and also as a natural environment independently of the uses made by Human.

Here are also legal texts specifically dedicated to soil protection. However, these texts are often not generally accompanied by binding measures. They are more of a declaratory nature. This is particularly the case for the European Soil Charter adopted under the aegis of the Council of Europe on 30 May 1972 (Council of Europe resolution 72/19 of 26 May 1972). In this text, soil is considered as 'one of humanity's most precious assets'" and 'an entity in itself' but it is also 'a limited resource which is easily destroyed'. The text stated that it should be protected from various threats such as erosion, contamination and urbanisation. This declaration invited the Member States to put in place a soil conservation policy. In a Revised European Charter for the Protection and Sustainable Management of Soil (28 May 2003), the soil functions are specified while recalling that 'Soil in itself is a living medium with intrinsic value'. The text also stresses that among fundamental principles, 'Soil protection should be systematically taken into consideration in all other policies, in particular farming, forestry, mining, industrial, tourism, transport, town planning and spatial development policy'.

Since then, a number of initiatives have emerged that bear witness to the global awareness of soil-related issues. In the early 1980s, on 26 November 1981, the World Soil Charter was adopted by the member countries of the FAO. This document treats soil as an essential resource and establishes principles and guidelines to help the various stakeholders to combat soil degradation and ensure its better conservation. The World Soil Charter also recalled that soil is a limited resource requiring specific protection with sufficient incentives or constraints to ensure the effective protection of this natural element recognised as the most precious asset of humanity. The revised Soil Charter of the World in 2015 confirms the terms of the Charter while emphasising that 'Soils are a key enabling resource, central to the creation of a host of goods and services integral to ecosystems and human well-being. The maintenance or enhancement of global soil resources is essential if humanity's overarching need for food, water, and energy security is to be met in accordance with the sovereign rights of each state over their natural resources'.

Furthermore, several reports on the state of the environment have since reiterated the message that soils are among the main resources to be considered and protected for the development of human societies. However, these messages remain largely declaratory even though they converge, whether on the scale of the reports of the intergovernmental bodies IBPES (IPBES, 2018) and IPCC (IPCC, 2019) or that of the Rio United Nations conventions (the convention on climate change, biological diversity or the one devoted to combating desertification). In particular, in 2018, IPBES supported a combined approach to land degradation and restoration. This report conveys the powerful message that land degradation, exacerbated by climate change, tends to narrow the range of possible solutions. In particular, it indicates that the later land restoration solutions are developed, the less effective they will be and the more costly and difficult they will be to implement. Shortly afterwards, in 2019, land, the medium of all activity, was again in the international arena with a special IPCC report focusing this time on climate change, desertification, land degradation, sustainable land management, food security and GHG fluxes in terrestrial ecosystems. The report states that soils are under unsustainable pressure: deforestation, intensive agriculture, soil depletion and extinction of life. It is argued that the way land is managed can be an accelerator or a solution to the climate crisis, which can have direct consequences for food security. Also aware of the role of soils in achieving climate and biodiversity targets, the European Commission has envisaged a roadmap from its 2030 Biodiversity Strategy on 'healthy soil for healthy life'.

However, a turning point seems to have been reached with the United Nations Convention of 17 June 1994 on Desertification, which it defines as 'land degradation in arid, semi-arid and dry sub-humid areas, resulting from various factors, including climatic variations and human activities' (Article 1a). It is the only international text that specifically addresses soil and its various forms of degradation but does not cover all soils. Although it is an international convention, the provisions to protect soils have not had the desired results. Land protection has recently found support in the Sustainable Development Goal[3] (SDG) 15 (UNEP, 2015), which aims to preserve and restore terrestrial ecosystems, ensure sustainable use and management of forests, combat desertification, halt and reverse land degradation, and halt the loss of biodiversity. It is specified by target 15.3, which aims to combat desertification, restore degraded land and soil, including land affected by desertification, drought and floods, and strive for a land degradation neutral world by 2030 (Keesstra et al. 2018). This neutrality originates (Safriel, 2017; Gilbey et al., 2019) from the UN Conference on Sustainable Development (Rio+20) (20-22 June 2012 in Brazil). To take into account the SDGs, land degradation neutrality was explicitly considered by the Convention to Combat Desertification. The intergovernmental working group (IWG) mandated to work on land degradation neutrality and therefore submitted a definition of land degradation to the 12th Conference of the Parties on this convention held on 12 and 23 October 2015 in Ankara, which was approved by decision (ICCD/COP(12)/20/Add., 2015): 'Land degradation neutrality is a state whereby the amount and quality of land resources necessary to support ecosystem functions and services and enhance food security remain stable or increase within specified temporal and spatial scales and ecosystems'.[4] Land degradation neutrality has become a central element of the 2028-2030 strategic framework of the United Nations Convention to Combat Desertification. Furthermore, achieving land degradation neutrality by 2030 a national goal (Chasek et al., 2019) was the decision taken by the States Parties to the United Nations Convention to Combat Desertification (UNCCD), meeting from 2 to 13 September 2019 in New Delhi (India) within the framework of their 14th conference (Executive secretary report, ICCD/COP(14)/2, 2019). This provision marks a move towards action to conserve a certain amount of quality soil. However, success remains dependent on its ratification by the signatory states. At an international level, the importance of global soil governance was recognised and led to the establishment of the Global Soil Partnership (GSP) at the FAO (Montanarella and Vagas, 2012) and recently to the launch of the RECSOIL programme dedicated to increasing SOC globally (Chapter 25 of this book).

3 The overall goals of the SDGs are to eradicate poverty in all its forms and in all countries, to protect the planet and to ensure prosperity for all, which correspond to the main pillars of sustainable development. Seventeen SDGs were defined by the UN member states in 2015 and constitute the 2030 Agenda.

4 The IWG definition of LDN is included here as contained in document ICCD/COP(12)/4 as amended and with the deletion of the text in square brackets.

Development of national law to implement global strategies

In addition to the development of global environmental strategies to enhance carbon sequestration in agricultural soils, such as the 4/1000 initiative, their implementation also requires a two-pronged normative framework, as it requires global orchestration and the guarantee of national implementation at the state level. National efforts may have been made. However, the relevance of a global strategy relies on the involvement of a wider community. It is precisely the role of international environmental law to respond to environmental threats through global international cooperation, which requires rules of law and international institutions to develop them and facilitate and monitor implementation.

To do this, several difficulties, not exhaustive, exist:

The international texts specifically devoted to soils are, for the most part non-legally binding texts. One exception, however, is the 1994 United Nations Convention to Combat Desertification (UNCCD).

Like any treaty, it implies compliance with its obligations by the States Parties, i.e. the sovereign States in charge of the legislative process. However, this international treaty has a limited geographical scope as long as the soils targeted are those threatened by desertification and the convention states that it concerns 'Countries Experiencing Serious Drought and/or Desertification, Particularly in Africa'. However, this is not a criterion for joining the Convention, as all countries can join.

At the crossroads of several issues, soil can also benefit from the binding effect of other international agreements and treaties. The issue of carbon sequestration in agricultural soils was mentioned in Article 3.4 of the 1998 Kyoto Protocol to decide which additional anthropogenic activities related to changes in GHG emissions could be added to or subtracted from the amounts assigned to Parties to the Protocol. The categories covered by this article are 'agricultural land and land-use change and forestry'. Carbon sequestration in agricultural soils also appears in the 2015 Paris Agreement, notably in the guise of Nationally Determined Contributions (NDCs) (previously only Intended Nationally Determined Contributions (INDCs) as preparatory documents for COP21). These NDCs are the materialization of the national efforts envisaged in the framework of the fight against climate change. These NDCs are at the heart of the Paris Agreement as they are a condition for its ratification. Article 4.2 states that '[e]ach Party shall prepare, communicate and maintain successive nationally determined contributions that it intends to achieve'. In addition, '[t]he Parties shall pursue domestic mitigation actions, with the aim of achieving the objectives of such contributions'. While several authors have welcomed the inclusion of soil and carbon in NDCs, few quantitative targets are included (Wiese et al., 2021).

More broadly, although many texts dedicated to soil protection and management do not have immediate legal effects, they contribute to shaping a common interest in soil quality but also constitute the basis for more binding texts in the future. Furthermore, although this does not concern all states, the European Union is an important normative relay for ensuring that the member states take account of soils and their carbon sequestration capacity (see below tab on agricultural soils, carbon sequestration and European Union law).

Soil and state sovereignty

Soil is a particular element of state sovereignty, a component of the state and a support for all economic activities. This is the case, for example, with forests, the management and protection of which may give rise to a certain amount of resistance from States. Considering the soil as an environmental element may be even more sensitive, as the soil is also a support for all activities. As a result, states may feel completely deprived of their national sovereignty.

For these reasons, it is not surprising that texts on soil protection and management are drafted in a declaratory, advisory and very general way. However, to legally implement soil carbon sequestration at scale, a more precise orchestration may be needed, requiring quantified targets as well as a soil typology associated with appropriate agricultural soil management practices.

The encroachment of international environmental law on soil into other areas.

In the case of agricultural activity, food sovereignty is a particularly crucial issue for states, although it is becoming increasingly clear that in the long term it will only be maintained by maintaining soil quality. This raises the question of arbitration between equally legitimate objectives that obey different time frames.

The interdependence of the issues to be addressed in order to guarantee the carbon sequestration capacity of the soil.

The intimate soil/climate/biodiversity relationship involves areas of interdependence that are entrusted in isolation to international institutions and obey sectoral normative logics. However, recent efforts are being made.

Maintaining the availability of quality land, particularly agricultural land, is an essential prerequisite for ensuring effective carbon storage. The capacity to sequester carbon, which will determine its achievable quality improvements, can also constitute an interesting legal lever since it is one of the intrinsic soil properties determining the ability of the soil to support plant growth. Such an approach can lead to a stronger legal interest in soil quality in general and the amount of SOC in particular.

Agricultural soil, carbon sequestration and European law

The European soil strategy (European Commission, 2021a), which is an offshoot of the Green Pact and, more specifically, of the biodiversity strategy, intends to pay particular attention to soil protection. Carbon sequestration in agricultural soils plays an important role in this respect. The strategy plans to act with existing tools and with a specific legislative proposal for 2023 on soil health. The strategy aims for healthy soils by 2050. The strategy is in line with several existing objectives, including that of achieving a climate-neutral Europe.

The achievement of the Zero Net GHG Emissions by 2050 target depends in particular on the carbon removals that will be achieved through soil restoration and better soil management to absorb the remaining emissions at the end of the decarbonisation pathway.

Concerning carbon sequestration in agricultural soils, the strategy focuses on 'developing a long-term vision for sustainable carbon cycles (including capture, storage and use of CO_2) in a climate-neutral EU economy'. This initiative was the subject of a Commission Communication on restoring sustainable carbon cycles (European Commission, 2021c). It aims to support the development of sustainable carbon absorption solutions to achieve climate neutrality by 2050. In particular, it takes up the challenge of launching an EU initiative on carbon storage in agricultural soils (carbon farming), announced in the 'farm to fork' strategy European Commission, 2020) and reiterated in the EU Forest Strategy 2030 European Commission, 2021b). Agricultural carbon storage is presented as a green business model that rewards land managers who adopt better land management practices. The funding logic is based on a double logic, practices-based or results-based, as it targets management practices that increase atmospheric carbon storage or the actual amount of carbon sequestered.

To launch this EU initiative in favour of carbon storage in agricultural soils, the European Commission relied on a report it recently published (European Commission, Directorate-General for Climate Action, Radley, G., Keenleyside, C., Frelih-Larsen, A., et al., 2021d). The study looked at existing climate change schemes in five promising areas: peatland restoration, agroforestry, maintenance and enhancement of SOC on mineral soils, SOC management on grasslands and carbon balance on livestock farms. The study concludes that agriculture, if result-oriented, 'can potentially make a significant contribution to climate change mitigation in the EU' and offer co-benefits such as increased biodiversity and ecosystem preservation. The same study highlighted that soil management practices that sequester carbon are already known, effective and low-cost practices.

Public funding is envisaged as a boost to carbon storage in agricultural land. The common agricultural policy, because of its link with agricultural land, is one of the preferred policies for supporting the deployment of agricultural carbon storage.

More broadly, to develop a market for these carbon credits, the initiative stresses the need to promote a standardised system of rules and benchmarks based on methodological standardisation. This is a prerequisite for a regulatory framework for the certification of carbon removals, which is the second ambition of this Commission initiative, alongside the promotion of carbon storage in agricultural soils.

For the time being, among the key actions planned to promote this storage, it is planned to create, in the framework of the 'A Soil Deal for Europe' mission, living laboratories that test and showcase agricultural carbon storage practices in various locations across Europe (https://ec.europa.eu/info/research-and-innovation/funding/funding-opportunities/funding-programmes-and-open-calls/horizon-europe/eu-missions-horizon-europe/soil-health-and-food_en).

2.2 Thinking about the legal protection of soil through the prism of its intrinsic qualities such as carbon sequestration

To strengthen the protection of soil, international legal texts have shifted the focus from indicating that soil is a finite resource to highlighting all soil functions and services. The interest in ES was heightened by the release of a synthesis report based on the work of the Millennium Ecosystem Assessment. In 2000, Kofi Annan, Secretary-General of the United Nations, published his report 'We the Peoples: A UN for the Twenty-First Century' (Annan, 2014). Among its recommendations is the 'ecosystem assessment'. At its request, a working group was set up in 2001 with more than 1360 experts from all over the world, whose main objectives were to assess the consequences of ecosystem change on human well-being and to establish the scientific basis for implementing the actions needed to improve the conservation and sustainable use of these systems, as well as their contribution to human well-being. The results of this evaluation were disseminated in a report, in particular in the form of a synthesis (MEA, 2005). This report is at the origin of the popularisation of the concept of ES, which were defined as 'benefits people obtain from ecosystem'. A typology of ES has been proposed by the MEA distinguishing between provisioning services (food, building materials, etc.), regulating services (water cycle, air renewal, etc.), cultural services (amenities, recreation, etc.) and support services (genetic diversity, habitats for species, etc.). In 2010, ES were again in the spotlight after the presentation of the synthesis report of the TEEB study (The Economics of Ecosystems and Biodiversity) (TEEB, 2010a) at the COP 10 of the Convention on Biological Diversity (CBD) in Nagoya (TEEB, 2010a). This global study launched by the G8 and five other countries provided guidelines for the economic value of natural capital and the 'dividend that society receives from natural capital' (TEEB, 2010a, p. 9), with dividends being ES. To make the economic value of nature visible (TEEB, 2010a), now referred to as natural capital, the work carried out in the TEEB study aimed to put into perspective 'the global economic benefit of biological diversity, the costs of biodiversity loss and the failure to take protective measures in relation to the costs of effective conservation' (TEEB, 2010b, back cover).

In this MEA, soils were recognised as the basis for many essential ES (Montanaro et al., 2017). One of these services is carbon sequestration, which has become increasingly important recently (see above). The educational dimension of the ES concept, which allows for a new appreciation of soil, has been negatively tainted by the anthropocentric and utilitarian character now attributed to ES. It is indeed crucial not to reduce soil to the services it provides; otherwise, a new form of land use conflict will emerge. Nevertheless, ES offer an opportunity to strengthen the legal protection of soil by recognising it as an ecosystem. It is therefore a question of recognising and protecting the qualities of this ecosystem and not just considering soil as a support for activities.

This recognition has two purposes. The first is to recall the natural capacities of the soil. In this case, it is the one concerning carbon sequestration. This prism reminds us that without the proper functioning of the soil, ES cannot be provided, or at least not in an optimal manner. It follows that any degradation of soils leads to an impairment of the proper functioning of the land and to a weakened overall performance of the terrestrial biosphere. In contrast, it also implies that ES cannot be understood in complete isolation from each other. For example, biodiversity in agricultural areas can promote carbon sequestration. Similarly, if the storage of organic matter plays a role in climate regulation, this same organic matter contributes to increasing the physical and chemical properties of the soil and thus its fertility. In other words, a holistic approach is necessarily required to think about the protection of the soils' carbon sequestration capacity.

A second virtue of recognising soil as an ecosystem can be highlighted. It leads to the recognition of soil as an entity in its own right and therefore as an entity that must be dealt with and not just as the sole receptacle of activities. In other words, the interests of soil as an ecosystem should be considered in the same way as other forms of land use, for example, building a house and farming the land. Such an approach therefore invites long-term thinking and a different hierarchy of expectations with regard to the soil and its uses.

The legal effects of an ES approach, which is still in its infancy, should be carefully monitored, as this concept has largely penetrated the law and is therefore likely to modify the applicable legal obligations. Indeed, a more direct consideration of the legal protection of soil ES, including that of carbon sequestration, may be envisaged, with the corollary of liability in the event of non-compliance with obligations to protect these services.

In addition to ensuring the availability of quality agricultural soils and thus their capacity to sequester carbon, it is also necessary to look at the conditions under which this sequestration is carried out so that the means implemented achieve the expected results.

3 Legal requirements for the implementation of soil carbon sequestration

The examination of the legal conditions related to the implementation of management techniques favouring soil carbon sequestration makes it possible both to question the very feasibility of carbon sequestration and to allow for its legal security for the actors involved. This is also a condition for its development.

This examination requires, first of all, clarifying what is meant by soil carbon sequestration. In particular, it is necessary to identify the practices that lead to increased soil carbon stocks and that will be the subject of legal intervention (Section 3.1). Secondly, insofar as practices that contribute to carbon sequestration are to be encouraged, they need to be based on a contractual mechanism involving the voluntary farmer in the use of practices that favour carbon sequestration. To achieve the desired result, however, this contractual mechanism faces difficulties inherent either in the carbon sequestration itself or in the contractual technique itself (Section 3.2).

3.1 Legal issues related to agricultural practices at the heart of carbon sequestration

While the term carbon sequestration is sufficient in itself to indicate the purpose of this technique in terms of the fight against climate change, it nevertheless needs to be clarified in terms of expectations for agricultural soils and, therefore, in terms of the agricultural practices required to achieve them. This effort at reclassification thus makes it possible to better define the associated legal issues.

Firstly, it is important to emphasise or recall that carbon sequestration implies increasing the amount of organic matter in soils by means of plant and animal residues returned to the soil. The current priority given to the fight against climate change is therefore expressed even in the choice of terminology. Indeed, there is 'the current trend is to use the term "soil carbon" instead of "soil organic matter". However, most soil and land management practices conducive to carbon sequestration actually also favour organic matter storage in the soil'[5] (Feller and Gallali, 2014). To act on the quantity of carbon sequestered in soils, it

5 But it could be different for biochar, for example.

is, therefore, through the prism of practices that promote the storage of organic matter in soils that law can intervene.

These agricultural practices can take the form of both actions and inactions to preserve carbon sequestration in agricultural soils. For example, it may involve not ploughing grasslands or favouring the use of natural fertilizers in cultivated soils. These actions and inactions are partly covered by the law and tend to be developed under the banner of agro-ecological practices.

However, in view of the challenges of soil protection, particularly to preserve and increase their organic matter content, soil protection should be strengthened. To ensure the effective preservation of the carbon content in agricultural soils, a targeted protection strategy combining a voluntary approach and an approach imposed unilaterally should be considered.

In addition to a currently perfectible approach of encouraging an increase in soil carbon content through a change in practices, the establishment of protection zones for soils that are already rich in carbon would make it possible not to erode this existing stock. Indirect protection measures for agricultural soils exist via, for example, the 1971 Ramsar Convention protecting wetlands of international importance, via the European 'habitats' directive of 21 May 1992 (OJ L 206, 22.7.1992) or via national protection measures such as nature reserves. However, the protection is not, or not directly, aimed at preserving the carbon content of agricultural soils or soils in general, but often the exceptional character of a site. The 'ordinary' character of many agricultural soils such as grasslands, which are naturally rich in carbon, may not benefit from this enhanced protection until the risks associated with changes in land use and agricultural practices are fully considered (Desrousseaux, 2016). In this respect, ecological zones specific to the protection of sequestered carbon could be envisaged, following the example of drinking water catchment protection zones. In this direction, measures should be taken to avoid GHG emissions from carbon-rich soils, e.g. during peatland cultivation (Chapter 21 of this book). So, in other words, the potential loss of carbon storage through this change of use should be fully considered under the environmental assessment procedure.

Secondly, although the fight against climate change is a global issue, carbon sequestration actions are carried out on a local scale (Amelung et al., 2020). The sum of these local actions nevertheless has an impact on a global scale and is therefore likely to justify an international approach to soil, which is still largely lacking.[6] However, control over agricultural land is a matter of national sovereignty. It is, therefore, on a national scale that rules, particularly in terms of land regulation or taxation, will be established. It is also at a more local level that the legal relationship between farmers and agricultural land is played

6 See above.

out.[7] All these legal rules contribute to the concrete implementation of carbon sequestration actions.

This local scale of intervention is also often justified by the pedological, climatic, economic and ecological diversity of soils and their use. This diversity at the local scale would therefore require specific and local measures to promote carbon sequestration, which depend on the scientific data available on the state of the soil. Taking into account this soil diversity could be materialised by soil mapping. To consider this diversity with regard to the objective of carbon sequestration, it could be possible to target the soils with the greatest potential for carbon sequestration. In the interests of efficiency, law can therefore prioritise the most relevant soils in terms of carbon sequestration and to be effective, the law can therefore prioritise the most relevant soils in terms of carbon sequestration and thus encourage only certain farmers to implement specific agricultural practices. Indeed, several criteria can be used to identify the most suitable soils (Virto et al., 2012; Kurganova et al., 2014). For example, soils in cold climates are more conducive to carbon storage because the degradation of organic matter by soil organisms is slowed down. Soil type is also important. In particular, Histosols or organic soils must be protected as they store about one-third of all terrestrial carbon (Chapter 21 of this book). Moreover, Chernozems and Andosols are rich in organic matter because different mechanisms protect their organic matter from biodegradation. Finally, the amount of carbon stored in soil depends on its use with grassland or forest soils containing higher carbon stocks than those with annual vegetation, i.e. under cropping systems (Bispo et al., 2016, p. 123). With regard to the large variety of soils and land use systems concerned by carbon sequestration potential, the contract may appear to be the ideal tool to provide the necessary flexibility.

Thirdly, in addition to the place, where practices are implemented, the nature of practices beneficial to carbon sequestration constitutes crucial information for the elaboration of legal aspects. Among the agricultural practices considered to favour soil carbon sequestration are the extension of crop rotations, the use of cover crops and/or organic fertilizers and the reduction of tillage (Chapter 16 of this book). They also include the management of grasslands to optimise carbon sequestration or the use of agroforestry systems (Smith et al., 2008, Chapters 18 and 20 of this book). These practices can be included in a specification and justify public aid when they are carried out. They can also be included in a contract between private contractors and implement these practices conditional on financial compensation.

Fourthly, recognising practices that increase carbon sequestration in soils must not hide the environmental risks associated with these different practices (Chapter 6 of this book). In other words, this implies that an environmental

7 See below.

assessment of agricultural practices, including for climate change purposes, should be foreseen and legally regulated. Without such an integrated approach, the sustainability of agricultural land management may be compromised. Indeed, the increase in organic matter in soils may be achieved with an increase in the application of livestock manure. The latter can be a source of water pollution by nitrates and/or phosphates. Indeed, the mineralisation of organic nitrogen leads to the production of nitrates and/or phosphates, which are both a nutrient and a potential water contaminant. For this reason, legal provisions have been introduced to control plant fertilisation and, more specifically, that of livestock effluents due to soil saturation. The spreading of effluents can also be a source of GHGs (CO_2, NO_2 and CH_4) (Thompson et al., 2019). These are released during the mineralisation of organic matter and should be accounted for and deducted from the quantity of carbon sequestered in the soil(s) concerned. Moreover, no-tillage is one of the most popular practices for sequestering carbon in soils. However, this practice is frequently associated with the use of plant protection products, which also tend to reduce the environmental impact (Chapter 22 of this book). Assessing the risks associated with carbon sequestration practices has the merit of not considering agriculture solely in terms of its contribution to the fight against climate change. Moreover, requiring this environmental assessment of carbon sequestration practices does not imply an adaptation of agriculture but a transformation of the dominant agricultural model into a more sustainable one.

To combat climate change, sustainable agricultural practices leading to soil carbon sequestration are encouraged by policymakers. To ensure their implementation, they must be supported by farmers, who wish to contribute to the fight against climate change, but they must also be in line with the soil being considered. The contract is the most common way to move the farmer towards carbon sequestration-friendly practices, as it is based on the farmer's willingness to do so. The contract also offers the flexibility to adapt to local situations. However, the meeting between the contractual approach and carbon sequestration practice may be a challenge in many ways.

3.2 The challenge of meeting carbon sequestration practice implementations by a contractual approach

To encourage agricultural practices that are beneficial for carbon sequestration to replace intensive practices that have the opposite effect, one of the challenges is to contract farmers who would like to adopt these practices, potentially time-consuming, labour intensive and risky. To encourage virtuous land practices and combat land degradation, the 2018 IPBES report recommends that 'Voluntary or regulation-based incentive mechanisms for safeguarding biodiversity and ecosystem services can help avoid, reduce and reverse land degradation. Such mechanisms include both market and non-market based approaches. Examples of market-based approaches include credit lines, insurance policies and future contracts that reward adoption of more sustainable land management practices, payments for ecosystem services and conservation tenders, as applied in some countries' (IPBES, 2018, p. XXV). The financial incentive can be achieved through both public and private funding. The contractual principle is quite simple as it consists in granting financial compensation to farmers who commit to sequestering carbon. Behind this apparent simplicity, however, there are several obstacles to be identified.

Main advantages and disadvantages of the contractual approach

Advantages

- Flexibility and adaptation to the different types of soils and also to the expectations and needs of the actors involved (landowners, farmers, private investors, etc.).
- Social acceptability and stronger incentive to change agricultural practices as the contract is usually associated with a financial counterpart in exchange for specific environmental actions (environmentally friendly agricultural practices and/or environmental results).

Disadvantages

- Environmental action is guaranteed for the duration of the contract. However, what happens afterwards? Permanence may be an issue but many national laws, as is the case in France, set a term for the agreements concluded. One cannot therefore be engaged indefinitely.
- The contract, by its very nature, is a voluntary act. So, it presupposes a choice by the contracting parties and therefore by the land concerned. There is a risk of disparity between contractors depending on the land chosen. In addition to this relative effect of contracts, which are therefore only binding on the contracting parties, it is crucial to create a minimum standard of measurement to provide some contractual uniformity.
- Sequestration of carbon may be favoured to maximise the short-term carbon benefits over the protection of carbon-rich soils.
- The effectiveness of these contracts requires the development of real monitoring, reporting and verification tools.

Firstly, contractual commitment is not just a matter of contractual technique. The contract is only the outcome of a process of knowledge, negotiation and trust. There is also a strong cultural dimension to this process, which can be either supportive or hostile to a commitment to carbon sequestration. From an objective point of view, the amount of carbon in agricultural soils can be increased by spreading livestock manure or, for example, by avoiding ploughing the soil. However, moving towards these practices or resorting to them again may face cultural obstacles. Indeed, the use of these practices is not without consequences for the exercise of the agricultural activity or even the agricultural production model itself. Using no-till means, for example, changing the way you sow, managing compacted soils and controlling the spread of weeds. It also requires crop and soil management using winter cover. As far as the spreading of livestock manure is concerned, this also requires new technical knowledge, particularly in view of the variability of the composition of livestock manure, and sometimes knowledge of the rules to be respected concerning authorised spreading periods or the required existence of a storage area and its management to avoid other GHG emissions. In addition to its technical constraints, the recent shift from livestock manure to mineral fertilisers can also be explained by a paradigm shift in agriculture, as explained by Feller and Gallali (2014). They explain that soil organic matter (or 'humus') has long been identified as a factor of fertility. According to the 'humus theory', a large part of plant dry matter comes from soil humus. Therefore, ensuring plant productivity implied managing organic returns to the soil as well as possible. This theory was replaced by the research of J. Liebig who, in 1840, demonstrated that plant nutrition is exclusively relying on mineral nutrients. From his findings, it became clear that the source of fertility comes essentially from the return of mineral elements to the soil. Organic matter management was therefore no longer considered essential for agricultural production. This led to specialisation at the farm level and the separation of cereal production from animal husbandry due to economic concerns.

However, enrichment with organic matter is generally known to improve soil quality and farmers are generally concerned with their soils' organic matter content. It has recently been exuberated by the necessity to adapt to climate change.

In these two examples, it appears that changing practices often are not isolated from the rest of the system. Therefore, an agro-ecological transition must consider all components of the entire food system.

Secondly, beyond the possible cultural obstacles associated with a change in agricultural practices, the purpose of the contract raises questions about the contractual risks it entails. Indeed, from the point of view of climate change mitigation, contracting is only of interest when a certain quantity of carbon is effectively and permanently sequestered. In other words, the appropriate formula would be a contract based on results, i.e. a given quantity of carbon sequestered in soil for a given time. Only in this way can carbon sequestration be considered a negative emissions technology, as the amount of carbon actually stored can be deducted from the amount of CO_2 in the earth's atmosphere. However, while in terms of environmental efficiency, such an approach is justified and tends to develop, it does present difficulties relating to the calculation of the quantity of carbon sequestered, the harmonisation of this calculation, the cost of this calculation as well as a risk for the contractors. Indeed, it requires the farmer to justify a given increase in the amount of carbon sequestration in his soils, which also requires the establishment of a valuable monitoring, reporting and verification scheme and beforehand an inventory. Contractually, if he does not fulfil his commitments, he cannot benefit from the financial compensation. On the one hand, monitoring carbon sequestration is not an easy task, making it difficult to obtain results (Chapter 10 of this book). On the other hand, this difficulty is accentuated by certain volatility of carbon stocks due to their non-permanence following change in practices and climate change, which leads to great uncertainty in terms of timescale and amounts of carbon sequestered. Moreover, climatic events such as natural hazards over which no one has control may affect soil carbon sequestration. However, in view of the now constant and progressive nature of the deleterious impact of climate change on agricultural systems, these climatic events would hardly qualify as 'force majeure' in law, i.e. exceptional and unforeseeable climatic events, justifying the non-fulfilment of contractual obligations. Rather, they should be included as a certainty to be taken into account before any contract is concluded. This would imply measuring the degree of impact of climate change on the quantity of carbon sequestered in order not to overestimate the reality of the quantities sequestered.

This way, the amount of carbon sequestered that the farmer is contractually obliged to respect would thus be based on a more realistic assessment.

Less risky for the farmer and avoiding the SOC monitoring issue, the contractual obligation can also relate solely to compliance with agricultural practices. The agricultural practices selected are those identified as favouring carbon sequestration in agricultural soils. However, the amount of SOC sequestered by these practices will rely on estimation rather than actual measurements. The question arises, if under these conditions, it will still be possible to claim meeting the criteria of NET? Furthermore, insofar as the results of agricultural practices identified as acting in favour of carbon sequestration are only assumed, this also tends to complicate the already porous boundary between mitigation measures, among which are carbon sequestration as NET, and climate change adaptation measures. Agricultural practices that are beneficial for carbon sequestration can also be seen as nature-based solutions (Langlais, 2020). In other words, they could be considered not as remedial measures but as adaptive measures.

Thirdly, in the face of carbon sequestration, the contractual logic should also respond to another challenge: that of not only optimising the quantity of carbon in soils but also maintaining this stock over the long term to become NET. However, carbon sequestration appears vulnerable to a change in agricultural practices. In other words, if an optimal quantity of carbon in agricultural soils has been obtained thanks to beneficial agricultural practices, the phenomenon is reversible in the event of a return to previous practices. In particular, this is the case for grassland transformed into cropland. This is also the case for unploughed land that becomes so. The contractual commitment cannot guarantee the maintenance of a particular practice on agricultural land, or at least not in the long term. Indeed, some legal contract regimes are time-limited, thus preventing a lifetime contractualisation. Moreover, a contract is not immutable in time, giving the contracting parties the possibility of breaking their commitment. Finally, these contracts are largely concluded intuitu personae. This means that after retirement, illness, accident, death, etc., the new farmer is free to sign a contract with the same obligations or not. Finally, carbon sequestration contracts are concluded between the farmer and a given operator. In case the farmer is under contract with a landowner, it may be advisable to include the landowner in the contractual negotiation, in particular, to ensure a temporal match between the contracts (Bodiguel, 2022).

In addition to the vulnerability of carbon sequestration in carbon-rich soils following or not following a change in practices should be considered. To foster carbon sequestration as NET, it is important to consider the limit of carbon storage in agricultural soils and to reduce their GHG emission potential. Indeed, beyond a certain limit, soils may no longer be able to store additional carbon. In general, a new equilibrium is reached a few decades after practice change depending on the initial stocks. Thereafter, the soil carbon stock needs to be protected. This requires knowledge of the carbon saturation thresholds or carrying capacity of a specific system to identify and prioritise soils according to their storage potential. Contractual choices must take these elements into account. So contracts for carbon sequestration in agricultural soils should not be concluded solely to increase carbon sequestration.

Indeed, it seems important not to consider all incentives to contract from the point of view of ecological additionality. The maintenance of this stock should therefore also be considered.

4 Conclusion

The possibility of sequestering carbon in agricultural soils is, in fact, conditioned by the availability of soils and their effective capacity to sequester carbon.

However, particularly in the face of rapid urbanisation, it is increasingly difficult to maintain a sufficient quantity of soil that meets this criterion. Moreover, the degradation of agricultural soils, linked in particular to intensive agricultural practices, tends to undermine the performance of soils, including their capacity to sequester carbon. While hopes for carbon sequestration to combat climate change are based on the soil as an essential substrate, their legal protection is insufficient, particularly in the face of conflicts of use and strong competition over these coveted areas (agricultural production, biofuels, urbanisation, biodiversity conservation, etc.).

To favour soil carbon sequestration itself, contracts may be the most appropriate instruments as they are flexible, allowing to adapt to specific local situations. Contracts also require several legal adjustments to consider the contractual risks and difficulties for the farmer as well as the environmental risks that may be associated with practices that favour carbon sequestration. Moreover, duration and amounts of carbon to be sequestered in a given situation are issues to be resolved before contracts can be established. All of these challenges to be met through the prism of legal issues relating to carbon sequestration reveal, however, a much larger task: that of the legal support methods for the agro-ecological transition.

5 Where to look for further information

Carbon sequestration in agricultural soils is a cross-cutting issue in full development. Its expansion is conditioned in particular by a strong demand for law to guarantee its concrete implementation. Moreover, this topic has the particularity of directly affecting agricultural production methods in addition to soil preservation as such or the fight against climate change. Consequently, the attention paid to carbon sequestration in agricultural soils raises legal questions that may be specific to each of these research areas. This characteristic makes it necessary to follow the research carried out by various agricultural, environmental and climate organisations. This is the case in particular at the FAO (Food and Agriculture Organisation) where new tools or initiatives are being launched to promote carbon sequestration in soils: for example, https://www.fao.org/soils-portal/soil-management/soil-carbon-sequestration/en/; https://www.fao.org/newsroom/detail/practical-tools-to-encourage-soil-organic-carbon-sequestration/en). This is also a major concern of the intergovernmental scientific and political platforms (IPBES and IPCC), which emphasise the link between carbon sequestration and the preservation of soil biodiversity and the fight against climate change in the case of the IPBES (in particular in its report on land degradation and restoration (https://ipbes.net/resource-file/18160)), or between carbon sequestration in soils and the fight against climate change in the case of the IPCC. All the reports, including recent IPCC reports, mention it as a means of achieving carbon neutrality. Measuring the role of agriculture in the fight against climate change, the IPCC has dedicated a special report to this issue: Report on climate change, desertification, land degradation, sustainable land management, food security, and greenhouse gas fluxes in terrestrial ecosystems (https://www.ipcc.ch/site/assets/uploads/sites/4/2020/02/SPM_Updated-Jan20.pdf).

All of these incentives for the development of sequestration in agricultural soils, to which the 4/1000 initiative (https://4p1000.org) actively contributes, are part of the construction of a law intended to accompany any action in this direction or, more specifically, a change in agricultural practices. Numerous legal texts promote carbon sequestration in soils (following the example of the United Nations Convention on Climate Change and its Paris Agreement adopted on 12 December 2015) or at the European Union level, as evidenced by the texts relating to the Common Agricultural Policy or the European Commission's initiatives in favour of carbon farming, which herald binding legal texts (https://ec.europa.eu/clima/eu-action/forests-and-agriculture/sustainable-carbon-cycles/carbon-farming_en). However, it should be emphasised that for a detailed understanding of the implementation of carbon sequestration in agricultural soils, it is at a local level and therefore domestic law that should be referred to. Property law, in particular, is specific to each country, as is the law applicable to contracts concluded with farmers. For example, in France, the legifrance.fr website lists all the applicable legal texts.

In parallel, the market associated with carbon sequestered in soils will also be an issue to be closely monitored, as it largely conditions the success of any legal incentive for carbon sequestration in soils. Moreover, this market implies thinking about and legally framing the harmonisation of the reference system for the tonne of carbon sequestered. This market, which is currently being developed, is an issue for many state actors. This is particularly the case in the United States (https://www.csis.org/analysis/soil-carbon-sequestration-myths-realities-and-biden-administrations-proposals). This is also the case in the European Union, which has described carbon farming as a 'business model.' Recent studies are helping to clarify this model (https://www.ecologic.eu/sites/default/files/publication/2021/70301-Carbon-farming-Making-agriculture-fit-for-2030.pdf). In addition, the European Commission launched a call for proposals entitled 'Research and Innovation and other actions to support the implementation of a mission in the area of Soil health and Food (HORIZON-MISS-2021-SOIL-02),' which closed in March 2022, was launched by the European Commission. This call has in particular the mission to explore the potential of carbon farming as a business model.

6 References

Amelung, W., Bossio, D., Vries, W. de, Kögel-Knabner, I., Lehmann, J., Amundson, R., Bol, R., Collins, C., Lal, R., Leifeld, J.,Minasny, B., Pan, G., Paustian, K., Rumpel, C., Sanderman, J., Groenigen, J. W. van, Mooney, S., van Wesemael, B., Wander, M. and Chabbi, A. (2020). Towards a global-scale soil climate mitigation strategy, *Nature Communications* 11(1), 5427, https://doi.org/10.1038/s41467-020-18887-7.

Annan, K. A. (2014). *We the Peoples: A UN for the Twenty-First Century*, Routledge, London, UK, 264 p.

Balesdent, J. and Arrouays, D. (1999). Usage des terres et stockage du carbone dans les sols du territoire français (1900–1999) (An estimate of the net annual carbon storage in french soils induced by land use change from 1900–1999), *Comptes Rendus de l'Académie d'Agriculture de France* 85(6), 265–277.

Bispo, A., Guellier, C., Martin, E., Sapijanskas, J., Soubelet, H., et al. (2016). *Les sols- Intégrer leur multifonctionnalité pour une gestion durable (Soils - Integrating Their Multifunctionality for Sustainable Management)*, Editions Quae, France, 384 p, Savoir Faire (Quae), 978-2-7592-2392-3.

Bodiguel, L. (2022). CO_2 vert et bail rural: les interactions entre le contrat de séquestration de carbone et le bail rural (Green CO_2 and rural leases: the interactions between carbon sequestration contracts and rural leases). In J.-B. Millard and H. Bosse-Platière (dir.) *Le CO2 vert "capturé" par le droit*, Lexis Nexis, Paris, pp. 132–143.

Chasek, P., Akhtar-Schuster, M., Orr, B. J., Luise, A., Rakoto Ratsimba, H. and Safriel, U. (2019). Land degradation neutrality: the science-policy interface from the UNCCD to national implementation, *Environmental Science and Policy* 92, 182–190.

Desrousseaux, M. (2016). *La protection juridique de la qualité des sols (Legal Protection of Soil Quality)*, LGDP, Paris, 484 p.

European Commission (2020). *Communication from the Commission to the European Parliament and the Council, the European Economic and Social Committee of the Regions, A Farm to Fork Strategy for a Fair, Healthy and Environmentally-Friendly Food System*, COM (2020) (vol. 381), 20.5.2020.

European Commission (2021a). *Communication from the Commission to the European Parliament and the Council, the European Economic and Social Committee of the Regions, EU Soil Strategy for 2030 Reaping the Benefits of Healthy Soils for People, Food, Nature and Climate*, COM (2021) (vol. 699), 17.11.2021.

European Commission (2021b). *Communication from the Commission to the European Parliament and the Council, the European Economic and Social Committee of the Regions, New EU Forest Strategy for 2030*, COM (2021) (vol. 572), 16.7.2021.

European Commission (2021c). Communication from the commission to the European Parliament and the council, sustainable carbon cycles, COM (2021) 800(15), 12.

European Commission (2021d). Directorate-general for climate action. In G. Radley, C. Keenleyside, A. Frelih-Larsen, et al. (eds.) *Setting Up and Implementing Result-Based Carbon Farming Mechanisms in the EU: Technical Guidance Handbook*, Publications Office. https://data.europa.eu/doi/10.2834/056153.

Feller, Ch. and Gallali, T. (2014). Preface in French scientific committee on desertification. In M. Bernoux and T. Chevallier (eds.) *Soil Carbon: Multiple Functions Benefiting Societies and the Environment*, CSFD, les dossiers thématiques, vol. 10, p. 2.

Fuss, S., Lamb, W. F., Callaghan, M. W., Hilaire, J., Creutzig, F., Amann, T., Beringer, T., de Oliveira Garcia, W., Hartmann, J., Khanna, T., Luderer, G., Nemet, G. F., Rogelj, J., Smith, P., Vicente, J. L. V., Wilcox, J., del Mar Zamora Dominguez, M. and Minx, J. C. (2018). Negative emissions–part 2: costs, potentials and side effects, *Environmental Research Letters* 13(6), https://doi.org/10.1088/1748-9326/aabf9f.

Gilbey, B., Davies, J., Metternicht, G. and Magero, C. (2019). Taking land degradation neutrality from concept to practice: early reflections on LDN target setting and planning, *Environmental Science and Policy* 100, 230–237.

ICCD/COP(12)/20/Add.1 (2015). Decision 3/COP.12, integration of the sustainable development goals and targets into the implementation of the United Nations convention to combat desertification and the intergovernmental working group report on land degradation neutrality, https://knowledge.unccd.int/sites/default/files/inline-files/COP.12.pdf, p. 8.

ICCD/COP(14)/2 (2019). Report by the executive secretary, integration of sustainable development goal 15 and related target 15.3 into the implementation of the United Nations Convention to combat desertification and land degradation neutrality, https://www.unccd.int/sites/default/files/sessions/documents/2019-07/ICCD_COP%2814%29_2-1910284E.pdf.

IPBES (2018). The IPBES assessment report on land degradation and restoration. L. Montanarella, R. Scholes, and A. Brainich (eds.) *Secretariat of the Intergovernmental Science-Policy Platform on Biodiversity and Ecosystem Services*, Bonn, Germany. 744 pages. https://doi.org/10.5281/zenodo.3237392.

IPCC (2018). Global warming of 1.5°C. An IPCC special report on the impacts of global warming of 1.5°C above pre-industrial levels and related global greenhouse gas emission pathways, in the context of strengthening the global response to the threat of climate change, sustainable development, and efforts to eradicate poverty. V. Masson-Delmotte, P. Zhai, H.-O. Pörtner, D. Roberts, J. Skea, P.R. Shukla, A. Pirani, W. Moufouma-Okia, C. Péan, R. Pidcock, S. Connors, J. B. R. Matthews, Y. Chen, X. Zhou, M. I. Gomis, E. Lonnoy, T. Maycock, M. Tignor, and T. Water eld (eds.).

IPCC (2019). Climate change and land, an IPCC special report on climate change, desertification, land degradation, sustainable land management, food security, and green house gas fluxes in terrestrial ecosystems, IPCC, www.ipcc.ch/report/srccl.

Juerges, N. and Hansjürgens, B. (2016). Soil governance in the transition towards a sustainable: bioeconomy - a review, *Journal of Cleaner Production* 170, 1628–1639.

Keesstra, S., Mol, G., de Leeuw, J., Okx, J., Molenaar, C., de Cleen, M. and Visser, S. (2018). Soil-related sustainable development goals: four concepts to make land degradation neutrality and restoration work, *Land* 7(4), 133.

Kurganova, I., Lopes de Gerenyu, V., Six, J. and Kuzyakov, Y. (2014). Carbon cost of collective farming collapse in Russia, *Global Change Biology* 20(3), 938–947. https://doi.org/10.1111/gcb.12379.

Langlais, A. (2020), Solutions fondées sur la nature : levier ou frein pour la préservation de la biodiversité ? Réflexions juridiques (Nature-based solutions: lever or brake for biodiversity conservation? Legal reflections). In M. Torre-Schaub (ed.) *Changement Climatique et Normes, Regards Interdisciplinaires*, Mare et Martin éd. pp. 181–196.

Langlais, A. and Lemoine-Schonne, M. (2022). *Construire le droit des ingénieries climatiques- Pour une approche croisée des enjeux climatiques et écosystémiques, UGA Editions, à paraître 2021* (building the law of climate engineering - for a cross-cutting approach to climate and ecosystem issues).

Millennium Ecosystem Assessment (MEA) (2005). *Ecosystems and Human Well-Being: Synthesis*, Island Press, Washington, DC, 137 p.

Minx, J. C., Lamb, W. F., Callaghan, M. W., Fuss, S., Hilaire, J., Creutzig, F., Amann, T., Beringer, T., de Oliveira Garcia, W., Hartmann, J., Khanna, T., Lenzi, D., Luderer, G., Nemet, G. F., Rogelj, J., Smith, P., Vicente Vicente, J. L., Wilcox, J. and del Mar Zamora Dominguez, M. (2018). Negative emissions: part 1 - research landscape, ethics and synthesis, *Environmental Research Letters* 13(6), https://doi.org/10.1088/1748-9326/aabf9b.

Montanarella, L. and Vargas, R. (2012). Global governance of soil resources as a necessary condition for sustainable development, *Current Opinion in Environmental Sustainability* 4(5), 559-564, https://doi.org/10.1016/j.cosust.2012.06.007.

Montanaro, G., Xiloyannis, C., Nuzzo, V. and Dichio, B. (2017). Orchard management, soil organic carbon and ecosystem services in Mediterranean fruit tree crops, *Scientia Horticulturae* 217, 92-101. https://doi.org/10.1016/j.scienta.2017.01.012.

National Academies of Sciences, Engineering, and Medicine (2019). *Negative Emissions Technologies and Reliable Sequestration: A Research Agenda*, The National Academies Press. Washington, DC. https://doi.org/10.17226/25259.

Nemet, G. F., Callaghan, M. W., Creutzig, F., Fuss, S., Hartmann, J., Hilaire, J., Lamb, W. F., Minx, J. C., Rogers, S. and Smith, P. (2018). Negative emissions - part 3: innovation and upscaling, *Environmental Research Letters* 13(6). https://doi.org/10.1088/1748-9326/aabff4

OFEV-Office fédéral de l'environnement (Suisse) (no year). https://www.bafu.admin.ch/bafu/fr/home/themes/climat/info-specialistes/reduction-emissions/technologies-d-emission-negative.html (Federal Office for the Environment (Switzerland)).

Rumpel, C., Amiraslani, F., Chenu, C., Garcia Cardenas, M., Kaonga, M., Koutika, L. S., Ladha, J., Madari, B., Shirato, Y., Smith, P., Soudi, B., Soussana, J. F., Whitehead, D. and Wollenberg, E. (2020). The 4p1000 initiative: opportunities, limitations and challenges for implementing soil organic carbon sequestration as a sustainable development strategy, *Ambio* 49(1), 350-360.

Safriel, U. (2017). Land Degradation Neutrality (LDN) in drylands and beyond-where has it come from and where does it go, *Silva Fennica* 51(1B), 20-24.

Smith, P., Martino, D., Cai, Z., Gwary, D., Janzen, H., Kumar, P., McCarl, B., Ogle, S., O'Mara, F., Rice, C., Scholes, B., Sirotenko, O., Howden, M., McAllister, T., Pan, G., Romanenkov, V., Scheinder, U., Towprayoon, S., Wattenbach, M. and Smith, J. (2008). Greenhouse gas emission in agriculture, *Philosophical Transactions of the Royal Society of London Series B* 363, 789-813.

Tanzer, S. E. and Ramírez, A. (2019). When are negative emissions negative emissions?, *Energy and Environmental Science* 12(4), 1210-1218.

TEEB (2010a). The economics of ecosystems and biodiversity: mainstreaming the economics of nature: a synthesis of the approach, conclusions and recommendations of TEEB. Retrieved 12 May 2018, www.teebweb.org/.

TEEB (2010b). The economics of ecosystems and biodiversity for business - executive summary 2010. Retrieved 12 May 2018, www.teebweb.org/.

Thompson, R. L., Lassaletta, L., Patra, P. K., Wilson, C., Wells, K. C., Gressent, A., Koffi, E. N., Chipperfield, M. P., Winiwarter, W., Davidson, E. A., Tian, H. and Canadell, J. G. (2019). Acceleration of global N_2O emissions seen from two decades of atmospheric inversion, *Nature Climate Change* 9(12), 993-998, https://doi.org/10.1038/s41558-019-0613-7.

UNEP (2015). Resolution n°70/1 adopted by the general assembly on 25 September 2015, "Transforming our world: the Sustainable Development Agenda to 2030".

Virto, I., Barré, P., Burlot, A. and Chenu, C. (2012). Carbon input differences as the main factor explaining the variability in soil organic C storage in no-tilled compared to inversion tilled agrosystems, *Biogeochemistry* 108(1-3), 17-26. https://doi.org/10.1007/s10533-011-9600-4.

Wiese, L., Wollenberg, E., Alcántara-Shivapatham, V., Richards, M., Shelton, S., Hönle, S. E., Heidecke, C., Emoke Madari, B. E. and Chenu, C. (2021). 'Countries' commitments to soil organic carbon in nationally determined contributions, *Climate Policy* 21(8), 1005-1019, https://doi.org/10.1080/14693062.2021.1969883.

Index